MATHEMATICS IN INDUSTRY　　**12**

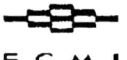

Luis L. Bonilla
Miguel Moscoso
Gloria Platero
Jose M. Vega

Editors

Progress in Industrial Mathematics at ECMI 2006

With 361 Figures, 52 in color and 53 Tables

 Springer

Editors

Luis L. Bonilla

Universidad Carlos III de Madrid
Avenida de la Universidad 30
28911 Leganes, Spain
Email: bonilla@ing.uc3m.es

Miguel Moscoso

Universidad Carlos III de Madrid
Avda. Universidad, 30
28911 Leganes (Madrid), Spain
Email: moscoso@math.uc3m.es

Gloria Platero

Instituto de Ciencia de Materiales
  de Madrid, CSIC
28049 Madrid, Spain
Email: Gloria.platero@icmm.csic.es

José M. Vega

Universidad Politécnica
  de Madrid
Plaza Cardenal Cisneros 3
28040 Madrid, Spain
Email: vega@fmetsia.upm.es

Library of Congress Control Number: 2007932402

Mathematics Subject Classification (2000): 00B20, 35-xx, 60-xx, 62-xx, 65-xx, 70-xx, 74-xx, 76-xx, 80-xx, 81-xx, 82-xx, 91-xx, 92-xx, 97-xx

ISBN 978-3-540-71991-5 Springer Berlin Heidelberg New York

Springer is a part of Springer Science+Business Media

springer.com

© Springer-Verlag Berlin Heidelberg 2008

Typeset by the editors and SPi using a Springer LaTeX macro-package
Cover design: *design & production* GmbH, Heidelberg
Printed on acid-free paper   SPIN: 12049349   46/2244/YL - 5 4 3 2 1 0

To our beloved ones

Preface

Increasing computing power in the last decades has given mathematical modeling an ever greater impulse and made it a very important tool to solve problems coming from industry. The European Consortium for Mathematics in Industry (ECMI) was founded 20 years ago by mathematicians from ten European universities to foster the use of mathematics to help European industry and commerce to pose and solve their problems. The aims of ECMI are to (a) promote the use of mathematical models and mathematics in industry, (b) form applied mathematicians capable of working effectively in industry and (c) work for these goals at the European scale. Efficient problem solving often requires the use of results in different mathematical fields, yet no single applied mathematician may be able to cover the whole subject. By providing a European research network, ECMI can bring together experts from a wide geographical range.

Since 1986, ECMI has incorporated many more institutions and industries throughout Europe and it has been consolidated as a brand name for Industrial Mathematics. Twenty years later, the biannual ECMI conference was celebrated for the first time in Spain, at the Universidad Carlos III de Madrid. This is a young university created in 1989. Technological studies and departments are located at the Leganés campus where the conference was held. Moreover, University Carlos III participates in the Leganés Scientific and Technological Park, together with the Autonomous Region of Madrid and the city of Leganés. They contribute to place Madrid at the forefront of research and development in Spain.

The scientific program covered a wide variety of topics related to technological sectors (aerospace and automotive industry, materials and electronics, information and telecommunication technologies, energy and environment, biology, biotechnology, life sciences, imaging) and to finances and economics. The different origin of participants helped making the conference multi-disciplinary. Active participation of industry was intended, with reasonable success. The present volume includes a part of the contributions to the conference, selected after a refereeing process. It is a pleasure to see that six plenary speakers have submitted papers for this volume. Vincenzo Capasso in his "Alan Tayler" lecture, besides presenting his scientific work on statistical geometric measure applied to medicine and materials science, recalls some of the challenges for Mathematics in Industry listed in the first ECMI brochure produced by Alan Tayler and himself in 1994, relates them to the present situation of an enlarged Europe, and tells us how these challenges remain important and pressing for us today. Antonio Barrero (Seville), Alfredo Bermúdez (Santiago), Russel Caflisch (UCLA), Luis Campos (Lisbon) and Pierre Degond (Tolouse) illustrate with their contributions the breadth of applications and variety of techniques that are embraced by ECMI. ECMI's commitment to educating students in Industrial Mathematics is reflected in the fact that many papers were given by students. The Wacker Prize, offered for a Master's Level thesis on an industrial problem was awarded to Filippo Terragni, in line with the tradition of excellent work by previous winners. Many of the minisymposia and special sessions included the activities of ECMI Special Interest Groups. Of the 35 minisymposia organized for the conference, many are gathered in this book, usually preceded by a short explanation about their contents. A number of contributed papers complete the volume. I hope that these proceedings will contribute both to show interesting and relevant mathematical problems and methods, and to strengthen cooperation between academia and industry, the absence of which is a major weakness of the European Science-Technology system.

As President of ECMI and on behalf of the ECMI Council, I wish to thank all those who have contributed to the success of the Conference. Among them the participants, the speakers, the International Scientific Committee and the National and Local Organizing Committees. Organizing this meeting has been possible thanks to the efforts of many people both at the Spanish national and local level to whom we are very grateful. In particular all the members of the Modeling, Simulation and Industrial Mathematics Group at Universidad Carlos III worked hard to run a smooth and successful conference which would not have been possible without their help. The dedication of our university congress bureau, Congrega, was also essential for the conference success. Ms. Bárbara Tapiador's help was very important to process the manuscripts that are gathered in the present book. I am grateful to my co-editors, Gloria Platero, Miguel Moscoso and José Manuel Vega for their invaluable help.

Lastly, the support of our sponsors is gratefully acknowledged: Ministerio de Educación y Ciencia (grant MTM-2005-24569-E), Comunidad de Madrid (grant S-0505/ENE/0229), Universidad Carlos III de Madrid, Universidad Politécnica de Madrid, Consejo Superior de Investigaciones Científicas (CSIC), Instituto de Tecnológico de Química y Materiales "Álvaro Alonso Barba", Ayuntamiento de Leganés and Springer.

Madrid, May 2007   *Luis L. Bonilla, President of ECMI*

Contents

Part II Minisymposia

Part III Contributed Papers

Part I

Plenary Lectures

On the Mean Geometric Densities of Random Closed Sets, and Their Estimation: Application to the Estimation of the Mean Density of Inhomogeneous Fibre Processes

Vincenzo Capasso and Alessandra Micheletti

Department of Mathematics, Università degli Studi di Milano, Via C. Saldini, 50, 20133 Milano, Italia
Vincenzo.Capasso@unimi.it, Alessandra.Micheletti@unimi.it

Dedicated to Alan Tayler

Preface [VC]

It has been a great honour for me to deliver the "Alan Tayler Lecture" in this ECMI Conference, to honour one of the leading founders and Presidents of ECMI. I have collaborated with Alan for many years, especially during my term as Chairman of the Educational Committee, and later during the first ECMI-HCM Project. While he was already very ill, he found the way to participate (even though only for a couple of days) in a workshop in Milan, opening ECMI to the Italian academic and industrial community, and highly supported the birth of MIRIAM (the Milan Research Centre for Industrial and Applied Mathematics).

I had a rewarding experience around the early 1990s producing, in a strict collaboration with Alan, the first ECMI Brochure [CT94] (see the ECMI web site) in order to advertise the specific role of ECMI within academia and industry in Europe.

It was clear to me that he had a vision of how to establish in Europe a cooperative action by the most active groups in the applications of mathematics to real world problems; I wish to remind the key issues stated in the brochure, since I may claim that these are still update.

"Realising the need of interaction between universities and research groups in industry, the European Consortium for Mathematics in Industry (ECMI) was founded in 1986 by mathematicians from ten European universities.

. . .

Mathematics, as the language of the sciences, has always played an important role in technology, and now is applied also to a variety of problems in commerce and the environment.

European industry is increasingly becoming dependent on high technology and the need for mathematical expertise in both research and development can only grow.

. . .

These new demands on mathematics have stimulated academic interest in Industrial Mathematics and many mathematical groups world-wide are committed to interaction with industry as part of their research activities.

In 1986 ten of these groups in Europe founded ECMI with the intention of offering their collective knowledge and expertise to European Industry.

The experience of ECMI members is that similar technical problems are encountered by different companies in different countries. It is also true that the same mathematical expertise may often be used in differing industrial applications.

If European industry is to compete in world markets it should take advantage of the competitive edge which may be gained from using European mathematical expertise.

No single European country is likely to have sufficient expertise of mathematical knowledge whereas ECMI can provide a comprehensive coverage of mathematical skills and their diverse applications." [CT94]

We are now facing the challenge of a larger European Union.

Alan had anticipated this by promoting an ECMI "patronage", financially supported by the EU, of those countries usually called "Central Europe", such as Čekia, Hungary, Poland, Romania, Slovakia.

I am sure that he would have liked to participate in the process of complete integration of all the new entries in the ECMI system.

Going back to the ECMI Brochure, a major scope of ECMI was identified as follows.

"C. TO OPERATE ON A EUROPEAN SCALE

Academic resources in Mathematics for Industry are also scarce and distributed across Europe; industrial needs are widely spread. Exchange and interactive programmes are necessary in training, research and industrial collaboration if there is to be an effective transfer of knowledge and skills. The EC is encouraging ECMI to involve relevant groups in Eastern Europe as Associate members."

As part of this encouragement, the EC provided funds to ECMI for organising a series of workshops in those countries, in collaboration with recognised colleagues at the local level. Thus anticipating the enlargement of the political Europe.

In my opinion, having the EC approved a significant enlargement of Europe towards East, listing soon 27 member states, ECMI, as an enlarged Consortium, should find new ways to exploit the best of the scientific resources of the old and the new member states together, to actively participate in the building up of a common competitive Europe. As far as scientific competence is concerned, there are excellencies in all regions of Europe, some of them well identifiable also in the new member states; a genuine will to sustain competence of Europe should go through ways to exploit all of them, with the usual ECMI cooperative attitude.

Another anticipation envisaged by Alan has been the shift of meaning of the key word "Industry" in the ECMI system.

"This collaboration may also be extended to developing mathematical models for the environment, earth sciences, biology and finance." [CT94]

We have already achieved the inclusion of what we call **Economathematics**, and today we are facing a further shift of attention towards **Medicine and Biotechnology**.

All over the world leading experts of **Mathematics for/in Industry**, are participating actively in the development of **Mathematics for/in Medicine**, thus undertaking the further challenge of contributing to the development of innovative methods for diagnosis and treatment of relevant diseases, from cancer to infectious diseases.

My own presentation here is aimed to showing an example of how mathematics, originally developed for mining industry or more in general for material science and chemical industry, is now moving to deal with problems of interest in medicine.

At first this research was motivated by polymer industry in Europe, and constitutes one of the most important success stories of collaborative research within ECMI, that was supported within the first HCM Project coordinated by Alan Tayler. As a documentation of the cooperation between different research teams in Europe within the ECMI Special Interest Group on "Polymers", the volume "Mathematical Modelling for Polymer Processing. Polymerization, Crystallization, Manufacturing", edited by myself, was published as Volume 2 in the **ECMI Series on Mathematics in Industry by Springer-Verlag, Heidelberg 2002**, showing an additional success story of ECMI: the start of the **Springer Series on Mathematics in Industry**.

1 Introduction

Many processes of biomedical or material science interest may be modelled as birth-and-growth processes (germ–grain models), which are composed of two processes, birth (nucleation, branching, etc.) and subsequent growth of spatial structures (cells, vessel networks, etc.), which, in general, are both stochastic in time and space. These structures induce a random division of the relevant spatial region, known as random tessellation (see Fig. 1). A quantitative description of the spatial structure of a tessellation can be given, in terms of the mean densities of interfaces (n-facets).

In applications to material science a main industrial interest is controlling the quality of the relevant final product in terms of its mechanical properties; as shown, e.g. in [FC98], these are strictly related to the final morphology

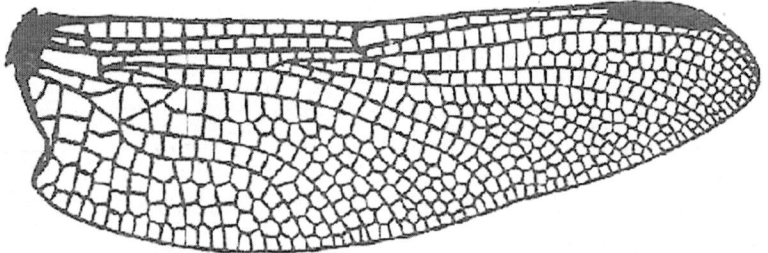

Fig. 1. The spatial tessellation generated by vessels in a dragonfly wing

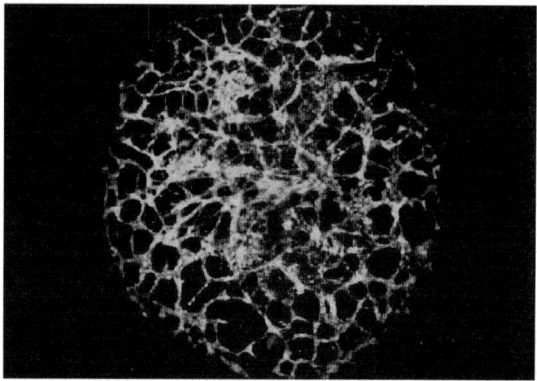

Fig. 2. Vascularization of an allantoid [Credit: Dejana et al. 2005]

of the solidified material, so that quality control in this case means optimal control of the final morphology.

In medicine, an important area of application of birth-and-growth processes and other models of stochastic geometry is tumour-induced angiogenesis. It can be modelled as a fibre process of Hausdorff dimension 1 in the relevant 2D or 3D space.

Tumour-induced angiogenesis is believed to occur when normal tissue vasculature is no longer able to support growth of an avascular tumour. At this stage the tumour cells, lacking nutrients and oxygen, become hypoxic. This is assumed to trigger cellular release of tumour angiogenic factors (TAFs) which start to diffuse into the surrounding tissue and approach endothelial cells (ECs) of nearby blood vessels. ECs subsequently respond to the TAF concentration gradients by forming sprouts, dividing, and migrating towards the tumour. A summary of these mechanisms can be found in the recent paper by Carmeliet [JK01] (see also Figs. 2–4 where examples of real or simulated vascular networks are depicted).

Initially, the sprouts arising from a parent vessel grow essentially parallel to each other. It is observed that once the finger-like capillary sprouts have

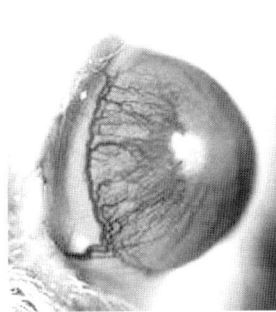
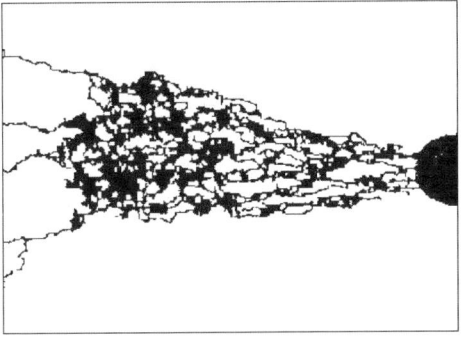

Fig. 3. *Left*: Angiogenesis on a rat cornea [Credit: Dejana et al. 2005]. The white spot is a pellet implanted in the cornea containing an angiogenetic substance, emulating the effect of a tumour. *Right*: A simulation of an angiogenesis due to a localized tumour mass (*black* region on the *right*) (from [CA99])

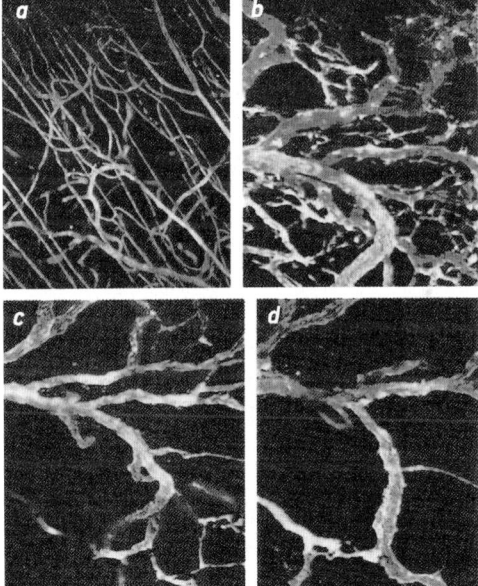

Fig. 4. Response of a vascular network to an antiangiogenic treatment (from [JK01])

reached a certain distance from the parent vessel, they tend to incline towards each other, leading to fusions called anastomoses. Such fusions lead to a network of vessels. On the other hand the sprout branching dramatically increases while approaching the tumour mass, eventually resulting in vascularization.

The coupling of the branching and growth process to the underlying chemical gradients is limited by the local density of the existing capillary network, thus leading to a mathematical strong coupling of this density and the kinetic parameters of the branching and growth process.

The study of angiogenesis has such potential for providing new therapies that it has received enthusiastic interest from the pharmaceutical and biotechnology industries. Indeed, dozens of companies are now pursuing angiogenesis-related therapies, and approximately 20 compounds that either induce or block vessel formation are being tested in humans. Although such drugs can potentially treat a broad range of disorders, many of the compounds now under investigation inhibit angiogenesis and target cancer. Intriguingly, animal tests show that inhibitors of vessel growth can boost the effectiveness of traditional cancer treatments (chemotherapy and radiation). Preliminary studies also hint that the agents might one day be delivered as a preventive measure to block malignancies from arising in the first place in people at risk for cancer.

In developing mathematical models of angiogenesis, the hope is to be able to provide a deeper insight into the underlying mechanisms which cause the process. It is therefore essential that predictive mathematical models are developed, capable of producing precise quantitative morphological features of developing blood vessels. Such models might be used for predicting the evolution of tumours (prognosis), and identifying optimal control strategies (medical treatment).

Unfortunately, a satisfactory modelling of angiogenesis requires a theory of stochastic fibre processes, evolving in time, and strongly coupled with underlying fields. In this case the theory of birth-and-growth processes (or branching-and-growth processes), developed for volume growth, cannot be applied to analyse realistic models, due to intrinsic mathematical difficulties, coming from the dependence of the kinetic parameters from the geometric spatial densities of the existing tumour, or capillary network itself [CM05, McDou06].

All these aspects induce stochastic time and space heterogeneities, thus motivating a more general analysis of the stochastic geometry of the process. The formulation of an exhaustive evolution model which relates all the relevant features of a real phenomenon dealing with different scales, and a stochastic domain decomposition at different Hausdorff dimensions, is a problem of high complexity, both analytical and computational.

Anyway statistical methods for the estimation of geometric densities may offer significant tools for diagnosis and dose/response analysis in medical treatments.

In the modelling of the above-mentioned systems it is of great importance to handle random closed sets of different (even though integer) Hausdorff dimensions. Following a standard approach in geometric measure theory, such sets may be described in terms of suitable measures. For a random closed set of lower dimension with respect to the environment space, the relevant measures induced by its realizations are singular with respect to the Lebesgue measure, and so their usual Radon–Nikodym derivatives are zero almost everywhere.

In Sect. 2 an original approach is reported, recently proposed by the research group of the authors, who have suggested to cope with these difficulties by introducing generalized densities (distributions) *á la Dirac–Schwartz*, for both the deterministic case and the stochastic case. In this last one, mean generalized densities are of interest.

These instruments may then help to formulate stochastic models (that is solving direct problems) for the over-mentioned applications; they also suggest methods for the solution of the related inverse problems, including methods of statistical analysis for the estimation of geometric densities of a stochastic fibre process that characterize the morphology of a real system. We apply such methods to real data, taken from the literature, and to simulated data, obtained by existing computational models of tumour-induced angiogenesis.

These methods can be used for validating computational models, and for monitoring the efficacy of possible medical treatment.

1.1 Nomenclature

We remind that a *random closed set (RACS)* Ξ in $\mathbb{R}^d$ is a measurable map

$$\Xi : (\Omega, \mathcal{F}, \mathbb{P}) \longrightarrow (\mathbb{F}, \sigma_{\mathbb{F}}),$$

where $\mathbb{F}$ denotes the class of the closed subsets in $\mathbb{R}^d$, and $\sigma_{\mathbb{F}}$ is the σ-algebra generated by the so-called *hit-or-miss topology* (see [Mat75]).

The theory of Choquet–Matheron shows that it is possible to assign a unique probability law associated with a *RACS* Ξ in $\mathbb{R}^d$ on the measurable space $(\mathbb{F}, \sigma_{\mathbb{F}})$ by assigning its *hitting functional* T_{Ξ}.

This is defined as

$$T_{\Xi} : K \in \mathcal{K} \longmapsto P(\Xi \cap K \neq \emptyset),$$

where $\mathcal{K}$ denotes the family of compact sets in $\mathbb{R}^d$.

Actually we may consider, equivalently, the restriction of T_{Ξ} to the family of closed balls $\{B_{\varepsilon}(x); x \in \mathbb{R}^d, \varepsilon \in \mathbb{R}_+ - \{0\}\}$.

In dependence of its regularity, a random closed set Θ_n with Hausdorff dimension n (i.e. $\dim_{\mathcal{H}} \Theta_n(\omega) = n$ for a.e. $\omega \in \Omega$), may induce a random Radon measure

$$\mu_{\Theta_n}(\cdot) := \mathcal{H}^n(\Theta_n \cap \cdot)$$

on $\mathbb{R}^d$ ($\mathcal{H}^n$ is the n-dimensional Hausdorff measure), and, as a consequence, an *expected measure*

$$\mathbb{E}[\mu_{\Theta_n}](\cdot) := \mathbb{E}[\mathcal{H}^n(\Theta_n \cap \cdot)]$$

(for a discussion about measurability of $\mathcal{H}^n(\Theta_n)$ we refer to [BM97, Z82]).

In several real applications, it is of interest to study the density (said *mean density*) of the measure $\mathbb{E}[\mu_{\Theta_n}]$ [BR04], and, in the dynamical case, its evolution in time [Mol92, Mol94]. Here we present a synthesis of a theory of random distributions as generalized densities of random measures, and mean geometric densities as expected values of random generalized densities, as proposed in [CV06c]. In particular we introduce a *Delta formalism*, á la Dirac–Schwartz, for the description of random measures associated with random closed sets of lower dimensions, such that the well known usual Dirac delta at a point follows as a particular case (see, for instance, [Jones82,KF70,Vlad79]).

In dealing with mean densities, a concept of *absolutely continuous random closed set* arises in a natural way in terms of the expected measure; indeed, an interesting property of a random set in $\mathbb{R}^d$ is whether the expected measure induced by the random set is absolutely continuous or not with respect to the d-dimensional Lebesgue measure ν^d. Thus, it is of interest to distinguish between random closed sets which induce an absolutely continuous expected measure, and random closed sets which induce a singular one. To this aim we introduce definitions of *discrete*, *continuous*, and *absolutely continuous* random closed set, coherently with the classical 0-dimensional case, in order to propose an extension of the standard definition of discrete, continuous, and absolutely continuous random variable, respectively (see also [CV06a, CV06b]).

2 Generalized Densities

In the sequel we will refer to a class of sufficiently regular random closed sets in the Euclidean space $\mathbb{R}^d$, of integer dimension n.

Definition 1 (n-regular set). *Given an integer $n \in [0, d]$, we say that a closed subset S of $\mathbb{R}^d$ is n-regular, if it satisfies the following conditions:*

(i) $\mathcal{H}^n(S \cap B_R(0)) < \infty$ *for any* $R > 0$

(ii) $\lim\limits_{r \to 0} \dfrac{\mathcal{H}^n(S \cap B_r(x))}{b_n r^n} = 1$ *for* $\mathcal{H}^n$*-a.e.* $x \in S$

Here b_n denotes the volume of the unit ball in $\mathbb{R}^n$.

Remark 1. Note that condition (ii) is related to a characterization of the $\mathcal{H}^n$-rectifiability of the set A ([Fal85], p. 256, 267, [AFP00], p. 83).

We may observe that if A_n is an n-regular closed set in $\mathbb{R}^d$, we have

$$\lim_{r \to 0} \frac{\mathcal{H}^n(A_n \cap B_r(x))}{b_n r^n} = \begin{cases} 1 & \mathcal{H}^n\text{-a.e. } x \in A_n, \\ 0 & \forall x \notin A_n; \end{cases}$$

as a consequence (by assuming $0 \cdot \infty = 0$), for $0 \leq n < d$ we have

$$\lim_{r \to 0} \frac{\mathcal{H}^n(A_n \cap B_r(x))}{b_d r^d} = \lim_{r \to 0} \frac{\mathcal{H}^n(A_n \cap B_r(x))}{b_n r^n} \frac{b_n r^n}{b_d r^d}$$

$$= \begin{cases} \infty & \mathcal{H}^n\text{-a.e. } x \in A_n, \\ 0 & \forall x \notin A_n. \end{cases}$$

It is well known that every positive Radon measure μ on $\mathbb{R}^d$ can be decomposed as

$$\mu = \mu_{\ll} + \mu_{\mathbb{P}\text{erp}},$$

where $\mu_{\ll}$ and $\mu_{\mathbb{P}\text{erp}}$ are the absolutely continuous, and the singular parts of μ, respectively, with respect to ν^d, the usual Lebesgue measure on $\mathbb{R}^d$.

It then follows that $\mu_{\ll}$ admits a (nontrivial) Radon–Nikodym derivative with respect to ν^d, which is known as its density; while the Radon–Nikodym derivative of $\mu_{\mathbb{P}\text{erp}}$, with respect to ν^d, would be zero ν^d- a.e.

Anyhow in analogy with the usual Dirac delta function $\delta_{x_0}(x)$ associated with a point $x_0 \in \mathbb{R}^d$ (a 0-regular closed set), a density can be introduced also for $\mu_{\mathbb{P}\text{erp}}$, in a generalized sense, according to Definition 2 [KF70].

Definition 2 (Generalized density). *We call $\delta_{\mu_{\mathbb{P}\text{erp}}}$, the generalized density (or, briefly, the density P) of $\mu_{\mathbb{P}\text{erp}}$, the quantity*

$$\delta_{\mu_{\mathbb{P}\text{erp}}}(x) := \lim_{r \to 0} \frac{\mu_{\mathbb{P}\text{erp}}(B_r(x))}{b_d r^d},$$

finite or not.

Clearly, if A_n is an n-regular closed set in $\mathbb{R}^d$ with $n < d$, then the measure

$$\mu_{A_n}(\cdot) := \mathcal{H}^n(A_n \cap \cdot)$$

is a singular measure with respect to ν^d. Based on Definition 1, the quantity

$$\delta_{A_n}(x) := \lim_{r \to 0} \frac{\mathcal{H}^n(A_n \cap B_r(x))}{b_d r^d},$$

(finite or not), can now be introduced as the (generalized) density associated with A_n.

With an abuse of notations, we may introduce the linear functional δ_{A_n} associated with the measure μ_{A_n}, as follows:

$$(\delta_{A_n}, f) := \int_{\mathbb{R}^d} f(x) \mu_{A_n}(dx),$$

for any $f \in C_c(\mathbb{R}^d, \mathbb{R})$, having denoted by $C_c(\mathbb{R}^d, \mathbb{R})$ the space of all continuous functions from $\mathbb{R}^d$ to $\mathbb{R}$ with compact support. In accordance with the usual representation of distributions in the theory of generalized functions, we formally write

$$\int_{\mathbb{R}^d} f(x) \delta_{A_n}(x) \, dx := (\delta_{A_n}, f).$$

Define the function

$$\delta_{A_n}^{(r)}(x) := \frac{\mathcal{H}^n(A_n \cap B_r(x))}{b_d r^d},$$

and correspondingly the associated measure

$$\mu_{A_n}^{(r)}(B) := \int_B \delta_{A_n}^{(r)}(x)\,\mathrm{d}x, \qquad B \in \mathcal{B}_{\mathbb{R}^d}.$$

As above, we may introduce the linear functional $\delta_{A_n}^{(r)}$ associated with the measure $\mu_{A_n}^{(r)}$, as follows:

$$\left(\delta_{A_n}^{(r)}, f\right) := \int_{\mathbb{R}^d} f(x)\mu_{A_n}^{(r)}(\mathrm{d}x),$$

It can be proven (see [CV06c]) that the sequence of measures $\mu_{A_n}^{(r)}$ weakly* converges to the measure μ_{A_n}; in other words, the sequence of linear functionals $\delta_{A_n}^{(r)}$ weakly* converges to the linear functional δ_{A_n}, i.e. $(\delta_{A_n}^{(r)}, f) \to (\delta_{A_n}, f)$ for any $f \in C_c(\mathbb{R}^d, \mathbb{R})$.

Consider now random closed sets.

Definition 3 (n-regular random set). *Given an integer n, with $0 \le n \le d$, we say that a random closed set Θ_n in $\mathbb{R}^d$ is n-regular, if it satisfies the following conditions:*

(i) For almost all $\omega \in \Omega$, $\Theta_n(\omega)$ is an n-regular set in $\mathbb{R}^d$
(ii) $\mathbb{E}[\mathcal{H}^n(\Theta_n \cap B_R(0))] < \infty$ for any $R > 0$

If Θ_n is a random closed set in $\mathbb{R}^d$, the measure

$$\mu_{\Theta_n}(\cdot) := \mathcal{H}^n(\Theta_n \cap \cdot)$$

is a random measure, and consequently δ_{Θ_n} is a *random linear functional* (i.e. (δ_{Θ_n}, f) is a real random variable for any test function f).

By extending the definition of expected value of a random operator à la Pettis (or Gelfand–Pettis) [AG80, Bosq00], we may define the *expected linear functional* $\mathbb{E}[\delta_{\Theta_n}]$ associated with δ_{Θ_n} as follows:

$$(\mathbb{E}[\delta_{\Theta_n}], f) := \mathbb{E}[(\delta_{\Theta_n}, f)] \qquad (1)$$

and the *mean generalized density* $\mathbb{E}[\delta_{\Theta_n}](x)$ of $\mathbb{E}[\mu_{\Theta_n}]$ by the formal integral representation:

$$\int_A \mathbb{E}[\delta_{\Theta_n}](x)\,\mathrm{d}x := \mathbb{E}[\mathcal{H}^n(\Theta_n \cap A)],$$

with

$$\mathbb{E}[\delta_{\Theta_n}](x) := \lim_{r \to 0} \frac{\mathbb{E}[\mathcal{H}^n(\Theta_n \cap B_r(x))]}{b_d r^d}.$$

It can be shown [CV06c] that an equivalent definition of (1) can be given in terms of the expected measure $\mathbb{E}[\mu_{\Theta_n}]$ by

$$(\mathbb{E}[\delta_{\Theta_n}], f) := \int_{\mathbb{R}^d} f(x)\mathbb{E}[\mu_{\Theta_n}](\mathrm{d}x),$$

for any f such that the above integral makes sense.

By using the integral representation of (δ_{Θ_n}, f) and $(\mathbb{E}[\delta_{\Theta_n}], f)$, (1) becomes

$$\int_{\mathbb{R}^d} f(x)\mathbb{E}[\delta_{\Theta_n}](x)\,\mathrm{d}x = \mathbb{E}\left[\int_{\mathbb{R}^d} f(x)\delta_{\Theta_n}(x)\,\mathrm{d}x\right];$$

so that, formally, we may exchange integral and expectation.

Remark 2. When $n = d$, integral and expectation can be really exchanged by Fubini's theorem. Since in this case $\delta_{\Theta_d}(x) = \mathbf{1}_{\Theta_d}(x)$, ν^d-a.s., it follows that $\mathbb{E}[\delta_{\Theta_d}](x) = \mathbb{P}(x \in \Theta_d)$. In particular, in material science, the density $V_V(x) := \mathbb{P}(x \in \Theta_d)$ is known as the (*degree of*) *crystallinity*.

If $n = 0$ and $\Theta_0 = X_0$ is an absolutely continuous random point with p.d.f. p_{X_0}, then $\mathbb{E}[\mathcal{H}^0(X_0 \cap \cdot)] = \mathbb{P}(X_0 \in \cdot)$ is absolutely continuous, and its density $\mathbb{E}[\delta_{X_0}](x)$ is just the probability density function $p_{X_0}(x)$.

Thus, for any lower dimensional random closed set Θ_n in $\mathbb{R}^d$, while it is clear that $\mu_{\Theta_n(\omega)}$ is a singular measure, when we consider the expected measure $\mathbb{E}[\mu_{\Theta_n}]$, it may happen that it is absolutely continuous with respect to ν^d, thus having a classical Radon–Nikodym derivative, so that $\mathbb{E}[\delta_{\Theta_n}](x)$ is a classical real-valued integrable function on $\mathbb{R}^d$ (see [CV06c], and [CV06a]). It is then of interest to say whether or not a classical mean density can be introduced for sets of lower Hausdorff dimensions, with respect to the usual Lebesgue measure on $\mathbb{R}^d$. In order to respond to this further requirement, in [CV06a] we have proposed a concept of absolute continuity for random closed sets.

To avoid pathologies, as discussed in [ACaV06] (see also [CV06d]), we introduce now a class of random sets, which, in particular, include all random sets we are interested in the sequel.

Definition 4 ($\mathcal{R}$ class). *We say that a random closed set Θ in $\mathbb{R}^d$ belongs to the class $\mathcal{R}$ if*

$$\dim_{\mathcal{H}}(\mathrm{Partial}\Theta) < d \quad and \quad \mathbb{P}(\mathcal{H}^{\dim_{\mathcal{H}}(\mathrm{Partial}\Theta)}(\mathrm{Partial}\Theta) > 0) = 1.$$

Definition 5 (Absolute continuity). *We say that a random closed set $\Theta \in \mathcal{R}$ is* (strongly) *absolutely continuous if*

$$\mathbb{E}[\mu_{\mathrm{Partial}\Theta}] \ll \nu^d \tag{2}$$

on $\mathcal{B}_{\mathbb{R}^d}$.

Remark 3. Note that, if $\Theta \in \mathcal{R}$ with $\dim_{\mathcal{H}}(\Theta) = d$ is sufficiently regular so that $\dim_{\mathcal{H}}(\mathrm{Partial}\Theta) = d - 1$, then it is absolutely continuous if

$$\mathbb{E}[\mathcal{H}^{d-1}(\mathrm{Partial}\Theta \cap \cdot)] \ll \nu^d(\cdot).$$

Remark 4. In the particular case that $\Theta = X$ is a random variable, Definition 5 coincides with the usual definition of absolute continuity of a random variable.

In fact, $\dim_{\mathcal{H}} X = 0$, $\mathbb{Partial} X = X$, and $\mathbb{E}[\mathcal{H}^0(X)] = \mathbb{P}(X \in \mathbb{R}^d) = 1$, so $X \in \mathcal{R}$ and then Condition (2) is equivalent to

$$\mathbb{E}[\mathcal{H}^0(X \cap \cdot)] = \mathbb{P}(X \in \cdot) \ll \nu^d.$$

To conclude this section, we may then claim that, if Θ_n, with $0 < n < d$, is an absolutely continuous random closed set, then $\mathbb{E}[\mu_{\Theta_n}] \ll \nu^d$, so that its local mean density $\mathbb{E}[\delta_{\Theta_n}](x)$ is a classical real-valued integrable function on $\mathbb{R}^d$.

3 Approximation of Mean Densities

In many real applications, it is of interest the estimation of the local mean density $\mathbb{E}[\delta_{\Theta_n}]$ of an absolutely continuous lower dimensional random closed set such as a fibre process of dimension $n = 1$ in a space of dimension $d > 1$ (see, e.g. [BR04] and [SKM95]).

For facing the problem of the zero ν^2-measure for points or lines in $\mathbb{R}^2$ it is natural to make use of a 2-D box approximation of points or lines. As a matter of fact, a computer graphic representation of them is anyway provided in terms of pixels, which can only offer a 2-D box approximation of points in $\mathbb{R}^2$. This is the motivation of this and the following sections, which tend to suggest estimators for local mean densities of absolutely continuous random closed sets of lower dimensions in a given d-dimensional space [ACaV06].

Given a random closed set Θ_n with Hausdorff dimension n, we consider the enlarged set $\Theta_{n_{\oplus r}}$, which is now of dimension d, and hence of nontrivial measure ν^d. We observe that $\mathbb{P}(x \in \Theta_{n_{\oplus r}}) = T_{\Theta_n}(B_r(x))$.

Proposition 1. *[ACaV06] Let Θ_n be a random closed set with Hausdorff dimension n, and $A \in \mathcal{B}_{\mathbb{R}^d}$ such that $\mathbb{P}(\mathcal{H}^n(\Theta_n \cap \mathbb{Partial} A) > 0) = 0$. If*

$$\lim_{r \to 0} \frac{\mathbb{E}[\nu^d(\Theta_{n_{\oplus r}} \cap A)]}{b_{d-n} r^{d-n}} = \mathbb{E}[\mathcal{H}^n(\Theta_n \cap A)], \tag{3}$$

then

$$\mathbb{E}[\mathcal{H}^n(\Theta_n \cap A)] = \lim_{r \to 0} \int_A \frac{T_{\Theta_n}(B_r(x))}{b_{d-n} r^{d-n}} \, \mathrm{d}x.$$

Sufficient conditions for (3) have been given in [ACaV06].

As a consequence of Proposition 1, if we denote by $\mu^{\oplus r}$ the measure on $\mathcal{B}_{\mathbb{R}^d}$ defined by

$$\mu^{\oplus r}(A) := \int_A \frac{T_{\Theta_n}(B_r(x))}{b_{d-n} r^{d-n}} \, \mathrm{d}x,$$

then it follows that $\mu^{\oplus r}$ weakly* converges to $\mathbb{E}[\mu_{\Theta_n}]$.

For every fixed $r > 0$, the measure $\mu^{\oplus r}$ is absolutely continuous with respect to the d-dimensional Lebesgue measure with density

$$\delta_n^{\oplus r}(x) := \frac{T_{\Theta_n}(B_r(x))}{b_{d-n}r^{d-n}}.$$

Such a function defines a linear functional, say $\delta_n^{\oplus r}$, associated with the measure $\mu^{\oplus r}$ as follows

$$(\delta_n^{\oplus r}, f) := \int_{\mathbb{R}^d} f(x)\mu^{\oplus r}(\mathrm{d}x).$$

Note that many kinds of random closed sets satisfy the proposition above, like fibre processes, line and segment processes, Boolean models, etc. (see [ACaV06]). As a consequence, estimating the probability that the random set Θ_n intersects the ball $B_r(x)$ may suggest (global) estimators of $\mathbb{E}[\mu_{\Theta_n}]$, and possibly (local) estimators of the mean density $\mathbb{E}[\delta_{\Theta_n}]$ (see, e.g. [BR04]).

If Θ_n is absolutely continuous, then there exists an integrable function λ_{Θ_n} (the Radon–Nikodym derivative) such that, for all $A \in \mathcal{B}_{\mathbb{R}^d}$,

$$\mathbb{E}[\mathcal{H}^n(\Theta_n \cap A)] = \int_A \lambda_{\Theta_n}(x)\,\mathrm{d}x.$$

So, in this case, we have that

$$\lim_{r \to 0} \int_A \frac{T_{\Theta_n}(B_r(x))}{b_{d-n}r^{d-n}}\mathrm{d}x = \int_A \lambda_{\Theta_n}(x)\,\mathrm{d}x. \qquad (4)$$

If Θ_n is a stationary random closed set, then $\delta_n^{\oplus r}(x)$ is independent of x and the expected measure $\mathbb{E}[\mu_{\Theta_n}]$ is motion invariant, i.e. it is absolutely continuous with density $\lambda_{\Theta_n}(x) = L \in \mathbb{R}_+$ for ν^d-a.e. $x \in \mathbb{R}^d$. It follows that

$$\lim_{r \to 0} \int_A \frac{T_{\Theta_n}(B_r(x))}{b_{d-n}r^{d-n}}\mathrm{d}x = \lim_{r \to 0} \frac{T_{\Theta_n}(B_r(0))}{b_{d-n}r^{d-n}}\nu^d(A),$$

and

$$\int_A \lambda(x)\,\mathrm{d}x = L\nu^d(A);$$

and so, by (4),

$$\lim_{r \to 0} \frac{T_{\Theta_n}(B_r(0))}{b_{d-n}r^{d-n}} = L.$$

Remark 5. When it is possible to exchange limit and integral in (4), by Proposition 1 we may claim that

$$\lim_{r \to 0} \frac{T_{\Theta_n}(B_r(x))}{b_{d-n}r^{d-n}} = \lambda_{\Theta_n}(x) \qquad \nu^d\text{-a.e. } x \in \mathbb{R}^d.$$

In the particular case $n = d$, we know that the measure $\mathbb{E}[\mu_{\Theta_d}]$ is always absolutely continuous with density $\lambda_{\Theta_d}(x) = \mathbb{P}(x \in \Theta_d)$. We may notice that $\delta_d^{\oplus r} = T_{\Theta_n}(B_r(x))$ and by Monotone Convergence Theorem we can exchange limit and integral, and so we have, as expected,

$$\lim_{r \to 0} T_{\Theta_d}(B_r(x)) = \mathbb{P}(x \in \Theta_d) = \lambda_{\Theta_d}(x).$$

Further, for $n = 0$, if $\Theta_0 = X$ is a random point in $\mathbb{R}^d$, we have $\mathbb{E}[\mathcal{H}^0(X \cap \cdot)] = \mathbb{P}(X \in \cdot)$. So, if X is absolutely continuous with probability density function f, we know that $\mathbb{E}[\mu_X] = \mathbb{P}_X$ is absolutely continuous with density f. In this case it can be shown that (3) holds, so that the sequence $\{\delta^{\oplus r}(x)\}$ converges to $f(x)$, as expected, which leads to the usual histogram estimation of $f(x)$ [ACaV06].

Example 1. As an additional example of applicability of the results above, let us consider the case in which Θ_n is given by a random union of absolutely continuous random closed sets of dimension $n < d$:

$$\Theta_n = \bigcup_{i=1}^{\Phi} E_i,$$

where Φ is a nonnegative discrete random variable with $\mathbb{E}[\Phi] < \infty$, and the E_i's are IID as E and independent of Φ. Then it follows that [ACaV06]

$$\lim_{r \to 0} \frac{T_{\Theta_n}(B_r(x))}{b_{d-n} r^{d-n}} = \mathbb{E}[\Phi] \lim_{r \to 0} \frac{T_E(B_r(x))}{b_{d-n} r^{d-n}},$$

provided that at least one of the two limits exists.

As a consequence, when it is possible to exchange limit and integral in (4), and so in particular when E is a stationary random closed set (which implies Θ_n stationary as well), we have

$$\lambda_{\Theta_n}(x) = \mathbb{E}[\Phi] \lim_{r \to 0} \frac{T_E(B_r(x))}{b_{d-n} r^{d-n}} = \mathbb{E}[\Phi] \lambda_E(x),$$

where λ_{Θ_n} and λ_E are the Radon–Nikodym derivatives of μ_{Θ_n} and μ_E, respectively. The above model may be used as a preliminary one for angiogenesis [CM05], but also for the earthworm burrow system in a soil [BR04, p.73].

4 Statistical Methods for Fibre Systems

We will here consider random fibre systems generated by Boolean models having a fibre as primary grain, that is a RACS Γ such that

$$\Gamma = \cup_{i \in \mathbb{N}} \Gamma_i \oplus x_i,$$

where

- $\{x_i\}_{i \in \mathbb{N}}$ is a spatial Poisson point process, possibly inhomogeneous, with intensity $\alpha(x), x \in \mathbb{R}^2$

- $\{\Gamma_i\}_{i\in\mathbb{N}}$ is a family of i.i.d. random fibres (i.e. random, a.s. bounded, 1-regular sets), passing a.s. through the origin

The resulting Boolean model is thus in general nonstationary and non-isotropic. The source of nonstationarity comes essentially from the nonstationarity of the germ process, i.e. from the *location* of the fibres, and not from intrinsic geometric irregularities of the fibres themselves. In fact the grains are assumed geometrically regular (1-regular) and with "good" statistical properties (i.i.d.). The main source of anisotropy instead comes from the distribution of fibres (grains) orientation, which may be nonuniform.

Note now that

$$T_{\Theta_n}(B_r(x)) = \mathbb{P}(x \in \Theta_{n\oplus r}) = \mathbb{P}(\Theta_n \cap B_r(x) \neq \emptyset)$$

thus we may rewrite Equality (4) in the following way

$$\int_A \lambda_n(x)\mathrm{d}x = \lim_{r\to 0} \int_A \frac{\mathbb{P}(\Theta_n \cap B_r(x) \neq \emptyset)}{b_{d-n}r^{d-n}}\,\mathrm{d}x \tag{5}$$

$$= \lim_{r\to 0} \int_A \frac{T_{\Theta_n}(B_r(x))}{b_{d-n}r^{d-n}}\,\mathrm{d}x \tag{6}$$

$$= \lim_{r\to 0} \int_A \frac{\mathbb{P}(x \in \Theta_{n\oplus r})}{b_{d-n}r^{d-n}}\,\mathrm{d}x. \tag{7}$$

Equalities (5)–(7) provide a way to introduce estimators of $\lambda_n(x)$ when Θ_n is a random fibre, or fibre system Γ, provided that the limit and the integrals in the right-hand terms of (5)–(7) can be exchanged, by estimating the quantities

$$\frac{T_\Gamma(B_r(x))}{b_{d-1}r^{d-1}} = \frac{\mathbb{P}(x \in \Gamma_{\oplus r})}{b_{d-1}r^{d-1}} = \frac{\mathbb{P}(\Gamma \cap B_r(y) \neq \emptyset)}{b_{d-1}r^{d-1}}.$$

We will call them *histogram-like* estimators, since the "enlargement" $\Gamma_{\oplus r}$ of the set Γ via the Minkowski addition of a d-dimensional ball, which approximates the fibre with a d-dimensional set, imitates the procedure used when we estimate the p.d.f. of a real random variable from an i.i.d. sample using moving histograms (see [Hard91, Pest98] for details), where we "enlarge" the Dirac-delta's measures concentrated on the sample points, approximating them with classical and sufficiently regular functions.

In the following we will provide two estimators for the mean geometric density of length, also called *intensity*, of the random fibre system Γ. The intensity can be used to characterize the mean geometric properties of the fibre system. Accordingly with the definitions introduced in the previous sections, the intensity of Γ is defined by

$$\lambda(x)\colon\ = \mathbb{E}[\delta_\Gamma](x) = \lim_{r\to 0} \frac{\mathbb{E}(\mathcal{H}^1(\Gamma \cap B_r(x)))}{r^d b_d}.$$

4.1 Basic Assumptions for the Estimation Procedure

Suppose to have one or more images of the random fibre system Γ under study and that the window $W \subseteq \mathbb{R}^d$ where Γ is observed can be divided in a partition of subwindows $\{A_k\}_{k=1,\ldots,K}$ such that:

A1 $A_j \cap A_k = \emptyset, \quad \forall j \neq k$

A2 $\bigcup_{k=1}^{K} A_k - W$

A3 in each window A_k limit and integral in (4) can be exchanged when $\Theta_n = \Gamma$. This is the case for example if in A_k the fibre system is (*locally*) stationary

A4 the intensity $\lambda(x)$ is sufficiently "smooth" to be locally well approximated by piecewise constant functions, assuming different constant values in each window A_k

We will now introduce a (nonstationary!) example where the previous assumptions are satisfied. The example will be used in the following as a case study for the properties of our estimators.

4.2 An Example of Inhomogeneous Poisson Segment Process

Let $d = 2$ and consider the Boolean model Γ formed by:

- Germs: A spatial nonhomogeneous Poisson point process $\{x_i\}_{i \in \mathbb{N}}, x_i \in \mathbb{R}^2$ having intensity $\alpha(x) = \alpha(x_1, x_2) = c x_1^2$, and c is a constant.
- Grains: A family $\{S_i\}_{i \in \mathbb{N}}$ of (deterministic) closed sets all distributed like the segment $S = [0, l] \times \{0\}$ of fixed length l.

The resulting Boolean model is

$$\Gamma = \bigcup_{i \in \mathbb{N}} S_i \oplus x_i$$

(see Fig. 5 where a realization is depicted). Note that since the germ intensity $\alpha(x)$ is a function of class C^∞, Assumption A4 is trivially satisfied.

Let us assume that the following equality holds

$$\lambda(x) := \lim_{r \to 0} \frac{\mathbb{E}\left(\nu^1(\Gamma \cap B_r(x))\right)}{2r} = \lim_{r \to 0} \frac{\mathbb{E}\left(\nu^1(\Gamma \cap Q_r(x))\right)}{2r}, 0 \qquad (8)$$

where $Q_r(x)$ is a square centred at x with side $2r$. This assumption is reasonable, since both cubes and spheres form a system of generators of the Borel σ-algebra in $\mathbb{R}^d$. Then Assumption A3 is satisfied thanks to the following.

Proposition 2. *Let Γ be the random segment system described above, $\lambda(x)$ be the mean intensity of length of the system, and suppose that Assumption (8) is satisfied, i.e.*

$$\lambda(x) = \lim_{r \to 0} \frac{\mathbb{E}\left(\nu^1(\Gamma \cap Q_r(x))\right)}{2r}.$$

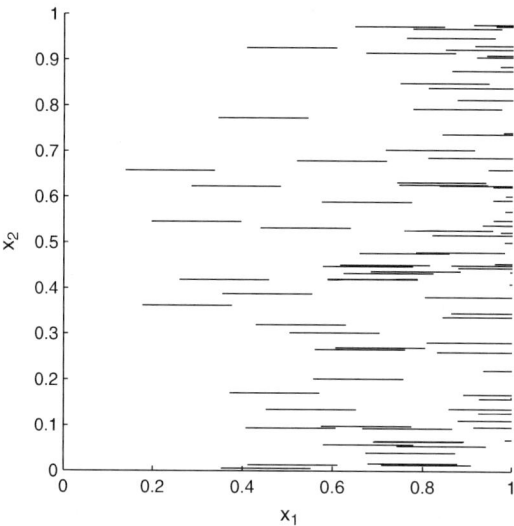

Fig. 5. A realization of the Boolean model in the example. The point process $\{x_i\}$ of germs is formed by the left-hand extremes of the segments

Then, the quantity

$$\bar{\lambda}(x): \ = \lim_{r \to 0} \frac{\mathbb{P}(\Gamma \cap Q_r(x) \neq \emptyset)}{2r}. \tag{9}$$

exists and is finite and we have, a.s.

$$\bar{\lambda}(x) = \lambda(x)$$

for ν^2- almost all $x \in \mathbb{R}^2$.

For the proof of this proposition see [CM06]. Let us remark that, in the proof, the particular functional form of $\bar{\lambda}(x)$ is not relevant.

5 Estimators of the Intensity

In the assumptions stated in Sect. 4.1, for all $x \in A_k$, let us denote by

$$
\begin{aligned}
\lambda_k : &= \lim_{r \to 0} \frac{\mathbb{E}(\nu^1(\Gamma \cap B_r(x)))}{b_d r^d} \\
&= \lim_{r \to 0} \frac{\mathbb{P}(x \in \Gamma_{\oplus r} \cap A_k)}{2r} \\
&= \lim_{r \to 0} \frac{T_\Gamma(B_r(x))}{2r}
\end{aligned}
$$

the (constant) intensity of the random fibre system in the subwindow A_k. We have explicited all the previous equalities since we will obtain different estimators, based on the estimate of the quantities:

1. $\mathbb{P}(x \in \Gamma_{\oplus r} \cap A_k)$
2. $T_\Gamma(B_r(x))$,

respectively.

Let us build first an estimator based on the estimate of $\mathbb{P}(x \in \Gamma_{\oplus r} \cap A_k)$. Let us overlap to A_k a grid of points $z_1, \ldots, z_p \in A_k$ and build the set $\Gamma_{\oplus r} \cap A_k$. Then a first estimator of λ_k is

$$\hat{\lambda}_{k,r,p}^1 := \frac{1}{2rp} \sum_{i=1}^{p} \mathbf{1}_{z_i \in \Gamma_{\oplus r} \cap A_k}, \tag{10}$$

where $\mathbf{1}_{z_i \in \Gamma_{\oplus r} \cap A_k}$ are i.i.d. Bernoulli random variables assuming value one with probability $\mathbb{P}(x \in \Gamma_{\oplus r} \cap A_k)$ which is independent of $x \in A_k$ in our assumptions. Since estimator (10) is the arithmetic mean of these variables, by applying the strong law of large numbers (SLLN) and Slutsky Theorem (see, e.g. [Pest98]) we obtain

$$\mathbb{E}(\hat{\lambda}_{k,r,p}^1) = \frac{\mathbb{P}(x \in \Gamma_{\oplus r} \cap A_k)}{2r} \longrightarrow \lambda_k, \text{ for } r \to 0 \tag{11}$$

$$\text{Var}(\hat{\lambda}_{k,r,p}^1) = \frac{(\mathbb{P}(x \in \Gamma_{\oplus r} \cap A_k))(1 - \mathbb{P}(x \in \Gamma_{\oplus r} \cap A_k))}{4r^2 p} \longrightarrow 0, \tag{12}$$
$$\text{for } r \to 0, p \to \infty, rp \to \infty$$

that is the asymptotic unbiasedness and weak consistency of the estimator, when $r \to 0, p \to \infty$ with $rp \to \infty$.

Note that this estimator is not much affected by edge effects, if the "enlargement" of Γ is performed correctly. If the fibres go across the whole window or have extremes internal to the window but far from the window border, edge effects are not present. For fibres having extremes close to the window border, edge effects can be reduced by reducing also the width r of the enlargement (see Fig. 6).

Let us now introduce an estimator based on the estimate of $T_\Gamma(B_r(x)), x \in A_k$. Let us again consider a grid of points $z_1, \ldots, z_p$ overlapped on the window A_k, such that $B_r(z_i) \subseteq A_k$ for all $i = 1, \ldots, p$ (this assumption has again the aim of reducing the edge effects). We then define

$$\hat{\lambda}_{k,r,p}^2 := \frac{1}{2rp} \sum_{i=1}^{p} \mathbf{1}_{\Gamma \cap B_r(z_i) \neq \emptyset}. \tag{13}$$

where again $\mathbf{1}_{\Gamma \cap B_r(z_i) \neq \emptyset}$ is a Bernoulli random variable assuming value 1 with probability

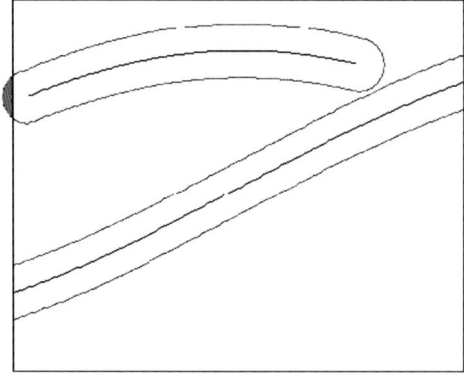

Fig. 6. Examples of edge effects: If the extreme of the fibre is internal and too close to the window border, a piece of enlargement is not considered in the estimator $\hat{\lambda}^1_{k,r}$

$$\mathbb{P}(\Gamma \cap B_r(z_i) \neq \emptyset) = T_\Gamma(B_r(z_i)) = T_\Gamma(B_r(x)),$$

$\forall x \in A_k$. Thus again by using the SLLN and Slutsky Theorem we obtain the asymptotic unbiasedness and weak consistency of this estimator, in fact, for any $x \in A_k$,

$$\mathbb{E}(\hat{\lambda}^2_{k,r,p}) = \frac{T_\Gamma(B_r(x))}{2r} = \frac{\mathbb{P}(x \in \Gamma_{\oplus r} \cap A_k)}{2r} \longrightarrow \lambda_k, \text{ for } r \to 0 \qquad (14)$$

$$\mathrm{Var}(\hat{\lambda}^2_{k,r,p}) = \frac{(T_\Gamma(B_r(x)))(1 - T_\Gamma(B_r(x)))}{4r^2 p} \qquad (15)$$

$$= \frac{(\mathbb{P}(x \in \Gamma_{\oplus r} \cap A_k))(1 - \mathbb{P}(x \in \Gamma_{\oplus r} \cap A_k))}{4r^2 p} \to 0, \qquad (16)$$

$$\text{for } r \to 0, p \to \infty, rp \to \infty.$$

6 Application of the Estimators to the Simulated Inhomogeneous Poisson Segment Process

In this section we will apply the estimators $\hat{\lambda}^1_{k,r,p}$ and $\hat{\lambda}^2_{k,r,p}$ introduced in Sect. 5 to the working example introduced in Sect. 4.2 and we will also derive the rate of convergence to 0 of the variance, in order to assess a method for choosing an "optimal bandwidth" of enlargement r, depending on p. Since the true intensity of this process is known, we use first this example to test empirically the properties of our estimators. In Sect. 7 we will apply the estimators to real or simulated processes where the true intensity is unknown.

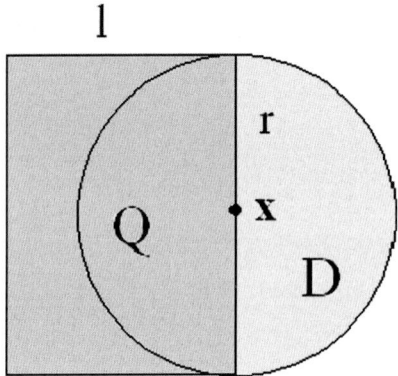

Fig. 7. The regions where a germ must appear in order that a segment hits $B_r(x)$

Assume that $r < l$, where l is the length of the segments forming the Boolean model, consider the quantities which have been estimated in Sect. 5 in the specific case of our example, where $x = (x_1, x_2) \in \mathbb{R}^2$ and refer to Fig. 7 for the definition of the regions Q and D.

$$
\begin{aligned}
\mathbb{P}(x \in \Gamma_{\oplus r}) &= \mathbb{P}(B_r(x) \cap \Gamma \neq \emptyset) \\
&= 1 - \mathbb{P}(B_r(x) \cap \Gamma = \emptyset) \\
&= 1 - \mathbb{P}(\text{no germs fall in } Q \cup D) \\
&= 1 - \exp\left[-\int_Q \alpha(x)\mathrm{d}x - \int_D \alpha(x)\mathrm{d}x \right] \\
&= 1 - \exp\left[-2r\int_{x_1-l}^{x_1} c\bar{x}_1^2\mathrm{d}\bar{x}_1 - \int_{x_1}^{x_1+r}\int_{x_2-\sqrt{r^2-(\bar{x}_1-x_1)^2}}^{x_2+\sqrt{r^2-(\bar{x}_1-x_1)^2}} c\bar{x}_1^2\mathrm{d}\bar{x}_1\mathrm{d}\bar{x}_2 \right] \\
&= 1 - \exp\left[-\frac{2}{3}rc\left(x_1^3 - (x_1-l)^3\right) \right. \tag{17}
\end{aligned}
$$

$$
\left. -\int_{x_1}^{x_1+r} 2c\bar{x}_1^2\sqrt{r^2-(\bar{x}_1-x_1)^2}\mathrm{d}\bar{x}_1 \right]. \tag{18}
$$

Now by computing a Taylor series expansion of (17)–(18) in a right neighborhood of $r = 0$, we obtain

$$
\mathbb{P}(x \in \Gamma_{\oplus r}) = 2r\lambda(x) - 2r^2(\lambda(x))^2 + o(r^2).
$$

By substituting this expansion in the expressions of the expected value and variance of the estimators $\hat{\lambda}_{k,r,p}^1$ and $\hat{\lambda}_{k,r,p}^2$, which are the same, given by (11), (12), and (14), (16), we get, for all $x \in A_k$ and $i = 1, 2$

$$\mathbb{E}(\hat{\lambda}_{k,r,p}^i) = \lambda(x) - r\lambda^2(x) + o(r) \tag{19}$$

$$\mathrm{Var}(\hat{\lambda}_{k,r,p}^i) = \frac{\lambda(x)}{2rp} - \frac{3\lambda^2(x)}{2p} + o\left(\frac{1}{p}\right). \tag{20}$$

The optimal enlargement or bandwidth r can then be computed by minimizing the mean square error, which (by neglecting infinitesimal terms of higher order) is given by

$$\mathrm{MSE}(\hat{\lambda}_{k,r,p}^i) = \mathrm{Var}(\hat{\lambda}_{k,r,p}^i) + \mathrm{Bias}^2(\hat{\lambda}_{k,r,p}^i)$$
$$= \frac{\lambda(x)}{2rp} - \frac{3\lambda^2(x)}{2p} + r^2\lambda^4(x).$$

By minimization one obtains

$$r_{\mathrm{optimal}} = \arg\min_r \mathrm{MSE}(\hat{\lambda}_{k,r,p}^i) = [4p\lambda(x)]^{-1/3}.$$

Note that the optimal bandwidth can be computed only if the true intensity $\lambda(x)$ is known, which is obviously not the case in general. The problem can be overcome in various ways, for example assuming that λ belongs to a given family of functions depending on parameters which can be estimated from the data, or with iterative methods, via the use of an initial guess for $\lambda(x)$ or for r_{optimal}. A discussion for the case of kernel density estimators of the p.d.f. of real-valued random variables can be found in [Hard91, Chap. 4].

7 Experimental Results

We applied estimators $\hat{\lambda}_{k,r,p}^1$ and $\hat{\lambda}_{k,r,p}^2$ to simulated data coming from the model described in Sect. 4.2. The simulation has been performed in the window $[0,1] \times [0,1]$; the constant c appearing in the intensity of germs has been assumed $c = 400$, and the length of segments was fixed to $l = 0.2$. The window $[0,1] \times [0,1]$ was divided into ten vertical stripes of equal width. The two estimators have been computed on each subwindow both by using a deterministic grid of p points z_i, coinciding with the grid of pixels of the image, and by overlapping a random grid of p uniformly distributed points $z_i, i = 1, \ldots, p$. The second method is less affected by correlation problems which may arise from points which have a spatially close location, but has higher computational costs. The optimal bandwidth r has been computed via the true value of $\lambda(x)$ in the centroid of each subwindow. The results are reported in Fig. 8. Since the estimators are biased, with first order bias given in (19), we corrected the estimators by subtracting $-r\lambda^2(x)$. The corrected estimators are reported in Fig. 9, and show a good agreement with the true value of the intensity of the process.

We also computed confidence bands for the estimators, both corrected and uncorrected for bias, by simulating 100 processes with the same intensity, performing on each simulated pattern the estimation procedure and taking the

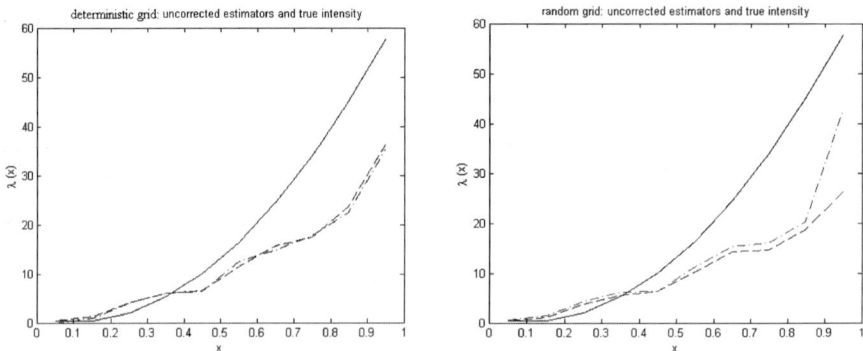

Fig. 8. *Left*: estimate with a deterministic grid; *right*: estimate with a random grid. *Dashed line* $= \hat{\lambda}^1_{k,r,p}$, *dotted-dashed line* $= \hat{\lambda}^2_{k,r,p}$, *continuous line* $=$ true value of $\lambda(x)$. The random grid used for the right-hand picture was formed by $p = 2,000$ uniformly distributed points. The number of pixels in each subwindow, used for the deterministic grid, is $11,628$

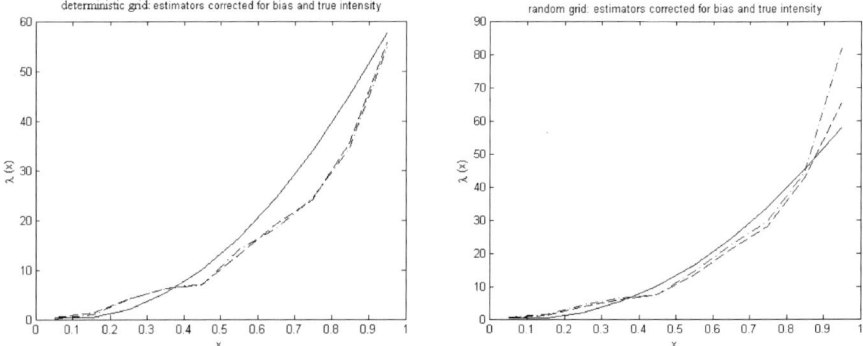

Fig. 9. *Left*: estimate with a deterministic grid; *right*: estimate with a random grid. *dashed line* $= \hat{\lambda}^1_{k,r,p}$, *dotted-dashed line* $= \hat{\lambda}^2_{k,r,p}$, *continuous line* $=$ true value of $\lambda(x)$. The estimators have been corrected for bias using the true value of the intensity. The random grid used for the bottom picture was formed by $p = 2,000$ uniformly distributed points. The number of pixels in each subwindow, used for the deterministic grid, is $11,628$

minimum and maximum values of the estimated intensity in each subwindow. The results are reported in Figs. 10 and 11. From the experimental results the estimators obtained by overlapping to the subwindows a deterministic equally spaced grid seem not to be equivalent to the ones obtained by overlapping a random grid of uniformly distributed points. The deterministic ones seem to have a larger variance than the random ones, and the random ones still show some negative bias, even after the correction, probably due to the terms of higher order which we neglected. The difference in the variance is due to the fact that in the derivation of the expected value and variance of the two esti-

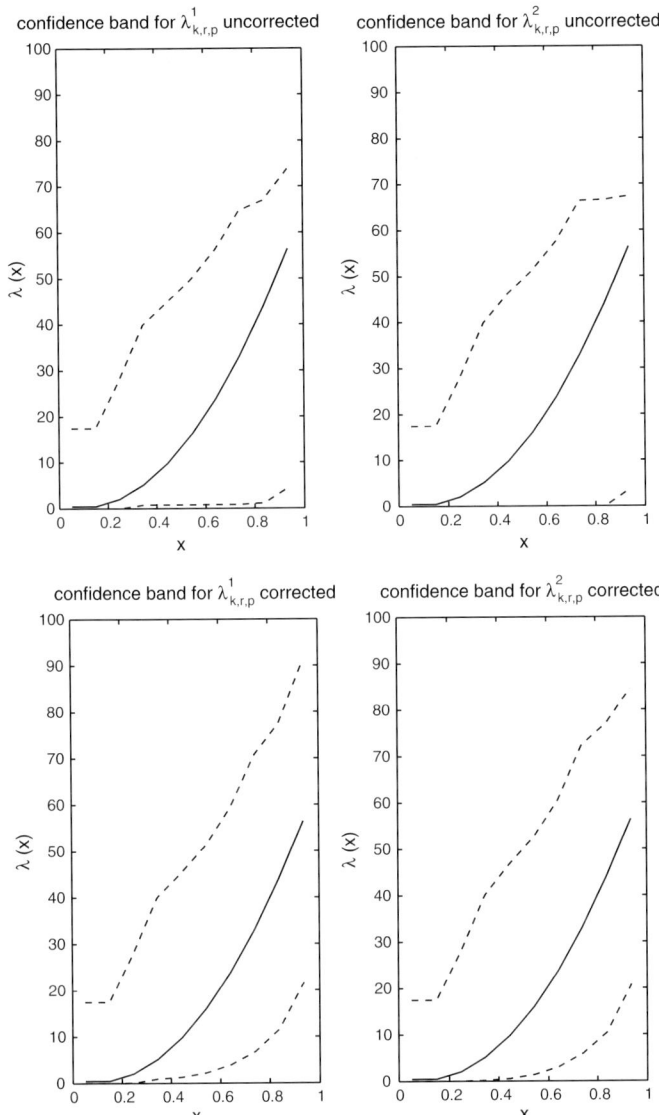

Fig. 10. *Dashed line* min–max confidence band computed by estimating the intensity over 100 simulations of the process; *continuous line* true value of $\lambda(x)$. For the estimation we used a *deterministic* equally spaced grid of points (coinciding with the pixels of the image) overlapped to each subwindow

mators, we assumed that the indicator functions appearing in their definition were i.i.d. Unfortunately the indicators are not independent if the points z_i are located on a regular grid, of width dx comparable with the length l of the segments of the Boolean model or with the "enlargement bandwidth" r. Note

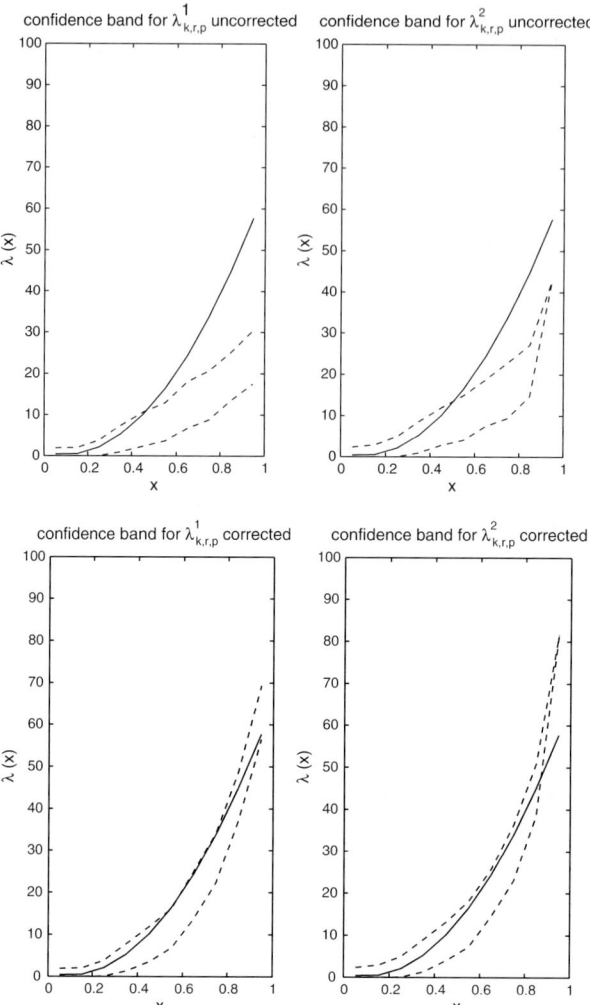

Fig. 11. *Dashed line* min–max confidence band computed by estimating the intensity over 100 simulations of the process; *continuous line* true value of $\lambda(x)$. For the estimation we used a *random* grid of $2,000$ uniformly distributed points overlapped to each subwindow

that the results obtained using a random grid could be improved by augmenting the number of random points of the grid, with a consequent increase of the computational costs.

Since in real applications the true intensity of the fibre process is unknown, we also applied an iterative method to compute the intensity. The method starts by enlarging of the same quantity r_{start} (initial guess) the fibres in all the subwindows; then the estimate procedure is applied and the estimate of

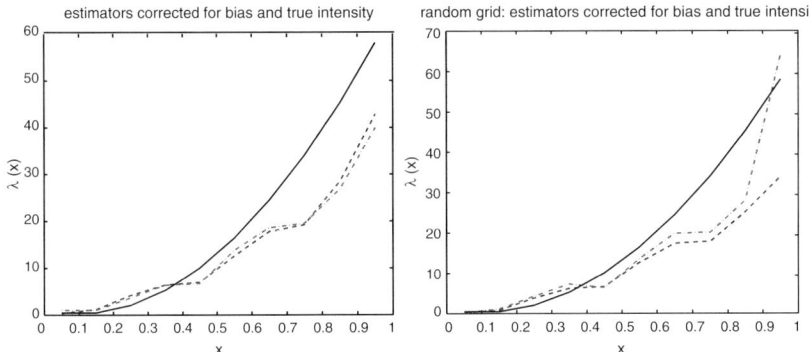

Fig. 12. Estimates with an iterative method. *Left figure*: estimate with a determin-istic grid. The number of pixels in each subwindow, used for the deterministic grid, is $11,628$. *Right figure*: Estimate with a random grid. The random grid used for the estimate was formed by $p = 2,000$ uniformly distributed points. The estimators have been corrected for bias; *dashed line* $= \hat{\lambda}^1_{k,r,p}$, *dotted-dashed line* $= \hat{\lambda}^2_{k,r,p}$, *continuous line* $=$ true value of $\lambda(x)$

$\lambda(x)$ is computed in each subwindow. The estimated intensity is then used to compute the optimal enlargement bandwidth r in the next iteration and an update of $\hat{\lambda}_{k,r,p}$ is computed. A given tolerance constant tol is fixed and the procedure is iterated up to when

$$\sup_x |\hat{\lambda}^i(x)_{m+1} - \hat{\lambda}^i(x)_m| < \text{tol},$$

where $\hat{\lambda}^i(x)_m$ is the intensity function estimated at iteration m ($i = 1, 3$ for the two considered estimators). The study of the termination of the iterative procedure is left to subsequent papers.

The results are reported in Fig. 12. Also in this case the estimators have been corrected for bias, using the estimated intensity for the correction instead of the true value of $\lambda(x)$. From the experimental results the termination and the results of the algorithm does not seem to depend strongly on the initial guess. The convergence looks faster for $\hat{\lambda}^1_{k,r,p}$ if we use a deterministic grid, and for $\hat{\lambda}^2_{k,r,p}$ if we use a random grid.

Min–max confidence bands have been computed over 100 simulations of the process also with the iterative method, using both a deterministic and a random grid; the results are reported in Figs. 13 and 14.

Remark 6. In this case estimator $\hat{\lambda}^1_{k,r,p}$ computed with a random grid seems to behave badly with respect to the others, in particular when the true intensity is high. Nevertheless this estimator has many computational advantages when applied to subwindows which have not a rectangular shape, since overlapping a random grid of points to a window having any shape and counting what points are falling inside the enlarged fibres, is much easier than selecting ran-dom points which have a spherical neighbourhood of fixed width r entirely

28   V. Capasso and A. Micheletti

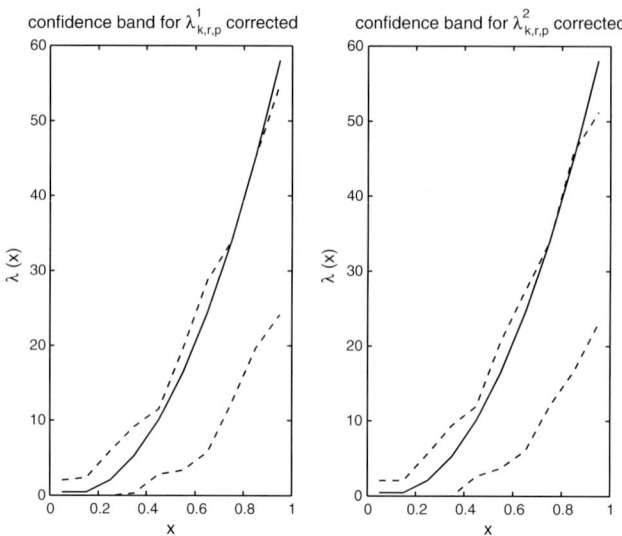

Fig. 13. Confidence bands for $\lambda(x)$ using $\hat{\lambda}^1_{k,r,p}$ (*left figure*) and $\hat{\lambda}^2_{k,r,p}$ (*right figure*). *Dashed line* = min–max confidence band computed by estimating the intensity over 100 simulations of the process; *continuous line* = true value of $\lambda(x)$. For the estimation we used a *deterministic* equally spaced grid of points (coinciding with the pixels of the image) overlapped to each subwindow

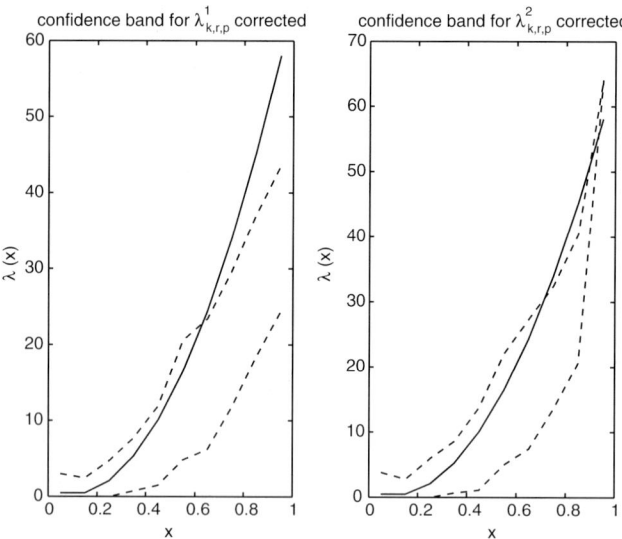

Fig. 14. Confidence bands for $\lambda(x)$ using $\hat{\lambda}^1_{k,r,p}$ (*left figure*) and $\hat{\lambda}^2_{k,r,p}$ (*right figure*). *Dashed line* = min–max confidence band computed by estimating the intensity over 100 simulations of the process; *continuous line* = true value of $\lambda(x)$. For the estimate we used a random grid formed by $p = 2,000$ uniformly distributed points

contained in the subwindow. Thus estimator $\hat{\lambda}^1_{k,r,p}$ will be more often used in the real applications which need a nonrectangular division in subwindows for a good analysis.

The estimators have then been applied to some simulations of real fibre processes, where the true intensity is not known. In Fig. 15 a simulation of the generation and branching of vessels driven by a chemotactic field generated by a tumour is reported. The tumour is located on the right-hand side of the window and the vessels start growing and branching from the left-hand side of the window in the right direction. The chemotactic field has a gradient in the x direction and influences both the speed of growth and the branching of the vessels. The intensity has been estimated both with a deterministic and a random grid, by dividing the observation window into ten vertical stripes of the same width. The estimators have been corrected for bias. The results are reported in Fig. 15. In Fig. 16 two simulations are reported where the

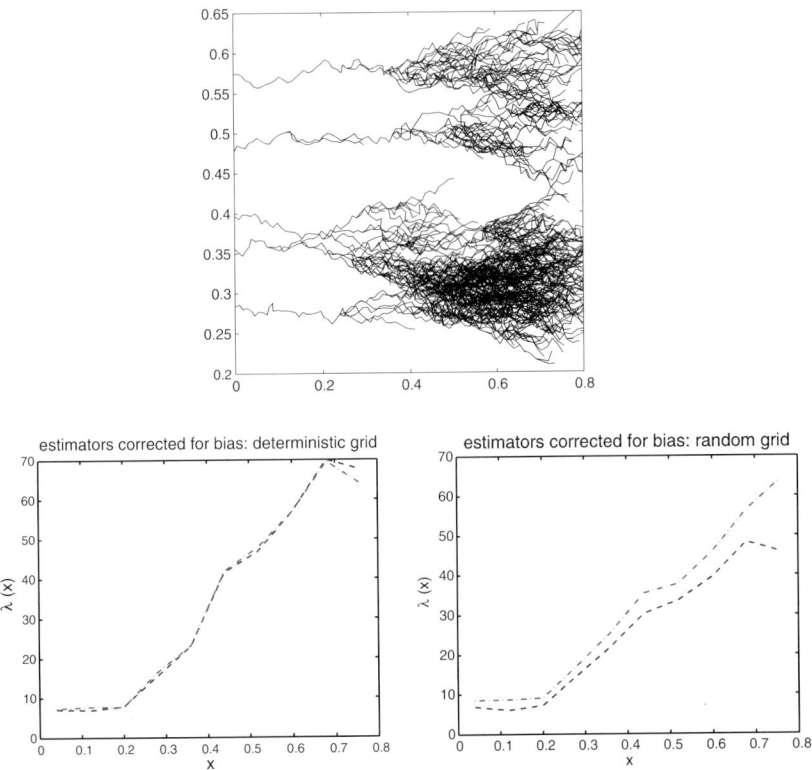

Fig. 15. Estimate of the fibre intensity of an angiogenetic process with a chemotactic field having a gradient in the x direction. *Bottom left*: estimate with a deterministic grid; *bottom right*: estimate with a random grid of 2,000 points. The estimators have been corrected for bias. *Dashed line* = $\hat{\lambda}^1_{k,r,p}$, *dotted-dashed line* = $\hat{\lambda}^2_{k,r,p}$

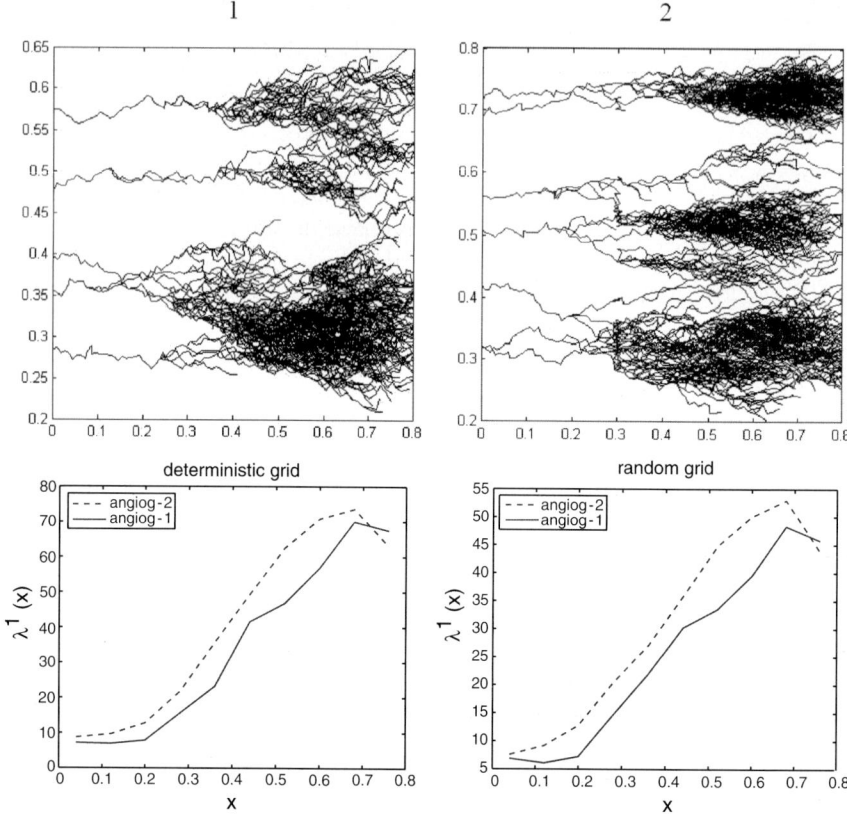

Fig. 16. Comparison between the two simulated angiogenetic processes depicted in the *top line*. *Bottom left*: comparisons of $\hat{\lambda}^1_{k,r,p}$ for the two processes estimated with a deterministic grid; *bottom right*: comparisons of $\hat{\lambda}^1_{k,r,p}$ for the two processes estimated with a random grid. In both cases the first process reveals an intensity lower than the second, and this was really the case in the performed simulation

intensities of branching where different. The difference is not much evident by simply looking at the patterns, but the estimate of the intensity reveals that the pattern on the left has a lower intensity than the pattern on the right for any value of x, and this was really the case, since the frequency of branching and speed of growth was settled higher in the right-hand pattern. This is thus an example where quantitative analysis is essential for the characterization and differentiation of the geometry.

In Fig. 17 an analogous process but driven by a chemotactic field with a spherical symmetry around a point-shaped tumour is reported. Because of the observed symmetry, in this case the window of observation has been divided

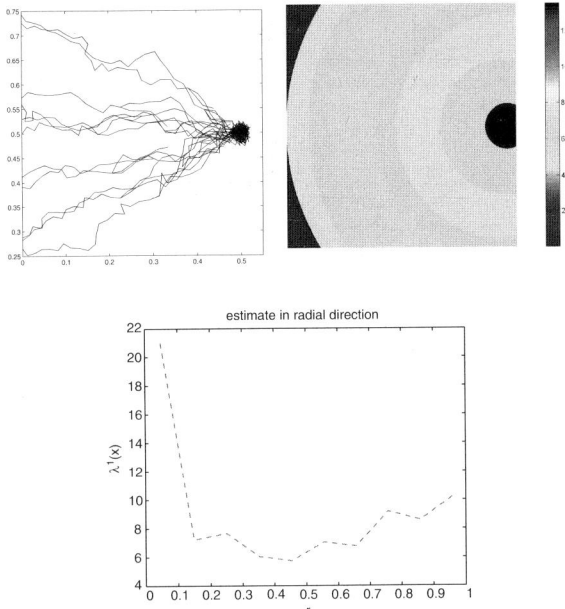

Fig. 17. Estimate of the fibre intensity of an angiogenetic process driven by a chemotactic field with a spherical symmetry. *Top line*: the fibre process and an estimate of $\hat{\lambda}^1_{k,r,p}$ using a random grid and dividing the window into ten spherical shells centred at the tumour; *bottom*: plot of $\hat{\lambda}^1_{k,r,p}$ with respect to the radial coordinate, centred at the tumour

into 10 spherical shells centred at the tumour location. Both the estimated values in each subregion in a 2D visualization and the plot of the estimated intensity with respect to the radial coordinate are reported. In this case, since the subwindows are not rectangular, only estimator $\hat{\lambda}^1_{k,r,p}$ has been computed (see Remark 6).

In Fig. 18 an estimator $\hat{\lambda}^1_{k,r,p}$ has been computed on three images of a vascular networks generated in allantoids (see [CM05] for a discussion of the relevance of these studies in tumour treatment). Two of the three allantoids have been treated with two different doses of an antiangiogenic substance, which should inhibit the formation of vessels. The figure on the left refers to an untreated control allantoid. Because of the spherical symmetry of the images, also in this case the observation window has been divided into spherical shells centred at the centroid of the allantoid. The results of the estimate reveal, in a quantitative way, that the increase of the dose of the substance results in a less widespread network and in a lower intensity of length of the vessels.

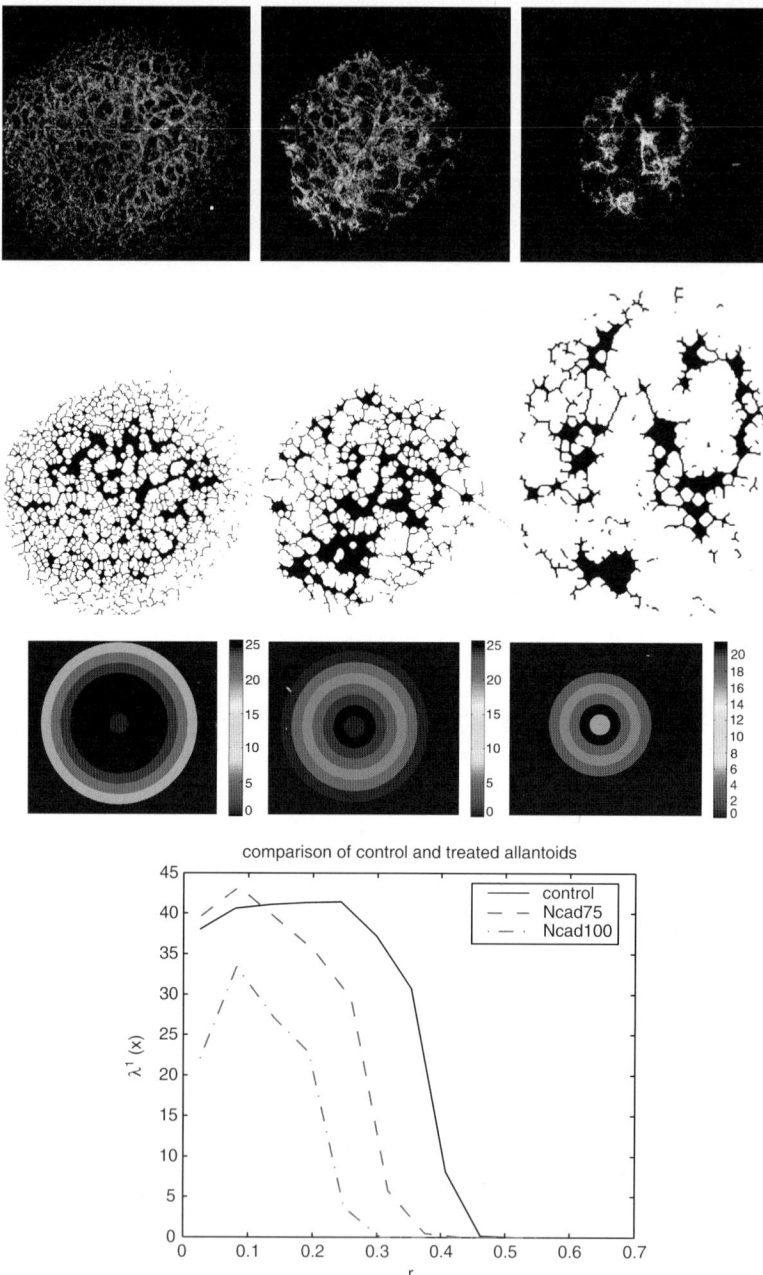

Fig. 18. Vascularization in allantoids. *First line*, from *left to right*: control experiment (untreated), treated with 0.75 mg of antiangiogenetic substance, treated with 1 mg of antiangiogenic substance. *Second line*: scheletonization of the upper images. *Third line*: 2D representation of the intensity estimate of the fibres in the skeletons; the space has been divided into ten spherical concentric shells. *Bottom line*: comparison of the radial estimates of the intensities of the three allantoids

Acknowledgements

It is a pleasure to acknowledge useful discussions with many Colleagues, at different Universities and Research Centres. Particular thanks are due to Professor Luigi Ambrosio (Scuola Normale Superiore, Pisa), Professor Elisabetta Dejana (IFOM, Milan), and Dr. Elena Villa (Department of Mathematics, University of Milan).

References

[ACaV06]   Ambrosio, L., Capasso, V., Villa, E.: On the approximation of geometric densities of random closed sets. RICAM Report N. 2006–14, 2006.

[ACoV06]   Ambrosio, L., Colesanti, A., Villa, E.: First order Steiner formulas for some classes of closed sets. An application to stochastic geometry. In preparation, 2006.

[AFP00]   Ambrosio, L., Fusco, N., Pallara, D.: Functions of Bounded Variation and Free Discontinuity Problems. Clarendon Press, Oxford (2000)

[AG80]   Araujo, A., Giné, E.: The Central Limit Theorem for Real and Banach Valued Random Variables. John Wiley & Sons, New York (1980)

[BM97]   Baddeley, A.J., Molchanov, I.S.: On the expected measure of a random set. In: Proceedings of the International Symposium on Advances in Theory and Applications of Random Sets (Fontainebleau, 1996). World Sci. Publishing, River Edge, NJ, 1, 3–20 (1997)

[BR04]   Benes, V., Rataj, J.: Stochastic Geometry: Selected Topics. Kluwer Academic Publishers, Norwell (2004).

[Bosq00]   Bosq, D.: Linear Processes in Function Spaces. Theory and Applications. Lecture Notes in Statistics 149. Springer-Verlag, New York (2000).

[Ca03]   Capasso, V. (ed): Mathematical Modelling for Polymer Processing. Polymerization, Crystallization, Manufacturing. Mathematics in Industry Vol 2. Springer Verlag, Heidelberg (2003)

[CM00a]   Capasso, V., Micheletti, A.: The local mean volume and surface densities for inhomogeneous random sets. Rend Circ. Mat. Palermo Suppl., 65, 49–66 (2000).

[CM00b]   Capasso, V., Micheletti, A.: Local spherical contact distribution function and local mean densities for inhomogeneous random sets. Stochastics and Stoch. Rep., 71, 51–67, (2000).

[CM05]   Capasso, V., Micheletti, A.: Stochastic geometry and related statistical problems in biomedicine. In: A. Quarteroni et al. (Eds), Complex Systems in Biomedicine. Springer, Milano (2005).

[CM06]   Capasso, V., Micheletti, A., Kernel-like estimators of the intensity of inhomogeneous fibre processes. Preprint (2006).

[CT94]   Capasso, V., Tayler, A. (Eds): ECMI Brochure, Bari (1994)

[CV06a]   Capasso, V., Villa, E.: On the continuity and absolute continuity of random closed sets. Stoch. An. Appl. 24, 381–397, (2006)

[CV06b]   Capasso, V., Villa, E.: Some remarks on the continuity of random closed sets. In: Lechnerová R., Saxl I., Beneš V. (eds), Proceedings of the Intenational Conference in Stereology, Spatial Statistics and Stochastic Geometry, UCMP, Prague, 69–74, (2006)

[CV06c]   Capasso, V., Villa, E.: On the geometric densities of random closed sets. RICAM Report 13/2006, Linz, (2006).

[CV06d]   Capasso, V., Villa, E.: On mean densities of inhomogeneous geometric processes arising in material science and medicine. Preprint (2006)

[CA99]   Chaplain, M.A.J., Anderson, A.R.A.: Modelling the growth and form of capillary networks. In: Chaplain, M.A.J. et al. (eds) On Growth and Form. Spatio-temporal Pattern Formation in Biology. John Wiley & Sons, Chichester (1999).

[Fal85]   Falconer, K.J.: The Geometry of Fractal Sets. Cambridge University Press, Cambridge (1985).

[Fed96]   Federer, H.: Geometric Measure Theory. Springer, Berlin (1996)

[FC98]   Friedman, L.H., Chrzan, D.G.: Scaling theory of the Hall-Petch relation for multilayers. Phys. Rev. Letters, **81**, 2715-2718 (1998).

[Hard91]   Hardle, W.: Smoothing techniques. With Implementation in S, Springer-Verlag, New York (1991).

[KF70]   Kolmogorov, A.N., Fomin S.V.: Introductory Real Analysis. Prentice-Hall, Englewood Cliffs (N.J.), (1970).

[JK01]   Jain, R.K., Carmeliet, P.F.: Vessels of Death or Life. Scientific American **285**, 38-45 (2001).

[Jones82] Jones, D.S.: The Theory of Generalised Functions. Cambridge University Press, Cambridge (1982)

[Mat75]   Matheron, G.: Random sets and integral geometry. John Wiley & Sons, New York (1975)

[McDou06] McDougall, S.R., Anderson, A.R.A., Chaplain, M.A.J.: Mathematical modelling of dynamic tumour-induced angiogenesis: Clinical implications and therapeutic targeting strategies. J. Theor. Biology, **241**, 564–589 (2006)

[Mol92]   Møller, J.: Random Johnson-Mehl tessellations. Adv. Appl. Prob., **24**, 814–844 (1992)

[Mol94]   Møller, J.: Lectures on Random Voronoi Tessellations. Lecture Notes in Statistics 87. Springer-Verlag, New York, Berlin, Heidelberg (1994)

[Pest98]   Pestman, W.R.: Mathematical Statistics. An Introduction. Walter de Gruyter, Berlin (1998).

[SKM95]   Stoyan, D., Kendall, W.S., Mecke, J.: Stochastic Geometry and its Application, John Wiley & Sons, New York (1995)

[Vlad79]   Vladimirov, V.S.: Generalized Functions in Mathematical Physics. Mir Publishers, Moscow (1979).

[Z82]   Zähle, M.: Random processes of Hausdorff rectifiable closed sets. Math. Nachr. **108**, 49–72 (1982)

Synthesis of Micro and Nanoparticles from Coaxial Electrified Jets

A. Barrero[1] and I.G. Loscertales[2]

[1] Escuela Técnica Superior de Ingenieros, Universidad de Sevilla, 41092 Sevilla, Spain
[2] Escuela Técnica Superior de Ingenieros Industriales, Universidad de Málaga, 29013 Málaga, Spain

Summary. The use of electrohydrodynamic (EHD) forces to generate highly charged coaxial jets of immiscible fluids, with diameters in the micro and nanoregime, has unravel itself as a quite interesting choice for producing complex nanostructures from a vast variety of precursors, provided they can solidify, polymerize or gel, in times comparable or shorter than the living time of the coaxial nanojet. For time ratios larger than one, the result of the process are micro or nanocapsules, while for time ratios smaller than one coaxial nanofibres are produced. We show examples of both situations, with organic and inorganic precursors. On the other hand, realization of the process in a liquid bath opens the door to production of controlled micro and nanosized complex emulsions.

1 Introduction

It is well known that the physical properties of a piece of a given substance (thermal and electrical conductivity, strength, toughness, etc.) depend not only on the substance itself but also on its characteristic size. In effect, let us consider an ideal experiment consisting of a material piece whose characteristic length L can be shortened in a controlled way by an external observer. The observer would find out that the values of the physical properties of the material piece undergone a dramatic change when L reaches values sufficiently small. The explanation for such an anomalous behaviour lays on the fact that the surface of a piece of matter decreases with L much more slowly than its volume does and, contrarily to what happens in our familiar macroscopic world, the atomic and molecular interactions of the surface becomes dominant compared to those in the volume once the nanoscopic limit is reached. The length at which the change of properties takes place is, roughly speaking, of the order of 100 nm so this length may be thought as the boundary below which nanotechnology and nanoscience apply. Therefore, its application domain ranges from isolated atoms/molecules to bulk materials, where length and timescales of the phenomena become comparable to those of the structure. Nanotechnology implies the ability to generate and to use structures, components, and devices with a size range from about 0.1 nm (atomic and molecular scale) to about 100 nm (or larger in some situations) by control at atomic, molecular, and macromolecular levels. Nanotechnology is a major breakthrough that will yield new tools for fundamental discoveries with broad impact on technology, materials, biomedical, energy, and environment. Moreover, their interdisciplinary character allows for unparalleled synergy between previously unrelated fields and therefore their applications are extremely diverse. Some few examples of potential applications that are being actively investigated are: advanced drug delivery via nanoparticles in medicine and pharmaceutics fields; chemical and biodetectors for security and other civilian uses; nanostructured catalysts in chemical and fuel industries; metallic and ceramic nanostructured materials with engineered properties, molecular manipulation of polymeric macromolecules, and nanostructured coatings, among others, in material science; nanofabrication of electronic products in electronics, etc. Commercially viable technologies are already available for some ceramic, metallic, and polymeric nanoparticles, nanostructured alloys, colorants and cosmetics, tissue engineering, electronic components such as those for media recording, and hard-disk reading, to name a few. In biomedicine, tissue engineering, for example, applies to regeneration of bones, arteries, and other organs by using biocompatible polymers: polycaprolactone (PCL) and polylactide-co-glycolide acid (PLGA). Basically, it is based on the fact that cells get together and rearrange faster around fibres with smaller diameters (500 nm) than the cells. These scaffolds made of woven fibres, which have proved to be a very efficient growing environment, are being used as biocompatible films to cover prostheses to avoid rejection. This stimulating tissue growth also applies in the cicatrization of wounds and burns. Another example of synergy between nanotechnology and medicine is the use of nanoparticles in drug delivery. The technique involves binding a therapeutic compound to a nanoparticle, or encapsulating it within a nanoshell. A key advantage of nanoshells is that they can be targeted to specific cell populations through conjugation with a monoclonal antibody. When the nanoshells reach the target site, their therapeutic contents are released by breaking them using a low intensity light source such as a laser; shells with controlled porous wall could be also used for the appropriate outflow of the drug. Drug delivery using nanoparticles provides high target specificity, with high potential for treatment of localized neurological disorders and cancer with therapeutic compounds which have side effects in the rest of the body. An alternative, noninvasive procedure for tumour ablation, which has been tested in mice, consists in the intravenous injection of nanoparticles with a dielectric core coated by a thin gold shell, Loo (2005). Based on the relative dimensions of the shell thickness and core radius (typical diameter of the shell is in 100 nm range), nanoshells may be designed to scatter and/or absorb light over a broad spectral range including the near-infrared (NIR), a wavelength region that provides maximal penetration of light through tissue. Immunotargeted nanoshells are engineered to absorb light, allowing selective destruction of targeted carcinoma cells through photothermal therapy. Production of micrometer- or even nanometer-sized particles and fibres can be tackled from two different approaches: bottom-up and top-down methods. Bottom-up refers to methods where materials and devices are built from molecular components which assemble themselves chemically using principles of molecular recognition. Bottom-up should broadly speaking be able to produce devices in parallel and much cheaper than top-down methods, but getting control over the methods is difficult when nanostructures become larger and more bulky than what is normally made by chemical synthesis. On the contrary, in top down methods, nanoobjects are obtained from the appropriate splitting of much larger physical systems without atomic level control along the process.

Top-down methods to produce micro- and nanoparticles require the division of a macroscopic (i.e. millimetric) piece of matter, generally a liquid, into tiny offsprings of micro- or nanometric size. Surface tension strongly opposes the huge increase of area inherent to this dividing process. Thus, to produce such small particles, energy must be properly supplied to the interface. This energy is the result of a mechanical work done on the interface by any external force field, i.e. hydrodynamic forces, electrical forces, etc. Two kinds of approaches can be distinguished, depending on how the energy is supplied. In one approach, such as in the mechanical emulsification techniques, the force fields (extensional and shear flows) employed to break up the interface between two immiscible fluids are so inhomogeneous that, in general, the offspring droplets present a very broad size distribution. Nevertheless, a good degree of monodispersity might be achieved for a particular combination of the emulsification parameters (shear rate, rotation speeds, temperature, etc.) and a given combination of substances. However, such a desirable condition might not exist if one of the substances is changed, if a new one is added, or if a different size is desired. The same occurs if capsules must be formed. Furthermore, in many instances, the formation of the structure depends on chemical interactions, usually preventing the process from being applicable to a broad combination of substances.

In the other approach, which has the advantage of being based on purely physical mechanisms, the force field stretches, steadily and smoothly, the fluid interface without breaking it until at least one of its radii of curvature reaches a well-defined micro or nanoscopic dimension d; at this point, the spontaneous break up of the stretched interface by capillary instabilities yields monodisperse particles with a size of the order of d, Barrero and Loscertales (2007). These types of flows are known as capillary flows due to the paramount role of the surface tension. For example, the formation and control of single and coaxial jets with diameters in the micrometer/nanometer range, and their eventual varicose breakup, lead to particles without structure (single jets) or compound droplets (coaxial jets), with the outer liquid encapsulating the inner one. On the other hand, if the liquid solidifies before the jet breaks, one obtains fibres (single jet) or coaxial nanofibres or hollow nanofibres (coaxial jets). The mean size of the particles obtained with these methods ranges from hundreds of micrometers to several nanometers, although the nanometric range is generally reached when electric fields are employed. The particles obtained using this approach are, in general, nearly monodisperse and its employment enables, in the case of capsules, a precise tailoring of both the capsule size and the shell thickness.

2 Capillary Flows Driven by Electrical Forces

2.1 Electrospray

The interaction of an intense electrical field with the interface between a conducting liquid and a dielectric medium has been known to exist since William Gilbert (1600) reported the formation of a conical meniscus when an electrified piece of amber was brought close enough to a water drop. The deformation of the interface is caused by the force that the electrical field exerts on the net surface charge induced by the field itself. Experiments show that the interface reaches a motionless shape if the field strength is below a critical value, whereas for stronger fields the interface becomes conical, issuing mass and charge from the cone tip in the form of a thin jet of diameter d. In the latter case, the jet becomes steady if the mass and charge it emits are supplied to the meniscus at the same rate. Taylor (1964) explained the conical shape of the meniscus as a balance between electrostatic and surface tension stresses; since then the conical meniscus has been referred to as the Taylor cone. The thin jet eventually breaks up into a stream of highly charged droplets with a diameter of the order of d. This electrohydrodynamic (EHD) steady-state process is the so-called steady cone-jet electrospray after Cloupeau and Prunet-Foch (1989), or just electrospray, see Fig. 1, Pantano et al. (1994).

The electrospray has been applied for bioanalysis (Fenn et al. 1989), fine coatings (Siefert 1984), synthesis of powders (Rullison and Flagan 1994), and electrical propulsion (Martinez-Sanchez et al. 1999), among other technological applications. Recently, the electrosprays in cone–jet mode were also stabilized inside dielectric liquid baths, Barrero et al. 2004; hence, the technique could be applied to the production of simple and double emulsions of the type water in oil, oil in water, and oil–water–oil.

Although the equations (Navier–Stokes and Maxwell equations) and boundary conditions governing the electrospray are known, the numerical simulation of the electrospray is quite complex due to (a) the disparity of length scales between the diameter of the jet and the needle diameter, or aperture, through which liquid is being injected, which can vary more than three orders of magnitude, (b) the existence of one (or more) free surface that must be consistently determined as part of the solution of the problem, and (c) the fact that the region where the interface breaks is time dependent in

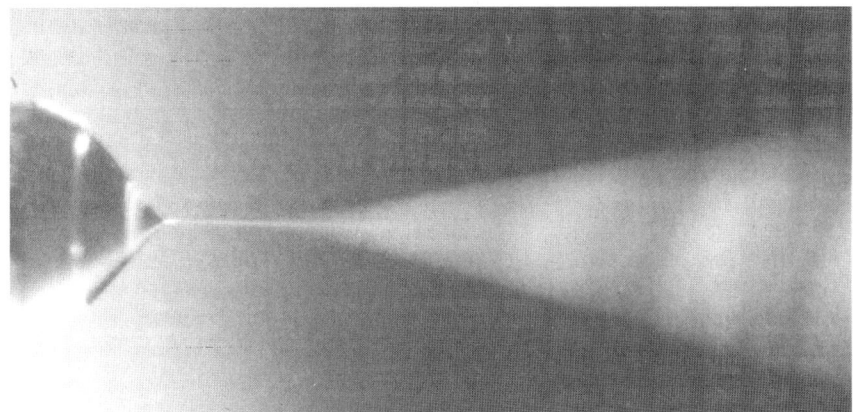

Fig. 1. Cone, jet, and spray in an electrospray; the electrosprayed liquid was methanol. The size of the charged droplets ranged between 380 and 720 nm, which are the wavelength of the blue and red radiation. As shown in the picture, droplets scatter the blue component avoiding its pass throughout the spray while the other components of the white light pass through the droplet cloud

spite of the steady character of the flow upstream of the breaking zone. For these reasons most works on electrospray have focused on experiments, which under the guide of the dimensional analysis have provided the widely accepted relationship between the current I and the flow rate q transported through the jet, Fernández de la Mora and Loscertales (1994),

$$\frac{I}{I_0} = g(\beta) \left(\frac{q}{q_0} \right)^{1/2} \quad \text{with} \quad I_0 = \left(\frac{\epsilon_0}{\rho} \right)^{1/2} \quad q_0 = \frac{\gamma \epsilon_0}{\rho K}, \tag{1}$$

where γ is the surface tension, ρ, K, and $\beta \epsilon_o$ are density, electrical conductivity and permittivity of the liquid respectively, ϵ_o is the vacuum permittivity, and $g(\beta) \sim \beta^{-1/4}$ is a dimensionless function that has been experimentally determined (Gañán-Calvo et al. 1997). However, the scaling law for the jet diameter d is still controversial because experimental errors in the reported measurements of the mean droplet diameter do not allow one to distinguish between the different proposed size laws. The scaling size laws that appear most frequently in the literature can be cast in the form

$$\frac{d}{d_0} = f(\beta) \left(\frac{q}{q_0} \right)^n \quad \text{with} \quad d_0 = \frac{\gamma \epsilon_0^2}{\rho K^2}, \tag{2}$$

where $f(\beta)$ is a dimensionless function of order of unity and exponent n takes the values $1/3$, $1/2$, and $2/3$ depending on the authors. For electrosprays, the minimum flow rate at which it can operate in steady-state conditions is approximately given by $q_{\min} \sim q_0$, which for liquids with electrical conductivities of the order of $1 \, \mathrm{S \, m^{-1}}$, the minimum jet diameter becomes of the order of a few nanometers.

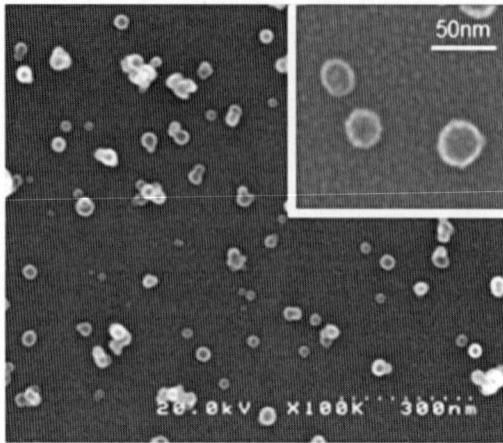

Fig. 2. Nonagglomerated spherical titanium oxide nanoparticles were prepared using an electrospray assisted chemical vapor deposition (ES-CVD) process. From Nakaso et al. (2003)

Numerical simulation of the cone jet electrospray has been considered in a recent paper, Higuera (2003); details of the equations and boundary conditions can be found there. To avoid the numerical difficulty of dealing with two highly disparate length scales, which appears in the case of liquids of relatively high electrical conductivity. Higuera did not consider the full problem from the needle to the final jet region (before breakup) but the cone-to-jet transition region and used the cone and the jet as asymptotic boundary conditions. The numerical analysis included the effect of the liquid viscosity, which had been neglected in prior experiments, and he approximately recovered the $I \sim Q^{1/2}$ law. An excellent review on the physics of electrosprays may be found in Fernandez de la Mora (2007).

The electrospray technique has proved its ability for the production of single nanoparticles; the ones shown in Fig. 2 are an example.

2.2 Electrospinning

The EHD flow described above can be also used to obtain very thin fibres if the jet solidifies before breaking into charged droplets. This process, known as electrospinning, occurs when the working fluid is a complex fluid, such as the melt of polymers of high molecular weight dissolved in volatile solvent, Doshi and Reneker (1995), Fridrikh et al. (2003). The rheological properties of these melts, sometimes enhanced by the solvent evaporation from the jet, slowdown, and even prevent the growth of varicose instabilities. As is well known, large values of liquid viscosity delay the jet breakup by reducing the growth rate of axisymmetric perturbations, so longer jets may be obtained. However, nonsymmetric perturbation modes can grow due to the net charge

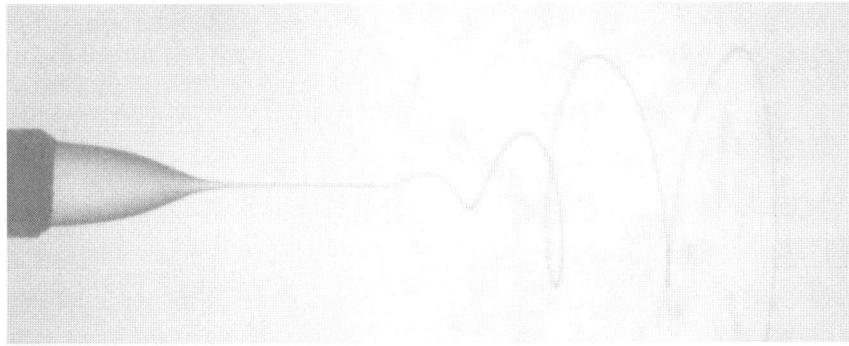

Fig. 3. Whipping instability in an electrified jet of glycerine in a bath of hexane. Courtesy of Mr. A. Gomez-Marín

carried by the jet. Indeed, if a small portion of the charged jet moves slightly off axis, the charge distributed along the rest of the jet will push that portion farther away from the axis, thus leading to a lateral instability known as whipping or bending instability. A picture capturing the development of the whipping instability in a jet of glycerine in a hexane bath is shown in Fig. 3.

The chaotic movement of the jet under this instability gives rise to very large tensile stresses, which lead to a dramatic jet thinning. The solidification process, and thus the production of micro- or nanofibres, is enhanced by the spectacular increase of the solvent evaporation rate due to the thinning process. This technique is very competitive to produce nanofibres as compared with other existing ones (i.e. phase separation, self-assembly, and template synthesis, among others), and it is therefore the subject of intense research.

2.3 Electrified Coaxial Jets

A new technique, which also uses EHD forces to generate coaxial jets of immiscible liquids, with diameters in the nanometer range, has been recently reported, Loscertales et al. 2002. The method is being used to synthesize nanoparticles with core-shell structure. Basically, the technique consists of the injection at appropriate flow rates of two immiscible liquids through two concentrically located needles. The inner diameter of the inner needle ranges from the order of 1 mm to tens of micrometers, whereas its outer diameter sets limits to the cross-section of the outer needle.

The outer needle is connected to an electrical potential of several kilovolts relative to a ground electrode. The inner needle is kept to an electrical potential that, depending on the conductivity of the outer liquid, can be varied from that of the outer needle to that of the extractor. For a certain range of values of the electrical potential and flow rates, a structured Taylor cone

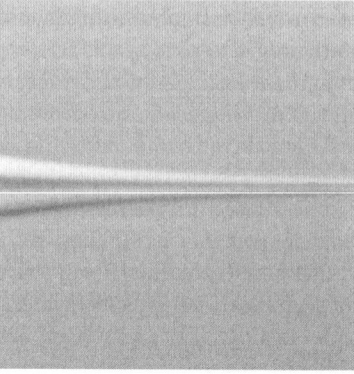

Fig. 4. Picture on the *left* shows a structured liquid Taylor cone; a downstream detail of the two coaxial jets emitted from the vertexes of the two menisci is given in picture on the *right*

is formed at the exit of the needles with an inner meniscus surrounded the inner one, see picture on the left in Fig. 4. A liquid thread is issued from the vertex of each one of the two menisci, giving rise to a compound jet of two coflowing liquids see picture on the right (Fig. 4). At the minimum jet section, the two-layered jet has an outer diameter of 4 µm.

To obtain this compound Taylor cone, at least one of the two liquids must be sufficiently conductive. Similarly to simple electrosprays, the electrical field pulls the induced net electric charge located at the interface between the conducting liquid and a dielectric medium and sets this interface into motion; because this interface drags the bulk fluids, it may be called the driving interface. The driving interface may be either the outermost or the innermost one; the latter happens when the outer liquid is a dielectric. When the driving interface is the outermost, it induces a motion in the outer liquid that drags the liquid–liquid interface. When the drag overcomes the liquid–liquid interfacial tension, a steady-state coaxial jet may be formed. On the other hand, when the driving interface is the innermost, its motion is simultaneously diffused to both liquids by viscosity, setting both in motion to form the coaxial jet. Scaling laws showing the effect of the flow rates of both liquids on the current transported by these coaxial jets and on the size of the compound droplets were recently investigated (Lopez-Herrera et al. 2003).

3 Core-Shell Nanoparticles

3.1 Nanocapsules and Hollow Nanospheres

The last technique has been applied, upon coaxial jet breakup, to microencapsulate aqueous solutions. An outer jet of Somos 6120, a Du Pont photopolymer

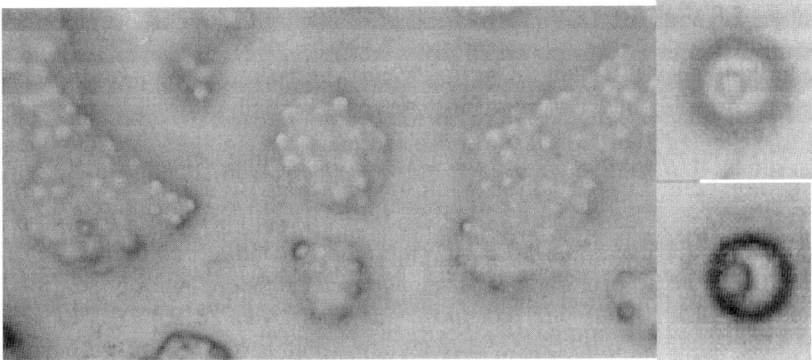

Fig. 5. Collection of near monodisperse capsules. Magnified views of two capsules formed under different parametrical conditions are also given in the two pictures on the *right*. In the *upper* one picture, the outer diameter is 10 µm, whereas the diameter of the capsule shown in the *lower* one is 8 µm

and a coflowing water inner jet were generated as described before. Compound droplets of water coated by Somos resulted from the jet breakup, so that a spray of compound droplets was formed and collected on a plate damped with water. In this case, the outer shell of the droplets was hardened with an ultraviolet light reactor. Before the hardening process, the charged aerosol was neutralized by corona discharge, so that losses were minimized. The liquid flow rates in this experiment were selected to obtain capsules in the micrometer range, because capsules in this range can be optically recorded to allow for visual observation, Fig. 5. Capsules of olive oil surrounding water of 150 nm of mean diameter have been also obtained with this technique. Some examples of applications of this approach to produce capsules include the encapsulation of water-based flavours within oil-based substances, and the opposite (oil-based flavours within water-based polymers) for food enrichment applications, Bocanegra et al. (2005).

Also, combination with sol–gel chemistry has proven fruitful, Larsen et al. (2003). In this case, the outer liquid was a sol–gel formulation, while the inner one was a regular nonstructured or "regular" liquid (like oil, water, glycerine, etc.). By adjusting the sol properties and the operating parameters, we have been able of producing hollow spheres, with mean diameters ranging from 10 µm down to 0.4 µm, and with shell thickness between 1 µm and less than 50 nm. Some of these results are shown in Fig. 6. Although the capsules were initially filled with the "regular" liquid, since the polymerization or gel transition forms porous solids, the inner liquid was easily solvent-extracted, so that after solvent evaporation a void cavity was left. In any of the above cases, the time of flight of the liquid capsules (that is the time from their formation up to their collection on a collector) was controlled to allow for either phase transition or polymerization (gelation). This can be easily done by reducing

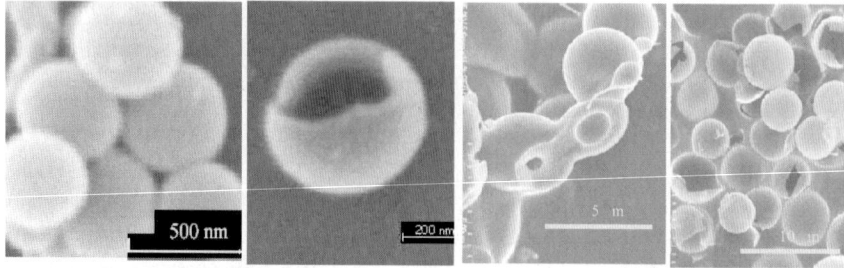

Fig. 6. Hollow spheres of SiO_2. Both diameter and shell thickness can be controlled by adjusting the flow rates

the charge level on the freshly formed capsules; this was accomplished in our lab by setting up a corona discharge of opposite polarity in the surrounding atmosphere.

3.2 Hollow Nanofibres and Coaxial Nanofibres

Another recent application of this technique relates to the production of nanofibres, compound nanofibres and hollow nanotubes. There are many procedures to build nanotubes of different materials, other than the popular carbon nanotubes. In general, the vast majority of these procedures resort to templates, Cepak and Martín (1999). A solid nanotemplate (i.e. a nanofibre or a pore membrane) is formed, around which nanotubes are grown. This growth usually happens in liquid phase, and it resorts to self-assembly of the proper molecules onto the surface of the template. The first complexity is due to this self-assembly process, which unfortunately appears to be very chemistry dependant. Usually, the recipe that works for one particular precursor does not work for another, even for very similar molecules. Once the shell is built around the template, still the template itself must be removed. This is typically done by degrading or decomposing the template thermally, or chemically, etc. This necessarily requires the shell to be more "resistant" than the template. In brief, the procedure is a multistep process, apart of the restriction imposed by the chemistry. One of the advantages of using compounds nanojets to produce nanotubes resorts on the fact that self-assembly is not a limiting step since the shape of the jet itself already constrain the material to the proper cylindrical shape. But during the same process, the inner liquid, which is also stretched to a cylindrical shape, plays the role of the template, thus limiting the inner surface of the nanotube. Furthermore, the template is not solid, but liquid, so that removing the template is much easier and much less energy consuming. Therefore, if solidification (or polymerization, or gelation) of the outer liquid occurs prior to the jet break up, then the nanotube is form in just one step, Loscertales et al. (2004).

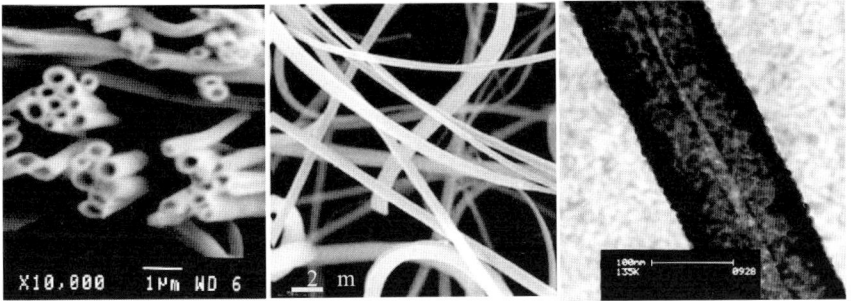

Fig. 7. Examples of hollow nanotubes and compound nanofibres

Figure 7 shows some examples of hollow nanotubes and compound nanofibres produced from electrified coaxial jets. Pictures have been taken with a scanning electron microscope, except the right one, which was taken with transmission electron microscope. The picture on the left shows nanotubes of SiO_2, with diameters of the order of 500 nm, and shell thickness of the order of 70 nm. The one in the middle shows ZnO_2 nanotubes, with diameters from 1 µm down to 400 nm; the wall thickness was of the order of 80 nm. Finally, the right one shows a coaxial character compound nanofibre of poly-ethylene-oxide (PEO) on the outside, and stained PEO in the inside. The outer and inner diameters are of 100 and 15 nm, respectively.

3.3 Simple and Double Emulsions

Finally, another extension of the EHD atomization is that when the surrounding atmosphere is not a gas nor vacuum, but a liquid insulator, Barrero et al. (2004). The same atomization process is possible within a liquid, which opens up the possibility of producing monodisperse micro- and nanoemulsions, Marín et al. (2007). Although work is still on its way, we have investigated the scaling laws for both the current and the size of the droplets. In this new situation, the role of surfactants, emulsifiers, and polymers in solution may be essential in order to stabilize such nanoemulsions. On top of that, the process may be executed with a compound Taylor cone instead (Fig. 8), so that double emulsions of nanometric size can be directly formed, still with a well controlled mean size and small size dispersion. Finally, the charged nature of the dispersed phase can be an advantage to control their trajectories and to select where to deposit them; this could be used to generate well controlled layers of nanoparticles on top of macroscopic objects to emulate colloidosomes, see for instance (Dinsmore et al. 2002).

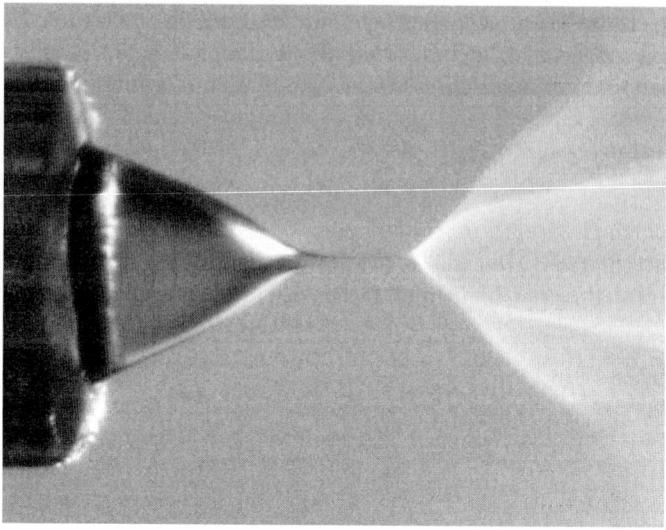

Fig. 8. Taylor cone of glycerol in a bath of hexane. The needle OD is 0.8 mm. The hydrosol in this case is formed by droplets of two different sizes: the main droplets, of 2 μm in diameter, and the satellite droplets, of about 0.8 μm in diameter

4 Conclusions

Some topdown methods to produce micro- and nanoparticles require one to divide a macroscopic (i.e. millimetric) piece of liquid into tiny offsprings of micro- or nanometric size. One of them uses electrical forces to generate coaxial jets with a diameter in the micro- and nanometric size ranges. Micro- or nanocapsules are formed upon jet breakup, whereas if the jet solidifies coaxial nanofibres or hollow nanofibres are obtained. This method can produce micro- and nanoparticles with or without inner structure. Generally, it enables both a precise control of the particle size and a narrow size distribution, which makes it attractive and competitive with other existing techniques. A noticeable feature of the method lies in the fact that the core-shell particles may be obtained in just one step; this is a clear advantage over multistep processes such as the emulsification techniques. However, the throughput of this EHD method is usually too small for many industrial purposes, restricting their use to some analytical applications. Increasing the production rate requires the operation of parallel devices. The main problems when trying to operate in parallel come from the shielding effect of the space charge created by the highly charged aerosol and from the electric crosstalk between neighbouring devices. Accordingly, the design of efficient approaches for operating parallel devices will probably become a very active area of research in the near future.

Acknowledgements

This work has been founded by the Spanish Ministry of Science and Technology under contract BFM2001-3860-C02-01 and by Yflow SL. The help provided by M. Lallave (Yflow), J.E. Díaz (Yflow) and D. Galán (Yflow), and A. Gómez (University of Seville) is also acknowledged.

References

1. Barrero A. and Loscertales I.G. *Annual Rev. Fluid Mech.* **39**, 89-106, 2007.
2. Barrero A., López-Herrera J., D. Boucard A. and Loscertales I.G. *J. Colloid Interf. Sci.* **272**, 104-108, 2004.
3. Bocanegra R., Gaonkar A.G., Barrero A., Loscertales I.G., Pechack D., Marquez M., *J. Food Sci.* **70** 492-497, 2005.
4. Cepak V.M. and Martín, C.R. *Chem. Mater.* **11**, 1363, 1999.
5. Cloupeau M. and Prunet-Foch B. *J. Electrost.* **22** 135-59, 1989.
6. Dinsmore A.D., Hsu, M.F., Nikolaides M.G., Márquez M., Bausch A.R., and Weitz D.A. *Science* **298**, 1006-1009, (2002).
7. Doshi J. and Reneker D.R. *J. Electrost.* **35**, 151-60, 1995.
8. Fenn J.B., Mann M., Meng C.K., and Wong S.F. *Science* **246**, 64-71 (1989).
9. Fernández de la Mora J. and Loscertales I.G. *J. Fluid Mech.* **260** 155-84, 1994. Fernández de la Mora J. *Annual Review of Fluid Mech.*, **39**, 217-244, 2007.
10. Fridrikh S.V., Yu J.H., Brenner M.P., Rutledge G.C. *Phys. Rev. Lett. 90*, 144502, 2003.
11. Gañán-Calvo A.M., Dávila J., Barrero A. *J. Aerosol Sci.* **28**, 249-75 1997.
12. Gilbert W. De Magnete (1600). Transl. P.F. Mottelay. Dover, UK. (1958)
13. Higuera F.J. *J. Fluid Mech. 484*, 303-327, 2003.
14. Larsen G., Velarde-Ortiz R., Minchow K., Barrero A., Loscertales I.G. *J. Am. Chem. Soc.* **125**, 1154-55, 2003.
15. Loo C., Lowery A., Halas N., West J., and Drezek R. *Nanoletters* **5**, 4, 709-711, (2005).
16. López-Herrera J., Barrero A., López A., Loscertales I.G., Márquez M. *J. Aerosol Sci.* **34**, 535-552, 2003.
17. Loscertales I.G., Barrero A., Guerrero I., Cortijo R., Márquez M. *Science* **295**, 1695-98, 2002.
18. Loscertales I.G., Barrero A , Márquez M., Spretz R., Velarde-Ortiz R., Larsen G. *J. Am. Chem. Soc.* **126**, 5376-77, 2004.
19. Marín A.G., Loscertales I.G., Márquez, M., Barrero A. *Phys. Rev. Lett.* **98**, 014502, 2007.
20. Martínez-Sanchez M., Fernandez de la Mora J., Hruby V., Gamero-Castano M., Khayms, V. *Proc. 26th Int. Electr. Propuls. Conf.*, Kitakyushu, Japan, pp. 93-100. Electr. Rocket Propuls. Soc. 1999.
21. K. Nakaso, B. Han, K.H. Ahn, M. Choi, and K. Okuyama. *J. Aerosol Sci.* **34**, 869-881, 2003.
22. Pantano C., Gañán-Calvo A.M. and Barrero A. *J. Aerosol Sci.* **25**, 1065-77, 1994.
23. Rulison A.J. and Flagan R.C. *J. Am. Ceramic Soc.* **77**, 3244-50, 1994.
24. Siefert W. *Thin Solid Films* **120**, 267-74, 1984.
25. Taylor G.I., *Proc. Royal Soc London* **A280**, 383-397, 1964.

Numerical Simulation of Induction Furnaces for Silicon Purification

A. Bermúdez[1], D. Gómez[1], M.C. Muñiz[1], P. Salgado[2], and R. Vázquez[1]

[1] Departamento de Matemática Aplicada, Universidade de Santiago
de Compostela, 15782 Santiago de Compostela, Spain
{mabermud, malola, mcarmen, marafa}@usc.es
[2] Departamento de Matemática Aplicada, EPS Lugo, Universidade de Santiago
de Compostela, 27002 Lugo, Spain
mpilar@usc.es

Summary. This paper deals with mathematical modelling and numerical simulation of induction heating furnaces for axisymmetric geometries. The mathematical model presented consists in a coupled thermo-magneto-hydrodynamic problem with phase change. We propose a finite element method and an iterative algorithm to solve the equations. Some numerical results for an industrial furnace used for silicon purification are shown.

1 Introduction

Silicon (Si) is the second most abundant element in the earth crust after oxygen. In natural form, it can be found mainly as silicon dioxide (Silica, SiO_2) and silicates. In particular, quartz and sand are two of the most common forms. Silicon is produced industrially by reduction of silicon dioxide, as quartz or quartzite, with carbon by a reaction which can be written in a simple way as follows:

$$Si\ O_2 + 2C = Si + 2CO. \tag{1}$$

Silicon has a wide variety of applications depending on its purity. Indeed, silicon is referred to by the approximate percentage of silicon contained in the material and the maximum amount of trace impurities present. Thus, *silicon metal* (or metallurgical grade silicon) refers to the silicon which contains about 1% of other elements. Its main application is as alloying of other metals like aluminum to produce cast parts, mainly for automotive industry. It is also a basic material in chemical industry for silicones. *Ferrosilicon* can contain more than 2% of other materials and represents the largest application of silicon. Almost all ferrosilicon products are consumed by the iron and steel industries.

Pure elementary silicon when doped with traces of elements such as boron and phosphorus is one of the best semiconductors. These substances have a myriad of applications in modern technology, because they are the core of any analog or digital electronic circuit. The use of silicon in semiconductor devices demands a much greater purity than afforded by metallurgical grade silicon. In fact, it is the purest silicon used in industry; it is known as the *9-nines* silicon (99,9999999% of purity).

With the growing of the photovoltaic industry, there is a great request of *solar silicon*, name given to the silicon suitable for use in photovoltaic applications, such as solar cells. Solar silicon must be extremely pure, even if the specifications of purity are less strict than for semiconductor silicon.

Induction heating techniques have been widely applied in the last years in the metallurgical and semiconductor industry for the purification of silicon ingots. Figure 1 illustrates the basic components of an induction heating system: a power supply, an induction coil and a workpiece, which is the piece to be heated. The power supply sends alternating current through the coil that circulates around the coil generating a magnetic field. When the workpiece is placed in the coil, the magnetic field induces eddy currents in it that, by the Joule effect, produce heat. It is this heat which warms up the workpiece.

Based on the induction heating technique, various kinds of induction furnaces are employed for different purposes, such as metal smelting ([CETAL, CRST93]), metal hardening ([CSL04, WKN94]) or crystal growing ([MR97, KP03]). In this work we consider an induction melting furnace as the one represented in Fig. 2. It consists of a cylindrical vessel (usually called the crucible)

Fig. 1. Induction system

Fig. 2. Induction furnace

made from a material such as graphite which is surrounded by an inductor coil made of a very conductive material (copper, for instance). Silicon is placed inside the crucible and the coil is supplied with an alternating current. The goal is to melt the silicon that initially is introduced in solid state.

The idea for the purification process is based on the fact that if silicon is melted and resolidified, the last parts of the mass to solidify contain most of the impurities. Thus, in zone melting, the first silicon purification method to be widely used industrially, rods of metallurgical grade silicon are heated to melt at one end. Then, the heater is slowly moved down the length of the rod, keeping a small length of the rod molten as the silicon cools and resolidifies behind it. Since most impurities tend to remain in the molten region, when the process is complete most of the impurities in the rod will have been moved into the end that was the last to be melted. This end is then cut off and discarded, and the process repeated if a still higher purity is needed. Usually, these methods are combined with chemical ones which involve the injection of gasses into (or onto) a molten silicon bath and are chosen to remove undesirable elements through formation of solid of gaseous reaction products.

An important advantage of induction heating is that the melt is very well stirred, since the Lorentz forces generated by the induced fields cause a movement in the liquid material.

The inductive system can be designed to maintain the silicon in a liquid state, control the shape of its free surface and to provide a strong electromagnetic stirring, ensuring a rapid transfer of pollutants from the bulk liquid to its surface. This stirring also aids in melting the charge since the moving fluid transfers heat from the crucible wall to the solid. The numerical simulation is used to control the design of the induction system, discussing, for instance, the effect of the power and the frequency on the process. One of the important items is the crucible.

From the mathematical point of view, the overall process is rather complex, involving thermal, electromagnetic, hydrodynamic and mechanical phenomena. In order to perform a numerical simulation of the furnace, the physical process is expressed as a coupled nonlinear system of partial differential equations arising from the thermo-magneto-hydrodynamic problem. In the last years several papers have been published which deal with the thermo-electromagnetic problem ([BGMS1, BGMS2, CETAL, CRST93, KP03]), with the magneto-hydrodynamic problem ([HSSH93, NEK99]) or with the thermo-magneto-hydrodynamic problem, but not fully coupled ([HO94, KHT96]). The authors have already dealt with the thermo-electromagnetic problem with phase change, using a finite element method [BGMS1]. The present work starts from the problem and the algorithms proposed in [BGMS1] and introduces the hydrodynamic problem and the convective heat transfer in the heat equation.

The outline of this chapter is as follows. In Sect. 2 we present the coupled mathematical model, assuming cylindrical symmetry. The equations of the electromagnetic model are expressed in terms of the magnetic vector potential. Moreover, the heat equations are written in terms of the enthalpy, to take into account the phase change. The hydrodynamic model is described by the incompressible Reynolds-averaged Navier–Stokes equations, to handle the effects of turbulence. In Sect. 3 we propose an iterative algorithm to solve the coupled problem. Finally, in Sect. 4 we present some numerical results for an industrial furnace devoted to the purification of silicon.

2 Statement of the Problem: Mathematical Modelling

We consider an induction furnace consisting of an induction coil surrounding a workpiece as the one sketched in Fig. 3. The goal is to compute the distribution of heat in the workpiece caused by the eddy currents, considering phase change and convective heat transfer.

Let Ω_0 be the radial section of the workpiece, and $\Omega_1, \Omega_2, \ldots, \Omega_m$ the radial sections of the windings of the coil. In fact, to be able to consider the problem in an axisymmetric setting, the induction coil is replaced by m rings with toroidal geometry. Moreover, Ω_a will denote the air around the conductors, so that $\Omega = \Omega_a \cup \Omega_0 \cup \Omega_1 \cup \cdots \cup \Omega_m$ will denote the two dimensional domain of the model (see Fig. 4). In principle, Ω is a half-plane and we should impose "boundary conditions" at infinity. For the sake of simplicity, we cut the domain far from the conductors and impose boundary conditions on the artificial boundary (see [BGMS2] for a BEM–FEM method to deal with the unbounded domain).

2.1 The Electromagnetic Model

Since we are considering alternating currents, all of the fields have the form:

$$\mathcal{F}(\mathbf{x}, t) = \mathrm{Re}\left[\mathrm{e}^{\mathrm{i}\omega t}\,\mathbf{F}(\mathbf{x})\right], \tag{2}$$

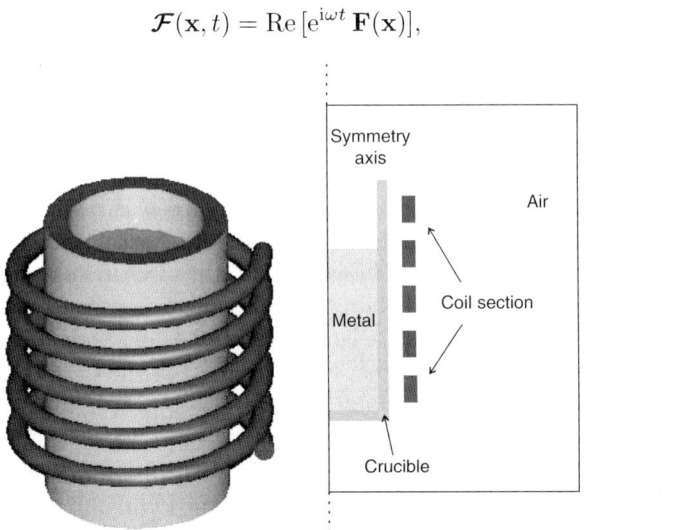

Fig. 3. Sketch of the induction furnace and diametral section

where t is time, $\mathbf{x} \in \mathbb{R}^3$ is the space position, ω is the angular frequency, i the imaginary unit and $\mathbf{F}(\mathbf{x})$ is the complex amplitude of the field. Moreover, as the induction furnace we are interested in works in a low-frequency regime, the Maxwell's equations can be reduced to the so-called *eddy current* model:

$$\operatorname{curl} \mathbf{H} = \mathbf{J}, \tag{3}$$

$$i\omega \mathbf{B} + \operatorname{curl} \mathbf{E} = \mathbf{0}, \tag{4}$$

$$\operatorname{div} \mathbf{B} = 0, \tag{5}$$

$$\operatorname{div} \mathbf{D} = \varrho, \tag{6}$$

to which we have to add the equation imposing the intensity current, I, flowing along the coil. In (3)–(6) $\mathbf{H}$, $\mathbf{J}$, $\mathbf{B}$, $\mathbf{E}$ and $\mathbf{D}$ are the complex amplitudes associated with the magnetic field, the current density, the magnetic induction, the electric field and the electric displacement, respectively, while ϱ denotes the charge density.

The system (3)–(6) needs to be completed by the constitutive relations

$$\mathbf{B} = \mu \mathbf{H}, \tag{7}$$
$$\mathbf{D} = \varepsilon \mathbf{E}, \tag{8}$$

where μ is the magnetic permeability and ε is the electric permittivity. We also need the Ohm's law

$$\mathbf{J} = \begin{cases} \sigma \mathbf{E} & \text{inside conductors,} \\ 0 & \text{in air,} \end{cases} \tag{9}$$

where σ is the electric conductivity.

Remark 1. In fact, the current density in the conductors is given by

$$\mathbf{J} = \sigma(\mathbf{E} + \mathbf{u} \times \mathbf{B}), \tag{10}$$

where $\mathbf{u}$ is the velocity field. In our problem the second term is only important when the furnace works at low frequencies and very high intensities, so we are neglecting it, for the sake of simplicity.

Due to the symmetry of the problem, we are interested in using a cylindrical coordinate system (r, θ, z), with the z-axis coinciding with the symmetry axis of the domain. Hereafter we denote $\mathbf{e}_r$, $\mathbf{e}_\theta$ and $\mathbf{e}_z$ the local orthonormal basis associated with this system of coordinates. Now we assume cylindrical symmetry, which means that no field depends on the angular variable θ. We further assume that the current density field has nonzero component only in the tangential direction $\mathbf{e}_\theta$, namely

$$\mathbf{J}(r, \theta, z) = J_\theta(r, z)\mathbf{e}_\theta.$$

A well-known result allows us to conclude from (5) that $\mathbf{B}$ is the curl of a magnetic vector potential, denoted by $\mathbf{A}$:

$$\mathbf{B} = \operatorname{curl} \mathbf{A}. \tag{11}$$

For the sake of uniqueness we take $\mathbf{A}$ to be divergence-free (Coulomb gauge), and we can also conclude that $\mathbf{A}$ is of the form

$$\mathbf{A}(r, \theta, z) = A_\theta(r, z)\mathbf{e}_\theta. \tag{12}$$

From (3), (4), (9) and (11) we deduce that there exist constants $C_k \in \mathbb{C}$, $k = 0, \ldots, m$, such that

$$i\omega A_\theta + \sigma^{-1} J_\theta = \frac{C_k}{r} \quad \text{in } \Omega_k, \tag{13}$$

recalling that Ω_k, $k = 1, \ldots, m$ denotes each connected component of the conductor, and that Ω_0 is the workpiece (see [BGMS1] or [BGMS2] for details).

The expression of the curl of a vector field in cylindrical coordinates and equations (3), (7), (12) and (13) combined together yield

$$-\left(\frac{\partial}{\partial r} \left(\frac{1}{\mu r} \frac{\partial (r A_\theta)}{\partial r} \right) + \frac{\partial}{\partial z} \left(\frac{1}{\mu} \frac{\partial A_\theta}{\partial z} \right) \right) + i\omega \sigma A_\theta = \frac{\sigma}{r} C_k, \tag{14}$$

in any connected component of the conducting domain, and

$$-\left(\frac{\partial}{\partial r} \left(\frac{1}{\mu r} \frac{\partial (r A_\theta)}{\partial r} \right) + \frac{\partial}{\partial z} \left(\frac{1}{\mu} \frac{\partial A_\theta}{\partial z} \right) \right) = 0 \tag{15}$$

in the air.

To be able to solve equations (14)–(15) we assume that the current intensities flowing in each ring are given data. Thus we add to the model the following equations

$$\int_{\Omega_k} J_\theta \, \mathrm{d}r \mathrm{d}z = \mathrm{I}_k, \quad k = 1, \ldots, m,$$

I_k being the intensity traversing Ω_k. For a further discussion about the model one can see [BGMS1], [BGMS2] or [CETAL]. An explanation about the physical meaning of the constants C_k can be seen in [CETAL] or [RS96]. An important result is that C_k must be zero in any simply connected region, in particular $C_0 = 0$ in the workpiece. From the mathematical point of view, these constants can be considered as Lagrange multipliers associated with the intensity constraints above.

Electromagnetic Boundary Conditions

As we have already said, the unbounded domain is cut far from the conductors to have a bounded domain. We shall denote by Γ^A the boundary of this

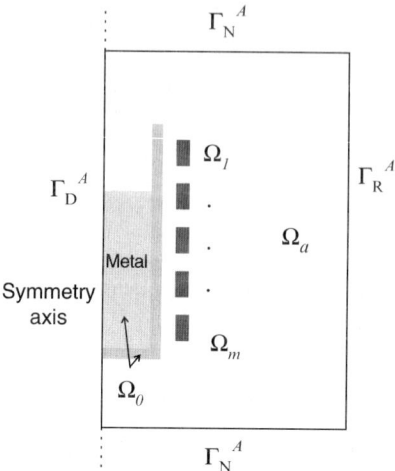

Fig. 4. Computational domain for the electromagnetic problem

computational domain and set $\Gamma^A = \Gamma_{\mathrm{R}}^A \cup \Gamma_{\mathrm{N}}^A \cup \Gamma_{\mathrm{D}}^A$ (see Fig. 4). Following [CETAL] the boundary conditions we impose are

$$\frac{\partial(r A_\theta)}{\partial r} + A_\theta = 0 \quad \text{on } \Gamma_{\mathrm{R}}^A, \tag{16}$$

$$\frac{\partial(r A_\theta)}{\partial z} = 0 \quad \text{on } \Gamma_{\mathrm{N}}^A, \tag{17}$$

$$A_\theta = 0 \quad \text{on } \Gamma_{\mathrm{D}}^A. \tag{18}$$

2.2 The Thermal Model

The above model must be coupled with the heat equation to study the thermal effects of the electromagnetic fields in the workpiece. As the furnace is designed to reach temperatures higher than the melting point of the metal we shall use the heat transfer equation in transient state with change of phase. Furthermore, since the molten metal is subject to electromagnetic and buoyancy forces, we also need to consider convective heat transfer. Let us suppose that we already know the velocity field $\mathbf{u}$ which is null in the solid part of the workpiece, then the equation for energy conservation is

$$\left(\frac{\partial e}{\partial t} + \mathbf{u} \cdot \mathbf{grad}\, e\right) - \mathrm{div}(k_{\mathrm{eff}}(\mathbf{x}, T)\, \mathbf{grad}\, T) = \frac{|\mathbf{J}|^2}{2\sigma} \quad \text{in } \Omega_0, \tag{19}$$

where e is the enthalpy, T is the temperature and k_{eff} is the effective thermal conductivity, which is the sum of the turbulent and molecular conductivities,

$k_{\text{eff}} = k + k_{\text{t}}$. The turbulent thermal conductivity is computed by using the formula

$$k_{\text{t}} = \frac{\eta_{\text{t}}}{\sigma_{\text{t}}}, \tag{20}$$

where η_{t} is the turbulent dynamic viscosity given by (39) below, and σ_{t} is the turbulent Prandtl's number, which is taken to be equal to 0.9.

We remark that the thermal conductivity k depends on temperature. We also assume that other material properties as the electric conductivity σ, the magnetic permeability μ and the dynamic viscosity η may depend on temperature.

The coupling between the thermal and the electromagnetic submodels is made by the heat released in the workpiece due to the Joule effect. This heat is represented in (19) by the term on the right-hand side, involving $\mathbf{J}$ which is obtained from (13). In fact, since the electromagnetic equations are expressed in the frequency domain, the heat source is determined by taking the mean value in a cycle (see [BGMS1]).

In (19) the terms between parenthesis on the left-hand side can be rewritten as the material time derivative of enthalpy, which we shall denote by $\dot{e}$. Moreover, assuming cylindrical symmetry and the fact that T does not depend on the angular coordinate θ, the heat equation becomes

$$\dot{e} - \frac{1}{r}\frac{\partial}{\partial r}\left(rk(r,z,T)\frac{\partial T}{\partial r}\right) - \frac{\partial}{\partial z}\left(k(r,z,T)\frac{\partial T}{\partial z}\right) = \frac{|J_\theta|^2}{2\sigma}. \tag{21}$$

Notice that, from (13), we obtain

$$J_\theta = -\mathrm{i}\omega\sigma A_\theta \quad \text{in } \Omega_0, \tag{22}$$

because $C_0 = 0$ in Ω_0.

Thermal Boundary Conditions

The computational domain for the thermal problem is the workpiece, i.e. Ω_0. We shall denote its symmetry axis by Γ_{S}, and by $\Gamma_{\text{R}}^{\text{T}}$ the part of the boundary that is not on the symmetry axis (see Fig. 5). Then, (21) is completed with the following radiation–convection condition on the boundary $\Gamma_{\text{R}}^{\text{T}}$:

$$k(\mathbf{x}, T)\frac{\partial T}{\partial \mathbf{n}} = \alpha(T_{\text{c}} - T) + \gamma(T_{\text{r}}^4 - T^4), \tag{23}$$

where α is the coefficient of convective heat transfer, T_{c} and T_{r} are the external convection and radiation absolute temperatures, respectively, the coefficient γ is the product of emissivity by Stefan–Boltzmann constant, and $\mathbf{n}$ is the outward unit normal vector to the boundary. Besides, on the axis Γ_{S} we set

$$k(\mathbf{x}, T)\frac{\partial T}{\partial \mathbf{n}} = 0.$$

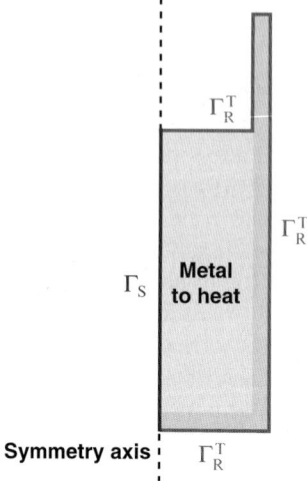

Fig. 5. Computational domain for the thermal problem

2.3 The Hydrodynamic Model

Let $\Omega_1(t)$ be the radial section of the molten metal, and $\Gamma_x(t)$, $\Gamma_d(t)$ and $\Gamma_n(t)$ the different parts of the boundary at time t (depicted in Fig. 6). We assume that the fluid motion is governed by the incompressible Navier–Stokes equations:

$$\rho(\mathbf{x}, T)\left(\frac{\partial \mathbf{u}}{\partial t} + \mathbf{u} \cdot \nabla \mathbf{u}\right) - \operatorname{div}(\eta(\mathbf{x}, T)D(\mathbf{u})) + \nabla p = \mathbf{f} \quad \text{in } \Omega_1(t), \qquad (24)$$

$$\operatorname{div} \mathbf{u} = 0 \quad \text{in } \Omega_1(t), \qquad (25)$$

where ρ denotes the density, $\mathbf{u}$ is the velocity field, η is the dynamic viscosity, p is the pressure and $D(\mathbf{u})$ denotes the symmetric part of $\operatorname{grad} \mathbf{u}$, namely

$$D = \frac{\operatorname{grad} \mathbf{u} + \operatorname{grad} \mathbf{u}^t}{2}.$$

We remark that the hydrodynamic domain is the molten region of the metal, which varies as the metal melts or solidifies, so it depends on time. Moreover, both density and viscosity are material properties which depend on temperature, so for the solution of the thermal problem is essential to solve the hydrodynamic problem.

The right-hand side term $\mathbf{f}$ contains the forces supported by the fluid due to natural convection (buoyancy forces) and those due to the electromagnetic field (Lorentz force):

$$\mathbf{f} = \rho(\mathbf{x}, T)\mathbf{g} + \mathbf{J} \times \mathbf{B}, \qquad (26)$$

where $\mathbf{g}$ is the acceleration of gravity.

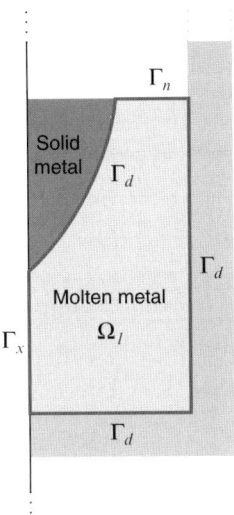

Fig. 6. Computational domain for the hydrodynamic problem

The term representing the Lorentz force is obtained from the solution of the electromagnetic problem. Since in the electromagnetic model we work in the frequency domain, the Lorentz force is determined by taking the mean value in a cycle, namely

$$\frac{\omega}{2\pi} \int_0^{2\pi/\omega} \mathcal{J}(\mathbf{x}, t) \times \mathcal{B}(\mathbf{x}, t)\, dt, \tag{27}$$

where $\mathcal{J}$ and $\mathcal{B}$ denote the current density and the magnetic induction, respectively, and ω is the angular frequency.

In (24) we can rewrite the terms into the parenthesis as the material time derivative of the velocity. If we do so, and use cylindrical coordinates, we obtain the equations we will use in our model:

$$\rho \dot{u}_r - \frac{1}{r}\left[\frac{\partial}{\partial r}\left(\eta r \frac{\partial u_r}{\partial r}\right) + \frac{r}{2}\frac{\partial}{\partial z}\left(\eta\left(\frac{\partial u_r}{\partial z} + \frac{\partial u_z}{\partial r}\right)\right) - \frac{u_r}{r^2}\right] + \frac{\partial p}{\partial r} = \mathbf{f}_r, \tag{28}$$

$$\rho \dot{u}_z - \frac{1}{r}\left[\frac{\partial}{\partial r}\left(\eta\frac{r}{2}\left(\frac{\partial u_r}{\partial z} + \frac{\partial u_z}{\partial r}\right)\right) + r\frac{\partial}{\partial z}\left(\eta\frac{\partial u_z}{\partial z}\right)\right] + \frac{\partial p}{\partial z} = \mathbf{f}_z, \tag{29}$$

$$\frac{1}{r}\frac{\partial}{\partial r}(r u_r) + \frac{\partial u_z}{\partial z} = 0, \tag{30}$$

where we recall that $\rho = \rho(r, z, T)$ and $\eta = \eta(r, z, T)$.

Initial and Boundary Conditions for the Hydrodynamic Model

Equations (28)–(30) are completed with the following initial and boundary conditions

$$\mathbf{u} = \mathbf{0} \quad \text{on } \Gamma_d(t), \tag{31}$$
$$S\mathbf{n} = 0 \quad \text{on } \Gamma_n(t), \tag{32}$$
$$S\mathbf{n} = 0 \quad \text{on } \Gamma_x(t), \tag{33}$$
$$\mathbf{u} = \mathbf{0} \quad \text{in } \Omega_1(0), \tag{34}$$

where S denotes the Cauchy stress tensor, $S = 2\eta D(\mathbf{u}) - pI$, and $\mathbf{n}$ is the outward unit normal vector to the boundary.

An Algebraic Turbulence Model: Smagorinsky's Model

We recall that the Reynolds number is a dimensionless quantity which gives the ratio of inertial forces to viscosity forces. It is given by

$$Re = \frac{\rho \mathbf{V} L}{\mu}.$$

When this number goes beyond a threshold the flow becomes turbulent, and it makes practically impossible to model its behaviour using the Navier–Stokes equations, due to the extremely fine required computational mesh. For numerical simulation purposes the Navier–Stokes equations are replaced with the so-called Reynolds-averaged Navier–Stokes equations (see [MP94]):

$$\rho(\mathbf{x}, T)\left(\frac{\partial \bar{\mathbf{u}}}{\partial t} + \bar{\mathbf{u}} \cdot \nabla \bar{\mathbf{u}}\right) - \operatorname{div}(\eta(\mathbf{x}, T)D(\bar{\mathbf{u}})) - \operatorname{div} R + \nabla \bar{p} = \mathbf{f} \quad \text{in } \Omega_1(t), \tag{35}$$

$$\operatorname{div} \bar{\mathbf{u}} = 0 \quad \text{in } \Omega_1(t), \tag{36}$$

where $\bar{\mathbf{u}}$ denotes the mean velocity and $\bar{p}$ the mean pressure. The tensor R is called the Reynolds stress tensor, and it represents the contribution of the turbulent part to the mean flow.

The Boussinesq assumption consists in taking the Reynolds tensor as

$$R = -\frac{1}{3}tr(R)I + 2\eta_t D(\bar{\mathbf{u}}), \tag{37}$$

where I is the identity tensor and η_t is the turbulent viscosity. Using this assumption we can now rewrite equation (35) as

$$\rho(\mathbf{x}, T)\left(\frac{\partial \bar{\mathbf{u}}}{\partial t} + \bar{\mathbf{u}} \cdot \nabla \bar{\mathbf{u}}\right) - \operatorname{div}(\eta_{\text{eff}}(\mathbf{x}, T)D(\bar{\mathbf{u}})) + \nabla \bar{p}^* = \mathbf{f} \quad \text{in } \Omega_1(t), \tag{38}$$

where $p^* = p - \frac{1}{3}tr(R)$ and η_{eff} is the effective viscosity, which is given by $\eta_{\text{eff}} = \eta + \eta_t$. Different models are obtained depending on the way in which the turbulent viscosity η_t is computed. A very simple and easy to implement model is the one proposed by Smagorinsky (see [MP94]), which consists in taking

$$\eta_t = \rho c h^2 |D(\bar{\mathbf{u}})|, \quad c \cong 0.01 \tag{39}$$

where $h(x)$ is the mesh size of the numerical method around point x.

3 Numerical Approximation

To obtain a suitable discretization of the material time derivative in (21) and (28) we have used the characteristics method (see [PIR82]).

Electromagnetic and thermal problems have been spatially discretized by a piecewise linear finite elements associated with a triangular mesh. The electromagnetic problem is solved in the workpiece, the inductors and the air, while the heat transfer equation is only solved in the workpiece.

The hydrodynamic problem has been spatially discretized by the finite element couple P_1-bubble/P_1, which is known to satisfy the *inf–sup* condition (see [BF91]). We remark that the hydrodynamic problem is only solved in the liquid domain Ω_1, which must be determined at each time step.

We also notice that, at each time step, the three problems form a coupled nonlinear system. Indeed, in the thermal problem the heat source depends on the solution of the electromagnetic problem, while the convective heat transfer needs from the hydrodynamic problem. Moreover, the Lorentz force in the hydrodynamic problem needs the solution of the electromagnetic problem. On the other hand, parameters k, σ, μ, ρ and η depend on temperature, and so does enthalpy. Furthermore, the radiation–convection boundary condition in the thermal problem depends on T^4. To handle the coupling between the three problems we propose a fixed point algorithm which is schematized in Fig. 7 below.

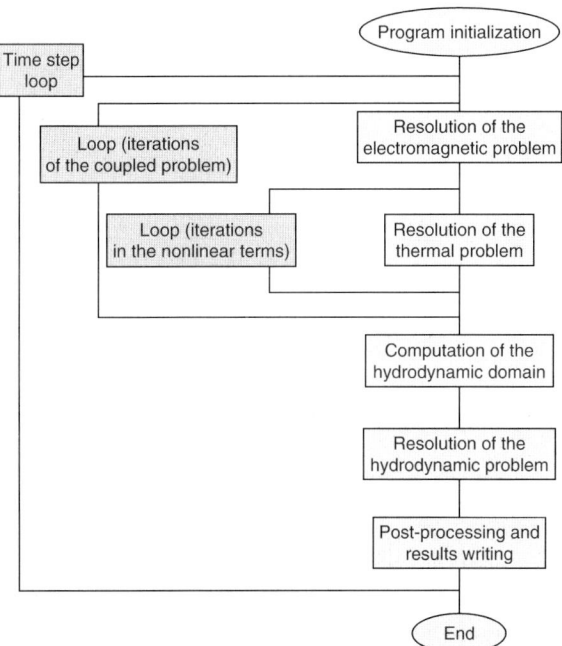

Fig. 7. Scheme of the algorithm

Remark 2. As we have seen in Remark 1 the velocity term in the Ohm's law is not considered, so the electromagnetic problem does not need the solution of the hydrodynamic problem. Moreover, in the thermal problem, the velocity field comes from the solution at the previous time step. Thus, we are allowed to solve the hydrodynamic problem segregated from the two other problems, which saves much computational time.

4 Numerical Results

In this section we present some numerical results obtained by using the algorithm introduced above, which has been implemented in a computer `Fortran` program. More precisely we have applied the algorithm to simulate an industrial furnace used for silicon purification.

We consider a workpiece consisting of a graphite crucible surrounded by an alumina layer and containing silicon. Since solid silicon is not very conductive, a graphite susceptor is required to heat the silicon charge; heating the silicon is then done by conduction and radiation from the graphite until the silicon melts and it conducts electric current. All of materials are initially at 30°C. The induction coil is made of water-cooled copper. The geometrical data of this furnace are summarized in Fig. 8 and Table 1. A detail of the computational

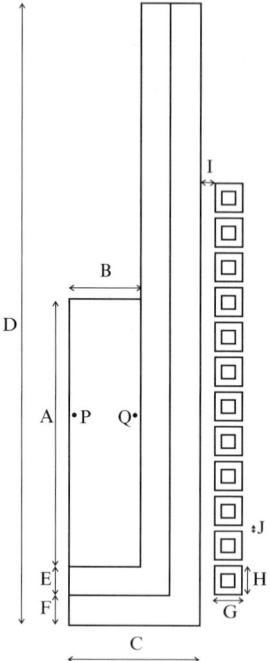

Fig. 8. Sketch of the geometry

Table 1. Geometrical data

A – Height of silicon:	0.45 m
B – Inner radius of crucible:	0.125 m
C – Outer radius of crucible:	0.225 m
D – Crucible height:	1.05 m
E – Crucible width:	0.05 m
F – Alumina layer width:	0.05 m
G – Turn diameter:	0.05 m
H – Turn height:	0.05 m
I – Distance between coil and crucible:	0.025 m
J – Distance between the turns:	0.01 m
Number of coil turns:	12
P, Q – Measure points	

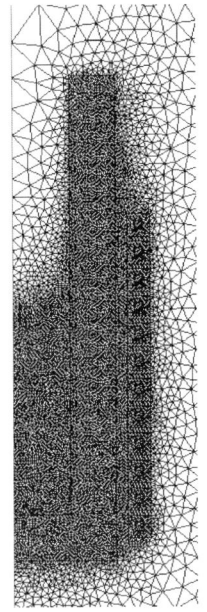

Fig. 9. Detail of the mesh

mesh can be seen in Fig. 9. The physical properties of the three materials in the workpiece depend on temperature and have been obtained from literature. Since we are not considering the thermal model in the coil, the electromagnetic properties of copper are supposed to be constant. Several simulations have been carried out, considering values of 100 Hz for the frequency and 5,500 Å for the intensity.

Figure 10 shows the temperature field in the workpiece, for 30 min and 180 min, respectively. In Fig. 11 we represent the temperature in the silicon

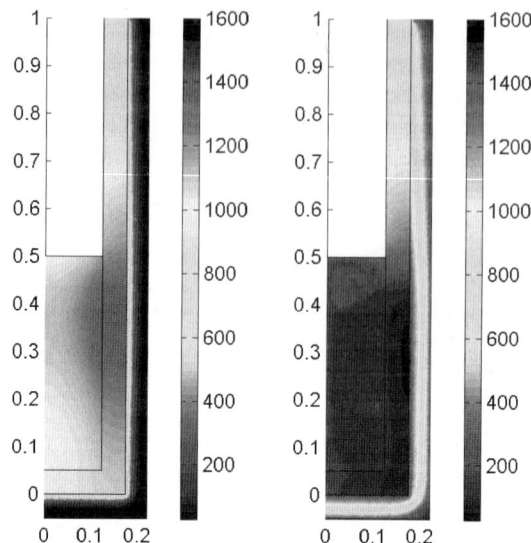

Fig. 10. Temperature field for $t = 30\,\text{min}$ (*left*) and $t = 180\,\text{min}$ (*right*)

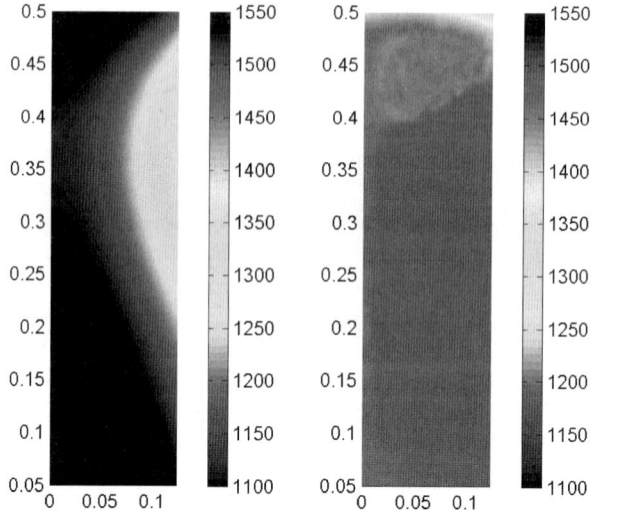

Fig. 11. Silicon temperature for $t = 30\,\text{min}$ (*left*) and $t = 180\,\text{min}$ (*right*)

for the same times. During the first 30 min, the temperature of the workpiece increases and the silicon begins to melt (the melting point is 1,412 °C) and after 180 min the silicon is completely liquid. Figure 12 shows the modulus of current density also for 30 min and 180 min, respectively. Notice that, since solid silicon is not an electric conductor, the induced current density concentrates in the graphite. As silicon temperature increases, so does its electrical conductivity and the induced current density on its surface.

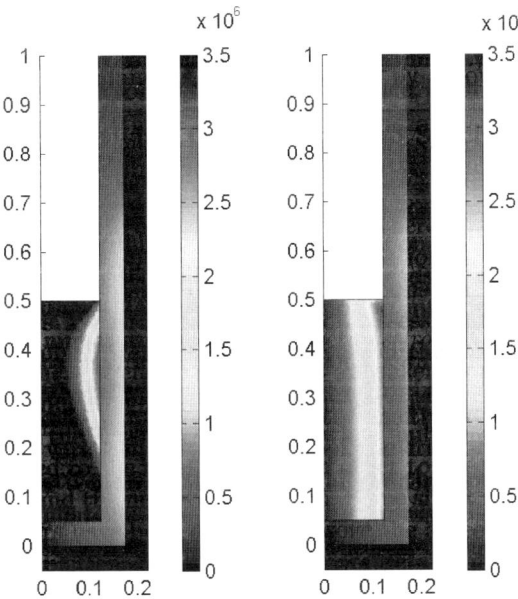

Fig. 12. Modulus of current density for $t = 30 \, \text{min}$ (*left*) and $t = 180 \, \text{min}$ (*right*)

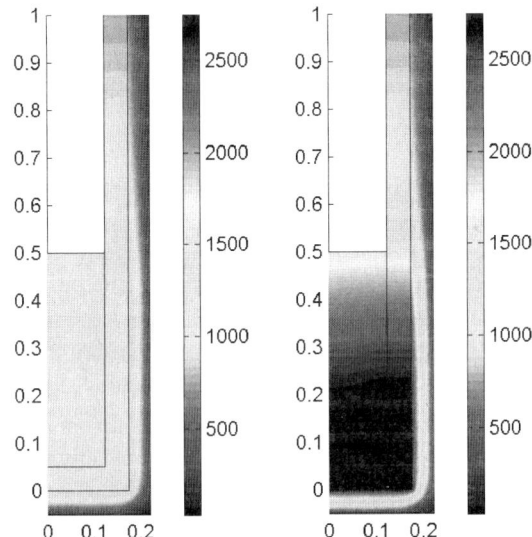

Fig. 13. Temperature with and without convection term ($t = 180 \, \text{min}$)

Figures 13 and 14 illustrate the importance of considering convective heat transfer when computing the temperature field. In Fig. 13 one can check how neglecting the convection term in the heat equation could cause the materials to reach very high and unrealistic temperatures that, in particular, would

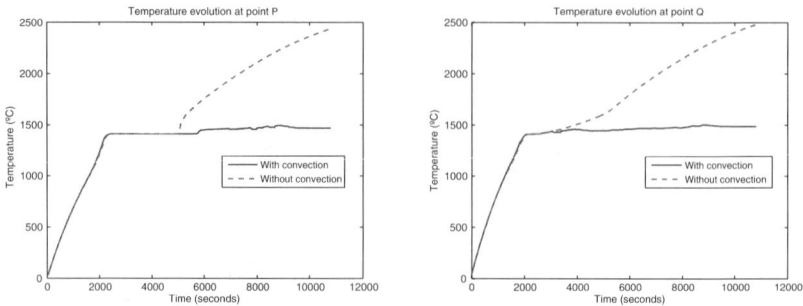

Fig. 14. Evolution of temperature at points P and Q

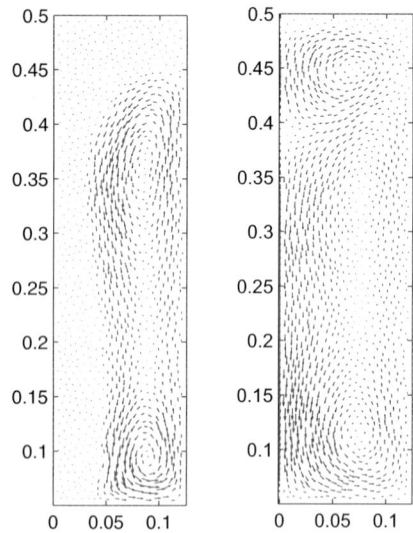

Fig. 15. Velocity field $t = 90\,\mathrm{min}$ (*left*) and $t = 180\,\mathrm{min}$ (*right*)

cause the crucible to melt. The same conclusions can be obtained from Fig. 14, that shows the evolution in time of the temperature of two different points in the silicon: a point P close to the symmetry axis and another point Q close to the graphite crucible, considering or not the convection term.

We complete these results by representing, in Fig. 15, the velocity field for times $t = 90$ and $t = 180\,\mathrm{min}$, respectively. We can appreciate the swirls due to Lorentz forces.

Acknowledgements

This work has been partly supported by MEC-FEDER (Spain) through research project DPI2003-01316 and Ferroatlántica I+D company under contract. The last author has received financial support by a predoctoral grant

FPI from the Spanish Ministry of Education and Science, cofinanced by the European Social Fund.

References

[BGMS1]   Bermúdez, A., Gómez, D., Muñiz, M.C., Salgado, P: Transient numerical simulation of a thermoelectrical problem in cylindrical induction heating furnaces. Adv. Comput. Math., To appear

[BGMS2]   Bermúdez, A., Gómez, D., Muñiz, M.C., Salgado, P: A FEM/BEM for axisymmetric electromagnetic and thermal modelling of induction furnaces, Syst. Int. J. Numer. Meth. Engng, To Appear.

[BF91]   Brezzi, F., Fortin, M.: Mixed and hybrid finite element methods. Springer Verlag, New York, (1991)

[CSL04]   Cajner, F., Smoljan, B., Landek, D.: Computer simulation of induction hardening, J. Mater. Process. Technol. **157-158**, 55–60 (2004)

[CETAL]   Chaboudez, C., Clain, S., Glardon, R., Mari, D., Rappaz, J., Swierkosz, M.: Numerical Modeling in Induction Heating for Axisymmetric Geometries, IEEE Trans. Magn., **33**, 739–745 (1997)

[MR97]   Chen, Q.S., Gao, P., Hu, W.R.: Effects of induction heating on temperature distribution and growth rate in large-size SiC growth system, J. Cryst. Growth, **266**, 320–326 (2004)

[CRST93]   Clain, S., Rappaz, J., Swierkosz, M., Touzani, R.: Numerical modelling of induction heating for two-dimensional geometries, Math. Models Appl. Sci., **3**, 805–822 (1993)

[HO94]   Henneberger, G., Obrecht, R.: Numerical calculation of the temperature distribution in the melt of industrial crucible furnaces In: Second International Conference on Computation in Electromagnetics. (1994)

[HSSH93]   Henneberger, G., Sattler, Ph. K., Shen, D., Hadrys, W.: Coupling of magnetic and fluid flow problems and its application in induction melting apparatus, IEEE Trans. Magn., **29**, 1589–1594 (1993)

[KHT96]   Katsumura, Y., Hashizume, H., Toda, S.: Numerical Analysis of fluid flow with free surface and phase change under electromagnetic force, IEEE Trans. Magn., **32**, 1002–1005 (1996)

[KP03]   Klein, O., Philip, P.: Transient numerical investigation of induction heating during sublimation growth of silicon carbide single crystals, J. Cryst. Growth, **247**, 219–235 (2003)

[MP94]   Mohammadi, B., Pironneau, O.: Analysis of the k-epsilon turbulence model. Wiley/Masson, New York (1994)

[NEK99]   Natarajan, T.T., El-Kaddah, N.: A methodology for two-dimensional finite element analysis of electromagnetically driven flow in induction stirring systems, IEEE Trans. Magn., **35**, n.3, 1773–1776 (1999)

[PIR82]   Pironneau, O.: On the transport-diffusion algorithm and its applications to the Navier-Stokes equations. Numer. Math., **38**, n.3, 309–332 (1982)

[RS96]   Rappaz, J., Swierkosz, M.: Mathematical modelling and numerical simulation of induction heating processes. Appl. Math. Comput. Sci., **6**, n.2, 207–221 (1996)

[WKN94]   Wanser, S., Krähenbühl, L., Nicolas, A.: Computation of 3D induction hardening problems by combined finite and boundary elements methods. IEEE Trans. Magn., **30**, 3320–3323 (1994)

Growth and Pattern Formation for Thin Films

Russel E. Caflisch

Department of Mathematics and Department of Material Science & Engineering,
University of California, 405 Hilgard Avenue, Los Angeles, CA 90095-1555, USA
caflisch@math.ucla.edu

Summary. Epitaxy is the growth of a thin film by attachment to an existing substrate in which the crystalline properties of the film are determined by those of the substrate. In heteroepitaxy, the substrate and film are of different materials, and the resulting mismatch between lattice constants can introduce stress into the system. We have developed an island dynamics model for epitaxial growth that is solved using a level set method. This model uses both atomistic and continuum scaling, since it includes island boundaries that are of atomistic height, but describes these boundaries as smooth curves. The strain in the system is computed using an atomistic strain model that is solved using an algebraic multigrid method and an artificial boundary condition. Using the growth model together with the strain model, we simulate pattern formation on an epitaxial surface.

1 Introduction

Epitaxy is the growth of a thin film on a substrate in which the crystal properties of the film are inherited from those of the substrate. Since an epitaxial film can (at least in principle) grow as a single crystal without grain boundaries or other defects, this method produces crystals of the highest quality.

The geometry of an epitaxial surface consists of step edges and island boundaries, across which the height of the surface increases by one crystal layer, and adatoms which are weakly bound to the surface. Epitaxial growth involves deposition, diffusion, and attachment of adatoms on the surface. Deposition is from an external source, such as a molecular beam. The principal dimensionless parameter (for growth at low temperature) is the ratio $D/(a^4 F)$, in which a is the lattice constant and D and F are the adatom diffusion coefficient and deposition flux. It is conventional to refer to this parameter as D/F, with the understanding that the lattice constant serves as the unit of length. Typical values for D/F are in the range of 10^4–10^8.

2 Island Dynamics

Burton, Cabrera, and Frank [2] developed the first detailed theoretical description for epitaxial growth. In this "BCF" model, the adatom density solves a diffusion equation with an equilibrium boundary condition ($\rho = \rho_{eq}$), and step edges (or island boundaries) move at a velocity determined from the diffusive flux to the boundary. Modifications of this theory were made, for example in [11], to include line tension, edge diffusion, and nonequilibrium effects. These are "island dynamics" models, since they describe an epitaxial surface by the location and evolution of the island boundaries and step edges. They employ a mixture of coarse graining and atomistic discreteness, since island boundaries are represented as smooth curves that signify an atomistic change in crystal height.

Adatom diffusion on the epitaxial surface is described by a diffusion equation of the form

$$\partial_t \rho - D\nabla^2 \rho = F - 2dN_{nuc}/dt \tag{1}$$

in which the last term represents loss of adatoms due to nucleation, and desorption from the epitaxial surface has been neglected. Attachment of adatoms to the step edges and the resulting motion of the step edges are described by boundary conditions at an island boundary (or step edge) Γ for the diffusion equation and a formula for the step-edge velocity v. The simplest of these is

$$\rho = \rho_* \tag{2}$$
$$v = D[\partial \rho / \partial n]$$

in which the brackets indicate the difference between the value on the upper side of the boundary and the lower side. Two choices for ρ_* are $\rho_* = 0$, which corresponds to irreversible aggregation in which all adatoms that hit the boundary stick to it irreversibly, and $\rho_* = \rho_{eq}$ for reversible aggregation. For the latter case, ρ_{eq} is the adatom density for which there is local equilibrium between the step and the terrace [2]. Numerical details on implementation of the level set method for thin film growth are provided in [5].

2.1 Nucleation

For the case of irreversible aggregation, a dimer (consisting of two atoms) is the smallest stable island, and the nucleation rate is

$$\frac{dN_{nuc}}{dt} = D\sigma_1 \langle \rho^2 \rangle, \tag{3}$$

where $\langle \cdot \rangle$ denotes the spatial average of $\rho(\mathbf{x}, t)^2$ and

$$\sigma_1 = \frac{4\pi}{\ln[(1/\alpha)\langle \rho \rangle D/F]} \tag{4}$$

is the adatom capture number as derived in [1]. The parameter α reflects the island shape, and $\alpha \simeq 1$ for compact islands. Expression (3) for the nucleation rate implies that the time of a nucleation event is chosen deterministically. Whenever $N_{\mathrm{nuc}}L^2$ passes the next integer value (L is the system size), a new island is nucleated. Numerically, this is realized by raising the level set function to the next level at a number of grid points chosen to represent a dimer.

The choice of the location of the new island is determined by probabilistic choice with spatial density proportional to the nucleation rate ρ^2. This probabilistic choice constitutes an atomistic fluctuation that must be retained in the level set model for faithful simulation of the epitaxial morphology. For growth with compact islands, computational tests have shown additional atomistic fluctuations can be omitted [16].

Additions to the basic level set method, such as finite lattice constant effects and edge diffusion, are easily included [17]. The level set method with these corrections is in excellent agreement with the results of kinetic Monte Carlo (KMC) simulations.

2.2 The Level Set Method

Within the level set approach, the union of all boundaries of islands of height $k + 1$, can be represented by the level set $\varphi = k$, for each k. For example, the boundaries of islands in the submonolayer regime then correspond to the set of curves $\varphi = 0$. The function ϕ is the level set function that evolves according to

$$\frac{\partial \phi}{\partial t} + v|\nabla \phi| = 0. \tag{5}$$

All the physical information is in the normal component v of the velocity function. Islands grow because atoms diffuse toward and attach to island boundaries, and shrink because they can detach from an island boundary.

3 Discrete Elasticity

In heteroepitaxy, strain is introduced into the epitaxial system due to the lattice mismatch between the two constituents of the material. Because of the strain, atoms are displaced by a vector $\mathbf{u}$ from their lattice position. The following discussion of atomistic strain and stress follows that in [19].

To describe the strain energy at each atom, $\mathbf{i} = (i, j, k)$, introduce the translation operators, $T_k^{\pm}$, and the discrete difference operators, $D_k^{\pm}$, D_k^0, defined as follows:

$$T_k^{\pm} f(\mathbf{i}) = f(\mathbf{i} \pm \mathbf{e_k}),$$

$$D_k^+ f(\mathbf{i}) = \frac{(T_k^+ - 1)f(\mathbf{i})}{h},$$

$$D_k^- f(\mathbf{i}) = \frac{(1 - T_k^-) f(\mathbf{i})}{h},$$

$$D_k^0 f(\mathbf{i}) = \frac{(T_k^+ - T_k^-) f(\mathbf{i})}{2h},$$

where h is the lattice constant and $\mathbf{e}_k$ is the vector in the kth direction for $k = 1, 2, 3$ with $\|\mathbf{e}_k\| = h$. Throughout this paper, we assume the lattice constant $h = 1$ for simplicity. We use i for the depth-like index, with $-\infty < i \leq n$. Here n is the maximum height of the material. An ABC is sought at $i = 0$, assuming that there is no force for $i < 0$.

Let $\mathbf{u}(\mathbf{i}) = (u_k(\mathbf{i}))_{k=1,\ldots,d}$ be the displacement at the discrete point $\mathbf{i}$ relative to an equilibrium lattice. The discrete strain components defined below ((6) and (7)) can be used to describe the discrete elastic energy. For $k, \ell = 1, 2, 3$ and $p, q = \pm$,

$$S_{k\ell}^{\pm}(\mathbf{u}(\mathbf{i})) = D_\ell^{\pm} u_k(\mathbf{i}), \tag{6}$$

$$S_{k\ell}^{pq}(\mathbf{u}(\mathbf{i})) = \frac{1}{2}(D_\ell^q u_k(\mathbf{i}) + D_k^p u_\ell(\mathbf{i})). \tag{7}$$

The discrete energy density at a point $\mathbf{i}$ is then given by

$$E(\mathbf{i})(\mathbf{u}, \mathbf{u}) = \sum_{k,p} \alpha_k^p (S_{kk}^p(\mathbf{u}))^2 + \sum_{k \neq \ell, p, q} \left\{ 2\beta_{k\ell}^{pq} (S_{k\ell}^{pq}(\mathbf{u}))^2 + \gamma_{k\ell}^{pq} S_{kk}^p(\mathbf{u}) S_{\ell\ell}^q(\mathbf{u}) \right\}.$$

The total energy is the sum

$$\mathcal{E} = \sum_{\mathbf{i}} E(\mathbf{i}). \tag{8}$$

The atomistic strain is determined by minimizing this energy with respect to variations in $\mathbf{u}$.

An effective numerical method for solving the atomistic strain equations using an algebraic multigrid method was developed in [4]. Moreover an artificial boundary condition can be imposed in the substrate close to the interface with the film, to greatly accelerate the computation [10].

4 Directed Self-Assembly

Regular patterns of nanoscale features, such as quantum dots [6, 7, 12], on an epitaxial surface are of considerable interest for possible applications, ranging from memory and logical devices to lasers. Features of this size are difficult to obtain by standard "top-down" approaches, such as lithography. The spontaneous growth of quantum dot arrays is a promising "bottom-up" approach, but it has proved difficult to control the size and spacing of quantum dots obtained in this way. Directed self-assembly is an intermediate approach, in which formation of the desired patterns is guided by prepatterning of the epitaxial system. For example, subsurface dislocation arrays have been suggested as a prepatterning method [8,18]. These buried dislocations introduce a long-range strain field, which alters the potential energy surface (PES) of the system. Similarly, islands that are capped by a buffer layer of a different material introduce a long-range strain field. It has been shown by density-functional theory (DFT) calculations for metal systems [15] and semiconductor systems [14] that both the adsorption energy E_{ad} and the transition energy E_{trans} of the PES change upon strain.

We model epitaxial growth on a surface with a spatially varying, anisotropic PES, using the following modification of the adatom diffusion equation (1)

$$\frac{\partial \rho}{\partial t} = F + \nabla \cdot (\mathbf{D}\nabla\rho) - 2\frac{dN}{dt} + \nabla \cdot \left(\frac{\rho}{k_B T}\mathbf{D}(\nabla E_{ad})\right). \qquad (9)$$

In (9), $\mathbf{D}$ is a diffusion tensor where the diagonal entries are labeled $D_i(\mathbf{x})$ and $D_j(\mathbf{x})$, and correspond to diffusion along the two directions i and j. For simplicity no other direction for diffusion is included (but could easily be incorporated). The last term is the thermodynamic drift, where k_B is the Boltzmann constant, and T is the temperature. We enforce a boundary condition $\rho(\mathbf{x}) = \rho_{eq}(D_{det}(\mathbf{x}), \mathbf{x})$, where $D_{det}(\mathbf{x})$ is a (spatially varying) detachment rate [3].

We assume a simple sinusoidal variation of E_{ad} and E_{trans}. Figure 1 shows the resulting patterns for PES with spatial variation that is one dimensional (left) and two dimensional (right). These simulation results bear a striking resemblance to the quantum dot patterns obtained in the experimental results of [8].

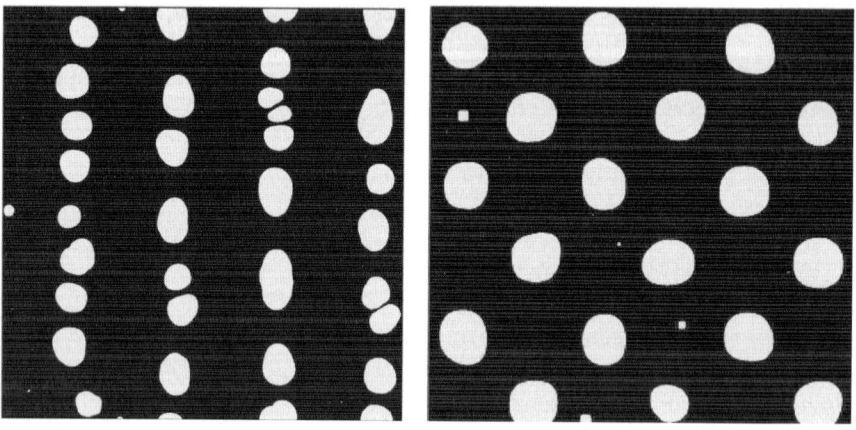

Fig. 1. Pattern formation for monolayer height islands due to a spatially varying PES, with sinusoidal variation in 1D (*left*) and 2D (*right*)

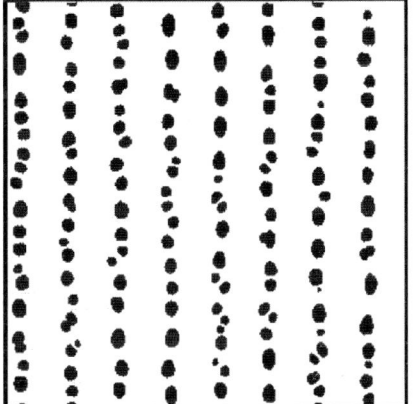

Fig. 2. Morphologies at coverages $\Theta = 0.1$ ML (*left*) and $\Theta = 0.3$ ML (*right*) obtained with a PES that has a much narrower variation

The morphologies shown so far were all obtained at a submonolayer precoalescence coverage of $\Theta = 0.2$ ML and with a PES that varies sinusoidally. Figure 2 shows the patterns that are obtained by a function that has sharper peaks that those of a sine function. The resulting islands at coverage $\Theta = 0.1$ monolayer (ML) are highly aligned. Moreover, at $\Theta = 0.3$ ML, all the islands that are aligned along the j-direction have coalesced in this direction, forming monolayer height "wires." For more details on these computations, see [13].

5 Conclusions

The island dynamics/level set method is capable of simulating epitaxial growth with processes such as adatom detachment from islands that would slow down other approaches. It can also be effectively combined with an atomistic strain code to simulate heteroepitaxial growth. The combined method can be used to study pattern formation due to strain in self-assembly and directed self-assembly.

Acknowledgements

This research was supported in part by the MARCO Center on Functional Engineered NanoArchitectonics (FENA) and by the NSF through grant DMS-0402276.

References

1. G.S. Bales and D.C. Chrzan. Dynamics of irreversible island growth during submonolayer epitaxy. *Phys. Rev. B*, **50**, 6057–6067, 1994.
2. W.K. Burton, N. Cabrera and F.C. Frank. The growth of crystals and the equilibrium structure of their surfaces. *Phil. Trans. Roy. Soc. London Ser. A*, **243**, 299–358, 1951.
3. R.E. Caflisch, W.E., M.F. Gyure, B. Merriman, and C. Ratsch, Phys. Rev. E. **59**, 6879 (1999).
4. R.E. Caflisch, Y.-J. Lee, S. Shu, Y. Xiao and Jinchao Xu. An application of multigrid methods for a discrete elastic model for epitaxial systems JCP, 219 (2006) 697–714.
5. S. Chen, M. Kang, B. Merriman, R.E. Caflisch, C. Ratsch, R. Fedkiw, M.F. Gyure and S. Osher. Level set method for thin film epitaxial growth. *Journ. Comp. Phys.*, **167**, 475–500, 2001.
6. D.J. Eaglesham and M. Cerullo, Phys. Rev. Lett. **64**, 1943 (1990).
7. S. Guha, A. Madhukar, and K.C. Rajkumar, Appl. Phys. Lett. **57**, 2110 (1990).
8. H.J. Kim, Z.M. Zhao, and Y.H. Xie, Phys. Rev. B **68**, 205312 (2003).
9. L.D. Landau and E.M. Lifshitz, *Theory of Elasticity*, Butterworth-Heinemann, Oxford, UK, 1986.
10. R.E. Sunmi Lee, Caflisch and Y.-J. Lee Artificial Boundary Conditions for Discrete Elasticity SIAM J. Applied Math. 66 (2006) 1749–1775.
11. B. Li and R.E. Caflisch. Analysis of island dynamics in epitaxial growth. *Multiscale Model. Sim.*, **1**, 150–171, 2002.
12. Y.-W. Mo, D.E. Savage, B.S. Swartzentruber and M.G. Lagally, Phys. Rev. Lett. **65**, 1020 (1990).
13. X. Niu, R. Vardavas, R.E. Caflisch and C. Ratsch. A Level Set Simulation of Directed Self-Assembly during Epitaxial Growth. Phys. Rev. B (2006) to appear.
14. E. Penev, P. Kratzer and M. Scheffler, Phys. Rev. B **64**, 085401 (2001).
15. C. Ratsch, A.P. Seitsonen and M. Scheffler, Phys. Rev. B **55**, 6750 (1997).
16. C. Ratsch, M.F. Gyure, S. Chen, M. Kang and D.D. Vvedensky. Fluctuations and scaling in aggregation phenomena. *Phys. Rev. B*, **61**, 10598–10601, 2000.
17. C. Ratsch, M.F. Gyure, R.E. Caflisch, F. Gibou, M. Petersen, M. Kang, J. Garcia and D.D. Vvedensky. Level-set method for island dynamics in epitaxial growth. *Phys. Rev. B*, **65**, #195403, U697–U709, 2002.
18. A.E. Romanov, P.M. Petroff and J.S. Speck, Appl. Phys. Lett. **74**, 2280 (1999).
19. A.C. Schindler, M.F. Gyure, D.D. Vvedensky, R.E. Caflisch, C. Connell and G. D. Simms. Theory of Strain Relaxation in Heteroepitaxial Systems Phys. Rev. B 67 (2003): art. no. 075316

On Waves in Fluids: Some Mathematical, Physical and Engineering Aspects

L.M.B.C. Campos

Centro de Ciências e Tecnologias Aeronáuticas e Espaciais (CCTAE)
and Secção de Mecânica Aeroespacial (SMA), Instituto Superior Técnico (IST),
1049-001 Lisboa, Portugal
lmbcampos.aero@mail.ist.utl.pt

Summary. The subject of waves in fluids is addressed from three complementary points-of-view: (Sect. 2) 60 mathematical forms of the acoustic wave equation in fluids, applying to linear and non-linear, non-dissipative and dissipative, sound waves in homogeneous or inhomogeneous, steady or unsteady media, at rest or in motion, e.g. potential and vortical flows; (Sect. 3) the physical interactions between (i) sound waves due to pressure fluctuations in a compressible fluid, with (ii) magnetic waves in an ionized fluid under external magnetic fields, (iii) internal waves in a stratified fluid under gravity and (iv) inertial waves due to Coriolis forces on a rotating fluid, viz. magneto-acoustic-gravity-inertial waves; (Sect. 4) some engineering problems in the area of aerocoustics, which has applications to aircraft, helicopters, rockets and other aerospace vehicles, including acoustic fatigue, sonic boom, interior noise and airport noise, concentrating on the last aspect.

1 Introduction

The classical wave equation describes the propagation of (i) linear (ii) non-dissipative sound waves in a (iii) homogeneous and steady medium (iv) at rest. There are many practical situations in which one or more of the assumptions (i)–(iv) do not hold, hence the importance to extend the acoustic wave equation to (i) inhomogeneous and unsteady media, for which mean state properties (such as mass density and sound speed), may depend, respectively, on position and time; (ii) moving media, e.g. potential mean flows, or vortical mean flows, such as shear flows or swirling flows; (iii) dissipation by thermal conduction and bulk and shear viscosity and (iv) non-linear effects, either weak or strong, depending on whether only second-order or also higher-order non-linearities are included.

Acoustic waves occur in the low atmosphere and in the ocean, are important in speech, hearing, music and high-fidelity sound reproduction and have applications in ultrasonics (e.g. crack detection), as well as unwanted effects (noise and acoustic fatigue). They are one (i) of the four types of waves in fluids, viz.: (ii) internal waves, in a stratified fluid under gravity occur in the ocean and in the atmosphere of the earth and other planets; (iii) inertial waves associated with the Coriolis force on rotating fluids, affect weather and climate on the earth and occur on other rotating celestial bodies like planets and stars and (iv) magnetic waves in an ionized fluid under an external magnetic field occur in fusion reactors and magnetohydrodynamic generators, in the earth's molten core and high atmosphere (ionosphere) and in the plasma which constitutes stars and permeates the interstellar medium. Their coupling leads to magneto-acoustic-gravity-inertial waves.

Aeroacoustics is a major area of application of acoustics, since it is relevant to many problems of aeronautics and astronauts e.g. (i) the noise of jet and propeller engines at take-off and climb is a major contributor to airport noise; (ii) at approach to land, with the engines at idle, the aerodynamic noise may by comparable; (iii) the sonic boom of supersonic aircraft has so far restricted commercial flight to subsonic speeds overland; (iv) the noise level of rockets is high enough to cause acoustic fatigue of launcher structures and satellite payloads and (v) the helicopter, due to the rotor and gearbox mechanisms it uses, poses noise and vibration problems which limit the exploitation of its ability to hover and fly low and slow near populated areas.

2 Sixty Acoustic Wave Equations

There are at least 60 forms of the acoustic wave equation in fluids (thus excluding solids), which may be grouped in nine classes. The derivation of the most general wave equation in each class can be made by elimination among the equations of fluid mechanics; in some cases variational and other methods can be used as alternatives. Thus, together with overlaps between different classes, there may be several derivations of the same wave equation and multiple cross-checks. In the present account one wave equation in each class is indicated, often but not always the most general [1, 2]. The acoustic wave equation has the same form for all acoustic variables (e.g. potential, gas pressure, mass density and velocity perturbations) for linear non-dissipative sound in an homogeneous steady medium, e.g. for the classical wave equation in a medium at rest or convected wave equation in a uniform flow. In more general conditions this is not usually the case and different acoustic variables satisfy different wave equations, so it is reasonable to aim for the simplest. Note also that non-linear waves are those with steep waveforms, viz. large amplitude waves are non-linear, but small amplitude waves with steep wavefronts ('ripples') are also non-linear.

2.1 Nine Classes of Acoustic Wave Equations

The classical wave equation

$$c_0^{-2}\ddot{\phi} - \nabla^2\phi = 0, \tag{1}$$

where ϕ is the acoustic potential, c_0 the sound speed and dot denotes time derivative $\ddot{\phi} = \partial^2\phi/\partial t^2$ assumes (i) an homogeneous and steady medium; (ii) medium at rest; (iii) linear perturbations and (iv) no dissipation. Next will be presented nine classes of acoustic wave equations, which generalize the classical wave equation.

2.2 Class I: Linear, Non-dissipative Sound in a Potential Mean Flow

The medium in assumed to be a potential flow of velocity $\mathbf{v}_0$, gas pressure p_0, mass density ρ_0 and sound speed c_0 which may depend on position (inhomogeneous medium) and/or on time (unsteady medium). Note that a potential flow is homentropic; in this case there is an acoustic potential ϕ. The wave equation can be deduced from equations of fluid mechanics [3, 4] or a variational method [5, 6]. The variational method uses the acoustic velocity and pressure perturbations:

$$\mathbf{v} = \nabla\phi, \tag{2a}$$

$$p = -\rho_0 \mathrm{d}\phi/\mathrm{d}t, \tag{2b}$$

where $\mathrm{d}/\mathrm{d}t$ is the material derivative for the mean flow:

$$\mathrm{d}/\mathrm{d}t = \partial/\partial t + \mathbf{v}_0 \cdot \nabla. \tag{3}$$

The difference of the kinetic energy per unit volume (4a) and compression energy (4b) in the quadratic approximation:

$$E_v = \rho_0 v^2 = \frac{1}{2}\rho_0(\nabla\phi)^2, \tag{4a}$$

$$E_p = \frac{p^2}{2\rho_0 c_0^2} = \frac{1}{2}\rho_0 c_0^{-2}(\mathrm{d}\phi/\mathrm{d}t)^2, \tag{4b}$$

specifies the acoustic Lagrangian:

$$\pounds(\phi, \dot{\phi}, \nabla\phi; \mathbf{x}, t) = \frac{1}{2}\left[(\nabla\phi)^2 - c_0^2(\dot{\phi} + \mathbf{v}_0 \cdot \nabla\phi)^2\right], \tag{5}$$

which satisfies the principle of stationary action:

$$0 = \delta \int \mathrm{d}^3\mathbf{x} \int \mathrm{d}t \; \pounds(\phi, \dot{\phi}, \nabla\phi; \mathbf{x}, t), \tag{6}$$

leading to the Euler–Lagrange equation:

$$\frac{\partial}{\partial t}\frac{\partial\pounds}{\partial\dot{\phi}} + \nabla \cdot \left[\frac{\partial\pounds}{\partial(\nabla\phi)}\right] = 0. \tag{7}$$

The substitution of (5) in the latter (7) specifies the wave equation in a potential flow (W1–W9 – there are nine particular cases):

$$\frac{d}{dt}\left(\frac{1}{c_0^2}\frac{d\phi}{dt}\right) - \frac{1}{\rho_0}\nabla \cdot (\rho_0 \nabla \phi) = 0. \tag{8}$$

In the case of an homogeneous, steady uniform flow it reduces to the convected wave equation:

$$c_0^{-2}d^2\phi/dt^2 - \nabla^2\phi = 0 \tag{9}$$

and in the general case it has ten terms

$$\ddot{\phi} - c_0^2\nabla^2\phi - c_0^2\nabla\phi \cdot \nabla(\log\rho_0) - 2\dot{\phi}c_0^{-1}\dot{c}_0 + 2(\mathbf{v}_0 \cdot \nabla\dot{\phi})$$
$$+(\dot{\mathbf{v}}_0 \cdot \nabla\phi) - 2\dot{\phi}\mathbf{v}_0 \cdot \nabla(\log c_0) - 2(\dot{\mathbf{v}}_0 \cdot \nabla\phi)c_0^{-1}\dot{c}_0$$
$$+(\mathbf{v}_0 \cdot \nabla)(\mathbf{v}_0 \cdot \nabla\phi) - 2(\mathbf{v}_0 \cdot \nabla\phi)\mathbf{v}_0 \cdot \nabla(\log c_0) = 0 \tag{10}$$

as follows (i) the first two terms form the classical wave equation (1); (ii) the third and fourth terms correspond to an inhomogeneous, unsteady medium at rest; (iii) the fifth term accounts for uniform low Mach number convection; (iv) the sixth to eighth terms includes inhomogeneous, unsteady low Mach number mean flow; (v) the ninth term represents uniform high Mach number convection and (vi) the tenth term includes non-uniform high Mach number mean flow.

2.3 Class II: Non-linear, Non-dissipative Sound in a Potential Mean Flow

The starting point is the exact continuity equation:

$$\nabla^2\Phi = \nabla \cdot \mathbf{V} = \frac{1}{\Gamma}\frac{D\Gamma}{dt} = \frac{1}{\Gamma C^2}\frac{DP}{dt}, \tag{11}$$

where is Φ total potential, $\mathbf{V}$ the total velocity, P the total pressure, Γ the total mass density and C the total sound speed:

$$C^2 = c_*^2 - (\gamma - 1)\left[\dot{\Phi} + \frac{1}{2}(\nabla\Phi)^2\right], \tag{12}$$

where c_* denotes the stagnation sound speed. For homentropic flow with enthalpy H:

$$\frac{1}{\Gamma}\frac{DP}{dt} = \frac{DH}{dt} = \dot{H} + \mathbf{V} \cdot \nabla H = \dot{H} + \nabla\Phi \cdot \nabla H, \tag{13}$$

the Bernoulli equation

$$H + \dot{\Phi} + \frac{1}{2}(\nabla\Phi)^2 = \text{const}, \tag{14}$$

leads to the exact potential equation

$$\ddot{\Phi} - \left\{c_*^2 - (\gamma - 1)\left[\dot{\Phi} + (\nabla\Phi)^2/2\right]\right\}\nabla^2\Phi \quad + 2\Phi \cdot \nabla\dot{\Phi} + \nabla\Phi \cdot [(\nabla\Phi \cdot \nabla)\nabla\Phi] = 0, \tag{15}$$

which includes non-linear terms up to the fourth-order. Assuming that the mean flow is non-uniform but steady (it is not possible to distinguish non-linear waves from a non-uniform, unsteady mean state), the exact wave equation is:

$$\frac{\mathrm{D}}{\mathrm{d}t}\left(\frac{\delta\phi}{\delta t}\right) - \left[c_*^2 - (\gamma - 1)\frac{\delta\phi}{\delta t}\right]\nabla^2\phi - c_*^2\nabla\phi \cdot \nabla(\log\rho_0) = 0, \qquad (16)$$

where the exact (17a) and self-convected (17b) material derivatives are used:

$$\frac{\mathrm{D}}{\mathrm{d}t} = \frac{\partial}{\partial t} + \mathbf{v}_0 \cdot \nabla + \nabla\phi \cdot \nabla, \qquad (17a)$$

$$\frac{\delta}{\delta t} = \frac{\partial}{\partial t} + \mathbf{v}_0 \cdot \nabla + \frac{1}{2}\nabla\phi \cdot \nabla. \qquad (17b)$$

There are six particular cases (W10–W15) of the wave equation (16); it has 15 terms:

$$\begin{aligned}
0 = &\ \ddot{\phi} - c_0^2\nabla\phi - c_0^2\nabla\phi \cdot \nabla(\log\rho_0) + 2(\mathbf{v}_0 \cdot \nabla\dot{\phi}) - 2\dot{\phi}\mathbf{v}_0 \cdot \nabla(\log c_0) \\
&+ (\mathbf{v}_0 \cdot \nabla)(\mathbf{v}_0 \cdot \nabla\phi) - 2(\mathbf{v}_0 \cdot \nabla\dot{\phi})\mathbf{v}_0 \cdot \nabla(\log c_0) + (\gamma - 1)\nabla^2\phi + 2\nabla\dot{\phi} \cdot \nabla\phi \\
&+ \nabla\phi\left[(\mathbf{v}_0 \cdot \nabla)\nabla\phi\right] + \mathbf{v}_0\left[(\nabla\phi \cdot \nabla)\nabla\phi\right] + \nabla\phi\left[(\nabla\phi \cdot \nabla)\mathbf{v}_0\right] \\
&- (\nabla\phi)^2\mathbf{v}_0 \cdot \nabla(\log c_0) + \frac{\gamma - 1}{2}(\nabla\phi)^2\nabla^2\phi + \nabla\phi \cdot \left[(\nabla\phi \cdot \nabla)\nabla\phi\right] \qquad (18)
\end{aligned}$$

(i) the first three coincide with the classical wave equation for linear waves in a steady inhomogeneous medium at rest; (ii) the fourth to seventh terms apply to a linear waves, in a moving medium (10); (iii) the eighth to eleventh terms account for quadratic non-linearities in a homogeneous medium; (iv) the twelfth and thirteenth terms include quadratic non-linearities in an inhomogeneous medium and (v) the 14th and 15th terms show that the highest-order non-linearities are cubic [7,8].

2.4 Class III: Linear, Non-Dissipative Sound in a Quasi-one-Dimensional Duct

Consider (Fig. 1) a straight duct with longitudinal coordinate x, non-uniform cross-section $A(x)$, steady shape (no coupling to elastic walls), containing a one-dimensional mean flow (which is always potential).

The Lagrangian per unit length:

$$\mathcal{L}^* = A\mathcal{L} = \frac{1}{2}\rho_0 A\left[\phi'^2 - c_0^{-2}(\dot{\phi} + v_0\phi')^2\right], \qquad (19)$$

where prime denotes derivative with regard to x, viz. $\phi' \equiv \partial\phi/\partial x$, leads to the high-speed wave equation

$$\frac{\mathrm{d}}{\mathrm{d}t}\left(\frac{1}{c_0^2}\frac{\mathrm{d}\phi}{\mathrm{d}t}\right) - \frac{1}{\rho_0 A}(\rho_0 A\phi')' = 0, \qquad (20)$$

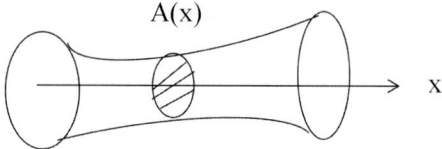

Fig. 1. Quasi-one-dimensional propagation in duct of varying cross-section

which is similar to a one-dimensional form of Class I, replacing in (10) the mass density ρ_0 per unit volume by the mass of fluid per unit length $\rho_0 A$; it has nine particular cases (W16–W24). It consists of 11 terms:

$$0 = \ddot{\phi} - c_0^2 \left[\phi'' + \phi'(A'/A + \rho_0'/\rho_0) \right] - 2\dot{\phi}\dot{c}_0/c_0 + 2v_0\dot{\phi}' + \dot{v}_0\phi'$$
$$-2\dot{\phi}v_0 c_0'/c_0 - 2v_0\phi' c_0'/c_0 + v_0(v_0\phi') - 2v_0^2\phi' c_0'/c_0, \tag{21}$$

namely (i) the first five apply to a horn [9–12] i.e. a duct of non-uniform cross-section without flow and (ii) the last six to a nozzle [13–17], i.e. a duct of non-uniform cross-section with mean flow.

2.5 Class IV: Non-Linear, Non-Dissipative Sound in a Quasi-One-Dimensional Duct

The combination of non-linearity (Class II) with a duct of non-uniform cross-section (Class III), leads to (Class IV) which has six particular cases (W25–W30). The most general is non-linear high-speed nozzle wave equation:

$$\frac{\mathrm{D}}{\mathrm{d}t}\left(\frac{\delta\phi}{\delta t}\right) - \left[c_0^2 - (\gamma - 1)\frac{\delta\phi}{\delta t}\right]\left(\phi'' + \phi'\frac{A'}{A}\right) - c_0^2\phi'\frac{\rho_0'}{\rho}, \tag{22}$$

where (i) the first term involves the non-linear (17a) and self-convected (17b) material derivatives; (ii) the second term has as a factor (12) the non-linear sound speed; (iii) the remaining factor in the second term is the Laplacian replaced by duct wave operator [9–12] and (iv) the last term applies to an inhomogeneous medium. The most general wave equation (W30) of Class IV is

$$0 = \ddot{\phi} - c_0^2\phi'' - c_0^2\phi' A'/A - c_0^2\phi' \rho_0'/\rho_0 - 2\dot{\phi}v_0 c_0'/c_0 + v_0(v_0\phi')'$$
$$-2v_0^2\phi' c_0'/c_0 + 2\dot{\phi}\dot{\phi}' + (\gamma - 1)\phi'\phi'' + (\gamma - 1)\dot{\phi}\phi' A'/A$$
$$+\phi'(v_0\phi')'v_0 + \phi'\phi'' + (\gamma - 1)v_0\phi'^2 A'/A - \phi'^2 v_0 c_0'/c_0$$
$$+\frac{\gamma - 1}{2}\dot{\phi}\phi'^2 + \frac{\gamma - 1}{2}\phi'^2\phi'' + \frac{\gamma - 1}{2}\phi'^3\frac{A'}{A}, \tag{23}$$

has 17 terms (i) the first four apply to linear waves in an inhomogeneous horn; (ii) the terms five to seven concern linear waves in an inhomogeneous nozzle; (iii) the terms eight to ten specify quadratic non-linearities in a horn; (iv) the terms 11–14 represent quadratic non-linearities in a nozzle and (v) the terms 15–17 represent cubic non-linearities. Note that all all cases of potential flows have been covered as shown in Table 1.

Table 1. Acoustics of potential flows

Waves	Free-space	1-D ducts
Linear	Class I	Class III
Non-linear	Class II	Class IV

2.6 Class V: Acoustic Waves in a Unidirectional Shear Flow

The acoustics of vortical flows is considered next, in the particular cases of (Sect. 2.5) shear flows [18–24]; (Sect. 2.6) rotating flows [25–27]. In both cases, since the mean flow is vortical, there is no acoustic potential; the scalar wave equation is obtained for the acoustic pressure perturbation. In a potential mean flow the are two acoustic modes plus decoupled vorticity (by Kelvin's theorem); in a vortical mean flow the sound couples to vorticity leading to a third-order wave equation. The simplest shear flow is unidirectional (24a) and leads to a material derivative (24b):

$$\mathbf{v}_0 = U(y, z)\mathbf{e}_x, \tag{24a}$$

$$\mathrm{d}/\mathrm{d}t = \partial/\partial t + U(y, z)\partial/\partial x. \tag{24b}$$

The acoustic wave equation, for acoustic pressure in unidirectional shear flow, has four (W31–W34) particular cases and consist of four terms:

$$0 = \frac{\mathrm{d}}{\mathrm{d}t}\left[\frac{1}{c_0^2}\frac{\mathrm{d}^2 p}{\mathrm{d}t^2} - \nabla^2 p\right] + \frac{\mathrm{d}}{\mathrm{d}t}\left[\nabla p \cdot \nabla(\log \rho_0)\right]$$

$$+2\rho_0\left(\frac{\partial U}{\partial y}\frac{\partial^2 p}{\partial x \partial y} + \frac{\partial U}{\partial z}\frac{\partial^2 p}{\partial x \partial y}\right) \tag{25}$$

as follows (i) the first two coincide with the convected wave equation (9) for homentropic flow without shear; (ii) the third term applies to isentropic, non-homentropic flow [compare with (10)] without shear and (iii) the fourth term shows that the presence of shear in the mean flow leads to a third-order wave equation.

2.7 Class VI: Acoustics of Sheared and Swirling Axisymmetric Mean Flow

For a rotating fluid, assuming an axisymmetric mean flow and using cylindrical coordinates, the mean flow velocity:

$$\mathbf{v}_0(r) = U(r)\mathbf{e}_z + r\Omega(r)\mathbf{e}_\theta, \tag{26}$$

consists of an axial shear and azimuthal rotation. There are 12 forms of the acoustic wave equation (W36–W60); the most general (W50) reduces in the low Mach number swirl and shear approximation:

$$(\Omega + \mathrm{d}U/\mathrm{d}r)^2 \ll r^2[c_0(r)]^2 \tag{27}$$

to the form:

$$0 = \frac{d}{dt} \left[\frac{1}{c_0^2} \frac{d^2 p}{dt^2} - \frac{1}{r} \frac{\partial}{\partial r} \left(r \frac{\partial p}{\partial r} \right) - \frac{1}{r^2} \frac{\partial^2 p}{\partial \theta^2} - \frac{\partial^2 p}{\partial z^2} \right]$$
$$+ 2 \frac{dU}{dr} \frac{\partial^2 p}{\partial z \partial r} + 2 \frac{d\Omega}{dr} \left(\frac{\partial^2 p}{\partial \theta \partial r} - \frac{1}{r} \frac{\partial p}{\partial \theta} \right)$$
$$+ \frac{d}{dt} \left[\frac{\partial p}{\partial r} \frac{\partial}{\partial r} (\log \rho_0) \right] + \frac{2\Omega}{r} \frac{\partial p}{\partial \theta} \frac{\partial p}{\partial r} \frac{\partial}{\partial r} (\log \rho_0) \qquad (28)$$

consisting of five terms (i) the first term corresponds to the convected wave operator (9) in cylindrical coordinates; (ii) the second term accounts for sheared mean flow [compare with (25)]; (iii) the third term corresponds to swirling mean flow and (iv) the fourth and fifth terms represent isentropic, non-homentropic mean flow.

2.8 Class VII: Viscous and Resistive Dissipation of Linear Sound

The magnitude of the viscous dissipation of sound is comparable to that for thermal conduction so both must be considered. The vorticity decouples and satisfies diffusion equation

$$\mathbf{\Omega} \equiv \nabla \times \mathbf{v}, \qquad (29\text{a})$$
$$\dot{\mathbf{\Omega}} = \nu \nabla^2 \mathbf{\Omega}, \qquad (29\text{b})$$

showing that it is dissipated only by shear viscosity ν; thus it is sufficient to consider a wave equation only for the dilatation $\Psi = \nabla \cdot \mathbf{v}$; the latter is dissipated by shear viscosity ν, bulk viscosity β and thermal conductive diffusivity α. There are two particular cases (W51 and W52) of the linear dissipative acoustic wave equation in an homogeneous medium at rest; it consists of five terms:

$$\ddot{\Psi} - c_0^2 \nabla \dot{\Psi} = (4\nu/3 + \beta + \alpha)\nabla^2 \ddot{\Psi} - c_0^2(\alpha/\gamma)\nabla^4 \Psi - \alpha(4\mu/3 + \beta)\nabla^4 \dot{\Psi} \qquad (30)$$

(i) the first two correspond to the classical wave equation (1) differentiated to the third-order in time; (ii) the third term corresponds to small diffusities and is of the second-order in space and time; (iii) the fourth term is of the fourth-order in space and involves the adiabatic exponent γ and (iv) the last term involves the product of diffusivities, so it applies to large diffusivities and is of the fourth-order in space and first-order in time.

2.9 Class VIII: One-dimensional Viscous Non-Linear Waves in a Quasi-One-Dimensional Duct of Variable Area

The quadratic non-linearities are sufficient to lead to wave front steeping and shock formation; the linear dissipation opposes this, leading to shock widening and decay. This is most readily demonstrated for simple wave (one Riemann invariant zero) with viscous damping using as variable the group velocity. There are four particulars cases (W53–W56) of viscous non-linear simple waves [28–30] in a quasi-one-dimensional duct of variable cross-section. In the most general case the wave equation is

$$\partial W/\partial t + W \partial W/\partial x = (2\nu/3 + \beta/2)\partial^2 W/\partial x^2 - [(\gamma+1)/4]\,CD(\log A)/dt, \quad (31)$$

where exact material derivative (32a) is applied to the group velocity (32b):

$$\frac{D}{dt} = \frac{\partial}{\partial t} + V\frac{\partial}{\partial x}, \tag{32a}$$

$$W \equiv V + C; \tag{32b}$$

there are four terms is (31) (i) the first two are the exact material derivative at group velocity and correspond to non-linear, non-dissipative simple waves (ii) the third term leads to Burger's equation involving shear and bulk viscosities (iii) the fourth term is associated with variable cross-section, e.g. includes cylindrical and spherical non-linear waves.

2.10 Class IX: Three-dimensional Non-linear Beam with Thermoviscous Dissipation

The thermoviscous dissipation coefficient

$$\vartheta \equiv \beta + 4\nu/3 + (\zeta/\rho_0)(1/C_v - 1/C_p), \tag{33}$$

involves the bulk β and shear ν viscosities and the thermal diffusivity ζ and specific heats at constant volume C_v and pressure C_p. It appears in the equation of a non-linear acoustic beam with thermoviscous dissipation:

$$\nabla^2 p - \frac{1}{c_0^2}\frac{\partial^2 p}{\partial t^2} + \frac{\vartheta}{c_0^4}\frac{\partial^3 p}{\partial t^3} = \frac{\gamma+1}{2\rho_0 c_0^4}\frac{\partial^2}{\partial t^2}(p^2). \tag{34}$$

The particular cases are (i) without dissipation the Westerwelt equation [31] and (ii) with viscous dissipation but no thermal dissipation the KZK-equation [32, 33]. The wave equation (34) consists of four terms (i) the first two are the classical wave equation (1); (ii) the third corresponds (33) to weak thermoviscous dissipation with no products of diffusivities and (iii) the fourth accounts for a weak quadratic non-linearity with no cubic terms.

In conclusion, without including chemical reaction or multi-phase flow, 60 forms of acoustic wave equation have been obtained in nine classes; this is still far from an exhaustive combination of all effects of (i) non-linearity, (ii) dissipation, (iii) mean flow and (iv) unsteady and inhomogeneous media. Thus there is plenty of scope for further developments.

3 Generation of Magneto-Acoustic-Gravity-Inertial (MAGI) Waves

There are four types of volume waves in fluids indicated in Table 2 plus all possible interactions, several of which occur in nature and/or have technological applications as shown in Table 3. Whereas for acoustic waves (Sect. 2) only propagation and dissipation was considered, for coupled magneto-acoustic-gravity-inertial waves generation and radiation will be considered as well [34–39].

3.1 Five Fundamental Equations of Fluids Under External Forces

The five fundamental equations of fluid mechanics including external forces are (i) the momentum equation for the velocity

$$\rho \left[\frac{d\mathbf{v}}{dt} + 2\mathbf{\Omega} \times \mathbf{V} + \mathbf{\Omega} \times (\mathbf{\Omega} \times \mathbf{X}) + \dot{\mathbf{\Omega}} \times \mathbf{X} \right] + \nabla p$$

$$= \rho \mathbf{g} + \frac{\mu}{4\pi} \mathbf{H} \times (\nabla \times \mathbf{H}) + \frac{\partial \sigma_{ij}}{\partial x_j}, \tag{35}$$

including the inertia force (equal to mass density times acceleration), the forces associated with rotation (Coriolis and centrifugal forces and unsteady rotation), the pressure gradient and gravity and magnetic forces (μ is the magnetic permeability); (ii) the continuity equation for the mass density

Table 2. Types of waves in fluids

Type of wave	Restoring force	Medium
Acoustic	Pressure	Compressible
Magnetic	Magnetic	Ionized
Internal	Gravity	Stratified
Inertial	Coriolis	Rotating

Table 3. Types of waves in fluids

Waves	Relevance
Acoustic	Sound, noise, speech, hearing, ultrasonics
Magnetic	Cold plasmas, fusion reactors
Internal, inertial	Ocean, atmosphere, weather
Acoustic-gravity	Ocean, atmosphere, weather
Magneto-inertial	Earth molten core, dynamo effect, Magnetic field generation in planets and stars
Magneto-acoustic	Cold plasmas, interstellar space
Magneto-acoustic-gravity	Earth ionosphere, stars
Magneto-acoustic-gravity-inertial	Most general (pulsars)

$$0 = \mathrm{d}\rho/\mathrm{d}t + \rho \nabla \cdot \mathbf{v} = \partial \rho / \partial t + \nabla \cdot (\rho \mathbf{v}); \tag{36}$$

(iii) the induction equation for the magnetic field

$$\partial \mathbf{H}/\partial t + \nabla \times (\boldsymbol{\nabla} \times \mathbf{H}) = -\nabla \times [\chi (\boldsymbol{\nabla} \times \mathbf{H})], \tag{37}$$

including convection of the magnetic field and Ohmic diffusity χ; (iv) the energy equation for entropy

$$\rho T \mathrm{d}s/\mathrm{d}t = -\nabla \cdot (k \nabla T) + \sigma_{ij} \partial V_i / \partial x_j + (\chi/4\pi)(\boldsymbol{\nabla} \times \mathbf{H})^2 + f(\rho, T), \tag{38}$$

including heat conduction, viscous dissipation, electrical resistance and thermal radiation and (v) the equation of state for gas pressure

$$\mathrm{d}p/\mathrm{d}t = c^2 \mathrm{d}\rho/\mathrm{d}t + \alpha \mathrm{d}s/\mathrm{d}t, \tag{39}$$

involving the sound speed and entropy.

3.2 Perturbation of Non-uniform, Steady Mean State

The fluid velocity, mass density, gas pressure, magnetic field and displacement vector

$$\{\mathbf{V}, \rho, p, \mathbf{H}, \mathbf{X}\}(\mathbf{x}, t) = \{\mathbf{0}, \rho_0, p_0, \mathbf{B}, \mathbf{r}\} + \{\mathbf{v}, \rho', p', \mathbf{h}, \boldsymbol{\xi}\}(\mathbf{x}, t) \tag{40}$$

consist of (i) a mean state, which is a medium at rest stratified $\rho_0(\mathbf{x}), p_0(\mathbf{x})$ under a non-uniform external magnetic field $\mathbf{B}(\mathbf{x})$ and (ii) an unsteady and non-uniform perturbation. The mean state is given by (i) the equations of continuity and state are trivially satisfied and (ii) the equations of induction (41a) and momentum (41b):

$$0 = \nabla \times [\chi (\boldsymbol{\nabla} \times \mathbf{B})]; \tag{41a}$$

$$\nabla p_0 = \rho_0 \mathbf{g} - (\mu/4\pi) \mathbf{B} \times (\boldsymbol{\nabla} \times \mathbf{B}) - \rho_0 [\boldsymbol{\Omega} \times (\boldsymbol{\Omega} \times \mathbf{r})]; \tag{41b}$$

the latter specifies magneto-rotating-hydrostatic equilibrium. Subtracting out the mean state, (41) from (35)–(39) the exact non-linear dissipative perturbation equations are obtained.

3.3 Exact Non-linear Dissipative Perturbation Equations

The linear, non-dissipative terms are separated on the l.h.s., from non-linear and/or dissipative terms on the r.h.s. in the Equations of (i) continuity (36):

$$\partial \rho'/\partial t + \nabla \cdot (\rho_0 \mathbf{v}) = -\nabla \cdot (\rho' \mathbf{v}); \tag{42}$$

(ii) state (39)

$$\partial p'/\partial t + \mathbf{v} \cdot \nabla p_0 - c_0^2 \rho_0 (\nabla \cdot \mathbf{v}) = -\mathbf{v} \cdot \nabla p' + (\rho_0 c_0^2 - \rho c^2)(\nabla \cdot \mathbf{v}) + \alpha \, \mathrm{d}s/\mathrm{d}t \equiv Z, \tag{43}$$

(iii) induction (37)

$$\partial \mathbf{h}/\partial t + \nabla \times (\mathbf{v} \times \mathbf{B}) = -\nabla \times (\mathbf{v} \times \mathbf{h}) - \nabla \times [\chi (\boldsymbol{\nabla} \times \mathbf{h})] \equiv \mathbf{Y}, \tag{44}$$

(iv) momentum (35)

$$\rho_0 \partial \mathbf{v}/\partial t + 2\mathbf{\Omega} \times \mathbf{v} + \mathbf{\Omega} \times (\mathbf{v} \times \boldsymbol{\xi}) + \nabla p' - \rho' \mathbf{g}$$
$$+ (\mu/4\pi) \left[\mathbf{B} \times (\nabla \times \mathbf{h}) + \mathbf{h} \times (\nabla \times \mathbf{B}) \right]$$
$$= \rho' \partial \mathbf{v}/\partial t - \rho(\mathbf{v} \cdot \nabla)\mathbf{v} - (\mu/4\pi) \left[\mathbf{h} \times (\nabla \times \mathbf{h}) \right] + \partial \sigma_{ij}/\partial x_j \equiv \mathbf{X}. \ (45)$$

The elimination between the linear, non-dissipative terms is performed by applying $\partial/\partial t$ to momentum equation (45) and substituting $\partial p'/\partial t$ from the continuity equation (42), $\partial p'/\partial t$ from the equation of state (43) and $\partial \mathbf{h}/\partial t$ from the induction equation (44). The result is the MAGI wave equation with sources.

3.4 Linear, Non-dissipative MAGI Wave Operator in an Inhomogeneous Medium

The MAGI equation in an inhomogeneous medium at rest under rotation and non-uniform gravity and non-uniform external magnetic field is specified by the vector wave equation for the velocity perturbation:

$$\mathbf{Q} \equiv \partial^2 \mathbf{v}/\partial t^2 - \rho_0^{-1} \nabla \left[\rho_0 c_0^2 (\nabla \cdot \mathbf{v}) \right] + \rho_0^{-1} \mathbf{g} \nabla \cdot (\rho_0 \mathbf{v})$$
$$+ \rho_0^{-1} \nabla \left[\rho_0 (\mathbf{v} \cdot \mathbf{g}) \right] + 2\mathbf{\Omega} \times (\partial \mathbf{v}/\partial t) + \mathbf{\Omega} \times (\mathbf{\Omega} \times \mathbf{v})$$
$$+ \rho_0^{-1} \nabla \left\{ \rho_0 \mathbf{v} \cdot \left[\mathbf{\Omega} \times (\mathbf{\Omega} \times \mathbf{r}) \right] \right\} + (\mu/4\pi\rho_0)\mathbf{B} \times \left\{ \nabla \times \left[\nabla \times (\mathbf{B} \times \mathbf{v}) \right] \right\}$$
$$+ (\mu/4\pi\rho_0) \left\{ \mathbf{v} \cdot \left[\mathbf{B} \times (\nabla \times \mathbf{B}) \right] + (\nabla \times \mathbf{B}) \times \left[\nabla \times (\mathbf{v} \times \mathbf{B}) \right] \right\}, \qquad (46)$$

which involves eight terms (i) the first two correspond to sound waves [compare (1)]; (ii) the third and fourth terms correspond to gravity waves; (iii) the fifth to seventh terms apply to inertial waves and (iv) eight and ninth terms concern hydromagnetic waves respectively in uniform and non-uniform external magnetic fields. The remaining terms are non-linear and/or dissipative and act as wave sources:

$$\mathbf{Q} \equiv \rho_0^{-1} \partial \mathbf{X}/\partial t + \mathbf{\Omega} \left\{ \mathbf{\Omega} \times \left[(\mathbf{v} \cdot \nabla)\boldsymbol{\xi} \right] \right\} - \rho_0^{-1} \nabla Z$$
$$+ \rho_0^{-1} \mathbf{g} \left[\nabla \cdot (\rho' \mathbf{v}) \right] - (\mu/4\pi\rho_0) \left[\mathbf{B} \times (\nabla \times \mathbf{Y}) + \mathbf{Y} \times (\nabla \times \mathbf{B}) \right]. \qquad (47)$$

forcing the wave equation.

3.5 Interpretation of MAGI Wave Operator

The MAGI wave operator represents wave propagation in (46), and is interpreted readily in the case of (i) an medium at rest, i.e. no mean flow; (ii) isothermal perfect gas, i.e. constant sound speed c_0; (iii) an uniform external magnetic field, i.e. no mean state electric current $\mathbf{B} = B\mathbf{1}$ and (iv) slow rotation $\Omega^2 \ll \omega^2$ relative to wave frequency ω. Note that the atmospheric mass density and pressure are non-uniform $\rho_0(\mathbf{x}), p_0(\mathbf{x})$, and hence the Alfvén speed:

$$a(\mathbf{x}) = B/\sqrt{4\pi\rho_0(\mathbf{x})} \tag{48}$$

is not constant, in the MAGI wave operator:

$$
\begin{aligned}
0 &= \partial^2\mathbf{v}/\partial t^2 - c_0^2\nabla(\nabla\cdot\mathbf{v}) - a^2\left[(\mathbf{l}\cdot\nabla)^2\mathbf{v} - (\mathbf{l}\cdot\nabla)\nabla(\mathbf{v}\cdot\mathbf{l})\right] \\
&\quad + 2\mathbf{\Omega}\times(\partial\mathbf{v}/\partial t) - \nabla(\mathbf{v}\cdot\mathbf{g}) - (\gamma-1)\mathbf{g}(\nabla\cdot\mathbf{v}) \\
&\quad - a^2\left[\nabla(\nabla\cdot\mathbf{v}) - \mathbf{l}(\mathbf{l}\cdot\nabla)(\nabla\cdot\mathbf{v})\right],
\end{aligned}
\tag{49}
$$

which is interpreted as follows (i) the first two terms correspond to sound waves

$$0 = \partial^2(\nabla\times\mathbf{v})/\partial t^2, \tag{50a}$$

$$0 = \left\{\partial^2/\partial t^2 - c_0^2\nabla^2\right\}(\nabla\cdot\mathbf{v}) \tag{50b}$$

since the vorticity is conserved (50a) and the dilatation satisfies (1) classical wave equation (50b); (ii) the third and fourth terms correspond to Alfvén waves, since the velocity perturbation along magnetic field conserved (51a):

$$0 = \partial^2(\mathbf{v}\cdot\mathbf{l})/\partial t^2, \tag{51a}$$

$$0 = \left\{\partial^2/\partial t^2 - a^2(\mathbf{l}\cdot\nabla)^2\right\}\left[\mathbf{v} - (\mathbf{v}\cdot\mathbf{l})\mathbf{l}\right], \tag{51b}$$

and the transverse velocity perturbation (51b) propagates along magnetic field lines at Alfvén speed; (iii) the fifth term corresponds to inertial waves involving the angular velocity of rotation; (iv) the sixth term accounts for internal waves, involving acceleration of gravity; (v) the seventh term concerns acoustic-gravity waves by coupling gravity to the dilatation; (iv) the eighth and ninth terms concern magneto-acoustic waves, coupling the dilatation and the Alfvén speed.

3.6 MAGI Wave Sources I: Hydrodynamic Tensor (Sound Waves)

In the case of acoustics alone the source term is (52a):

$$Q_i^{(1)} = -\rho_0^{-1}\partial^2 T_{ij}/\partial t\partial x_j, \tag{52a}$$

$$T_{ij} = \rho v_i v_j + (p' - c_0^2\rho')\delta_{ij} + \sigma_{ij}, \tag{52b}$$

where (52b) is the Lighthill tensor [8, 40, 41]. Using the notation:

$$A_i * B_j = A_iB_j + A_jB_i - \frac{1}{2}(\mathbf{A}\cdot\mathbf{B})\delta_{ij}, \tag{53}$$

its three terms are interpreted as follows (i) the first term is a quadrupole representing turbulence as a source of waves:

$$\overset{1}{T}_{ij} = \frac{1}{2}\rho v_i * v_j = \rho v_i v_j - \frac{1}{2}\rho v^2\delta_{ij}, \tag{54}$$

and consists of the Reynolds stresses minus the dynamic pressure; (ii) the second terms is a dipole representing fluid inhomogeneities, i.e. regions of different temperature or chemical composition:

$$\overset{2}{T}_{ij} = (\bar{p} - c_0^2 \rho')\delta_{ij} = \alpha s \delta_{ij};$$

(55a)

$$\bar{p} = p - \frac{1}{2}\rho v^2$$

(55b)

and consist of non-isentropic terms in equation of state, i.e. entropy inhomogeneities and (iii) the third term is the viscous stresses $\overset{3}{T}_{ij} = \sigma_{ij}$ representing viscous dissipation.

3.7 MAGI Wave Sources II: Hydromagnetic Tensor

The hydromagnetic tensor acts as source of magneto-acoustic waves [34–39] and contains all terms involving magnetic field:

$$Q_i^{(2)} = -\rho_0^{-1}\partial^2 R_{ij}/\partial t \partial x_j$$

(56)

and also has three analogous terms (i) the first term is a quadrupole representing wave generation by hydromagnetic turbulence

$$\overset{1}{R}_{ij} = (\mu/8\pi)h_i * h_j = (\mu/4\pi)h_i h_j - (\mu/8\pi)h^2\delta_{ij},$$

(57)

viz. the magnetic stresses are analogous to Reynolds stresses (54); (ii) the second term is a dipole modelling wave generation by ionized inhomogeneities:

$$\overset{2}{R}_{ij} = (\mu/4\pi)B_i * [\nabla \times (\mathbf{v} \times \mathbf{h})]_j$$

(58)

and is non-zero if velocity and magnetic field perturbations are non-parallel [compare (55)] and (iii) the third term accounts for dissipation by Joule effect:

$$\overset{3}{R}_{ij} = (\chi/4\pi)B_i * (\nabla^2 \mathbf{h}))_j$$

(59)

and corresponds to the viscous stresses.

3.8 MAGI Wave Sources III and IV: Hydrorotation and Hydrogravity Tensors

The forcing term involving angular velocity:

$$\mathbf{Q}^{(3)} = \mathbf{\Omega} \wedge \{\mathbf{\Omega} \wedge [(\mathbf{v} \cdot \nabla)]\,\boldsymbol{\xi}\},$$

(60)

is quadratic in rotation velocity, and represents the hydrorotation tensor. All the remaining source terms form the hydrogravity tensor, which has three terms, like the Lighthill and hydromagnetic tensors (i) The first is a dipole term representing wave generation by turbulence:

$$\mathbf{Q}_{41} = -\partial^2[(\rho'/\rho_0)\mathbf{v}]/\partial t^2. \tag{61}$$

(ii) the second is a dipole term representing wave generation by fluid inhomogeneities:

$$\mathbf{Q}_{42} = \rho_0^{-1}\nabla^2(p'\mathbf{v}) + \nabla[(\gamma-1)p'(\nabla\cdot\mathbf{v})] - (\mathbf{g}/\rho_0)\nabla\cdot(\rho'v) + [\gamma/(\gamma-1)]T_0\nabla s; \tag{62}$$

(iii) the third is a dissipative term which involves the entropy:

$$\mathbf{Q}_{43} = -\rho_0^{-1}\nabla(\alpha ds/dt) = -\rho_0^{-1}\nabla\{[\rho T/(\gamma-1)]ds/dt\}, \tag{63}$$

where could be substituted from the energy equation (38).

3.9 Dispersion Relation and Possible Decouplings

For wavelength small compared to scale of non-uniformity of the medium, the Fourier decomposition is used for the wave variable the viz. velocity $\tilde{v}_i$ and wave source i.e. the forcing Q_i:

$$v_i, Q_i(\mathbf{x}, t) = \int\int\int\int_{-\infty}^{+\infty} \tilde{v}_i, \tilde{Q}_i(\mathbf{k}, \omega)e^{i(\mathbf{k}\cdot\mathbf{x}-\omega t)}d^3k\,d\omega, \tag{64}$$

leading to the dispersion relation:

$$D_{ij}(\mathbf{k}, \omega)\tilde{v}_j = \tilde{Q}_i(\mathbf{k}, \omega) \tag{65}$$

involving the dispersion matrix:

$$D_{ij}(\mathbf{k}, \omega) = -\omega^2\delta_{ij} + c_0^2 k_i k_j + a^2\left[k_i k_j + (\mathbf{k}\cdot\mathbf{l})^2\delta_{ij} - (\mathbf{k}\cdot\mathbf{l})(k_i l_j + k_j l_i)\right]$$
$$+i\omega e_{ijk}\Omega_k + k_i g_j - i(\gamma-1)g_i k_j. \tag{66}$$

The possible decouplings depend on four directions (i) the external magnetic field $\mathbf{l}$; (ii) gravity $\mathbf{g}$; (iii) rotation $\mathbf{\Omega}$ and (iv) wavevector $\mathbf{k}$. The dispersion matrix (66) consists of five terms (i) the first two correspond to sound waves, which are isotropic and non-dispersive; (ii) the third corresponds to hydromagnetic waves which are non-dispersive and anisotropic and (iii) the fourth and fifth–sixth terms correspond, respectively, to inertial and gravity waves, which are both dispersive and anisotropic.

3.10 Radiation of MAGI Waves

Inverting the dispersion relation (65) and substituting in the first equation of (64) leads to:

$$v_i(\mathbf{x}, t) = \int \int \int \int_{-\infty}^{+\infty} \left[\bar{D}_{ij}(\mathbf{k}, \omega)/D_{ij}(\mathbf{k}, \omega) \right] \tilde{Q}_j(\mathbf{k}, \omega) e^{i(\mathbf{k}\cdot\mathbf{x}-\omega t)} d^3k \, d\omega,$$

(67)

where $\bar{D}_{ij}$ are the cofactors of dispersion matrix, $\tilde{Q}_j$ the source spectrum, and $D(\mathbf{k}, \omega)$ the determinant of dispersion matrix; its roots specify six modes or three pairs of wave modes:

$$D(\mathbf{k}, \omega) = - \prod_{n=1}^{6} [\omega - \omega_n(\mathbf{k})]$$

(68)

and correspond to poles in $d\omega$ integral in (67), which can be evaluated by residues:

$$v_i(\mathbf{x}, t) = \sum_{m=1}^{M} \int \int \int \int_{-\infty}^{+\infty} F_i^m(\mathbf{k}, t) e^{i\mathbf{k}\cdot\mathbf{x}} d^3k,$$

(69)

where the radiation vector

$$F_i^m(\mathbf{k}, t) = 2\pi i D_{ij}(\mathbf{k}, \omega(\mathbf{k})) \tilde{Q}_j(\mathbf{k}, \omega(\mathbf{k})) \left\{ \prod_{\substack{n=1 \\ n \neq m}}^{6} [\omega_m(\mathbf{k}) - \omega_n(\mathbf{k})] \right\}^{-1} e^{-i\omega_m(\mathbf{k})t},$$

(70)

is valid at arbitrary distance.

3.11 Radiation to Observer in the Far-field

An observer in the far-field receives waves from points in the wavenumber surface where the group velocity points to him; if these are regular points, a quadratic approximation is valid:

$$k_3 - k_{30}^m = \frac{1}{2} \sum_{s=1}^{2} \varepsilon_s^m (k_3 - k_{30}^m),$$

(71)

where $\varepsilon_1^m, \varepsilon_2^m$ are the principal curvatures of wave of wavenumber of wavenumber surface of mode m. The $dk_1 dk_2$ integrals can be evaluated by method of stationary phase, leading to:

$$\mathbf{v}(\mathbf{x}, t) = \sum_{m=1}^{6} \mathbf{G}_m(k_{01}, k_{02}, \mathbf{x}, t)(2\pi/x_3) |\varepsilon_1^m \varepsilon_2^m|^{-1/2}$$

$$\times \exp\left\{ -i \left[-(x_1)^2 \varepsilon_1^m + (x_2)^2/\varepsilon_2^m \right]/2x_3 \right\} \exp\left\{ -\frac{\pi}{4} \sum_{s=1}^{2} \operatorname{sgn}(\varepsilon_s^m) \right\} \tag{72}$$

which may be interpreted as follows (i) the first factor is the radiation vector at the stationary point:

$$\mathbf{G}_m(k_{01}, k_{02}, \mathbf{x}, t) \equiv \int_{-\infty}^{+\infty} \mathbf{F}^m(k_0, t) \exp(i\mathbf{k}_0 \cdot \mathbf{x}) dk_{03} \left\{1 + O(|\mathbf{x}|)^{-1}\right\}; \quad (73)$$

(ii) the second factor is the decay with distance; (iii) the third factor shows that the Gaussian curvature $\varepsilon = \sqrt{\varepsilon_1 \varepsilon_2}$ of wavenumber surface determines the beam aperture so that the energy flux $\varepsilon v^2 \sim$ const. is constant; (iv) the fourth factor shows that the phase decays away from beam direction and (v) the fifth factor is the phase, viz. $-\pi/2$ for a synclastic wavenumber surface [two positive principal curvatures $1 = \text{sgn}(\varepsilon_1) = \text{sgn}(\varepsilon_2)$], and $\pi/2$ for an anti-clastic surface [negative principal curvatures of $\text{sgn}(\varepsilon_1) = \text{sgn}(\varepsilon_2) = -1$], so that there is a phase jump of π when crossing a caustic.

3.12 Law of Intensity of Radiation of MAGI Wave

From an order-of-magnitude evaluation it follows that the intensity of radiation of MAGI waves scales as

$$Q_0 \sim \rho_0 \ell^2 c^{-1} Q^2, \quad (74)$$

where ρ_0 is mass density, ℓ is the lengthscale of source region, c is the phase velocity of waves and Q is the source strength. In aeroacoustics [8, 40–44] the source strength for a monopole, dipole (55a) or quadripole (54) source, respectively:

$$Q_0 \sim \rho_0 u^2 \left\{1, u/c, (u/c)^2\right\} \quad (75)$$

leads to the Lighthill's fourth, sixth and eighth-power laws of velocity:

$$I_0 \sim \rho_0 l^2 c^{-1} u^4 \left\{1, (u/c)^2, (u/c)^4\right\}. \quad (76)$$

For MAGI waves the sources are the hydrodynamic, hydromagnetic, hydrorotation and hydrogravity tensors and lead to the radiation intensity:

$$I \sim \rho_0 l^2 c^{-1} u^4 \left[u^2 + \mu h^2/4\pi\rho_0 + \Omega^2 l^2 + gl\right]^2. \quad (77)$$

For example the law (77) agrees with the solar radiation flux 3×10^{33} erg cm^{-2} s^{-1}, and applies to stellar luminosities [34–39].

4 Some Problems in Aeroacoustics

Table 4 indicates some problems in aeroacoustics [8, 44–47], of which we single ou for more detailed consideration that marked with an asterisk [48, 49] in the Table 4.

Table 4. Some problems in aeroacoustics

Problems in aeroacoustics

- Large amplitude sound
  - Structural fatigue
  - Shock noise
  - Engine surge
  - Sonic boom
- Linear acoustics
  - Internal noise
    - Passive reducion
    - Active reduction
    - Systems (air conditioning, etc...)
  - External noise
    - Propulsion
      - Jets
      - Propellers
      - Rotors
      - Fans/compressors
    - Airframe
      - Undercarriadge
      - Flaps
      - Boundary layer
- Operations
  - Design issues
    - Duct liners*
    - Combustion
    - Shielding
  - Environmental aspects
    - Flight path optimization
    - Atmospheric propagation
    - Ground effects

4.1 Cylindrical or Coaxial Nozzles with Non-Uniform Wall Impedance

Consider an annular (Fig. 2) or cylindrical (Fig. 3) nozzle with an azimuthally varying wall impedance distribution. The reflection of a mode by a the wall creates other modes, e.g. an axisymmetric mode $m = 0$ is reflected non-axisymmetrically by an azimuthally varying wall impendence distribution. Thus the attenuation of sound due to an uniform impedance distribution is modified by intermodal scattering, which can be used to enhance noise absorption.

4.2 Cylindrical Waves in an Axial Nozzle Flow

The convected wave equation in cylindrical coordinates (28) in an uniform axial flow of velocity U with the acoustic pressure as variable is given by:

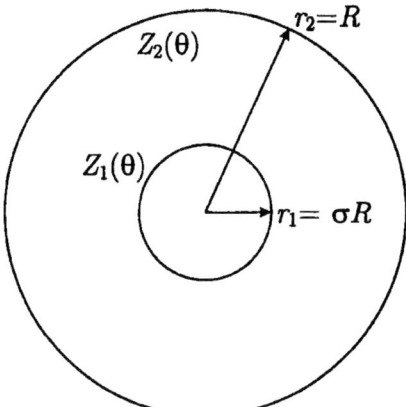

Fig. 2. Annular nozzle with circumferential non-uniform impedance distributions $Z_1(\theta)$ and $Z_2(\theta)$, respectively, over the inner and outer surfaces of the duct

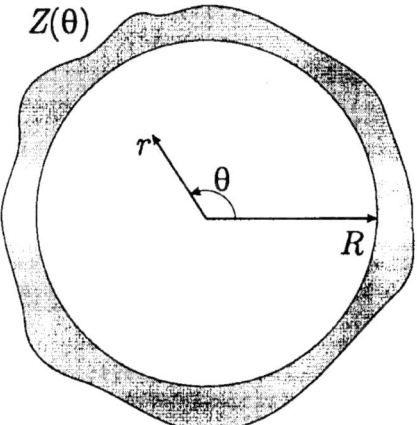

Fig. 3. In the particular case of the cylindrical nozzle of radius $r = R$ the circomferentially varying impedance $Z(\theta)$ causes axisymmetric modes to be reflected non-axisymmetricaly

$$\left\{ \frac{1}{c^2}\left(\frac{\partial}{\partial t} + U\frac{\partial}{\partial z} \right)^2 - \left[\frac{1}{r}\frac{\partial}{\partial r}\left(r\frac{\partial}{\partial r} \right) + \frac{1}{r^2}\frac{\partial^2}{\partial \theta^2} + \frac{\partial^2}{\partial z^2} \right] \right\} p(r,\theta,z,t) = 0, \tag{78}$$

where

$$0 \le r \le R, \quad 0 \le \theta \le 2\pi, \quad 0 \le z \le L, \quad -\infty < t < +\infty. \tag{79}$$

Using a Fourier representation in θ, z, t

$$p(r,\theta,z,t) = \sum_{m=-\infty}^{+\infty} e^{im\theta} \sum_{l=-\infty}^{+\infty} e^{i2\pi lz/L} \int_{-\infty}^{+\infty} e^{-i\omega t} P_{lm}(r,\omega)d\omega, \tag{80}$$

where m is the azimuthal wavenumber, ω the frequency and $2\pi\ell/L$ the axial wavenumber, it follows that the radial dependence is specified by a Bessel equation

$$r^2 P'' + r P' + (k^2 r^2 - m^2)P = 0; \tag{81}$$

the solution for cylindrical nozzle which is finite at the centre is specified by a Bessel function of first kind (82a):

$$P_{lm}(r,\omega) = A_{lm} J_m(kr), \tag{82a}$$

$$k^2 \equiv \frac{1}{c^2}\left(\omega - \frac{2\pi lU}{L}\right)^2 - \left(\frac{2\pi l}{L}\right)^2, \tag{82b}$$

with radial wavenumber (82b).

4.3 Rigid Wall Boundary Condition

The simplest boundary condition is taken for a start, viz. a rigid wall for which the normal velocity is zero:

$$0 = \partial p/\partial r|_{r=R} \quad \Rightarrow \quad 0 = J'_m(kR) \tag{83}$$

so that the eigen-values are specified by the roots of the derivative of the Bessel function:

$$J'_m(j_{mn}) = 0, \tag{84a}$$

$$k_{mn} = j_{mn}/R. \tag{84b}$$

The radial wavenumber and eigen-functions

$$P_m(r; k, \omega) = J_m(j_{mn}r/R) \tag{85}$$

together with the frequencies of natural modes

$$\omega_{lmn} = 2\pi U/L \pm c\sqrt{(2\pi l/L)^2 + (j_{mn}/R)^2}, \tag{86}$$

appear in the total wave field as a superposition of eigen-functions:

$$p(r,\theta,z,t) = \sum_{m=-\infty}^{+\infty} e^{im\theta} \sum_{l=-\infty}^{+\infty} e^{i2\pi lz/L} \sum_{n=1}^{+\infty} e^{-i\omega_{lmn}t} A_{lmn} J_m(k_{mn}r), \tag{87}$$

with amplitudes determined:

$$A_{lmn} = \frac{1}{\pi L}\left[\left(1 - \frac{m^2}{k_{lmn}^2}\right)\{J_m(k_{lmn}R)\}^2\right]$$

$$\times \int_0^{2\pi} e^{-im\theta}\mathrm{d}\theta \int_0^L e^{i2\pi lz/L}\mathrm{d}z \int_0^R r J_m(k_{lmn})p(r,\theta,z,0)\mathrm{d}r \tag{88}$$

from initial wave field.

4.4 Uniform Impedance Boundary Condition

The total wave field (87, 88) applies to other boundary conditions, provided that the eigen-values (84b) and hence the eigen-functions (85) and natural frequencies (86) be modified. The eigen-values are determined next for uniform (Sect. 4.4) and non-uniform (Sect. 4.5) wall impedance. The impedance relates pressure and velocity spectra at wall (89a):

$$\tilde{p}(R, \theta, z; \omega) = \bar{Z}(\omega)\tilde{v}_r(R, \theta, z; \omega), \tag{89a}$$

$$\bar{Z}(\omega) = Z(\omega)/\rho c, \tag{89b}$$

where the specific impedance, i.e. impedance divided by that of a plane wave may be used (89b). The r-component of momentum equation

$$\left(\frac{\partial}{\partial t} + U\frac{\partial}{\partial z}\right)v_r(r, \theta, z, t) + \frac{1}{\rho_0}\frac{\partial}{\partial r}p(r, \theta, z, t) = 0, \tag{90}$$

in terms of eigen-functions

$$\rho_0(\omega - 2\pi l U/L)iV_{lm}(r; \omega) = \mathrm{d}P_{lm}/\mathrm{d}r. \tag{91}$$

specifies the eigen-values or radial wavenumbers as the roots of

$$iZJ'_m(kR) = \alpha_l J_m(kR), \tag{92a}$$

$$\alpha_l \equiv \sqrt{1 + (2\pi l/kL)^2}. \tag{92b}$$

For a rigid wall, i.e. infinite impedance this leads to the roots of $J'_m(kR) = 0$ as before (83, 84a,b); for an impedance wall i.e. complex Z, this leads to complex radial wavenumbers k_{lmn} and complex natural frequencies ω_{lmn}:

$$\omega_{lmn} = 2\pi l U/L \pm c\sqrt{(2\pi l/L)^2 + k_{lmn}^2}, \tag{93}$$

implying mode decay in time

$$\exp(-i\omega_{lmn}t) = \exp\left[-i\Re(\omega_{lmn})t\right]\exp\left[\Im(\omega_{lmn})t\right], \tag{94}$$

for $\Im(\omega_{lmn}) < 0$.

4.5 Circumferentially Non-Uniform Wall Impedance

A circumferentially non-uniform wall impedance can be represented by Fourier series:

$$Z(\theta; \omega) = \sum_{-\infty}^{+\infty} Z_{m'}(\omega)e^{-im'\theta} \tag{95}$$

with 'impedance harmonics' of amplitude $Z'_m(\omega)$. The acoustic pressure perturbation spectrum

$$\tilde{p}(r,\theta,z;\omega) = \sum_{m=-\infty}^{+\infty} e^{im\theta} \sum_{l=-\infty}^{+\infty} e^{i2\pi lz/L} \sum_{n=1}^{+\infty} A_{lmn} J_m(kr) \qquad (96)$$

and acoustic radial velocity perturbation spectra

$$\tilde{v}_r(r,\theta,z;\omega) = \sum_{l=-\infty}^{+\infty} e^{i2\pi lz/L} \sum_{m=-\infty}^{+\infty} e^{im\theta} V_{lm}(r;\omega) \qquad (97)$$

are related by the wall boundary condition (91), viz.:

$$\sum_{l=-\infty}^{+\infty} e^{i2\pi lz/L} \left[\sum_{m=-\infty}^{+\infty} A_m e^{im\theta} J_m(kr) \mp \sum_{m,m'=-\infty}^{+\infty} \frac{1}{i\alpha_l} A_m Z_{m'} e^{i(m+m')\theta} J_m'(kr) \right] = 0,$$

$$(98)$$

re-arrangement of (4.5) shows that

$$\sum_{m'=-\infty}^{+\infty} \left[J_{m'}(kr)\delta_{mm'} \mp \frac{Z_{m-m'}}{i\alpha_l} J_{m'}'(kr) \right] A_{m'} = 0 \qquad (99)$$

and thus for the amplitudes to be not all zero the determinant of the term is square brackets must vanish.

4.6 Coupling of Azimuthal Modes

From the preceding it follows that the radial wavenumbers k_{lmn} the roots of infinite determinant:

$$D \equiv \begin{vmatrix} \ddots & \vdots & \vdots & \vdots & \vdots \\ \cdots & [J_{-1}(kR) \mp \frac{Z_0}{i\alpha_l} J_{-1}'(kR)] & \mp\frac{Z_{-1}}{i\alpha_l} J_0'(kR) & \mp\frac{Z_{-2}}{i\alpha_l} J_1'(kR) & \cdots \\ \cdots & \mp\frac{Z_1}{i\alpha_l} J_{-1}'(kR) & [J_0(kR) \mp \frac{Z_0}{i\alpha_l} J_0'(kR)] & \mp\frac{Z_{-1}}{i\alpha_l} J_1'(kR) & \cdots \\ \cdots & \mp\frac{Z_2}{i\alpha_l} J_{-1}'(kR) & \mp\frac{Z_1}{i\alpha_l} J_0'(kR) & [J_1(kR) \mp \frac{Z_0}{i\alpha_l} J_1'(kR)] & \cdots \\ & \vdots & \vdots & \vdots & \ddots \end{vmatrix} = 0.$$

$$(100)$$

In the case I of uniform impedance the determinant is diagonal thus the eigenvalues k_{lmn} are roots of each diagonal element (92), and there is no coupling of azimuthal modes. In the case II of non-uniform impedance

$$Z(\theta;\omega) = Z_0(\omega) + Z_1(\omega)e^{i\theta} + Z_{-1}(\omega)e^{-i\theta} + \dots \qquad (101)$$

with 'impedance harmonics' $Z_{\pm s}$ up to order s, the determinant (100) has $(2s+1)$ non-zero bands around the diagonal, implying the coupling of different azimuthal modes by the non-uniform impedance. For example a symmetric mode $m = 0$ is reflected by wall with non-uniform azimuthal impedance $Z(\theta;\omega)$ as a superposition of modes $m = 0, \pm1, \dots, \pm s$.

4.7 Axially Varying Wall Impedance

An axially varying impedance is also represented by a Fourier series:

$$Z(z;\omega) = \sum_{l'=-\infty}^{+\infty} Z^{l'}(\omega)\mathrm{e}^{\mathrm{i}2\pi l'z/L}, \tag{102}$$

a function of z instead of θ in (102) instead of z in (95). Substituting the acoustic velocity and pressure perturbation at the wall

$$\sum_{m=-\infty}^{+\infty} \mathrm{e}^{\mathrm{i}m\theta} \left[\sum_{l=-\infty}^{+\infty} \mathrm{e}^{\mathrm{i}2\pi lz/L} A_{lm} J_m(kr) - \sum_{l,l'=-\infty}^{+\infty} \mathrm{e}^{\mathrm{i}2\pi(l+l')z/L} \frac{Z^{l-l'}}{\mathrm{i}\alpha_{l'}} A_{lm} J'_m(kr) \right] = 0, \tag{103}$$

leads after simplification to

$$\sum_{l'=-\infty}^{+\infty} A_{l'm} \left[\delta_{l'l} J_m(kr) \mp \frac{Z^{l-l'}}{\mathrm{i}\alpha_{l'}} J'_m(kr) \right] = 0 \tag{104}$$

showing that of for the amplitudes to be not all zero the determinant must vanish

$$E_m = \det \left\{ \delta_{l'l} J_m(kr) \mp \frac{Z^{l-l'}}{\mathrm{i}\alpha_{l'}} J'_m(kr) \right\}; \tag{105}$$

the radial wavenumbers are roots of this single infinite determinant.

4.8 Two-dimensionally Varying Wall Impedance

In the most general case the wall impedance varies both axially and azimuthaly and is represented by double Fourier series:

$$Z(\theta, z;\omega) = \sum_{l=-\infty}^{+\infty} \mathrm{e}^{\mathrm{i}2\pi lz/L} \sum_{-\infty}^{+\infty} \mathrm{e}^{-\mathrm{i}m'\theta} Z^{l'}_{m'}(\omega). \tag{106}$$

Substitution in the wall boundary condition leads to eigen-values k_{lmn} which are roots of 'double' infinite determinant, i.e. an infinite determinant whose elements are infinite determinants, viz. (i) the inner determinant is

$$D^{l,l'} \equiv \begin{vmatrix} \ddots & \vdots & \vdots & \vdots & \vdots \\ \cdots [J_{-1}(kR) + \frac{Z_0^{l-l'}}{\mathrm{i}\alpha_{l'}} J'_{-1}(kR)] & \frac{Z_{-1}^{l-l'}}{\mathrm{i}\alpha_{l'}} J'_0(kR) & \frac{Z_{-2}^{l-l'}}{\mathrm{i}\alpha_{l'}} J'_1 & \cdots \\ \cdots \frac{Z_1^{l-l'}}{\mathrm{i}\alpha_{l'}} J'_{-1}(kR) & [J_0(kR) + \frac{Z_0^{l-l'}}{\mathrm{i}\alpha_{l'}} J'_0(kR)] & \frac{Z_{-1}^{l-l'}}{\mathrm{i}\alpha_{l'}} J'_1 & \cdots \\ \cdots \frac{Z_2^{l-l'}}{\mathrm{i}\alpha_{l'}} J'_{-1}(kR) & \frac{Z_1^{l-l'}}{\mathrm{i}\alpha_{l'}} J'_0(kR) & [J_1(kR) + \frac{Z_0^{l-l'}}{\mathrm{i}\alpha_{l'}} J'_1(kR)] \cdots \\ \vdots & \vdots & \vdots & \vdots & \ddots \end{vmatrix} \tag{107}$$

(ii) the outer determinant is

$$F(kR, \alpha, Z_m^l) = \det \left[D^{l-l'}(kR, \alpha, Z_m^l) \right].\tag{108}$$

In all cases the eigen-values are computed by truncating determinants, and the accuracy checked comparing 'smaller' and 'bigger' truncations.

4.9 Eigen-Values for Uniform and Non-Uniform Liners

The comparison is made between a uniform wall impedance Z_0 and a non-uniform wall impedance $Z_{-1} = Z_1$, viz.

$$Z(\theta) = Z_0 + Z_1 e^{i\theta} + Z_{-1} e^{-i\theta} = Z_0 + 2Z_1 \cos\theta,\tag{109}$$

for the values indicated in Table 5.

A cylindrical nozzle without flow is chosen, for four values of the ratio of length to radius $\mu = L/R = 1, 2, 5, 10$. The plots of the real and imaginary parts of the dimensionless radial wavenumber concern (i) the weakly non-uniform (case A) wall impedance in Fig. 4 and (ii) the stronger non-uniformity (case B) in Fig. 5.

Table 5. Uniform and non-uniform wall impedance

Impedances	Z_0	$Z_1 = Z_{-1}$
Case A	$1 + i$	$0.1 + i0.1$
Case B	$1 + i$	$0.1 + i0.1$

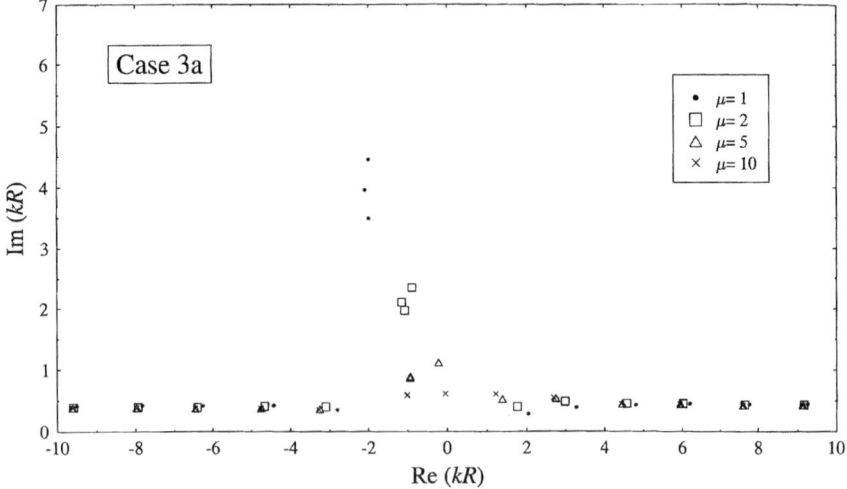

Fig. 4. Dimensionless radial wavenumbers for weakly non-uniform acoustic liner

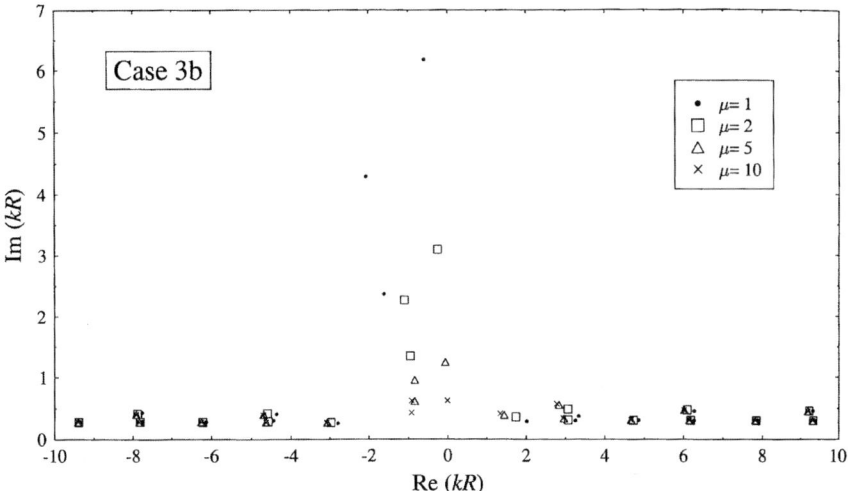

Fig. 5. Dimensionless radial wavenumbers for strongly non-uniform acoustic liner

4.10 Optimization of Non-Uniform Acoustic Liners

One method of optimization is to (i) calculate all eigen-values i.e. complex radial wavenumbers k_{mn}; (ii) for a given frequency calculate corresponding axial wavenumbers K_{mn}

$$(1 - M^2)K_{mn} = -M\omega/c \pm \left|(\omega/c)^2 - (1 - M^2)(k_{mn})^2\right|^{1/2}, \qquad (110)$$

where $M = U/c$ is the Mach number of the mean flow; (iii) since the axial wavenumber are complex:

$$\exp(iK_{lmn}z) = \exp[iz\Re(K_{mn})]\exp[-z\Im(K_{mn})], \qquad (111)$$

the imaginary part determines acoustic wave decay along duct axis; (iv) the slowest decaying mode is selected as the main contributor to noise and (v) the non-uniform wall impedance is chosen so as to maximize the decay of this mode.

4.11 Impedance Which Maximizes Decay of Slowest Decaying Mode

Taking as example a wave of frequency $f = 1\,\mathrm{kHz}$ in a duct of radius $R = 1\,\mathrm{m}$, without mean flow for a sound speed $c_0 = 340\,\mathrm{m\,s^{-1}}$, the slowest decaying mode is: $kR = 8.03438 + i3.020068$. For an azimuthally varying impedance distribution:

$$Z(\theta) = Z_0(1 - \varepsilon\cos\theta), \qquad (112)$$

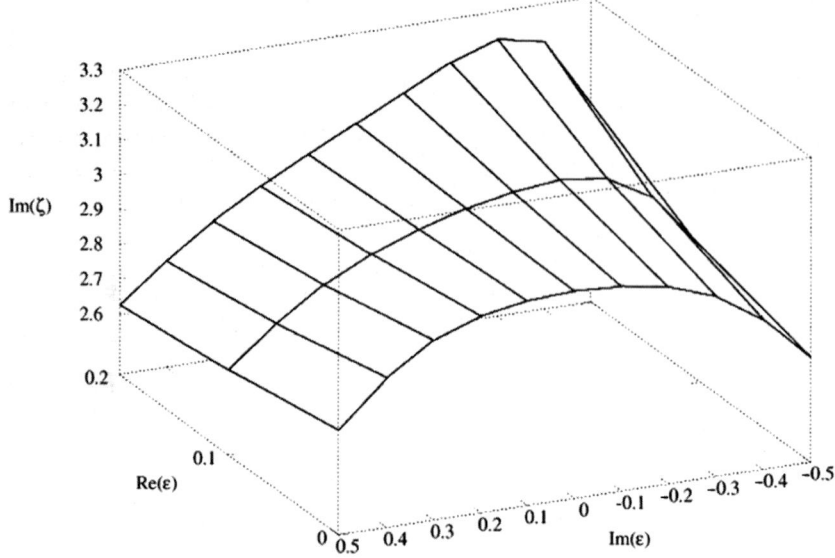

Fig. 6. Spatial decay $\Im(\zeta)$ is slowest decaying mode plotted as a function of the real $\Re(\varepsilon)$ and imaginary $\Im(\varepsilon)$ parts of the non-uniform impedance term ε, based on the 3×3 determinant centred on mode 14

with constant term $Z_0 = 1 + i$, the complex value of ε is chosen to maximize the decay of the mode. The optimal ε turns out to be $\varepsilon = 0.2 - i\,0.3$, as shown in Fig. 6.

A more general impedance distribution:

$$Z(\theta) = Z_0 \left[1 - \sum_{n=1}^{N} \varepsilon_n \cos(n\theta) \right] , \tag{113}$$

allows choice of the N parameters to maximize the decay of N modes.

4.12 Minimization of Total Acoustic Energy

Optimizing for fastest decay of one or more modes can lead to energy transfer to other modes. Thus it may be best to minimize total acoustic energy

$$E_m(z) = \sum_n E_{nm} e^{-2z\Im(k_{mn})}, \tag{114}$$

where the amplitude of modes can be obtained by (a) an equipartition hypothesis that all modes have same energy, which is usually true for m (azimuthal) and less so for n (radial) modes; (b) calculating the amplitudes from initial wave field as in (88) and (c) calculating the amplitudes by forcing, i.e. wave generation, leading to

$$\left[\frac{1}{r}\frac{\partial}{\partial r}r\frac{\partial}{\partial r} + \frac{1}{r^2}\frac{\partial^2}{\partial\theta^2} + \frac{\partial^2}{\partial z^2} - \frac{1}{c^2}\left(\frac{\partial}{\partial t} + U\frac{\partial}{\partial z}\right)^2 \right] p(r,\theta,z,t) = S(r,\theta,z)\mathrm{e}^{\mathrm{i}\omega t},$$

$$(115)$$

which is the forced convected wave equation in cylindrical coordinates (78) with sound source of frequency ω and arbitrary spatial distribution.

4.13 Wave Forcing, Without or with Single and Double Resonance

The sound source is decomposed into cylindrical harmonics:

$$S(r,\theta,z) = \sum_{l=-\infty}^{+\infty} \mathrm{e}^{\mathrm{i}2\pi lz/L} \sum_{m=-\infty}^{+\infty} \mathrm{e}^{\mathrm{i}m\theta} \sum_{n=1}^{+\infty} J_m(k_{mn}r)S_{lmn},\qquad(116)$$

with known amplitudes

$$S_{lmn} = \frac{1}{2\pi L}\frac{2}{R^2}\left[\left(1 - \frac{m^2}{(k_{lmn}R)^2}\right)\{J_m(k_{mn}R)\}^2 \right]$$
$$\times \int_0^{2\pi} \mathrm{e}^{-\mathrm{i}m\theta}\mathrm{d}\theta \int_0^L \mathrm{e}^{\mathrm{i}2\pi lz/L}\mathrm{d}z \int_0^R r J_m(k_{mn})S(r,\theta,z)\mathrm{d}r. \quad(117)$$

The forced acoustic field is then given (i) the absence of resonance by

$$P_{lmn}(t) = \mathrm{e}^{\mathrm{i}(m\theta-2\pi lz/L-\omega t)} J_m(k_{mn}r)\frac{c^2}{(\omega-\omega_{lmn})(\omega+\omega_{lmn})-4\pi lU/L},$$

$$(118)$$

(ii) for single resonance of type 1, i.e. $\omega = \omega_{lmn}$ by

$$P_{lmn}(t) = -\frac{1}{2}\mathrm{i}t\mathrm{e}^{\mathrm{i}\omega_{lmn}t}\mathrm{e}^{\mathrm{i}(m\theta-2\pi lz/L)} J_m(k_{mn}r)\frac{c^2 S_{lmn}}{\omega_{lmn}-2\pi lU/L},\qquad(119)$$

(iii) for single resonance of type 2, i.e. $\omega + \omega_{lmn} = 4\pi lU/L$ by

$$P_{lmn}(t) = -\frac{1}{2}\mathrm{i}t\mathrm{e}^{\mathrm{i}(\omega_{lmn}-4\pi lU/L)t}\mathrm{e}^{\mathrm{i}(m\theta-2\pi lz/L)} J_m(k_{mn}r)\frac{c^2 S_{lmn}}{2\pi lU/L-\omega_{lmn}},$$

$$(120)$$

(iv) for double resonance i.e. coincidence of types $2 \equiv 1$, by

$$\omega=\omega_{lmn} = 2\pi lU/L : P_{lmn}(t) = -\frac{1}{2}t^2\mathrm{e}^{-\mathrm{i}2\pi lU/L}\mathrm{e}^{\mathrm{i}(m\theta-2\pi lz/L)} J_m(k_{mn}r)c^2 S_{lmn}.$$

$$(121)$$

Note that impedance wall lead to complex natural frequencies excluding resonance for real forcing frequencies. Resonance is possible only for forcing with time decaying amplitude, but then the acoustic pressure would decay exponentially in time, instead of increasing linearly in time for single (119) and (120) or quadratically in time for in double (121) true resonance.

4.14 Acoustic Modes in Non-Uniform Lined Nozzle

The ten acoustic modes in Table 6 are illustrated in Fig. 7 for a wave of frequency $f = 1\,\text{kHz}$, in a nozzle of length $L = 5\,\text{m}$ and radius $R = 1\,\text{m}$, in the absence of flow (solid line) and for a mean flow Mach number $M = 0.3$ (dotted line). The wall impedance is non-uniform (109) with $Z_0 = 1 + \text{i}$ and $\varepsilon = 0.3 + \text{i}\,0.2$.

4.15 Mode Radiation Out-of-the Duct to the Far-field

The sound radiation out of a duct can be calculated in four cases (i) in the case of a flanged duct the source distribution on duct exit disk

Table 6. Ten acoustic modes with $m = 0$

N	1	2	3	4	5	6	7	8	9	10
e	0	0	1	1	1	1	2	2	2	2
n	1	2	1	2	3	4	1	2	3	4

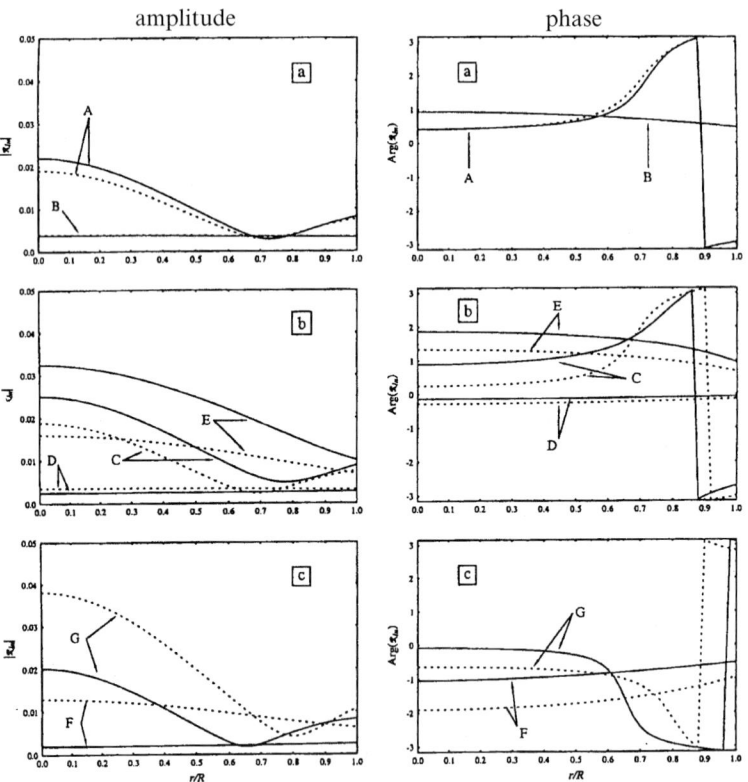

Fig. 7. Acoustic modes in a non-uniformly lined duct with and without mean flow

Table 7. Mode radiation out of duct

Panel	a	b	c	d
m	0	0	0	0
n	1	2	3	4

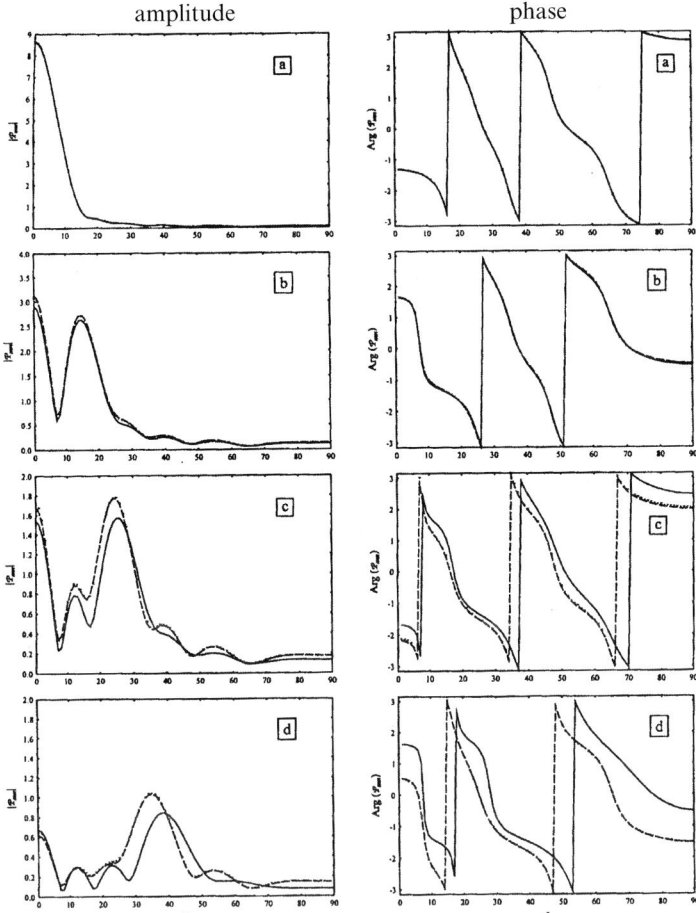

Fig. 8. Mode radiation out of a nozzle with weakly or strongly non-uniform wall impedance

$$p(r, \theta, t) = \frac{e^{i\omega(r/c - t)}}{4\pi r} \sum_{m=-\infty}^{+\infty} \sum_{n=1}^{+\infty} P_{mn}(r, \theta) A_{mn} \tag{122}$$

radiates to an observer in far-field, and the received duct modes are given by

$$P_{mn}(r,\theta) = \frac{-2\mathrm{i}k_{mn}}{\alpha}\left\{1+\mathrm{i}\frac{\mathrm{d}}{\mathrm{d}(\omega r/c)}\right\}$$

$$\times \int_0^{2\pi}\mathrm{d}\alpha\int_0^{\alpha}\mathrm{d}R\,Re^{\mathrm{i}(\omega R/c)\sin\theta\cos\alpha}\mathrm{e}^{\mathrm{i}m\alpha}J_m(k_{mn}R); \quad (123)$$

(ii) in the case of an unflanged duct the Wiener–Hopf technique is used to account for diffraction by the nozzle lip [50]; (iii) in the case of a nozzle with flow, the vortex sheet issuing from lip modifies sound diffraction [51,52] and (iv) in the case of a turbulent and irregular shear issuing from nozzle lip, there is spectral and directional broadening of sound [53,54].

For the four modes in Table 7 the case 1 is chosen for illustration in Fig. 8, by plotting the amplitude (left) and phase (left) vs. the angle θ of reception with the duct axis. The duct has non-uniform wall impedance (109), with $Z_0 = 2.5 - \mathrm{i}\,0.4$ and two cases (i) weak non-uniformity $\varepsilon = 0.1 + \mathrm{i}\,0.1$ (solid line) and (ii) stronger non-uniformity $\varepsilon = 0.2 - \mathrm{i}\,0.3$ (dotted line).

5 Conclusion

The presentation has emphasized mathematical results, as comparison with experiments was made elsewhere for aeroacoustics [8,45,46] and astrophysics [36,38,39]. The mathematical, physical and engineering aspects of waves in fluids were addressed each for one general issue: (Sect. 2) the classical and convected wave equations have significant restrictions so there are many applications which need more general forms; (Sect. 3) fluids support not only acoustic waves, but also magnetic, internal and inertial waves (water waves, instability waves and chemical reactions were omitted), and there are analogies and differences, relevant to mutual interactions; (Sect. 4) there is a wide variety of problems in aeroacoustics, with practical motivations. One example of the latter is the use of acoustic liners, which absorb noise but cause (i) a weight penalty and (ii) a drag (or fuel consumption) increase. Non-uniform liners try to maximize sound absorption for a given weight and drag penalty. They are still a subject of discussion since the optimum impedance distribution depends on the combination of modes to be attenuated. The three parts of this talk have also concentrated on three aspects of the physics of waves: (Sect. 2) wave propagation for sound; (Sect. 3) wave generation for MAGI waves; (Sect. 4) sound attenuation for aeroacoustics.

Acknowledgements

I would like to thank professor Luis Bonilla and the organizers for the invitation, and to professors H. & J.R. Ockendon for some remarks after the talk.

References

1. Campos, L.M.B.C.: On 36 forms of the acoustic wave equation in potential flows and inhomogeneous media. Reviews of Applied Mechanics, to appear, (2007).
2. Campos, L.M.B.C.: On 24 forms of the acoustic wave equation in vortical flows and dissipation media. Reviews of Applied Mechanics, to appear, (2007).
3. Howe, M.S.: Contributions to the theory of aerodynamic sound, with application to excess jet noise and the theory of the flute. J. Fluid Mech., **71**, 625–673 (1975).
4. Campos, L.M.B.C.: On the emission of sound by an ionized inhomogeneity. Proc. Roy. Soc., **A351**, 65–91 (1977).
5. Campos, L.M.B.C.: On linear and non-linear wave equations for the acoustics of high-speed potential flows. J. Sound Vib., **110**, 41–57 (1986).
6. Campos, L.M.B.C.: On generalizations of the Doppler factor, local frequency, wave invariant and group velocity. Wave Motion, **10**, 193–207 (1988).
7. Crighton, D.G.: Model equations of non–linear acoustics. Ann. Rev. Fluid Mech., **11**, 11–33 (1979).
8. Campos, L.M.B.C.: On waves in gases. Part I: Acoustics of jets, turbulence, and ducts. Rev. Mod. Phys., **58**(1), 117–182 (1986).
9. Webster, A.G.: Acoustical impedance, and the theory of horns and of the phonograph. Proc. Nat. Acad. Sci., **5**, 275–282 (1919).
10. Rayleigh, J.W.S.: On the propagationog sound in narrow tubes of variable section. Phil. Mag., **31**, 89–96 (1916).
11. Lagrange, J.L.: Nouvelles recherches sur la nature et la propagation du son. Misc. Turin., **2**, 11–170 (1760).
12. Campos, L.M.B.C.: Some general properties of the exact acoustic fields in horns and baffles. J. Sound Vib., **95**(2), 177–201 (1984).
13. Campos, L.M.B.C.: On the propagation of sound in nozzles of varying cross-section containing a low Mach number mean flow. Zeitz. Flugwis. Weltraumf., **8**, 97–109 (1984).
14. Campos, L.M.B.C., Lau, F.J.P.: On sound in an inverse sinusoidal nozzle with low Mach number mean flow. J. Acoustic. Soc. Am., **100**, 355–363 (1996).
15. Campos, L.M.B.C., Lau, F.J.P.: On the acoustics of low Mach number bulged, throated and baffled nozzles. J. Sound Vib., **196**, 611–633 (1996).
16. Campos, L.M.B.C., Lau, F.J.P.: On the convection of sound in inverse catenoidal nozzles. J. Sound Vib., **244**, 195–209 (2001).
17. Lau, F.J.P., Campos, L.M.B.C.: On the effect of wall undulations on the acoustics of ducts with flow. J. Sound Vib., **270**, 361–379 (2003).
18. Haurwitz, W.: Zur theorie der Wellenbewegungen in Luft und Wasser. Veroff. Geophys. Inst. Univ. Leipz, **6**, 334–364 (1932).
19. Kuchemann, D.: Storungbewegungen in einer Gasstromung mit Grenzschicht. Z. Angew. Math Mech., **31**, 79–84 (1983).
20. Pridmore-Brown, D.C.: Sound propagation in a fluid through an attenuating duct. J. Fluid Mech., **4**, 393–406 (1958).
21. Mohring, W., Muller, E.A., Obermeier, F.: Problems in flow acoustics. Rev. Mod. Phys, **55**, 707–723 (1983).
22. Campos, L.M.B.C., Oliveira, J.M.G.S., Kobayashi, M.H.: On sound propagation in a linear flow. J. Sound Vib., **219**(5), 739–770 (1999).
23. Campos, L.M.B.C., Serrão, P.G.T.A.: On the acoustics of an exponential boundary layer. Phil. Trans. Roy. Soc. London, **A356**, 2335–2378 (1998).

24. Campos, L.M.B.C., Kobayashi, M.H.: On the reflection and transmmission of sound in a thick shear layer. J. Fluid Mech., **420**, 1–24 (2000).
25. Howe, M.S.: The generation of sound by vorticity waves in swirling flows. J. Fluid Mech., **81**, 369–383 (1977).
26. Tam, C.K.W., Auriault, L.: The wave modes in ducted swirling flows. J. Fluid Mech., **371**, 1–20 (1998).
27. Campos, L.M.B.C., Serrão, P.G.T.A.: On the acoustics of unbounded and ducted vortex flows. SIAM J. Appl. Math., **65**, 1353–1368 (2004).
28. Burgers, J.M.: A mathematical model illustrating the theory of turbulence. Adv. Appl. Math., **10**, 171–179 (1948).
29. Burgers, J.M.: The non-linear diffusion equation. D. Reidel (1964).
30. Campos, L.M.B.C., Leitão, J.F.P.: On the computation of special functions with applications to non-linear and inhomogeneous waves. J. Comp. Mech., **3**, 343–360 (1988).
31. Westerwelt, P.J.: Parametric acoustic array. J. Acoust. Soc. Am., **35**, 335–337 (1963).
32. Zabalotskaya, E.A., Khoklov. Quasiplane waves in non-linear acoustics of confined beams. Sov. Phys. Acoust., **26**, 217–220 (1969).
33. Kuznetsov, V.P.: Equation of non-linear acoustics. Akust. Zh., **16**, 548–551 (1970).
34. Campos, L.M.B.C.: On the generation and radiation of magnetoacoustic waves. J. Fluid Mech., **81**, 529–549 (1977).
35. Campos, L.M.B.C.: Modern trends in research on waves in fluids. Part II: Propagation and dissipation in compressible and ionized atmospheres. Portugaliæ Phys., **14**, 145–173 (1983).
36. Campos, L.M.B.C.: On waves in gases. Part II: Interaction of sound with magnetic and internal modes. Rev. Mod. Phys., **59**(1); 363–462 (1987).
37. Campos, L.M.B.C.: On oscillations in sunspot umbrae and wave radiation in stars. Mont. Not. Roy. Astron. Soc., **241**, 215–229 (1989).
38. Campos, L.M.B.C.: On hydromagnetic waves in atmospheres with application to the sun. Theor. Comp. Fluid Dyn., **10**, 37–70 (1998).
39. Campos, L.M.B.C.: On the generation, propagation and radiation of magneto-acoustic-gravity waves. Geophys. Astrophys. Fluid Mec., to appear, (2007).
40. Lighthill, M.J.: On sound generated aerodynamically. I: General theory. Proc. Roy. Soc. London, A**211**, 564–587 (1952).
41. Lighthill, M.J.: On sound generated aerodynamically. II. turbulence as a source of sound. Proc. Roy. Soc. London, A**222**, 1–32 (1954).
42. Lighthill, M.J.: Bakerian lecture: On sound generated aerodynamically. Proc. Roy. Soc. London, A**267**, 147–182 (1954).
43. Lighthill, M.J.: Waves in Fluids. Cambridge University Press (1978).
44. Campos, L.M.B.C.: On the fundamental acoustic mode in variable area, low mach number nozzles. Prog. Aerosp. Sci., **22**, 1–27 (1986).
45. Campos, L.M.B.C.: On the spectra of aerodynamic noise and aeroacoustic fatigue. Prog. Aerosp. Sci., **33**, 353–389 (1997).
46. Campos, L.M.B.C.: On some recent advances in aeroacoustics. Int. J. Sound Vib., **11**, 27–45 (2006).
47. Campos, L.M.B.C., Oliveira, J.M.G.S. On the optimization of non-uniform acoustic liners on annular nozzles. J. Sound Vib., **275**(3–5), 557–576 (2004).

48. Campos, L.M.B.C., Oliveira, J.M.G.S.: On the acoustic modes in a cylindrical nozzle with an arbitrary impedance distribution. J. Acoust. Soc. Am., **116**(6); 3336–3347 (2004).

49. Noble, B.: The Wiener-Hopf Technique. Oxford University Press (1958).

50. Munt, R.M.: The interaction of sound with a subsonic jet issuing from a semi-infinite pipe. J. Fluid Mech., **83**, 609–640 (1977).

51. Rienstra, S.W.: Acoustic radiation from a semi-infinite duct in a uniform subsonic mean flow. J. Sound Vib., **94**, 267–288 (1984).

52. Campos, L.M.B.C.: The spectral broadening of sound in turbulent shear layers. Part 1: The transmission of sound through turbulent shear layers. J. Fluid Mech., **89**, 723–749 (1978).

53. Campos, L.M.B.C.: The spectral broadening of sound in turbulent shear layers. Part 2: The spectral broadening of sound and aircraft noise. J. Fluid Mech., **89**, 751–783 (1978).

54. Campos, L.M.B.C.: Sur la propagation du son dans les écoulements non-uniformes et non-stationaires. Rev. d'Acoust., **67**, 217–233 (1984).

Quantum Diffusion Models Derived from the Entropy Principle

P. Degond[1], S. Gallego[1], F. Méhats[2], and C. Ringhofer[3]

[1] MIP, UMR 5640 (CNRS-UPS-INSA), Université Paul Sabatier, 118 route de Narbonne, 31062 Toulouse Cedex, France
degond@mip.ups-tlse.fr
[2] IRMAR (UMR CNRS 6625), Universit de Rennes, Campus de Beaulieu, 35042 Rennes Cedex, France
florian.mehats@univ-rennes1.fr
[3] Department of Mathematics, Arizona State University, Tempe, Arizona 85284-184, USA
ringhofer@asu.edu

Summary. In this chapter, we review the recent theory of quantum diffusion models derived from the entropy minimization principle. These models are obtained by taking the moments of a collisional Wigner equation and closing the resulting system of equations by a quantum equilibrium. Such an equilibrium is defined as a minimizer of the quantum entropy subject to local constraints of given moments. We provide a framework to develop this minimization approach. The results of numerical simulations show that these models capture well the various features of quantum transport.

1 Introduction

The goal of this paper is to give an introduction to the theory of quantum diffusion models derived from the entropy principle. These lecture notes report on previously published works [10–12, 14, 15].

These models are obtained by taking the moments of a collisional Wigner equation and closing the resulting system of equations by a quantum equilibrium. Such an equilibrium is defined as a minimizer of the quantum entropy subject to local constraints of given moments. We provide a framework to develop this minimization approach and apply it to quantum diffusion models. We also give some preliminary numerical results.

More precisely we consider a collisional Quantum Liouville equation

$$i\hbar\partial_t\rho = [\mathcal{H}, \rho] + i\hbar\mathcal{Q}(\rho), \tag{1}$$

where ρ is the density operator (i.e., a hermitian nonnegative trace-class operator of trace unity representing the statistical state of the quantum system). $\hbar$ is the Planck constant and $\mathcal{H}$ is the Hamiltonian operator defined by

$$\mathcal{H}\psi = -\frac{\hbar^2}{2}\Delta\psi + V(x,t)\psi, \tag{2}$$

where V is an external potential (the case of mean-field Hartree potential can be considered as well). The operator $\mathcal{Q}(\rho)$ is an unspecified collision operator which describes the interaction of the particles with themselves and with their environment and accounts for dissipation mechanisms. The only assumption that will be used is that this operator dissipates entropy (see below).

Let $W[\rho](x,p)$ denote the Wigner transform of ρ

$$W[\rho](x,p) = \int \underline{\rho}\left(x - \frac{1}{2}\xi, x + \frac{1}{2}\xi\right) e^{i\xi \cdot p/\hbar}\, d\xi, \tag{3}$$

where $\underline{\rho}(x, x')$ is the distribution kernel of ρ

$$\rho\psi = \int \underline{\rho}(x, x')\psi(x')\, dx'.$$

We recall that the inverse Wigner transform (or Weyl quantization) is given by the following formula:

$$W^{-1}(w)\psi = \frac{1}{(2\pi)^d}\int w\left(\frac{x+y}{2}, \hbar k\right)\psi(y)e^{ik(x-y)}\, dk\, dy \tag{4}$$

and defines $W^{-1}(w)$ as an operator acting on the element ψ of L^2. The function w is also called the Weyl symbol of ρ. W and W^{-1} are Isometries between $\mathcal{L}^2$ (the space of operators such that the product $\rho\rho^\dagger$ is trace-class, where $\rho^\dagger$ is the Hermitian conjugate of ρ) and $L^2(\mathbb{R}^{2d})$:

$$\mathrm{Tr}\{\rho\sigma^\dagger\} = \int W[\rho](x,p)\overline{W[\sigma](x,p)}\,\frac{dx\, dp}{(2\pi\hbar)^d}. \tag{5}$$

Taking the Wigner transform of (1), we get the following collisional Wigner equation for $w = W[\rho]$

$$\partial_t w + p \cdot \nabla_x w + \Theta^\hbar[V]w = Q(w) \tag{6}$$

with

$$\Theta^\hbar[V]w = -\frac{i}{(2\pi)^d\hbar}\int \left(V\left(x + \frac{\hbar}{2}\eta\right) - V\left(x - \frac{\hbar}{2}\eta\right)\right)$$

$$\times w(x, q)\, e^{i\eta \cdot (p-q)}\, dq\, d\eta \tag{7}$$

and $Q(w)$ is the Wigner transform of $\mathcal{Q}(\rho)$.

In Sect. 2, we are going to make use of the entropy dissipation properties of Q to derive quantum hydrodynamic models.

2 Quantum Energy-Transport Model

In this section, we report on the work [11, 12].

In order to derive quantum diffusion model, we need to specify the collisions operator $\mathcal{Q}$ in Quantum Liouville equation

$$i\hbar\partial_t\rho = [\mathcal{H}, \rho] + i\hbar\mathcal{Q}(\rho) \tag{8}$$

or in the Wigner equation

$$\partial_t w + p \cdot \nabla_x w + \Theta^h[V]w = Q(w). \tag{9}$$

In the absence of a precise definition of the physical collision mechanism, the most simple choice is a relaxation operator also called BGK operator. The collision operator expresses the relaxation of the collision operator to the Local Thermodynamical Equilibrium, in our case, the quantum Maxwellian. We want to investigate a case where this collision operator is written

$$Q(w)(p) = -\nu(w - \mathcal{E}\mathrm{xp}(A + C|p|^2/2)), \tag{10}$$

where we recall that $\mathcal{E}\mathrm{xp}\, w = W\,(\exp{(W^{-1}w)})$ and exp is the exponential of operators. The functions $A(x)$ and $C(x)$ are such that the operator Q locally conserves mass and energy. More precisely, let us write

$$\mathcal{M}_{n,\mathcal{W}} = \mathcal{E}\mathrm{xp}(A + C|p|^2/2), \tag{11}$$

the Quantum Maxwellian whose local mass at point x is $n(x)$ and local energy is $\mathcal{W}(x)$. Then, (A, C) is such that

$$\int \mathcal{E}\mathrm{xp}(A + C|p|^2/2) \begin{pmatrix} 1 \\ |p|^2/2 \end{pmatrix} \widetilde{\mathrm{d}p} = \begin{pmatrix} n \\ \mathcal{W} \end{pmatrix}. \tag{12}$$

In density operator form, the Quantum Maxwellian is written

$$\rho_{n,\mathcal{W}} = W^{-1}(\mathcal{M}_{n,\mathcal{W}}) = \exp(W^{-1}(A + C|p|^2/2)) \tag{13}$$

with, for all test functions ϕ

$$\mathrm{Tr}\{\rho_{n,\mathcal{W}}\,\phi\} = \int n\phi\,\mathrm{d}x, \quad \mathrm{Tr}\{\rho_{n,\mathcal{W}}\,\phi|p|^2/2\} = \int \mathcal{W}\phi\,\mathrm{d}x. \tag{14}$$

The Quantum Maxwellian $\rho_{n,\mathcal{W}} = \exp(W^{-1}(A + C|p|^2/2))$ is a solution of the entropy minimization principle: to find

$$\min\{H[\rho] = \mathrm{Tr}\{\rho(\ln\rho - 1)\}\}, \quad \text{subject to}$$
$$\mathrm{Tr}\{\rho_{n,\mathcal{W}}\,\phi\} = \int n\phi\,\mathrm{d}x, \quad \mathrm{Tr}\{\rho_{n,\mathcal{W}}\,\phi|p|^2/2\} = \int \mathcal{W}\phi\,\mathrm{d}x\}. \tag{15}$$

In Wigner form, the quantum entropy $H[\rho]$ has the expression

$$H[\rho] = \mathrm{Tr}\{\rho(\ln\rho - 1)\} = \int w(\mathcal{L}\mathrm{n}\,w - 1)\,\mathrm{d}x\,\widetilde{\mathrm{d}p}, \tag{16}$$

where the quantum logarithm is defined according to $\mathcal{L}\mathrm{n}\,w = W[\ln(W^{-1}(w))]$.

For a given Wigner distribution w, let us denote by $\mathcal{M}_w := \mathcal{M}_{n,\mathcal{W}}$ the Quantum Maxwellian which possesses the same density n and energy $\mathcal{W}$ as w:

$$\int \mathcal{M}_w \begin{pmatrix} 1 \\ |p|^2/2 \end{pmatrix} \mathrm{d}p = \int w \begin{pmatrix} 1 \\ |p|^2/2 \end{pmatrix} \mathrm{d}p. \tag{17}$$

Then, the quantum BGK operator is written

$$Q(w) = -\nu(w - \mathcal{M}_w). \tag{18}$$

In density operator form, we shall denote the quantum Maxwellian which has the same mass and energy as ρ by $\mathcal{M}_\rho$. Then the quantum BGK operator is written

$$\mathcal{Q}(\rho) = -\nu(\rho - \mathcal{M}_\rho). \tag{19}$$

The physical situation modeled by $Q(w)$ is typically when the energy exchanges among the particles themselves are more efficient than with the surrounding and that a different temperature than that of the background is possible. In short channel transistors, the electron typical energy exceeds the phonon energy by almost two orders of magnitude. Then, the phonon collisions can be viewed as quasielastic and most of the energy exchanges are with the other electrons via Coulomb interaction. In plasmas, a similar situation arises between electrons and ions because of the very small electron to ion mass ratio.

We observe that we need the two sets of variables: the conservative variables $(n, \mathcal{W})$ and the entropic variables (A, C). The passage between $(n, \mathcal{W})$ and (A, C) is a functional change of variable which is done through the entropy and its Legendre dual (see reference given above).

Let us now summarize the properties of Q:

(i) Mass and energy conservation

$$\int Q(w) \begin{pmatrix} 1 \\ |p|^2 \end{pmatrix} \mathrm{d}p = 0 \tag{20}$$

(ii) Null set of Q (equilibria)

$$Q(w) = 0 \Longleftrightarrow \exists\, (A, C) \text{ such that } w = \mathcal{E}\mathrm{xp}(A + C|p|^2/2) \tag{21}$$

(iii) Entropy decay

$$\int Q(w) \, \mathcal{L}\mathrm{n}w \, \mathrm{d}x \, \widetilde{\mathrm{d}p} = \mathrm{Tr}\{\mathcal{Q}(\rho) \ln \rho\} \leq 0 \tag{22}$$

Properties (i) and (ii) are obvious from definition (18) and the conservation relations (12). The only delicate point is entropy decay (iii). In the classical case, the proof uses that the logarithm is an increasing function. This is no more true here in the case of the quantum logarithm. Indeed, because the dependence between w and $\mathcal{L}\mathrm{n}(w)$ is functional, the statement that $\mathcal{L}\mathrm{n}(w)$ is increasing w.r.t w is meaningless. So another proof must be developed. It uses convexity argument [11].

Now, we consider a diffusion scaling of the collisional Wigner equation

$$\eta^2 \frac{\partial w^\eta}{\partial t} + \eta(v \cdot \nabla_x w^\eta - \Theta(w^\eta)) = Q(w^\eta). \tag{23}$$

This scaling is obtained through the change $t \to t/\eta$ and $Q \to Q/\eta$ which means that the collision operator is large and that we are looking at long time scales.

The limit $\eta \to 0$ of (23) is the so-called quantum Energy-Transport model. Indeed, as $\eta \to 0$, $w^\eta \longrightarrow \mathcal{E}\mathrm{xp}(A + C|p|^2/2)$ where (A, C) satisfy the Energy-Transport model which consists of the mass and energy conservation equations

$$\frac{\partial n}{\partial t} + \nabla_x \cdot j_n = 0, \tag{24}$$

$$\frac{\partial \mathcal{W}}{\partial t} + \nabla_x \cdot j_\mathcal{W} + \nabla_x V \cdot j_n = 0, \tag{25}$$

where $(n, \mathcal{W})$ is related with (A, C) through

$$\int \mathcal{E}\mathrm{xp}(A + C|p|^2/2) \begin{pmatrix} 1 \\ |p|^2/2 \end{pmatrix} \widetilde{dp} = \begin{pmatrix} n \\ \mathcal{W} \end{pmatrix} \tag{26}$$

and the fluxes $(j_n, j_\mathcal{W})$ are given by

$$j_n = -\nu^{-1}[\nabla \Pi + n\nabla V], \tag{27}$$

$$j_\mathcal{W} = -\nu^{-1}[\nabla \mathbb{Q} + (\mathcal{W}\,\mathrm{Id} + \Pi)\nabla V - \frac{\hbar^2}{8} n\nabla(\Delta V)] \tag{28}$$

with the tensors $\Pi(A, C)$ and $\mathbb{Q}(A, C)$ given by

$$\Pi(A, C) = \int \mathcal{E}\mathrm{xp}(A + C|p|^2/2)\, p \otimes p\, \widetilde{dp}, \tag{29}$$

$$\mathbb{Q}(A, C) = \int \mathcal{E}\mathrm{xp}(A + C|p|^2/2)\, p \otimes p\, |p|^2/2\, \widetilde{dp}. \tag{30}$$

Like in the classical case (see, e.g., [4,5,9]), the system consists of balance equations for the conservative variables $(n, \mathcal{W})$, the fluxes of which are expressed in terms of the gradients of the entropic variables (A, C). The passage $(n, \mathcal{W})$ to (A, C) can be done through the use of the entropy functional or its Legendre dual. However, by contrast with the classical case, there is no clear symmetric positive-definite matrix structure relation between the fluxes and the gradients of the entropic variables.

Let us now consider entropy decay. The fluid entropy is given by the kinetic entropy evaluated for the equilibrium: $S(n, \mathcal{W}) = H(\mathcal{M}_{n,\mathcal{W}})$ and has the following expressions

$$S(n, \mathcal{W}) = \int \mathcal{M}_{n,\mathcal{W}} (\mathcal{L}\mathrm{n}\mathcal{M}_{n,\mathcal{W}} - 1) \, dx \, \widetilde{dp}$$

$$= \int \mathcal{E}\mathrm{xp}(A + C|p|^2/2)(A + C|p|^2/2 - 1) \, dx \, \widetilde{dp}$$

$$= \int (n(A - 1) + C\mathcal{W}) \, dx. \tag{31}$$

The quantum Energy-Transport model decreases the entropy:

$$\frac{\mathrm{d}}{\mathrm{d}t}S(n,\mathcal{W}) \le 0. \tag{32}$$

The proof follows exactly the same arguments as for the hydrodynamic model and is omitted.

We close this section about quantum Energy-Transport models by a few remarks. The first one is that there is no rigorous proof neither for the existence of solutions nor for its derivation from the collisional Wigner equation. Numerical simulations have not been performed yet either. In the literature, quantum Energy-Transport models can be found but their derivation (and the model itself) are different. For instance, we refer to the Energy-Transport extension of the DG (Density-Gradient) model by Chen and Liu [8].

2.1 Quantum Drift-Diffusion Model

This section summarizes a series of works [10–12, 14].

In the classical setting, the Drift-Diffusion model is a simplification of the Energy-Transport model when the assumption of constant temperature is made. To derive a Quantum-Drift-Diffusion model, we start by a discussion of the appropriate BGK operator.

This operator will be defined as a relaxation to a quantum Maxwellian with a fixed temperature, and can be expressed by

$$Q(w)(v) = -\nu(w - \mathcal{E}\mathrm{xp}(A - |p|^2/2)), \tag{33}$$

where the function $A(x)$ is such that the operator conserves mass. Here again, we take a constant temperature equal to unity for the sake of simplicity.

For a given density $n(x)$, the Quantum Maxwellian which has density n in Wigner form is given by

$$\mathcal{M}_n = \mathcal{E}\mathrm{xp}(A - |p|^2/2), \tag{34}$$

$$\int \mathcal{E}\mathrm{xp}(A - |p|^2/2)\,\widetilde{\mathrm{d}p} = n. \tag{35}$$

In density operator form it is written

$$\rho_n = W^{-1}(\mathcal{M}_n) = \exp(W^{-1}(A - |p|^2/2)), \tag{36}$$

with, for all test function ϕ

$$\mathrm{Tr}\{\rho_n\,\phi\} = \int n\phi\,\mathrm{d}x. \tag{37}$$

This Quantum Maxwellian satisfies the free energy minimization principle: $\rho_n = \exp(W^{-1}(A - |p|^2/2))$ is a solution of the problem: to find

$$\min \{G[\rho] = \mathrm{Tr}\{\rho(\ln\rho - 1) + \mathcal{H}\rho\}$$

$$\text{subject to: } \mathrm{Tr}\{\rho_n \ \phi\} = \int n\phi \, \mathrm{d}x \, , \quad \forall \text{ test fct } \phi\}, \tag{38}$$

where $\mathcal{H} = |p|^2/2 + V$ is the system Hamiltonian.

In Wigner form, the free energy is written

$$G[\rho] = \mathrm{Tr}\{\rho(\ln\rho - 1) + \mathcal{H}\rho\} = \int [w(\mathcal{L}\mathrm{n}\, w - 1) + \mathcal{H}w] \, \mathrm{d}x \, \widetilde{\mathrm{d}p} \tag{39}$$

with the quantum logarithm $\mathcal{L}\mathrm{n}\, w = W[\ln(W^{-1}(w))]$.

For a given Wigner distribution w, we denote $\mathcal{M}_w := \mathcal{M}_n$ the Quantum Maxwellian which has the same density n as w:

$$\int \mathcal{M}_w \, \mathrm{d}p = \int w \, \mathrm{d}p. \tag{40}$$

Then quantum BGK operator is finally written

$$Q(w) = -\nu(w - \mathcal{M}_w). \tag{41}$$

In density operator formulation, we denote by $\mathcal{M}_\rho$ the Quantum Maxwellian associated with ρ, and the BGK operator is written:

$$\mathcal{Q}(\rho) = -\nu(\rho - \mathcal{M}_\rho). \tag{42}$$

The situation modeled by $Q(w)$ is that of a system where energy exchanges between the particles and the surrounding relax the temperature to the background temperature.

Again, two variables appear, the conservative variable n and the entropic variable A, with a functional change of variable between these two variables which can be expressed through the free energy and its Legendre dual.

We now list the properties of Q

(i) Mass conservation

$$\int Q(w) \, \mathrm{d}p = 0, \tag{43}$$

(ii) Null set of Q (equilibria)

$$Q(w) = 0 \iff \exists A \text{ such that } w = \mathcal{E}\mathrm{xp}(A - |p|^2/2), \tag{44}$$

(iii) Free energy decay

$$\int Q(w)(\mathcal{L}\mathrm{n}w + \mathcal{H}) \, \mathrm{d}x \, \widetilde{\mathrm{d}p} = \mathrm{Tr}\{\mathcal{Q}(\rho)(\ln\rho + \mathcal{H})\} \leq 0. \tag{45}$$

We now look at the Wigner equation under diffusion scaling

$$\eta^2 \frac{\partial w^\eta}{\partial t} + \eta(v \cdot \nabla_x w^\eta - \Theta(w^\eta)) = Q(w^\eta), \tag{46}$$

The limit $\eta \to 0$ leads to the Quantum Drift-Diffusion model: More precisely, as $\eta \to 0$, $w^\eta \longrightarrow \mathcal{E}\mathrm{xp}(A - |p|^2/2)$ where A satisfies the Energy-Transport model which consists of the mass conservation equation

$$\frac{\partial n}{\partial t} + \nabla_x \cdot j_n = 0 \tag{47}$$

with

$$\int \mathcal{E}\mathrm{xp}(A - |p|^2/2)\,\widetilde{\mathrm{d}p} = n \tag{48}$$

and the flux j_n given by

$$j_n = -\nu^{-1}[\nabla\Pi + n\nabla V] \tag{49}$$

with

$$\Pi(A) = \int \mathcal{E}\mathrm{xp}(A - |p|^2/2)\,p \otimes p\,\widetilde{\mathrm{d}p}. \tag{50}$$

Now, the fluid free energy is the kinetic free energy evaluated on the equilibrium $\mathcal{G}(n) = G(\mathcal{M}_n)$ and is given by

$$\begin{aligned}
\mathcal{G}(n) &= \int \mathcal{M}_{n,\mathcal{W}}\left(\mathrm{Ln}\mathcal{M}_{n,\mathcal{W}} - 1 + \mathcal{H}\right)\mathrm{d}x\,\widetilde{\mathrm{d}p} \\
&= \int \mathcal{E}\mathrm{xp}(A - |p|^2/2)(A - |p|^2/2 - 1 + \mathcal{H})\,\mathrm{d}x\,\widetilde{\mathrm{d}p} \\
&= \int n(A + V - 1)\,\mathrm{d}x.
\end{aligned} \tag{51}$$

Then if either V is independent of t or V is given by Poisson's equation

$$\Delta V = n, \tag{52}$$

then

$$\frac{\mathrm{d}}{\mathrm{d}t}\mathcal{G}(n) \leq 0 \tag{53}$$

(in the latter case, we have to multiply the term nV by a factor $1/2$).

We now give a more tractable expression of the pressure tensor Π than (50), given by

$$\nabla\Pi(A) = n\nabla A. \tag{54}$$

This leads to an equivalent formulation of the QDD model:

$$\frac{\partial n}{\partial t} + \nabla_x \cdot j_n = 0, \tag{55}$$

$$j_n = -\nu^{-1}(n\nabla(A + V)), \tag{56}$$

$$\int \mathcal{E}\mathrm{xp}(A - |p|^2/2)\,\widetilde{\mathrm{d}p} = n. \tag{57}$$

The moment reconstruction problem (57) has also a simpler expression if we suppose that the Hamiltonian $H(A) = |p|^2/2 - A$ has a discrete spectrum with eigenvalues $\lambda_p(A)$ and eigenfunctions $\psi_p(A)$, $p = 1, \ldots, \infty$. Indeed, we have

$$n(A)(x) = \sum_{p=1}^{\infty} \exp(-\lambda_p(A)) |\psi_p(A)(x)|^2. \qquad (58)$$

The "final" expression of the Quantum Drift-Diffusion model is therefore

$$\frac{\partial n}{\partial t} + \nabla_x \cdot j_n = 0, \qquad (59)$$

$$j_n = -\nu^{-1}(n\nabla(A + V)), \qquad (60)$$

$$n(A)(x) = \sum_{p=1}^{\infty} \exp(-\lambda_p(A)) |\psi_p(A)(x)|^2, \qquad (61)$$

with $\lambda_p(A)$ and $\psi_p(A)$ the eigenvalues and eigenvectors associated with the modified Hamiltonian $H(A) = |p|^2/2 - A$.

Now, we would like to consider the equilibrium states of the QDD model, defined by $j_n = 0$. This obviously implies $A = -V$ (up to a constant that we take equal to zero). Therefore, the moment reconstruction problem becomes

$$n(x) = \sum_{p=1}^{\infty} \exp(-\lambda_p) |\psi_p(x)|^2, \qquad (62)$$

with λ_p, ψ_p the eigenvalue and eigenvector associated with the "true" system Hamiltonian $H(-V) = |p|^2/2 + V$. If additionally, n is related with V through Poisson's equation (52), this leads to the well-known Schrödinger–Poisson problem which characterizes equilibrium states.

Now, if we assume that we are close to equilibrium, we can make the approximation $A \approx -V$ and replace A by $-V$ in the moment reconstruction problem (57), which leads to the following system

$$\frac{\partial n}{\partial t} + \nabla_x \cdot j_n = 0, \qquad (63)$$

$$j_n = \nu^{-1}(n\nabla(A + V)), \qquad (64)$$

$$n(A)(x) = \sum_{p=1}^{\infty} \exp(A + V - \lambda_p(-V)) |\psi_p(-V)(x)|^2, \qquad (65)$$

in which case, the spectral problem to be solved is associated with the "true" system Hamiltonian $H(-V) = |p|^2/2 + V$. This system is known as the Schrödinger–Poisson-Drift-Diffusion and has been investigated by Sacco and coauthors in [13, 17, 20].

We now investigate $\hbar$ expansions of the QDD model. Up to $O(\hbar^2)$ terms, the QDD model reads

$$\partial_t n + \nabla \cdot j_n = 0, \tag{66}$$

$$j_n = -\nu^{-1}[\nabla n - n\nabla(V + V_{\mathrm{B}}[n])), \tag{67}$$

$$V_{\mathrm{B}}[n] = -\frac{\hbar^2}{6}\frac{1}{\sqrt{n}}\Delta(\sqrt{n}). \tag{68}$$

This model is called the Density-Gradient model and has first been proposed by Ancona and coauthors [1–3]. We note that this is just the classical Drift-Diffusion model with the addition of the Bohm potential (divided by a factor 3 as compared with the Bohm potential of the single-particle hydrodynamics). Usually, this factor is treated as a fitting parameter in the simulation codes.

It is a remarkable fact that the Density-Gradient model has an entropy, which is nothing but the free energy of the QDD model expanded up to $O(\hbar^2)$ terms:

$$\mathcal{G}_2(n) = \int_{\mathbb{R}^d} n(\ln n - 1 + V + V_{\mathrm{B}}[n])\,\mathrm{d}x. \tag{69}$$

If V is independent of t it can be shown that

$$\frac{\mathrm{d}}{\mathrm{d}t}\mathcal{G}_2(n) = -\int_{\mathbb{R}^d} \frac{1}{\nu n}|\nabla n + n\nabla(V + V_{\mathrm{B}}[n])|^2\,\mathrm{d}x \leq 0. \tag{70}$$

A similar expression would hold if V is solved through Poisson's equation (52). The proof can be found in [11].

The Density-Gradient model has been widely investigated in the literature. The mathematical theory has been settled first by Ben Abdallah and Unterreiter in [6] and later by Pinnau [18]. Numerical methods have been developed by Pinnau and Unterreiter [19] and Jngel and Pinnau [16]. The present approach provides a derivation of the DG model from first principles and proves (for the first time) that DG model is compatible with free energy decay.

About the full QDD model (i.e., with no $\hbar$ expansion), there is no rigorous proof, neither of existence nor of convergence.

We now present some numerical simulations. We look at open boundary conditions. We first analyze the influence of the effective mass on the shape of the current–voltage characteristic. The temperature is chosen equal to $77\,\mathrm{K}$ and the mobility is supposed to be constant and equal to $0.85\,\mathrm{m}^2\,\mathrm{V}^{-1}\mathrm{s}^{-1}$. The permittivity is also supposed to be constant and equal to $11.44\,\epsilon_0$. Figure 1 shows four different IV curves with different values of the effective mass inside and outside the double barriers. These curves show a certain sensitivity of the model to the value of the effective mass inside the barrier.

Figure 2 shows the time evolution of the density from the peak to the valley when the effective mass is $m_2 = 1.5 \times 0.092 m_\mathrm{e}$ inside the barriers and $m_1 = 1.5 \times 0.067 m_\mathrm{e}$ outside it (corresponding to the IV curve at the bottom right of Fig. 1). To obtain this figure, we apply a voltage of $0.25\,\mathrm{V}$ and wait

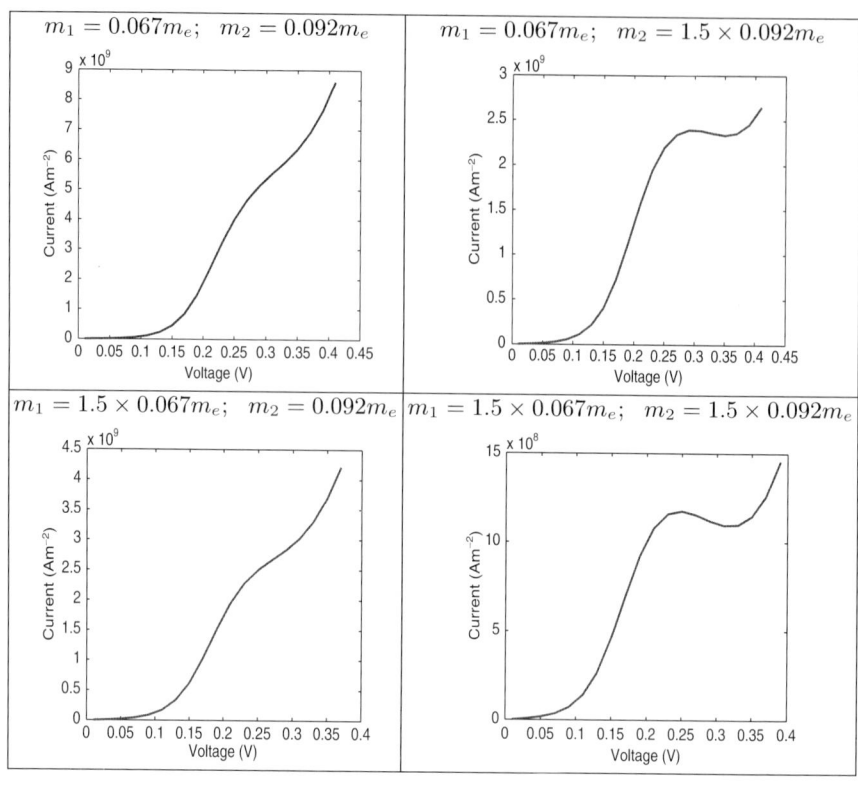

Fig. 1. Influence of the effective mass on the IV curve, m_1 being the mass outside the barriers, and m_2 being the mass inside

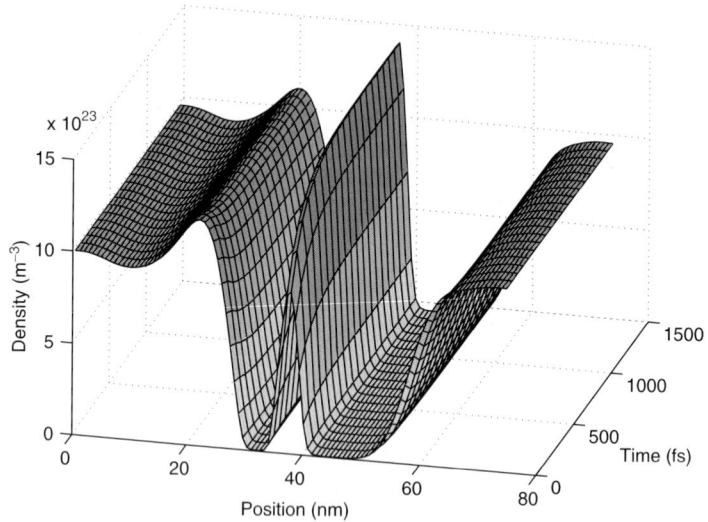

Fig. 2. Evolution of the density from the peak (applied bias: 0.25 V) to the valley (applied bias: 0.31 V)

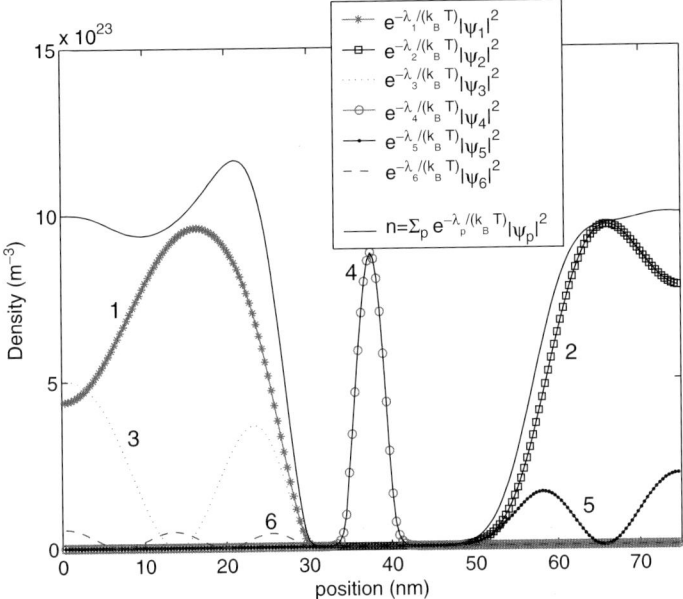

Fig. 3. Density at the peak (applied bias: 0.25 V)

for the electrons to achieve the stationary state. Then we suddenly change the value of the applied bias to 0.29 V and we record the evolution of the density. As expected, the density inside the well grows significantly and the stationary state is achieved at about 1,500 fs.

The next two figures (Figs. 3 and 4) display the details of the reconstruction of the density from the eigenstates ψ_p (for $p = 1, \ldots, 6$) of the modified Hamiltonian $H[A]$. The density $e^{-\lambda_p}|\psi_p|^2$ corresponding to each eigenstate is plotted for two values of the applied bias, respectively, corresponding to the current peak (Fig. 3) and to the valley (Fig. 4). Table 1 shows the values of the corresponding energies λ_p. Last, Fig. 5 shows the transient current at the left contact ($x = 0$). A detailed discussion of these results can be found in [10].

In Fig. 6, we show the results obtained with the Density Gradient model using the same parameters as defined for the QDD model. As we can see, results are qualitatively similar but differ significantly. Even with a smoother external potential (replacing the two step functions by two gaussians), it appears that the current–voltage characteristics are still different for the two models as suggested by Fig. 7. To finish, Fig. 8 shows the role of the temperature on the current for an applied bias of 0.2 V and for the three models QDD, DG, and CDD with a constant mass equal to $0.067m_e$.

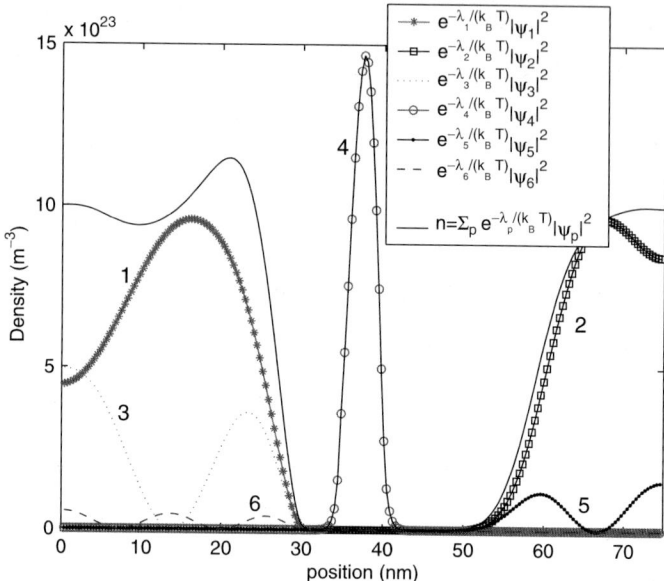

Fig. 4. Density at the valley (applied bias: 0.31 V)

Table 1. Eigenvalues (energies [eV]) of the modified Hamiltonian $H[A]$ at the peak and at the valley

	λ_1	λ_2	λ_3	λ_4	λ_5	λ_6	λ_7
Peak	0.87	1.05	1.56	2.03	2.28	3.03	4.47
Valley	0.87	1.11	1.57	1.70	2.54	3.05	5.03

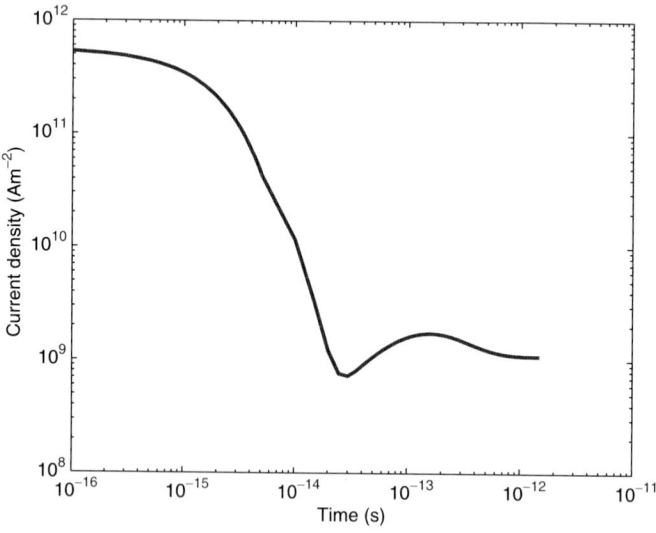

Fig. 5. Transient current density

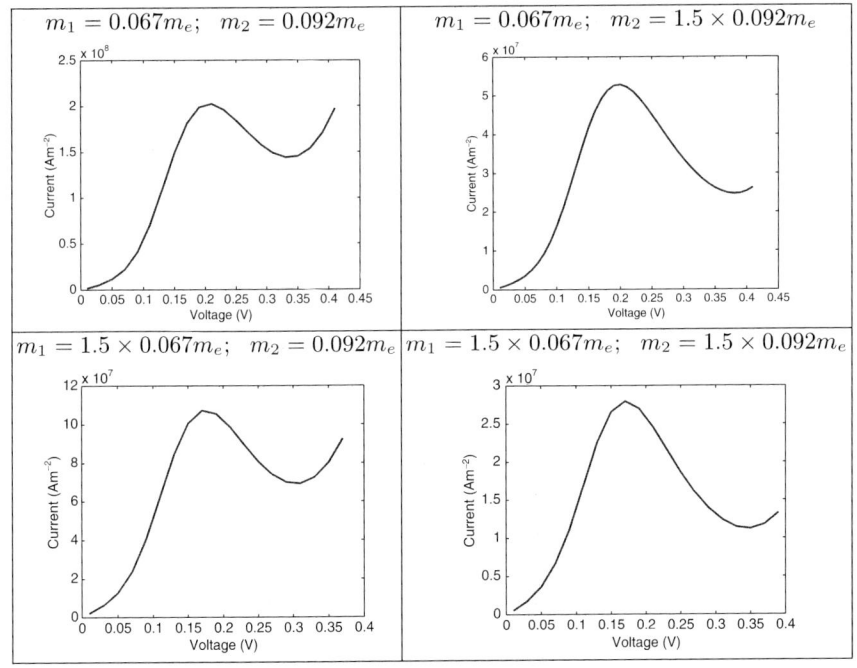

Fig. 6. IV curves obtained with the DG model (m_1 being the mass outside the barriers, and m_2 being the mass inside)

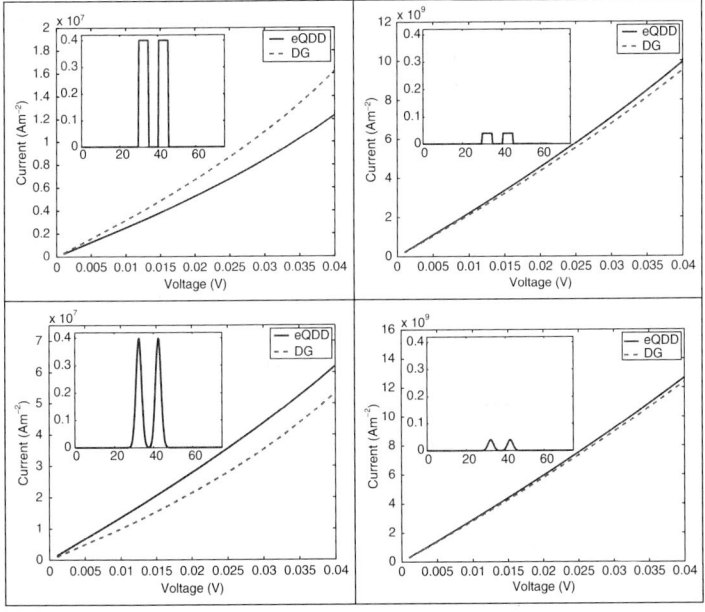

Fig. 7. Influence of the shape and the height of the double barrier on the current–voltage characteristics for the QDD and DG models

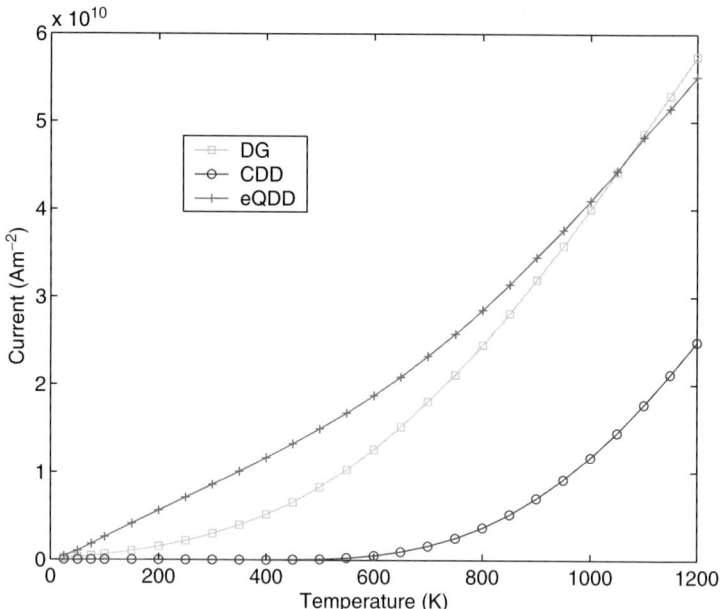

Fig. 8. Current–temperature curve (applied bias: 0.2 V)

3 Summary and Conclusion

In these notes, diffusion models have been derived by means of the entropy minimization approach. We have first proposed a formulation of a quantum BGK operator (which models a relaxation of the Wigner distribution function toward a quantum equilibrium). Then, we have performed a diffusion approximation of the resulting Quantum Kinetic Equation and provided new Quantum Energy-Transport or Drift-Diffusion models. The Quantum Drift-Diffusion model has been analyzed in more detail. This model differs from classical models by the reconstruction of the density from the chemical potential (through an eigenvalue problem). We can recover the Density-Gradient (DG) model of Ancona, and Iafrate [2] as an $O(\hbar^2)$ approximation, and the Schrödinger–Poisson Drift-Diffusion (SPDD) model of Sacco et al. [13] in situations close to equilibrium. A large set of numerical simulations have been realized and show that the qualitative behavior of the model is fairly satisfactory, while a certain sensitivity to some physical parameters still needs to be understood.

The quantum Energy-Transport model needs to be analyzed in the same way. The first step would be to find a simplified expression of the model (having local values of the pressure tensors in terms of the conservative and entropic variables).

Multidimensional simulations will require more computing power but are within reach. A better account of the continuous spectrum of the operators would certainly improve the results, notably close to the boundaries.

Of the overall approach, some other extensions and applications will require further developments. One would wish to introduce many particle effects more accurately than through the use of the BGK collision operator. Using this approach, phonon–electron collision operators for electrons in crystals could be derived. Also, the introduction of confinement in one or more directions would lead to subband models which could be applied to systems such as quantum wires or quantum dots. Following the same lines, Born–Oppenheimer approximations in quantum chemistry could also be used in the framework in these models and would lead to hybrid quantum-classical models, in the spirit of [7]. Applications could span from reaction dynamics in chemistry to biology problems such as ionic channels in cell membrane physiology.

References

1. Ancona, M. G., Diffusion-Drift modeling of strong inversion layers, COMPEL *6* 11–18 (1987)
2. Ancona, M. G., Iafrate, G. J., Quantum correction of the equation of state of an electron gas in a semiconductor, Phys. review B, *39*, 9536–9540 (1989)
3. Ancona, M. G., Tiersten, H. F., Macroscopic physics of the silicon inversion layer, Phys. review B, *35*, 7959–7965 (1987)
4. Ben Abdallah, N., Degond, P., On a hierarchy of macroscopic models for semiconductors, J. Math. Phys., *37*, 3306–3333 (1996)
5. Ben Abdallah, N., Degond, P., Gnieys, S., An energy-transport model for semiconductors derived from the Boltzmann equation, J. Stat. Phys., *84*, 205–231 (1996)
6. Ben Abdallah, N., Unterreiter, A., On the stationary quantum drift-diffusion model, Z. Angew. Math. Phys., *49*, 251–275 (1998)
7. Burghardt, I., Parlant, G., On the dynamics of coupled Bohmian and phase-space variables, a new hybrid quantum-classical approach, Journal of Chemical Physics, *120*, 3055–3058 (2004)
8. Chen, R-C., Liu, J-L., A quantum corrected energy-transport model for nanoscale semiconductor devices, J. Comput. Phys., *204*, 131–156 (2005)
9. Degond, P., Mathematical modelling of microelectronics semiconductor devices, AMS/IP Studies in Advanced Mathematics, AMS Society and International Press, 77–109, (2000)
10. Degond, P., Gallego, S., Mhats, F., An entropic quantum drift-diffusion model for electron transport in resonant tunneling diodes, J. Comp. Phys., to appear
11. Degond, P., Mhats, F., Ringhofer, C., Quantum energy-transport and drift-diffusion models, J. Stat. Phys., *118*, 625–667 (2005)
12. Degond, P., Mhats, F., Ringhofer, C., Quantum hydrodynamic models derived from the entropy principle, Contemp. Math., *371*, 107–131 (2005)
13. de Falco, C., Gatti, E., Lacaita, A. L., Sacco, R., Quantum-Corrected Drift-Diffusion Models for Transport in Semiconductor Devices, J. Comput. Phys., *204*, 533–561 (2005)

14. Gallego, S., Mhats, F., Entropic discretization of a quantum drift-diffusion model, SIAM J. Numer. Anal., *43*, 1828–1849 (2005)
15. Gallego, S., Mhats, F., Numerical approximation of a quantum drift-diffusion model, C. R. Math. Acad. Sci. Paris, *339*, 519–524 (2004)
16. Jngel, A., Pinnau, R., A positivity preserving numerical scheme for a fourth order parabolic equation, SIAM J. Num. Anal., *39*, 385–406 (2001)
17. Micheletti, S., Sacco, R., Simioni, P., Numerical Simulation of Resonant Tunnelling Diodes with a Quantum-Drift-Diffusion Model, Scientific Computing in Electrical Engineering, Lecture Notes in Computer Science, Springer-Verlag, pp. 313–321 (2004)
18. Pinnau, R., The Linearized Transient Quantum Drift Diffusion Model - Stability of Stationary States, Z. Angew. Math. Mech., *80*, 327–344 (2000)
19. Pinnau, R., Unterreiter, A., The Stationary Current-Voltage Characteristics of the Quantum Drift Diffusion Model, SIAM J. Numer. Anal., *37*, 211–245 (1999)
20. Pirovano, A., Lacaita, A., Spinelli, A., Two-Dimensional Quantum effects in Nanoscale MOSFETs, IEEE Trans. Electron Devices, *47*, 25–31 (2002)

Statistical Aspects of Size Functions for the Description of Random Shapes: Applications to Problems of Lithography in Microelectronics

Alessandra Micheletti[1], Filippo Terragni[1], and Mauro Vasconi[2]

[1] ADAMSS and Department of Mathematics, Università degli Studi di Milano, Via C. Saldini, 50, 20133 Milano, Italia
Alessandra.Micheletti@mat.unimi.it
[2] STMicroelectronics Srl, Via C. Olivetti, 2, 20041 Agrate Brianza, Italia
mauro.vasconi@st.com

Summary. Here the theory of size functions is introduced and joined to some statistical techniques in order to build confidence regions for a family of random shapes. An algorithm for the computation of the discrete counterpart of the size functions is also introduced. The method is applied to the quality control of shapes impressed with a laser on a silicon wafer, in microelectronics. The robustness of the size functions in the description of random shapes has led to good experimental results, and thus to the possibility of enclosing this method into an automatic procedure for the quality control of electronic devices.

1 Introduction

In real applications, objects rarely have exactly the same shape within measurement error; hence the randomness of shapes need to be taken into account. Thanks to the development of information technologies, the last decade has seen a considerable growth of interest in the statistical shape theory and its application to various scientific areas.

The solution of the problem of describing a "shape" via functions taking values in a finite dimensional space, without loosing important information, is essential for a mathematical and statistical approach. Recently new geometrical descriptors of shapes, called *size functions*, have been proposed [4]. These functions are able to capture "globally" the topological and geometrical features of an object, differently from landmarks [2,6] (which usually are specific points, angles, distances, etc. on the object, chosen by an expert) which are widely used in literature but whose results in a statistical context are strongly dependent on their choice, leading to a sort of subjective quantitative analysis.

Size functions depend on the choice of a *measuring function* and usually only a small number of choices can lead to different statistical results. A measuring function takes into account the most relevant shape features of the object in the considered application and it is chosen on the basis of the invariance properties that the geometrical descriptors must satisfy (e.g., invariance with respect to rotations, translations, scaling, etc.).

The theory of size functions has been developed mainly in a deterministic framework. A first attempt to join this theory with randomness is here presented. In particular, we show how to combine size functions with some well-known statistical techniques in order to obtain good results in random shape recognition and classification [9].

Since in most of applications data are provided as digital images of the shapes under study, preprocessing of such images (which should filter out the undesired noise, perform edge detection, etc.) is a fundamental step toward the computation of a discrete approximation of the size functions associated with such shapes. Suitable algorithms may compute a graphical representation of a (discrete) size function. Then, thanks to the robustness of this descriptor and by applying some *cluster analysis techniques* (based on a suitable distance between size functions), it is possible to find 2D confidence regions for a family of shapes and to detect the presence of outliers, i.e., of shapes not belonging to the family under study.

We applied this technique to some specific problems arising in microlithography of electronic devices, like:

1. Introducing a suitable distance to compare the shapes of the impressed structures
2. Specifying confidence regions for the geometries impressed using standard process parameters
3. Testing the effects of changing some process parameters on the resulting geometry
4. Looking for the most critical points (if any) in the impressed structures

The application of this methodology to experimental data has led to the definition of a procedure that could be implemented to control in a powerful and automatic way the quality of the devices. For other applications, to biomedical problems, see [7].

2 Size Functions and Shape Description

Let $\mathscr{M}$ be a finite union of compact arcwise connected and locally arcwise connected subsets of an Euclidean space and let $\varphi : \mathscr{M} \to \mathbb{R}$ be a continuous function, called *measuring function*. The pair $(\mathscr{M}, \varphi)$ denotes in a formal way the shape of the object $\mathscr{M}$. For every $x \in \mathbb{R}$ let $\mathscr{M}\langle \varphi \leqslant x \rangle$ denote the set $\{P \in \mathscr{M} : \varphi(P) \leq x\}$. Thus we can introduce the following definition [4].

Definition 1. *Consider the function* $l_{(\mathscr{M},\varphi)} : \mathbb{R} \times \mathbb{R} \to \mathbb{N} \cup \{+\infty\}$ *defined by setting* $l_{(\mathscr{M},\varphi)}(x,y)$ *equal to the number of equivalence classes into which the set* $\mathscr{M}\langle \varphi \leqslant x \rangle$ *is divided by the relation of* $\langle \varphi \leqslant y \rangle$-*homotopy, where two points* $P, Q \in \mathscr{M}$ *are* $\langle \varphi \leqslant y \rangle$-*homotopic if and only if either* $P = Q$ *or a continuous path* $\gamma : [0,1] \to \mathscr{M}$, *joining* P *and* Q, *exists in* $\mathscr{M}$ *such that* $\varphi(\gamma(t)) \leq y$ *for*

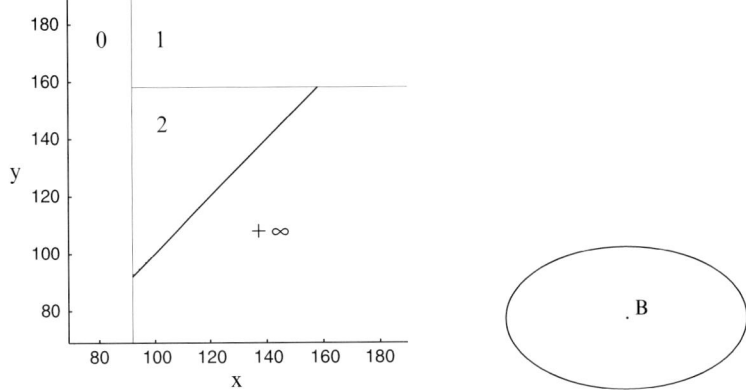

Fig. 1. Size function of an ellipse contour with respect to the distance from its center of mass

every $t \in [0,1]$. *We shall call* $l_{(\mathcal{M},\varphi)}$ *the* size function *associated with the pair* $(\mathcal{M},\varphi)$.

The size function $l_{(\mathcal{M},\varphi)}$ describes the shape of $\mathcal{M}$ through information given by φ, whose choice depends on the specific application problem we are interested in. An important property of size functions is that they inherit the invariance properties, if any, of the chosen measuring functions. Thus it is sufficient to take measuring functions with the desired invariance to obtain invariant size functions.

The size function $l_{(\mathcal{M},\varphi)}$ conveys relevant information about the pair under study only in the half-plane $x < y$. Thus in the following we will consider only this region. We point out that, for $x < y$, size functions have a simple geometric interpretation: in such a case $l_{(\mathcal{M},\varphi)}(x,y)$ is equal to the number of arcwise connected components of $\mathcal{M}\langle \varphi \leqslant y \rangle$ containing at least one point of $\mathcal{M}\langle \varphi \leqslant x \rangle$.

An example of size function is illustrated in Fig. 1. We show the size function of an ellipse contour $\mathcal{M}$ with respect to the measuring function $\varphi(z)$ which associates to each point $z \in \mathcal{M}$ its distance from the center of mass of $\mathcal{M}$. More precisely, we represent the domain of $l_{(\mathcal{M},\varphi)}$ with its discontinuities: the number displayed in each region of the domain denotes the value of the size function in that region.

The discontinuities of size functions are related to specific points and vertical lines in the real plane, each one with a multiplicity, called *cornerpoints* and *cornerlines* respectively. We refer to [3,4] for further details.

The abscissa of every cornerline corresponds to the global minimum taken by φ on an arcwise connected component of $\mathcal{M}$. Moreover, when $\mathcal{M}$ and φ are sufficiently regular, it can be shown that the coordinates of each cornerpoint (in the half-plane $x < y$) are couples of critical values for φ.

It can be proven that all and only the discontinuity points of a size function are generated by its cornerpoints and cornerlines, and the domain of the size function (for $x < y$) is divided by its discontinuities into overlapping triangular regions (possibly of infinite area), each one of them related to a cornerpoint or a cornerline (see previous Fig. 1). Moreover for $x < y$, cornerpoints and cornerlines with their multiplicities uniquely determine the value of $l_{(\mathcal{M},\varphi)}$ almost everywhere, so that they contain all information conveyed by the size function about the shape under study. This result has a fundamental consequence: size functions can reduce the analysis of a shape to a finite dimensional problem, since its main features are described by a finite number of cornerpoints and cornerlines.

3 Size Functions and Shape Comparison

The problem of comparing shapes can be dealt with by defining a suitable distance between the size functions describing the considered shapes. An idea is to compare two size functions by measuring the *cost* of moving and overlapping the cornerpoints and cornerlines of one size function to those of the other one, by minimizing the longest movement. Since, in general, the number of cornerpoints of two size functions is different, we also enable the cornerpoints to be transported onto the points of the diagonal Δ with equation $y = x$. This leads to the definition of the *matching distance* between size functions (see [1] for further details).

In order to introduce the matching distance we need some new definitions. In this section, for simplicity, we will assume that $\mathcal{M}$ is also arcwise connected and the only cornerline $x = \bar{x}$ for $l_{(\mathcal{M},\varphi)}$ will be formally identified with the point $(\bar{x}, \infty)$. Let then $\mathscr{S}$ be the set $\{(x,y) \in \mathbb{R}^2 : x \leq y\} \cup \{(k, \infty) : k \in \mathbb{R}\}$.

Definition 2. *Let $l_{(\mathcal{M},\varphi)}$ be the size function associated with the pair $(\mathcal{M}, \varphi)$. We shall call* representative sequence *for $l_{(\mathcal{M},\varphi)}$ any sequence of points $a :$ $\mathbb{N} \to \mathscr{S}$, briefly denoted by (a_i), with the following properties:*

1. *a_0 is the cornerline for $l_{(\mathcal{M},\varphi)}$*
2. *For each $i > 0$, either a_i is a cornerpoint for $l_{(\mathcal{M},\varphi)}$ or a_i Belongs to Δ*
3. *If p is a cornerpoint for $l_{(\mathcal{M},\varphi)}$ with multiplicity $\mu(p)$, then the cardinality of the set $\{i \in \mathbb{N} : a_i = p\}$ is equal to $\mu(p)$*
4. *The set of indices for which a_i belongs to Δ is countably infinite*

We now introduce the following pseudodistance d in order to assign a cost to each displacement of cornerpoints and cornerlines.

Definition 3. *Let d be the pseudodistance on $\mathscr{S}$ such that for every $p = (x, y)$ and $\bar{p} = (\bar{x}, \bar{y})$*

$$\mathrm{d}(p, \bar{p}) := \min \left\{ \max\{|x - \bar{x}|, |y - \bar{y}|\}, \max \left\{ \frac{y - x}{2}, \frac{\bar{y} - \bar{x}}{2} \right\} \right\}, \qquad (1)$$

with the convention about ∞ that $\infty - y = y - \infty = \infty$ for $y \neq \infty$, $\infty - \infty = 0$, $\infty/2 = \infty$, $|\infty| = \infty$, $\min\{\infty, c\} = c$, $\max\{\infty, c\} = \infty$.

In other words, the pseudodistance d between two points p and $\bar{p}$ compares the cost of moving p to $\bar{p}$ and the cost of moving p and $\bar{p}$ onto the diagonal and takes the smaller.

Definition 4. *Let l_1 and l_2 be two size functions. If (a_i) and (b_i) are two representative sequences for l_1 and l_2, respectively, then the* matching distance *between l_1 and l_2 is the number*

$$d_{\mathrm{match}}(l_1, l_2) := \inf_{\sigma} \sup_{i} \; \mathrm{d}(a_i, b_{\sigma(i)}),$$

where $i \in \mathbb{N}$ and σ varies among all the bijections from $\mathbb{N}$ to $\mathbb{N}$.

Theorem 1 states the fundamental property of stability of the matching distance between size functions.

Theorem 1. *Let us consider a pair $(\mathscr{M}, \varphi)$. For every real number $\varepsilon \geq 0$ and for every measuring function $\psi : \mathscr{M} \to \mathbb{R}$ such that $\max_{P \in \mathscr{M}} |\varphi(P) - \psi(P)| \leq \varepsilon$, the matching distance between $l_{(\mathscr{M},\varphi)}$ and $l_{(\mathscr{M},\psi)}$ is smaller than or equal to ε.*

This result allows us to use size functions as *robust* shape descriptors in presence of random perturbations on the considered shapes, often arising in real applications due to noise, errors or intrinsic randomness. Thanks to this property of robustness, the presence of randomness on a shape is revealed in the domain of the corresponding size function by small displacements of its cornerpoints and cornerlines and by the presence of small triangles near the diagonal Δ. As an example, in Fig. 2 we show an ellipse contour perturbed with noise and its size function with respect to the distance from its center of mass.

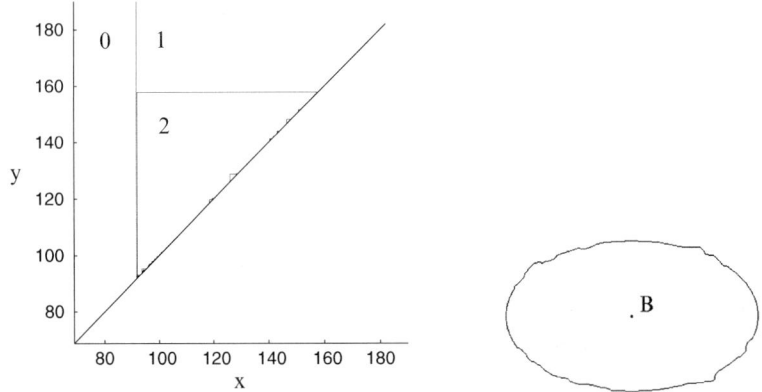

Fig. 2. Size function of a deformed ellipse contour with respect to the distance from its center of mass

Thus the cornerpoints which describe (with the cornerlines) the main characteristics of the shape under study are those standing "sufficiently far" from Δ. Since they are always in a finite number [4], *shape analysis and classification are then reduced to the statistical study of the location of finite sets of points and lines in the real plane.*

4 Analysis of Random Shapes Impressed on Integrated Devices

In this section, we will show how we applied the mathematical tools previously described for the recognition and classification of random shapes in the field of microlithography of electronic devices [8, 9].

During the lithographic process, structures with particular geometries are impressed on silicon wafers on several overlapped levels (they show particular shapes if looked at from above). In order to guarantee a correct working of the final device, structures on different levels must be perfectly aligned. It is then essential to control that their shapes are always well-impressed, that is showing features satisfying the desired specifications. Size functions are powerful tools which can be very useful to deal with these problems. We stress that, since the shapes printed on wafers have an intrinsic randomness, as the lithographic process often involves factors with unpredictable effects, size functions need to be used in a statistical context.

We concentrated our analysis on four structures whose lithographic printing seemed to be particularly crucial. Figure 3 shows SEM images of these structures impressed with optimal process conditions. As an example we now consider the first of them and we explain how its shape could be analyzed.

Let us take a sample of SEM images of this structure impressed with standard process parameters. The first step to be performed consists of a preprocessing in order to filter out the undesired noise and to detect the edges of the displayed structures. An example of the patterns we get from these preliminary operations is shown in Fig. 4. We then describe the shapes of these patterns by computing the corresponding size functions with respect to the distance from the center of mass.

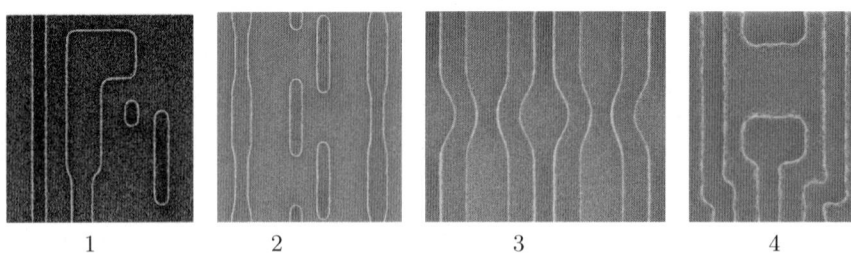

Fig. 3. SEM images of the structures considered in our statistical shape analysis

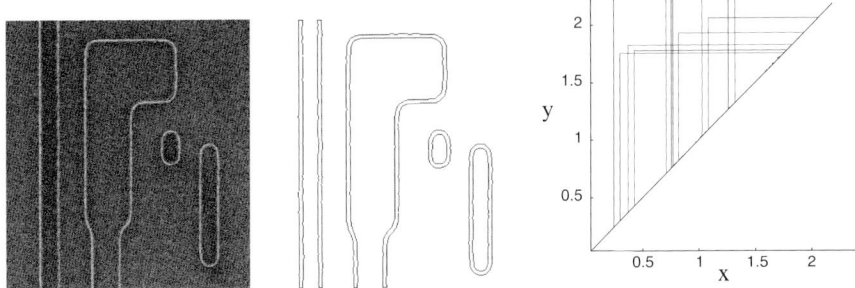

Fig. 4. Pattern of edges detected from a SEM image of the first structure (*in the center*) and the corresponding size function with respect to the distance from its center of mass (*on the right*)

4.1 Computation of Size Functions

Before studying our sample of size functions, it is interesting to understand how we got it. Therefore, we briefly hint at how a discrete approximation of a size function could be defined and how an efficient algorithm for its computation could be implemented (see [3, 5, 9]).

Let us consider a pair $(\mathcal{M}, \varphi)$. We shall assume that $\mathcal{M}$ is a compact connected and locally connected subset of $\mathbb{R}^2$ and the measuring function φ is the restriction to $\mathcal{M}$ of a continuous function $g : \mathbb{R}^2 \to \mathbb{R}$, with modulus of continuity $\omega(\delta)$, $\delta > 0$. Our first purpose is that of approximating the considered pair $(\mathcal{M}, \varphi)$ with a new pair $(G, \bar{\varphi})$, where G will be a finite graph and $\bar{\varphi}$ will be a (discrete) function defined on the set of its vertices.

Definition 5. *Let* $\mathscr{P} = \{P_0, P_1, \ldots, P_h\}$ *be a finite set of points of* $\mathbb{R}^2$ *and let us denote by* $\mathscr{B}_\delta$ *the set of the* $h + 1$ *open balls* $B(P_i, \delta)$ *of radius* $\delta > 0$ *with center at the points of* $\mathscr{P}$. *Let us assume that* $\mathscr{B}_\delta$ *verifies the following properties:*

1. $\mathcal{M}$ *is contained in* $\bigcup_{i=0}^{h} B(P_i, \delta)$
2. *for every* $i = 0 \ldots h$, $B(P_i, \delta) \cap \mathcal{M}$ *is a nonempty connected set*

We shall call $\mathscr{B}_\delta$ *a* δ-*covering of* $\mathcal{M}$.

Let us assume that a δ-covering $\mathscr{B}_\delta$ of $\mathcal{M}$ is given, with $\{P_0, P_1, \ldots, P_h\}$ as the set of its centers.

Definition 6. *We shall call size graph* $(G, \bar{\varphi})$ *associated with* $\mathscr{B}_\delta$ *the (finite) labeled graph so that:*

1. *The set of vertices is* $V = \{P_0, P_1, \ldots, P_h\}$
2. *Two vertices* P_i, P_j *in* V *are adjacent if and only if the set* $(B(P_i, \delta) \cup B(P_j, \delta)) \cap \mathcal{M}$ *is connected*

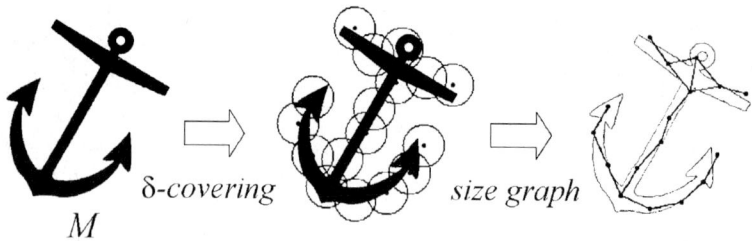

Fig. 5. A δ-covering and the corresponding graph associated with the object $\mathcal{M}$ on the *left*

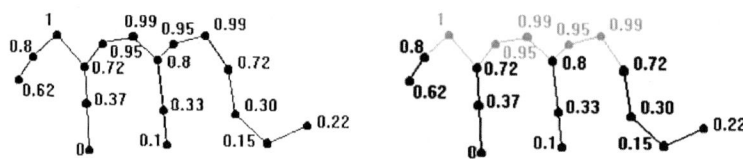

Fig. 6. On the *left*, an example of size graph $(G, \bar{\varphi})$, where $\bar{\varphi}$ is the height function with respect to the lowest vertex. On the *right*, it is shown the subgraph $G\langle\bar{\varphi} \leqslant 0.8\rangle$ of G. According to Definition 7, we have $l_{(G,\bar{\varphi})}(0.5, 0.8) = 3$

3. *We label each vertex P_i in V by the real number $\bar{\varphi}(P_i) := g(P_i)$ (thus $\bar{\varphi}$ is a discrete measuring function on V)*

The size graph $(G, \bar{\varphi})$ δ-approximates (Fig. 5) the considered pair $(\mathcal{M}, \varphi)$. In our application, G will be the finite graph whose vertices are the centers of the (ordered) pixels describing the edges of the considered structures displayed in 2D digital images.

Now we have all the tools to introduce the definition of *discrete size function*. For every $x \in \mathbb{R}$, let $G\langle\bar{\varphi} \leqslant x\rangle$ be the subgraph of G obtained by erasing all the vertices of G at which $\bar{\varphi}$ takes a value strictly greater than x and all the edges connected to these vertices.

Definition 7. *We shall call* discrete size function *associated with the size graph $(G, \bar{\varphi})$ the function $l_{(G,\bar{\varphi})} : \{x \leq y\} \rightarrow \mathbb{N}$ that associates to each point (x, y) the number of connected components of $G\langle\bar{\varphi} \leqslant y\rangle$ containing at least one vertex of $G\langle\bar{\varphi} \leqslant x\rangle$.*

Thus, the discrete size function $l_{(G,\bar{\varphi})}$ (Fig. 6) approximates the size function $l_{(\mathcal{M},\varphi)}$ when the considered object $\mathcal{M}$ is approximated by a finite set of points. Furthermore the better this approximation is, the more precise is the information about the size function we can get on the basis of its discrete counterpart, as stated in Theorem 2.

Theorem 2. *Assume that a size graph* $(G, \bar{\varphi})$ *is given,* δ-*approximating the pair* $(\mathcal{M}, \varphi)$. *Then, for every* $x, y \in \mathbb{R}$ *and every* $\bar{\omega} \geq \omega(\delta)$ *with* $x + \bar{\omega} \leq y - \bar{\omega}$, *the following inequalities hold:*

(1) $l_{(G,\bar{\varphi})}(x - \bar{\omega}, y + \bar{\omega}) \leq l_{(\mathcal{M},\varphi)}(x, y) \leq l_{(G,\bar{\varphi})}(x + \bar{\omega}, y - \bar{\omega})$

(2) $l_{(\mathcal{M},\varphi)}(x - \bar{\omega}, y + \bar{\omega}) \leq l_{(G,\bar{\varphi})}(x, y) \leq l_{(\mathcal{M},\varphi)}(x + \bar{\omega}, y - \bar{\omega})$

Using the good properties of the discrete size functions, we succeeded in implementing an efficient algorithm for computing the size function associated with a shape. Note that size graphs are usually very big, thus increasing the costs involved in the computation of the discrete size functions. So the problem of reducing the size graphs without changing the associated discrete size functions is very important in order to use these tools for a statistical shape analysis. We tackled this problem in our algorithm implementation, using the $\mathcal{L}$-*reduction* method (see [5]).

4.2 Statistical Shape Analysis

Now let us go back to the sample of size functions we computed from the patterns whose shape we want to study. Thanks to robustness of size functions, we expect that the location of the cornerpoints and cornerlines of each pattern will be slightly different. Thus, after removing the small triangles near the diagonal Δ, which are due to noise, we obtain clusters of points in the half-plane $x < y$, each one related to a cornerpoint aside from Δ of the computed size functions. Similarly, if we identify each cornerline with its abscissa, we get clusters of points on the x-axis.

Let us consider only the clusters of cornerpoints in the two-dimensional plane, since the one-dimensional clusters of the abscissas of cornerlines can be treated in a similar and even simpler way. Remember that cornerpoints are linked to local critical values of the measuring function, thus bringing more information than the location of cornerlines, which is only related to the global minimum of the measuring function on each arcwise connected component of $\mathcal{M}$.

The clusters of cornerpoints must be identified, that is each point must be assigned to one specific cluster. We identified them by applying a suitable *cluster analysis technique*, based on the matching distance between size functions. A statistical analysis of such clusters can give precious information about the shapes under study.

We built confidence regions for the clusters of cornerpoints obtained from our family of shapes impressed with optimal process conditions, that is for shapes which should have good features in order to guarantee the correct working of the final device. Such regions can be constructed by specifying a confidence region to detect the presence of outliers (i.e., of observations belonging to the queues of the distribution of the sample under study) in each cluster of cornerpoints (see [9]). Figure 7 shows confidence regions at significance level 0.95 for the clusters of cornerpoints, which can be used for testing

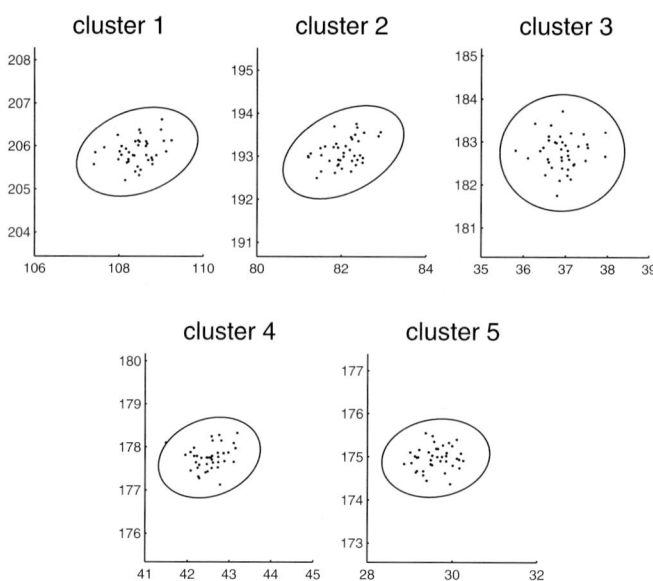

Fig. 7. Confidence regions at significance level 0.95 for the clusters of cornerpoints

the effects of changing some process parameters on the resulting geometry. More precisely, if most of the cornerpoints of the size function describing a new observed shape (for the same structure) fall outside the regions, we shall say that such a shape is "well-impressed" only with a probability of 5%. On the other hand if (almost) all of them stand inside the regions, the chosen shape could not be regarded as "far" from a shape printed in standard process conditions.

4.3 Results

We applied the above procedure to the shapes of the first structure on a wafer exposed with different values of exposure energy and focus offset. The results we got are shown in Fig. 8: Positions in which the computed size functions have all the cornerpoints (except for one at most) standing inside the specified regions (which have been colored) are close to the central part of the wafer, where standard process parameters have been used. This is consistent with our expectations, that is with the existence of a quite large process window which guarantees structures satisfying the desired specifications. Moreover, the process window as obtained by this method is the same as deduced by an expert engineer through current practices. Our results seem also to point out that the impressed shapes are more sensitive to the focus offset changes (along the horizontal direction on the wafer). Similar results were obtained for the other structures depicted in Fig. 2.

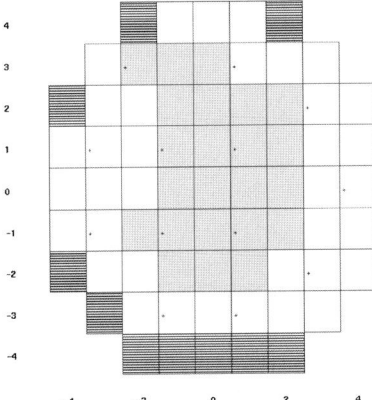

Fig. 8. Layout of a wafer exposed with variable process parameters. Positions in which the computed size functions have all the cornerpoints (except for one at most) standing inside the specified regions have been colored in *light gray* (the *black* positions have not been exposed)

The analysis we carried out had the aim of improving the quality control process of integrated devices. We point out that, differently from the local measurements which are usually done on wafers (the location or distance of specific crucial points or lines is usually measured), it takes into account the whole topology of the structures under study and it is independent from the subjective choice of the points to be measured. Thus the quality control process can be improved, since it would automatically take into account also parts on a structure which are generally regarded as less critical and therefore ignored.

Moreover the application of the described methodology to experimental data has led to the definition of a procedure that could be implemented to automatically recognize and classify, from a probabilistic point of view, structures on wafers showing or not well-impressed shapes.

It is interesting to see that the implemented algorithm for computing size functions allows us to identify which points on the considered pattern can be regarded as the most critical ones, with respect to the shape description given by the chosen measuring function. The choice of the measuring function, or the choice of the part on the pattern to be analyzed, is crucial in the correct identification of the critical points. Figure 9 shows an example. On the left a pattern associated with the first structure is displayed with its critical points, when it is described by the related size function with respect to the distance from its center of mass B. Instead if we limit the analysis to the pattern on the right and we use the distance from the fixed point P as a measuring function, some (but not all!) of the critical points correspond to parts of the structure which are known to be particularly sensitive to variations during the exposition. Thus the other points can be regarded as other critical points,

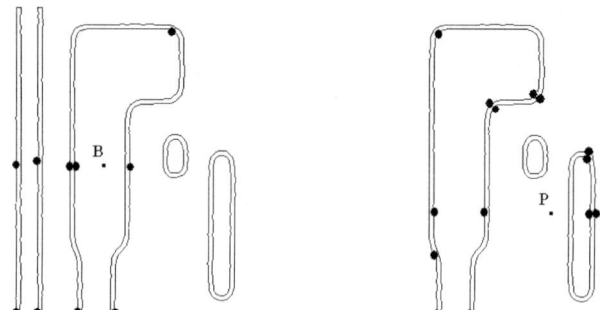

Fig. 9. Critical points on two patterns when using the distance from the center of mass B (*on the left*) and the distance from the fixed point P (*on the right*) as a measuring function

whose influence in the shape description may be not much evident to an expert, which probably should be taken into account for a correct check of the goodness of the impressed shape under study.

References

1. M. D'Amico, P. Frosini, C. Landi, Natural pseudo-distance and optimal matching between reduced size functions, Preprint, 2005.
2. I. L. Dryden, K. V. Mardia, *Statistical Shape Analysis*, Wiley, New York, 1998.
3. P. Frosini, C. Landi, Size theory as a topological tool for computer vision, *Pattern Recognition and Image Analysis* **9**, pp. 596–603, 1999.
4. P. Frosini, C. Landi, Size functions and formal series, *AAECC* **12**, pp. 327–349, 2001.
5. P. Frosini, M. Pittore, New methods for reducing size graphs, *Intern. J. Computer Math.* **70**, pp. 505–517, 1999.
6. A. Micheletti, Statistical shape analysis applied to automatic classification, *Industry Days 2003-2004*, D. Aquilano et al. Eds. The Miriam Project Series, Società Editrice Esculapio, Bologna, pp. 91–100, 2005.
7. A. Micheletti, Size functions applied to the statistical shape analysis and classification of tumor cells. Minisymposium Shape and Size in Medicine and Biotechnology. In this volume, 2007.
8. A. Micheletti, E. Severgnini, F. Terragni, M. Vasconi, Statistical shape analysis applied to microlithography, *Proceedings of SPIE International Symposium on Microlithography, Data Analysis and Modeling for Pat-terning Control III*, 2006.
9. F. Terragni, *Statistical Aspects of Size Functions for the Description of Random Shapes. Applications to Lithography Problems in Microelectronics*, MSc. Thesis, University of Milan, 2005. In Italian.

Part II

Minisymposia

Minisymposium "Flow Control in Aircrafts"

A. Abbas[1] and J.M. Vega[2]

[1] Airbus, Spain
[2] Universidad Politécnica de Madrid, Spain

Delaying laminar-turbulent transition in boundary layers attached to commercial aircrafts has a significant impact in drag reduction, which in turn contributes to reducing both fuel consumption and environmental impact. This is a classical subtle problem in Fluid Mechanics that has received a continued attention during the last decades from both the experimental and the theoretical points of view, and involves some fascinating open problems in Applied Mathematics. The flavour of current European efforts in this direction can be appreciated in the various contributions below, which have been selected to illustrate the multidisciplinary character of this field.

Carlo Cossu, from the École Polytechnique, Palaiseau, summarizes some recent results on stabilizing Tollmien-Schlichting waves in a Blasius boundary layer using nearly optimal streaks of moderate amplitude (larger streaks would enhance transition). The streaks are effectively generated using appropriate roughness elements placed near the leading edge of the plate. This provides an effective and quite promising passive method to delay transition, as has been experimentally checked.

Xuesong Wu, from Imperial College, describes the mathematical theory involved in the analysis of nonlinear interaction between various planar and oblique Tollmien-Schlichting modes. The associated perturbative scheme is quite subtle and requires to consider several sublayers in the boundary layer (five, in the example considered in the paper) and allows to uncover the catalytic role of some of the modes in enhancing/suppressing spatial growth of the remaining modes. Leading order results are given in terms of small fractional powers (such as one tenth in the example considered in the paper) of the Reynolds number R, which could make quantitative predictions problematic at moderately large R. But the associated qualitative prediction provides physical insight and can be extremely useful in the search for effective means of controlling transition.

Eusebio Valero, from the Polytechnic University of Madrid, describes some recent results in using mode interaction processes to stabilize Tollmien-Slichting waves through parametric coupling of unstable modes with stable ones, and coupling to the associated mean flow. The idea is not new, but is connected with related methods to suppress classical hydrodynamic instabilities, such as the Rayleigh-Taylor instability. Effective stabilization is obtained that in addition is fairly robust. The method is also connected with the mode interaction processes involved in the two previous papers in this minisymposium.

Paolo Luchini, from the University of Salerno, discusses some subtle connections between flow topology and drag. In particular, he tells us three stories in connection to (a) the possibility of obtaining a lower-than-laminar drag through a zero-mean blowing and suction; (b) the effect of wavy blowing and suction in generating large scale vortices in an already turbulent flow, and the effect of these on both drag and skin friction; and (c) the role of the steady streaming flow produced by oscillatory blowing and suction in changing the overall drag. When the three stories are put together, it becomes clear that a distinction must be made between whether we are reducing drag or creating trust when we act on the flow in an oscillatory manner. We believe that these simple concepts can open new ways of looking for effective devices to reduce the overall drag.

Peter Carpenter, from the University or Warwick, describes some recent results on the effect of free stream turbulence on transition in a boundary layer attached to a compliant wall. The conclusion is that compliant walls are effective devices in flow control, even in environments with relatively high free stream turbulence levels. Although, as he points out compliant walls are not practical in aeronautical applications, they are of interests in related fields, such as drag reduction in marine vehicles. Also, compliant walls appeared from early studies of the way in which dolphins seem to manage to decrease skin friction. This was the beginning of the intent of translating to industrial devices some solutions that nature seems to have found to reduce friction drag in swimming and flying animals.

Using Non-Normality for Passive Laminar Flow Control

C. Cossu[1], L. Brandt[2], J.H.M. Fransson[2], and A. Talamelli[3]

[1] LadHyX, CNRS/Ecole Polytechnique, F-91128 Palaiseau, France
[2] KTH Mechanics, SE-10044, Stockholm, Sweden
[3] II Facoltà di Ingegneria, Università di Bologna. I-47100 Forlì, Italy

1 Introduction

In the absence of external perturbations, the boundary layer developing on a flat plate is uniform in the spanwise direction and is well described by the Blasius and Falkner-Skan similarity solutions (see, e.g. [Sch79]). These solutions become linearly unstable when the Reynolds number $Re = \sqrt{U_\infty x/\nu}$ based on the freestream velocity U_∞, the streamwise distance from the leading edge of the plate x and the kinematic viscosity of the fluid ν exceeds a critical value Re_c ($Re_c = 304$ for the Blasius solution). This primary linear instability appears in the form of two-dimensional Tollmien–Schlichting (TS) waves localized inside the boundary layer. As TS waves grow to amplitudes of the order of 1% of the free-stream velocity, secondary instability sets in (for a review see [Her88]), eventually leading to breakdown and transition to turbulence. This scenario is today well understood and is often referred to as the 'classical' transition scenario of boundary layers in low noise environments.

In the two-dimensional boundary layer, however, small amounts of streamwise vorticity are very effective in pushing low momentum fluid away from the wall and high momentum fluid towards the wall eventually leading to large elongated spanwise modulations of the streamwise velocity called streamwise streaks. The mechanism of streak generation, described above and known as the 'lift-up effect' is based on an inviscid process and applies to shear flows in general. The effect of viscosity eventually dominates rendering the growth of the streaks only transient. The transient growth, that can be of the order of Re^2 [Gus91, RH93], can however, be very large and is related to the non-normal nature of the linearised stability operator (for a review the reader may refer to [SH01]). The most dangerous perturbations, leading to the 'optimal transient growths', have been found to consist of streamwise vortices and have been computed for a number of shear flows. In the presence of streaks, the streamwise velocity profiles develop inflection points which may support inviscid instabilities for sufficiently large streak amplitudes (26% of the freestream velocity for optimal streaks in the Blasius boundary layer [ABBH01]). These secondary streak instabilities typically lead to 'bypass' transition to turbulence, i.e. transition without the primary TS instability. Optimal streamwise vortices are therefore usually seen as 'dangerous' perturbations for the boundary layer stability because they are capable of initiating the transition to turbulence with extremely low initial energy.

In this contribution we will summarize the main results of recent studies that have shown that the artificial forcing of streaks of moderate amplitude, stable to secondary inflectional instabilities, has a stabilizing effect on the TS waves instability [CB02, CB04]. The choice of an optimal forcing minimizes the actuator input energy to levels of $O(1/Re^2)$. In this context non-normality is therefore used as an efficient amplifier in the control protocol. Careful experiments have validated this concept by demonstrating that stable moderate amplitude nearly optimal streaks can be generated using appropriate roughness elements [FBTC04] that these streaks have a stabilizing effect on low amplitude Tollmien–Schlichting waves [FBTC05] and that they are indeed able to effectively delay transition [FTBC06, Cho06].

2 Basic Flows

The basic flows considered in the two early theoretical studies, consist in zero pressure gradient boundary layers with steady, nonlinearly saturated, spanwise periodic streaks of different amplitudes. Optimal perturbations, consisting of vortices aligned in the streamwise direction [ABH99, Luc00], are used to generate the streaks with minimum input energy. Following [ABBH01], the optimal perturbation computed by Andersson *et al.* [ABH99] is used as inflow condition close to the leading edge and its downstream evolution is followed till nonlinear saturation for different initial amplitudes. Direct numerical integrations are used to compute the basic flows and the evolution of the perturbations in the presence of forcing. The incompressible 3D Navier–Stokes equations are integrated using a pseudospectral code described in Lundbladh *et al.* [LBS+99]. The code uses Fourier expansions in the streamwise and spanwise directions and Chebyshev polynomials in the wall-normal direction. The time stepping scheme is a low storage third-order Runge–Kutta method for the nonlinear terms and a second-order Crank–Nicolson method for the linear terms. Dealiasing is used in the streamwise and spanwise directions. A fringe region is employed to enforce inflow and outflow boundary conditions in a periodic domain; in the case of 'temporal' simulations a volume force is used to keep the basic flow parallel. For the computations presented below we have used use a box with inlet at Re $= 272$ and dimensions of $1128\,\delta_{*0} \times 20\,\delta_{*0} \times 12.83\,\delta_{*0}$, in the streamwise, wall-normal and spanwise directions, respectively, where $576 \times 65 \times 32$ collocation points are used. We denote by δ_{*0} the boundary layer displacement thickness at the inlet. The spanwise extension of the domain corresponds to one wavelength of the optimally growing streaks. Denoting by x, y and z the streamwise, wall-normal and spanwise

Table 1. Streak amplitude for the computed basic flows

Case	Inlet A_{ST}	Maximum A_{ST}	A_{ST} at $Re = 609$
A	0.0000	0.0000	0.0000
B	0.0618	0.1400	0.1396
C	0.0927	0.2018	0.2017
D	0.1235	0.2558	0.2558

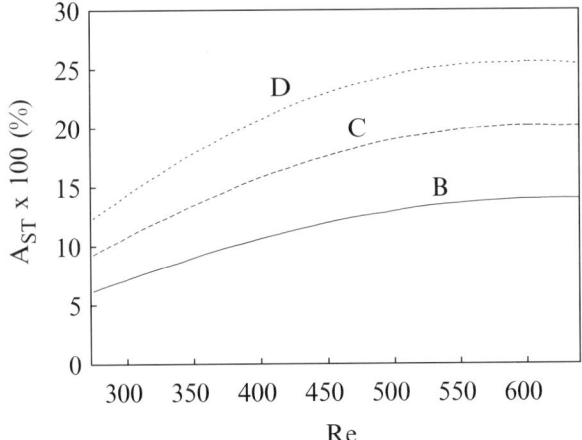

Fig. 1. Streamwise spatial evolution of the amplitude of streaks B, C and D (adapted from [CB02])

coordinates, respectively, we use the following definition of the streak amplitude [ABBH01]: $A_{ST}(x) = [\max_{y,z}(U - U_B) - \min_{y,z}(U - U_B)]/2U_\infty$, where U_∞ is the free stream velocity, $U_B(x, y)$ is the Blasius solution, and $U(x, y, z)$ is the streamwise velocity of the streak. The four different cases considered are listed in Table 1. Case A is nothing but the Blasius boundary layer without streaks. In Fig. 1 the evolution of the amplitude of the streaks B,C,D vs. the Reynolds number is displayed. Only streaks with amplitude $A_{ST} > 0.26$ are subject to secondary inflectional instabilities [ABBH01], and therefore all the considered basic flows are stable to this kind of instability.

3 Linear Stability of the Streaky Basic Flows

The spatial stability of the computed basic flows to the TS waves is tested by forcing two-dimensional harmonic perturbations of dimensionless frequency $F = 2\pi 10^6 f\nu/U_\infty^2$ into the boundary layer. The same computational parameters adopted in the evaluation of the basic flows are used in this type of simulations. The perturbation is induced by a two-dimensional time periodic volume force localized at the inlet position, extending up to $Re = 279$, of

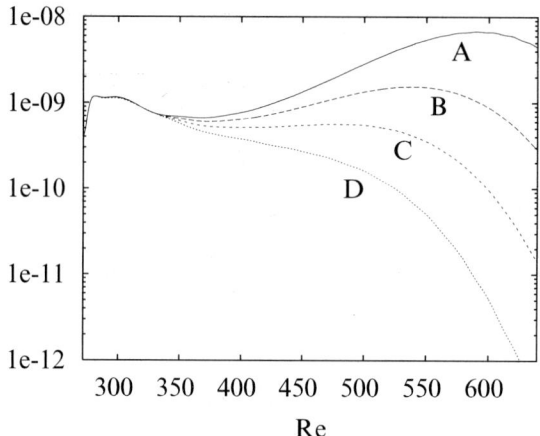

Fig. 2. Spatial evolution of the amplitude of 2D perturbations in the Blasius boundary layer without streaks (case A) and with streaks of increasing amplitude (cases B to D) (adapted from [CB02])

amplitude small enough to ensure a linear evolution of the perturbations. The computations were carried on for sufficiently large times to achieve converged time periodic solutions in all the computational domain. In Fig. 2 we show the downstream development of the amplitude, based on the energy density norm of two-dimensional waves at the frequency $F = 131.6$ of the forcing.

In the Blasius boundary layer (case A) the perturbations decay until they reach branch I of the linear neutral stability curve situated at $Re = 369$ in the parallel flow approximation. After, they begin to grow till branch II is reached at $Re = 581$. When the basic flow contains a low amplitude streak (case B) an unstable domain still exists but the growth of the TS waves is attenuated. Case C presents a region of marginal stability around $Re = 436$, where the streak amplitude is about 0.17. In the case of largest amplitude streaks (case D), the forced TS-waves are stable. Similar results apply to forcing frequencies $F = 160$ and $F = 200$. The results of these spatial numerical simulations have been confirmed by a modal stability analysis of the local streaky velocity profiles [CB04]. The observed stabilization of the TS waves in the Blasius boundary layer has been attributed to the modification of the main flow due to growth of the finite amplitude streaks. The basic flow distortion $\Delta U(y,z) = U(y,z) - U_B(y)$ can be separated into its spanwise averaged part $\overline{\Delta U}(y)$ and its spanwise varying part $\widetilde{\Delta U}(y,z) = \Delta U(y,z) - \overline{\Delta U}(y)$. Note that nonlinear effects are essential to generate $\overline{\Delta U}(y)$. In Fig. 3a, b we reproduce the spanwise averaged velocity $\overline{U}(y) = U_B(y) + \overline{\Delta U}(y)$ of the basic solutions at $Re = 609$ and the corresponding $\overline{\Delta U}(y)$. It can be seen how the increase of the streak amplitude leads to fuller $\overline{U}$-profiles having a stabilizing effect on the TS-waves.

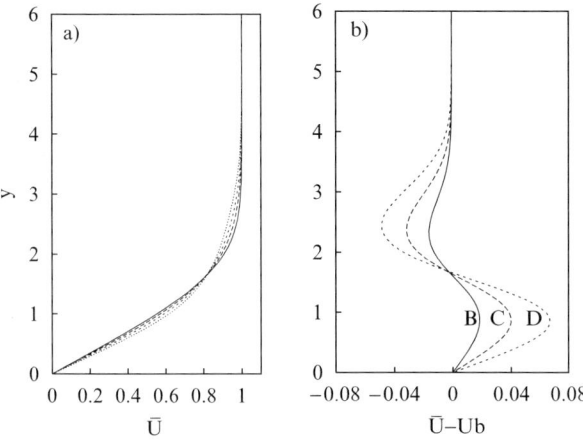

Fig. 3. (a) Spanwise averaged streamwise velocity profiles $\overline{U}(y)$, at $Re = 609$, of the Blasius boundary layer (*solid line*) and of the streaky boundary layers B, C and D. (b) Corresponding spanwise averaged basic flow distortion $\overline{\Delta U}(y) = \overline{U}(y) - U_{\rm B}(y)$ (adapted from [CB02])

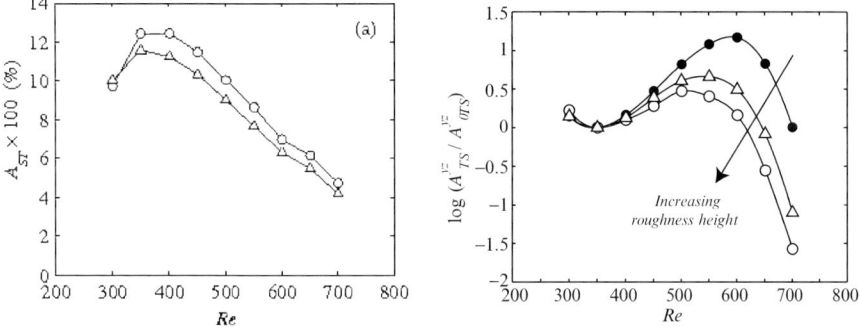

Fig. 4. (a) Streak amplitude of experimental streaks for successively increasing roughness heights $h_1=13$ mm and $h_2=14$ mm (*triangle, big circle*) (b) corresponding TS amplification curves (*triangle, big circle*) compared to the curve obtained for the reference 2D boundary layer (*filled circle*) $F = 130$ (adapted from [FBTC05])

4 Experimental Results

Experiments have been carried on in the Minimum-Turbulence-Level (MTL) wind tunnel at KTH Mechanics in Stockholm. Nearly optimal streaks have been generated using a spanwise array of equispaced cylindrical roughness elements placed on the wall of a flat plate, near the leading edge. The reader is referred to [FBTC04,FBTC05] for more details about the experimental apparatus. These studies have confirmed that streaks of increasing amplitude have an increasingly stabilizing effect on the TS waves (see Fig. 4). Experiments have also shown that even with moderately stabilizing streaks like the ones

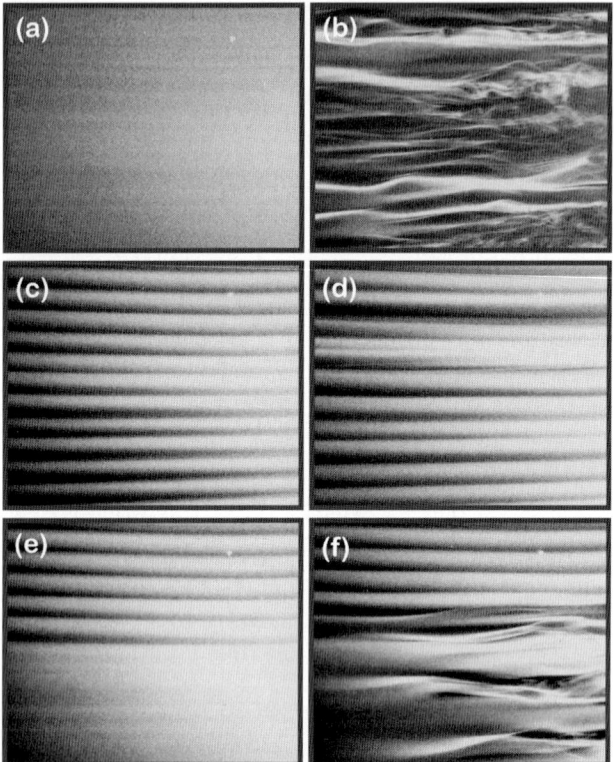

Fig. 5. Smoke flow visualizations from above with flow from left to right. (a) and (b) show the two-dimensional boundary layer, without streaks, with no excitation and with excitation of 201 mV, respectively. The flow in (b) is turbulent. (c) shows the streaky base flow with no excitation. In the presence of streaks with excitation of 450 mV (d), the flow remains laminar. (e) shows a half-streaky boundary layer obtained removing half the roughness elements and without forcing. With a forcing at 157 mV (f) the streaky part of the boundary layer remains laminar while the uncontrolled part undergoes transition (adapted from [FTBC06])

studied in [FBTC05] and that are roughly like the case B considered in the numerical simulations, it is possible to delay transition to turbulence on a flat plate [FTBC06] (see Fig. 5).

5 Conclusions

The scope of this paper was to show how optimal transient growth due to non-normality can be used as an effective amplifier of a passive laminar flow control. Optimal perturbations, usually seen as the most dangerous for transition are here seen as the most effective for control. It is likely that the use of optimal perturbations to modify the basic state at leading order with low energy could be applied to other strongly non-normal systems on different physical applications.

References

[ABBH01] P. Andersson, L. Brandt, A. Bottaro, and D. Henningson. On the break-down of boundary layers streaks. *J. Fluid Mech.*, 428:29–60, 2001.

[ABH99] P. Andersson, M. Berggren, and D. Henningson. Optimal disturbances and bypass transition in boundary layers. *Phys. Fluids*, 11(1):134–150, 1999.

[CB02] C. Cossu and L. Brandt. Stabilization of Tollmien–Schlichting waves by finite amplitude optimal streaks in the Blasius boundary layer. *Phys. Fluids*, 14:L57–L60, 2002.

[CB04] C. Cossu and L. Brandt. On Tollmien–Schlichting waves in streaky boundary layers. *Eur. J. Mech./B Fluids*, 23:815–833, 2004.

[Cho06] K. S. Choi. The rough with the smooth. *Nature*, 440:754, 2006.

[FBTC04] J. Fransson, L. Brandt, A. Talamelli, and C. Cossu. Experimental and theoretical investigation of the non-modal growth of steady streaks in a flat plate boundary layer. *Phys. Fluids*, 16:3627–3638, 2004.

[FBTC05] J. Fransson, L. Brandt, A. Talamelli, and C. Cossu. Experimental study of the stabilisation of Tollmien–Schlichting waves by finite amplitude streaks. *Phys. Fluids*, 17:054110, 2005.

[FTBC06] J. Fransson, A. Talamelli, L. Brandt, and C. Cossu. Delaying transition to turbulence by a passive mechanism. *Phys. Rev. Lett.*, 96:064501, 2006.

[Gus91] L. H. Gustavsson. Energy growth of three-dimensional disturbances in plane Poiseuille flow. *J. Fluid Mech.*, 224:241–260, 1991.

[Her88] Th. Herbert. Secondary instability of boundary-layers. *Annu. Rev. Fluid Mech.*, (20):487–526, 1988.

[LBS+99] A. Lundbladh, S. Berlin, M. Skote, C. Hildings, J. Choi, J. Kim, and D. S. Henningson. An efficient spectral method for simulation of incompressible flow over a flat plate. Technical Report KTH/MEK/TR-99/11-SE, KTH, Department of Mechanics, Stockholm, 1999.

[Luc00] P. Luchini. Reynolds-number independent instability of the boundary layer over a flat surface. part 2: Optimal perturbations. *J. Fluid Mech.*, 404:289–309, 2000.

[RH93] S. C. Reddy and D. S. Henningson. Energy growth in viscous channel flows. *J. Fluid Mech.*, 252:209–238, 1993.

[Sch79] H. Schlichting. *Boundary-Layer Theory*. Mc Graw-Hill, New York, 1979.

[SH01] P. J. Schmid and D. S. Henningson. *Stability and Transition in Shear Flows*. Springer, New York, 2001.

On the Catalytic Effect of Resonant Interactions in Boundary Layer Transition

Xuesong Wu[1], Philip A. Stewart[2], and Stephen J. Cowley[2]

[1] Department of Mathematics, Imperial College London, UK
   `x.wu@ic.ac.uk`
[2] DAMTP, Cambridge University, UK
   `ps57@damtp.ac.uk`, `s.j.cowley@damtp.ac.uk`

Summary. This paper is concerned with a fascinating phenomenon in boundary layer transition, namely, three-dimensional disturbances undergo rapid amplification despite that they have smaller linear growth rates than two-dimensional ones. Physical mechanisms are sought by considering two types of nonlinear interactions between oblique and planar instability modes. The first is the well-known subharmonic resonance. The relevant mathematical theory and its main predictions are briefly summarised. This mechanism, however, operates only among a very restrictive set of modes, and hence is unable to explain the broadband nature of the amplifying disturbances observed in experiments. The second mechanism involves the interaction between a planar and a pair of oblique Tollmien–Schlichting (T–S) waves which are phase-locked in that they travel with (nearly) the same phase speed. It is a more general type of interaction than subharmonic resonance since no further restriction is imposed on the frequencies. Yet similar to subharmonic resonance, this interaction also leads to super-exponential growth of the oblique modes, while the planar mode remains to follow linear stability theory. The dominant planar mode therefore plays the role of a catalyst, the implications of which for the e^N-method and for transition control are discussed.

1 Introduction

Laminar-turbulent transition is one of the unsolved fundamental problems in classical physics since it in essence is concerned with how a simple system becomes disordered and chaotic in both time and space through a sequence of nonlinear processes. It is also a problem of technological importance in industries, where accurate prediction and effective control of transition are crucial. For example, transition occurs in the flow over the wing and fuselage of an aircraft, with the turbulent state exerting a much greater drag. Considerable saving may be achieved if a laminar flow can be maintained in the whole or a large portion of the wing. Turbulence in the boundary layer also emits considerable amount of noise. In such circumstances, it is desirable to suppress turbulence.

At high speeds, turbulence crucially affects surface heat transfer, which is of great concern in the re-entry phase of a space shuttle mission. In chemically reacting flows, it is necessary to enhance turbulence in order to achieve complete reaction. This is especially important for the projected scramjet engines, where effective mixing is vital for overcoming the difficulty of ignition and combustion caused by short residence time of reactants.

Laminar-turbulent transition in boundary layers and other shear layers are often initiated by amplification of two-dimensional disturbances. This is well understood on the basis of linear stability theory, which predicts that planar instability modes have larger growth rates. However, prior to onset of turbulence, three-dimensional disturbances are invariantly observed to undergo rapid development to overtake the initially dominant two-dimensional perturbation, contradicting the linear stability theory. The mechanisms must necessarily be inherently nonlinear. Yet the primary method of transition prediction, the so-called e^N-method, is based on the calculation of the accumulated amplification as predicted by linear stability theory.

The present paper is concerned with three related questions:

1. What are the dominant mechanisms for inducing the rapid amplification of three-dimensional disturbances?
2. Given that transition is caused by nonlinear mechanisms, can the current e^N-method be possibly justified?
3. How can an improved understanding of transition mechanisms guide control strategy?

As we shall see, central to these questions is that the dominant planar mode plays the role of a *catalyst* in the sense that it promotes three-dimensional disturbances while itself is little affected by nonlinearity. We shall examine relevant experimental evidence before presenting mathematical theories.

2 Experimental Evidence of Catalytic Effect

The three-dimensional nature of boundary layer transition has long been recognised since the experiments of Klebanoff, Tidstrom & Sargent (1962), where a planar mode and oblique modes with the same frequency were excited simultaneously in a controlled fashion. While the former dominates the early stage of the development, three-dimensional disturbances eventually prevail to form alternating valleys and peaks along the spanwise direction. The preferential amplification of the three-dimensional disturbances has been customarily attributed to the so-called fundamental type of interaction between a pair of oblique mode $(\alpha, \pm\beta, \omega)$ and the planar mode $(\alpha_0, 0, \omega_0)$, where α and α_0 are the streamwise numbers, β spanwise wavenumber, and ω the frequency.

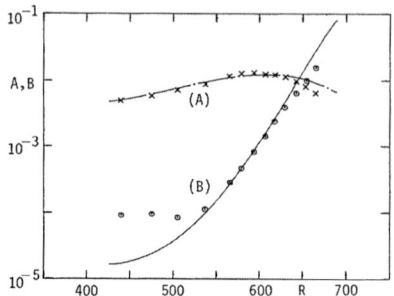

Fig. 1. Development of planar (*cross*) and oblique subharmonic T–S waves (*circle*) (Herbert 1988)

Raetz (1959) was the first to show that a more powerful interaction, a subharmonic resonance involving a triad of a planar wave $(\alpha_0, 0, \omega_0)$ and a pair of oblique waves $(\alpha, \pm\beta, \omega)$ at the subharmonic frequency $\omega = \omega_0/2$, may take place in boundary layer. Its importance to transition was demonstrated by Craik (1971), who derived the amplitude equations for three resonating T–S waves. His work prompted the landmark experiments of Kachanov & Levchenko (1984), who introduced both the planar and subharmonic oblique waves in a controlled manner and measured their subsequent development. Figure 1 is a typical measurement. It shows that the planar mode amplifies according to local linear stability theory, whilst oblique modes hardly exhibits any growth at all in the linear regime. However, they start to grow at a faster rate to become dominant once the planar mode reaches a threshold amplitude. Kachanov & Levchenko (1984) also mapped the downstream development of the disturbance spectrum (Fig. 2). While the seeded planar and oblique subharmonics appear as distinct peaks, a striking feature is that a broad peak in the low frequency portion of the spectrum emerges downstream. A similar phenomenon was observed by Corke & Mangano (1989), as is shown in Fig. 3 taken from their paper. These evidences suggest that a broad band of relatively low frequency waves undergo substantial amplification to attain magnitudes comparable with the subharmonic modes.

Borodulin, Kachanov & Koptsev (2002) recently investigated the effect of a harmonic forcing on the development of background disturbances. Unlike previous experiments, where both planar and oblique modes are excited, they simply excited a planar mode. Their key results are shown in Fig. 4, where the response is compared with that without forcing. As is illustrated, the harmonic forcing precipitated the growth of a broad band disturbances, which would otherwise acquire minimal amplitudes. Figure 5 shows the development of several selected components. They reach amplitudes which are about one order of magnitude larger compared with the unforced case. The seeded planar mode itself more or less follows local linear instability theory despite its considerable size. In this sense, it acts as a catalyst. What is the most remarkable is that

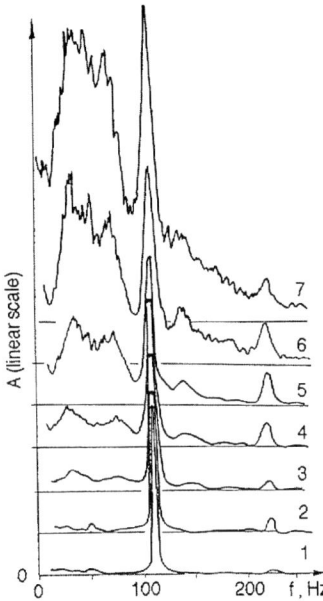

Fig. 2. Disturbance spectrum at different streamwise locations (Kachanov & Levchenko 1984)

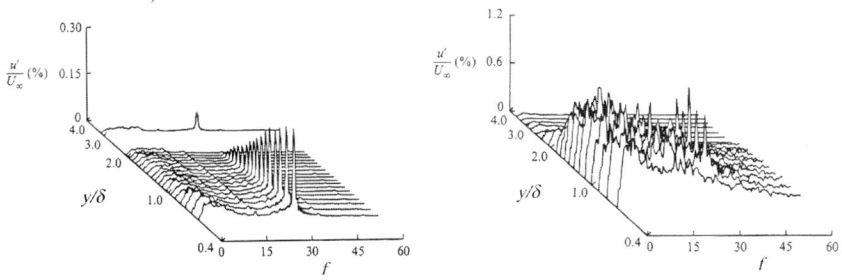

Fig. 3. Development of disturbance spectra at different y positions (Corke & Mangano 1989). *Left*: upstream; *right*: downstream

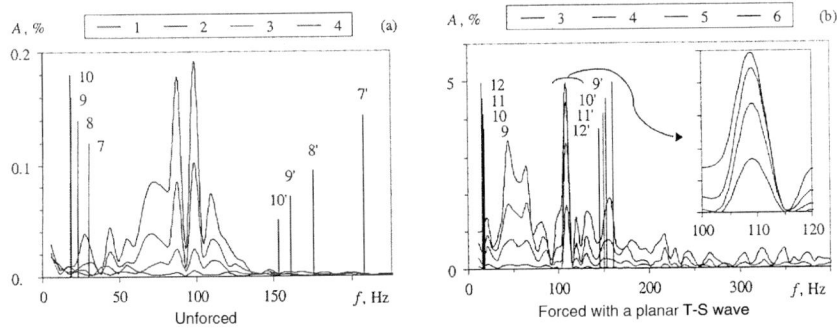

Fig. 4. Comparison of the disturbance spectra with and without forcing (Borodulin, Kachanov & Koptsev 2002)

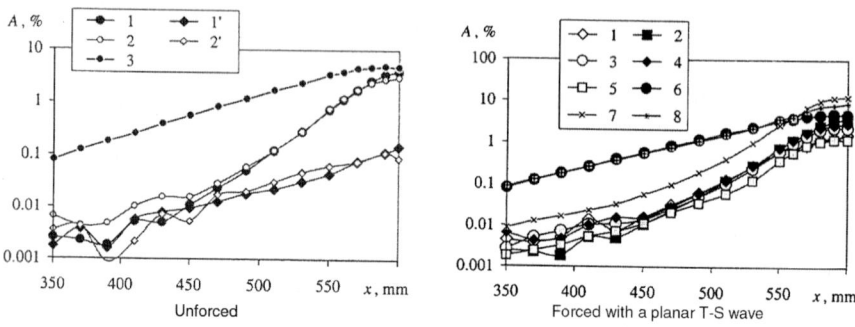

Fig. 5. Development of selected three-dimensional disturbances (Borodulin et al. 2002)

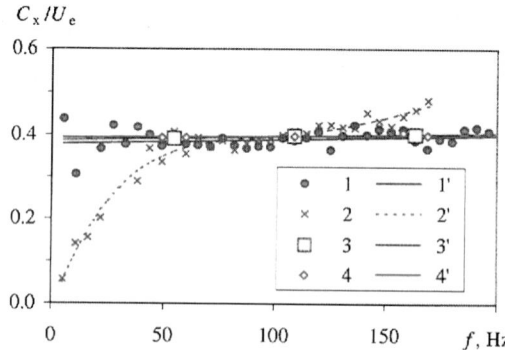

Fig. 6. Phase speeds of selected rapidly amplifying three-dimensional disturbances with planar wave forcing (*filled circle*) (Borodulin et al. 2002)

the rapidly amplifying disturbances all share nearly the same phase speed (Fig. 6), i.e. their phases are almost locked in a coordinate system moving with the common phase speed.

We now present relevant mathematical descriptions of two mechanisms, subharmonic resonance and phase-locked interaction, and demonstrate that through these interactions a planar mode indeed plays a catalytic role.

3 Mathematical Formulation

The theory is based on consideration of mutual interactions between a planar mode (which is usually taken to be the most unstable mode) and a pair of suitable oblique waves. They may be expressed as

$$\epsilon A_0(x_1)\,e^{i(\alpha_0 x - \omega_0 t)} + \delta A(x_1)\,e^{i(\alpha x - \omega t)}\big(e^{i\beta z} + e^{-i\beta z}\big), \qquad (1)$$

where ϵ and δ stand for the magnitudes of the respective waves, and the dependence on the transverse coordinate y is suppressed for brevity. Depending on the relation between the wavenumbers α_0, α and β, two particularly effective interactions may occur.

We shall derive appropriate evolutions equations by using asymptotic analysis based on the assumption that the Reynolds number

$$R = U_\infty \delta^* / \nu \gg 1 \tag{2}$$

where the reference length δ^* is the boundary layer thickness at the location of interest. The analysis is conducted for the upper-branch scaling regime (Bodonyi & Smith 1981), in which

$$\alpha_0 \sim \alpha \sim \sigma, \quad \beta \sim \sigma, \quad \omega_0 \sim \omega \sim \sigma^2,$$

where $\sigma = R^{-1/10}$. The viscosity-induced growth rate is of $O(\sigma^4)$, and hence the slow variable x_1 is defined as (Goldstein & Durbin 1986)

$$x_1 = \sigma^4 x.$$

The linear instability acquires an asymptotic structure consisting of five layers. Dominant nonlinear interactions takes place in the critical layer centred at y_c, where the base flow velocity $U_B(y_c) = c = \omega/\alpha$ (to leading order).

4 Subharmonic resonant triad

The resonant-triad interaction is a universal mechanism taking place in a three-wave system, when the sums of the wavenumbers and frequencies of the two waves equal to those of the third wave respectively. In the case of upper branch T–S waves, it follows from Squire's transformation that the resonance condition is satisfied for (Smith & Stewart 1987, Mankbadi, Wu & Lee 1993)

$$\alpha = \tfrac{1}{2}\alpha_0, \quad \beta = \tfrac{\sqrt{3}}{2}\alpha_0. \tag{3}$$

A self-consistent description was given by Mankbadi et al. (1993), who showed that the resonant interaction starts to affect the oblique modes when the planar mode has reached the threshold magnitude

$$\epsilon \sim \sigma^{10}. \tag{4}$$

On assuming that the oblique modes have a sufficiently small amplitude, an analysis of the interaction within the critical layer, which is viscosity dominated and equilibrium at this stage, leads to evolution equations

$$A'(x_1) = \kappa A + l B A^*, \quad B'(x_1) = \kappa_0 B. \tag{5}$$

These equations indicate that the planar mode affects the oblique modes through their mutual interaction at quadratic level, but the latter produce no back effect on the former, which continues to follow linear stability theory. This is referred to as parametric resonance stage. It has been shown that

$$A \sim \exp\left(\kappa x_1 + a_\infty e^{\kappa_0 x_1}\right), \quad \text{as} \quad x_1 \to \infty, \tag{6}$$

where $\Re(a_\infty) > 0$, suggesting that while the planar mode remains linear, it causes the oblique modes to amplify super-exponentially. This is a manifestation of catalytic effect.

However, the super-exponential growth of the form (6) cannot continue forever. As was pointed out by Goldstein (1994, 1995), the rapid growth of the oblique modes renders the critical layer non-equilibrium when $x_1 \sim O(\kappa_0^{-1} \ln \sigma^{-1})$, suggesting the introduction of the shifted coordinate

$$\tilde{x} = x_1 - \kappa_0^{-1} \ln \sigma^{-1}.$$

After this new effect is included, the evolution switches to super-exponential growth of a different form, namely (Goldstein 1994)

$$A \sim \exp(b_\infty e^{\kappa_0 \tilde{x}/4} / \sigma), \quad \text{as} \quad \tilde{x} \to \infty. \tag{7}$$

The continued rapid growth of the oblique modes will eventually lead to a fully-coupled stage, where the oblique waves react back on the planar wave (Wu 1995). Depending on the size of δ, that might happen before or after the planar wave becomes nonlinear.

5 Phase-Locked Interaction

Subharmonic resonance is rather restrictive in that it operates among a particular triadic set specified by (3). It follows that one planar mode can only promote the growth of a specific pair of modes, or a narrow band of modes close to that pair when the concept is generalised to include detuning. To explain the experimental observation that a single planar enhances a broad band of three-dimensional disturbances, we consider phase-locked interaction between a planar and a single (or a pair of) oblique modes, which have nearly the same phase speed. The concept of phase-locked modal interaction was first proposed by Wu & Stewart (1996) for inviscid Rayleigh instability modes. We shall show that the mechanism operates as well for viscous T–S waves, albeit some new features arise due to different nature of the instability. From Squire's transformation, it can be inferred that the phase-locking requirement,

$$\omega/\alpha = \omega_0/\alpha_0, \tag{8}$$

is met by *any* oblique mode provided its wavenumbers α and β satisfy

$$\alpha^2 + \beta^2 = \alpha_0^2. \tag{9}$$

This is a much less restrictive condition than (3). As a generalization, we allow the phase speeds to differ by $O(\sigma^3)$, i.e.

$$c_0 = c + \sigma^2 \Delta$$

with $\Delta = O(1)$ being the scaled phase-speed mismatch, the inclusion of which is important, as will be shown shortly. The disturbance evolves through several distinct nonlinear stages.

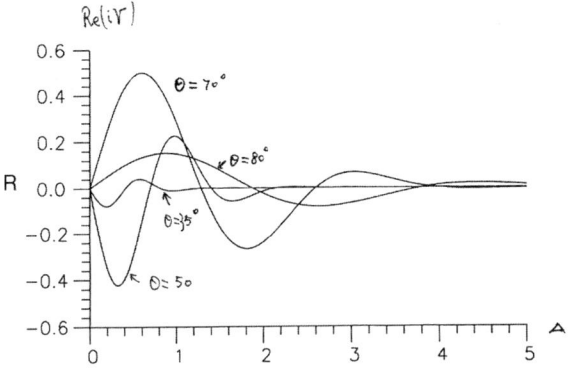

Fig. 7. Real part of Landau coefficient l_r, vs. phase-speed mismatching Δ. Curves (i)–(iv) correspond to $\theta = 80°$, $70°$, $50°$ and $35°$, respectively

5.1 Stage I: Viscous (Equilibrium) Critical Layer

The first nonlinear stage commences when the planar mode reaches a threshold amplitude

$$\epsilon = \sigma^{17/2}.$$

The critical layer is equilibrium and dominated by viscous effect. Analysis yields the amplitude equations

$$A'(x_1) = \kappa A + l|A_0|^2 A, \quad A'_0(x_1) = \kappa_0 A_0, \tag{10}$$

where l is a function of Δ. The amplitude A has the solution

$$A = \exp\left(\kappa x_1 + l(2\kappa_0)^{-1} e^{2\kappa_0 x_1}\right). \tag{11}$$

In this equilibrium regime, the property of the solution is dictated by l_r, and only for $l_r > 0$ do the oblique modes grow super-exponentially. The variation of l_r with Δ is shown in Fig. 7. As is expected, $l_r \to 0$ as $\Delta \to \infty$, implying a diminishing effect when the phase is sufficiently de-locked. It is noted that there exists an optimal mismatch at which l_r attains its global maximum. Interestingly, for perfect matching ($\Delta = 0$), $l_r = 0$.

5.2 Stage II: Non-Equilibrium Critical Layer

Owing to the super-exponential growth, the oblique waves evolve at an increasingly rapid rate, and eventually the non-equilibrium effect becomes important in the critical layer when $\sigma^4 A'/A \sim \sigma^4 e^{2\kappa_0 x_1} \sim \sigma^3$ i.e. when $x_1 \sim \kappa_0^{-1} \log \sigma^{-1/2}$, and hence we introduce variable $\tilde{x}$ by writing

$$x_1 = \kappa_0^{-1} \log \sigma^{-1/2} + \tilde{x} \quad \text{with} \quad \tilde{x} = O(1).$$

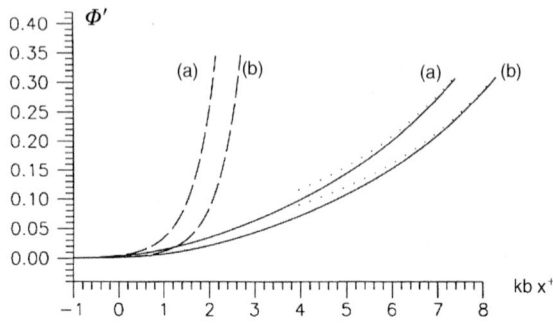

Fig. 8. Oblique-modes growth rate Φ' vs. $\kappa_b x^\dagger \equiv \kappa_0 \tilde{x}$ for $\theta = 40°$ (curves (a)), $70°$ (curves (b)), showing the evolution from the equilibrium to non-equilibrium stages. The *dashed* and *dotted lines* represent, respectively, the upstream (14) and downstream (15) super-exponential growth

The oblique modes now evolve over a shorter length scale than the planar wave, and its amplitude A takes the WKBJ form:

$$A = \bar{A}\, e^{\Phi(\tilde{x})/\sigma} . \tag{12}$$

It is found that the function Φ' satisfies a transcendental equation

$$\Phi'(\tilde{x}) = l \left[\int_0^\infty Q(\xi, \lambda)\, e^{-\Phi'(\tilde{x})\xi}\, d\xi \right]^2 e^{2\kappa_0 \tilde{x}} \tag{13}$$

with the rather complex expression for $Q(\xi, \lambda)$ being omitted. In the upstream limit,

$$\Phi \rightarrow l(2\kappa_0)^{-1}\, e^{2\kappa_0 \tilde{x}} \quad \text{as} \quad \tilde{x} \rightarrow -\infty. \tag{14}$$

In the downstream limit $\tilde{x} \rightarrow \infty$,

$$\Phi \rightarrow a_\infty\, e^{2\kappa_0 \tilde{x}/7}, \quad A \rightarrow A_\infty \exp(a_\infty\, e^{2\kappa_0 \tilde{x}/7}/\sigma), \tag{15}$$

indicating that the oblique waves undergo a super-exponential growth, which differs from that in the previous stage (11), but is the same as that for the Rayleigh waves (cf. Wu & Stewart 1996). Figure 8 displays such a switch of from the equilibrium to non-equilibrium dynamics.

5.3 Stage III: Fully-Coupled Stage

Owing to their fast super-exponential growth, the oblique modes eventually react back on the planar wave as their amplitude reaches $O(\delta)$, i.e. at $\tilde{x}_s$ where $\delta \exp\{(l_r + \Phi(\tilde{x}_s))/\sigma\} = \sigma^7$. The evolution then occurs on the $O(\sigma^{-3})$ shorter length scale and so we introduce

$$\bar{x} = (\tilde{x} - \tilde{x}_s)/\sigma.$$

The interaction is similar to that for the Rayleigh instability waves, and the analysis leads to fully coupled amplitude equations

$$A' = \int_0^\infty K_p(\xi, \eta; \Delta) B(\bar{x} - \hat{\sigma}_d \xi) B^*(\bar{x} - \xi - \hat{\sigma}\eta) A(\bar{x} - \xi - \eta) \, \mathrm{d}\xi \, \mathrm{d}\eta$$

$$+ \int_0^\infty K_a(\xi, \eta; \Delta) A(\bar{x} - \xi) A(\bar{x} - \xi - \eta) A^*(\bar{x} - 2\xi - \eta) \, \mathrm{d}\xi \, \mathrm{d}\eta, \qquad (16)$$

$$B' = \int_0^\infty K_b(\xi, \eta; \Delta) A(\bar{x} - \xi) B(\bar{x} - \xi - \eta) A^*(\bar{x} - \nu_s \xi - \nu_0 \eta) \, \mathrm{d}\xi \, \mathrm{d}\eta$$

$$+ \int_0^\infty K_c(\xi, \eta; \Delta) B(\bar{x} - \xi) A(\bar{x} - \xi - \eta) A^*(\bar{x} - \nu_s \xi - \eta) \, \mathrm{d}\xi \, \mathrm{d}\eta, \qquad (17)$$

where the expression for K_p, K_a, K_b and K_c are omitted here for brevity.

Numerical solutions of above equations suggest that both amplitudes terminate within a finite-distance at a singularity of the form

$$A(\bar{x}) \rightarrow \tilde{a}_s(\bar{x}_s - \bar{x})^{-(3 + \mathrm{i}\,\psi_a)}, \quad B(\bar{x}) \rightarrow \tilde{b}_s(\bar{x}_s - \bar{x})^{-(7/2 + \mathrm{i}\,\psi_b)}. \qquad (18)$$

While the singularity is non-physical, its occurrence signals a rapid evolution on the short scale comparable with the wavelength.

6 Concluding Remarks

We have presented two types of powerful interactions which promote rapid amplification of certain three-dimensional disturbances in the form of super-exponential growth. In particular, through phase-locked interaction, forcing a planar mode can simultaneously enhance *all* oblique modes which have the nearly the same phase speed. The planar mode on the other hand remains linear. A fundamental physical insight that this theoretical result offers is that the dominant planar mode acts as a catalyst.

Once three-dimensional disturbances evolve on a very short length scale, transition is likely to complete within a relatively small distance. As a first approximation the transition point may be taken to be at the streamwise location where the dominant planar mode has acquired the threshold to activate the catalytic effect. Since it remains essentially linear, the required amplification factor to reach the threshold may be calculated by integrating the linear growth rate. This may explain why the e^N-method, which is based entirely on linear instability concept, works quite well despite that transition is a highly nonlinear process. The usual criticism that this method does not take into account nonlinearity seems not entirely warranted. The primary limitation of the e^N-method is rather that predicting transition based on amplification factor alone without considering receptivity means that it is applicable only to situations where the ambient disturbance level is about the same.

The catalytic effect implies suppressing/enhancing planar T–S modes can be a simple and yet effective means for delaying/hastening transition.

References

1. Bodonyi, R.J. & Smith, F.T. The upper-branch stability of the Blasius boundary layer, including non-parallel flow effects. *Proc. R. Soc. Lond.* **A375**, 65–92 (1981).
2. Borodulin, V.I., Kachanov, Y.S. & Koptsev, D.B. Experimental study of resonant interactions of instability waves in a self-similar boundary layer with an adverse pressure gradient: III. Broadband disturbances. *J. Turbulence* **3**(64), 1-19 (2002).
3. Craik, A.D.D. Non-linear resonant instability in boundary layers. *J. Fluid Mech.* **50**, 393-413 (1971).
4. Corke, T.C. & Mangano, R.A. Resonant growth of three-dimensional modes in transitioning Blasius boundary layers. *J Fluid Mech.* **209**, 93-150 (1989).
5. Goldstein, M.E. & Durbin, P.A. Nonlinear critical layers eliminate the upper branch of spatially growing Tollmien-Schlichting waves. *Phys. Fluids* **29**, 2344-2345 (1986).
6. Goldstein, M.E. Nonlinear interactions between oblique instability waves on nearly parallel shear flows. *Phys. Fluids* **A6**, 724-735 (1994).
7. Goldstein, M.E. The role of nonlinear critical layers in boundary-layer transition. *Phil. Trans. R. Soc. Lond.*, **A 352**, 425-442 (1995).
8. Herbert, T. Secondary instability of boundary layers. *Ann. Rev. Fluid Mech.* **20**, 487-526 (1988).
9. Kachanov, Yu. S. & Levchenko, V. Ya. The resonant interaction of disturbances at laminar-turbulent transition in a boundary layer. *J. Fluid Mech.* **138**: 209–247 (1984).
10. Klebanoff, P.S., Tidstrom, K.D. & Sargent, L.M. The three-dimensional nature of boundary layer instability. *J. Fluid Mech.* **12**, 1–34 (1962).
11. Mankbadi, R.R., Wu, X. & Lee, S.S. A critical-layer analysis of the resonant triad in Blasius boundary-layer transition: nonlinear interactions. *J. Fluid Mech.* **256**, 85-106 (1993).
12. Raetz, G.S. A new theory of the cause of transition in fluid flows. *Norair Rep. NOR-59-383*, Hawthorne, California (1959).
13. Smith, F.T. & Stewart, P.A. The resonant-triad nonlinear interaction in boundary-layer transition. *J. Fluid Mech.* **179**, 227-252 (1987).
14. Wu, X. Viscous effects on fully coupled resonant–triad interactions: an analytical approach. *J. Fluid Mech.* **292**, 377-407 (1995).
15. Wu, X. & Stewart, P.A. Interaction of phase-locked modes: a new mechanism for the rapid growth of three-dimensional disturbances. *J. Fluid Mech.* **316**, 335-372 (1996).

Stabilization of Tollmien–Schlichting Waves by Mode Interaction

Carlos Martel, Eusebio Valero, and José M. Vega

E.T.S.I. Aeronáuticos, Universidad Politécnica de Madrid, Plaza Cardenal Cisneros, 3. 28040-Madrid, Spain
vega@fmetsia.upm.es

1 Introduction

Decreasing skin friction in boundary layers attached to aircraft wings can have an impact in both fuel consumption and pollutant production, which are becoming crucial to reduce operation costs and meet environmental regulations, respectively. Skin friction in turbulent boundary layers is about ten times that of laminar boundary layers. Thus, an obvious method to reduce friction drag is to delay transition to turbulence, which is a fairly involved process in real aircraft wings [J98]. Transition sis promoted either by *Tollmien–Schlichting* (TS) and *Klebanov* (K) modes [K94], with the former playing an essential role. Various methods (e.g., suction [SG00, ZLB04], wave cancellation [WAA01, LG06]) have been proposed to reduce TS modes in laminar boundary layers. Mode interaction methods have been successfully used in fluid systems to control related instabilities, such as the Rayleigh–Taylor instability [LMV01]. Here, we present some recent results on using these methods to control TS modes in a compressible, 2D boundary layer over a flat plate at zero incidence. A given unstable TS mode can be stabilized by coupling its spatial evolution with that of a second selected stable TS mode, in such a way that the stable mode takes energy from the unstable one and gives a stable coupled evolution of both modes. The coupling device is a wavetrain in the boundary layer, with appropriate wavenumber and frequency, which can be created by an array of oscillators on the wall, and promotes both (i) parametric coupling between the stable and unstable TS modes and (ii) a mean flow that is also stabilizing. Three differences with wave cancelation methods are relevant. Namely, (a) nonlinear terms play an essential role in the process; (b) the unstable TS mode is stabilized (its growth rate is decreased), not just canceled; and (c) stabilization does not depend on the phase of the incoming wave, which implies that active control is not necessary.

This paper is devoted to analyzing the effect and is organized as follows. After formulating the problem in Sect. 2, the stabilizing process is described in

Sect. 3, where the relevant mode interactions are described. A short description
of the numerical tool used to calculate the non stationary flow in a boundary
layer attached to a flat plate, and the post processing tool developed to filter
the amplitudes of the various marginal modes involved is described in Sect. 4.
The numerically obtained results are described and discussed in Sect. 4. The
paper ends with some concluding remarks, in Sect. 5.

2 Compressible Navier–Stokes Equations

The continuity, momentum, and energy equations are nondimensionalized us-
ing the streamwise length $\hat{x}_0$, outer velocity $\hat{u}_0$, density $\hat{\rho}_0$, and temperature T_0
as characteristic length, velocity, density, and temperature, respectively; time
and the modified pressure $(=\text{pressure} - \hat{\rho}_0 R \hat{T}_0)$ are nondimensionalized with
$\hat{x}_0/\hat{u}_0$ and $\hat{\rho}_0 \hat{u}_0^2$, respectively. With the usual notation, the resulting equations
are

$$\frac{\partial \rho}{\partial t} + \boldsymbol{\nabla} \cdot (\rho \boldsymbol{v}) = 0, \quad 1 + \mathcal{B}_1 p = \rho T, \tag{1}$$

$$\frac{\partial \boldsymbol{v}}{\partial t} + (\boldsymbol{v} \cdot \boldsymbol{\nabla})\boldsymbol{v} = -\boldsymbol{\nabla} p + \frac{1}{Re}\left[\boldsymbol{\nabla} \cdot [M(\boldsymbol{v} + \boldsymbol{v}^\top)] - \frac{2}{3}\boldsymbol{\nabla}(M\boldsymbol{\nabla} \cdot \boldsymbol{v})\right], \tag{2}$$

$$\rho\frac{DT}{Dt} = \boldsymbol{\nabla} \cdot \left[\frac{M}{PrRe}\boldsymbol{\nabla} T\right] + \mathcal{B}\frac{Dp}{Dt} +$$

$$\frac{\mathcal{B}M}{Re}\{\boldsymbol{\nabla} \cdot [\boldsymbol{v} \cdot (\boldsymbol{\nabla} \boldsymbol{v} + \boldsymbol{\nabla} \boldsymbol{v}^\top)] - \boldsymbol{v} \cdot \boldsymbol{\nabla} \cdot (\boldsymbol{\nabla} \boldsymbol{v} + \boldsymbol{\nabla} \boldsymbol{v}^\top)\}, \tag{3}$$

in terms of the velocity $\boldsymbol{v} = (u, v)$, the density ρ, the modified pressure p,
and the temperature T, where the Re and Ma are the Reynolds and Mach
numbers based on the conditions at the outer flow, and Pr is the Prandtl
number, which can be considered as constant $Pr = 0.72$ for air.

$$\mathcal{B} = (\gamma - 1)Ma^2 \quad \text{and} \quad \mathcal{B}_1 = \gamma Ma^2 \tag{4}$$

are nondimensional measures of *viscous dissipation* and *compressibility*, respectively, where $\gamma = 1.4$ for air. The function M, results from the dependence of viscosity on temperature, assumed to obey a *Sutherland formula*,
which gives

$$M(T) = \frac{T^{3/2}(1 + s)}{T + s}, \tag{5}$$

with $s = \hat{s}/\hat{T}_0$ and $\hat{s} = 110\,\text{K}$. Thus, e.g., $s = 0.37$ if $\hat{T}_0 = 300\,\text{K}$. The
computational domain and the boundary conditions depend on the various
approximations made below.

3 Mode Interaction; Parametric Forcing

In a boundary layer above a plate at zero incidence, with a local thickness

$$\delta = \sqrt{x/Re} \ll 1, \tag{6}$$

a self-similar, approximated steady state solution exists,

$$(\rho_S, u_S, v_S, T_S, p_S) = (\rho(\zeta), u(\zeta), \delta V(\zeta), p, t(\zeta), \tag{7}$$

where

$$\zeta = y/\delta. \tag{8}$$

Replacing these into (1)–(3) and neglecting $O(1/R)$-terms, where

$$R = Re\delta \equiv \sqrt{xRe} \ll 1 \tag{9}$$

is the Reynolds number based on the local boundary layer thickness, we obtain the following ODE system

$$- (\rho u)' + 2(\rho V)' = 0, \quad \rho T = 1, \tag{10}$$

$$\rho(-\zeta u + 2V)u' = 2[M(T)u']', \tag{11}$$

$$\rho(-\zeta u + 2V)T' = 2Pr^{-1}[M(T)T']' + 2\mathcal{B}M(T)(u')^2. \tag{12}$$

The boundary conditions are

$$u = v = T' = 0 \quad \text{at } \zeta = 0, \qquad u_S = 1, \quad T = T_e \quad \text{at } \zeta = \infty, \tag{13}$$

where the outer flow temperature T_e is predetermined, and we are assuming a thermally insulated wall, namely assuming that the steady state is reached after a transient in which thermal equilibrium between the solid and the air is reached.

The stability of this self-similar steady state is analyzed considering two kinds of modes. K modes are, in some sense, the natural modes of the boundary layer because they exhibit the same scaling (7)–(8) as the basic steady state. These modes are nearly marginal (namely, exhibit a zero growth rate), exhibit a power law growth along the streamwise coordinate [L00], and play a secondary role in the transition process: they can either enhance [R01] or delay [CB01] transition. TS modes, instead, exhibit a streamwise wavelength comparable to the boundary layer thickness, and are analyzed setting

$$(\rho, u, v, T, p) = (\tilde{\rho}, \tilde{u}, \tilde{v}, \tilde{T}, \tilde{p}) + A(r, \phi, \delta\psi, \theta, \pi)e^{i(\int_0^x \varphi \, dx - \omega t)/\delta} + \text{c.c.}, \tag{14}$$

where c.c. stands for the complex conjugate, the complex wavenumber $\alpha = \varphi'$, and the complex amplitude A is allowed to depend slowly on x and t. Substituting these into (1)–(3), linearizing and retaining $O(1/R)$-terms, we obtain

$$i(\alpha\tilde{u} - \omega)r + (\tilde{\rho}\psi)' + i\alpha\tilde{\rho}\phi = 0, \quad \mathcal{B}_1\pi = \tilde{T}r + \tilde{\rho}\theta, \tag{15}$$

$$3i\tilde{\rho}(\alpha\tilde{u} - \omega)\phi + 3\tilde{\rho}\tilde{u}'\psi = -3i\alpha\pi + R^{-1}\left[3(\tilde{M}\phi')' + \tilde{M}\left(i\alpha\psi' - 4\alpha^2\phi\right)\right.$$
$$\left. +3i\alpha\tilde{M}'\tilde{T}'\psi + 3(\tilde{M}'\tilde{u}'\theta)'\right], \tag{16}$$

$$3i\tilde{\rho}(\alpha\tilde{u} - \omega)\psi = -3\pi' + R^{-1}\left[4(\tilde{M}\psi')' + \tilde{M}\left(i\alpha\phi' - 3\alpha^2\psi\right)\right.$$
$$\left. +i\alpha\tilde{M}'\left(-2\tilde{T}'\phi + 3\tilde{u}'\theta\right)\right], \tag{17}$$

$$i\tilde{\rho}(\alpha\tilde{u} - \omega)\theta + \tilde{\rho}\tilde{T}'\psi = (PrR^{-1})\left[(\tilde{M}\theta')' - \alpha^2\tilde{M}\theta + (\tilde{M}'\tilde{T}'\theta)'\right]$$
$$+ i\mathcal{B}(\alpha\tilde{u} - \omega)\pi + \mathcal{B}R^{-1}[2\tilde{M}\tilde{u}'(\phi' + i\alpha\psi) + \tilde{M}'(\tilde{u}')^2\theta], \tag{18}$$

where $\tilde{M}$ denotes $M(\tilde{T})$. The appropriate boundary conditions are

$$\phi = \psi = \theta = 0 \quad \text{at } \zeta = 0, \infty. \tag{19}$$

Note that now (cf. (13)) we are assuming that the wall is isothermal, with the steady state temperature. This is because (a) the heat capacity and the thermal conductivity of the plate are both much larger than those of the air, and (b) the characteristic time of the nonsteady flow (essentially, the period of the Tollmien–Schlichting waves) is much smaller than the conductive time in the plate. Also note that we are retaining small $O(1/R)$-terms, which account for viscous effects and are essential to trigger the instability that promotes TS waves; an asymptotic analysis as $R \to \infty$ of (15)–(19) leads to a triple-deck problem [S82], which requires to consider fractional powers of R^{-1} and yields a poor approximation. Thus, the usual strategy is to retain $O(R^{-1})$-terms, as we do here, and solve numerically the resulting stiff problem. This can be done either discretizing the boundary value problem or using a shooting method combined with a continuous orthonormalization method [D83]; we have done the latter. In either case, the boundary conditions at $\zeta = \infty$ must be imposed at a (large but) finite distance, treated conveniently [K76] to avoid large errors due to wave reflection. Solving (15)–(19) yields marginal instability curves in the planes α vs. R and ω vs. R that are tongues like those shown in Fig. 1, where instability sets in when entering the tongues. Instability is convective and thus it can be seen as a spatial instability [Ch05], which develops with a fixed frequency (ω/δ=real=constant, see (15)). This means invoking (6)–(9) moving along a straight line passing through the origin in the ω vs. Re pane in Fig. 1.

Some remarks are now in order:

– As anticipated above, the real parts of α and ω (wavenumber and frequency) are much larger than the imaginary parts (spatial and temporal growth rates) inside the tongues in Fig. 1. This is illustrated in Fig. 2,

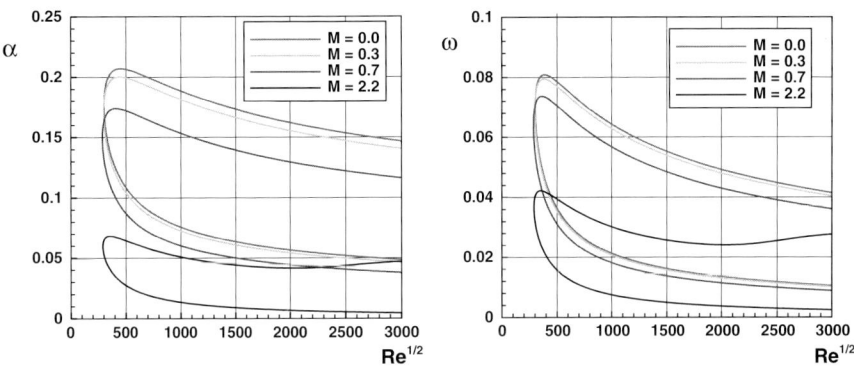

Fig. 1. Neutral stability curves

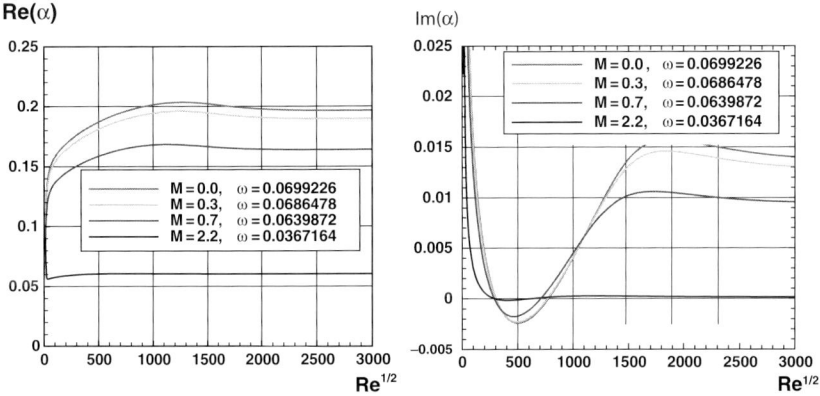

Fig. 2. Wavenumber and damping rate vs. $R = Re^{1/2}$ for fixed ω

where the real and imaginary parts of α are plotted vs. R for the indicated values of ω and M. This means, in particular, that the spatial evolution of these modes exhibit two well separated scales and can be described as slowly modulated wavetrains, of the form (14), with φ and ω real and the complex amplitude A satisfying a linear equation of the form

$$A' = \mu A, \tag{20}$$

where the growth rate μ is such that $|\mu| \ll \varphi'/\delta$.

– Parametric coupling between two TS modes, with frequencies and wavenumbers ω_j and κ_j for $j = 1$ and 2, is promoted through (quadratic) nonlinear terms by a wavetrain in the boundary layer, with a frequency ω and wavenumber κ, such that

$$\omega = \omega_1 + \omega_2, \quad \kappa = \kappa_1 + \kappa_2. \tag{21}$$

The complex amplitudes of both TS modes, A_1 and A_2 (which would obey equations of the type (20) in the absence of parametric coupling) evolve according to a system of coupled equations, of the form

$$A_1' = \mu_1 A_1 + \beta_1 a \bar{A}_2, \quad A_2' = \mu_2 A_2 + \beta_2 a \bar{A}_1, \quad (22)$$

where μ_1 and μ_2 are the growth rates of the TS modes (both small), a is the (small) amplitude of the wavetrain, and the complex coefficients β_1 and β_2 are of order one. Thus, effective parametric forcing requires that $|a| \sim |\mu_j|$, which according to our comment above requires that a be small, of order 0.001 (see Fig. 2). This coupled evolution can be either more stable or more unstable than the original uncoupled evolution, depending on the coupling coefficients β_1 and β_2. Since the coupling coefficients depend on x, this cannot be elucidated analytically, but can be illustrated in the constant coefficient case, in which coupling stabilizes provided that the real part of $\beta_1 \bar{\beta}_2$ be negative. It turns out that the wavetrain is stabilizing in all situations that have been checked.

- The parametric forcing described above is not the end of the mode interaction story. Once the two original TS modes and the wavetrain are present, nonlinearities force infinitely many new modes, with frequencies and wavenumbers $(j - k)\omega + (j_1 - k_1)\omega_1 + (j_2 - k_2)\omega_2$ and $(j - k)\kappa + (j_1 - k_1)\kappa_1 + (j_2 - k_2)\kappa_2$, for any natural numbers $j, k, j_1, k_1, j_2,$ and k_2; the amplitudes are $\gamma a^j \bar{a}^k A_1^{j_1} \bar{A}_1^{k_1} A_2^{j_2} \bar{A}_2^{k_2}$, where a, A_1, and A_2 are the (small) complex amplitudes of the wavetrain and the TS modes, and the coefficient γ is $O(1)$ if the excited mode is not a nearly marginal mode, but can be large otherwise. The latter case is that in which the mode interaction process is effective. For instance, in the parametric interaction case above, $j = 1$, $j_1 = -1$, $k = k_1 = j_2 = k_2 = 0$ and the excited mode is the second TS mode, which is nearly marginal. Some additional, not so strong resonances can also appear, see Fig. 5 in Sect. 4. But there is an additional resonance that is always present and is associated with a *mean flow*. If

$$|A_1| \sim |A_2| \ll a \ll 1, \quad (23)$$

the mean flow is produced mainly by the wavetrain, and exhibits an amplitude that is of the order of $Ra^2 = a^2$, which can affect the stability of the TS waves. It turns out that the effect of the mean flow is taken into account replacing (22) by

$$A_1' = (\mu_1 + \beta_3 Ra^2)A_1 + \beta_1 a \bar{A}_2, \quad A_2' = (\mu_2 + \beta_4 Ra^2)A_2 + \beta_2 a \bar{A}_1, \quad (24)$$

with the complex coefficients β_3 and β_4 of order one. The coefficients in this equation can be obtained via weakly nonlinear analysis, which is omitted here. Instead, we use (24) to guess the order of magnitude of the various amplitudes for these mode interaction be effective, namely

$$|\mu_1| \sim |\mu_2| \sim |a| \sim Ra^2 \ll 1. \quad (25)$$

This will be used in the DNS analysis that is considered next.

4 Direct Numerical Simulation

In a flat plate at zero incidence boundary layer, (1)–(3) should be integrated in a domain close to the plate and extending streamwise to a position beyond transition, which occurs at a fairly high Reynolds number ($\sim 5 \times 10^6$). Since the smallest scale associated with the viscous sublayer must be described, this is quite costly numerically. Thus, we take a computational domain in the streamwise direction covering only a portion of the boundary layer and impose the steady profile at the entrance. Namely, the computational domain is

$$x_0 < x < x_0 + L, \quad 0 < y < y_0, \tag{26}$$

with $L \ll x_0$ but somewhat large as to include several wavelengths of the relevant waves, and y_0 somewhat large compared to the boundary layer thickness. In this region, (1)–(3) apply, with the Reynolds number R based on the distance from the leading edge to the entrance of the computational domain. For convenience, we first calculate the steady state solution, $(\boldsymbol{v}_s, \rho_s, T_s)$, with boundary conditions

$$\boldsymbol{v}_s = \tilde{\boldsymbol{v}}(y), \quad \rho_s = \tilde{\rho}(y), \quad T_s = \tilde{T}(y) \quad \text{at } x = x_0, \tag{27}$$

$$u = 1, \quad \partial v/\partial y = \partial \rho/\partial y = \partial T/\partial y = 0 \quad \text{at } y = y_0, \tag{28}$$

$$v = 0, \quad \partial T/\partial y = 0 \quad \text{at } y = 0, \tag{29}$$

$$\partial \boldsymbol{v}/\partial x = \boldsymbol{0}, \quad \partial T/\partial x = 0 \quad \text{at } x = x_0 + L, \tag{30}$$

where the plate is assumed to be thermally insulated, and $(\tilde{\boldsymbol{v}}, \tilde{\rho}, \tilde{T})$ is the Blasius self-similar steady state solution at $x = x_0$, given by (7).

The boundary conditions for the nonstationary problem are assumed to be such that:

- The temperature of the plate (and the air just above the plate) is assumed to remain at its steady state value. This is because the characteristic time in the air (associated with TS oscillations) is much shorter than the heat conduction time in the solid.
- The solution should match with the uniform flow outside the boundary layer. When imposing boundary conditions at a finite distance from the plate, spurious reflection of both acoustic and hydrodynamic waves must be avoided.
- In order to generate a TS wave in the boundary layer, a vibrating membrane can be used with a vibrating frequency equal to that of the TS wave; see (33) and (35) below. The size of the membrane is not essential (because the spatially parabolic character of the boundary layer). A size similar to the wavelength of the TS wave is nevertheless convenient to facilitate generation.
- In order to generate a wavetrain in the air, a wavetrain-like boundary condition for the vertical velocity is imposed in the plate; see (33)–(36). This can be achieved in practice using a periodic array of oscillators, one at each period of the wavetrain, with a frequency ω and appropriate phase shifts.

– The boundary conditions at the exit are unessential (again, because of the parabolic character of the boundary layer), but a buffer near the exit (where the solution depends on the selected boundary conditions) must be excluded from post processing.

With these ideas in mind, the boundary conditions for the nonstationary problem are (cf. (27)–(30))

$$v_s = \tilde{v}(y), \quad \rho_s = \tilde{\rho}(y), \quad T_s = \tilde{T}(y) \quad \text{at } x = x_0, \tag{31}$$

$$u = 1, \quad \partial v/\partial y = \partial \rho/\partial y = \partial T/\partial y = 0 \quad \text{at } y = y_0, \tag{32}$$

$$u = 0, \quad v = v_0(x,t) T = T_s(x,0) \quad \text{at } y = 0, \tag{33}$$

$$\partial v/\partial x = \mathbf{0}, \quad \partial T/\partial x = 0 \quad \text{at } x = x_0 + L, \tag{34}$$

where

$$v_0(x,t) = \varepsilon \sin \kappa_1(x - x_1) \sin \omega_1 t \quad \text{if } |x - x_1| < \pi/\kappa_1, \tag{35}$$

$$v_0(x,t) = a \sin(\kappa x - \omega t) \quad \text{if } x_2 < x < x_0 + L, \tag{36}$$

$$v_0(x,t) = 0 \quad \text{if either } x_0 < x \le x_1 - \pi/\kappa_1, \text{ or } x_1 + \pi/\kappa_1 \le x < x_0 + L. \tag{37}$$

Now, the numerical tool must be sufficiently precise as to give a precise description of the TS modes involved, which exhibit quite small amplitude and growth rates; in particular, numerical viscosity must be quite small. Also, in order to isolate the contribution of the various modes in the complete flow field provided by the numerical tool, a temporal fast Fourier transform tool is used that gives the components of the flow at various frequencies.

Now, we are in a position to simulate the mode interaction process explained in Sect. 3 and illustrated in Fig. 3.

To this end, we perform the following simulations at $Ma = 0.3$, $Re = 1.96 \times 10^7$ (which gives $Re \simeq 4,430$ using (9)), $\Omega_1 = \Omega_{\text{TSu}} = 90$, $\kappa_1 = \kappa_{TSu} = 410$ (frequency and wavenumber of an unstable TS mode, with growth rate $d = 29$), $\Omega_2 = \Omega_{\text{TSs}} = 150$, $\kappa_2 = \kappa_{TSs} = 440$ (stable TS mode, with $d = -42$), $\varepsilon = 2 \times 10^{-3}$, and various values of a.

If $a = 0$, then only the unstable TS mode is forced (Fig. 3 top). In order to check the numerical approximation, the horizontal velocity profile of the TS mode is compared in Fig. 4 top with its counterpart obtained from the linear approximation. (15)–(19). If $\varepsilon = 0$, $a = 0.001$, $\Omega = 240$, and $\kappa = 850$, then we only generate a wavetrain (Fig. 3 middle). The numerical approximation is now checked in Fig. 4 bottom, where the linear approximation of the wavetrain is calculated using (15)–(19), except the boundary condition for the vertical velocity at $\zeta = 0$, which is replaced by $\psi = a$.

If both $\varepsilon \ne 0$ and $a \ne 0$, then since the frequencies and wavenumbers satisfy the resonance relations (21), the stable TS is also forced parametrically

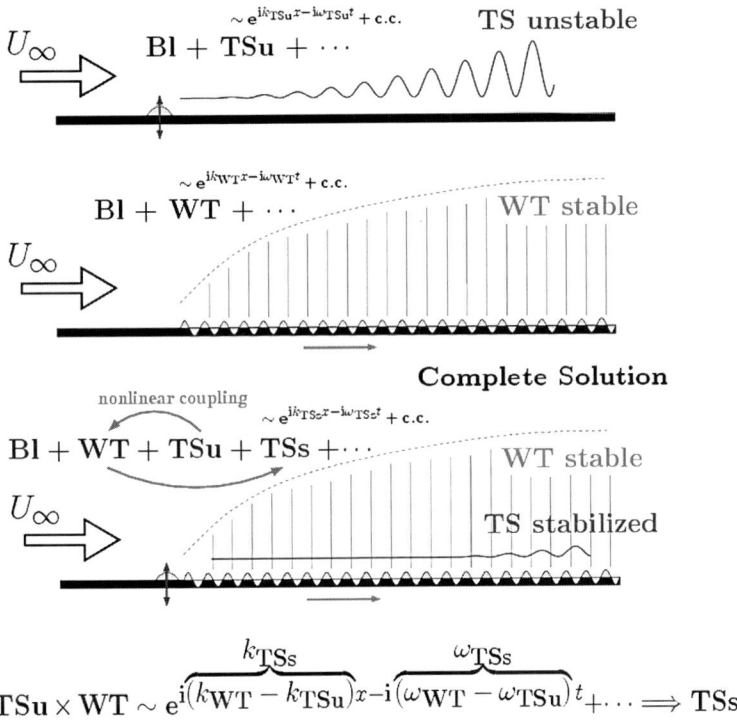

Fig. 3. The mode interaction process. *Top*: unstable TS wave, TSu, generated with an actuator. *Middle*: generation of a wavetrain, WT, at the bottom of the plate (red). *Bottom*: the WT induces a nonlinear coupling between a stable TS (TSs) and unstable TS, reducing spatial growth

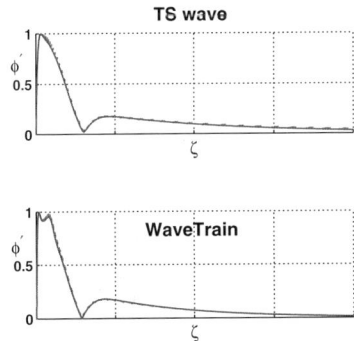

Fig. 4. Comparison of the horizontal velocity profiles provided by DNS and linear stability for the unstable TS mode with $\Omega = 90$ (*top*) and the wavetrain obtained with $\Omega = 240$ and $\kappa = 850$ (*bottom*)

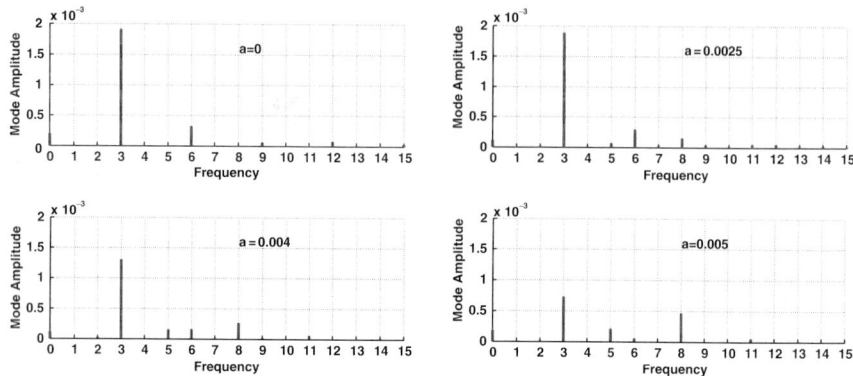

Fig. 5. Temporal fast Fourier transform (FFT), with frequency $= \Omega/30$ in abscissa. Thus the unstable and stable TSs, and the wavetrain correspond to frequencies 3, 5, and 8, respectively

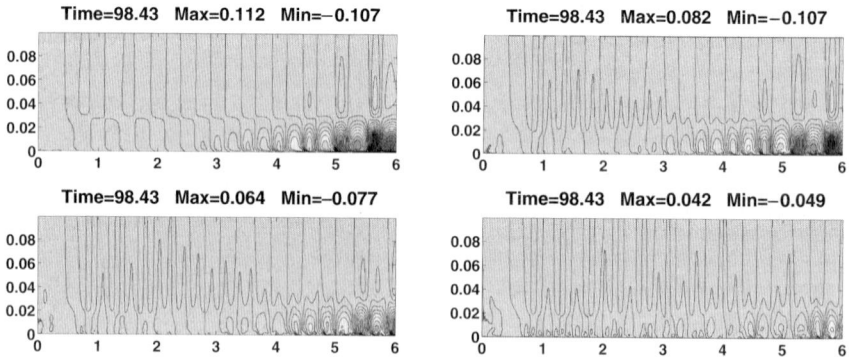

Fig. 6. Streamwise perturbation velocity contours for the WT wave amplitudes corresponding to the dots marked in Fig. 5 ($a = 0, 0.0025, 0.004$, and 0.005 from *left* to *right* and *top* to *bottom*)

and its spatial evolution is coupled with that of the unstable TS, as explained in Sect. 3. This is illustrated in the FFT plots in Fig. 5, where it is seen that if $a = 0$ the unstable TS mode (with $\Omega = 90$ and its harmonics appear while for increasing values of a, both the wavetrain (with $\Omega = 240$) and the stable TS mode (with $\Omega = 150$) are also present with increasing amplitude, while the amplitude of the unstable TS mode decreases, meaning that the latter is stabilized. In particular, at $a = 0.005$ the unstable TS mode has been divided by three. Note that the mean flow corresponds to $\Omega = 0$ and remains small for all considered values of a. This is because since $Re = 4,430$, $a^2 Re \ll 1$ for the considered values of a, and that term accounting for the mean flow in (24) is much smaller than that term accounting for parametric forcing. The mean flow seems to give additional stabilization at $a = 0.005$, but this should be checked. This stabilization process is further illustrated in the snapshots plotted in Fig. 6, where it is clearly appreciated that the strength of the unstable TS mode at $a = 0$ is strongly reduced by the presence of the wavetrain. Thus, this mode interaction process is quite effective.

5 Concluding Remarks

We have applied a mode interaction process to stabilize TS waves in a compressible, 2D boundary layer attached to a flat plate at zero incidence. In order to stabilize a given unstable TS wave, a wavetrain is created in the boundary layer that couples parametrically the spatial evolution of the unstable TS wave with the evolution of a stable TS wave, in such a way that the stable takes energy from the unstable and stabilizes it. The process has been explained qualitatively in Sect. 3, where the required order of magnitude of the amplitude, frequency, and wavenumber of the stabilizing wavetrain was anticipated. In addition, we also anticipated that the mean flow produced by the wavetrain can also play a role. All these has been confirmed in Sect. 4, where a battery of numerical simulation was performed. Additional DNS simulations, not presented here, show that the process is quite robust in the sense that it is fairly insensitive to both the frequency and wavenumber of the wavetrain. This must be because of the stalibilizing effect of the wavetrain, which does not depend on any resonance relation and seems to also contribute to stabilizing the system. But the analysis of this is beyond the scope of this paper.

References

[Ch05]   Chomaz, J.M.: Global instabilities in spatially developing flows: non-normality and nonlinearity, Annu. Rev. Fluid Mech., **37**, 357–392 (2005)

[CB01]   Cossu, C. and Brant, L.: Stabilization of Tollmien–Schlichting waves by finite amplitude optimal streaks in a Blasius boundary layer. Phys. Fluids, **14**, 016318-1-17 L.57–L.60(2002)

[D83]   Davey, A.: An authomatic orthonormalization method for solving stiff boundary-value problems, J. Compt. Phys., **51**, 343–356 (1983)

[J98]   Joslin, R.D.: Aircraft laminar flow control. Annu. Rev. Fluid Mech., **30**, 1–29 (1998)

[K94]   Kachanov, Y.S.: Physical mechanisms of laminar boundary layer transition. Annu. Rev. Fluid Mech., **26**, 411–482 (1994)

[K76]   Keller, H.: Numerical Solutions of Two Point Boundary Value Problems, Vol.24 of CBMS-NSF Regional Conferences in Applied Mathematics, SIAM, 1976.

[LG06]   Li, Y., Gaster, M.: Active control of boundary layer instabilities. J. Fluid Mech., **550**, 185–205 (2006)

[LMV01] Lapuerta, V., Mancebo, F.J. and Vega, J.M.: Control of Rayleigh–Taylor instability by vertical vibration in large aspect ratio containers. Physical Review E, **64**, 016318-1-17 (2001)

[L00]   Luchini, P.: Reynolds number independent instability of the boundary
           layer over a flat plate: optimal perturbations. J. Fluid Mech., **404**, 289–
           309 (2000)
[R01]   Reshotko, E.: Transient growth: A factor in bypass transition. Phys. Flu-
           ids, **13**, 1067–1075 (2001)
[SG00]   Schlichting, H., Gersten, K.: Boundary layer theory. Springer, Berlin
           (2000)
[S82]   Smith, F.T.: On the high Reynolds number theory of laminar flows. IMA
           J. Appl. Math., **28**, 207–281 (1982)
[WAA01]   Walther, S., Airiau, C., Bottaro, A.: Optimal control of Tollmien–
           Schlichting waves in a developing boundary layer. Phys. Fluids, **13**, 2087–
           2096 (2001)
[ZLB04]   Zuccher, S, Luchini, P. Bottaro, A.: Algebraic growth in a Blasius bound-
           ary layer: optimal and robust control by mean of suction in the nonlinear
           regime. J. Fluid Mech., **513**, 135–160 (2004)

Acoustic Streaming and Lower-than-Laminar Drag in Controlled Channel Flow

P. Luchini

Department of Mechanical Engineering, Università di Salerno, Italy

1 Introduction

This contribution is about an unforeseen connection that arose while studying three seemingly unrelated research problems. For this reason I thought it appropriate to be presented at a meeting on the applications of mathematics to industry. I will follow the outline of the oral presentation and expose the three stories first, to later comment about their connection.

1.1 Story #1: Lower-than-Laminar Drag

In "Flow Control: new challenges for a new Renaissance", Bewley [1] discussed among others the feasibility of a *thought experiment* in which roller-like spanwise vortices, induced by external forces, "lubricate" a channel flow and reduce friction at the wall. After examining various aspects of this problem, he concluded that it was unlikely that a skin-friction reduction below laminar level could be obtained if actuation to generate the vortices was applied in the form of zero-mean blowing-and-suction at the wall and formulated the following

> **Conjecture.** The lowest sustainable drag of an incompressible constant mass-flux channel flow, in either 2D or 3D, when controlled via a distribution of zero-net mass-flux blowing/suction over the channel walls, is exactly that of the laminar flow.

Several arguments make this conjecture reasonable. For instance:

- A zero-mean blowing and suction (v-velocity component) has zero momentum flux ($\langle uv \rangle = 0$, just because $u = 0$).
- It seems at first sight counter-intuitive that a pressure drop reduction (or flow-rate increase) could be realized without any momentum transfer from the wall.
- In a subsequent paper, Bewley could mathematically prove that *heat transfer* cannot be reduced to lower-than-laminar levels by zero-mean blowing and suction.

The Numerical Counter-Example

Nonetheless the conjecture is untrue. A contribution presented at the 2005 Meeting of the APS – Division of Fluid Dynamics [2] (and later published as [3]) contained a *numerical experiment* showing that below-laminar drag could be sustained in a two-dimensional channel flow with surface blowing/suction in the form of upstream travelling waves.

At the same meeting Bewley (from the audience) immediately recognized that, just as not all conjectures are bound to be true, this one was not. Here we shall work with our own version of the counter-example (Fig. 1). To exaggerate the effect, we have chosen a large amplitude of 1 (i.e. a sinusoidal normal velocity with an amplitude equal to the mean velocity of the longitudinal flow) and a relatively low Reynolds number. The mean profile shown in the figure corresponds to a flow-rate increase by 40% for the same pressure gradient and leaves no doubt that a friction reduction has occurred.

We provisionally name this the
Puzzle #1:

- Pure blowing/suction ($v \neq 0$, $u = 0$) has zero momentum flux at the wall ($\langle uv \rangle = 0$); shear stress is unchanged.
- Flow rate is larger, pressure gradient is unchanged.
- How can a substantial flow-rate increase be sustained without any *reaction on the wall?*

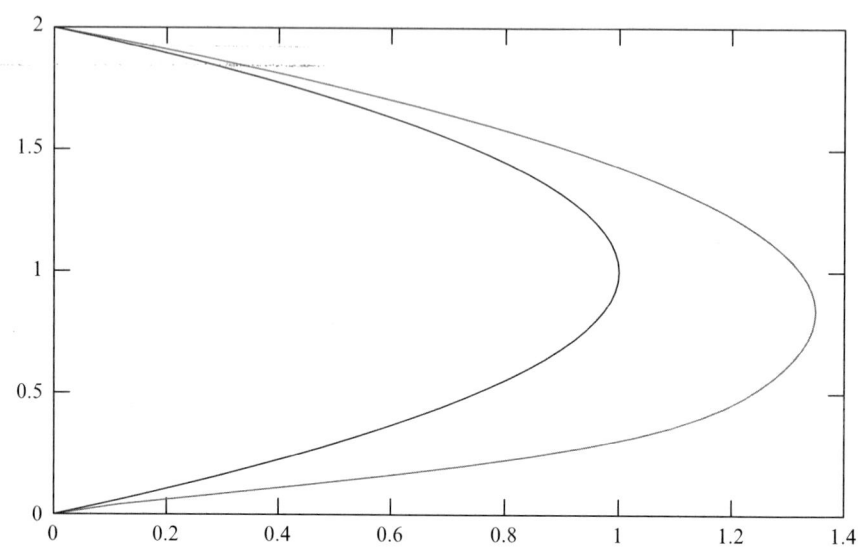

Fig. 1. A reconstructed numerical counter-example: Channel flow, $Re = 100$, wavelength $= 2$, wave amplitude $= 1$, wave speed $= 0.3$. Flow-rate increase: 40%

1.2 Story #2: Turbulent Flow over a Wavy Wall

In a totally separate line of thought, it was observed years ago that a wavy blowing and suction at the wall can *destabilize* channel flow to longitudinal vortices/streaks [4]. Stimulated by this idea, we started a research project (still ongoing, actually) to investigate whether similar large-scale vortices/streaks can be excited in an already turbulent flow. The signal we expected from the generation of such vortices was a *mixing* and *drag* increase. Our first effort [5] produced the table shown in Fig. 1.

In these direct numerical simulations of turbulence, the formation of vortices and an *increase* in turbulent mixing and drag was expected. The vortices were not unequivocally observed (higher Reynolds number may be needed), but both drag increase and an unexpected skin-friction *reduction* were observed depending on wave amplitude and wavelength. We shall name this the

Puzzle #2.

Table 1. Change in friction coefficient induced by different combinations of transpiration intensity A and wavenumber α_t. Q_t is the resulting transpiration flow rate over half wavelength. The values in wall units are computed with the friction velocity of the reference case.

Case	α_t	A	$10^2 Q_t/Q_x$	λ_t^+	A^+	$10^3 C_f$	$\%\Delta C_f$
0	0	0	0.00	∞	0	8.15	0.0
1	0.5	0.03	9.00	2262	0.71	25.03	207.1
2	1.0	0.002	0.30	1131	0.05	8.23	1.0
3	1.0	0.005	0.75	1131	0.12	8.68	6.6
4	1.0	0.01	1.50	1131	0.24	10.48	28.6
5	1.0	0.02	3.00	1131	0.47	15.18	86.3
6	1.0	0.03	4.50	1131	0.71	19.50	139.2
7	1.0	0.04	6.00	1131	0.94	23.08	183.2
8	1.5	0.03	3.00	753	0.71	15.43	89.3
9	2.0	0.02	1.50	565	0.47	10.11	24.0
10	2.0	0.03	2.20	565	0.71	12.40	52.2
11	2.5	0.02	1.20	452	0.47	8.61	5.6
12	2.5	0.03	1.80	452	0.71	9.98	22.4
13	3.0	0.03	1.50	377	0.71	8.64	6.0
14	3.5	0.03	1.30	323	0.71	8.05	−1.2
15	4.0	0.03	1.10	283	0.71	7.91	−3.0
16	4.5	0.03	1.00	251	0.71	7.87	−3.5
17	5.0	0.005	0.15	226	0.12	8.12	−0.4
18	5.0	0.01	0.30	226	0.24	8.06	−1.2
19	5.0	0.02	0.60	226	0.47	7.95	−2.5
20	5.0	0.025	0.75	226	0.59	7.91	−3.0
21	5.0	0.03	0.90	226	0.71	7.85	−3.7
22	5.0	0.04	1.20	226	0.94	7.90	−3.0
23	5.0	0.05	1.50	226	1.18	7.93	−2.7

Table 1. (Continued)

Case	α_t	A	$10^2 Q_t/Q_x$	λ_t^+	A^+	$10^3 C_f$	$\%\Delta C_f$
24	5.0	0.08	2.40	226	1.89	7.85	−3.7
25	5.0	0.10	3.00	226	2.36	7.71	−5.4
26	5.0	0.12	3.60	226	2.83	7.55	−7.3
27	5.0	0.16	4.80	226	3.77	7.26	−11.0
28	5.0	0.20	6.00	226	4.72	7.02	−13.9
29	5.5	0.03	0.80	206	0.71	7.90	−3.0
30	6.0	0.03	0.75	188	0.71	7.92	−2.8
31	6.5	0.03	0.69	174	0.71	7.91	−3.0
32	7.5	0.03	0.60	151	0.71	8.03	−1.4
33	9.0	0.03	0.50	126	0.71	8.11	−0.5
34	11.0	0.03	0.41	103	0.71	8.17	0.2
35	22.0	0.03	0.21	51	0.71	8.23	1.0

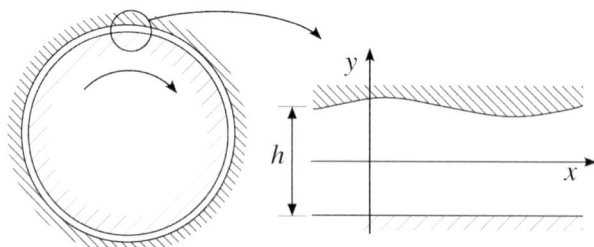

Fig. 2. Fluid-based ultrasonic motor

1.3 Story #3: Acoustic Streaming

The third, originally unrelated, line of thought arose as a study of acoustic streaming as applied to ultrasonic motors (see Fig. 2).

Solid piezoelectric ultrasonic motors are a technological reality. They are based on the nonlinearity of contact friction between two solid discs. Fluid-based ultrasonic motors, instead, are experimental devices based on the non-linearity of wave propagation in a fluid and more precisely on the phenomenon known as *acoustic streaming*. In an effort to study this phenomenon, we [6] adapted the theory of acoustic streaming to the situation where gap width is comparable to boundary-layer thickness and compressibility effects cannot be neglected. In doing so, particular attention was also paid to the effect of temperature oscillations and of the elastic impedance of the vibrating wall.

In fact, streaming is not at all limited to compressible flow, and some of its most important applications are in the domain of surface waves (see, e.g. [7]). For this reason, it is also known as *steady streaming* or *Rayleigh streaming*.

2 Not Drag but Thrust

The unexpected connection among the three stories came about when we realized that *streaming can explain the sub-laminar drag reduction*. In other

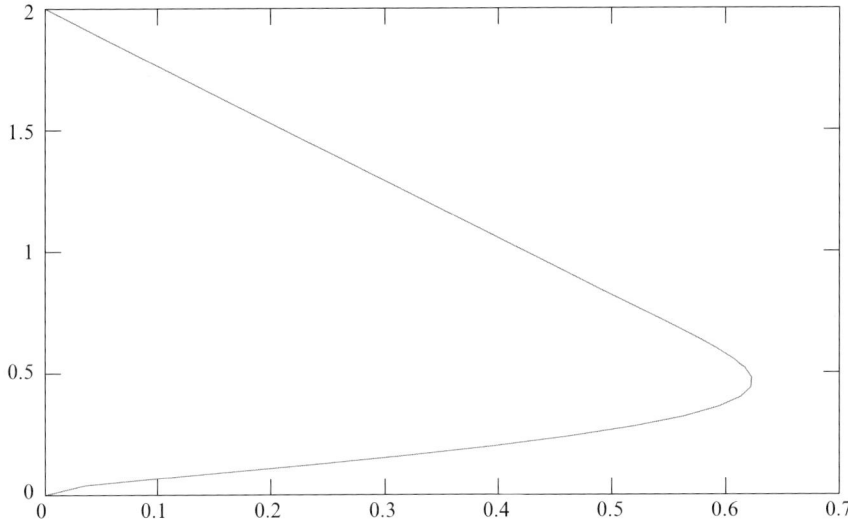

Fig. 3. Pressure gradient off! Channel flow, $Re = 100$, wavelength $= 2$, wave amplitude $= 1$, wave speed $= 0.3$ *upstream*

words, the effect of a zero-mean blowing-and-suction at the wall is not a (multiplicative) influence on the drag-producing mechanism driven by the pressure gradient, but an (additive) pumping that pushes the fluid in one direction or other independent of any mean pressure gradient applied. The easiest way to verify this in our little numerical experiment was to turn the pressure gradient off (Fig. 3).

The flow rate thus obtained is still as much as 50% of what was produced by the pressure gradient in the absence of blowing/suction. If it is remembered that the combination of the two had produced 140%, it can be seen that the effect is nearly additive. Of course, if one were to judge Fig. 3 in terms of drag reduction, he would have to admit that drag reduction can be 100% (non-zero flow with zero pressure gradient). The only asymmetry that determines the direction of flow is the direction of propagation of the applied travelling wave, which here is upstream. Just as a check, one can reverse this direction, and a specular reverse flow is obtained (Fig. 4).

Of course, pumping costs energy, and one has to realize that zero pressure gradient does not imply zero energy expenditure. The work rate spent to produce the blowing and suction alone, for the case of Figs. 3 and 4 is *500 times* as much as necessary to produce Poiseuille flow. Although the balance becomes less unfavourable at larger Reynolds number, it does not look like a technique you want to use for practical drag reduction.

But, as far as proof of principle is concerned, the phenomenon is interesting, and we must still illustrate the connection to our story #2, so stick with us and let us see whether an effect is still there when the wave speed is

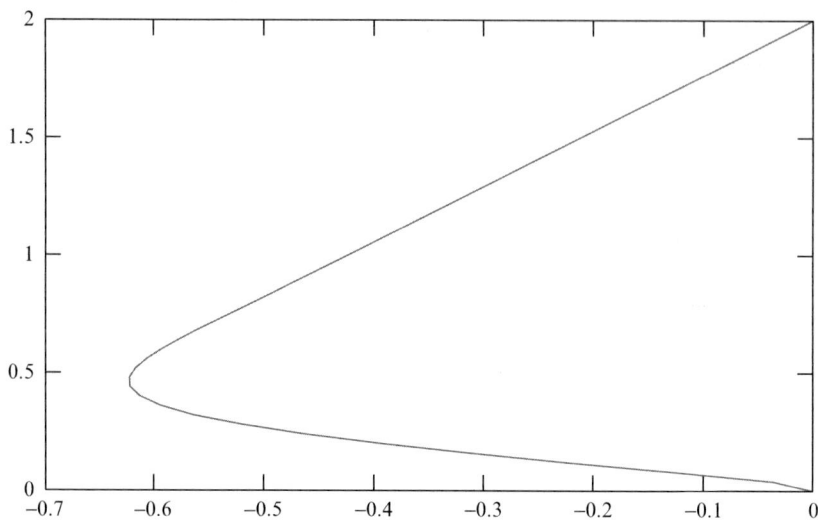

Fig. 4. Downstream wave speed. Channel flow, $Re = 100$, wavelength $= 2$, wave amplitude $= 1$, wave speed $= -0.3$ *downstream*

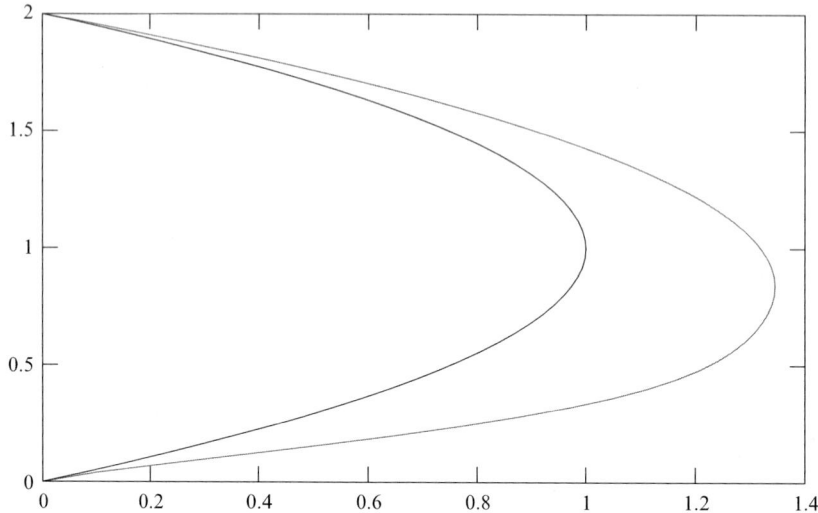

Fig. 5. Wave speed off! Channel flow, $Re = 100$, wavelength $= 2$, wave amplitude $= 1$, wave speed $= 0$

zero (i.e. the pattern of blowing and suction is fixed to the wall). This case is illustrated by Fig. 5.

We still see a flow-rate increase, precisely by 38% more w.r.t. Poiseuille flow (and a work rate 145 times as large as Poiseuille flow). So was the travelling

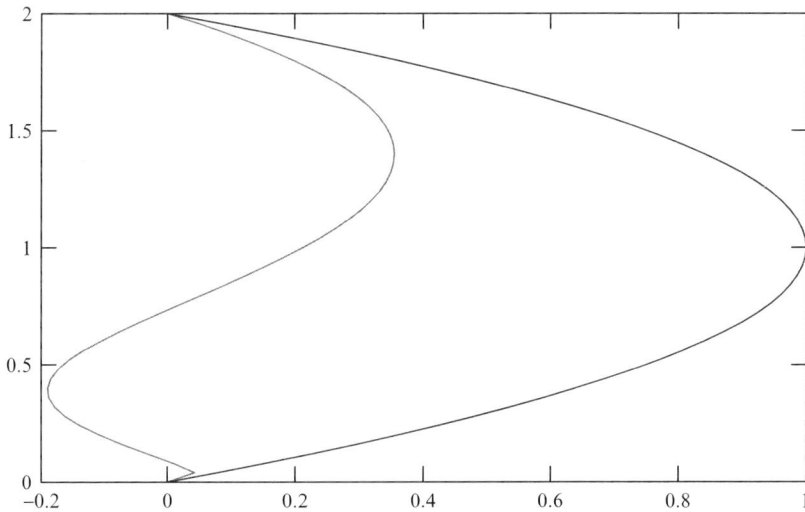

Fig. 6. Downstream wave speed. Channel flow, $Re = 100$, wavelength $= 2$, wave amplitude $= 1$, wave speed $= -0.3$. Flow rate: 91% *less* w.r.t. Poiseuille flow. Work rate: 5,500 times as much w.r.t. Poiseuille flow

wave an essential factor? The answer is that, since now we do have a pressure gradient, the relative speed between wave and fluid is non-zero even if the absolute wave speed is. In the previous examples, the wave had to move *up-stream* in order to produce a *downstream* flow, and this is precisely the relative direction in which the wave moves when fixed to the wall. In fact, if the wave is made to move downstream (Fig. 6), a flow-rate reduction is effected.

So, part of the effect observed in [5] can be ascribed to streaming. (Another part, as explained in that paper, is actual drag reduction due to the fact that, in a turbulent flow, the outflow at the suction locations is turbulent while the inflow at the blowing sites is laminar.)

3 Is the Absence of Reaction on the Walls Counter-Intuitive?

It remains to be discussed what is there that is wrong with the arguments in favour of the no-drag-reduction conjecture presented at the beginning of this paper. Particularly counter-intuitive (but nevertheless true) is that the streaming effect of a travelling wave can be produced without any reaction force on the walls (nor on any other supporting pillar). In other words, it is in principle possible, although not necessarily economically convenient, to push oil through a pipeline without exerting any force on the pipeline itself. In order to simplify matters the most and see how this is possible, let us suppose the fluid immersed in an external force field (to which the streaming-producing

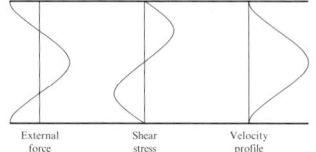

Fig. 7. A simpler example: *zero* net force − *zero* wall shear − *non-zero* flow rate

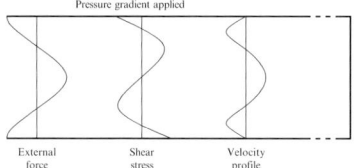

Fig. 8. An alternate view

average stress is equivalent). By a simple static balance, the reaction force exerted by the fluid on the walls must equal the resultant of the external forces. If, however, the force field has the distribution depicted in Fig. 7, the resultant is zero. Yet, the flow velocity, produced by integrating the force field twice according to the laws of Newtonian viscosity, has the distribution reported in the same figure and clearly a non-zero integral (flow rate).

An alternate view of the same phenomenon is depicted in Fig. 8. Here a stirring force in a closed container produces a wall stress with no net flow rate. The force on the wall is balanced by a pressure gradient. Nobody will probably find counter-intuitive this latter picture. If a fluid is stirred in a pipe closed at one end, so as to produce a flow towards the closed end at the centre and a backflow flush against the walls, the net effect will be a non-zero force on the lateral walls and a non-zero pressure gradient, with zero flow rate. Yet if this flow is linearly superposed (as allowed by the one-dimensional version of the Navier–Stokes equations) to a Poiseuille flow with an equal and opposite pressure gradient, the effect will be a non-zero flow rate with a zero pressure gradient and wall friction. This is exactly what the streaming does.

4 Conclusion

A non-zero mean flow rate in a channel or pipe can persist in the absence of any friction force on the wall (or other source of waves or force field). Through this effect, lower-than-laminar skin friction *can* indeed be achieved through zero-mean blowing and suction. However, the appropriate way to describe this phenomenon is not as *drag reduction* but as additional *thrust* due to Rayleigh streaming. The thrust can point in either direction independently of the pressure gradient. The thrust costs energy, and in terms of energy dissipation the budget is a definite *loss*.

In connection with the latter statement, it was pointed out to me by Bewley that the power dissipation of Poiseuille flow is indeed an absolute minimum and cannot be reduced further by any sort of zero-mean blowing and suction. A very neat proof of this is provided in [8]. It should be noted, however, that this proof does not exclude the possibility for blowing-and-suction to lower the energy dissipation of turbulent flow below the level of the same turbulent flow on an impermeable wall. Some results of [3] point in this direction.

Acknowledgements

An invitation to visit the University of California, San Diego during the Summers of 2004 and 2005 and several prolonged discussions with T. Bewley were of stimulus to this paper.

References

1. Bewley, T.R., Flow Control: new challenges for a new Renaissance, *Prog. Aero. Sci.* **37**, 21-58 (2001).
2. Sung Kang, Taegee Min, Jason Speyer & John Kim, Sublaminar skin-friction drag in controlled channel flow, *2005 58th Annual Meeting of the APS Division of Fluid Dynamics*, Chicago 20-22 Nov. 2005. Abstract: AT.00004.
3. Min, T., Kang, S.M., Speyer, J.L. & Kim, J., Sustained sub-laminar drag in a fully developed channel flow, *J. Fluid Mech.* **558**, 309-318 (2006)
4. Cabal, A., Szumbarsky, J. & Floryan, J.M., Stability of flow in a wavy channel, *J.Fluid Mech.* **475**, 191 (2002).
5. Quadrio, M., Floryan, J.M. & Luchini, P., Effect of streamwise-periodic wall transpiration on turbulent friction drag, 50th CASI Conference, Montreal, April 2003; accepted for publication in J. Fluid Mech.
6. Luchini, P., & Charru, F., Acoustic streaming past a vibrating wall, *Phys. Fluids* **17**, 122106(1-7), (2005).
7. Riley N., Steady Streaming, *Annu. Rev. Fluid Mech.* **3 3**, 43-65 (2001).
8. Bewley, T.R. & Aamo, O.M., A 'win-win' mechanism for low-drag transients in controlled two-dimensional channel flow and its implications for sustained drag reduction, *J. Fluid Mech.* **499**, 183–196 (2004).

Recent Progress in the Use of Compliant Walls for Laminar Flow Control

Peter W. Carpenter

School of Engineering, University of Warwick, Coventry, CV4 7AL, UK
P.W. Carpenter@warwick.ac.uk, http://www.warwick.ac.uk/fac/sci/eng

Summary. It has been known for some time that an appropriately designed compliant wall (artificial dolphin skin) is highly effective for laminar flow control in low-disturbance environments. Unfortunately, compliant walls are not really practical for aeronautical applications. Accordingly, we focus here on marine applications. The marine environment tends to have much higher levels of freestream turbulence than found in flight conditions typical of cruise. Herein, we explore the effects of freestream turbulence on laminar–turbulent transition. In particular, we investigate the velocity streaks generated in the boundary layer by freestream turbulence. Furthermore, we carry out a numerical-simulation study of the effects of wall compliance on the velocity streaks. We find that boundary layers over compliant walls are much less receptive to streaks than those over a rigid surface. This implies that compliant walls should be effective at laminar flow control even in environments with relatively high levels of freestream turbulence.

1 Introduction

Interest in the use of compliant walls or artificial dolphin skins for laminar flow control dates back to the seminal papers of Kramer [1, 2]. He achieved drag reductions of up to 60% in experimental tests on bodies of revolution covered with his specially designed compliant coatings. Kramer, himself, was inspired by a belief that the flow over the dolphin (the bottle-nosed dolphin *Tursiops truncatus* is the most commonly studied) remained laminar despite the high Reynolds numbers. Accordingly, the design of his compliant coatings was based closely on his interpretation of the structure of the dolphin epidermis. His views on dolphin hydrodynamics arose partly from personal observation and partly from the earlier seminal paper of Gray [3]. Gray, a zoologist, estimated the specific power required from the dolphin's propulsive muscles to allow it to swim at the, then accepted, maximum sustained speed. He found that the resulting estimated specific power, assuming aerobic metabolism, exceeded the mammalian norm by about sevenfold. This result became known as *Gray's Paradox* (see Babenko and Carpenter [4] for a recent review of dolphin hydrodynamics).

In theory, compliant walls could be used in air as well as in water. Unfortunately, the large difference between the densities of air and typical elastomeric materials leads to a mismatch in inertias except in the case of impractically flimsy compliant walls. Accordingly, as explained in more detail by Carpenter et al. [5] the use of compliant walls is not practical in aeronautical applications. It is entirely possible that understanding the flow physics that underlie the good laminar-flow-control capability of appropriately designed compliant walls will lead to alternative devices with similar benefits that can function in air flow. One example of this is the passive porous wall [6]. However, for the remainder of the paper we shall assume that the applications of compliant walls lie in the marine environment. In this case, the principal sources of environmental disturbances that are responsible for creating the boundary-layer perturbations leading to laminar–turbulent transition are freestream turbulence, but at somewhat higher levels than seen in typical aeronautical applications, and suspended particulate matter. Here we shall concentrate on the effects of freestream turbulence. Little is known about the effects of suspended particles, particularly when the particles are small, i.e. of a dimension that is a small fraction of the boundary-layer thickness.

2 The Effects of Freestream Turbulence on Transition

The creation of perturbations within the boundary layer by vortical and acoustic perturbations is commonly known as *receptivity* [7]. Here we shall confine our attention to vortical perturbations, in other words freestream turbulence. There are many ways of modelling freestream vortical perturbations. Our approach is based on the velocity-vorticity formulation of the Navier–Stokes equations due to Davies and Carpenter [8]. This formulation is based on three primary variables (ω_x, ω_y, w), namely the perturbation vorticity components in the streamwise, x, and spanwise, y, directions and the wall-normal velocity perturbation. These primary variables are governed by three equations: two vorticity transport equations and a Poisson equation for w. For the present study the perturbations are assumed to be small, so the governing equations have been linearized. Furthermore, for simplicity, we have confined attention to quasi-two-dimensional boundary layers without a streamwise pressure gradient. To simplify further we have assumed a constant boundary-layer thickness. In other words, we have made the well-known parallel-flow approximation. This implies that the undisturbed velocity and vorticity profiles correspond to the Blasius profile with wall-normal velocity set to zero. The effects of freestream turbulence are modelled by adding vorticity sources to the right-hand sides of the governing equations for ω_x and ω_y. The vorticity sources take the form of derivatives of a fictitious body force $\mathbf{F} = (F_x, F_y, 0)$

$$-\frac{\partial F_y}{\partial z} = G_x \mathrm{e}^{\mathrm{i}\beta y}\delta(x - x_f)\delta(z - z_f)H(t)\mathrm{e}^{\mathrm{i}ft} \qquad (1)$$

$$\frac{\partial F_x}{\partial z} = G_y \mathrm{e}^{\mathrm{i}\beta y}\delta(x - x_f)\delta(z - z_f)H(t)\mathrm{e}^{\mathrm{i}ft} \qquad (2)$$

where z is the wall-normal coordinate, t is time, and H(t) is the Heaviside step function. Thus, the vorticity sources have amplitudes G_x and G_y and vary sinusoidally in the spanwise direction with wavenumber β and vary, in general, harmonically in time with frequency f. The sources are located at (x_f, z_f) where z_f is located at the boundary-layer edge. All variables are non-dimensionalized with respect to the boundary-layer displacement thickness δ^*, the kinematic viscosity ν and the freestream velocity U_∞. The Reynolds number is defined as $R = U_\infty \delta^*/\nu$.

When the frequency f is non-zero then both types of vorticity source generate waves in the free stream with a phase speed equal to U_∞. These waves have been seen in some other experimental and computational studies and are mentioned by Saric et al. [7]. By a receptivity mechanism, as yet not fully explained, in both cases these generate three-dimensional Tollmien–Schlichting waves in the boundary layer. As the Tollmien–Schlichting waves grow exponentially with propagation downstream the freestream waves decay. The receptivity is much greater for the spanwise vorticity source. As expected from hydrodynamic stability theory, the two-dimensional ($\beta = 0$) Tollmien–Schlichting waves grow more rapidly than three-dimensional ones. It is often argued [7] that there is a mismatch in wavelength between the freestream disturbances and the Tollmien–Schlichting waves. It is thereby inferred that a local geometric feature with the appropriate dimension, be it strong boundary-layer growth near the leading edge or some sort of large-scale roughness element, is required for the generation of Tollmien–Schlichting waves. This is certainly a sound argument for acoustic disturbances, but for the freestream vortical disturbances the ratio of the wavelength to that of the Tollmien–Schlichting waves is at most three. And it turns out that all that is needed to generate Tollmien–Schlichting waves is some sort of spatial inhomogeneity – in this case the sudden onset of the freestream waves at $x = x_s$. The results discussed above were presented by Kudar et al. [9].

When the frequency of the vorticity source is zero or very low, the boundary layer is not receptive to spanwise vorticity. The streamwise vorticity source of (1) with $f = 0$, however, generates a sheet of vorticity that extends downstream and varies sinusoidally in magnitude in the spanwise direction [9, 10] (see Fig. 1). This can be considered as modelling the very low-frequency components of streamwise vorticity seen in the free stream. These freestream vortices of alternating sign generate elongated streak-like structures within the boundary layer. Typical simulation results taken from Kudar [10] are depicted in Fig. 2. These streak-like flow structures are often termed Klebanoff modes after the researcher who first observed them experimentally [11, 12]. Low-frequency sources corresponding to (1) also produce streak-like structures

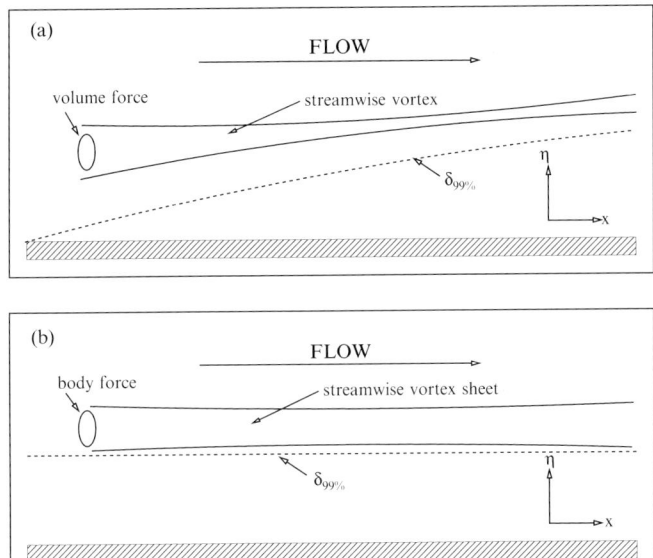

Fig. 1. Schematic sketches comparing the location of the vorticity source relative to (**a**) the real growing boundary layer and (**b**) our simplified constant-thickness boundary layer

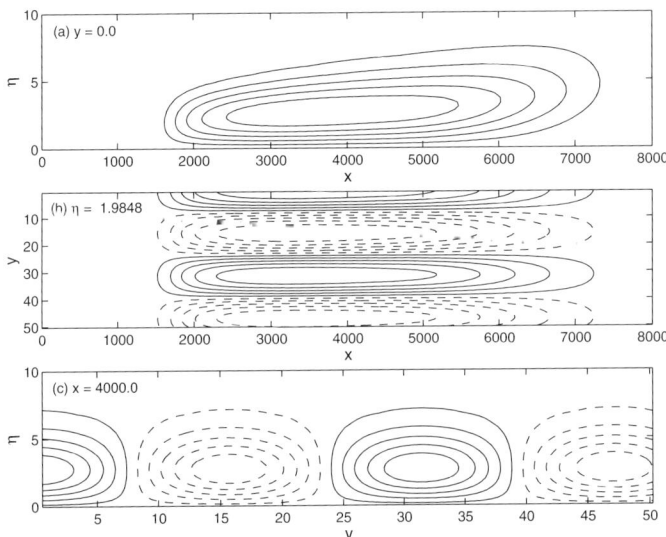

Fig. 2. Contour plots of streamwise velocity perturbation showing the Klebanoff mode, as would be seen in a Blasius boundary layer, at $t = 6,400$ viewed (**a**) in the spanwise direction, (**b**) from above, and (**c**) as a cross-sectional view in the streamwise direction. In views (**a**) and (**b**) the flow is from left to right, in (**c**) the flow is going into the page. *Dashed lines* indicate negative contours. η is the Blasius similarity variable, $\beta = 0.34$, $f = 0$, $R = 1,720$ (taken from Kudar [10])

but the receptivity drops sharply as frequency increases from zero [9, 12]. It should be mentioned that our use of vorticity source of the form (1) is similar to Fasel [12] except that he used a body force, rather than vorticity source, of form (1), thereby generating a more complex dipole-like vortex sheet in the free stream.

The results presented in this paper concerning the streak-like flow structures correspond to a Blasius boundary layer. The source strength, G_x, in (1) is kept fixed, and the frequency is zero throughout. The vorticity source of (2) is absent. Figure 3 shows how the maximum velocity perturbation varies with time for cases where the forcing is only applied for a certain period of time and for the case of constant forcing. The resulting flow structures at a particular instant that were generated by constant forcing are illustrated in Fig. 2.

It can be seen from Fig. 3 that for the case of constant forcing the amplitude of the velocity streak grows with time until it asymptotes to a constant value. This steady-state value varies with the dimensionless spanwise wavenumber β of (1), as illustrated in Fig. 4. It can be seen that the optimum value of dimensionless wavenumber corresponds closely with the experimental value measured by Klebanoff [11]. This is as expected because one would expect to see only the strongest, most receptive, streaks in an experiment. Furthermore, there is also close agreement between our simulation results and the previous theoretical results of Bertolotti [13].

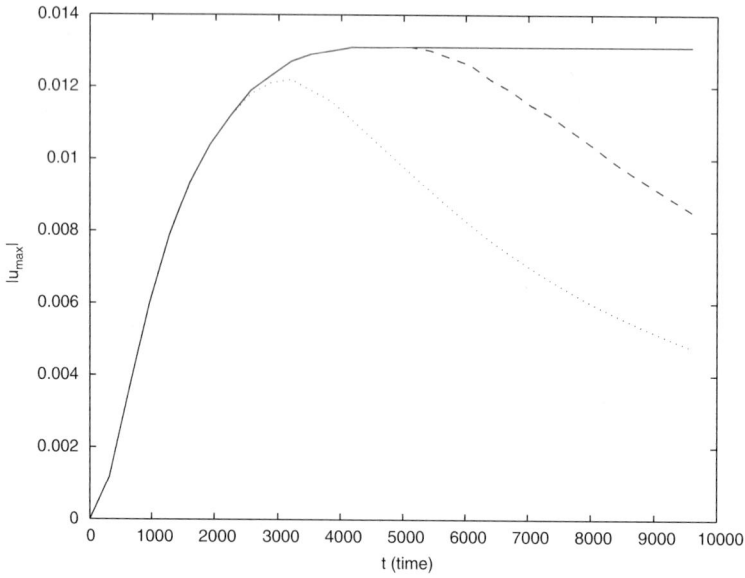

Fig. 3. Variation of the maximum velocity amplitude of the streak ($|u_{\mathrm{max}}|$) with time for a continuously forced streak (*solid line*), and for ones for which the forcing is discontinued after $t = 3,200$ (*dashed line*) and $t = 1,600$ (*dotted line*), $\beta = 0.34$, $f = 0$, $Re = 1,720$ (taken from Kudar [10])

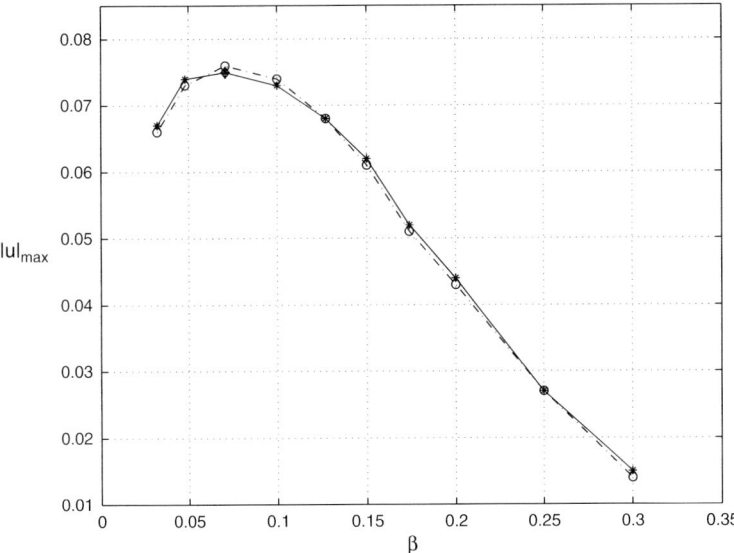

Fig. 4. Maximum streak amplitude as a function of the dimensionless spanwise wavenumber of the forcing for $R = 1,000$. Our results (*circle*) are compared with Bertolotti's theory [13] (*asterisk*) and Klebanoff's experiment [11] (*diamond*) (taken from Ali [14])

Figure 2 gives an idea of the form of the Klebanoff mode after it has reached its 'steady state'. It can be seen that it takes the form of alternating high- and low-speed velocity streaks in the spanwise direction. The maximum steady-state streak velocity amplitude increases algebraically with Reynolds number. This is evident to some extent from Fig. 5a. In the real spatially developed flow this corresponds to algebraic growth with x, as found by Fasel [12]. Even though, we have assumed a constant boundary-layer thickness, it is still possible to predict this algebraic growth. All the dimensions in our simulations scale with displacement thickness. In our simulations we find that the maximum streak velocity amplitude varies as Re^μ where μ is slightly greater than 1. Bearing in mind that $\delta^* \propto \sqrt{x}$, the variation found in our simulations compares reasonably well with the experimental variation of $\sqrt{x}$ found by Boiko et al. [15].

For the low levels of freestream turbulence typically seen in flight conditions at cruise, the streak-like flow structures are very weak. Under these conditions, owing to their exponential growth, Tollmien–Schlichting waves are the dominant route to transition. As the level of freestream turbulence rise, the streak-like flow structures become stronger and become more evident in experimental studies. They now play an important role in transition. Initially, this is through nonlinear interaction with the Tollmien–Schlichting waves [12]. However, at even higher levels of turbulence they 'bypass' the Tollmien–Schlichting waves completely and act as an independent route to transition.

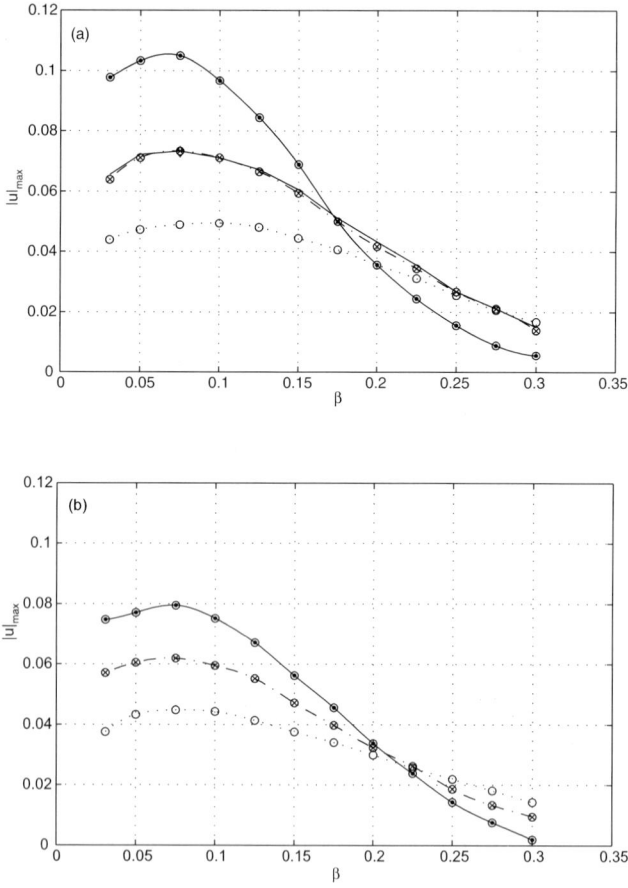

Fig. 5. Variation of maximum streamwise velocity perturbation with dimensionless spanwise wavenumber β of the forcing over (**a**) rigid surface and (**b**) Kramer-type compliant wall at three values of Reynolds number; (*solid line*), $R = 1,414$; (*dashed-dotted line*), $R = 1,000$; (*dotted line*), $R = 707$. In (**a**) for the rigid surface, data points correspond to our simulations, continuous curves to Bertolotti's theory [13] (taken from Ali [14])

3 Effects of Wall Compliance on Tollmien–Schlichting Waves

The stabilizing effects of compliant walls on Tollmien–Schlichting waves – the precursors to laminar–turbulent transition in quasi-two-dimensional boundary layers – were demonstrated theoretically by Benjamin [16] shortly after Kramer's papers appeared. Benjamin also identified the presence of other instabilities of a hydroelastic nature. Kramer's concept of how his compliant coatings damped Tollmien–Schlichting waves was inconsistent with

Benjamin's theory. And this, combined with the failure of other experimentalists to corroborate his high values of drag reduction, led to a certain amount of disillusionment with Kramer's ideas. However, Carpenter and Garrad [17] later reviewed the experimental efforts aimed at replicating Kramer's results, and found them to be faulty. Furthermore, they developed a theoretical model of Kramer's compliant coatings and used it to show that they were capable of postponing laminar–turbulent transition. Shortly afterwards Gaster [18] (see also [19]) carried out a careful experimental study in which the growth of instabilities over a compliant wall were measured and compared with theory. The agreement was very good. In the case of two of his three compliant walls transition was actually caused by the amplification of one of the flow-induced wall instabilities – travelling-wave flutter (see [20]). Even in this case the experimental data for growth rate agreed well with theory (see [19]). This good agreement between theory and experiment, and subsequent studies based on theory and numerical simulation, established pretty much beyond doubt that very effective laminar flow control could be achieved using compliant walls.

4 Effects of Wall Compliance on Velocity Streaks

We shall now turn to the effects of wall compliance on the velocity streaks or Klebanoff modes. To model the dynamics of the compliant walls we use

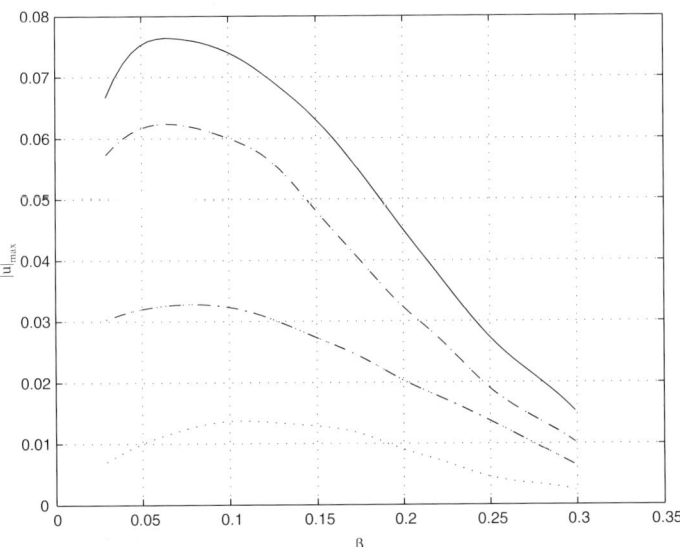

Fig. 6. Variation of maximum streamwise velocity perturbation with the dimensionless spanwise wavenumber β of the forcing over surfaces of various levels of compliance. $R = 1,000$. (*solid line*), rigid; (*dashed line*), Kramer-type wall; (*dashed-dotted line*), 0.5× stiffness of Kramer-type wall; (*dotted line*), 0.333× stiffness of Kramer-type wall (taken from Ali [14])

the plate-spring model introduced by Carpenter and Garrad [17] to model Kramer's [1, 2] compliant coatings. Figure 5 compares the variation of maximum streak velocity amplitude with spanwise wavenumber for a rigid wall and a compliant wall with dimensionless wall parameters identical to the case presented in [17] for a Kramer compliant coating having an elastic modulus of 500 kPa. In practice, at a given Reynolds number only streaks with the optimum value of β, i.e. the value corresponding to the maximum streak velocity amplitude, are likely to be seen experimentally or in real applications. It can be seen that, measured by this maximum amplitude, the receptivity of the compliant wall to Klebanoff modes is much reduced at all Reynolds numbers. This is even more apparent in Fig. 6, where the maximum streak velocity amplitude is plotted against spanwise wavenumber for walls of varying degrees of compliance.

5 Conclusions

We have investigated the effects of freestream turbulence on laminar–turbulent transition. We have carried out a numerical-simulation study of the effects of wall compliance on the velocity streaks (or Klebanoff modes) generated in boundary layers by freestream turbulence. The results clearly demonstrate that compliant walls are much less receptive to Klebanoff modes than rigid walls and therefore should remain capable of postponing or suppressing laminar–turbulent transition even in environments with relatively high levels of freestream turbulence.

References

1. Kramer, M.O.: Boundary-layer stabilization by distributed damping. J. Aeronautical Sciences, **24**, 459 (1957)
2. Kramer, M.O.: Boundary-layer stabilization by distributed damping. J. American Society of Naval Engineers, **72**, 25-33 (1960)
3. Gray, J.: Studies in animal locomotion VI: The propulsive power of the dolphin. J. Experimental Biology, **13**, 192-199 (1936)
4. Babenko, V.V., Carpenter, P.W.: Dolphin hydrodynamics. In: P.W. Carpenter, T.J. Pedley (eds.) Flow past highly compliant boundaries and in collapsible tubes. Kluwer Academic Publishers, Dordrecht (2003)
5. Carpenter, P.W., Lucey, A.D., Davies, C.: Progress on the use of compliant walls for laminar flow control. J. Aircraft, **38**, 504-512 (2001)
6. Carpenter, P.W., Porter, L.J.: Effects of passive porous walls on boundary-layer instability. AIAA J., **39** (4), 597-604 (2001)
7. Saric, W.S., Reed, H.L., Kerschen, E.J.: Boundary-layer receptivity to freestream disturbances. Annual Review of Fluid Mech., **34**, 291-319 (2002)
8. Davies, C., Carpenter, P.W.: A novel velocity-vorticity formulation of the Navier–Stokes equations with applications to boundary-layer disturbance evolution. J. Comp. Phys., **172**, 119-165 (2001)

9. Kudar, K.L, Carpenter, P.W., Davies, C.: Generation of boundary-layer disturbances by freestream forcing. Bull. Amer. Phys. Soc., **50** (9), 180 (2005)
10. Kudar, K.L.: Flow control using pulsed jets. PhD thesis, Univ. of Warwick,
U.K., (2004)
11. Klebanoff, P.S.: Effects of freestream turbulence on a laminar boundary layer.
Bull. Amer. Phys. Soc., **10**(11), 1323 (1971)
12. Fasel, H.F.: Numerical investigation of the interaction of the Klebanoff mode
with a Tollmien-Schlichting wave. J.Fluid. Mech., **450**, 1-33 (2002)
13. Bertolotti, F.P.: Response of the Blasius boundary layer to free-stream vorticity.
Phys. Fluids, **9**, 2286-2299 (1997).
14. Ali, R.: Receptivity and transition in boundary layers over rigid and compliant
surfaces. PhD thesis, Univ. of Warwick, U.K., (2003)
15. Boiko, A.V., Westin, K.J.A., Klingmann, B.G.B., Kozlov, V.V., Alfredsson,
P.H.: Experiments in a boundary layer subjected to free-stream turbulence.
J. Fluid Mech., **281**, 219-245 (1994)
16. Benjamin, T.B.: Effects of a flexible boundary on hydrodynamic stability. J.
Fluid Mech., **9**, 513-532 (1960)
17. Carpenter, P.W., Garrad, A.D.: The hydrodynamic stability of flow over
Kramer-type compliant surfaces. Pt. 1. Tollmien-Schlichting instabilities. J.
Fluid Mech, **155**, 465-510, (1985)
18. Gaster, M.: Is the dolphin a red herring? In: Liepmann, H.W., Narasimha, R
(eds) IUTAM Symp. on Turbulence management and Relaminarisation.
Springer, Berlin, Heidelberg, New York (1987)
19. Lucey, A.D., Carpenter, P.W.: Boundary layer instability over compliant walls:
Comparison between theory and experiment. Phys. Fluids, **7**, 2355-2363 (1995)
20. Carpenter, P.W., Garrad, A.D.: The hydrodynamic stability of flow over
Kramer-type compliant surfaces. Pt. 2. Flow-induced surface instabilities. J.
Fluid Mech., **170**, 199-232 (1986)

Minisymposium "Global Flow Instability"

V. Theofilis

School of Aeronautics, Universidad Politécnica de Madrid, Spain
vassilis@aero.upm.es

Summary

Linear stability theory is concerned with the evolution of small-amplitude disturbances superimposed upon a steady- or time-periodic so-called basic flow. The vast majority of investigations during the second half of the last century has dealt with the analysis of one-dimensional ("parallel") basic flows. On the other hand, *Global flow instability* deals with essentially non-parallel (as well as with weakly non-parallel) flows [1] and is an emerging and highly active area of research, to which a Minisymposium has been dedicated. Four invited contributions from three countries were presented, one summarizing experimental work and the rest presenting alternative numerical methodologies to solve the large eigenvalue problem resulting in the context of BiGlobal instability analysis. Applications addressed ranged from laminar and turbulent separation control (Avi Seifert, Tel-Aviv University), vortex instabilities (Michael Broadhurst, Imperial College London), and cavity flow hydrodynamic (Leo González, School of Naval Engineering, UP Madrid) and aeroacoustic (Javier de Vicente, School of Aeronautics, UP Madrid) instabilities. With the exception of the first author, whose contribution is outlined below, papers were submitted describing in detail the contents of the talks delivered.

Avi Seifert explored the *Relationship of global flow instability and flow control,* based on experimental results in a wide variety of external aerodynamics configurations. He discussed possible relationships between global instability modes and control of separated regions. It was repeatedly found that the effective frequencies for control of separated flow on numerous configuration results in a Strouhal number of order unity. This Strouhal number is based on the length of the baseline separated flow region. Regardless of the turbulence level upstream of separation, the curvature and the history of the boundary layer, the effective frequency for reattaching a separated flow remains of order unity. It was hypothesized that a feed-back loop exists between the reattaching flow, sending upstream an acoustic wave that when coincides with the frequency of the actuator causes enhanced effectiveness and receptivity of the excitation introduced by the actuator. It is hoped that global flow instability analysis of a properly measured or computed baseline flow will enable to reproduce, explain and eventually predict this type of resonance.

Michael Broadhurst, in collaboration with Spencer J. Sherwin, used BiGlobal linear theory and Direct Numerical Simulation to discuss *Helical instability and breakdown of a Batchelor trailing vortex*. The new perspective offered concerned the relaxation of the restrictive assumption of azimuthal homogeneity, invariably used in earlier analyses of the same phenomenon. Their main conclusions were that (a) helical instability is responsible for the onset of spiral-type vortex breakdown and (b) pressure gradients were shown to exert a strong influence on the evolution of vortex breakdown. In the latter respect, they presented an extension of the Parabolized Stability Equations technique, which is capable of addressing the issue of pressure-gradients in the axial flow direction.

Leo González discussed *A finite-element alternative for BiGlobal instability analysis*, as an alternative to well-established spectral methods for the spatial discretization of the BiGlobal eigenvalue problem. Motivation for this approach is provided by the desire to address instability in flows over or within complex geometries. Low-order elements have been used (by contrast to the high-order spectral-element used by the previous authors) and several validation cases in closed and open flows have established the ability of the methodology presented to address the problem at hand. The flexibility of the method was exploited by analyzing, for the first time, a lid-driven cavity of triangular shape. From a numerical point of view, the key conclusion has been the need for a high-order extension of the finite-element method, that is presently unavailable for this class of stability problems.

Finally, Javier de Vicente, in collaboration with E. Valero and V. Theofilis, presented *Numerical considerations in spectral multi-domain methods for BiGlobal instability analysis of open cavity configurations*. They mainly focused on results of their parallelization efforts associated with the solution of the sparse-matrix based BiGlobal eigenvalue problem. Both incompressible and compressible flows can be addressed by the algorithms developed, respectively corresponding to hydrodynamic and aeroacoustic instabilities. Target application in this work has been the open cavity configuration, in the presence of model stores placed inside the cavity. The authors presented some distributed-memory solutions to the eigenvalue problem; on the basis of the associated convergence rates and CPU timings they concluded that shared-memory solutions were an alternative worthy of exploration.

Discussion of key issues presented during the Minisymposium followed, with a good degree of interaction between the speakers and the audience. Flow control was singled out as a promising direction [2], in which the tools presented in the Minisymposium may be applied. In this context, and at the invitation of the organizer, the Minisymposium was concluded by a short exposition by P. Luchini (a world-leading expert in the subject of global flow instability) of the connection between adjoint-based flow-control methodologies and global flow instability. Readers interested in further information on recent developments on the topic of global flow instability, may also visit: http://www.aero.upm.es/es/departamentos/crete05/Home.html

References

1. Theofilis, V., *Advances in global linear instability analysis of nonparallel and three-dimensional flows*, Prog. Aero. Sci., **39**, (2003), pp. 249–315.
2. S. S. Collis, R. D. Joslin, A. S. and Theofilis, V., *Issues in active flow control: theory, control, simulation and experiment*, Prog. Aero. Sciences **40**, (2004), pp. 237–289.

Helical Instability and Breakdown of a Batchelor Trailing Vortex

Michael S. Broadhurst and Spencer J. Sherwin

Department of Aeronautics, Imperial College London, UK
michael.broadhurst@imperial.ac.uk, s.sherwin@imperial.ac.uk

1 Introduction

A particular feature of swirling flows with a strong core vorticity is the phenomenon of vortex breakdown. For vortices with an appreciable axial velocity component, [Hall (1972)] defines vortex breakdown as, "an abrupt change in the "vortex" structure with a very pronounced retardation of the flow along the axis". One factor that is known to influence vortex breakdown, reviewed by [Leibovich (1984)], is the role of instability. This was also recognised by [Ash and Khorrami (1995)], who describe a possible mechanism of breakdown as, 'a final outcome of vortex instability, with the caveat that vortex breakdown can also be produced by external means'. External influences might include pressure gradients. Consequently, the aim of the current research is to demonstrate the relationship between instability and spiral-type breakdown of a Batchelor vortex, and to assess the influence of pressure gradients on vortex stability using the parabolised stability equations.

2 Numerical Methods

A combination of linear stability analysis and direct numerical simulation has been used to investigate the role of instability in the incipience of vortex breakdown. The stability analysis of flows governed by the Navier–Stokes equations is based upon the decomposition of all flow variables into a steady basic state solution of the equations upon which small amplitude disturbances are permitted to develop (i.e. $\mathbf{q} = \bar{\mathbf{q}} + \mathbf{q}'$). By allowing a mild dependence of the base flow on the streamwise spatial coordinate z, an eigenmode Ansatz is introduced, according to which

$$\mathbf{q}'(x, y, z, t) = \hat{\mathbf{q}}(x, y, z) \exp \mathrm{i}\Theta + \text{c.c.} \tag{1}$$

$$\Theta = \Theta_{3\mathrm{D}} = \int_{z_0}^{z} \beta(\xi)\mathrm{d}\xi - \Omega t. \tag{2}$$

Applied to the linearised Navier–Stokes equations, this leads to the three-dimensional parabolised stability equations (3D-PSE):

$$\hat{u}_x + \hat{v}_y + \mathrm{i}\beta\hat{w} = -\hat{w}_z, \tag{3}$$

$$\{\mathcal{L} - \bar{u}_x\}\,\hat{u} - \bar{u}_y\hat{v} - \hat{p}_x + \mathrm{i}\Omega\hat{u} = \bar{w}\hat{u}_z + \bar{u}_z\hat{w} - \frac{2\mathrm{i}\beta}{\mathrm{Re}}\hat{u}_z - \mathrm{i}\frac{\mathrm{d}\beta}{\mathrm{d}z}\hat{u}, \tag{4}$$

$$-\bar{v}_x\hat{u} + \{\mathcal{L} - \bar{v}_y\}\,\hat{v} - \hat{p}_y + \mathrm{i}\Omega\hat{v} = \bar{w}\hat{v}_z + \bar{v}_z\hat{w} - \frac{2\mathrm{i}\beta}{\mathrm{Re}}\hat{v}_z - \mathrm{i}\frac{\mathrm{d}\beta}{\mathrm{d}z}\hat{v}, \tag{5}$$

$$-\bar{w}_x\hat{u} - \bar{w}_y\hat{v} + \mathcal{L}\hat{w} - \mathrm{i}\beta\hat{p} + \mathrm{i}\Omega\hat{w} = \bar{w}\hat{w}_z + \bar{w}_z\hat{w} - \frac{2\mathrm{i}\beta}{\mathrm{Re}}\hat{w}_z - \mathrm{i}\frac{\mathrm{d}\beta}{\mathrm{d}z}\hat{w} + \hat{p}_z. \tag{6}$$

where $\mathcal{L} = \frac{1}{\mathrm{Re}}\left\{\partial_{xx} + \partial_{yy} - \beta^2\right\} - \bar{u}\partial_x - \bar{v}\partial_y - \mathrm{i}\beta\bar{w}$, and $\beta(\xi) = \bar{\beta}(\xi) + \mathrm{i}\sigma(\xi)$ is a complex wavenumber. Implicit in this derivation is that the disturbance takes the form of a rapidly varying phase function and a slowly varying shape function, for which second derivatives along with products of first derivatives (with respect to z) can be neglected. For homogeneous flows in z, (3)–(6) form a matrix system corresponding to the linearised incompressible Navier–Stokes equations known as a BiGlobal stability analysis. This can be solved using a suitable method, such as an exponential power method (see, for example, [Barkley and Tuckerman (2000)]). For inhomogeneous flows, such as a vortex developing in an external pressure gradient, (3)–(6) form a parabolic system, that can be solved using a suitable marching procedure, analogous to the one proposed by [Bertolotti et al. (1992) Bertolotti, Herbert, and Spalart] for the two-dimensional parabolised stability equations.

3 Vortex Stability and Breakdown

Evaluated using a BiGlobal stability analysis, the most unstable perturbation mode of a Batchelor vortex ([Batchelor (1964)]), with a swirl value of $q = 0.8$, and a Reynolds number based on the vortex core radius of $Re = 1,000$ is illustrated in Fig. 1, for jet-like and wake-like axial velocity profiles, respectively. Direct numerical simulation (DNS) using $\mathcal{N}\epsilon\kappa\tau\alpha r$[1] with a Fourier series approximation in the axial direction has been used to investigate the non-linear development of the various modes of instability. Helical modes of instability were found to cause a lateral expansion of the cross-section of the vortex core, and a corresponding drop in axial velocity. Enforcing an axial periodicity in the solution restricts how the streamwise component of the velocity can change, limiting the extent of axial deceleration. To resolve this problem, 3D-DNS on the same Batchelor vortex has been conducted (see Fig. 2): as the helical instability evolves, an abrupt axial deceleration develops – indicative

---

[1] A spectral/hp-element solver developed by [Karniadakis and Sherwin (2005)].

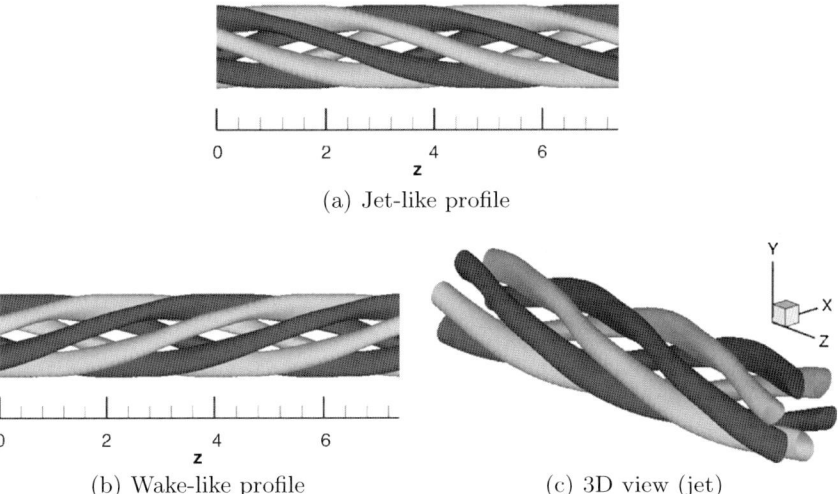

(a) Jet-like profile

(b) Wake-like profile   (c) 3D view (jet)

Fig. 1. Most unstable perturbation mode of a Batchelor trailing vortex; with $Re = 1,000$, $q = 0.8$, $\beta = 1.7$. Visualised using iso-surfaces of axial vorticity magnitude. The *dark* and *light grey* surfaces correspond to values of $+0.2$ and -0.2, respectively

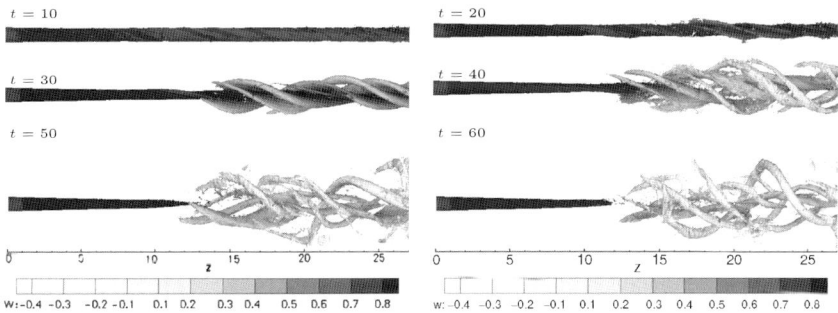

Fig. 2. Evolution of vortex breakdown of an isolated Batchelor vortex; with $q = 0.8$ and $Re = 1,000$. Visualised using iso-surfaces of $\lambda_2 = -0.4$, shaded by the axial velocity component

of vortex breakdown. This suggests a causal relationship between helical instability and spiral-type breakdown. The mechanism by which this occurs is discussed by [Broadhurst et al. (2006) Broadhurst, Theofilis, and Sherwin].

4 Influence of Pressure Gradients

A novel approach to analysing vortex stability, called the parabolised stability equations in three-dimensions (3D-PSE), has recently been validated, and – in conjunction with 3D-DNS – is currently being used to investigate the influence of pressure gradients on vortex breakdown. The PSE concept, discussed by [Herbert (1997)] for boundary layers, has been extended to analyse three-dimensional vortical flows. Preliminary results, in agreement with 3D-DNS, have indicated that an adverse pressure gradient is destabilising, whereas a favourable pressure gradient is stabilising. A pressure gradient is applied by considering the potential flow around a circular cylinder. As the flow approaches a circular cylinder, there is an associated adverse pressure gradient caused by the stagnation point at the leading edge. Alternatively, a favourable pressure gradient is obtained downstream of the trailing edge of the cylinder. These pressure gradients can be realised by superimposing the Batchelor vortex profile onto the relevant section of the potential flow around a circular cylinder, with associated far field boundary conditions. The results in an adverse pressure gradient are illustrated in Fig. 3, which demonstrates how a pressure gradient modifies the growth rate of the most unstable helical modes. For reference, the bold lines illustrate the growth rate of a vortex in the absence of a pressure gradient, equivalent to a BiGlobal stability analysis.

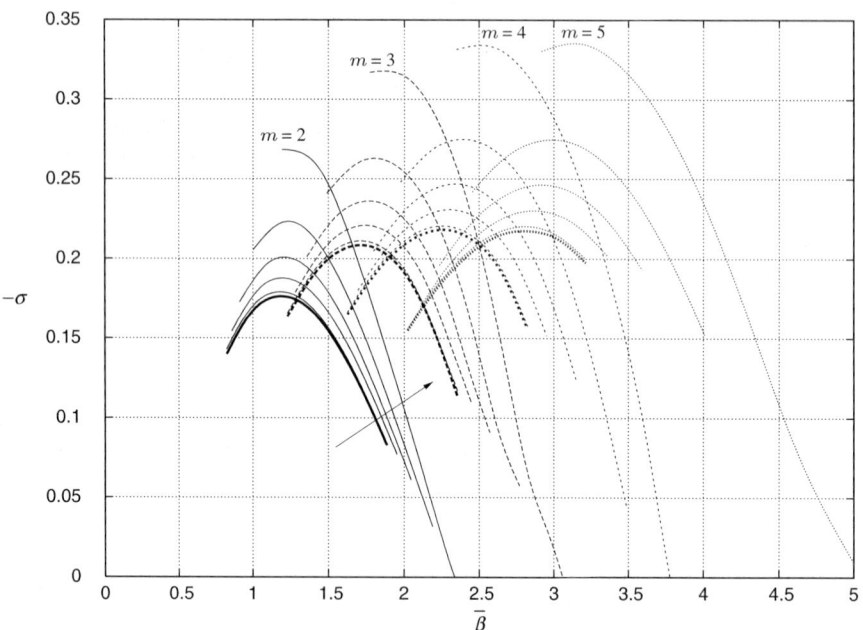

Fig. 3. Influence of an adverse pressure gradient on the stability of an isolated Batchelor vortex. The *bold lines* illustrate the growth rate without a pressure gradient, and the *arrow* indicates the direction of increasing downstream distance: results are illustrated for $z = 0, 10, 20, 30,$ and 40. m is the azimuthal wavenumber

5 Conclusions

The relationship between vortex instability and breakdown has been introduced, and it is suggested that helical modes of instability are responsible for the onset of spiral-type vortex breakdown. External pressure gradients are known to significantly influence the onset and evolution of vortex breakdown (see, for example, [Sarpkaya (1974)]), with adverse pressure gradients promoting vortex breakdown. A suitable technique to analyse the influence of pressure gradients on vortex stability is a parabolised stability analysis, which permits flows with a mild variation in the axial direction. The results agree with experimental and numerical observations, and suggest that adverse pressure gradients are destabilising, whereas favourable pressure gradients suppress vortex breakdown.

References

[Ash and Khorrami (1995)] R. L. Ash and M. R. Khorrami. Vortex stability. In S. I. Green, editor, *Fluid Vortices*, volume 30 of *Fluid Mechanics and its Applications*, chapter 8, pages 317–372. Kluwer Academic Publishers, 1995. ISBN 0-7923-3376-4.

[Barkley and Tuckerman (2000)] D. Barkley and L. S. Tuckerman. Bifurcation analysis for timesteppers. In E. Doedel and L. S. Tuckerman, editors, *Numerical Methods for Bifurcation Problems and Large-Scale Dynamical System*, volume 119 of *IMA Volumes in Mathematics and its Application*, pages 466–543. Springer, New York, 2000.

[Batchelor (1964)] G. K. Batchelor. Axial flow in trailing line vortices. *Journal of Fluid Mechanics*, 20(4):645–658, 1964.

[Bertolotti et al. (1992) Bertolotti, Herbert, and Spalart] F. P. Bertolotti, T. Herbert, and P. R. Spalart. Linear and nonlinear stability of the Blasius boundary layer. *Journal of Fluid Mechanics*, 242:441–474, 1992.

[Broadhurst et al. (2006) Broadhurst, Theofilis, and Sherwin] M. S. Broadhurst, V. Theofilis, and S. J. Sherwin. Spectral element stability analysis of vortical flows. In Rama Govindarajan, editor, *Sixth IUTAM Symposium on Laminar-Turbulent Transition*, Fluid Mechanics and its Applications. Springer, 2006.

[Hall (1972)] M. G. Hall. Vortex breakdown. *Annual Review of Fluid Mechanics*, 4:195–218, 1972.

[Herbert (1997)] T. Herbert. Parabolized stability equations. *Annual Review of Fluid Mechanics*, 29:245–283, 1997.

[Karniadakis and Sherwin (2005)] G. Em. Karniadakis and S. Sherwin. *Spectral/hp Element Methods for Computational Fluid Dynamics*. Numerical Mathematics and Scientific Computation. Oxford University Press, 2nd edition, 2005. ISBN 0-19-852869-8.

[Leibovich (1984)] S. Leibovich. Vortex stability and breakdown: Survey and extension. *AIAA Journal*, 22:1192–1206, September 1984.

[Sarpkaya (1974)] T. Sarpkaya. Effect of the adverse pressure gradient on vortex breakdown. *AIAA Journal*, 12(5):602–607, May 1974.

A Finite-Element Alternative for BiGlobal Linear Instability Analysis

Leo M. González

School of Naval Arquitecture, Universidad Politécnica de Madrid, Avd. Arco de la Victoria s/n, 28040 Madrid, Spain
leo.gonzalez@upm.es

Summary. Viscous linear three-dimensional BiGlobal instability analyses of incompressible flows have been performed using finite-element numerical methods, with a view to extend the scope of application of this analysis methodology to flows over complex geometries.

1 Mathematical Formulation

The two-dimensional equations of motion are solved in the laminar regime at appropriate Re regions, in order to compute steady basic flows $(\bar{u}_i, \bar{p})$ whose stability will subsequently be investigated.

1.1 Eigenvalue Problem (EVP) Formulation and Solution Methodology

The basic flow is perturbed by small-amplitude velocity $\tilde{u}_i$ and kinematic pressure $\tilde{p}$ perturbations, as follows

$$u_i = \bar{u}_i + \varepsilon \hat{u}_i(x,y)\, \mathrm{e}^{\mathrm{i}\beta z} \mathrm{e}^{\omega t} + \text{c.c.} \tag{1}$$

$$p = \bar{p} + \varepsilon \hat{p}(x,y)\, \mathrm{e}^{\mathrm{i}\beta z} \mathrm{e}^{\omega t} + \text{c.c.}, \tag{2}$$

where $\varepsilon \ll 1$, c.c. denotes conjugate of the complex quantities $(\tilde{u}_i, \tilde{p})$, β is a real wavenumber parameter, while ω is the complex eigenvalue sought. Introducing the ansatz into the linearized Navier–Stokes equations, the system is transformed into a (complex) generalized eigenvalue problem for the determination of ω,

$$A \begin{pmatrix} \hat{u}_1 \\ \hat{u}_2 \\ \hat{u}_3 \\ \hat{p} \end{pmatrix} = -\omega B \begin{pmatrix} \hat{u}_1 \\ \hat{u}_2 \\ \hat{u}_3 \\ \hat{p} \end{pmatrix}. \tag{3}$$

The complex generalized eigenvalue problem (3) has either real or complex solutions, corresponding to stationary ($\omega_i = 0$) or traveling ($\omega_i \neq 0$) modes.

2 Results

2.1 The Instability Analyses

The Rectangular Duct Flow

First, a square duct at low Reynolds number value $Re = 100$ is considered, which is known to permit a relatively coarse resolution [Th04], such that numerical experimentation is straightforward. The complex EVP (3) is solved on $O(10^4)$ nodes and varying the Krylov subspace dimension, m; the results are presented in Table 1. Convergence of the leading eigenvalue is achieved at a moderate Krylov subspace dimension, $m = 20$, using a well-acceptable 600 Mb of in-core memory. The relative error of the eigenvalue obtained compared with the spectral collocation result [Th04] of the same complex EVP is of $O(10^{-6})$.

Also worth noting is that, on account of the increase of the (serial) computational time as the Reynolds number increases, it becomes increasingly inefficient to attempt a solution of the complex BiGlobal EVP at Reynolds numbers beyond $Re = O(10^3)$. This is to be expected, given the low formal order of accuracy of the method.

On the other hand, computational efficiency considerations aside, once sufficient resolution is provided, the method is capable of providing results in very good agreement with the established spectral computations. The predictions of the leading eigenmode frequency at critical conditions as function of the duct aspect ratio is shown in Table 2, where the relative error in this quantity, compared with the spectral computations of [Th04], can be seen to vary between 4×10^{-3} at the lower two Reynolds numbers and 1.5×10^{-2} at the highest Re value. The eigenfunctions pertinent to the least-damped mode at $(Re, \beta) = (100, 1)$ at $AR = 3.5$ are presented in Fig. 1.

Rectangular and Triangular Regularized Lid-Driven Cavities

The instability problem in the regularized rectangular lid-driven cavity has been solved employing the EVP methodology based on numerical solution

Table 1. Grid-dependence of eigenvalue results in square duct flow

Nodes	Memory (Mb)	Time (min)	ω_r	ω_i
		$Re = 100$		
5 129	680	17	-0.140498	0.594178
11 605	702	49	-0.140503	0.594177
60 465	1 950	280	-0.140507	0.594177
		$Re = 1 000$		
5 129	648	13	-0.078650	0.868472
11 605	642	43	-0.072671	0.862796
60 465	2 037	442	-0.070679	0.865575

Parameters used are $\beta = 1$; Krylov subspace dimension, $m = 60$.

Table 2. Critical parameter (Re, β) values of the four most significant modes as a function of duct aspect ratio [Th04] $(m = 40)$

AR	Re	β	Nodes	ω_i
3.5	36 600	0.71	13 279	0.121660885
4	18 400	0.80	29 725	0.161186414
5	10 400	0.91	57 657	0.210532778

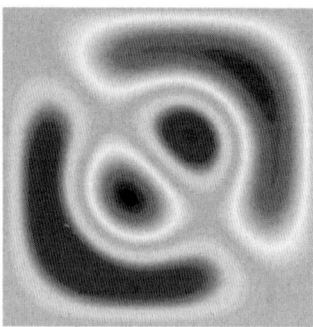

Fig. 1. Eigenfunction $\hat{w}$ pertaining to the least-damped mode of a square duct flow at $Re = 100, \beta = 1.0$

of (3). In addition, a spectral collocation algorithm [Th00, Th04] has been used for comparisons. Attention is first focused on the stable test case $(Re, \beta) = (200, 2)$. The grid used for the basic flow calculations, comprising $O(2 \times 10^4)$ (quadratic) velocity nodes, has been used for the instability analyses. Interestingly, at this Reynolds number the regularization condition results in a general stabilization of the global eigenmodes, especially at large β values, when compared with the standard lid-driven cavity (LDC) flow, in which the singular boundary condition $\bar{u}(x, y = 1) = 1$ is used. This result is in line with the analogous prediction of Theofilis [Th00], who analyzed a family of regularized profiles of the class discussed here.

A consequence of the difference in amplification/damping rates between the two cavity configurations is the increase of the linear critical Reynolds number pertinent to all known modes of the singular lid-driven cavity, S1, T1, T2, and T3 [Th00]. The effect of the aspect ratio on the instability of the regularized LDC has been examined. Four cases have been considered, $AR = 0.5, 1, 2$, and 4, in order to be able to draw qualitative conclusions on the effect of AR on the stability of the three-dimensional flow. The results for $A = 4$ are shown in graphical form in Fig. 2.

The triangular cavity flow has been substantially less investigated, and only from a basic flow point of view. However, it is clear that should a linear instability be present in the triangular cavity, the corresponding critical Reynolds number will define the upper limit beyond which two-dimensional numerical solutions of the basic flow problem will only be of academic value.

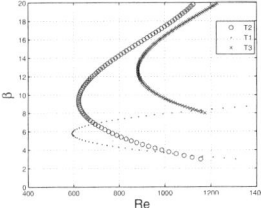

Fig. 2. Neutral curves of the first four eigenmodes in the regularized square lid-driven cavity at aspect ratio AR = 4. T: travelling modes

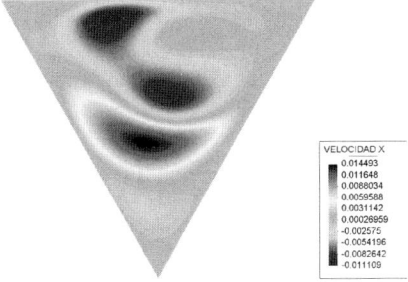

Fig. 3. Eigenfunction $\hat{u}$ pertaining to the least-damped mode of the triangular cavity flow at $Re = 1870, \beta = 6.73$

Resolutions comprising up to $O(2 \times 10^4)$ nodes were found to be adequate in order to provide reliable amplification rate information.

A zero-crossing has been found to occur at the (near-) critical parameters

$$(Re, \beta) \approx (1870, 6.73). \tag{4}$$

The linearly unstable mode discovered is stationary; the amplitude functions of its components have been found to comprise only real parts, shown in Fig. 3. In contrast to the rectangular cavity examined earlier, within the parameter range examined, no traveling (or other stationary) modes have been found in the triangular cavity.

A Batchelor Vortex

Also results were obtained in the well-studied Batchelor vortex instability problem [MP92]. The second-order FEM method discussed by González et al. [LG06] is employed on this analytically-constructed basic flow and the system describing linear perturbations from a BiGlobal point of view is solved in a large domain of $[-40, 40]^2$ (scaled with the Batchelor vortex radius), such that homogeneous Dirichlet boundary conditions may be imposed on all amplitude functions at the boundary of the domain. Two conclusions have been drawn, first that the (low-order) FEM is capable of delivering accurate results, in a

200   L.M. González

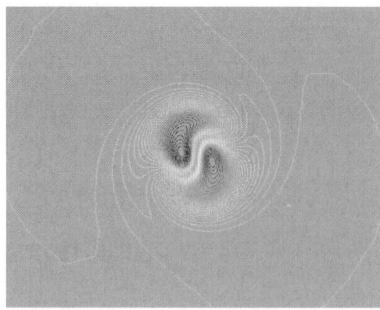

Fig. 4. Eigenfunction $\hat{u}$ pertaining to the least-damped mode of a single Batchelor vortex at $Re = 100$, $\beta = 0.418$

manner analogous with that demonstrated in the closed and open flows discussed in reference [LG06]. Second, the mesh required in order for such results to be obtained is prohibitively large; $O(8 \times 10^4)$ nodes were used to resolve the cases presented, resulting in typical memory and runtime requirements of $O(4Gb)$ and $O(10)$ CPU hours on an Intel P-IV at 3 GHz. The mesh used and the amplitude function corresponding to the axial disturbance velocity component, $\hat{w}$, are respectively shown in Fig. 4. In view of these results, two possible approaches have been followed, parallelization of the low-order accurate method, or implementation of a novel high-order hp-FEM algorithm for the solution of the BiGlobal eigenvalue problem (3).

References

[Th00]   V. Theofilis: Globally unstable basic flows in open cavities. In: 6th AIAA Aeroacoustics Conference and Exhibit. AIAA-2000-1965

[LG07]   L.M. González, J. de Vicente, V. Theofilis: High-order finite element methods for global viscous linear instability analysis of internal flows. In: 18th AIAA Computational Fluid Dynamics Conference. AIAA-2007

[Th04]   V. Theofilis, P.W. Duck, J. Owen: Viscous linear stability analysis of rectangular duct and cavity flows. J. Fluid. Mech., **505**, 249–286 (2004)

[LG06]   L.M. González, V. Theofilis, R. Gomez Blanco: Finite-element numerical methods for viscous incompressible BiGlobal linear instability analysis on unstructured meshes. AIAA Journal, accepted (2006)

[MP92]   E.W. Mayer and K.G. Powell: Viscous and inviscid instabilities of a trailing vortex. J. Fluid. Mech., **245**, 91–114 (1992)

Numerical Considerations in Spectral Multidomain Methods for BiGlobal Instability Analysis of Open Cavity Configurations

J. de Vicente, E. Valero, and V. Theofilis

School of Aeronautics, Universidad Politcnica de Madrid
javier@dmae.upm.es

Summary. A novel approach for the solution of the viscous incompresible and/or compressible BiGlobal eigenvalue problems (EVP) in complex open cavity domains is discussed. The algorithm is based on spectral multidomain spatial discretization, decomposing space into rectangular subdomains which are resolved by spectral collocation based on Chebyshev polynomials. The eigenvalue problem is solved by Krylov subspace iteration. Here particular emphasis is placed on aspects of the parallel developments that have been necessary, on account of the high computing demands placed on the solver, as ever more complex *"T-store"* configurations are addressed.

1 Theory

1.1 Spectral Collocation Approximation

A Chebyshev spectral expansion of the function $u(x)$ is considered on the Gauss–Lobatto nodes,

$$u_N(x) = \sum_{j=0}^{N} h_j(x) u(x_j) \tag{1}$$

being $h_j(x)$ the Lagrange interpolation functions. The unknowns become the values of $u(x)$ at the grid points. Differentiation is introduced by using the interpolation polynomial which permits expressing derivatives as

$$U = \mathcal{D}U, \quad U^{(p)} = \mathcal{D}^{(p)}U. \tag{2}$$

Using this technique, the solution of an eigenvalue problem is reduced to constructing the matrix $\mathcal{L}$ of the differential operator and manipulate it properly in order to impose boundary conditions. The extension of this one-dimensional idea to several Cartesian spatial dimensions is straightforward.

1.2 BiGlobal Theory

Linear stability analysis in the BiGlobal framework involves the substitution of a decomposition of any of the independent flow variables into the equations of motion [Th03]. All quantities are considered to be composed of an $O(1)$ steady two-dimensional basic state and $O(\epsilon)$ unsteady three-dimensional perturbations, according to the BiGlobal Ansatz

$$\mathbf{q}(x, y, z, t) = \bar{\mathbf{q}}(x, y) + \epsilon \hat{\mathbf{q}}(x, y) \exp i(\beta z - \Omega t) \qquad (3)$$

for the determination of the complex eigenvalue Ω. $\Omega_r \equiv \Re\{\Omega\}$ represents a frequency and $\Omega_i \equiv \Im\{\Omega\}$ is the amplification/damping rate of the disturbance, while barred and hatted quantities denote basic and disturbance flow, respectively. Discretization of the linearized equations of motion lead to the two-dimensional BiGlobal eigenvalue problem (EVP)

$$Ax = \Omega M x. \qquad (4)$$

1.3 Spectral Multidomain Discretization

Multidomain spatial discretization divides the space into rectangular subdomains each resolved by spectral collocation. Once the two-dimensional BiGlobal eigenvalue problem has been formed for each domain boundary and interface conditions are imposed in order to form the global matrix discretizing the eigenvalue problem. There are several choices in the eigenvalue recovering algorithm. The storage of the matrix elements and the use of different parallel machines determines the final algorithm, as schematically shown in Fig. 1.

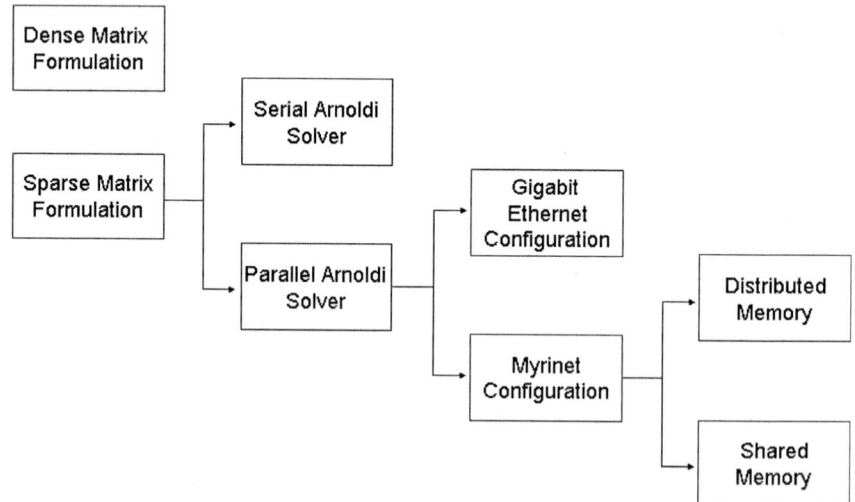

Fig. 1. Computational choices for the solution of the BiGlobal EVP

2 Results

2.1 Dense vs. Sparse Format

A key requirement for the numerical solution of the BiGlobal EVP is the availability of sufficient resolution, for the eigenvector structures to be resolved adequately, which translates into large memory requirements. In addition, the multidomain spectral collocation technique generates matrices in which sparsity is a special feature. This property may be exploited by making use of specialized formats for storing and handling the data Compressed Column Format (CCF) that has been chosen, due to its compatibility with most of the libraries employed for parallel solution of linear systems. CCF stores the sparse matrix in three arrays:

$val[1:N]$ = Value of the nonzeros in column order. N, is the number of nonzero elements.

$rowindex[1:N]$ = Integer array containing the row index of the element.

$sumcol[1:M+1]$ = Integer array that contains pointers to the beginning of each column in $rowindex$ and val. M is the number of columns in the matrix.

So, let

$$A = \begin{pmatrix} 1 & 0 & 0 & 2 & 0 \\ 0 & 1 & -1 & 3 & 1 \\ 1 & 0 & 1 & 2 & 0 \\ 0 & 0 & 0 & 2 & -1 \\ 2 & 0 & 1 & 0 & 0 \end{pmatrix} \tag{5}$$

then
$$\begin{aligned} val(A) &= 1, 1, 2, 1, -1, 1, 1, 2, 3, 2, 2, 1, -1 \\ rowindex(A) &= 1, 3, 5, 2, 2, 3, 5, 1, 2, 3, 4, 2, 4 \\ sumcol(A) &= 1, 4, 5, 8, 12, 14 \end{aligned}$$

Figure 2 shows a typical "T-store" configuration, in which a relatively complex object is placed in the floor of the open cavity. Accordingly, spectral multidomain is used to refine selective areas of the flow. The memory requirement associated with such an EVP, if stored in dense format, is 8 Gb. By contrast, using CCF only 0.5 Gb is required; this has been a strong motivation for the sparse-path of EVP solution to be followed, as shown in Fig. 1.

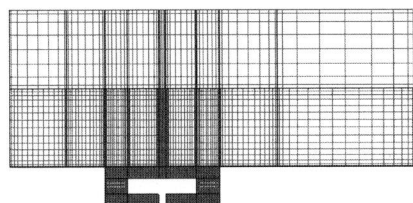

Fig. 2. Example T-store configuration in the open cavity

2.2 Parallel Approach

Once the matrix discretizing the EVP is stored in the CCF, it is solved by Krilov subspace iteration in two stages. First the matrix is LU-decomposed using SuperLU [LU], which is a general purpose library for the direct solution of large, sparse, nonsymmetric systems of linear equations on high performance parallel machines. The library routines also perform an LU decomposition with partial pivoting, and triangular system solves through forward and backward substitution. This LU-decomposition is fed into an Arnoldi iteration to recover the leading eigenvalues. However, storing the LU-decomposition is itself demanding in memory; that associated with the T-store configuration shown in Fig. 2 requires an additional 2 Gbytes. Consequently, in all but the lowest Reynolds numbers[1], the LU decomposition must be performed in parallel.

2.3 Hardware Considerations

Parallel performance broads in one side the scope of tackled problems, involving, however, some new difficulties: new programming strategy compatible with MPI structure and also a greater dependency on hardware in the sense of not only processors features but net architecture and communications among processors. Example results for the solution of the (large) linear system $Ax = b$ have been obtained on two different clusters, one denoted "*Gigabit Ethernet*," and one denoted "*Myrinet*." The characteristics of these distributed-memory machines are summarized in Table 1; results obtained, shown in Table 2, show the dependence of the performance of the SuperLU algorithm on the architecture chosen. Analogous results have been obtained on the Myrinet cluster, using one of the matrices from the validation suite of the CCF format. Results shown in Table 3 show that the theoretical linear speed-up is achieved as long as sections of the matrix are kept in the (shared) memory of each computing node. This tendency stops as long as the number of processors increases beyond two and communication amongst nodes becomes the determining factor; Fig. 3 shows this result in graphical form.

These results for the solution of the linear system have led to the conclusion that use of sparse linear solvers on distributed-memory machines may not be as competitive as their counterparts on shared-memory architectures. Efforts are currently underway to develop a new version of the EVP solver for the BiGlobal instability problem in shared-memory parallel machines, such as the one indicated in Table 1.

---

[1]Such cases have been presented elsewhere [JV06]

Table 1. Computing cluster characteristics: *Gigabit* and *Myrinet* clusters feature distributed-memory

Gigabit Cluster	*Myrinet Cluster*
12 Intel Xeon compute nodes, each featuring:	256 IBM BladeCenter JS20 compute nodes, each featuring:
• 2 single-core 32-bit Pentium	• 2 single-core, 2.20 Ghz 64-bit PowerPC 970FX
• 2 GB DDR memory	• 4 GB PC2700 ECC DDR memory
• Gigabit Ethernet	• Dual Gigabit Ethernet with Myrinet interface

Shared-memory cluster
2 HP Integrity Superdome compute nodes, each featuring:
- 64 Itanium2, 1.5 Ghz
- 384 GB DDR memory

Table 2. Performance of the parallel LU-decomposition $Ax = b$, in which A features $O(10^{10})$ nonzero elements

# proc	Processor Distribution	*Gigabit* Cluster		*Myrinet* Cluster	
		LU-decomposition (s)	LU-decomposition (Flops)	LU-decomposition (s)	LU-decomposition (Flops)
1	1×1	0.14	56.14	0.11	57.14
2	1×2	10.14	5.19	0.24	25.49
2	2×1	12.29	2.19	0.13	48.16
4	1×4	13.11	0.42	1.90	3.17
4	2×2	23.69	0.50	1.34	2.36
4	4×1	11.13	0.42	0.41	14.57
8	1×8	–	–	3.36	1.80
8	2×4	–	–	2.56	2.36
8	4×2	–	–	1.39	4.33
8	8×1	–	–	0.34	17.62

Table 3. Performance of the Myrinet cluster on the parallel solution of $Ax = b$, with $\dim(A) = 7 \times 10^5$ elements

# proc	Processor Distribution	*Myrinet* Cluster			
		LU-decomposition (s)	Solve Time (s)	LU-decomposition (Flops)	Solve (Flops)
1	1×1	4.15	0.25	1159.12	86.56
2	1×2	2.65	0.24	1815.61	88.70
2	2×1	2.43	0.18	1981.04	117.86
4	1×4	1.96	0.55	2451.85	39.13
4	2×2	1.87	0.17	2570.00	127.75
4	4×1	2.20	0.19	2184.66	113.46
6	1×6	1.89	0.63	2548.34	34.57
6	2×3	1.85	0.35	2599.90	6.80
6	3×2	1.92	0.16	2506.08	13.17
6	6×1	2.62	0.20	1834.86	110.03
8	1×8	1.98	0.83	2434.88	25.98
8	8×1	2.80	0.20	1722.45	108.89

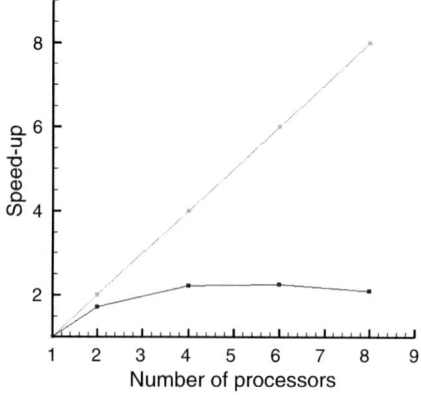

Fig. 3. Speed-up for the LU-decomposition in several processors. Theoretical (achieved in the case of two processors) result is shown in *green*; results of Table 3 shown in *red*

References

[Th03] V. Theofilis: Advances in global linear instability analysis of nonparallel and three-dimensional flows. Prog. Aero. Sci., Vol. **39**, 249–315 (2003).

[LU]   Xiaoye S. Li and James W. Demmel: SuperLU-DIST: A Scalable Distributed-Memory Sparse Direct Solver for Unsymmetric Linear System. ACM Trans. Mathematical Software, **29**, 110–140 (2003).

[JV06] J. de Vicente, E. Valero, L. Gonzalez and V. Theofilis: Spectral multi-domain methods for BiGlobal instability analysis of complex flows over open cavity configurations. No. AIAA Paper 2006-2877 (2006).

Minisymposium "Analysis of Dynamical Problems in Turbomachinery"

Carlos Martel[1] and Roque Corral[2]

[1] Universidad Politécnica de Madrid, Spain
[2] Industria de Turbo Propulsores, S.A., Spain

The ordinary day to day operation of current aircraft turbomachines involves geometrically highly complex parts (bladed disks and vanes) working under extreme mechanical and thermal conditions. The mathematical modeling of these systems typically requires a compressible unsteady aerodynamic description of the fluid flow coupled to linear and nonlinear elastic models for the solid structure. The idea of this minisymposium is to give some insight into various interesting problems of industrial relevance associated with the modeling and analysis of the dynamics and vibration of Turbomachinery structures. The chapter by Berthillier et al. deals with a central problem in Turbomachinery vibration: the problem of blade mistuning. Bladed disks are cyclic structures in which a sector is repeated many times but, because of small unavoidable imperfections (mistuning), all sectors are not identical and these small sector-to-sector variations can give rise to very dangerous localized amplifications of the vibration response that result in a severe increase of blade fatigue. On the other hand, the chapter by Petrov presents some recent advances in the numerical study of structures with friction contact interfaces, which give rise to nonlinear elastic models for the vibration of the structure that can exhibit multiplicity of solutions, hysteresis, sub and superharmonic resonances, etc. And, finally, the last chapter by Corral and Gallardo describes a methodology for the estimation of the vibration levels for aerodynamically unstable Low-Pressure Rotor blades, where the saturation of the vibration amplitude results from the friction at the fir-tree attachment.

Modal Identification of Mistuned Bladed Discs

Marc Berthillier, Bendali Salhi, and Joseph Lardiès

FEMTO-ST 24 chemin de l'Epitaphe, 25000 Besançon, France
`marc.berthillier@univ-fcomte.fr`

1 Introduction

For cyclic structures with aeroelastic coupling, coriolis, and other rotational effects, the different eigen modes are traveling modes with constant interblade phase angle. However, because of small imperfections, as manufacturing tolerances, bladed discs are only quasi-cyclic structures. As a consequence, the dynamic behavior of actual bladed discs may be tremendously modified compared to their cyclic idealization. These small imperfections are called detuning or mistuning depending if they are or not deliberate. The eigen values are usually slightly affected, but the modes shapes could become localized. The vibrating energy is no more distributed along all the blades, but confined to a limited number of blades. From an industrial point of view, the effects of localization could be positive or negative. For example, in aeroelasticity, unstable rotors could be stabilized by the introduction of a judicious detuning pattern. In contrary, mistuning can greatly increase the forced response level, usually when localization occurs.

On the experimental side, the problem is that it is nearly impossible to know in advance with enough accuracy which blade will exhibit the largest vibration amplitude. As rotating instrumentation is necessarily limited, it is a challenge for the monitoring of real machines with strain gages. For that reason, techniques that allow the survey of all blades as Tip Timing methods have been developed in the recent years. The exploitation methods for Tip Timing signals are still in progress [LaI05], [Tuy83]. We present in this chapter a method to identify the modal properties of mistuned bladed disc with time domain signals, reconstructed from Tip Timing measurements. The method for the signal reconstruction is not addressed here. The main difficulties that we have to overcome are the unknown excitation forces and the very high modal density.

2 Mechanical Model

To validate the modal identification method, we developed a simple but realistic mistuned bladed disc model with 22 blades. This model will provide time domain responses for identification and exact modal properties to compare with. The structural part of the model is represented Fig. 1.

The equations of motion are of the form

$$M\ddot{q} + C\dot{q} + Kq = F_{\text{aeroelastic}} + F_{\text{aerodynamic}}$$

where M, C, and K represent the structural, mass, damping and stiffness matrices, and q the displacement vector composed of 2×22 degrees of freedom. The stiffness coefficients have been chosen in order that the blades modes are well separated from the disc modes (the blade modes family has frequencies around 60 Hz, the lowest frequency for a disc mode is close to 500 Hz). This tuned model is mistuned by the introduction of a different stiffness for each blade $K_a(j) = K_a(1 + \delta_j)$ with $j = 1, 22$. Aeroelastic coupling has been introduced, in the way of circulant matrices coefficients, considering only five different coefficients [GGQ96]. The eigen values of the tuned and mistuned model are presented in Fig. 2. These computed eigensolutions will be called

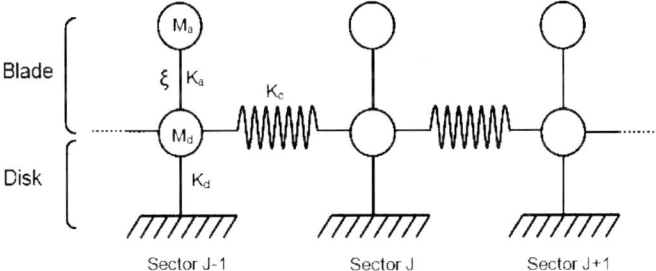

Fig. 1. Mechanical model

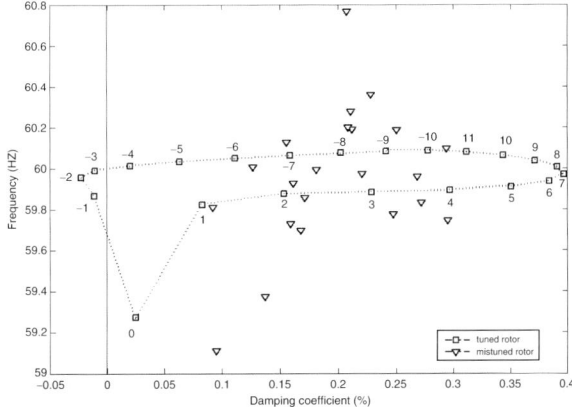

Fig. 2. Stability loop for the tuned and mistuned model

exact solutions in the rest of the paper. We can note that the tuned system is unstable and that the mistuned system is stable. Mistuning couple all the tuned modes stabilizing the unstable ones by the others.

To generate the time forced response for identification, an asynchronous white noise excitation $F_{\text{aerodynamic}}$ has been used. It is representative of turbulent excitation of the blades. The time response is generated for all blades using the convolution integral after diagonalization of the system.

3 Identification Algorithm

The continuous space state and observation equations of a second order dynamical system with stochastic excitations can be written for discrete time series in the following subspace form [Nat86], [La200], [V99]:

$$z_{k+1} = Az_k + Be_k \quad \text{(discrete stochastic state equation)}$$
$$y_k = Cz_k + e_k \quad \text{(discrete observation equation)}$$

where the subscript k is for the time $k\Delta t$, y_k is the vector of observation, e_k is the vector of stochastic innovations, z_k is the vector of states, A is the transition matrix, B is the Kalman gain matrix, and C is the observation matrix.

The eigen values λ and eigen vectors Ψ of the transition matrix A are related to the modal properties (frequency, critical damping ratio, and mode shape) of the underlying second order dynamic system by the relations

$$\omega_i = \frac{1}{2\Delta t}\sqrt{[\ln(\lambda_i\lambda_i^*)]^2 + 4\left[\text{Arc cos}\left(\frac{\lambda_i + \lambda_i^*}{2\sqrt{\lambda_i\lambda_i^*}}\right)\right]^2}$$

$$\zeta_i = \sqrt{\frac{[\ln(\lambda_i\lambda_i^*)]^2}{[\ln(\lambda_i\lambda_i^*)]^2 + 4\left[\text{Arc cos}\left(\frac{\lambda_i+\lambda_i^*}{2\sqrt{\lambda_i\lambda_i^*}}\right)\right]^2}}$$

$$\Phi = C\Psi$$

The major step of the identification process is to determine the matrices A and C from the measurements y_k. For that we define the bloc Hankel matrix H as the covariance matrix between the future and the past.

$$H = E\left[y_k^+, y_{k-1}^{-T}\right]$$

where E denotes the expectation operator, $y_k^+ = (y_k^T, y_{k+1}^T, ..., y_{k+f-1}^T)^T$ the future data vector, and $y_k^- = (y_k^T, y_{k-1}^T, ..., y_{k-p+-1}^T)^T$ the past data vector.

The bloc Hankel matrix H can be estimated from the blades time responses. To estimate the matrices A and C we factorize the bloc Hankel matrix into its observability and controllability matrices, O and K:

$$H = \begin{bmatrix} C \\ CA \\ \vdots \\ CA^{f-1} \end{bmatrix} \begin{bmatrix} G & AG & \cdots & A^{p-1}G \end{bmatrix} = OK$$

The orthogonal-triangular (QR) decomposition of $H = QR$ with R an upper triangular matrix and Q a unitary matrix provides an estimation of the observability matrix O. Finally, to estimate A, it is necessary to introduce the following shifted observability matrices

$$O^{\downarrow} = \begin{bmatrix} C \\ CA \\ \vdots \\ CA^{f-2} \end{bmatrix} \quad \text{and} \quad O^{\uparrow} = \begin{bmatrix} CA \\ CA^2 \\ \vdots \\ CA^{f-1} \end{bmatrix}$$

obtained by removing, respectively, the last bloc line and the first bloc line of the matrix O. The following relation is obtained: $O^{\uparrow} = O^{\downarrow}A$, consequently $A = O^{\downarrow+}O^{\downarrow}$ where $O^{\downarrow+}$ is the pseudo inverse of $O^{\downarrow}$.

4 Evaluation of the Algorithm

The subspace modal identification algorithm presented in the preceding section has been applied to the time responses generated from the model presented on Sect. 2. Time responses of 30 min 27 s with a time step of 710^{-3} s have been used. The stability diagram plotting the evolution of the identified eigen values for various orders of the system is presented in Fig. 3.

We find in the frequency band of the blades, 22 frequencies and damping ratios. We can see that only one frequency and two damping ratios are not stable. The identified modes shapes can be compared with the exact ones in the following way: for each complex mode Φ, we define a real mode $\tilde{\Phi}$ as the deflection for the instant of time where the maximum amplitude is obtained. Four of these modes as well as the MAC matrix for 22 modes are shown in Fig. 4. The MAC matrix is defined as:

$$\mathrm{MAC}_{ij} = \frac{\left| \tilde{\Phi}_i^{\mathrm{T}} \tilde{\Phi}_j \right|}{\sqrt{\left(\tilde{\Phi}_i^{\mathrm{T}} \tilde{\Phi}_i \right) \left(\tilde{\Phi}_j^{\mathrm{T}} \tilde{\Phi}_j \right)}}$$

with $\tilde{\Phi}_i$ an exact mode shape and $\tilde{\Phi}_j$ an identified mode shape. We can see in Fig. 4 (MAC matrix) that two exact modes have not been found by the identification process (modes number 6 and 17).

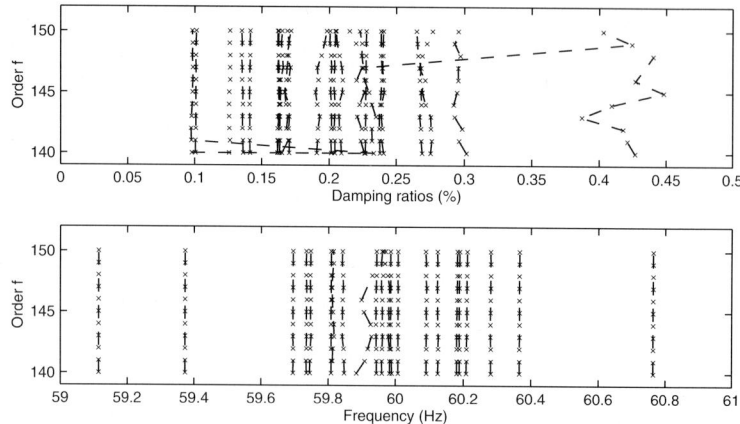

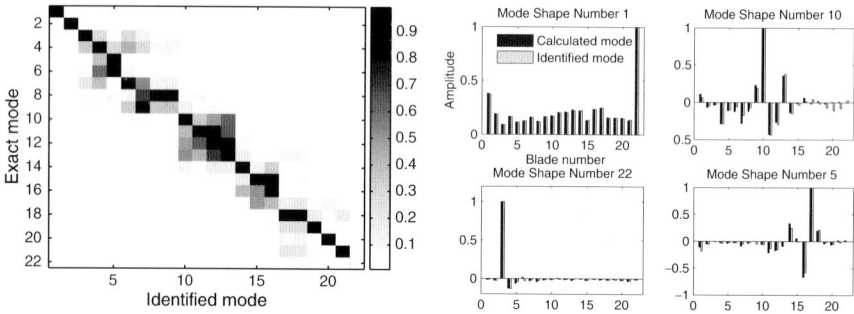

Fig. 3. Stability diagram

Fig. 4. Mode shape comparison

Consequently, the identification found two spurious modes (number 9 and 18). One could be easily identify in the stability diagram as unstable, the other duplicate another mode with nearly the same frequency. For the remaining 20 modes, the correlation between exact and identified modes is acceptable to excellent. The relative differences between the exact and identified frequencies are below 0.07%, and the relative difference for the damping ratios is below 10% except for mode number 8 for which it is of 27%.

5 Conclusion

We present a rather promising method to identify modal properties of bladed discs in operation from time response measurements obtained for example through a Tip Timing procedure. This method is a real improvement to the traditional FFT that provides only a poor estimation of the modal frequencies in the case of mistuned rotors. The main drawback is that rather long time responses have to be recorded, nearly 30 min for the case presented here. Further work is to be done to reconstruct time signals with appropriate time step from Tip Timing measurement. This work is a first step in the identification of aeroelastic coupling, aerodynamic excitation, and mistuning pattern.

References

[LaI05]   Lawson, C.P., Ivey, P.C.: Turbomachinery blade vibration amplitude measurement through tip timing with capacitance tip clearance probes. Sensors ans Actuators, **A 118**, 14-24 (2005)

[DCW02] Dimitriadis, G., Carrington, I.B, Wright, J.R., Cooper, J.E.: Blade-Tip Timing Measurement of Synchronous Vibrations of Rotating Bladed Assemblies, Mechanical Systems and Signal Processing, **16(4)**, 599-622 (2002)

[Cra87]   Crawley, E.F.: Aeroelastic formulation for tuned and mistuned rotors, AGARD-AG-298, Volume 2 (1987)

[La100]   Lardiès, J.: Estimation of parameters and model order in state space innovation forms, Inverse Problem in Engineering, **8**, 75-92 (2000)

[La200]   Lardiès, J.: A stochastic realisation algorithm with application to modal parameter estimation, mechanical Systems and Signal Processing, **15**, 275-285 (2000)

[Tmn05] Ta, M.N.: Analyse modale par sous espaces et par la transformée en ondelettes. PhD Thesis, Université de Franche-Comté, Besançon (2005)

Aeroelastic Instability of Low-Pressure Rotor Blades

Roque Corral[1,2] and Juan Manuel Gallardo[1]

[1] Industria de Turbopropulsores S. A. 28830 Madrid, Spain
[2] Adjunct professor at the School of Aeronautics, UPM, Madrid, Spain

Summary. A methodology for the prediction of the vibration levels of welded-in-pairs low-pressure-turbine rotors is presented. It combines three-dimensional viscous linear aerodynamic analyses with a simple friction model for the fir-tree attachment. Results are presented for an existing rotor and compared with experimental data.

1 Introduction

Modern low pressure turbines (LPTs) are made up of very slender and thin airfoils due to steady trend to design very efficient, low cost, low weight turbomachinery. This is specially true for LPTs since the continuous trend to increase engine's by-pass ratio poses extraordinary difficulties to the LPT design.

Cost and weight reductions in LPTs are obtained by reducing the part count, increasing the lift per airfoil, and designing light, high aspect ratio airfoils. The latter lowers the natural frequencies of the assembly, and therefore, the reduced frequency k up to a point in which airfoils may become aerodynamically unstable, giving rise to the onset of flutter. Nowadays flutter may become a dominant constrain on the design of modern LPTs, precluding the use of more efficient aerodynamic and structural configurations.

When the rotor airfoils are aerodynamically unstable, the blade vibration grows up to a point in which the motion is nonlinearly saturated. Typically, this is due to the dry friction that takes place in the fir-tree attachment, although in principle other devices as under-platform dampers or cover plates may contribute to increase the damping of the rotor. In any case what is important at this point is to estimate the vibration level of the rotor, which ultimately will determine the life of the component. The effect of friction on aerodynamically unstable rotors from a conceptual point of view was first studied by Sinha and Griffin [3, 4].

To clarify some of the basic issues addressed in the present investigation, it is interesting to review the main results obtained for a single degree of freedom (SDOF) problem. If the unsteady aerodynamics is linear, then the unsteady pressure scales with the amplitude of vibration, δ, and hence the aerodynamic work per cycle scales as the square of the amplitude, $W_{\text{aero}} \propto \delta^2$. Dry friction scales with the amplitude in different ways, depending on the vibration amplitude. For very small vibration amplitudes ($\delta < \delta_{\text{off-set}}$) the blade is stuck on the attachment and the work dissipated is null. For large vibration amplitudes there is a macro displacement of the rotor in the attachment. The tangential force is constant and therefore the work dissipated per cycle is proportional to the displacement, $W_{\text{macro}} \propto \delta$. Between both situations there is a regime known as microslip where only a fraction of surface in contact is sliding. Different models exist to describe this behavior (see for instance the Mindlin's model [5] to describe the contact between two elastic spheres). At this stage what is important to highlight is that the work dissipated per cycle is of the form $W_{\text{micro}} \propto \delta^n$, with $n > 2$, typically $\delta = 3$.

The situation is depicted in Fig. 1, the balance between the aerodynamic self-excitation and the dry friction provides either one or three situations, depending on the relative value between them. The trivial solution, $\delta = 0$, is unstable and any small perturbation from $\delta = 0$ moves the system towards the solution 2, which is a stable cyclic limit. Solution 2 is an attractor and any perturbation of the cyclic limit comes back to the solution 2 unless we reach the amplitude δ_3, which is an absolute stability limit, since the solution 3 is unstable. Alternatively, if the aerodynamic self-excitation is too large, or the friction work too low, the only solution is the trivial one and the system is unstable.

There are computational [2] and experimental evidences (see Fig. 2) that indicate that some LPT welded-in pairs rotors are aerodynamically unstable. Under these circumstances we are interested in predicting the vibration amplitude of the rotor to estimate the alternate stress field. Figure 2 shows

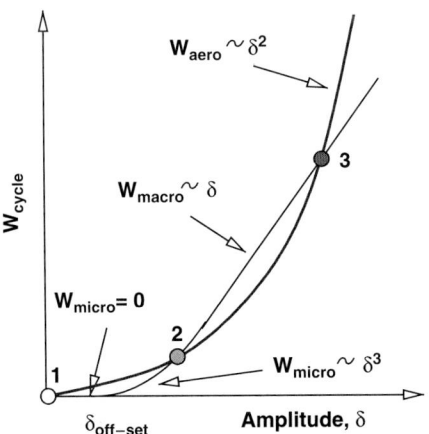

Fig. 1. Energy balance for the SDOF model

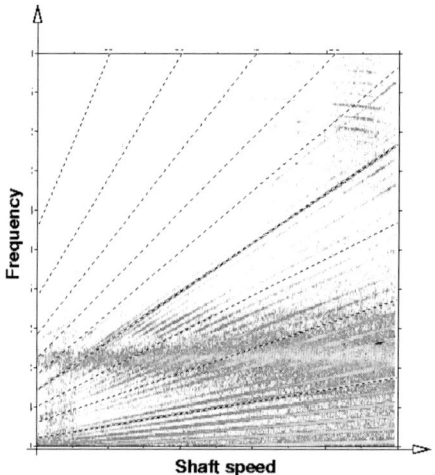

Fig. 2. SG readings for a aerodynamically unstable LPT rotor blade

engine measurements in a strain gauge located in the shank of a LPT rotor. Apart from the synchronous excitation a nonsynchronous excitation of a certain mode is clearly seen in the whole range of shaft speed. This is an indication that this mode is probably aerodynamically unstable.

In this paper we will first describe the analysis methodology. Then we will discuss the results for several LPT rotor blades and will compare them with experimental data.

2 Analysis Methodology

2.1 Aerodynamic Damping

The aerodynamic damping is computed using a Navier–Stokes linearized unstructured code known as MusT-L [6]. The main hypothesis is that the flow may be decomposed into two parts: a steady or mean background flow, plus a small and periodic unsteady perturbation, which in turn may be expressed as a Fourier series in time. The approach assumes that the effect of the aerodynamics on the mode-shape is negligible and the structural model may be precomputed, injected, and interpolated in the aerodynamic solver. The mode displacements are imposed on the blade surface and a moving mesh approach is used in the aerodynamic solver. When several blades are required to describe the structural mode-shape (such as in welded-in-pairs rotors), several passages are employed as well in the simulation. Phase-lagged boundary conditions are used to analyze traveling waves with arbitrary spatial wave numbers (nodal diameters) in a single passage.

Because of manufacturing tolerances there are slight variations in the characteristics of the different rotor blades. This may involve a significant alteration of the vibratory behavior of the row and its aeroelastic stability. To account for this effect analyses based on the fundamental mistuning model (FMM) [2] are performed.

Since the aerodynamics is linear the aerodynamic work per cycle is quadratic with the vibration amplitude δ_a, i.e.,

$$W_a = W'_a \delta_a^2.$$

2.2 Friction Damping

Centrifugal forces give rise to very high contact pressure in the fir-tree contact faces. On the other hand, the alternating stresses due to the vibration motion are comparatively much smaller than the normal steady load. This implies that dissipation in fir-trees is caused by microslip phenomena, where both deformation and sliding of the contacting surfaces play a significant role. The processes involved are not completely understood and different approaches may be found in the literature to model this behavior. Most recent theories are based on the presence of micro-asperities on the contacting surfaces. Here, we will consider a simplified version of the model proposed by Sellgren and Olofsson [9], where the hypothesis that the tangential displacements are small is used. Similar results may be obtained however with other models.

By analyzing the hysteresis loop of the friction force we can relate the local energy dissipation in a cycle with the (macroscopic) local tensional state in the contact surface via an equation such as

$$W \simeq K(\sigma) A_c \tau^3.$$

Here, $K(\sigma)$ is a function of the *static* tensional state and the material properties, A_c is the contact area, and τ is the *alternate* shear stress.

Following the hypothesis of small displacements in the contact surfaces, this alternate shear stress may be related to the blade macroscopic vibration amplitude δ_a by means of analyses with clamped boundary conditions at the contact surfaces. Taking into account all the aforementioned considerations the energy dissipated in a friction loop is

$$W_F = W'_F \delta_a^3.$$

By balancing the aerodynamic work added to the blade and the energy dissipated by the friction, the equilibrium vibration amplitude (point 2 in Fig. 1) may be determined. For a more detailed description of the methodology the reader is referred to [10].

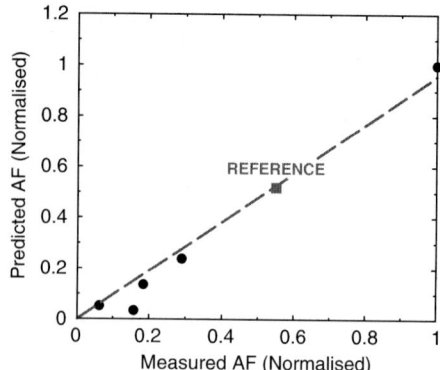

Fig. 3. AF results from the methodology

3 Results

The methodology has been applied to several rotor rows from different stages of several engines. Results can be seen in Fig. 3. One of the cases is used to calibrate the contact properties for the friction model (red square). Dashed line represents the ideal case where the predictions are coincident with the experimental data.

The agreement between the simulations and the experiments is fairly good in a wide range of vibration amplitudes. It is important to highlight that, although all the blades are aerodynamically unstable, several of them exhibit acceptably small vibration amplitudes and should be considered as admissible designs.

4 Concluding Remarks

An attempt to predict strain-gauge readings associated to self-excited vibrations has been presented. The aerodynamic damping is computed using a 3D linearized viscous method while the damping associated to dry friction is estimated by means of a simple model.

The vibration amplitudes associated to the strain-gauge readings are reproduced with the present method within engineering accuracy, what makes it very attractive from a practical point of view.

References

1. Corral, R., Lpez, C. and Vasco, C., "Linear Stability Analysis of LPT Rotor Packets - Part I: Methodology and Two-Dimensional Results". ASME Paper 2004-GT-54119, Proceedings of the 49th ASME Gas Turbine and Aero Engine Congress, Exposition and Users Symposium, Vienna, Austria. June 14-17, 2004

2. Corral, R., Gallardo, J. M. and Vasco, C., "Linear Stability Analysis of LPT Rotor Packets - Part I: Methodology and Two-Dimensional Results". ASME Paper 2004-GT-54120, Proceedings of the 49th ASME Gas Turbine and Aero Engine Congress, Exposition and Users Symposium, Vienna, Austria. June 14-17, 2004

3. Sinha, A., and Griffin, J. H., "Friction Damping of Flutter in Gas Turbine Engine Airfoils", AIAA Journal of Aircraft, Vol 20, No. 4, April 1983.

4. Sinha, A. and Griffin, J. H. "Effects of friction dampers on Aerodynamically Unstable Rotor Stages", AIAA Journal Vol. 23, No. 2, pp 262-270, Feb. 1985.

5. Midlin, R. D. and Deresiewicz, H., "Elastic Spheres in Contact Under Varying Oblique Forces", Journal of Applied Mechanics, pp. 327-344, 1953

6. Corral, R., Escribano, A., Gisbert, F., Serrano, A., and Vasco, C., "Validation of a Linear Multigrid Accelerated Unstructured Navier-Stokes Solver for the Computation of Turbine Blades on Hybrid Grids", AIAA Paper 2003-3326, presented at the 9th AIAA/CEAS Conference & Exhibit, Hilton Head, South Caroline, 12-14, 2003

7. Corral, R., Crespo, J., and Gisbert, F., "Parallel Multigrid Unstructured Method for the Solution of the Navier-Stokes Equations", AIAA Paper 2004-0761, presented at the 42nd AIAA Aerospace Science Meeting and Exhibit, Reno, Nevada, January 5-8, 2004

8. Sellgren, U., and Olofsson, U., "Application of a Constitutive Model for Micro-Slip in Finite Element Analysis", Computer Methods in Applied Mechanics and Engineering Vol. 170/1-2, pp. 65-77, 1999

9. Kielb, J., and Abhari, R. S., "Experimental Study of Aerodynamic and Structural Damping in a Full-scale Rotating Turbine", ASME Paper 2001-GT-0262, presented at the 46th International Gas Turbine & Aero Engine Congress & Exhibition, New Orleans, Louisiana, June 4-7, 2001

10. Corral, R. and Gallardo, J.M., "A Methodology for the Vibration Amplitude Prediction of Self-Excited Rotors Based on Dimensional Analysis", ASME Paper 2006-GT-90668, presented at the 51st International Gas Turbine & Aero Engine Congress & Exhibition, Barcelona, Spain, May 8-11, 2006

Recent Advances in Numerical Analysis of Nonlinear Vibrations of Complex Structures with Friction Contact Interfaces

E.P. Petrov

Imperial College London, Mechanical Engineering Department, South Kensington Campus, London, SW7 2AZ, UK
y.petrov@imperial.ac.uk

1 Introduction

Vast majority of machinery structures are assembled structures: they consist of two or, usually, more components assembled together and these joined components interact with each other at friction contact interfaces. The forces acting at friction contact interfaces are generally strongly nonlinear. Among many sources of the nonlinearity of the interaction are (1) unilateral contact of interaction along directions normal to contact surfaces, when compression normal stresses can act at these surfaces but tension stresses are not allowed; (2) variation of contact areas during loading, including closing and opening clearances and interferences resulting in contact-separation transitions over a whole interface surface or over its some parts; (3) friction forces with their magnitude and stick-slip transitions affected by contact-separation and normal stress variation. In this chapter recent developments in modeling and numerical analysis of nonlinear vibration of structures with friction interfaces are discussed.

This chapter is not intended to make a thorough review of the state of art in this vast scientific field but is focused on developments recently made by the author. Three major research directions are developed in order to provide effective tools for analysis of practical structures with friction contact and other nonlinear interfaces, namely (1) contact interaction modeling, which include development of new friction theories and special contact interface elements facilitating contact interaction modeling; (2) efficient methods for analysis of nonsmooth dynamics of realistic structure, containing, possibly, millions of degrees of freedom (DOFs); and (3) advanced tools for effective design of nonlinear structures, which include such capabilities as sensitivity analysis of nonlinear forced response, direct parametric analysis, optimization of design parameters. A scheme of the capabilities developed is shown in Fig. 1.

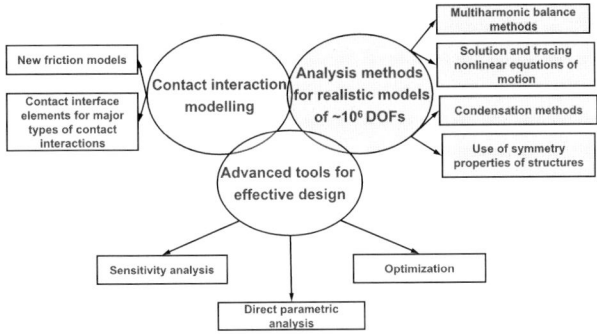

Fig. 1. Capabilities developed for analysis of nonsmooth dynamics of structures with friction contact interfaces

2 New Friction Constitutive Laws and Modeling of Friction Contact Interfaces

Friction is one of the most important sources of damping in structures with joints. The problem of developing friction models is one of the oldest in mechanics and the model developed by Coulomb in 1785 is widely used in structural dynamics owing to its simplicity. However, Coulomb's friction model has inherent limitations when attempting to capture experimentally observed friction effects found after Coulomb. For example, it does not take into account effect of normal load variation on slip-stick transitions, cannot make adequate modeling when a trajectory of relative motion is more complex than a simple line and has a difficulty in microslip modeling.

New efficient models have been developed in [1]. These friction models can model friction forces occurring at contact interfaces for arbitrary trajectory of relative motion and under time-varying normal load, including cases of separation. They allow for also time-varying friction contact parameters, such as friction coefficient and contact stiffness coefficients; anisotropy and variation of the friction characteristics over the contact surfaces. The capabilities of the new friction models are demonstrated and effects of trajectory of motion, anisotropy of friction characteristics and of variation in off-time friction characteristics, and normal load are discussed.

Special friction contact elements have been developed in [2] and [3] for a general case of multiharmonic steady-state vibrations. These friction contact elements can be spread over area where friction contact interactions are expected and actual contact area, possibly varying in time, is determined as a result of calculation together with forced response levels, contact interaction forces, and contact interfaces stiffness characteristics corresponding to the response levels. Expressions for interaction forces and stiffness matrix are derived analytically. Owing to this, exact and very fast calculation can be

222 E.P. Petrov

made for any types of relative motion of contacting surfaces, including clos-
ing/opening gaps/interferences.

In [4] analytical expressions for first and second order sensitivity coefficients for the contact forces and tangent stiffness matrices with respect to
parameter of the friction contact such as gap value, friction coefficient, static normal stress value, stiffness coefficients of the contact surface have been
derived for the first time.

3 Methods for Analysis of Nonsmooth Nonlinear Dynamics of Large-Scale Finite Element Models of Structures

3.1 Frequency-Domain Analysis of Steady-State Forced Response

A steady-state, periodic vibration response is often of major interest, and
therefore, displacements' variation in time can be represented by a restricted
Fourier series. The total number of harmonics kept in such multiharmonic expansion of displacements and selection of harmonics numbers are dependent
on accuracy of calculation required. It should be noted that all major types
of periodic vibration, which are possible for strongly nonlinear structures,
can be found, including (1) major resonances; (2) superharmonic resonances;
(3) subharmonic resonances; and (4) combination resonances. The multiharmonic representation allows transformation of nonlinear equations of motion
from time domain to frequency domain; therefore, instead of time-consuming
integration of differential equation of motion the algebraic nonlinear equations
are obtained with respect to coefficients of harmonics.

Methods for solution of such equations, for a case of realistic models comprising possible millions of degrees of freedom, continuation approaches and
condensation methods are developed and discussed in [2,5,6].

3.2 Use of Cyclic Symmetry in Analysis of Strongly Nonlinear Structures

Many practical structures are designed to be cyclically symmetric, i.e., to
have a repetitive, "cyclic" part of the structure, which can form the whole
structure by simple rotations of this part (called also "a sector") around its
symmetry axis. For strongly nonlinear structures a general method allowing
for cyclic symmetry has been proposed and validated for different types of non-
linearities in [5]. The method allows exact calculation of steady-state forced
response for a whole structure using its single sector, with special boundary
conditions applied at surfaces where this sector is attached to the rest of the
structure. Types of excitation and boundary conditions which allow use of
cyclic symmetry in nonlinear forced response analysis are formulated. For a

case of structures with violations of the cyclic symmetry, so-called "mistuned" structures, efficient analysis methods are proposed in [7].

3.3 Analysis of Sensitivity, Uncertainty, and Stochastic Characteristics of the Nonlinear Forced Response

To make a justified choice of design parameters there is a need to determine how sensitive predicted response levels are to variations of the design parameters. Sensitivity characteristics of the forced response facilitate choosing the optimal parameters for a structure and allow assessment of robustness and fidelity of the calculated forced response levels in the presence of inevitable variability of the design parameters

In [4] a method is proposed to calculate sensitivity of nonlinear forced response levels to variation of parameters of the friction contact interfaces and gaps. The effectiveness of the method allows the first and second order sensitivity coefficients to be calculated simultaneously with calculation of the forced response in wide frequency ranges. The method is based on analytical derivation of the friction contact elements that provide highly accurate and extremely fast calculations of the forced response sensitivity.

Methods of calculations of uncertainty, stochastic characteristics, and probability density function of the nonlinear forced response of structure with stochastic friction contact parameters are proposed in [8].

3.4 Concept and Methods for Direct Parametric Analysis of Strongly Nonlinear Structures

In design practice there is usually a need to understand how forced response levels are dependent on the choice of design parameter values when they can vary in wide ranges. Customary multivariant calculations of the forced response calculated for discrete, a priori selected sets of the contact parameters values require significant computation expense and in many cases do not provide information required.

An effective method for direct parametric analysis of nonlinear forced response for structures with friction contact interfaces has been developed in [6]. The method allows, for the first time, forced response levels to be calculated directly as a function of contact interface parameters such as the friction coefficient, contact surface stiffness, clearances, interferences, and normal stresses. As a result of the calculation, the functional dependency of the forced response level on each design parameter separately, or, on their simultaneous variation, is determined for wide ranges of parameter variation. In [9] the method has been extended to allow the direct parametric analysis for resonance peak amplitudes and frequencies. The method provides a unique capability to calculate dependencies of resonance frequencies and amplitudes for strongly nonlinear structures.

4 Conclusions

A methodology has been developed for efficient analysis of forced response for strongly nonlinear structure with friction, gap, and impact interfaces. The methodology allows using realistic, large-scale models of practical structures with millions of DOFs.

Development of the analytical formulation for friction contact interfaces elements ensures unprecedented speed and accuracy of calculation and provides a breakthrough in analysis of the nonlinear forced response for structures with friction and gaps. Original methods for advanced analysis of nonlinear forced response have been proposed, including allowing for cyclic symmetry, sensitivity analysis of the forced response, direct parametric analysis, and determination of response uncertainty and probabilistic characteristics.

References

[1]   Petrov, E.P. and Ewins, D.J., "Generic friction models for time-domain vibration analysis of bladed discs," Trans. ASME: J. of Turbomachinery, 2004, Vol.126, January, pp. 184-192

[2]   Petrov, E.P. and Ewins, D.J. "Analytical formulation of friction interface elements for analysis of nonlinear multi-harmonic vibrations of bladed discs", Trans. ASME: J. of Turbomachinery, 2003, Vol.125, April, pp. 364-371

[3]   Petrov, E.P. and Ewins, D.J., "Effects of damping and varying contact area at blade-disc joints in forced response analysis of bladed disc assemblies", Trans. ASME: J. of Turbomachinery, 2006, Vol.128, January, pp. 403-410

[4]   Petrov, E.P., "Sensitivity analysis of nonlinear forced response for bladed discs with friction contact interfaces", Proc. of ASME Turbo Expo 2005, June 6-9, 2005, Reno-Tahoe, USA, GT2005-68935, 12pp

[5]   Petrov, E.P., "A method for use of cyclic symmetry properties in analysis of nonlinear multiharmonic vibrations of bladed discs," Trans. ASME: J. of Turbomachinery, 2004, Vol.126, January, pp. 175-183

[6]   Petrov, E.P., "Method for direct parametric analysis of nonlinear forced response of bladed discs with friction contact interfaces", Trans. ASME: J. of Turbomachinery, 2004, Vol.126, October, pp. 654-662

[7]   Petrov, E.P. and Ewins, D.J., "Method for analysis of nonlinear multiharmonic vibrations of mistuned bladed discs with scatter of contact interface characteristics", Trans. ASME: J. of Turbomachinery, 2005, Vol. 127, January, pp. 128-136

[8]   Petrov, E.P., "A sensitivity-based method for direct stochastic analysis of nonlinear forced response for bladed discs with friction interfaces," Proc. of ASME Turbo Expo 2007, May 14-17, 2007, Montreal, Canada, GT2007-27981, 11pp

[9]   Petrov, E.P., "Direct parametric analysis of resonance regimes for nonlinear vibrations of bladed discs", Proc. of ASME Turbo Expo 2006, June 8-11, 2006, Barcelona, Spain, GT2006-90147, 10pp

Minisymposium "Numerical Methods for Conservation Laws"

G. Russo

University of Catania, Italy

In this minisymposium, Gabriella Puppo (Politecnico di Torino) and Giovanni Russo (U. Catania) present "Central Runge-Kutta schemes for stiff balance laws," Susana Serna (UCLA) presents "Flow calculations using shock capturing schemes based on Power limiters," and Fausto Cavalli, Giovanni Naldi, Matteo Semplice (U. Milano) and Gabriella Puppo (Politecnico di Torino) present "A comparison between relaxation and Kurganov–Tadmor schemes."

Central Runge–Kutta Schemes for Stiff Balance Laws

Gabriella Puppo[2] and Giovanni Russo[1]

[1] Dipartimento di Matematica ed Informatica, Università di Catania, Viale
Andrea Doria 6, 95125 Catania, Italy
russo@dm.unict.it

[2] Dipartimento di Matematica, Politecnico di Torino, Corso Duca degli Abruzzi,
24, 10129 Torino, Italy
gabriella.puppo@polito.it

Summary. In this work, we propose a new family of high order finite volume methods for stiff balance laws. These methods are characterized by an explicit integration of the nonlinear convective terms, while the possibly stiff source is computed implicitly with a novel approach that avoids cell coupling. For this reason, the methods enjoy a favorable stability restriction, without requiring the solution of large nonlinear systems of equations.

1 Introduction

Consider the system of balance laws

$$u_t + f_x(u) = \frac{1}{\varepsilon} g(u), \tag{1}$$

where $\varepsilon > 0$ is a stiffness parameter and f is hyperbolic, i.e., the Jacobian of f has real eigenvalues and a complete system of eigenvectors, for all u. Systems of the form (1) arise in several applications, for instance dynamics of gas mixtures with chemical reactions or change of phases, multiphase flows, kinetic systems for rarefied gases, extended thermodynamics, hydrodynamical models for semiconductor devices, continuous models for traffic flow, and granular flow, to mention just a few (see, for example, [4] and references therein. In several applications, the stiffness parameter ε is very small, so that the source term can be very large and the system becomes stiff. On the other hand, the convective term $f_x(u)$ has to be discretized by a nonlinear method, even in the case of linear flow, because the space discretization calls for nonlinear algorithms to prevent the onset of spurious oscillations. Thus, it is convenient to use implicit time integrators on the stiff term to avoid the use of very small time steps forced by the stability restriction, but it is also convenient to integrate explicitly the convective term to avoid the need to solve large nonlinear algebraic systems of equations.

Several applications require a nonuniform unstructured grid in order to concentrate degrees of freedom where the solution exhibits a complex structure, and/or to fit the grid to a computational domain Ω which may not be accurately covered with a uniform, cartesian mesh. In both cases, a finite difference discretization is not possible: high order finite differences in fact require a regular of a smoothly varying grid [6], and are difficult to use onto a domain with a generic shape. For such reasons finite volume methods are sometimes preferred, since they can be adapted to unstructured grids.

A finite volume formulation of (1) can be written as

$$\frac{\mathrm{d}}{\mathrm{dt}} \bar{u}_k = -\frac{1}{m(T_k)} \int_{\partial T_k} f(u)n + \frac{1}{\varepsilon} \langle g(u) \rangle_k, \qquad (2)$$

where T_k is the generic element of the mesh, i.e., a triangle or quadrilateral with measure $m(T_k)$, $\bar{u}_k$ denotes the cell average of the solution u on the element T_k, and $\langle g(u) \rangle_k \equiv (\int_{T_k} g(u))/m(T_k)$ denotes the average of the source on the element T_k.

We note that an implicit treatment of the source term couples the elements T_k together since, for a high order scheme, $\langle g(u) \rangle_k \neq g(\bar{u})_k$.

In this work, we propose a family of finite volume methods based on a staggered grid which overcomes the difficulty inherent in the construction of high order finite volume methods with an implicit source term. The key ingredient is to adapt to the present framework the construction of Central Runge Kutta (CRK) schemes proposed in [3].

2 IMEX CRK Schemes

For the sake of simplicity, from now on we will consider a one-dimensional problem on a uniform grid, with nodes x_j and uniform mesh spacing h. The finite volume formulation now reads

$$\frac{\mathrm{d}}{\mathrm{dt}} \bar{u}(x) = -\frac{1}{h} \left(f(u(x + h/2, t)) - f(u(x - h/2, t)) \right) + \frac{1}{\varepsilon} \langle g(u) \rangle|_x . \qquad (3)$$

Starting from the cell averages $\bar{u}_j^n$ at time t^n, we reconstruct the piecewise polynomial function $U(x, t^n)$, which is discontinuous at the cell edges located in $x_{j \pm 1/2} = x_j \pm h/2$. Next, (3) is integrated in time with a ν stage IMEX Runge–Kutta scheme, as in [4], on the staggered grid $[x_j, x_{j+1}]$. Let $\tilde{b}_i, \tilde{a}_{i,l}$ and $b_i, a_{i,l}$, with $i, l = 1, \ldots \nu$ be the coefficients in the Butcher tableaux of the explicit and implicit Runge Kutta schemes forming the IMEX pair, respectively. Then the numerical cell averages at time t^{n+1} are given by

$$\bar{u}_{j+1/2}^{n+1} = \bar{u}_{j+1/2}^n - \frac{\Delta t}{h} \sum_{i=1}^{\nu} \tilde{b}_i \left(f_{j+1}^i - f_j^i \right) + \frac{\Delta t}{\varepsilon} \sum_{i=1}^{\nu} b_i \langle g(U^{(i)}) \rangle_{j+1/2}, \qquad (4)$$

where $f_j^i \equiv f(U_j^{(i)})$, while the stage values $U_j^{(i)}$ are computed pointwise with the following equations:

$$U_j^{(i)} = U_j^n - \frac{\Delta l}{h} \sum_{l=1}^{i-1} \tilde{a}_{i,l} \partial_x f(U_j^{(l)}) + \frac{\Delta t}{\varepsilon} \sum_{l=1}^{i} a_{i,l} g(U_j^{(l)}). \tag{5}$$

Since the reconstructed function $U^{(i)}$ is smooth at the grid points x_j, all quantities appearing in the equation above are well defined. We remark that the evaluation of the stage values is performed pointwise on the primitive grid.

A discrete approximation of the space derivative of the flux appearing in (5) is performed by a suitable Central WENO reconstruction [2]. A similar reconstruction is used to compute the pointwise value $U_j^{(1)}$ from cell averages in (5) and the staggered cell average $\bar{u}_{j+1/2}^n$ in (4).

In the computation of the new stage value, (5), at each stage one has to solve one nonlinear equation for $U_j^{(i)}$, which can be solved cell by cell, since at this level all cells are decoupled. The solution is then updated in the sense of cell averages in (4), and here the cell averages of the source term can be explicitly computed by quadrature, using the stage values $U^{(i)}(x)$ found through a suitable reconstruction technique. For more details, see [2, 3, 5].

Results

We demonstrate the performance of this class of schemes on the Broadwell model, which is a toy model for kinetic problems. The system is of the form (1) with [4]

$$u = (\rho, m, z)^T, \quad f = (m, z, m)^T, \quad g = (0, 0, (\rho^2 + m^2)/2 - \rho z)^T. \tag{6}$$

As $\varepsilon \to 0$, the system relaxes to a 2×2 system of conservation laws for ρ and m, while the variable z becomes a function of the other variables. We test schemes of order 1 (CRK1: the IMEX scheme is simply formed by the explicit and the implicit Euler schemes, with piecewise constant reconstruction), order 2 (CRK2: second order IMEX scheme used in [5], piecewise linear reconstruction with MinMod limiter), and order 4 (CRK4: third order IMEX scheme as in [5], with 4th order central WENO reconstruction, see [2]).

First we show the convergence of the schemes for various values of the stiffness parameter ε on the smooth test problem with periodic initial data: $\rho(x,0) = \rho_0 = 1 + a_\rho \sin(2\pi x/L), v(x,0) = v_0 = 1/2 + a_v \sin(2\pi x/L), z(x,0) = a_z \frac{1}{2}(\rho_0 + \rho_0 v_0^2)$, where v is the velocity, $v = m/\rho$. The computational region is $L = 20$. The system has eigenvalues $\mu = 0, \mu = \pm 1$, thus the mesh ratio λ was chosen as $\lambda = 0.9/2$. The computation was stopped at $T = 20$. Figure 1 shows the order of accuracy of the second and fourth order schemes as a function of ε, computed against a reference solution. The figure shows clearly that full accuracy is reached only for very small values of ε or for $\varepsilon = O(1)$. However, it is noteworthy that the higher order scheme has always a much higher accuracy than the lower order IMEX-CRK2 scheme.

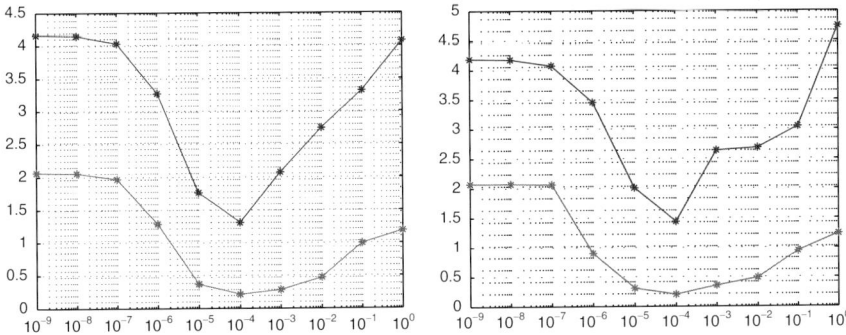

Fig. 1. Accuracy of CRK-IMEX schemes of order 2 (*lower curve*) and 4 (*upper curve*) in the L^1 (*left*) and L^∞ (*right*) norms

Table 1. Error in the L^1 norm for the second order CRK2 scheme (left table) and CRK4 (right table)

ε	$N = 50$	$N = 100$	$N = 200$	$N = 400$	$N = 50$	$N = 100$	$N = 200$	$N = 400$
1.	0.556e−2	0.307e−2	0.149e−2	0.656e−3	0.118e−4	0.577e−6	0.280e−7	0.166e−8
0.1	0.234e−1	0.177e−1	0.107e−1	0.541e−2	0.139e−4	0.767e−6	0.573e−7	0.575e−8
0.01	0.301e−1	0.269e−1	0.216e−1	0.156e−1	0.104e−4	0.168e−5	0.283e−6	0.423e−7
1e−3	0.172e−1	0.163e−1	0.149e−1	0.123e−1	0.712e−5	0.175e−5	0.485e−6	0.115e−6
1e−4	0.236e−2	0.195e−2	0.177e−2	0.153e−2	0.360e−5	0.399e−6	0.130e−6	0.527e−7
1e−5	0.794e−3	0.264e−3	0.177e−3	0.137e−3	0.328e−5	0.181e−6	0.195e−7	0.576e−8
1e−6	0.681e−3	0.152e−3	0.457e−4	0.187e−4	0.326e−5	0.163e−6	0.918e−8	0.949e−9
1e−7	0.670e−3	0.142e−3	0.363e−4	0.929e−5	0.326e−5	0.161e−6	0.817e−8	0.499e−9

A comparison between left and right tables in Table 1 show clearly that the higher order scheme is by far superior even on coarse grids and that the decay of accuracy on moderately stiff values of ε does not translate in large errors. Both second and fourth order scheme show a pronounced degradation of the accuracy for intermediate values of the stiffness parameter ε. The main reason is that we used initial conditions which are not "well prepared," i.e., the parameter $a_z = 0.2$ is not 1. With this choice we want to show that the IMEX numerical schemes that we used are robust enough to recover full accuracy for both very small and very large values of the stiffness parameter, even in presence of an initial layer.

The effect of degradation of accuracy for intermediate values of ε and the techniques to overcome it obtaining uniform accuracy are studied in detail in the PhD thesis [1].

Note that CRK4 is actually third order accurate in time. However, fourth order accuracy appears at the resolution used in the test, because space error is usually dominant than time discretization error at such resolutions.

We end with a Riemann problem, with data $[\rho, m, z] = [2, 1, 1]$ for $x < 0.2$ and $[\rho, m, z] = [1, 0.13962, 1]$ for $x > 0.2$, on the computational domain $[-1, 1]$,

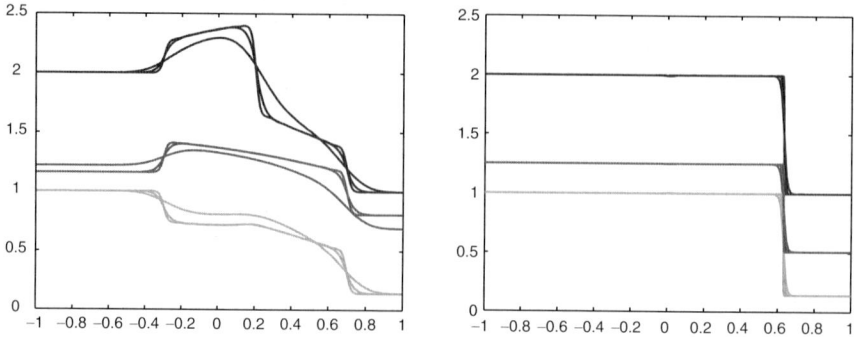

Fig. 2. Riemann problem for the Broadwell model. *Left*: comparison of schemes of order 1, 2, and 4 for $\varepsilon = 1$. *Right*: comparison of several grid spacings for the high order CRK4 scheme with $\varepsilon = 10^{-8}$

with free flow boundary conditions. Figure 2 shows a comparison of several schemes of the CRK family (orders 1, 2, and 4) for $\varepsilon = 1$, while on the right the solution of the CRK4 scheme with $\varepsilon = 10^{-8}$ is shown. In the figure, all three components of the solution are shown in different shades of grey. We note the increase in accuracy as the order is increased, and the high and nonoscillatory resolution obtained with the high order scheme on the discontinuities resulting from the stiff problem.

References

1. S. BOSCARINO, *Uniform accuracy of Implicit-Explicit Runge Kutta methods*, PhD thesis, University of Catania, 2006.
2. D. LEVY, G. PUPPO, AND G. RUSSO, *Central WENO Schemes for Hyperbolic Systems of Conservation Laws*, Math. Model. and Numer. Anal., 33 (1999), pp. 547–571.
3. L. PARESCHI, G. PUPPO, AND G. RUSSO, *Central Runge Kutta schemes for Conservation Laws*, SIAM J. Sci. Comp., 26 (2005), pp. 979–999.
4. L. PARESCHI AND G. RUSSO, *Implicit-explicit Runge-Kutta methods and applications to hyperbolic systems with relaxation*, J. Scientific Computing, 25 (2005), pp. 129–155.
5. G. PUPPO AND G. RUSSO, *Staggered Finite Difference Schemes for Balance Laws*, in Hyperbolic Problems: Theory, Numerics, Applications, Yokohama publishers, 2006, pp. 243–250.
6. C. W. SHU, *Essentially Non-Oscillatory and Weighted Essentially Non-Oscillatory Schemes for Hyperbolic Conservation Laws*, in Advanced Numerical Approximation of Nonlinear Hyperbolic Equations, Lecture Notes in Mathematics (A. Quarteroni editor), Springer, Berlin, 1998.

Flow Calculations using Shock Capturing Schemes Based on Power Limiters

Susana Serna

Institute of Geophysics and Planetary Physics, University of California,
Los Angeles, CA, USA
serna@math.ucla.edu

1 Introduction

In this research work we address the issue of the use of slope limiters to design high order reconstruction procedures when combined with shock capturing schemes for the approximation of hyperbolic conservation laws.

We compare WENO ([JS96]) reconstruction procedure with the Weighted PowerENO one introduced in [SM04] defined as a result of applying a weaker slope limiter (*powereno* limiter) on second order differences than the one (*mineno* limiter) used by WENO. We compute with both methods the solution of the compressible Rayleigh–Taylor instability where complex structure appear. The growth of this instability is sensitive to numerical diffusion; therefore, reduced viscosity and high resolution of the contact discontinuity is important [MOS92, SZS03]. Weighted PowerENO resolves fine structure with reduced viscosity compared with WENO.

2 Power Limiters and Weighted PowerENO Method

The main goal of high order methods is to reduce smearing at discontinuities with high accuracy along smooth regions of the flow avoiding Gibbs phenomena.

ENO procedures [HEOC] use the smoothest polynomial interpolation by choosing the divided differences of smallest size following a tree-like algorithm. This selection procedure consists of a limiter function (*mineno* limiter) acting on the successive divided differences of the data.

For methods of order of accuracy larger than two, as parabolic ENO methods, second order differences need to be limited to ensure local total variation bounded property [HEOC, SM04]. However, when limiting second order differences, small scales may be destroyed if a very strong limiter, like the one used for ENO methods, is applied. The main effect on the numerical solution is the increasingly smearing across contact discontinuities as time advances.

In Serna and Marquina [SM04] a new class of limiters, the so-called *power* limiters, were introduced to be applied on neighboring second order differences to define new five point stencil ENO parabolic reconstruction procedures.

Power limiters are functions of two variables based on an average (power mean) of two nonnegative numbers. The $\text{power}_p(x, y)$ mean for $x > 0$, $y > 0$, and p a positive integer is defined in [SM04] as

$$\text{power}_p(x, y) = \frac{(x + y)}{2} \left(1 - \left| \frac{x - y}{x + y} \right|^p \right). \tag{1}$$

The corresponding limiters are defined for any x and y as

$$\text{powermod}_p(x, y) = \frac{(\text{sgn}(x) + \text{sgn}(y))}{2} \text{power}_p(|x|, |y|), \tag{2}$$

$$\text{powereno}_p(x, y) = \text{minsign}(x, y) \text{power}_p(|x|, |y|), \tag{3}$$

where $\text{sgn}(x)$ is the sign function, and

$$\text{minsign}(x, y) = \begin{cases} \text{sgn}(x), & |x| <= |y|, \\ \text{sgn}(y) & \text{otherwise.} \end{cases}$$

The following inequalities are satisfied for any $x > 0$ and $y > 0$ and for $0 < p < q$ [SM04]:

$$\min(x, y) = \text{power}_1(x, y) \leq \text{power}_p(x, y) \leq \text{power}_q(x, y) \leq \frac{x + y}{2}.$$

The *minimum* is the strongest average among all power_p for $p > 1$. A limiter based on power_p mean is weaker as larger p is chosen.

Power ENO reconstruction procedure [SM04] is a third order accuracy method defined as a correction of third order ENO method by applying the powereno_3 limiter on second order differences. Weighted PowerENO is a fifth order accuracy reconstruction procedure written as a convex combination of the third order accurate Power ENO parabolas. In a similar way as WENO procedure is designed in [JS96], Weighted PowerENO results uniformly fifth order accurate in smooth regions.

Weighted PowerENO presents substantially reduced smearing near discontinuities and good resolution of corners and local extrema compared with WENO [SM04]. The reduced viscosity of Weighted PowerENO is due to the use of a weaker limiter on second order differences. This behavior is important to capture fine structure generated at contact discontinuities along the evolution of complex flows.

3 Numerical Method

Let us consider a system of conservation laws of the form

$$\mathbf{u}_t + (\mathbf{f}(\mathbf{u}))_x + (\mathbf{g}(\mathbf{u}))_y = \mathbf{S}(\mathbf{u}), \tag{4}$$

with the initial data $\mathbf{u}(x, y, 0) := \mathbf{u}_0(x, y)$.

We introduce the computational grid $x_j = j\Delta x$, $y_k = k\Delta y$, (Δx and Δy are the spatial steps), $t_n = n\Delta t$, the time discretization, (Δt is the time step), $I_j^x = [x_{j-\frac{1}{2}}, x_{j+\frac{1}{2}}]$, and $I_k^y = [y_{k-\frac{1}{2}}, y_{k+\frac{1}{2}}]$ are the spatial cells, and $C_{jk}^n = I_j^x \times I_k^y \times [t_n, t_{n+1}]$ is the computational cell, where $x_{j+\frac{1}{2}} = x_j + \frac{\Delta x}{2}$ and $y_{k+\frac{1}{2}} = y_k + \frac{\Delta y}{2}$.

Let $\mathbf{u}_{jk}^n$ be an approximation of the mean value in $I_j^x \times I_k^y$

$$\frac{1}{\Delta x \Delta y} \int_{x_{j-\frac{1}{2}}}^{x_{j+\frac{1}{2}}} \int_{y_{k-\frac{1}{2}}}^{y_{k+\frac{1}{2}}} \mathbf{u}(x, y, t_n) \, dx dy \tag{5}$$

of the exact solution $\mathbf{u}(x, y, t_n)$ of the initial value problem (4) obtained from a finite difference scheme in conservation form:

$$\mathbf{u}_{j,k}^{n+1} = \mathbf{u}_{j,k}^n - \frac{\Delta t}{\Delta x}(\tilde{\mathbf{f}}_{j+\frac{1}{2},k} - \tilde{\mathbf{f}}_{j-\frac{1}{2},k}) - \frac{\Delta t}{\Delta y}(\tilde{\mathbf{g}}_{j,k+\frac{1}{2}} - \tilde{\mathbf{g}}_{j,k-\frac{1}{2}}) + \Delta t \; \mathbf{S}(\mathbf{u}_{j,k}^n), \tag{6}$$

where the numerical fluxes, $\tilde{\mathbf{f}}$ and $\tilde{\mathbf{g}}$, are functions of $2l$ variables

$$\tilde{\mathbf{f}}_{j+\frac{1}{2},k} = \tilde{\mathbf{f}}(\mathbf{u}_{j-l+1,k}^n, \cdots \mathbf{u}_{j+l,k}^n) \quad \text{and} \quad \tilde{\mathbf{g}}_{j,k+\frac{1}{2}} = \tilde{\mathbf{g}}(\mathbf{u}_{j,k-l+1}^n, \cdots \mathbf{u}_{j,k+l}^n)$$

that are consistent with the fluxes of (4), $\tilde{\mathbf{f}}(\mathbf{u}, \cdots, \mathbf{u}) = \mathbf{f}(\mathbf{u})$ and $\tilde{\mathbf{g}}(\mathbf{u}, \cdots, \mathbf{u}) = \mathbf{g}(\mathbf{u})$.

To approximate the numerical fluxes at cell interfaces, we use the Marquina flux formula (MFF) [DM96]. MFF prescribes the appropriate viscosity to be stable and to develop the physically consistent features of the shock wave phenomena. This flux formula computes the numerical flux $\mathbf{F}^M(\mathbf{u}_l, \mathbf{u}_r)$ by performing a characteristic field decomposition at $\mathbf{u}_l$ and $\mathbf{u}_r$, using Godunov's method for nontransonic local characteristic fields and local Lax–Friedrichs method for transonic ones. Thus, the first order scheme based on MFF is

$$\mathbf{u}_{jk}^{n+1} = \mathbf{u}_{jk}^n - \frac{\Delta t}{\Delta x}\left(\mathbf{F}^M(\mathbf{u}_{jk}^n, \mathbf{u}_{j+1,k}^n) - \mathbf{F}^M(\mathbf{u}_{j-1,k}^n, \mathbf{u}_{j,k}^n)\right)$$

$$\tag{7}$$

$$- \frac{\Delta t}{\Delta y}\left(\mathbf{G}^M(\mathbf{u}_{jk}^n, \mathbf{u}_{j,k+1}^n) - \mathbf{G}^M(\mathbf{u}_{j,k-1}^n, \mathbf{u}_{j,k}^n)\right) + \Delta t \; \mathbf{S}(\mathbf{u}_{jk}^n),$$

with $\mathbf{u}_{jk}^0 = \mathbf{u}(x_j, y_k, 0)$, $\mathbf{F}^M(\mathbf{u}, \mathbf{u}) = \mathbf{f}(\mathbf{u})$, and $\mathbf{G}^M(\mathbf{u}, \mathbf{u}) = \mathbf{g}(\mathbf{u})$.

Higher order of accuracy in space is obtained by applying fifth order accurate reconstruction procedures on local characteristic variables and local characteristic fluxes extrapolating them to the left and right states of cell interfaces following the so-called Shu–Osher "flux formulation," [SO89]. We integrate in time using the third order Runge–Kutta method [SO89]. The scheme is stable under a CFL restriction.

3.1 Rayleigh–Taylor Instability

Rayleigh–Taylor instability is generic to a wide range of physical phenomena, ([MOS92] and references there in). This type of instability is generated as a consequence of a heavy fluid driven into a light one under the acceleration of a gravitational field. In the initial state of the Rayleigh–Taylor instability an unstable interface separates two fluids of different densities. A persistent acceleration (gravity field) causes the perturbation to grow as the heavier fluid pushes through the perturbation. Long spikes of the heavier fluid fall into the lighter fluid at the same time as bubbles of the lighter fluid rise into the heavier one. The perturbation grows exponentially in time.

Rayleigh–Taylor instability can be described by means of the two-dimensional Euler equations of compressible gas dynamics, (4), with

$$
\mathbf{u} = \begin{bmatrix} \rho \\ \rho u \\ \rho v \\ E \end{bmatrix} \quad \mathbf{f}(\mathbf{u}) = \begin{bmatrix} \rho u \\ P + \rho u^2 \\ \rho uv \\ u(E + P) \end{bmatrix} \quad \mathbf{g}(\mathbf{u}) = \begin{bmatrix} \rho v \\ \rho uv \\ P + \rho v^2 \\ v(E + P) \end{bmatrix} \quad \mathbf{S}(\mathbf{u}) = \begin{bmatrix} 0 \\ 0 \\ \rho g \\ \rho v g \end{bmatrix},
$$

where ρ is the density, u and v are the components of the velocity in both directions, P is the pressure, g is the constant gravity field, and E is the total energy, $E = \frac{P}{(\gamma-1)} + \frac{1}{2}\rho(u^2 + v^2)$. We consider the ideal gas equation of state to compute the pressure $P = (\gamma - 1)\rho\epsilon$.

The numerical experiments are performed on a rectangular domain $[0, 0.25] \times [0, 1]$, with the initial data

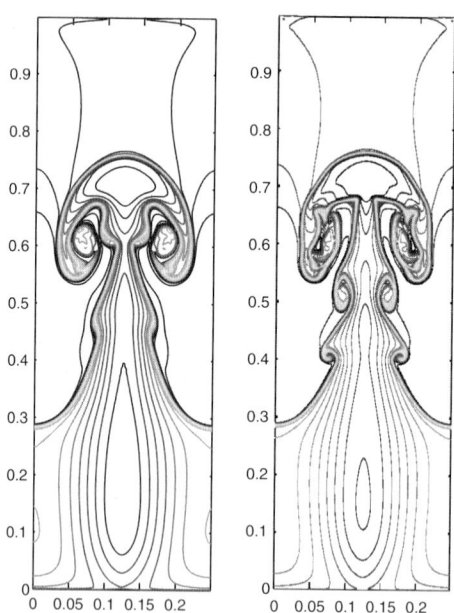

Fig. 1. *Left*: MFF-WENO. *Right*: MFF-Weighted PowerENO

$$(\rho, u, v, P) = \begin{cases} (2, 0, -0.25 \cdot c \cdot \cos(8\pi x), 2y + 1) & 0 \le y < 0.5 \\ (1, 0, -0.25 \cdot c \cdot \cos(8\pi x), y + 1.5) & 0.5 \le y < 1, \end{cases}$$

where $c = \sqrt{\frac{\gamma P}{\rho}}$ is the sound speed. In this experiment we consider $\gamma = \frac{5}{3}$, $g = 1$ and evolve until time $t = 1.95$.

Reflective boundary conditions are imposed for the left and right boundaries. At the top boundary, $(y = 1)$, the flow values are set as $(\rho, u, v, P) = (1, 0, 0, 2.5)$ and at the bottom boundary, $(y = 0)$, as $(\rho, u, v, P) = (2, 0, 0, 1)$.

Both experiments have been implemented using a grid of 200×800 points and a CFL factor of 0.6. Along the evolution, the small initial sinusoidal perturbation grows into a mushroomshaped object and develops side rolls as displayed in Fig. 1. We observe much better resolution for the Weighted PowerENO method in complicated solution structure than WENO method. In particular the Kelvin–Helmholtz vortex structure appearing in the evolution is better resolved for the Weighted PowerENO method.

Acknowledgements

Grant MTM2005-07708 is acknowledged.

References

[DM96]   Donat, R. and Marquina, A. Capturing Shock Reflections: An improved Flux Formula, J. Comput. Phys. 125 (1996) 42–58.

[HEOC]   Harten, A., Engquist, B., Osher, S. Chakravarthy, S: Uniformly high order accurate essentially non-oscillatory schemes III, J. Comput. Phys. 71 (2) 1987, 231–303.

[JS96]   Jiang, G.S., and Shu, C.W.: Efficient implementation of weighted ENO schemes, J. Comput. Phys. 126 (1996) 202–228.

[MOS92]   Mulder, W., Osher, S., Sethian, J.A.: Computing interface motion in compressible gas dynamics. J. Comput. Phys., 100, 209–228, (1992)

[SM04]   Serna, S., Marquina, A.: Power ENO methods: A fifth-order accurate Weighted Power ENO method. J. Comput. Phys., 194, 632–658 (2004)

[SZS03]   Shi, J., Zhang, YT., Shu, CW.: Resolution of high order WENO schemes for complicated flow structures. J. Comput. Phys., 186, 690–696 (2003)

[SO89]   Shu, CW, Osher, SJ: Efficient Implementation of Essentially Non-Oscillatory Shock Capturing Schemes II, J. Comput. Phys. 83, 1989, 32–78.

A Comparison Between Relaxation and Kurganov–Tadmor Schemes

Fausto Cavalli[1], Giovanni Naldi[1], Gabriella Puppo[2], and Matteo Semplice[1]

[1] Dipartimento di Matematica, Universit di Milano, Via Saldini 50, 20133 Milano, Italy
{cavalli, naldi, semplice}@mat.unimi.it
[2] Dipartimento di Matematica, Politecnico di Torino, Corso Duca degli Abruzzi, 24, 10129 Torino, Italy
gabriella.puppo@polito.it

Summary. In this work we compare two semidiscrete schemes for the solution of hyperbolic conservation laws, namely the relaxation [JX95] and the Kurganov Tadmor central scheme [KT00]. We are particularly interested in their behavior under small time steps, in view of future applications to convection diffusion problems. The schemes are tested on two benchmark problems, with one space variable.

1 Motivation

We are interested in the solution of systems of equations of the form

$$u_t + f_x(u) = D\,p_{xx}(u), \qquad (1)$$

where $f(u)$ is hyperbolic, i.e., the Jacobian of f is provided with real eigenvalues and a basis of eigenvectors for each u, while $p(u)$ is a nondecreasing Lipschitz continuous function, with Lipschitz constant μ and $D \geq 0$.

We continue the study of convection diffusion equations with the aid of high order relaxation schemes started in [CNPS06] for the case of the purely parabolic problem.

In many applications, such as multiphase flows in porous media, $p(u)$ is nonlinear and possibly degenerate. In these conditions, an implicit solution of the diffusion term can be computationally very expensive: in fact it may be necessary to solve large nonlinear algebraic systems of equations which, moreover, can be singular at degenerate points, i.e., where $p(u) = 0$. For this reason, it is of interest to consider the *explicit* solution of (1). This in turn poses one more difficulty. An explicit solution of (1) requires a parabolic CFL condition, that is, stability will restrict the possible choice of the time step Δt to $\Delta t \leq C(\Delta x)^2$, where Δx is the grid spacing. In other words, it may be necessary to choose very small time steps. But conventional solvers for convective operators typically work at their best for time steps close to a *convective* CFL, i.e., $\Delta t \leq C\Delta x$. When the time step is much smaller, they exhibit a very large artificial diffusion of the form $O((\Delta x)^{2r}/\Delta t)$, where r is the accuracy of the scheme, see for instance [KT00]. Clearly in these conditions artificial diffusion becomes very large for $\Delta t \to 0$.

As a first step to the numerical solution of problem (1), we concentrate on semidiscrete schemes for the solution of the convective part of (1). Such schemes enjoy an artificial diffusion which depends weakly on Δt, and are therefore particularly suited for the solution of convection–diffusion equations.

We will compare two semidiscrete methods for the integration of systems of hyperbolic equations. We are interested in the representation of solutions which can be characterized by strong gradients, and in the degenerate case, even by discontinuities. Moreover, we are interested in comparing the behavior of the schemes for small values of Δt, and for such small values of the time step, we will investigate the resolution of discontinuous solutions and the behavior of the error in a few test problems.

The schemes analyzed in this work are the Kurganov Tadmor central scheme proposed in [KT00], and the relaxation scheme proposed in [JX95]. These methods discretize the equations starting from very different ideas; however, they share some interesting properties. First of all, they are both semidiscrete schemes. Therefore, they require separate discretizations in space and time, which is the key to the fact that artificial diffusion depends mainly on space discretization. Secondly, they are both Riemann solver free methods. The Kurganov–Tadmor scheme is based on a central approach: the solution of the Riemann problem is computed on a staggered cell, before being averaged back on the standard grid. In this fashion, the numerical solution is updated on the edges of the staggered grid, where it is smooth, and can be computed via a Taylor expansion, with no need to solve the actual Riemann problem. The relaxation scheme instead moves the nonlinearities of the convective term to a stiff source term, and the transport part of the system becomes linear, with a fixed and well known characteristic structure. Thus again there is no need to use approximate or exact Riemann solvers.

For these reasons both schemes can be applied as black-box methods to a fairly general class of balance laws.

2 Results

For the Kurganov–Tadmor (KT) scheme we have followed the componentwise implementation of the method described in [KT00]. The scheme is written in conservation form, with numerical flux

$$F_{j+1/2}(t) = \tfrac{1}{2} \left[f(u^+_{j+1/2}(t)) + f(u^-_{j+1/2}(t)) \right.$$
$$\left. -a_{j+1/2}(t)\left(u^+_{j+1/2}(t) - u^-_{j+1/2}(t)\right)\right], \tag{2}$$

where $u^+_{j+1/2}(t)$ and $u^-_{j+1/2}(t)$ are the boundary extrapolated data, computed at the edges of each cell with a piecewise linear reconstruction at time t, and

$a_{j+1/2}(t)$ is a measure of the maximum propagation speed at the cell edge. For the case of systems of equations, in particular in the nonconvex case, this value must be carefully tuned, and it is the same for all components, when the scheme is implemented componentwise.

On the other hand, the relaxation scheme requires an accurate choice of the subcharacteristic velocities A^2. The relaxation system is

$$
\begin{cases}
\dfrac{\partial \mathbf{u}}{\partial t} + \dfrac{\partial \mathbf{v}}{\partial x} = 0 \\[2mm]
\dfrac{\partial \mathbf{v}}{\partial t} + A^2 \dfrac{\partial \mathbf{u}}{\partial x} = -\dfrac{1}{\varepsilon}\left(\mathbf{v} - f(\mathbf{u})\right).
\end{cases}
\tag{3}
$$

As $\varepsilon \to 0$, the system (3) formally relaxes to the original conservation laws, provided the subcharacteristic condition holds, namely that $(A^2 - (f'(u))^2)$ is positive-definite.

For a scalar conservation law, we take $A^2 = \max(|f'(u)|)$ as in [JX95], while for the Euler system of gas-dynamics we take A^2 to be the diagonal matrix with entries $\max_j(|u_j - c_j|)$, $\max_j(|u_j|)$, and $\max_j(|u_j + c_j|)$. Here u is the velocity and c is the speed of sound. We update these quantities at each time step, so that A^2 can be chosen as small as possible (in the paper [JX95] A^2 was chosen as a constant diagonal matrix but this results in a larger numerical diffusion).

Because of the diagonal form of A^2, the convective operator is block diagonal with 2×2 blocks. Each block is independently diagonalized and we compute the numerical fluxes using a second order ENO reconstruction [HEOC87].

We use the second order Heun Runge–Kutta method for the time integration of both the KT and the relaxation schemes.

Table 1 shows the errors in the L^1 norm for the linear advection equation $u_t + u_x = 0$ with initial data $u(x,0) = \sin(2\pi x)$. We use the standard convective CFL condition $\Delta t = C\Delta x$ and the parabolic CFL, $\Delta t = C(\Delta x)^2$. We note that the errors are almost the same for the two schemes for the convective CFL, while the relaxation scheme seems superior for the parabolic CFL.

Table 1. Linear advection of a sine function

	Convective CFL		Parabolic CFL	
	KT	Relax	KT	Relax
20	2.03E−1	2.16E−1	6.19E−1	1.02E−1
40	7.58E−2	7.66E−2	2.04E−1	4.58E−2
80	2.71E−2	2.73E−2	9.10E−2	1.34E−2
160	8.22E−3	8.25E−3	2.67E−2	3.82E−3
320	2.29E−3	2.29E−3	7.62E−3	1.03E−3
640	6.11E−4	6.12E−4	2.06E−3	2.77E−4
1280	1.61E–4	1.61E–4		

Errors in L^1 at $t = 1$.

A key requirement for a numerical scheme for conservation laws is the ability to pick the entropy solution in nonconvex problems. Here we show a Riemann problem for the nonconvex flux $f(u) = (u^2 - 1)(u^2 - 4)/4$, as in [KT00]. The Riemann problem breaks into two shocks connected by a rarefaction wave. The results are shown in Fig. 1. Clearly both schemes are able to resolve the correct discontinuities and they have approximately the same resolution, the KT scheme being slightly less diffusive.

Figure 2 shows the density component of the Lax Riemann problem in gas dynamics. The accurate choice suggested above for the matrix A^2 in the relaxation system yields a slightly higher resolution than KT.

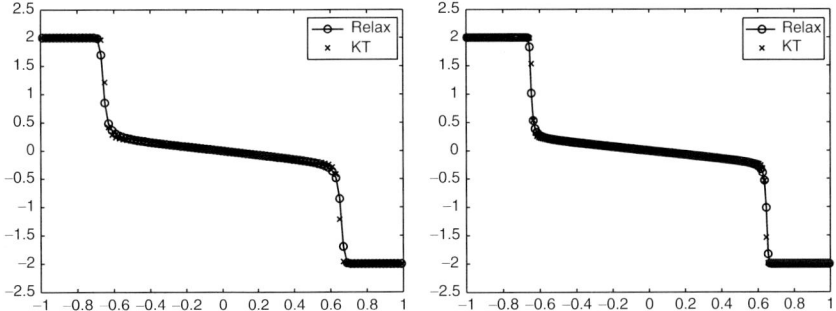

Fig. 1. Nonconvex flux. Kurganov–Tadmor and relaxation schemes, with $n = 100$ (*left*) and $n = 200$ (*right*)

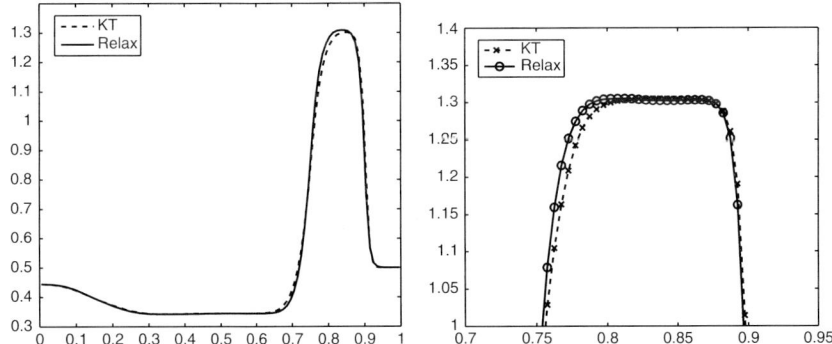

Fig. 2. Lax Riemann problem, density component. KT (*dashed*) and relaxation (*solid line*) with $n = 100$ (*left*) and $n = 200$ (*right*), where a detail of the density peak is shown

3 Concluding Remarks

We have compared two semidiscrete schemes for conservation laws. We find that although the schemes are constructed with very different philosophies, they yield comparable results on some significant test problems. We think that the relaxation scheme is slightly more robust, since it results from the relaxation of a viscous profile, provided the subcharacteristic condition is satisfied. Also, the actual errors obtained with a parabolic CFL in Table 1 seem to favor the relaxation scheme.

We also wish to mention higher order extensions of the schemes studied in this work: namely the third order central upwind scheme described in [KNP01], endowed with a more carefully crafted artificial diffusion with respect to [KT00] and the third order extension of the relaxation scheme proposed in [Sea06].

References

[CNPS06] F. Cavalli, G. Naldi, G. Puppo, and M. Semplice. High order relaxation schemes for non linear diffusion problems. 2006. Submitted to SINUM.

[HEOC87] Ami Harten, Björn Engquist, Stanley Osher, and Sukumar R. Chakravarthy. Uniformly high-order accurate essentially nonoscillatory schemes. III. *J. Comput. Phys.*, 71(2):231–303, 1987.

[JX95] S. Jin and Z. Xin. The relaxation schemes for systems of conservation laws in arbitrary space dimensions. *Comm. Pure and Appl. Math.*, 48:235–276, 1995.

[KNP01] Alexander Kurganov, Sebastian Noelle, and Guergana Petrova. Semidiscrete central-upwind schemes for hyperbolic conservation laws and Hamilton-Jacobi equations. *SIAM J. Sci. Comput.*, 23(3):707–740 (electronic), 2001.

[KT00] Alexander Kurganov and Eitan Tadmor. New high-resolution central schemes for nonlinear conservation laws and convection-diffusion equations. *J. Comput. Phys.*, 160(1):241–282, 2000.

[Sea06] Mohammed Sead. High-resolution relaxation scheme for the two-dimensional Riemann problems in gas dynamics. *Numer. Methods Partial Differential Equations*, 22(2):397–413, 2006.

Minisymposium "Multibody Dynamics"

B. Simeon

Technische Universität München, Germany

In this minisymposium, Christoph Lunk, Bernd Simeon (TU München) present "The Reverse Method of Lines in Flexible Multibody Dynamics," J. Linn, T. Stephan (Fraunhofer ITWM, Kaiserslautern), J. Carlsson, and R. Bohlin (Fraunhofer–Chalmers Research Center FCC, Göteborg) present "Fast simulation of quasistatic rod deformations for VR applications," and Michael Speckert and Klaus Dreßler (Fraunhofer ITWM, Kaiserslautern) present "Simulation and optimization of suspension testing systems."

The Reverse Method of Lines in Flexible Multibody Dynamics

Christoph Lunk and Bernd Simeon

Zentrum Mathematik, TU München, D-85748 Garching, Germany
{simeon|lunk}@ma.tum.de

Summary. Adaptivity is a crucial prerequisite for efficient and reliable simulations. In multibody dynamics, adaptive time integration methods are standard today, but the treatment of elastic bodies is still based on an a priori fixed spatial discretization. This contribution introduces a basic algorithm in the fashion of the reverse method of lines that is able to adapt both the spatial grid and the time step size from step to step. The example of a catenary with a moving pantograph head illustrates the approach.

1 Introduction

Flexible multibody systems are aimed at the growing simulation demands in vehicle dynamics, robotics, and in air- and spacecraft development. These mixed systems contain both rigid and elastic bodies as well as the usual interconnections like joints that constrain the motion of pairs of bodies or springs and dampers that act as compliant elements.

The mathematical model of a flexible multibody system consists of a set of ordinary differential equations (ODEs) or differential-algebraic equations (DAEs), which are coupled with some partial differential equations (PDEs). The PDEs are the equations of elasto dynamics for the deformation of bodies while the ODE or DAE part describes the so-called gross motion, i.e., spatial translations and rotations. Beams are the most frequent elastic members, but plates and shells and even full 3D structures have also become widespread in multibody formalisms. We refer to [S98, S06] for an extensive survey on the underlying mathematical models.

Simulation methods for flexible multibody systems typically employ first a space discretization of the PDEs. This reduces the overall equations to an extended system of ODEs or DAEs. Standard interfaces between finite element and multibody codes facilitate this step considerably. The resulting semidiscretized system involves two types of state variables, namely, those for the gross motion and those for the deformations.

From the point of numerical analysis, flexible multibody systems lead to two key problems that deserve particular attention:

1. Good approximation of elastic deformation and low system dimension are contradictory goals. In particular, in complex applications, the result of the semidiscretization in space depends strongly on engineering judgement since adaptive grids are not available so far.
2. Gross motion and elastic deformation may have widely different time scales, and this turns the time integration into a challenging problem.

This chapter addresses mainly problem (1). We propose an algorithm in the fashion of the reverse method of lines in order to combine adaptivity in time and space. The time integrator in this approach, however, has to cope with problem (2), which means that stability and numerical dissipation are important features.

This chapter is organized as follows: As a starting point, Sect. 2 summarizes the equations of motion both in the rigid and elastic case. Section 3 introduces the time integration scheme along with a sketch of the reverse method of lines, and finally in Sect. 4 simulation results for a catenary system with moving pantograph head are presented.

2 Equations of Motion

As a point of departure, we consider a mechanical system composed of rigid bodies only and denote by the vector $q(t) \in \mathbb{R}^{n_q}$ the position coordinates of all bodies depending on time t. According to Euler and Lagrange, the equations of constrained mechanical motion read

$$M\,\ddot{q} = f(q, \dot{q}, t) - G^T(q)\lambda, \qquad (1a)$$

$$0 = g(q). \qquad (1b)$$

Here, M stands for the mass matrix, f for the applied forces, g for the holonomic constraints, and $G = \partial g/\partial q$ for the constraint Jacobian. Besides the position coordinates q, the Lagrange multipliers $\lambda(t) \in \mathbb{R}^{n_\lambda}$, $n_\lambda < n_q$, are also unknowns. The equations of motion (1) form a system of index 3 if the constraint Jacobian G has full rank, which is equivalent to the min–max condition

$$\min_{\lambda} \max_{v} \frac{\lambda^T G v}{\|\lambda\|_2 \|v\|_2} = \sigma_{\min}(G) > 0$$

with minimum singular value $\sigma_{\min}$.

In case of elastic bodies, the mathematical model involves a coupling of the above equations of motion with the PDEs that govern the deformation. However, in the engineering literature on flexible multibody systems, mathematical modeling and numerical treatment are often intertwined. The elastic body is first discretized in space, which results in a finite dimensional structure. Thereafter, the equations of motion and the coupling conditions are formulated in terms of certain shape functions or finite element displacements.

If we consider a single elastic body occupying the domain $\Omega \subset \mathbb{R}^3$ and neglect the gross motion for the moment, i.e., assume the setting of linear elasticity, Cauchy's equations are

$$\rho \ddot{u} = \operatorname{div} \sigma(u) + \beta \quad \text{in } \Omega \tag{2}$$

with boundary conditions $u = u_0$ on Γ_0 and $\sigma(u)n = \tau$ on Γ_1. The displacement field $u(x,t) \in \mathbb{R}^3$ satisfies thus a generalized wave equation where ρ denotes the mass density, $\beta(x,t)$ the density of body forces, $\sigma(u) \in \mathbb{R}^{3\times3}$ the stress tensor, $u_0(x,t)$ the Dirichlet boundary conditions, and $\tau(x,t)$ the surface tractions with normal vector $n(x)$. Hooke's law relates stress tensor σ and strain tensor $\varepsilon = 1/2(\nabla u + \nabla u^T)$ via $\sigma(u) = C \cdot \varepsilon(u)$ with elasticity tensor C. Note that the bounday condition $u = u_0$ on Γ_0 is a constraint in the multibody context since $u_0 = u_0(q)$ may depend on the motion of neighboring bodies. For a more detailed discussion of this aspect and the corresponding saddle point problem formulation see [S06].

Using a floating frame of reference approach, we take rotation and translation in space into account via

$$\varphi(x,t) = y(t) + A(\alpha(t))(x + u(x,t)). \tag{3}$$

Here, $\varphi(x,t) \in \mathbb{R}^3$ is the motion of a material point of the elastic body, $y(t) \in \mathbb{R}^3$ is the translation between the inertial and the floating frame, and $A(\alpha) \in SO(3)$ is a rotation matrix that depends on the angles α.

A space discretization is introduced by the Galerkin projection

$$u(x,t) \doteq N_u(x) \cdot q_e(t) \tag{4}$$

where $N_u(x) \in \mathbb{R}^{3\times n_{qe}}$ is a matrix of known global shape functions and $q_e(t) \in \mathbb{R}^{n_{qe}}$ is the vector of corresponding displacement coefficients. The equations of motion of a flexible multibody system follow now from the same variational principle as in the rigid body case, and we obtain a differential-algebraic system that has the same structure as (1). The unknowns q split into $q = (q_r, q_e)$ with rigid motion variables q_r and elastic displacements q_e.

3 Time Integration and Reverse Method of Lines

Our choice of the time integration scheme for the equations of motion (1) is inspired by the following requirements:

- Both position constraint $0 = g(q)$ and velocity constraint $0 = G(q)\dot{q} = G(q)v$ are used to stabilize the discretization.
- The acceleration constraint need not be evaluated.

- The computational effort is comparable to a BDF method for the stabilized equations of motion.
- Adjustable numerical dissipation is available.

As described in [LS06], the following method meets these requirements and is particularly suited for applications in flexible multibody dynamics. One time step of the α-RATTLE method for position $\boldsymbol{q}_{n+1}$, velocity $\boldsymbol{v}_{n+1}$, and acceleration $\boldsymbol{a}_{n+1}$ is given by

$$\boldsymbol{M}\frac{\boldsymbol{p}_{n+1} - \boldsymbol{p}_n}{h} = \boldsymbol{M}\left(\boldsymbol{v}_n + h(\frac{1}{2} - \beta)\boldsymbol{a}_n + h\beta\boldsymbol{a}_{n+1}\right) - \frac{h}{2}\boldsymbol{G}_{n+1}^T\boldsymbol{\lambda}_{n+1}, \quad (5a)$$

$$\boldsymbol{M}\frac{\boldsymbol{v}_{n+1} - \boldsymbol{v}_n}{h} = \boldsymbol{M}\left((1-\gamma)\boldsymbol{a}_n + \gamma\boldsymbol{a}_{n+1}\right) - \frac{1}{2}\boldsymbol{G}_n^T\boldsymbol{\lambda}_{n+1} - \frac{1}{2}\boldsymbol{G}_{n+1}^T\boldsymbol{\tau}_{n+1}, \quad (5b)$$

$$(1 - \alpha_m)\boldsymbol{M}\boldsymbol{a}_{n+1} = \alpha_f\boldsymbol{f}_n + (1 - \alpha_f)\boldsymbol{f}_{n+1} - \alpha_m\boldsymbol{M}\boldsymbol{a}_n, \quad (5c)$$

$$0 = \boldsymbol{G}_{n+1}\boldsymbol{v}_{n+1}, \quad (5d)$$

$$0 = \boldsymbol{g}_{n+1}. \quad (5e)$$

Both position and velocity constraints are thus enforced at each step. The method coefficients should satisfy the conditions $\gamma = 1/2 - \alpha_m + \alpha_f$ for second order and $-1 \le \alpha_m \le \frac{1}{2}$ for zero-stability. Thus, the remaining parameters α_m, α_f, and β can be used to adapt the method to special requirements, in particular to specify numerical dissipation.

Though the time integration method (5) has been formulated for the semi-discretized equations of motion, we may also apply it formally to the corresponding infinite dimensional problem and reverse the order of time and space discretization. In this way, the spatial grid can be adapted from time step to time step as proposed in [BS98]. The main challenge in our setting here is the saddle point structure and the presence of constraints.

In short, one time step $n \rightsquigarrow n + 1$ of the reverse method starting with mesh $\mathcal{T}_n$ and given data $\boldsymbol{M}, \boldsymbol{f}_n, \boldsymbol{G}_n$ reads

1. solve (5) for $\boldsymbol{q}_{n+1}, \boldsymbol{v}_{n|1}, \boldsymbol{a}_{n+1}, \boldsymbol{\lambda}_{n+1}, \boldsymbol{\mu}_{n+1}$;
2. compute embedded solution $\tilde{\boldsymbol{q}}_{n+1}$;
3. estimate spatial error err_x of $\boldsymbol{q}_{n+1}$;
4. if $\mathrm{err}_x > \mathrm{tol}_x$: adapt mesh $\mathcal{T}_n$; project $\boldsymbol{q}_n$; go to 1.);
5. estimate time error err_t;
6. if $\mathrm{err}_t > \mathrm{tol}_t$:
   decrease time stepsize h; project $\boldsymbol{q}_n$ if necessary; go to 1.)
   else accept step;

Clearly, this algorithm is only a rough sketch of an actual implementation, and several details like the combination of space and time errors require a more elaborate discussion that will be published in a forthcoming paper.

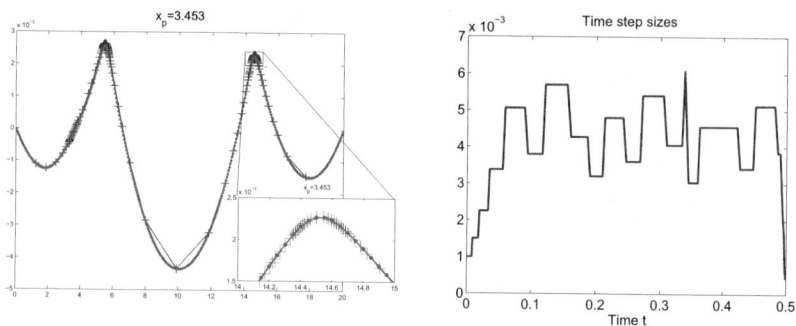

Fig. 1. Snapshot of contact wire with adapted grid (*left*) and time stepsizes (*right*)

4 Simulation Example

Finally, we present first results of the above reverse method of lines for the simulation of a catenary system with moving pantograph. We take the setting of the simple benchmark problem with two droppers as described in [AS00] and apply the α-RATTLE method (5) as time integrator. The contact wire of the catenary is discretized by cubic beam elements, and the averaging error estimator [C05] is used to adapt the spatial grid. Figure 1 shows a snapshot of the dynamic simulation and the time stepsize history for this simulation.

In conclusion, we would like to stress that the reverse method of lines represents a promising approach in flexible multibody dynamics. However, the interplay of time and space error control requires further investigations, and the implementation in a more general setting needs still to be done.

References

[AS00] Arnold, M., Simeon, B.: Pantograph and catenary dynamics: a benchmark problem and its numerical solution. Appl. Numer. Math. **34**, 345-362 (2000)

[BS98] Bornemann, F., Schemann, M.: An Adaptive Rothe Method for the Wave Equation. Comput. Visual Sci. **1**, 137-144 (1998)

[C05] Carstensen, C.: A unifying theory of a posteriori finite element error control. Numer. Math. **100** 617–637 (2005)

[LS06] Lunk, Ch., Simeon, B.: Solving constrained mechanical systems by the family of Newmark and α-methods. ZAMM **86**, 772–784 (2006)

[S98] Shabana, A.: Dynamics of Multibody Systems. Cambridge University Press, Cambridge 1998

[S06] Simeon, B.: On Lagrange Multipliers in Flexible Multibody Dynamics. Comp. Methods Appl. Mech. Eng. **195**, 6993–7005 (2006)

Fast Simulation of Quasistatic Rod Deformations for VR Applications

J. Linn[1], T. Stephan[1], J. Carlsson[2], and R. Bohlin[2]

[1] Fraunhofer ITWM, Kaiserslautern
    (linn@itwm.fraunhofer.de)
[2] Fraunhofer–Chalmers Research Center FCC, Göteborg
    (www.fcc.chalmers.se)

Summary. We present a model of flexible rods – based on *Kirchhoff's* geometrically exact theory – which is suitable for the fast simulation of quasistatic deformations within VR or functional DMU applications. Unlike simple models of *"mass & spring"* type typically used in VR applications, our model provides a proper *coupling of bending and torsion*. The computational approach comprises a *variational formulation* combined with a *finite difference discretization* of the continuum model. Approximate solutions of the equilibrium equations for sequentially varying boundary conditions are obtained by means of *energy minimization* using a nonlinear CG method. The computational performance of our model proves to be sufficient for the interactive manipulation of flexible cables in assembly simulation.

1 Introduction

The handling of flexible objects in multibody simulation (MBS) models is both a long term research topic [1, 2] as well as an active area of current research within the MBS community [3–5]. A standard approach supported by most commercial software packages represents flexible bodies by means of vibrational modes (e.g. of *Craig–Bampton* type [6, 7]) computed by modal analysis within the framework of linear elasticity. The modal representation of a flexible structure usually yields a drastic reduction of the degrees of freedom and thereby provides a reduced model. However, such methods are suitable (as well as by definition restricted) to model forced oscillations effecting *small* deformations within a flexible structure.

If the flexible bodies of interest possess special geometrical properties characterising them as *slender* (or *thin*) structures (i.e. rods, plates or shells), their overall deformation in response to moderate external loads may become *large*, although locally the stresses and strains remain *small*. Therefore, models suitable to describe such *large deformations of slender structures* must be capable to account for *geometric nonlinearities*. Compared to object geometries that require fully three-dimensional volume modelling, the reduced dimensionality of rod or shell models is accompanied by a considerable reduction in the number of degrees of freedom, which makes the inclusion of appropriately discretised versions of the *full* models (in contrast to modally reduced ones) into a MBS framework [4] computationally feasible even for time critical simulation applications.

Modelling of Flexible Structures in VR Applications

The application aimed at within the framework of this article is the modelling of flexible cables or tubes (e.g. those externally attached to manufacturing robots) such that *quasistatic* deformations occurring during sufficiently slow motions of these cables can be simulated in *real time*. This capability is crucial for the seamless integration of a cable simulator module within VR (*virtual reality*) or FDMU (*functional digital mock up*) software packages used for *interactive simulation* (e.g. of assembly processes).

Although the dominant paradigm to assess the quality of an animation or simulation within these application areas – as well as related ones like computer games or movies – seems to be "...*It's good enough if it looks good* ..." [8], such that a mere "*fake*" [9] of structure deformation is considered to be acceptable (at least for those applications were "...*fooling of the eye* ..." [8] is the main issue), the need for "physics based" approaches increases constantly, and the usage of models that are more [12] or less [10,11,13] based on ideas borrowed from classical structural and rigid body mechanics is not uncommon, especially if the primary concern is not visual appearance but physical information (see e.g. [14]).

2 Cosserat and Kirchhoff Rod Models

In structural mechanics *slender objects* like cables, hoses, etc. are described by one-dimensional *beam* or *rod* models which utilise the fact that, due to the relative smallness of the linear dimension D of the cross-section compared to the length L of a rod, the local stresses and strains remain small and the cross sections are almost unwarped, even if the overall deformation of the rod relative to its undeformed state is large. This justifies kinematical assumptions that restrict the cross sections of the deformed rod to remain *plane and rigid*.

In the following we give brief introduction to rod models of *Cosserat* and *Kirchhoff* type, the latter being a special case of the former. We do not present the most general versions of these models, which are discussed at length in the standard references [16] and [18]. The approach we finally use as a basis for the derivation of a generalized "*mass & spring*" type model by finite difference discretisation is an extensible variant of Kirchhoff's original theory [15] as presented in [17] (see part II, 16–19) for a hyperelastic rod with symmetric cross-section subject to a constant gravitational body force.

2.1 Kinematics of Cosserat and Kirchhoff Rods

A *(special) Cosserat rod* [18] is a *framed curve* [21] formally defined as a mapping $s \mapsto (\varphi(s), \hat{\mathbf{F}}(s))$ of the interval $\mathcal{I} = [0, L]$ into the configuration space $\mathbb{R}^3 \times SO(3)$ of the rod, where L is the length of the undeformed rod. Its constituents are *(i)* a space curve $\varphi : \mathcal{I} \to \mathbb{R}^3$ that coincides with the *line of centroids* piercing the cross sections along the deformed rod at their geometrical center, and *(ii)* an "curve of frames" $\hat{\mathbf{F}} : \mathcal{I} \to SO(3)$ with the origin of each frame $\hat{\mathbf{F}}(s)$ attached to the point $\mathbf{x}_s = \varphi(s)$. The matrix representation of the frame $\hat{\mathbf{F}}(s)$ w.r.t. a fixed global coordinate system $\{\mathbf{e}^{(1)}, \mathbf{e}^{(2)}, \mathbf{e}^{(3)}\}$ of $\mathbb{R}^3$ may be written as a triple of column vectors, i.e. $\hat{\mathbf{F}}(s) = \left(\mathbf{d}^{(1)}(s), \mathbf{d}^{(2)}(s), \mathbf{d}^{(3)}(s)\right)$, obtained as $\mathbf{d}^{(k)}(s) = \hat{\mathbf{F}}(s) \cdot \mathbf{e}^{(k)}$. By definition $\mathbf{d}^{(3)}$ coincides with the unit *cross section normal* vector located at $\varphi(s)$.

For simplicity we assume the undeformed rod to be *straight and prismatic* such that its intial geometry relative to $\{\mathbf{e}^{(1)}, \mathbf{e}^{(2)}, \mathbf{e}^{(3)}\}$ is given by the direct product $\mathcal{A} \times \mathcal{I}$ with a constant cross section area $\mathcal{A}$ parallel to the plane spanned by $\{\mathbf{e}^{(1)}, \mathbf{e}^{(2)}\}$. Introducing coordinates (ξ_1, ξ_2) in the plane of the cross section $\mathcal{A}$ relative to its geometrical center, we may parametrise the material points $\mathbf{X} \in \mathcal{A} \times \mathcal{I}$ of the undeformed rod geometry by $\mathbf{X}(\xi_1, \xi_2, s) = \sum_{k=1,2} \xi_k \mathbf{e}^{(k)} + s\, \mathbf{e}^{(3)}$, and the deformation mapping $\mathbf{X} \mapsto \mathbf{x} = \boldsymbol{\Phi}(\mathbf{X})$ is given by the formula $\mathbf{x}(\xi_1, \xi_2, s) = \varphi(s) + \sum_{k=1,2} \xi_k \mathbf{d}^{(k)}(s)$. The kinematics of a framed curve as presented above determine the possible deformations of a *Cosserat* rod. These are *stretching* (in the direction of the curve tangent), *bending* (around an axis in the plane of the cross section), *twisting* (of the cross section around its normal) and *shearing* (i.e. tilting of the cross section normal w.r.t. the curve tangent).

Following Chouaieb and Maddocks [21] we denote a frame $\hat{\mathbf{F}}(s)$ as *adapted* to the curve $\varphi(s)$ if $\mathbf{d}^{(3)}(s)$ coincides with the *unit tangent* vector $\mathbf{t}(s) = \partial_s \varphi(s)/\|\partial_s \varphi(s)\|$ along the curve. An adapted frame satisfies the *Euler–Bernoulli* hypothesis, which states that the cross sections remain always orthogonal to the centerline curve in a deformed state also. Curves with adapted frames describe the possible deformations of (extensible) *Kirchhoff* rods. Compared to Cosserat rods the kinematics of Kirchhoff rods are further restricted, as they do not allow for shear deformations. The *inextensibiliy* condition $\|\partial_s \varphi\| = 1$ constitutes an additional kinematical restriction.

Measuring the slenderness of a rod of cross-section diameter D and length L in terms of the small parameter $\varepsilon = D/L$ and assuming (hyper)elastic material behaviour one may show that the potential energy terms corresponding to bending and torsion are of the order $\mathcal{O}(\varepsilon^4)$, while the energy terms corresponding to stretching and shearing scale are of the order $\mathcal{O}(\varepsilon^2)$. In this way the latter effectively act as *penalty terms* that enforce the kinematical restrictions $\|\partial_s \varphi\| = 1$ and $\mathbf{d}^{(3)}(s) = \mathbf{t}(s)$. This explains how Kirchhoff rods appear as a natural limit case of Cosserat rods subject to moderate deformations provided ε is sufficiently small.

2.2 Hyperelastic Kirchhoff Rods with Circular Cross-Section

As we are interested in a rod model that is suitable for the simulation of moderate cable deformations, both the Cosserat as well as the Kirchhoff approach would fit for our purpose. A characteristic feature of the Cosserat model consists in the description of the bending and torsion of the rod in terms of the frame variables, while the bending of the centerline curve $\varphi(s)$ is produced only indirectly via shearing forces that try to align the curve tangent to the cross section normal. In contrast to that, Kirchhoff's model [20] encodes bending strain directly by the curvature of $\varphi(s)$ and therefore provides a direct pathway to mass & spring type models formulated in terms of (discrete) dof of the centerline.

Averaging the normal Piola–Kirchhoff tractions and corresponding torques over the cross-section surface of the deformed rod located at $\varphi(s)$ yields *stress resultants* $\mathbf{f}(s)$ and *stress couples* $\mathbf{m}(s)$, i.e. resultant force and moment vectors per unit reference length [19]. If the rod is in a *static equilibrium* state, these vectors satisfy the differential balance equations of forces and moments

$$\partial_s \mathbf{f} + \mathbf{G} = \mathbf{0} , \quad \partial_s \mathbf{m} + \partial_s \varphi \times \mathbf{f} = \mathbf{0} , \tag{1}$$

where $\mathbf{G}$ represents a (not necessarily constant) body force acting along the rod, and we assumed that no external moment is applied in between the rod boundaries.

In the case of an extensible Kirchhoff rod which in its undeformed state has the form of a straight cylinder with circular cross section, the assumption of a hyperelastic material behaviour yields the expression [17]

$$\mathbf{m}(s) = EI\, \mathbf{t}(s) \times \partial_s \mathbf{t}(s) + GJ\, \Omega_t\, \mathbf{t}(s) \tag{2}$$

for the stress couple, where E is Young's modulus, G is the shear modulus, I measures the geometrical moment of inertia of the cross section ($I = \frac{\pi}{4} R^4$ for a circular cross section of radius R) and $J = 2I$. The quantities EI and GJ determine the stiffness of the rod w.r.t. bending and torsion. The strain measure related to the bending moment is given by the vector

$$\mathbf{t} \times \partial_s \mathbf{t} = \frac{\partial_s \varphi \times \partial_s^2 \varphi}{\|\partial_s \varphi\|^2} = \|\partial_s \varphi\|\, \kappa\, \mathbf{b} \tag{3}$$

which is proportional to the *Frenet curvature* $\kappa(s)$ of the centerline and (if $\kappa > 0$) points in the direction of the binormal vector $\mathbf{b}(s)$. The strain measure related to the torsional moment is determined by the *twist*

$$\Omega_t(s) = \mathbf{t}(s) \cdot [\mathbf{d}(s) \times \partial_s \mathbf{d}(s)] , \tag{4}$$

where $\mathbf{d}(s)$ is any unit normal vector field to the centerline given as a fixed linear combination $\mathbf{d} = \cos(\alpha_0)\, \mathbf{d}^{(1)} + \sin(\alpha_0)\, \mathbf{d}^{(2)}$ of the frame vectors $\mathbf{d}^{(1)}(s)$ and $\mathbf{d}^{(2)}(s)$ for some constant angle α_0. Note that the special constitutive relation (2) implies that in equilibrium the twist Ω_t is constant.

As a Kirchhoff rod is (by definition) unshearable, only the *tangential* component of the stress resultant $\mathbf{f}(s)$ is constitutively determined by the *tension*

$$\mathbf{t}(s) \cdot \mathbf{f}(s) =: T(s) = EA \left(\|\partial_s \varphi\| - 1 \right) \tag{5}$$

related to the elongational strain ($\|\partial_s \varphi\| - 1$). The resistance of the rod w.r.t. stretching is determined by EA where $A = |\mathcal{A}|$ is the size of the cross-section area (in our case $A = \pi R^2$). The *shearing force* acting parallel to the cross section is given by $\mathbf{f}_{sh}(s) = \mathbf{f}(s) - T(s)\,\mathbf{t}(s)$. It is not related to any strain measure but has to be determined from the equilibrium equations a Lagrange parameter corresponding to the internal constraint $\mathbf{d}(s) \cdot \mathbf{t}(s) = 0$.

To determine the deformation of the rod in static (or likewise quasistatic) equilibrium one has to solve the combined system of the equations (1)–(5) for a suitable set of *boundary conditions*, e.g. like those discussed in [20]. (This issue will not be discussed here.) Equivalently, one may obtain the centerline $\varphi(s)$ and the unit normal vector field $\mathbf{d}(s)$ that represents the adapted frame of the rod by *minimization of the potential energy*

$$W_{\text{pot}}[\varphi, \mathbf{d}] = \int_0^L w_{\text{el}}(s)\,ds - \int_0^L \mathbf{G}(s) \cdot \varphi(s)\,ds. \tag{6}$$

According to (2)–(5) the *elastic energy density* is a quadratic form in the various strain measures given by

$$w_{\text{el}}(s) = \frac{EI}{2} \left(\mathbf{t} \times \partial_s \mathbf{t} \right)^2 + \frac{GJ}{2} \, \Omega_t^2(s) + \frac{EA}{2} \left(\|\partial_s \varphi\| - 1 \right)^2 \tag{7}$$

and determines the *stored energy function* $W_{\text{el}}[\varphi, \mathbf{d}] = \int_0^L w_{\text{el}}(s)\,ds$ containing the internal part of W_{pot}. A specific choice of boundary conditions may be accounted for by modified expressions for $w_{\text{el}}(0)$ and $w_{\text{el}}(L)$, which are obtained from (7) by fixing combinations of the kinematical variables $\varphi(s)$ and $\mathbf{d}(s)$ and their derivatives at prescribed values (as required by the b.c.) and substituting these into $w_{\text{el}}(s)$.

3 Discrete Rod Models of "Mass & Spring" Type

The final step of our approach towards a model of flexible rods suitable for the fast computation of quasistatic rod deformations is the discretisation of the potential energy by applying standard (e.g. central) finite difference stencils to the elastic energy density (7) and corresponding quadrature rules (e.g. trapezoidal) to the energy integrals (6). (Boundary conditions are treated in the way described at the end of the previous section.)

This procedure results in a discrete model of an extensible Kirchhoff rod that has a similar structure like the simple "mass & spring" type models presented in [11] and [13]. However, as a benefit of the systematic derivation procedure on the basis of a proper continuum model, our discrete rod model is able to capture the rather subtle coupling of bending and torsion deformation.

We compute approximate solutions of the equilibrium equations for sequentially varying boundary conditions by a minimization of the discrete potential energy using a nonlinear CG method [22]. The computational efficiency of our approach is illustrated by the typical results shown in Fig. 1 above. As the calculation times are comparable to those mentioned in Gregoire and Schomer [13], we estimate that our model is suitable for the interactive manipulation of flexible cables in assembly simulation (as indicated by preliminary tests with a software package developped at FCC.)

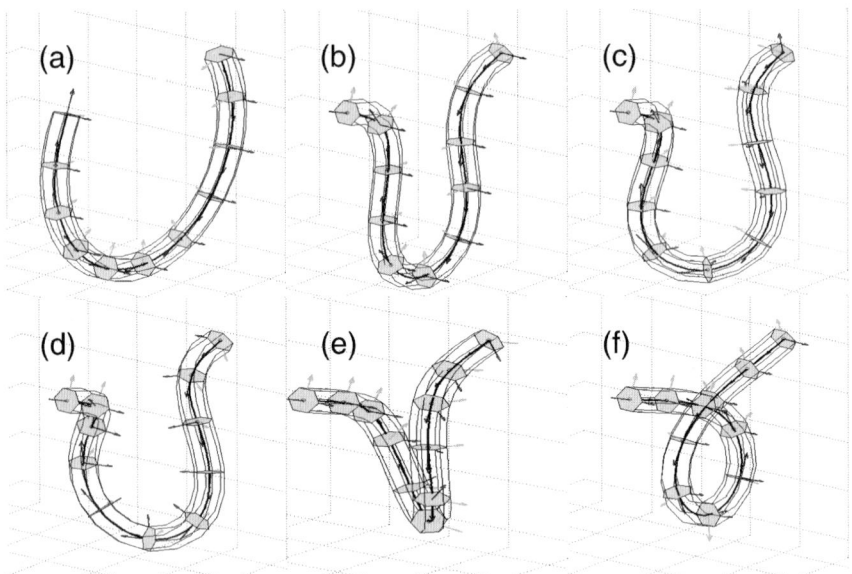

Fig. 1. Sequential deformation of a discrete, hyperelastic Kirchhoff rod of symmetric cross section: (**a**) Starting from a circle segment, the tangents of the boundary frames are bent inward to produce (**b**) an (upside down) Ω-shaped deformation of the rod at zero twist. To demonstrate the effect of mutual coupling of bending and torsion in the discrete model, the boundary frame at $s = L$ is twisted counterclockwise by an angle of 2π while the other boundary frame at $s = 0$ is held fixed. The pictures (**c**)–(**f**) show snapshots of the deformation state taken at multiples of $\pi/2$. The overall deformation from (**a**)–(**f**) was split up into a sequence of 25 consecutive changes of the boundary conditions defined by the terminal frames of the rod. For a discretization of the cable into 10 segments, the simulation took 150 ms on 1 CPU of an AMD 2.2 GHz double processor PC, which amounts to an average computation time of 6 ms per step

References

1. W.O. Schiehlen: Multibody system dynamics: Roots and perspectives, *Multibody System Dynamics* **1**, p. 149–188 (1997)
2. A.A. Shabana: Flexible multibody dynamics: Review of past and recent developments, *Multibody System Dynamics* **1**, p. 189–222 (1997)
3. B. Simeon: Numerical Analysis of Flexible Multibody Systems, *Multibody System Dynamics* **6**, p. 305–325 (2001)
4. P. Betsch: *Computational Methods for Flexible Multibody Dynamics*, Habilitationsschrift (2002)
5. A.A. Shabana: *Dynamics of Multibody Systems (Third edition)*, Cambridge University Press (2005)
6. R.R. Craig Jr. and M.C.C. Bampton: Coupling of Substructures for Dynamic Analysis, *AIAA Journal* Vol. **6**, No. 7, July 1968
7. R.R. Craig: *Structural Dynamics*, Wiley (1981)
8. D. Roble and T. Chan: Math in the Entertainment Industry, p. 971–990 in B. Engquist and W. Schmidt (ed.): *Mathematics Unlimited — 2001 and Beyond*, Springer (2001)
9. R. Barzel: Faking dynamics of ropes and springs, *IEEE Comput. Graph Appl.* **17**(3), p. 31–39 (1996)
10. D. Baraff and A. Witkin: Large steps in cloth simulation, p. 43–54 in *Proceedings of SIGGRAPH 98*, Computer Graphics Proceedings ed. by M. Cohen, Addison Wesley (1998)
11. A. Look and E. Schömer: A virtual environment for interactive assembly simulation: from rigid bodies to deformable cables, in *5th World Multiconference on Systemics, Cybernetics and Informatics (SCI'01)*, Vol. 3 (Virtual Engineering and Emergent Computing), p. 325–332 (2001)
12. D.K. Pai: Strands: Interactive simulation of thin solids using Cosserat models, *Comput. Graph. Forum* **21**(3), p. 347–352 (2002)
13. M. Gregoire and E. Schömer: Interactive simulation of one-dimensional felxible parts, *ACM Symposium on Solid and Physical Modeling (SPM'06)*, p. 95–103 (2006)
14. H. Baaser: Längenoptimierung von Bremsschläuchen und deren FE–Analyse auf Mikroebene, *VDI–Berichte* Nr. **1967**, p. 13–22 (2006)
15. G. Kirchhoff: Über das Gleichgewicht und die Bewegung eines unendlich dünnen elastischen Stabes, *J. Reine Angew. Math. (Crelle)* **56**, p. 285–343 (1859)
16. A.E.H. Love: *A Treatise on the Mathematical Theory of Elasticity (4th edition)*, Cambridge University Press (1927), reprinted by Dover (1944)
17. L.D. Landau and E.M. Lifshitz: *Theory of Elasticity (Course of Theoretical Physics, Vol. 7, 3rd edition)*, Butterworth Heinemann (1986)
18. S. Antman: *Nonlinear Problems of Elasticity (2nd Edition)*, Springer (2005)
19. J.C. Simo: A finite strain beam formulation: the three dimensional dynamic problem – Part I, *Comp. Meth. Apl. Mech. Eng.* **49**, p. 55–70 (1985)
20. J.H. Maddocks: Stability of nonlinearly elastic rods, *Arch. Rational Mech. & Anal.* **85**, p. 311–354 (1984)
21. N. Chouaieb and J.H. Maddocks: Kirchhoff's problem of helical equilibria of uniform rods, *J. Elasticity* **77**, p. 221–247 (2004)
22. J.R. Shewchuck: An Introduction to the Conjugate Gradient Method Without the Agonizing Pain, `www.cs.berkeley.edu/jrs/` (1994)

Simulation and Optimization of Suspension Testing Systems

M. Speckert and K. Dreßler

Fraunhofer ITWM, Kaiserslautern, Germany
`speckert@itwm.fraunhofer.de, dressler@itwm.fraunhofer.de`

1 Introduction

In automotive industry complex multi channel servo-hydraulic test rigs are used for physical testing of suspensions. Typically, wheel forces measured on a test track (target loads) are to be reproduced on the test rig. Most of the traditional rigs use one hydraulic actuator for one DOF, i.e. an actuator for the vertical force, one for the longitudinal force, etc. In [Wie02], a new concept for suspension test rigs based on the hexapod technology has been proposed. (see Fig. 2 in Sect. 3). Here six actuators are driving a platform (parallel kinematics) which is attached to the wheel hub.

In 2004 Volkswagen decided to introduce this concept into its suspension testing environment. The hexapods have been developed by MOOG-FCS in the Netherlands. A project has been initiated to set up a model and a simulation environment for the new testing system in order to accompany the introduction of the system, optimise the design and give support during the preparation of future suspension tests on the system.

The model of the system should be capable of simulating an entire physical test including the hydraulics and the control mechanisms. It is divided into three subsystems, namely the suspension model (M1), the mechanics of the hexapod (M2) and a model for the hydraulics and the controlling (M3). These subsystems are assembled to the entire testing environment as sketched below.

Hexapod mech. M2	$\leftrightarrow$	Susp. model M1	$\leftrightarrow$	Hexapod mech. M2
$\updownarrow$				$\updownarrow$
Hydr./Control M3	$\longleftrightarrow$	Target loads	$\longleftrightarrow$	Hydr./Control M3

In this paper, the application of the model to the derivation of the requirements for the actuators and the improvement of the first design of the hexapod is described. More technical details about the model, some applications and the implementation (multibody simulation code, additional subroutines, etc.) can be found in [Spe06].

2 The Suspension and Test Rig Models

2.1 The Suspension Model M1

A front and a rear suspension taken from an elasto-kinematically validated database at Volkswagen have been chosen for this project. Both suspension models contain rigid bodies for most of the suspension components, flexible beam elements for the stabilizers and joints and non-linear bushing elements to connect the different parts. The entire model leads to a system of differential algebraic equations (DAE). See [Sch99] for an introduction to multibody simulation and its numerical aspects.

The loads for the suspension simulations are taken from measurements on a test track containing rough road profiles, curves and braking events. Vertical displacements and the accelerations of the wheel hub during the simulations have been compared to the measurements in order to check the quality of the suspension model and the numerical solution. Some slight modifications of the damping parameters have been done to improve the agreement between measurements and simulation. See [Spe06] for more details about the suspension models and the measurement data and [Eic00, Kap02] for more details about the general application of multibody simulation in the vehicle development.

2.2 The Multibody Model M2

As can be seen in Fig. 2, one hexapod consists of a base and a top platform, which are connected via six identical actuators. The joints between the actuators and the platforms have two rotational degrees of freedom (DOFs). One actuator is composed of the piston and the cylinder, which in turn are connected using a cylindrical joint. This construction has six DOFs, namely the displacements Δl_i of the pistons or equivalently the distances l_i of the joints. They uniquely define the position x_R and the orientation α_R of the wheel centre (called tool centre point TCP). This relation cannot be expressed in closed form, however, the inverse relation $l_i = l_i(x_R, \alpha_R)$ can be written down explicitly.

In [Spe06], the equations of motion of the hexapod are derived. Since there are six DOFs, they can be written as six coupled second order differential equations for the TCP variables x_R, α_R. Besides the inertia terms for the bodies, they contain the forces and moments at the TCP and the hydraulic or actuator forces f_i. These equations are linear in the scalar actuator forces f_i. Thus, the forces can be calculated easily if the motion of the top platform (x_R, α_R as functions of time) is given. Usually, the motion is unknown and a result of the simulation. However, in this context, the wheel motion has been predicted using only the suspension models and the linear equations have been used to calculate the actuator forces as described in Sect. 3.

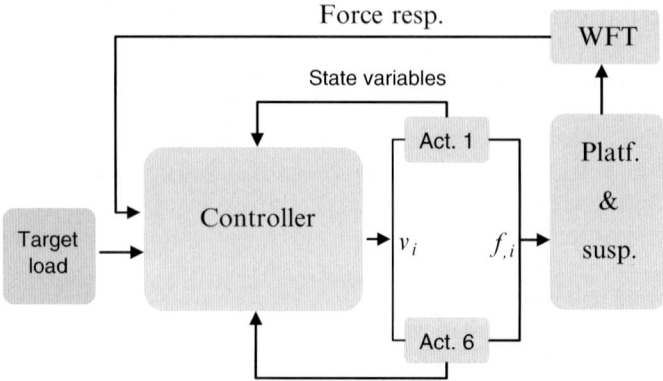

Fig. 1. Sketch of the data flow

2.3 The Model M3 for Hydraulics and Control

Figure 1 roughly shows the data flow of the complete system. The target
loads, i.e. the wheel forces as measured on the test track are fed into the controller together with the actual wheel forces (response) during the simulation.
Based on their deviation, a control signal (valve setpoint v_i) is calculated for
each actuator. The controller is modelled as a system of differential algebraic
equations relating the actuator state variables, the target loads, the response
forces and the valve setpoints. See [Gla00] or [Fri96] for more about controlling
strategies.

The valve setpoints v_i are input for the hydraulic model, which consists out
of the valve model (often a first or second order linear differential equation)
and the hydraulic model (see [Dro97] for a simple example). Again, this gives
a more or less complex system of differential algebraic equations.

2.4 The Complete Model

The suspension model M1, the mechanical model M2 for the hexapod and
the model M3 for the hydraulics and control can be combined to a model of
the complete testing system (see [Spe06] for implementation details). Figure 2
shows an assembly for a rear suspension. With this simulation environment
all steps, which have to be performed during a physical suspension test, can
be simulated.

3 Improving the Design of the Hexapod

Since the purpose of the test rig is to simulate test track driving including
rough road profiles, there will be high accelerations needed at the wheel hub.

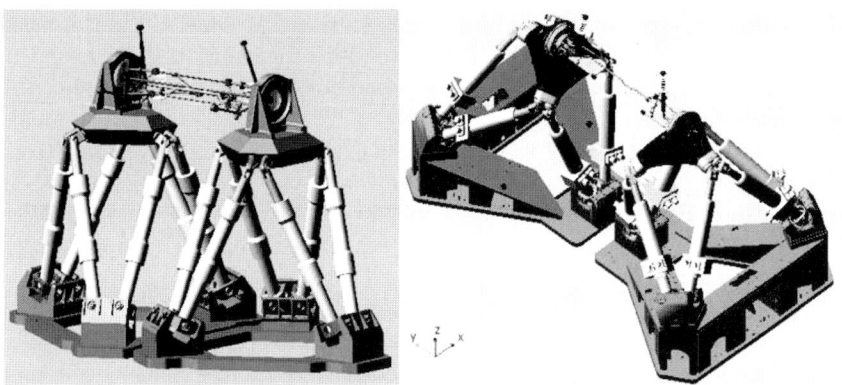

Fig. 2. Old (*left*) and new (*right*) design of the hexapod

The actuators have to excite the suspension accordingly, leading to high actuator forces. The models M1 and M2 have been used to calculate these actuator forces. To this end, the suspension models M1 have been excited with the target loads and the wheel displacements (reference displacements) have been logged.

Using the explicit model of the hexapod mechanics (Sect. 2.2) the actuator forces have been calculated from the target loads and the reference displacements. They contain very sharp peaks, which are implied by high accelerations. Using smoothed reference displacements (low pass filtering) strongly decreases the accelerations and thus the actuator forces. However, it has been shown in [Spe06], that the actuator forces calculated from the smoothed reference displacements lead to almost similar response of the suspension with respect to displacements and the forces at the wheel hub. This fact has been taken into account during the formulation of the requirements and the design of the hexapod.

The force calculation has been performed for several configurations. The radius of the base and top platform, the neutral actuator length and the distance of the joints at the top platform were modified. In addition, the actuators have been strengthened, the mass of the top platform has been reduced and the whole base platform has been tilted, leading to the configuration as shown in Fig. 2 which meanwhile is operational at Volkswagen in Wolfsburg.

4 Summary and Future Work

The goal of the project, namely the development of a complete simulation environment for suspension testing on the hexapod concept, has been reached. All models are integrated into an ADAMS/Car environment and can thus be used in the development and testing process at Volkswagen.

The control mechanism has been redesigned at MOOG-FCS in the meantime and the new concept will be integrated into the simulation environment. The hydraulic model used in this project has not been finally validated. Both topics are subject to ongoing work.

The multibody model of the hexapod has successfully been applied to the improvement of the hexapod design. The separation of the complete assembly into the suspension and the hexapod model and the consideration of the reference displacements has proven to be an effective way to optimise the configuration.

The mathematical framework for the system simulation is DAE solving. In this case, the commercial tool MSC/ADAMS has been used for modelling as well as for solving. The stabilized index 2 formulation has been used, see [ADA03] and [Sch99] for details. Since the valve input signals, which are needed for the execution of a certain test, are not known in advance, system identification (typically frequency based) and iterative learning control algorithms (see [Lju99] and [Moo93]) will play an important role in the future.

References

[Wie02]   Wierda, H.: Hexapods for Automotive Testing, Testing Technology Conference, May 2002
[Spe06]   Speckert, M., Dreßler, K.: MBS Simulation of a hexapod based suspension test rig, NAFEMS Seminar Virtual Testing, Wiesbaden, May 2006
[Sch99]   Von Schwerin, R.: MultiBody System SIMulation, Numerical methods, Algorithms and Software, Lecture Notes in Computational Science and Engineering, Springer 1999
[Eic00]   Eichler, M., Lion, A.: Gesamtfahrzeugsimulationen auf Prüfstrecken zur Bestimmung von Lastkollektiven, VDI Bericht 1559, Berechnung und Simulation im Fahrzeugbau, 2000, 369–398.
[Kap02]   Kaps, L., Lion, A. Stolze, F. Zhang, G.: Ganzheitliche Analyse von Fahrzeugprototypen mit Hilfe von virtuellen Fahrzeugmodellen und virtuellen Prüfgeländen, VDI Bericht 1701, Berechnung und Simulation im Fahrzeugbau, 2002, 653–678.
[Gla00]   Glad, T., Ljung, L.: Control Theory: Multivariable and Nonlinear Methods, Taylor & Francis, London, 2000
[Fri96]   Friedland, B.: Advanced Control System Design, Prentice Hall, Englewood cliffs, N.J., 1996
[Dro97]   Dronka, S.: Zur Einbindung hydraulischer Elemente in SIMPACK, Simpack-News, 2. Jahrgang, 2. Ausgabe, 1997
[ADA03] MSC-Software GmbH: ADAMS 2003 User Manual
[Lju99]   Ljung, L., System Identification - Theory For the User, 2nd ed, PTR Prentice Hall, Upper Saddle River, N.J., 1999
[Moo93] Moore, K.L.: Iterative Learning Control for Deterministic Systems. London: Springer-Verlag 1993

Minisymposium "Some Topics in Astrodynamics and Space Geodesy"

J.M. Gambi[1] and P. Romero[2]

[1] Universidad Carlos III de Madrid
[2] Universidad Complutense de Madrid, Spain

The first presentation in this minisymposium *Optimal Station Keeping for Geostationary Satellites with Electric Propulsion Systems Under Eclipse Constraints* by P. Romero discusses the possible implementation of an optimal strategy to satisfy the constraints imposed by the occurrence of eclipses on the geostationary orbit. The second presentation by M. Folgueira and coworkers, the *International Reference Systems for Astrodynamics and Space Geodesy*, discusses different Earth rotation models in the proposal recently adopted by the IAU and IUGG. Finally, the new post-Newtonian covariant measurement formulations for SLR, SST and GPS and their possible implementation are discussed by J.M. Gambi and coworkers in the third presentation.

Optimal Station Keeping for Geostationary Satellites with Electric Propulsion Systems Under Eclipse Constraints

P. Romero[1], J.M. Gambi[2], E. Patiño[3], and R. Antolin[4]

[1] Instituto de Astronomía y Geodesia (UCM-CSIC), Facultad de Matemáticas, Universidad Complutense de Madrid, E-28040 Madrid, Spain
[2] M.S.M.I., Universidad Carlos III , E-28911 Leganés, Spain
[3] Dep. Matemática Aplicada, E.T.S. Arquitectura, Universidad Politécnica de Madrid, E-28040 Madrid, Spain
[4] DIIAR Polo Regionale di Como, Politecnico di Milano, I-22100 Como, Italy

Summary. In order to keep geostationary satellites within the prescribed boundaries to satisfy mission requirements, orbital station keeping manoeuvres are performed periodically to compensate natural perturbations on the satellites.

The propulsion systems currently used to modify the orbit are of chemical nature (usually, hydrazine) but new trends in spatial propulsion point towards the use of electric systems. The use of these systems introduces new problems such as the impossibility to perform manoeuvres at eclipse epochs.

A procedure is proposed here to analyze the implementation of optimal strategies in terms of electric energy consumption to satisfy the additional constraints imposed by the use of these kind of systems.

1 Introduction

A satellite in geostationary orbit is subjected to various forces which tend to move it from its assigned orbital position. Station keeping manoeuvres are therefore required and implemented by on-board thrusters. The use of electric propulsion systems for station keeping [GFJ00, Klu04] is lead to achieve important reductions of the total amount of the satellite's masses. But the implementation of these systems makes it necessary the revision of the strategies for the station keeping due to two reasons: first, the limitations in magnitude of the impulses provided by these systems; and second, the impossibility to perform manoeuvres during long time periods at eclipse epochs because of the high electric energy consumption.

In this paper, in order to check if optimal strategies minimizing the magnitude of impulses (needed to control geostationary satellites) satisfy the additional constraints imposed by the use of electric propulsion systems, a simulation of the station keeping process is carried out and the numerical results are analyzed. The station keeping technique implemented uses analytical expressions to determine long-term variations in the orbit evolution, as well as linearized equations to compute the correction manoeuvres (see, e.g. [Soo94]).

2 Problem Specification and Modelling

A satellite in geostationary orbit is intended to be at rest with respect to the rotating Earth. But natural perturbations tend to shift the satellite from its assigned position at a nominal longitude l_s (see Fig. 1).

The different models needed to describe the orbit evolution to compute manoeuvres, as well as the eclipses determination, are described in this Section.

2.1 Geostationary Orbit Evolution

The perturbed satellite motion is determined by the differential equations $\ddot{x}_i + \frac{\mu x_i}{x_i{}^3} = F_i$, $i = 1 - 3$, where F_i are the perturbing forces. The method of variation of the constants of the unperturbed motion ($F_i = 0$) leads to the Lagrange equations. Since the orbit is geostationary, to avoid numerical indetermination, the following set of orbital elements are considered (see, e.g. [EP03]): the semi major axis, a; the two-dimensional inclination vector, $\mathbf{i} = (i_x = i \cos \Omega, i_y = i \sin \Omega)$; the eccentricity vector, $\mathbf{e} = (e_x = e \cos(\Omega+\omega), e_y = e \sin(\Omega+\omega))$ and the mean longitude, $l = \Omega+\omega+M-\vartheta_G$; i being the orbital inclination with respect to the equatorial plane, Ω the right ascension of the ascending node, e the dimensionless eccentricity, ω the argument of the perigee, M the mean anomaly and ϑ_G the Greenwich sidereal time.

The evolution of these elements is obtained by means of the following linearized Lagrange equations [Cne80]

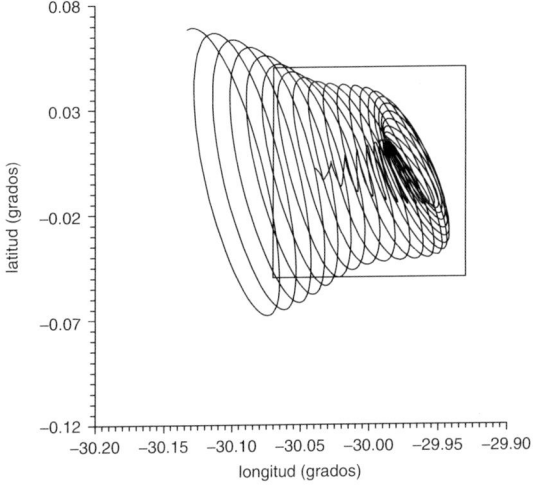

Fig. 1. Monthly evolution of a satellite at a nominal longitude of $l_s = 30°\mathrm{W}$

$$\frac{dl}{dt} = n - \Omega_{\oplus} - \frac{2}{na} \frac{\partial R_1}{\partial a}, \quad \frac{da}{dt} = \frac{2}{na} \frac{\partial R_1}{\partial l}, \tag{1}$$

$$\frac{de_x}{dt} = -\frac{1}{na^2} \frac{\partial R_2}{\partial e_y}, \quad \frac{de_y}{dt} = \frac{1}{na^2} \frac{\partial R_2}{\partial e_x}, \tag{2}$$

$$\frac{di_x}{dt} = -\frac{1}{na^2} \frac{\partial R_3}{\partial i_y}, \quad \frac{di_y}{dt} = \frac{1}{na^2} \frac{\partial R_3}{\partial i_x}, \tag{3}$$

where n is the mean motion; R_1 is the terrestrial perturbing potential; R_2 is the potential due to the lunisolar attraction and R_3 is the potential due to the solar radiation pressure.

To analyze the strategies satisfying the eclipse constraints, we have considered the evolution of mean orbital elements when the perturbing function only contains those terms causing long period perturbations. Thus, the evolution of the mean longitude is a parabola, the annual evolution of the mean eccentricity vector can be approximated by a circle, and the evolution of the mean inclination vector has a secular drift in a direction, Ω_{sec}, with periodic components superimposed (for a detailed description, see [RGP06]).

2.2 Linear Equations for the Station Keeping Manoeuvres and Optimal Strategies

In the east/west station keeping (EWSK) an impulse tangent to the orbit modifies both the longitude drift and the eccentricity. Two tangential burns ΔV_1 and ΔV_2, separated half a sidereal day, have to be implemented to maintain the satellite within the specified limits in longitude. On the other hand, in the north/south station keeping (NSSK) the inclination correction is performed by means of a normal impulse ΔV_n. The effects on the corresponding elements are given by the following equations [Soo94]:

$$\Delta \mathbf{e} = \mathbf{e}^+ - \mathbf{e}^- = \frac{2}{V} (\Delta V_1 - \Delta V_2) \begin{pmatrix} \cos s_b \\ \sin s_b \end{pmatrix}, \tag{4}$$

$$\Delta l = l(t) - l^- = -\frac{3}{V} (\Delta V_1 + \Delta V_2) \left[\Omega_{\oplus} (t - t_b) - 90° \right], \tag{5}$$

$$\Delta \mathbf{i} = \mathbf{i}^+ - \mathbf{i}^- = -\Delta V_n / V \begin{pmatrix} \cos s_b \\ \sin s_b \end{pmatrix}, \tag{6}$$

where s_b is the sidereal mean time at the satellite that corresponds to the thrust time t_b; $\Omega_{\oplus}$ is the Earth's angular velocity and V is the geosynchronous velocity.

The challenge in planning the manoeuvres is to design optimal strategies to minimize the magnitude of the impulses. The secular mean line (SML) strategy is usually chosen to plan the NSSK. With this strategy the correction is applied in the direction of the secular drift for the long-term evolution of the inclination vector [SDB88]. For the EWSK manoeuvres the optimal direction

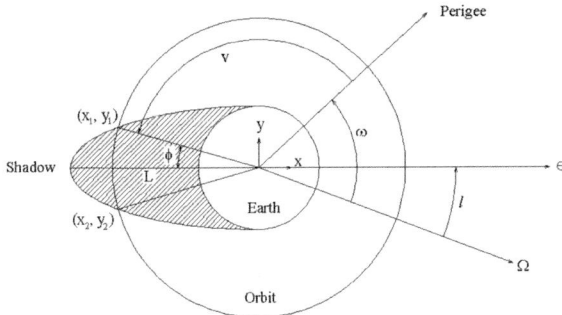

Fig. 2. Shadow ellipse and satellite orbit

is obtained by pointing the perigee of the satellite orbit towards the Sun, i.e. according to the strategy called Sun pointing perigee (SPP) [KW82]. It is characteristic of these optimization strategies that they define the direction of the manoeuvres (which, in turn, determines the time of the day for the thrusts).

2.3 Eclipses

To determine the time and duration of eclipses, we have modelled the shadow of the Earth as the semi-ellipse obtained by the intersection of the shadow cone with the orbital plane of the satellite (see Fig. 2). Then, the initial and final points of an eclipse correspond to the solutions, (x_1, y_1) and (x_2, y_2), of the system describing the satellite orbit and the semi-ellipse of shadow,

$$x^2 + y^2 = a^2, \quad \frac{x^2}{L^2} + \frac{y^2}{R^2} = 1, \tag{7}$$

where $L = R/\sin b$ (with R, the Earth radius and b, the Sun elevation angle above the orbit plane).

3 Numerical Results

Numerical simulations have been carried out to evaluate the cited manoeuvres required in the NSSK and EWSK process. In order to ensure simplicity in the operations, weekly manoeuvres cycles of 14 days have been chosen to check the possibility of their implementation. To this end, a satellite located at $l_s = 30°\,W$ has been considered.

For 2006, this satellite is under eclipse during 90 days, being the starting and final dates of eclipses $02/25/2006 - 04/10/2006$ for the spring epoch, and $08/31/2006 - 10/14/2006$ for the autumn epoch. The maximum durations are $1^h 9^m 24^s$ (spring) and $1^h 9^m 25.5^s$ (autumn) plus 4^m of penumbra. Figure 3 shows the results for the spring eclipse epoch for 2006 at $30°\,W$ with the time of implementation for the four NSSK manoeuvres determined with the SML strategy and the eight EWSK manoeuvres determined with the SPP. As a result, it can be seen why those optimal strategies can be implemented (Fig. 3).

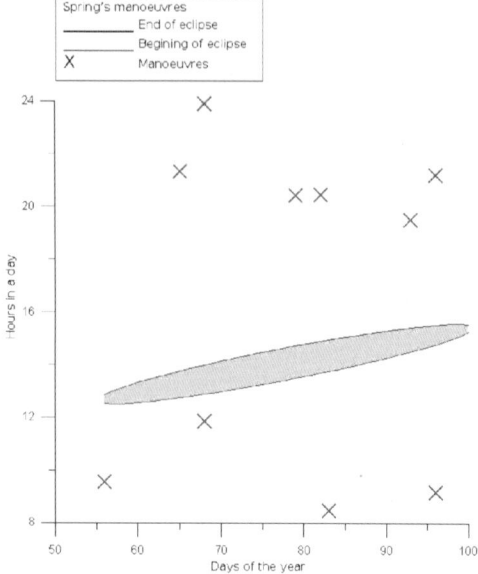

Fig. 3. Manoeuvres at eclipse epoch in spring

References

[Cne80] G. Legendre, Le maintien à poste de satellites géostationnaries I. Cours de Technologie Spatiale. CNES Toulouse, pp. 583–609 (1980)

[EP03] B.P. Emma, H.J. Pernicka, Algorithm for autonomous longitude and eccentricity control for geostationary spacecraft, Journal of Guidance Control and Dynamics **26**, **(3)**, 483–490 (2003)

[GFJ00] E. Gottzein, W. Fichter, A. Jablonsky, O. Juckenhofel, M. Mittnacht, C. Muller, M. Surauer, Challenges in the control and autonomy of communications satellites, Control Engineering Practice, **8**, 409–427 (2000)

[KW82] A.A. Kamel, C.A. Wagner, On the Orbital Eccentricity Control of Syncronous Satellites, J.Astronutical Sci., **XXX**, 61–73 (1982)

[Klu04] C.A. Kluever, Geostationary orbit transfers using solar electric propulsion with specific impulses modulation, Journal of Spacecraft and Rockets, **41**, 461–466 (2004)

[RGP06] P. Romero, J.M. Gambi, E. Patino, Stationkeeping manoeuvres for geostationary satellites using feedback control techniques, Aerospace Science and Technologie, doi.10.1016/j.ast.2006.08.003 (2006)

[SDB88] D. Slavinskas, H. Daddaghi, W. Bendent, G. Johnson, Efficient Inclination Control for Geostationary Satellites, Journal of Guidance, Control and Dynamics **11**, **(6)**, 584–589 (1988)

[Soo94] E.M. Soop, Handbook of Geostationary Orbits, Space Technology Library, Kluwer Academic Publishers, Dordrecht (1994)

International Reference Systems for Astrodynamics and Space Geodesy

M. Folgueira[1,2], N. Capitaine[2], and J. Souchay[2]

[1] Instituto de Astronomía y Geodesia (UCM-CSIC), Facultad de CC,
   Matemáticas, Universidad Complutense de Madrid, Plaza de Ciencias,
   3. Ciudad Universitaria, 28040 Madrid, Spain
   martafl@mat.ucm.es
[2] Observatoire de Paris, SYRTE, UMR 8630/ CNRS,
   61 avenue de l'Observatoire, 75014 Paris, France
   n.capitaine@obspm.fr, jean.souchay@obspm.fr

1 Introduction

The fields of Astrodynamics and Space Geodesy research are experiencing continuous growth. Advancements in science and technology are enabling missions with much more challenging goals. In response, many new techniques have been introduced to solve these demanding new mission design problems with a high precision.

Many physical, astronomical and geodetic models related to such problems assume the availability of a unique reference system to establish the equations of the problem. In practice, and in accordance with the advance of physical theories, observational methods and measuring devices, one faces a multitude of historical concepts, which have undergone continued revisions up to the present date.

Following this introduction, Sect. 2 is devoted to the review of the different international reference systems, the related conventions and the Earth rotation models adopted by international organizations (IAU and IUGG) which are employed in the fields of Geodesy and Astrodynamics. Section 3 deals with the modelling of the rotation undertaken in the European DESCARTES-NUTATION sub-projet entitled: *"Advances in the integration of the equations of the Earth's rotation in the framework of the new parameters adopted by the IAU 2000 Resolutions"*. It is followed in the last section by a brief discussion about the state of art of this problem.

2 International Reference Systems in Geodesy and Astrodynamics

Geodesy and Astrodynamics frequently need a precise positioning of points on and outside the Earth's surface. Such positions are determined using two reference frameworks: the terrestrial frame, fixed in relation to the Earth's crust and rotating synchronously with the planet, and the celestial frame, which is immobile in space. Traditionally, celestial reference frames have been tied to the rotational and translational motion of the Earth. In 1997, the IAU decided to establish a new *International Celestial Reference System* (ICRS):

1. Origin: the solar system barycentre within a relativistic framework
2. Their axes fixed with respect to distant extragalactic radio objects
3. The fundamental plane: closely aligned with the mean equator at J2000
4. The origin of right ascension: defined by an adopted right ascension of the quasar 3C 273B

The practical realization of the ICRS is the *International Celestial Reference Frame* (ICRF). The IAU has charged the IERS with the responsibility of monitoring the International Celestial Reference System (ICRS) and maintaining its current realization, the International Celestial Reference Frame (ICRF), and links with other celestial reference frames. Starting in 2001, these activities are run jointly by the ICRS Centre (Paris Observatory and US Naval Observatory) of the IERS and the International VLBI Service for Geodesy and Astrometry (IVS), in coordination with the IAU Working Group on Reference Systems. Complementary to the ICRS, the *International Terrestrial Reference System* (ITRS) provides the conceptual definitions of an Earth-fixed reference system:

1. It is geocentric, the centre of mass being defined for the whole earth, including oceans and atmosphere.
2. The unit of length is the metre SI. This scale is consistent with the TCG time coordinate for a geocentric local frame, in agreement with IAU and IUGG (1991) Resolutions. This is obtained by appropriate relativistic modelling.
3. Its orientation was initially given by the BIH orientation at 1984.0.
4. The time evolution of the orientation is ensured by using a no-net-rotation condition with regards to horizontal tectonic motions over the whole Earth.

The ITRS is realized by estimates of the coordinates and velocities of a set of stations observed by Laser Ranging (SLR), Lunar Laser Ranging (LLR), Global Positioning System (GPS) and Very Long Baseline Interferometry (VLBI) and DORIS. Its name is *International Terrestrial Reference Frame* (ITRF). General documentation on terrestrial reference systems and frames is available at the ITRS Centre of the IERS: http://hpiers.obspm.fr/ and also in [McCarthy & Petit, 2003].

The transformation from the International Celestial Reference System to the International Terrestrial Reference System is expressed in this way:

$$[\mathsf{ITRS}] = W(t)\,R(t)\,C(t)\ [\mathsf{ICRS}] = M(t)\ [\mathsf{ICRS}], \tag{1}$$

which represents a dynamical tic between the International Celestial and Terrestrial Reference Systems. This relation includes, by means of the matrix $M(t)$, precession, nutation -$C(t)$-, sidereal time -$R(t)$- and polar motion parameters -$W(t)$-. In the representation using the 'non-rotating origin', $C(t)$ equals to [Capitaine, 1990]:

$$C(t) = R_3(-s)\cdot\begin{pmatrix} 1-aX^2 & -aXY & -X \\ -aXY & 1-aY^2 & -Y \\ X & Y & 1-a(X^2+Y^2) \end{pmatrix} \tag{2}$$

(X,Y) being the rectangular coordinates of the Celestial Intermediate Pole (CIP) unit vector in the ICRS, which have the advantage of being directly related to VLBI observations. s is a small quantity: $\dot{s} = a(Y\dot{X} - X\dot{Y})$, $a = 1/(1+Z)$ and $Z = \sqrt{1-(X^2+Y^2)}$. The CIP is defined by the motions of Tisserand mean axis of the Earth with periods greater than two days in the celestial reference system [IAU Transactions, 2000].

From the expressions (1) and (2), the accuracy in the transformation between International Celestial and Terrestrial Reference Systems will depend directly on the precision of the algorithms for computing the Celestial Intermediate Pole's position. There are now algorithms for computing CIP position, in the form of new expressions for precession and nutation. That is to say, in a indirect way. The new nutation has been generated by [Mathews et al., 2002] by the convolution of the MHB 2000 transfer function with the rigid Earth nutation series REN 2000 of [Souchay et al., 1999]. It has provided the components of the nutation in longitude and in obliquity: $\Delta\psi$ and $\Delta\varepsilon$. These new nutation series used together the P03 precession development of [Capitaine et al., 2003] yield the computed path of the Celestial Intermediate Pole.

The main goal in the above project, mentioned in Sect. 1, is to provide a new precise theoretical development for computing directly the instantaneous motion of the Celestial Intermediate Pole. For this purpose, our approach will consist in developing the equations for Earth rotation using explicitly the (X,Y) coordinates with the aim of providing the precession-nutation model directly in the form recommended by the IAU 2000 Resolutions.

3 Modelling

3.1 Approaches for the Rigid Earth

The choice of variables is an important point to be taken into account in the description of the Earth's rotation as well as in many different problems of Space Geodesy and Astrodynamics. In this problem, the equations of the rotational motion of the rigid Earth can be formulated in terms of angular variables either canonical or non-canonical:

- The rectangular components $(\omega_1, \omega_2, \omega_3)$ of the angular velocity vector $\overrightarrow{\omega}$ along the principal axes of inertia, or alternatively the Euler angles between the figure axes and a fixed reference plane, are basic non-canonical variables that are classically used for writing the Euler dynamical equations. These variables can be expressed as functions of the rectangular coordinates of CIP what will provide us the rotational equations of the Earth as functions of (X, Y) [Capitaine et al., 2006]

$$\begin{cases} F_1\ddot{X} + F_2\ddot{Y} + F_3\dot{X}^2 + F_4\dot{Y}^2 + F_5\dot{X}\dot{Y} + F_6\dot{X} + F_7\dot{Y} = \frac{L}{\bar{A}} \\ G_1\ddot{X} + G_2\ddot{Y} + G_3\dot{X}^2 + G_4\dot{Y}^2 + G_5\dot{X}\dot{Y} + G_6\dot{X} + G_7\dot{Y} = \frac{M}{\bar{A}} \,, \end{cases}$$
$$(3)$$

where (L, M) are the components of the external torque in the celestial reference system, $\bar{A} = \frac{A+B}{2}$ with A, B are the Earth's equatorial principal moments of inertia and F_i and G_i are functions of (X, Y).

- Two sets of canonical variables can be used in the Hamiltonian approach. These variables are represented by the amplitude of the angular-momentum vector $(\overrightarrow{L})$, the X- and Y- components of this vector with respect to the inertial reference system, the x- and y-components of $\overrightarrow{L}$ with respect to the figure axes and their canonically conjugate variables [Folgueira et al., 2006].

3.2 Methods of Integration

We have investigated the appropriate methods of integration to this study and the solutions for the X and Y variables in the axially symmetric case:

- Non-canonical variables:
    - Numerical integration $\rightarrow$ Fifth-order adaptive step size Runge-Kutta-Fehlberg algorithm: numerical solution
    - Semi-analytical integration $\rightarrow$ Method of variation of constants: semi-analytical solution
- Canonical variables:
    - Analytical integration$\rightarrow$ Hori-Deprit's averaging perturbation method: analytical solution

4 Conclusions

The achieved objectives of our work may be summarized as follows:

1. The selection of the set of variables appropriated to our study.
2. The obtention of the equations of the Earth rotation problem considering directly the Earth Rotation Parameters.

3. The study of the different integration methods to be carried out, which include analytical, semi-analytical and numerical approaches, in order to obtain the solution with microarcsecond accuracy.
4. To test the efficiency of these methods of integration.
5. To obtain the results according to the level of accuracy of IAU 2000 Resolutions.
6. The comparison between the new solutions and those obtained indirectly from the classical solutions for $\Delta\psi$ and $\Delta\varepsilon$.
7. The discussion about the step: rigid Earth $\longrightarrow$ elastic Earth.

Acknowledgements

The research was carried out in the Department of *"Systèmes de Référence Temps Espace" (SYRTE)* of Observatoire de Paris and received financial support from Descartes Prize Allowance (M. Folgueira).

References

[Capitaine, 1990] Capitaine, N.: The celestial pole coordinates. Celest. Mech., **48**, 127–143 (1990)

[Capitaine et al., 2003] Capitaine, N., Wallace, P.T. and Chapront, J.: Expressions for IAU 2000 precession quantities. Astron. Astrophys. **412**, 567–586 (2003)

[Capitaine et al., 2006] Capitaine, N., Folgueira, M. and Souchay, J.: Earth rotation based on the celestial coordinates of the celestial intermediate pole I. The dynamical equations. Astron. Astrophys. **445**, 347–360 (2006)

[Folgueira et al., 2006] Folgueira, M., Souchay, J. and Capitaine, N.: On the appropriate sets of variables for the rigorous study of the Earth's rotation in the framework of IAU 2000 Resolutions. Proceedings of Journées 2005 Systèmes de Référence Spatio-temporels (2006) *(in press)*

[IAU Transactions, 2000] IAU Transactions 2000, Vol. XXIVB; in the Proceedings of the Twenty-Fourth General Assembly; Manchester, Ed. H. Rickman, Astronomical Society of the Pacific, Provo, USA, 34 (2001)

[Mathews et al., 2002] Mathews, P.M., Herring, T.A. and Buffett B.A.: Modeling of nutation and precession: New nutation series for nonrigid Earth and insights into the Earth's interior. J. Geophys. Res., 10.1029/2001JB000390 (2002)

[McCarthy & Petit, 2003] McCarthy D. and Petit, G., eds.: IERS Conventions 2003. IERS Technical Note 32, Publ. Frankfurt am Main: Verlag des Bundesamts für Kartographie und Geodäsie (2003)

[Souchay et al., 1999] Souchay, J., Loysel, B., Kinoshita, H. and Folgueira, M.: Corrections and New Developments in Rigid Earth Nutation Theory: III. Final Tables "REN-2000" including Crossed-Nutation and Spin-Orbit Coupling Effects. Astron. Astrophys. Suppl. Ser. **135**, 111–131 (1999)

Post-Newtonian Covariant Measurement Formulations in Space Geodesy

J.M. Gambi[1], M.L. García del Pino[1], M.C. Rodriguez[1], M. Salas[1], and P. Romero[2]

[1] M.S.M.I., Universidad Carlos III , E-28911 Leganés, Spain
[2] Instituto de Astronomía y Geodesia (UCM-CSIC), Facultad de Matemáticas, Universidad Complutense de Madrid, E-28040 Madrid, Spain

1 Introduction

Most spatial current high precision geodetic techniques, like those used in the Global Positioning System, have led to widely consider the assumption of a slightly curved space-time in the vicinity of the Earth (according to the general theory of relativity) as the essential basis to build geometric models for measurements formulations that allow correct interpretations of the results, at least up to the level of accuracy required at the present and near-future time. In this contribution, Synge's world function for the local geometric models associated to a global model of that space-time is used to give a flexible and structured set of covariant two-way local formulations for the four basic kind of measurements involved in Space Geodesy. Both local and global models are made compatible by using local and global Fermi coordinates, respectively. The measurements formulations are one-to-one general (weak) relativistic versions of the local classical formulations currently implemented in altimetry, satellite laser ranging and satellite-to-satellite tracking on the one hand, and on the other, of the classical version of the ballistic problem.

2 Modelling Assumptions

The local geometric models correspond to local Fermi coordinates, $X^{(\alpha)}$, associated to reference frames given by observers, O, and Fermi transported tetrads, $\lambda^i_{(a)}$, along O [Syn60]. (Latin index runs from 1 to 4, and Greek, from 1 to 3). Up to the second order of approximation in v and m_E/r ($c = 1$), these metrics are given by

$$g_{(\alpha\beta)} = \delta_{\alpha\beta} + \frac{1}{2}S_{O(\alpha\beta\mu\nu)}X^{(\mu)}X^{(\nu)}$$

$$g_{(\alpha 4)} = S_{O(\alpha 4\mu\nu)}X^{(\mu)}X^{(\nu)}$$

$$g_{(44)} = -1 - 2b_O B_{O(\gamma)}X^{(\gamma)} + \frac{3}{2}S_{O(44\mu\nu)}X^{(\mu)}X^{(\nu)}, \tag{1}$$

where $b_O B_{O(\gamma)}$ is the first curvature of O, and $S_{O(ab\mu\nu)} = -\frac{1}{3}\left(R_{O(a\mu b\nu)} + R_{O(a\nu b\mu)}\right)$ are the relevant terms of the symmetrized Riemann tensor calculated at O up to this order of approximation. Essentially, it may be said that $b_O B_{O(\gamma)}$ is, from the classical point of view, $-\vec{g}$, and $S_{O(ab\mu\nu)}$ the tidal potential of the external bodies (e.g. Sun and Moon) calculated at O. $S_{O(ab\mu\nu)}$ is derived with the global geometric model for the vicinity of the Earth, which here is built similarly to the model by Ashby and Bertotti is constructed [AB86].

In terms of global Fermi coordinates, x^α (with classical standard notations), the global metric at any point of the vicinity of the Earth results to be

$$
\begin{aligned}
g_{\alpha\beta} = \delta_{\alpha\beta}&\left\{1 + 2\frac{m_E}{r}\left[1 - \left(\frac{a_1}{r}\right)^2 J_2 P_2(\cos\theta)\right]\right\}\\
&-\frac{1}{3}\left[\frac{M_\odot}{R_\odot^3}\left(2\delta_{\alpha\beta}\delta_{\mu\nu} - 2\delta_{\alpha\nu}\delta_{\beta\mu} + 3\delta_{\alpha\nu}\frac{R_\odot^\mu R_\odot^\beta}{R_\odot^2}\right.\right.\\
&\qquad\left.+3\delta_{\beta\mu}\frac{R_\odot^\alpha R_\odot^\nu}{R_\odot^2} - 3\delta_{\alpha\beta}\frac{R_\odot^\mu R_\odot^\nu}{R_\odot^2} - 3\delta_{\mu\nu}\frac{R_\odot^\alpha R_\odot^\beta}{R_\odot^2}\right)\\
&\quad+\frac{M_L}{R_L^3}\left(2\delta_{\alpha\beta}\delta_{\mu\nu} - 2\delta_{\alpha\nu}\delta_{\beta\mu} + 3\delta_{\alpha\nu}\frac{R_L^\mu R_L^\beta}{R_L^2}\right.\\
&\qquad\left.\left.+3\delta_{\beta\mu}\frac{R_L^\alpha R_L^\nu}{R_L^2} - 3\delta_{\alpha\beta}\frac{R_L^\mu R_L^\nu}{R_L^2} - 3\delta_{\mu\nu}\frac{R_L^\alpha R_L^\beta}{R_L^2}\right)\right]x^\mu x^\nu
\end{aligned}
$$

$$g_{\alpha 4} = -\Omega_{\alpha\mu}x^\mu$$

$$
\begin{aligned}
g_{44} = &-1 + 2\frac{m_E}{r}\left[1 - \left(\frac{a_1}{r}\right)^2 J_2 P_2(\cos\theta)\right] + \Omega^2\left[(x^1)^2 + (x^2)^2\right]\\
&-\left[\frac{M_\odot}{R_\odot^3}\left(\delta_{\mu\nu} - 3\frac{R_\odot^\mu R_\odot^\nu}{R_\odot^2}\right) + \frac{M_L}{R_L^3}\left(\delta_{\mu\nu} - 3\frac{R_L^\mu R_L^\nu}{R_L^2}\right)\right]x^\mu x^\nu.
\end{aligned}
\tag{2}
$$

Therefore, since taking into account the expressions for $\lambda^i_{(a)}$ at O (see, for example, [Sof89]) it results that $R_{O(ab\mu\nu)} = R_{Oijkm}\lambda^i_{(a)}\lambda^j_{(b)}\lambda^k_{(\mu)}\lambda^m_{(\nu)} = R_{Oab\mu\nu}$ up to the second order, then we have that

$$
\begin{aligned}
R_{O(\alpha\mu\beta\nu)} = -\frac{m_E}{r^3}&\left(2\left[\delta_{\alpha\nu}\delta_{\mu\beta} - \delta_{\alpha\beta}\delta_{\mu\nu}\right]\left[1 - 3\left(\frac{a_1}{r}\right)^2 J_2 P_2(\cos\theta)\right]\right.\\
&-\frac{3}{r^2}\left[\delta_{\alpha\nu}x^\mu x^\beta + \delta_{\mu\beta}x^\alpha x^\nu - \delta_{\alpha\beta}x^\mu x^\nu - \delta_{\mu\nu}x^\alpha x^\beta\right]\\
&\left.\times\left[1 - 5\left(\frac{a_1}{r}\right)^2 J_2 P_2(\cos\theta)\right]\right)
\end{aligned}
$$

$$-\frac{M_\odot}{R_\odot^3}\left(2\left[\delta_{\alpha\nu}\delta_{\mu\beta}-\delta_{\alpha\beta}\delta_{\mu\nu}\right]\right.$$

$$\left.-\frac{3}{R_\odot^2}\left[\delta_{\alpha\nu}R_\odot^\mu R_\odot^\beta+\delta_{\mu\beta}R_\odot^\alpha R_\odot^\nu-\delta_{\alpha\beta}R_\odot^\mu R_\odot^\nu-\delta_{\mu\nu}R_\odot^\alpha R_\odot^\beta\right]\right)$$

$$-\frac{M_L}{R_L^3}\left(2\left[\delta_{\alpha\nu}\delta_{\mu\beta}-\delta_{\alpha\beta}\delta_{\mu\nu}\right]\right.$$

$$\left.-\frac{3}{R_L^2}\left[\delta_{\alpha\nu}R_L^\mu R_L^\beta+\delta_{\mu\beta}R_L^\alpha R_L^\nu-\delta_{\alpha\beta}R_L^\mu R_L^\nu-\delta_{\mu\nu}R_L^\alpha R_L^\beta\right]\right)$$

$$R_{O(\alpha\mu4\nu)}=0 \tag{3}$$

$$R_{O(4\mu4\nu)}=\frac{m_E}{r^3}\left(\delta_{\mu\nu}\left[1-3\left(\frac{a_1}{r}\right)^2 J_2 P_2(\cos\theta)\right]\right.$$

$$\left.-3\frac{x^\mu x^\nu}{r^2}\left[1-5\left(\frac{a_1}{r}\right)^2 J_2 P_2(\cos\theta)\right]\right)$$

$$-\delta_{\mu\nu}\Omega^2(\mu,\nu\neq3)+\frac{M_\odot}{R_\odot^3}\left(\delta_{\mu\nu}-3\frac{R_\odot^\mu R_\odot^\nu}{R_\odot^2}\right)+\frac{M_L}{R_L^3}\left(\delta_{\mu\nu}-3\frac{R_L^\mu R_L^\nu}{R_L^2}\right).$$

The world function $\Omega(X^{(i_1)},X^{(i_2)})$ is defined as half the square of the measure of the geodesic joining the points P_1 $(X^{(i_1)})$ and P_2 $(X^{(i_2)})$ [Syn60], and for the metrics (1) it takes the form

$$\frac{1}{2}\Delta X^{(\alpha)}\Delta X^{(\alpha)}$$

$$-\frac{1}{2}\Delta X^{(4)}\Delta X^{(4)}+\frac{1}{4}S_{O(\alpha\beta\mu\nu)}X^{(\mu_1)}X^{(\nu_1)}\Delta X^{(\alpha)}\Delta X^{(\beta)}+S_{O(\alpha4\mu\nu)}$$

$$X^{(\mu_1)}X^{(\nu_1)}\Delta X^{(\alpha)}\Delta X^{(4)}$$

$$+\frac{1}{2}\left(-2b_O B_{O(\gamma)}X^{(\gamma_1)}+\frac{3}{2}S_{O(44\mu\nu)}X^{(\mu_1)}X^{(\nu_1)}\right)(\Delta X^{(4)})^2. \tag{4}$$

3 Relative Distance

In terms of the world function (4), the relative position of the particle O_2 with respect to O_1 for the observer O at P is given by

$$r_{O_1O_2|O_P(\alpha)}(s_O)=-\Omega_{i_1}(P_1,P_2)\lambda^{i_1}_{(\alpha)}, \tag{5}$$

where Ω_{i_1} is the covariant derivative of $\Omega(X^{(i_1)},X^{(i_2)})$ with respect to $X^{(i_1)}$ calculated with the metric (1); $P_1(X^{(i_1)})$ is the event of O_1 with the same O-proper time, s_O, than P; $P_2(X^{(i_2)})$ is the event of O_2 with the same O_1-proper time than P_1 and $\lambda^{i_1}_{(\alpha)}$ are given by the finite expressions of any Fermi transported triad along O_1 calculated with the metric (1) [Gar04]. The final result is

$$r_{O_1 O_2 | O_P(\alpha)}(s_O) = \Delta X^{(\beta)} \left(\delta_{\alpha\beta} + \frac{1}{4} S_{O_P(\alpha\beta\mu\nu)} X^{(\mu_1)} X^{(\nu_1)} - \frac{1}{2} v^{\alpha_1} v^{\beta_1} \right.$$

$$\left. - \frac{1}{4} S_{O_P(\alpha\mu\beta\gamma)} X^{(\mu_1)} \Delta X^{(\gamma)} \right), \tag{6}$$

where v^{α_1} is the relative velocity of O_1 with respect to O at P. Therefore, the corresponding relative distance is

$$\sigma_{O_1 O_2 | O_P}(s_O) = \left(r_{O_1 O_2 | O_P(\alpha)}(s_O) r^{(\alpha)}_{O_1 O_2 | O_P}(s_O) \right)^{1/2} = \left(\Delta X^{(\alpha)} \Delta X^{(\beta)} \right)^{1/2}.$$

$$\left[\delta_{\alpha\beta} - v^{\alpha_1} v^{\beta_1} + \frac{1}{2} S_{O_P(\alpha\beta\mu\nu)} X^{(\mu_1)} X^{(\nu_1)} - \frac{1}{2} S_{O_P(\beta\gamma\alpha\mu)} X^{(\mu_1)} \Delta X^{(\gamma)} \right]^{1/2}. \tag{7}$$

4 Local Measurement Procedures

Let $s^1_{O_1}$ be the O_1-proper time corresponding to the instant Q_1 at which an electromagnetic signal is emitted from O_1 and let $s^2_{O_1}$ the O_1-proper time corresponding to the instant Q_2 at which the signal is received by O_1 after the signal is bounced at $P_2 \in O_2$. Then, by straightforward calculations, it can be deduced the expression that gives the distance (7) in terms of $s^1_{O_1}$ and $s^2_{O_1}$. This is

$$\sigma_{O_1 O_2 | O_P}(s_O) = \frac{s^2_{O_1} - s^1_{O_1}}{2} \left[1 + \frac{1}{2} b_{O_1} B^{(\alpha)}_{O_1} \Delta X^{(\alpha)} - \frac{1}{4} S_{O_P(44\alpha\beta)} \Delta X^{(\alpha)} \Delta X^{(\beta)} \right]. \tag{8}$$

In particular, when $b_{O_1} = 0$ the following expression corresponds to the local measurement model for Satellite-to-Satellite Tracking (SST)

$$\sigma = \frac{s^2_{O_1} - s^1_{O_1}}{2} \left[1 - \frac{1}{4} S_{O_P(44\alpha\beta)} \Delta X^{(\alpha)} \Delta X^{(\beta)} \right]. \tag{9}$$

When $O \equiv O_1$, the following expression can be applied to Satellite Laser Ranging (SLR)

$$\sigma = \frac{s^2_O - s^1_O}{2} \left[1 + \frac{1}{2} b_O B^{(\alpha)}_O X^{(\alpha_2)} - \frac{1}{4} S_{O(44\alpha\beta)} X^{(\alpha_2)} X^{(\beta_2)} \right]; \tag{10}$$

and, finally, for measurements in Altimetry we have

$$\sigma = \frac{s^2_{O_1} - s^1_{O_1}}{2} \left[1 - \frac{1}{4} S_{O_P(44\alpha\beta)} X^{(\alpha_1)} X^{(\beta_1)} \right]. \tag{11}$$

For the general formulae (8), the directions of emission and reception with respect to $\lambda^{i_1}_{(\alpha)}$ at Q_1 and Q_2, respectively, result to be

$$
\begin{aligned}
\theta_{(\alpha_1)} = {} & \mu_{(\alpha)} \left[1 + \frac{1}{4} S_{O_P(44\beta\gamma)} \Delta X^{(\beta)} \Delta X^{(\gamma)} \right] \\
& - \frac{1}{2} S_{O_P(44\alpha\beta)} \Delta X^{(\beta)} \left(\Delta X^{(\gamma)} \Delta X_{(\gamma)} \right)^{1/2} \\
& + \frac{1}{2} \Big[S_{O_P(\alpha4\beta\gamma)} + S_{O_P(\alpha\delta\beta\gamma)} v^{\delta_1} - \frac{1}{2} S_{O_P(44\alpha\beta)} v^{\gamma_1} \\
& \qquad - \frac{1}{2} S_{O_P(44\alpha\gamma)} v^{\beta_1} + S_{O_P(44\beta\gamma)} v^{\alpha_1} \Big] \Delta X^{(\beta)} \Delta X^{(\gamma)} \qquad (12)
\end{aligned}
$$

$$
\begin{aligned}
\theta_{(\alpha_2)} = {} & \mu_{(\alpha)} \left[1 + \frac{1}{4} S_{O_P(44\beta\gamma)} \Delta X^{(\beta)} \Delta X^{(\gamma)} \right] \\
& - \frac{1}{2} S_{O_P(44\alpha\beta)} \Delta X^{(\beta)} \left(\Delta X^{(\gamma)} \Delta X_{(\gamma)} \right)^{1/2} \\
& - \frac{1}{2} \Big[S_{O_P(\alpha4\beta\gamma)} + S_{O_P(\alpha\delta\beta\gamma)} v^{\delta_2} - \frac{1}{2} S_{O_P(44\alpha\beta)} v^{\gamma_2} \\
& \qquad - \frac{1}{2} S_{O_P(44\alpha\gamma)} v^{\beta_2} + S_{O_P(44\beta\gamma)} v^{\alpha_2} \Big] \Delta X^{(\beta)} \Delta X^{(\gamma)}. \qquad (13)
\end{aligned}
$$

5 The Ballistic Problem

The following formulae, which correspond to the targeting directions of an object O_2 at $P_1 \in O_1$ are, in a certain sense, the inverse of (12) and (13). The direction of the line-of-sight with respect to O at P is

$$
\theta_{(\alpha_1)} = -r_{O_2 O_1 | O_P(\alpha)} - \sigma_{O_2 O_1 | O_P} (v^{\alpha_2} - v^{\alpha_1}) - \frac{1}{2} \sigma^2_{O_2 O_1 | O_P} S_{O_P(\alpha 4 4 \gamma)} r^{(\gamma)}_{O_2 O_1 | O_P};
$$
$$(14)$$

the pointing direction is

$$
\theta_{(\alpha_2)} = -r_{O_2 O_1 | O_P(\alpha)} + \sigma_{O_2 O_1 | O_P} (v^{\alpha_2} - v^{\alpha_1}) - \frac{1}{2} \sigma^2_{O_2 O_1 | O_P} S_{O_P(\alpha 4 4 \gamma)} r^{(\gamma)}_{O_2 O_1 | O_P},
$$
$$(15)$$

and the angle of advance is

$$
\cos \theta' = 1 - 2(v^{\alpha_2} - v^{\alpha_1})^2 + 2 \frac{r_{O_2 O_1 | O_P(\gamma)} r_{O_2 O_1 | O_P(\delta)}}{\sigma^2_{O_2 O_1 | O_P}} (v^{\gamma_2} - v^{\gamma_1})(v^{\delta_2} - v^{\delta_1}), \quad (16)
$$

so that when $O \equiv O_1$ we have that

$$
\theta_{(\alpha_1)} = X_{(\alpha_2)} - \sigma v^{\alpha_2} + \frac{1}{2} \sigma^2 S_{O_1(\alpha 4 4 \gamma)} X^{(\gamma_2)} \qquad (17)
$$

$$
\theta_{(\alpha_2)} = X_{(\alpha_2)} + \sigma v^{\alpha_2} + \frac{1}{2} \sigma^2 S_{O_1(\alpha 4 4 \gamma)} X^{(\gamma_2)} \qquad (18)
$$

$$
\cos \theta' = 1 - 2(v^{\alpha_2})^2 + 2 \frac{X_{(\gamma_2)} X_{(\delta_2)}}{\sigma^2} v^{\gamma_2} v^{\delta_2}, \qquad (19)
$$

where $X^{(\alpha_2)}$ is the relative position of O_2 with respect to O_1 at P_1 (see (6)), and σ is the relative distance of O_2 with respect to O_1 at P_1 (see (7)).

These magnitudes, relative position and distance, will be the result of a chain of iterations between (9), and (17) and (18) so that, at some step they must be predicted by using another formulation giving at least the same order of approximation (for example, those given in [ZGR02] and [Bah01]).

References

[AB86]   N. Ashby, B. Bertotti, Relativistic Effects in Local Inertial Frames, Phys. Rev. D, **34**, 2246–2259 (1986).

[Bah01]   T.B. Bahder, Navigation in Curved Space-Time, Am. J. Phys., **69**, 315–321 (2001).

[Gar04]   M.L. García del Pino, Descripción covariante relativista de observaciones en Astrometría y Geodesia Espacial, Ph.D. Thesis, University Carlos III de Madrid (2004).

[Sof89]   M.H. Soffel, Relativity in Astrometry, Celestial Machanics and Geodesy, Springer Verlag, Berlin Heidelberg (1989).

[Syn60]   J.L. Synge, Relativity: The General Theory, North Holland, New York (1960).

[ZGR02]   P. Zamorano, J.M. Gambi, P. Romero, M.L. García del Pino, Description of orbits with polar Gaussian coordinates in the post-Newtonian approximation of a spherically symmetric field, Il Nuovo Cimento **117 B (4)**, 441–448 (2002).

Minisymposium "Clean Coal Conversion Technologies"

C. Dopazo[1] and P.L. Garcia-Ybarra[2]

[1] Fluid Mechanics Group, Univeristy of Zaragoza, Spain
dopazo@unizar.es
[2] Dept. Fisica Matematica y de Fluidos, UNED, Spain
pgybarra@ccia.uned.es

Among fossil fuels (coal, oil and gas), world proved reserves of coal are the largest and, at the current consumption rates, they would last for over two hundred years. Then, all along the current century, coal is expected to continue playing a key role, as a vital primary fuel for energy generation purposes. At present, coal combustion is a rather mature technology; nevertheless, advanced conversion processes are continuously being developed in order to reduce gaseous atmospheric emissions and other pollutants from coal power plants. Additional environmental concerns have recently emerged, as some health impacts are correlated to trace metal emissions and submicronic airborne particles. Moreover, the former indications of a potential global impact of emissions of greenhouse gases on climate change are becoming evidences. Scientific research and technological development on coal energy utilization are mainly focused on improving the understanding of the combustion and gasification underlying basic processes, as well as designing thermal cycles at higher pressures and temperatures. Efforts are aimed at increasing efficiency, reducing emissions and at separating the carbon dioxide from the flue gases for its possible subsequent storage.

The proceedings of this minisymposium include a summary of the talks presented by A. Linan and J.L. Ferrin devoted to the mathematical simulation of coal particles in pulverised coal furnaces; an analytic combustion model is incorporated into the governing equations, leading to a convenient Eulerian–Lagrangian mixed formulation, and to numerical integration via an efficient algorithm. The second contribution, by J.L. Castillo and P.L. Garcia-Ybarra, reviews post-combustion processes related to the transport and deposition of ash and soot particles which generate problems of fouling and emissions. The third contribution, by N. Fueyo et al., develops a comprehensive – fully Eulerian – mathematical description of a multiphase flow model to cope with the topic of flue-gas desulphurization by wet scrubbers. Finally, the work by J. Jimenez and J. Ballester describes experimental techniques in a drop tube furnace for correlating the data on kinetic parameters of solid fuel particle combustion with an Arrhenius rate.

Mathematical Modelling of Coal Particles Combustion in Pulverised Coal Furnaces

A. Bermúdez[1], J.L. Ferrín[1], A. Liñán[2], and L. Saavedra[1]

[1] Universidad de Santiago de Compostela, Spain
  `mabermud@usc.es`, `maferrin@usc.es`, `laurasl@usc.es`
[2] Universidad Politécnica de Madrid, Spain
  `alinan@aero.upm.es`

1 Introduction

The purpose of this paper is to contribute to the mathematical modelling of the combustion of coal particles in pulverised coal furnaces, and also to propose an algorithm for its numerical solution. The mathematical model includes two coupled phases: the solid phase, for the coal particles, where a Lagrangian description is used and an Eulerian description for the gas phase, where the effects of the combustion of coal particles are homogenised.

2 Mathematical Model

The mathematical model take into account the simultaneous processes of moisture evaporation and devolatilisation together with the heterogeneous gasification reactions of the char. These processes can take place in a kinetically or diffusion-controlled way. For the gas phase reactions, the Burke–Schumann analysis for very fast reactions will be generalised to account for the competition for oxygen of CO, H_2 and the volatiles.

The validity of the model is dependent on the inequalities $L \gg l_c \gg l_p \gg a$, between the length scales, L of the burner, l_c of the computational cell, l_p of the interparticle distance and a the radius of the coal particle. A detailed derivation can be seen in [BFL].

2.1 The Combustion Model

The simplified kinetic model we are going to consider consists of the following physico-chemical processes within the porous particles, expressed by the heterogeneous reactions

$\underline{1}$ $C(s) + CO_2 \rightarrow 2CO \quad + (q_1)$

$\underline{2}$ $C(s) + \frac{1}{2}O_2 \rightarrow CO \quad + (q_2)$

$\underline{3}$ $C(s) + H_2O \rightarrow CO + H_2 \quad + (q_3)$

$\underline{4}$ $V(s) \rightarrow V(g) \quad + (q_4)$

$\underline{5}$ $H_2O(s) \rightarrow H_2O(g) \quad + (q_5)$

and the gas phase reactions

$\underline{6}$ $CO + \frac{1}{2}O_2 \rightarrow CO_2 \quad + (q_6)$

$\underline{7}$ $V(g) + \nu_1 O_2 \rightarrow \nu_2 CO_2 + \nu_3 H_2O + \nu_4 SO_2 \quad + (q_7)$

$\underline{8}$ $H_2 + \frac{1}{2}O_2 \rightarrow H_2O \quad + (q_8)$,

where index s denotes the solid phase and index g the gas phase, whereas q_i is the heat released by reaction i per unit of gasified mass. For simplicity, all the volatiles are represented by a single molecule $V(g) = C_{\kappa_1} H_{\kappa_2} O_{\kappa_3} S_{\kappa_4}$ of molecular mass M_{vol}, with coefficients deduced from the ultimate analysis of the coal. The molar stoichiometric coefficients ν_i are given in terms of the composition of the volatile molecule.

For the generation of volatiles and moisture evaporation, we shall use a simple kinetic model given by

$$w_4 = B_4 e^{-E_4/\mathcal{R}T} \rho_V, \tag{1}$$
$$w_5 = B_5 e^{-E_5/\mathcal{R}T} \rho_{H_2O}, \tag{2}$$

where ρ_V and ρ_{H_2O} are the local values within the coal particle of the density of volatiles and H_2O remaining in condensed form. For the char gasification reactions, as well as for the gas phase reactions, we could adopt similar expressions for the overall reaction rates per unit volume. However, in our analysis we shall consider that these reactions are either frozen or that the limit of infinite reaction rates applies.

2.2 Gas Phase Model

The coal particle combustion model to be developed in the following section has to be coupled with a gas phase model, which establishes the local average conditions of the gas where the coal particles are burnt. They are represented by mean field values, denoted by the subscript or superscript g, of the mass fractions, temperature and velocity of the gaseous mixture.

In our analysis we shall consider that the limit of infinite reaction rates applies to reactions 6, 7 and 8, leading to the non-coexistence with O_2 of CO, V and H_2, independently of the detailed form of the rates. Therefore, in order to obtain equations without the gas phase reaction terms we consider the following conserved scalars:

$$\beta_1^g = Y_{O_2}^g - \frac{4}{7}Y_{CO}^g - \frac{32\nu_1}{M_{vol}}Y_V^g - 8Y_{H_2}^g, \tag{3}$$

$$\beta_2^g = Y_{CO_2}^g + \frac{11}{7}Y_{CO}^g + \frac{44\nu_2}{M_{vol}}Y_V^g, \tag{4}$$

$$\beta_3^g = Y_{H_2O}^g + \frac{18\nu_3}{M_{vol}}Y_V^g + 9Y_{H_2}^g, \tag{5}$$

$$\beta_4^g = Y_{SO_2}^g + \frac{64\nu_4}{M_{vol}}Y_V^g, \tag{6}$$

$$H^g = h_T^g + q_6 Y_{CO}^g + q_7 Y_V^g + q_8 Y_{H_2}^g. \tag{7}$$

Then, the conservation equations of the gaseous species and energy are

$$\mathcal{L}_g(\beta_1^g) = f_{O_2}^m - \frac{4}{7}f_{CO}^m - \frac{32\nu_1}{M_{vol}}f_V^m - 8f_{H_2}^m, \tag{8}$$

$$\mathcal{L}_g(\beta_2^g) = f_{CO_2}^m + \frac{11}{7}f_{CO}^m + \frac{44\nu_2}{M_{vol}}f_V^m, \tag{9}$$

$$\mathcal{L}_g(\beta_3^g) = f_{H_2O}^m + \frac{18\nu_3}{M_{vol}}f_V^m + 9f_{H_2}^m, \tag{10}$$

$$\mathcal{L}_g(\beta_4^g) = f_{SO_2}^m + \frac{64\nu_4}{M_{vol}}f_V^m, \tag{11}$$

$$\mathcal{L}_g(H^g) = f^e + q_6 f_{CO}^m + q_7 f_V^m + q_8 f_{H_2}^m - \nabla \cdot \mathbf{q}_{rg}, \tag{12}$$

with $\mathcal{L}_g$ being the differential operator defined by

$$\mathcal{L}_g(u) = \frac{\partial(\rho_g u)}{\partial t} + \nabla \cdot (\rho_g u \mathbf{v}_g) - \nabla \cdot (\rho_g \mathcal{D}\nabla u), \tag{13}$$

where $\mathcal{D}$ is a gas phase diffusion coefficient which, for simplicity, will be considered to be the same for all species and equal to the thermal diffusivity. The effects of the particles gasification and combustion appear in the right-hand side of those equations as homogenised sources which will be calculated later after analysing the distribution of temperature and concentrations within the individual particles and in the gaseous neighbourhood of each particle.

The solution of the previous conservation equations provides the temperature and the mass fractions of the species in the gas mixture, with the gaseous domain in two regions: Ω_O, defined by $\beta_1^g > 0$, where there is oxygen in the gaseous environment of the particles with zero concentration of the volatiles, H_2 and CO generated by the gasification, because they will react with the oxygen in a diffusion flame sheet inside the particle, or outside in its vicinity. In a second region Ω_F, defined by $\beta_1^g \leq 0$, the mass fraction $Y_{O_2}^g$ is zero, so the particles are gasifying in an oxygen free environment. In the first region the particles do not represent sources of CO, volatiles or H_2 for the mean values of the bulk interstitial gas. In the second region the volatiles, CO and H_2 join, without locally burning, the homogenised gas phase and, if we assume combustion reactions 6, 7 and 8 to be infinitely fast, these species will burn, in the form of group combustion, where they meet the oxygen, in a gaseous diffusion flame.

2.3 Particle Gasification Model

We shall deal with coal particles that contain a significant fraction of ashes not lost during the devolatilisation or char oxidation stages. Thus we shall consider in our model that the apparent radius of each particle remains constant, although the density of H_2O, volatiles and char will change with time.

The density of the coal particle is given by $\rho_p = \rho_{H_2O} + \rho_V + \rho_C + \rho_{ash}$. In order to model the particle gasification, we must provide equations for the evolution of ρ_{H_2O}, ρ_V and ρ_C and also for the temperature of the particle, T_p.

The evolution of ρ_{H_2O}, ρ_V and ρ_C, with the radial coordinate r and time t, will be given by

$$\frac{\partial \rho_V}{\partial t} = -w_4, \qquad \frac{\partial \rho_{H_2O}}{\partial t} = -w_5, \qquad \frac{\partial \rho_C}{\partial t} = -w_C, \qquad (14)$$

in terms of the mass rates, per unit volume and time, of generation of volatiles w_4, water vapour w_5 and char gasification w_C, given by (1)–(2) and $w_C = w_1 + w_2 + w_3$. For the following description of the char gasification reactions, they are considered to be infinitely fast.

Depending on the region where particle burns, the analysis of its gasification changes considerably. Thus, if the particle lies in Ω_F there is no oxygen inside the particle and the gas phase reactions terms w_6, w_7 and w_8 disappear from mass conservation equations. In that case, the corresponding analysis, which is given later in this section, simplifies considerably. However, analysing the situation where the particle lies in Ω_O and, in particular, when the oxygen reaches the particle surface ($\beta_1^s > 0$), is more complicated; the details are given in [BFL].

When the particle lies in region Ω_F, and therefore the oxygen does not reach the particle surface, the mass conservation equations describing the radial distribution of the gas phase mass fractions Y_{CO_2} and Y_{H_2O}, within the pores of the particles, are

$$\frac{1}{r^2}\frac{\partial}{\partial r}(r^2 \rho_g v_g Y_{CO_2}) - \frac{1}{r^2}\frac{\partial}{\partial r}\left(r^2 \rho_g \mathcal{D}_e \frac{\partial Y_{CO_2}}{\partial r}\right) = -\frac{11}{3}w_1, \qquad (15)$$

$$\frac{1}{r^2}\frac{\partial}{\partial r}(r^2 \rho_g v_g Y_{H_2O}) - \frac{1}{r^2}\frac{\partial}{\partial r}\left(r^2 \rho_g \mathcal{D}_e \frac{\partial Y_{H_2O}}{\partial r}\right) = -\frac{3}{2}w_3 + w_5. \qquad (16)$$

These equations can be integrated once, in the limit of Damköhler numbers $Da_i = (a^2/\mathcal{D}_e)B_i e^{-E_i/\mathcal{R}T_p} \gg 1$, for $i = 1, 3$, (which is applicable at high particle temperatures) to give

$$r^2 \rho_g v_g Y_{CO_2} - r^2 \rho_g \mathcal{D}_e \frac{\partial Y_{CO_2}}{\partial r} = -\frac{11}{3}m_1'' r_c^2, \qquad (17)$$

$$r^2 \rho_g v_g Y_{H_2O} - r^2 \rho_g \mathcal{D}_e \frac{\partial Y_{H_2O}}{\partial r} = -\frac{3}{2}m_3'' r_c^2 + w_5 \frac{r^3}{3}, \qquad (18)$$

for $r_c < r < a$, with the boundary conditions

$$Y_{CO_2} = Y_{H_2O} = 0 \text{ at } r = r_c, \tag{19}$$

leading to

$$\frac{11}{3}\frac{\lambda_1}{\lambda} = \left\{ Y^s_{CO_2} + \frac{11}{3}\frac{\lambda_1}{\lambda} \right\} e^{\lambda \frac{\mathcal{D}}{\mathcal{D}_e}\left(1-\frac{a}{r_c}\right)}, \tag{20}$$

$$\frac{3}{2}\frac{\lambda_3}{\lambda} - \frac{\lambda_5}{\lambda} = \left\{ Y^s_{H_2O} + \frac{3}{2}\frac{\lambda_3}{\lambda} - \frac{\lambda_5}{\lambda} \right\} e^{\lambda \frac{\mathcal{D}}{\mathcal{D}_e}\left(1-\frac{a}{r_c}\right)}, \tag{21}$$

written in terms of the nondimensional reaction rates defined by

$$\lambda = \sum_{i=1}^{5} \lambda_i, \quad \lambda_i = \frac{\dot{m}_i}{\rho_g a \mathcal{D}}, \tag{22}$$

where $4\pi \dot{m}_i$ is the mass gasification reaction rate, due to the heterogeneous i reaction. Moreover, $Y^s_{CO_2}$ and $Y^s_{H_2O}$ are the surface values of Y_{CO_2} and Y_{H_2O} to be calculated later using the gas phase analysis.

The time evolution of r_c, the radius of the shrinking core, is determined from the time evolution equation of the char density as

$$\frac{\rho^0_C}{\rho_g a \mathcal{D}} r_c^2 \frac{dr_c}{dt} = -(\lambda_1 + \lambda_3), \tag{23}$$

where we have neglected the changes in ρ_C within the char core, during the first stage of the kinetically controlled char gasification, so we have approximated ρ_C by its initial value ρ^0_C for $r < r_c$.

Finally, the time evolution of the temperature of the particle, considered to be uniform, is given by the equation

$$\frac{4}{3}\pi a^3 \rho_p c_s \frac{dT_p}{dt} = 4\pi a^2 (q''_p + q''_r) + 4\pi \rho_g a^2 \mathcal{D}(q_1\lambda_1 + q_3\lambda_3 + q_4\lambda_4 + q_5\lambda_5), \tag{24}$$

where $4\pi a^2 q''_p$ and $4\pi a^2 q''_r$ are the rates of heat reaching the particle by conduction and radiation.

In order to determine the values of the mass fractions and the heat flux at the surface of the particle, we need to model the gas environment outside the particle. The mass and energy conservation equations are of the form (15) and (16), with the exception that now v_g is the true velocity of the gas phase and $\mathcal{D}_e$ must be replaced by the gas phase diffusion coefficient $\mathcal{D}$, and no sources in their right-hand sides. Again, we solve the corresponding system of equations in the limit of infinite Damköhler numbers for the gas phase reactions by introducing the same Schvab–Zeldovich combinations defined by (3)–(7).

Thus, as an example, if particle lies in the region Ω_F by integrating the conservation equations with appropriate boundary conditions we obtain

$$Y_{CO_2}^s = Y_{CO_2}^g e^{-\lambda} - \frac{11}{3}\frac{\lambda_1}{\lambda}(1 - e^{-\lambda}), \tag{25}$$

$$Y_{H_2O}^s = Y_{H_2O}^g e^{-\lambda} + \left(\frac{\lambda_5}{\lambda} - \frac{3}{2}\frac{\lambda_3}{\lambda}\right)(1 - e^{-\lambda}), \tag{26}$$

$$q_p'' = \frac{k}{ac_p}(h_T^g - h_T^s)\frac{\lambda}{e^\lambda - 1}. \tag{27}$$

In order to close the model, expressions for the homogenised sources in the gas phase per unit volume and time, at each point x of the boiler, can be obtained from the individual sources of one particle by

$$f^\alpha(x) = \sum_{j=1}^{N_e}\sum_{i=1}^{N_p} \tilde{q}_j \frac{p_{ij}}{100} \int_0^{t_f^{ij}} F_{ij}^\alpha(t)\delta(x - x_s^{ij}(t))dt, \tag{28}$$

where $F_{ij}^\alpha(t)$ denotes the contribution of one individual particle of type i introduced through inlet j, at instant t, $x_s^{ij}(t)$ is the position occupied by this particle at instant t, $\delta(x)$ is the Dirac measure at point 0, t_f^{ij} is the time needed for the particle to be completely burned or to leave the furnace, $\tilde{q}_j$ is the mass flow of coal through inlet j, p_{ij} is the percentage of particles of type i through inlet j, and N_e and N_p are the number of inlets and types of particles, respectively.

Expressions for the mass and energy sources in the case of an individual particle burning in region Ω_F are

$$F_{O_2}^m = F_{SO_2}^m = 0, \tag{29}$$

$$F_{CO_2}^m = \frac{4\pi ak}{c_p}\left(-\frac{11}{3}\lambda_1\right), \tag{30}$$

$$F_{H_2O}^m = \frac{4\pi ak}{c_p}\left(\lambda_5 - \frac{3}{2}\lambda_3\right), \tag{31}$$

$$F_{CO}^m = \frac{4\pi ak}{c_p}\left(\frac{14}{3}\lambda_1 + \frac{7}{3}\lambda_3\right), \tag{32}$$

$$F_V^m = \frac{4\pi ak}{c_p}\lambda_4, \tag{33}$$

$$F_{H_2}^m = \frac{4\pi ak}{c_p}\frac{1}{6}\lambda_3, \tag{34}$$

$$F^e = 4\pi ak\left(\frac{c_s}{c_p}T_p - T_g\right)\frac{\lambda}{e^\lambda - 1} - c_s T_p \frac{dm_p}{dt}. \tag{35}$$

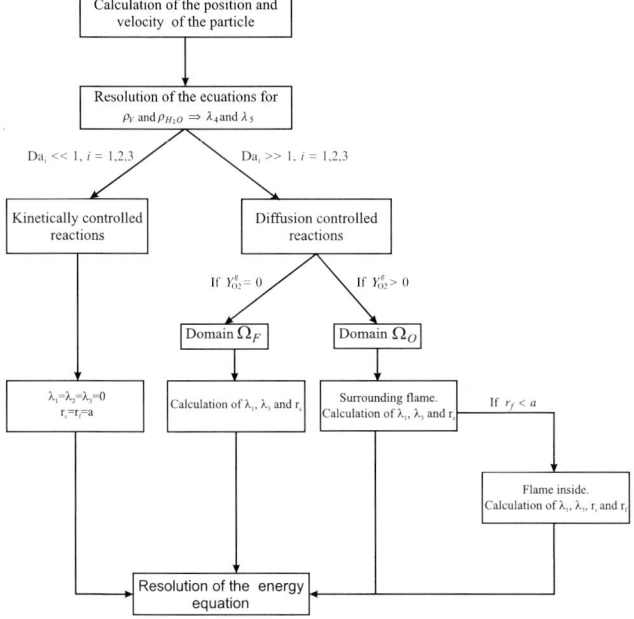

Fig. 1. Algorithm

3 Numerical Solution

Once given the ambient gas conditions, the algorithm proposed for solve the mathematical model introduced in the previous section (and detailed in [BFL]) can be seen in Fig. 1.

The validation of the numerical algorithm and some numerical results can be seen in [Saa06].

References

[BFL]   Bermúdez, A., Ferrín, J.L., Liñán, A.: The modelling of the generation of volatiles, H_2 and CO, and their simultaneous diffusion controlled oxidation, in pulverised coal furnaces. *Submitted to* Combustion Theory and Modelling.

[Saa06] Saavedra, L.: Simulación numérica de la combustión de partículas de carbón y simulación en Mecánica de Fluidos. Trabajo de Investigación Tutelado, Universidad de Santiago de Compostela, Spain (2006)

Transport of Particles and Vapors in Flue Gases and Deposition on Cold Surfaces

José L. Castillo and Pedro L. Garcia-Ybarra

Dept. Fisica Matematica y Fluidos, Facultad de Ciencias, UNED, Madrid, Spain
jcastillo@ccia.uned.es, pgybarra@ccia.uned.es

1 Introduction

In coal combustion processes, a large amount of nonvolatile material is emitted as particular matter carried by the gas stream. Moreover, some condensable vapors (usually sulfates and nitrates) are formed by reaction in the flue gases. The control of these particles and vapors is a key factor in clean coal conversion technologies. Thus, the formation of soot and fly ash deposits and the condensation of vapors over heat exchanger tubes and exhaust lines reduce the heat transfer efficiency and promote corrosion problems, leading to shorter lifetimes of the equipment and increasing the production and maintenance costs. Also, the emission of submicron particles to the ambient air is an environmental issue of capital importance. Moreover, the bulk (porosity, hardness) and surface (roughness) properties of the formed deposit depend on the particle arrival dynamics. Therefore, the analysis of particle and vapor transport under controlled conditions and the study of deposit formation from particle laden gases are problems of wide practical implications in coal combustion. In particular, there is a need of theoretical analysis on the dynamics of particles in gases under strong temperature differences and intense radiative fluxes, as well as on the behavior of particles near obstacles to evaluate the deposition rates. Some model problems linked to the behavior of particles and vapors in gases and deposit formation will be discussed here.

2 Dynamics of Particle in Gases

The transport properties of particles in gas streams (soot and flying ash in the case of combustion environments) determine their distribution in the gas, as well as their resident time and thus, the coagulation, condensation, and deposition processes undergone for the particles in the system. Thus, particle accumulation regions may appear in the gas (where the particles concentrate around prescribed trajectories inducing a local enhancement of the coagulation rate) due to the competition of different transport mechanisms. In general, small particles in gases do not follow the gas streamlines [16], the main causes being: inertia of particles in rapidly varying flows [11], thermophoresis (drift of particles in a gas down a temperature gradient) [8, 10, 15, 16, 18], photophoresis (caused by inhomogeneous surface temperatures induced by radiative fluxes) [2], buoyancy, electrophoresis, external forces, and Brownian diffusion. In general, the mean particle velocity $\mathbf{v}_p$ can be written as

$$\mathbf{v}_p = \mathbf{v} + \mathbf{v}_{rel}, \tag{1}$$

here $\mathbf{v}$ is the local gas velocity and $\mathbf{v}_{rel}$ is the average particle velocity relative to the gas which may have several contributions. Due to the high temperature gradients involved in coal combustion, thermophoresis may become the leading diffusive transport for intermediate soot and flying ash particle sizes (from submicron to micron sizes) pushing the particles away from the hotter regions. This thermophoretic velocity is

$$\mathbf{v}_T = -\alpha_T D \frac{\nabla T}{T}, \tag{2}$$

where α_T is the thermal diffusion factor and D the diffusion coefficient. The particle dynamics depends mainly on two dimensionless parameters, the thermal diffusion strength α, and the Schmidt number Sc

$$\alpha = \frac{\alpha_T D}{\nu} \qquad Sc = \frac{\nu}{D} \tag{3}$$

with ν the gas kinematic viscosity. For sufficiently large particle, Sc takes on very large values and Brownian diffusion may become negligible whereas for a wide range of particle sizes, the value of α is quite insensitive to particle size and form [15, 16]. Moreover, the presence of intense radiative fluxes may lead to a photophoretic drift of the particles in the gas [2].

Usually the particles form large aggregates which can be simulated as continuous porous particles. Recently [9], we have obtained the drag on an aggregate composed by a large number of unitary spheres N of radius R_1, distributed in a fractal manner with a cumulative monomer number distribution given by

$$N(r) = \beta \left(\frac{r}{R_1} \right)^{D_f}, \tag{4}$$

where r is the distance to the aggregate center, D_f is the fractal dimension, and β is a the fractal prefactor of order unity. The aggregated was modeled as a fractal porous sphere with variable porosity, using a low permeability asymptotics; that is, in the limit

$$k^{1/2} \equiv \frac{[\chi(R)]^{1/2}}{R} \ll 1 \tag{5}$$

where χ is the particle permeability and R is the aggregate radius. The drag on a moving aggregate can be estimated by matching the solutions in three different regions: An outer region (the fluid Stokes region) and an inner region (the porous particle core governed by a Darcy law), both connected through a transition region at the surface (governed by a Brinkman equation). The solution of this multilayered problem leads to the drag force on the aggregate

$$F = -6\pi\mu R U \left\{ 1 - k^{1/2} + k \left[\frac{(\chi'/\chi)_{r=R}}{4} - \frac{3+\delta}{2+\delta} \right] + O\left(k^{3/2}, \right) \right\} \quad (6)$$

χ' denotes the derivative of the particle permeability with the distance to the particle center, being a negative quantity. Moreover,

$$\delta \equiv \frac{\sqrt{D_f^2 + 8(3 - D_f)} - D_f}{2}. \quad (7)$$

Therefore, the Stokes drag is reduced by the particle permeability due to the possibility for the fluid to pass through the aggregate.

3 Behavior of Particles Near Obstacles Under Strong Temperature Differences

The knowledge of the particle dynamics is needed to analyze the distribution of these particles around the surfaces confining the gas stream and to obtain the particle deposition rates on these surfaces. In a previous work [8], we studied the behavior of particles around obstacles when there exists a large temperature difference between the solid and the gas stream. For cold surfaces in hot gases, the thermophoretically induced mass flux of a dilute aerosol toward the body surface was obtained and the results were compared with some available experimental measurements [11]. In most practical cases, a simple empirical law may correlate the deposition rate

$$J \approx b_1 \left[1 - \exp\left(-b_2 \, \alpha \frac{T_\infty - T_w}{T_w} \right) \right], \quad (8)$$

with T_∞ and T_w denoting the mainstream temperature and the wall temperature, respectively, and b_1, b_2 are correlation constants of order unity.

On the other hand, for hot plates in cold streams, thermophoresis pushes the particles away from the plate. In the absence of Brownian diffusion a dust free region appears above the hot surface and the thickness of the dust free region increases with the temperature difference. However, Brownian diffusion induces a leakage of the particles across the dust free region and leads to a deposition rate on the solid surface

$$J \approx \frac{a}{Sc^{1/2}} \left[\alpha \frac{T_w - T_\infty}{T_w} \right]^{1/4} \exp\left[-b \, Sc \left(\alpha \frac{T_w - T_\infty}{T_w} \right)^{3/2} \right], \quad (9)$$

which is exponentially small for large values of the Schmidt number Sc.

In some applications involving particle laden gases, there is a need to avoid the particle deposition to confining surfaces. The particle transport properties may be used to reduce the arrival and deposition of particles to surfaces and the efficiency of these methods lies on the generation of a particle rejection field near the surface, but one has to account for Brownian diffusion which is always present, and produces a diffusive transport of the particles against this rejection field. Due to thermophoretic effects that drifts the particles away from heated bodies, heating the surface may be used as a repulsion method to reduce the particle arrival to worthy surfaces [1, 7, 17]. Some other repulsion methods can be devised; as for instance, blowing through the surface which generates a local gas flow field away from the surface opposing the arrival of particle to the wall [1, 13].

4 Monte Carlo Simulation of Deposit Growth Dynamics

Once the particle become in contact with the surface, they may form deposits that evolve with time. The bulk properties (hardness, porosity, permeability, effective thermal conductivity) and the surface structure (roughness, rigidity, reactivity) of the generated deposits will affect the temperature field and the velocity field around them. These deposit properties are primarily controlled by the way the particle reach the deposit. The particle motion near the deposit can be split into two additional contributions; namely, a mean particle velocity, U, and a random motion with a characteristic diffusion coefficient, D. Thus, the motion is characterized by a Peclet number, $Pe = Ua/D$, where a is the particle diameter. The Peclet number provides the relative importance of the deterministic motion to the Brownian motion for the particles. In a recent work [14], a Monte Carlo model for the simulation of particle deposit growth by advection and diffusion toward a flat surface has been proposed. The model allows to follow the evolution of the deposit and to determine the main morphological and structural features of the generated deposits, depending on the transport properties of the arriving particles. The deposit structure is characterized by its interface (mean height and surface roughness) and bulk (density) properties. Numerical correlations, fitted by simple expressions for these magnitudes were obtained, relating them to time (number of deposited particles), and Peclet number. The density profiles inside the deposit show three different layers: a near wall region (affected by the presence of the – flat and smooth – initial solid body), a plateau region (with a constant mean density), and an active growth layer (with decreasing density) at the surface. Decreasing Peclet numbers lead to deposits characterized by a lower mean density in the plateau region and a larger width of the active growth region at the surface, presenting more open structures which are less compact and easier to remove.

5 Vapor Deposition

On the other hand, vapors present in the gas streams may condensate on cold surfaces or on preexisting particles, or even nucleate in cold regions. The deposition of vapors and particles on surfaces has been studied focusing on some model problems. Thus, in [3–6, 12] the deposition of vapors on cold surfaces was studied allowing for vapor condensation within the thermal boundary layer. When the dew point is reached inside this layer, vapors either nucleate and form new particles or condensate on already existing particles. Then, the total deposition of material on the surface is obtained as the addition of the transport of vapors by diffusion and the vapor condensated on the arriving particles.

6 Final Remarks

Problems related to transport of particles and vapors, vapor nucleation, condensation of vapors on particles, deposition of vapors and particles on surfaces, and formation of structured deposits have wide practical implications in combustion processes. A few simple model problems have been discussed here. In actual combustion systems, the particle cloud is formed by condensable vapors and a broad distribution of particles sizes, interacting between each other. Inertia will play a key role for the larger particle sizes whereas the dynamics of the smaller particles will be controlled by diffusive processes (as thermophoresis or Brownian diffusion). Further advances on these lines require the availability of experimental results (under well controlled conditions), together with analytical studies of typical flow configurations and the use of integrated computational tools to study the long time evolution of deposits large enough to modify the flow field around them and the heat exchange processes between the gas and the deposits.

Acknowledgement

Work supported by DGES-MEC (Spain) under projects ENE2005-09190-C04-02 and DPI2005-04601 and by Comunidad de Madrid project S-505/ENE/0229.

References

1. Castillo, J.L., Garcia-Ybarra, P.L.: Diffusive leakage of small particles towards blowing surfaces. J. Aerosol Sci., **29**, S1107 (1998)
2. Castillo, J.L., Mackowski, D., Rosner, D.E.: Photophoretic modification of the transport of absorbing particles across combustion gas boundary layers. Prog. Energy and Comb. Sci., **16**, 253–260 (1990)

3. Castillo, J.L., Rosner, D.E.: A nonequilibrium theory of surface deposition from particle-laden, dilute condensible vapor-containing laminar boundary layers. Int. J. Multiphase Flow, **14**, 99–120 (1988)
4. Castillo, J.L., Rosner, D.E.: Theory of surface deposition from a unary dilute vapor-containing stream, allowing for condensation within the laminar boundary layer. Chem. Engng. Sci., **44**, 925–937 (1989)
5. Castillo, J.L., Rosner, D.E.: Equilibrium theory of surface deposition from particle-laden dilute, saturated vapor containing laminar boundary layers. Chem. Engng. Sci., **44**, 939–956 (1989)
6. Castillo, J.L., Rosner, D.E.: Theory of turface deposition from a binary dilute vapor-containing stream, allowing for equilibrium condensation within the laminar boundary layer. Int. J. Multiphase Flow, **15**, 97–118 (1989)
7. Friedlander, S.K., Fernandez de la Mora, J., Gokoglu, S.A.: Diffusive leakage of small particles across the dust-free layer near a hot wall. J. Colloid Interface Sci., **125**, 351–355 (1988)
8. Garcia-Ybarra, P.L., Castillo, J.L.: Mass transfer dominated by thermal diffusion in laminar boundary layers. J. Fluid Mech., **336**, 379–409 (1997)
9. Garcia-Ybarra, P.L., Castillo, J.L., Rosner, D.E.: Drag on a large spherical aggregate with self-similar structure: an asymptotic analysis. J. Aerosol Sci., **37**, 413–428 (2006)
10. Goren, S.L.: Thermophoresis of aerosol particles in the laminar boundary layer on a flat plate, J. Colloid Interface Sci., **61**, 77–85 (1977)
11. Konstandopoulos, A.G., Rosner, D.E.: Inertial effects on thermophoretic transport of small particles to walls with streamwise curvature-I: Theory. Intl J. Heat Mass Transfer, **38**, 2305–2315 (1995), -II: Experiment. Intl J. Heat Mass Transfer, **38**, 2317–2327 (1995)
12. Liang, B., Gomez, A., Castillo, J.L., Rosner, D.E.: Experimental studies of nucleation phenomena within thermal boundary layers - influence on chemical vapor deposition rate processes. Chem. Engng. Comm., **85**, 113–133 (1989)
13. Perea, A., Castillo, J.L., Garcia-Ybarra, P.L.: Fickian leakage on boundary layers with blowing. J. Aerosol Sci., **34**, S117–118 (2003)
14. Rodriguez-Perez, D., Castillo, J.L., Antoranz, J.C.: Relationship between particle deposit characteristics and the mechanism of particle arrival. Phys. Rev. E, **72**, 021403 (2005)
15. Rosner, D.E., Mackowski, D.W., Garcia-Ybarra, P.L.: Size- and structure-insensitivity of the thermophoretic transport of aggregated soot particles in gases. Comb. Sci. Tech., **80**, 87–101 (1991)
16. Rosner, D.E., Mackowski, D.W., Tassopoulos, M., Castillo, J.L., Garcia-Ybarra, P.L.: Effects of heat transfer on the dynamics and transport of small particles suspended in gases. I&EC Res., **31**, 760–769 (1992)
17. Stratmann, F., Fissan, H., Papperger, A., Friedlander, S.: Suppression of particle deposition to surfaces by the thermophoretic force, Aerosol Sci. Tech., **9**, 115–121 (1988)
18. Talbot, L., Cheng, R.K., Schefer, R.W., Willis, D.R.: Thermophoresis of particles in a heated boundary layer. J. Fluid Mech., **101**, 737–758 (1980)

A Comprehensive Mathematical Model of Flue-gas Desulfurization

N. Fueyo[1], A. Gomez[1], and J.F. Gonzalez[2]

[1] Fluid Mechanics Group, University of Zaragoza, Spain
   Norberto.Fueyo@unizar.es
[2] Endesa Generacion SA, Spain

1 Flue-gas Desulfurization

When burned to produce energy, sulfur-containing fossil-fuels, such as coal or oil, often generate sulfur dioxide (SO_2). SO_2 is known to be damaging to humans (at high concentration levels) and to the environment, being one of the main precursors of acid rain. Coal is the most abundant fossil fuel, with reserves estimated to be in excess of 150 years at current consumption rates. Thus, technologies aiming at minimising the environmental impact of coal utilization are subject of vigourous research worldwide. Among these, flue-gas cleanup, such as Flue-gas desulfurization (FGD), is perhaps the one offering at present the lowest technological risk, and the fastest route to implementation. FGD can be achieved using a number of technologies [3], but the vast majority of currently-installed capacity is for wet scrubbers. Wet scrubbers combine high SO_2 removal efficiency, high reagent utilization, and compact designs.

Wet scrubbing is a complex mathematical problem. It involves multiphase flow, with the contact between the scrubbing agent and the flue gas being the driving force of the process. There is mass-transfer between the phases, and chemical reactions within each phase. Limestone is the most most-widely used scrubbing agent. It consists mostly of calcium carbonate ($CaCO_3$), and is plentiful and inexpensive. Further, wet scrubbers using limestone as the absorber can produce gypsum as a by-product. This can be sold to the construction industry, thus generating revenue and avoiding landfilling fees. The production of gypsum requires the addition of an oxidation tank to the scrubber, with adds to the complexity of the mathematical model.

The integration of chemistry and fluid dynamics in a single, comprehensive, mathematical model is the aim of the present paper. The model equations for the multiphase flow can be solved only numerically, and in this paper this is done using Computational Fluid Dynamics techniques. Further, we model simultaneously both the scrubber and the oxidation tank as two separate domains, with similar physicochemical models, linked through boundary conditions.

2 Multiphase, Multidomain Model

As outlined above, the mathematical model required is multidomain and multiphase. The two domains are the scrubber (or absorber) and the tank. In the scrubber, the flue gases are showered with limestone slurry, often in the form of a spray. SO_2 in the flue gases is transferred to (absorbed into) the limestone slurry. The clean flue gases are directed to the plant stack, while the limestone slurry with the dissolved SO_2 falls by gravity onto the oxidation tank; here, additional air is injected, in the form of bubbles, to completely oxidise the gypsum (which is usually later de-watered and sold). The process is outlined in Fig. 1.

The flow, either in the scrubber or the tank, is also made up of two distinct phases. In the scrubber, the flue gas is a continuous phase exchanging properties with the disperse slurry-phase. In the tank, the situation is the reciprocal one, with the liquid slurry being the continuous phase and the air injected through the spargers being the disperse one. An Eulerian–Eulerian multiphase model has been employed to simulate each domain. This model treats both phases (disperse and continuum) as Eulerian continua. They can coexist at each point with a certain volume fraction (or probability of presence) and each will generally have distinct properties which are accounted for by their own set of equations. The model considers conservation equations for the local amount of phase, its momentum and the within-phase concentration (mass fraction) of all the relevant chemical species.

The conservation of mass for each phase α is governed by its volume-fraction equation

$$\frac{\partial (\rho_\alpha r_\alpha)}{\partial t} + \nabla \cdot (\rho_\alpha r_\alpha \mathbf{v}_\alpha) - \nabla \cdot (\Gamma_\alpha \nabla r_\alpha) = \dot{m}_{\beta\alpha} , \tag{1}$$

where ρ_α is the density of phase α, r_α is the volume fraction of the phase, $\mathbf{v}_\alpha = (v_{\alpha_1}, v_{\alpha_2}, v_{\alpha_3})$ is the velocity vector, Γ_α is the phase-diffusion coefficient

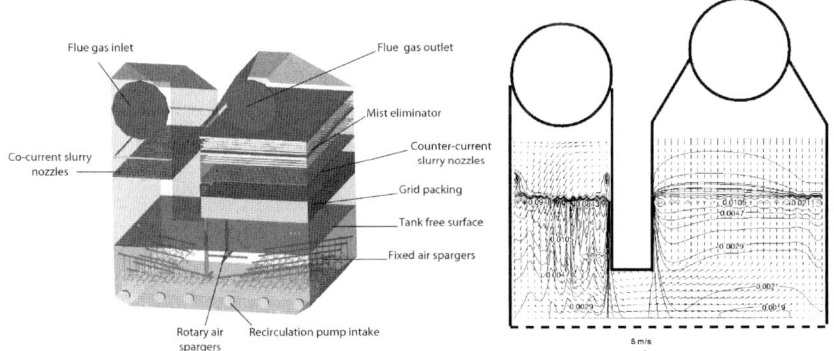

Fig. 1. (*Left*) Plant schematic; (*Right*) gas-phase velocity-vectors and slurry volume-fraction in the scrubber

(accounting for turbulent mixing in the present model), and $\dot{m}_{\beta\alpha}$ represents the mass-transfer rate from phase β into phase α. The volume fractions add to unity: $r_\alpha + r_\beta = 1$.

The momentum equation for the ith component of the phase velocity-vector $\mathbf{v}_\alpha$ is

$$\frac{\partial}{\partial t}(\rho_\alpha r_\alpha v_{\alpha i}) + \nabla \cdot (\rho_\alpha r_\alpha v_{\alpha i} \mathbf{v}_\alpha) - \nabla \cdot (\Gamma_\alpha v_{\alpha i} \nabla r_\alpha) - \nabla \cdot (\mu_\alpha r_\alpha \nabla v_{\alpha i}) = -r_\alpha \frac{\partial P}{\partial x_i} + S_\alpha, \quad (2)$$

where S_α represents different source terms, such as drag or body forces.

Finally, the species conservation equation is, for species A in the phase α

$$\frac{\partial (\rho_\alpha r_\alpha Y_\alpha^A)}{\partial t} + \nabla \cdot (\rho_\alpha r_\alpha Y_\alpha^A \mathbf{v}_\alpha) - \nabla \cdot (\Gamma_\alpha Y_\alpha^A \nabla r_\alpha) - \nabla \cdot (\Gamma_{Y_\alpha^A} r_\alpha \nabla Y_\alpha^A) = S_\alpha^A, \quad (3)$$

where Y_α^A is the mass fraction, $\Gamma_{Y_\alpha^A}$ is the diffusion coefficient for the species, and S_α^A represents different source terms, such as absorption or chemical reaction.

Turbulence is represented with a two-equation k–ϵ model in the continuous phase (i.e., gas in the scrubber and slurry in the tank). The model is a standard one [4], with turbulence-modulation corrections to account for the effect of the disperse phase [7].

3 Chemical Model

The chemical model implemented calculates the local mass fractions of the main species involved in the desulfurization process, in both the scrubber and the tank, through transport equations such as (11). For the flue gas in the scrubber, these species are SO_2, O_2, and CO_2. For the slurry phase, the relevant species are SO_2, HSO_3^-, and SO_3^{2-}, which are modeled together as a single variable named SS (for sulfurous species); CO_2, HCO_3^-, and CO_3^{2-} which are treated similarly as a single variable CS (for carbon species); Ca^{2+}; SO_4^{2-}; $CaCO_3(s)$; and $CaSO_4 \cdot 2H_2O(s)$. Both limestone and gypsum ($CaCO_3(s)$ and $CaSO_4 \cdot 2H_2O(s)$) are treated as species in the liquid phase, as in [5], and not as third, solid phase. The main difference in the equation for each species is the source term, which represents absorption, desorption, and chemical reaction. The corresponding source terms for each species are indicated in Table 1; their rationale is briefly discussed below. (In the expressions shown, the equilibrium constants and solubility products are taken from [1], and Henry's law constants from [8]. A_{int} is the interface area between gas and slurry per unit volume, k are mass-transfer coefficients, E are enhancement factors due to chemical reaction, H is Henry's constant, W are molecular weights, and C are molar concentrations.)

Table 1. Source terms for the chemical species

Source	Value
$S_g^{SO_2}$	$W_{SO_2} A_{int} \left(1/k_g^{SO_2} + H_{SO_2}/k_l^{SO_2} E_{SO_2}\right)^{-1} \left(C_g^{SO_2} - H_{SO_2} C_l^{SO_2}\right)$
$S_g^{O_2}$	$W_{O_2} E_{O_2} A_{int} k_l^{O_2} \left(C_g^{O_2}/H_{O_2} - C_l^{O_2}\right)$
$S_g^{CO_2}$	$W_{CO_2} E_{CO_2} A_{int} k_l^{CO_2} \left(C_g^{CO_2}/H_{CO_2} - C_l^{CO_2}\right)$
S_{SS}	$S_g^{SO_2} - S_g^{O_2} 2 W_{SO_2}/W_{O_2}$
$S_l^{CaCO_3}$	$W_{H^+} \left(S_g^{SO_2}/W_{SO_2} + S_g^{O_2}/W_{O_2}\right)/2$
S_l^{CS}	$S_l^{CaCO_3} - S_g^{CO_2}$
$S_l^{CaSO_4}$	$1.1 \, 10^{-4} A_{gyp} r_l W_{CaSO_4 \cdot 2H_2O} \, (RS - 1)$
$S_l^{Ca^{2+}}$	$S_l^{CaCO_3} - S_l^{CaSO_4}$
$S_l^{SO_4^{2-}}$	$S_g^{O_2} 2 W_{SO_2}/W_{O_2} - S_l^{CaSO_4}$

$SO_2(g)$ is absorbed by the slurry in the scrubber, and the source term in the equation is, therefore, the absorption rate [6]. The mass-transfer coefficients k are calculated from the Sherwood number-correlations. $O_2(g)$ is absorbed in the tank and in the scrubber. The source term is similar to the source term for $SO_2(g)$, but it is considered that the absorption or desorption rate is controlled by mass transfer on the liquid side [6]. $CO_2(g)$ can be absorbed or desorbed in the scrubber and in the tank. The source term is similar to that for $O_2(g)$, because the absorption or desorption rate is controlled by the mass transfer on the liquid side [6]. The compound-species SS is formed in the scrubber, where the absorption of $SO_2(g)$ takes place, and it is consumed in the tank with the oxidation of HSO_3^- to SO_4^{2-}. The respective molar concentrations of SO_2, HSO_3^-, and SO_3^{2-} depend on the pH (pH $= -\log(C^{H^+})$).

4 Results

The mathematical model outlined above has been used to simulate an actual FGD plant, viz that in operation at the ENDESA Teruel powerstation in Spain (schematic in Fig. 1, left). The FGD plant has been designed for a high-sulfur coal (4.5%) and for a desulfurization efficiency better than 90%. The combined limestone consumption is approximately $100 \, \text{Ton} \, \text{h}^{-1}$, for a gypsum production of $180 \, \text{Ton} \, \text{h}^{-1}$.

The plant is of the two-pass kind, with a co-current and a counter-current section. The flue gas enters the plant through the co-current section, where it is showered with the limestone slurry injected through the limestone nozzles. At the bottom of the scrubber there is a slurry oxidation and neutralisation tank. The co-current slurry droplets disengage from the gas flow and fall to the tank, while the gases turn $180°$ toward the counter-flow section of the scrubber, where it is further sprayed with limestone slurry.

The appropriate operational data have been obtained from the plant, both for the definition of the scenarios simulated in the model and for model

N. Fueyo et al.

validation. The main operational parameters for the plant for a flue-gas mass-flow-rate of $1,257,364\,\mathrm{N\,m^3\,h^{-1}}$, a coal sulfur contents of 3.5%, and a design desulfurization efficiency of 95% are: fresh-slurry mass-flow-rate, $112,886\,\mathrm{kg\,h^{-1}}$; fresh slurry limestone concentration, $333\,\mathrm{g\,l^{-1}}$; recycled-slurry mass-flow-rate, $56,073,600\,\mathrm{kg\,h^{-1}}$; oxidation-air flow-rate, $35,300\,\mathrm{N\,m^3\,h^{-1}}$.

Because of space limitations, only a fraction of the results is presented. Figure 1 shows the flue-gas velocity pattern with superimposed slurry volume-fraction. Thus the flue gas enters the absorber through the co-current section, where it is showered with the slurry, and then performs a U-turn toward the counter-current section, where it is again sprayed.

Figure 2 displays contours of SO_2 concentration in the flue gas on a vertical plane in the scrubber. SO_2 levels decrease as it is absorbed into the slurry. Most of the absorption takes place in the co-current section. This agrees with observations at the plant, and also with results from other double-loop scrubbers [2]. In the counter-current section, the SO_2 concentration in the flue gas is smaller, and a grid is installed to enhance the contact between the phases and therefore the mass-transfer. The figure also indicates that the concentration of SO_2 is not uniform across the exit plane, reflecting the effect of the different residence times: the flue gas exiting close to the outer wall has a lower SO_2 level. The predicted overall flue-gas desulfurization efficiency is 90.1%, which can be compared with the actual efficiency of 95%.

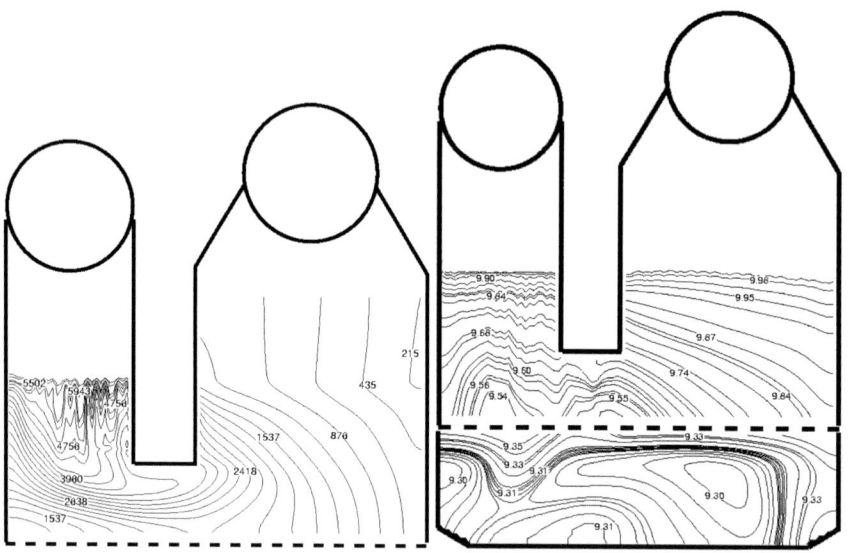

Fig. 2. (*Left*) Contours of SO_2 (*ppmv*, dry basis) in the scrubber; (*Right*) contours of limestone concentration ($\mathrm{g\,l^{-1}}$) in the slurry on a vertical plane across the scrubber and tank

References

1. L. Brewer. Thermodynamic values for desulfurization processes. In ACS Symposium Series, volume 188. American Chemical Society, 1982.
2. DOE. Milliken clean coal demonstration project: A DOE assessment. Technical report, U.S. Department of Energy, 2001.
3. IEA. Flue gas desulfurization (FGD) for SO2 control. http://www.ieacoal. org.uk/, 2006.
4. W. P. Jones and B. E. Launder. The prediction of laminarization with a two-equation model of turbulence. International Journal Heat Mass Transfer, 5:301314, 1972.
5. K. I. Keskinen, V. Alopaeus, J. Koskinen, T. Kinnunen, H. Pitkänen, J. Majander, and U. Wärnström. CFD simulation of the oxidation tank reactor of a wet flue gas desulfurization process. 2002 AIChe Annual Meeting, Indianapolis, IN, Nov. 3-8, 2002.
6. O. Levenspiel. The Chemical reactor Omnibook. Wiley, 1993.
7. M. Lopez de Bertodano, S. T. Lee, R. T. Lahey, and D. A. Drew. The prediction of 2-phase turbulence and phase distribution phenomena using a reynolds stress model. ASME Journal of Fluids Engineering, 112:107, 1990.
8. R. Sander. Compilation of Henrys law constants for inorganic and organic species of potential importance in environmental chemistry. http://www.mpchmainz. mpg.de/sander/res/henry.html, 1999.

Determination of the Kinetic Parameters of a Pulverized Fuel from Drop Tube Experiments

Santiago Jiménez[1] and Javier Ballester[2]

[1] LITEC-CSIC
   yago@litec.csic.es
[2] LITEC-University of Zaragoza
   ballester@unizar.es

1 Introduction

The correct simulation of industrial plants firing pulverized fuels (pf: coal, biomass, etc.) by means of commercial CFD codes relies on a number of sub-models for the various processes, including, e.g. heat transfer (radiation, conduction through deposits, etc.) and particle combustion. The latter is of major importance in the design of the combustion chamber and the selection of the mills or, conversely, regarding the feasibility of burning a new fuel in an existing boiler. In the last decade, the introduction of new, internationally traded coals and alternative fuels into the power market has motivated renewed interest in the experimental and theoretical characterization of the combustion of these fuels. Regarding experimentation, it is generally accepted that the 'reactivity' of a fuel can not be determined in desktop analytical instruments; instead, drop tube furnaces or entrained flow reactors (EFR) must be used in order to reproduce the high temperature, high heating rate conditions found in a real pf combustion chamber [1]. Several alternative experimental procedures have been developed in the past and are still used (see, e.g. [2,3]). On the other hand, two general approaches are used in the literature to model pulverized coal/biomass char combustion: one intends to characterize the evolution of the pores inside the burning particle, and considers both internal and external diffusion, whereas the kinetics for the basic homogeneous and heterogeneous reactions are taken from low temperature analysis or fundamental knowledge of the chemistry involved (e.g. [4]); the other one, followed here, makes use of an apparent kinetics based on the outer particle surface, and includes external diffusion [5]. In the latter case, two parameters governing an Arrhenius-like kinetics are the main unknowns to be determined from the experiments performed in an EFR. The aim of this paper is to discuss some aspects of the mathematical procedure for the determination of those parameters.

2 Char Oxidation Model

The 'single film' model first proposed by Field et al. [5], the most extended among commercial codes, describes the oxidation of a char particle by means of apparent oxidation kinetics based on the outer surface of the particle. The equality of oxygen diffused from the bulk atmosphere surrounding the particle and that consumed in the oxidation results in the following equation [2, 5]

$$\frac{D_{O2} \cdot W_{O2}}{R \cdot T_g}(P_{O2,g} - P_{O2,s}) = \frac{2}{3}d_p \cdot Ac \cdot P_{O2,s}^n \cdot e^{-\frac{Ec}{R \cdot T_p}} \tag{1}$$

where the subscripts g refers to the bulk gas and s to the particle surfaces, n is the reaction ratio, and Ac and Ec represent the frequency factor and the activation energy of the apparent kinetics, respectively. These parameters are, essentially, the only ones to be determined from experiments. This pseudo-empirical formulation has proved its applicability to a wide range of fuels and combustion conditions in the pulverized range, and overcomes the great difficulties found by other approaches which intend to model the internal diffusion of gases through the pores of the particle [4].

3 Deconvolution Procedures

Figure 1 presents the experimental data obtained at the EFR, in terms of unburnt fraction as a function of the length travelled by the particles along the reactor and the combustion conditions (gas and wall temperature, oxygen concentration) [2]. The curves shown correspond to a Spanish anthracite, thoroughly sieved in the range 53–63 μm. The traditional procedure for obtaining the kinetic parameters from these data includes the representation of the burnout rates in Arrhenius plots (i.e. in logarithmic scale, as a function of $1/T$), and the fit of the results to linear regressions, whose slope should be Ec. In order to calculate the particle temperature, quasi-stationary conditions must be assumed [5], which is a reasonable hypothesis in a wide range of situations, including those corresponding to Fig. 1. However, due to the evident 'curvature' of the burnout curves, very different burnout rates for roughly similar particle temperatures are obtained depending on the range of burnout levels considered along the curves. Figure 2 explicitly shows how the experimental rates span over more than one order of magnitude for each combustion condition considered. This fact poses serious problems for the correct adjustment of the data to a linear fit. Little information has been traditionally given in the literature on the details of the fitting, which nevertheless may lead to significantly different kinetic parameters depending on the number of points (or range of burnouts) considered from each curve. For example, Fig. 2 shows two different fits to the data of Fig. 1, considering alternatively only the first two points in each curve, and all of them. The activation energies so derived

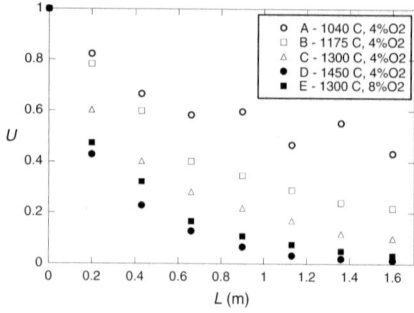

Fig. 1. Unburnt fraction of the coal along the EFR length, for the five combustion conditions used

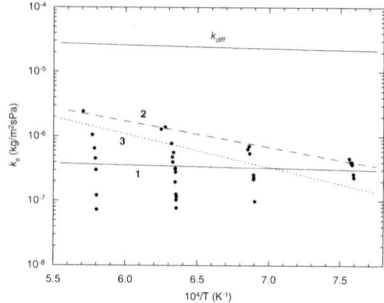

Fig. 2. Arrhenius plot derived from the experimental results, and different lines $ks = Ac\exp(-Ec/RTp)$: Fit 1 – $Ac = 7 \times 10^{-4}\,\mathrm{g\,m^{-2}\,s^{-1}\,Pa^{-1}}$, $Ec = 9.2\,\mathrm{kJ\,mol^{-1}}$ (all data points); Fit 2 – $Ac = 0.43$, $Ec = 76.4$ (*two upper points*); Fit 3 – $Ac = 1.4$, $Ec = 99$ (*best fit with the new procedure*)

vary from 76.4 to $9.2\,\mathrm{kJ\,mol^{-1}}$, which illustrates the relevance of this choice in the result obtained. The best fit with the new procedure is also included in Fig. 2 for comparison ($Ec = 99\,\mathrm{kJ\,mol^{-1}}$).

Moreover, the calculation of the combustion curves corresponding to a monosized sample of fuel particles results in straight lines; in order to explain the reduction of burnout rate observed in the experimental results, several 'passivation factors' that affect the 'reactivity' of the particle at the last stages of combustion have been proposed in the past (see, e.g. [6]).

Most of the uncertainties associated to this procedure are avoided if the particle size distribution (PSD) is recognised as polidisperse, as it always is (even after iterative sieving), and the measured distribution used in the deconvolution procedure [2]. As a drawback, an analytical treatment of the data is impossible, and ab initio calculations which simulate the combustion history of the whole distribution must be accomplished. This new method has been applied to the data of Fig. 1: the fuel PSD was measured by means of laser diffractometry, and the combustion of the particles in the EFR was simulated from injection to collection at the different sampling points for a certain pair of parameters Ec, kc. The difference between simulations and experimental data was computed for each curved and stored. By iterating the simulations for a wide range of Ec, Ac values, an optimum pair is found (i.e. those values for which the 'error' is minimal). In this case, $Ec = 99\,\mathrm{kJ\,mol^{-1}}$, $Ac = 1.4\,\mathrm{g\,m^{-2}\,s^{-1}\,Pa^{-1}}$.

The simulations correctly predict the combustion curves, and specifically most of their 'curvature' is reproduced with this approach. Figure 3 illustrates the effect of considering the multiple size classes included in the fuel PSD, by comparing an experimental curve with the predictions corresponding to hypothetical monosized samples and to the actual PSD used. As expected, carbon consumption is nearly linear with residence time (or length travelled) for each size class, but the weighed sum correctly fits the experimental data for most of the burnout range studied. This is of practical importance, since the prediction of the burnout fractions at lower end of the combustion curve is crucial for the minimization of carbon losses in real systems.

Several minimization algorithms could be applied to the search of the optimal kinetic parameters in the framework of this procedure (e.g. genetic algorithms); however, the massive ('square boxes') method used provides some insight into the dependence of the error with both parameters. A 'valley' of minima is always found, with steep side slopes but considerably flat along its centreline, so that a continuum of pairs Ec, Ac is found to adequately fit the experimental data; nevertheless, as mentioned above, an optimal pair is found in this case.

Compared to the traditional method, which is essentially based on the first derivate of the data and thus considerably sensitive to experimental uncertainties, the new approach has a greater tolerance to noise. Also, it is most suitable for the introduction of variations in the sub-models embedded in the char oxidation model (e.g. evolution of particle diameter with burnout, etc.).

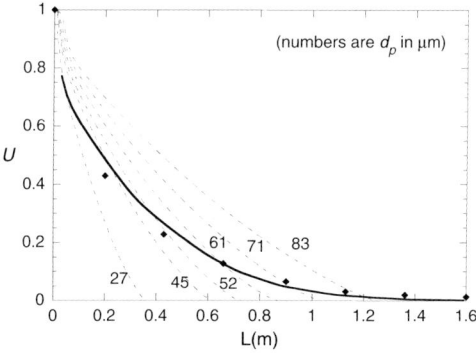

Fig. 3. Predictions of U vs. path along the EFR, for different particle sizes (conditions of Run D, $Ac = 1.4\,\mathrm{g\,m^{-2}\,s^{-1}\,Pa^{-1}}$, $Ec = 99\,\mathrm{kJ\,mol^{-1}}$) (experimental results for Run D are also presented, as well as the best fit)

4 Conclusions and Further Work

It has been shown that the traditional procedure, based on the representation of the data in Arrhenius plots, entails significant errors essentially due to the inherent assumption of monodispersity of the fuel tested. If, on the contrary, the distribution of sizes is measured and considered, the 'analytical' treatment of the data is impossible, and a direct calculation of the combustion history of the whole distribution must be done. A simple (massive) algorithm is used to find the 'optimal' kinetic parameters (i.e. those for which the fitting error is minimal) [2]. This general procedure is found to better fit the experimental data available, and its tolerance to noisy data is much higher, compared to the traditional method.

Several research lines for ongoing and future work are indicated in the following:

- The extension of the searching algorithm to the devolatilization process, also described by Arrhenius kinetics, and to other fuels, such as pulverized biomass. In this case, the suitability of the char oxidation sub-model used must be confirmed, since biomass particles are typically in the range of hundreds of microns, where the condition of thermally thin particles, assumed for pulverized coal, might not be applicable.
- The introduction of variations in the fuel combustion model to account for phenomena such as fragmentation, which could be of great practical relevance in the prediction of carbon losses in real boilers.
- Further analysis of the numerical procedure would be needed in order to establish the minimal number and characteristics of the tests required to allow deriving reliable kinetic parameters from experimental data. This is also of practical importance, since a reduction of the number of tests would imply a decrease in the experimental effort and cost to characterize a particular fuel.

References

[1] Carpenter, A.N., Skorupska, N.M.: Coal combustion - Analysis and Testing, IEACR/64, IEA Coal Research, London, 1993.
[2] Ballester J., Jimnez S.: Kinetic parameters for the oxidation of pulverized coal as measured from drop tube tests. Combustion and Flame, 142:210-222 (2005).
[3] Mitchell, R.E.: Experimentally determined overall burning rates of coal chars. Combustion Science and Technology, 53:165-186 (1987).
[4] Smith, I.W.: The combustion rates of coal chars: a review. Proc. Combustion Institute, 19:1045-1065 (1982).
[5] Field, M.A., Gill, D.W., Morgan, B.B., Hawksley, P.G.W.: Combustion of Pulverised Coal, The British Coal Utilization Research Association, Leatherhead, UK, 1967.
[6] Hurt, R., Sun, J.-K., Lunden, M.: A kinetic model of carbon burnout in pulverized coal combustion. Combustion and Flame, 113:181-197 (1998).

Minisymposium "Mathematical Problems in Oil Industry"

A. Fasano

University of Florence, Italy

In recent years the need of exploiting reservoirs of oil of lesser quality has pushed the research of the chemical, physical and rheological behaviour of oils particularly rich in heavy hydrocarbons. The latter category includes a large class of n-alkanes, collectively termed "wax", up to the so-called asphaltenes. Asphaltenes may develop a tendency to aggregate and to precipitate. Wax can segregate at sufficiently low temperatures and also give rise to the phenomenon of molecular diffusion induced by thermal gradients. The outcome of all such phenomena is the formation of deposits, which can reduce the lumen of pipelines, possibly leading to obstruction. Therefore it is quite obvious that the possibility of predicting the rate of precipitation of asphaltenes or the rate of wax deposition has a great economic impact.

This minisymposium was dealing precisely with this subject. Three talks were presented, concerning the problem of asphaltene precipitation (S. Correra) and problems of wax migration (M. Primicerio) and deposition in pipelines (L. Fusi). They well represent the state of the art in the mathematical modelling of these phenomena. Starting from a sound theoretical background, the models are formulated and the corresponding theories are developed to the point of producing simulations. Validation is obtained by comparison with laboratory or field data. Of course the models are based on approximations, sometimes limiting their range of applicability. The situation in these processes is so complicated that this research area is still in full expansion and a lot of theoretical and experimental work has already been planned. As a matter of fact, the relative importance of several simultaneous processes (e.g. wax migration, segregation, gelification, ablation, ageing of deposits, etc.) going through different stages is presently not completely understood, so that there is a strong expectation that the global view of the physical picture may greatly improve in the next future.

An Asphaltene Precipitation Model Using a Lattice Approach

S. Correra

EniTecnologie S.p.A, Via F. Maritano 26 - 20097 S. Donato M.se (MI)
sebastiano.correra@eni.it

1 Introduction

Asphaltenes constitute the heaviest, most polar fraction of crude oil [1]; they form heavy organic deposits in oil production ducts, inducing flow rate reductions. The principle of economy (Ockham's razor) was employed to develop onset-constrained colloidal asphaltene model (OCCAM) [2]. It is a particularisation of the Flory–Huggins model [3]; a binary system is considered, constituted by the solvent mixture (pseudocomponent 1), grouping together components and solvents (possibly) added, and the asphaltene (pseudocomponent 2). At the onset of asphaltene precipitation, the following relationship is fulfilled:

$$\frac{V_1}{RT} \left(\delta_1 - \delta_2 \right)^2 = \chi_{\mathrm{cr}}, \tag{1}$$

where

V_1 = molar volume of solvent mixture
R = gas constant
T = absolute temperature
δ_1 = solubility parameter of the solvent mixture
δ_2 = asphaltene solubility parameter
χ_{cr} = critical value of the interaction parameter.

The parameters of the model are tuned by fitting experimental data of onset of asphaletene separation and the model can be employed to predict asphaltene instability conditions [4], [5]. It only needs a few, physically well-defined parameters, easy to estimate with relatively cheap measurements. Here, the lattice approach is re-examined in order to take into account self-aggregation.

2 The Lattice Description

For a polymer the lattice description leads to the well-known Flory–Huggins expression for the entropy of mixing:

$$\frac{\Delta S^M}{k} = -n_2 \ln \phi_2 - n_1 \ln \phi_1, \tag{2}$$

in which

n_1 = number of solvent molecules (each one is able to fill a single site of the lattice);

n_2 = number of polymer molecules (each one is able to fill r sites of the lattice);

ϕ_1 = volume fraction occupied by the solvent;

ϕ_2 = volume fraction occupied by the solute

In (2) the dominant term is $n_1 \ln \phi_1$; this is why expression (2) was adopted without modifications.

3 Enthalpy of Mixing

A honeycomb lattice is considered, of order $z = 6$; asphaltene molecules are mixed with the solvent, and each solvent molecule occupies a single site. The asphaltene molecule is constituted by an aromatic plate, which occupies m sites, and c paraffinic chains, each of them occupying q sites. The total number of sites of the lattice is:

$$n_o = n_1 + rn_2, \tag{3}$$

in which:

n_o = total number of lattice sites;

n_1 = total number of solvent molecules;

n_2 = total number of asphaltene molecules;

r = number of sites occupied by an asphaltene molecule.

It is considered that a single asphaltene molecule occupies r sites:

$$r = m + cq, \tag{4}$$

in which:

m = number of sites occupied by the aromatic "core" of the asphaltene molecule;

c = number of paraffinic chains;

q = length (in number of sites) of a single paraffinic chain.

The mixing can be considered as a quasi-chemical reaction between solvent and asphaltene, in which two solvent-asphaltene contacts replace a couple of solvent–solvent and asphaltene-asphaltene contacts. If p of these contacts are formed and neglecting all volume changes on mixing, the enthalpy of mixing is obtained:

$$\Delta H^M = \Delta U^M + \Delta(PV) \approx \Delta U^M = p_{\text{sol-asp}} \Delta u_{\text{sol-asp}}.$$

Here, separate contributions from paraffinic chains and aromatic core are considered.

Paraffinic chains. For a single chain, the contacts are $(z-2)q+1 \approx zq$; therefore, for c chains, there are czq contacts.

Aromatic core. An approximate expression is used for aromatic plate-solvent bonds:

$$2\frac{mz}{2} + \frac{z}{2}\nu z \approx mz.$$

Overall change. The probability that a given cell is occupied by the solvent is just ϕ_1. This leads to:

$$\Delta H^M = n_2\phi_1(cqz\Delta u_{\text{sol-chains}} + mz\Delta u_{\text{sol-core}})$$
$$= n_2\phi_1(cq\chi_{\text{sol-chains}} + m\chi_{\text{sol-core}})kT. \tag{5}$$

Two interaction parameters (χ) have been introduced. Now, an asphaltene aromaticity factor ξ is defined

$$\xi = \frac{m}{m+cq}. \tag{6}$$

Then

$$\Delta H^M = n_1\phi_2[(1-\xi)\chi_{\text{sol-chains}} + \xi\chi_{\text{sol-core}}]kT = n_1\phi_2\chi kT. \tag{7}$$

Expressing the two χ in terms of solubility parameters, the following relationship is obtained:

$$\xi = \frac{(\delta_2-\delta_1)^2 - (\delta_{chains}-\delta_1)^2}{(\delta_{core}-\delta_1)^2 - (\delta_{chains}-\delta_1)^2}. \tag{8}$$

4 The Free Energy of Mixing

The free energy of mixing is obtained from (2) and (7); it allows to calculate phase equilibrium compositions. With typical values of the parameters (Tables 1 and 2) and $\xi = 0.65$, a mass of heptane equal to 2.28 times the mass of oil is able to destabilize a $50-50\%w$ mixture of oil and toluene.

5 Asphaltene Aggregation

Aromatic cores of asphaltene molecules form stacks, with the aliphatic tails that make a sort of paraffinic boundary. If asphaltene molecules form a stack, the following replacements are to be considered:

Table 1.

Property	Mw	ρ	V	δ
	$(\mathrm{g\,mol^{-1}})$	$(\mathrm{g\,cm^{-3}})$	$(\mathrm{cm^3\,mol^{-1}})$	$((\mathrm{cal\,cm^{-3}})^{1/2})$
oil	300	0.85	352.9	8.60
toluene	92	0.86	106.8	8.89
n-pentane	72	0.61	116.2	7.09
asphaltene	2,000	1.25	1,600.0	9.70

Table 2.

Parameter	δ	χ
	$((cal\ cm^{-3})^{1/2})$	$-$
chain	6.5	0.34
core	10	1.14

$$\begin{cases} n_2 \longrightarrow \frac{n_2}{s}, \\[2mm] \xi \longrightarrow \frac{m}{m+scq} = \frac{1}{1+s(cq/m)} = \frac{1}{1+s\lambda}, \qquad \lambda = \frac{cq}{m}, \\[2mm] \phi_1,\ \phi_2,\ n_1 \ \text{not varied.} \end{cases}$$

In this way:

$$\Delta G^M = kT \left\{ n_1 \ln \phi_1 + \frac{n_2}{s} \ln \phi_2 + + n_1 \phi_2 \right.$$
$$\left. \left[\frac{s\lambda}{1+s\lambda} \chi_{sol-chains} + \frac{1}{1+s\lambda} \chi_{sol-core} \right] \right\} \tag{9}$$

6 Results and Conclusions

By employing (9) it is possible to show that aggregation tends to lower the free energy of mixing. With the same values of parameters, but varying the aggregation level, the model was employed to describe the stabilising effect of aggregation. Results are shown in Fig. 1, in which the ratio (weight of paraffin at the onset)/(weight of oil) is reported versus the stack aggregation numbers. Clearly, the more the system aggregates, the more stable it becomes, as a limit to aggregation is not yet present in the model.

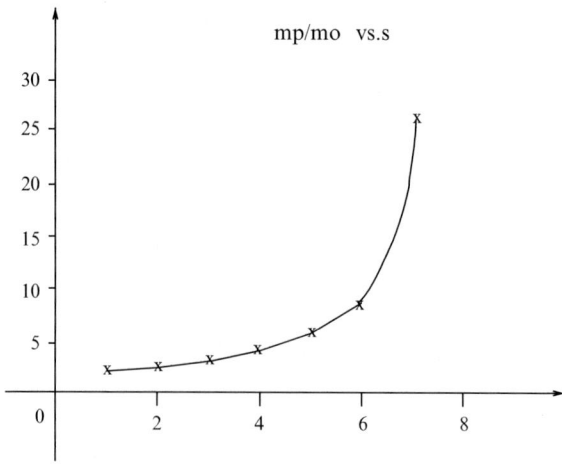

Fig. 1. Ratio (weight of paraffin)/(weight of oil) at the onset vs. the aggregation number

References

1. T.F. Yen and G.V. Chilingarian (editors), *Asphaltenes and asphalts*, Developments in Petroleum Science, Elsevier Science B.V., The Netherlands, 1994.
2. S. Correra, F. Donaggio, *OCCAM: Onset-Constrained Colloidal Asphaltene Model*. In Proceedings International Symposium on Formation Damage, Lafayette, Louisiana, Feb. 23-24, 2000. SPE Paper 58724.
3. D. H. Napper, *Polymeric stabilization of colloidal dispersions*, Academic Press, London, U.K., 1983.
4. S. Correra, *Stepwise Construction of an Asphaltene Precipitation Model*, Petroleum Science and Technology, 22(7-8), 943-959 (2004).
5. S. Correra, D. Merino-Garcia, *Simplifying the Thermodynamic Modelling of Asphaltene Behaviour*, The 7th International Conference on Petroleum Phase Behaviour and Fouling, June 25-29, 2006, Biltmore Estate, Asheville, North Carolina, 2006.

Formation and Growth of Wax Deposit in the Pipelining of Crude Oils

S. Correra[1], D. Merino-Garcia[1], A. Fasano[2], and L. Fusi[2]

[1] Eni S.p.A, Via F. Maritano 26 - 20097 S. Donato Milanese (MI)
   sebastiano.correra@eni.it, daniel.merino.garcia@eni.it
[2] Dipartimento di Matematica "U. Dini", Viale Morgagni 67/a, 50134 Firenze
   fasano@math.unifi.it, fusi@math.unifi.it

Summary. This work presents a model for the turbulent flow of a waxy crude oil in a pipeline, in which deposition is taken into account. Waxy crude oils (WCO's) are mineral oils with high content of heavy molecular weight compounds, usually called waxes. When a sufficiently low temperature is reached (*cloud point*, T_{cloud}) waxes begin to solidify, entrapping the oil in a gel-like structure.

The presence of solid waxes may lead to the formation of a deposit layer on the pipe walls during transportation at low temperatures. This phenomenon has important consequences, such as the increase of pressure requirements and, in the worst scenario, the blockage of the line. Deposition can be due to different mechanisms (see [1]), although there is a general agreement on considering that molecular diffusion is the dominant one. Diffusion refers to the radial mass flow of dissolved waxes towards the pipe wall due to a concentration gradient.

In the model presented herein, molecular diffusion is taken as the only deposition mechanism. The model is also based on the assumption that the deposit thickness is small compared to the pipe radius, as explained elsewhere [2]. Moreover, the effects of ablation, ageing and desaturation are also addressed in the model. Ablation refers to the removal of part of the deposit by the fluid shearing. On the other hand, ageing is a phenomenon that decreases the oil fraction in the deposit. Finally, desaturation takes into account the fact that the fluid is being depleted of waxes.

1 The Thermal Field

Let us analyze the following case: a fluid is circulating in a cylindrical pipe of radius R and length L in turbulent regime. The thermal field is homogeneized over cross sections except in a thin thermal boundary layer where it can be written as

$$T(r,z) = (T_o - T_e) \exp\left\{-\frac{2\pi hR}{\rho c Q} z\right\} \cdot \left\{1 - \frac{hR}{k} \ln\left(\frac{r}{R}\right)\right\} + T_e, \qquad (1)$$

with

$$\frac{\partial T(r,z)}{\partial r} = -\left(T_{\mathrm{o}} - T_{\mathrm{e}}\right)\exp\left\{-\frac{2\pi hR}{\rho cQ}z\right\}\cdot\frac{hR}{kr}. \tag{2}$$

In (1) h is the heat transfer coefficient, k is the heat conductivity, ρ is the density, c is the heat capacity, Q is the volumetric flow rate, T_{o} is the temperature of the oil at the inlet and T_{e} the temperature of the surroundings. Expression (1) is compatible with the assumption that heat transfer is quasi-steady and takes place mainly in the radial direction. The thickness of the boundary layer can be obtained from the momentum boundary layer by means of classical correlations. The momentum boundary layer is determined imposing balance between the drag and propulsive forces in a unit length portion of the pipe. Indeed, denoting with σ_{m} the momentum boundary layer thickness and introducing the ratio $\epsilon_{\mathrm{m}} = \sigma_{\mathrm{m}}/R$, such a balance is expressed by

$$F_{\mathrm{D}} = \frac{2\eta Q}{\epsilon_{\mathrm{m}}R} = F_{\mathrm{P}} = \Delta P\pi R^2, \tag{3}$$

where η is the viscosity of the oil and P is pressure. Relation (3) allows us to calculate ϵ_{m}. Then the correlation (see [3]) $\epsilon_{\mathrm{T}} = \epsilon_{\mathrm{m}} \times 0.41$ is used to evaluate the thermal boundary layer thickness $\sigma_{\mathrm{T}} = \epsilon_{\mathrm{T}}R$.

The temperature of the turbulent core T_{c} is obtained writing the energy balance

$$\rho c\pi R^2 V\left(T_{\mathrm{c}}(z) - T_{\mathrm{c}}(z+\mathrm{d}z)\right) = -\int_z^{z+\mathrm{d}z} 2\pi Rk\frac{\partial T}{\partial r}(R,z)\mathrm{d}z, \tag{4}$$

where $V = Q/R^2\pi$ is the velocity in the turbulent core. In the limit $\mathrm{d}z \to 0$

$$\frac{\mathrm{d}T_{\mathrm{c}}}{\mathrm{d}z} = \frac{2\pi hR(T_{\mathrm{o}} - T_{\mathrm{e}})}{\rho cQ}\exp\left\{-\frac{2\pi hR}{\rho cQ}z\right\}. \tag{5}$$

Integrating with $T_{\mathrm{c}}(0) = T_{\mathrm{o}}$ the following solution is obtained

$$T_{\mathrm{c}}(z,t) = (T_{\mathrm{o}} - T_{\mathrm{e}})\exp\left\{-\frac{2\pi hR}{\rho cQ}z\right\} + T_{\mathrm{e}}. \tag{6}$$

2 The Deposition Equation

The reduced pipe radius is denoted by $\nu = R - \sigma_{\mathrm{d}}$[1] and we assume $\nu \approx R$. When the bulk is saturated, the wax solubility is an increasing function of temperature $C_{\mathrm{s}}(T)$ (assumed linear in T, i.e. $\mathrm{d}C_{\mathrm{s}}/\mathrm{d}T = \beta = \mathrm{const.} > 0$). The deposition rate is

$$j_{\mathrm{dep}} = -\frac{D}{\psi}\frac{\mathrm{d}C_{\mathrm{s}}}{\mathrm{d}T}\frac{\partial T}{\partial r}\bigg|_{r=\nu} =: -\frac{D\beta}{\psi}\frac{\partial T}{\partial r}\bigg|_{r=\nu}, \tag{7}$$

---

[1] σ_{d} is the deposit thickness.

where ψ is the solid fraction of the deposit and D is the dissolved wax diffusivity. At this stage ψ is assumed to be constant; in Sect. 4 an expression for the evolution of ψ with time is proposed. The rate of removal by ablation is proportional to the shear stress at the deposit front

$$\dot{j}_{abl} = -\frac{A\tau}{\psi}, \tag{8}$$

where A is the ablation coefficient. Since $\tau = (\eta Q)/(\pi R^2 \epsilon_m \nu)$,

$$\dot{j}_{abl} = -\frac{A\eta Q}{\pi R^2 \psi \epsilon_m \nu}, \tag{9}$$

The deposition equation is obtained by writing

$$\frac{d}{dt} \int_z^{z+dz} \rho\pi(R^2 - \nu^2(\xi,t))d\xi = \int_\Sigma (\mathbf{j}_{dep} \cdot \mathbf{n} - \mathbf{j}_{abl} \cdot \mathbf{n})dS \tag{10}$$

where Σ is the deposition surface and $\mathbf{n}$ its outward normal. Supposing enough regularity for the function ν (10) becomes

$$\rho\frac{\partial \nu}{\partial t} = \frac{D\beta}{\psi}\frac{\partial T}{\partial r}(\nu,z) + \frac{A\eta Q}{\pi\epsilon_m \psi\nu^3}. \tag{11}$$

Recalling (2) it is possible to integrate the above equation with $\nu(z,0) = R$. This leads to

$$\sigma_d = R - \nu = \frac{D\beta t}{\psi\rho R}\left[\frac{hR(T_o - T_e)}{k}\exp\left\{-\frac{2\pi hR}{\rho cQ}z\right\} - \frac{A\eta Q}{\pi\epsilon_m D\beta R^2}\right]_+$$

$$\cdot H(T_{cloud} - T_w), \tag{12}$$

where $[..]_+$ is the positive part, T_w is temperature at the wall and H is the Heaviside function (which guarantees that deposition is effective only after temperature at the wall has fallen below T_{cloud}). For deposition to occur the quantity in square brackets must be nonnegative, i.e. the molecular flux by diffusion must exceed the ablation rate.

3 The Deposition Segment

It can be proved that deposition occurs only in the interval $[z_f, z_e]$, where

$$z_f = \frac{Q\rho c}{2\pi hR}\ln\left[\frac{T_o - T_e}{T_{cloud} - T_e}\right], \tag{13}$$

$$z_e = \frac{Q\rho c}{2\pi hR}\ln\left(\frac{(T_o - T_e)\pi\epsilon_m hD\beta R^3}{kA\eta Q}\right). \tag{14}$$

In fact z_f (obtained imposing $T_w = T_{cloud}$) is the axial position from which deposition may start and z_e is the axial position from which ablation is dominant over deposition (the quantity in the square brackets in (12) is negative). Obviously the model makes sense only if $z_f < z_e$.

4 Ageing

The phenomenon of ageing consists in the gradual release of oil from the deposit. This causes an increase of ψ according to some kinetics. For the sake of simplicity the following is herein proposed

$$\frac{\partial \psi}{\partial t} = \frac{1}{t_a}(1 - \psi), \tag{15}$$

where t_a is a characteristic consolidation time and ψ depends only on time. Such an equation can be integrated to get

$$\psi = 1 - (1 - \psi_o) \exp\left(-\frac{t}{t_a}\right), \tag{16}$$

where ψ_o is the initial deposit wax fraction.

5 The Total Mass of Deposit

The total mass of deposit (with oil inclusion) at a certain time t will be given by

$$M_{tot} = \int_{z_f}^{z_e} \rho\pi(R^2 - v^2(z,t))dz. \tag{17}$$

Assuming that $z_f < z_e$ and that ψ is given by (15), from (12) we get

$$M_{tot} = 2t_a D\beta \ln\left\{\frac{1}{\psi_o}\left[\exp\left(\frac{t}{t_a}\right) - 1\right] + 1\right\}$$

$$\cdot\left[\frac{A\eta Q}{\pi\varepsilon_m D\beta R^2}(z_f - z_e) + (T_o - T_e)\frac{Q\rho c}{2\pi k}\right.$$

$$\left.\cdot\left(\exp\left\{-\frac{2\pi h R z_f}{\rho c Q}\right\} - \exp\left\{-\frac{2\pi h R z_e}{\rho c Q}\right\}\right)\right]. \tag{18}$$

6 Desaturation

Denoting with $G(z,t)$ the concentration of segregated solid wax in the bulk it is possible to write the balance

$$\left\{\frac{\partial G}{\partial t} + V\frac{\partial G}{\partial z}\right\}\pi v^2 = 2\pi v\left\{D\beta\frac{\partial T}{\partial r} + \frac{A\eta Q}{\pi\varepsilon_m v^3}\right\}. \tag{19}$$

Imposing the steady state solution $\hat{G}(z)$ we get

$$\hat{G}(z) = G_o + \frac{2\pi}{Q}\int_{z_f}^{z}\left\{vD\beta\frac{\partial T}{\partial r} + \frac{A\eta Q}{\pi\varepsilon_m v^2}\right\}dz', \tag{20}$$

where G_o denotes the value of G at $z = z_f$. Integrating (20) we get

$$\hat{G}(z) = G_o + \left[\frac{2A\eta}{\varepsilon_m R^2}(z - z_f) \right.$$
$$\left. + \frac{D\beta(T_o - T_e)}{\alpha} \left(\exp\{-\frac{2\pi\alpha z}{\mu Q}\} - \exp\{-\frac{2\pi\alpha z_f}{\mu Q}\} \right) \right], \quad (21)$$

which represents the concentration of the segregated phase for $z > z_f$. Desaturation may be achieved at a distance z_{des} such that $\hat{G}(z_{des}) = 0$, that is

$$G_o + \left[\frac{2A\eta}{\varepsilon_m R^2}(z_{des} - z_f) \right.$$
$$\left. + \frac{D\beta(T_o - T_e)}{\alpha} \left(\exp\{-\frac{2\pi\alpha z_{des}}{\mu Q}\} - \exp\{-\frac{2\pi\alpha z_f}{\mu Q}\} \right) \right] = 0.$$

Obviously the interesting case is when $z_{des} < z_e$. For $z > z_{des}$ deposition continues with the same rate as long as wax concentration $c(z, t)$ in the oil stays above the value of saturation concentration corresponding to the wall temperature. Thus, when $C_s(T_w(z)) < C_s(T_c(z))$, we write

$$\left\{ \frac{\partial c}{\partial t} + V\frac{\partial c}{\partial z} \right\} \pi\nu^2 = 2\pi\nu \left\{ D\beta\frac{\partial T}{\partial r} + \frac{A\eta Q}{\pi\varepsilon_m \nu^3} \right\}, \quad (22)$$

Referring once more to the steady state $\hat{c}$ we have

$$\hat{c}(z) = C_s(T_c(z_{des})) + \frac{2\pi}{Q}\int_{z_{des}}^{z} \left\{ \nu D\beta\frac{\partial T}{\partial r} + \frac{A\eta Q}{\pi\varepsilon_m \nu^2} \right\} dz', \quad (23)$$

The oil will be completely depleted of solid wax at z_s. This is given by

$$\hat{c}(z_{des}) = C_s(T_w(z_s)), \quad (24)$$

i.e. there is no concentration gradient of dissolved wax and, consequently, no deposition.

References

1. E.D. Burger, T.K. Perkins, J.H. Striegler, *Studies of wax deposition in the trans Alaska pipeline*, Journal of Petroleum Technology, 1075-1086, (June 1981).
2. S. Correra, A. Fasano, L. Fusi, D. Merino-Garcia, *Calculating deposit formation in the pipelining of waxy crude oils*, to appear on Meccanica.
3. Kays W.M., Crawford M.E., *Convective Heat and Mass Transfer*, McGraw Hill (1980).

Simulations of the Spurt Phenomenon for Suspensions of Rod-Like Molecules

Christiane Helzel

Department of Applied Mathematics, Bonn University, Germany
`helzel@iam.uni-bonn.de`

Summary. We simulate the Doi model for suspensions of rigid rod-like molecules. This model couples a microscopic Fokker–Planck type equation (the Smoluchowski equation) to a macroscopic Stokes equation. The Smoluchowski equation describes the evolution of the distribution of the rod orientation. It is a drift-diffusion equation on the sphere in every point of physical space. The drift term in the microscopic equation depends on the local macroscopic velocity gradient. Furthermore, the microscopic orientation of the rods leads to elastic effects which affect the rheological properties of the macroscopic flow.

For sufficiently high macroscopic shear rates the coupled problem shows the spurt phenomena, which describes a sudden increase in the volumetric flow rate. In this regime the drift term in the Smoluchowski equation is dominant and thus a numerical method appropriate for transport dominated PDEs is used.

1 Introduction

There is a fast growing literature on the physics of liquid suspensions, since these kind of materials occur in a large variety of applications. The trend is away from ad-hoc macroscopic models to models with a more detailed description of the microscopic behavior. The reason is that there is no single macroscopic model which may capture the entire wealth of potentially relevant phenomena on the micro-scale. A detailed mathematical model requires a description of the microscopic molecular orientations and the macroscopic rheological response. Such a micro-macro model, the so-called Doi model, is considered here (see Sect. 2).

The coupled micro-macro system shows interesting phenomena, in particular the spurt phenomenon. Spurt describes a sudden increase of the volumetric macroscopic flow rate at a critical stress. This phenomenon had long been observed experimentally. For the Doi model, the occurrence of spurt was recently analyzed by Otto and Tzavaras in [5]. In [3], we presented a numerical method for the approximation of the micro-macro model. Using this numerical method we further investigated the spurt phenomenon in different regimes.

In order to simulate the Smoluchowski equation, we use a finite volume type discretization which is motivated by methods for transport dominated problems. The Smoluchowski equation is a drift-diffusion equation on the sphere. In [3], the sphere was discretized by a longitude-latitude grid. This is appropriate for flow situations where the drift term vanishes at the poles. For general flow situations this is not the case and we therefore now use a discretization of the sphere (introduced in [1]) which avoids the pole singularity.

2 The Mathematical Model for Suspensions of Rod-Like Molecules in the Dilute Regime

Doi and Edwards [2] derived kinetic models for suspensions of rod-like molecules in different regimes (dilute, semidilute, and concentrated). Here we restrict our considerations to the *dilute regime*, where the rods are well separated.

The *microscopic model* is described by a local probability distribution $\psi(t, \mathbf{x}, \mathbf{n})d\mathbf{n}$. It gives the time dependent probability that a rod with center of mass at $\mathbf{x}$ has an axis in the area element $d\mathbf{n}$. The evolution of ψ is given by the Smoluchowski equation

$$\partial_t \psi(t, \mathbf{x}, \mathbf{n}) + \mathbf{u}(t, \mathbf{x}) \cdot \nabla_{\mathbf{x}} \psi(t, \mathbf{x}, \mathbf{n}) + \nabla_{\mathbf{n}} \cdot (P_{\mathbf{n}\perp} \nabla_{\mathbf{x}} \mathbf{u}(t, \mathbf{x}) \mathbf{n} \psi(t, \mathbf{x}, \mathbf{n}))$$
$$= D_r \Delta_{\mathbf{n}} \psi(t, \mathbf{x}, \mathbf{n}). \tag{1}$$

Here the second term describes advection of the centers of mass by the macroscopic velocity $\mathbf{u}$, the third term describes the rotation of the axis due to a macroscopic velocity gradient $\nabla_{\mathbf{x}} \mathbf{u}$ and the term on the right hand side models rotational diffusion. Gradient, divergence, and Laplacian on the sphere are denoted by $\nabla_{\mathbf{n}}$, $\nabla_{\mathbf{n}}\cdot$ and $\Delta_{\mathbf{n}}$, while gradient and divergence in macroscopic physical space are denoted by $\nabla_{\mathbf{x}}$ and $\nabla_{\mathbf{x}}\cdot$. The term $P_{\mathbf{n}\perp} \nabla_{\mathbf{x}} \mathbf{u}\mathbf{n} := \nabla_{\mathbf{x}} \mathbf{u}\mathbf{n} - (\mathbf{n} \cdot \nabla_{\mathbf{x}} \mathbf{u}\mathbf{n}) \mathbf{n}$ denotes the projection of the vector $\nabla_{\mathbf{x}} \mathbf{u}\mathbf{n}$ on the tangent space in $\mathbf{n}$.

A velocity gradient $\nabla_{\mathbf{x}} \mathbf{u}$ distorts an isotropic distribution ψ which leads to an increase in entropy. Thermodynamic consistency requires that this is balanced by a stress tensor $\boldsymbol{\sigma}(t, \mathbf{x})$ given by

$$\boldsymbol{\sigma}(t, \mathbf{x}) = \int_{S^2} (3\mathbf{n} \otimes \mathbf{n} - \mathbf{id}) \, \psi(t, \mathbf{x}, \mathbf{n})d\mathbf{n}. \tag{2}$$

Here, $\boldsymbol{\sigma}$ plays the role of an elastic stress arising as additional term in the Stokes equation that models the macroscopic flow.

The *macroscopic equation* has the form

$$\nabla_{\mathbf{x}} \cdot \left(\left(\nabla_{\mathbf{x}} \mathbf{u}(t, \mathbf{x}) + \nabla_{\mathbf{x}}^t \mathbf{u}(t, \mathbf{x}) - p(t, \mathbf{x}) \mathrm{id} + \boldsymbol{\sigma}(t, \mathbf{x}) \right) \right) = -\mathbf{F}_{\mathrm{ext}}$$
$$\nabla_{\mathbf{x}} \cdot \mathbf{u}(t, \mathbf{x}) = 0, \tag{3}$$

where $\mathbf{u}$ is the macroscopic velocity, p the pressure and $\mathbf{F}_{\mathrm{ext}}$ is an externally imposed volume force.

3 A Numerical Method for the Smoluchowski Equation

Numerical approximations of the coupled micro-macro system were discussed in [3]. There the Smoluchowski equation was discretized on a longitude-latitude grid. Here, we present an alternative discretization of the Smoluchowski equation which avoids time step restrictions due to a pole singularity. We now consider

$$\partial_t \psi(t, \mathbf{n}) + \nabla_{\mathbf{n}} \cdot (P_{\mathbf{n}^{\perp}} \nabla_{\mathbf{x}} \mathbf{u}_{\mathrm{ext}} \mathbf{n} \psi(t, \mathbf{n})) = D_r \varDelta_{\mathbf{n}} \psi(t, \mathbf{n}) \tag{4}$$

with suitable initial values $\psi(t_0, \mathbf{n})$ and a fixed externally imposed macroscopic velocity gradient $\nabla_{\mathbf{x}} \mathbf{u}_{\mathrm{ext}}$. Note that we also require $\int_{S^2} \psi(t, \mathbf{n}) d\mathbf{n} = 1$ for all times t, if it is satisfied initially.

We use a logically rectangular quadrilateral grid where a single rectangular computational domain is mapped to the sphere in such a way that the ratio of the largest to the smallest grid cell is about 1.7, i.e., the pol singularity of a longitude-latitude mesh is avoided. This sphere grid was introduced in [1] and is indicated in Fig. 1.

We discretize (4) by using an operator splitting approach in which we solve the subproblems

$$\partial_t \psi(t, \mathbf{n}) + \nabla_{\mathbf{n}} \cdot (P_{\mathbf{n}^{\perp}} \nabla_{\mathbf{x}} \mathbf{u}_{\mathrm{ext}} \mathbf{n} \psi(t, \mathbf{n})) = 0 \tag{5}$$

and

$$\partial_t \psi(t, \mathbf{n}) = D_r \varDelta_{\mathbf{n}} \psi(t, \mathbf{n}) \tag{6}$$

separately during each time step.

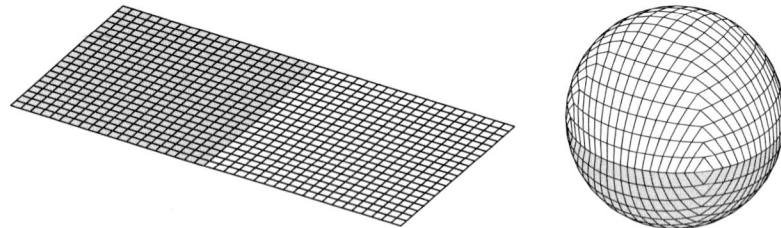

Fig. 1. Computational mesh and the sphere grid from [1]

There are two general ways to discretize partial differential equations on mapped grids. We can either transform the PDE into an equation in computational coordinates and discretize the transformed equation, or we can discretize the PDE directly in physical space and work with reference to a Cartesian frame. We use the latter approach to discretize (5) and the former approach to discretize (6). We use LeVeques wave propagation algorithm for curvilinear grids [4, Sect. 23] which is implemented in the CLAWPACK software. Note that in [3], we discretized the Smoluchowski equation on a longitude-latitude grid. For this we transformed the equation to computational coordinates and discretized the transformed equations. For several important flow situations the drift term in the Smoluchowski equation vanishes at the poles. In such cases (to which we restricted our considerations in [3]) a longitude-latitude grid is appropriate. The method outlined here is more general since it can efficiently be used for any externally imposed velocity gradient.

A finite volume method for (5) has the general form

$$\Psi^{n+1} = \Psi^n - \frac{\Delta t}{|C|} \sum_{j=1}^{N} h_j \breve{F}_j^n, \tag{7}$$

where $\breve{F}_j^n$ represents the average normal flux across the j-th side of the grid cell C, h_j is the length of the j-th interface and N is the number of sides. A finite volume method on a quadrilateral mesh cell can be written in the form

$$\Psi_{ij}^{n+1} = \Psi_{ij}^n - frac\Delta t \kappa_{ij} \Delta x_c \left(F_{i+\frac{1}{2},j} - F_{i-\frac{1}{2},j} \right) - \frac{\Delta t}{\kappa_{ij} \Delta y_c} \left(G_{i,j+\frac{1}{2}} - G_{i,j-\frac{1}{2}} \right) \tag{8}$$

with

$$\kappa_{ij} = |C_{ij}|/\Delta x_c \Delta y_c, \quad F_{i-\frac{1}{2},j} = \gamma_{i-\frac{1}{2},j} \breve{F}_{i-\frac{1}{2},j}, \quad \gamma_{i-\frac{1}{2},j} = h_{i-\frac{1}{2},j}/\Delta y_c, \tag{9}$$

where Δx_c and Δy_c describes the length and the width of a grid cell in computational space. To obtain first order accurate fluxes we calculate

$$s_{i-\frac{1}{2},j} = (\nabla_{\mathbf{x}} \mathbf{u} \mathbf{n} - \mathbf{n} \cdot (\nabla_{\mathbf{x}} \mathbf{u} \mathbf{n}) \mathbf{n})_{i-\frac{1}{2},j} \cdot \boldsymbol{\nu}_{i-\frac{1}{2},j},$$

where $\boldsymbol{\nu}_{i-\frac{1}{2},j}$ is the normal vector at the interface $(i-\frac{1}{2},j)$ in the local tangent plane. This gives us the fluxes

$$\breve{F}_{i-\frac{1}{2},j} = \begin{cases} s_{i-\frac{1}{2},j} \Psi_{i-1,j}^n : s_{i-\frac{1}{2},j} \geq 0 \\ s_{i-\frac{1}{2},j} \Psi_{i-1,j}^n : s_{i-\frac{1}{2},j} < 0 \end{cases}$$

Furthermore, second order correction terms are included in the update as described in [4].

In order to approximate the heat equation (6), we transform the equation to computational coordinates and discretize the transformed equation in the form of a finite volume method (8) using analytical formulas for the metric

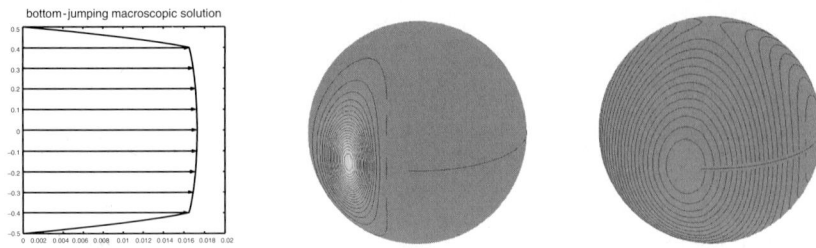

Fig. 2. Solution of the coupled flow problem showing the spurt phenomenon (*left*). The microscopic solution of the Smoluchowski equation for small and large Deborah number is also indicated (we show contour lines for the density ψ of the probability distribution of rod orientation)

terms, see [1]. For the time discretization the RKC-method, an explicit solver for parabolic PDEs, is used.

We now consider a macroscopic shear flow of the form $\mathbf{u} = (u(y), 0, 0)^T$ for $y \in [-\frac{1}{2}, \frac{1}{2}]$ with no slip boundary conditions $u(-1/2) = u(1/2) = 0$. The first plot in Fig. 2 shows a macroscopic velocity profile in the spurt regime. Along the boundary layer the microscopic solution structure corresponds to solutions of the Smoluchowski equation in the large Deborah number regime ($De := |\nabla_x \mathbf{u}_{\text{ext}}|/D_r$) where the molecules strongly align in flow direction (see the second plot in Fig. 2). In this regime the drift term in the Smoluchowski equation is dominant and thus the finite volume approach presented here leads to accurate results. We can also use the method for the Smoluchowski equation in the diffusion dominated regime of small Deborah number, which we observe in the bulk of the flow domain. In this regime the orientation of the rod-like molecules is almost isotropic with a slight preference of orientations in a 45° angle to the flow direction (see the last plot in Fig. 2).

References

1. D.A. Calhoun, C. Helzel, and R.J. LeVeque, Logically rectangular grids and finite volume methods for pdes in circular and spherical domains, 2006, submitted.
2. M. Doi and S.F. Edwards, *The Theory of Polymer Dynamics*, Oxford University Press, 1986.
3. C. Helzel and F. Otto, Multiscale simulations for suspensions of rod-like molecules, *J. Comput. Phys.*, 216:52–75, 2006.
4. R.J. LeVeque, *Finite Volume Methods for Hyperbolic Problems*, Cambridge University Press, 2002.
5. F. Otto and A.E. Tzavaras, Continuity of velocity gradients in suspensions of rod-like molecules, submitted to Comm. Math. Phys.

Minisymposium "Flow in Porous Media"

Nils Svanstedt

Chalmers University, Sweden

In this minisymposium we consider computational and theoretical modeling of a variety of problems with complicated microstructure. As the title of this minisymposium says we present modeling of filtration problems in porous media. In his talk I. S. Pop derives upscaled Buckley–Leverett equations for two-phase flow in a porous medium with application to oil recovery. The upscaled equation is derived by the use of classical homogenization techniques. He also presents numerical results on the effective saturation. In the talk by N. Neuss he considers the Dirichlet problem for the Poisson equation with homogeneous boundary data in a domain Ω_ϵ with rapidly varying boundary $\partial\Omega_\epsilon$. He uses homogenization to derive an approximate solution u to the solution u_ϵ. Here u solved an effective Dirichlet problem for the Poisson equation with homogeneous boundary data in a domain Ω with nonoscillatory boundary $\partial\Omega$. He also presents numerical simulations of this very nice approximation technique. The third talk is by C. Timofte and concerns thermal diffusion with nonlinear lower order terms and nonlinear flux laws. The application in mind is the thermal transmission between two substances embedded in complicated microstructures. Using mathematical homogenization theory she derives effective laws for systems of nonlinear thermal diffusion equations. In particular she points out two cases of practical importance, the Langmuir and the Freundlich kinetics. In the fourth presentation N. Svanstedt considers a convection-diffusion model with possibly highly oscillatory random convection field and diffusion matrix. By combining tools from stochastic homogenization and reiterated homogenization he derives an effective convection enhanced diffusion model. He also presents some numerical simulations of the cell solutions for some two-dimensional convection-diffusion examples.

Multiscale Stochastic Homogenization of Convection-Diffusion Equations

Nils Svanstedt

Department of Computational Mathematics, Chalmers University

1 Introduction

In this short communication we consider the homogenization problem for the following initial-boundary value problem:

$$
\begin{cases}
\dfrac{\partial u_\varepsilon^\omega}{\partial t} + \dfrac{1}{\varepsilon_3} B(T_3(\dfrac{x}{\varepsilon_3})\omega_3) \cdot D(u_\varepsilon^\omega) \\
\quad -\operatorname{div}\left(a\left(T_1(\tfrac{x}{\varepsilon_1})\omega_1, T_2(\tfrac{x}{\varepsilon_2})\omega_2, t\right) Du_\varepsilon^\omega \right) = f \text{ in } Q, \\
\operatorname{div} B = 0 \text{ in } Q, \\
u_\varepsilon^\omega(x,0) = u_0(x) \text{ in } \Omega, \\
u_\varepsilon^\omega(x,t) = 0 \text{ in } \partial\Omega \times (0,T),
\end{cases}
\tag{1}
$$

where Ω is an open bounded set in $\mathbb{R}^n$, T is a positive real number and $Q = \Omega \times (0, T)$. We also assume that ε_1 and ε_1 are two well separated functions (scales) of $\varepsilon > 0$ which converge to zero as ε tends to zero. Well separatedness means

$$
\lim_{\varepsilon \to 0} \frac{\varepsilon_2}{\varepsilon_1} = 0.
$$

We also assume that the scale ε_3 is well separated from one of the scales ε_1 and ε_2 but might coincide with the other. The conditions on the field B and the map a are given in Definition 1. With these conditions it is well-known that for given data $f \in L^2(0,T;H^{-1}(\Omega))$ and $u_0 \in L^2(\Omega)$ there exists a unique solution $u_\varepsilon^\omega \in L^2(0,T;H_0^1(\Omega))$ to (1) with time derivative $\frac{\partial u_\varepsilon}{\partial t} \in L^2(0,T;H^{-1}(\Omega))$ for every fixed $\varepsilon > 0$ and almost all random variables $(\omega_1, \omega_2, \omega_3) \in X_1 \times X_2 \times X_3$.

The multiscale stochastic homogenization problem for (1) consists in studying the asymptotic behavior of the solutions u_ε^ω as ε tends to zero [1,2]. The main result is that the sequence of solutions $\{u_\varepsilon^\omega\}$ to (1) converges in the sense of G-convergence [3] to the solution u to a homogenized problem of the form

$$\begin{cases} \dfrac{\partial u}{\partial t} - \operatorname{div}\left(\mathcal{B}\left(t, Du\right)\right) = f \ \text{ in } \Omega \times (0, T), \\ u(x, 0) = u_0(x) \ \text{ in } \Omega, \\ u(x, t) = 0 \ \text{ in } \partial\Omega \times (0, T), \end{cases} \tag{2}$$

where the *convection enhanced effective diffusion matrix* $\mathcal{B}$ depends on t but is no longer oscillating in space with ε. For a detailed version of the result in this communication we refer to [5].

2 Homogenization of the Convection-Diffusion Equation

Let $\{(X_k, \mathcal{F}_k, \mu_k)\}_{k=1}^M$ denote a family of probability spaces, where each $\mathcal{F}_k$ is a complete σ-algebra and each μ_k is the associated probability measure. We assume that for each $x \in \mathbb{R}^n$, X_k is acted on by the dynamical system

$$T_k(x) : X_k \rightarrow X_k$$

We are interested in the asymptotic behavior (as $\varepsilon_i \rightarrow 0$, $i = 1, 2, 3$) of the sequence (1) of initial-boundary value problems. Since $\operatorname{div} B = 0$ there exists a skew-symmetric matrix S such that $\operatorname{div} S = B$. In space dimension two we get:

$$S\left(T_3\left(\frac{x}{\varepsilon_3}\right)\omega_3\right) = \begin{pmatrix} 0 & s\left(T_3(\frac{x}{\varepsilon_3})\omega_3\right) \\ -s\left(T_3(\frac{x}{\varepsilon_3})\omega_3\right) & 0 \end{pmatrix}. \tag{3}$$

where s is the stream function corresponding to the field B. We define the map

$$\mathcal{A}_\varepsilon^\omega(x, t) = \mathcal{A}\left(T_1\left(\frac{x}{\varepsilon_1}\right)\omega_1, T_2\left(\frac{x}{\varepsilon_2}\right)\omega_2, T_3\left(\frac{x}{\varepsilon_3}\right)\omega_3, t\right)$$
$$= \begin{pmatrix} a(T_1(\frac{x}{\varepsilon_1})\omega_1, T_2(\frac{x}{\varepsilon_2})\omega_2, t) & -s\left(T_3(\frac{x}{\varepsilon_3})\omega_3\right) \\ s\left(T_3(\frac{x}{\varepsilon_3})\omega_3\right) & a(T_1(\frac{x}{\varepsilon_1})\omega_1, T_2(\frac{x}{\varepsilon_2})\omega_2, t) \end{pmatrix}.$$

Before we state the main theorem we also define the appropriate class of coefficients:

Definition 1 *We say that* $C(\omega_1, \omega_2, \omega_3, t) = (C(\omega, t)_{ij}) \in S^2$ *if for* $0 < \alpha < \beta < \infty$ *we have*

$$\alpha|\xi|^2 \le C(\omega, t)_{ij}\xi_j\xi_i \le \beta|\xi|^2$$

for all $\xi \in \mathbb{R}^n$ *a.e. in* $X_1 \times X_2 \times X_3 \times (0, T)$.

We can now write (1) as a diffusion problem and state the following theorem:

Theorem 1 *Consider the sequence of diffusion equations:*

$$\begin{cases} \dfrac{\partial u_\varepsilon^\omega}{\partial t} - \mathrm{div}(\mathcal{A}_\varepsilon^\omega(x,t)Du_\varepsilon^\omega) = f_\varepsilon \ \ in \ \ Q, \\ u_\varepsilon^\omega(0) = u_0^\omega, \ \ in \ \ \Omega, \\ u_\varepsilon^\omega \in L^2(0,T;H_0^1(\Omega)), \end{cases}$$

Assume that $\mathcal{A}_\varepsilon^\omega \in S^2$ and that

$$|\mathcal{A}_\varepsilon^\omega(x,t) - \mathcal{A}_\varepsilon^\omega(x,s)| \le \eta(t-s)$$

where η is the modulus of continuity function. Also assume that the underlying dynamical systems $T_1(x)$, $T_2(x)$ and $T_3(x)$ are ergodic and that the scale ε_3 is the fastest (strong convection). Then

$$u_\varepsilon^\omega(\cdot,t) \rightharpoonup u \ \ in \ \ L^2(0,T;H_0^1(\Omega))$$

and

$$\mathcal{A}_\varepsilon^\omega(\cdot,t)Du_\varepsilon^\omega \rightharpoonup \mathcal{B}(t)Du \ \ in \ \ L^2(0,T;[L^2(\Omega)]^n)$$

where u is the solution to the homogenized problem

$$\begin{cases} \dfrac{\partial u}{\partial t} - \mathrm{div}(\mathcal{B}(t)Du) = f \ \ in \ \ Q, \\ u \in L^2(0,T;H_0^1(\Omega)). \end{cases} \tag{4}$$

For a fixed $\xi \in \mathbb{R}^n$ the operator $\mathcal{B}(t)$ is defined as

$$\mathcal{B}(t)\xi = \int_{X_1} \mathcal{B}_1(\omega_1,t)(\xi + z_1^\xi(\omega_1,t))\,d\mu_1(\omega_1)$$

where $z_1^\xi(\omega_1,t) \in V_{pot}(X_1)$ is the solution to the ϵ_1-scale local problem

$$\langle \mathcal{B}_1(\omega_1,t)(\xi + z_1^\xi(\omega_1,t)), \Phi_1(\omega_1) \rangle = 0$$

for all $\Phi_1(\omega_1) \in V_{pot}(X_1)$, $t \in [0,T]$. The operator $\mathcal{B}_1(\omega_1,t)$ is defined as

$$\mathcal{B}_1(\omega_1,t)\xi = \int_{X_2} \mathcal{B}_2(\omega_1,\omega_2,t,\xi + z_2^{\omega_1,\xi}(\omega_2,t))\,d\mu_2(\omega_2)$$

where $z_2^{\omega_1,\xi}(\omega_2,t) \in V_{pot}(X_2)$ is the solution to the ϵ_2-scale local problem

$$\langle \mathcal{B}_2(\omega_1,\omega_2,t)(\xi + z_2^{\omega_1,\xi}(\omega_2,t)), \Phi_2(\omega_2) \rangle = 0$$

for all $\Phi_2(\omega_2) \in V_{pot}(X_2)$ a.e. $\omega_1 \in X_1$, $t \in [0,T]$. The operator $\mathcal{B}_2(\omega_1,\omega_2,t)$ is defined as

$$\mathcal{B}_2(\omega_1,\omega_2,t)\xi = \int_{X_3} \mathcal{A}(\omega_1,\omega_2,\omega_3,t,\xi + z_3^{\omega_1,\omega_2,\xi}(\omega_3,t))\,d\mu_3(\omega_3)$$

where $z_3^{\omega_1,\omega_2,\xi}(\omega_3,t) \in V_{pot}(X_3)$ is the solution to the ϵ_3-scale local problem

$$\langle \mathcal{A}(\omega_1,\omega_2,\omega_3 t)(\xi + z_3^{\omega_1,\omega_2,\xi}(\omega_3,t)), \Phi_3(\omega_3) \rangle = 0$$

for all $\Phi_3(\omega_3) \in V_{pot}(X_3)$ a.e. $\omega_1 \in X_1$, $\omega_2 \in X_2$, $t \in [0,T]$.

Proof The proof is completely analogous to the proof of Theorem 7 in [4]. Just choose $p = 2$ and perform one more reiterated homogenization.

3 Some Cell Solutions of the Convection Field

As a simple illustration we solve two cell problems for periodic diffusion with and without oscillating convection fields, see Figures 1 and 2. The numerical computations are as well as the theoretical modeling in 2D. We have chosen the diffusion map

$$a(x,y) = 2 + \sin(2\pi x)\sin(2\pi y).$$

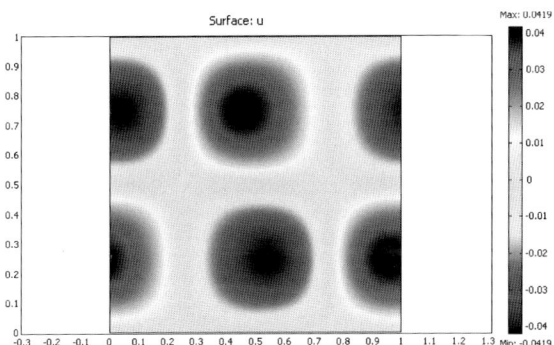

Fig. 1. Periodic cell solution (no convection)

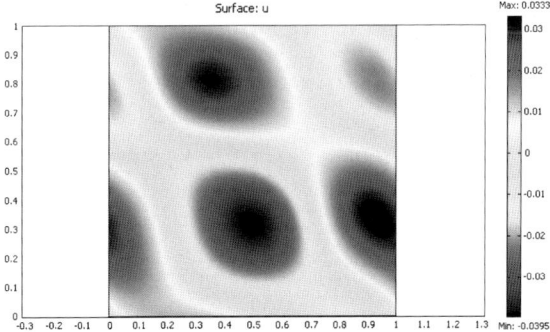

Fig. 2. Periodic cell solution (large convection)

The forcing f and the convection field B are chosen as analogous periodic functions and fields, respectively. According to the definition of the classical cell problem we choose

$$f(x, y) = 2\pi \cos(2\pi x) \sin(2\pi y).$$

References

1. Y. Efendiev and A. Pankov, *Numerical homogenization of nonlinear random parabolic operators*, SIAM Multiscale Model. Simul., Vol. 2 (2004), 237-268.
2. A. Fannjiang and G. Papanicolaou, *Convection enhanced diffusion for periodic flows*, SIAM J. Appl. Math., Vol. 54 (1994), 333-408.
3. N. Svanstedt, *G-convergence of parabolic operators*, Nonlinear Analysis, 36, No. 7, (1999), 807-843.
4. N. Svanstedt, *Multiscale stochastic homogenization of monotone operators*, Networks and Heterogeneous Media, Vol. 2, No. 1, (2007), 181-192.
5. N. Svanstedt, *Multiscale stochastic homogenization of convection-diffusion problems*, submitted.

Numerical Approximation of Boundary Layers for Rough Boundaries

Nicolas Neuss

Universität Kiel, Christian-Albrechts-Platz 4, 24098 Kiel
Since 1.8.2006: Universität Karlsruhe (TH), Engler Str. 2, 76189 Karlsruhe
neuss@mathematik.uni-karlsruhe.de

1 Introduction

In physical problems, interesting phenomena often occur at boundaries or interfaces between different media. Often these phenomena are complicated due to the nature of the process or due to the intricate geometry of the interface. Therefore, they are usually described by effective boundary or interface laws.

In this contribution, we will discuss a model cases in a quasi-periodic setting, where the parameter function in an effective boundary condition can be calculated from the microscopic setting. Theoretically, this case was treated in [5]. Practical computations can be found in treatment of this case was done. First, we construct a suitable approximation and give a priori estimates for the error. Second, we consider the efficient numerical calculation of the effective law, and its use for approximating the solution to the original problem.

2 The Model Problem

We consider the model problem described in [5]. Let $\Omega \subset I\!\!R^n$ be a domain with smooth boundary $\partial\Omega \subset I\!\!R^n$, which has a tubular neighborhood T_δ of width $\delta > 0$. Let Ω^ε be another domain such that its boundary $\partial\Omega^\varepsilon$ lies inside the tubular neighborhood and is described in local charts $\{\varphi_i : U_i \to \partial\Omega\}$ with open sets $U_i \subset I\!\!R^{n-1}$ as a graph of the form

$$\gamma^\varepsilon(x) = \varepsilon\gamma_i(\varphi_i^{-1}(x), \frac{\varphi_i^{-1}(x)}{\varepsilon}) \tag{1}$$

with smooth functions $\gamma_i : U_i \times I\!\!R^{n-1} \to I\!\!R$ which are 1-periodic in the second variable where ε is assumed to be small.

On Ω^ε, we are given the problem

$$-\Delta u^\varepsilon(x) = f(x), \ x \in \Omega^\varepsilon, \quad u^\varepsilon(x) = 0, \ x \in \partial\Omega^\varepsilon. \tag{2}$$

Now, this problem is difficult to handle numerically due to the intricate structure of $\partial\Omega^\varepsilon$, such that it is very desirable to find an approximation which is easier to compute. For achieving this, we assume that the right-hand side f is extended to $\Omega \cup \Omega^\varepsilon$ (either by 0, or such that $\|f\|_{L^\infty(\Omega^\varepsilon\setminus\Omega)}$ is bounded). Then the solution u to the Poisson problem

$$-\Delta u(x) = f(x)\,,\ x \in \Omega\,, \qquad u(x) = 0\,,\ x \in \partial\Omega \tag{3}$$

approximates u^ε up to order $O(\varepsilon)$ in the $L^2(\Omega)$-norm. This is a sufficiently good approximation only for small ε.

However, it is possible to obtain a better approximation by computing a corrector $\eta \in H^1(\Omega)$ satisfying

$$-\Delta\eta = 0\,,\qquad x \in \Omega\,,$$
$$\eta(x) = c^{bl}(x)\frac{\partial}{\partial\nu}u(x)\,,\ x \in \Gamma\,, \tag{4}$$

Then $u + \eta$ satisfies the interior estimate

$$\|u^\varepsilon - (u + \varepsilon\eta)\|_{L^2(\Omega')} \le C(f, \Omega')\varepsilon^2 \tag{5}$$

for every domain Ω' which is compactly embedded in Ω.

3 The Two-Dimensional Case

For $\Omega \subset \mathbb{R}^2$, the situation simplifies considerably, see, e.g. [3], [1]. Especially, it is a reasonable simplification that we have to use only one chart $\phi:(0,L)\to\partial\Omega$ which is a parametrization of $\partial\Omega$ by arclength, and $\gamma^\varepsilon : (0,L) \to \mathbb{R}$ is a periodic map with $\gamma^\varepsilon = \gamma(s,\frac{s}{\varepsilon})$ where $\gamma : (0,L) \times (0,1)$ is L-periodic in the first variable and 1-periodic in the second variable. Here, for evaluating the function c^{bl} at a position $x \in \partial\Omega$, it is necessary to solve for each $x \in \partial\Omega$ the cell problem:
    Find $\beta_x \in H^1(\bar{Z})$ with

$$Z = \{y \in (0,1) \times \mathbb{R} : y_2 < \gamma(x,y_1)\} \tag{6}$$

such that

$$-\Delta_y\beta_x(y_1,y_2) = 0 \qquad 0 \ne y_2 \le \gamma(x,y_1) \tag{7}$$
$$\beta_x(y) = 0 \qquad y_2 = \gamma(x,y_1) \tag{8}$$
$$\beta_x(y) \to 0 \qquad y_2 \to -\infty \tag{9}$$
$$\beta_x(0,y_2) = \beta_x(1,y_2) \qquad y_2 \in \mathbb{R} \tag{10}$$
$$\left[\frac{\partial\beta_x}{\partial n}\right](y) = 1 \qquad y_2 = 0\,. \tag{11}$$

In [5] it is shown that the solution β_x exists, that it decays exponentially fast, and that its norm and the norm of its derivatives are bounded in terms of the smoothness of γ. Consequently, also the function

$$c^{bl} : \partial\Omega \to \mathbb{R}, \quad x \mapsto \int_0^1 \beta_x(y_1, 0)\, dy_1 \tag{12}$$

satisfies

$$\|D^k c^{bl}\|_\infty \le C \tag{13}$$

where C depends only on the smoothness of γ. The evaluation of c^{bl} at a point $x \in \partial\Omega$ involves solving the elliptic cell problem (7)–(11) together with computing the average (12).

4 Numerical Approximation of the Cell Problems

Since β_x is smooth for $y \in Z \setminus (\{y : y_2 = 0\} \cup \{y : y_2 = \gamma_x(y_1)\}$, an efficient numerical approximation requires a method of high discretization order. Additionally, the method must be capable of dealing with curved boundaries and the unbounded cell domain Z.

Our method of choice is a finite element method of order p on a conforming mesh which is adaptively refined. The coarsest mesh covers only a finite part $Z_{-1} = Z \cap \{y : y_2 > -1\}$ and is conforming, i.e. it uses nonlinear cell mappings to fit the curved boundary given by $y_2 = \gamma(x, y_1)$ precisely. Solving the cell problem on this truncated domain yields an approximation β_h, and the application of a residual type error estimator where the dual problem is solved with a method of order $p + 1$ computes an error distribution which is used to guide mesh refinement. If the bottom cell with boundary $y_2 = l$ is refined, the computational domain is automatically extended to become Z_{l-1}. This is an admissible approach, because the fast decay of β and β_h guarantees that the error $\|\nabla(\tilde{\beta}_h - \beta)\|_{L^2(Z \setminus Z_l)}$ of a suitable extension $\tilde{\beta}_h$ of β_h to $Z \setminus Z_l$ can be shown to be much smaller than the error $\|\nabla(\tilde{\beta}_h - \beta)\|_{(0,1) \times (l, l+1)}$.

Remark 1. In the model situation considered here, the use of distorted meshes could be avoided by transforming the domain Z_l itself into a rectangular domain. However, this would not be possible for more complicated shapes of the boundary, e.g. the case where $\gamma(x, \cdot)$ is not a function graph (cf. [1]).

5 Numerical Approximation of c^{bl}

Inside a discretization of (4), it is necessary to evaluate the function c^{bl} which can be computed solving the cell problems as described in the previous section. However, we still have to establish how accurate this approximation should be.

It is reasonable to connect the approximation accuracy of c^{bl} to the accuracy with which u and η are approximated. Thus, if the mesh T_H for the discretization of (3) is such that an a-posteriori error estimate shows

$$\|u_H - u\|_\infty = E(H) \tag{14}$$

we compute $\tilde{c}^{bl}$ such that

$$\varepsilon \|c^{bl} - \tilde{c}^{bl}\|_\infty \left\| \frac{\partial u_H}{\partial n} \right\|_{L^\infty(\partial\Omega)} \le E(H) \tag{15}$$

which ensures (using a discrete maximum principle) that we have on the same mesh T_H the estimate

$$\|\varepsilon\eta - \varepsilon\eta_H)\| \le CE(H) \tag{16}$$

where C is a moderate constant depending on the mesh quality and the finite element degree.

Remark 2. Since the calculation of c^{bl} involves solving the system (7)–(11), it can be worthwhile to separate its calculation from its use in discretizing the corrector problem (4). For example, if c^{bl} is very smooth, an interpolation of c^{bl} from relatively few sample points is already good enough for obtaining an approximation $\tilde{c}^{bl}$ satisfying (15).

6 Numerical Test

We consider the specific example of Ω being the unit circle $B_1(0)$, and Ω^ε being described by

$$\varphi \mapsto \gamma\left(\varphi, \frac{\varphi}{\varepsilon}\right) = 6\varepsilon \sin^2(\varphi) \sin\left(\frac{\varphi}{\varepsilon}\right) \tag{17}$$

for $\varepsilon = \frac{1}{40}$ in the tubular neighborhood

$$(\varphi, r) \mapsto r \begin{pmatrix} \cos\varphi \\ \sin\varphi \end{pmatrix}. \tag{18}$$

Now we consider problem (2) with right-hand side $f \equiv 1$. For a prescribed meshsize h, we obtain a mesh T_h^ε for Ω^ε using the program Triangle, see [6]. For discretizing and solving on T_h^ε, the finite element library FEMLISP was used, see [2, 4]. We obtain an approximate solution u_h^ε using a discretization of quadratic finite elements of Lagrange type, see Figure 1. For studying point-wise convergence, we look at the value at the origin. On a mesh with about 10000 cells, we obtain a value $u_h^\varepsilon(0) = 0.22044\ldots$.

Of course, using $f \equiv 1$ in (3), we have $u = \frac{1}{4}(1 - \|x\|^2)$ from which we see that $u(0) = 0.25$. This smooth function can be easily approximated to very high accuracy on a coarse mesh T_H. On the other hand, 0.25 is only a

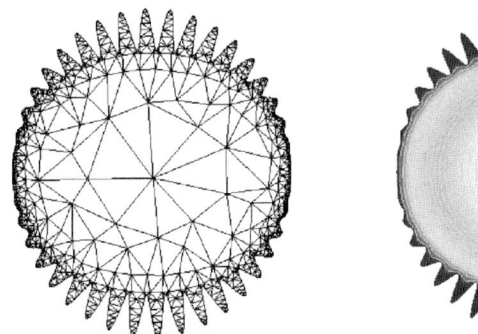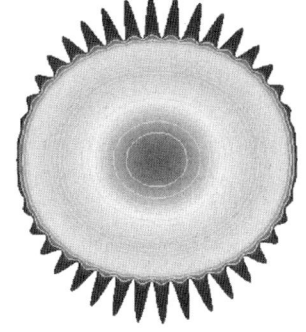

Fig. 1. Ω^ε with the initial mesh T_h^ε and u_h^ε

rather bad approximation to $u_h^\varepsilon(0)$. The corrector problem (4) now reads as follows: find some function η which solves a Laplace problem with a boundary condition $\eta(x) = \frac{\partial u}{\partial n}(x) c^{bl}(x) = \frac{1}{2} c^{bl}(x)$ for $x \in \partial \Omega$ where c^{bl} is computed as described in Sect. 5. Solving this Dirichlet problem we obtain $\eta_H(0) = -0.028$ such that in total we obtain $u_H(0) + \varepsilon \eta_H(0) = 0.222$. Thus, it becomes obvious that the inclusion of the first order corrector η_H significantly improves the approximation while the numerical effort is much smaller than the effort for computing u_h^ε. It is clear that the effect would become even more pronounced for smaller ε.

References

1. T. Abboud and H. Ammari. Diffraction at a curved grating: TM and TE cases, homogenization. J. Math. Anal. Appl. 202:995–1026, 1996.
2. Femlisp. http://www.femlisp.org.
3. A. Madureira and F. Valentin. Analysis of curvature influence on effective boundary conditions. C. R. Acad. Sci. Paris, Ser I, 335:499–504, 2002.
4. N. Neuss. Femlisp — a multi-purpose tool for solving partial differential equations. Technical report, IWR, Universität Heidelberg, 2003.
5. N. Neuss, M. Neuss-Radu, and A. Mikelić. Effective laws for the poisson equation on domains with curved oscillating boundaries. Appl. Anal., pages 479–502, 2006.
6. Jonathan Richard Shewchuk. Triangle: Engineering a 2D Quality Mesh Generator and Delaunay Triangulator. In Ming C. Lin and Dinesh Manocha, editors, Applied Computational Geometry: Towards Geometric Engineering, volume 1148 of Lecture Notes in Computer Science, pages 203–222. Springer-Verlag, May 1996. From the First ACM Workshop on Applied Computational Geometry.

Upscaling in Nonlinear Thermal Diffusion Problems in Composite Materials

Claudia Timofte

Faculty of Physics, University of Bucharest, P.O. Box MG-11, Bucharest, Romania
claudiatimofte@yahoo.com

1 Introduction

The general question which will make the object of this paper is the homogenization of some nonlinear problems arising in the modelling of thermal diffusion in a two-component composite. We shall consider, at the microscale, a periodic structure formed by two materials with different thermal properties. We shall deal with two situations: in the first one, we assume that we have some nonlinear sources acting in both components and that at the interface between our two materials the temperature and the flux are continuous, while in the second problem we shall address here, we assume that the flux is still continuous, but depends in a nonlinear way on the jump of the temperature field. In both cases, since the characteristic sizes of these two components are small compared with the macroscopic length-scale of the flow domain, we can apply an homogenization procedure.

As usual in homogenization, we shall be interested in obtaining a suitable description of the asymptotic behavior, as the small parameter which characterizes the sizes of our two regions tends to zero, of the temperature field in the periodic composite.

Using the so-called energy method introduced by L. Tartar (see [Tar77]), in the first case we can prove that the limit problem will be a new nonlinear elliptic boundary-value problem, with extra zero-order terms capturing the effect of the nonlinear sources acting in our two parts of the domain. The asymptotic behavior of the solution of the second problem will be governed by a new nonlinear system, similar to the famous Barenblatt's model (see [BZK60]), with extra zero-order terms capturing the effect of the interfacial barrier and of the nonlinear sources.

Our results constitute a generalization of those obtained in [BZK60], [3] and [EP02], by considering nonlinear sources and nonlinear transmission conditions. For detailed proofs of the results of this paper, we refer to [Tim06].

The structure of our paper is as follows: in Sect. 2 we analyze the case of classical transmission boundary conditions. The case of partially fissured media with interfacial thermal barrier is addressed in Sect. 3.

2 Classical Transmission Boundary Conditions

Let Ω be a bounded domain in $\mathbb{R}^n$ ($n \geq 3$), having the boundary of class C^2. In the first problem we address in this paper, we consider that Ω is an ε-periodic structure, consisting of two parts: a fluid phase Ω^ε and a solid skeleton (grains), $\Omega \setminus \overline{\Omega^\varepsilon}$, ε representing a small parameter related to the characteristic size of the grains. We consider periodic structures obtained by removing from Ω, with period εY, where $Y = (-\frac{1}{2}, \frac{1}{2})^n$, an elementary hole T, with boundary Γ of class C^2, which has been appropriated rescaled and for which $\overline{T} \subset Y$. More precisely, for each ε and for any $k \in \mathbb{Z}^n$, set T_k^ε the translated image of εT by εk and denote by T^ε the set of all the holes contained in Ω, $T^\varepsilon = \bigcup \{T_k^\varepsilon \mid \overline{T_k^\varepsilon} \subset \Omega, \ k \in \mathbb{Z}^n \}$. Let $S^\varepsilon = \bigcup \{\partial T_k^\varepsilon \mid \overline{T_k^\varepsilon} \subset \Omega, \ k \in \mathbb{Z}^n \}$. So, $\partial \Omega^\varepsilon = \partial \Omega \cup S^\varepsilon$. Also, let $\Omega^\varepsilon = \Omega \setminus \overline{T^\varepsilon}$, $Y^* = Y \setminus T$ and $\theta = |Y^*|$.

Let us consider a family of inhomogeneous media occupying the region Ω, parameterized by ε and represented by $n \times n$ matrices $A^\varepsilon(x)$ of real-valued coefficients defined on Ω. The positive parameter ε will also define a length scale measuring how densely the inhomogeneities are distributed in Ω. We shall deal with periodic structures, defined by $A^\varepsilon(x) = A(\frac{x}{\varepsilon})$. Here $A = A(y)$ is a smooth matrix-valued function on $\mathbb{R}^n$ which is Y-periodic. We use the symbol # to denote periodicity properties. We shall assume that

$$\begin{cases} A \in L_\#^\infty(\Omega)^{n \times n}, \\ A \text{ is a symmetric matrix}, \\ \text{For some } 0 < \alpha < \lambda, \ \alpha |\xi|^2 \leq A(y)\xi \cdot \xi \leq \lambda |\xi|^2 \quad \forall \xi, \ y \in \mathbb{R}^n \end{cases}$$

and we shall denote the matrix A by A_1 in Y^* and by A_2, respectively, in T.

As already mentioned in Introduction, in such a domain we shall study a thermal diffusion problem modelling the transmission of temperature between our two components, with an unknown flux on the boundary of each grain. A simplified version of this kind of models can be formulated as follows:

$$\begin{cases} -\text{div } (A_1^\varepsilon \nabla u^\varepsilon) + \beta(u^\varepsilon) = f & \text{in } \Omega^\varepsilon, \\ -\text{div } (A_2^\varepsilon \nabla v^\varepsilon) + ag(v^\varepsilon) = 0, & \text{in } \Pi^\varepsilon \\ A_1^\varepsilon \nabla u^\varepsilon \cdot \nu = A_2^\varepsilon \nabla v^\varepsilon \cdot \nu & \text{on } S^\varepsilon, \\ u^\varepsilon = v^\varepsilon & \text{on } S^\varepsilon, \\ u^\varepsilon = 0 & \text{on } \partial \Omega. \end{cases} \tag{1}$$

Here, $\Pi^\varepsilon = \Omega \setminus \overline{\Omega^\varepsilon}$, ν is the exterior unit normal to Ω^ε, $a > 0$, $f \in L^2(\Omega)$ and β and g are continuous functions, monotonously non-decreasing and such that $\beta(0) = 0$ and $g(0) = 0$. We shall suppose that there exist a positive constant C and an exponent q, with $0 \leq q < n/(n-2)$, such that

$$|\beta(v)| \le C(1 + |v|^q) \tag{2}$$

$$|g(v)| \le C(1 + |v|^q). \tag{3}$$

These two general situations are well illustrated, for instance, by the following important practical examples:

(a) $g(v) = \dfrac{\delta v}{1 + \gamma v}$, $\delta, \gamma > 0$ (Langmuir kinetics)

and

(b) $\beta(v) = |v|^{p-1}v$, $0 < p < 1$ (Freundlich kinetics).

Let us consider the functional spaces

$$V^\varepsilon = \left\{ v \in H^1(\Omega^\varepsilon) \mid v = 0 \text{ on } \partial\Omega \right\},$$

$$H^\varepsilon = \left\{ w^\varepsilon = (u^\varepsilon, v^\varepsilon) \mid u^\varepsilon \in V^\varepsilon, v^\varepsilon \in H^1(\Pi^\varepsilon),\ u^\varepsilon = v^\varepsilon \text{ on } S^\varepsilon \right\},$$

with $\|w^\varepsilon\|_{H^\varepsilon}^2 = \|\nabla u^\varepsilon\|_{L^2(\Omega^\varepsilon)}^2 + \|\nabla v^\varepsilon\|_{L^2(\Pi^\varepsilon)}^2$.

The variational formulation of problem (1) is the following one:

$$
\begin{cases}
\text{Find } w^\varepsilon \in H^\varepsilon \text{ such that} \\[4pt]
\int_{\Omega^\varepsilon} A_1^\varepsilon \nabla u^\varepsilon \cdot \nabla \varphi dx + \int_{\Pi^\varepsilon} A_2^\varepsilon \nabla v^\varepsilon \cdot \nabla \psi dx + \int_{\Omega^\varepsilon} \beta(u^\varepsilon)\varphi dx \\[4pt]
+a \int_{\Pi^\varepsilon} g(v^\varepsilon)\psi dx = \int_{\Omega^\varepsilon} f\varphi dx \quad \forall(\varphi, \psi) \in H^\varepsilon.
\end{cases}
\tag{4}
$$

By classical existence and uniqueness results (see [Bre72]), we know that (4) is a well-posed problem. Then, the main result of this section is given by:

Theorem 1. *One can construct an extension* $P^\varepsilon u^\varepsilon$ *of the solution* u^ε *of the variational problem (4) such that* $P^\varepsilon u^\varepsilon \rightharpoonup u$, *weakly in* $H_0^1(\Omega)$, *where* u *is the unique solution of*

$$
\begin{cases}
-\mathrm{div}(A^0 \nabla u) + \beta(u) + a\dfrac{|T|}{|Y^*|}g(u) = f & \text{in } \Omega, \\[6pt]
u = 0 & \text{on } \partial\Omega.
\end{cases}
$$

Here, $A^0 = ((a_{ij}^0))$ *is the homogenized matrix, whose entries are defined by:*

$$a_{ij}^0 = \int_Y \left(a_{ij} + a_{ik}\frac{\partial \chi_j}{\partial y_k} \right) dy,$$

in terms of the functions χ_j, $j = 1, ..., n$, *weak solutions of the cell problems*

$$
\begin{cases}
-\mathrm{div}(A\nabla(y_j + \chi_j)) = 0 & \text{in } Y, \\[4pt]
\chi_j - Y \text{ periodic.}
\end{cases}
$$

Let us notice, too, that if we denote by $\widetilde{v^\varepsilon}$ the extension by zero of v^ε to the whole of Ω, then $\widetilde{v^\varepsilon} \rightharpoonup |T|u$, weakly in $L^2(\Omega)$.

In (1) we took the ratio of our diffusion coefficients to be of order one. However, a much more interesting problem would arise if we consider different orders for the diffusion in the "obstacles" and in the "pores". If one takes the ratio of the diffusion coefficients to be of order ε^2, then the limit model will be the so-called *double-porosity model*. This scaling preserves the physics of the flow inside the grains, as $\varepsilon \to 0$. The effective limit model includes two equations, one in T and another one in Ω, the last one containing an extra-term which reflects the influence of the grains (see [1], [2]).

3 Diffusion in Partially Fissured Media

For describing the second problem, we consider that the domain Ω is a periodic structure formed by two connected components representing two materials with different thermal features. So, we assume this time that both Ω^ε and Π^ε are connected, but only Ω^ε reaches the external fixed boundary of the domain Ω. In such a domain, we shall analyze the asymptotic behavior of the solutions of the following nonlinear system (for the linear case, see [3] and [EP02]):

$$\begin{cases} -\operatorname{div}\,(A_1^\varepsilon \nabla u^\varepsilon) + \beta(u^\varepsilon) = f & \text{in } \Omega^\varepsilon, \\ -\operatorname{div}\,(A_2^\varepsilon \nabla v^\varepsilon) = f, & \text{in } \Pi^\varepsilon \\ A_1^\varepsilon \nabla u^\varepsilon \cdot \nu = A_2^\varepsilon \nabla v^\varepsilon \cdot \nu & \text{on } S^\varepsilon, \\ A_1^\varepsilon \nabla u^\varepsilon \cdot \nu = a\varepsilon g(v^\varepsilon - u^\varepsilon) & \text{on } S^\varepsilon, \\ u^\varepsilon = 0 & \text{on } \partial\Omega. \end{cases} \qquad (5)$$

Hence, we assume that at the interface between our two materials the flux is continuous and depends in a nonlinear way on the jump of the temperature field. Here, $f \in L^2(\Omega)$, $a > 0$, β is a continuous function, monotonously non-decreasing and such that $\beta(0) = 0$ and g is a continuously differentiable function, monotonously non-decreasing, with $g(0) = 0$. Also, β and g satisfy the conditions (2)–(3) In fact, due to the compactness injection theorems in Sobolev spaces, it would be enough to assume, in both problems we address here, that β satisfies the growth condition (2) for some $0 \le q < (n+2)/(n-2)$.

If, in this case, we take $H^\varepsilon = \{w^\varepsilon = (u^\varepsilon, v^\varepsilon) \mid u^\varepsilon \in V^\varepsilon, v^\varepsilon \in H^1(\Pi^\varepsilon)\}$, with $\|w^\varepsilon\|_{H^\varepsilon}^2 = \|\nabla u^\varepsilon\|_{L^2(\Omega^\varepsilon)}^2 + \|\nabla v^\varepsilon\|_{L^2(\Pi^\varepsilon)}^2 + \varepsilon\|u^\varepsilon - v^\varepsilon\|_{L^2(\Gamma^\varepsilon)}^2$, the variational formulation of problem (5) is the following one:

$$\begin{cases} \text{Find } w^\varepsilon \in H^\varepsilon \text{ such that} \\ \int\limits_{\Omega^\varepsilon} A_1^\varepsilon \nabla u^\varepsilon \cdot \nabla\varphi \mathrm{d}x + \int\limits_{\Pi^\varepsilon} A_2^\varepsilon \nabla v^\varepsilon \cdot \nabla\psi \mathrm{d}x + \int\limits_{\Omega^\varepsilon} \beta(u^\varepsilon)\varphi \mathrm{d}x \\ +a\varepsilon \int\limits_{\Gamma^\varepsilon} g(u^\varepsilon - v^\varepsilon)(\varphi - \psi)\mathrm{d}\sigma = \int\limits_{\Omega^\varepsilon} f\varphi \mathrm{d}x + \int\limits_{\Pi^\varepsilon} f\psi \mathrm{d}x \quad \forall(\varphi, \psi) \in H^\varepsilon. \end{cases} \qquad (6)$$

By classical existence and uniqueness results (see [Bre72]) we know that (6) is a well-posed problem. Then, we can prove the following result, which represents a generalization to the nonlinear case of Barenblatt's model (see [BZK60]):

Theorem 2. *One can construct two extensions $P^\varepsilon u^\varepsilon$ and $P^\varepsilon v^\varepsilon$ of the solutions u^ε and u^ε of problem (6) such that $P^\varepsilon u^\varepsilon \rightharpoonup u$, $P^\varepsilon v^\varepsilon \rightharpoonup v$, weakly in $H_0^1(\Omega)$, where*

$$\begin{cases} -div\ (\overline{A}^1\nabla u) + \theta\beta(u) - ag(v-u) = \theta f & in\ \Omega, \\ -div\ (\overline{A}^2\nabla v) + ag(v-u) = (1-\theta)f & in\ \Omega. \end{cases}$$

Here, $\overline{A}^1$ and $\overline{A}^2$ are the homogenized matrices, defined by:

$$\overline{A}_{ij}^1 = \int_{Y_1} \left(a_{ij} + a_{ik}\frac{\partial\chi_{1j}}{\partial y_k} \right) dy,$$

$$\overline{A}_{ij}^2 = \int_{Y_2} \left(a_{ij} + a_{ik}\frac{\partial\chi_{2j}}{\partial y_k} \right) dy,$$

in terms of the functions $\chi_{1k} \in H_{per}^1(Y_1)/\mathbb{R}$, $\chi_{2k} \in H_{per}^1(Y_2)/\mathbb{R}$, $k = 1, ..., n$, weak solutions of the cell problems

$$\begin{cases} -\nabla_y \cdot ((A(y)\nabla_y\chi_{1k}) = \nabla_y A(y)e_k, & y \in Y_1, \\ (A(y)\nabla_y\chi_{1k}) \cdot \nu = -A(y)e_k \cdot \nu, & y \in \Gamma, \end{cases}$$

$$\begin{cases} -\nabla_y \cdot ((B(y)\nabla_y\chi_{2k}) = \nabla_y B(y)e_k, & y \in Y_2, \\ (B(y)\nabla_y\chi_{2k}) \cdot \nu = -B(y)e_k \cdot \nu, & y \in \Gamma. \end{cases} \quad\blacksquare$$

References

[BZK60]   Barenblatt, G.I., Zheltov, Y.P. and Kochina, I.N.: On basic conceptions of the theory of homogeneous fluids seepage in fractured rocks (in Russian). Prikl. Mat. Mekh., **24**, 852–864 (1960).

[BLM96]   Bourgeat, A., Luckhaus, S. and Mikelić, A.: Convergence of the homogenization process for a double-porosity model of immiscible two-phase flow. SIAM J. Math. Anal., **27**, 1520–1543 (1996).

[Bre72]   Brézis, H.: Problèmes unilatéraux. J. Math. Pures et Appl., **51**, 1–168 (1972).

[CDT03]   Conca, C., Díaz, J.I. and Timofte, C.: Effective chemical processes in porous media. Math. Models Methods Appl. Sci. (M3AS), **13**(10), 1437–1462 (2003).

[Ene01]   Ene, H.I.: On the microstructure models of porous media. Rev. Roumaine Math. Pures Appl., **46**, 289–295 (2001).

[EP02]   Ene, H.I. and Polisevski, D.: Model of diffusion in partially fissured media. Z. angew. Math. Phys., **53**, 1052–1059 (2002).

[Tar77]   Tartar, L.: Problèmes d'homogénéisation dans les équations aux dérivées partielles. In: Cours Peccot, Collège de France (1977).

[Tim06]   Timofte, C.: Upscaling in nonlinear diffusion problems in composite materials (in preparation).

Effective Two-Phase Flow Models Including Trapping Effects at the Micro Scale

C.J. van Duijn[1], H. Eichel[2], R. Helmig[2], and I.S. Pop[1]

[1] Universität Stuttgart, Institut für Wasserbau, Pfaffenwaldring 61, 70569 Stuttgart, Germany {eichel, rainer.helmig}@iws.uni-stuttgart.de
[2] Technische Universiteit Eindhoven, Department of Mathematics and Computer Science, P.O. Box 513, 5600 MB Eindhoven, The Netherlands {C.J.v.Duijn, I.Pop}@tue.nl

Summary. We consider a two-phase flow model in a heterogeneous porous column. The medium consists of many homogeneous layers that are perpendicular to the flow direction and have a periodic structure resulting in a one-dimensional flow. Trapping may occur at the interface between a coarse and a fine layer. An effective (upscaled) model is derived by homogenization techniques.

1 Introduction

We consider the flow of water and oil in a heterogeneous porous medium. The model studied here is relevant for the water-drive oil recovery, when water is injected into reservoirs to drive oil towards production well. The presence of rock heterogeneities will decrease the efficiency of the recovery process. This becomes obvious if paths of high permeability are encountered from injection to production wells. Then water will flow essentially through these paths, leaving much of the oil behind in the reservoir. Furthermore, if the heterogeneities are perpendicular to flow, oil may be trapped at the interfaces separating high and low permeability layers. This situation was analyzed mathematically in [DM95], and studied experimentally in [vL98].

In case of small scale heterogeneities, the complexity of the model rules out computations that take into account the model in full detail. Therefore it becomes necessary to have effective parameters or constitutive relationships, accounting for the averaged behavior of the system on a larger scale. Such models can be derived in various ways, and we refer to [WG96] and [Far02] for overviews. Here we consider a periodic medium consisting of alternating homogeneous layers of high an low permeability. These layers are assumed transversal to the flow direction, so the problem can be reduced to one spatial dimension. We employ homogenization techniques (see [Hor97]) to derive the effective flow equations. As shown for example in [NC05], this technique gives good results in the case of unsaturated flows. The present work is strongly related to [DM02]. It gives an alternative effective model to the one derived there for the case when the capillary forces dominate the viscous ones. Having in mind the presentation in [DM02], in Sect. 2 we shortly discuss the model and give the governing equations and interface conditions, and derive the effective equations. We conclude the paper with a numerical experiment. More details will appear in a forthcoming work (see also [DE05]).

2 The Model

In this section we briefly describe the heterogeneous model, and proceed with the derivation of the corresponding macro scale model. In doing so we refer to the detailed discussion in the first two sections of [DM02]. The medium consists of homogeneous layers of constant thickness. For simplicity we assume that all the characteristics of the medium are homogeneous, excepting the absolute permeability. This is location dependent, with values that jump between $K^+ K_{\text{ref}}$ (in a highly permeable layer) and $K^- K_{\text{ref}}$ (in a low permeable one). K_{ref} is a constant and dimensional reference absolute permeability.

By u we denote the reduced oil saturation: $0 \leq u \leq 1$. Then the water saturation is $1 - u$. We assume that the injection flow rate is constant in time. Since the model is one dimensional, the total specific discharge $q > 0$ is constant. The underlying mass and momentum equations can be combined into a single transport equation for one saturation only (see [DM95] or [DM02] for details). The typical nonlinearities of the model are k_w and k_o, the relative permeabilities of the fluid phases, as well as p_c, the capillary pressure. For these characteristics we use the Corey and Leverett expressions:

$$k_w = k_w(u) = (1-u)^2, \quad k_o = k_o(u) = u^2, \quad \text{and}$$
$$p_c(x,u) = \sigma \sqrt{\Phi/(K(x)K_{\text{ref}})} J(u) \quad \text{with} \quad J(u) = (1-u)^{-1/2}. \tag{1}$$

Here Φ is the porosity of the medium, σ the interfacial tension between the phases in the pores, and J the Leverett function. The extension to other cases is straightforward. The x-dependence of p_c is induced by the variable dimensionless component of the absolute permeability, $K(x) = K^{\pm}$. Note that $J(0) = 1 > 0$, implying the existence of an oil entry pressure: a pressure $p_c(x,0)$ has to be exerted on the oil before it can enter a fully water saturated medium (see [Hel97]).

In dimensionless form the oil-transport equation can be brought to

$$\partial_t u + \partial_x F(u) = \partial_t u + \partial_x \left(f(u) - N_c K(x)\lambda(u)\partial_x p_c(x,u) \right) = 0, \quad \text{with}$$
$$p_c(x,u) = \frac{J(u)}{\sqrt{K(x)}}, \quad f(u) = \frac{k_o(u)}{k_o(u)+Mk_w(u)}, \quad \text{and} \quad \lambda(u) = k_{rw}(u)f(u). \tag{2}$$

Notice the occurrence of the two dimensionless numbers: the capillary number N_c and the viscosity ratio M. Both are assumed to be of moderate order, $O(1)$.

2.1 The Interface Conditions

The absolute permeability K is discontinuous at any interface separating two homogeneous layers. This makes the definition of (2) across such interfaces impossible, where instead matching conditions are defined. Without loss of generality, we consider the interface located at $x = 0$, with the coarse material (K^+) inside the halfspace $x < 0$, and the fine material ($K^- < K^+$) appearing for $x > 0$. Then the continuity of flux reads

$$F(t) = F(0+, t) = F(0-, t), \quad \text{for all} \quad t > 0. \tag{3}$$

This condition is completed by the extended pressure condition (see [DM95]):

$$
\begin{cases}
u(0-, t) < u^* \text{ implies } u(0+, t) = 0, \\
u(0-, t) \geq u^* \text{ implies } \dfrac{J(u(0-,t))}{\sqrt{K^+}} = \dfrac{J(u(0+,t))}{\sqrt{K^-}}.
\end{cases}
\tag{4}
$$

The threshold saturation u^* in (4) is uniquely defined by:

$$J(u^*)/\sqrt{K^+} = J(0)/\sqrt{K^-}. \tag{5}$$

The capillary pressure is continuous only if oil is present on both sides of the interface. If the fine medium contains no oil, the entry pressure model implies a discontinuous capillary pressure. This is the mechanism for oil trapping in the coarse ($K = K^+$) material, see [DM95] for details.

2.2 The Upscaling

We turn now to the medium consisting of periodically repeating homogeneous layers, having the thickness L_y. This represents the micro scale. With L_x being the (macroscopic) column length, we define $\varepsilon = L_y/L_x$ and seek for the effective equations as $\varepsilon \searrow 0$. To do so we assume all quantities depending on two spatial variables: the macro scale x, and the micro scale $y = \frac{x}{\varepsilon}$. Then we expand all quantities asymptotically in ε and equate terms of the same order in ε. In this way, the multiscale oil saturation u_ε reads:

$$u_\varepsilon(x,t) = u^0(x,y,t) + \varepsilon u^1(x,y,t) + \varepsilon^2 u^2(x,y,t) \ldots, \tag{6}$$

where each u^k is periodic in $y = x/\varepsilon$. To make the periodicity explicit, consider two adjacent layers. We have K^+ for $y \in (-1, 0)$ (coarse layer), and K^- for $y \in (0, 1)$ (fine layer). For the multiscaled flux $F_\varepsilon = F(u_\varepsilon)$ we assume the continuity of fluxes on all scales at $y = 0$, as well as at the end points $y = \pm 1$:

$$F^k|_{y=0+} = F^k|_{y=0-}, \text{ and } F^k|_{y=1-} = F^k|_{y=-1+} \text{ for all } k > 0. \tag{7}$$

For the second condition we first notice that for $p_c(u_\varepsilon)$ we have:

$$p_c^0 = J(u^0)/\sqrt{K}, \text{ and } p_c^1 = u^1 J'(u^0)/\sqrt{K}. \tag{8}$$

Then the condition (4) is imposed periodically for p_c^0, in the spirit of (7). As shown in [DM02], this is only possible if the lowest order saturation u^0 is constant in each layer. To be specific, if $u^0 > 0$ in both layers, we have:

$$u^0(y) = \begin{cases} c > u^*, & \text{for } -1 < y < 0, \\ \bar{c}, & \text{for } 0 < y < 1, \end{cases} \qquad (9)$$

where c and $\bar{c}$ in $(0, 1]$ are related by

$$J(c)\sqrt{K^+} = J(\bar{c})/\sqrt{K^-}. \qquad (10)$$

Further, if $u^0(y > 0) = 0$, then $u^0(y < 0) = c \leq u^*$. Then the oil cannot flow into the fine layer, and trapping occurs at the interface $y = 0$.

To derive the effective equation, we need an additional condition at the interface between layers. In [DM02] the continuity of u^1 was assumed:

$$u^1(y = 0+) = u^1(y = 0-), \text{ and } u^1(y = 1-) = u^1(y = -1+). \qquad (11)$$

Notice that this assumption refers to the $O(\varepsilon)$ quantity, u^1. In the present work we consider another possibility, the continuity of p_c^1. By (8) this gives:

$$u^1(0-)\frac{J'(c)}{\sqrt{K^+}} = u^1(0+)\frac{J'(\bar{c})}{\sqrt{K^-}}, \text{ and } u^1(-1+)\frac{J'(c)}{\sqrt{K^+}} = u^1(1-)\frac{J'(\bar{c})}{\sqrt{K^-}}. \qquad (12)$$

Defining the effective oil saturation as $U(x, t) = (c + \bar{c})/2$, we can use (10) to express c and $\bar{c}$ in terms of U whenever $\bar{c} > 0$. The case $\bar{c} = 0$ is trivial, since then $U = c/2 \leq U^* := u^*/2$. With (12), the effective model becomes ([DE05])

$$\partial_t U + \partial_x \left[\mathcal{F}(U) - N_c \Lambda(U)\partial_x \mathcal{P}_C(U)\right] = 0. \qquad (13)$$

Whenever $0 \leq U \leq U^* = u^*/2$, $\mathcal{F}(U) = \Lambda(U) = 0$. Further, $\mathcal{F}(1) = 1$ and $\Lambda(1) = 0$. For $U^* < U < 1$, implying $0 < \bar{c} < 1$ and $u^* < c < 1$, we have

$$\mathcal{F}(U) = \frac{\frac{1}{K^+ k_w(c)} + \frac{1}{K^- k_w(\bar{c})}}{\frac{1}{K^+ k_w(c) f(c)} + \frac{1}{K^- k_w(\bar{c}) f(\bar{c})}}, \text{ and } \Lambda(U) = \frac{2}{\frac{1}{K^+ \lambda(c)} + \frac{1}{K^- \lambda(\bar{c})}}. \qquad (14)$$

By (10) the upscaled capillary pressure is continuous in the nontrivial case:

$$\mathcal{P}_c(U) = J(c)/\sqrt{K^+} = J(\bar{c})/\sqrt{K^-}. \qquad (15)$$

Remark 1. Note that the above effective model is degenerate for $0 \leq U \leq U^* = \frac{u^*}{2}$, showing that oil cannot flow oil unless its effective saturation U does exceed U^*. Therefore up to the threshold saturation U^* the oil will be trapped in the reservoir.

For neither of the choices (11) and (12) we have a rigorous mathematical proof of the convergence of the homogenization process. We refer to [Sch06] for first results in this sense. There the micro scale oil saturation is assumed strictly positive. In this way the capillary pressure becomes continuous across interfaces. The additional $O(\varepsilon)$ assumption in [Sch06] involves the averaged slopes for u in the adjacent homogeneous layers, leading to an upscaled model that is equivalent to the one derived here.

In the absence of a rigorous mathematical proof for the convergence of the homogenization process, the choice between (11) and (12) is suggested by numerical experiments. Though referring strictly to terms that are $O(\varepsilon)$, the two assumptions (11) and (12) lead to different effective models, as follows by comparing the effective coefficients in (14) – (15) with the ones in (2.20) of [DM02]. For the effective equation obtained in the former case, a numerical evidence of convergence was difficult to achieve unless capillary effects are becoming strongly dominant, e.g. $N_c > 1$. As revealed by the experiment below, the present approach yields an effective solution that agrees well with the averaged solution of the micro scale model also for moderate values of N_c.

In Figure 1 below, we display the two different effective solutions and the solution of the full (heterogeneous) model. The computations are performed in MUFTE-UG [BH99]. For comparing the results, the fine scale solution is averaged over the micro scale cells consisting of two adjacent layers, coarse and fine. In all computations we assume that the the reservoir is initially fully oil saturated $u(t = 0) = 1$ and we set $u = 0$ and $u = 1$ at the left and the right boundary. Further we take $N_c = 1$, and the dimensionless interval of computation is $(0, 2)$. The fine scale solutions are computed for different numbers of layers, yielding $\varepsilon = 1/20, 1/40$, or $1/80$. We use the nonlinear functions given in (1), with $K^+ = 1$ and $K^- = 0.5$.

In the left figure we present the solutions of the two effective models, computed at $t = 0.8$. Obviously, the two solutions are not identical. To show

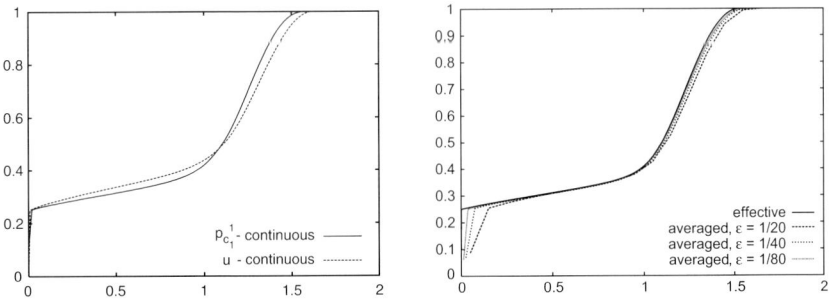

Fig. 1. *Left*: Effective solutions, obtained at $t = 0.8$ for the present approach (*solid*) and the approach in [DM02] (*dashed*). *Right*: Effective saturation (*solid*) computed assuming p_c^1 - continuous, and averaged saturations at $t = 0.8$.

the appropriateness of the present approach, in the right figure we compare the effective solution obtained assuming (12) with the averaged small-scale solution for different values of ε. A clear convergence is encountered as $\varepsilon \searrow 0$.

Conclusions and Acknowledgment

We have employed homogenization techniques to derive an effective two-phase flow model for a periodically layered medium. Trapping effects are occurring due to the difference in the micro scale entry pressures. The model strongly depends on the assumed behavior of the higher order terms at the interfaces separating the homogeneous layers. The resulting effective model is a nonlinear parabolic problem of degenerate type, incorporating convection, where the effective functions are weighted harmonic means of the corresponding small scale ones. These functions vanish whenever the effective saturation is below a threshold saturation, representing the maximal amount of trapped oil.

Acknowledgements

We thank Dr. Insa Neuweiler (Stuttgart) for suggesting us to consider the alternative micro scale continuity assumption. The work of H. Eichel was supported through a European Community Marie Curie Fellowship (HPMT-CT-2001-00422). We also acknowledge the support of the DFG (FIMOTUM He 2531/5-1), and of the BSIK-BRICKS project (theme MSV1), funded by the Dutch government.

References

[BH99]   Bastian, P., Helmig, R.: Efficient fully-coupled solution techniques for two-phase flow in porous media. Adv. Water Resour., **23**, 199–216 (1999)

[DE05]   van Duijn, C.J., Eichel, H., Helmig, R., Pop, I.S.: Effective equations for two-phase flow in porous media: the effect of trapping at the micro scale. CASA Report, **05-34**, Eindhoven University of Technology (2005)

[DM02]   van Duijn, C.J., Mikelić, A., Pop, I.S.: Effective equations for two-phase flow with trapping on the micro scale. SIAM J. Appl. Math., **62**, 1531–1568 (2002)

[DM95]   van Duijn, C.J., Molenaar, J., de Neef, M.J., Effects of capillary forces on immiscible two-phase flow in heterogeneous porous media. Transp. Porous Media, **21**, 71–93 (1995)

[Far02] Farmer, C.L.: Upscaling: a review. Int. J. Numer. Meth. Fluids, **40**, 63-78 (2002)

[Hel97] Helmig, R.: Multiphase Flow and Transport Processes in the Subsurface. Springer-Verlag, Heidelberg (1997)

[Hor97] Hornung, U.: Homogenization and Porous Media. Interdisciplinary Applied Mathematics. Springer Verlag, New York (1997)

[vL98] van Lingen, P.: Quantification and reduction of capillary entrapment in cross-laminated oil reservoirs, PhD Thesis, TU Delft (1998).

[NC05] Neuweiler, I., Cirpka, O.A.: Homogenization of Richards equation in permeability fields with different connectivities. Water Resour. Res., **41**, W02009, (2005)

[Sch06] Schweizer, B.: Homogenization of degenerate two-phase flow equations with oil-trapping. *Preprint 2006-04, Dept. of Math., University of Basel* (2006).

[WG96] Wen, X.H., Gómez-Hernández, J.J.: Upscaling hydraulic conductivities in heterogeneous media: An overview. J. Hydrol., **183**, ix–xxxii (1996)

Minisymposium "Shallow Water and Simulation of Environmental Flows"

C. Parés

Universidad de Málaga, Spain

In this minisymposium, C.E. Castro and E.F. Toro present "ADER DG and FV schemes for shallow water flows", M. Castro Díaz, E.D. Fernández Nieto and A. Ferreiro Ferreiro present "Numerical simulation of bedload sediment transport using finite volume schemes", L. Ferrer, Ad. Uriarte and M. González present "New Trends and Applications in Oceanographic Numerical Modelling", M. J. Castro, A. M. Ferreiro, J. A. García, J. M. González and C. Parés present "Study and Development of Numerical Models for the simulation of geophysical flows: the DamFlow Project", and I. Gejadze, M. Honnorat, F.X. Le Dimet and J. Monnier present "On variational data assimilation for 1D and 2D fluvial hydraulics".

ADER DG and FV Schemes for Shallow Water Flows

Cristóbal E. Castro and Eleuterio F. Toro

Laboratory of Applied Mathematics, Faculty of Engineering
University of Trento, Italy
cristobal.castro@ing.unitn.it, toro@ing.unitn.it

1 Introduction

We are concern with ADER [3] high-order numerical methods for the time-dependent two-dimensional non-linear shallow water equations [2] in the framework of finite volumes (FV) and discontinuous Galerkin (DG) finite elements methods using non-structured triangular meshes.

High order in space and time is obtained by (1) a high-order spatial distribution of the solution in each element, (2) the solution of the derivative Riemann problem (DRP) [4] and (3) an accurate computation of the numerical flux and volume integrals. Regarding the high-order spatial distribution of the solution, in the FV method one requires cell average reconstructions [1] at each time step; in the case of DG the high-order representation of the data is built into the scheme to the desired order and no reconstruction is needed. However, in the presence of high gradients numerical oscillations arise in the DG case, which requires the implementation of special reconstruction.

We assess the methods by comparing numerical solutions with exact solutions. The rest of the chapter is organized as follows: Sect. 2 describes the governing equations and the hyperbolic character. In Sect. 3 we construct the numerical method. In Sect. 4 we describe the ADER approach. Test problems are presented in Sect. 5 and conclusions are drawn in Sect. 6.

2 The Two-Dimensional Shallow Water System

The two-dimensional shallow water equations with source term due to bottom variation can be written as follows:

$$
\begin{aligned}
&\partial_t \left(h \right) + \partial_x \left(hu \right) + \partial_y \left(hv \right) = 0 \,, \\
&\partial_t \left(hu \right) + \partial_x \left(hu^2 + \tfrac{1}{2} g h^2 \right) + \partial_y \left(huv \right) = -ghb_x \,, \\
&\partial_t \left(hv \right) + \partial_x \left(huv \right) + \partial_y \left(hv^2 + \tfrac{1}{2} g h^2 \right) = -ghb_y \,,
\end{aligned}
\tag{1}
$$

where $h(x, y, t)$ is the depth, $u(x, y, t)$ and $v(x, y, t)$ are the component of the particle velocity in the x and y direction, respectively, $b(x, y)$ is the bathymetry and g is the gravity force. Here we consider a reference level $z = 0$ with $z = b(x, y) > 0$ and $z = b(x, y) + h(x, y, t)$ the free surface.

The system (1) can be written in vector form, where the unknown vector $\mathbf{Q}$ contains the conservative variables.

$$\mathbf{Q}_t + \mathbf{F}(\mathbf{Q})_x + \mathbf{G}(\mathbf{Q})_y = \mathbf{S}(\mathbf{Q}) , \tag{2}$$

with

$$\mathbf{Q} = \begin{bmatrix} h \\ hu \\ hv \end{bmatrix} , \quad \mathbf{F}(\mathbf{Q}) = \begin{bmatrix} hu \\ hu^2 + \frac{1}{2}gh^2 \\ huv \end{bmatrix} , \tag{3}$$

$$\mathbf{G}(\mathbf{Q}) = \begin{bmatrix} hv \\ huv \\ hv^2 + \frac{1}{2}gh^2 \end{bmatrix} , \quad \mathbf{S}(\mathbf{Q}) = \begin{bmatrix} 0 \\ -ghb_x \\ -ghb_y \end{bmatrix} . \tag{4}$$

In order to verify the hyperbolic character of the system we express (1) in quasi-linear form expanding the Jacobian matrices $\mathbf{A}(\mathbf{Q}) = \partial \mathbf{F}(\mathbf{Q})/\partial \mathbf{Q}$ and $\mathbf{B}(\mathbf{Q}) = \partial \mathbf{G}(\mathbf{Q})/\partial \mathbf{Q}$.

$$\mathbf{Q}_t + \mathbf{A}(\mathbf{Q})\mathbf{Q}_x + \mathbf{B}(\mathbf{Q})\mathbf{Q}_y = \mathbf{S}(\mathbf{Q}) . \tag{5}$$

System (5) is said to be hyperbolic if for any vector $\mathbf{n} = [n_x, n_y]$ the matrix $\mathbf{C} = n_x\mathbf{A} + n_y\mathbf{B}$ has real eigenvalues and a complete set of linear independent eigenvectors. This can be easily verified for (1). For more details see [2].

In the next section we construct the numerical method.

3 Numerical Method

The numerical method is constructed for a computational domain obtained by a conforming triangulation of the physical domain $\Omega \in R^2$. The control volume considered is a triangular element $T_m \in \Omega$. In what follows we will use the divergence form of (2).

$$\partial_t \mathbf{Q} + \nabla \cdot \mathbf{H}(\mathbf{Q}) = \mathbf{S}(\mathbf{Q}) , \tag{6}$$

where $\nabla = [\partial_x, \partial_y]$ and $\mathbf{H}(\mathbf{Q}) = [\mathbf{F}(\mathbf{Q}), \mathbf{G}(\mathbf{Q})]^T$.

Finite element discontinuous Galerkin numerical methods are constructed considering that the unknown vector $\mathbf{Q}(\mathbf{x}, t)$ is approximated numerically by the vector $\mathbf{Q}_h(\mathbf{x}, t)$ given by a linear combination of spatial polynomial basis functions $\phi_l(\mathbf{x}) \in R^2$ of order $o - 1$, where o is the order of the method, and temporal scalar degrees of freedom $\hat{q}_l(t) \in R$. In contrast with FV schemes, where cell averages are evolved, Discontinuous Galerkin schemes evolve the degrees of freedom $\hat{q}_l(t)$.

$$q(\mathbf{x}, t) \approx q_h(\mathbf{x}, t) \equiv \sum_{l=0}^{N} \hat{q}_l(t) \phi_l(\mathbf{x}) . \tag{7}$$

Considering one component of (6), we multiply by the basis function $\phi_k(\mathbf{x})$ and integrate over the control volume T_m. Using the product rule, the Gauss's divergence theorem and introducing (7), we project the continuous function q into the discrete space q_h. Considering the orthogonality of the basis functions, we obtain,

$$\partial_t \hat{q}_k \int_{T_m} \phi_k \phi_k \, \mathrm{d}\mathbf{x} + \int_{\partial T_m} \phi_k \, H(q) \cdot \hat{n} \, \mathrm{d}\mathbf{x} - \int_{T_m} \nabla \phi_k \cdot H(q) \, \mathrm{d}\mathbf{x} = \int_{T_m} \phi_k \, S(q) \, \mathrm{d}\mathbf{x}. \tag{8}$$

Integrating (8) within time interval $[t^n, t^{n+1}]$ and rearranging the terms, we obtain the expression for the time evolution for the degrees of freedom $\hat{q}_k$:

$$\hat{q}_k^{n+1} = \hat{q}_k^n - \frac{1}{|J| \, m_k} \left[\int_{t^n}^{t^{n+1}} \int_{\partial T_m} \phi_k \, H(q) \cdot \hat{n} \, \mathrm{d}\mathbf{x} \, \mathrm{d}t \right.$$
$$\left. - \int_{t^n}^{t^{n+1}} \int_{T_m} \nabla \phi_k \cdot H(q) \, \mathrm{d}\mathbf{x} \, \mathrm{d}t - \int_{t^n}^{t^{n+1}} \int_{T_m} \phi_k \, S(q) \, \mathrm{d}\mathbf{x} \, \mathrm{d}t \right]. \tag{9}$$

Equation (9) give us an explicit one-step evolution equation for the degrees of freedom $\hat{q}_k$ inside the triangle T_m from time t^n to time t^{n+1}. The particular case where only one basis function $\phi_0 = 1$ is used in (7), (9) gives the finite volume numerical scheme.

Depending on the order of the approximation $q \approx q_h$ in (7) for DG methods, and the order of a spatial reconstruction used for FV, the space integrals in (9) are computed using a quadrature rule of suitable order. In the presence of strong discontinuities, non-oscillatory spatial reconstructions are used. Here we use the one presented in [1].

4 ADER Schemes

The ADER approach originally proposed by Toro et al. [3] allows us to construct arbitrary high order accuracy numerical methods in the framework of FV and DG. This is obtained by constructing a time-dependent function that approximates the time evolution of the vector $\mathbf{Q}(\mathbf{x}_h, \tau)$ for a particular spatial point $\mathbf{x}_h$, following the Cauchy–Kowalewski method

$$\mathbf{Q}(\mathbf{x}_h, \tau) = \mathbf{Q}(\mathbf{x}_h, 0) + \sum_{k=1}^{r-1} \frac{\tau^k}{k!} \partial_t^{(k)} \mathbf{Q}(\mathbf{x}_h, 0) . \tag{10}$$

This expansion is a Taylor's time series expansion around the leading term $\mathbf{Q}(\mathbf{x}_h,0)$ with high order derivatives terms with coefficients $\partial_t^{(k)}\mathbf{Q}(\mathbf{x}_h,0)$.

Assuming that a continuous and differentiable polynomial space distribution of the unknown vector $\mathbf{Q}(\mathbf{x},\mathbf{0})$ is available in the element T_m at local time $\tau = 0$, for a given spatial integration point $\mathbf{x}_h \in T_m/\partial T_m$ we can evaluate the function $\mathbf{Q}(\mathbf{x}_h,\mathbf{0})$ and all space derivatives. Using the balance laws (2) and the Cauchy–Kowalewski procedure we can transform the space derivatives into time derivatives. If the spatial point $\mathbf{x}_h \in \partial T_m$ the function $\mathbf{Q}(\mathbf{x}_h,\mathbf{0})$ and its derivatives are discontinuous, considering that the neighbour triangle has a different polynomial representation of the data, then we use derivative Riemann problem (DRP) solvers in order to compute the leading term and all high order derivatives. Here we implement the DRP solver presented in [4].

Once the time evolution function (10) is obtained, time integrals in (9) are computed numerically with the required order, evaluating the function (10) at the designed integration point.

5 Convergence Test

Normally, exact solutions for two-dimensional problems with source term are unknown. It is possible to construct an exact solution as follow. We set the solution of the problem $\tilde{\mathbf{Q}}(x,y,t)$, then we evaluate (1) finding a new source term $\tilde{\mathbf{S}}(x,y,t)$. Defining the exact solution $\tilde{\mathbf{Q}}$ as $b(x,y) = \exp(-8(x^2+y^2))/5$, $u(x,y,t) = (1+\sin(x\pi))/10$, $v(x,y,t) = (1+\sin(y\pi))/10$ and $H(x,y,t) = h(x,y,t)+b(x,y) = \exp(t/10)$, we evaluate (2) to find $\tilde{\mathbf{S}}(x,y,t)$.

$$\tilde{\mathbf{S}}(x,y,t) = \partial_t\tilde{\mathbf{Q}} + \partial_x\mathbf{F}(\tilde{\mathbf{Q}}) + \partial_y\mathbf{G}(\tilde{\mathbf{Q}}) - \mathbf{S}(\tilde{\mathbf{Q}}) , \tag{11}$$

Solving the following initial value problem we can measure the error of the numerical solution.

$$\left.\begin{array}{ll} \text{PDEs: } \partial_t\mathbf{Q} + \nabla\cdot\mathbf{H}(\mathbf{Q}) = \mathbf{S}(\mathbf{Q}) + \tilde{\mathbf{S}} , & \mathbf{x}\in[-1,1]\times[-1,1] , \quad \mathbf{t}>\mathbf{0} , \\ \text{IC:} \qquad \mathbf{Q}(x,y,0) = \tilde{\mathbf{Q}}(x,y,0) \end{array}\right\} \tag{12}$$

We set periodic boundary conditions and final time $t = 1$ s. The errors and convergences order are presented in Tables 1–3 for the second component of $\mathbf{Q}$. The expected order of convergence is reached for all methods.

Table 1. Convergence rates test: second-order method

Mesh	L_1 Error	O_1	L_2 Error	O_2	L_∞ Error	O_∞
h/2	1.34×10^{-2}	1.80	9.76×10^{-3}	1.77	3.79×10^{-2}	1.17
h/4	2.45×10^{-3}	2.48	1.94×10^{-3}	2.35	1.45×10^{-2}	1.40
h/8	4.66×10^{-4}	2.41	3.59×10^{-4}	2.45	4.30×10^{-3}	1.77

Table 2. Convergence rates test: third-order method

Mesh	L_1 Error	O_1	L_2 Error	O_2	L_∞ Error	O_∞
h/2	4.07×10^{-3}	2.62	3.07×10^{-3}	2.58	2.50×10^{-2}	1.41
h/4	4.13×10^{-4}	3.33	3.17×10^{-4}	3.31	1.07×10^{-3}	4.60
h/8	5.59×10^{-5}	2.90	4.20×10^{-5}	2.93	1.16×10^{-4}	3.23

Table 3. Convergence rates test: fourth-order method

Mesh	L_1 Error	O_1	L_2 Error	O_2	L_∞ Error	O_∞
h/2	1.37×10^{-3}	3.58	8.96×10^{-4}	3.64	2.13×10^{-3}	3.75
h/4	5.31×10^{-5}	4.74	4.44×10^{-5}	4.38	2.13×10^{-4}	3.36
h/8	3.96×10^{-6}	3.78	3.27×10^{-6}	3.79	2.05×10^{-5}	3.40

6 Conclusions and Further Work

We have solved the non-linear shallow water system on a two-dimensional unstructured grid with a high order ADER-type numerical method. Derivative Riemann Problem solutions are used to evaluate accurately the numerical fluxes. High order space reconstruction was used to reconstruct cell average values. The expecting convergence rate was reached for the test presented. Further developments in the treatment of the friction terms and wet/dry fronts are subject of current investigation by the authors.

References

1. M. Dumbser and M. Käser. Arbitrary high order non-oscillatory finite volume schemes on unstructured meshes for linear hyperbolic systems. *J. Comp. Phys. (In Press)*, 00:000–000, 2006.
2. E.F. Toro. *Shock-Capturing Methods for Free-Surface Shallow Flows.* John Wiley and Sons, Chichester, 2001.
3. E.F. Toro, R.C. Millington, and L.A.M. Nejad. Towards very high order Godunovs schemes. In E.F. Toro, editor, *Godunov Methods. Theory and Applications*, pages 907–940. Kluwer/Plenum Academic Publishers, 2001.
4. E.F. Toro and V.A. Titarev. Derivative Riemann Solvers for Systems of Conservation Laws and ADER Methods. *J. Comp. Phys.*, 212:150–165, 2006.

Numerical Simulation of Bedload Sediment Transport Using Finite Volume Schemes

M. Castro Díaz[1], E.D. Fernández Nieto[2], and A. Ferreiro Ferreiro[3]

[1] Dpto. Análisis Matemático, Universidad de Málaga
castro@anamat.cie.uma.es
[2] Dpto. de Matemática Aplicada I, Universidad de Sevilla
edofer@us.es
[3] Dpto. de Ecuaciones Diferenciales y Análisis Numérico, Universidad de Sevilla
anafefe@us.es

1 Introduction

In this work we study high order finite volume methods by state recon-structions. We apply it to beload sediment transport problems. For the hydrodynamical component we consider Shallow Water Equations, for the morphodynamical component we consider a continuity equation. The final system of equations can be re-written as a 2D non-conservative system

$$\frac{\partial W}{\partial t} + \mathcal{A}_1(W)\frac{\partial W}{\partial x_1} + \mathcal{A}_2(W)\frac{\partial W}{\partial x_2} = 0, \tag{1}$$

where $W(x,t) : \mathcal{O} \times (0,T) \to \Omega \subset \mathbb{R}^N$, $\mathcal{O}$ is a bounded domain of $\mathbb{R}^2$, Ω is a convex subset of $\mathbb{R}^N$, $\mathcal{A}_i : \Omega \to M_{N \times N}$ regular and locally bounded matricial functions.

We define, for a given vector $\eta = (\eta_1, \eta_2) \in \mathbb{R}^2$: $\mathcal{A}(W, \eta) = \mathcal{A}_1(W)\eta_1 + \mathcal{A}_2(W)\eta_2$. And we suppose that system (1) is strictly hyperbolic, that is, for all $W \in \Omega \subset \mathbb{R}^N$ and $\forall \eta \in \mathbb{R}^2$, matrix $\mathcal{A}(W, \eta)$ has N real and different eigenvalues: $\lambda_1(W, \eta) < \cdots < \lambda_N(W, \eta)$, being $R_j(W, \eta)$, $j = 1, \ldots, N$ the associated eigenvectors. Consequently $\mathcal{A}(W, \eta)$ is diagonalizable: $\mathcal{A}(W, \eta) = \mathcal{K}(W, \eta)\mathcal{L}(W, \eta)\mathcal{K}^{-1}(W, \eta)$, where $\mathcal{L}(W, \eta)$ is the diagonal matrix whose coeffi-cients are the eigenvalues of $\mathcal{A}(W, \eta)$ and $\mathcal{K}(W, \eta)$ is the matrix which columns are the vectors $R_j(W, \eta)$, $j = 1, \ldots, N$.

2 Shallow Water Equations with Sediment Transport

We consider the system,

$$
\begin{cases}
\dfrac{\partial h}{\partial t} + \dfrac{\partial q_1}{\partial x} + \dfrac{\partial q_2}{\partial x} = 0, \\[2mm]
\dfrac{\partial q_1}{\partial t} + \dfrac{\partial}{\partial x_1}\left(\dfrac{q_1^2}{h} + \dfrac{1}{2}gh^2\right) + \dfrac{\partial}{\partial x_2}\left(\dfrac{q_1 q_2}{h}\right) = gh\dfrac{\partial S}{\partial x_1} - ghS_{f,x_1}, \\[2mm]
\dfrac{\partial q_2}{\partial t} + \dfrac{\partial}{\partial x_1}\left(\dfrac{q_1 q_2}{h}\right) + \dfrac{\partial}{\partial x_2}\left(\dfrac{q_2^2}{h} + \dfrac{1}{2}gh^2\right) = gh\dfrac{\partial S}{\partial x_2} - ghS_{f,x_2}, \\[2mm]
\dfrac{\partial S}{\partial t} - \xi\dfrac{\partial q_{b,x_1}}{\partial x_1} - \xi\dfrac{\partial q_{b,x_2}}{\partial x_2} = 0,
\end{cases}
\tag{2}
$$

where the unknowns are the height of the water column $h(\mathbf{x}, t)$, the discharge $q(\mathbf{x}, t) = (q_1(\mathbf{x}, t), q_2(\mathbf{x}, t))$ and $S(\mathbf{x}, t) = H(\mathbf{x}) - z_b(\mathbf{x}, t)$, where H is the bathimetry of the fixed bottom and z_b is the height of the sediment layer column. By q_{b,x_1} and q_{b,x_2} we denote the solid transport discharge for the two components x_1 and x_2, respectively. The definition of q_b depends on the considered model. In this work we present Grass' model and Meyer-Peter and Müller's model. S_{f,x_1} and S_{f,x_2} are the Manning's friction laws.

The system of equations (2) can be rewriten under the structure of a 2D non-conservative hyperbolic system (1).

The definition of q_{b,x_1} and q_{b,x_2} for the particular cases of Grass and Meyer-Peter & Müller are:

1. *Grass' model*:

$$
\begin{aligned}
q_{b,x_1} &= A_g u_1 (u_1^2 + u_2^2), \\
q_{b,x_2} &= A_g u_2 (u_1^2 + u_2^2);
\end{aligned}
\tag{3}
$$

where $u_1 = q_1/h$ and $u_2 = q_2/h$. Being A_g the interaction constant between the fluid and the sediment $(0 \leq A_g \leq 1)$.

2. *Meyer-Peter & Müller's model*:

$$
\begin{aligned}
q_{b,x_1} &= 8\sqrt{(G-1)gd_i^3}\ \mathrm{sgn}\,(u_1)\,(\tau_{*,x_1} - \tau_{*,c}), \\
q_{b,x_2} &= 8\sqrt{(G-1)gd_i^3}\ \mathrm{sgn}\,(u_2)\,(\tau_{*,x_2} - \tau_{*,c});
\end{aligned}
\tag{4}
$$

where τ_* and τ_{*c} are the non-dimensional shear stress and the critical shear stress. The definition of τ_* is

$$
\tau_{*,x_1} = \frac{\gamma\eta^2|u_1|\sqrt{u_1^2 + u_2^2}}{(\gamma_s - \gamma)d_i R_h^{1/3}}, \quad
\tau_{*,x_2} = \frac{\gamma\eta^2|u_2|\sqrt{u_1^2 + u_2^2}}{(\gamma_s - \gamma)d_i R_h^{1/3}}.
$$

Where γ denotes the specific weight of the fluid, γ_s denotes the specific sediment weight, η is the Manning's coeffient corresponding to the sediment layer, d_i is the mean diameter. τ_{*c} is usually set to 0.047.

3 2D High Order Finite Volume Methods by State Reconstructions

We consider the 2D non-conservative system

$$\frac{\partial W}{\partial t} + \mathcal{A}_1(W)\frac{\partial W}{\partial x_1} + \mathcal{A}_2(W)\frac{\partial W}{\partial x_2} = 0. \tag{5}$$

Our objective is to use a high order finite volume method based on state reconstructions for (5).

We suppose that the computational domain is subdivided in cells or control volumes, $V_i \subset \mathbb{R}^2$, that we suppose to be defined by a closed poligone. We use the following notation: for a given control volume V_i, $\mathcal{N}_i$ is the set of index j such that V_j is a neighbour of V_i, E_{ij} is the common edge between the two control volumes V_i and V_j, and $|E_{ij}|$ is its length, $\eta_{ij} = (\eta_{ij,1}, \eta_{ij,2})$ is the unitary normal vector to the edge E_{ij} outward to V_i.

For a given control volume V_i we denote by P_i the state reconstruction operator over it. And for a given vector η_{ij} from V_i to V_j and normal to the edge E_{ij}, we denote by $W_{ij}^-(t, \mathbf{x}_{ij})$ and $W_{ij}^+(t, \mathbf{x}_{ij})$ $\forall \mathbf{x}_{ij} \in E_{ij}$, the limit of P_i (P_j respectively) when $\mathbf{x}$ tends to $\mathbf{x}_{ij}$ by the interior of V_i (V_j respectively).

We suppose that P_i is order p over the boundary of V_i, it is order q at the interior of the control volume V_i and ∇P_i is an approximation of order m of the gradient of the solution (see [3]).

3.1 High Order 2D Finite Volume Method

We consider the following numerical scheme (see [3]):

$$W_i'(t) = -\frac{1}{|V_i|}\left[\sum_{j \in \mathcal{N}_i} |E_{ij}| \sum_{l=1}^{n(\bar{r})} w_l \mathcal{A}_{ij,l}^-(W_{ij,l}^+, W_{ij,l}^-, \eta_{ij})(W_{ij,l}^+ - W_{ij,l}^-)\right.$$

$$\tag{6}$$

$$\left. + \int_{V_i}\left(\mathcal{A}_1(P_i(\mathbf{x}))\frac{\partial P_i}{\partial x_1}(\mathbf{x}) + \mathcal{A}_2(P_i(\mathbf{x}))\frac{\partial P_i}{\partial x_2}(\mathbf{x})\right)d\mathbf{x}\right],$$

where w_l, $l = 1, \ldots, n(\bar{r})$, are the weights of a quadrature formula associate to the 1D integral over the edge E_{ij}. If by x_l we denote the points over the edge E_{ij} of the quadrature formula then $W_{ij,l}^\pm = W_{ij}^\pm(\mathbf{x}_l)$. In practice this formula is chosen in function of the order of the state reconstruction: if by $\bar{r}$ we denote the order of the quadrature formula, then $\bar{r} > p$.

By $\mathcal{A}_{ij}$ we denote Roe matrix associated to the 1D non-conservative projected problem over η_{ij}. The non-conservative product $\mathcal{A}(W)W_x$ of the projected 1D problem makes difficult the definition of weak solutions for this kind of systems. After the theory developed by Dal Maso, LeFloch and Murat, a definition of non-conservative products as Borel measures is introduced, which is based on the selection of a family of paths in the phases space. Also a family of paths in this case must be chosen in order to define Roe matrix (see [Pares04]).

4 Numerical Test

In this section we present an experiment where we simulate sediment layer evolution over a soil which is not eroded. The experiment has been performed in a channel of length 15 m and width 0.5 m, of Hydraulic Laboratory from Escuela Superior de Ingenieros de Caminos, Canales y Puertos of A Coruña University. The experiment presented in this chapter was made by E. Peña González (see [Peña02]).

The experimental test was developed introducing a sand layer in central part of laboratory channel, and inducing hydrodynamic conditions to erode the sand layer, until to get steady state. The channel has a very small slope of 0.052%. Sand layer was situated in the interval [4.5 m, 9 m], with a thickness of 4.5 cm; being media diameter grain equal to 1 mm.

For numerical simulation we have used a mesh with 6,008 finite volumes. As CFL condition we consider 0.8. For boundary conditions, upstream we impose a discharge equal to $0.0285\,\mathrm{m^2\,s^{-1}}$, and downstream it is fixed the thickness of column water to 0.129 m. Sediment porosity is equal to 0.4. Friction between fluid and bed is modeled using a Manning's law with coefficient equal to 0.0125, that is discretized semi-implicitly. Friction coefficient between fluid and sediment layer is 0.0196. As boundary conditions we impose an incoming mass-flow equal to $0.0285\,\mathrm{m^2\,s^{-1}}$ upstream, while downstream is fixed thickness fluid equal to 0.129 m. We use a state reconstruction operator of second order of MUSCL type (see [3]).

As this test is basically one-dimensional, so we compare 2D solution with solutions obtained by 1D generalized Roe scheme and Roe-Weno2 scheme (see [Castro06], [CFF06]). In Fig. 1 is shown comparison at instant $t = 120\,\mathrm{min}$, of 2D solution (continuous line with stars), with solution obtained using generalized Roe scheme (continuous line) and Roe-Weno2 scheme (dotted line). We also compare in Fig. 2 with experimental data. The model that we are using does not include pressure forces, so the sediment does not fall by its own weight due to gravity effects, but it must reproduce at least, the downstream sand slope and the median profile of the sediment layer. This behavior is reflected in Fig. 2.

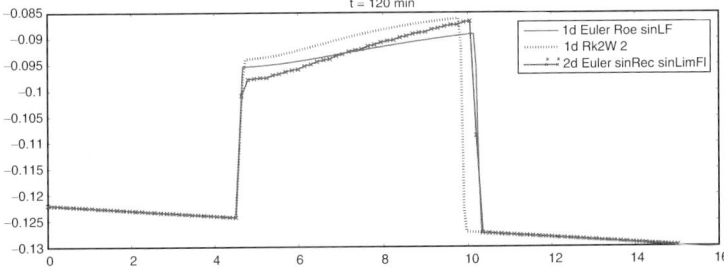

Fig. 1. Sediment layer evolution at $t = 120\,\mathrm{min}$. Euler-Roe 1d (*continuous line*). Weno2-Rk2 1d (*dotted line*). Euler-Roe 2d (*continuous line with stars*)

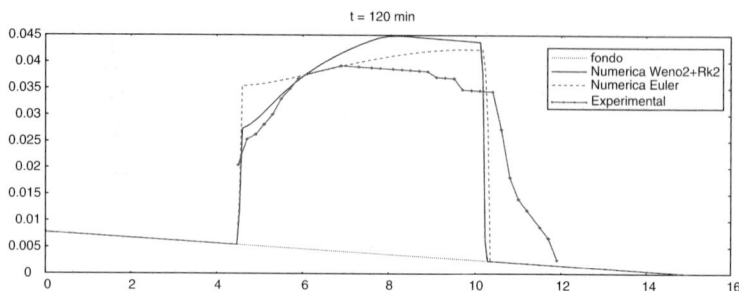

Fig. 2. Comparison with experimental data

References

[CFF06]   M.J. Castro, E.D. Fernández-Nieto, A.M. Ferreiro. Sediment transport
             models in Shallow Water equations and numerical approach by high
             order finite volume methods. Submitted (2006).

[Castro06]   M.J. Castro, J. M. Gallardo and Carlos Pars. Finite volume schemes
             based on weno reconstruction of states for solving nonconservaative
             hyperbolic systems. Applications to shallow water systems. Mathemat-
             ics of Computation. Accepted for publication. (2006)

[Ferreiro06] A.M. Ferreiro. Desarrollo de técnicas de post-proceso de flujos
             hidrodinámicos, modelización de problemas de transporte de sedimen-
             tos y simulación numérica mediante técnicas de volúmenes finitos. The-
             sis, University of Sevilla, Spain (2006).

[Pares04]   C. Parés, M.J. Castro. On the well-balanced property of Roe's method
             for nonconservative hyperbolic systems. Applications to shallow water
             systems. ESAIM: M2AN **38**(5): 821-852, (2004)

[Peña02]   Enrique Peña González. Estudio numérico y experimental del trans-
             porte de sedimentos en cauces aluviales. Tesis Doctoral. Universidade
             da Coruña. Grupo de Ingeniería del agua y del medio ambiente (2002).

New Trends and Applications in Oceanographic Numerical Modelling

L. Ferrer, Ad. Uriarte, and M. González

Unidad de Investigación Marina, AZTI-Tecnalia, Herrera Kaia - Portu aldea z/g, 20110, Pasaia - Gipuzkoa, Spain
lferrer@pas.azti.es

1 Introduction

The development of numerical models for the simulation of marine hydro-dynamics together with the increment in computational capacity of the new computers and the collaboration amongst interdisciplinary research groups, is slowly allowing to catch up with the delay of oceanic forecasting in comparison to other research areas like meteorology. Thanks to the knowledge gained in relation to oceanographic processes and the interaction between the ocean and atmosphere, new forecasting tools are being developed. These, comprise models of currents and waves, pollutant drift and aging, interaction between physical and biological processes with application to the management of fishing resources, or models for environmental impact assessment of coastal activities and uses (submarine outfalls, aquaculture cages, spill of dredged material, etc.). This chapter shows some examples of several applications of numerical models to regional and local scale areas.

2 Sinking of the *Prestige*

The sinking of the oil tanker *Prestige* off the coast of Galicia (in the north-western part of Spain), on the 19th November 2002, was the base to establish in the Bay of Biscay an *Operational Oceanography System*. This system included data analysis of in situ tracked buoys and oceanic and meteorological stations, satellite and visual observations of the sea, and numerical model predictions (Fig. 1). At regional scale, the drifting of the *Prestige* pollutant at sea surface from the sinking area to the affected regions was the final result from the combination of a large number of interacting factors: the local wind stress, acting directly on the fuel oil patches or indirectly through the Ekman layer, the density-driven circulation and the development of mesoscale structures such as surface eddies and fronts [Alv06], [Gon06].

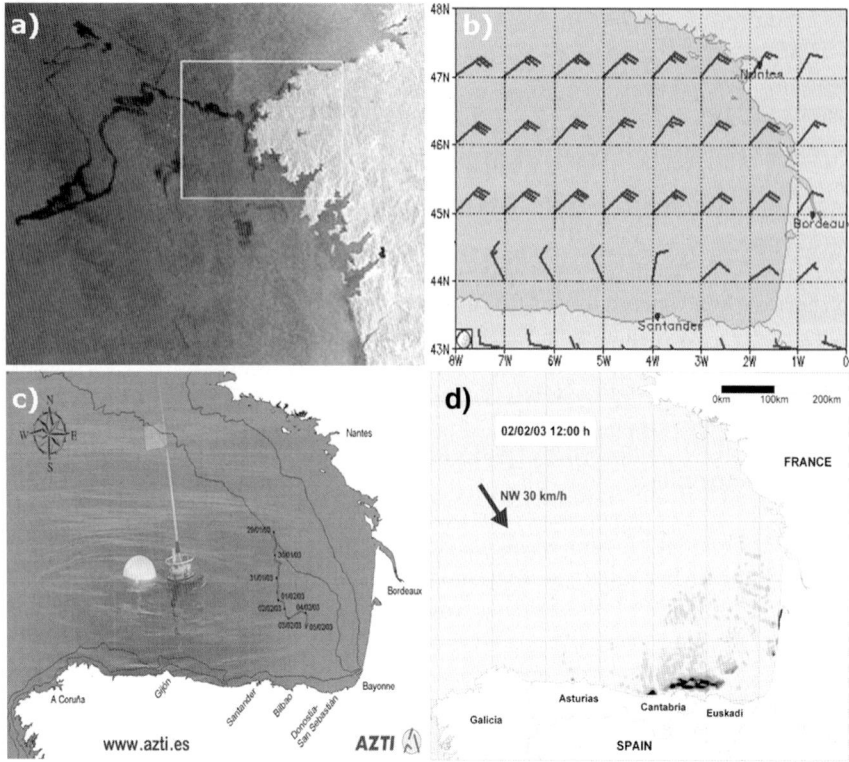

Fig. 1. Operational Oceanography System established during the *Prestige* event to follow the pollutant: (**a**) Satellite observation of the oil spill. (**b**) Wind fields obtained from meteorologic numerical modelling. (**c**) Trajectory of a drifting buoy released at the time of the oil spills. (**d**) Oil location derived by the *TRIMODENA* modelling system in February 2003

During the *Prestige* event, hydrodynamic and particle dispersion models included in the *TRIMODENA* modelling system [Gon01], were used to predict the current fields and the trajectories of the visualised oil slicks which were not recovered or controlled at sea. On the basis of the data sets obtained and the derived predictions, daily reports were provided to the local administrations, for decision-making regarding the deployment of the fishing fleet at sea (190 vessels, ranging in length from 9 to 30 m, and approx. 1,100 fishermen) and experts and volunteers on land. In total 21,000 tonnes of fuel, approximately, were retrieved at sea by the Basque fishing fleet, representing a ratio of 6.6 tonnes recovered at sea per tonne on land. The recovey patterns of the oil were consistent with the numerical model predictions.

3 Fish Recruitment

An important effort has been directed by marine research institutes to analyse the physics of the ecosystem in the Bay of Biscay; this is due to the large economical importance of the fisheries sector in this area. Physical variables as temperature, salinity, currents at the upper layers of the water column, and main river runoff have a fundamental role in the selection of the spawning areas by the different species, in the fish dispersion, retention and growth, especially during the early life stages (eggs and larvae), and in the future recruitment. Numerous studies have been undertaken in the last years, involving physical aspects like: hydrological sampling, current meters and drifter deployments, air and sea surface temperature measurements, and satellite data analysis (e.g. [Alv06], [Gon06], [KL96], [PL92]).

Nevertheless, the complexity of the system showed by the drifting buoys at sea surface during the *Prestige* event, was such, that numerous questions are still open. The buoy data provided evidence that wind was the most important mechanism affecting the surface water movement [Gon06]. Nowadays, high order and resolution hydrodynamic models, as *ROMS* model (Regional Ocean Modelling System, [SM05]), are the direction chosen to explore the physical processes affecting fish recruitment. The results of *ROMS* simulations (Fig. 2) confirm the variability and weakness of the general oceanic circulation in the Bay of Biscay, and the frequent presence of eddies [KL96]. These physical aspects, combined with the biological ones, determine the egg distribution of anchovy in the months of May and June in the Bay of Biscay.

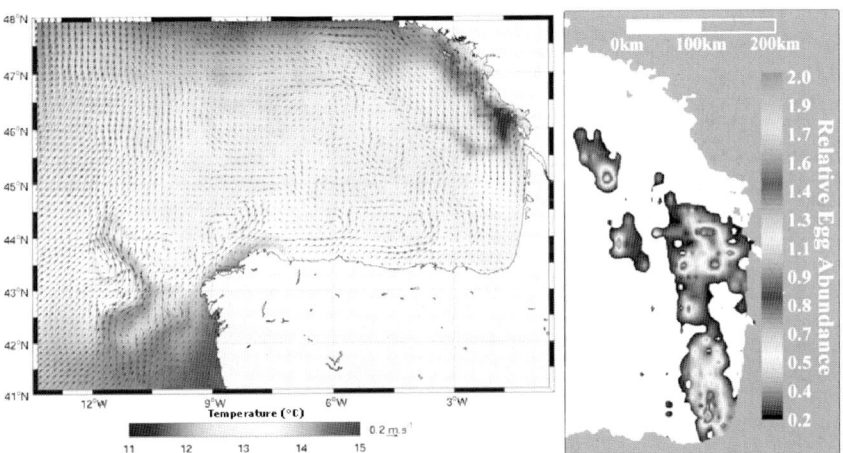

Fig. 2. Current and temperature fields derived by the *ROMS* model for 15th April 2003 (*left*), and spatial relative egg abundance of anchovy (*Engraulis encrasicolus*) in May–June obtained from field data between the years 1999 and 2002 (*right*)

4 Coastal Regions

The local authorities have a series of responsibilities and functions on their coastal regions which are defined by the Law of the different countries. Amongst their main responsibilities is the control of the environment pollution caused by the high anthropogenic pressure, especially that caused by economic and commercial activities, which can modify the water quality and its natural characteristics. Generally, these water changes affect the adjacent areas which use to have public or fishing interests. These activities of negative environmental impact use to occur inside harbours or estuarine zones. The control on these areas can be done through an exhaustive data recovery (water quality, hydrodynamic, sediment analysis, etc.) and numerical modelling.

Another kind of impacts are those generated by extreme natural events, as can be high waves in heavy sea and strong wind conditions. In these cases, the simulation of the wave propagation, as well as the induced currents and sediment transport can be done with softwares as *SMC* (Coastal Modelling System). This was developed by the Coastal Engineering and Oceanography Group (University of Cantabria) with the Coastal and Environmental Department of the Spanish Government. The model uses the parabolic solution of the mild-slope equations for the wave propagation [KD83]. On Fig. 3, it can be seen a wave propagation image inside La Concha bay estimated by the *SCM* model, with the results of the wave impacts on Ondarreta beach [Lir06].

Fig. 3. Extreme wave conditions on the coast of San Sebastin (Spain): (**a**) High wave impact at the promenade around Urgull mount. (**b**) Remains of artificial structures at Ondarreta beach. (**c**) Gravel and rocks at Ondarreta beach during the low tide. (**d**) Monochromatic wave fronts inside La Concha bay estimated by *SMC* model, with the Urgull mount on the right side and Ondarreta beach on the left side

5 Conclusions

Operational Oceanography can be defined as the activity of systematic and long-term routine measurements of the seas, oceans, and atmosphere, and their rapid interpretation and dissemination. Following this definition, the new trends in oceanographic modelling are the development and use of high order numerical models in order to generate data products, including warnings (coastal floods, ice and storm damage, harmful algal blooms and contaminants), hydrodynamics, climate variability, living resources, etc. These models, together with the increment in computational capacity of the new poweful computers and an exhaustive data acquisition, must be prepared to work in real-time in an operational way.

References

[Alv06] Álvarez-Salgado, X.A., Herrera, J.L., Gago, J., Otero, P., Soriano, J.A., Pola, C.G., Garca-Soto, C.: Influence of the oceanographic conditions during spring 2003 on the transport of the Prestige tanker fuel oil to the Galician coast. Mar. Pollut. Bull., 53 (5-7), 239–249 (2006)

[Gon06] González, M., Uriarte, Ad., Pozo, R., Collins, M.: The Prestige crisis: operational oceanography applied to oil recovery by the Basque fishing fleet. Mar. Pollut. Bull., 53, 369–374 (2006)

[Gon01] González, M., Espino, M., Comerma, E., Gyssels, P., Hernáez, M., Uriarte, Ad., García, M.A.: A numerical tool for hydrocarbon pollution forecasting in the autonomous port of Bilbao. Oil & Hydrocarbon Spills II, WIT Press, C.A. Brebbia & G.R. Rodrguez (Eds.), 95–104 (2001)

[KD83] Kirby, J.T., Darlrymple, R.A.: A parabolic equation for the combined refraction-diffraction of Stokes waves by mildy varying topography. J. Fluid Mech., 136, 543–566 (1983)

[KL96] Koutsikopoulos, C., Le Cann, B.: Physical processes and hydrologycal structures related to the Bay of Biscay anchovy. Sci. Mar., 60 (Supl. 2), 9–19 (1996)

[Lir06] Liria, P., Gyssels, P., Galparsoro, I., Santiago, Z., González, M., Uriarte, Ad.: Morphodynamic study of the Ondarreta Beach in San Sebastián (Spain). Environmental Problems in Coastal Regions VI, WIT Transactions on Ecology and the Environment, 88, 183–192 (2006)

[PL92] Pingree, R.D., Le Cann, B.: Three anticyclonic Slope Water Oceanic eDDIES (SWODDIES) in the southern Bay of Biscay in 1990. Deep Sea Res., 39, 1147–1175 (1992)

[SM05] Shchepetkin, A.F., McWilliams, J.C.: The regional oceanic modeling system (ROMS): a split-explicit, free-surface, topography-following-coordinate oceanic model. Ocean Model., 9, 347–404 (2005)

Study and Development of Numerical Models for the Simulation of Geophysical Flows: The DamFlow Project

M.J. Castro, A.M. Ferreiro, J.A. García, J.M. González, and C. Parés

Universidad de Málaga
castro@anamat.cie.uma.es

Summary. The goal of this chapter is to describe briefly the DamFlow project whose goal is the efficient implementation of a numerical parallel solver to simulate geophysical flows. This chapter focuses on the numerical solution of shallow water systems by means of finite volume methods. A technique to develop a high level C++ small matrix library that takes advantage of SIMD registers of modern processors is introduced. A visualization toolkit specifically designed for the pre and post-process of the simulated problems is also presented. Finally, some numerical results are shown.

1 Equations and Numerical Scheme

We consider the shallow water system:

$$\frac{\partial W}{\partial t} + \frac{\partial F_1}{\partial x_1}(W) + \frac{\partial F_2}{\partial x_2}(W) = S_1(W)\frac{\partial H}{\partial x_1} + S_2(W)\frac{\partial H}{\partial x_2} + S_{\mathrm{f}}(W), \quad (1)$$

where

$$W = \begin{bmatrix} h \\ q_x \\ q_y \end{bmatrix}, \quad (2)$$

$$F_1(W) = \begin{bmatrix} q_x \\ \dfrac{q_x^2}{h} + \dfrac{1}{2}gh^2 \\ \dfrac{q_x q_y}{h} \end{bmatrix}, \quad F_2(W) = \begin{bmatrix} q_y \\ \dfrac{q_x q_y}{h} \\ \dfrac{q_y^2}{h} + \dfrac{1}{2}gh^2 \end{bmatrix},$$

$$S_1(W) = \begin{bmatrix} 0 \\ gh \\ 0 \end{bmatrix}, \quad S_2(W) = \begin{bmatrix} 0 \\ 0 \\ gh \end{bmatrix}.$$

Here, $D \subset \mathbb{R}^2$ represents the plane projection of the volume occupied by the flow; $H(\mathbf{x})$ is the depth function measured from a fixed level of reference; g is the gravity; $h(\mathbf{x}, t)$ and $\mathbf{q}(\mathbf{x}, t) = (q_x(\mathbf{x}, t), q_y(\mathbf{x}, t))$ are, respectively, the thickness and the mass-flow of the water layer at the point $\mathbf{x}$ at time t, and they are related to velocity $\mathbf{u}(\mathbf{x}, t) = (u_x(\mathbf{x}, t), u_y(\mathbf{x}, t))$ through the equation:

$$\mathbf{q}(\mathbf{x}, t) = \mathbf{u}(\mathbf{x}, t) h(\mathbf{x}, t).$$

The term $S_{\mathrm{f}}(W)$ parameterize the friction, wind and Coriolis effects.

The discretization of (1) is performed by means of a finite volume scheme. The computational domain is divided into discretization cells or finite volumes. At the edges of each cell, a projected one-dimensional Riemann Problem in the normal direction to the edge is considered, whose solution is approached by a Roe-type first order well-balance numerical scheme (see [1], [2] for details).

2 A Brief Description of the Implementation

This scheme is parallelized in a PC cluster by using MPI. A domain decomposition technique is used to break the domain into pieces that are sent to each node of the cluster. For each piece of the domain, the computations are performed in its corresponding node of the cluster, and a communication procedure among the nodes is performed at the end of each computational time step. In order to reduce the data transfer, a specific buffer structure has been developed. The implementation has been carried out with MPI and C++.

Even if very good speed-up results are obtained, the nature of our targeted problems demands more computing power: for example, 2D tidal simulations of months or years in the Strait of Gibraltar can lead to days or even weeks of CPU time in a PC cluster. Most of this CPU time is spent on performing a huge number of small matrix computations, similar to those carried out in multimedia software and hardware. Modern CPU's (for example Pentium IV processors) are provided with specific SIMD units devoted to these purposes. We introduce a technique to develop a high level C++ small matrix library that takes advantage of SIMD registers, hiding the difficulties related to the use of very low level coding (mostly assembler) needed to develop an efficient SIMD implementation. This has been implemented in a 8 dual Intel Xeon EM64T cluster. Each Xeon EM64T processor has 16 SSE2 registers that provide to the processors with an SIMD parallel architecture. To develop a high level C++ library without loosing the efficiency of a low level SSE implementation, several high level techniques have been used: templates, operator overloading, function inlining, etc. (see [4] for details).

3 Numerical Performance of the Matrix Library

We consider a shallow layer of water flowing in a rectangular channel of 1 m width and 10 m long with a bump placed at the middle of the domain given by the depth function $H(x_1, x_2) = 1 - 0.2\, e^{-(x_1 - 5)^2}$. Three meshes of the domain are constructed with 2,590, 5,162 and 10,832 volumes, respectively. The initial condition is $\mathbf{q}(x_1, x_2) = \mathbf{0}$, and:

$$
h(x_1, x_2) = \begin{cases} H(x_1, x_2) + 0.7 & \text{if } 4 \le x_1 \le 6, \\ H(x_1, x_2) + 0.5 & \text{other case.} \end{cases} \tag{3}
$$

The numerical scheme is run in the time interval $[0, 10]$ with CFL = 0.9. Wall boundary conditions $\mathbf{q} \cdot \eta = 0$ are considered. Table 1 shows the CPU time for each run. As it can be seen in Fig. 1, the speed-up of the parallelization using SSE noticeably diminishes for meshes 1 and 3 in the one layer case, with respect to the case in which they are not used. The reason of this phenomena is that, due to the great efficiency of the SSE parallelization, the calculus time for each iteration in each node is very small compared to the time spent in communications. The efficiency of mixing both kinds of parallelism increases with the mesh size. To show this behaviour, we consider a fourth mesh much finer than mesh number 3 (mesh 4, with 244.163 volumes) to compute again test 1 and compare the speed-up (see Table 2).

Table 1. Speed-up: meshes 3 and 4

CPUs	Mesh 1		Mesh 2		Mesh 3	
	SSE	NON-SSE	SSE	NON-SSE	SSE	NON-SSE
1	0 m 18.507 s	4 m 52.201 s	0 m 51.764 s	14 m 16.735 s	3 m 5.985 s	50 m 21.319 s
2	0 m 10.685 s	2 m 32.606 s	0 m 29.066 s	7 m 6.800 s	1 m 38.830 s	25 m 25.037 s
4	0 m 6.876 s	1 m 17.556 s	0 m 17.078 s	3 m 38.655 s	0 m 53.459 s	12 m 43.717 s
8	0 m 4.340 s	0 m 40.120 s	0 m 10.032 s	1 m 51.360 s	0 m 29.315 s	6 m 26.135 s

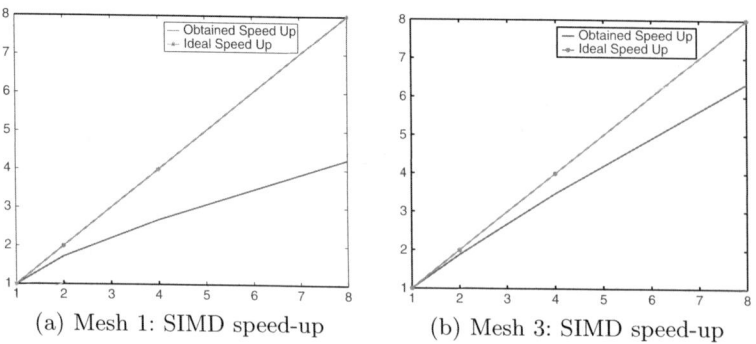

(a) Mesh 1: SIMD speed-up   (b) Mesh 3: SIMD speed-up

Fig. 1. Speed-up for meshes 1 and 3: one layer model

Table 2. Speed-up: meshes 3 and 4

N. CPUs.	1	2	4	8
Time for mesh 4	25 m 26.436 s	12 m 53.427 s	6 m 34.203 s	3 m 24.476 s
Speed-up for mesh 3	1	1.8818	3.4790	6.3443
Speed-up for mesh 4	1	1.9736	3.8722	7.4651

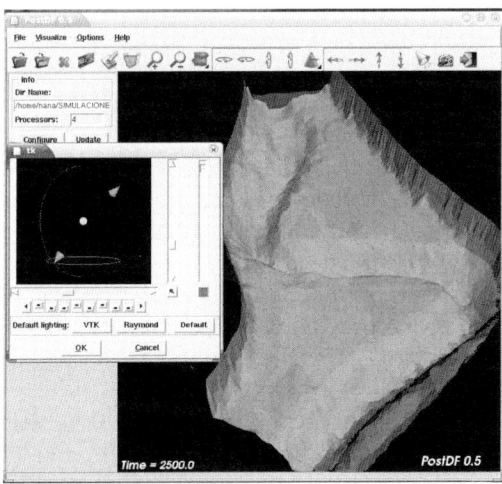

Fig. 2. Screenshot of the post-processing tool

4 A Brief Description of the Visualization Toolkit

A specific post-processing tool has been developed in order to analyze the results of our models. The main characteristics of this tool are (more details in [3]):

- Ability to visualize results in non-structured finite volume meshes.
- Flexibility for the treatment of very large computational files (several Gigabytes files).
- Visualization of each computed physical magnitude (velocities, fluxes, sediments...).
- Ability to join in real time results coming from different processors of the cluster in order to visualize the complete domain.
- Ability to visualize results from a remote machine avoiding the transfer of data to the visualization computer.

This post-processing tool has been developed using Python as basis language. The 2D and 3D visualization kernels are based in MatplotLib and VTK, respectively. Finally some functionalities are implemented in C++ in order to get better performance. A screenshot is shown in Fig. 2.

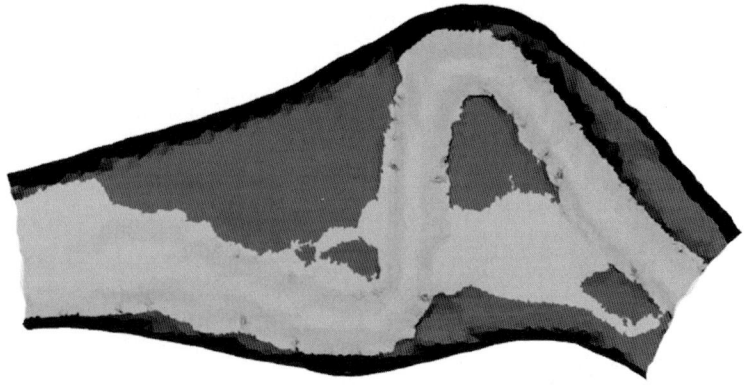

Fig. 3. Mero river flood simulation

5 Numerical Examples

The model has been applied to the simulation of the Mero river floods
(A Coruña – Spain), see Figure 3. We have validated the model by comparing our computational results with experimental data provided from a model
of the Mero river made to scale by CITEEC in the University of A Coruña
(Spain). We have been provided with experimental data of velocities, flows,
water depth, etc. in different sections of the computational domain.

Experimental and computational data has been compared obtaining a very
good agreement (see [4] for more details).

References

[1] M. J. Castro, C. Parés. "On the Well-Balance Property of Roe's Method for
     Nonconservative Hyperbolic Systems. Applications to Shallow-Water Systems".
     ESAIM-Math. Model. Num., 38(5): 821-852, 2004.
[2] M. J. Castro, J. A. García, J. M. González, C. Parés. "A parallel 2D finite
     volume scheme for solving systems of balance laws with nonconservative prod-
     ucts: application to shallow flows". Computer Methods in Applied Mechanics
     and Engineering,196: 2788-2815,2006.
[3] Ana M. Ferreiro. "Desarrollo de técnicas de post-proceso de flujos
     hidrodinámicos, modelización de problemas de transporte de sedimentos y sim-
     ulación numérica mediante técnicas de volúmenes finitos". PhD Thesis, Univer-
     sidad de Sevilla, March 2006.
[4] José A. García Rodríguez. "Paralelización de esquemas de volúmenes finitos:
     aplicación a la resolución de sistemas de tipo aguas someras". PhD Thesis,
     Universidad de Málaga, June 2005.

On Variational Data Assimilation for 1D and 2D Fluvial Hydraulics

I. Gejadze[2], M. Honnorat[1], F.X. Le Dimet[1], and J. Monnier[1]

[1] LJK - MOISE project-team, Grenoble, France
   Jerome.Monnier@imag.fr
[2] Civil Engineering Dept, Univ. Strathclyde, UK

1 Introduction

We address two problems related to variational data assimilation (VDA) [1] as applied to river hydraulics (1D and 2D shallow water models). In real cases, available observations are very sparse (especially during flood events). Generally, they are very few measures of elevation at gauging stations. The first goal of the present study is to estimate accurately some parameters such as the inflow discharge, manning coefficients, the topography and/or the initial state. Since the elevations measures (eulerian observations) are very sparse, we develop a method which allow to assimilate extra lagrangian data (trajectory particles at the surface, e.g. extracted from video images). The second goal aims to develop a joint data assimilation - coupling method. We seek to couple accurately a 1D global net-model (rivers net) and a local 2D shallow water model (zoom into a flooded area), while we assimilate data. This "weak" coupling procedure is based on the optimal control process used for the VDA. Numerical twin experiments demonstrate that the present two methods makes it possible to improve on one hand the identification of river model parameters (e.g. topography and inflow discharge), on the other hand an accurate 1D–2D coupling combined with the identification of inflow boundary conditions.

1.1 The 2D Forward Model

The 2D forward model considered rely on the shallow water equations (SWE) (h is the water elevation, $\mathbf{q} = h\mathbf{u}$ the discharge, $\mathbf{u}$ the depth-averaged velocity):

$$
\begin{cases}
\partial_t h + \operatorname{div}(\mathbf{q}) = 0 & \text{in } \Omega \times]0, T] \\
\partial_t \mathbf{q} + \operatorname{div}(\frac{1}{h}\mathbf{q} \otimes \mathbf{q}) + \frac{1}{2}g\nabla h^2 + gh\nabla z_b + g\frac{n^2\|\mathbf{q}\|}{h^{7/3}}\mathbf{q} = 0 & \text{in } \Omega \times]0, T]
\end{cases}
\tag{1}
$$

with initial conditions $(h_0, \mathbf{q}_0)$ given, g the magnitude of the gravity, z_b the bed elevation, n the Manning roughness coefficient. Boundary conditions are:

at inflow, the discharge $\bar{q}$ is prescribed; at outflow, either the water elevation $\bar{z}_s$ is prescribed or incoming characteristics are prescribed; and walls conditions. Given the control vector $\mathbf{c} = (h_0, \mathbf{q}_0, n, z_b, \bar{q}, \bar{z}_s)$, the state variable $(h, \mathbf{q})$ is determined by solving the forward model.

2 Assimilation of Lagrangian Data

Lagrangian DA consists in using observations described by lagrangian coordinates in the DA process. Here, we consider observations of particles transported by the flow (e.g. extracted from video images). The link between the lagrangian data made of N particle trajectories denoted by $X_i(t)$ and the classical eulerian variables of the shallow water model is made by the following equations, see [4]:

$$\begin{cases} \frac{d}{dt} X_i(t) = \gamma \mathbf{u}\left(X_i(t), t\right) & \forall t \in]t_i^0, t_i^f[\\ X_i(t_i^0) = x_i^0 , \end{cases} \quad \text{for } i = 1, \ldots, N \quad (2)$$

where t_i^0 and t_i^f are the time when the particle enter and leave the observation domain, γ is a multiplicative constant. We consider two kinds of observations (classical eulerian observations $h^{\text{obs}}(t)$ and trajectories of particles transported by the flow $X_i^{\text{obs}}(t)$). Then, we build the following composite cost function:

$$j(\mathbf{c}) = \frac{1}{2} \int_0^T \left\| Ch(t) - h^{\text{obs}}(t) \right\|^2 dt + \frac{\alpha_t}{2} \sum_{i=1}^N \int_{t_i^0}^{t_i^f} \left| X_i(t) - X_i^{\text{obs}}(t) \right|^2 dt \quad (3)$$

where α_t is a scaling parameter, C the observation operator.

2.1 Numerical Results

Particle trajectories associated with local water depth measurements are used for the joint identification of local bed elevation z_b and initial conditions $(h^0, \mathbf{u}^0)$. A constant discharge $\bar{q}$ is prescribed at inflow, Fig. 1a. A vertical cut of the fluid domain in the longitudinal plane in Fig. 1b shows the bed and the free surface elevation for this configuration.

Twin DA experiments are carried out: observations are created by the model from the reference steady flow described above. Water depth is recorded continuously in time at the abscissae $x_1 = 15$ m and $x_2 = 70$ m, for the whole width of the domain. These measurements are used as observations denoted by $h_i^{\text{obs}}(y; t)$ for $i = 1, 2$. With regards to the creation of trajectories observations, virtual particles are dropped in the reference steady flow and transported by a turbulent surface velocity $\mathbf{u}^t = \gamma \mathbf{u} + \mathbf{u}^p$, where $\gamma = 1$ and $\mathbf{u}^p$ is a Gauss–Markov process. A total of $N_{\text{obs}} = 640$ particles is released in the flow.

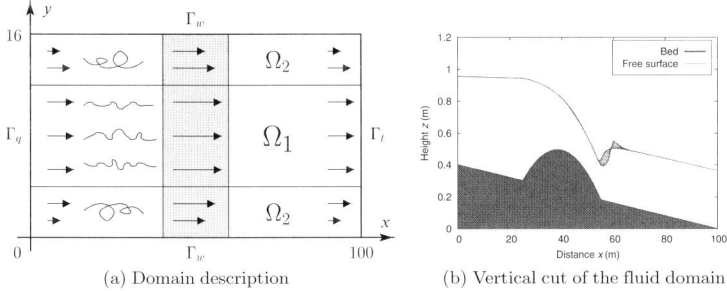

(a) Domain description   (b) Vertical cut of the fluid domain

Fig. 1. Flow configuration

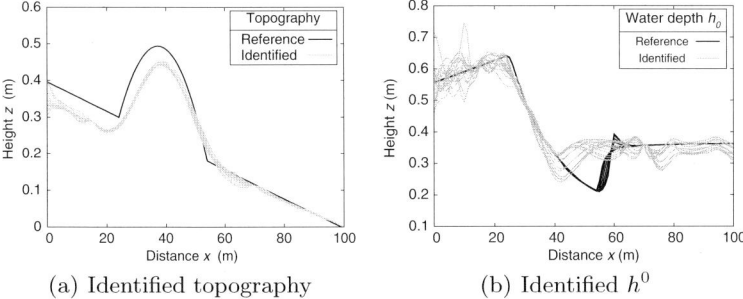

(a) Identified topography   (b) Identified h^0

Fig. 2. Joint identification of the topography and the initial conditions using water depth measurements and particle trajectories with $\alpha_t = 1 \times 10^{-4}$

We seek to identify jointly the reference topography and the reference initial conditions (water depth h^0 and velocity $\mathbf{u}^0$) used to create the observations, from the a priori hypothesis that the bed is made of a longitudinal slope of without bump and the initial conditions correspond to the steady state obtained with the modified topography. To that purpose, we carry out DA using cost function (3). As shown in Fig. 2 a, the identified topography is close to the reference, with a good recovery of the bump. As for the initial conditions, we can see in Fig. 2 c,d that it reproduces the same main features as the reference.

3 A Joint Assimilation-Coupling Procedure

Operational models used in hydrology are generally net-models based on the 1D Saint–Venant equations with storage areas. Here we shortly describe a method which superpose locally the previous 2D SWE along the 1D channel, see [3]. The first issue is to nest properly the local 2D model into the 1D global

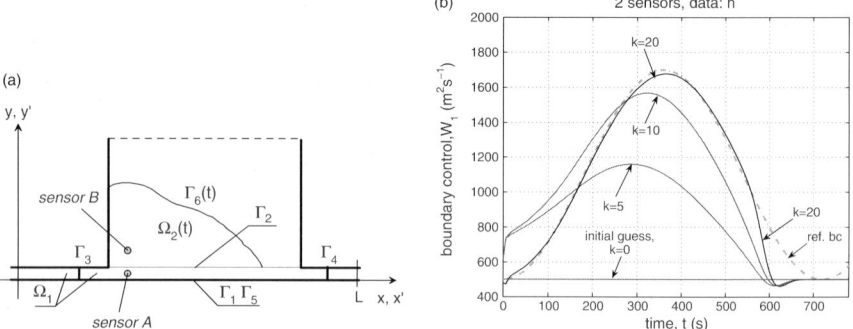

Fig. 3. (a) Problem layout (b) Assimilation of data (h) by the JAC algorithm: $W_1(t)$ after k iterations

model. To this end, we specify incoming characteristics at lateral boundary conditions (BC), Fig. 3:

$$(x,y) \in \Gamma_3 : \quad \begin{aligned} q + (c - u)h &= w_1(x,y,t) \\ p - vh &= w_3(x,y,t), \ \forall u > 0, \end{aligned} \qquad (4)$$

$$(x,y) \in \Gamma_4 : \quad \begin{aligned} q - (c + u)h &= w_2(x,y,t) \\ p - vh &= w_3(x,y,t), \ \forall u < 0, \end{aligned} \qquad (5)$$

where the coefficients in (4) and (5) are: $c = (gh|_{t-\tau})^{1/2}$, $u = (q/h)|_{t-\tau}$, $v = (p/h)|_{t-\tau}$ and τ is a time shift, which is taken equal to the time integration step in the numerical implementation; w_1, w_2 and w_3 (depending on the sign of u) are the incoming characteristic variables that must be specified based on their counterparts W_1, W_2 from the global model defined as follows:

$$Q + (c - u)H = W_1(x,t), \quad Q - (c + u)H = W_2(x,t), \qquad (6)$$

where H, Q are variables of the "dimensional" 1D SWE problem (i.e. scaled by the main channel width assuming the rectangular cross-section) and $c = (gH|_{t-\tilde{\tau}}/b)^{1/2}$, $u = (Q/H)|_{t-\tilde{\tau}}$.

The feedback from local to global model is achieved by computing a generalized defect correction term, which will be a source term to the global model equations, see [3] for more details. Another issue is that we couple two different models (1D and 2D). The problem is formulated as a DA problem, while the local model boundary conditions are considered as unknown controls. The coupling conditions in this formulation become penalty terms of the extended objective function $J = (\gamma J^* + J_1 + J_2)$ with $J^* = \sum_i \int_0^T (U_i - \hat{U}_i)^2 \, dt$ and $J_k = \int_0^T (\int \Gamma_{k+1} w_k d\Gamma - W_k|_{\Gamma_{k+1}})^2$, while as constraints we consider the one-way relaxed model described by the following steps: (a) given current approximation w_k solve the 2D SWE local problem; (b) compute "defect correction"; (c) given current approximation (or known values) $W_1(0), W_2(L)$ and the "defect correction" computed at previous stage solve the 1D SWE problem; (d) compute extended objective function J. We refer to this method as to a "joint assimilation-coupling method" (JAC).

In the numerical examples [2] given below we solve DA problem for the 1D section (main channel) looking for the unknown inflow BC (characteristic $W_1(t, 0)$), while data is measured in the area covered by the 2D local model and is assimilated into this model correspondingly. The control problem for J is solved using the adjoint [4, 5] of the one-way relaxed model described by steps (a)–(c). Data is collected in two points as shown in Fig. 3a. The reference value is chosen to cause a "flooding event", i.e. massive overflowing of the main channel in the area where the 2D local model is superposed. (Rem. Under these conditions assimilation of measurements from sensor A into the 1D model alone may fail to produce meaningful results because this model is not adequate. Data from sensor B cannot be assimilated in principle).

In the following assimilation examples, Fig. 3b, we can see the reference BC (in dashed line) and the retrieved value after k iterations of the JAC algorithm (in sharp solid lines). A line that corresponds to $k = 0$ is the initial guess. This example shows that the JAC method converges and allows retrieving the unknown BC of the 1D model, while data is assimilated into the weakly connected local 2D model.

References

1. F.-X. Le Dimet and O. Talagrand. Variational algorithms for analysis and assimilation of meteorogical observations. *Tellus*, 38A:97–110, 1986.
2. I. Gejadze, M. Honnorat, X. Lai, J. Marin, and J. Monnier. DASSFLOW v2.0 : a variational data assimilation software for 2d shallow water river flows. To appear, INRIA reasearch report, 2007.
3. I. Gejadze and J. Monnier. On a 2d zoom for 1d shallow-water model: coupling and data assimilation. *Submitted*, 2007.
4. M. Honnorat, J. Monnier, and FX Le Dimet. Lagrangian data assimilation for river hydraulics simulations. *Submitted*, 2007.
5. J.-L. Lions. *Optimal Control of Systems Governed by Partial Differential Equations*. Springer-Verlag, 1971.

Minisymposium "Multiscale Problems in Materials"

A. Carpio

Universidad Complutense de Madrid, Spain

Over the past years, Mathematical Materials Science has become an important discipline, with a significant impact both on mathematics and materials science. Modeling many properties of materials involves describing phenomena taking place in a wide range of scales, from the nanoscale up to the macroscale. One of the big challenges nowadays is to develop the tools to handle multiscale problems and understand how dynamics on one scale affect the others. Experimental, computational and theoretical advances provide a better understanding of materials, making it easier to design new materials with desired properties.

This minisimposium is focused on mechanical properties, defects and growth in Materials Science. It contains selected contributions containing modeling, analysis, computation and experimental results for some cutting edge problems.

Y. Farjoun addresses growth phenomena in a model of aggregation which combines surface activated and diffusion limited growth. A continuum approximation produces an asymptotic solution showing three stages: nucleation, growth and coarsening. Late coarsening in this model selects the discontinuous similarity solution of the Lifshitz–Slyozov equation.

O. Rodríguez de la Fuente presents some experimental and computational results on nanoindentation tests in gold crystals. Atomistic simulations reproduce the experimentally observed plastic mechanisms of dislocation nucleation and motion. The strengths of surfaces are compared.

I. Plans discusses formation of misfist dislocations in heteroepitaxial growth using a simple nearest neighbour model to predict the number of film layers needed for an edge misfit dislocation network to appear in InAs/GaAs heteroepitaxy. Numerical results agree reasonably well with experiments.

P. Ariza presents a study of dislocation newtworks in BCC crystals using "discrete crystal elasticity" models derived through discrete differential calculus and algebraic geometry applied to crystal lattices. A method to compute dislocated structures solving integer optimization problems is discussed.

J. Ockendon describes an asymptotic method for the analysis of dislocation pile-ups. The dislocation density for particular geometries is easily found in the continuum limit. This method allows to recover the dislocation positions knowing the dislocation density.

An Asymptotic Solution of Aggregation Dynamics

Yossi Farjoun[1] and John Neu[2]

[1] MIT, Cambridge, Massachusetts, USA
  yfarjoun@mit.edu

[2] UC Berkeley, Berkeley, California, USA

Summary. We present a model of aggregation, whereby clusters are created according to the Zeldovich nucleation rate and subsequently undergo diffusion limited growth as in the classic Lifshitz–Slyozov (LS) model. The mathematical formulation of this model as an advection PDE signaling problem is singular in the small super-saturation limit. Using singular perturbation methods, we find three successive eras: Nucleation, growth and coarsening. The long-term limit of the coarsening era solution converges to the discontinuous similarity solution of the LS PDE.

1 Introduction

Aggregation of identical particles (monomers) into large clusters is a universal phenomenon throughout physics, chemistry, and biology. Two classical models have been proposed and they are well suited for their respective domains. The Becker–Döring (BD) model [1] of surface activated reactions compares the rate at which particles leave the surface of a cluster with the rate at which particles surrounding the cluster join it. The BD model is used primarily for deriving the Zeldovich rate [11] of *nucleation*, which describes the rate at which clusters overcome a free energy barrier. Lifshitz and Slyozov (LS) proposed a model [5] in which the growth of the clusters is controlled by the diffusive flux of monomers. The LS model is used to describe the *growth* of clusters after they are nucleated.

After setting up the aggregation signaling problem using a continuum approximation, we present its asymptotic solution. It exhibits three successive and distinct eras: Nucleation, growth, and coarsening, with increasingly larger time-scales. During the nucleation era, the clusters are created and they start growing. No more clusters are being formed during the growth era, but the existing ones grow and deplete the monomer around them. In the coarsening era, the monomer density is so depleted, that the growth of the large clusters is only possible by the evaporation of the smaller ones. The $t \to \infty$ limit of the coarsening era solution converges to the discontinuous similarity solution of the LS PDE.

The current chapter only skims through a partial set of the results and refers the interested reader to a future article by the authors for the full derivation and more results.

2 Aggregation Model

The standard derivation of the nucleation and growth models suffices for our purposes. Thus, we cite the results needed for the description of the model, and forgo any explanation or derivation. For a complete description of the aggregation model see, for example [2, 7–10].

2.1 Super-Saturation and the Nucleation Rate

The cluster size distribution has an equilibrium solution when the density of monomers is smaller than the *saturation density*, f_s. When f_1, the monomer density is greater than f_s an equilibrium solution does not exist. In this case the creation and growth rates of the clusters are controlled by the *super-saturation*, η,

$$\eta \equiv \frac{f_1 - f_s}{f_s},\tag{1}$$

The small cluster-sizes have a quasi-equilibrium distribution, however, due to thermal fluctuations, a small fraction of clusters grow beyond a free energy barrier, i.e. *nucleate*. Zeldovich showed that the rate of nucleation per unit volume is

$$j = \Omega e^{-\frac{\sigma^3}{2\eta^2}}, \quad \Omega \equiv w f_s \sqrt{\frac{\sigma}{6\pi}},\tag{2}$$

for $0 < \eta \ll 1$. Here, w is the rate constant for any particular "surface" particle to evaporate from the cluster, and σ is the energy per unit area of the surface of a cluster (in units of $k_B T$). It is assumed that w and σ are asymptotically constant for large clusters.

2.2 The Growth Rate of Clusters

Large clusters, grow according to the Lifshitz–Slyozov growth rate

$$\dot{n} = d\left(\eta n^{1/3} - \sigma\right), \quad d = (3 \cdot 16\pi^2)^{1/3}\frac{D f_s}{v^{2/3}},\tag{3}$$

where D is the diffusion constant of monomers outside the clusters and v is the volume per particle in the bulk of the cluster.

3 The Signaling Problem and the Three Eras

To follow the changing clusters sizes, a continuous function $r(n, t)$ is introduced. It specifies the density of clusters of size n at time t. The aggregation dynamics are then rewritten as a PDE for $r(n,t)$,

$$\partial_t r + \partial_n(ur) = 0 \text{ in } n > 0, \quad u = \mathrm{d}(n^{\frac{1}{3}}\eta - \sigma), \tag{4}$$

$$\eta = \frac{f - f_s}{f_s} - \frac{1}{f_s} \int_{n_*}^{\infty} n\, r(n, t)\, \mathrm{d}n, \tag{5}$$

$$r(n, 0) \equiv 0. \tag{6}$$

The super-saturation is determined via (5) which states that particles are either monomers, or part of large clusters, this is an approximation we use here. PDE (4) is the translation of the growth rate of an individual cluster to the advection of the cluster-size density, r. Equation (6) states the initial condition: no clusters and monomer of density f.

Throughout the analysis we use the initial value of η, $\frac{f-f_s}{f_s}$ as a gauge parameter, ε. By analyzing the small ε limit we find that the signaling problem (4)–(6) has three distinct "eras", or scalings. In the following subsections we present the three eras, their reduced equations, and solutions.

3.1 Nucleation

The characteristic cluster size n in the nucleation era has $n \gg n_* = \left(\frac{\sigma}{\eta}\right)^3$ (as will be shown shortly). Thus, (3) reduces to

$$\dot{n} = \mathrm{d}\eta n^{\frac{1}{3}},$$

and the size distribution $r(n, t)$ satisfies advection PDE

$$\partial_t r + \partial_n(\mathrm{d}\eta n^{\frac{1}{3}} r) = 0 \tag{7}$$

in $n > 0$. *Outgoing characteristics* from $n = 0$ require boundary conditions, corresponding to the rate nucleation rate of clusters. For this we use the Zeldovich rate:

$$\mathrm{d}\eta n^{1/3} r \to j \equiv \Omega f_s e^{-\sigma^3/2\eta^3}, \text{ as } n \to 0. \tag{8}$$

Since the nucleation rate (8) depends strongly on the super-saturation, the *change* in super-saturation, $\Delta\eta \equiv \varepsilon - \eta$, is introduced. The nucleation era lasts until the rate of nucleation diminishes to a small fraction of the original rate. The dominant balance relations in (5), (4, and (8) for the nucleation era results in the scales in the following table.

Variable	η	n	t	r	$\Delta\eta$
Unit	ε	$\left(\dfrac{\mathrm{d}f_s\varepsilon^4}{\Omega\sigma^3}e^{\frac{\sigma^3}{2\varepsilon^2}}\right)^{3/5}$	$\left(\dfrac{\varepsilon}{\mathrm{d}\sigma^2}\right)^{3/5}\left(\dfrac{\Omega}{f_s}\right)^{-2/5}e^{\sigma^3/5\varepsilon^2}$	$\left(\dfrac{\Omega^2\sigma}{d^2 f_s^{1/3}\varepsilon^3}e^{-\frac{\sigma^3}{\varepsilon^2}}\right)^{3/5}$	$\dfrac{\varepsilon^3}{\sigma^3}$

We note that, as expected, the characteristic cluster size is much larger that n_*.

Using these scales we find the reduced equations for the nucleation era:

$$\partial_t r + \partial_n(n^{\frac{1}{3}}r) = 0, \quad \text{in } n > 0, \tag{9}$$

$$j = n^{\frac{1}{3}}r \rightarrow e^{\Delta\eta}, \quad \text{as } n \rightarrow 0^+, \tag{10}$$

$$\Delta\eta = -\int_0^\infty nr\, dn, \tag{11}$$

$$r(n,0) = 0. \tag{12}$$

By using the characteristics of (9), we obtain an integral equation for $\Delta\eta$:

$$\Delta\eta(t) = -\int_0^t \left(\frac{2}{3}(t-\tau)\right)^{\frac{3}{2}} e^{\Delta\eta(\tau)}\, d\tau, \tag{13}$$

A solution to (13) is found by numerical integration, and $r(n,t)$ is reconstructed using the characteristics. Figure 1 shows a succession of "snap-shots" of the cluster-size density $r(n,t)$ at various values of t. The (scaled) total density of clusters created during the nucleation era emerges from the numerical solution:

$$R \equiv \int_0^\infty r\, dn = \int_0^\infty j(\varphi)\, d\varphi \approx 1.34.$$

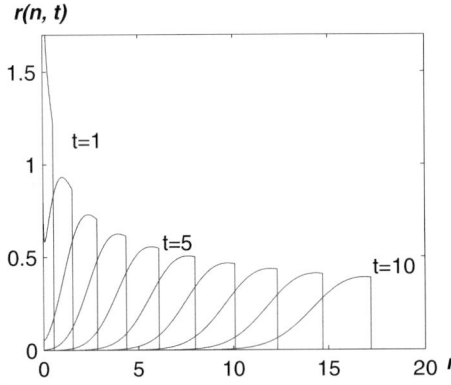

Fig. 1. The cluster-size density, $r(n,t)$, for various values of t

3.2 Growth

At the tail-end of the nucleation era, clusters are being nucleated at a vanishingly small rate. The characteristic cluster size $[n]$ here is still much larger than n_*, and so the simple growth rate, $\dot{n} = d\eta n^{\frac{1}{3}}$, is still a good approximation. The leading order dynamics of the cluster-size distribution is described by its width and the size of the largest clusters, a and N. Using N and a, we define an "inner variable" x, and a corresponding "inner density" $q(x, t)$:

$$x = \frac{n - N(t)}{a(t)}, \qquad q(x, t) = \frac{a(t)}{R}r(N(t) + a(t)x, t). \qquad (14)$$

Choosing N and a so that,

$$\dot{N} = N^{\frac{1}{3}}\eta, \qquad \dot{a} = \frac{1}{3}N^{-\frac{2}{3}}a\eta, \qquad (15)$$

causes the dynamics of $q(x, t)$ to be trivial to first order. That is $\partial_t q \equiv 0$. The remaining problem is an ODE system for N. A dominant balance of the equations finds the growth era scales:

Variable	N	η	t	a
Unit	τ^{-3}	1	τ^{-2}	τ^{-1}

$, \tau = \dfrac{R^{\frac{1}{3}}\varepsilon^{\frac{2}{3}}}{\sigma}$

These scales are relative to the nucleation era. So, for example, the largest cluster, N, is scaled with $\frac{\sigma^3}{R\varepsilon^2}\left(\frac{df_s\varepsilon^4}{\Omega\sigma^3}e^{\frac{\sigma^3}{2\varepsilon^2}}\right)^{3/5}$, and η remains scaled with ε. Using the scales above and taking the $\varepsilon, \tau \to 0$ limit, results in an ODE for N:

$$\dot{N} = N^{\frac{1}{3}}(1 - N), \qquad \partial_t q \equiv 0, \qquad \eta = 1 - N, \qquad a = CN^{\frac{1}{3}}. \qquad (16)$$

The solution to (16) is shown in Fig. 2. The function q and constant C are determined by asymptotic matching with the nucleation era solution:

$$q(x, t) = \frac{C}{R}j(-Cx), \qquad C = \frac{1}{R}\int_0^\infty t\, j(t)\, dt \approx 0.81.$$

Here, $j(t)$ is the nucleation flux found during the nucleation era.

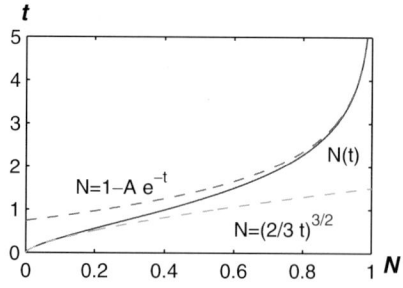

Fig. 2. The size of the largest cluster, $N(t)$, during the growth era. The *two dashed lines* show asymptotic behaviors of $N(t)$. $A \approx 2.1$

3.3 Coarsening

The end of the growth era is marked by a vanishing super-saturation and a stationary distribution, $\dot{N} = 0$. For such a small super-saturation, the approximation made in the two previous eras, that $\sigma \ll \eta n^{\frac{1}{3}}$ is no longer valid. Another way of looking at it is that the critical cluster size, n_*, "catches up" with the distribution during the coarsening era. Thus, we use the *full* advection velocity from (4)

$$u = d(n^{\frac{1}{3}}\eta - \sigma) \tag{17}$$

in the advection PDE. At $n = 0$ the characteristics are incoming, and so no boundary conditions are required (the incoming characteristics correspond to the final breakup of clusters into monomers). Since the initially narrow distribution widens during the coarsening era, we use a new independent variable $y = \frac{n}{N}$, and define $q(y, t)$ appropriately.

The scales of the coarsening era are listed in the following table.

Variable	N	η	t	w	q		
Unit	τ^{-3}	$s\tau$	$\frac{1}{s\tau^3}$	$s\tau^3$	1	$\tau \equiv \frac{R^{\frac{1}{3}}\varepsilon^{\frac{2}{3}}}{\sigma}$,	$s \equiv \left(\frac{d^6 f_s \varepsilon^9}{\Omega \sigma^8} e^{\frac{\sigma^3}{2\varepsilon^2}}\right)^{-\frac{1}{5}}$

These scales are with respect to the *nucleation* scaling, thus for example, the resulting scale of N, the largest cluster size, is the same in the coarsening and growth eras. The reduced equations for the coarsening era are disentangled with some algebra:

$$0 = q_t + (wq)_y, \text{ for } 0 \le y \le 1, \qquad \eta = \frac{M^{1/3} M_0}{M_{1/3}}, \tag{18}$$

$$w = M\left(\frac{M_0}{M_{1/3}}\left(y^{1/3} - y\right) + (y - 1)\right), \qquad N = \frac{1}{M}. \tag{19}$$

The moments M, M_0, and $M_{1/3}$ are defined by

$$M = \int_0^1 y\, q\, dy, \qquad M_0 = \int_0^1 q\, dy, \qquad M_{1/3} = \int_0^1 y^{1/3} q\, dy.$$

We solved PDE (18) and (19) for q numerically using `clawpack` [4]. Effective initial conditions were found by linearizing the advection velocity and following the widening of the narrow distribution over a time period of $-6\log\tau$. Shifting the origin of time by this amount causes the coarsening era solutions for different values of ε to collapse onto a single solution, which is used as the effective initial condition. As $t \to \infty$, the distribution q converges to the discontinuous similarity solution of the PDE. Figure 3 shows snapshots of the solution and the similarity solution. In Fig. 4, the original distribution $r(n, t)$ is plotted instead of $q(y, t)$.

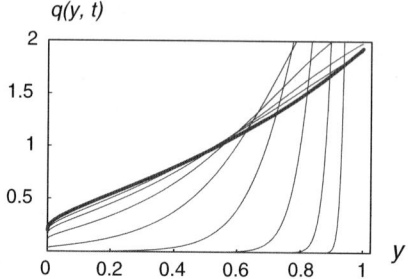

Fig. 3. The (*normalized*) numerical solution $q(y,t)$. On the right at $t = -10$, and indistinguishable from the discontinuous similarity solution (*dark line*) at $t \approx 4$

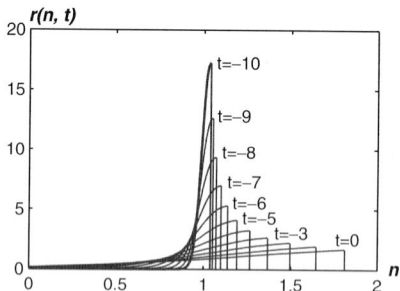

Fig. 4. The numerical solution $r(n,t)$. The *dark curve* is the effective initial condition, at $t = -10$

4 Conclusions

Three eras — nucleation, growth, and coarsening — emerge from the asymptotic analysis of the aggregation model. In the small super-saturation limit they are characterized by increasingly larger times scales. Starting with pure monomer, the distribution relaxes to the discontinuous similarity solution. The solution is accompanied by previously unknown physical scales, and amount of clusters generated, R. We predict that other phenomena, such as the ignition transient [3, 6] will smooth the discontinuity in a larger characteristic time.

References

1. R. Becker and W. Döring, *Kinetishe Behandlung der Keimbildung in übersättigten Dämpfen*, Ann. Phys. **24** (1935), 719.
2. K. F. Kelton, *Crystal nucleation in crystals and glasses*, Solid State Phys. (H. Ehrenreich and D. Turnbull, eds.), vol. 45, Academic, New York, 1991, pp. 75–177.
3. K. F. Kelton, A. L. Greer, and C. V. Thompson, *Transient nucleation in condensed systems*, J. Chem. Phys. **79** (1983), 6261.

4. R. J. LeVeque, *Finite-volume methods for hyperbolic problems*, Cambride University Press, 2002, http://www.amath.washington.edu/~claw/.
5. I. M. Lifshitz and V. V. Slyozov, *The kinetics of precipitation from supersatu- rated solid solutions*, J. Phys. Chem. Solids **19** (1961), 35–50.
6. J. C. Neu, J. A. Cañizo, and L. L. Bonilla, *Three eras of micellization*, Phys. Rev. E **66** (2002), no. 061406, 1–15.
7. B. Niethammer, *On the evolution of large clusters in the Becker–Döring model*, J. Nonlin. Sci. **13** (2003), 115–155.
8. O. Penrose, *The Becker–Döring equations at large times and their connection with the LSW theory of coarsening*, J. Stat. Phys. **89** (1997), 305–320.
9. J. J. L. Velazquez, *The Becker–Döring equations and the Lifshitz–Slyozov theory of coarsening*, J. Stat. Phys. **92** (1998), 198–236.
10. D. T. Wu, *Nucleation theory*, Solid State Physics, vol. 50, Academic Press, San Diego, CA, 1996, pp. 37–187.
11. J. B. Zeldovich, *On the theory of new phase formation; cavitation*, Acta Physiochim, URSS **18** (1943), 1–22.

Atomistic Simulations of the Incipient Plastic Deformation Mechanisms on Metal Surfaces

O. Rodríguez de la Fuente

Departamento de Física de Materiales, Universidad Complutense, Madrid-280400, Spain
oscar.rodriguez@fis.ucm.es

1 Introduction

We know since decades that, to exploit and control mechanical properties of materials, a profound knowledge of the defects generated during deformation is required. The advent of nanotechnologies and the maturity of surface science have catalysed the development of nanoindentation techniques. Nanoindentation experiments, with both high spatial and load resolution, allow to determine defects emerging at the surface, and their relationship with discontinuities in the load vs. penetration curve. But sub-surface defect morphology generally remains hidden, and quite often only indirect conclusions can be inferred. Simulations are then a very valuable tool to unveil defect configurations. Among the present challenges regarding nanoindentation, one is the origin and correct interpretation of the piled-up material, i.e. the material displaced from the indentation point. Another problem is the role of the surface roughness on the mechanical properties at the nanoscale.

2 Methodology

All simulations in the present work are atomistic. The interatomic potential used is the embedded atom method [DB84] for gold, which belongs to a wide and extensively used class of potentials, suitable for metallic systems. Simulation cells, with lateral periodic boundary conditions and (001) or (111) top surface orientations contain up to a few million atoms, depending on the specific simulation. The bottom layer is kept rigid, so that the whole system is not displaced downwards during indentation. Cell lateral dimensions are continuously scaled to keep the diagonal components of the stress tensor equal to zero. The nanoindentor is simulated as a spherical purely repulsive potential, ignoring all possible attractive forces during approach and contact. The indenter radius varies between 30 and 300 nm. In every indentation step the nanoindentor position is lowered in steps of 0.01 Å. The whole system is then fully relaxed in this new configuration with a conjugate gradient minimization algorithm until system energy is minimized. To visualize the simulated system the software AtomEye [LJ03] is used.

3 Results

In previous publications [OR02] we have reported the results of experimental nanoindentations performed on gold surfaces. Although we just observe, to be more precise, the intersection with the surface of emerging dislocation lines in different configurations, we have proposed models explaining the complete sub-surface structure, as well as the generation and displacement mechanisms of these defects. Atomistic simulations are as well helpful to interpret some of these defects and their generation mechanisms. In very general terms, and referring to a Au(001) surface, we can classify the defect configurations in two large classes: dislocation half-loops with total Burgers vector (a) parallel to the surface and (b) with a component perpendicular to the surface. The former were called *mesas* and the later *screw loops*. We have recently performed more extensive atomistic simulations, which we present here. Apart from *mesas*, other types of dislocations are nucleated below the indenter. Simulations generally show the nucleation of a tiny half-loop with screw character, which expands ands grows under the stress applied by the indenter. These very incipient events are associated with an abrupt fall in the applied force. Dislocation lines glide following compact crystallographic [011] directions, leaving a step on the surface. In fact, this step is a trace on the surface of the subsurface trajectory of the dislocation as it glides. As penetration increases, dislocations grow and new ones are nucleated.

Figure 1 shows a view of the dislocation network formed under the indentation point. Some surface steps created during the indentation process are visible. One of the steps ends at the emergence point of a screw dislocation. In previous publications we have proposed that, at a given distance, the dislocation line cross-slips to another glide plane (this is due to the anisotropic stress distribution around the indentation point and is explained in detail in a previous publication [CE04]). Present simulations support previous proposals about the origin and nature of the piled-up material around nanoindentations. In the meanwhile, other recent works strongly suggest, for different kinds of solids (gold thin films [AA06], barite [AA] and KBr [FT06]), the same kind of mechanism. It can be inferred that the formation of the piled-up material around a nanoindentation by means of nucleation, glide and cross-slip of screw dislocations (giving rise to crystallographic superimposed terraces) can be considered to be a quite general mechanism.

An analogous study has been carried out on defective surfaces. As a first approach we can consider a stepped surface, with regularly separated monoatomic steps. Several simulations have been performed on surfaces with different density of steps. Figure 2 shows the force vs. penetration curve for

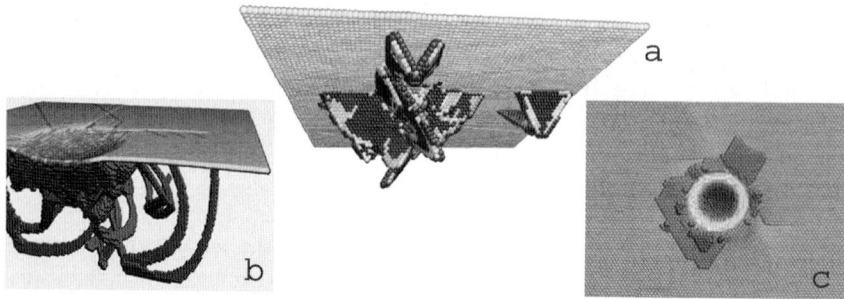

Fig. 1. (**a**), (**b**) Sub-surface views of defect configurations generated during nanoindentation in Au(001). Just surface and defective atoms are shown. Most of the defects are *mesas* in **a**, and screw dislocations intersecting the surface in **b** (where just a setion of the whole simulation cell is visualized). (**c**) Top view of a Au(111) surface, in false grey scale, showing the formation of small terraces around the indentation point (piled-up material) due to a dislocation mechanism

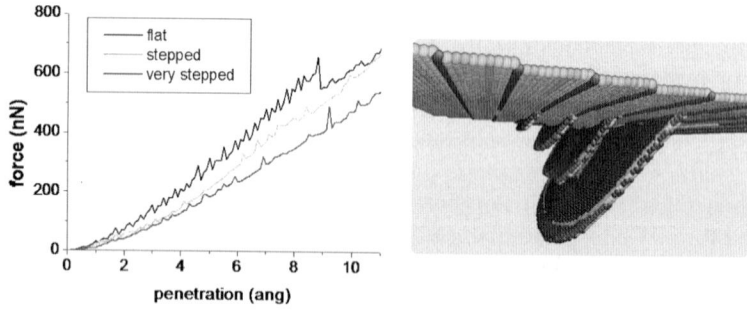

Fig. 2. Force vs. penetration curve on one flat and two stepped surfaces (with different step separations). On the right, a sub-surface lateral view of the simulation cell during nanoindentation is shown. Dislocation loops stem from the steps along the same set of (111) slip planes

three of the simulations carried out. It is clear that a lower force is necessary to penetrate the same distance into the material when it is stepped, and that the higher the defect density, the lower the surface stiffness. An analysis of the intermediate states of the crystal shows that the first dislocations below the flat surface do not appear before the first drop in the force (around 8.8 Å for the case shown in the figure). But for the case of stepped surfaces, dislocations nucleate below the indenter shortly after mechanical contact, and it does not necessarily produce a drop in the force, as shown in the curve. This kind of discontinuities are usually considered a clear sign of formation of the first plastic defects in high force resolution nanoindentation experiments. In apparent contradiction, we note here that, in stepped surfaces, plastic defects can be generated with no apparent drop in the indenter force. Discontinuities in the force curve thus should not be considered as a good indicator of plastic activity in rough surfaces. Additionally, stepped surfaces show two peculiarities with respect to the nucleation site and geometry of the dislocations. In a fcc solid, there exist four equivalent (111) slip planes. For flat surfaces, the exact region where dislocations are generated and their orientation (slip plane) are not completely determined, and any of these planes can be activated as stress is applied. But, according to our results, dislocation loops in a stepped surface are generated just at the step, and they slip and glide along the same family of slip planes: the ones parallel to the step line and lying below the higher part of the step (see Fig. 2). Just one of the four slip planes is activated, and the result is that all generated loops are parallel. There exist two main reasons explaining these peculiarities. The indenter pressure on a step is higher due to the reduced contact area, as shown in [ZJ01] (the indenter, initially, does not contact the lower part of the step). This fact favours dislocation nucleation at these sites. But, additionally, atoms in a step has a reduced number of neighbours. To create a dislocation, atomic bonds need to be broken before the displacement and rebonding of the atoms involved in the slip process. According to this, atoms with a reduced number of neighbours should be easier to displace (a fewer number of bonds have to be broken and less energy is required). And it is because of the asymmetric distribution of bonds around an atom in a step that displacement is favoured along one specific direction (below the upper part of the step), and dislocations are activated in the same slip plane. Moreover, and speaking in terms of final energy balance, less energy is required to form a dislocation loop below a step (since this step segment is partially eliminated) than in a flat surface (since a new step segment is created). Finally, we have performed nanoindentations on surface three-dimensional structures like the one shown in Fig. 3. They are stepped pyramids and simulate the real asperities or mounds present at the nanoscale in rough surfaces. Like in the case of steps, one can tune the roughness by varying the distance between the steps. Simulations show that results on the steps can be extended to the nanopyramids. The force curves (not shown)

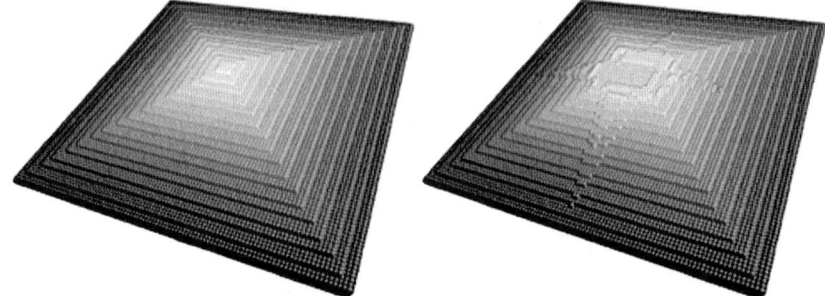

Fig. 3. Simulated protusion before (*left*) and after (*right*) nanoindentation. After plastic deformation, the upper levels have disappeared. The displaced material has been split aside (*traces are visible*) and no dislocations remain below

reveal that, the steeper the pyramid, the less stiff it is. And, due to the particular morphology of these structures, the very initial dislocation structures glide in the directions parallel to the surface, meet adjacent steps and finally disappear. They are annihilated by the sides, which act as dislocation sinks, and do not penetrate into the bulk. Thus, these kind of structures are able to absorb very incipient plastic deformation through self-deformation, leaving the material below with no defect trace. This could have an influence on subsequent deformations, since no dislocations are present.

4 Conclusions

Atomistic nanoindentation simulations on gold surfaces reproduce and extend previous experimental observations. Incipient plasticity consists of the emission of dislocation loops, some of which glide and cross-slip giving rise to the piled-up material around the indentation point. Simulations on stepped surfaces show a peculiar plastic behaviour, revealing an active role of surface defects at the onset of plasticity. Surface steps lower the stress threshold for nucleating dislocations, and determine their orientation. Indentations on nanoscale surface protrusions show analogous properties as for steps, with an additional characteristic: their stepped faces act as dislocation sinks, leaving the subsurface region undefective after plastic deformation.

References

[DB84] Daw M.S. *et al.*: Embedded-atom method: Derivation and application to impurities, surfaces and other defects in metals. Phys. Rev. B, **29**, 6443–6453 (1984)
[LJ03]   J. Li, Modelling Simul. Mater. Sci. Eng. **11** 173 (2003)
[OR02] Rodríguez de la Fuente, O. *et al.*: Dislocation Emission around Nanoindentations on a (001) fcc Metal Surface Studied by Scanning Tunneling Microscopy and Atomistic Simulations. Phys. Rev. Lett., **88**, 036101 (2002); Carrasco, E. *et al.*: Dislocation cross slip and formation of terraces around nanoindentations in Au(001). Phys. Rev. B, **68**, 180102 (2003)
[CE04] Carrasco, E. *et al.*: Analysis at atomic level of dislocation emission and motion around nanoindentations in gold. Surf. Sci. **572**, 467–475 (2004)
[AA06] Asenjo, A. *et al.*: Dislocation mechanisms in the first stage of plasticity of nanoindented Au(111) surfaces, Phys. Rev. B, **73**, 075431 (2006)
[AA]   Asenjo, A. *et al.*: to be published
[FT06]   Filleter, T. *et al.*: Atomic-scale yield and dislocation nucleation in KBr, Phys. Rev. B, **73**, 155433 (2006)
[ZJ01]   Zimmerman, J.A. *et al.*: Surface Step Effects on Nanoindentation, Phys. Rev. Lett., **87**, 165507 (2001)

Critical Thickness for Misfit Dislocation Formation in InAs/GaAs(110) Heteroepitaxy

I. Plans[1], A. Carpio[2], L.L. Bonilla[1], and R.E. Caflisch[3]

[1] MSMI, Universidad Carlos III, Ave. Universidad 30, 28911 Leganés, Spain
   ignacio.plans@uc3m.es
[2] Dept. Matemática Aplicada, Universidad Complutense, 28040 Madrid, Spain
[3] Dept. Mathematics, University of California, Los Angeles, CA 90095-1555, USA

Summary. A two-dimensional discrete elasticity model is used to compute the critical thickness at which interfacial pure edge dislocations are energetically preferred to form in the InAs/GaAs(110) heteroepitaxial system. The calculated critical thickness of six monolayers, is fairly close to the measured value in experiments, five.

1 Introduction

Heteroepitaxial growth of InAs on GaAs(110) [1] has been examined in detail using different techniques, including scanning tunneling microscopy (STM) and transmission electron microscopy (TEM). The strain relief mechanisms depend strongly on the orientation of the substrate over which layers are being grown. In the case of a (110) substrate, growth of the InAs film occurs in three different stages, depending on the film thickness, h_f:

- $1 \leq h_f \leq 3$ monolayers (ML). A uniform network of very small 1 ML high InAs islands is formed.
- $3 \leq h_f \leq 200$ ML. The islands formed during the first stage coalesce and the film grows layer by layer so that no three-dimensional (3D) structures on top of the last layer are created. Early on, for $3 \leq h_f \leq 5$ ML, an array of pure edge misfit dislocations (90° MDs) is formed at the interface, and it induces a lattice distortion that is visible at the surface. The dislocation lines have the [100] direction, whereas their Burgers vectors lay along the [1$\bar{1}$0] direction. This stage is called the [1$\bar{1}$0]-relaxed stage.
- $h_f \geq 200$ ML. The [001] direction is also relaxed by an array of 60° MDs directed along [1$\bar{1}$0]. This array is perpendicular to the previous one, and thus a complete dislocation network has been formed at the interface [2].

In this chapter, we consider the second stage of film growth. According to experiments [3,4], the first interfacial dislocations are found for $h_f = 3$ ML, and a complete array of 90° MD has been formed when $h_{f,\mathrm{exp}}^* \sim 5$ ML. By using energy arguments and a discrete elasticity model of heteroepitaxial growth in InAs/GaAs(110), we compute the critical film thickness h_f^* necessary to create the MD array. We find $h_f^* = 6\,\mathrm{ML}$.

2 The Model

We model atoms on a plane perpendicular to the dislocation lines (parallel along [001]) that contains the Burgers vector of MDs (along [1$\bar{1}$0]). Each layer of the resulting 2D square lattice comprises two layers of the 3D zincblende structure of the material. We choose the slice of InAs/GaAs (110) so that indium atoms are grown on a Ga substrate, both on square lattices that have different lattice constants. Cartesian axes on our 2D lattice will be chosen along [1$\bar{1}$0] and [110], as indicated in Fig. 1. The elastic constants and the lattice constant referred to these axes are related to those in the [100] direction according to the formulas [(13)–(43) in [5], $H = 2C_{44}^{[100]} + C_{12}^{[100]} - C_{11}^{[100]}$ is the anisotropy factor]:

$$C_{11}^{[1\bar{1}0]} = C_{11}^{[100]} + H/2, \quad C_{12}^{[1\bar{1}0]} = C_{12}^{[100]} - H/2, \quad C_{44}^{[1\bar{1}0]} = C_{44}^{[100]} - H/2, \quad (1)$$

$$a^{[1\bar{1}0]} = a^{[100]}/\sqrt{2}. \quad (2)$$

Our computational domain comprises M layers: layers $1, ..., p$, correspond to the substrate, and layers $p+1, ..., M$, to the film. Each computational layer

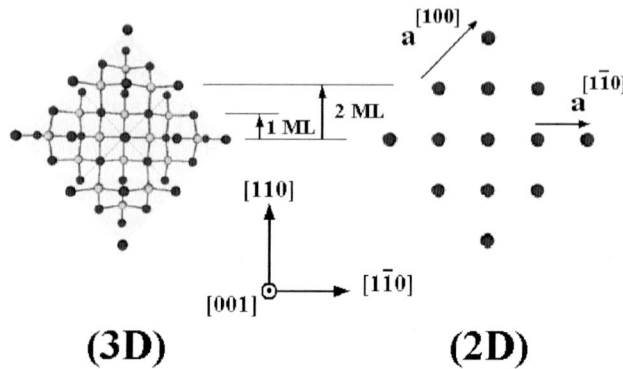

Fig. 1. Extracting a plane of atoms from the three-dimensional (3D) zincblende lattice results in a two-dimensional (2D) square lattice. Each layer of the 2D lattice represents 2 ML of the real crystal

Table 1. Elastic constants in $10^9\,\mathrm{N\,m^{-2}}$ and lattice parameters in Å

	$C_{11}^{[100]}$	$C_{12}^{[100]}$	$C_{44}^{[100]}$	$a^{[100]}$	$C_{11}^{[1\bar{1}0]}$	$C_{12}^{[1\bar{1}0]}$	$C_{44}^{[1\bar{1}0]}$	$a^{[1\bar{1}0]}$
InAs	83	45.0	39.5	6.05	$C_{11}^+ = 103.50$	$C_{12}^+ = 24.50$	$C_{44}^+ = 19.00$	$a^+ = 4.28$
GaAs	118	53.5	59.0	5.65	$C_{11}^- = 144.75$	$C_{12}^- = 26.75$	$C_{44}^- = 32.25$	$a^- = 4.00$

represents two physical monolayers, as shown in Fig. 1. A cell in the square lattice, labeled by indices (l, m), contains one atom located at coordinates $x_i(l, m; t)$, $i = 1, 2$ at time t. The side of a square cell containing a substrate atom is $a(l, m) = a^-$ $(m \leq p)$, and it is $a(l, m) = a^+$ for a cell containing a film atom $(m > p)$. The lattice misfit is $\epsilon = (a^+ - a^-)/a^+ = 7\%$.

In our model, we consider a discrete energy $V(\{x_i(l, m; t)\})$,

$$V = \sum_{l,m} a^3(l, m) W(l, m; t) = \frac{1}{2} \sum_{l,m,i,j,r,s} a^3(l, m)\, c_{ijrs} g_{ij} g_{rs}, \tag{3}$$

in which the strain energy density W depends on the tensor of elastic constants c_{ijrs} and on displacement differences that become $\partial u_i / \partial x_j$ in the continuum limit (u_i is the displacement field):

$$g_{ii} = g\left(\frac{D_i^+ x_i(l, m; t)}{a(l, m)} - 1\right) \sim \frac{\partial u_i}{\partial x_i}, \quad g_{ij} = g\left(\frac{D_j^+ x_i(l, m; t)}{a(l, m)}\right) \sim \frac{\partial u_i}{\partial x_j}, i \neq j, \tag{4}$$

$i, j = 1, 2$. Here $D_1^+ x_i(l, m; t) = x_i(l+1, m; t) - x_i(l, m; t)$ and $D_2^+ x_i(l, m; t) = x_i(l, m + 1; t) - x_i(l, m; t)$ and $g(x) \sim x$ for small x. Note that the x_i are absolute coordinates, not displacements from equilibrium positions. The atoms at the top layer do not have any other ones above them, so that $D_2^+ x_i(l, M; t) = 0$, which represents a free surface boundary condition. In the continuum limit, W agrees with anisotropic linear elasticity:

$$W(l, m; t) \rightarrow \frac{C_{11}^\alpha}{2}\left(\frac{\partial u_1}{\partial x_1}\right)^2 + \frac{C_{11}^\alpha}{2}\left(\frac{\partial u_2}{\partial x_2}\right)^2 + C_{12}^\alpha \frac{\partial u_1}{\partial x_1} \frac{\partial u_2}{\partial x_2} + \frac{C_{44}^\alpha}{2}\left(\frac{\partial u_2}{\partial x_1} + \frac{\partial u_1}{\partial x_2}\right)^2 \tag{5}$$

Here $\alpha = -, +$ depending on whether cell (l, m) belongs to the substrate $(m \leq p)$ or to the film $(m > p)$, respectively.

3 Methodology

The potential energy V yields a force $-\partial V / \partial x_i$ on the atom located at x_i and Newton's second law provides the equations of motion for our model. Local equilibrium configurations can be found from stationary solutions or from energy minima. It is computationally more efficient to seek for stationary configurations of the equations of motion by solving the overdamped equations:

$$\beta \frac{dx_i(l, m; t)}{dt} = -\frac{\partial V}{\partial x_i(l, m; t)}, \quad i = 1, 2, \tag{6}$$

where $\beta = 1$ is the damping coefficient. The relaxation method starts from an initial guess $\{x_i(l, m; t_0)\}$ and (6) are solved until a stationary configuration $\{x_i(l, m; t_\infty)\}$ is reached. Then its *energy density* is the corresponding energy V in (3) divided by the sample volume. We will compare the energy densities of the coherent (i.e., without dislocations) and dislocated configurations.

Initial conditions are chosen as close as possible to a stationary solution to which the system is observed to relax. The substrate is set so as to have its equilibrium lattice constant a^- as bond length. In the coherent configuration, epilayer atoms are set to be vertically aligned to those in the substrate, and they have their own lattice spacing a^+ in the dislocated configuration. The difference between the substrate and film lattice constants causes the formation of a MD array at the interface. At both sides of the domain boundary conditions are periodic, atoms in the substrate lower layer ($m = 1$) do not move, and the top layer is a free surface; cf. Sect. 1.

To ensure that the depressions at surfaces are found right above dislocation cores, we need to relabel atoms in dislocated configurations. The right-hand side of (6) includes the coordinates of first and second neighbors of the atom (l, m). These neighbors determine a stencil of dependence, which may be updated. When dislocations are present at the interface, we compute the lists of neighbors, *upper-neighbor(l)* and *lower-neighbor(l)*, for atoms at layers p and $p + 1$, respectively. The upper neighbor of a given atom (l, p) will be that minimizing $|x_1(upper\text{-}neighbor(l), p+1; t) - x_1(l, p; t)|$. Similarly, an atom $(l, p+1)$ will find its first neighbor at the layer p by minimizing $|x_1(l, p+1; t) - x_1(lower\text{-}neighbor(l), p; t)|$. This allows to update the stencil of dependence. The corresponding change in the energy V is introduced by redefining the gradients at layer p: $D_2^+ x_i(l, p; t) = x_i(upper\text{-}neighbor(l), p + 1; t) - x_i(l, p; t)$. The critical thickness h_f^* is the minimum number of film layers for which the dislocated configuration has lower stationary energy density than the coherent one. Recall that each computational layer in $h_{f,2D}$ represents 2 ML of the 3D crystal (Figs. 1 and 2).

In our simulations, we inserted eight dislocations ($N_x = 121$ columns of atoms for the substrate, only 113 for the film) and 15 substrate layers. This value is similar to the inter-dislocation distance, hence the substrate may be considered infinitely extended. No dependence of h_f^* on the system size was observed.

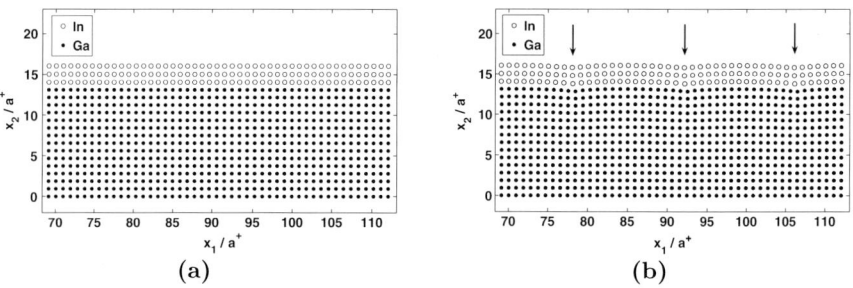

Fig. 2. Partial view of **(a)** coherent and **(b)** dislocated relaxed configurations with $g_2(x) = \tan^{-1}(\pi x)/\pi$. The *arrows* point towards the valleys formed above dislocation core regions. Three computational layers $h_{f,2D} = 3$ represent six 3D MLs

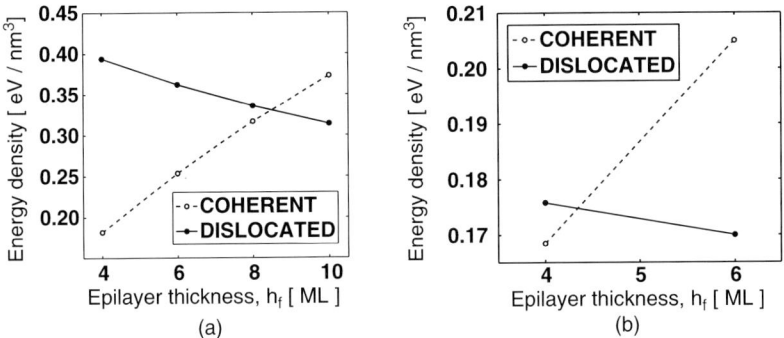

Fig. 3. Energy density vs. film thickness for (**a**) the linear case, $g_1(x) = x$, and (**b**) for $g_2(x) = \tan^{-1}(\pi x)/\pi$

4 Results and Conclusions

Figure 3 shows our results. We inserted two different functions: the first one, $g_1(x) = x$, is purely linear, and $g_2(x) = \tan^{-1}(\pi x)/\pi$, is anharmonic. The simulations with both functions yield qualitatively similar results: the energy density of the dislocated configuration increases as the epilayer thickness increases, whereas the energy density of the coherent configuration decreases. Besides, in the relaxed dislocated configurations (see Fig. 2), valleys are formed above the dislocation lines, in agreement with the experimental observations.

In the linear case, the energy density of the dislocated configuration is lower than that of the coherent one starting from a critical thickness $h^*_{f,2D,1} = 5$, that corresponds to $h^*_{f,1} = 10\,\text{ML}$. For the other function, we have $h^*_{f,2D,2} = 3$, $h^*_{f,2} = 6\,\text{ML}$. The latter is closer to the experimental value, $h^*_{f,\exp} \sim 5$, as shown in Fig. 2b.

To summarize, we used a simple 2D discrete elasticity model to compute the critical thickness at which it is energetically preferred for the InAs/GaAs(110) heteroepitaxial system to form interfacial pure edge dislocations. Despite its simplicity, the model provides qualitatively correct critical thickness (6 ML to experimentally observed 5 ML), and coherent and dislocated energy densities as functions of film thickness.

Acknowledgments

This work has been supported by the Spanish Ministry of Education grants MAT2005-05730-C02-01 (LLB and IP), MAT2005-05730-C02-02 (AC) and FPI grant BES-2003-1610 (IP), and by the Universidad Complutense grants Santander/UCM PR27/05-13939 and CM/UCM 910143 (AC). I. Plans acknowledges the Materials Modeling in Applied Mathematics group at UCLA for their hospitality and fruitful discussions.

References

[1] Joyce, B. A., Vvedensky, D. D., Mater. Sci. Eng., R. **46**, 127 (2004).
[2] Zhang, X., Pashley, D. W., Hart, L., Neave, J. H., Fawcett P. N., Joyce, B. A., J. Cryst. Growth **131**, 300 (1993).
[3] Belk, J. G., Sudijono, J. L., Zhang, X. M., Neave, J. H., Jones, T. S., Joyce, B. A., Phys. Rev. Lett. **78**, 475 (1997).
[4] Shiraishi, K., Oyama, N., Okajima, K., Miyagishima, N., Takeda, K., Yamaguchi, H., Ito, T., Ohno.,T., J. Crystal Growth **237-239**, 206 (2002).
[5] Hirth, J. P., and Lothe, J.: Theory of Dislocations. Wiley, New York (1982).

Discrete Dislocation Dynamics in Crystals

M.P. Ariza[1], A. Ramasubramaniam[2], and M. Ortiz[3]

[1] Escuela Superior de Ingenieros, Universidad de Sevilla, 41092-Sevilla, Spain
   mpariza@us.es
[2] Department of Mechanical & Aerospace Engineering, Princeton University, Princeton, NJ 08544, USA
   aramasub@princeton.edu
[3] Division of Engineering and Applied Science, California Institute of Technology, Pasadena, CA 91125, USA
   ortiz@aero.caltech.edu

1 Introduction

We present a study of 3D dislocation dynamics in BCC crystals based on discrete crystal elasticity. Ideas are borrowed from discrete differential calculus and algebraic geometry to construct a mechanics of discrete lattices. The notion of lattice complexes provides a convenient means of manipulating forms and fields defined over the crystal. Atomic interactions are accounted for via linearized embedded atom potentials thus allowing for the application of efficient fast Fourier transforms. Dislocations are treated within the theory as energy minimizing structures that lead to locally lattice-invariant but globally incompatible eigendeformations. The discrete nature of the theory automatically eliminates the need for core cutoffs. The quantization of slip to integer multiples of the Burgers vector along each slip system leads to a large integer optimization problem. We suggest a new method for solving this NP-hard optimization problem and the simulation of large 3D systems.

2 Stored Energy of Discrete Dislocations

In order to provide a point of reference, before presenting the details of the discrete theory, we recall first some well-known notions from the classical continuum theories. Within the harmonic approach, the energy of a crystal is a convex function of the displacement field. The underlying crystalline structure however allows for displacements that leave the lattice invariant. The total energy of a crystal is thus a non-convex function of the displacements when crystallographic slip is allowed. This deficiency can be remedied by recourse to the theory of eigendeformations [Mur87, OP99]. The plastic distortion in a slipped crystal is constrained by crystallography and may be written as

$$\beta_{ij} = \sum_{\alpha=1}^{N} \gamma^{\alpha} s_i^{\alpha} m_j^{\alpha}, \qquad (1)$$

the sum in α running over all the available slip systems in the crystal. Thus the plastic distortion is built from lattice preserving deformations such as crystallographic slip and is referred to as an eigendeformation. The elastic energy of the crystal is a functional of the displacement field and the eigendeformation and is given by

$$E[u, \beta] = \int_V \frac{1}{2} C_{ijkl}(u_{i,j} - \beta_{ij})(u_{k,l} - \beta_{kl}) dV , \qquad (2)$$

which is now quadratic in the elastic distortion field $\beta_{ij}^e = u_{i,j} - \beta_{ij}$ and piecewise quadratic in $u_{i,j}$ where C_{ijkl} are the usual elastic moduli. This approach will be employed in an analogous manner to accommodate crystallographic slip within the discrete formulation.

2.1 Eigendeformations in Discrete Lattices

The energy of a harmonic crystal admits the representations [AO05]

$$E(\boldsymbol{u}) = \frac{1}{(2\pi)^n} \int_{[-\pi,\pi]^n} \frac{1}{2} \langle \hat{\boldsymbol{\Psi}}(\boldsymbol{\theta}) \widehat{d\boldsymbol{u}}(\boldsymbol{\theta}), \widehat{d\boldsymbol{u}}^*(\boldsymbol{\theta}) \rangle \, d^n\theta, \qquad (3a)$$

$$E(\boldsymbol{u}) = \frac{1}{(2\pi)^n} \int_{[-\pi,\pi]^n} \frac{1}{2} \langle \hat{\boldsymbol{\Phi}}(\boldsymbol{\theta}) \hat{\boldsymbol{u}}(\boldsymbol{\theta}), \hat{\boldsymbol{u}}^*(\boldsymbol{\theta}) \rangle \, d^n\theta, \qquad (3b)$$

where $\boldsymbol{\Psi}$ and $\boldsymbol{\Phi}$ are the force-constant fields of the lattice. We write

$$\langle \hat{\boldsymbol{\Psi}}(\boldsymbol{\theta}) \widehat{d\boldsymbol{u}}(\boldsymbol{\theta}), \widehat{d\boldsymbol{u}}^*(\boldsymbol{\theta}) \rangle \equiv \sum_{\alpha=1}^N \sum_{\beta=1}^N \hat{\Psi}_{ik} \begin{pmatrix} \boldsymbol{\theta} \\ \alpha \ \beta \end{pmatrix} \widehat{du}_i(\boldsymbol{\theta}, \alpha) \widehat{du}_k^*(\boldsymbol{\theta}, \beta) \qquad (4a)$$

$$\langle \hat{\boldsymbol{\Phi}}(\boldsymbol{\theta}) \hat{\boldsymbol{u}}(\boldsymbol{\theta}), \hat{\boldsymbol{u}}^*(\boldsymbol{\theta}) \rangle \equiv \hat{\Phi}_{ik}(\boldsymbol{\theta}) \hat{u}_i(\boldsymbol{\theta}) \hat{u}_k^*(\boldsymbol{\theta}) \qquad (4b)$$

for shorthand. The preceding representations show that the force-constant fields are related as

$$\hat{\boldsymbol{\Phi}} = \boldsymbol{Q}_1^\dagger \hat{\boldsymbol{\Psi}} \boldsymbol{Q}_1 \qquad (5)$$

$\boldsymbol{Q}_1$ being the matrix of the Fourier representation of the differential of a 1-form (see [AO05]).

   In the spirit of the eigendeformation theory, the elastic energy may be assumed to be of the form

$$E(\boldsymbol{u}, \boldsymbol{\xi}) = \frac{1}{2} \langle \boldsymbol{B}(d\boldsymbol{u} - \boldsymbol{\beta}), d\boldsymbol{u} - \boldsymbol{\beta} \rangle, \qquad (6)$$

$\boldsymbol{\xi} \in \mathbb{Z}$ is the integer-valued slip field corresponding to every slip system. Equation (6) replaces (3b) in the presence of crystallographic slip. Clearly, if $\boldsymbol{\beta} = d\boldsymbol{v}$, i.e., if the eigendeformations are compatible, then the energy-minimizing displacements are $\boldsymbol{u} = \boldsymbol{v}$ and $E = 0$. However, because slip is crystallographically constrained, $\boldsymbol{\beta}$ is not compatible in general. By virtue of this lack of compatibility, a general distribution of slip induces residual stresses in the lattice and a nonvanishing elastic energy, or stored energy.

If the distribution of eigendeformations is known, and in the absence of additional constraints, the energy of the lattice can be readily minimized with respect to the displacement field. Suppose that the crystal is acted upon by a distribution of forces $\boldsymbol{f} : E_0 \to \mathbb{R}^n$. The total potential energy of the lattice is then

$$F(\boldsymbol{u}, \boldsymbol{\xi}) = E(\boldsymbol{u}, \boldsymbol{\xi}) - \langle \boldsymbol{f}, \boldsymbol{u} \rangle. \tag{7}$$

Minimization of $F(\boldsymbol{u}, \boldsymbol{\xi})$ with respect to $\boldsymbol{u}$ yields the equilibrium equation

$$\boldsymbol{A}\boldsymbol{u} = \boldsymbol{f} + \delta\boldsymbol{B}\boldsymbol{\beta}, \tag{8}$$

where $\delta\boldsymbol{B}\boldsymbol{\beta}$ may be regarded as a distribution of eigenforces corresponding to the eigendeformations $\boldsymbol{\beta}$. The equilibrium displacements, are, therefore,

$$\boldsymbol{u} = \boldsymbol{A}^{-1}(\boldsymbol{f} + \delta\boldsymbol{B}\boldsymbol{\beta}) \equiv \boldsymbol{u}_0 + \boldsymbol{A}^{-1}\delta\boldsymbol{B}\boldsymbol{\beta}, \tag{9}$$

where $\boldsymbol{u}_0 = \boldsymbol{A}^{-1}\boldsymbol{f}$ is the displacement field induced by the applied forces in the absence of eigendeformations. Conditions under which the minimum problem just described is well-posed and delivers a unique energy-minimizing displacement field have been given in [AO05]. The corresponding minimum potential energy is

$$\begin{aligned} F(\boldsymbol{\beta}) &= \frac{1}{2}\langle \boldsymbol{B}\boldsymbol{\beta}, \boldsymbol{\beta} \rangle - \frac{1}{2}\langle \boldsymbol{A}^{-1}(\boldsymbol{f} + \delta\boldsymbol{B}\boldsymbol{\beta}), \boldsymbol{f} + \delta\boldsymbol{B}\boldsymbol{\beta} \rangle \\ &= \frac{1}{2}\langle \boldsymbol{B}\boldsymbol{\beta}, \boldsymbol{\beta} \rangle - \frac{1}{2}\langle \boldsymbol{A}^{-1}\delta\boldsymbol{B}\boldsymbol{\beta}, \delta\boldsymbol{B}\boldsymbol{\beta} \rangle - \langle \boldsymbol{A}^{-1}\delta\boldsymbol{B}\boldsymbol{\beta}, \boldsymbol{f} \rangle - \frac{1}{2}\langle \boldsymbol{A}^{-1}\boldsymbol{f}, \boldsymbol{f} \rangle \\ &= \frac{1}{2}\langle \boldsymbol{B}\boldsymbol{\beta}, \boldsymbol{\beta} \rangle - \frac{1}{2}\langle \boldsymbol{A}^{-1}\delta\boldsymbol{B}\boldsymbol{\beta}, \delta\boldsymbol{B}\boldsymbol{\beta} \rangle - \langle \boldsymbol{B}\boldsymbol{\beta}, \mathrm{d}\boldsymbol{u}_0 \rangle - \frac{1}{2}\langle \boldsymbol{A}\boldsymbol{u}_0, \boldsymbol{u}_0 \rangle. \end{aligned} \tag{10}$$

The first two terms

$$E(\boldsymbol{\beta}) = \frac{1}{2}\langle \boldsymbol{B}\boldsymbol{\beta}, \boldsymbol{\beta} \rangle - \frac{1}{2}\langle \boldsymbol{A}^{-1}\delta\boldsymbol{B}\boldsymbol{\beta}, \delta\boldsymbol{B}\boldsymbol{\beta} \rangle \tag{11}$$

in (10) give the self-energy of the distribution of lattice defects represented by the eigendeformation field $\boldsymbol{\beta}$, or stored energy; the third term in (10) is the interaction energy between the lattice defects and the applied forces; and the fourth term in (10) is the elastic energy of the applied forces.

Recall that the crystal under consideration possesses M slip systems and its eigendeformations admit a representation in terms of an integer-valued slip field $\boldsymbol{\xi} \equiv \{\xi^s, s = 1, \ldots, M\}$. Then, the stored energy (11) can be written in the form

$$E(\boldsymbol{\xi}) = \frac{1}{2}\langle \boldsymbol{H}\boldsymbol{\xi}, \boldsymbol{\xi} \rangle, \tag{12}$$

where the operator $\boldsymbol{H}$ is defined by the identity

$$\langle \boldsymbol{H}\boldsymbol{\xi}, \boldsymbol{\xi}\rangle = \langle \boldsymbol{B}\boldsymbol{\beta}, \boldsymbol{\beta}\rangle - \langle \boldsymbol{A}^{-1}\delta \boldsymbol{B}\boldsymbol{\beta}, \delta \boldsymbol{B}\boldsymbol{\beta}\rangle, \tag{13}$$

$\boldsymbol{H}\boldsymbol{\xi}$ represents the resolved shear-stress field resulting from a slip distribution $\boldsymbol{\xi}$, and therefore $\boldsymbol{H}$ may be regarded as an atomic-level hardening matrix.

Finally, we require to find the slip distribution $\boldsymbol{\xi}$ that minimizes the potential energy $F(\boldsymbol{\xi})$. This is an integer optimization problem and is known to be NP complete – solving it is thus an entirely non-trivial task. In Sect. 3 we illustrate the minimization procedure with a concrete example.

3 Applications

In order to verify the efficacy and computational advantages of this approach we have conducted numerical tests on a variety of BCC metals. Details about these applications can be found in [RAO].

3.1 Core Structure of BCC Screw Dislocations

The core-structure of dislocations is widely believed to be responsible for plastic deformation at low temperatures in BCC metals. Figure 1 shows the DD map of the core structure that has the threefold rotational point-group symmetry about the [111] axis. It compares well with the easy core structure obtained by [IA00] among others.

3.2 Point of Dilatation

We have applied the solution procedure presented in [RAO] to the case of a point of dilatation. In Fig. 1 we show the slip distribution as a result of the equilibration process.

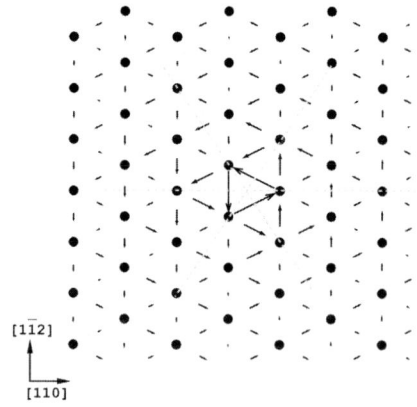

Fig. 1. Differential displacement map of Mo screw dislocation using a quadrupolar cell

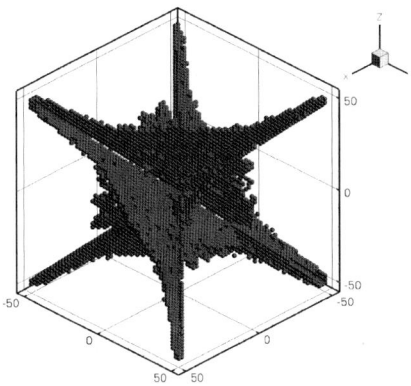

Fig. 2. Simulation cell containing one million Vanadium atoms. Slipped atoms after equilibration

Acknowledgements

We gratefully acknowledge the support of the Department of Energy through Caltech's ASC Center for the Simulation of the Dynamic Response of Materials.

References

[Mur87] Mura, T.: Micromechanics of defects in solids. Kluwer Academic Publishers, Boston (1987)

[OP99]   Ortiz, M. and Phillips, R.: Nanomechanics of defects in solids. Advances in Applied Mechanics, **36**, 1–79 (1999)

[AO05]   Ariza, M.P. and Ortiz, M.: Discrete Crystal Elasticity and Discrete Dislocations in Crystals. Archive for Rational Mechanics and Analysis, **178**, 149–226 (2005)

[IA00]   Ismail-Beigi, S. and Arias, T.A.: Ab initio study of screw dislocations in Mo and Ta: a new picture of plasticity in bcc transition metals. Physical Review Letters, **84**, 1499–1502 (2000)

[RAO]   Ramasubramaniam, A., Ariza, M.P. and Ortiz, M.: A Discrete Mechanics Approach to Dislocation Dynamics in BCC Crystals. Journal of the Mechanics and Physics of Solids (in press)

Interconnection of Continuum and Discrete Models of Dislocation Pile-ups

R.E. Voskoboinikov, S.J. Chapman, and J.R. Ockendon

Oxford Centre for Industrial and Applied Mathematics, Mathematical Institute,
24-29 St Giles', Oxford OX1 3LB, UK
voskoboynikov@maths.ox.ac.uk

Summary. A new asymptotic approach for analysing pile-ups of large numbers of dislocations is described. As an example, the pile-up of n identical screw or edge dislocations on a single slip plane under the action of an external loading in the direction of a locked dislocation in that plane is considered. As $n \to \infty$ the continuum number density of the dislocations can be easily obtained whereas direct evaluation of the discrete dislocation positions from the set of force balance equations is not straightforward. However, in the framework of our method these positions can be revealed using the corresponding dislocation density.

Introduction

The development of coarse-grain models that do not lose the essential details is a real challenge in the scope of new computational techniques to study and understand the behaviour of dislocations in structural materials. Because the relevant number density can run up to $10^9 - -10^{13} \, \mathrm{cm}^{-2}$, the accurate discrete consideration of dislocation networks at linear scales suitable for engineering applications can only be done using high-performance computing facilities. A conventional way to treat dislocations in structural materials is to work in terms of a continuum dislocation density that can be obtained relatively easy. Although the latter approach provides an adequate insight concerning the macroscopic stress and displacements, it fails at a scale at or below the separation between neighbouring dislocations.

Alternatively, by reducing the system of nonlinear force balance equations to an ordinary differential equation (ODE) in the vicinity of each dislocation, Eshelby et al. [1] found the equilibrium positions of dislocations in certain particular cases. However, this powerful technique did not develop further because it required the use of an indeterminate *"force balance-to-ODE"* procedure, which is the main issue of this chapter. As a model system, we will consider the simplest configuration of a pile-up of screw or edge dislocations on a single slip plane stressed against a locked dislocation by a constant applied stress. Using the corresponding continuum number density, we obtain the equilibrium distribution of dislocations in the pile-up and compare it to the results of [1].

1 Governing Equations

We aim to find the equilibrium configuration of n identical straight dislocations in the slip plane $y = 0$ located at positions $x = x_i > 0$, $i = 1, \ldots, n$, piling up against a locked dislocation with the same Burgers vector at the origin by an constant stress, σ_{ext}. The normalised stress on $y = 0$ due to such a configuration is

$$\sigma(x) = \sum_{j=1}^{n} \frac{1}{x - x_j} + \sigma_0(x),$$

where $\sigma_0(x) = 1/x - \sigma_{\text{ext}}$. In equilibrium the regular part of σ, obtained by subtracting the dislocation self-stress, must be zero at each x_i, giving the set of n equations of equilibrium

$$\sum_{j=1, j \neq i}^{n} \frac{1}{x_i - x_j} + \sigma_0(x_i) = 0. \tag{1}$$

This problem has been previously considered by Eshelby et al. [1], who introduce the polynomial

$$f(x) = \prod_{i=1}^{n} (x - x_i), \tag{2}$$

whose zeros, x_i, $i = 1, \ldots, n$, correspond to the dislocation positions. In terms of $f(x)$ the force balance (1) can thus be written

$$\lim_{x \to x_i} \left(\frac{f'(x)}{f(x)} - \frac{1}{x - x_i} + \sigma_0(x_i) \right) = 0 \quad \text{or} \quad \frac{f''(x_i)}{2 f'(x_i)} + \sigma_0(x_i) = 0, \tag{3}$$

on expanding $f(x)$ in a Taylor series near each x_i. Eshelby et al. [1] proceed by considering the ordinary differential equation

$$f''(x) + 2\sigma_0(x) f'(x) + q(x, n) f(x) = 0, \tag{4}$$

where q is to be determined. If q can be chosen so that this equation has a polynomial solution, and q is not singular at the zeros x_i of $f(x)$, then the force balance (3) is satisfied, and the problem is solved.

2 The Continuum Approximation

Finding the function q for a general dislocation pile-up problem is nontrivial. For the moment let us assume that q is known and use (4) to determine the continuum dislocation density in the limit $n \to \infty$. The substitution

$$f(x) = v(x)e^{-\int^x \sigma_0(x')\,dx'} \tag{5}$$

reduces (4) to its normal form

$$v''(x) + \kappa^2 v(x) = 0, \tag{6}$$

where

$$\kappa^2 = q(x,n) - \sigma_0'(x) - \sigma_0^2(x) \tag{7}$$

and the zeros of $v(x)$ coincide with those of $f(x)$. Rescaling $\xi = x/n$ reduces (6) to

$$v''(\xi) + n^2\kappa^2(\xi,n)v(\xi) = 0. \tag{8}$$

With the assumed expansion

$$\kappa^2(\xi,n) \sim \kappa_0^2(\xi) + 2\kappa_0(\xi)\kappa_1(\xi)n^{-1} + \cdots \tag{9}$$

as $n \to \infty$, we can make the WKB expansion, see [2],

$$v(\xi) \sim \Re\left\{ e^{in\phi(\xi)} \sum_{k=0}^{\infty} \frac{A_k(\xi)}{n^k} \right\}. \tag{10}$$

Substituting (10) in (8) and equating coefficients as $n \to \infty$ gives the eikonal equation

$$-[\phi'(\xi)]^2 + \kappa_0^2(\xi) = 0 \tag{11}$$

and leading-order amplitude equation

$$2A_0'(\xi)\phi'(\xi) + A_0(\xi)\phi''(\xi) - 2i\kappa_0(\xi)\kappa_1(\xi)A_0(\xi) = 0. \tag{12}$$

Hence

$$v(\xi) \sim C\kappa_0(\xi)^{-\frac{1}{2}} \exp\left(in \int_0^\xi \kappa_0(\xi')\,d\xi' + i\int_0^\xi \kappa_1(\xi')\,d\xi' \right) + \text{c.c.}, \tag{13}$$

where c.c. denotes complex conjugate, and $C = Re^{i\chi}$ is a complex constant, with R and χ real.

3 Dislocation Density in WKB Region

The total stress in the WKB region can be expressed in terms of $f'(x)/f(x)$ as well as in terms of $v'(x)/v(x)$. However, $v(x)$ is a linear combination of two exponentials in the dislocation region where $\kappa^2 > 0$, whereas in the dislocation-free zone the same WKB analysis holds but $\kappa(\xi,n)$ is imaginary, and the exponents of exponentials are real. In order to avoid exponential growth as $\xi \to \infty$, we have to keep the decaying term only. So, in dislocation-free zone we have

$$\frac{f'}{f} = \sum_{j=1}^{n} \frac{1}{\xi - \xi_j} \sim -n\sigma_0 + in\kappa_0 - i\kappa_1 - \frac{\kappa_0'}{2\kappa_0} + \ldots = \frac{v'}{v} - n\sigma_0. \qquad (14)$$

In order to use the Euler–Maclaurin approximation formula for the sum we define $g(\xi, \xi') = 1/(\xi - \xi')$ and introduce the transformation $z = z(\xi')$ such that the dislocations are equally spaced in z, that is $z_i = z(\xi_i) = i/n$. Then, setting $g(\xi, \xi') = G(\xi, z)$ we have

$$\sum_{i=1}^{n} G(\xi, z_i) = n \int_0^1 G(\xi, z)\, dz - \frac{G(\xi, 0)}{2} + \frac{G(\xi, 1)}{2} + \cdots .$$

Returning to the original variable we have

$$\sum_{i=1}^{n} \frac{1}{\xi - \xi_i} = n \int_0^{\xi^*} \frac{\rho(\xi')}{\xi - \xi'} d\xi' - \frac{1}{2\xi} + \frac{1}{2(\xi - \xi^*)} + \cdots ,$$

where $\xi^* = \xi_n$ is the position of the last dislocation, and $\rho = dz/d\xi'$ is the dislocation density, which clearly satisfies

$$\int_0^{\xi^*} \rho\, d\xi' = [z]_0^{\xi^*} = z(\xi_n) = 1. \qquad (15)$$

Hence, in the dislocation free zone, $\xi > \xi^*$, we get

$$n \int_0^{\xi^*} \frac{\rho(\xi')}{\xi - \xi'} d\xi' - \frac{1}{2\xi} + \frac{1}{2(\xi - \xi^*)} + \cdots \sim -n\sigma_0 + in\kappa_0 - i\kappa_1 - \frac{\kappa_0'}{2\kappa_0} + \ldots . \qquad (16)$$

Analytical continuation to the region $\xi < \xi^*$ gives

$$n \int_0^{\xi^*} \frac{\rho(\xi')\, d\xi'}{\xi - \xi'} + n\pi i \rho(\xi) - \frac{1}{2\xi} + \frac{1}{2(\xi - \xi^*)} + \cdots \sim -n\sigma_0 + in\kappa_0 - i\kappa_1 - \frac{\kappa_0'}{2\kappa_0} + \ldots . \qquad (17)$$

Expanding $\sigma_0(\xi)$, $\rho(\xi)$ and ξ^* as $n \to \infty$

$$\sigma_0(\xi) \sim \sigma_{00} + \frac{\sigma_{01}}{n} + \ldots ; \quad \rho(\xi) \sim \rho_0 + \frac{\rho_1}{n} + \ldots ; \quad \xi^* \sim \xi_0^* + \frac{\xi_1^*}{n^{2/3}} + \ldots \qquad (18)$$

and equating real and imaginary parts of coefficients of powers of n we obtain the leading order equations for dislocation density

$$\int_0^{\xi_0^*} \frac{\rho_0(\xi')}{\xi - \xi'} d\xi' = -\sigma_{00}, \qquad (19)$$

$$\pi \rho_0 = \kappa_0. \qquad (20)$$

Equation (19) is the well-known singular integral equation for continuum dislocation density, see [3], whereas (20) provides the relationship between the continuum dislocation density and, from (7), the leading order approximation for the unknown function q. In the case of the pile-up stressed against a locked dislocation $\sigma_{00} = \sigma_{\text{ext}}$ and the solution of 19), in accordance with the inversion theorem [4], is given by

$$\rho_0 = \frac{1}{\pi}\sqrt{\frac{2\sigma_{\text{ext}}}{\xi} - \frac{2\sigma_{\text{ext}}}{\xi_0^*}}, \tag{21}$$

where $\xi_0^* = 2/\sigma_{\text{ext}}$ in accordance with (15). Taking into account (7) $q = 2\sigma_{\text{ext}}/\xi$ to leading order as $n \rightarrow \infty$. Hence, in terms of the new variable $x = t/(2\sigma_{ext})$ (4) becomes

$$tf''(t) + (2 - t)f'(t) + nf(t) = 0, \tag{22}$$

which is a particular case of the associated Laguerre differential equation, with the required polynomial solution

$$f(t) = L_n^1(t); \tag{23}$$

hence $L_n^1(t)$ is the associated Laguerre polynomial (the other independent solution is $U(n, 2, t)$, a confluent hypergeometric function of the second kind, which is not polynomial). The positions of the dislocations in a pile-up against a lock to leading order are given by the zeros of $L_n^1(t)$ in agreement with the solution obtained by Eshelby et al. in [1]. However, our approach has not required us to guess information concerning q.

4 Conclusions

We have described the procedure for finding the dislocation density in a pile-up against a lock when $\xi = O(1)$. As a byproduct, our method gives a systematic derivation of the lowest order approximation of the q function arising in [1] as $n \rightarrow \infty$.

References

1. Eshelby, J.D., Frank, F.C., Nabarro, F.R.N., The equilibrium of linear arrays of dislocations, Philosophical Magazine **42** (1951) 351–364.
2. Hinch, E.J., Perturbation Methods, Cambridge University Press 1991.
3. Hirth, J.P., Lothe, J., Theory of dislocations, 2nd Ed., Krieger Publishing, Malabar Florida, 1992.
4. Muskhelishvili, N.I., Singular Integral Equations, Groningen, 1953.

Simplified P_N Models and Natural Convection–Radiation

René Pinnau and Mohammed Seaïd

Fachbereich Mathematik, Technische Universität Kaiserslautern, D–67663
Kaiserslautern, Germany
pinnau@mathematik.uni-kl.de, seaid@mathematik.uni-kl.de

Summary. In this chapter we examine the accuracy and efficiency of the simplified
P_N approximations of radiative transfer for natural convection problems in a square
enclosure. A Boussinesq approximation of the Navier–Stokes equations is employed
for the fluid subject to combined natural convection and radiation. Coupled with the
simplified P_N models, the system of equations results into a set of partial differential
equations independent of the angle variable. Numerical results for different Rayleigh
numbers are presented.

1 Introduction

Due to the high numerical complexity of simulations including radiative effects
there is presently a whole hierarchy of approximate models available which
allows to reduce the numerical costs significantly and still reproduces the main
physical phenomena. The reduced models range from half space moment approximations over full space moment systems to the diffusive-type simplified
$P_N(SP_N)$ systems [3,4,8]. The latter were developed recently and tested extensively for various radiative transfer problems [7]. They were successfully used
to simulate many high temperature applications, like glass cooling, the design
of combustion chambers for gas turbines, or crystal growth processes [1, 2, 6].

Another interesting application is the simulation of a glass melting furnace, where one needs to include conduction, convection and radiation into
the model. This additional radiative heat exchange appearing in the energy
balance poses severe numerical problems for CFD simulations due to the
enhanced complexity of the model leading to a high dimensional discrete

R. Pinnau and M. Seaïd

phase space. Without radiation the melting furnace might be modelled by natural convection in a differentially heated cavity (see [5] and the references therein). There, the main focus is on the understanding of the Rayleigh–Bérnard convection due to temperature gradients and on its adequate numerical resolution.

Here, we will present first results on the applicability of the SP_N hierarchy in the context of coupled radiation, convection and diffusion problems in a square enclosure. The vertical walls of the enclosure are heated with uniform different temperatures and the other walls are adiabatic. A Boussinesq approximation of the Navier–Stokes equations is employed for the fluid subject to combined natural convection and radiation. Coupled with the SP_N models, the system of equations results into a set of partial differential equations independent of the angle variable.

The chapter is organized as follows. In the next section we state the model equations consisting of a coupled PDE system and in Sect. 3 numerical results are presented.

2 The Model for Natural Convection–Radiation

The physical system consists of a square enclosure with sides of length L subject to a thermal variation $(T_H - T_C)$, where T_H and T_C are temperatures of the hot and cold boundary walls. The enclosure consists of a gray, absorbing, emitting, and non-scattering fluid surrounded by rigid black walls. The fluid is Newtonian and all the thermophysical properties are assumed to be constant, except for density in the buoyancy term that can be adequately modelled by the Boussinesq approximation [5] and that compression effects and viscous dissipation are neglected. The system we want study reads in dimensionless form

$$\nabla \cdot \mathbf{u} = 0,$$

$$\frac{D\mathbf{u}}{Dt} + \nabla p - Pr\nabla^2\mathbf{u} = RaPrT\mathbf{e}, \qquad (1)$$

$$\frac{DT}{Dt} - \nabla^2 T = -\frac{\kappa}{Pl}\nabla \cdot Q_R,$$

where $\mathbf{e}$ is the unit vector and Dw/Dt is the material derivative. The variables are the velocity vector $\mathbf{u}$, the temperature T and the pressure p. The parameters are the Prandtl number Pr, the Planck number Pl and the Rayleigh number Ra as well as the absorption coefficient κ. The dimensionless radiative heat flux is given by

$$\nabla \cdot Q_R = \frac{1}{\tau^2}\left(\varphi - B(T)\right), \qquad (2)$$

where φ is the total incident radiation, which will be computed via the SP_N approximations, and τ is the optical thickness. The scaled Planck function is given by $B(T) = (T + 1)^4$. To formulate a well-posed problem, equations (1) have to be solved in a bounded domain Ω with smooth boundary $\partial\Omega$ and subject to given initial and boundary conditions. We have $\partial\Omega = \Gamma_1 \cup \Gamma_2 \cup \Gamma_3 \cup \Gamma_4$, where Γ_1 and Γ_2 represent the hot and cold walls, respectively, whereas Γ_2 and Γ_4 are adiabatic walls. Hence, the boundary conditions are

$$\mathbf{u}(t, \hat{\mathbf{x}}) = 0, \qquad\qquad \hat{\mathbf{x}} \in \partial\Omega, \qquad\qquad (3)$$

for the flow, and

$$\begin{aligned}
T(t, \hat{\mathbf{x}}) &= T_\text{H}, & \hat{\mathbf{x}} &\in \Gamma_1, \\
T(t, \hat{\mathbf{x}}) &= T_\text{C}, & \hat{\mathbf{x}} &\in \Gamma_3, \\
\mathbf{n}(\hat{\mathbf{x}}) \cdot \nabla T(t, \hat{\mathbf{x}}) &= 0, & \hat{\mathbf{x}} &\in \Gamma_2 \cup \Gamma_4,
\end{aligned} \qquad (4)$$

for the temperature. In (4), $\mathbf{n}(\hat{\mathbf{x}})$ denotes the outward unit normal in $\hat{\mathbf{x}}$ with respect to $\partial\Omega$. Now, we shortly present the first models in the SP$_N$ hierarchy, which are used in the upcoming simulations. For details of their derivation we refer to [3]. In the present work, we consider only the SP$_1$ and SP$_3$ approximations and our techniques can be straightforwardly extended to other SP$_N$ approximations.

The SP$_1$ approximation is given by

$$B(T) = \varphi - \frac{\tau^2}{3\kappa^2} \nabla^2 \varphi + \mathcal{O}(\tau^4),$$

yielding

$$-\frac{\tau^2}{3\kappa} \nabla^2 \varphi + \kappa\varphi = \kappa B(T). \qquad\qquad (5)$$

Further, the SP$_3$ approximation is given by

$$B(T) = \left(1 - \frac{\tau^2}{3\kappa} \nabla^2 - \frac{4\tau^4}{45\kappa^4} \nabla^4 - \frac{44\tau^6}{945\kappa^6} \nabla^6\right) \varphi + \mathcal{O}(\tau^8),$$

and its associated equations are

$$\begin{aligned}
-\frac{\tau^2}{\kappa} \mu_1^2 \nabla^2 \varphi_1 + \kappa\varphi_1 &= \kappa B(T), \\
-\frac{\tau^2}{\kappa} \mu_2^2 \nabla^2 \varphi_2 + \kappa_k \varphi_2 &= \kappa B(T).
\end{aligned} \qquad (6)$$

The new variables φ_1 and φ_2 in (6) are related to the total incident intensity by

$$\varphi = \frac{\gamma_2 \varphi_1 - \gamma_1 \varphi_2}{\gamma_2 - \gamma_1}. \qquad\qquad (7)$$

Once the mean intensity φ is obtained from the above SP$_N$ approximations the radiative heat flux is formulated as in (2). The boundary conditions for the SP$_1$equation (5) are

$$\frac{\tau}{3\kappa}\mathbf{n}(\hat{\mathbf{x}}) \cdot \nabla\varphi(t,\hat{\mathbf{x}}) + \varphi(t,\hat{\mathbf{x}}) = B\left(T_{\mathrm{H}}\right), \qquad \hat{\mathbf{x}} \in \Gamma_1,$$

$$\frac{\tau}{3\kappa}\mathbf{n}(\hat{\mathbf{x}}) \cdot \nabla\varphi(t,\hat{\mathbf{x}}) + \varphi(t,\hat{\mathbf{x}}) = B\left(T_{\mathrm{C}}\right), \qquad \hat{\mathbf{x}} \in \Gamma_3, \qquad (8)$$

$$\mathbf{n}(\hat{\mathbf{x}}) \cdot \nabla\varphi(t,\hat{\mathbf{x}}) = 0, \qquad \hat{\mathbf{x}} \in \Gamma_2 \cup \Gamma_4.$$

For the detailed boundary conditions for the SP_3equations (6) we refer to [9].

3 Numerical Results

The model system is discretized in space and time using a characteristic-Galerkin method and a splitting algorithm is used to advance from one time step to the next one (for details on the discretization and the splitting algorithm we refer to [9]). In the following computations we used a grid size $\Delta x = 1/64$, the time step Δt is fixed to 0.05 and steady-state solutions are displayed. We used the following criteria

$$\left\|T^{n+1} - T^n\right\|_{L^2} \leq 10^{-6},$$

to stop the time integration process. All the linear systems of algebraic equations are solved using the conjugate gradient solver with incomplete Cholesky decomposition (ICCG). For the simulations we used the following parameters:

$$T_{\mathrm{C}} = -0.5, \quad T_{\mathrm{H}} = 0.5, \quad Pr = 0.71, \quad \kappa = 1, \quad Pl = 1, \quad \tau = 1.$$

In Fig. 2 we depict the computed temperature along the line $y = 0$ for the system without radiation and with radiation modelled by the SP_1 and the SP_3 approximation. Note the larger temperature in the interior due to the additional radiative energy. As expected this effect is more pronounced for a lower Rayleigh number. To get an expression how the flow field is affected, we depict the corresponding streamlines for $Ra = 10^7$ in Fig. 2.

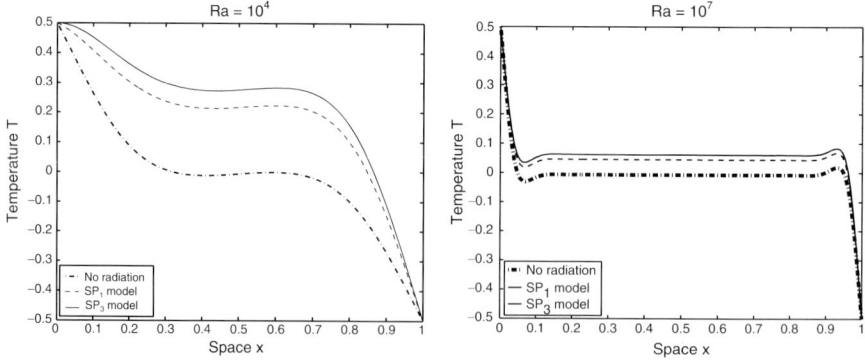

Fig. 1. Cross-section of the temperature

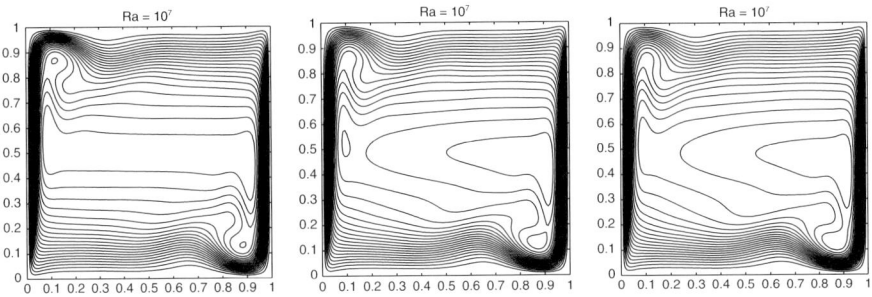

Fig. 2. Results for no radiation (*first column*), SP₁ (*second column*) and SP₃ (*right column*)

References

1. R. Backofen, T. Bilz, A. Ribalta and A. Voigt: SP_N-Approximations of Internal Radiation in Crystal Growth of Optical Materials. J. Crystal Growth., **266** 264–270 (2004)
2. M. Frank, M. Seaïd, J. Janicka, A. Klar, R. Pinnau and G. Thömmes: A Comparison of Approximate Models for Radiation in Gas Turbines. Int. J. Progress in CFD, **3** 191–197 (2004)
3. E. Larsen, G. Thömmes, A. Klar, M. Seaïd and T. Götz: Simplified P_N Approximations to the Equations of Radiative Heat Transfer and Applications. J. Comp. Phys., **183** 652–675 (2002)
4. D. Levermore: Moment closure hierachies for kinetic theories. J. Stat. Phys., **83** (1996).
5. Y. Jaluria: Natural Convection Heat and Mass Transfer. Pergamon Ress, Oxford (1980)
6. R. Pinnau and G. Thömmes: Optimal boundary control of glass cooling processes. M2AS, **120** 1261–1281 (2004)
7. M. Seaïd, M. Frank, A. Klar, R. Pinnau and G. Thömmes: Efficient Numerical Methods for Radiation in Gas Turbines. J. Comp. Applied Math, **170** 217–239 (2004)
8. M. Schäfer, M. Frank, and R. Pinnau: A hierarchy of approximations to the radiative heat transfer equations: Modelling, analysis and simulation. Math. Mod. Meth. Appl. Sci., **15** 643–665 (2005)
9. R. Pinnau and M. Seaïd: Simplified P_N solutions for natural convection–radiation in enclosures. In preparation (2006)

Minisymposium "Nonlinear Charge and Spin Transport in Semiconductor Nanostructures"

G. Platero

Instituto de Ciencia de Materiales de Madrid, CSIC, Spain

In this minisymposium the electronic and transport properties of different low-dimensional nanodevices have been discussed. In these devices, the interplay between charge, spin and vibrational degrees of freedom determines their main electronic and transport features. Moreover, the number of atoms in the system determines the more suitable theoretical framework and numerical techniques for each particular system.

In the first contribution, Prof. A.P. Jauho reviews different mathematical and computational tools useful to study different low-dimensional systems. He discusses first principles electronic structure methods that are appropriate for systems with a large number of atoms, such as semiconducting nanowires. Other theoretical frameworks based in the density matrix renormalization group can be used to describe strongly interacting systems.

Quantum dots in the Kondo regime belong to the class of strongly correlated systems. Dr. R. López studies the transport properties of a double quantum dot inserted in an Aharonov–Bohm interferometer, where interactions play a main role. Two limits are analyzed: low interdot Coulomb interaction, where spin fluctuations play the main role, and strong interdot Coulomb interaction.

In the case where the quantum dot is attached to superconducting contacts, the interplay between the Josephson and the Kondo effects has to be considered. Prof. A. Martín-Rodero has developed a comprehensive analysis of the interplay between Josephson effect, Kondo and antiferromagnetic coupling in a double quantum dot system attached to superconducting leads.

Nanoelectromechanical devices, as movable single electron transistors, are systems where the interplay between the charge and the mechanical degree of freedom determines the nonlinear transport properties. In particular, in the shuttle regime, a quantum dot oscillates, transfering one electron per cycle from one contact to another. Dr. A. Donarini presents a numerical technique for solving the generalized master equation and described the different operating regimes of the shuttle devices.

Spintronics is a very alive ramification of electronics, where the spin instead of the charge plays the main role. Recent transport experiments in double quantum dots show the important role played by the Pauli exclusion principle in current rectification. Spin blockade is observed at certain regions of dc voltages, and the interplay between Coulomb and spin blockade can be used to block the current in one bias direction while allowing it to flow in the opposite one. Then these devices could behave as externally controllable spin-Coulomb rectifiers with potential application in spintronics as spin memories and transistors. Spin decoherence and relaxation induced by hyperfine interaction have shown to reduce spin blockade producing a leakage current. Dr. J. Iñarrea presents a theoretical model based in rate equations for the charge occupations and nuclei polarizations which accounts for hyperfine and electron–phonon interaction and which allows to describe the spin blockade regime and the nuclear and electron spin dynamics.

In double quantum dots, it is possible to pump spin polarized electrons by means of external ac voltages. This is described by Dr. R. Sánchez using a theoretical model based in the density matrix formalism and the Markov approximation. He shows how to control the spin current polarization by tuning the ac frequency and intensity. He also shows how the spin blockade could be removed by photoassisted tunneling through the system and how the spin decoherence time could be inferred from the tunneling current.

In a clean semiconductor quantum wire the Rashba interaction affects the energy bands, modifying the wire magnetization and the linear conductance curves. Prof. Ll. Serra has calculated the spectral and transport properties of ballistic quasi-one dimensional systems in the presence of spin–orbit (Rashba) coupling. For a wire with local spin–orbit coupling, he predicts the occurrence of Fano lineshapes. He also shows that the local Rashba interaction acts in a strictly one-dimensional channel as an attractive impurity, leading to the formation of purely bound states. In a quasi-one dimensional system these bound states couple to the conduction ones through the Rashba intersubband mixing, giving rise to pronounced dips in the linear conductance plateaus.

Diluted magnetic semiconductors are very suitable materials to be incorporated as compounds in normal semiconductor nanodevices. They can be used for instance, as spin injectors and therefore, they are frequently used for spintronic purposes. Dr. D. Sánchez presents a theoretical model for analyzing spin-dependent transport in magnetically doped II–VI resonant tunneling diodes. He discusses spin transport for different diode configurations: magnetic or normal contacts and magnetically doped or normal semiconductor quantum wells.

Dr. Rossier has analyzed transport through a CdTe magnetically doped (with Mn) quantum dot doped. Using the density matrix formalism, he has shown that, under certain conditions, single electron transport through a single atom magnet can result in hysteretic behavior of the linear conductance versus gate voltage.

Electronic Transport in Nanowires at Different Length Scales

Antti-Pekka Jauho

MIC – Department of Micro and Nanotechnology, NanoDTU, Technical University of Denmark, 2800 Kgs. Lyngby, Denmark
antti@mic.dtu.dk

Summary. Nanowires, i.e., systems with a diameter of the order of 1–10 nm, and length up to microns, form a subclass of modern nanoscale systems, which hold a great promise for future technologies. For example, they could be used as interconnects in future's nanoelectronics, or they could form the basis of extremely sensitive sensors. In addition to their possible practical applications, nanowires exhibit a wide range of physical properties, which are of their own intrinsic interest. The theoretical scientist attempting to model charge transport in these systems faces many challenges. The number of atoms or active charge carriers requiring a microscopic treatment may vary from a few to several millions. The transport may be coherent, or dominated by interaction effects. No single formalism can capture all the different facets, and in this article a review of a few selected modern techniques, operative at different length scales, is given. Specifically, we shall be considering four different physical systems: (1) semiconducting nanowires; (2) gold atomic wires; (3) molecular electronics, and (4) one-dimensional strongly correlated chains.

1 Semiconducting Nanowires

1.1 Introduction

Semiconductor nanowires can be grown by a number of methods, and present technology allows one to change the chemical composition of the nanowire essentially within on lattice spacing. Figure 1 shows an example of a nanowire fabricated at Lund University. Since the different chemical compounds comprising the sample of Fig. 1 have different band-gaps, it is possible fine-tune the functionality of a given nanowire by a judicious choice of the potential energy landscape in which the carriers move. The narrowest wires have cross-sections of the order of 10 nm × 10 nm, and their lengths can reach to micrometers, and hence they consist of tens of thousands or millions of atoms. Even though the quality of these wires is very high, they nevertheless contains imperfections, such as defects, surface roughness, or intentionally introduced dopants. A fully first-principles approach, for example Density Functional Theory, is not

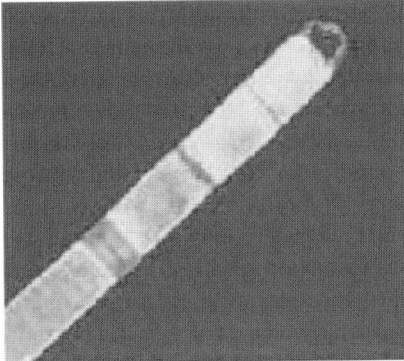

Fig. 1. A semiconductor nanowire grown at Lund University. The different colors indicate different materials, with different band-gaps. From [Sam03]

feasible, because the computational effort scales as $\mathcal{O}(N^3)$. Thus one needs some approximative methods which less stringent computational demands, but which nevertheless use parameters which are determined microscopically.

1.2 Theory

One possible way to proceed is to use a first-principles electronic structure method, which outputs some effective tight-binding parameters, thereby yielding an effective Hamiltonian

$$\hat{H} = \sum_i \left[\epsilon_i |i\rangle\langle i| + \sum_{j=n.n} \left(t_{ij}|j\rangle\langle i| + t_{ij}^*|i\rangle\langle j| \right) \right], \tag{1}$$

It should be emphasized that despite of the simple appearance of (1), its parameters contain information about a self-consistent first-principles calculation, thereby allowing a detailed, microscopic study of the effects of various dopants, impurities, or defects. There are a number of methods of how to evaluate the transport properties of a system described by a Hamiltonian such as (1). In our recent work [MRBJ06] we have used the SIESTA-code [SAG02] to evaluate the tight-binding parameters for a number of different disordered Si-nanowires, and then studied transport by either using a quantum diffusion approach (which we also call the Kubo method, see below) [Roc97], or the recursive Green function technique [Tod96], and in what follows we give some representative results for both of these methods, and discuss their relative merits and drawbacks. The basic set-up for the two methods is sketched in Fig. 2.

The idea behind the method of [Roc97] is to evaluate the energy resolved conductance $G(E, L)$ of a device of length L via the diffusion constant, $D_E(\tau)$, obtained from the Kubo formula (which explains the chosen nomenclature):

$$G(E, L) = 2e^2 \pi n(E) \frac{D_E(\tau)}{L}$$

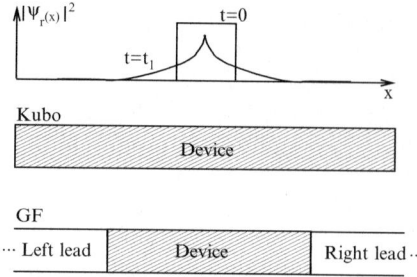

Fig. 2. *Top*: Schematic time evolution of a random phase state initially located in the central region of the nanowire. *Middle*: the geometry in the Kubo method consists only of a large device region. *Bottom*: in the Green function approach a device region is connected to two semi-infinite leads

$$D_E(\tau) = \frac{\sum_i \langle i(\tau)|\hat{X}\delta(E - \hat{H})\hat{X}|i(\tau)\rangle}{\mathrm{Tr}\{\delta(E - \hat{H})\}} \frac{1}{\tau}, \tag{2}$$

where $\hat{X}$ is the position operator, and the $|i(t)\rangle$ is the time-propagated state:

$$|i(t)\rangle = \mathrm{e}||^{-\mathrm{i}\hat{H}t/\hbar}|i(0)\rangle. \tag{3}$$

In order to evaluate the trace, one must generate a set of states which samples the available space of all states sufficiently densely. This can be carried out quite effectively by choosing a relatively modest number of random-phase states, all localized in the central part of the device, and then allowing them evolve in time to compute $D_E(\tau)$ via (2) (see Fig. 2). The procedure is repeated until convergence is achieved; however there is no a priori knowledge of how many times this must done, and it is a matter of trial and error to find the optimal parameter values. The computationally most demanding task consists of the repeated evaluations of the time-propagated states $|i(t)\rangle$. An efficient way to carry this out is to expand the time-evolution operator in terms of the Chebyshev polynomials, which have very appealing computational properties because of the recurrence relations they obey. Figure 3 compares the convergence of the Chebyshev method and a straightforward Taylor expansion of the time-evolution operator.

The recursive Green function method, our second approach in this section, is described in detail in many references, see. e.g. [Tod96]; we summarize it only very briefly here. In short, one first generates the "surface Green functions" for the isolated leads, and then uses these to "grow" the device, one unit cell at a time, making efficient use of the previously calculated results. Once the sample has reached the desired length L, the resulting conductance (obtained via Landauer formula) is recorded. In case of a disordered system, one needs to generate several realizations of the disorder, and then perform an ensemble averaging.

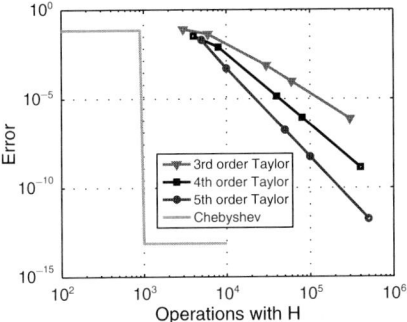

Fig. 3. Convergence test for a one-dimensional toy model, for which the exact solution is known. The error, defined as $\max\{||\psi\rangle_i - |\tilde{\psi}\rangle_i|\}$, where ψ and $\tilde{\psi}$ are the exact and approximate time-evolved states, is plotted as a function of the number of evaluations of the Hamiltonian. The simulation time-interval was 500 fs

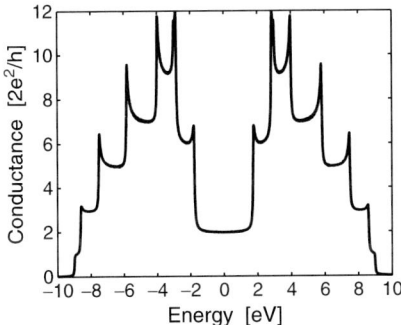

Fig. 4. Conductance of an ideal (5,5) nanotube calculated with the quantum diffusion method

1.3 Physical Examples

As a first bench-mark we compute the conductance an ideal carbon nanotube with the Kubo method. For a pristine nanotube the conductance should have a strict staircase form: each one-dimensional subband contributes to the conductivity by $2e^2/h\times$(degeneracy). The overshoots seen in Fig. 4 at the transition edges are spurious and indicate that Kubo method does not yield accurate results for one-dimensional ballistic systems close to an opening or closing of a conduction channel. This problem can be traced to the singular behavior of the one-dimensional density of states at the band-edge, which should be exactly canceled by the vanishing band-velocity; the numerics however have difficulties in achieving this. On the other hand, for disordered or higher-dimensional systems, which are more relevant from the practical point of view, and are also the focus of our study, these spurious effects do not play a significant role.

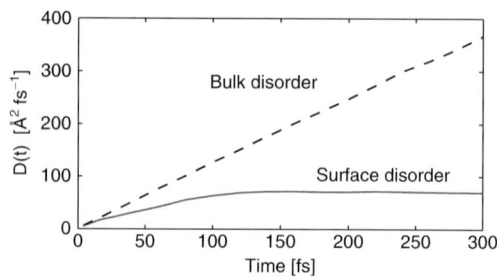

Fig. 5. Comparison of bulk and edge disorder

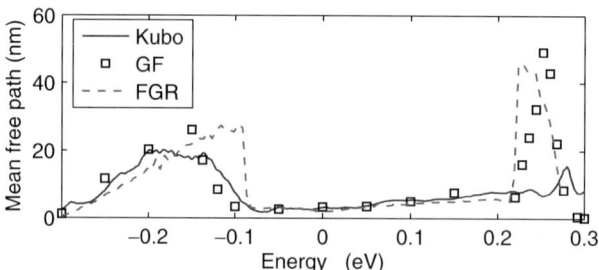

Fig. 6. Mean free path computed with three methods: *solid line*: Kubo method; boxes: recursive Green functions; *dashed*: Fermi Golden rule (not discussed here, see, however, [MRBJ06]). The Green function results are averaged over 200 different realizations of the disorder, while the Kubo results are mean values of 10 different samples

In Fig. 5 we show the time-dependent diffusion constant, (2), for two different type of Anderson disorders: the on-site energies of either the bulk or the surface atoms are given a random increment. As is seen from the figure, the bulk disorder has hardly any effect – the wire stays ballistic with a diffusion constant that increases linearly with time, while the edge disorder leads to an ohmic behavior. This is easily understood because the extended states, which carry the current, lie close to the surface, and are therefore much more sensitive to disorder.

In Fig. 6 we show a comparison of the Kubo method and the recursive Green function method. As is seen, the two methods are in general in reasonable agreement; the main discrepancy at $E = 0.24\,\text{eV}$ is due to the inherent difficulty of the Kubo method to deal with sharp densities of states, discussed in connection with Fig. 4. Based on a large number of calculations similar to those reported in Fig. 6, and described in detail in [MRBJ06], we conclude that the Kubo method is advantageous if one needs the transport properties at many different energies, since the energy resolved results are obtained in just one calculation, while the Green function method requires a full calculation for each energy. On the other hand, if one needs only a few energies, such as in metallic systems where only energies close to the Fermi energy are

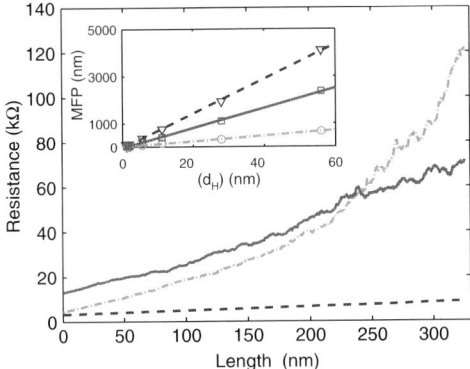

Fig. 7. Effect of Hydrogen adatoms on the length dependence of the resistance, at three different energies, $E = -0.15\,\text{eV}$ (*dash-dotted line*); $E = -0.3\,\text{eV}$ (*dashed line*), and $E = -0.03\,\text{eV}$ (*solid line*). Energies are measured from the Fermi level

relevant, then the Green function method with its higher inherent accuracy is to be preferred.

We have also studied the effect of Hydrogen adatoms on the conductance of a Si-wire. Depending on the density of the adatoms one may observe either a ballistic behavior (the resistance is a linear function of the system length), or localized behavior (the resistance grows exponentially with length). An example is shown in Fig. 7; many other disorder effects are considered in our recent paper [MRBJ06], to which the reader is referred.

2 The Density Functional: Nonequilibrium Green Function Paradigm

2.1 Formalism

In this section we give a brief introduction to the computational schemes that combine some ab initio electron structure theory and the nonequilibrium Green function theory. In the rapidly growing literature one can find several implementations; here we use the code developed in [BMOTS02] as an illustrative example.

Most electronic structure calculations are restricted in the sense that the geometry must be finite, or periodic, and that the electronic system is in equilibrium. The present situation is very different: now a small subsystem, i.e., the nanowire, lacking translational invariance couples to semi-infinite leads *and* the electronic subsystem can be far from equilibrium. Ideally one should describe the whole system (the central region and electrodes) on equal footing. As is well-known, the Density Functional Theory gives the exact electronic density and total energy, if the exact exchange-correlation functional

was known. Since this is not the case, one must resort to approximate forms
of the functional, such as the local-density approximation (LDA), or the generalized gradient approximation (GGA), or something else. There is no theory
to say which (approximate) functional is the best, rather the choice is made
based on painstaking tests, and comparisons in some limits where alternative
methods, or experiments, can give benchmarks. In an attempt to extend DFT
to nonequilibrium situations one must go one step further: the Kohn–Sham
single-particle wave-functions $\psi_{KS}(x)$ are used when calculating the current.
This implies a leap of faith: as is well-known, the ψ_{KS} are useful mathematical
objects used in the construction for the ground-state density, but which have
no immediate physical interpretation. Nonperturbative many-particle effects,
such as the Kondo effect, are excluded from the treatment. On the other hand,
inelastic effects can be included, as discussed below. A further development
of the present approach could conceivably be reached by the current-density
formalism [VK96], or time-dependent density-functional formalism [RG84].

At the core of the DFT-NEGF implementation described in [BMOTS02]
is the SIESTA code [SAG02] for calculating the electronic properties for large
numbers of atoms. This approach has many technical advantages because
of the employed finite range orbitals for the valence electrons: not only do
the numerics get faster but also the system partitioning into leads and the
central region becomes unambiguous. The SIESTA approach can be extended
to nonequilibrium by using a nonequilibrium electron density as an input.
In nonequilibrium Green function theory [HJ96] the nonequilibrium density
readily follows from the lesser Green function,

$$n(x) = -\mathrm{i}G^<(x = x', t = t') = \int \frac{\mathrm{d}\epsilon}{2\pi\mathrm{i}} G^<(x = x', \epsilon). \tag{4}$$

$G^<$ follows directly from the Keldysh equation, because the self-energy is a
known function for mean-field theories (such as DFT): $\Sigma^< = \mathrm{i}(\Gamma^L f_L + \Gamma^R f_R)$,
and consequently $G^< = \mathrm{i}G^r(\Gamma^L f_L + \Gamma^R f_R)G^a$. Hence, all that one needs
are the retarded and advanced Green functions, and these are obtained by
evaluating

$$\mathbf{G}^{r,a}(E) = [E\mathbf{I} \pm \mathrm{i}\eta - \mathbf{H}]^{-1}, \tag{5}$$

where

$$\mathbf{H} = \begin{pmatrix} \mathbf{H}_L + \mathbf{\Sigma}_L & \mathbf{V}_L & 0 \\ \mathbf{V}_L^\dagger & \mathbf{H}_C & \mathbf{V}_R \\ 0 & \mathbf{V}_R^\dagger & \mathbf{H}_R + \mathbf{\Sigma}_R \end{pmatrix} \tag{6}$$

The semi-infinite left and right leads are accounted for by the self-energies
$\mathbf{\Sigma}_{L/R}$, see, e.g., [Datta]. The matrices $\mathbf{V}_{L/R}$ give the coupling of the leads
to the central region, described by the Hamiltonian $\mathbf{H}_C$. Importantly, to
determine $\mathbf{V}_R$, $\mathbf{V}_L$, or $\mathbf{H}_C$ one does not need to evaluate the density matrix outside the $L - C - R$ region, if the $L - C - R$ region is defined so large
that all screening takes place inside it.

Summarizing, and somewhat simplifying, the TRANSIESTA iterative loop consists of the steps

$$\text{initial } n(x) \Rightarrow \text{SIESTA} \Rightarrow \psi_{\text{KS}}(x) \Rightarrow \text{NEGF} \Rightarrow \text{new } n(x), \qquad (7)$$

and the iteration is repeated until convergence is achieved for the desired quantity, such as the current for a given voltage difference. For a detailed description of many of the technical details suppressed here we refer to the paper by Brandbyge et al. [BMOTS02]. The scheme outlined above, and similar parallel implementations, have been applied by a large group of researchers to many specific physical systems. Occasionally the agreement with experiments reaches a quantitative level, which is indeed very satisfying, while sometimes the predicted current can be orders of magnitude too large. At present, there is no consensus of whether the discrepancies are due to poorly controlled experiments, bad implementations of the DFT-NEGF scheme, or due to an inadequacy of the entire concept. A possible cause for the discrepancy has very recently been identified in [TFSB05], who suggest that self-interaction corrections (which are not included in the GGA-LDA underlying most theoretical work) could remedy some of the problems. Nevertheless, a lot of research remains to be done.

2.2 Vibrational Effects in Atomic Gold Wires

The issue of vibrational effects in molecular electronics has recently drawn a lot of interest because inelastic scattering and energy dissipation inside atomic-scale conductors are of paramount importance for device characteristics, working conditions, and their stability [KLP04, WLKR04, SUR04, F04]. Inelastic effects are important, not only because of their potentially detrimental influence on device functioning, but also because they can open up new possibilities and operating modes. Vibrational effects are often visible in the measured conductances of nanoscopic objects; here we focus on recent experimental studies on free standing atomic gold wires. Agraït and co-workers [AURV02] used a cryogenic STM tip to first create an atomic-scale gold wire (lengths up to seven gold atoms have been achieved), and then measured its conductance as a function of the displacement of tip, and the applied voltage. The data showed clear drops of conductance at a certain voltage, and the interpretation was that an excitation of an inelastic mode was taking place, leading to enhanced back-scattering, and hence drop in the conductance. It should be pointed out that opening a new vibrational mode in the atomic scale conductor does not necessarily lead to a decrease in conductance (one can envisage various assisted processes), and a proper theory should be able to predict conductance enhancement as well, whenever the physics dictates so.

Here I will briefly describe our recent work [FBLJ04, FPBJ06] on inelastic effects in the kind of wires studied by Agraït et al. [AURV02]. An important aspect of our work is that we go beyond lowest order perturbation theory in the electron-vibration coupling, and therefore polaronic effects can be included. However, as we shall discuss below, this approach is computationally very expensive, and physically motivated approximation schemes are of paramount importance if one wishes to describe more complicated systems than few-atom metallic nanowires. We also address the issue of phonon heating, albeit within a phenomenological model, as discussed below.

The calculational method consists of three steps. (1) The mechanical normal modes and frequencies of the gold chain are evaluated. (2) The electronic structure and electron-vibration coupling elements are evaluated in a localized atomic-orbital basis set. (3) The inelastic transport is evaluated using NEGF by using a self-consistent Born approximation self-energy in the Dyson and Keldysh equations for the respective Green functions. The electrical current and the power transfer are then evaluated with (here, for the left lead; for a detailed derivation, see [FPBJ06])

$$I_L = \frac{e}{h} \int \mathrm{d}\epsilon \, t_L(\epsilon) \tag{8}$$

$$P_L = \int \frac{\mathrm{d}\epsilon}{2\pi\hbar} \epsilon t_L(\epsilon) \tag{9}$$

$$t_L(\epsilon) = \mathrm{Tr}\left\{ \boldsymbol{\Sigma}_L^<(\epsilon)\mathbf{G}^>(\epsilon) - \boldsymbol{\Sigma}_L^>(\epsilon)\mathbf{G}^<(\epsilon) \right\}, \tag{10}$$

where Hartree and Fock parts of self-energy components are

$$\boldsymbol{\Sigma}^{H,r} = \mathrm{i} \sum_\lambda \frac{2}{\Omega_\lambda} \int \frac{\mathrm{d}\epsilon'}{2\pi} \mathbf{M}^\lambda \mathrm{Tr}[\mathbf{G}^<(\epsilon')\mathbf{M}^\lambda] \tag{11}$$

$$\boldsymbol{\Sigma}^{H,<} = 0 \tag{12}$$

$$\boldsymbol{\Sigma}^{F,r}(\epsilon) = \mathrm{i} \sum_\lambda \int \frac{\mathrm{d}\epsilon'}{2\pi} \mathbf{M}^\lambda [D_0^r(\epsilon - \epsilon')\mathbf{G}^<(\epsilon')$$

$$+ D_0^r(\epsilon - \epsilon')\mathbf{G}^r(\epsilon') + D_0^<(\epsilon - \epsilon')\mathbf{G}^r(\epsilon')]\mathbf{M}^\lambda \tag{13}$$

$$\boldsymbol{\Sigma}^{F,<}(\epsilon) = \mathrm{i} \sum_\lambda \int \frac{\mathrm{d}\epsilon'}{2\pi} \mathbf{M}^\lambda D_0^<(\omega - \omega')\mathbf{G}^<(\epsilon')\mathbf{M}^\lambda. \tag{14}$$

Here the vibrational modes are labeled by λ, and Ω_λ is the corresponding eigenfrequency. It is worth noting that the lack of translational invariance makes the retarded Hartree term non-zero, and potentially important. Also, at this stage the phonon propagators are undamped – an approximation that merits further investigation. The coupled equations are iterated until convergence is achieved, and in the following we give some representative results.

We have considered a number of atomic gold wires under different states of strain, as shown in Fig. 8. We calculate the phonon signal in the nonlinear

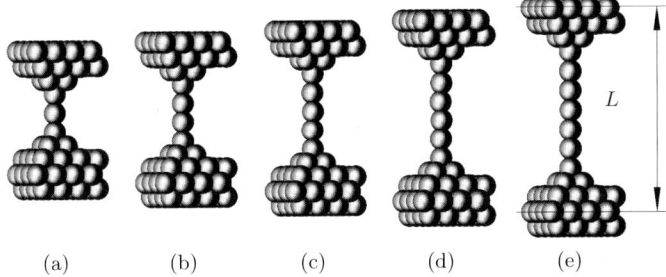

Fig. 8. Generic gold wire supercells containing 3–7 atoms bridging pyramidal bases connected to stacked Au(100) layers

differential conductance vs. bias voltage for two extremal cases: the energy transferred from the electrons to the vibrations is either (1) instantaneously absorbed into an external heat bath, or (2) accumulated and only allowed to leak via electron–hole pair excitations. These limits are referred to as the externally damped and externally undamped cases, respectively.

Since a typical experiment is done at low temperatures, the mode occupation in the externally damped case vanishes, $N_\lambda \approx 0$. In the externally undamped case the mode occupation N_λ is an unknown parameter entering the electron–phonon self-energy, and additional physical input is necessary to determine this parameter. We argue as follows. Since the system is in a steady state, the net power transferred from the electrons to the device must vanish, i.e., $P_L + P_R = 0$. Using (9) one then obtains the required constraint on N_λ. This procedure works in a straightforward way if there is only a single active mode, but if several modes are present, a more detailed theory of how the phonon modes equilibriate would be needed.

When comparing to the experiments of Agraït et al. [AURV02] (Figure 9), one sees that the externally undamped model is in near quantitative agreement with the data: the conductance drop at the onset of inelastic scattering, and the slope after the drop are very well reproduced. We view this as strong evidence of the presence of heating in the experiment, but at the same time recognize the need for a detailed microscopic theory including phonon–phonon interactions.

2.3 More Complicated Systems: The Lowest Order Expansion (LOE)

The numerical task of solving the SCBA equations (10)–(14) is prohibitive for all but the simplest systems. However, for systems where the electron–phonon coupling is weak, and the density of states varies slowly with energy, a much more efficient scheme has been developed recently by Paulsson and co-workers [PFB05], to be referred to as the LOE method. The idea here is to get rid of the time-consuming numerical energy-integrations in (10)–(14). This

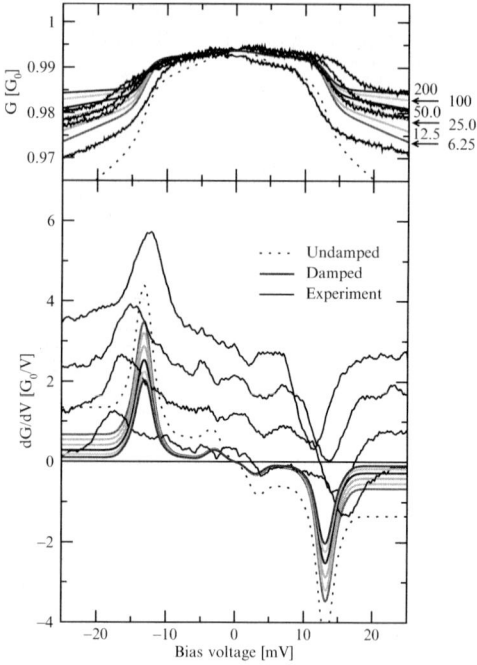

Fig. 9. Comparison between theory and experiment [AURV02] for the inelastic conductance of an atomic gold wire. The measured charactersitics correspond to different states of strain of the wire (around 7 atoms long). The calculations are for the 7 atom wire at $L = 29.20\,\text{Å}$. (Reproduced from [FPBJ06])

can be achieved by appealing to the above-mentioned weak energy dependence and the weakness of the electron–phonon interaction, which allows an analytic integration after an expansion in the electron–phonon matrix element has been carried out. We do not reproduce the rather lengthy expressions here (in fact, Mathematica was essential in obtaining them), and merely state that they allow a very efficient numerical evaluation, cutting the computational time down by orders of magnitude. For a fuller discussion, we refer to [PFB05] and [FPBJ06]. As an illustrative example, we reproduce here results from [PFB06], where the transport properties of a number of conjugated and saturated hydrocarbon molecules between gold electrodes were evaluated. The motivation for this work was provided by the recent experiments of [KLP04], who reported detailed experimental data for these systems. In particular, Fig. 10 compares the experimental and theoretical inelastic electron tunneling spectroscopy spectra for one of these molecules. Quite interestingly, the main features are in good agreement, and with the help of the theory one thus can identify the few specific vibrational modes that are important for the transport properties.

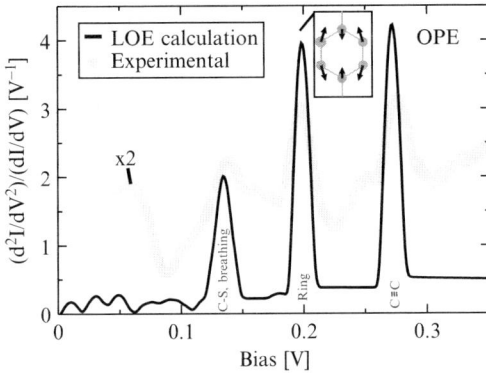

Fig. 10. Calculated IETS spectrum for an OPE molecule compared to the experimental data from [KLP04]. Each of the three inelastic scattering peaks arise from different kinds of vibrations localized on the molecule. (Reproduced from [PFB06])

3 The Density Matrix Renormalization Group Applied to Transport

In the sections above we have discussed systems which can be treated with perturbation theory methods. However, a very interesting subclass of systems does not yield to this approach: systems where the correlations are so strong that the actual ground state is not adiabatically connected to the noninteracting ground states. Examples of such systems include the Kondo phenomenon, Luttinger liquids, charge–density waves, and several others. To evaluate the transport properties of strongly correlated systems is one of the most challenging and most active fields of research in condensed matter theory. A variety of methods are available, and here we describe one of them – our recent attempts to bring the Density-Matrix Renormalization Group (DMRG) out of equilibrium. The DMRG method was introduced by S. White about 15 years ago [White92, White93], and it has been proven to be an extremely successful method for determining ground state properties of strongly correlated systems. In fact, in the few cases where exact solutions are known by analytical means, such as the Bethe Ansatz method, the results obtained by DMRG are extremely accurate. The method is quite subtle and a proper discussion is not appropriate in the present context. Very shortly, the DMRG algorithm systematically optimizes the basis set used to describe the interacting system; the optimization process can be formulated as a variational principle for the system density matrix. Depending on the problem under investigation, one can choose a number of physical properties for which optimal solution is sought, these are called the target states. The DMRG method has a number of limitations, though. It becomes numerically very expensive, if one tries to move away from one dimension. It is essentially a ground-state formalism (in this sense it is plagued by similar problems as one faces when

applying DFT to nonequilibrium). Also, it is very numerical in its nature –
the final results are not always easy interpret in physical terms. Nevertheless,
the potential rewards in having (numerically) exact results for transport properties for (some) strongly interacting systems are so large, that many groups
are presently working to achieve this goal.

The ultimate goal would be to calculate the full nonlinear IV-curve, for
example with the Meir–Wingreen formula [MW92, HJ96],

$$I = \frac{e^2}{h} \int d\epsilon \frac{\Gamma^L \Gamma^R}{\Gamma^R + \Gamma^L} A(\epsilon)[n_F(\epsilon - \mu_L) - n_F(\epsilon - \mu_R)], \tag{15}$$

where $A(\epsilon)$ is the interacting spectral function for the central region.
A straightforward DMRG evaluation of this formula, however, is not possible,
because DMRG finds the lowest energy state(s), and in a biased structure
all particles would accumulate in the low-bias region. How to overcome this
problem is one of the outstanding issues. On the other hand, linear response
conductance can be evaluated with DMRG, and here we describe some recent
results obtained by D. Bohr and co-workers [BSW06].

In linear response one may use the Kubo formula to obtain formal expressions for the electrical conductance. Bohr et al. [BSW06] have recently shown
that the conductance calculation boils down to the evaluation of either of the
following correlation functions:

$$g_{J_i N} = -\frac{e^2}{h} \langle \Psi_0 | J_{n_i} \frac{4\pi i\eta}{(H_0 - E_0)^2 + \eta^2} N | \Psi_0 \rangle$$

$$g_{JJ} = \frac{e^2}{h} \langle \Psi_0 | J_{n_1} \frac{8\pi\eta(H_0 - E_0)}{[(H_0 - E_0)^2 + \eta^2]^2} J_{n_2} | \Psi_0 \rangle. \tag{16}$$

Here, Ψ_0 is the exact interacting ground-state, J_{n_i} is the current operator at
site n_i, N is the occupation number operator, and η is cut-off parameter that
must have finite, yet small value in the numerical calculations. Equations (16)
are in a form that can be evaluated with the DMRG algorithm. They are
equivalent, and they have both their merits and dismerits in the numerical
work (see [BSW06] for a fuller discussion). In what follows we shall display
some numerical results to illustrate the utility of these formulas.

As explained above, due to the numerical nature of the DMRG approach it
is always very important to bench-mark it against some known results. Since
analytical results for an interacting result are rare, we have carried out a number of comparisons against exactly solvable noninteracting systems. Figure 11
shows the conductance of a noninteracting single-level system, computed both
with the DMRG (the curves denoted by g), and via an exact diagonalization
(the curves denoted by f). Several interesting conclusions can be drawn from
Fig. 11.

The figure clearly demonstrates the numerical exactness of the DMRG
approach: the DMRG results (crosses) are essentially indistinguishable from

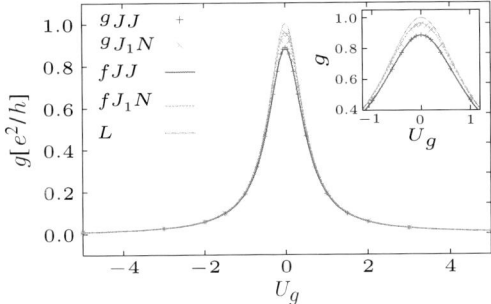

Fig. 11. The conductance of a single-level system as a function of the gate voltage. The curves labeled with g are DMRG results, while those labeled with f are exact diagonalization results. Also shown (label L) is the Lorentzian conductance for a system with semi-infinite leads. *Inset*: conductance close to zero gate voltage

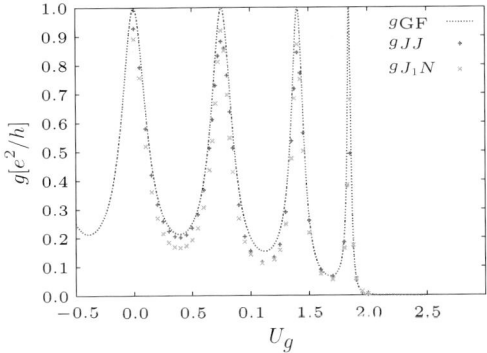

Fig. 12. Conductance of a noninteracting seven-site system as a function of the gate voltage. *Continuous line*: exact results obtained with Green functions; *crosses*: numerical results obtained with DMRG

the exact diagonalization results (continuous lines). Also, we conclude that for this particular case the current–density correlation function performs better than the current–current correlation function. Neither method obtains at resonance the full unit conductance of a system with semi-infinite leads. This is related to subtle finite-size effects which always are present in a numerical approach. It is a matter of considerable difficulty to optimize the various cutoffs so that the finite-size effects are minimized, and typically one is forced to proceed with a trial-and-error approach. Another bench-mark against an exactly solvable model is shown in Fig. 12, where we compare the DMRG results against a Green function calculation for a system, where the central region consists of seven sites. Again, the general agreement is very good, except for conductance peaks, where finite-size effects lead to a slight suppression.

Very importantly one should bear in mind that the DMRG calculation could also have been carried out at the same numerical cost for an interacting

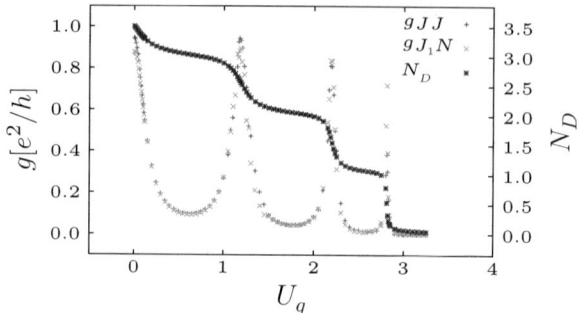

Fig. 13. Conductance as a function of gate voltage for a seven site system, with a relatively weak nearest-neighbor Hubbard interaction (the Luttinger Liquid regime). The total system length is 150 sites. Reproduced from [BSW06]

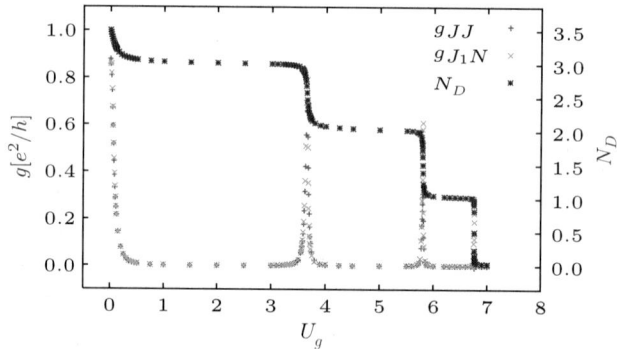

Fig. 14. Conductance as a function of gate voltage for a seven-site system, with a strong nearest-neighbor Hubbard interaction (the charge–density wave regime). Reproduced from [BSW06]

system, for which exact results cannot be obtained. We conclude by displaying results obtained for an interacting system (Figures 13 and 14), where the central region consists of seven sites, and the spinless fermions interact with the neighboring sites with a Hubbard interaction [BSW06]. We see that both the resonance widths and positions are strongly affected by the interaction effects (see Fig. 12). The results shown in these figures form an important proof-of-principle demonstration. However, much work remains to be done in order to extend these calculations to more realistic situations. Special attention should be devoted to spinfull systems, more realistic description of the leads, and to Aharonov–Bohm geometries, and work is in progress along this direction.

4 Conclusion

We have reviewed a number of methods for computing the conduction properties of nanowires. Depending on the goals of the investigation one may choose mean field models capable of treating millions of atoms, in one extreme, or numerically exact methods restricted to one-dimensional and rather small systems, in the other extreme. In our presentation we have also tried to identify open problems and unresolved issues, and areas for further improvement. We have not discussed experimental issues in detail, but emphasize that in our opinion the real driving force is provided by new experiments which place the theoretical models under stringent tests.

Acknowledgements

All the heavy numerical work underlying the present review has been carried out by Dan Bohr, Mads Brandbyge, Thomas Frederiksen, Troels Markussen, Magnus Paulsson, and Riccardo Rurali, and the author wishes to express his thanks for the pleasant collaborations.

References

[Sam03]   L. Samuelson, Mater. Today **6**, 22 (2003).
[MRBJ06]   T. Markussen, R. Rurali, M. Brandbyge, and A. P. Jauho: Phys. Rev. B **74**, 245313 (2006).
[SAG02]   J. M. Soler, E. Artacho, J. D. Gale, A. García, J. Junquera, P. Ordejn, and D. Snchez-Portal, J. Phys.: Condens. Matter **14**, 2745 (2002).
[Roc97]   S. Roche and D. Mayou, Phys. Rev. Lett. **79**, 2519 (1997).
[Tod96]   T. N. Todorov, Phys. Rev. B **54**, 5801 (1996).
[BMOTS02]   M. Brandbyge, J. L. Mozos, P. Ordejn, J. Taylor, and K. Stokbro, Phys. Rev. B **65**, 165401 (2002).
[VK96]   G. Vignale and W. Kohn, Phys. Rev. Lett. **77**, 2037 (1996).
[RG84]   E. Runge and E. K. U. Gross, Phys. Rev. Lett. **52**, 997 (1984).
[HJ96]   H. Haug and A. P. Jauho: Quantum kinetics in transport and optics in semiconductors. Springer Series in Solid State Sciences **123**, Berlin (1996).
[Datta]   S. Datta: Electronic transport in mesoscopic systems. Cambridge University Press, Cambridge (1995).
[TFSB05]   C. Toher, A. Filippetti, S. Sanvito, K. Burke: Phys. Rev. Lett. **95**, 146402 (2005).
[KLP04]   J. G. Kushmerick, J. Lazorcik, C. H. Patterson, R. Sahshidhar, D. S. Seferos, G. C. Bazan: Nano Lett. **4**, 639 (2004).
[WLKR04]   W. Y. Wang, T. Lee, I. Kretzschmar, M. A. Reed: Nano Lett. **4**, 643 (2004), 643 (2004).
[SUR04]   R. H. M. Smit, C. Untiedt, J. M. van Ruitenbeek: Nanotechnology **15**, 472 (2004).
[F04]   K. Flensberg: Phys. Rev. B **68**, 205323 (2004).

[AURV02]   N. Agraït, C. Untiedt, G. Rubio-Bollinger, S. Vieira: Phys. Rev. Lett. **88**, 216803 (2002).

[FBLJ04]   T. Frederiksen, M. Brandbyge, N. Lorente, A. P. Jauho: Phys. Rev. Lett. **93**, 256601 (2004).

[FPBJ06]   T. Frederiksen, M. Paulsson, M. Brandbyge, A. P. Jauho: cond-mat/0611562 (submitted to Phys. Rev. B).

[PFB05]   M. Paulsson, T. Frederiksen, M. Brandbyge: Phys. Rev. B **72**, 201101(R) (2005).

[PFB06]   M. Paulsson, T. Frederiksen, and M. Brandbyge: Nano Letters **6**, 258 (2006).

[White92]   S. R. White: Phys. Rev. Lett. **96**, 2863 (1992).

[White93]   S. R. White: Phys. Rev. B **48**, 10345 (1993).

[MW92]   Y. Meir, N. S. Wingreen: Phys. Rev. Lett **68**, 2512 (1992).

[BSW06]   D. Bohr, P. Schmitteckert, P. Wölfle: Europhys. Lett. **73**, 246 (2006).

SU(4) Kondo Effect in a Mesoscopic Interferometer

Rosa López

Departament de Física, Universitat de les Illes Balears, E-07122 Palma
de Mallorca, Spain
rosa.lopez-gonzalo@uib.es

Summary. We investigate theoretically the transport properties of a closed
Aharonov–Bohm interferometer containing two quantum dots in the Kondo limit.
We find two distinct physical scenarios depending on the strength of the interdot
Coulomb interaction. For negligible interdot interaction, transport is governed by
the interference of two Kondo resonances, whereas for strong interdot interaction
transport takes place via simultaneous correlations in both spin and orbital sectors.

1 Introduction

Progressive advance in nanofabrication technology has achieved the realization of tiny droplets of electrons termed quantum dots (QDs) with a high-
precision tunability of the transport parameters [1]. One of the most exciting
features of a QD is its ability to behave as a quantum impurity with spin $1/2$.
At temperatures lower than the Kondo temperature ($T_{\rm K}$), the localized spin
becomes strongly correlated with the conduction electrons and consequently
is screened [2]. Experimentally, the formation of the resulting singlet state
[having SU(2) symmetry] is demonstrated by a narrow peak at zero bias in
the differential conductance [3].

A natural step forward is the understanding of the magnetic interactions
of two artificial Kondo impurities [4]. The study of double QDs is mainly motivated by the possibility that they may represent the key stone to implement
a tunable spin two-qubit circuit [5]. When the two QDs are interacting, the
orbital degrees of freedom come into play as a pseudo-spin, as shown experimentally in [6], which may give rise to exotic physical scenarios. In particular,
it was recently found that when the interdot Coulomb interaction (U_{12}) is
large, there arises a Kondo correlated state possessing a SU(4) symmetry.
This unusual Kondo state involves entanglement between the (real-)spin and
pseudo-spin [7] leading to a great enhancement of $T_{\rm K}$. In this work, we consider
a double quantum dot (DQD) embedded in a prototypical mesoscopic interferometer threading a magnetic flux Φ. Our motivation is twofold: (1) striking

effects such as Fano resonances arise already in the noninteracting case and, more interestingly, (2) as the interdot interaction gets stronger, the local density of states (DOS) on the DQD changes drastically [8]. Here, we provide *a unified picture of the combined influence of wave interference, Kondo effect, and interdot interaction on the electronic transport through a DQD in and out of equilibrium.* Of particular interest is the behavior of the transmission through this system. When the interdot Coulomb energy is negligible each QD can accomodate one electron and both spins become screened. We find that each QD develops a Kondo resonance at the Fermi level $E_F = 0$. Their interference causes a very narrow *dip* in $\mathscr{T}$ except at $\phi = \Phi/\Phi_0 \approx 0 \pmod{2\pi}$ with $\Phi_0 = h/e$ the flux quantum. For large U_{12} only one electron lies in the DQD system. Remarkably, for $\phi = \pi \pmod{2\pi}$ a single Kondo state emerges with total SU(4) symmetry [7]. The screened magnitude is now the *hyperspin* $\mathbf{M} \equiv \sum_{i,j}(S^i + 1/2)(T^j + 1/2)$, where S^i (T^j) is the ith (jth) component of the real- (pseudo-) spin. As a result, $\mathscr{T}$ shows a peak instead of a dip.

2 Theoretical Approaches: Scaling Analysis and Numerical Renormalization Group

In this section we present the different theoretical approaches used to treat this system. Firstly, we employ the "scaling approach" and then we apply a more sophisticated technique, the numerical renormalization group (NRG) to confirm our previous predictions. When U_{12} is vanishingly small (and yet $U_1, U_2 \to \infty$), the two dots are both singly occupied: $\langle n_1 \rangle = \langle n_2 \rangle \approx 1$. Each dot can thus be regarded as a magnetic impurity with spin 1/2. A Schrieffer-Wolff transformation [2] gives a Kondo-like Hamiltonian (the total Hamiltonian is $\mathscr{H} = \mathscr{H}_0 + \mathscr{H}_K$):

$$\mathscr{H}_K = \frac{J_1}{4}\mathbf{S} \cdot \left[\psi_1^\dagger \boldsymbol{\sigma}\psi_1 + \psi_2^\dagger \boldsymbol{\sigma}\psi_2\right] + \frac{J_2}{4}\mathbf{S} \cdot \left[\psi_1^\dagger \boldsymbol{\sigma}\psi_2 + h.c.\right]$$
$$+ \frac{J_3}{4}(\mathbf{S}_1 - \mathbf{S}_2) \cdot \left[\psi_1^\dagger \boldsymbol{\sigma}\psi_1 - \psi_2^\dagger \boldsymbol{\sigma}\psi_2\right] - I\mathbf{S}_1 \cdot \mathbf{S}_2 , \qquad (1)$$

where $\boldsymbol{\sigma}$ denotes the Pauli matrices and $\psi_\mu = \sum_k \psi_{\mu,k}$ with $\psi_{\mu,k} = [c_{\mu,k,\uparrow}\ c_{\mu,k,\downarrow}]$ ($\mu = 1, 2$) the spinor. Here we have taken the canonical transformation $c_{1(2),k,\sigma} = \left(e^{\pm i\pi/4}c_{L,k,\sigma} + e^{\mp i\pi/4}c_{R,k,\sigma}\right)/\sqrt{2}$ for clearer interpretation. In (1), $\mathbf{S}_i = \psi_i^\dagger \boldsymbol{\sigma}\psi_i$ is the spin operator on the dot i, where $\psi_i = [d_{i,\uparrow}\ d_{i,\downarrow}]$ ($i = 1, 2$), and $\mathbf{S} = \mathbf{S}_1 + \mathbf{S}_2$. The physical values for the coupling constants in (1) are: $J_1 = 2|V|^2/|\varepsilon_d|$, $J_2 = J_1\cos(\phi/2)$ and $J_3 = J_1\sin(\phi/2)$. The normalized $I = \rho_0 J_1^2/2$ corresponds to a ferromagnetic RKKY coupling (ρ_0 the DOS in the leads). Under the renormalization group (RG) transformation (or scaling analysis) [2], the Kondo couplings scale to strong coupling according to the scaling equation $dJ_1/d\ell = 2\rho_0 J_1^2$, where $\ell = -\log D$ and D is the bandwidth. Other Kondo couplings keep the relations $J_2/J_1 = \cos(\phi/2)$, $J_3/J_1 = \sin(\phi/2)$

under RG transformations(details are provided in [9]). For $\phi \gtrsim 0$ the Kondo temperature reads $T_K^{SU(2)} = D \exp(-1/2\rho_0 J_1)$. When $I \gg T_K^{SU(2)}$, the strong RKKY interaction makes $\mathbf{S}$ a triplet, which is eventually screened following a two-stage procedure [10]. An exception is for ϕ exactly 0, where $\mathbf{S}_1$ and $\mathbf{S}_2$ are coupled only to a single conduction band through the RKKY interaction and therefore they are underscreened at $T \to 0$ [10]. Nevertheless, in an actual experimental situation [11] the QDs are far apart and the RKKY interaction may be negligible. Then, for $\phi \gtrsim 0 \pmod{2\pi}$ and $T = 0$ the ground state is always a Fermi liquid. In the limit of $U_{12} \to \infty$ the system properties change completely. Now, only one electron is acommoded in the whole DQD system, i.e., $\langle n_1 + n_2 \rangle \approx 1$ having either spin $\uparrow$ or $\downarrow$. The orbital degrees of freedom (pseudo-spin) play as significant a role as the spin, and the DQD behaves as an impurity with four degenerate levels with different tunneling amplitudes depending on the applied flux. Due to the orbital degrees of freedom involved in the interference, the symmetry of the wavefunction is crucial. Therefore, in this limit, it is more useful to work with a representation in terms of the symmetric (even) and antisymmetric (odd) combinations of the localized and delocalized orbital channels. Then the field operators are $\psi_d^\dagger = [d_{e,\uparrow}^\dagger \; d_{e,\downarrow}^\dagger \; d_{o,\uparrow}^\dagger \; d_{o,\downarrow}^\dagger]$ for the DQD and $\psi_k^\dagger = [c_{e,k,\uparrow}^\dagger \; c_{e,k,\downarrow}^\dagger \; c_{o,k,\uparrow}^\dagger \; c_{o,k,\downarrow}^\dagger]$ for the leads. To examine the low-energy properties of the system, we obtain for all values of ϕ the following effective Hamiltonian:

$$\mathcal{H}_K = \frac{J_1}{4}\left[\mathbf{S}\cdot(\psi^\dagger\boldsymbol{\sigma}\psi) + \mathbf{S}\cdot(\psi^\dagger\boldsymbol{\sigma}\tau^z\psi)\overline{T}^z\right] - J_5\overline{T}^z + \frac{J_2}{4}\left[\mathbf{S}\cdot(\psi^\dagger\boldsymbol{\sigma}\tau^\perp\psi)\cdot\mathbf{T}^\perp\right.$$

$$+ (\psi^\dagger\tau^\perp\psi)\cdot\overline{\mathbf{T}}^\perp\left] + \frac{J_3}{4}(\psi^\dagger\tau^z\psi)\overline{T}^z + \frac{J_4}{4}\left[\mathbf{S}\cdot(\psi^\dagger\boldsymbol{\sigma}\tau^z\psi) + \mathbf{S}\cdot(\psi^\dagger\boldsymbol{\sigma}\psi)\overline{T}^z\right], \quad (2)$$

where $\overline{\mathbf{T}} = \psi_d^\dagger\boldsymbol{\tau}\psi_d$ (with $\boldsymbol{\tau}$ being the Pauli matrices in the pseudo-spin space) is the pseudo-spin operator on the DQD and $\psi = \sum_k \psi_k$. Notice that in the even/odd basis the DQD pseudo-spin is rotated: $T^x \to \overline{T}^z$, $T^y \to -\overline{T}^y$ and $T^z \to \overline{T}^x$. The effective coupling constants are: $J_1 = J_3 = 2|V|^2/|\varepsilon_d|$, $J_2 = J_1\sin(\phi/2)$ and $J_4 = J_1\cos(\phi/2)$. The bare value of $J_5 = 4\rho_0|V|^2\cos(\phi/2)\ln|(\varepsilon_d+D)/(\varepsilon_d-D)|$ gives almost zero when $|\varepsilon_d| \ll D$. Importantly, we show now that the system exhibits a crossover from 0-flux to π-flux. Near the 0-flux [$\phi \approx 0 \pmod{2\pi}$], the DQD odd orbital is completely decoupled from the odd-symmetric lead. The Kondo-like model involves only the spin in the even orbital: $\mathcal{H}_K = J\mathbf{S}_e \cdot (\psi_e^\dagger\boldsymbol{\sigma}\psi_e)(1 + \overline{T}^z) + (J/4)(\psi_e^\dagger\psi_e)\overline{T}^z - J_5\overline{T}^z$, where $J = 2|V|^2/|\varepsilon_d|$. For this model the ground state corresponds to a Fermi liquid with a greatly enhanced Kondo temperature $T_K = D \exp(-1/4\rho_0 J)$ and a frozen orbital pseudo-spin due to the second term (J_5 does not flow to the strong coupling regime) [9]. Near the π-flux [$\phi \approx \pi \pmod{2\pi}$], one obtains the Kondo-like Hamiltonian

$$\mathcal{H}_K = \frac{J}{4}\left[\mathbf{S}\cdot(\psi^\dagger\boldsymbol{\sigma}\psi) + (\psi^\dagger\boldsymbol{\tau}\psi)\cdot\overline{\mathbf{T}} + \mathbf{S}\cdot(\psi^\dagger\boldsymbol{\sigma}\boldsymbol{\tau}\psi)\cdot\overline{\mathbf{T}}\right]. \quad (3)$$

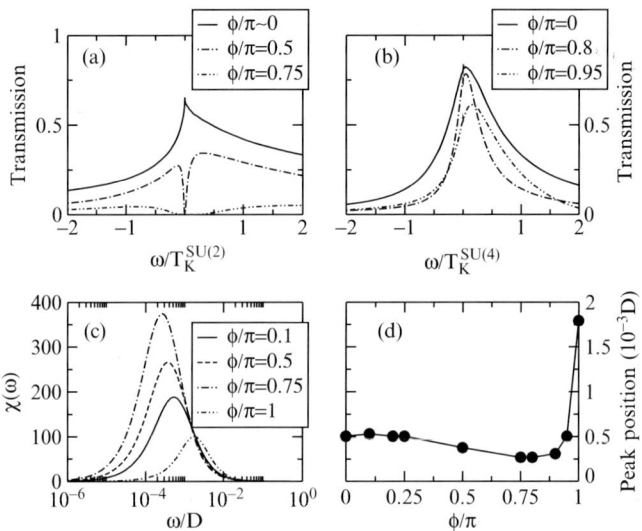

Fig. 1. *NRG results: Top panel*: Transmission probability vs. flux for **(a)** both dot levels are at $\varepsilon_d = -7\Gamma$ (with Γ being the lead-dot tunneling coupling, we set the same for both dots) and their intradot Coulomb interactions are respectively $U_1 = U_2 = 5D$. Case for negligible interdot Coulomb interaction $U_{12} = 0$, and **(b)** $\varepsilon_d = -14\Gamma$, $U_1 = U_2 = 5D$, with a very strong Coulomb interaction $U_{12} = 5D$. We set $\Gamma = D/60$. Bottom panel ($U_{12} = 5D$): **(c)** Spin susceptibility (in an arbitrary unit) in the limit of strong interdot interaction. **(d)** The peak position of the susceptibility as a function of the flux ϕ

This is the SU(4) Kondo model, the spin and the orbital degrees of freedom being entangled due to the third term. The RG equation reads $dJ/d\ell = 4\rho_0 J^2$, leading to $T_K^{SU(4)} = D \exp(-1/4\rho_0 J)$. As the flux departs from π, the degeneracy of the even and odd orbitals is lifted and the SU(4) symmetry is broken. The crossover from the SU(4) to the SU(2) Kondo model occurs at a given critical flux ϕ_c. From our NRG calculation (see below) we estimate $\phi_c \approx 0.75\pi$. Our results for the NRG calculations are plotted in Fig. 1. The case of $U_{12} \to 0$ is considered in Fig. 1a where the transmission is shown. Here, $\mathscr{T}$ displays always a dip except for $\phi = 0 \pmod{2\pi}$. When $\phi = 0 \pmod{2\pi}$ the DOS of each dot has a resonance exactly at E_F leading to a constructive interference. However, for $\phi \neq 0 \pmod{2\pi}$, $\mathscr{T}$ can be written as a combination of a Breit-Wigner resonance at E_F plus a Fano antiresonace [12]. A *dip* in the transmission is then obtained. We focus now on Fig. 1b, where $U_{12} \to \infty$. We find that the transmission consists of a peak for any flux value. Besides, the peak does not change appreciably for some $\phi < \phi_c$ [corresponding to SU(2) Kondo physics], and for $\phi > \phi_c$ [corresponding to SU(4) Kondo physics] the peak decreases very rapidly. The value of ϕ_c is the last ingredient we have to explain. Fortunately, ϕ_c can be extracted from the position of the spin suceptibility $\chi(\omega)$ peak, which yields a reasonable value of the Kondo temperature. Figure 1c shows the evolution of χ when ϕ increases. Remarkably, when the flux enhances, at some point the position of the peak moves toward higher frequencies. By tracing the peak position as a function of ϕ we plot Fig. 1d. We observe that $T_K(\phi)$ is almost constant when ϕ goes from zero to $\phi_c \approx 0.75\pi$. This fact allows us to establish a criterium for the crossover between the SU(2) and SU(4) Kondo states in the DQD system around the critical value of $\phi_c \approx 0.75\pi$.

3 Conclusions

We have analyzed the transport properties of a DQD inserted in a Aharonov–Bohm interferometer when interactions play a dominant role. We have demonstrated that crucial differences arise in the limits of negligible and large interdot Coulomb interaction, and that they can be measured directly in a transport experiment. In the former case, only spin fluctuations matter and the transmission shows a dip. For large interdot Coulomb interaction, this is quenched with increasing flux. Here, the Kondo state changes its symmetry, from SU(2) to SU(4) as ϕ approaches π.

Acknowledgements

We thank David Sánchez for useful discussions. This research was supported by Spanish Grant No. FIS2005-02796 (MEC) and the "Ramón y Cajal program".

References

1. For a review, see L.P. Kouwenhoven *et al.*, in *Mesoscopic Electron Transport*, edited by L.L. Sohn *et al* (Kluwer, Dordrecht, 1997).
2. A.C. Hewson, *The Kondo Problem to Heavy Fermions* (Cambridge University Press, Cambridge, UK, 1993).
3. D. Goldhaber-Gordon, *et al.*, Nature **391**, 156 (1998); S.M. Cronenwett *et al.*, Science **281**, 540 (1998); J. Schmid *et al.*, Physica B **256-258**, 182 (1998).
4. H. Jeong *et al.*, Science **293**, 2221 (2001).
5. D. Loss and E.V. Sukhorukov, Phys. Rev. Lett. **84**, 1035 (2000).
6. U. Wilhelm and J. Weis, Physica E **6**, 668 (2000).
7. L. Borda *et al.*, Phys. Rev. Lett. **90**, 026602 (2003).
8. D. Boese *et al.*, Phys. Rev. B **66**, 125315 (2002).
9. R. López *et al.*, Phys. Rev. B **71**, 115312 (2005). M-S Choi *et al.*, Phys. Rev. Lett. **95**, 067204 (2005).
10. C. Jayaprakash *et al.*, Phys. Rev. Lett. **47**, 737 (1981).
11. A.W. Holleitner *et al.*, Phys. Rev. B **70**, 075204 (2004).
12. T.V. Shahbazyan and M.E. Raikh, Phys. Rev. B **49**, 17 123 (1994).

Josephson Effect and Magnetic Interactions in Double Quantum Dots

F.S. Bergeret, A. Levy Yeyati, and A. Martín-Rodero

Departamento de Física Teórica de la Materia Condensada C-V, Universidad
Autónoma de Madrid, E-28049 Madrid, Spain
fs.bergeret@uam.es

Summary. Double quantum dot (DQD) structures provide a good system for
studying the competition between the Kondo effect and the antiferromagnetic coupling of the electrons in the dots [2]. Such a system can be realized not only in
semiconducting heterostructres but also in structures consisting of molecules and
nanotubes attached to metallic electrodes [1, 4]. If the latter are superconducting
then the interplay between the Josephson and the Kondo effects has to be taken
into account. In the case of a single quantum dot placed between two superconductors, the energy gap in the density of the states may lead to a suppression of the
Kondo effect and the appearance of an unscreened magnetic moment. This leads to
the so-called π-phase with a reversal of the sign of the Josephson current [3]. In the
case of a DQD the situation is more complicated since besides the Josephson and the
Kondo effect one should also take into account the magnetic interaction between the
dots. Here we provide a comprehensive analysis of the interplay between Josephson
effect, Kondo and antiferromagnetic coupling in a S-DQD-S systems. We analyze
the phase diagram and the appropriate correlation functions for a broad range of
parameters. Like in the single S-QD-S system we identify phases in which the sign
of the Josephson coupling is reversed.

The system under consideration is depicted in Fig. 1 and consists of
two coupled quantum dots in series placed between two superconducting
electrodes. The electronic degrees of freedom are represented by a double
Anderson model associating a single spin degenerate level with each QD. The
corresponding Hamiltonian is given by

$$H_{el} = \sum_{i,\sigma} \epsilon_{i\sigma}\hat{n}_{i\sigma} + U \sum_i \hat{n}_{i\uparrow}\hat{n}_{i\downarrow}$$

$$+H_{12} + H_{1L} + H_{2R} + H_{\text{L}} + H_{\text{R}} \tag{1}$$

The index i denotes the dot ($i = 1, 2$). The terms H_{L} and H_{R} describe
the uncoupled leads as BCS superconductors; H_{12} is the coupling term be-
tween the dots given by $H_{12} = \sum_\sigma t c_{1\sigma}^\dagger c_{2\sigma} + \text{h.c.}$, and the last two terms

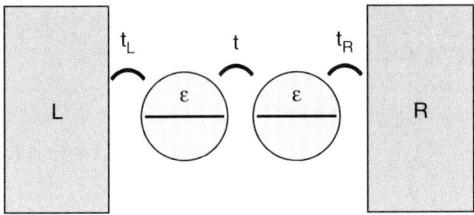

Fig. 1. Schematic view of the DQD structure

describe the coupling between dot 1 (2) and the left (right) electrode, *i.e.* $H_{1L} = \sum_{k\sigma} t_L c_{1\sigma}^\dagger c_{kL\sigma} + \text{h.c.}$ and $H_{2R} = \sum_{k\sigma} t_R c_{2\sigma}^\dagger c_{kR\sigma} + \text{h.c.}$. In the last expressions t, t_L and t_R are the corresponding hopping amplitudes. The Coulomb interaction within each dot is described by the U term.

Our aim is to determine the ground state of the system for a given set of parameters. We consider the most interesting case of strong coupling between the QD's, i.e. $t \gg \Delta$ and distinguish four different states: the pure 0 and π states for which the energy as a function of the superconducting phase ϕ has a minimum at $\phi = 0, \pi$ respectively; and two mixed phases, which are designed as $0'$ and π' depending of the relative stability of the minima [3]. In a first approach we consider the zero bandwidth limit (ZBWL) for the superconducting electrodes [3,5]. In this case we can diagonalize numerically H_{el}. In a second approach we use the slave-boson mean field (SBMF) approximation [Col93,7]. It is worth mentioning that in order to describe the main features of the DQD system, *i.e.* different superconducting phases and competition between AF and Kondo regimes, it is convenient to use the more general representation of [7], which is valid for finite values of U. We will see that both the ZBWL and the SB lead to similar results.

1 Exact Diagonalization

The main results in the ZBWL are sumarized in Fig. 2 which shows the (U, ϵ) phase diagram for $t = 10\Delta$ and two different values of $t_{R,L} = 2\Delta$ (panel (a)) and $t_{R,L} = 2.5\Delta$ (panel (b)). The range of ϵ is chosen to show only the region where the transition occurs, which corresponds to a charge per dot between 0 and 1. Due to the particle-hole symmetry an identical picture is obtained in the region of lower ϵ when the charge per dot is between 1 and 2 (not shown here).

One can see from Fig. 2 that for $t_L = t_R = 2$ all phases 0, $0'$, π' and π, represented by different colors, appear at the transition region. It is interesting to analyze the spin correlations functions $<\boldsymbol{\sigma}_1\boldsymbol{\sigma}_2>$ and for example $<\boldsymbol{\sigma}_L\boldsymbol{\sigma}_1>$. We choose the line in the phase diagram which corresponds to $U = 800\Delta$

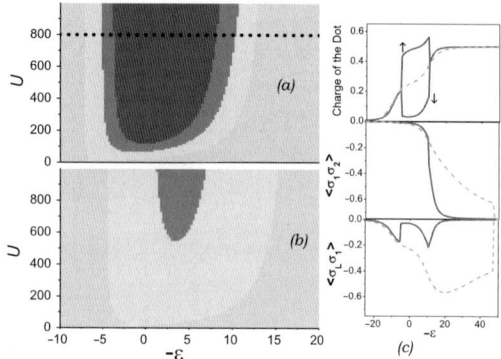

Fig. 2. Phase diagram obtained by exact diagonalization (from [8])

and show these correlation functions in panel (c) of Fig. 2 together with the occupation number $n_{\uparrow,\downarrow}$ for each spin projection per dot. The main conclusions are:

- Charge evolution along this line shows an overall a jump $0 \to 1/2$ per spin in the π region.
- The appearance of a magnetic moment $S = 1/2$ for the full S-DQD-S system is signaled by the broken symmetry $n_{\downarrow} \neq n_{\uparrow}$. The function $<\sigma_1\sigma_2>$ measures the spin correlations between the electrons in the two dots. As one can see it evolves continuously from 0 to $-3/4$. The latter value corresponds to a complete antiferromagnetic (AF) correlation.
- The superconducting state leads to a partial suppression of the Kondo correlations. The latter are best described in this simple model by the functions $<\sigma_L\sigma_1>$ or $<\sigma_R\sigma_2>$ which correspond to spin correlations between dots and leads. The Kondo regime corresponds for example to $<\sigma_L\sigma_1> \to -3/4$, i.e. to the formation of a singlet state between the lead and the dot. As can be observed in the lower panel of Fig. 2c the Kondo correlations are strongly suppressed by superconductivity compared to the normal case (dashed line).
- If the coupling between the leads and the dots is larger then it is more difficult for the superconductivity to suppress the Kondo correlations. As one can see from Fig. 2b already for $t_L = t_R = 2.5$ the system shows only three phases (0, $0'$ and π'). Further increase of the coupling with the electrodes will lead to a complete suppression of the π' state and so on.

2 Slave Boson Mean Field

So far we have presented the results within the ZBWL for the electrodes. We now go beyond this approach and consider a finite bandwidth W. An exact solution of the problem described by the Hamiltonian (1) cannot be obtained easily. We will use the slave-boson representation in its general form valid for a finite U [7], which allows the possibility of magnetic solutions. The auxiliary Bose fields are designed by e_i (empty state), $p_{i\alpha}$ (single occupied state corresponding to spin σ) and d_i (double occupied state) and we define the operator $z_{i\sigma} = (1 - d_i^2 - p_{i\sigma}^2)^{-1/2}(e_i p_{i\sigma} + p_{i\bar\sigma} d_i)(1 - e_i^2 - p_{i\sigma}^2)^{-1/2}$, where $i = 1, 2$ denotes dot 1 or 1. In the enlarged space the Hamiltonian (1) has the form

$$H = H_L + H_R + \sum_{i\sigma} \epsilon_i z_{i\sigma}^\dagger z_{i\sigma} n_{i\sigma} + \sum_i U d_i^\dagger d_i$$

$$+ \sum_\sigma t(z_{1\sigma}^\dagger z_{2\sigma} f_{1\sigma}^\dagger f_{2\sigma} + \text{h.c}) + \sum_{k,\sigma} t_{L(R)} (z_{1(2)\sigma}^\dagger f_{1(2)\sigma}^\dagger c_{kL(R)\sigma} + h.c.)$$

$$- \sum_i \alpha_i (e_i^\dagger e_i + d^\dagger d_i + \sum_\sigma p_{i\sigma}^\dagger p_{i\sigma} - 1)$$

$$- \sum_{i\sigma} \beta_{i\sigma} (f_{i\sigma}^\dagger f_{i\sigma} - p_{i\sigma}^\dagger p_{i\sigma} - d^\dagger d_i) \qquad (2)$$

where the $f_{i\sigma}$ are fermionic operators and α_i, $\beta_{i\sigma}$ are the Lagrange multipliers corresponding to the constrains $e_i^\dagger e_i + d^\dagger d_i + \sum_\sigma p_{i\sigma}^\dagger p_{i\sigma} = 1$ and $f_{i\sigma}^\dagger f_{i\sigma} = p_{i\sigma}^\dagger p_{i\sigma} + d^\dagger d_i$.

In the mean field approximation we replace the Bose operators by their expectation values. Thus, in the mean field Hamiltonian the parameters should be renormalized according to $\epsilon_{\sigma i} = \epsilon_i - \beta_{i\sigma}$, $\tilde t_\sigma = t z_{1\sigma} z_{2\sigma}$ and $\tilde t_{L(R)\sigma} = t_{L(R)} z_{1(2)\sigma}$. The expectation values of the Bose operators must be determined self-consistently from the equations obtained by variation of the effective action minimizing with respect to the fields e, d and p_σ [7]. As a first check of this approximation we have compared the results obtained by exact diagonalization and the SBMF in the ZBWL limit. As can be seen in Fig. 4, the SBMF technique provides an excellent estimation of the ground state

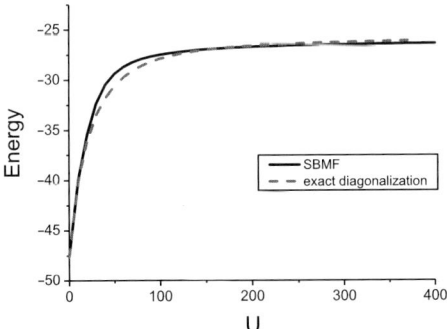

Fig. 3. Comparison of the exact diagonalization (*dashed line*) and the SBMF (*solid line*) in the ZBWL limit. We have taken $t_R = t_L = 3$, $t = 10$ and $\epsilon = -10$. All energies are given in units of Δ

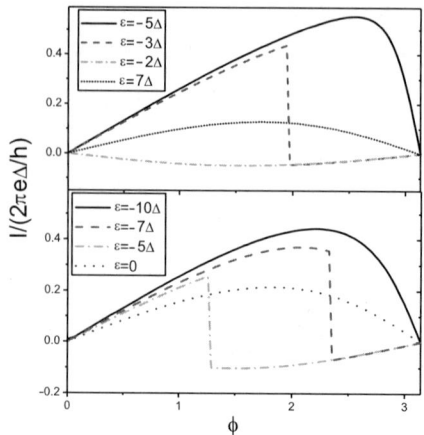

Fig. 4. The current-phase relation for $t = 10\Delta$, $U = 800\Delta$, $\Gamma_L = \Gamma_R = 2.25$ (*upper panel*) and $\Gamma_L = \Gamma_R = 4$ (*lower panel*) (from [8])

energy both in the $U \to 0$ and $U \to \infty$ limit. It also provides a satisfactory description of the energy evolution in the intermediate regime.

We have solved numerically the mean field equations and computed the Josephson current through the DQD system for a certain set of parameters. In Fig. 4 we show the current-phase dependence for $U = 800$, $t = 10$ and two different values of the parameter $\Gamma_{L,R} = t_{L,R}^2/W$. We see that for $\Gamma_L = \Gamma_R = 2.25$ and $\epsilon = -5$ the system is in the 0 state. By increasing the value of ϵ the system passes through the mixed and the π state, until the ground state corresponds again to a 0 junction ($\epsilon = 7$). In the case of larger coupling, $\Gamma_{L,R} = 4$, the π state never takes place. These results are, at least qualitatively, in agreement with those obtained in the ZBWL.

References

1. M. R. Buitelaar, A. Bachtold, T. Nussbaumer, M. Iqbal and C. Schönenberger, Phys. Rev. Lett. **88**, 156801 (2002)
2. H. Jeong, A. M. Chang and M. R. Melloch, Science (Washington D.C., U.S.) **293**, 2221 (2001).
3. E. Vecino, A. Martín-Rodero and A. Levy Yeyati, Phys. Rev. B **68**, 035105 (2003).
4. A. Yu. Kasumov *et al.*, Phys. Rev. B **72**, 033414 (2005).
5. I. Affleck, J.-S. Caux, and A. M. Zagoskin, Phys. Rev. B **62**, 1433 (2000).
6. P. Coleman, Phys. Rev. B **29**, 3035 (1984).
7. G. Kotliar and A. E. Ruckenstein, Phys. Rev. Lett. **57**, 1362(1986).
8. F. S. Bergeret, A. Levy Yeyati and A. Martín-Rodero, Phys. Rev. B **74**, 132505 (2006).

Quantum Shuttle: Physics of a Numerical Challenge

Andrea Donarini

Institut für Theoretische Physik, Universität Regensburg,
93040 - Regensburg, Germany

Summary. Shuttle devices are a class of nanoelectromechanical systems generically described as movable single electron transistors. They exhibit an electromechanical instability from the standard tunnelling regime to the shuttling regime in which the quantum dot oscillates and transfer one electron per cycle. I present a theory for the device in which both the electrical and mechanical degrees of freedom are quantized. The different operating regimes are detected by analyzing current, noise and Wigner function distributions. The calculation of the stationary solution for the Generalized Master Equation which describes the system dynamics is the starting point for the evaluation of these quantities and represents a numerically challenging problem due to the size of the Hilbert space necessary to capture the tunnelling to shuttling transition.

1 The Archetypal Model

The archetypal shuttle device (SD) consists of a movable quantum dot (QD) suspended between source and drain leads. One can imagine the dot attached to the tip of a cantilever or connected to the leads by some soft ligands or embedded into an elastic matrix. In the model the nanoparticle is confined to an harmonic potential. We give a schematic visualization of the device in Fig. 1.

Due to its small diameter, the QD has a very small capacitance and thus a charging energy that exceeds the thermal energy $k_B T$. For this reason we assume that only one excess electron can occupy the device (Coulomb blockade) and we describe the electronic state of the oscillating dot as a two-level system (empty/charged). Electrons can tunnel between leads and dot with tunnelling amplitudes which are exponentially dependent on the position of the central island. This is due to the exponentially decreasing/increasing overlapping of the electronic wave functions.

The Hamiltonian of the model reads:

$$H = H_{\text{sys}} + H_{\text{leads}} + H_{\text{bath}} + H_{\text{tun}} + H_{\text{int}}, \tag{1}$$

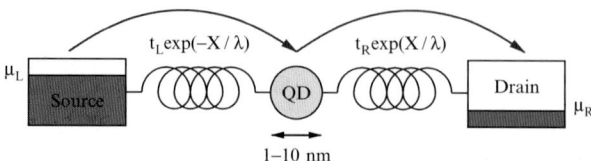

Fig. 1. Schematic representation of a shuttle device: electrons tunnel from the left lead at chemical potential (μ_L) to the quantum dot and eventually to the right lead at lower chemical potential μ_R. The position dependent tunnelling amplitudes are indicated. X is the displacement from the equilibrium position. The springs represent the harmonic potential in which the central dot can move

where

$$H_{\text{sys}} = \frac{\hat{p}^2}{2m} + \frac{1}{2}m\omega^2\hat{x}^2 + (\varepsilon_1 - e\mathcal{E}\hat{x})c_1^\dagger c_1$$

$$H_{\text{leads}} = \sum_k (\varepsilon_{l_k}c_{l_k}^\dagger c_{l_k} + \varepsilon_{r_k}c_{r_k}^\dagger c_{r_k})$$

$$H_{\text{tun}} = \sum_k [T_l(\hat{x})c_{l_k}^\dagger c_1 + T_r(\hat{x})c_{r_k}^\dagger c_1] + h.c. \tag{2}$$

$$H_{\text{bath}} + H_{\text{int}} = \text{generic heat bath}$$

Using the language of quantum optics we call the movable grain alone *the system*. This is then coupled to two electric baths (the leads) and a generic heat bath. The system is described by a single electronic level of energy ε_1 and a harmonic oscillator of mass m and frequency ω. When the dot is charged the electrostatic force ($e\mathcal{E}$) acts on the grain and gives the *electrical influence* on the mechanical dynamics. The electric field $\mathcal{E}$ is generated by the voltage drop between left and right lead. In our model, though, it is kept as an external parameter, also in view of the fact that we will always assume the potential drop to be much larger than any other energy scale of the system (with the only exception of the charging energy of the dot). The operator form $\hat{x}, \hat{p}$ for the mechanical variables is due to the quantum treatment of the harmonic oscillator. The leads are Fermi seas kept at two different chemical potentials (μ_L and μ_R) by the external applied voltage ($\Delta V = (\mu_L - \mu_R)/e$). The oscillator is immersed into a dissipative environment that we model as a collection of bosons and is coupled to that by a weak bilinear interaction:

$$H_{\text{bath}} = \sum_q \hbar\omega_q d_q^\dagger d_q$$

$$H_{\text{int}} = \sum_q \hbar g(d_q + d_q^\dagger)(d + d^\dagger) \tag{3}$$

where the bosons have been labelled by their wave number q. The coupling to the electric baths is introduced by the tunnelling Hamiltonian H_{tun}. The tunnelling amplitudes $T_l(\hat{x})$ and $T_r(\hat{x})$ depend exponentially on the position operator $\hat{x}$ and represent the *mechanical feedback* on the electrical dynamics:

$$T_{l,r}(\hat{x}) = t_{l,r} \exp(\mp \hat{x}/\lambda), \qquad (4)$$

where λ is the tunnelling length.

2 The Dynamics: Generalized Master Equation

The Hamiltonian for the shuttle device includes terms describing (1) the electronic part of the movable QD, (2) its mechanical motion (which is quantized), (3) the position dependent coupling of the QD and the leads, (4) the leads (treated as noninteracting fermions), and (5) coupling to environment, which damps the mechanical motion [1–4]. Since we are only interested on the dynamics of the quantum dot, we integrate out the environmental degrees of freedom (the lead electrons, and a generic heat bath) to obtain a Generalized Master Equation for the "system" ($=$ QD + quantized oscillator) density operator:

$$\dot{\sigma}(t) = \mathcal{L}\sigma(t) = (\mathcal{L}_{\text{coh}} + \mathcal{L}_{\text{driv}} + \mathcal{L}_{\text{damp}})\sigma(t). \qquad (5)$$

Here $\mathcal{L}_{\text{coh}}, \mathcal{L}_{\text{driv}}$ and $\mathcal{L}_{\text{damp}}$ are superoperators corresponding to the coherent evolution, coupling to leads, and damping of the QD. In the spirit of the classical master equation, it is sufficient to consider the diagonal electronic components (i.e., an empty and an occupied QD, respectively), since the electrical coherences are rapidly damped by the macroscopic leads. Nevertheless a generalization of the original concept that maintain the mechanical coherences is necessary to capture the electromechanical correlation characterizing the shuttle instability. The resulting equation of motion for the reduced density matrix reads:

$$
\begin{aligned}
\dot{\sigma}_{00}(t) &= \frac{1}{i\hbar}[H_{\text{osc}}, \sigma_{00}(t)] - \frac{\Gamma_{\text{L}}}{2}(e^{-\frac{2x}{\lambda}}\sigma_{00}(t) \\
&\quad + \sigma_{00}(t)e^{-\frac{2x}{\lambda}}) + \Gamma_{\text{R}}e^{\frac{x}{\lambda}}\sigma_{11}(t)e^{\frac{x}{\lambda}} \\
&\quad + \mathcal{L}_{\text{damp}}\,\sigma_{00}(t) \;, \\
\dot{\sigma}_{11}(t) &= \frac{1}{i\hbar}[H_{\text{osc}} - eEx, \sigma_{11}(t)] + \Gamma_{\text{L}}e^{-\frac{x}{\lambda}}\sigma_{00}(t)e^{-\frac{x}{\lambda}} \\
&\quad - \frac{\Gamma_{\text{R}}}{2}(e^{\frac{2x}{\lambda}}\sigma_{11}(t) + \sigma_{11}(t)e^{\frac{2x}{\lambda}}) \\
&\quad + \mathcal{L}_{\text{damp}}\,\sigma_{11}(t). \qquad (6)
\end{aligned}
$$

where

$$\mathcal{L}_{\text{damp}}\sigma = -\frac{i\gamma}{2\hbar}[x, \{p, \sigma\}] - \frac{\gamma m\omega}{\hbar}(\bar{N} + 1/2)[x, [x, \sigma]] \;.$$

The physical parameters defining the quantum shuttle are thus the (bare) tunnelling rates between QD and leads $\Gamma_{L/R}$, the oscillator frequency ω, the damping rate of the oscillator γ, the temperature T, and the tunnelling length λ.

3 Stationary State: A Mathematical Challenge

The master equation generally describes the irreversible dynamics due to the coupling between the system and the infinite number of degrees of freedom of the environment. It is reasonable to require that in absence of a time dependent driving mechanism the system tends asymptotically to a stationary condition defined by the equation:

$$\mathcal{L}\sigma^{\text{stat}} = 0. \tag{7}$$

3.1 A Matter of Matrix Sizes

We calculate the stationary matrix σ^{stat} numerically: we have to find the null vector of the matrix representation for the Liouvillean super-operator $\mathcal{L}$. The challenge arises from the matrix size. If N represents the size of the truncated Hilbert space of the harmonic oscillator that we consider in our calculation, $2N \times 2N$ is the size of the reduced density matrix σ and $2N^2 \times 2N^2$ the corresponding size of the Liouvillean matrix (remember we neglect electrical coherences).

The description of the SD dynamics requires (especially in the shuttling regime) amplitude oscillations of the vibrating dot between 5 and 10 times larger than the zero point fluctuations. For this reason, we are left to study the null space of matrices of typical size of $2 \cdot 10^4 \times 2 \cdot 10^4$. We solved this numerical problem using the iterative Arnoldi scheme.

3.2 The Arnoldi Scheme

The Arnoldi scheme is an efficient numerical method for the calculation of the null space, since (1) it allows to work with operators on the system Hilbert space only, and (2) it requires to look for the best approximation to the null vector in spaces which are typically much smaller that the Liouville space.

The central rôle in the Arnoldi scheme is played by Krylov spaces. For a given Liouvillean $\mathcal{L}$ and a temptative matrix σ of the Liouville space we define the Krylov space as:

$$\mathcal{K}_j(\mathcal{L}, \sigma) \equiv \text{span}(\sigma, \mathcal{L}\sigma, \ldots, \mathcal{L}^{j-1}\sigma), \tag{8}$$

where j is a small natural number. It is important to note that for the construction of the Krylov space all what we need are the matrices $\sigma, \mathcal{L}\sigma, \mathcal{L}^2\sigma, \ldots$ and not explicitly the superoperator $\mathcal{L}$. The method proceeds by looking for the best approximation of the null vector for the matrix representation of the Liouvillian within the Krylov space $\mathcal{K}_j(\mathcal{L}, \sigma)$. We call this vector σ'. This minimization is operated into a space of size $j + 1$, is much less demanding than the one required in the original problem and is performed using singular value decomposition (SVD) [5]. If the criterion $\|\mathcal{L}\sigma' < \epsilon$ for a given threshold value ϵ is not satisfied, we restart the procedure with the improved guess $\sigma = \sigma'$, otherwise we accept the solution. A schematic representation of the Arnoldi iterative method is presented in Fig. 2.

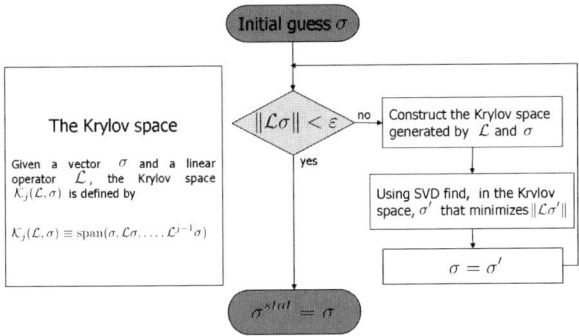

Fig. 2. Schematic representation of the iterative Arnoldi scheme

3.3 Preconditioning

The Arnoldi scheme is iterative and can suffer from convergence problems. It is not a priori clear how many iterations one needs to converge and fulfill the convergence criterion. A possible answer to a non-convergent code is, though, to reformulate the problem into an equivalent and (hopefully) convergent form. The basic idea is to find a regular operator $\mathcal{M}$ on the Liouville space, invertible, easy to implement, such that the original problem $\mathcal{L}[\sigma^{\text{stat}}] = 0$ can be recast into the form:

$$\mathcal{M}[\mathcal{L}[\sigma^{\text{stat}}]] = 0 \tag{9}$$

and that the finite version of the operator $\mathcal{ML}$ gives rise to a (fast) convergent iteration scheme. The operator $\mathcal{M}$ is also known as the *preconditioner*.

The Arnoldi scheme is particularly efficient in finding the best approximation of the eigenvalues and corresponding eigenvectors for those eigenvalues that are separated from the rest of the spectrum. Since we want to calculate the null vector it is important that the preconditioner moves the non-vanishing part of the spectrum far from the origin. For the problem (7) of the stationary solution of our GME, a good preconditioner is represented by the operator $\mathcal{M} = \mathcal{L}_{\text{Sylv}}^{-1}$ where

$$\mathcal{L}_{\text{Sylv}} = \mathbf{A}\sigma + \sigma\mathbf{A}^{\dagger} = \begin{bmatrix} A_{00}\sigma_{00} + \sigma_{00}A_{00}^{\dagger} & 0 \\ 0 & A_{11}\sigma_{11} + \sigma_{11}A_{11}^{\dagger} \end{bmatrix} \tag{10}$$

and

$$
\begin{aligned}
A_{00} &= -\frac{i}{\hbar}H_{\text{osc}} - \frac{\Gamma_{\text{L}}}{2}e^{-\frac{2x}{\lambda}} - \frac{i\gamma}{2\hbar}xp - \frac{\gamma m\omega}{\hbar}\left(n_B + \frac{1}{2}\right)x^2 \\
A_{11} &= -\frac{i}{\hbar}(H_{\text{osc}} - e\mathcal{E}x) - \frac{\Gamma_{\text{R}}}{2}e^{\frac{2x}{\lambda}} - \frac{i\gamma}{2\hbar}xp - \frac{\gamma m\omega}{\hbar}\left(n_B + \frac{1}{2}\right)x^2
\end{aligned}
\tag{11}
$$

where n_B is the average occupation number of the energy states of the harmonic oscillator in equilibrium with the bath.

4 The Three Regimes

Once the static density matrix is found, the current is readily calculable from

$$
\begin{aligned}
I^{\text{stat}} &= e\text{Tr}_{\text{osc}}\{\Gamma_R e^{2x/\lambda}\sigma_{11}^{\text{stat}}\}\\
&= e\text{Tr}_{\text{osc}}\{\Gamma_L e^{-2x/\lambda}\sigma_{00}^{\text{stat}}\}.
\end{aligned}
\tag{12}
$$

Also the noise [2], and even the higher cumulants [4], can be calculated with similar methods. In particular, we find that the Fano factor $F = S(0)/2eI$ (here $S(0)$ is the zero-frequency component of the noise spectrum) can be expressed as

$$
F = 1 - \frac{2e\Gamma_R}{I}\text{Tr}_{\text{osc}}\left\{e^{2x/\lambda}\left[\mathcal{Q}\mathcal{L}^{-1}\mathcal{Q}\times\left(\Gamma_R e^{x/\lambda}\rho_{11}^{\text{stat}}e^{x/\lambda}0\right)\right]_{11}\right\}.
\tag{13}
$$

Here $\mathcal{Q}$ is a projection operator that projects away from the stationary state. Very importantly, the pseudoinverse $\mathcal{R}$ of the Liouvillean, defined as $\mathcal{Q}\mathcal{L}^{-1}\mathcal{Q}\equiv\mathcal{R}$ is tractable by similar numerical methods as used in the evaluation of the current (we use the generalized minimum residual method (GMRes)). Before showing results for the current and noise, we discuss an important visualization tool.

We have found that Wigner functions are an excellent investigation tool for the numerical results obtained for the stationary density matrix. The intuitive picture comes from the well-known results in the classical limit: the Wigner representation (or, equivalently, the phase-space representation) of a regularly moving harmonic oscillator is a circle. On the other hand, irregular motion under the influence of external noise gives rise to a Gaussian probability distribution centered at the origin. Since the QD can be either empty or occupied, it is advantageous to introduce *charge resolved* Wigner functions ($n = 0$ corresponds to an empty dot, while $n = 1$ represents the occupied dot), defined as

$$
W_{nn}^{\text{stat}}(q,p) = \int_{-\infty}^{\infty}\frac{d\xi}{2\pi\hbar}\left\langle q-\frac{\xi}{2}\Big|\sigma_{nn}^{\text{stat}}\Big|q+\frac{\xi}{2}\right\rangle\exp\left(i\frac{p\xi}{\hbar}\right).
\tag{14}
$$

The behavior of the total Wigner distribution as a function of the mechanical damping shows precisely a smooth transition between the dot and the circular structure at high and low damping, respectively. The threshold for this transition is given by the effective tunnelling rates of the electrons.

The following picture arises: every time an electron jumps on the movable grain the latter is subject to the electrostatic force $e\mathcal{E}$ that accelerates it towards the right. Energy is pumped into the mechanical system and the dot starts to oscillate. If the damping is high compared to the tunnelling rates the oscillator dissipates this energy into the environment before the next tunnelling event: on average the dot remains in its ground state. On the contrary for very small damping the relaxation time of the oscillator is long and multiple "forcing events" happen before the relaxation takes place.

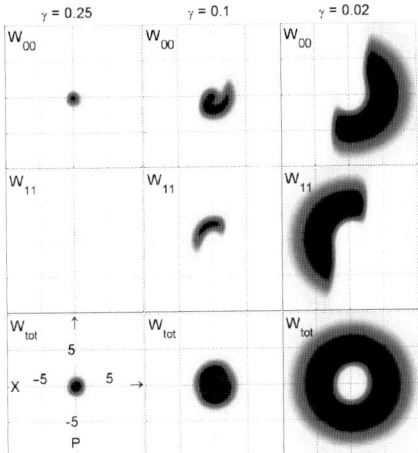

Fig. 3. Phase space picture of the tunnelling-to-shuttling transition. The respective rows show the Wigner distribution functions for the discharged (W_{00}), charged (W_{11}), and both (W_{tot}) states of the oscillator in the phase space. ($\Gamma = 0.05, \lambda = 1$)

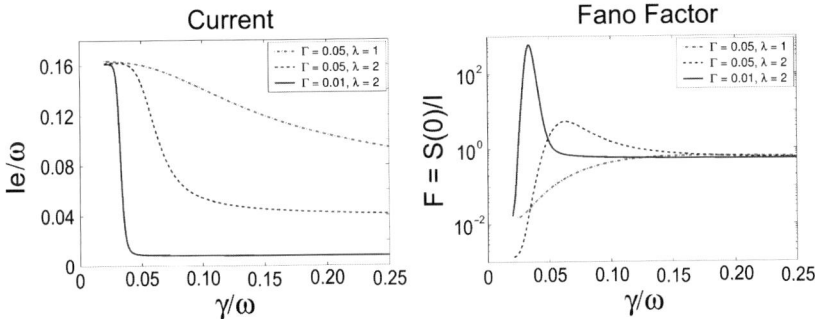

Fig. 4. Particle stationary current (*left*) and Fano function (*right*) of the SD plotted as a function of the damping rate

This continuously drives the oscillator far from equilibrium and a stationary state is reached only when the energy pumped per cycle into the system is dissipated during the same cycle in the environment.

It is not difficult to realize that, like for a macroscopic swing, in order to sustain the motion one needs coordination between forcing (here related to the electrical dynamics) and oscillations. This coordination is revealed by the charge resolved Wigner distributions W_{00} and W_{11}. The ring that appears in the total distribution is asymmetrically shared by the empty and charged-dot distributions (Fig. 3).

We study also the current as a function of the mechanical damping (left panel of Fig. 4). At low damping the current saturates, at the "magic" value $I \approx 0.16\omega$ (the frequency of the harmonic oscillator $\omega/2\pi$). This value is independent of the others parameters of the model. Increasing the damping the stationary current drops, more or less rapidly, to a plateau, dependent this time on both the bare tunnelling rate and length. Further increase of the damping does not change the scenario.

If we compare the current results with the Wigner function distribution we can recognize a correspondence between the shuttling charge-position (momentum) correlation and the saturation point, as well as a progressive disappearing of the ring structure in correspondence with the current transition. The high damping plateau in the current sets in when the mechanical oscillator lays into its ground state and the Wigner distribution function is reduced to a fuzzy spot close to the origin of the phase space.

In the right panel of Fig. 4 we present the Fano factor as a function of the mechanical damping γ for different values of the bare injection rate Γ and tunneling length λ. We recognize common features in the three curves. At high damping the Fano factor is of order 1 and (at least for the "most classical" set of parameters $\lambda = 2x_0$ and $\Gamma = 0.05\omega, 0.01\omega$) close to what we expect from a resonant tunneling system. The discrepancies from the symmetric double barrier are due to the quantum fuzziness in the position of the dot that influences the injection and ejection rates and to the charge dependent equilibrium position of the dot. Diminishing the damping the Fano factors encounter a more or less pronounced maximum and drop finally to very low values for small damping. The maximum at intermediate damping rates is more pronounced and sharper the more classical are the parameters and can reach values of $F \approx 600$ for the most classical case.

A comparison with the current curves (Fig. 4) shows that the peak in the Fano factor corresponds to the transition region from the tunneling to the shuttling current. Similarly the correspondence can be established also with the Wigner function distribution: the region of damping in which tunneling (dot in W_{tot}) and shuttling (ring in W_{tot}) features coexist is associated with a super-poissonian Fano factor. The very low ($F \approx 0.01$) Fano factors for low damping are a signature of the deterministic transport that takes place in the shuttling regime. It is interesting to note that this regularity persists also deep in the quantum regime as can be seen for $\Gamma = 0.05$ and $\lambda = 1$. The relative uncertainty in the amplitude of the oscillation (see Figure 4) does not seem to influence the current noise.

5 Conclusions

In conclusion, we have presented a numerical technique for solving the generalized Master equation governing an archetypal model of shuttle device. The obtained numerical results are interpreted with the help of phase space representations and allow to identify three operating regime of the SD. We believe that the methods discussed here are also applicable to many other quantum transport situations, where the matrix representations of the relevant operators are very large, but where only certain extremal eigenvalues are important.

References

1. T. Novotný, A. Donarini and A.-P. Jauho, Phys. Rev. Lett. **90**, 256801 (2003).
2. T. Novotný, A. Donarini, C. Flindt and A.-P. Jauho, Phys. Rev. Lett. **92**, 248302 (2004).
3. C. Flindt, T. Novotný and A.-P. Jauho, Phys. Rev. B **70**, 205334 (2004).
4. C. Flindt, T. Novotný and A.-P. Jauho, Europhys. Lett. **69**, 745 (2004).
5. T. Eirola and O. Nevanlinna, *Numerical linear algebra; iterative methods.* Lecture notes, Helsinki University of Technology, 2003. `http://www.math.hut.fi/teaching/175/ex03`

Microscopical Model for Hyperfine Interaction in Electronic Transport Through Double Quantum Dots: Spin Blockade Lifting

Jesús Iñarrea[1,2], Gloria Platero[2], and Allan H. MacDonald[3]

[1] Escuela Politécnica Superior, Universidad Carlos III, Leganes, Madrid, Spain
[2] Instituto de Ciencia de Materiales, CSIC, Cantoblanco, Madrid 28049, Spain
[3] Department of Physics, University of Texas at Austin, Austin, Texas 78712

1 Introduction

Recent transport experiments in vertical double quantum dots (DQDs) show that Pauli exclusion principle plays an important role [1, 2] in current rectification. In particular, spin blockade (SB) is observed at certain regions of dc voltages, and the interplay between Coulomb and SB can be used to block the current in one direction of bias while allowing it to flow in the opposite one. Then DQDs could behave as externally controllable spin-Coulomb rectifiers with potential application in spintronics as spin memories and transistors. Spin de-coherence and relaxation processes [3, 4] induced by spin–orbit (SO) scattering [5] or hyperfine (HF) interaction [6], have shown to reduce SB producing a leakage current in the voltage region where the blockade occurs. We theoretically analyze recent experiments of transport through two weakly coupled vertical QDs [1]. In these experiments current flow is allowed when the electrons in each QD have antiparallel spins and a finite gate voltage allows one electron in the left dot to tunnel sequentially to the right one. However, there is a similar probability for the electron coming from the left lead to be parallel or antiparallel to the electron spin occupying the right dot. In the first case, the electron cannot tunnel to the right dot due to Pauli exclusion principle and SB takes place, presenting a plateau in the I/V_{DC} curve.

2 Theoretical Model

The theoretical model presented here has been carried out in the frame of rate equations for the electronic charge occupations and for nuclei polarizations in both QDs which are coupled via electron and nuclei spin interaction and which we solve self-consistently. The electron and nuclei spin interactions (HF) brings to the Overhauser effect, which is also called flip-flop interaction. In our theoretical model we include a microscopic description of the spin electron and nuclei interaction and its effect on the electron dynamics. According to measurements on QDs by Fujisawa et al. [3] the sf time, $\tau_{sf} > 10^{-6}$s, is much longer than the typical tunnelling time, $\tau_{tun} = 1 - 100$ ns, or the momentum relaxation time, $\tau_{mo} = 1 - 10$ ns, meaning that sf processes due to HF interaction are important mostly in the SB region. Our system consists of a vertical DQD under a DC voltage in the presence of a magnetic field parallel to the current. We consider a hamiltonian: $H = H_L + H_R + H_T^{LR} + H_{\text{leads}} + H_T^{l,D}$ where $H_L(H_R)$ is the hamiltonian for the isolated left (right) QD and is modelled as one-level (two-level) Anderson impurity, $H_T^{LR}(H_T^{l,D})$ describes tunnelling between QDs (leads and QDs) and H_{leads} is the leads hamiltonian. The basis considered has 20 states: $|1\rangle = |0,\uparrow\rangle; |2\rangle = |0,\downarrow\rangle; |3\rangle = |\uparrow,\uparrow\rangle; |4\rangle = |\downarrow,\downarrow\rangle$

$\rangle; |5\rangle = |\uparrow,\downarrow\rangle; |6\rangle = |\downarrow,\uparrow\rangle;$
$|7\rangle = |0,\uparrow\uparrow^*\rangle; |8\rangle = |0,\downarrow\downarrow^*\rangle; |9\rangle = |0,\uparrow\downarrow\rangle; |10\rangle = |\uparrow,0\rangle; |11\rangle = |\downarrow,0\rangle;$
$|12\rangle = |0,\uparrow^*\rangle; |13\rangle = |0,\downarrow^*\rangle; |14\rangle = |\uparrow,\uparrow^*\rangle; |15\rangle = |\downarrow,\downarrow^*\rangle; |16\rangle = |\uparrow,\downarrow^*\rangle;$
$|17\rangle = |\downarrow,\uparrow^*\rangle; |18\rangle = |0,0\rangle; |19\rangle = |0,\downarrow,\uparrow^*\rangle; |20\rangle = |0,\uparrow\downarrow^*\rangle$

We have considered two levels, the ground state and the first excited level, in the right QD. Those states marked with (*) correspond to the excited state in the right QD. The time evolution equations for the electron charge occupations are: $\dot{\rho}(t)_{ss} = \sum_{m\neq s} W_{sm}\rho_{mm} - \sum_{k\neq s} W_{ks}\rho_{ss}$ where ρ_{ss} consists of the charge occupation of the electronic s-state. $W_{i,j}$ is the transition rate from the j-state to the i-state due to different mechanisms: coupling with the electric baths (contacts), phonon scattering or spin flip (sf) scattering due to HF interaction. We calculate $W_{i,j}$ by means of the Fermi Golden Rule (FGR). The inter-dot transition rate should account for both the elastic and inelastic current between the dots. For inelastic transitions, energy is exchanged with phonons in the environment. This contribution has been measured by Fujisawa et al. [7] and theoretically analyzed by Brandes et al. [8]. In order to calculate the inelastic transition rate $W_{1,2}^{ph}$ we have considered the theoretical model in [8]: $W_{1,2}^{ph} = \frac{\pi T_{12}^2}{\hbar}\left[\frac{\alpha_{\text{pie}}}{\varepsilon} + \frac{\varepsilon}{\hbar^2 w_\xi^2}\right]\left[1 - \frac{w_d}{w}\sin\frac{w}{w_d}\right]$ where α_{pie} is a piezoelectric coupling parameter, $\varepsilon = \hbar w = \mu_1 - \mu_2$, $w_d = c/d$ being c the sound velocity and d the inter-dot distance. Finally, $\frac{1}{w_\xi^2} = \frac{1}{\pi^2 c^3}\frac{\Xi^2}{2\rho_M c^2 \hbar}$ where ρ_M is the mass density and Ξ is the deformation potential.

In order to calculate the electronic spin-flip scattering rate $W_{i,j}^{sf}$ we have developed a microscopical model starting from the HF hamiltonian in the presence of an external magnetic field (B) plus the electronic Zeeman term: $\hat{H} = g_e\mu_B\mathbf{S}\cdot\mathbf{B} + \frac{A}{N}\sum_{i=1}^{N}\left[S_z I_z^i + \frac{1}{2}(S_+ I_-^i + S_- I_+^i)\right]$ where we have considered that the hyperfine constants for all the nuclei are equal: $A_i = A/N$, A_i being the i-nuclei hyperfine constant, N is the number of nuclei within the electronic envelope function ($A = 90\,\mu$eV for GaAs). $\hat{H}$ can also be written as: $\hat{H} = \hat{H}_z + \hat{H}_{sf}$ where $\hat{H}_z = g_e\mu_B\mathbf{S}\cdot\mathbf{B} + \frac{A}{N}\sum_{i=1}^{N}\left[S_z I_z^i\right]$ is the part of the hamiltonian responsible for the electronic Zeeman splitting due to external B and the induced nuclear magnetic field B_N and $\hat{H}_{sf} = \frac{A}{N}\sum_{i=1}^{N}\left[\frac{1}{2}(S_+ I_-^i + S_- I_+^i)\right]$ is responsible for the sf of a nucleus and of an electron. The sf time is calculated with the FGR: $\frac{1}{\tau_{sf}} = \frac{2\pi}{\hbar}|<\hat{H}_{sf}>|^2\frac{\gamma}{E_{\rm ST}^2+\gamma^2}$ γ being the electronic state broadening, $E_{\rm ST} = J - \Delta Z_e$, for the triplet state $|3\rangle = |\uparrow,\uparrow\rangle$ and $E_{\rm ST} = J + \Delta Z_e$ for the triplet state $|4\rangle = |\downarrow,\downarrow\rangle$. $\Delta Z_e = g_e\mu_B B + \frac{A}{2}(P_{1/2} + 3P_{3/2})$ is the total electronic Zeeman splitting, including the one produced by the effective nuclear field. $P_{1/2}$ and $P_{3/2}$ are the corresponding nuclear spin polarizations for the nuclear spin $I_z = 1/2$ and $I_z = 3/2$: $P_{1/2} = \frac{N_{1/2}-N_{-1/2}}{N}$; $P_{3/2} = \frac{N_{3/2}-N_{-3/2}}{N}$

For example, for the sf process $|\downarrow,\downarrow\rangle \rightarrow |\uparrow,\downarrow\rangle$ there are three different sf rates: $W^{sf}_{5,4;(I_Z:3/2\rightarrow1/2)} = \frac{1}{\tau_{sf}(3/2\rightarrow1/2)}\left[\frac{N_{3/2}}{N}\right]$ $W^{sf}_{5,4;(I_Z:1/2\rightarrow-1/2)} = \frac{1}{\tau_{sf}(1/2\rightarrow-1/2)}\left[\frac{N_{1/2}}{N}\right]$ $W^{sf}_{5,4;(I_Z:-1/2\rightarrow-3/2)} = \frac{1}{\tau_{sf}(-1/2\rightarrow-3/2)}\left[\frac{N_{1/2}}{N}\right]$ where $N_{3/2}$ $(N_{1/2})$ is the number of nuclei with $I_Z = 3/2$ $(I_Z = 1/2)$. There are similar equations for sf in the right QD. The equations that describe the time evolution of the nuclei spin polarization, for example for $P_{1/2}$ for the left QD have the form:

$$\dot{P}_{1/2} = 2W^{sf}_{6,3(-1/2\rightarrow1/2)}\rho_3 - 2W^{sf}_{5,4(1/2\rightarrow-1/2)}\rho_4 + W^{sf}_{5,4(3/2\rightarrow1/2)}\rho_4$$

$$-W^{sf}_{6,3(1/2\rightarrow3/2)}\rho_3 - W^{sf}_{6,3(-3/2\rightarrow-1/2)}\rho_3 + W^{sf}_{5,4(-1/2\rightarrow-3/2)}\rho_4 - \frac{P_{1/2}}{\tau_{\rm relax}} \quad (1)$$

where $\tau_{\rm relax}$ is the nuclear spin relaxation time ($\approx$ ms [9]). The rate equation for the charge occupation of the state $|3\rangle = |\uparrow,\uparrow\rangle$ is:

$$\dot{\rho}_3 = W_{3,1}\rho_1 + W_{3,7}\rho_7 + W_{3,11}\rho_{11} - (W_{1,3} + W_{7,3} + W_{11,3} + W^{sf}_{5,3(1/2\rightarrow3/2)}$$

$$+W^{sf}_{5,3(-1/2\rightarrow1/2)} + W^{sf}_{5,3(-3/2\rightarrow-1/2)} + W^{sf}_{6,3(1/2\rightarrow3/2)} + W^{sf}_{6,3(-1/2\rightarrow1/2)}$$

$$+W^{sf}_{6,3(-3/2\rightarrow-1/2)})\rho_3 \quad (2)$$

The system consisting of time evolution equations for the electronic states occupations ρ_i and for nuclei polarization for each QD is self-consistently solved and from that we calculate the stationary current through the system.

3 Results

Experiments [1] show, for $B \neq 0$, stationary $I/V_{\rm DC}$ curve for different B. We observe an additional peak at finite B which moves to lower $V_{\rm DC}$ as B increases. A finite B parallel to I, produces an energy shift experienced by the Fock-Darwin states due to its coupling with the electronic orbital momentum (B couples with the electronic orbital angular momentum of the first excited state of the right QD ($l = 1$)). Increasing B the first excited state of the right QD $(0, \uparrow\uparrow^*)$ enters in the transport window and comes into resonance with the ground state of the left QD, opening a new transport channel, and thus I flows through the device and SB is lifted. In Fig. 1 we present the stationary $I/V_{\rm DC}$ curve calculated for different B. We observe an additional peak at finite B

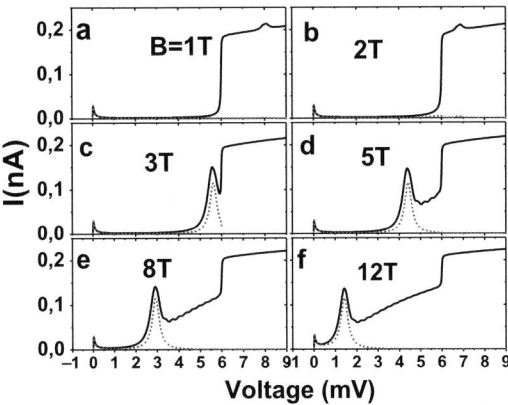

Fig. 1. Stationary I/V_{DC} curve for different B. *Single line*: both elastic (direct tunneling) and inelastic (phonon-assisted) contributions. *Dotted line*: only elastic transitions

which moves to lower V_{DC} as B increases, as in the experiments by Ono [1]: for different values of B, the resonance condition: $|3\rangle = |\uparrow, \uparrow\rangle \Rightarrow |7\rangle = |0, \uparrow \uparrow^*\rangle$ and $|4\rangle = |\downarrow, \downarrow\rangle \Rightarrow |8\rangle = |0, \downarrow \downarrow^*\rangle$, occurs at different values of V_{DC}. In this figure, the single line corresponds to the situation where both elastic (direct tunneling) and inelastic (phonon-assisted) contributions are considered and the dotted line means that only elastic inter-dot transitions are taken into account. The results presented in Fig. 1 are in good agreement with the experimental curve by Ono et al. [1].

Work supported by the MCYT (Spain), grant MAT2005-06444 (J.I. and G.P.), by the Ramón y Cajal program (J.I.) and by the EU Human Potential Programme: HPRN-CT-2000-00144.

References

1. K. Ono et al., Science **297** 1313 2002.
2. A.C. Johnson et al., cond-mat/0410679.
3. T. Fujisawa et al., Nature (London) **419**, 278 (2002); J.M. Elzerman et al., Nature, **430**, 431 (2004).
4. O. Gywat et al., Phys. Rev. B, **69**, 205303 (2004).
5. V. N. Golovach et al., Phys. Rev. Lett., **93**, 016601 (2004).
6. S. I. Erlingsson et al., Phys. Rev. B, **64**, 195306 (2001); *ibid*, Phys. Rev. B, **66**, 155327 (2002).
7. T. Fujisawa et al., Science **282**, 932 (1998).
8. T. Brandes et al., Phys. Rev. Lett., **83**, 3021, (1999); T. Brandes, Phys. Rep., **408**, 315-408, (2005); T. Brandes et al., Phys. Rev. B, **67**, 125323 (2003).
9. I.A. Merkulov et al., Phys. Rev. B, **65**, 205309 (2002); W.A. Coish et al., Phys. Rev. B, **70**, 195340 (2004).
10. K. Ono and S. Tarucha, Phys. Rev. Lett., **92**, 256803 (2004).

Rabi Dynamics in Driven Tunneling Devices

Rafael Sánchez and Gloria Platero

Instituto de Ciencia de Materiales, CSIC, Cantoblanco, Madrid 28049, Spain
rafael.sanchez@icmm.csic.es

Summary. We study the dynamics of the reduced density matrix for coupled quantum dots systems under the influence of time-dependent ac potentials. By disregarding non-resonant processes, we find analytical expressions for the stationary charge current in different regimes.

Solid state quantum dot (QD) systems have become probes for different and very assorted phenomena predicted for quantum systems. Concretely, two-level systems have been successfully developed and manipulated in double quantum dots (DQD) with one level each, coupled one to each other and to fermionic leads by tunnel barriers. Thus, by transport measurements, it is possible to access the quantum behaviour of electrons.

The probability of finding the electron of a closed two-level system shows a sinusoidal behaviour strongly dependent on the energy difference between the two states [1] and maximum when they have the same energy. A spacially resolved version of these *Rabi oscillations* appear in a DQD [2] when a level of each QD is in resonance with the other. Here, the interdot tunnel barrier plays the role of the coupling between the two levels and the electron is delocalized between both sites while the coupling to the fermionic environment causes the *damping* of the oscillations.

The introduction of a time dependent potential *connects* non-resonant levels if the ac frequency fits the energy diference between them [3,4]. Thus, as we will see, the interaction with the field gives the electron enough energy to jump to the other level performing *photon-assisted Rabi oscillations* [5] in a mechanism reminiscent of ESR experiments.

1 Model

We will consider a system consisting in two QDs connected in series to two electron reservoirs which can be described by the Hamiltonian $\hat{H} = \hat{H}_0 + \hat{H}_{L\Leftrightarrow R} + \hat{H}_T + \hat{H}_{ac}(t)$, where $\hat{H}_0$ describes the uncoupled DQD plus leads system. The inter-dot coupling, $\hat{H}_{L \Leftrightarrow R} = \sum_{l,r} T_{lr} \hat{c}_l^\dagger \hat{c}_r + $ h.c., connects coherently two states, $|l\rangle$ and $|r\rangle$, in the left and right dots, respectively, while the coupling to the leads, $\hat{H}_T = \sum_{l \in \{L,R\},k} \gamma_l \hat{d}_{lk}^\dagger \hat{c}_l + $ h.c., is considered weak and will be treated perturbatively. The time-dependent ac potential, is introduced as an oscillation of opposite phase on the energy of the levels of the QDs: $\hat{H}_{ac}(t) = \frac{V_{ac}}{2} \cos \omega t (\hat{n}_L - \hat{n}_R)$ where V_{ac} and ω (in units where $\hbar = 1$, $e = 1$) are the amplitude and frequency of the applied field, respectively. For simplicity, we do not consider the spin of the electron.

The dynamics of the system can be described by the reduced density operator, $\hat{\rho} = \text{tr}_R \hat{\chi}$, obtained by tracing all the reservoir states in the density operator of the whole system, $\hat{\chi}$. The Liouville equation, $\dot{\hat{\rho}}(t) = -i[\hat{H}(t), \hat{\rho}(t)]$, gives us the time evolution of the system.

1.1 Photon-Assisted Tunneling

In order to derive the master equation for the density matrix elements, it is convenient to remove the time dependence from the energy of the QD levels. This is made by applying a unitary transformation, $\hat{U}(t) = e^{i \frac{V_{ac}}{2\omega} \sin \omega t (\hat{n}_L - \hat{n}_R)}$ to the Hamiltonian: $\hat{H}'(t) = \hat{U}(t) \left(\hat{H} - i\partial_t \right) \hat{U}^\dagger(t) = \hat{H}_0 + \hat{H}'_{L \Leftrightarrow R}(t) + \hat{H}'_T(t)$. The time dependence is now included in the coupling terms:

$$\hat{H}'_{L \Leftrightarrow R}(t) = \sum_{\nu=-\infty}^{\infty} \sum_{l,r} T_{lr}(-1)^\nu J_\nu \left(\frac{V_{ac}}{\omega} \right) e^{i\nu\omega t} \hat{c}_l^\dagger \hat{c}_r + \text{h.c.} \tag{1}$$

where $J_\nu(x)$ is the νth order Bessel function of the first kind and, similarly:

$$\hat{H}'_T(t) = \sum_{\nu=-\infty}^{\infty} \sum_{l \in \{L,R\} k} \gamma_l(-1)^\nu J_\nu \left(\frac{V_{ac}}{2\omega} \right) e^{i\nu\omega t} \hat{d}_{lk}^\dagger \hat{c}_l + \text{h.c.} \tag{2}$$

Note that the argument of the Bessel function is twice in the interdot term than in the coupling to the leads. This is because the expected value of $\hat{n}_L - \hat{n}_R$ changes in ± 2 when an electron tunnels from one QD to the other and in ± 1 when it tunnels through the contact barriers.

Therefore, the influence of the ac potential, though applied only to the levels of the DQD, affects not only the interdot transitions but also the tunneling of electrons through the contact barriers [6–10]. As we will see, the transformed tunneling term (2) gives a modification in the tunneling rates, which now include the possibility that the electron get from the absorption or emission of ν photons with frequency ω enough energy to tunnel through the contact barriers. In the hopping term, responsible for the coherent dynamics inside the DQD, we can keep only the terms that put in resonance the states of both dots, an approximation reminiscent of the rotating wave approximation used in Quantum Optics [11–14], disregarding non-resonant oscillating terms. For instance, if the energy difference between the states, $|l\rangle$ and $|r\rangle$ fits the energy absorbed from n photons: $\omega_{rl} \approx n\omega$, the hopping Hamiltonian can be simplified to[1]:

$$\hat{H}'_{L\leftrightarrow R} = \sum_{l,r}(-1)^n \frac{\Omega_{lr}^{(n)}}{2}e^{in\omega t}\hat{c}_l^\dagger \hat{c}_r + \text{h.c..} \tag{3}$$

We have defined the Rabi frequency of the *n-photon-assisted delocalization process*: $\Omega_{lr}^{(n)} = 2J_n\left(\frac{V_{ac}}{2\omega}\right)\tau_{lr}$.

The time evolution of the system is given by the Liouville equation: $\dot{\rho}(t) = -i[\hat{H}, \hat{\rho}(t)]$. Assuming the Markov and Born approximations [15], we obtain, after some algebra [16,17], the master equation for the relevant density matrix elements:

$$\dot{\rho}_{jj}(t) = \sum_{k\neq j}\Gamma_{jk}\rho_{kk}(t) - \sum_{k\neq j}\Gamma_{kmj}\rho_{jj}(t) \tag{4}$$

$$\dot{\rho}_{ll}(t) = (-1)^{n+1}\Omega_{lr}^{(n)}\Im\rho_{lr}(t) + \sum_{k\neq l}\Gamma_{lk}\rho_{kk}(t) - \sum_{k\neq l}\Gamma_{kl}\rho_{ll}(t) \tag{5}$$

$$\dot{\rho}_{rr}(t) = (-1)^{n}\Omega_{lr}^{(n)}\Im\rho_{lr}(t) + \sum_{k\neq r}\Gamma_{rk}\rho_{kk}(t) - \sum_{k\neq r}\Gamma_{kr}\rho_{rr}(t) \tag{6}$$

$$\dot{\rho}_{lr}(t) = i(\omega_{rl} - n\omega)\rho_{lr}(t) - (-1)^{n}i\frac{\Omega_{lr}^{(n)}}{2}(\rho_{rr}(t) - \rho_{ll}(t)) - \Lambda_{lr}\rho_{lr}(t). \tag{7}$$

Γ_{mn} are the *photon-assisted tunneling* rates for transitions through the contact barriers:

$$\Gamma_{jk} = \sum_{\nu=-\infty}^{\infty}J_\nu^2\left(\frac{V_{ac}}{2\omega}\right)\xi_{jk}(\omega_{jk} + \nu\omega), \tag{8}$$

where

$$\xi_{jk}(\varepsilon) = 2\pi|\gamma_l|^2\left\{f_l(\varepsilon)\delta_{N_j,N_k+1}\delta_{N_j^l,N_k^l+1} + \bar{f}_l(\varepsilon)\delta_{N_j,N_k-1}\delta_{N_j^l,N_k^l-1}\right\} \tag{9}$$

are the *non-driven* tunneling rates. $N_k = \sum_j N_k^j$ is the number of electrons in the DQD in state $|k\rangle$, $\bar{f}_l(\varepsilon) = 1 - f_l(-\varepsilon)$, $f_l(\varepsilon) = 1/\left(1 + e^{(\varepsilon-\mu_l)\beta}\right)$ is the Fermi distribution function of the lead l involved in the transition, with $\beta = 1/k_BT$ and chemical potential μ_l. The interaction with the reservoirs induces decoherence in the Rabi oscillations, Λ_{lr}, which can be written as a function of the transition rates: $\Re\Lambda_{lr} = \frac{1}{2}\left(\sum_{k\neq r}\Gamma_{kr} + \sum_{k\neq l}\Gamma_{kl}\right)$.

The current that flows through the right contact is obtained with the relation $I_R = \sum_{j,k}(\Gamma_{kj}\rho_{jj} - \Gamma_{jk}\rho_{kk})\delta_{N_j^R-1,N_k^R}$.

---

[1] In this approximation the renormalization of the resonant frequency is lost [3,9]. However, this is small if $\tau_{lr} \ll |\omega_{rl}|$. Our results are exact in strict resonance.

2 Example: Two Interacting Electrons

Let us study a DQD that can be occupied by up to two extra electrons. We consider a charging energy, U, due to the *interdot* Coulomb repulsion, while the *intradot* Coulomb repulsion is supposed to be high enough to avoid the double occupation of a single QD. The system can then be described by the four state basis: $|0\rangle, |L\rangle, |R\rangle, |2\rangle$, if it is empty, contains one electron in the left, in the right or in each QD, respectively.

After writing the master equation for this system, we can treat analytically different configurations by varying the parameters of the system as, for instance, chemical potentials of the leads.

2.1 Pumping Regime

If $\varepsilon_L + U < \mu_L$ and $\mu_R < \varepsilon_R + U$ [16], the transport will be blocked unless we introduce a resonant ac potential which breaks the equilibrium by the absorption of n photons ($\omega_{RL} = n\omega$). Then, PAT processes contribute to the transport through the interdot and the contact barriers. In the unbiased low temperature case, if $\gamma_L = \gamma_R$, the tunneling rates for the processes that contribute to the current from left to right are equal ($\Gamma_{0R} = \Gamma_{L0} = \Gamma_{L2} = \Gamma_{2R} = \Gamma_+$) and so are the ones that contribute to the current from right to left, which involve the absorption of one photon ($\Gamma_{R0} = \Gamma_{0L} = \Gamma_{2L} = \Gamma_{R2} = \Gamma_-$). Then, the stationary current can be written, by doing $\dot{\rho}(t) = 0$:

$$I_{pump} = \frac{\frac{1}{2}\Omega^2 \left(\Gamma_+ - \Gamma_-\right)}{\Omega^2 + \left(\Gamma_+ + \Gamma_-\right)^2 + (\omega_{RL} - \omega)^2} \tag{10}$$

which, since it is proportional to the Rabi frequency, $\Omega = 2\tau J_n(\frac{V_{ac}}{\omega})$, is quenched when $J_n\left(\frac{V_{ac}}{\omega}\right) = 0$ is satisfied (*n-photons-assisted dynamical localization* [5], see Fig. 1 for $n = 1$). Furthermore, in the ideal case of having infinite width conduction bands, $\Gamma_+ - \Gamma_-$ is proportional to $J_0^2\left(\frac{V_{ac}}{2\omega}\right)$, leading to a suppression of the net current when the Bessel function is zero.

The current is also strongly suppressed for high ac intensities.

2.2 High Bias Regime

If $\mu_L \gg \varepsilon_L + U$ and $\mu_R \ll \varepsilon_R + U$, the processes contributing to the current to the left would need the absorption of a large number of photons to occur, so we can neglect them ($\Gamma_{R0} = \Gamma_{0L} = \Gamma_{2L} = \Gamma_{R2} = 0$). On the other hand, from the normalization condition for the Bessel functions, we have: $\Gamma_{0R} = \Gamma_{L2} = 2\pi|\gamma_R|^2 \equiv \Gamma_R$ and $\Gamma_{L0} = \Gamma_{2R} = 2\pi|\gamma_L|^2 \equiv \Gamma_L$, obtaining a stationary current:

$$I_{h-b} = \frac{\Omega^2 \Gamma_L \Gamma_R \left(\Gamma_L + \Gamma_R\right)}{\left(\Gamma_L^2 + \Gamma_R^2\right)\Omega^2 + \left(2\Omega^2 + \left(\Gamma_L + \Gamma_R\right)^2 + 4(\omega_{RL} - \omega)^2\right)\Gamma_L \Gamma_R}, \tag{11}$$

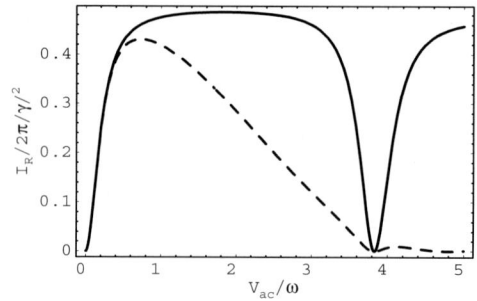

Fig. 1. Stationary current in resonance ($\omega = \omega_{RL}$) for the high bias (*solid*) and pumping (*dashed*) regimes in the symmetric case: $\gamma_L = \gamma_R = \gamma$ and $t_{LR} \sim 2\pi|\gamma|^2$.

that shows *photon-assisted dynamical localization*, cancelling the transport in spite of the high bias voltage applied to the contacts, see Fig. 1.

Work supported by the Ministerio de Educación y Ciencia of Spain through Grant No. MAT2005-00644. R.S. was supported by CSIC-Programa I3P, cofinanced by Fondo Social Europeo.

References

1. Cohen–Tannoudji, C., Diu, B. and Laloë, F., Quantum Mechanics. John Wiley & Sons, New York (1977)
2. van der Wiel, W.G. *et al*, Rev. Mod. Phys. 75, 1 (2003)
3. Stoof, T.H. and Nazarov, Yu.V., Phys. Rev. B 53, 1050 (1996)
4. Cota, E., Aguado, R., Creffield, C.E. and Platero, G., Nanotechnology 14, 152 (2003)
5. Platero, G., and Aguado, R., Phys. Rep. 395, 1 (2004)
6. Tien, P.K. and Gordon, J.P., Phys. Rev. 129, 647 (1963)
7. Kouwenhoven, L.P. et al., Phys. Rev. B 50, 2019 (1994)
8. Flensberg, K., Phys. Rev. B 55, 13118 (1997)
9. Strass, M., Hänggi, P. and Kohler, S., Phys. Rev. Lett. 95, 130601 (2005)
10. Sánchez, R., Platero, G., Aguado, R. and Cota, E., phys. stat. sol. (a) 203, 1154 (2006)
11. Bloch, F., Siegert, A., Phys. Rev., 57, 522 (1940)
12. Agarwal, G.S., Phys. Rev. A, **4**, 1778 (1971)
13. Agarwal, G.S., Phys. Rev. A, **7**, 1195 (1973)
14. Knight, P.L. and Allen, L., Phys. Rev. A, 7, 368 (1973)
15. Blum, K. Density Matrix Theory and Applications. Plenum, New York (1996)
16. Hazelzet, B.L., Wegewijs, M.R., Stoof, T.H. and Nazarov, Yu.V., Phys. Rev. B 63, 165313 (2001)
17. Sánchez, R., Cota, E., Aguado, R. and Platero, G., Phys. Rev. B 74, 35326 (2006)

Quantum-Transmitting-Boundary Algorithm with Local Spin–Orbit Coupling

Llorenç Serra[1,2] and David Sánchez[1]

[1] Departament de Física, Universitat de les Illes Balears, E-07122 Palma de Mallorca, Spain
[2] Institut Mediterrani d'Estudis Avançats IMEDEA (CSIC-UIB), E-07122 Palma de Mallorca, Spain
llorens.serra@uib.es, david.sanchez@uib.es

Summary. We review the application of the quantum-transmitting-boundary algorithm to compute electronic currents in the presence of localized spin–orbit couplings. As specific physical realization we choose a semiconductor quantum wire containing a Rashba spin–orbit dot. The Rashba dot leads to the formation of quasi-bound states and to Fano profiles in the energy dependence of the wire conductance.

1 Introduction

The description of electronic transport at a microscopic scale requires solving the Schrödinger equation for current-carrying states. In the so-called ballistic transport regime this implies finding the single-electron wave functions impinging on and being scattered off a given potential inhomogeneity. The boundary conditions for these scattering states are considerably more involved than for bound states. Indeed, while the latter just require the wave function to vanish asymptotically, the former involve finding transmission and reflection coefficients that are unknown a priori and, thus, amount to selfconsistent boundary conditions [1, 2].

An efficient algorithm to obtain scattering states is the quantum-trasmitting-boundary method (QTBM) proposed by Lent and Kirkner [3]. As noted by these authors, in dimensions higher than one the scattering problem cannot be solved as an initial value problem, where one recursively finds the wave function at one point in terms of the wave function at the preceding ones, but has to be transformed into a coupled linear system of equations yielding the wave function at all points in a single step. This implicit scheme is the essence of the QTBM and we shall discuss below its application to compute the conductance of a semiconductor quantum wire containing a region of spin–orbit Rashba coupling.

Interest in spin–orbit effects in semiconductors is mostly due to the tunability of the Rashba coupling by means of external electric gates. This tunability is expected to allow the controlled manipulation of electron spins, a central requirement of any *spintronic* device. We shall prove that for specific energies the electron wavefunction resonates, manifesting a strong localization inside the Rashba dot – the region with Rashba spin–orbit coupling. These quasibound states interfere with the direct transmission along the wire to the extent that the conductance completely vanishes for some specific Rashba dot lengths and coupling intensities. In general, we obtain wire conductances whose energy dependence follows a generalized Fano profile, a mechanism we have named the Fano–Rashba effect in [4] and [5]. Also relevant to this work are the discussion in terms of matching methods in [6–8]. The Chapter is organized as follows: Sect. 2 presents the physical system, Sect. 3 details our implementation of the QTBM while the results are discussed in Sect. 4. Finally, Sect. 5 is devoted to conclusions and outlook.

2 Physical System

We assume the effective-mass model for the conduction band states of a two-dimensional electron gas in GaAs and consider a parabolic confinement in the y direction, yielding a ballistic quantum wire along x. The system Hamiltonian reads

$$H = \frac{p_x^2 + p_y^2}{2m} + \frac{1}{2}m\omega_0^2 y^2 + H_R .\tag{1}$$

The Rashba Hamiltonian H_R that we consider is characterized by an intensity $\alpha(x)$ essentially vanishing everywhere except for $-\ell/2 < x < \ell/2$ where it takes the value α_0. In detail, H_R reads

$$H_R = \frac{\alpha(x)}{2\hbar}(p_y\sigma_x - p_x\sigma_y) + \text{H.C.} ,\tag{2}$$

where σ_x and σ_y are the Pauli matrices and the Hermitian conjugation is used to ensure Hermiticity. In the numerical applications we take a smoothed variation $\alpha(x) = \alpha_0[f(x - \ell/2) - f(x + \ell/2)]$, with $f(x) = 1/(1 + e^{x/\sigma})$ and σ a small diffusivity. The results are not very sensitive to the precise value of σ provided $\sigma < \ell$. Figure 1 shows a simple sketch of the physical system under consideration.

3 The QTBM

We consider a uniform discretization of the xy plane in a grid of points as indicated in Fig. 1. The QTBM requires as many equations as grid points in order to obtain a closed system of equations determining the wave function $\Psi(x,y,\eta)$ (where $\eta = \uparrow, \downarrow$ is the spin variable) for all grid points. The grid is divided in a central subset (C) containing the Rashba dot and two lateral subsets, left (L) and right (R), which are in the translationally invariant asymptotic regions. For each point in subset C we use the Schrödinger equation

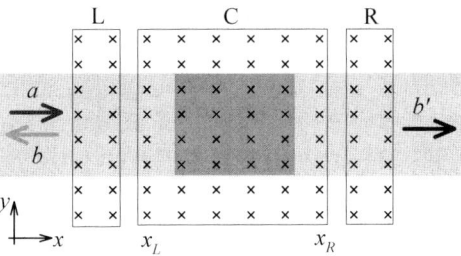

Fig. 1. Sketch of the physical system and the uniform grid. *Light gray* region corresponds to the quantum wire and the *dark gray square* indicates the position of the Rashba dot. The grid points are separated in subsets L, C, and R, for which different equations are applied. Incident (*a*), reflected (*b*), and transmitted (*b'*) waves are also indicated. x_L and x_R indicate the x value for the points in C closest to the left and right boundary, respectively

$H\Psi = E\Psi$, with H given by (1) and E the given energy of the incident electron. The width of the lateral regions L and R is determined by the number of points n_d used for the finite difference discretization of the Laplacian operator as $n_d/2$. We typically use $n_d = 5$ or 7, which corresponds to a width of 2 or 3 points, respectively. For points exceeding the grid in the y direction the wavefunction is assumed to vanish identically.

It remains now to specify the equations to be used for grid points in subsets L and R. In these regions it is particularly adequate to expand the full wave function in the oscillator transverse modes $\{\phi_n(y), \varepsilon_n\}$ and spin eigenstates $\chi_s(\eta)$, $s = \pm$. These are the solutions of the 1D problem $(-\frac{\hbar^2}{2m}\frac{d^2}{dx^2} + \frac{1}{2}m\omega_0^2 y^2)\phi_n(y) = \varepsilon_n\phi_n(y)$ with $\varepsilon_n = (n - 1/2)\hbar\omega_0$, $n = 1, 2, \ldots$, and the spin up and down eigenstates in a given arbitrary direction, respectively. For definiteness, the expansion for the points in subset L reads

$$\Psi(x, y, \eta) = \sum_{ns} a_{ns}e^{ik_n(x-x_L)}\phi_n(y)\chi_s(\eta) + \sum_{ns} b_{ns}e^{-ik_n(x-x_L)}\phi_n(y)\chi_s(\eta)$$

$$+ \sum_{ns}' b_{ns}e^{\kappa_n(x-x_L)}\phi_n(y)\chi_s(\eta). \quad (3)$$

The first two contributions in (3) are the incident and reflected propagating waves having $\varepsilon_n \leq E$ and wavenumbers $k_n = \sqrt{2m(E - \varepsilon_n)}/\hbar$. The third contribution corresponds to the sum of evanescent states, for which $\varepsilon_n > E$ and $\kappa_n = \sqrt{2m(\varepsilon_n - E)}/\hbar$, which is truncated for a large enough n. We have defined x_L as the x coordinate of the grid points of region C lying closest to region L (see Fig. 1). The set of incident amplitudes $\{a_{ns}\}$ is a known input to the problem and the $\{b_{ns}\}$ coefficients can be related to the wave function in region C:

$$b_{ns} = \sum_{\eta} \int dy\, \phi_n^*(y)\chi_s^*(\eta)\, \Psi(x_L, y, \eta) - a_{ns}\,. \quad (4)$$

The a_{ns} amplitude is obviously absent when applying (4) to evanescent states. Now, the combination of (3) and (4) gives a closed expression in terms of the grid that we can apply for each point of subset L. Similar expressions are easily found for the R subset noting that, in this case, the $\{a_{ns}\}$ coefficients vanish since we only consider incidence from the left, and making the replacements $k_n \rightarrow -k_n$, $\kappa_n \rightarrow -\kappa_n$, and $x_L \rightarrow x_R$ in (3) and (4). The final result is a set of linear equations whose only inputs are the energy E and left incident amplitudes $\{a_{ns}\}$. The system is highly sparse and can be effectively solved using standard numerical routines [9], yielding the wave function as well as the transmission and reflection amplitudes from which one obtains the linear conductance [1].

4 Results

Figure 2 shows the linear conductance of the wire as a function of E for a specific Rashba dot. Energies and lengths are given in units of the transverse oscillator energy $\hbar\omega_0$ and length $\ell_0 = \sqrt{\hbar/m\omega_0}$. A small diffusivity $\sigma = 0.2\ell_0$ has been used to smooth the Rashba coupling steps. In absence of Rashba dot the conductance is quantized with abrupt changes of e^2/h, the conductance quantum, every time a transverse mode becomes propagating. Focusing on the first plateau shown in Fig. 2a, we note that the Rashba inhomogeneity produces smooth conductance oscillations at the beginning of the plateau as well as a conspicuous asymmetric dip for $E \sim 1.32\hbar\omega_0$. While the oscillations are usual in quantum scattering, the existence of the dip with asymmetric line

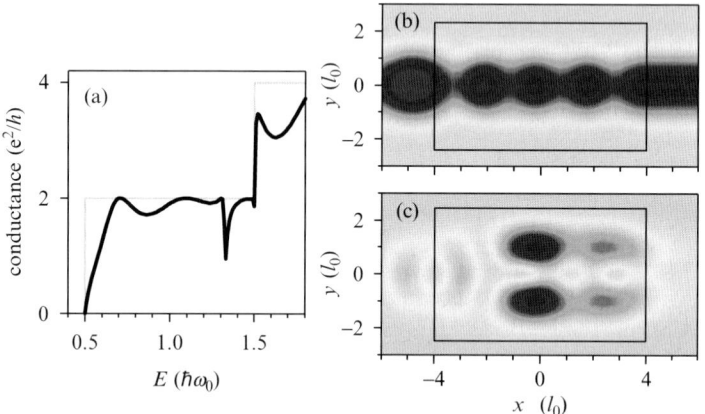

Fig. 2. (a) Conductance for a Rashba dot of $\ell = 8\ell_0$ and $\alpha_0 = 0.7\hbar\omega_0\ell_0$. The thin *gray line* is the result in absence of the Rashba dot. (b) and (c) Probability density for a left-incident spin-up wave along y of energy $E = \hbar\omega_0$ (b) and $E = 1.32\hbar\omega_0$ (c). *Darker color* means higher probability density and the *rectangle* indicates the position of the Rashba dot

shape is due to the formation of a Fano resonance. Indeed, the interference between a transmitting channel and a quasibound state can severely quench the conductance for specific energies.

The existence of a resonating quasibound state is proved in Figs. 2b, c by showing the probability density associated with a spin up wave, with spin along $+y$, in the lowest transverse mode impinging on the Rashba dot from the left. Indeed, at the energy of the conductance asymmetric dip the wave function is strongly localized to the Rashba dot [panel (c)] while in other cases it clearly extends to the asymptotic wire regions [panel (b)].

5 Conclusions

The QTBM allows to obtain the wave function and linear conductance of a quantum wire containing inhomogeneities. Here we have reviewed the general formulation of the QTBM for a wire with a Rashba dot, where a spin–orbit coupling of Rashba type is active. The scattering problem is formulated as an implicit system of linear equations, yielding the wave function for all the grid points as well as the transmission and reflection coefficients. The transmission coefficients for all incident modes at a given energy determine the wire linear conductance. The Rashba dot sustains quasibound states that interfere with the direct transmission path along the wire and lead to an energy-dependence of the linear conductance displaying oscillations and asymmetric Fano-resonance dips.

We acknowledge R. López for valuable discussions. This work was supported by the Grant No. FIS2005-02796 (MEC) and the "Ramón y Cajal" program.

References

1. Datta, S.: Electronic transport in mesoscopic systems. Cambridge University Press (1995)
2. Ferry, D.K., Goodnick, S.M.: Transport in nanostructures. Cambridge University Press (1997)
3. Lent, C.S., Kirkner, D.J.: J. Appl. Phys. **67**, 6353 (1990)
4. Sánchez, D.; Serra Ll.: Phys. Rev. B **74**, 153313 (2006)
5. Serra, Ll., Sánchez, D.: J. Phys.: Conf. Ser. in press (cond-mat/0610147)
6. Shelykh, I.A.; Galkin, N.G.; Phys. Rev. B **70**, 205328 (2004)
7. Cserti, J.; Csordás, A.; Zülicke, U.: Phys. Rev. B **70**, 233307 (2004)
8. Zhang, L.; Brusheim, P.; Xu, H.Q.: Phys. Rev. B **72**, 045347 (2005)
9. Harwell subroutine library.

Spintronic Transport in II–VI Magnetic Semiconductor Resonant Tunneling Devices

David Sánchez

Departament de Física, Universitat de les Illes Balears, E-07122 Palma de Mallorca, Spain
david.sanchez@uib.es

Summary. We investigate electron transport through resonant tunneling diodes doped with magnetic impurities. Due to exchange interaction between impurities and carriers, there arises a giant Zeeman splitting which dominates the I–V curves. We discuss a simple model which accounts for spin effects in these systems and examine its applicability in realistic samples.

1 Introduction

Spintronic devices make use of the electron's spin degree of freedom for information processing. A basic requirement of these devices is their all-electrical ability to create [1, 2] and detect spin-polarized currents. Possible candidate materials for full on-chip integration are diluted magnetic semiconductors (DMSs) which are doped with magnetic (Mn) impurities [3, 4]. These impurities decrease the sample mobility but recent progress in II–VI compounds (e.g., CdTe or ZnSe) [5] has reduced Mn scattering while, at the same time, spin effects are maximized. As a result, large spin-splittings in the electron conduction band are observed [1] in the presence of a relatively small magnetic field. Since carriers, which arise from n-doping, move in the conduction band spin-orbit effects can be neglected and spin relaxation times as large as 1 ns are possible to achieve.

We are here concerned with DMS resonant tunneling diodes (RTD) which show strong spin effects due to the combination of negative differential conductance, giant Zeeman splitting, and low dimensionality. These systems have recently received attention [6–11]. A RTD is a well known quantum semiconductor device [12] formed by two tunnel junctions in a series, see Fig. 1. The double barrier confines the electrons in the growth direction and their longitudinal energy become discretized in the quantum well comprising energy subbands. The well between the barriers couples to electron reservoirs and current flows when a dc voltage bias is applied. Since the bias rises the emitter band bottom relative to the well level ε_0, the RTD exhibits a resonance behavior in the current–voltage I–V characteristics.

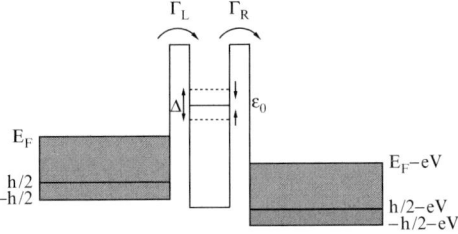

Fig. 1. Schematic representation of the energy landscape of a RTD with a single energy level in the quantum well, ε_0, attached to two electron reservoirs with Fermi energy E_F. The quantum level and the emitter band bottom may be spin split due to an external magnetic field

2 Theoretical Model

The giant Zeeman splitting is due to the exchange interaction between d Mn local moments and itinerant s electrons, favoring parallel alignment. Then, the exchange can be modeled as $\mathcal{H}_{\rm int}^{sd} = \tilde{J}_{sd} \sum_I \mathbf{S}_I \cdot \mathbf{s}(\mathbf{r}_I)$, with I labels the position of the Mn impurities and $\tilde{J}_{sd}$ is the coupling constant. We now make use of the virtual crystal approximation combined with a mean-field theory [4] to simplify the Hamiltonian: $\mathcal{H}_{\rm int}^{sd} = J_{sd}\langle S_z \rangle \sum_i s_z(\mathbf{r}_i)$, where $J_{sd} \equiv x\tilde{J}_{sd}$. Here, we have placed a Mn impurity at each lattice site and $\tilde{J}_{sd}$ is thus reduced by a factor x. Then, each Mn spin is replaced with its average along the direction of the field. Hence, the DMS quantum well is a paramagnetic system where spin–spin interaction gives rise to an effective magnetic field. The resulting spin splitting given by

$$\Delta = N_{\rm Mn} J_{sd} S_0 B_S \left(\frac{S g \mu_{\rm B} B}{k_{\rm B}(T + T_{\rm eff})} \right), \tag{1}$$

where $N_{\rm Mn}$ is the concentration of $S = 5/2$ Mn spins, $B_S(x) = (1 + 0.5/S)\,{\rm ch}[(S+0.5)x] - 0.5S\,{\rm ch}(0.5S)$ is the Brillouin function with g the Landé factor, B is the applied magnetic field, T is the temperature, and S_0 and $T_{\rm eff}$ are the Mn effective spin and temperature, respectively. At large B the splitting reaches saturation and the s-spins become fully polarized. When the Mn impurities are placed in the emitter, the giant Zeeman splitting is denoted with h to distinguish it from the splitting Δ in the well.

We neglect orbital effects due to B as spin effects are independent of the B direction. Further, we disregard spin-relaxation effects for the moment, thereby the current is carried by spins up ($\sigma = +$) and spins down ($\sigma = -$) in parallel. As a result, the resonant tunneling current $J = J_+ + J_-$ through the double barrier system depicted in Fig. 1 is determined within the transmission formalism [12]:

$$J_\sigma = \frac{em}{4\pi^2\hbar^3} \int_{eV - \sigma h/2}^{\infty} {\rm d}E_z {\rm d}E_\perp \, T_\sigma(E_z, \varepsilon_0, V)[f_{\rm L}(E_z + E_\perp) - f_{\rm R}(E_z + E_\perp)], \tag{2}$$

with V the bias voltage applied to the structure. The Fermi functions f_L and f_R describe the distribution of electrons with total energy $E_z + E_\perp$ in the left and right leads with electrochemical potentials $\mu_L = E_F + eV$ and $\mu_R = E_F$.

We first consider the case of a magnetic injector and a normal quantum well. To simplify (2) we consider an infinitely narrow resonance (δ-resonance), $T(E_z) = 2\pi(\Gamma_L\Gamma_R/\Gamma)\delta(E_z - \varepsilon_0)$ where $\Gamma_{L(R)}$ are the partial hybridization widths associated to the decay rate of electrons localized in the well. The total linewidth is $\Gamma = \Gamma_L + \Gamma_R$. When $E_0 > E_F$ we find at zero temperature $J_\sigma(V) = e\nu\Gamma_L\Gamma_R(E_F + eV - \varepsilon_0)/\hbar\Gamma$ for $\varepsilon_0 - E_F < eV < \varepsilon_0 + \sigma h/2$ and zero elsewhere ($\nu = m/2\pi\hbar^2$). We infer that the maximum value attained by the total current $J = J_\uparrow + J_\downarrow$ is independent of the splitting h. For $E_0 < E_F$, the spin-dependent current is $J_\sigma(V) = e^2\nu\Gamma_L\Gamma_R V/\hbar\Gamma$ for $0 < eV < \varepsilon_0 + \sigma h/2$. Hence, the peak current is also h-independent.

For more realistic modeling we use the Breit-Wigner approximation, which is a good approach for RTDs close to resonance [13]:

$$T(E_z) = \frac{\Gamma_L\Gamma_R}{(E_z - \varepsilon_0)^2 + \Gamma^2/4}. \tag{3}$$

In contrast to the δ-resonance, the dependence of the electric current on voltage and h is *nonlinear*. This leads to predictions that differ from the δ-resonance. For instance, consider for simplicity the case of a resonance dominated by Γ either because E_F and temperature are quite small or due to strong interface roughness or disorder. Then, we can expand (2) in powers of $1/\Gamma$. We find for small spin splittings that the resonance peaks at $eV_{res} = \varepsilon_0 - E_F/3 + h^2/3E_F$. Clearly, this expression shows a shift of V_{res} with increasing h. Inserting this result in (2) we find to leading order in $1/\Gamma$ that the current peak $J_p = (e\nu/\pi\hbar)E_F^2(1 + h^2/4E_F^2)$ is, in fact, an *increasing* function of the magnetic field. In Fig. 2 we show numerical simulations for the parameters $\varepsilon_0 = 21$ meV, $E_F = 10$ meV, $\Gamma_0 = 15$ meV, and $k_B T = 4$ K (consistent with a Zn$_{0.94}$Mn$_{0.06}$Se/Zn$_{0.7}$Be$_{0.3}$Se/ZnSe/Zn$_{0.7}$Be$_{0.3}$Se/ZnSe RTD), taking into account the energy dependence of Γ. We observe that the peak current increases with h in agreement with the discussion above. Thus, this effect allows to generate high peak current peaks, which arise from spin effects only and which can be tuned with an external magnetic field without changing the sample parameters.

We now focus on the case of a normal injector and a DMS well. In the δ-resonance limit, we find $J_\sigma(V) = e\nu\Gamma_L\Gamma_R(E_F + eV - E_0 + \sigma\Delta/2)/\hbar\Gamma$ for $\max(E_F, E_0 - E_F - \sigma\Delta/2) < eV < E_0 - \sigma\Delta/2$. This simple expression predicts a splitting of the $I-V$ curve as observed in the experiment [7]. The maximum current for each spin channel is $J_\sigma^{max} = (e\nu/\hbar)(\Gamma_L\Gamma_R/\Gamma)E_F$. The total current peak is given by $I_p = 2e\nu E_F(1 - \Delta/2E_F)/\hbar\Gamma$, which decreases with increasing Δ, as expected.

At low magnetic fields, the giant Zeeman splitting in the well is much larger than the Zeeman splitting in the normal injector, $\Delta \gg h$, where h is now $g\mu_B B$. However, at large B the well magnetization saturates and the

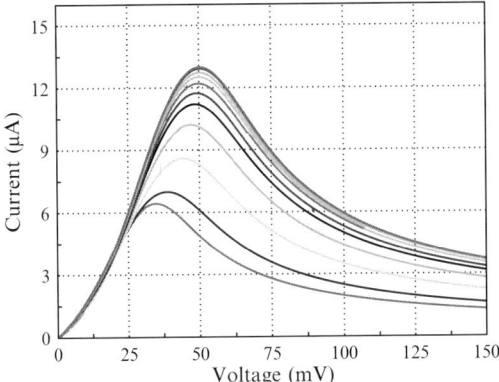

Fig. 2. Theoretical I–V curves at $4\,\mathrm{K}$ for a RTD with a spin-polarized injector increasing the spin splitting from $h = 0$ to $h = 2E_\mathrm{F}$ in steps of $h = 0.2E_\mathrm{F}$ (from bottom to top)

Zeeman splitting in the injector starts to play a role. In the I–V curves, which show two peaks corresponding to transport for each spin channel, there should be an increase (reduction) of the current amplitude for spin up (down) carriers due to the spin polarized population in the injector as B grows. (Here, we have taken spins up as the majority spins). Including this effect in the expressions above is easy. In the low $k_\mathrm{B}T$ limit one finds that the maximum current per spin, $J_\sigma^{\max} = (e\nu A/\hbar)(\Gamma_\mathrm{L}\Gamma_\mathrm{R}/\Gamma)(E_\mathrm{F} + \sigma h/2)$, increases (decreases) for spins up (down). Therefore, for small Γ, $J_\sigma^{\max}$ is a linear function of the applied field. Defining the relative current peak change, $\xi_\sigma(h) = [J_\sigma^{\max}(h) - J_\sigma^{\max}(0)]/I_\sigma^{\max}(0)$, we find that $\xi_\sigma(h) = \sigma h/2E_\mathrm{F}$. Now, for free electrons we approximate the spin polarization $N_\uparrow - N_\downarrow = \int_{-h/2}^{E_\mathrm{F}} D(E)\,\mathrm{d}E - \int_{h/2}^{E_\mathrm{F}} D(E)\,\mathrm{d}E \approx D(E_\mathrm{F})h$, where $D(E)$ is the density of states in the injector. Thus, the injector polarization, $p = (N_\uparrow - N_\downarrow)/(N_\uparrow + N_\downarrow) \approx \xi_\uparrow$, can be extracted from the increase or reduction of the I–V spin-splitting peaks.

In Fig. 3 we present numerical simulations of (2,3) for a magnetic RTD increasing both B and $k_\mathrm{B}T$ in such a way that Δ remains constant (1) and h increases. The I–V curves show the giant Zeeman splitting and an enhancement (reduction) of the majority (minority) spin resonance for increasing Zeeman splitting in the leads. The expected polarization detected by the RTD is plotted in the inset, which reinforces the idea that the polarization can be measured via the change of the peak height.

We note that our model deals with free electrons. In reality, electrostatic interactions induce bistable current solutions for a given V, which may lead to domain formation in extended systems [14]. In addition, spin relaxation, which tends to equilibrate the spin subsystems avoiding spin bottleneck effects, may become important in particular situations and it would be then desirable

458   D. Sánchez

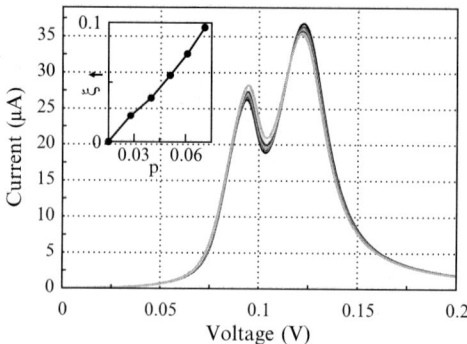

Fig. 3. Theoretical I–V curves for a II–VI DMS RTD with a 4%-Mn doped quantum well. The parameters are $E_{\rm F} = 10\,{\rm meV}$, $\varepsilon_0 = 54\,{\rm meV}$, $\Gamma_0 = 2\,{\rm meV}$. *Inset:* Relative change in the current amplitude of the majority spin peak as a function of the injector spin polarization

to include these effects in modeling RTD I–V curves. We follow Ref. [15] and establish rate equations for the spin-dependent density in the quantum well n_σ:

$$\frac{{\rm d}n_\sigma}{{\rm d}t} = \frac{J_{iw}^\sigma - J_{wc}^\sigma}{e} - \nu\frac{\mu_\sigma - \mu_{\bar\sigma}}{\tau_{\rm sf}} \tag{4}$$

where J_{iw}^σ and J_{wc}^σ are the well currents from the injector and toward the collector, μ_σ is the electrochemical potential in the well associated to spins σ, and $\tau_{\rm sf}$ is a phenomenological spin-flip time. This equation must be solved self-consistently with a mean-field approach for the potential drops along the RTD in terms of Poisson equations [15]. Thus, we see that there is considerable latitude for model improvements in future research.

3 Conclusions

We have discussed a theoretical model for spin-dependent transport in magnetically doped II–VI resonant tunneling diodes. Interestingly, we found that a magnetic injector increases the current peak and that this enhancement is tunable with a magnetic field. Further, when the quantum well is a magnetic semiconductor the I–V curves become split due to giant Zeeman splitting. We demonstrated that the splitting peaks change in amplitude when the field further increases and that this amplitude change can be used for detection of spin polarized currents.

Acknowledgements

I thank C. Gould, A.H. MacDonald, L.W. Molenkamp and G. Platero for collaborations in related work. This work was supported by the Spanish MEC Grant No. FIS2005-02796 and the "Ramón y Cajal" program.

References

1. R. Fiederling, M. Keimi, G. Reuscher, W. Ossau, G. Schmidt, A. Waag, and L.W. Molenkamp, Nature (London) **402**, 787 (1999).
2. Y. Ohno, D.K. Youn, B. Beschoten, F. Matsukura, H. Ohno, and D.D. Awschalom, Nature (London) **402**, 790 (1999).
3. J.K. Furdyna, J. Appl. Phys. **64**, R29 (1988).
4. T. Dietl, H. Ohno, F. Matsukura, J. Cibert, and D. Ferrand, Science **287**, 1019 (2000).
5. D.D. Awschalom and N. Samarth, J. Magn. Magn. Mat. **200**, 130 (1999).
6. T. Hayashi, M. Tanaka, and A. Asamitsu, J. Appl. Phys. **87**, 4673 (2000).
7. A. Slobodskyy, C. Gould, T. Slobodskyy, C.R. Becker, G. Schmidt, and L.W. Molenkamp, Phys. Rev. Lett. **90**, 246601 (2003).
8. P. Bruno and J. Wunderlich, J. Appl. Phys. **84**, 978 (1998).
9. N.N. Beletskii, G.P. Berman, and S.A. Borysenko, Phys. Rev. B **71**, 125325 (2005).
10. S. Ganguly, A.H. MacDonald, L.F. Register, and S. Banerjee Phys. Rev. B **73**, 033310 (2006).
11. C. Ertler and J. Fabian, cond-mat/0610617 (unpublished).
12. L.L. Chang, L. Esaki, and R. Tsu, Appl. Phys. Lett. **24**, 593 (1974).
13. M. Büttiker, IBM J. Res. Develop. **32**, 63 (1988).
14. L. L. Bonilla and H. T. Grahn, Rep. Prog. Phys. **68**, 577 (2005).
15. D. Sánchez, A.H. MacDonald, and G. Platero, Phys. Rev. B **65**, 035301 (2002).

Hysteretic Linear Conductance in Single Electron Transport through a Single Atom Magnet

J. Fernández-Rossier[1] and R. Aguado[2]

[1] Departamento de Física Aplicada, Universidad de Alicante, 03690 Alicante, Spain
jfrossier@ua.es
[2] Instituto de Ciencia de Materiales de Madrid (ICMM), CSIC, Spain

Summary. We consider single electron transport through a II-VI semiconductor quantum dot doped with a single Mn atom. The spin dynamics of the Mn atom is controlled by the carriers electrically injected in the dot. We find that the charge-vs.-gate curve can display hysteretic behaviour when the Mn-carrier interaction is anisotropic. We discuss the origin and implication of this result.

1 Introduction

The interplay between spin dynamics and spin-polarized transport in nanometric devices results in a variety of interesting physical phenomena, like magneto-resitance and spin transfer [1]. The fabrication of single electron transistors (SET) that permit current flow through a single molecule magnet opens new perspectives in this field [2,3]. A new direction in the single-electron control of nanomagnet comes from the recent fabrication of self-assembled CdTe quantum dots (QD) doped with a single Mn atom [4,5] embedded in an electrically active circuit [6]. Substitutional Mn in (Cd,Mn)Te has a +2 oxidation state with spin $S = 5/2$ associated to the localized d electrons [7]. Single exciton spectroscopy of a single Mn in a QD [4–6] provides a very good undertanding of the effective spin Hamiltonian for the Mn and the carriers in the dot [8] and makes it possible to model single electron transport through such interesting system [9]. These advances are promising steps towards the fabrication of a SET based on Mn doped II-VI QD, widely studied from the theory side [9–15].

We have recently studied the single electron transport through a CdTe QD, doped with a single Mn atom and charged by a gate voltage, V_G. We found that, in some instances, the charge Q vs. gate V_G curve was different depending on whether the dot was being charged or discharged. Consequently, the linear conductance $G_0(V_G)$ also presents hysteretic behaviour. Here we further explore this phenomenon. We argue that a hysteretic $G_0(V_G)$ curve can be a finger-print to characterize transport through single-molecule magnets.

2 Formalism

We consider a QD weakly coupled to two metallic and non-magnetic electrodes (source and drain). The dot can be gated so that the average charge can vary between 0 and +1 (injection of a single valence-band hole). Other charge states have been considered elsewhere [9]. The total Hamiltonian reads $\mathcal{H} = \mathcal{H}_{QD} + \mathcal{H}_C + \mathcal{H}_L + \mathcal{H}_R + \mathcal{V}_L + \mathcal{V}_R$. Here $\mathcal{H}_{QD}$ features a single orbital level, with twofold Kramers degeneracy, exchanged coupled to a spin $\mathbf{M}$. In this paper we take $M = 5/2$, adequate for a single Mn atom. The applied magnetic field is zero. In analogy with previous work [9, 10, 16–19], we make use of a quantum master equation for the dissipative dynamics of the *reduced density matrix* $\rho_{NM}(t)$ written in the basis of many-body states of the dot, $|N\rangle$. Importantly, this quantum master equation includes the combined dynamics of both populations and coherences. The second quantization Hamiltonian of the isolated dot reads

$$\mathcal{H}_{QD} = \epsilon_0 \sum_\sigma f_\sigma^\dagger f_\sigma + \sum_{a=x,y,z} J_a M_a S_{\sigma,\sigma'}^a f_\sigma^\dagger f_{\sigma'}. \tag{1}$$

Here $f_\sigma^\dagger$ injects a hole with (pseudo)spin σ in the quantum dot. Different choices of J_a reflect the interplay between spin–orbit interaction, shape of the dot and orbital origin of the valence band [5, 8, 9]. For the first hole level in a dot with cylindrical symmetry, we have $J_z >> J_x = J_y$. In the absence of light hole heavy hole mixing, we have $J_\perp \equiv J_x = J_y = 0$. We shall consider both the $J_\perp = 0$ and $J_\perp \neq 0$ cases. For $Q = 0$ H_{QD} has $2S + 1 = 6$ eigenstates describing the Mn spin. For $Q = +1$ H_{QD} has 12 eigenstates describing both the Mn spin and the hole (iso-)spin. We label them as $H_{QD}|N\rangle = E_N|N\rangle$.

The metallic electrodes are described by $\mathcal{H}_L = \sum_{\sigma,k} \epsilon_k a_{k\sigma}^\dagger a_{k\sigma}$ and $\mathcal{H}_{rmR} = \sum_{\sigma,p} \epsilon_p b_{p\sigma}^\dagger b_{p\sigma}$ whereas $\mathcal{V}_L = \sum_{\sigma,k,\alpha} V_{\sigma,k,\alpha} f_\alpha^\dagger a_{k\sigma} + \text{h.c.}$ and $\mathcal{V}_R = \sum_{\sigma,p,\alpha} V_{\sigma,k,\alpha} f_\alpha^\dagger b_{p\sigma} + \text{h.c.}$ are the standard *spin-conserving* tunneling Hamiltonian that couple the metallic reservoirs and the dot. Here α labels the isospin of the QD state. Assuming that the quantum dot is weakly coupled to the electronic reservoirs (sequential tunneling), the dissipative dynamics of the density matrix is governed by a Markovian kernel, $\dot{\rho}(t) = A\rho(t)$, fully characterized by the rates $\Gamma_{N,M} = \sum_{r \in L,R} \Gamma_r n_r(E_N - E_M) \sum_\sigma |\langle N|f_\sigma|M\rangle|^2$. Here $f_\sigma^+ \equiv f_\sigma^\dagger$, $f_\sigma^- \equiv f_\sigma$, n_r^+ is the Fermi function of reservoir r and $n_r^- = 1 - n_r^+$. The notation $\Gamma_{N,M}^\pm$ implies that states M with charge Q are connected with states N with charge $Q \pm 1$. The coupling to the leads is parameterized by $\Gamma_{L,R} = \frac{2\pi}{\hbar}|V_{L,R}|^2 N_{L,R}$, where $N_{L,R}$ is the DOS of the metallic reservoir. For a given value of gate, bias voltage and temperature we find the *steady state* density matrix $\tilde{\rho}$ (namely, $A\tilde{\rho} = 0$), from which we can compute the average charge, magnetization and current. Importantly, the current expression includes both diagonal and non-diagonal terms in the density matrix [9].

3 Results

We now consider the charging and discharging process of the Mn-doped QD. The V_G is varied so that the charge changes from 0 to +1 or vice-versa. We take the temperature equal to $0.05\,\text{meV} = k_B T = 5\Gamma_L = 5\Gamma_R$. In standard SET the $G_0(V_G)$ curve is independent of whether the dot is being charged or discharged and has a single peak at the V_G for which the ground state energies of the manifolds with 0 and 1 carriers are the same. In a previous work we reported that this is no longer the case for Mn-doped quantum dots [9]. Here we present some details to provide additional insight of this new phenomena. The initial condition for the solution of the master equation is a thermal density matrix. We choose the inital V_G so that the charge state is either 0 or +1. The master equation is solved and the steady state DM is obtained for each V_G. Importantly, as the gate is ramped the initial condition for the density matrix is the steady state of the previous run. Because the only Mn spin-relaxation mechanism included in our approach comes from the combined action of exchange interaction with the hole and single-hole tunneling events, sometimes our simulations reach a steady state different from the thermal equilibrium. This simulates an experiment in which the gate is ramped at a pace *faster* than the spin relaxation mechanisms not included in the calculation, i.e. those operating in (Cd,Mn)Te. Importantly, the relaxation time of Mn in CdTe reaches $\approx 10^{-3}\,\text{s}$ in the dilute limit [20] and it might be even longer in quantum dots. We expect that the $G(V_G)$ curve will be different depending on the pace at which the gate voltage is ramped. This resembles the $M(H)$ curves of single molecule magnets, which depend on the pace at which the external field is applied [21].

In our first simulation we consider the case $J_\perp = 0$ for which the charging curve shows three peaks, whereas in the isotropic case (not shown here), $J_z = J_\perp$, the results are identical to the standard case [9]. In the upper (lower) panels of the figure we plot the population (linear conductance) of the QD as a function of V_G for two cases (charging and discharging the QD) and two values of the transverse interaction $J_\perp$. In both cases we take $J_z = 0.6\,\text{meV}$ whereas $J_\perp = 0$ in panels (a–d). A very small bias voltage ($0.005\,\text{meV}$) is applied so that current flows. The eigenstates of the QD Hamiltonian with $Q = +1$ are split in six doublets. The eigenstates of the lowest energy doublet are $|M_z = \pm 5/2, \sigma = \mp\rangle$. The results in panels (b) and (d) are quite standard: the dot starts in an equilibrium situation, with the two states of the $Q = +1$ ground state doublet equally occupied and all the other states with either $Q = +1$ and $Q = 0$ with zero occupation. As the gate brings the states $Q + 1$ with $|M_z| = 5/2$ into resonance with the $Q = 0$ states, the

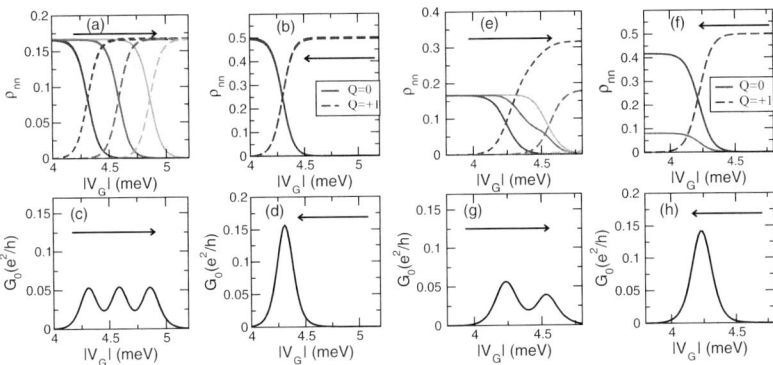

Fig. 1. (**a-d**) QD with $J_\perp = 0$. (**e-h**) QD with $J_\perp = 0.1 J_z$. *Upper panels*: Steady state values of the QD occupations as a function of gate voltage for charging (**a, e**) and discharging (**b, f**) the QDs. *Lower panels*: the corresponding linear conductances for charging (**c, g**) and discharging (**d, h**) the QDs

populations are transferred from the charged to the neutral states, always *conserving* the Mn spin. A single peak in the linear conductance occurs at the degeneracy point. The non-standard results are shown in panels (a) and (c), corresponding to charging the QD. At the initial gate, the six $\mathcal{Q} = 0$ states are equally populated (they are degenerate) and all the $\mathcal{Q} = 1$ states are high above in energy and their occupation is zero. As the gate is ramped so that the charged states are closer in energy, a first transfer of population from neutral to charged states occurs in the $M_z = \pm 5/2$, corresponding to the lowest energy doublet of the $\mathcal{Q} = +1$ states. Notice that the population of the two $\mathcal{Q} = 0$ doublets with $M_z \neq \pm 5/2$ remains unaffected. Only when the second charged doublet $\mathcal{Q} = +1, M_z = \pm 3/2$ becomes in resonance with the neutral states, population transfer occurs, resulting in a second conductance peak. Notice that the steady state occupations reached when the dot is charged (panel a) are different from those of equilibrium ($V_G = 5.2\,\mathrm{meV}$ in panel b).

The anomalous behaviour is related to the strict conservation of the Mn spin along the growth axis M_z. We now explore what happens when Mn spin flip terms are included in the Hamiltonian. To do that we take $J_\perp = 0.1 J_z$, with the same J_z than before. This choice of J_a corresponds to weak LH-HH mixing [9]. Because of $J_\perp \neq 0$, the states with $\mathcal{Q} = +1$ are not eigenstates of M_z anymore and instead of six doublets, the spectrum has five doublets and two singlets [9]. The results of our simulations are shown in the panels (e–h) of the figure, for integration times much larger than $1/\Gamma$. As in the preceeding case, the removal of the hole from the dot yields a single peak in the $G_0(V_G)$ curve. The injection of the hole, however, results in two peaks, different both from the standard one peak result and from the anomalous three peak result obtained for the $J_\perp = 0$ case. We see that strict conservation of M_z is not necessary to obtain hysteretic $G_0(V_G)$ curves. Panel (e) shows how the population of the initially occupied $\mathcal{Q} = 0$ states is transferred to the initially empty $Q = +1$ states. In contrast with the pure Ising case, the depletion of the $|M_z| = 3/2$ doublet occurs via transfer both to the ground state doublet and the first excited state of the $Q = +1$ manifold. The latter also depletes the $Q = 0$ $|M_z| = 1/2$ doublet. As a result $G_0(V_G)$ has only two peaks associated to the resonance condition for the $Q = +1$ first two doublets. A single peak is recovered only if the integration time goes to infinity. The fact that M_z is not conserved is also seen during the reverse process of hole ejection from the dot, in panel (f).

As a final remark, we note that according to our simulations we can *filter* electrically the Mn spin. For instance, if we start with an equilibrium distribution with $Q = 0$ the six Mn spins are equally populated. We now charge the dot (panels a and e) and wait until thermal equilibrum is reached. At that point, the occupied states have zero (or small) overlap with the $M_z \neq \pm 5/2$ sector. If the gate is reversed so that the hole is removed from the dot the final density matrix is different from the initial one and only the occupation of the $|M_z| = 5/2$ doublet is significantly larger than that of the others (panels b and f). This result shows a possible protocol of electrical manipulation of the quantum state of a single Mn spin.

4 Conclusions

We have shown that single electron transport through a single atom magnet can result in hysteretic behavior of the linear conductance vs. gate. The anomalous result occurs when the following conditions are met: (1) the effective Hamiltonian of the nanomagnet is *strongly* modified by the addition of a single carrier (2) The exchange interaction is anisotropic and (3) the charged nanomagnet can reach a (quasi)-steady state different from equilibrium at time scales shorter than the slow spin relaxation times of the neutral Mn [20]. Observability of the hysteretic linear conductance requires ramping the gate faster than this time scale.

Acknowledgements

Funding from Spanish Ministry of Education (Grants FIS200402356, MAT2005-07369-C03-03, Ramón y Cajal Program) and Generalitat Valenciana (GV05-152) is acknowledged.

References

1. J. C. Slonczewski, J. Magn. Magn. Mater. 159, L1 (1996).
2. H. B. Heersche *et al.*, Phys. Rev. Lett. **96**, 206801 (2006).
3. Moon-Ho Jo *et al.* Nano Lett. **6**, 2014 (2006).
4. L. Besombes *et al.* Phys. Rev. Lett. **93**, 207403 (2004).

5. Y. Léger Phys. Rev. Lett. **95**, 047403 (2005).
6. Y. Léger *et al.* Phys. Rev. Lett. **97**, 107401 (2006).
7. J. K. Furdyna, J. Appl. Phys **64** R29 (1988).
8. J. Fernández-Rossier, Phys. Rev. B. **73**, 045301 (2006).
9. J. Fernández-Rossier, and R. Aguado, Phys. Rev. Lett. **98**, 106805 (2007).
10. A. Efros, E. Rashba, M. Rosen, Phys. Rev. Lett. **87**, 206601 (2001).
11. J. Fernández-Rossier and L. Brey, Phys. Rev. Lett. **93** 117201 (2004).
12. A. O. Govorov, Phys. Rev. B **72**, 075359 (2005). A. O. Govorov, Phys. Rev. B **72**, 075358 (2005).
13. J. I. Climente *et al.*, Phys. Rev. B**71**, 125321 (2005).
14. F. Qu, P. Hawrylak, Phys. Rev. Lett. **95**, 217206 (2005). *ibid* **96**, 157201 (2006).
15. J.-M. Tang, J. Levy, and M. E. Flatte Phys. Rev. Lett. **97**, 106803 (2006).
16. X. Waintal, P. Brouwer, Phys. Rev. Lett. **91**, 247201 (2003).
17. X. Waintal, O. Parcollet, Phys. Rev. Lett. **94**, 247206 (2005).
18. C. Timm, F. Elste F. Elste and C. Timm Phys. Rev. B **71**, 155403 (2005); *ibid* Phys. Rev. B **73**, 235304 (2006).
19. M. Braun, J. König, J. Martinek, Phys. Rev. B **70**, 195345 (2004)
20. J. Lambe and C. Kikuchi, Physical Review **119**, 1256 (1960); D. Scalbert *et al.*, Solid State Commun., **66**, 571 (1988).
21. D. Gatteschi and. R. Sessoli, Angew. Chem. **42**, 268 (2003).

Minisymposium "Ferromagnetic Carbon Nanostructures"

T. Makarova[1] and M.A.H. Vozmediano[2]

[1] Umea University, Sweden
[2] Universidad Carlos III de Madrid, Spain

The discovery of nanostructured forms of molecular carbon has led to renewed interest in the varied properties of this element. Recent experiments and theoretical studies have suggested that electronic instabilities in pure graphite may give rise to superconducting and ferromagnetic properties, even at room temperature. Magnetic carbon could be used to make inexpensive, metal-free magnets for applications in medicine and biology, nanotechnology and telecommunications. The following topics in the invited presentations to the minisymposium have been included in these proceedings: The state of the art of magnetism in nanographite is reviewed in two papers from the point of view of experimentalists (T. Makarowa and M.A. Ramos). The theoretical support is presented in the report by M.P. Lopez-Sancho, M.A.H. Vozmediano, F. Stauber, and F. Guinea. A. Cortijo discusses the electronic properties of topological defects in graphene, and F. Guinea presents a transistor effect in bilayer graphene. An additional presentation by L. Pisani on numerical studies of graphene nanoribbons could not be included in this volume.

Ferromagnetic Carbon Nanostructures

Tatiana L. Makarova[1,2]

[1] Institute of Physics, Umeå University, 90187 Umeå, Sweden
[2] Ioffe PTI, 194021 St. Petersburg, Russia
tatiana.makarova@physics.umu.se

Introduction

Carbon nanostructures are regarded as all-carbon structures with the nanometer size. Building blocks of the future, building blocks of future information and energy technologies – here are the permanent epithets for carbon nanostructures. Scientific interest, sparked by the discovery of fullerenes [1], refocused on carbon nanotubes [2] and other exotic structures like nanofibers, nanoribbons, nanohoops, nanocones and nanohorns [3], toroids [4] and helicoidal tubes [5], onions and peapods [6], Schwarzites and Haeckelites [7]. More recently, it was discovered that the two-dimensional building block for creating the nanostructures of any other dimensionality, graphene, itself possesses unique electronic properties: ballistic electron transport, constant velocity for the electrons confined in the graphene sheets (massless particle behaviour), half-integer shift in the quantum Hall effect and quantized minimum conductivity [8–10]. The linear dependence of the energy on momentum in graphene leads to unusual features, not encountered in other materials [11].

Detailed understanding of the structure, electronic properties and potential applications of carbon nanostructures is the basis for a new technology, which will modify their properties in a targeted way. Carbon nanostructures offer record values of strength and flexibility, can exhibit ballistic conductivity, superconductivity and superlubricity. Magnetic properties are less investigated [12]. It is well known that small amounts of carbon are capable of destroying the strong exchange interactions between iron atoms in stainless steel. Can one expect exchange interactions in all-carbon structures?

Nanosized Carbon Structures

Carbon nanostructures exhibit a wide variety of unusual structural and electronic properties. An important feature of graphitic structures is 2D itinerant π- electron system, which is responsible for high conductivity and large diamagnetism. Nearly all (probably, excluding nanodiamonds and tetrahed-rally bonded amorphous carbon) carbon nanostructures could be constructed from an ordinary hexagonal graphene layer by cutting, bending, rolling and zipping. These carbon structures can be closed or open, singlewalled or mul-tiwalled, and can have zero, positive or negative Gaussian curvature.

Rolled graphene sheets, fullerenes and carbon nanotubes possess closed π-electron systems. These systems do not have open edges, or the influ-ence of the open edges is negligible. They are characterized by negative dia-magnetic susceptibility [13]. More complicated hypothetical close-shell pure carbon nanostructures such as corrugated nanotori constructed from coales-cent fullerenes [14] may exhibit positive magnetic susceptibility (paramag-netism). Carbon nanotori with magic numbers have colossal paramagnetic response [15], perforated fullerenes and nanoporous graphitic structures, ex-hibiting negative Gaussian curvature which behave as strong paramagnets experiencing large magnetic moments when an external magnetic field is ap-plied [16]. A giant magnetoconductance is predicted to occur in twisted nan-otubes in presence of an applied magnetic field [17].

Diamagnetism of nanographenes can be understood in terms of diamag-netic ring currents. Defects in graphite always reduce the diamagnetic signal. In a simplified picture, vacancies, adatoms, pores and bond rotations enhance local paramagnetic ring currents and produce local magnetic moments [18].

Nanographite, a stack of nanosized graphene layers, is a nanosized π-electron system with open edges. The periphery of a nanographite pattern can be described as a combination of zigzag and armchair edges (Fig. 1). In the open-edge systems, the edges around their boundary produce distinctive electronic features, namely, the zigzag edges produce strongly spin-polarized states, which are spatially localized around the edges. The presence of these states modifies the electronic structure of nanographite as a whole: It produces edge-inherited non-bonding π-electronic state (edge state) in addition to the

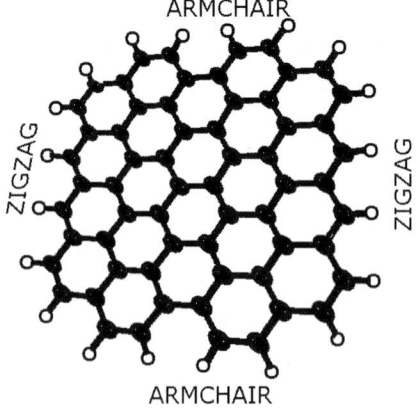

Fig. 1. Nanographene

π- and π^*-bands, giving entirely different electronic structure from bulk graphite. These so-called 'peculiar' states are extended along the edges but at the same time are localized at the edges [19]. These states produce large electronic density of states at the Fermi level and play an important role in the unconventional nano-magnetism.

An important difference exists between so-called graphitic and nongraphitic nanocarbon. Graphitic nanocarbon is characterized by the three-dimensional order in the direction perpendicular to the planes; the presence of the defects in the planes is not taken into account. In the case of random distribution of the packets of the graphite-like layers, the material is called 'nongraphitic carbon' [20]. Formally, the sp^2/sp^3 ratio is not relevant; however, general trend in the electronic properties of graphitic and non-graphitic nanocarbon is the following: The increase in the sp^2/sp^3 ratio leads to the clusterization of the sp^2 sites and to establishing the long range structural order. This favours the formation of the itinerant π-electron system; therefore, graphitic nanocarbon is usually characterized with larger diamagnetism than non-graphitic carbons. However, even sp^2-rich disordered carbon may remain non-graphitic, and the simplest examples of this are the fullerene solids or carbon nanofoam [21].

A particular case of a graphene modification is the Stone–Wales defects, or topological defects, caused by the rotation of carbon atoms which leads to the formation of five- or sevenfold rings. A novel class of curved carbon structures, Schwarzites and Haeckelites, has been proposed theoretically [22]. Schwarzite is a form of carbon containing graphite-like sheets with hyperbolic curvature, so far, periodic schwarzites have not been realized experimentally; however, there is experimental evidence that random Schwarzite structures are present in a cluster form in such carbon phases as spongy carbon [23] and carbon nanofoam [24]. Haeckelite is a theoretically predicted material in which pentagons, hexagons and heptagons are equally considered as regular building blocks. Calculations have shown that this structure is energetically more favourable than fullerenes [25]. Haeckelite structures (nanostructures containing non-hexagonal rings) were produced on the HOPG (highly oriented pyrolitic graphite) substrate: Y-branched carbon nanotubes and coiled carbon nanotubes [26]. It was shown theoretically that the schwarzite carbon structures which do not contain under-coordinated carbon atoms carry a net magnetic moment in the ground state. In the systems with negative Gaussian curvature, unpaired spins can be introduced by sterically protected carbon radicals [27].

Intrinsic Magnetic Defects in Nanocarbon

Electronic structure of nanocarbon is controlled by the defects [28]. It has been shown theoretically [29] that the interplay of disorder and interactions in a

2D graphene layer gives rise to a rich phase diagram where strong coupling phases can become stable. Local defects can lead to the magnetic ordering.

The examples of intrinsic carbon defects are the lattice defects, vacancies and voids in the graphene structure, which give rise to localized states at the Fermi energy, and the number of these states roughly scales with the defect perimeter [30]. The repulsive electron–electron interaction leads to spin polarization and to the formation of localized moments which interact ferromagnetically. However, under the reasonable assumptions on the defect concentration, the exchange integral is low, and the estimated Curie temperature does not exceed 1 K.

Topological defects in graphene lead to the presence of curvature in the samples of this material. The formation of the heptagon–pentagon defects introduces corrections to the local density of states, and the spatial extent of the correction is such that the relative intensity decays to 10% in approximately 20 unit cells [31].

The key problem in graphite magnetism is the nature and stability of the defects and the range of the magnetic interaction (J) between the localized spins of the defects. It is important that disorder increases J [32].

A specific case of defects is the presence of the first raw elements, although they cannot be unambiguously classified as intrinsic carbon defects. The most important defect is hydrogen: Unsaturated valence bonds at the boundaries of graphene flakes are filled with stabilizing elements; among these stabilizers hydrogen atoms are the common ones. The entrapment of hydrogen by dangling bonds at the nanographite perimeter can induce a finite magnetization. A theoretical study of a graphene ribbon in which each carbon atom is bonded to two hydrogen atoms at one edge and to a single hydrogen atom at the other edge shows that the structure has a finite total magnetic moment [33]. Combination of different edge structures (by hydrogenation, fluorination or oxidation) is proposed as a method to design magnetic nanographite [34].

Other elements that may strongly influence magnetic behaviour of carbon are boron and nitrogen. Border states in hexagonally bonded BNC heterosheets have been predicted to lead to a ferromagnetic ground state, a manifestation of flat-band ferromagnetism [35]. In heterostructured nanotubes, partly filled states at the interface of carbon and boron nitride segments, may acquire a permanent magnetic moment. Depending on the atomic arrangement, heterostructured C/BN nanotubes may exhibit an itinerant ferromagnetic behaviour owing to the presence of localized states at the zigzag boundary of carbon and boron nitride segments [36].

Magnetic Properties of Nanocarbon

Magnetic susceptibility χ_{total} of nanocarbon may be generally described by the following formula:

$$\chi_{\text{total}} = \chi_{\text{core}} + \chi_{\text{orb}} + \chi_{\text{vv}} + \chi_{\text{p}} + \chi_{\text{L}} + \chi_{\text{c}}$$

where χ_{core} is the diamagnetic contribution from the core electrons which can be estimated from the sum of Pascal's constants [37], χ_{orb} is the orbital diamagnetism arising from the inter-band transition between graphitic linear bands, χ_{vv} is the paramagnetic van Vleck term originating from virtual magnetic dipole transitions between the valence and conduction bands, χ_p is the Pauli paramagnetism of itinerant electrons which depends on the density of states around the Fermi energy, χ_L is the Landau diamagnetism of the itinerant electrons and χ_c is the temperature-dependent Curie–Weiss term $\chi_c = C/(T - \Theta)$ where C and Θ stand for the Curie constant and the Weiss temperature, respectively.

Experimentally, various types of magnetic behaviour were discovered in nanocarbon derived from graphite. Theoretical predictions have been confirmed by strong experimental evidence that the edge states in nanographite disordered network govern its magnetic properties. Normally the nanocarbons are diamagnets, and diamagnetism competes with the Curie–Weiss behaviour and the Pauli paramagnetic temperature-independent term. In highly disordered structures a paramagnetic behaviour was observed, and the presence of non-bonding π-electrons at the edges of nanosized graphene was invoked for the explanation of the paramagnetism [38]. Unusually strong paramagnetism was found in dense graphitic filaments formed via thermal decomposition of mesitylene in an applied electric field [39].

Unconventional magnetic properties, including anti-ferromagnetic interactions, spin–glass state, disordered magnetism, magnetic switching phenomenon have been described in nanocarbons. In non-graphitic but sp^2-rich disordered carbons prepared by pulsed laser deposition the presence of anti-ferromagnetic interactions between the localized spins has been identified from the negative Weiss temperature and from the magnetization curves, which did not follow the expected Brillouin curve for non-interacting spins with $S = 1/2$ at low temperatures [40]. The average distance between the localized spins has been calculated under the assumption of the homogeneous distribution of the spins. The estimated $17\,\text{Å}$ value sufficiently exceeds the one expected for the direct exchange interaction. Similar effects were observed on nanographite obtained from the heat treatment of nano-diamond particles [41]. Strong anti-ferromagnetic coupling has been found between the spins localized on the surface of the particles. Hydrothermal treatment sufficiently enhances the exchange interactions. The exchange coupling between the spins becomes appreciable in the range of $40\,\text{K}$, and after the hydrothermal treatment in supercritical water this range increases to $200\,\text{K}$ (Fig. 2) [42]. Again, the strength of anti-ferromagnetic interactions is significantly larger than that simply expected from the average spin–spin distance.

Another example of open-shell carbon nanostructures is ACF (activated carbon fibers), which can be considered as a three-dimensional random network of nanographitic domains with characteristic dimensions of several nanometers [43]. Temperature dependencies of the susceptibility taken in zero field cooled regime indicate the presence of a quenched disordered magnetic

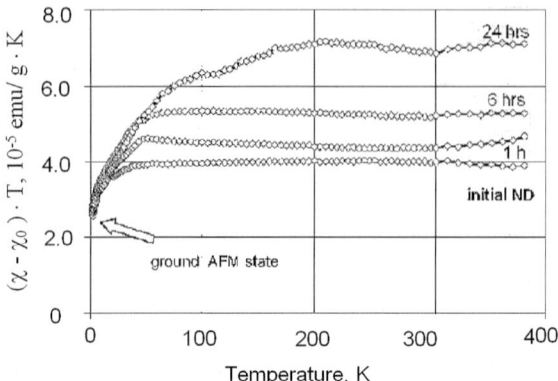

Fig. 2. The downturn of curves $(\chi - \chi_0) \cdot T$ vs. T manifests the onset of anti-ferromagnetic interactions between the spins. Reprinted with the permission from Osipov et al. [42]

structure like a spin glass state [44]. This effect appears in the vicinity of the metal–insulator transition, where the coexistence of the edge-state localized spins and the conduction π-electrons causes the magnetic state in which the exchange interactions between the localized spins are mediated by the conduction electrons. The range of exchange interaction is estimated as 2–3 nm, and such a long-range nature of the exchange interaction proves that the nanographite magnetism is *sui generis*.

An important proof for the edge-state inherited unconventional magnetism is the magnetic switching phenomenon, which has been found in the activated carbon fibers. Physisorption of water drastically changes magnetic properties, although water itself is nonmagnetic [45]. Water molecules compress the nanographite domains, reducing the interlayer distance in a stepwise manner. This leads to the enhancement of the anti-ferromagnetic exchange interaction of the edge-state localized spins at the adjacent nanographene layers [46].

Ferromagnetism in Carbon Nanostructures

Magnetic ordering at high temperatures in carbon-based compounds has been persistently reported since 1986. In some cases the reported experimental data give convincing proof for the intrinsic origin of the effect. This is the case of the proton-irradiated carbon structures, where the ultimate purity of the material is proved by simultaneous measurements of the magnetic impurities [47]. In other cases the carbonaceous materials do contain 0.001–0.1% iron, whereas the measured magnetization values several times exceed the value expected from the impurities assuming all iron is in its ferromagnetic form. This noticeable surplus as well as the absence of superparamagnetic behaviour, which is typical for iron–carbon composites with iron concentrations ranging up to several at. % [48–50] suggests that impurities either do not contribute to magnetic properties or their role is far from trivial. Nontrivial origin of magnetic behaviour in contaminated carbon-based materials may result from catalytic [51] or template [52] properties of transition metal atoms, and it was shown that ferromagnetic properties of carbon preserve after washing out the transition metal ions [53]. The effect of triggering carbon magnetism by the presence of transition metals was reported on graphite [54], carbon nanotubes [55] and C/Fe layers [56]; however, this effect was not found in graphite–magnetite composites [57] and fullerene-like Ni-C nanostructures [58]. Eventually, there is growing evidence that carbon does not need iron to become magnetic.

Figure 3 shows the magnetization loop for the ultra pure spectral graphite rod. The concentration of impurities in this sample is the following: B = 0.03 ppm, Si = 0.1 ppm, Fe = 0.03 ppm, Mg < 0.03 ppm, Ti = 0.03 ppm, Al < 0.03 ppm, V = 0.03 ppm, Ca < 0.03 ppm, Cu < 0.03 ppm. Maximum magnetization value for the extrinsic (iron-conditioned) magnetization could be 6.6×10^{-6} emu g^{-1}; the measured value is 1.2×10^{-3} emu g^{-1}, i.e. 180 times higher. Remanence magnetization is 1.2×10^{-4} emu g^{-1}, coercive force is 160 Oe. A diamagnetic contribution -6.59×10^{-6} emu (G g)$^{-1}$ was subtracted from the original data. At room temperature no signs of nonlinearity were detected (empty circles in Fig. 1). Although we operate with very small values of both magnetization and metal content, fairly large mass of the analyzed sample (156 mg) reduces the mistakes to minimum.

Similar signals were found in high purity HOPG samples from NTI-Europe, ZYA quality [59].The authors discard the possibility that this behaviour is due to magnetic impurities, since the PIXE (particle induced X-ray emission) experiments performed on some of these samples always gave concentrations several times less than needed for the measured magnetization. We believe that the occurrence of the weak signal of magnetic ordering in ultra-pure

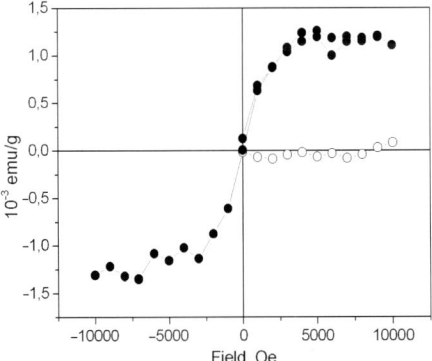

Fig. 3. $M(H)$ at 1.9 K (*full circles*) and at 298 K (*open circles*) for the ultra pure spectral carbon

metal-free graphite rod and high purity HOPG samples manifests a defect-related magnetism. However, the signals are very weak, and it is not possible to attribute these signals to any of defects described above: edge states, or bond defects, or interstitial or surface states. More importantly, the nature of the ordering mechanism remains to be determined.

Conclusions

Carbon materials that exhibit ferromagnetic behaviour have been predicted theoretically and reported experimentally in recent years. The initial surprising experiments were confirmed by the independent groups. The fact that carbon atoms can be magnetically ordered at room temperature was confirmed by the direct experiment: an element-sensitive method X-ray magnetic circular dichroism [60]. However, it is still a challenge to demonstrate experimentally the paramagnetic and ferromagnetic properties of bulk carbon materials.

We envisage that magnetism in different families of nanostructures will be playing a key role in the development of emerging technologies in the present century. Single-molecule transistors, all-carbon integrated circuits, molecular actuators, flat displays all could be produced on the basic of carbon nanostructures. An extension of these possibilities to spintronic devices is a tempting opportunity.

Acknowledgements

This work is supported by the European project 'Ferrocarbon' Contract 12881 (NEST), RFBR 05-02-17799 and Swedish Research Council.

References

1. H. Kroto, J. R. Heath, S. C. O'Brien, R. F. Curl and R. E. Smalley, Nature (London) **318**, 162 (1985).
2. S. Ijima, Nature (London), **354** 56 (1991).
3. M. Ge, K. Sattler, Chem. Phys. Lett. **220**, 192 (1994).
4. S. Itoh, S. Ihara, J. Kitakami, Phys. Rev B **47**, 1703 (1993).
5. S. Amelinckx, X. B. Zhangm D. Bernaerts, X. F. Zhang, V. Ivanov, J. B. Nagy, Science **265**, 635 (1994).
6. B.W. Smith, M. Monthioux, and D.E. Luzzi, Nature (London) **396**, 3239 (1998).
7. Mackay A. L. and Terrones H., Nature (London) **352**, 762 (1991).
8. K.S. Novoselov, A.K. Geim, S.V. Morozov, D. Jiang, Y. Zhang, S.V. Dubonos, I.V. Grigorieva and A.A. Firsov, Science **306**, 666 (2004).
9. K.S. Novoselov, A.K. Geim, S.V. Morozov, D. Jiang, M.I. Katsnelson, I.V. Grigorieva, S.V. Dubonos and A.A. Firsov, Nature (London) **438**, 197 (2005).

10. Y. Zhang, Y. Tan, H. L. Stormer, P. Kim Nature (London) **438**, 201 (2005).
11. F. Guinea. 14th European Conference on Mathematic for Industry, July 10th-14th, 2006, Leganés, Madrid, Spain. Book of Abstracts, p. 48.
12. P. Stamenov, J.M.D. Coey, J Magn Magn Mater **290**, 279 (2005).
13. J. Heremans, C.H. Olk and D.T. Morelli, Phys Rev B **49**, 15122 (1994).
14. J.A. Rodríguez-Manzo, F. López-Urías, M. Terrones and H. Terrones, Nano Lett. **4**, 2179 (2004).
15. L. Liu, G.Y. Guo, C.S. Jayanthi and S.Y. Wu, Phys. Rev. Lett. **88**, 217206 (2002).
16. H. Terrones, F. López-Urías, E. Muñoz-Sandoval, J.A. Rodríguez-Manzo, A. Zamudio, A.L. Elías and M. Terrones. Solid St. Sci. **8**, 303 (2006).
17. S. Steven, W.D. Bailey, D. Tomaanek, Y. Kwon, and C. Lambert, Europhys. Lett. **59**, 75 (2002).
18. F. Lopez-Urias, J. A. Rodriguez-Manzo, E. Munoz-Sandoval, M. Terrones, H. Terrones. Opt. Mat. **29**, 110 (2006).
19. M. Fujita, K. Wakabayashi, K. Nakada, and K. Kusakabe, J. Phys. Soc. Jpn. **65**, 1920 (1996).
20. R. E. Franklin, Acta Crystallogr. **4**, 253 (1951).
21. A.V. Rode, R.G. Elliman, E.G. Gamaly, A.I. Veinger, A.G. Christy and S.T. Hyde, Appl.Surf. Sci **197**, 644 (2002).
22. A. L. Mackay, H. Terrones, Nature **352**, 762 (1991).
23. E. Barborini, P. Piseri, P. Milani, G. Benedek, C. Ducati, J. Robertson. Appl. Phys. Lett. **81**, 3359 (2002).
24. A. V. Rode, E. G. Gamaly, A. G. Christy, J. G. F. Gerald, S. T. Hyde, R. G. Elliman, B. Luther-Davies, A. I. Veinger, J. Androulakis, J. Giapintzakis, Phys. Rev. B **70**, 054407 (2004).
25. H. Terrones, M. Terrones, E. Hernández, N. Grobert, J.-C. Charlier, and P. M. Ajayan, Phys. Rev. Lett. **84**, 1716 (2000).
26. L. P. Biro, R. Ehlich, Z. Osvath, A. Koos, Z. E. Horvath, J. Gyulai, J. B. Nagy, Mat. Sci. Eng. **19**, 3 (2002).
27. N. Park, M. Yoon, S. Berber, J. Ihm, E. Osawa, D. Tomanek. Phys. Rev. Lett. **91**, 237204 (2003).
28. M. A. H. Vozmediano, M. P. Lopez-Sancho, T. Stauber, F. Guinea, Phys. Rev. B **72**, 155121 (2005).
29. M. A. H. Vozmediano, F. Guinea, M. P. Lopez-Sancho, J. Phys. Chem. Sol. **67**, 562 (2006).
30. M. P. Lopez-Sancho. 14th European Conference on Mathematic for Industry, July 10th-14th, 2006, Leganés, Madrid, Spain. Book of Abstracts, p. 47. Submitted to the Proceeding volume.
31. Alberto Cortijo. 14th European Conference on Mathematic for Industry, July 10th-14th, 2006, Leganés, Madrid, Spain. Book of Abstracts, p. 47. Submitted to the Proceeding volume.
32. L. Pisani. 14th European Conference on Mathematic for Industry, July 10th-14th, 2006, Leganés, Madrid, Spain. Book of Abstracts, p. 47.
33. K. Kusakabe and M. Maruyama, Phys. Rev. B **67**, 092406 (2003).
34. M. Maruyama, and K. Kusakabe, J. Phys. Soc. Jpn. **73**, 656 (2004).
35. S. Okada and A. Oshiyama, Phys. Rev. Lett. **87**, 146803 (2001).
36. J. Choi, Y. Kim, K. J. Chang, and D. Tomanek, Phys. Rev., B **67** 125421 (2003).
37. O. Kahn, in Molecular Magnetism, VCH publishers, New York (1993).

38. B.L.V. Prasad, H. Sato, T. Enoki, Y. Hishiyama, Y. Kaburagi and A.M. Rao, Phys Rev B **62**, 11209 (2000).
39. J. M. Calderon-Moreno, A. Labarta, X. Batlle, D. Crespo, V. G. Pol, S. V. Pol, A. Gedanken, Carbon **44**, 2864 (2006)
40. K. Takai, M. Oga, T. Enoki and A. Taomoto. Diam. Rel. Mat. **13**, 1469 (2004)
41. O. E. Andersson, B. L. V. Prasad, H. Sato, T. Enoki, Y. Hishiyama, Y. Kaburagi, M. Yoshikawa, S. Bandow. Phys. Rev B **58**, 16387 (1998).
42. V. Osipov, M. Baidakova, K. Takai, T. Enoki, A. Vul'. Fuller. Nanot. Car. Nan. **14**, 565 (2006)
43. Y. Shibayama, H. Sato, T. Enoki, X.X. Bi, M.S. Dresselhaus, M. Endo, J. Phys. Soc. Jpn. **69**, 754 (2000).
44. Y. Shibayama, H. Sato, T. Enoki, M. Endo, Phys. Rev. Lett. **84**, 1744 (2000).
45. H. Sato, N. Kawatsu, T. Enoki, M. Endo, R. Kobori, S. Maruyama, and K. Kaneko, Solid State Commun. **125**, 641 (2003).
46. K. Harigaya and T. Enoki, Chem. Phys. Lett. **351**, 128 (2002)
47. P. Esquinazi, D. Spemann, R. Höhne, A. Setzer, K.-H. Han, and T. Butz, Phys. Rev. Lett. **91**, 227201 (2003).
48. D. Babonneau, J. Briatico, F. Petroff, T. Cabioc'h, A. Naudon. J. Appl. Phys. **87**, 3432 (2000).
49. T. Enz, M. Winterer, B. Stahl, S. Bhattacharya, G. Miehe, K. Foster, C. Fasel, H. Hahn, J. Appl. Phys **99**, 044306 (2006).
50. M. Schwickardi, S. Olejnik, E. L. Salabas, W. Schmidt, F. Schuth, Chem. Commun, **38**, 3987 (2006)
51. M. Lautens, W. Klute, W. Tam, Chem. Rev. **96**, 49 (1996).
52. B. M. Trost. Pure Appl. Chem, **60**, 1615 (1988).
53. H. Ueda, J. Mater. Sci. **36**, 5955 (2001).
54. J.M.D Coey, M. Venkatesan, C. Fitzgerald, A. Douvalis, and I. Sanders, Nature **420**, 156 (2002)
55. O. Cespedes, M.S. Ferreira, S. Sanvito, M. Kociak, and J.M.D. Coey, J. Phys.: Cond. Mat. **16**, L155 (2004).
56. H.-Ch. Mertings, S. Valencia, W. Gudat, P.M. Oppeneer, O. Zaharko, and H. Grimmer, Europhys. Lett. **66** 743 (2004).
57. R. Hohne, M. Ziese, and P. Esquinazi, Carbon **42**, 3109 (2004)
58. S. V. Pol, V. G. Pol, A. Frydman, G. N. Churilov, A. Gedanken. J. Phys. Chem. B **109**, 9495 (2005)
59. M. A. Ramos. 14th European Conference on Mathematic for Industry, July 10th-14th, 2006, Leganés, Madrid, Spain. Book of Abstracts, p. 48. Submitted to the Proceeding volume.
60. H. Ohldag, T. Tyliszczak, R. Hohne, D. Spemann, P. Esquinazi, M. Ungureanu, and T. Butz, cond-mat/0609478.

Looking for Ferromagnetic Signals in Proton-Irradiated Graphite

M.A. Ramos[1], A. Asenjo[2], M. Jaafar[2], A. Climent-Font[3],
A. Muñoz-Martín[3,4], J. Camarero,[5] M. García-Hernandez[2], and M. Vázquez[2]

[1] LBT-UAM, Depto. de Física de la Materia Condensada, C-III, Universidad
Autónoma de Madrid, Cantoblanco, 28049 Madrid
miguel.ramos@uam.es
[2] Instituto de Ciencia de Materiales, ICMM-CSIC, Cantoblanco, 28049 Madrid
[3] Centro de Microanálisis de Materiales (CMAM), Universidad Autónoma
de Madrid, Cantoblanco, 28049 Madrid
[4] Parque Científico de Madrid. Campus de Cantoblanco, 28049 Madrid
[5] LASUAM, Depto. de Física de la Materia Condensada, C-III, Universidad
Autónoma de Madrid, Cantoblanco, 28049 Madrid

1 Introduction

Pure graphite, the stable crystalline allotrope of carbon at room temperature
and ambient pressure, is known to exhibit a strong and anisotropic "textbook"
diamagnetism, due to its delocalized π electrons. Nevertheless, in the last
two decades several researchers have reported more or less clear evidences of
ferromagnetic behavior in carbon at room temperature. We might mention
the work by japanese groups [Mur91, Mur92], who observed it in amorphous
carbon (relatively rich in hydrogen). The origin of this ferromagnetic behavior
could be theoretically justified as arising from the mixture of sp^2 and sp^3
bonding in carbon structure [Ovc88]. More recently, new findings of this kind
have appeared in the literature, as the presence of ferromagnetic signals in
some polymerized fullerenes reported by Makarova et al. [Mak01], or that
found in proton-irradiated highly-oriented pyrolitic graphite (HOPG) by the
group led by Esquinazi [Esq02, Han03]. In the latter experimental work, the
analysis of possible magnetic impurities has been much more rigorous as to
overcome the natural skepticism arose by the former experiments. Moreover,
several theoretical works seem to support the importance of disorder [Voz04]
and/or of vacancy-hydrogen complexes [Leh04] for the appearance of magnetic
moments in graphite.

The interest in the possibility of producing organic materials with magnetic properties is obvious. Therefore, we have undertaken a joint research
line to study this subject, by making use of the 5 MV tandem ion-accelerator
hosted by the CMAM in the Universidad Autónoma de Madrid. At the same
time of the ion implantation, the PIXE technique allows to determine in

situ the amount of magnetic impurities in the sample, a crucial issue given the weakness of the reported ferromagnetic signals. The possible existence of the latter have been studied through SQUID magnetometry, Magnetic Force Microscopy (MFM) and magneto-optic Kerr effect (MOKE).

2 Experimental

High purity HOPG was used (NTI-Europe, ZYA quality, $0.4° \pm 0.1°$ rocking curve). Proton and/or carbon irradiations were conducted in a 5 MV tandem ion-accelerator (HVEE, using a Cockroft-Walton power supply system). Particle induced X-ray emission (PIXE) measurements allowed us to assess the amount of local concentration of heavier impurities. In some cases, we employed a fine square mesh of copper (G2000HS, SPI, with a pitch of 12.5 μm, with separating copper bars of 5 μm and squared holes of 7.5 μm each side) as a mask onto the irradiation area. Measurements of the total magnetic moment of the samples were performed with a SQUID magnetometer from quantum design. Possible ferromagnetic behavior at the surface of the irradiated samples was studied by means of a magnetic force microscope (MFM) from Nanotec Electronica S.L., operating under externally applied magnetic field [Ase00] at ambient temperature. In all experiments, a double-step procedure was employed: First, a simple topographic scan is taken by maintaining a constant amplitude of oscillation of the cantilever, very close to the surface. Afterwards, long-range interactions are measured in a second scan with no feedback, following the topography of the sample and measuring the frequency shift that is proportional to the magnetic force gradient. This second scan is performed by retracing the tip tens of nanometer from the sample in order to avoid the topographic interaction. We have also conducted magneto-optic measurements by using a high resolution vectorial Kerr set-up [Cam05].

3 Results and Discussion

First of all, we cut HOPG samples with a typical surface area of $5 \times 3.3 \, mm^2$ and 0.2–0.3 mm thick, using clean diamond wire. We conducted ion-beam irradiation of H^+ protons of 3 MeV energy, in high vacuum. Montecarlo SRIM simulations indicate a corresponding implantation depth of $75.3 \pm 1.3 \, μm$ for the H^+ ions. Spot size was here always about $1 \, mm^2$. In Fig. 1, we show the measured magnetization of the samples, with magnetic fields applied parallel to graphene planes, after subtracting the linear (negative) diamagnetic background, expected for bulk, pure graphite. Total irradiated doses ranged 40–1,000 μC. As can be seen, in all cases a (weak) ferromagnetic curve is observed, with the sample of intermediate irradiation dose, 200 μC exhibiting the higher ferromagnetic signal. Nevertheless, we found somewhat surprisingly that also a nonirradiated HOPG sample exhibit some ferromagnetic signal, comparable with the samples of lesser magnetization. We discard the possibility of all this behavior being due to magnetic impurities, since our

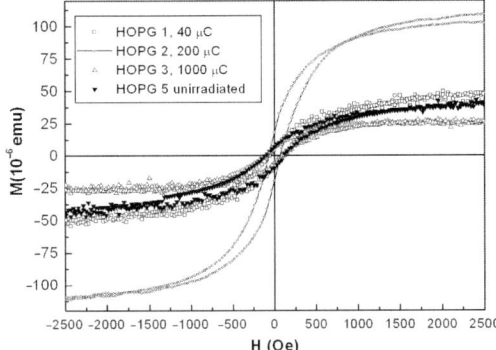

Fig. 1. SQUID measurements of the total magnetic moment of differently irradiated HOPG samples. See legend for ion-doses implanted. A linear diamagnetic background has been subtracted in all curves to show up the ferromagnetic contribution

PIXE experiments performed on some of these samples always gave concentrations below 10 ± 4 ppm of Fe element, and indetectable for other magnetic impurities. We believe that these observations simply confirm the findings of ferromagnetic behavior in many nonirradiated HOPG samples, with different kinds of preparation quality, structural vacancies or disorder, as found by Esquinazi and coworkers [Esq02]. However, our first experiments using MFM on these proton-irradiated samples, as on other ones irradiated in air with ion-beam spots smaller than $100 \, \mu m$, provided no clear evidence of magnetic behavior at the surface of proton-irradiated regions, in contrast to earlier reports [Esq02,Han03]. MOKE experiments performed in the same samples also gave negative results, always a pure linear diamagnetic curve was obtained. This is not too surprising, since the found ferromagnetic contributions superimposed on a large diamagnetic signal are very weak. Moreover, it is not clear what should be its relative strength at the superficial regions probed by these techniques.

In order to improve the layout of the samples surface to have a better contrast for MFM studies, we decided to use a grid or mask of copper. Thus, we put several grids of a fine copper mesh on one $1 \, cm^2$ HOPG sample to be irradiated. The squared holes were of $7.5 \times 7.5 \, \mu m^2$ with copper separating bars of 5 and $20 \, \mu m$ thick. In the experiments shown here, all studied sample regions were first irradiated with a dose of $150 \, \mu C$ of C^{4+} carbon ions impinging on the sample with an energy of $25 \, MeV$, the spot size being around $1 \, mm^2$. Hence the carbon ions were to stop at the middle of the depth of the copper separating bars. Through the holes of the grid, however, these carbon ions will penetrate the HOPG sample with a calculated implantation range of $20.3 \, \mu m$. In one region of the sample, a second irradiation was then conducted. A dose of $225 \, \mu C$ of H^+ protons was implanted, with an energy of $1.25 \, MeV$ chosen, so that the protons also stop at mid depth of the copper separating

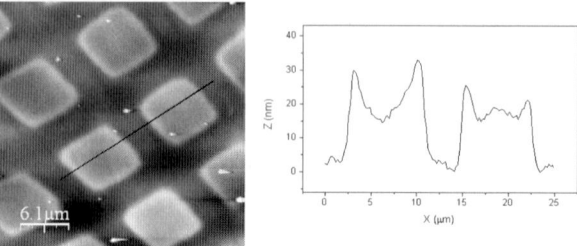

Fig. 2. Topographic profile of a HOPG sample irradiated with a copper mask with a dose of $150\,\mu C$ of C^{4+} carbon ions of 25 MeV, and $225\,\mu C$ of H^+ protons of 1.25 MeV (see text for details)

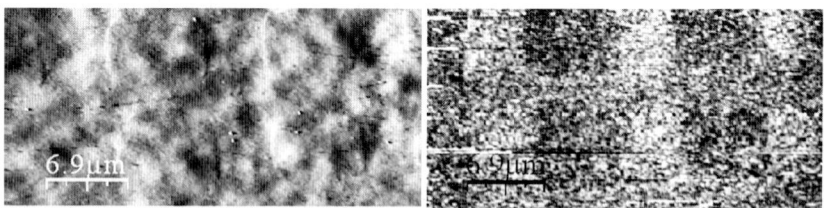

Fig. 3. Topographic image (*left picture*) and corresponding magnetic-contrast image (*right picture*) taken after applying an external magnetic field of 3 kOe to the same sample shown in Fig. 2

bars, whereas through the holes the penetration into the HOPG sample would be of about $20.0\,\mu m$. The surface topography of the latter sample after both consecutive ion irradiations and removing of the mask is shown in Fig. 2, where the topographic modifications induced by the irradiations are seen to clearly follow the pattern of the copper grid. Figure 3 shows a clear magnetic-contrast pattern in the ion-irradiated spots. In contrast to this case, in the region of the sample were only the first carbon irradiation was performed, no magnetic features were observed, as can be seen in Fig. 4. MFM probes with different magnetic moment [Ase06] have been used in order to enhance the tip–sample interaction. This seems to support the relevant role played by H^+ ions in promoting ferromagnetism in carbon.

4 Summary and Conclusions

We have presented some representative experimental results found in proton- and carbon-irradiated HOPG samples, aiming to confirm or disregard the existence of ferromagnetic behavior in pure carbon. To address that, we have combined macroscopic, bulk magnetic measurements (SQUID) with microscopic ones (MFM, MOKE). In brief, we have confirmed the existence of

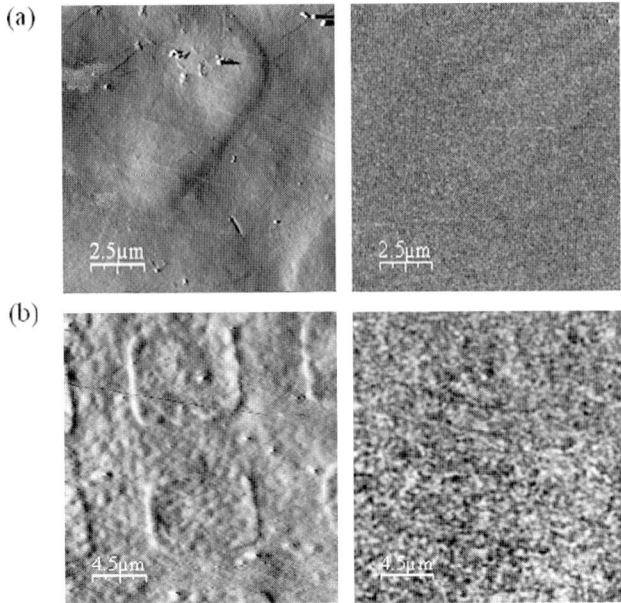

Fig. 4. Topographic images (*left picture*) and corresponding magnetic-contrast images (*right picture*) taken on a HOPG sample irradiated through a copper mask, with a dose of 150 μC of C^{4+} carbon ions of 25 MeV by using (**a**) a Mesp-LM MFM probe and (**b**) a MESP MFM probe from Veeco

ferromagnetic features, both at macroscopic and microscopic scales. Nevertheless, the produced effects are still so weak, and the variables involved so many and unknown, that much more systematic studies are needed.

References

[Ase00]   Asenjo A., Garcia D., Garcia J. M., Prados C., Vázquez, M.: MFM study of dense stripes domains in FeB/CoSiB multilayers and its evolution under an external applied field. Phys. Rev. B, **62**, 6538 (2000)

[Ase06]   Asenjo A., Jafaar M., Navas D., Vázquez M.: Quantitative Magnetic Force Microscopy analysis of the magnetization process in nanowire arrays. J. Appl. Phys., **100**, 023909 (2006)

[Cam05]   Camarero, J., Sort, J., Hoffmann, A., García-Martín, J. M., Dieny, B., Miranda, R., Nogués, J.: Origin of the Asymmetric Magnetization Reversal Behavior in Exchange-Biased Systems: Competing Anisotropies. Phys. Rev. Lett. **95**, 057204 (2005)

[Esq02]   Esquinazi, P., Setzer, A., Höhne, R., Semmelhack, C., Kopelevich, Y., Spemann, D., Butz, T., Kohlstrunk, B., Lösche, M.: Ferromagnetism in oriented graphite samples. Phys. Rev. B, **66**, 024429 (2002)

[Esq02]   Esquinazi, P., Spemann, D., Höhne, R., Setzer, A., Han, K.-H., Butz, T.:
           Induced magnetic ordering by proton irradiation in graphite. Phys. Rev.
           Lett., **91**, 227201 (2003)
[Han03]   Han, K.-H., Spemann, D., Esquinazi, P., Höhne, R., Riede, V., Butz, T.:
           Ferromagnetic spots in graphite produced by proton irradiation. Adv.
           Mater., **15**, 1719-1722 (2003)
[Mur91]   Murata, K., Ushijima, H., Ueda, H., Kawaguchi, K.: Magnetic properties
           of amorphous-like carbons prepared by tetraaza compounds by the chem-
           ical vapour deposition (CVD) method. J. Chem. Soc., Chem. Commun.,
           1265-6 (1991)
[Mur92]   Murata, K., Ushijima, H., Ueda, H., Kawaguchi, K.: A stable carbon-
           based organic magnet. J. Chem. Soc., Chem. Commun., 567-569 (1992)
[Ovc88]   Ovchinnikov, A. A., Spector, V. N.: Organic ferromagnets - new results.
           Synth. Metals, **27**, B615-B624 (1988)
[Mak01]   Makarova, T., Sundqvist, B., Höhne, R., Esquinazi, P., Kopelevich, Y.,
           Scharff, P., Davydov, V. A., Kashevarova, L. S., Rakhmanina, A. V.:
           Magnetic carbon. Nature (London), **413**, 716-718 (2001)
[Voz04]   Vozmediano, M. A. H., Guinea, F., Lopez-Sancho, M. P.: Interactions,
           disorder and local defects in graphite. cond-mat/0409567 (2004)
[Leh04]   Lehtinen, P. O., Foster, A. S., Ma, Y., Krasheninnikov, A.,
           Nieminen, R. M.: Irradiation-Induced Magnetism in Graphite: A Density
           Functional Study. Phys. Rev. Lett., **93**, 187202 (2004)

Ferromagnetism and Disorder in Graphene

M.P. López-Sancho[1], M.A.H. Vozmediano[2], T. Stauber[1], and F. Guinea[1]

[1] Instituto de Ciencia de Materiales de Madrid, CSIC, Cantoblanco, E-28049
    Madrid, Spain
    `pilar@icmm.csic,es`
[2] Unidad Asociada CSIC-UC3M, Universidad Carlos III de Madrid, E-28911
    Leganés, Madrid, Spain
    `geli@math.uc3m.es`

1 Introduction

Magnetic correlations in carbon-based materials have been reported for many years, but the lack of reproducibility have aroused scepticism about this topic. However, in recent years, the improvement of the characterization techniques has allowed the observation of ferromagnetism and the precise measurement of the impurity amounts in different samples of HOPG and Kish graphite [1]. Besides the ferromagnetic hysteresis loops [2] reported at room temperature, the enhancement of ferromagnetic behavior by proton bombardment of graphite has been observed in samples with an amount of impurities much lower than that needed to produce the saturation magnetization measured [3]. Irradiation induced magnetism in carbon nanostructures has been reported by N and C implantation [4]. A relation between the magnetic properties of pure bulk ferromagnetic graphite and the topographic defects introduced in the pristine material have been reported by comparison of atomic force microscopy (AFM) images and magnetic force microscopy (MFM) [5]. Soft X-ray dichroism spectromicroscopy has been used to analyze the magnetic order of metal free carbon films. A clear evidence of intrinsic ferromagnetic order at room temperature has been obtained in these carbon films that have been irradiated by a focused proton beam [6]. All these experiments suggest that there is a relation between topological defects in the lattice induced by irradiation and the ferromagnetic correlations [6]. We will study the ferromagnetism in a two-dimensional graphene plane by considering disorder, vacancies, and defects in the atomic network.

The graphene sheet is described by a tight binding model including the π orbitals, perpendicular to the planes. The bands of graphite are well described by this model, including only nearest neighbor coupling. The conduction and valence bands obtained are degenerate at the six corners of the hexagonal Brillouin zone (BZ). At half-filling the Fermi level lies at the mid point of the bands; therefore, the Fermi surface is reduced to the six K points located at the BZ corners, only two of then are inequivalent. In the proximity of these points, the dispersion relation is linear and the low energy excitations can be studied taking the continuum limit at these points. The effective Hamiltonian obtained turns out to be the Dirac operator [8, 9]. The density of states of these linear bands vanishes at the Fermi level.

2 Inclusion of Disorder

The disorder may have an important influence in the occurrence of ferromagnetism in 2D graphene. On the other hand, vacancies, dislocations, edges, or cracks are present in most of the samples specially after irradiation with protons. We will study the formation of local moments near extended defects in the continuum approximation. It is known that disorder significantly changes the states described by the two-dimensional Dirac equation [10], and usually, the density of states at low energies is increased. Lattice defects, such as pentagons and heptagons, or dislocations, can be included in the continuum model by means of a non-abelian gauge field [9, 11] that reproduces the effects of the curvature of the lattice and the possible exchange of Fermi points. Within the same theoretical scheme it has also been shown that certain types of disorder randomly distributed in the graphene lattice enhances the effect of the interactions [6] and can stabilize new phases. In addition, a graphene plane can show states localized at interfaces [5], which in the absence of other types of disorder lie at the Fermi energy.

The tight binding model defined by the π orbitals at the lattice sites can have edge states when the sites at the edge all belong to the same sublattice [5] (zigzag edge). These states lie at zero energy which for neutral graphene planes correspond to the Fermi energy. In the continuum model described earlier, these localized states are normalizable solutions $(\Psi_1(\mathbf{r}), \Psi_2(\mathbf{r}))$ of the Dirac equation for $\epsilon = 0$:

$$(i\partial_x \pm \partial_y)\Psi_1(\mathbf{r}) = i\partial_{z,\bar{z}}\Psi_1(z,\bar{z}) = 0$$

$$(i\partial_x \mp i\partial_y)\Psi_2(\mathbf{r}) = i\partial_{\bar{z},z}\Psi_2(z,\bar{z}) = 0, \qquad (1)$$

where $z, \bar{z} = x \pm iy$. These equations are satisfied if $\Psi_1(\mathbf{r})$ is an analytic function of z and $\Psi_2(\mathbf{r}) = 0$, or if $\Psi_1(\mathbf{r}) = 0$ and $\Psi_2(\mathbf{r})$ is an analytic function of $\bar{z}$.

We now consider a semi-infinite honeycomb lattice with an edge at $y = 0$ and which occupies the half plane $x > 0$. A possible solution that decays as $x \to \infty$ is

$$\Psi_1(x,y) \propto e^{-kz} = e^{iky}e^{-kx}, \Psi_2(\mathbf{r}) = 0 .$$

These solutions satisfy the boundary conditions at $y = 0$ if the last column of carbon atoms belong to the sublattice where the component Ψ_1 is defined. Then, the next column belongs to the other sublattice, where the amplitude of the state is, by construction, zero.

This kind of solutions can be generalized to describe other types of extended defects that will be produced in experiments where graphite samples

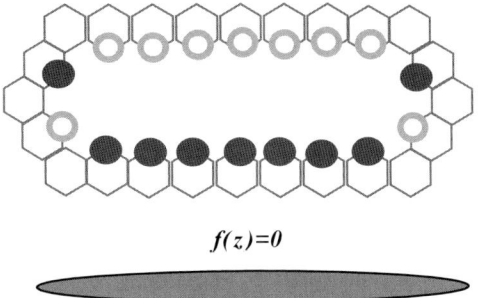

$$f(z)=0$$

Fig. 1. Elongated crack in the honeycomb structure. The crack is such that the sites in the upper edge belong to one sublattice, while those at the lower edge belong to the other. *Bottom*: approximate cut in the complex plane which can be used to represent this crack at long distances

are bombarded by protons. In a strongly disordered sample, large defects made up of many vacancies can exist. These defects will give rise to localized states, when the termination at the edges is locally similar to the surfaces discussed above. Only possible localized states can exist at zero energy, where the density of extended states vanishes. The wave functions obtained from the Dirac equations will be normalizable and analytic functions of the variables $z = x + iy$ or $\bar{z} = x - iy$ of the form $\Psi(z) \equiv [f(z), 0]$ obeying the boundary conditions imposed by the shape of the defect.

Extended vacancies with approximate circular shape can support solutions of the type $f(z) \propto z^{-n}, n > 1$. By using conformal mapping techniques, solutions can be found with the boundary conditions appropriate to the shapes of different defects.

A simple case is the elongated crack depicted in Fig. 1, which we assume to extend from $x = -a$ to $x = a$, and to have a width comparable to the lattice constant along the y axis. The analytic function $f(z)$ associated with localized states near a crack of this shape should satisfy $\mathrm{Re} f(z) = 0$ at the crack edges, because the boundaries of the crack include atoms from the two sublattices. Hence, the boundary of the crack leads to a branch cut in the complex function $f(z)$. Labeling edge states by a quantum number n, we find that the function Ψ can be written for these states as

$$\Psi_n \equiv \left\{ \mathrm{Re} \left[\frac{A}{z^n \sqrt{z^2 - a^2}} \right], 0 \right\} .$$

A similar solution is obtained by exchanging the upper by the lower spinor component and replacing $z \leftrightarrow \bar{z}$. Because of the discreteness of the lattice, the allowed values of n should be smaller than the number of lattice units spanned by the crack. We have checked numerically the existence of these localized states by diagonalizing the tight binding hamiltonian in finite lattices of different sizes. It is found that the states closest to $\epsilon = 0$ show a dependence

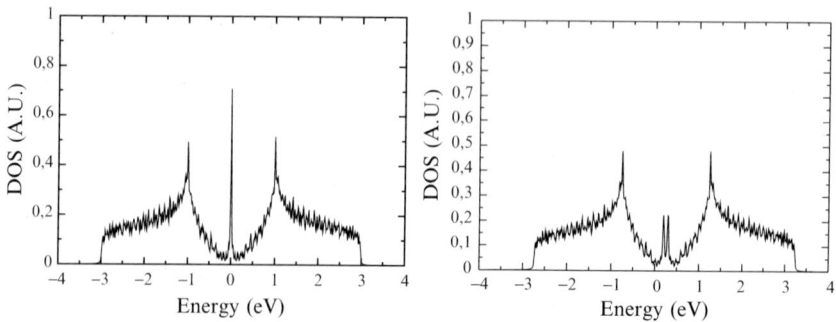

Fig. 2. Density of states of a 24×24 cluster with periodic boundary conditions and three contiguous vacancies: (*left*) the on-site interaction term is $U = 0$, (*right*) with an on-site interaction $U = 0.5t$

$\epsilon_{\text{loc}} \propto L^{-2}$, which suggest a power law localization, in agreement with the previous analysis. The total density of states of a given cluster is shown in Fig. 2.

In the presence of a finite local repulsion, the flat band of localized states will tend to become polarized, leading to a a ferromagnetic alignment of the electrons in these states. We have checked the formation of local moments near cracks and similar defects by performing Hartree–Fock calculations in finite clusters, and modeling the electron–electron interaction by an on site repulsive term U. Typical total density of states for the unpolarized state of a cluster with 24×24 unit cells, three contiguous vacancies, and periodic boundary conditions are shown in Fig. 2. A small, but finite repulsive term $U = 0.5t \simeq 1.4\,\text{eV}$ leads to the splitting of a central peak, as shown in the righthand side of Fig. 2. The total polarization of the cluster is also small, $S_z = 3/2$, indicating that only the electrons from the states around the impurity contribute to the formation of a local moment. These local moments interact with the extended states, which will mediate a RKKY like interaction between local moments at neighboring defects. The change in the wavefunction of the extended electrons can be calculated using perturbation theory, in terms of the spin susceptibility of a clean system. The susceptibility per unit area was calculated in [9], and can be written at small momenta as

$$\chi(\mathbf{q}) \propto \frac{|\mathbf{q}|}{v_F}. \tag{2}$$

The total potential induced around a defect is proportional to the length of the perimeter of the defect and U. This potential is distributed over an area comparable, or larger, to the surface of the defect, $A_d \sim L_d^2$. We obtain

$$J_{\text{RKKY}}(\mathbf{r}) \sim U^2 N_d^2 a^4 \int_{|\mathbf{k}| \ll L_d^{-1}} d^2\mathbf{k}\, e^{i\mathbf{k}\mathbf{r}} \frac{|\mathbf{k}|}{v_F} \sim U^2 N_d^2 \frac{a^4}{v_F |\mathbf{r}|^3}, \tag{3}$$

where a is the lattice constant.

Due to the absence of a finite Fermi surface, the RKKY interaction in (3) does not have oscillations and the magnetic moments will tend to be ferromagnetically aligned. The total polarization per unit area at low temperatures is proportional to $c \times N_d$, where c is the concentration of defects, and N_d is proportional to their average size [6].

3 Conclusions

We have shown that, under very general circumstances, lattice defects, vacancies, and voids in the graphene structure give rise to localized states at the Fermi energy. The number of these states scales roughly with the perimeter of the defect. Repulsive electron–electron interactions lead to the polarization of these states and to the formation of local moments. The RKKY interaction mediated by the valence electrons decays as r^{-3}, where r is the distance between defects, and shows no oscillations, due to the absence of a finite Fermi surface in a graphene layer. The interaction is ferromagnetic, and the system cannot show the frustration effects and spin glass features observed in other disordered systems with local moments. On the other hand, the Curie temperature estimated assuming a random distribution of local moments is low, $T_C \sim 1K$, for reasonable values of the defect concentration. It may happen that percolation effects and the finite extension of the localized states which give rise to the local moments will increase the value of T_C.

Acknowledgments

Funding from MCyT (Spain) through grant FIS2005-05478-C02-01 and European Union FERROCARBON Contract 12881 (NEST) is acknowledged.

References

1. 'Carbon Based Magnetism', edited by Makarova and F. Palacio, Elsevier B.V., Amsterdam (2006).
2. Kopelevich, Y. *et al* Phys. Rev. Lett. **90**, 156402 (2003).
3. Esquinazi, P., *et al*, Phys. Rev. Lett. **91**, 227201 (2003).
4. Talapatra, S. *et al* Phys. Rev. Lett. **95**, 097201 (2005).
5. A. W. Mombrú *et al*, Phys. Rev. B **71**, 100404(R) (2005).
6. Ohldag, H. *et al* cond-mat/0609478 (2006).
7. Vozmediano, M.A.H., López-Sancho, M.P., Stauber, T., and Guinea, F. Phys. Rev. B **72**, 155121 (2005).
8. McClure, J.W., Phys. Rev. **112**, 715 (1958).
9. González, J., Guinea, F., and Vozmediano, M.A.H., Phys. Rev. B **73**, 134421 (2001); Nucl. Phys. B **406** 771 (1993).
10. Horovitz, B. and La Doussal, P., Phys. Rev. B **65**, 125323 (2002).
11. J. González, F. Guinea and M. A. H. Vozmediano, Mod. Phys. Lett. **B7**, 1593 (1993).
12. Wakayabashi, K. Phys. Rev. B **64**, 125428 (2001).

Topological Defects and Electronic Properties in Graphene

Alberto Cortijo[1] and M.A.H. Vozmediano[2]

[1] Instituto de Ciencia de Materiales de Madrid, CSIC, Cantoblanco E28049
   Madrid, Spain
   `cortijo@icmm.csic.es`
[2] Grupo de Modelización y Simulación Numéricas, Universidad Carlos III
   de Madrid, E28913 Leganés, Madrid, Spain
   `vozmediano@icmm.csic.es`

Summary. In this work we will focus on the effects produced by topological disorder on the electronic properties of a graphene plane. The presence of this type of disorder induces curvature in the samples of this material, making quite difficult the application of standard techniques of many-body quantum theory. Once we understand the nature of these defects, we can apply ideas belonging to quantum field theory in curved space-time and extract information on physical properties that can be measured experimentally.

1 Introduction

Graphene is a two-dimensional material formed by isolated layers of carbon atoms arranged in a honeycomb-like lattice. Each carbon atom is linked to three nearest neighbors due to the sp^2 hybridization process, which leads to three strong σ bonds in a plane and a partially filled π bond perpendicular to the plane. These π bonds will determine the low energy electronic and transport properties of the system.

It is possible to derive a long wavelength tight binding hamiltonian for the electrons in these π bonds [3]. This hamiltonian is

$$H = -iv_{\mathrm{F}} \int \mathrm{d}^2\mathbf{r}\,\bar{\Psi}(\mathbf{r})\gamma^j \partial_j \Psi(\mathbf{r}), \tag{1}$$

where v_{F} being a constant with dimensions of velocity ($v_{\mathrm{F}} \sim 10^3\,\mathrm{m\,s^{-1}}$). The wave equation derived from the hamiltonian (1) is the Dirac equation in two dimensions with the coefficients γ^j being an appropriate set of Dirac matrices. We can set for instance, $\gamma^1 = \mathbf{1} \otimes \sigma_1$ and $\gamma^2 = \tau_3 \otimes \sigma_2$, where the σ, τ matrices are related to the sublattice and Fermi point degrees of freedom, respectively. The unexpected form of the tight-binding Hamiltonian comes from two special features of the honeycomb lattice: first, the unit cell contains two carbon atoms belonging to different triangular sublattices, and second, in the neutral system at half filling, the Fermi surface reduces to two nonequivalent Fermi points. We will study the low energy states around any of these two Fermi points. The dispersion relation obtained from (1) is $\varepsilon(\mathbf{k}) = \pm v_F|\mathbf{k}|$, leading to a constant density of states, $\rho^0(\omega) = \frac{8}{\pi}|\omega|$.

2 A First Model for the Topological Defects in Graphene

Several types of defects like vacancies, adatoms, complex boundaries, and structural or topological defects have been observed experimentally in the graphene lattice [2] and studied theoretically (see, for example, [3–5]).

Topological defects are produced by substitution of an hexagonal ring of the honeycomb lattice by an n-sided polygon with any n. Their presence impose nontrivial boundary conditions on the electron wave functions, which are difficult to handle. A proposal made in [6] was to trade the boundary conditions imposed by pentagonal defects by the presence of appropriate gauge fields coupled to the spinor wave function. A generalization of this approach to include various topological defects was presented in [7]. The strategy consists of determining the phase of the gauge field by parallel transporting of the state in suitable form along a closed curve surrounding all the defects.

$$\Psi(\theta = 0) = T_C \Psi(\theta = 2\pi) \Leftrightarrow \Psi(\theta = 0) = \exp\left(\oint_C \mathbf{A}_a T^a \mathrm{d}\mathbf{r}\right) \Psi(\theta = 2\pi), \quad (2)$$

where $\mathbf{A}_a$ are a set of gauge fields and T^a a set of matrices related to the pseudospin degrees of freedom of the system.

When dealing with multiple defects, we must consider a curve surrounding all of them, as the one sketched in Fig. 1:

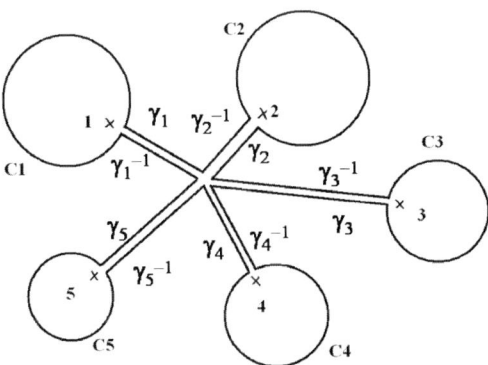

Fig. 1. Prototypical curve enclosing multiple defects in which the state will be parallel transported

The contour C is made of closed circles enclosing each defect and straight paths linking all the contours to a fixed origin. The parallel transport operator P_C associated to the closed path is thus a composition of transport operators over each piece:

$$P = P_{\gamma 1} \cdot P_1 \cdot P_{\gamma 1}^{-1} \cdots P_{\gamma N} \cdot P_N \cdot P_{\gamma N}^{-1}. \tag{3}$$

As explained in [7] the total holonomy turns out to be[1]

$$P = (\mathrm{i})^N (\tau_2)^N \exp\left(\frac{2\pi\mathrm{i}}{6}(N^+ - N^-)\sigma_3\right) \exp\left(\frac{2\pi\mathrm{i}}{3}\sum_{j=1}^{N}(n_j - m_j)\tau_3\right). \tag{4}$$

From (4) we see that we have in principle three different gauge fields to incorporate into the Dirac equation, which couple to the matrices σ_3, τ_2, and τ_3 and whose associated fluxes are adjusted from (2).

3 Generalization of the Model

In spite of its elegance, the model presented in the previous section does not contain the effects due to the curvature of the layer in the presence of these defects. The model can be generalized to account for curvature effects [6, 8] by coupling the gauge theory obtained from the analysis of the holonomy in a curved space.

The substitution of a hexagon by a polygon with $n < 6$ sides gives rise to a conical singularity with deficit angle $(2\pi/6)(6 - n)$, which is similar to the singularity generated by a cosmic string in general relativity. The Dirac Equation for a massless spinor in a curved spacetime is [9]

$$\mathrm{i}\gamma^\mu(x)(\partial_\mu - \Gamma_{j\mu}^{(T)})\psi = 0, \tag{5}$$

where $\Gamma_{j\mu}^{(T)}$ is a set of spin connections related to the pseudospin matrices in (4) and $\gamma^\mu(x)$ are generalized Dirac matrices satisfying the anticommutation relations

$$\{\gamma^\mu(x), \gamma^\nu(x)\} = 2g^{\mu\nu}(x). \tag{6}$$

The metric tensor in (6) corresponds to a curved spacetime generated by an arbitrary number of N parallel cosmic strings placed in (a_i, b_i) (here we will follow the formalism developed in [10]):

$$\mathrm{d}s^2 = -\mathrm{d}t^2 + e^{-A(x,y)}(\mathrm{d}x^2 + \mathrm{d}y^2), \tag{7}$$

---

[1]The usual chiral lattice real vector basis for the honeycomb lattice is used in this derivation.

with $\Lambda(x,y) = \sum_{i=1}^{N} 4\mu_i \log([(x-a_i)^2 + (x-b_i)^2]^{1/2})$. The parameters μ_i are related to the angle defect or surplus by the relationship $c_i = 1 - 4\mu_i$ in such manner that if $c_i < 1(> 1)$ then $\mu_i > 0(< 0)$.

From equation (5) we can write down the equation for the electron propagator, $S_F(x,x')$:

$$i\gamma^{\mu}(x)(\partial_{\mu} - \Gamma_{j\mu}^{(T)})S_F(x,x') = \frac{1}{\sqrt{-g}}\delta^3(x-x'). \tag{8}$$

The local density of states $N(\omega,\mathbf{r})$ is obtained from the solution of (8) by Fourier transforming the time component and taking the limit $\mathbf{r}' \to \mathbf{r}$:

$$N(\omega,\mathbf{r}) = \text{Im}\text{Tr}S_F(\omega,\mathbf{r},\mathbf{r}). \tag{9}$$

Provided that we only consider the presence of pentagons and heptagons, the parameters μ_i are all equal and small ($\mu_i \equiv \mu = 1/24$). We will solve equation (8) perturbatively in μ.

When dealing with (8) we will reduce the number of spin connections derived in the previous section by the following considerations: First, we will consider a scenario where the number of pentagonal and heptagonal defects is the same – so the total number of defects is even. This suppresses the contribution from the first exponential in (4). If we consider that pentagonal and heptagonal defects come in pairs as usually happens in the observations, we can neglect the effect of mixing of the the two sublattices that each individual odd-sided ring produces and hence eliminate the spin connection related to τ_2 from (8). Furthermore, we can disregard the spin connection related to τ_3 by the following argument: We will solve (8) perturbatively to first order of the parameter μ. In general, if S_F^0 is the unperturbed Dirac propagator and $\hat{V}(\omega,\mathbf{r})$ the perturbation potential, the first term of such solutions is

$$S_F^1(\omega,\mathbf{r},\mathbf{r}') = \mu \int d^2\mathbf{r}'' S_F^0(\omega,\mathbf{r},\mathbf{r}'')\hat{V}(\omega,\mathbf{r}'')S_F^0(\omega,\mathbf{r}'',\mathbf{r}'), \tag{10}$$

and we trace $S_F^0(\omega,\mathbf{r},\mathbf{r})$ in order to get the first contribution to the density of states $\delta N(\omega,\mathbf{r})$. The trace operation eliminates all the terms appearing in (10) which are proportional to a traceless matrix, including the matrix related to τ_2. In fact, up to this order in perturbation theory, the only term that survives will be the one proportional to γ^0. With all this in mind, the relevant spin connection terms are

$$\Gamma_1(\mathbf{r}) = -\frac{1}{2}\gamma^1\gamma^2\partial_y\Lambda, \quad \Gamma_2(\mathbf{r}) = -\frac{1}{2}\gamma^2\gamma^1\partial_x\Lambda. \tag{11}$$

After all these simplifications we can write (8) in a more suitable form. Expanding the terms in (11) in powers of μ we get the potential $\hat{V}(\omega,\mathbf{r})$

$$\hat{V}(\omega,\mathbf{r}) = -2\Lambda\gamma^0\omega + i\Lambda\gamma^j\partial_j + \frac{i}{2}\gamma^j(\partial_j\Lambda). \tag{12}$$

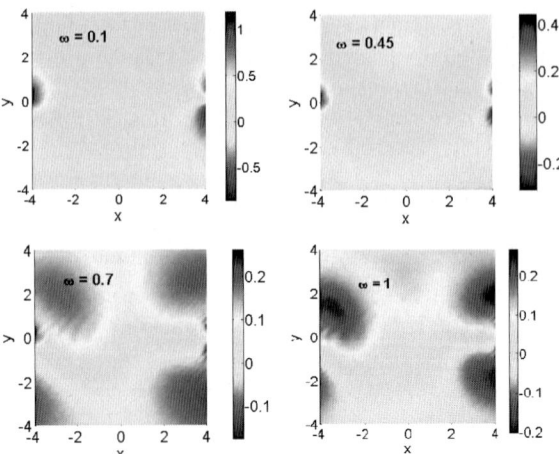

Fig. 2. First order correction to the local density of states in a region around two pairs of heptagon–pentagon defects located out of the image for increasing values of the energy

As we said, expression (10) gives us the first correction to the local density of states in real space. In Fig. (2) we present an example of the results obtained. We show the first order correction to the local density of states coming from two pairs of heptagon–pentagon defects located out of the image for increasing values of the energy. What we see is that as the frequency increases, the local density of states is enhanced and inhomogeneous oscillations are observed in a wide area around the defects. The spatial extent of the correction is such that the relative intensity decays to 10% in approximately 20 unit cells. The model described in this work can be applied to other configurations of defects, such as simple pairs or stone-Wales defects. These results can be found in [11].

Acknowledgements

Funding from MCyT (Spain) through grant FIS2005-05478-C02-01 and European Union FERROCARBON Contract 12881 (NEST) is acknowledged.

References

1. P. R. Wallace, Phys. Rev. **71**, 622 (1947).
2. A. Hashimoto et al., Nature **430**, 870 (2004).
3. P. O. Lehtinen et al., Phys. Rev. Lett. **91**, 017202 (2003).
4. M. A. H. Vozmediano et al., Phys. Rev. B **72**, 155121 (2005).
5. M. P. López-Sancho, T. Stauber, F. Guinea and M. A. H. Vozmediano, Contribution in this volume, 2007.

6. J. González, F. Guinea and M. A. H. Vozmediano, Phys. Rev. Lett. **69**, 172 (1992).
7. P. E. Lammert and V. H. Crespi, Phys. Rev. B. **69**, 035406 (2004).
8. E. A. Kochetov and V. A. Osipov, J. Phys. A**32**, 1961 (1999).
9. N. D. Birrell and P. C. W. Davies, Quantum Fields in Curved Space. Cambridge U. P., 1982.
10. A. N. Aliev, M. Hörtacsu and N. Ozdemir, Class. Quantum. Grav. **14**, 3215 (1997).
11. A. Cortijo and M. A. H. Vozmediano, LANL preprint cond-mat/0603717.

Transport Through a Graphene Transistor

F. Guinea[1], A.H. Castro Neto[2], and N.M.R. Peres[3]

[1] Instituto de Ciencia de Materiales de Madrid, Cantoblanco, 28049 Madrid, Spain
paco.guinea@icmm.csic.es
[2] Department of Physics, Boston University, 590 Commonwealth Avenue, Boston,
MA 02215, USA
neto@bu.edu
[3] Center of Physics and Departamento de Física, Universidade do Minho,
P-4710-057, Braga, Portugal
peres@fisica.uminho.pt

1 Introduction

Single layer graphene and stacks of graphene layers have recently attracted
a great deal of attention, because of their unusual electronic properties and
potential applications [1,2].

The low energy band structure is well described by the two-dimensional
Dirac equation [3]. This result leads to many anomalous electronic properties,
such as localized states at the energy of the Dirac point [4–9], new Landau
levels in the presence of a magnetic field [8,10], antilocalization effects [11–13],
or a pseudodiffusive behavior at zero energy [14–16], unusual channel quantization at finite energies and confinement properties [17].

We analyze here the electronic transport across a graphene potential step,
or a potential barrier (a graphene transistor). A related calculation can be
found in [18]. A more involved calculation, which analyzes transport from
single layer graphene to a double layer system is given in [19].

2 The Model

We discuss the transport properties of a clean graphene stripe where a gate
voltage is applied to a part of it. The two situations considered are shown in
Fig. 1: the transmission across a square potential step and the transmission
across a square potential barrier. In the following, we use units such that the
Fermi velocity $v_F = 1$.

2.1 Potential Step

We consider a Dirac quasiparticle with energy ϵ, reaching the potential barrier
at an angle θ. The incoming wavefunction can be written as

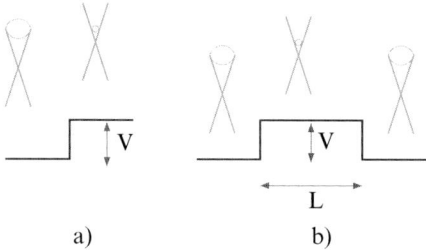

Fig. 1. Sketch of the models analyzed in the text. *Left*: step potential. *Right*: square barrier

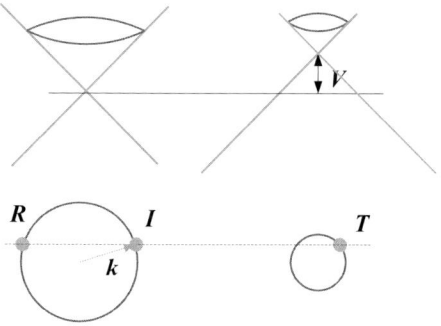

Fig. 2. Sketch of the matching conditions at the barriers described in the text

$$\Psi_{\text{inc}}(\mathbf{r}) \equiv e^{i\mathbf{kr}} \begin{pmatrix} 1 \\ e^{i\theta} \end{pmatrix}, \tag{1}$$

where $k = |\mathbf{k}| = \epsilon$. The current in the direction perpendicular to the barrier is $\langle \sigma_x \rangle = +\cos(\theta)$. The collision with the step leads to a reflected and a transmitted wave. If $\epsilon - V > 0$, the transmitted wave is electron like, and we can write

$$\Psi_{\text{ref}}(\mathbf{r}) \equiv Re^{-i\mathbf{kr}} \begin{pmatrix} 1 \\ -e^{-i\phi_k} \end{pmatrix},$$

$$\Psi_{\text{trans}}(\mathbf{r}) \equiv Te^{i\mathbf{k'r}} \begin{pmatrix} 1 \\ e^{i\phi_{k'}} \end{pmatrix}, \tag{2}$$

where $k' = |\mathbf{k'}| = \epsilon - V$, $\phi_{k'} = \arctan(k_y/k_x) = \theta$, $k'_y = k_y$ and $k'\phi_{k'} = \arctan(k'_y/k'_x) = \arcsin[(k\sin(\theta)/(k - V)]$. The matching conditions are sketched in Fig. 2.

The wavefunction has to be continuous at the position of the potential step, $x = 0$, leading to the equations

$$1 + R = T,$$

$$e^{i\phi_k} - Re^{-i\phi_k} = Te^{i\phi_{k'}}, \tag{3}$$

so that

$$T = \frac{2\cos(\phi_{\mathbf{k}})}{e^{-i\phi_{\mathbf{k}}} + e^{-i\phi_{\mathbf{k'}}}} \ . \tag{4}$$

The ratio of the transmitted current to the incoming current is

$$\frac{J_{\text{trans}}(\theta)}{J_{\text{inc}}(\theta)} = \frac{|T|^2 \cos(\phi_{\mathbf{k'}})}{\cos(\phi_{\mathbf{k}})} = \frac{2\cos(\phi_{\mathbf{k}})\cos(\phi_{\mathbf{k'}})}{1 + \cos(\phi_{\mathbf{k}} - \phi_{\mathbf{k'}})} \ . \tag{5}$$

It is worth noting that the transmission is 1 for normal incidence, $\theta = 0$, in a similar way to the lack of backscattering in nanotubes.

We set $J_{\text{inc}}(\theta) = 1$ and define the conductance of the step as

$$\sigma = \int_{-\pi/2}^{\pi/2} J_{\text{trans}}(\theta) d\theta \ . \tag{6}$$

The previous analysis has been performed for electron like transmitted quasiparticles. For $V \geq \epsilon$ the transmitted wave is hole like. Performing a similar analysis, we obtain

$$\left.\frac{J_{\text{trans}}(\theta)}{J_{\text{inc}}(\theta)}\right|_{\text{hole}} = \frac{|T|^2 \cos(\phi_{\mathbf{k'}})}{\cos(\phi_{\mathbf{k}})} = \frac{2\cos(\phi_{\mathbf{k}})\cos(\phi_{\mathbf{k'}})}{1 + \cos(\phi_{\mathbf{k}} + \phi_{\mathbf{k'}})} \ . \tag{7}$$

2.2 Potential Barrier

We now analyze the potential barrier of length L sketched in Fig. 1b. The matching conditions now are

$$
\begin{aligned}
1 + R &= T_1 + R_1, \\
e^{i\phi_{\mathbf{k}}} - Re^{-i\phi_{\mathbf{k}}} &= T_1 e^{i\phi_{\mathbf{k'}}} - R_1 e^{-i\phi_{\mathbf{k'}}}, \\
T_1 e^{ik'_x L} + R_1 e^{-ik'_x L} &= T, \\
T_1 e^{ik'_x L} e^{i\phi_{\mathbf{k'}}} - R_1 e^{-ik'_x L} e^{-i\phi_{\mathbf{k'}}} &= T e^{i\phi_{\mathbf{k}}},
\end{aligned}
\tag{8}
$$

which lead to

$$
\begin{aligned}
T_1 e^{ik'_x L}(e^{-i\phi_{\mathbf{k'}}} + e^{i\phi_{\mathbf{k'}}}) &= T(e^{-i\phi_{\mathbf{k'}}} + e^{i\phi_{\mathbf{k}}}), \\
R_1 e^{-ik'_x L}(e^{-i\phi_{\mathbf{k'}}} + e^{i\phi_{\mathbf{k'}}}) &= T(e^{i\phi_{\mathbf{k'}}} - e^{i\phi_{\mathbf{k}}}), \\
e^{i\phi_{\mathbf{k}}} + e^{-i\phi_{\mathbf{k}}} &= T_1(e^{-i\phi_{\mathbf{k}}} + e^{i\phi_{\mathbf{k'}}}) + R_1(e^{-i\phi_{\mathbf{k}}} - e^{-i\phi_{\mathbf{k'}}}).
\end{aligned}
\tag{9}
$$

The transmission coefficient across the barrier, T, can be written as

$$T = \frac{2\cos(\phi_{\mathbf{k}})\cos(\phi_{\mathbf{k'}})}{e^{-ik'_x L}[2 + 2\cos(\phi_{\mathbf{k}} + \phi_{\mathbf{k'}})] + e^{ik'_x L}[-2 + 2\cos(\phi_{\mathbf{k}} - \phi_{\mathbf{k'}})]}, \tag{10}$$

leading to the current

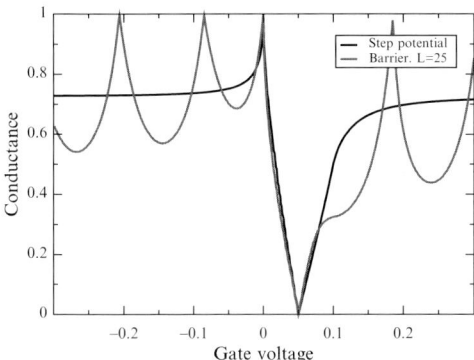

Fig. 3. Conductance of the two types of devices analyzed in the text. The energy of the incident wave is $\epsilon = 0.05$, and the length of the barrier is $L = 25$

$$\frac{J_{\text{trans}}(\theta)}{J_{\text{inc}}(\theta)} = \frac{2\cos^2(\phi_{\mathbf{k}})\cos^2(\phi_{\mathbf{k}'})}{[1 - \sin(\phi_{\mathbf{k}})\sin(\phi_{\mathbf{k}'})]^2 + \cos^2(\phi_{\mathbf{k}})\cos^2(\phi_{\mathbf{k}'}) + \cos(2k'_x L)}$$

$$\times \frac{1}{[1 - \sin(\phi_{\mathbf{k}})\sin(\phi_{\mathbf{k}'})]^2 - \cos^2(\phi_{\mathbf{k}})\cos^2(\phi_{\mathbf{k}'})}. \tag{11}$$

As in the case of a step barrier, the transmission is perfect for $\theta = 0$.

A similar derivation leads for the case of hole like transmission, $V \geq \epsilon$ to

$$\left.\frac{J_{\text{trans}}(\theta)}{J_{\text{inc}}(\theta)}\right|_{\text{hole}} = \frac{2\cos^2(\phi_{\mathbf{k}})\cos^2(\phi_{\mathbf{k}'})}{[1 + \sin(\phi_{\mathbf{k}})\sin(\phi_{\mathbf{k}'})]^2 + \cos^2(\phi_{\mathbf{k}})\cos^2(\phi_{\mathbf{k}'}) + \cos(2k'_x L)}$$

$$\times \frac{1}{[1 + \sin(\phi_{\mathbf{k}})\sin(\phi_{\mathbf{k}'})]^2 - \cos^2(\phi_{\mathbf{k}})\cos^2(\phi_{\mathbf{k}'})}. \tag{12}$$

3 Results

Results for the integrated current over angles are shown in Fig. 1. The conductance follows approximately the V-shaped density of states of the region where the gate potential is applied. The transmission is always finite, in agreement with the suppression of backscattering at normal incidence. In addition, we find Fabry–Perot interference patterns, with resonances where the *total* transmission is equal to one.

References

1. K. S. Novoselov *et al*, Science **306**, 666 (2004).
2. K. S. Novoselov *et al*, Proc. Nat. Acad. Sci. **102**, 10451, (2005).
3. P. R. Wallace, Phys. Rev. **71**, 622 (1947).

4. K. Wakayabashi and M. Sigrist, Phys. Rev. Lett. **84**, 3390 (2000).
5. K. Wakayabashi, Phys. Rev. B **64**, 125428 (2001).
6. M.A.H. Vozmediano *et al*, Phys. Rev. B **72**, 155121 (2005).
7. V. M. Pereira *et al*, Phys. Rev. Lett. **96**, 036801 (2006).
8. N. M. R. Peres, F. Guinea and A. H. Castro Neto, Phys. Rev. B **73**, 125411 (2006).
9. M. Titov, cond-mat/0611029 (2006).
10. V. P. Gusynin and S. G. Sharapov, Phys. Rev. Lett. **95**, 146801 (2005).
11. S.V. Morozov *et al*, Phys. Rev. Lett. **97**, 016801 (2006).
12. E. McCann *et al*, Phys. Rev. Lett. **97**, 146805 (2006).
13. A. F. Morpurgo and F. Guinea, Phys. Rev. Lett. **97**, 196804 (2006).
14. J. Tworzydlo, *et al*, Phys. Rev. Lett. **96**, 246802 (2006).
15. J. A. Vergés, F. Guinea, E. Chiappe and E. Louis, cond-mat/0610201 (2006).
16. E. Prada, P. San-José, B. Wunsch and F. Guinea, cond-mat/0611189 (2006).
17. N. M. R. Peres, A. H. Castro Neto and F. Guinea, Phys. Rev. B **73**, 241403 (2006).
18. M. I. Katsnelson, K. S. Novoselov and A. K. Geim, Nature Physics **2**, 620 (2006).
19. J. Nilsson, A. H. Castro Neto, F. Guinea and N. M. R. Peres, cond-mat/0607343 (2006).

Minisymposium "PDAE Modelling and Multiscale Simulation in Microelectronics and New Technologies"

G. Alì and R. Pulch

[1] Consiglio Nazionale delle Ricerche, Napoli, Italy
[2] Bergische Universität Wuppertal, Germany

Mathematical models of physical systems form the basis of numerical simulations used for industrial applications. The continuous advancement in technical design demands refined models, where more effects have to be included. Thus sophisticated analysis as well as efficient numerical methods have to be tailored to the arising complex systems. On the one hand, modelling dynamical systems by partial differential equations (PDEs) may involve singular matrices, which yield partial differential algebraic equations (PDAEs) in the sense of singular PDEs. On the other hand, a coupling of time-dependent systems of differential algebraic equations (DAEs) with PDEs describing spatial effects is called a system of PDAEs, too. Both cases represent concepts required in advanced simulation of technical processes. The systems often include a multirate behaviour with largely differing time scales. Hence, multiscale simulation using the underlying structure has to be performed for achieving an efficient technique. In particular, the design of electronic circuits is based on numerical simulation of DAE models resulting from a network approach, which specifies the evolution of node voltages and branch currents in time. Due to down-scaling, parasitic effects can not be neglected in the modelling any more. Thus, spatial physical phenomena like heat distribution, electromagnetic interaction or complex semiconductor behaviour are considered, where corresponding PDE models apply. The arising systems of PDAEs become more and more important in microelectronics to enable a realistic simulation. Furthermore, a specific signal model for oscillators yields singular PDEs in microelectronics. Nevertheless, PDAE models apply in other applications like chemical engineering, biology, mechanical engineering, hydrodynamics, etc., too. In the minisymposium, we have presented approaches based on PDAEs in the field of microelectronics and chemical engineering. Thereby, the emphasis has been on PDAEs in the sense of coupled systems of DAEs and PDEs. Mathematical modelling, analysis and numerical aspects have been addressed. Concerning the design of electronic circuits, the topics of PDAE modelling and multiscale simulation are present in the new Marie Curie Research Training Network COMSON (COupled Multiscale Simulation and Optimization in Nanoelectronics) supported by the European Commission.

Domain Decomposition Techniques for Microelectronic Modeling

G. Alì[1], M. Culpo[2], and S. Micheletti[2]

[1] Istituto per le Applicazioni del Calcolo "M. Picone", sez. di Napoli, via P. Castellino 111, I-80131 Napoli, and INFN-Gruppo c. Cosenza, Italy
ali@na.iac.cnr.it
[2] MOX - Modeling and Scientific Computing, Dipartimento di Matematica "F. Brioschi", Politecnico di Milano, via Bonardi 9, I-20133 Milano, Italy
{culpo, micheletti}@mate.polimi.it

1 Introduction

This paper is meant to be the continuation of the previous work [1] where a coupled ODE/PDE method for the simulation of semiconductor devices was introduced. From a strictly mathematical viewpoint, analytical results on coupled PDE/ODE systems (as arising in integrated circuit simulation) can be found in [2]. In particular, in the present paper, we investigate numerically new algorithms of Domain Decomposition type for the simulation of circuits containing distributed devices (Sect. 2) as well as semiconductors in which some part is modeled with lumped parameters (Sect. 3). The results presented here have been investigated in the seminal work [3], while a more extended analysis is ongoing [4].

2 Extra Device Approach

The approach that we present here is devoted to circuit analysis. Suppose we have to deal with a complex circuit network where, however, only few devices are critical with respect to the behavior of the network. On the one hand, the use of a complex PDE model to describe the whole network may be unnecessary and, even though very accurate, it would also require a lot of resources, thus undermining the overall efficiency. On the other hand, using some "black box" method to model the critical devices would possibly be quite inaccurate. What we propose here is to use the PDE model only where strictly needed, keeping the lumped circuit model for the other parts. In the literature, this kind of heterogeneity is usually addressed as mixed-mode device simulation, and a Newton-like numerical procedure is implemented in the simulator MINIMOS-NT [5]. Our approach is instead based on Domain

Decomposition techniques that suitably allows for the coupling of distributed devices, modeled by PDEs, with external circuits described by ODEs. With this aim we have employed a suitable extension of the Dirichlet/Neumann algorithm [6] to enforce the continuity of both currents and node potentials at the device–circuit interface.

For simplicity, suppose that there is only one device that requires a PDE description like, for example, a pn junction diode. We can consider the circuit viewed from the diode terminals as a generic bipole: thus the circuit is divided in two separate subdomains, i.e., the PDE-diode and the ODE-bipole. To apply the Dirichlet/Neumann algorithm it is necessary to figure out how to treat the boundary conditions (b.c.). We have chosen to use Neumann b.c. (i.e., current-operated) for the ODE-circuit and Dirichlet b.c. (voltage-operated) for the PDE-diode. To fix some notation, suppose that the distributed and the lumped models are described by the two problems

$$\frac{\partial u}{\partial t} + \mathcal{D}(u, V) = 0 \quad \text{in } \Omega \times (0, T], \quad \text{and} \quad \frac{dw}{dt} + \mathcal{L}(w, I) = 0 \quad \text{in } (0, T], \quad (1)$$

respectively, where $u = u(x, t)$ represents the internal (state) variable of the PDE part, with $x \in \Omega$ and $t \in (0, T]$, $w = w(t)$ those of the ODE, while $V = V(t), I = I(t)$ are the vectors of the potentials and of the currents, respectively, at the boundary of the device occupying the domain Ω. The quantities $\mathcal{D}, \mathcal{L}$ are suitable differential operators and proper boundary and initial conditions are understood. Let us first introduce a partition $\{t_n\}$ of $[0, T]$ into N subintervals such that $0 = t_0 < t_1 < \ldots < t_{N-1} < t_N = T$. We want to advance the solutions $u(x, t), w(t)$ from $t = t_n$ until $t = t_{n+1}$, for $n = 0, 1, \ldots, N - 1$. The Dirichlet/Neumann algorithm can be thought of as a fixed point iteration for the potentials $V(\cdot)|_{(t_n, t_{n+1}]}$. For this purpose, set the iteration counter $j \leftarrow 0$. Then given some initial guess $V^{(j)}$, the algorithm comprises the following steps:

1. Solve $\frac{\partial u}{\partial t}^{(j+1)} + \mathcal{D}(u^{(j+1)}, V^{(j)}) = 0$ in $\Omega \times (t_n, t_{n+1}]$ for $u^{(j+1)}(x, t)$;
2. Compute $I^{(j+1)} = \mathcal{I}(u^{(j+1)}, V^{(j)})$;
3. Solve $\frac{dw}{dt}^{(j+1)} + \mathcal{L}(w^{(j+1)}, I^{(j+1)}) = 0$ in $(t_n, t_{n+1}]$ for $w^{(j+1)}(t)$;
4. Compute $V^{(j+1)} = \theta \, \mathcal{V}(w^{(j+1)}, I^{(j+1)}) + (1 - \theta) \, V^{(j)}$;
5. Check for convergence: if $\|V^{(j+1)} - V^{(j)}\|_{(t_n, t_{n+1}]} < \varepsilon$ then finish, else $j \leftarrow j + 1$ and go to 1.

Note that the functions $I = \mathcal{I}(u, V), V = \mathcal{V}(w, I)$ return the output current of the device and the potentials at the circuit terminals, respectively, while $0 < \theta < 1$ is a suitable relaxation parameter, and ε a given tolerance. In practice, the fixed point mapping is understood with respect to the final value $V(t_{n+1})$ only, and spatial/temporal discretization schemes have to be employed as well. This algorithm admits also a very interesting circuit interpretation, see Fig. 1 (left). As we can identify Neumann b.c. with the currents

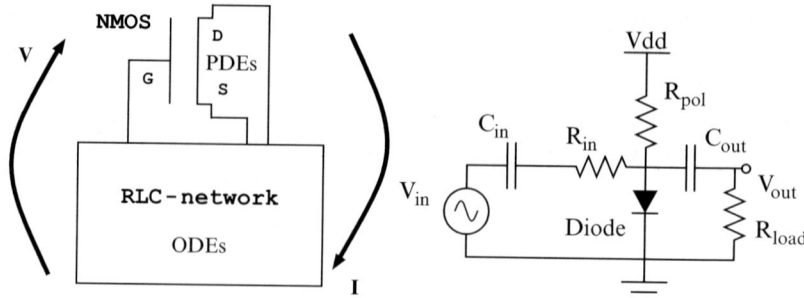

Fig. 1. Example of the Domain Decomposition approach for network analysis (*left*) and scheme for the attenuator of the model problem (*right*)

Table 1. Numerical data used for the extra device test case

Diode	P zone	N zone
Doping (uniform)	$N_A = 10^{17}\,\mathrm{cm}^{-3}$	$N_D = 10^{17}\,\mathrm{cm}^{-3}$
Length	$5\,\mu\mathrm{m}$	$5\,\mu\mathrm{m}$
Minority carrier lifetime	$10\,\mathrm{ns}$	$10\,\mathrm{ns}$
Simulation time	$1\,\mu\mathrm{s}$	
Time step	$5\,\mathrm{ns}$	

Circuit
$R_{in} = 100\,\mathrm{k\Omega}$ $R_{pol} = 100\,\mathrm{k\Omega}$ $R_{load} = 1\,\mathrm{k\Omega}$
$C_{in} = 10\,\mathrm{nF}$ $C_{out} = 10\,\mathrm{nF}$
$V_{DD} = 5\,\mathrm{V}$ $V_{in} = \sin(2\pi f t)\,\mathrm{mV}$ $f = 1\,\mathrm{MHz}$

at the diode terminals, and Dirichlet b.c. with the values of the corresponding potentials, we have to deal at each time step with a voltage-operated PDE device and with a current-operated ODE circuit. We point out that this procedure is implementable without any knowledge about the internal codes of both the PDE and the ODE solvers. Actually, one can use these particular solvers as building blocks for implementing the Domain Decomposition algorithm.

We carry out a sensitivity analysis with respect to the relaxation parameter of the Dirichlet/Neumann algorithm for the transient simulation of a small signal circuit, i.e., an attenuator (Fig. 1, right). The diode is treated as a 1D device and is described by the Drift-Diffusion (DD) transport model [7]. The data used in the simulation are gathered in Table 1. The circuit is solved via the Tableau analysis [8]. The sensitivity for both the Dirichlet and the Neumann b.c. are shown in Fig. 2, where the number of iterations vs. time and relaxation parameter are shown. Notice that under-relaxation is needed for convergence with an optimal parameter of about 0.15 in both cases.

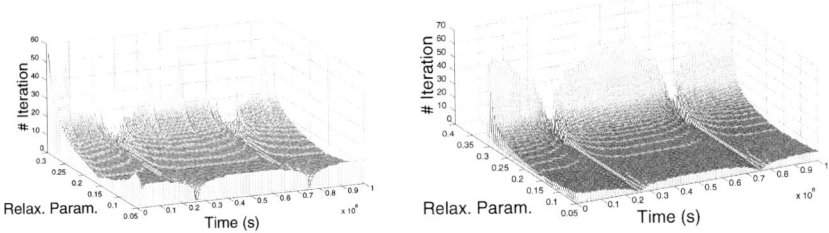

Fig. 2. Sensitivity for Dirichlet b.c. (*left*) and Neumann b.c. (*right*)

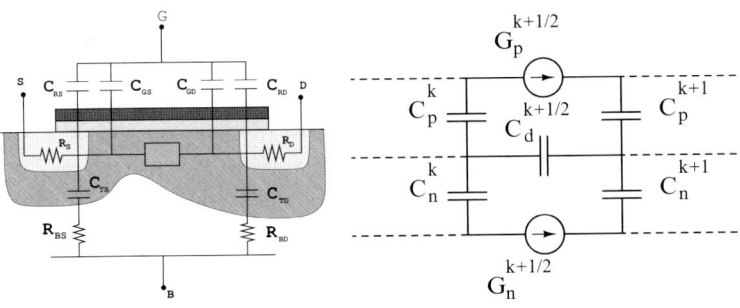

Fig. 3. Example of ODE/PDE coupling for a single device (*left*) and basic building-block for the circuit extraction technique (*right*)

3 Intra Device Approach

The second approach we present is focused on the simulation of a single distributed device. In this case we treat with a PDE model a particular region of interest of the device at hand and with an ODE model the other parts. As in the extra device case, the terminal current and voltage continuity is enforced via the Dirichlet/Neumann approach. In Fig. 3 (left) we see a possible application of this procedure to a MOSFET where only the channel (green box) is modeled with PDEs. For the derivation of a suitable ODE model we have used a technique proposed in [1,9] that allows for the extraction of a compact *physics-based* circuit model from dc device simulations. We have also improved the model, deriving the constitutive relations for the currents from a linearization of the very well-known Scharfetter-Gummel formulas [10]. This has the advantage of being mathematically consistent with the PDE model, where the current continuity equations are also discretized via the Scharfetter-Gummel formulas, and eliminates the degree of freedom in the choice of the conductances that characterizes the approach in [9]. In particular, the expressions of the conductances for the electrons read

$$G_{mn}^{k+1/2} = -\frac{q\mu_n}{h}\left[\dot{B}(\Delta_k\psi)\,n_{k+1} + \dot{B}(-\Delta_k\psi)\,n_k\right],$$

$$G_{rn}^{k+1/2} = \frac{q\mu_n}{h}B(\Delta_k\psi)\,n_{k+1}, \quad G_{fn}^{k+1/2} = -\frac{q\mu_n}{h}B(-\Delta_k\psi)\,n_k,$$

where q is the absolute value of the electron charge, μ_n the electron mobility, h the length of a typical grid element, n_k, ψ_k the electron concentration and electric potential at the k-th grid point, $\Delta_k\psi = \frac{\psi^{k+1}-\psi^k}{V_{\text{th}}}$, $B(\cdot), \dot{B}(\cdot)$ the Bernoulli function and its derivative, while V_{th} is the thermal voltage.

The algorithm that we propose is in some way similar to the one adopted in the extra device case, except that we have to guarantee that the ODE system be consistent with the PDE model. This is why we have used a physics based circuit extraction that calibrates the lumped model from dc device simulations only. Basically, this procedure allows us to describe a semiconductor region of finite size with the basic circuit block shown in Fig. 3 (right).

We have tested our algorithm on a model problem, i.e., a voltage-operated 1D diode. We use the lumped model only in the quasi neutral zones, while the PDE model (based on the DD equations) is employed for the depletion region. This choice is motivated by the consideration that this region is where most of the physically relevant processes take place, so that it represents the part of the device needing a more accurate description.

We carry out a sensitivity analysis with respect to the relaxation parameter for a transient simulation. The Gummel map [11] is employed for the solution of the DD equations. The data used in the simulation are collected in Table 2. The lumped circuits are solved with the MNA analysis [8, 12]. The sensitivity results are displayed in Fig. 4, through the number of iterations vs. time and

Table 2. Numerical data used in the simulation

Diode	P zone	N zone
Doping	$N_A = 10^{16}\,\text{cm}^{-3}$	$N_D = 10^{16}\,\text{cm}^{-3}$
Mobility	$1{,}000\,\text{cm}^2/(\text{V s})$	$1{,}000\,\text{cm}^2/(\text{V s})$
Length	$5\,\mu\text{m}$	$5\,\mu\text{m}$
Polarization	$0.6\,\text{V}$	No generation/recombination
AC signal	$\sin(2\pi ft)\,\text{mV}$	$f = 100\,\text{kHz}$

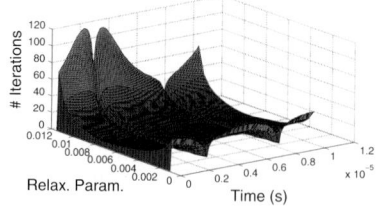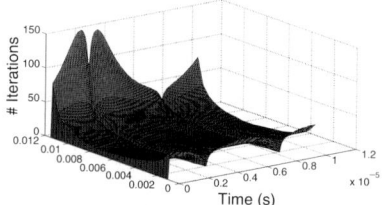

Fig. 4. Sensitivity analysis for the coupled model with fifth- (*left*) and tenth-order circuits (*right*)

relaxation parameter, and where the order of the circuits refers to the number of blocks used for modeling each quasi neutral zone. The relaxation refers only to the Neumann boundary conditions. As in the extra-device case, under-relaxation is required for convergence and the optimal parameter is about 0.007.

4 Conclusions

We have presented and validated a Domain Decomposition procedure for simulating extra-device as well as intra-device structures modeled by coupled PDE/ODE systems. So far, only simple devices and circuits have been tested. We plan to extend the numerical technique to more complex circuits and to multidimensional devices.

References

1. Alì, G., Micheletti, S.: Domain Decomposition techniques and coupled PDE/ODE simulation of semiconductor devices. In Anile, A., Alì, G., Mascali, G., eds.: Scientific Computing in Electrical Engineering. Volume 9 of Mathematics in Industry. Berlin, Springer (2006)
2. Alì, G., Bartel, A., Günther, M., Tischendorf, C.: Elliptic partial differential algebraic multiphysics models in electrical network design. Math. Models Methods Appl. Sci. **13** (2003) 1261–1278
3. Culpo, M.: Accoppiamento Eterogeneo ODE/PDE per il Trasporto di Carica nei Semiconduttori. Dissertation, Politecnico di Milano (2006)
4. Alì, G., Culpo, M., Micheletti, S.: In preparation. (2006)
5. Grasser, T.: Mixed-Mode Device Simulation. Dissertation, Technische Universität Wien (1999)
6. Quarteroni, A., Valli, A.: Domain Decomposition Methods for Partial Differential Equations. The Clarendon Press, Oxford University Press, New York (1999)
7. Selberherr, S.: Analysis and Simulation of Semiconductor Devices. Springer-Verlag, Wien, New York (1984)
8. Chua, L., Desoer, C., Kuh, E.: Linear and Nonlinear Circuits. McGraw-Hill (1987)
9. Pacelli, A., Mastrapasqua, M., Luryi, S.: Generation of equivalent circuits from physics based device simulation. IEEE Trans. Circuits Syst. **19** (2000) 1241–1250
10. Scharfetter, D., Gummel, H.: Large signal analysis of a silicon read diode oscillator. IEEE Trans. Electron Devices **16** (1969) 64–77
11. Gummel, H.: A self-consistent iterative scheme for one dimensional steady-state transistor calculations. IEEE Trans. Electron Devices **11** (1964) 455–465
12. Ho, C., Ruehli, A., Brennan, P.: The modified nodal approach to network analysis. IEEE Trans. Circuits Syst. **22** (1975) 504–509

A Concept for Classification of Partial Differential Algebraic Equations in Nanoelectronics

Andreas Bartel and Roland Pulch

Bergische Universität Wuppertal, Department of Mathematics and Sciences,
Chair of Applied Mathematics and Numerical Analysis, Gaußstr. 20,
D-42119 Wuppertal, Germany
[bartel, pulch]@math.uni-wuppertal.de

Summary. The design of electronic circuits is based on numerical simulation of corresponding mathematical models. Systems of differential algebraic equations (DAEs) reproduce the time behaviour of idealised electric networks. In nanoelectronics, miniaturisation causes parasitic effects, which can not be neglected any longer. These spatial phenomena yield models consisting of partial differential equations (PDEs). Thus the circuit's behaviour is given by partial differential algebraic equations (PDAEs), which couple DAEs in time and PDEs in time/space. We present a rough concept for classifying existing PDAE models in nanoelectronics. The categorisation rests primarily upon the physical background in each model.

1 Introduction

The mathematical model of dynamical systems often results from some network approach, which yields time-dependent systems of differential algebraic equations (DAEs). That is, we consider ideally joint lumped elements, without spatial coordinate, but with the topology information given by the incidences of these elements. In contrast, spatial physical effects are described by partial differential equations (PDEs) in space or time/space. Thus an enhanced model requires a coupling of DAEs and PDEs, which yields systems of so-called partial differential algebraic equations (PDAEs). Such systems of PDAEs arise in many technologies like mechanical engineering as coupled multibody systems with sole or flexible/plastic systems, see [5], in nanoelectronics (see below) and others.

Furthermore, the wording PDAE is also used for singular implicit PDEs, i.e. where singular matrices arise in front of partial derivatives, see [8], for example. In case of electronic circuits, a specific multivariate model yields an efficient representation of amplitude and/or frequency modulated signals including widely separated time scales. The introduction of different time variables (for the occurring scales) transforms the circuit's DAE into a PDAE in the sense of a singular PDE, see [9].

In this paper, we focus on PDAE models in nanoelectronics setting with PDE-enhancement of DAE models, rather than singular PDEs. Modified nodal analysis yields large systems of DAEs for ideal circuits, see [7]. We write such a system in the general form

$$\mathbf{f} : \mathbb{R}^k \times \mathbb{R}^k \times I \to \mathbb{R}^k, \quad \mathbf{f}(\mathbf{y}, \dot{\mathbf{y}}, t) = \mathbf{0}, \quad t \in I := [0, T], \tag{1}$$

where $\mathbf{y} : I \to \mathbb{R}^k$ denotes unknown node voltages and branch currents. A consistent initial value $\mathbf{y}(0) = \mathbf{y}_0$ completes the usual electric network model. In addition, we formulate schematically a system of PDEs corresponding to a parasitic effect via an operator

$$\mathcal{L} : D \times I \times V \to \mathbb{R}^m, \quad \mathcal{L}(\mathbf{x}, t, \mathbf{v}) = \mathbf{0}, \quad \mathbf{x} \in D \subset \mathbb{R}^d, \quad t \in I \tag{2}$$

with a solution $\mathbf{v} : D \times I \to \mathbb{R}^m$ in some function space V. Initial and/or boundary conditions have to be specified appropriately. Coupling the systems (1) and (2) using some variables/functions results in a PDAE. The coupling can be done via artificial variables, source terms, boundary conditions (BCs) or even more sophisticated constructions.

Since, the mathematical structure of PDAEs is rather complex, we can not derive a universal classification of all existing PDAE models. Alternatively, we introduce a rough concept to categorise some important models arising in ongoing research within the field of nanoelectronics.

2 PDAE Models

Each spatial physical phenomenon requires a corresponding modelling via a PDE. The following aspects arise due to miniaturisation in chip design. We distinguish two general types of coupling.

2.1 Refined Modelling

Usually semiconductors, transmission lines and other components with spatial distribution are given by subcircuits of lumped electric elements (companion models). To obtain a somewhat more precise model (also considering downscaling phenomena), we replace one or several of these subcircuit descriptions by a PDE model for the corresponding electric effect in the network. These can be one or several semiconductor elements, which behave critical in an electronic network, and where it makes sense to simulate these elements more detailed. Another possibility is to replace a transmission line model based on DAEs by an according PDE. This is a natural way, which bypasses a huge number of more or less artifical parameters of the companion model.

This approach is called refined modelling. It has a special type of coupling. Boundary conditions for the Ohmic contacts of the PDE model are the node potentials of the connect network nodes (Dirichlet condition). At the remaining boundaries in multiple dimensions, where there is no electric contract, one may have von-Neumann conditions with no flux or field conditions at insulated contacts. On the other hand, the output of the PDE model is an electric current, which is eventually a source term to the network's DAE. Abstractly, we obtain systems of the type

$$
\begin{aligned}
A\mathbf{u}_t + \mathcal{L}_D\mathbf{u} - \mathbf{h}(\mathbf{u}, t) &= \mathbf{0} \quad\text{(PDE in } I \times D) \\
\mathbf{u}|_{\Gamma_1} &= \mathbf{g}(\mathbf{y}) \text{ (Dirichlet BC)} \\
\tfrac{\partial}{\partial \mathbf{n}}\mathbf{u}|_{\Gamma_2} &= \mathbf{h}(\mathbf{y}) \text{ (von-Neumann BC)} \\
\mathbf{f}(\mathbf{y}, \dot{\mathbf{y}}, t) &= \boldsymbol{\lambda}(\mathbf{u}) \text{ (DAE in } I),
\end{aligned}
\tag{3}
$$

where $\mathcal{L}_D$ represents a differential operator with respect to space. The involved PDE can be of mixed type (elliptic, hyperbolic, parabolic). Thereby, the coupling is performed via the input $\boldsymbol{\lambda}$ and the boundary conditions $\mathbf{g}$ and $\mathbf{h}$ (where we have a decomposition of the boundary: $\partial D = \Gamma_1 \cup \Gamma_2$). Furthermore, analysing complex systems (3) may yield simpler but still highly accurate companion models for the underlying component.

In nanoelectronics, the PDAE systems, which have been considered in the literature or are part of ongoing research, can principally be categorised into the following cases:

Semiconductors: Here transistors are described by drift-diffusion or quantum mechanical equations coupled with the electric network. Existence and uniqueness results for nonstationary and stationary drift-diffusion network systems are found in [1,2], for an index analysis of the arising PDAE, we refer to [4]. Currently, efficient numerical codes are being developed.

Transmission line effects: Also down-scaling causes a decreasing distance of transmission lines and thus an undesired interaction arises. Telegrapher's equation describes the underlying physical effect. The coupling of PDEs and DAEs accords to the form (3). Now the involved PDE is exclusively of hyperbolic type, which implies a specific numerical treatment, see [6] for details.

Electromagnetic fields: The DAEs (1) result from a network approach to avoid a simulation of the complete circuit using Maxwell's equations. However, if some crucial parts of the circuit demand a refined model, a separation from the network can be done. Thus we apply Maxwell's equations to represent the small part, whereas we use the network DAEs for the major part.

2.2 Multiphysical Extension

This modelling is much more complex, since we do not add a physical dimension to the electric network, but have a distributed additional effect:

Thermal aspects: The increase of the clock rate in chips causes a higher power loss in the electronic network. Thus we have to consider heat distribution and conduction between the circuit's elements. In contrast to the effects described above, the heat evolution runs in parallel to the time-dependence of voltages and currents. Thus a thermal network can be associated to the electric network. In the thermal part, specific 0D elements can be refined into elements with spatial distribution or elements can be located in macro structures. Combining the heat equation for the spatial elements with the network yields

$$A\mathbf{u}_t + \mathcal{L}_D\mathbf{u} - \mathbf{h}(\mathbf{u}, t) = \mathbf{s}(\mathbf{x}) \quad (\text{PDE in } I \times D)$$
$$\mathbf{u}|_{\partial D} = \mathbf{g}(\mathbf{x}) \quad (\text{BC}) \tag{4}$$
$$\mathbf{f}(\mathbf{y}, \dot{\mathbf{y}}, t, \boldsymbol{\mu}(\mathbf{u})) = \mathbf{0} \quad (\text{DAE in } I).$$

In this case, the included PDE is of parabolic type (Fourier law). The coupling is present in the source terms and boundary conditions $\mathbf{s}$, $\boldsymbol{\mu}$, $\mathbf{g}$: Here dissipated power is not only entering the boundary conditions, but is also a source term for the evolution equation; on the other hand, the temperature enters the electric network as parameter and thus causes a more general dependence. For further details, we refer to [3].

Electromagnetics: In principle, one can interpret an electromagnetic field influencing the complete circuit as a multiphysical case, too. Consequently, the contribution of the field to each component has to be modelled appropriately.

3 Illustrative Example

We consider the electric circuit given in Fig. 1. In the refined description, where the diode is modelled by semiconductor equations, we have:
a) electric network: (current through voltage source j_V, node potential u_1, u_2)

$$j_V + \frac{u_1 - u_2}{R} = 0, \quad \frac{u_2 - u_1}{R} + j_d + C\frac{\mathrm{d}}{\mathrm{d}t}u_2 = 0, \quad u_1 - v(t) = 0. \tag{5a}$$

b) 1D drift-diffusion: (electron/hole density n/p, electrostatic potential V – currents j_p, j_n)

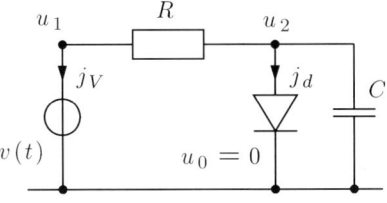

Fig. 1. Example circuit

$$-q\partial_t n + \partial_x j_n = qR, \qquad j_n = -q\left\{-D_n\partial_x n + \mu_n n \partial_x V\right\}, \quad (5b)$$

$$q\partial_t p + \partial_x j_p = -qR, \qquad j_p = q\left\{-D_p\partial_x p - \mu_p p \partial_x V\right\}, \quad (5c)$$

$$-\epsilon\partial_x^2 V - q(C + p - n) = 0, \qquad j_d = \frac{\epsilon}{l}\frac{\mathrm{d}}{\mathrm{d}t}u_2 + \frac{1}{l}\int_0^l \{j_n + j_p\}\mathrm{d}x, \quad (5d)$$

$$\begin{pmatrix} V(0,t) - V_{\mathrm{bi}} \\ V(l,t) - V_{\mathrm{bi}} \end{pmatrix} = \begin{pmatrix} u_2 \\ 0 \end{pmatrix}, \quad n(0,t) = n(l,t) = n_0, \quad p(0,t) = p(l,t) = p_0. \quad (5e)$$

This involves parameters: resistance R, capacity C, input $v(t)$ for the network equations; the diode is defined on a 1D-line segment ($[0,l]$) with mobilities μ_n, μ_p, diffusivities D_n, D_p, unit of charge q, permittivity ϵ, doping profile $C : [0,l] \to \mathbb{R}$ and according built-in potential $V_{\mathrm{bi}} : [0,l] \to \mathbb{R}$. Furthermore, j_d specifies the output current of the diode and therefore our coupling quantity.

The boundary conditions for the carrier densities (n_0, p_0) are obtained from equilibrium assumptions, see [1] for more details. As a further refinement of this example, one could image telegrapher's equation for the parallel connection of the diode and the capacitor.

4 Conclusions

In nanoelectronics, a sophisticated modelling of physical effects leads to systems of PDAEs. An elementary classification of some crucial models has been presented, which considers the type of the underlying PDEs and especially the physical nature of the coupling: models can be (i) refined, replacing simple (0D) descriptions by (more advanced) spatial models, or (ii) multiphysically extended, adding a new layer of effects.

Acknowledgements. This work is part of the MCA-RTN project COMSON (COupled Multiscale Simulation and Optimisation in Nanoelectronics) supported by the European Union.

References

1. Alì, G.; Bartel, A.; Günther, M.: Parabolic differential-algebraic models in electric network design. SIAM J. MMS 4 (2005) 3, pp. 813-838.
2. Alì, G.; Bartel, A.; Günther, M.; Tischendorf, C.: Elliptic partial differential-algebraic multiphysics models in electrical network design. Math. Model. Meth. Appl. Sc. 13 (2003) 9, pp. 1261-1278.
3. Bartel, A.: Partial Differential-Algebraic Models in Chip Design — Thermal and Semiconductor Problems. VDI-Verlag, Düsseldorf 2004.
4. Bodestedt, M.; Tischendorf, C.: PDAE models of integrated circuits and index analysis. to appear in: Math. Comput. Model. Dyn. Syst.
5. Büttner, J.; Simeon, B.: Numerical Treatment of Material Equations with Yield Surfaces. in: Hutter, K.; Baaser, H. (eds.); Deformations and failure of metallic continua, Lecture Notes in Appl. Comp. Mech. 10, Springer-Verlag, Berlin 2003.

6. Günther, M.: A joint DAE/PDE model for interconnected electrical networks. Math. Comput. Model. Dyn. Syst. 6 (2000), pp. 114-128.
7. Günther, M.; Feldmann, U.: CAD based electric circuit modeling in industry I: mathematical structure and index of network equations. Surv. Math. Ind. 8 (1999), pp. 97-129.
8. Lucht, W.; Strehmel, K.; Eichler-Liebenow, C.: Indexes and special discretization methods for linear partial differential algebraic equations. BIT 39 (1999) 3, pp. 484-512.
9. Pulch, R.: Multi time scale differential equations for simulating frequency modulated signals. Appl. Numer. Math. 53 (2005) 2-4, pp. 421-436.

Numerical Simulation of a Class of PDAEs with a Separation of Time Scales

Benoît Chachuat[1] and Paul I. Barton[2]

[1] Laboratoire d'Automatique, École Polytechnique Fédérale de Lausanne, EPFL-STI-LA, Station 9, CH-1015 Lausanne, Switzerland
`benoit.chachuat@epfl.ch`

[2] Department of Chemical Engineering, Massachusetts Institute of Technology, Cambridge, MA 02139, USA
`pib@mit.edu`

1 Introduction

Problems that exhibit multiple time scales arise naturally in many scientific and engineering fields. For transient, distributed process systems, the corresponding models consist of partial differential equations (PDEs), possibly coupled to ordinary differential equations (ODEs) or differential-algebraic equations (DAEs) that describe lumped processes or are used as boundary conditions.

Particularly popular for the numerical solution of time-varying problems in one spatial dimension is the so-called *method of lines* (MoL) [6], which yields a large-scale DAE system by discretizing one of the independent variables of the original system, usually the spatial variable. One of the main reasons for this popularity lies in its relative simplicity and the fact that state-of-the-art numerical solvers can be used to integrate the resulting DAEs accurately. But, when applied to problems with a wide range of time-scales, the MoL often results in prohibitive computational times because shocks and fronts can develop and move around the spatial domain, which need very fine spatial discretizations and give rise to stiff DAEs. These difficulties can be alleviated by moving mesh or mesh refinement techniques, possibly combined with shock capturing schemes, although the complexity of the MoL is then significantly increased. Reliability issues also arise from the challenging task of consistently initializing large-scale DAEs, which happen to be a serious obstacle for embedding the simulation into an optimization problem.

In this paper, we propose an alternative solution method for a particular class of one-dimensional PDAEs with a separation of time scales where (i) the fast variables are spatially distributed, and (ii) the slow variables are lumped. Recent studies in micro-chemical and micro power generation processes suggest that many systems of practical interest are well approximated by this formulation, yet such problems have remained largely unexplored in the literature. In Sect. 2, we formulate the problem and discuss its approximation in the light of the theory of singular perturbations. Then, we show how the special structure of these problems can be exploited to compute an approximate solution of the PDAEs, both reliably and efficiently, in Sect. 3. The proposed approach is demonstrated on an application related to the start-up of micro-scale chemical processes for portable power generation in Sect. 4. Finally, Sect. 5 concludes the paper.

2 Problem Formulation and QSS Approximation

Consider the system of one-dimensional, first-order, quasi-linear PDAEs

$$\mathbf{u}_t + \mathsf{A}(t, x, \mathbf{u})\mathbf{u}_x = \mathbf{r}(t, x, \mathbf{u}), \tag{1}$$

where t and x stand for the independent variables, $(t, x) \in \Omega := [t_0, t_f] \times [a, b]$; $\mathbf{u}(t, \cdot) \in \mathcal{H}([a, b], \mathbb{R}^{n_u})$ denotes the dependent variable; $\mathsf{A}(t, x, \mathbf{u})$ and $\mathbf{r}(t, x, \mathbf{u})$ are a $n_u \times n_u$ matrix and a n_u vector, respectively, whose elements are sufficiently smooth functions of t, x and $\mathbf{u}$.

Assumption 1. *A natural partition of* $\mathbf{u}$ *into* slow *and* fast *subsets of variables* $\left(\mathbf{u}^s(t), \mathbf{u}^f(t, x)\right) \in \mathbb{R}^{n_s + n_f}$ *exists for the PDAEs* (1). *Moreover, the slow variables* $\mathbf{u}^s(t)$ *are* lumped, *and the fast variables* $\mathbf{u}^f(t, x)$ *are hyperbolic, with all the characteristics pointing in the same direction.*

Under these assumptions, the PDAEs (1) may be rewritten in the form of a singularly perturbed model:

$$\mathbf{u}_t^s(t) = \mathbf{r}^s[t, \mathbf{u}^s(t), \mathbf{u}^f(t, b), \varepsilon] \tag{2}$$

$$\varepsilon\, \mathbf{u}_t^f(t, x) + \mathsf{A}^f[t, x, \mathbf{u}^s(t), \mathbf{u}^f(t, x), \varepsilon]\mathbf{u}_x^f(t, x) = \mathbf{r}^f[t, x, \mathbf{u}^s(t), \mathbf{u}^f(t, x), \varepsilon], \tag{3}$$

where ε stands for the perturbation parameter, $0 < \varepsilon \ll 1$. The initial and boundary conditions for the slow and fast variables are specified as

$$\mathbf{u}^s(t_0) = \boldsymbol{\eta}^f(\varepsilon) \tag{4}$$

$$\mathbf{u}^f(t_0, x) = \boldsymbol{\eta}^f(x, \varepsilon), \ \forall x \in (a, b]; \quad \boldsymbol{\xi}^f(\mathbf{u}^f(t, a), t, \varepsilon) = \mathbf{0}, \ \forall t \in [t_0, t_f], \tag{5}$$

where $\boldsymbol{\eta}^f$ and $\boldsymbol{\xi}^f$ are sufficiently smooth functions of their arguments. A number of remarks are in order:

Remark 1. For the sake of clarity, the equation (2) giving the slow variables shows dependence on the fast variables at $x = b$ only. However, the approach can be readily extended to the case where (2) depends on any functional of $\mathbf{u}^f(t, \cdot)$.

Remark 2. The fast variables being hyperbolic, the Cauchy problem should be well-posed for the fast subsystem [4]. Moreover, the eigenvalues of A^f should be all real and have the same sign since the characteristics point in the same direction. Here, the Cauchy data (5) assume that the eigenvalues of A^f are nonnegative.

On setting $\varepsilon = 0$, the fast model (3) reduces to the DAEs

$$A^f[t, x, u^s, \bar{u}^f(x), 0]\bar{u}^f_x(x) = r^f[t, x, u^s, \bar{u}^f(x), 0]; \;\; \xi^f(\bar{u}^f(a), t, 0) = 0, \quad (6)$$

for each t, u^s. By analogy to the classical theory of singular perturbations [3], we say that (2)(3) is in *standard form* if a solution $\bar{u}^f(x) := h^f(x; t, u^s)$ to the initial value problem (6) exists and is unique on $[a, b]$, for each t, u^s. The so-called *quasi-steady-state (QSS) model* is then obtained as

$$\bar{u}^s_t(t) = r^s\left[t, x, \bar{u}^s(t), h^f(b; t, \bar{u}^s(t)), 0\right]; \;\; \bar{u}^s(t_0) = \eta^s(0). \quad (7)$$

Note that the number of equations in the state model reduces from $n_s + n_f$ to n_s under the QSS approximation. Conditions under which the QSS solution provides a $O(\varepsilon)$ approximation of the PDAE solution can be obtained upon extending Tikhonov's theorem [3] to hyperbolic fast subsystems. In particular, a crucial requirement is that the boundary layer system be exponentially stable around 0 at $x = b$, uniformly in t, u^s.

3 Solution of ODEs with IVP-DAEs Embedded

For most practical applications, the fast subsystem (6) cannot be solved explicitly. A numerical procedure is therefore needed for calculating the right-hand side of the slow subsystem (7). That is, the QSS model corresponds to an initial value problem (IVP) in ODEs (in time), the right-hand side of which depends on the solution of an IVP in DAEs (in space),

$$\bar{u}^s_t(t) = r^s\left[t, x, \bar{u}^s(t), \bar{u}^f(b; t, \bar{u}^s(t)), 0\right]; \qquad \bar{u}^s(t_0) = \eta^s(0)$$

s.t. $A^f[t, x, \bar{u}^s(t), \bar{u}^f(x), 0]\bar{u}^f_x(x) = r^f\left[t, x, \bar{u}^s(t), \bar{u}^f(x), 0\right]; \;\; \xi^f(\bar{u}^f(a), t, 0) = 0.$

We shall refer to such problems as *ODEs with IVP-DAEs embedded* herein.

The numerical solution of ODEs with IVP-DAEs embedded proceeds by integrating the $n - s$ ODEs (outer system), forward in time; at each time step, a full integration of the n_f DAEs (inner system) is then performed, forward in space. In this work, we used the numerical solver DSL48S in the software package DAEPACK [5] for solving either subsystems, which implements a multistep BDF method [1] and has built-in sensitivity analysis capabilities. The Jacobian matrix $\nabla_{\bar{u}^s(t)} r^s$ can be calculated by forward sensitivity analysis as

$$\nabla_{\bar{\mathbf{u}}^s(t)} \mathbf{r}^s = \frac{\partial \mathbf{r}^s}{\partial \bar{\mathbf{u}}^s(t)} + \frac{\partial \mathbf{r}^s}{\partial \bar{\mathbf{u}}^f} \frac{\partial \bar{\mathbf{u}}^f}{\partial \bar{\mathbf{u}}^s(t)} \bigg|_{x=b}$$

$$\text{with: } \mathbf{A}^f \left(\frac{\partial \bar{\mathbf{u}}^f}{\partial \bar{\mathbf{u}}^s(t)} \right)_x = \left(\frac{\partial \mathbf{r}^f}{\partial \bar{\mathbf{u}}^f} - \bar{\mathbf{u}}_x^f{}^T \frac{\partial \mathbf{A}^f}{\partial \bar{\mathbf{u}}^f} \right) \frac{\partial \bar{\mathbf{u}}^f}{\partial \bar{\mathbf{u}}^s(t)} + \left(\frac{\partial \mathbf{r}^f}{\partial \bar{\mathbf{u}}^s(t)} - \bar{\mathbf{u}}_x^f{}^T \frac{\partial \mathbf{A}^f}{\partial \bar{\mathbf{u}}^s(t)} \right) ;$$

$$\frac{\partial \boldsymbol{\xi}^f}{\partial \bar{\mathbf{u}}^f} \frac{\partial \bar{\mathbf{u}}^f}{\partial \bar{\mathbf{u}}^s(t)} \bigg|_{x=a} = \mathbf{0}.$$

Note that, unlike MoL, the need for making an a priori discretization of either the time or the spatial coordinate is removed, for both the time and space steps are adapted by the numerical solver directly. Moreover, since the inner and outer systems can both be solved to a high accuracy, based on the error control mechanism of the numerical solver, one can guarantee a high accuracy for the solution of the QSS system. Further, this approach solves a system of n_f DAEs repeatedly, instead of $(n_f + n - s) \times N$ DAEs at once in the MoL approach (N being the number of lines). The issue of initializing a large number of DAEs is thus dramatically alleviated, which makes the solution process more reliable. We shall also see below that the proposed approach compares favorably with the MoL from a computational viewpoint.

4 Application to Micro Power Generation

We consider an application of the QSS approach to the start-up simulation of a micro power generation system employing a high-temperature fuel cell. The fuel cell stack is coupled to a small battery that heats up the fuel cell stack and meets the power demand until the fuel cell is fully operational.

Assuming a uniform temperature in the fuel cell stack (fabricated from silicon), the energy balance equations yield a set of ODEs

$$\frac{d\mathscr{E}^{bat}}{dt} = -\mathscr{P}_{out}^{bat} - \mathscr{P}_{heat}^{bat} + \mathscr{P}_{rech}^{fc} - \dot{Q}_{loss}^{bat}$$

$$\frac{dT}{dt} = \frac{1}{V^{dev} c_p^{dev}} \left[\dot{H}_{in}^{dev} - \dot{H}_{out}^{dev} - \mathscr{P}_{out}^{fc} - \mathscr{P}_{rech}^{fc} + \mathscr{P}_{heat}^{bat} - \dot{Q}_{loss}^{dev} \right],$$

where T and $\mathscr{E}^{bat}$ denote the fuel cell stack temperature and the remaining energy in battery, respectively. On the other hand, mass and species balance equations in each unit of the fuel cell stack correspond to one-dimensional, quasi-linear PDAEs:

$$\frac{1}{V} \frac{\partial F}{\partial x} = \frac{1}{T} \frac{dT}{dt} + \frac{1}{\rho} \sum_{j=1}^{n_r} \sum_{i=1}^{n_c} \nu_{i,j} r_j$$

$$\frac{\partial y_i}{\partial t} + \frac{F}{V} \frac{\partial y_i}{\partial x} = \frac{1}{\rho} \sum_{j=1}^{n_r} \left[\nu_{i,j} r_j - y_i \sum_{k=1}^{n_c} \nu_{k,j} r_j \right] \qquad i = 1 \ldots n_c, \qquad (8)$$

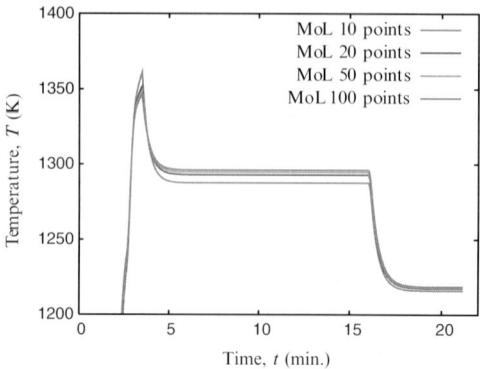

	CPU time[†]
MoL 10 lines	7.4
MoL 20 lines	17.9
MoL 30 lines	32.8
MoL 50 lines	77.5
MoL 75 lines	165.3
MoL 100 lines	248.4
QSS approach	171.2

[†]3.2GHz Athlon processor, 1Gb RAM.

Fig. 1. Comparison of the ODEs with IVP-DAEs embedded approach with the MoL approach. *Left*: Accuracy of MoL vs. number of lines; *Right*: CPU time comparison

where V, F, y_i and r_j stand for the volume, the molar flow rate, the molar fraction of gas species $i = 1, \ldots, n_c$, and the rate of reaction $j = 1, \ldots, n_r$, respectively; $\rho = \frac{P}{\mathcal{R}T}$ is the molar density. See [2] for details on the fuel cell stack and the kinetic rates. For this system, the state variables are naturally split up into slow $(T, \mathscr{E}^{\text{bat}})$ and fast (F, y_i) subsets. The QSS model is simply obtained by zeroing the time derivatives $\frac{\partial y_i}{\partial t}$ in (8).

The temperature profiles obtained with the MoL approach are shown in Fig. 1 (left), for a typical start-up scenario. Note that obtaining an accuracy of about 1 K requires that around 100 lines be considered (in each unit). On the other hand, it was found that the accuracy of the QSS approach is within 1 K of the PDAE solution. Computationally, the proposed approach takes about the same time as the MoL approach for about 75 lines (Fig. 1, right), and thus compares favorably for achieving the desired accuracy.

5 Conclusions

Modeling and simulation of dynamic systems with a separation of time-scales leads naturally to singular perturbation models. In this paper, we have considered a special class of PDAEs where (i) the slow variables are lumped, and (ii) the fast variables are hyperbolic with all the characteristics pointing in the same direction. Under these conditions, the QSS model yields a system of ODEs with IVP-DAEs embedded, and a numerical integrator can be used to solve for the spatial profile of the fast variables at each time step for the slow variables. The ability to use an adaptive spatial mesh is highly advantageous to the reliability and accuracy of the simulation, as was demonstrated on an application related to the start-up of micro-scale chemical processes for portable power generation.

References

1. Asher, U.K., Petzold, L.R.: Computer Methods for Ordinary Differential Equations and Differential-Algebraic Equations, SIAM, Philadelphia, PA (1998)
2. Chachuat, B., Mitsos, A., Barton, P.I.: Optimal Design and Steady-State Operation of Micro Power Generation Employing Fuel Cells. *Chem. Eng. Sci.*, **60**(16), 4535–4556 (2005)
3. Khalil, H.K.: Nonlinear Systems, Prentice Hall, 3rd Ed., Upper Saddle River, NJ (2002)
4. Jeffrey, A.: Quasilinear Hyperbolic Systems and Waves. Pitman Publishing, London, UK (1976)
5. Tolsma, J.E., Barton, P.I.: DAEPACK: An Open Modeling Environment for Legacy Models. *Ind. & Eng. Chem. Res.*, **39**(6), 1826–1839 (2000)
6. Vande Wouwer, A. and Saucez, P. and Schiesser, W.E.: Adaptive Method of Lines, Chapman & Hall, CRC Press (2001)

Model Order Reduction for Nonlinear Differential Algebraic Equations in Circuit Simulation

Thomas Voss[3], Arie Verhoeven[1,2], Tamara Bechtold[1], and Jan ter Maten[1]

[1] NXP Semiconductors
   tamara.bechtold@philips.com
[2] Eindhoven University of Technology
   averhoev@win.tue.nl
[3] Delft University of Technology
   t.voss@tudelft.nl

Summary. In this paper we demonstrate model order reduction of a nonlinear academic model of a diode chain. Two reduction methods, which are suitable for nonlinear differential algebraic equation systems are used, the trajectory piecewise linear approach and the proper orthogonal decomposition with missing point estimation.

1 Introduction

The dynamics of electrical circuits at time t can be generally described by the nonlinear, first order, differential-algebraic equation (DAE) system of the form:

$$\frac{\mathrm{d}}{\mathrm{d}t}\mathbf{q}(\mathbf{x}) + \mathbf{j}(\mathbf{x}) + B\mathbf{u}(t) = \mathbf{0}, \tag{1}$$

where $\mathbf{x} \in \mathbb{R}^n$ represents the unknown vector of circuit variables in time t, the vector-valued functions $\mathbf{q}, \mathbf{j} : \mathbb{R} \times \mathbb{R}^n \to \mathbb{R}^n$ represent the contributions of, respectively, reactive elements (such as capacitors and inductors) and of nonreactive elements (such as resistors) and $B \in \mathbb{R}^{n \times m}$ is the distribution matrix for the excitation vector $\mathbf{u} : \mathbb{R} \to \mathbb{R}^m$. There are several established methods, such as sparse-tableau, modified nodal analysis, etc. which generate the system (1) from the netlist description of electrical circuit. The dimension n of (1) is of the order of the number of elements in the circuit, which means that it can be extremely large, as today's VLSI circuits have hundreds of millions of elements.

Mathematical model order reduction (MOR) aims to replace (1) by a system of much smaller dimension, which can be solved by suitable DAE solvers within acceptable time. At present, however, only linear MOR techniques are well-enough developed and properly understood to be employed [1]. To that end, we either linearise the system (1) or decouple it into nonlinear and linear subcircuits (interconnect macromodeling or parasitic subcircuits [2]). The nonlinear MOR techniques are less developed and less understood than the linear ones. In this paper we present the application of two most promising nonlinear reduction methods on an academic diode chain model. These are the trajectory piecewise linear approach (TPWL) [3] and the proper orthogonal decomposition (POD) [4] supported by missing point estimation (MPE) technique [5].

2 Trajectory Piecewise Linear Model Order Reduction

The idea behind the TPWL method is to linearise (1) several times along a training trajectory (corresponding to some typical input). The local systems are then used to create a global reduced subspace. The final TPWL model is constructed as a weighted sum of all local linearised reduced systems.

2.1 Creating the Local Linearised Models

The disadvantage of standard linearisation methods is that they deliver good results, only in the surrounding of the chosen linearisation tuple (LT)$(x(t_i), t_i)$. To overcome this, in TPWL approach several linearised models are created. This guarantees the quality of the results whenever the solution stays close to one of the chosen LTs. The procedure for selection of LTs can be described by the following steps:

1. Set an absolute accuracy factor $\varepsilon > 0$, set $i = 1$.
2. Linearise the system around the i-th LT (x_i, t_i). This implies:

$$C_i \dot{\mathbf{x}} + G_i \mathbf{x} + B_i \mathbf{u}(t) = 0 \qquad (2)$$

   with $C_i = \frac{\partial}{\partial x} \mathbf{q}(t, \mathbf{x})\big|_{\mathbf{x}_i, t_i}$ and $G_i = \frac{\partial}{\partial x} \mathbf{j}(t, \mathbf{x})\big|_{\mathbf{x}_i, t_i}$, where x_i stays for $x(t_i)$. Save C_i, G_i and B_i.
3. Reduce the linearised system to dimension $r \ll n$ with an appropriate linear MOR method, like "Poor Man's TBR" [6] or Krylov-subspace methods [7]. This implies:

$$C_i^r \dot{\mathbf{z}} + G_i^r \mathbf{z} + B_i^r \mathbf{u}(t) = 0 \qquad (3)$$

   where $C_i^r = V_i^\top C_i V$, $G_i^r = V_i^\top G_i V_i$, $B_i^r = V_i^\top B$ with $V_i \in \mathbb{R}^{n \times r}$, $\mathbf{z} \in \mathbb{R}^r$ and $\mathbf{x} \approx V_i \mathbf{z}$. Save the local projection matrix V_i.
4. Integrate both, the reduced system (3) and the original system (1) choosing the same time-steps t_k. When $\frac{\|V_i \mathbf{z}(t_k) - \mathbf{x}(t_k)\|}{\|\mathbf{x}(t_k)\|} > \varepsilon$ chose $(\mathbf{x}(t_k), t_k)$ as $i + 1$-th LT. Set $i = i + 1$. Go to step 2.

The steps 2–4 are repeated until the end of the given trajectory has been reached. In this way, s local reduced subspaces with bases $V_1, \ldots, V_s$ are created.

2.2 Creating the Global Reduced Subspace

All local reduced subspaces are merged into the global reduced subspace and each local linearised system (2) is now projected onto this global subspace. The procedure can be described by the following steps:

1. Define $\tilde{V} = [V_1, \ldots, V_s]$.
2. Calculate the SVD of $\tilde{V}$: $\tilde{V} = U\Sigma W^\top$ with $U = [u_1, \ldots, u_n] \in \mathbb{R}^n$, $\Sigma \in \mathbb{R}^{n \times rs}$ and $W \in \mathbb{R}^{rs \times rs}$.
3. Define new global projection matrix V_g as $[u_1, \ldots, u_r]$.
4. Project each local linearised system (2) onto V_g.

2.3 Creating the TPWL Model by Weighting

All local reduced linearised reduced systems are combined in a weighted sum to build the global TPWL model:

$$\sum_{i=1}^{s} w_i V_g^\top C_i V_g \dot{\mathbf{z}} + \sum_{i=1}^{s} w_i V_g^\top G_i V_g \mathbf{z} + \sum_{i=1}^{s} w_i V_g^\top B_i \mathbf{u}(t) = 0. \tag{4}$$

A weight w_i determines the influence of the i-th local system to the global system. The weights can be chosen by making them distance depending, which means that w_i is chosen large if the solution $\mathbf{z}$ of (4) is close to the i-th LT, else the weight should be small. For more details on how to chose weights, see [8].

3 Proper Orthogonal Decomposition Combined with Missing Point Estimation

The idea behind POD is to directly project the original nonlinear system (1) onto some subspace with smaller dimension. As this, however, does not lead to the reduction of the computational time, MPE is used to speed up the simulation.

3.1 "Classical" POD

The POD projection basis V_{POD} is an orthonormal basis, which is derived from the collected "snapshots" at the time points t_i:

$$X = [\mathbf{x}(t_1) \ldots \mathbf{x}(t_s)] \tag{5}$$

The POD basis is found from the SVD of X: $X = U\Sigma W^\top$ with $U = [u_1, \ldots, u_n] \in \mathbb{R}^n$, $\Sigma \in \mathbb{R}^{n \times s}$ and $W \in \mathbb{R}^s$, as $V_{\text{POD}} = [u_1, \ldots, u_r]$ with $r \ll n$. Finally the original system is replaced by the following Galerkin projection

$$\frac{d}{dt} V_{\text{POD}}^\top \mathbf{q}(V_{\text{POD}}\mathbf{z}) + V_{\text{POD}}^\top \mathbf{j}(V_{\text{POD}}\mathbf{z}) + V_{\text{POD}}^\top B\mathbf{u}(t) = 0. \tag{6}$$

3.2 Missing Point Estimation

In the projection schemes, usually the original numerical model is projected onto the chosen subspace. In the case of linear systems, i.e. when $\mathbf{q}(\mathbf{x}) = C\mathbf{x}$ and $\mathbf{j}(\mathbf{x}) = G\mathbf{x}$ the projections $V_{\mathrm{POD}}^{\mathrm{T}}\mathbf{q}(V_{\mathrm{POD}}\mathbf{z})$ and $V_{\mathrm{POD}}^{\mathrm{T}}\mathbf{j}(V_{\mathrm{POD}}\mathbf{z})$ can be computed "in advance" and will deliver the matrices of the reduced system (as in (3)). For the nonlinear systems however, the projection requires the complete evaluations of $\mathbf{q}$ and $\mathbf{j}$ and hence, the solution of (6) will not be faster than the solution of (1). In order to speed it up, a so called missing point estimation can be applied. Assume that:

$$V_{\mathrm{POD}} \approx P^{\top}P\tilde{V}_{\mathrm{POD}}, \qquad (7)$$

where $P \in \{0,1\}^{g \times n}$ is a selection matrix with $PP^{\top} = I_g$. Now introduce the restricted basis $V_{\mathrm{MPE}} = P\tilde{V}_{\mathrm{POD}}$. Then $V_{\mathrm{POD}}^{\mathrm{T}}\mathbf{q}(V_{\mathrm{POD}}\mathbf{z}) \approx \tilde{V}_{\mathrm{POD}}^{\mathrm{T}}P^{\top}P\mathbf{q}$ $(P^{\top}P\tilde{V}_{\mathrm{POD}}\mathbf{z}) = V_{\mathrm{MPE}}^{\mathrm{T}}P\mathbf{q}(P^{\top}V_{\mathrm{MPE}}\mathbf{z})$ and similar for j. Hence, only g elements of $\mathbf{q}$ and $\mathbf{x}$ have to be evaluated, which is much cheaper than evaluating $\mathbf{q}$ and $\mathbf{j}$ if $g \ll n$.

We use an iterative version of the greedy algorithm [5] in order to find a selection matrix P with minimal dimension g, such that

$$\mathrm{cond}(V_{\mathrm{POD}}^{\top}P^{\top}PV_{\mathrm{POD}}) < \mathrm{TOL} \qquad (8)$$

is fullfiled.

4 Numerical Results

We considered the academic diode chain model shown in Fig. 1, which is described through the following equations:

$$V_1 - U_{\mathrm{in}}(10^9 t) = 0,$$
$$i_E - g(V_1, V_2) = 0,$$
$$g(V_1, V_2) - g(V_2, V_3) - C\dot{V}_2 - \tfrac{1}{R}V_2 = 0,$$
$$\vdots$$
$$g(V_{N-1}, V_N) - g(V_N, V_{N+1}) - C\dot{V}_N - \tfrac{1}{R}V_N = 0,$$
$$g(V_N, V_{N+1}) - C\dot{V}_{N+1} - \tfrac{1}{R}V_{N+1} = 0,$$

$$g(V_a, V_b) = \begin{cases} (I_s e^{\frac{V_a - V_b}{V_T}} - 1) & \text{if } V_a - V_b > 0.5 \\ 0 & \text{otherwise} \end{cases}$$

$$U_{\mathrm{in}}(t) = \begin{cases} 20 & \text{if } t \leq 10 \\ 170 - 15t & \text{if } 10 < t \leq 11 \\ 5 & \text{if } t > 11 \end{cases}$$

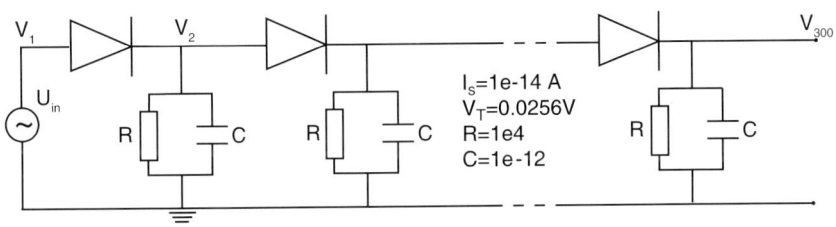

Fig. 1. Structure of the test circuit

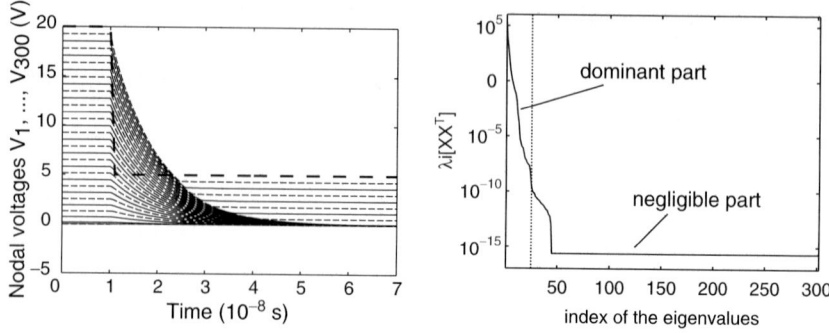

Fig. 2. Numerical solution of the full-scale nonlinear diode chain model (*left*) and the eigenvalues of the correlation matrix $\frac{1}{n}XX^{\mathrm{T}}$ (*right*)

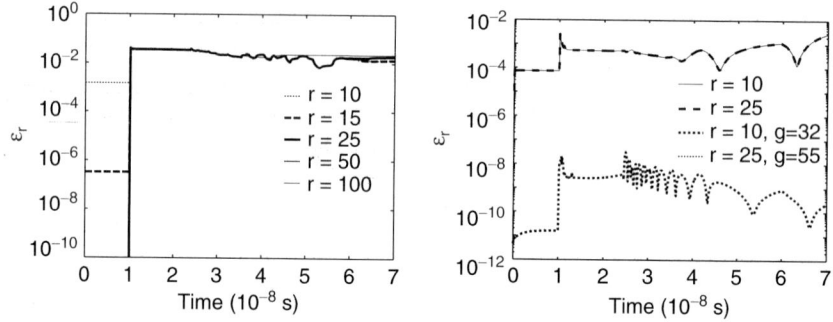

Fig. 3. Relative errors over all nodes for the reduced models created by TPWL (*left*) and by POD (*right*)

Figure 2 (left) shows the numerical solution (nodal voltage in each node) of the original model, computed by the Euler Backward method with fixed step sizes of 0.1 ns. It further indicates (right) the redundancy of the model, as most of the eigenvalues of the correlation matrix $\frac{1}{n}XX^{\mathrm{T}}$ can be neglected. Figure 3 shows the relative errors over all nodes in the time interval [0, 70 ns], defined as $\varepsilon_r = \frac{||Vz-x||}{||x||}$, for the reduced models of different orders constructed by TPWL (left) and POD (right). For TPWL the relative error is most of the time lower then the chosen error bound $\varepsilon = 0.025$. Furthermore, for higher order reduced models, a smaller number of LTs has been used than for the reduced models with lower order, as the local systems with higher orders are more accurate. For, e.g. a reduced model of order 100 we have used 42 LTs and for smaller reduced models 60 LTs. The POD models are, as expected, more accurate, but much slower to simulate than the TPWL models (see the corresponding extraction and simulation times in Table 1). A significant speed up has been achieved by combining the POD with MPE.

Table 1. Comparison of extraction and simulation times in seconds

Model	r	Extr. time	Sim. time	Model	r	g	Extr. time	Sim. time
Original	302	0	142	POD	10	302	142	168
TPWL	10	290	1.1	POD	25	302	142	182
TPWL	25	285	1.5	POD + MPE	10	32	146	74
TPWL	50	206	2.3	POD + MPE	25	55	151	123

5 Conclusion and Outlook

The TPWL method seems to be a promising technique to reduce the simulation time for nonlinear DAE systems. It's main advantage is the application of well-developed linear model reduction techniques. The POD method delivers reduced models which are more accurate but also much more expensive to compute. Hence, the missing point estimation is necessary to achieve a reduction of simulation time at all. Both techniques offer a good starting point for further research on MOR of non-linear dynamical systems.

Finally, we would like to thank Dr. B. Tasić for his help with the diode chain model and to acknowledge the EU support through the COMSON RTN project.

References

1. Antoulas, A. C.: Approximation of Large-Scale Dynamical Systems. Society for Industrial and Applied Mathematics, (2005)
2. Freund, R. W.: Krylov-subspace methods for reduced order modeling in circuit simulation. Journal of Computational and Applied Mathematics, Vol. 123, pp. 395-421, (2000)
3. Rewienski, M., White, J.: A Trajectory Piecewise-Linear Approach to Model Order Reduction and Fast Simulation of Nonlinear Circuits and Micromachined Devices. In: Proc. of the Int. Conf. on CAD, 252–7 (2001)
4. Astrid, P.: Reduction of process simulation models: a proper orthogonal decomposition approach. PhD dissertation, Eindhoven University of Technology, Department of Electrical Engineering, (2004)
5. Astrid, P., Verhoeven, A.: Application of Least Squares MPE technique in the reduced order modeling of electrical circuits. In: Proceedings MTNS, Kyoto (2006)
6. Phillips, J., Silvera, L.M.: Poor Man's TBR: A simple model reduction scheme, IEEE transactions on computer-aided design of integrated circuits and systems. Vol. 14 No. 1, (2005)
7. Odabasioglu, A., Celik, M., Pileggi, L.T.: PRIMA: Passive Reduced-order Interconnect Macromodeling Algorithm, IEEE Transactions on Computer-Aided Design of Integrated Circuits and Systems, Vol. 17, No. 8, 645-654, (1998)
8. Voss, T.: Model reduction for nonlinear differential algebraic equations, M.Sc. Thesis, University of Wupertal, (2005) Journal of Computational and Applied Mathematics, Vol. 123, pp. 395-421, (2000)

Minisymposium "Numerical Methods for Semiconductor Kinetic Equations (COMSON Minisymposium)"

A. Majorana

University of Catania, Italy

In this minisymposium, A. Majorana and V. Romano (U. Catania) present "Comparing kinetic and MEP model of charge transport in semiconductors", and M. Galler and F. Schürrer (TU Graz) present "A Deterministic Solver to the Boltzmann-Poisson System Including Quantization Effects for Silicon-MOSFETs".

Comparing Kinetic and MEP Model of Charge Transport in Semiconductors

A. Majorana and V. Romano

Department of Mathematics and Computer Science, University of Catania, viale
A. Doria 6, 95125 Catania, Italy
majorana@dmi.unict.it, romano@dmi.unict.it

Summary. The distribution function based on the maximum entropy principle
(MEP) in the case of 8 moments is compared with the direct solution of the
Boltzmann transport equation in typical one dimensional benchmark problems for
semiconductor silicon devices. The energy bands are assumed to be described by the
Kane dispersion relation.

1 Kinetic Model

In the semi-classical approximation the charge transport in semiconductors is
described by a Boltzmann equation for the one particle distribution function
$f(t, \boldsymbol{x}, \boldsymbol{k})$

$$\frac{\partial f}{\partial t} + \boldsymbol{v}(\boldsymbol{k}) \cdot \nabla_{\boldsymbol{x}} f - \frac{q}{\hbar} \boldsymbol{E} \cdot \nabla_{\boldsymbol{k}} f = \mathcal{C}[f]. \tag{1}$$

t and $\boldsymbol{x}$ are time and space coordinates and $\boldsymbol{k}$ the crystal momentum of the
electron.

The electron velocity $\boldsymbol{v}(\boldsymbol{k})$ is related to the electron energy $\mathcal{E}(\boldsymbol{k})$ by the
relation

$$\boldsymbol{v}(\boldsymbol{k}) = \frac{1}{\hbar} \nabla_{\boldsymbol{k}} \mathcal{E}(\boldsymbol{k}).$$

In general, the expression of $\mathcal{E}(\boldsymbol{k})$ (the so called band structure) depends
on the material and is very complicated. Reasonable approximation for the
applications is *Kane's dispersion relation* which takes into account the non-
parabolicity at high energies

$$\mathcal{E}(\boldsymbol{k}) = \frac{1}{1 + \sqrt{1 + 2\frac{\alpha}{m^*}\hbar^2|\boldsymbol{k}|^2}} \frac{\hbar^2|\boldsymbol{k}|^2}{m^*} = \sqrt{\frac{1}{4\alpha^2} + \frac{\hbar^2|\boldsymbol{k}|^2}{2\alpha m^*}} - \frac{1}{2\alpha}, \qquad \boldsymbol{k} \in \mathbb{R}^3$$

where m^* is the effective mass and α the non-parabolicity parameter.

The absolute value of the electron charge is denoted by q. The electric field
$\boldsymbol{E}$ is related to the electron distribution by Poisson's equation

$$\boldsymbol{E} = -\nabla\phi, \qquad \epsilon\Delta\phi = -q(N_D - N_A - n),$$

where ϕ is the electric potential, ϵ the permittivity of the semiconductor, N_D and N_A donor and acceptor density, and n the electron density

$$n(t, \boldsymbol{x}) = \int_{\mathbb{R}^3} f(t, \boldsymbol{x}, \boldsymbol{k}) \, d\boldsymbol{k}.$$

$\mathcal{C}[f]$ is the collision operator, which takes into account scattering of the electrons with acoustical and optical phonons and with impurities [JaLu89]. In the non degenerate case the collision term is written in the linear approximation

$$\mathcal{C}[f](\mathbf{k_A}) = \int_{\mathbb{R}^3} [P(\mathbf{k_B}, \mathbf{k_A})f(\mathbf{k_B}) - P(\mathbf{k_A}, \mathbf{k_B})f(\mathbf{k_A})] \, d\mathbf{k_B}$$

with $P(\mathbf{k_A}, \mathbf{k_B})$ transition rate from the state with wave-vector $\mathbf{k_A}$ to the state with wave-vector $\mathbf{k_B}$. In this article we will consider a silicon semiconductor.

The direct solution of the semiclassical Boltzmann transport equation (BTE) is a daunting computational task and mainly it is based on a stochastic approach (Monte Carlo simulations). Recently deterministic solutions of BTE have been obtained in [Ga05, CGMS06, GM06] by using a fifth order conservative finite difference WENO scheme combined with a third order TVD Runke-Kutta time discretization.

The results give a very accurate description of the electron dynamics in the device. However, the CPU time is still not adequate for CAD purposes in electronics engineering, although some attempts of parallelization [MCM06]. This has prompted the development of macroscopic models.

2 The Maximum Entropy System for Electrons in Semiconductors

Besides the electron density n, other physically relevant macroscopic quantities are

$$\boldsymbol{u} = \frac{1}{n} \int_{\mathbb{R}^3} \boldsymbol{v}(k)f \, d\boldsymbol{k} \qquad \text{the average electron velocity } \boldsymbol{u} \text{ relative to the crystal,}$$

$$W = \frac{1}{n} \int_{\mathbb{R}^3} \mathcal{E}(\boldsymbol{k})f \, d\boldsymbol{k} \qquad \text{the average electron energy,}$$

$$\boldsymbol{S} = \frac{1}{n} \int_{\mathbb{R}^3} \boldsymbol{v}(k)\mathcal{E}(\boldsymbol{k})f \, d\boldsymbol{k} \qquad \text{the flux of energy.}$$

Multiplying (1) with weight functions $\boldsymbol{a} = (a_1, \ldots, a_m)^T$ and integrating over $\boldsymbol{k}$, we obtain equations for the moments

$$\frac{\partial \rho}{\partial t} + \sum_{j=1}^{3} \frac{\partial}{\partial x_j} \langle f\, v_j\, \boldsymbol{a} \rangle = \langle (\mathcal{C}[f] + \gamma \boldsymbol{E} \cdot \nabla_{\boldsymbol{k}} f)\, \boldsymbol{a} \rangle \,, \quad \gamma = q/\hbar, \qquad (2)$$

where $\langle \cdot \rangle$ means $\boldsymbol{k}$ integration. The system would be closed if

$$f(t, \boldsymbol{x}, \boldsymbol{k}) = F(\rho(t, \boldsymbol{x}), \boldsymbol{k}) \,.$$

In the maximum entropy approach $F(\rho, \boldsymbol{k})$ is taken as solution of the problem

$$\text{maximize } H(f) = -\langle f\,(\log f - 1)\rangle \quad \text{with } f \geq 0 \text{ and } \langle f\,\boldsymbol{a}\rangle = \rho \qquad (3)$$

We introduce the Lagrange functional

$$L(f, \boldsymbol{\lambda}) := H(f) - \boldsymbol{\lambda} \cdot (\rho - \langle f\,\boldsymbol{a}\rangle)$$

where $\boldsymbol{\lambda}$ is the vector of Lagrange multipliers. The necessary condition that all directional derivatives vanish in the maximum $f_{\boldsymbol{\lambda}}$ leads to

$$0 = \delta L(f_{\boldsymbol{\lambda}}, \boldsymbol{\lambda}) = (-\log f_{\boldsymbol{\lambda}} + \boldsymbol{\lambda} \cdot \boldsymbol{a})\,\delta f_{\boldsymbol{\lambda}} \quad \text{so that} \quad f_{\boldsymbol{\lambda}} = \exp(\boldsymbol{\lambda} \cdot \boldsymbol{a}).$$

Finally, the Lagrange multipliers $\boldsymbol{\lambda}$ are chosen in such a way $\rho = \langle f_{\boldsymbol{\lambda}}, \boldsymbol{a}\rangle$ are satisfied and one can now close the moment system (2)

$$\frac{\partial \rho}{\partial t} + \frac{\partial}{\partial x_j} \boldsymbol{G}_j(\rho) = \boldsymbol{P}(\rho) \qquad (4)$$

where $\boldsymbol{G}_j(\rho) = \langle F(\rho)\, v_j \boldsymbol{a}\rangle \,, \quad \boldsymbol{P}(\rho) = \langle (\mathcal{C}[F(\rho)] + \gamma \boldsymbol{E} \cdot \nabla_{\boldsymbol{k}} F(\rho))\, \boldsymbol{a}\rangle.$

The solvability of the maximum entropy problem has been proved in [JR05] when the Kane dispersion relation is used.

Assuming $n, \boldsymbol{V}, W, \boldsymbol{S}$ as fundamental variables, in [AR99, RO00, RO01, AMR03] the MEP 8-moment model, which describes the electron as a heat-conducting gas, has been deduced and investigated obtaining the relations between Lagrange multipliers and basic moments by an expansions in terms of a small anisotropy parameter.

The corresponding MEP distribution reads

$$f_{ME} = \exp\left(-\lambda - \boldsymbol{\lambda}^V \cdot \boldsymbol{v} - \lambda^W \mathcal{E} - \mathcal{E}\,\boldsymbol{\lambda}^S \cdot \boldsymbol{v}\right) \qquad (5)$$

with the following characterisation of the moment cone

$$\Lambda = \left\{ \boldsymbol{\lambda} = \left(\lambda, \boldsymbol{\lambda}^V, \lambda^W, \boldsymbol{\lambda}^S\right) : \ \boldsymbol{\lambda} \in \mathbb{R}^8, \lambda^W > 0 \quad \text{and} \quad v_\infty |\boldsymbol{\lambda}^S| < \lambda^W \right\},$$

where $v_\infty = 1/\sqrt{2\alpha m^*}$ is the asymptotic value of $v(\mathcal{E})$. The MC results justify the *ansatz* of small anisotropy and f_{ME} is approximated as

$$f_{ME} \simeq \exp\left(-\frac{1}{k_B}\lambda - \lambda^W \mathcal{E}\right)\left[1 - \left(\boldsymbol{\lambda}^V \cdot \boldsymbol{v} + \mathcal{E}\,\boldsymbol{\lambda}^S \cdot \boldsymbol{v}\right)\right]$$

where λ^W, λ^V and λ^S are functions of W, $\boldsymbol{u}$ and $\boldsymbol{S}$ (see [AMR03] for the explicit relations and details).

After having expressed $\boldsymbol{k}$ in terms of its unit vector $\boldsymbol{l}$ and $\mathcal{E}$, the zeroth and first harmonics of f_{ME} are given by

$$f_{ME}^{(0)}(\mathcal{E}) = \int_{S^2} f_{ME}(\mathcal{E}, \boldsymbol{l}) \, g(\mathcal{E}) \, d\Omega, \quad \boldsymbol{f}_{ME}(\mathcal{E}) = \int_{S^2} \boldsymbol{v}(\mathcal{E}) f_{ME}(\mathcal{E}, \boldsymbol{l}) \, g(\mathcal{E}) \, d\Omega$$

where $g(\mathcal{E}) = \sqrt{2}(m^*)^{3/2} \, \hbar^{-3} \sqrt{\mathcal{E}(1 + \alpha\mathcal{E})}(1 + 2\alpha\mathcal{E})$ is the density of states and $d\Omega$ the element of solid angle (the same symbol f_{ME} has been used for the MEP function expressed in the new variable). The dependence on space and time is through the lagrangian multipliers that in turn depend on the macroscopic variables $n, \boldsymbol{V}, W, \boldsymbol{S}$. Therefore once the balance equations (4), coupled to the Poisson equation, are solved, f_{ME} is obtained as well.

3 Comparison Between the MEP Distribution Function and the Direct Solution of the Boltzmnn Equation

As first test we consider a one dimensional homogeneous silicon device under several values of the applied electric field. In the table we compare the results for the macroscopic velocity and energy for several electric fields.

E (kV cm^{-1})	BTE velocity (10^5 m sec^{-1})	MEP velocity (10^5 m sec^{-1})	BTE energy (eV)	MEP energy (eV)
10	0.059915	0.6579	0.060655	0.06037
20	0.079029	0.8546	0.087296	0.09069
30	0.087372	0.9355	0.11733	0.1237
40	0.091356	0.9744	0.14899	0.1581
50	0.093176	0.9964	0.18118	0.2222

The zeroth order harmonics of f_{ME} and of the direct solution of the Boltzmann equation are compared in Fig. 1 for the electric field $E = 10 \, kV \, \text{cm}^{-1}$ and $E = 50 \, kV \, \text{cm}^{-1}$.

As second test we consider a one dimensional $n^+ - n - n^+$ silicon diode, with total length 0.6 micron, channel length 0.4 micron, doped as follows

$$N_D(x) - N_A(x) = \begin{cases} 10^{18} \ \text{cm}^{-3} \ \text{in the n+ region} \\ 10^{16} \ \text{cm}^{-3} \ \text{in the n region} \end{cases}$$

under an applied voltage of 2 Volt.

The zeroth and first order harmonics of f_{ME} and the direct solution of the Boltzmann equation are compared in Figs. 2 and 3 along the device. A reasonable agreement is observed.

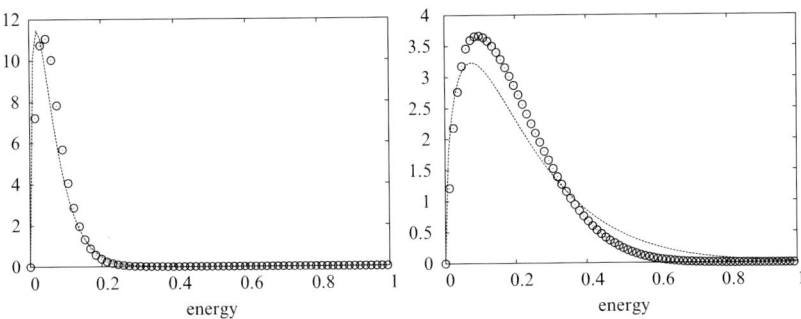

Fig. 1. The zeroth order harmonic for the solution of the BTE (-) and the MEP (o) solution in the bulk case for $E = 10$ kv cm^{-1} (*left*) and $E = 50$ kV cm^{-1} (*right*)

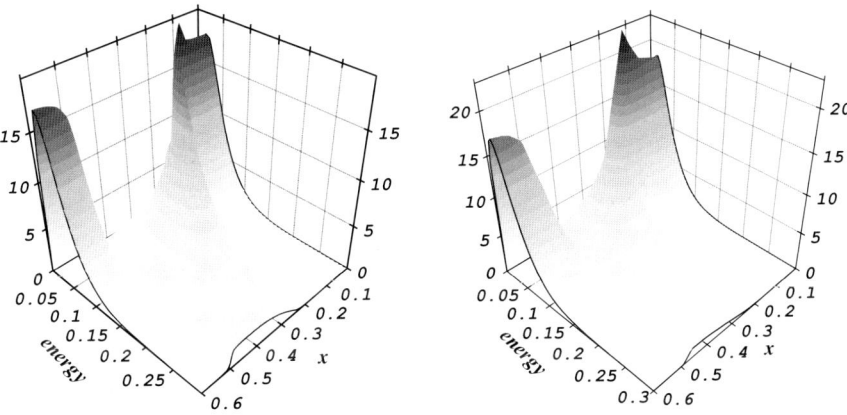

Fig. 2. The zeroth order harmonic for the solution of the BTE (*left*) and the MEP solution (*right*) along the device vs. $\mathcal{E}$

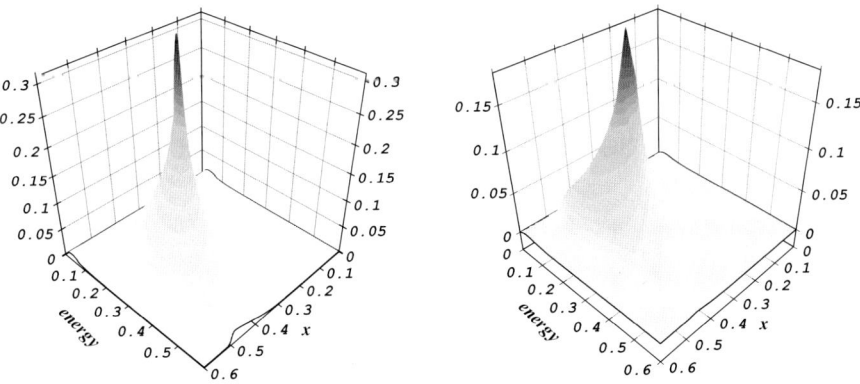

Fig. 3. First order harmonic for the solution of the BTE (*left*) and the MEP solution (*right*)

Acknowledgements

The authors acknowledge the financial support by the EU Marie Curie RTN project **COMSON** grant n. MRTN-CT-2005-019417.

References

[JaLu89]   Jacoboni C., Lugli P.: The Monte Carlo Method for Semiconductor Device Simulation. Springer (1989)

[Ga05]   Galler, M.: Multigroup equations for the description of the particle transport in semiconductors. In: Series on Advances in Mathematics for Applied Sciences Vol. 70, World Scientific Publishing (2005)

[CGMS06]   Carrillo, J.A., Gamba, I.M., Majorana, A., and Shu, C.-W.: 2D semiconductor device simulations by WENO-Boltzmann schemes: efficiency, boundary conditions and comparison to Monte Carlo methods. J. Comput. Phys. **214**, 55–80 (2006)

[GM06]   Galler, M., and Majorana, A.: Deterministic and stochastic simulation of electron transport in semiconductors. To appear in Transp. Theory and Stat. Phys., 6th MAFPD (Kyoto) special issue (2006).

[MCM06]   Mantas, J.M., Carrillo, J. A., and Majorana, A.: Parallelization of WENO-Boltzmann Schemes for Kinetic Description of 2D Semiconductor Devices. In: Scientific Computing in Electrical Engineering, Mathematical in Industry Vol. 9, Springer (2006)

[JR05]   Junk, M., and Romano, V.: Maximum entropy moment system of the semiconductor Boltzmann equation using Kane's dispersion relation. Cont. Mech. Thermodyn. **17**, 247–267 (2005)

[AR99]   A.M. Anile, A.M., and Romano, V.: Non parabolic band transport in semiconductors: closure of the moment equations. Continuum Mech. Thermodyn. **11**, 307–325 (1999).

[RO00]   Romano, V.: Non parabolic band transport in semiconductors: closure of the production terms in the moment equations. Continuum Mech. Thermodyn. **12**, 31–51 (2000).

[RO01]   Romano, V.: Nonparabolic band hydrodynamical model of silicon semiconductors and simulation of electron devices. Math. Meth. Appl. Sciences **24**, 439–471 (2001).

[AMR03]   Anile, A.M., Mascali, G., and Romano, V.: Recent developments in hydrodynamical modeling of semiconductors. In: Mathematical Problems in Semiconductor Physics, Lecture Notes in Mathematics 1832, Springer (2003).

A Deterministic Solver
to the Boltzmann-Poisson System Including
Quantization Effects for Silicon-MOSFETs

M. Galler and F. Schürrer

Institute of Theoretical and Computational Physics, Graz University
of Technology, Petersgasse 16, A-8010 Graz, Austria
galler@itp.tugraz.at, schuerrer@itp.tugraz.at

Summary. We present a deterministic solver to the Boltzmann-Poisson system
for simulating the electron transport in silicon MOSFETs. This system consists
of the Boltzmann transport equations (BTEs) for free electrons and for the two-
dimensional electron gas (2DEG) formed at the Si/SiO2 interface. Moreover, the
Poisson equation is coupled to the BTEs. Eigenenergies and wave functions of the
2DEG are dynamically calculated from the Schrödinger-Poisson system. Numerical
studies prove the applicability and the efficiency of the proposed numerical technique
for simulating ultrasmall semiconductor devices.

1 Introduction

The increasing miniaturization in semiconductor technology leads to semiconductor devices, which are strongly influenced by quantum mechanical effects.
In such devices, the motion of electrons is often confined in quantum wells.
Hence, these carriers are termed two-dimensional electron gas. The consideration of 2DEGs is very important, since they occur in modern heterostructure
devices as well as in conventional silicon MOSFETs [1].

In this paper, we simulate the electron transport in a silicon MOSFET
including the quantization effects. So far, such simulations on a kinetic level
have mainly been performed by means of Monte Carlo techniques [2,3]. However, our work concerning the 2DEG transport in homogeneous channels [4,5]
have shown that deterministic solvers to the transport equations are an interesting alternative to the usual stochastic schemes. Hence, we adopt these
schemes in order to properly describe the 2DEG dynamics in inhomogeneous
channels as they are found in realistic devices.

This paper is organized as follows. In Sect. 2, we present the Boltzmann–
Poisson-Schrödinger (BPS) system, on which our simulations are based.
Section 3 deals with the numerical scheme, which is used to solve this sys-
tem. Finally, some numerical results are given in Sect. 4.

2 Boltzmann-Poisson-Schrödinger System

In this section, we shortly summarize the basic equations of our Si-MOSFET simulation. A schematic illustration of the considered device is displayed in Fig. 1. We assume that notable quantum effects only occur in a domain between the suitably chosen positions x_1 and x_2. In this region, we solve the effective mass Schrödinger equation,

$$\left[-\hbar^2 \partial_{yy} + 2m^* U(t, x, y)\right] \varphi_\nu(t, x, y) = 2m^* \varepsilon_\nu(t, x) \varphi_\nu(t, x, y), \qquad (1)$$

in order to obtain the eigenvalues ε_ν and envelope wave functions φ_ν as functions of the position (x, y) at time t. The subband index ranges between $\nu = 1, 2, \ldots, \gamma$, where γ is a chosen integer. The boundary values of φ_ν are set to $\varphi_\nu(x, 0) = 0$ and $\varphi_\nu(x, l_y) = 0$ for $x \in [x_1, x_2]$. The potential energy U in $[0, l_x] \times [0, l_y]$ is related to the electrostatic potential V by $U = -eV$ with the elementary charge e. For determining V, we solve the Poisson equation in the Si and the SiO$_2$ region,

$$\partial_x[\epsilon_0 \epsilon_r \partial_x V(t, x, y)] + \partial_y[\epsilon_0 \epsilon_r \partial_y V(t, x, y)] = -e[N_D(x, y) - n(t, x, y)] \qquad (2)$$

with the dielectric constant $\epsilon_0 \epsilon_r$ and the donor density N_D. The electron density n is found as the zero-order moment of the distribution function of 3D electrons and those of 2D electrons f_ν. The evolution of the distribution functions is governed by Boltzmann equations. For 2D electrons, it reads

$$\partial_t f_\nu + v_x^\nu \partial_x f_\nu - \frac{e}{\hbar} E_x^\nu \partial_{k_x} f_\nu = \sum_\mu \mathcal{C}^{\nu, \mu}[f_\nu], \qquad (3)$$

where k_x is the x-component of the 2D wave vector $\mathbf{k}$. The group velocity of electrons in x-direction is given by $v_x^\nu(\mathbf{k}) = \partial_{k_x} E_\nu(\mathbf{k})/\hbar$ with the energy E_ν related to a 2D electron with wave vector $\mathbf{k}$ in the subband ν. The effective electric field acting on these electrons is obtained from $E_x^\nu(t, x) = \partial_x \varepsilon_\nu(t, x)/e$. The collision term $\mathcal{C}^{\nu, \mu}[f_\nu]$ involves intrasubband scattering ($\nu = \mu$) and intersubband scattering ($\nu \neq \mu$) by acoustic and optical phonons and it includes electron degeneracy. For more details, we refer to [1, 3, 4, 6].

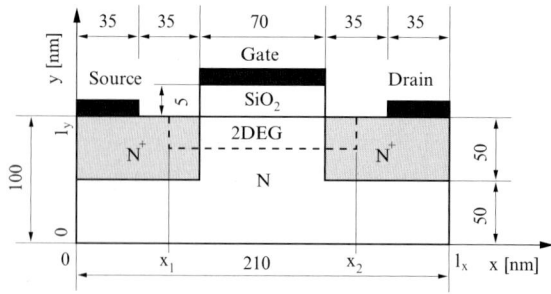

Fig. 1. Schematic illustration of the simulated Si-MOSFET

The equations (1), (2), (3) together with a Boltzmann equation for 3D electrons form the BPS system. For a realistic simulation of a MOSFET, strongly influenced by quantum effects, they must be solved self-consistently.

3 Numerical Scheme

To begin with, we specify the region of the phase space in which 2D transport takes place. Therefore, we introduce a border energy E_{b}, which is set to $E_{\mathrm{b}} = \varepsilon_\gamma$. All of the electrons with total energy lower than E_{b} are described by BTEs for 2DEG, the others by means of a 3D BTE. In real space, the 2D region is given by $[x_1, x_2] \times [E_{\mathrm{b}}(x) = U(x, y), l_y]$.

For numerically approximating the 2D BTEs, we transform them into a conservative form with the change of variables $\mathbf{k}^\nu(E, \varphi) = k^\nu(E)(\cos\varphi, \sin\varphi)$, where $k^\nu(E) = [2m^*(E - \varepsilon_\nu)]^{1/2}/\hbar$ and $E \in [\varepsilon_\nu, E_{\mathrm{b}}], \varphi \in [0, 2\pi]$. Hence, we express the 2D wave vector in terms of energy E and angle φ based on a parabolic energy-momentum dispersion relation. Rewriting Eq. (3) with $f_\nu = f_\nu(t, x, E, \varphi)$ leads to

$$\partial_t (Z_\nu f_\nu) + \partial_x (a_1 Z_\nu f_\nu) + \partial_E (a_2 Z_\nu f_\nu) + \partial_\varphi (a_3 Z_\nu f_\nu) = \sum_\mu \tilde{C}^{\nu,\mu}[Z_\nu f_\nu] \quad (4)$$

with $Z_\nu(E) = m^* \Theta(E - \varepsilon_\nu)/(2\pi\hbar)^3$ and the Heaviside step function Θ. The function a_2^ν is defined, for instance, by $a_2^\nu = -e\hbar E_x^\nu(t, x) k^\nu(E) \cos\varphi/m^*$. Next, we introduce the discretization of energy $E_i^\nu = \varepsilon_\nu + i\Delta E, i = 0, 1, \ldots, N^\nu$, and angle $\varphi_j = j\Delta\varphi, j = 0, 1, \ldots, R, \Delta\varphi = \pi/2R, R \in \mathbb{N}$ and the ansatz

$$Z_\nu(E) f_\nu(t, x, E, \varphi) \approx \sum_{i=1}^{N^\nu} \sum_{k=1}^{R} n_{ij}^\nu(t, x) Z_\nu(E) \chi_i^E(E) \chi_j^\varphi(\varphi). \quad (5)$$

Here, $N^\nu(x)$ are integer numbers so that $E_{N^\nu - 1} < E_{\mathrm{b}} \leq E_{N^\nu}$, while $\chi_i^E(E)$ and $\chi_j^\varphi(\varphi)$ are characteristic functions defined, for instance, by $\chi_i^E = \Lambda E^{-1}$ for $E \in [E_{i-1}, E_i]$ and $\chi_i^E = 0$ otherwise. The ansatz (5) is inserted in the BTE (4) and the result is successively integrated over the cells $\mathcal{Z}_{ij} = [E_{i-1}, E_i] \times [\varphi_{j-1}, \varphi_j]$. This procedure together with an upwind scheme and a MinMod slope limiter for determining the coefficients n_{ij}^ν at cell boundaries leads to a set of multigroup equations, which are partial differential equations in x and t. The spatial derivative is approximated by a fifth-order WENO scheme [7] and the resulting set of ordinary differential equations is solved by a TVD Runge-Kutta method with time steps so that the CFL condition is fulfilled.

The BTE of 3D electrons based on a non-parabolic dispersion law is solved in a similar way. For details, we refer to [6, 8]. Transitions between the 2D and 3D electrons by scattering and advection are handled as follows. We treat the transfer between them classically, although our model contains sharp borders between 2D and 3D region. In a real device, such sharp borders, which normally demand a quantum mechanical description, are not found, since 3D behaviour changes gradually into 2D behavior. Hence, we demand classical conversation of electron number ($\Delta n_{3D} = -\Delta n_{2D}$), momentum ($k_x^{3D} = k_x^{2D}, k_z^{3D} = k_z^{2D}$) and total energy $E_{3D} = E_{2D}$, when transferring the outflow of one region into the inflow of the other.

4 Results

In this section, we present some results of our simulation of a Si-MOSFET. The used geometry is found in Fig. 1. The donor densities are set to $N=10^{11}$ cm^{-3} and $N^+=10^{18}$ cm^{-3}. The applied voltages are set to $V_S=0$ at source, $V_G=1$ V at gate and $V_D=1$ V at drain. The material parameters used are the same as given in [8]. The 2DEG is built up by $\gamma = 10$ subbands. Starting from an initial equilibrium distribution, we solved the BPS system up to 5 ps, where the stationary state is almost reached.

In Fig. 2, we display the total electron density in the simulated Si-MOSFET. The high electron density at the Si/SiO$_2$ interface forming the conduction channel is clearly visible. An interesting detail can be seen in the given contour plot of the electron density. We observe that the maximum density is not found at the Si/SiO$_2$ interface, as it would be the case for a semiclassical simulation, but it is shifted inside the device.

Figure 3 shows the distribution function of 3D electrons averaged over one angle and that of electrons in the first 2DEG subband. In a pure 3D simulation, the 3D distribution would also be filled at low energies. This part of the 3D distribution function is zero in our case, since these low energy electrons are treated as 2D electrons. Hence, they are contained in f_ν. Finally, we note that the presented distribution functions are very smooth as they can only be obtained with a deterministic scheme at affordable computational costs.

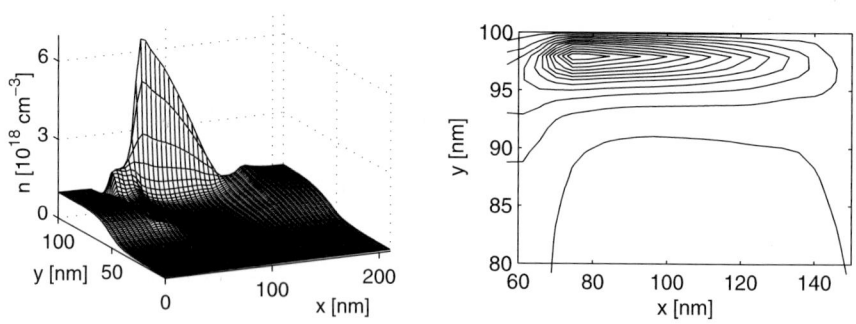

Fig. 2. Total electron density n (*left*) and a contour plot of n vs. position (x, y)

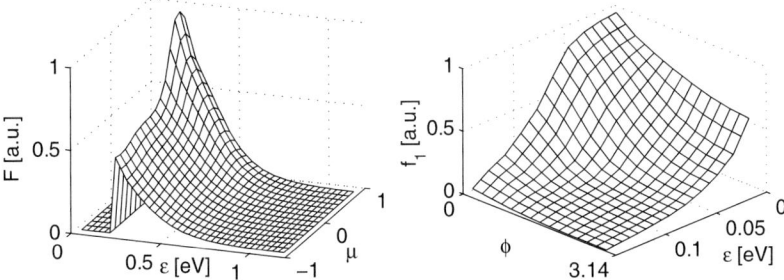

Fig. 3. Distributions of 3D electrons F (*left*) and 2D-electrons in the first subband f_1 (*right*) vs. kinetic energy ϵ and angles μ and φ at $x = 105$ nm and $y = 98$ nm

5 Conclusion

We simulate the electron transport in silicon MOSFETs by means of a deterministic solver to the BPS system. Numerical studies prove the applicability and the efficiency of the proposed numerical technique. In order to allow a more realistic device simulation, our scheme should be extended in the future. For instance, non-spherical energy bands and further scattering mechanisms like impurity scattering and surface roughness scattering must be incorporated. In addition, hole transport should also be regarded. However, the results we have obtained so far are promising, and we believe that our approach will become an important tool for sophisticated semiconductor device simulation.

Acknowledgements

This work has been supported by the Fonds zur Förderung der wissenschaftlichen Forschung, Vienna, under contract number P17438-N08.

References

1. Lundstrom, M.: Fundamentals of Carrier Transport. Cambridge University Press, Cambridge (2000)
2. Fischetti, M. V., Laux, S. E.: Monte Carlo study of electron transport in silicon inversion layers. Phys. Rev. B **48**, 2244-2274 (1993)
3. Tomizawa, K.: Numerical Simulation of Submicron Semiconductor Devices. Artech House, Boston (1993)
4. Galler, M., Schürrer F.: A deterministic solver for the transport of the AlGaN/ GaN 2D electron gas including hot-phonon and degeneracy effects. J. Comput. Phys., **210**, 519-534 (2005)
5. Ertler, C., Schürrer, F.: A deterministic study of hot phonon effects in a 2D electron gas channel formed at an AlGaN/GaN heterointerface. J. Comput. Electron., **5**, 15-26 (2006).

6. Galler, M.: Multigroup Equations for the Description of the Particle Transport in Semiconductors. World Scientific, Singapore (2005)
7. Jiang, G., Shu, C.-W.: Efficient implementation of weighted ENO schemes. J. Comput. Phys., **126**, 202–228 (1996)
8. Galler, M., Majorana, A., Schürrer, F.: A multigroup-WENO solver for the non-stationary Boltzmann-Poisson system for semiconductor devices. In: Anile, A. M., Ali, G., Mascali, G. (ed) Scientific Computing in Electrical Engineering. Springer, New York (2004)

Minisymposium of the ECMI SIG "Shape and Size in Medicine, Biotechnology and Material Sciences"

Alessandra Micheletti

Department of Mathematics, Università degli Studi di Milano

Thanks to the development of information technologies, the last decade has seen a considerable growth of interest in the statistical theory of shape and its application to many and diverse scientific areas.

In applications, bodies rarely have exactly the same shape within measurement error; hence randomness of shapes need to be taken into account.

Often the diagnosis of a pathology, or the description of a biological process mainly depend on the shapes present in images of cells, organs, biological systems, etc., and mathematical models which relate the main features of these shapes with the correct outcome of the diagnosis, or with the main kinetic parameters of a biological systems are still not present.

From the mathematical point of view, shape analysis uses a variety of mathematical tools from differential geometry, geometric measure theory, stochastic geometry, etc. As far as applications are concerned, we emphasize here topics which are relevant in medicine and biotechnology. We deal with direct and inverse problems.

Among direct problems, spatio-temporal pattern formation deals with the analysis of how patterns are created and developed in biology. An example of application to growth of plants is here presented.

Among inverse problems, some stochastic geometric techniques of shape analysis and mathematical morphology are here proposed to measure in a quantitative way the random variability of objects and perform automatic classification. Recent methods of image analysis include optical imaging of objects in turbid media, which can be used as a non-invasive technique for the detection of tumors in the body. An example of study of x-ray tomographic data (Schlieren data) is here also presented.

Size Functions Applied to the Statistical Shape Analysis and Classification of Tumor Cells

Alessandra Micheletti[1] and Gabriel Landini[2]

[1] Department of Mathematics, Università degli Studi di Milano, Via C. Saldini, 50, 20133 Milano, Italia
`Alessandra.Micheletti@unimi.it`
[2] Oral Pathology Unit, School of Dentistry, The University of Birmingham, St. Chad's Queensway, Birmingham, B4 6NN, England, U.K.
`G.Landini@bham.ac.uk`

Summary. Here the Theory of Size Functions is introduced and joined to some statistical techniques of discriminant analysis, to perform automatic classification of families of random shapes. The method is applied to the classification of normal and malignant tumor cell nuclei, described via their section profiles. The results here reported are compared with other techniques of shape analysis, already applied to the same data, showing some improvements.

1 Introduction

The solution of the problem of describing a "shape" via functions taking values in a finite dimensional space, without loosing important information, is essential for a mathematical and statistical approach. Recently new geometrical descriptors of shapes, called *size functions*, have been proposed [4]. These functions are able to capture "globally" the topological and geometrical features of an object, differently from landmarks [2, 8] (usually these are specific points, angles, distances, etc., on the object, chosen by an expert) which are widely used in literature but whose results in a statistical context are strongly dependent on their choice, leading to a sort of subjective quantitative analysis.

Size functions depend on the choice of a *measuring function* and usually only a small number of choices can lead to different statistical results.

The theory of size functions has been developed mainly in a deterministic framework. Here, we join this theory with randomness and with suitable statistical techniques in order to obtain good results in random shape recognition and classification.

This technique has been here applied to the automatic classification of normal and tumor cell nuclear profiles, as observed in electron microscope sections of human epithelial tissue samples.

2 Size Functions and Shape Description

Let $\mathcal{M}$ be a finite union of compact arcwise connected and locally arcwise connected subsets of an Euclidean space and let $\varphi : \mathcal{M} \to \mathbb{R}$ be a continuous function, called *measuring function*. The pair $(\mathcal{M}, \varphi)$ denotes in a formal way the shape of the object $\mathcal{M}$.

For every $x \in \mathbb{R}$ let $\mathcal{M}\langle \varphi \leqslant x \rangle$ denote the set $\{P \in \mathcal{M} : \varphi(P) \leq x\}$. Thus, we can introduce the following definition [4].

Definition 1. *Consider the function* $l_{(\mathcal{M},\varphi)} : \mathbb{R} \times \mathbb{R} \to \mathbb{N} \cup \{+\infty\}$ *defined by setting* $l_{(\mathcal{M},\varphi)}(x, y)$ *equal to the number of equivalence classes into which the set* $\mathcal{M}\langle \varphi \leqslant x \rangle$ *is divided by the relation of* $\langle \varphi \leqslant y \rangle$-*homotopy, where two points* $P, Q \in \mathcal{M}$ *are* $\langle \varphi \leqslant y \rangle$-*homotopic if and only if either* $P = Q$ *or a continuous path* $\gamma : [0, 1] \to \mathcal{M}$, *joining* P *and* Q, *exists in* $\mathcal{M}$ *such that* $\varphi(\gamma(t)) \leq y$ *for every* $t \in [0, 1]$. *We shall call* $l_{(\mathcal{M},\varphi)}$ *the* size function *associated with the pair* $(\mathcal{M}, \varphi)$.

The size function $l_{(\mathcal{M},\varphi)}$ describes the shape of $\mathcal{M}$ through information given by φ, whose choice depends on the specific problem we are interested in. An important property of size functions is that they inherit the invariance properties, if any, of the chosen measuring functions.

The size function $l_{(\mathcal{M},\varphi)}$ conveys relevant information about the pair under study only in the half-plane $x < y$. Furthermore for $x < y$ size functions have a simple geometric interpretation: in such a case $l_{(\mathcal{M},\varphi)}(x, y)$ is equal to the number of arcwise connected components of $\mathcal{M}\langle \varphi \leqslant y \rangle$ containing at least one point of $\mathcal{M}\langle \varphi \leqslant x \rangle$.

An example of size function is illustrated in Fig. 1. We show the size function of the contour of an ellipse $\mathcal{M}$ with respect to the measuring function $\varphi(z)$ which associates to each point $z \in \mathcal{M}$ its distance from the center of mass of $\mathcal{M}$. More precisely, we represent the domain of $l_{(\mathcal{M},\varphi)}$ with its discontinuities: the number displayed in each region of the domain denotes the value of the size function in that region.

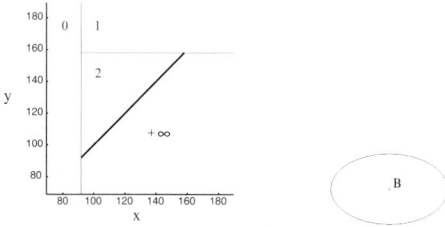

Fig. 1. Size function of an ellipse with respect to the distance from the center of mass

2.1 Cornerpoints and Cornerlines

The discontinuities of size functions are related to specific points and vertical lines in the real plane, each one with a multiplicity, called *cornerpoints* and *cornerlines*, respectively. We refer to [3,4] for further details.

The abscissa of every cornerline corresponds to the global minimum taken by φ on an arcwise connected component of $\mathcal{M}$, while the coordinates of the cornerpoints are couples of critical values for φ.

It can be proven that all and only the discontinuity points of a size function are generated by its cornerpoints and cornerlines (whose number is a.s. finite); viceversa, cornerpoints and cornerlines with their multiplicities uniquely determine the value of $l_{(\mathcal{M},\varphi)}$ almost everywhere, so that they contain all information conveyed by the size function about the shape under study.

2.2 Matching Distance

In order to compare different shapes a suitable distance between the representative size functions must be introduced. An idea is indeed to compare size functions by measuring the *cost* of moving and overlapping the cornerpoints and cornerlines of one size function to those of the other one, by minimizing the longest movement. Since, in general, the number of cornerpoints of the two size functions is different, we also enable the cornerpoints to be transported onto the points of the diagonal Δ with equation $y = x$. This leads to the definition of the *matching distance* between size functions (see [1] for the definition and further details).

The matching distance between size functions has the fundamental property of stability with respect to small perturbations of the shape. This result allows us to use size functions as *robust* shape descriptors in presence of randomness. The presence of randomness on a shape is revealed in the domain of the corresponding size function by small displacements of its cornerpoints and cornerlines and by the presence of small triangles near the diagonal Δ. As an example, in Fig. 2 we show an ellipse perturbed with noise and its size function with respect to the distance from the center of mass.

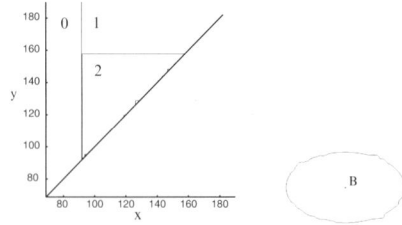

Fig. 2. Size function of a deformed ellipse with respect to the distance from the center of mass

Thus the cornerpoints which describe (with the cornerlines) the main characteristics of the shape under study are those standing "sufficiently far" from Δ. Since, for sufficiently regular shapes, they are (a.s.) in a finite number [4], shape analysis and classification are then reduced to the statistical study of the location of finite sets of points and lines in the real plane.

3 Application to the Classification of Tumor and Normal Cells

The nearest neighbor method for nonparametric discriminant analysis (see [6] for an introduction to the method) has been applied to the automatic recognition of tumor and normal cells, by analyzing the shape of their nuclear profiles via the size functions. Usually the pathologist classifies the cells by direct visual analysis of the nuclear profile, according to the symmetry, smoothness, and regularity of the contour, which are significantly different between the majority of the elements of the two classes, tumor and normal.

We used a data set consisting of 1,337 images of nuclear profiles of cells coming from electron microscope sections of oral epithelium samples; 637 out of the analyzed images came from seemingly normal tissues free of pathology, while 700 nuclear profiles came from invasive oral squamous cell carcinoma (the commonest malignant tumor in oral tissues). The same data had already been analyzed in [7] using three asymptotic fractal parameters as descriptors of nuclear shape. Discriminant analysis applied to these descriptors led to a correct classification of 78.8% of the cells (88% of the normal cells and 70.2% of the tumor cells). In [8] the data were analyzed using landmarks located on points of maximum curvature of the contour, leading to a correct recognition of the 76.6% of the cells.

Here we have applied the nearest neighbor method to the size functions, computed using the distance from the centroid of the nuclei as measuring function. We used the matching distance to build the discriminant function. The computation of the matching distance between two size functions having n cornerpoints involves $n!$ comparisons of couples of cornerpoints [4]; this leads rapidly to an explosion of the computational costs when the number of cornerpoints due to noise increases. Thus, we filtered out the noise by considering for each size function only the m cornerpoints which are the furthest apart from the diagonal. The constant m is a bandwidth which must be properly chosen. The nearest neighbor method is also based on the choice of another bandwidth, the number K of nearest neighbors which are considered for each new object that must be classified. The values of K and m have here been chosen empirically on the basis of the results obtained on our already classified data set.

The results obtained for different choices of the bandwidths are reported in Table 1. The results look better than the ones obtained in [8] using landmarks, and are comparable with the ones obtained in [7] using the asymptotic

542   A. Micheletti and G. Landini

Table 1. Percentages of correct classification with the nearest neighbors method

Chosen bandwidths	Tumor nuclei (%)	Normal nuclei (%)	Total (%)
$K = 17, m = 3$	71.21	80.79	76.00
$K = 15, m = 3$	70.53	83.28	76.90
$K = 11, m = 3$	71.62	82.55	77.08
$K = 15, m = 2$	70.53	83.72	77.12
$\mathbf{K = 11, m = 2}$	**72.31**	**82.55**	**77.43**
$K = 7, m = 2$	71.62	80.65	76.13

K = number of considered nearest neighbors; m = number of cornerpoints used for each size function. The leave one out method has been used to compute the percentage of correct recognition. The best result is reported in bold

fractal dimension; in particular an improvement can be observed in the correct recognition of tumor cells.

Further improvements could perhaps be obtained by changing the measuring function. The problem of finding a method for the automatic selection of the best measuring function from a given family is still open and under investigation. The solution of this problem is crucial and may lead to standard procedures for automatic diagnosis.

Acknowledgements

Fruitful discussions are acknowledged by A.M. with Patrizio Frosini and Massimo Ferri, University of Bologna, and with Vincenzo Capasso and Filippo Terragni, University of Milan.

References

1. M. D'Amico, P. Frosini, C. Landi, Natural pseudo-distance and optimal matching between reduced size functions, Preprint, 2005.
2. I. L. Dryden, K. V. Mardia, *Statistical Shape Analysis*, Wiley, New York, 1998.
3. P. Frosini, C. Landi, Size theory as a topological tool for computer vision, *Pattern Recognition and Image Analysis* **9**, pp. 596–603, 1999.
4. P. Frosini, C. Landi, Size functions and formal series, *AAECC* **12**, pp. 327–349, 2001.
5. P. Frosini, M. Pittore, New methods for reducing size graphs, *Intern. J. Computer Math.* **70**, pp. 505–517, 1999.
6. P.A. Lachenbruch, *Discriminant Analysis*, Hafner Press, New York, 1975.
7. G. Landini, J.W. Rippin, Quantification of nuclear pleomorphism using an asymptotic fractal model, *Anal. Quant. Cyt. Hist.*, **18**, pp. 167–176, 1996.
8. A. Micheletti, Statistical shape analysis applied to automatic recognition of tumor cells, *Fractals in Biology and Medicine IV*, G.A. Losa et al. Editors, Birkhauser Verlag, Basel, pp. 165–174, 2005.

A Mathematical Morphology Approach to Cell Shape Analysis

Jesus Angulo

Centre de Morphologie Mathématique - Ecole des Mines de Paris
35, rue Saint-Honoré, 77300 Fontainebleau, France
jesus.angulo@ensmp.fr, http://cmm.ensmp.fr/~angulo

1 Introduction: Context and Motivation

Morphological analysis of cells (size, shape, texture, etc.) is fundamental in quantitative cytology. Anomalies and variations from the typical cell are associated with pathological situations, e.g. useful in cancer diagnosis, in cell-based screening of new active molecules, etc.

Mathematical morphology is a nonlinear image processing technique based on minimum and maximum operations [SER82], i.e. the basic structure is a complete lattice [HEI94]. This contribution aims to apply mathematical morphology operators to quantify the shape of round-objects which present irregularities from an ideal circular pattern. More specifically we illustrate, on the one hand, the application of morphological granulometries for size/shape multi-scale description and on the other hand, the radial/angular decompositions using skeletons in polar-logarithmic representation. We discuss also the aspects related to the properties of invariance of these tools, which is important to describe cell shapes acquired under different magnifications, orientations, etc.

The performance of these mathematical shape descriptors is shown by means of examples from haematological cytology [ANG06] (to classify red blood cells) and from cell-based high-content screening assays [LEM06] (to quantify the populations of hepatocytes), see Fig. 1.

Let $E, \mathcal{T}$ be non-empty sets. We denote by $\mathcal{F}(E, \mathcal{T})$ the power set $\mathcal{T}^E$, i.e. the set of functions from E onto $\mathcal{T}$. Typically for the digital 2-D images $E \subset \mathbb{Z}^2$. Let f be a grey level image, $f(x) \in \mathcal{F}(E, \mathcal{T})$ ($x \in E$ is the pixel position), in the case of discrete image values $\mathcal{T} = \{t_{\min}, t_{\min} + 1, \cdots, t_{\max}\}$ (in general $\mathcal{T} \subset \mathbb{Z}$ or $\mathbb{R}$, or any compact subset of $\mathbb{Z}$ or $\mathbb{R}$) is an ordered set of grey-levels. We suppose here that a binary image f (sometimes denoted X) is a two-levels image, i.e. $t_{\min} = 0$ and $t_{\max} = 1$.

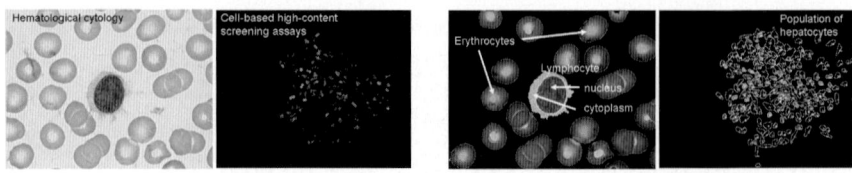

Fig. 1. Microscopic cell images (*left*), segmented cell shapes to be analysed (*right*)

2 Multi-Scale Shape Descriptors Using Granulometries

Given a grey level image $f \in \mathcal{F}(E, \mathcal{T})$, the two basic morphological operators are *dilation*: $\delta_{nB}(f(x)) = \{f(y) : f(y) = \sup[f(z)], z \in n(B_x)\}$, and *erosion*: $\varepsilon_{nB}(f(x)) = \{f(y) : f(y) = \inf[f(z)], z \in n(B_x)\}$, where B is a subset of $\mathbf{Z}^2$ and $n \in \mathbf{N}$ a scaling factor. $n(B_x)$ is called *structuring element* (shape probe) B of size n (homotetic of factor n) centred at point x. Here we suppose that B is plane, symmetric and compact convex. Typically, nB are families of disks (isotropic) or of segments (orientated). Note that $\delta(f), \varepsilon(f) \in \mathcal{F}(E, \mathcal{T})$. Erosion shrinks positive peaks. Peaks thinner than the structuring element disappear. As well, it expands the valleys and the sinks. Dilation produces the dual effects.

The two elementary operations of erosion and dilation can be composed together to yield a new set of operators having desirable feature extractor properties which are *opening*: $\gamma_{nB}(f) = \delta_{nB}(\varepsilon_{nB}(f))$, and *closing*: $\varphi_{nB}(f) = \varepsilon_{nB}(\delta_{nB}(f))$. Opening (closing) removes positive (negative) structures according to the predefined size and shape criterion of the structuring element (smooth in a nonlinear way).

A granulometry is a size distribution based on a pyramid of morphological operators. Formally, it can be defined as an one-parameter family of *openings* [MAT67] $\Gamma = (\gamma_\lambda)_{\lambda \geq 0}$ such that: (1) γ_0 is the identity mapping, i.e. $\gamma_0(f) = f$; (2) γ_λ is increasing, i.e. $f \leq g \Rightarrow \gamma_\lambda(f) \leq \gamma_\lambda(g), \forall \lambda \geq 0, \forall f, g$; (3) γ_λ is antiextensive, i.e. $\gamma_\lambda(f) \leq f, \forall \lambda \geq 0, \forall f$; (4) γ_λ follows the absorption law, i.e. $\forall \lambda \geq 0, \forall \mu \geq 0, \gamma_\lambda \gamma_\mu = \gamma_\mu \gamma_\lambda = \gamma_{\max(\lambda, \mu)}$. Moreover, granulometries by *closings* (or *anti-granulometry*) can also be defined as families of increasing closings $\Phi = (\varphi_\lambda)_{\lambda \geq 0}$.

The morphological openings, γ_{nB}, B compact convex, satisfies the four granulometric postulates. They also satisfy two fundamental properties: (5) the γ_{nB} are translation invariant; (6) $\gamma_n(f) = \gamma_1(\frac{1}{n}f)$, i.e. there is a unit sieve γ_1, and any other sieve in the process can be evaluated by first scaling the image by the reciprocal of the parameter, filtering by the unit sieve, and then rescaling.

Let $m(f)$ be the Lebesge measure of a discrete image f. Performing the granulometric analysis is equivalent to mapping each opening of size λ with

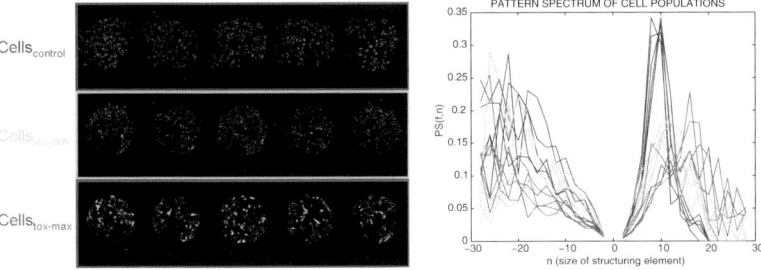

Fig. 2. *Left*, cell population based high content toxicity biosensor, three examples of toxic concentration. *Right*, pattern spectra, $PS(f, n)$, with openings (for size/shape description) and closing (for aggregation study) of size $n = -30$ to 30

a measure of the opened image $\gamma_\lambda(f)$. The granulometry curve or *pattern spectrum* [MAR89] of f with respect to Γ is defined as the following (normalised) mapping: $PS_\Gamma(f, n) = PS(f, n) = \frac{m(\gamma_n(f)) - m(\gamma_{n+1}(f))}{m(f)}$, $n \geq 0$. $PS_\Gamma(f)$ maps each size n to some measure of the bright image structures with this size: loss of bright image structures between two successive openings. $PS_\Gamma(f)$ is a probability density function (a histogram): a large impulse in the pattern spectrum at a given scale indicates the presence of many image structures at that scale. By duality, the concept of pattern spectra extends to anti-granulometry curve $PS_\Phi(f)$ by closings, $PS_\Phi(f, -n) = PS(f, -n) = (m(\varphi_n(f)) - m(\varphi_{n-1}(f)))/m(f)$, and is used to characterise the size of dark image structures. The pattern spectrum can be directly used to compare shapes. Moreover different parameters (moments, partial sums, etc.) can be derived from the pattern spectrum to measure the complexity, dispersion, etc., of the shape [SIV97] [BAT97].

Granulometric analysis is very useful to describe the shape of individual cells (see for instance in [ANG06] the cytoplasmic profile classification using the partial sums of $PS_\Gamma(f, n)$). Figure 2 shows an example of application of granulometric analysis to characterise three classes of cell populations (control and two values of toxicity). In this case, the pattern spectra $PS(f, n)$ of segmented cells allow us to classify the populations according to the size/shape of cells (with the family of openings) or with respect to their aggregation (closings).

3 Radial/Angular Decompositions Using Skeletons in Log-Polar Coordinates

It is difficult to take advantage of radial/angular properties of round-objects (definition of neighbourhood, adapted structuring elements, etc.) when mathematical morphology operators are defined in $\mathcal{F}(E, \mathcal{T})$ (space E corresponds to Cartesian coordinates). The conversion into logarithmic polar coordinates as well as the derived cyclic morphology, recently studied by [LUE05], appears to be a way that provides interesting results to obtain inclusion (extrusion) decompositions by means of angular/radial closings (openings) and to describe shape angularities by computing radial skeletons.

The *log-polar transformation* converts the cartesian image function $f(x, y)$: $E \to \mathcal{T}$ into another log-polar image function $f^\circ(\rho^{\log}, \theta) : E_{\rho^{\log}\theta} \to \mathcal{T}$, where the angular coordinates are placed on the vertical axis and the logarithmic radial coordinates are placed on the horizontal one. More precisely, with respect to a central point (x^c, y^c): $\rho = \sqrt{(x - x^c)^2 + (y - y^c)^2}$, $\rho^{\log} = \log(\rho)$, $0 \le \rho^{\log} \le R$; $\theta = \arctan\left(\frac{y - y^c}{x - x^c}\right)$, $0 \le \theta < 2\pi$. The support is the space $E_{\rho^{\log}\theta}$, $(\rho^{\log}, \theta) \in (\mathbb{Z} \times \mathbb{Z}_p)$ (discrete period of p pixels equivalent to 2π). A relation is established where the points at the top of the image ($\theta = 0$) are neighbours to the ones an the bottom ($\theta = p - 1$). The choice of (x^c, y^c) is relatively critical. We propose to use the maxima of the distance function or ultimate erosion [SER82]. The image $f^\circ \in \mathcal{F}(E_{\rho^{\log}\theta}, \mathcal{T})$ presents two properties useful for shape analysis: (1) rotations in the cartesian image $f(x, y)$ become vertical cyclic shifts in the transformed log-pol $f^\circ(\rho^{\log}, \theta)$; (2) the changes of size in f become horizontal shifts in f°.

The use of classical structuring elements in the log-pol image is equivalent to the use of *'radial–angular' structuring elements* in the original image, e.g. $g^\circ = \delta_B(f^\circ)$ where B is a vertical structuring element corresponds in g to the dilation by an arc. (a square in g° corresponds to a circular sector in g). This property yields a method for extracting inclusions/extrusions from the contour of a relatively rounded shape with vertical openings or closings. The proportion of the vertical size from the structuring element with respect to the whole vertical size represents the angle affected in the original cartesian image. With respect to a standard extraction in E, the choice of size in $E_{\rho^{\log}\theta}$ is not as critical.

The morphological *skeleton by homotopic thinning* of a binary image, $\mathrm{skel}(f)$, is a transformation which produces a connected medial axis of the shape [SER82] [HEI92]. However, the skeleton of a round object in E is usually biased and for these kind of shapes is more appropriate to work in $E_{\rho^{\log},\theta}$. Two definitions are possible. The *radial inner skeleton* $\mathrm{skel_{in}}(f^\circ)$ is the skeleton obtained by an homotopic thinning from the log-pol transformation of an object. In the invert transformation to cartesian coordinates, the branches of the radial inner skeleton have radial sense and tend to converge to the centre ($\rho = 0$). The *radial outer skeleton* $\mathrm{skel_{out}}(f^\circ)$ is obtained from the negative image of the log-pol image and in the corresponding cartesian image, the branches tend to diverge to an hypothetical circumference in the infinity ($\rho \longrightarrow \infty$).

Figure 3 gives an approach to extract the extrusions/intrusions of red blood cells which is used to classify them according to their shape. The approach for extrusions is composed of several steps: (1) $f \to f^\circ$, (2) residue of vertical opening $f_1^\circ = f^\circ - \gamma_{B_{\mathrm{vert}}}(f^\circ)$, (3) radial outer skeleton $f_2^\circ = \mathrm{skel_{out}}(f^\circ)$,

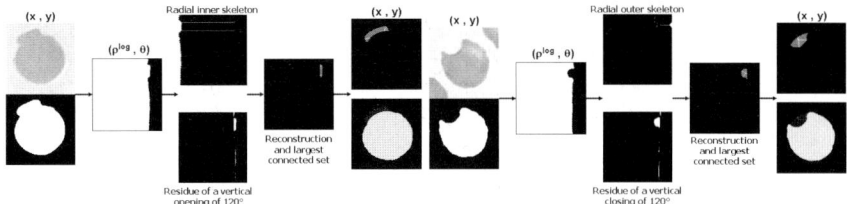

Fig. 3. Erythrocyte shape analysis: morphological algorithm for detecting extrusions (*left*) and intrusions (*right*)

(4) reconstruction to extract the connected components associated to the skeleton $f_3^\circ = \gamma^{\mathrm{rec}}(f_1^\circ, f_2^\circ)$ and (5) $f_3^\circ \to f_3$. The algorithm for the intrusions is the same changing the opening by a closing and the radial outer skeleton by a inner skeleton.

References

[ANG06]   Angulo, J., Klossa, J., Flandrin, G.: Ontology-based lymphocyte popula-
          tion description using mathematical morphology on colour blood images.
          Cellular and Molecular Biology, **52**(6), 3–16 (2006)
[LUE05]   Luengo-Oroz, M.A., Angulo, J., Flandrin, G., Klossa, J.: Mathematical
          morphology in polar-logarithmic coordinates. In: Proc. of the 2nd Iberian
          Conference on Pattern Recognition and Images Analysis (IbPRIA'05),
          Estoril, Portugal, Springer LNCS 3523, 199–206 (2005)
[HEI92]   Heijmans, H.J.A.M.: Mathematical Morphology as a Tool for Shape
          Description. In: Workshop Shape in Picture, Driebergen, The Netherlands
          (1992)
[HEI94]   Heijmans, H.J.A.M.: Morphological Image Operators. Academic Press,
          Boston (1994)
[LEM06]   Lemaire, F., Mandon, C., Reboud, J., Papine, A., Angulo, J., Pointu, H.,
          Diaz-Latoud, C., Sallette, J., Lajaunie, C., Chatelain, F., Arrigo, A.-P.,
          Schaack, B.: High content toxicity assays in nanodrops reveals individual
          cell behaviour. PLOS One, (to appear) (2006)
[MAR89]   Maragos, P.: Pattern Spectrum and Multiscale Shape Representation.
          IEEE Transactions on Pattern Analysis and Machine Intelligence, **11**(7),
          701–716, (1989)
[MAT67]   Matheron, G.: Elements pour une thèorie des milieux poreux. Masson,
          Paris (1967)
[SER82]   Serra, J.: Image Analysis and Mathematical Morphology, Vol. I., Vol. II
          Theoretical Advances. Academic Press, London (1982, 1988)
[SIV97]   Sivakumar, K., Goutsias, J.: Discrete morphological size distributions and
          densities: estimation techniques and application. Journal of Electronic
          Imaging, **6**, 31–53 (1997)
[BAT97]   Batman, S., Dougherty, E.R.: Size distributions for multivariate mor-
          phological granulometries: Texture classification and statistical propeties.
          Optical Engineering, **36**, 1518–1529 (1997)

Reconstruction of Transducer Pressure Fields from Schlieren Data

R. Kowar

Department of Computer Science, University of Innsbruck, Technikerstr. 21a,
A-6020, Austria
richard.kowar@uibk.ac.at

Summary. In order to ensure safety and optimal performance of medical ultrasound transducers it is necessary to measure the acoustic pressure fields of transducers. For the estimation of such pressure fields we use light intensity data that is obtained by a Schlieren system. Schlieren data corresponds mathematically to squared x-ray tomographic data. Acoustic pressure fields attain positive and negative values, but only the square of the line integrals are provided by the Schlieren system. Therefore the signs of the line integrals are not known, and Schlieren data cannot be reduced to data of classical x-ray CT. For the numerical estimation of pressure fields we used the loping Landweber–Kaczmarz method.

1 Schlieren Optical System and Data Acquisition

Given is a tank that is filled with water and that lies within a *Schlieren optical system* (see Fig. 1). A rotatable ultrasound transducer is mounted at the center of the top side of the tank with vertical rotation axis. The aim of a Schlieren optical system is to measure approximately the *first order diffraction pattern* of the laser light passing through the tank. In this case the light intensity measured by a Schlieren optical system is proportional to the square of the line integral of the pressure along the light path.

The problem of Schlieren tomography is to reconstruct the pressure field within the tank that was generated by the transducer from Schlieren data gathered from different angles σ_j, $j = 0, \ldots, N-1$ of rotation of the transducer.

For more details on Schlieren data we refer to [1–3, 7–10].

2 A Mathematical Model for Schlieren Tomography

Let q denote the pressure field within the water tank (at fixed altitude z and at fixed time T) generated e.g. by an ultrasound transducer.

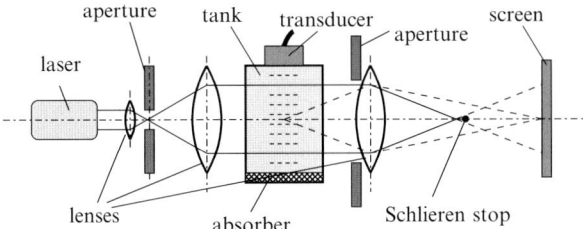

Fig. 1. Schlieren optical system. The *Schlieren stop* can be considered as a black dot on a transparent screen that reduces the diffraction pattern of order zero. The aim of the system is to reduce the diffraction patterns of order larger than one

Let $D := \{\mathbf{x} \in \mathbf{R}^2 \,|\, |\mathbf{x}| < 1\}$ and $J := \{0, 1, \ldots, N-1\}$. For $j \in J$ let $\boldsymbol{\sigma}_j \in S^1$ and I_j^δ denote the *recording angle* of the Schlieren system and the corresponding measured intensity function, respectively. The goal of Schlieren tomography is to reconstruct the pressure field $q : D \to \mathbf{R}$ for each altitude z at fixed time T from the Schlieren data $(I_j)_{j \in J}$.

We define the *parameter-to-data map* of our problem by

$$F : H_0^1(D) \to L^2([-1,1])^N, \quad q \mapsto (F_0, \ldots, F_{N-1})^{\mathrm{T}} \tag{1}$$

with

$$F_j(q)(s) := \left(\int_{\mathbf{R}} q(s\,\boldsymbol{\sigma}_j + r\,\boldsymbol{\sigma}_j^\perp)\,\mathrm{d}r \right)^2, \quad s \in [-1,1]. \tag{2}$$

Here, s denotes the signed normal distance of the line $L_j(s) := s\,\boldsymbol{\sigma}_j + \mathbf{R}\,\boldsymbol{\sigma}_j^\perp$ from the origin of D and q is considered as the function continued from D to $\mathbf{R}^2$ by zero. *Schlieren tomography* is concerned with the solution of the system of equations

$$F_j(q) = I_j^\delta \quad (j \in J) \tag{3}$$

for Schlieren data $(I_j^\delta)_{j \in J}$.

2.1 Properties of the Parameter-to-Data Map

Let $q \in H_0^1(D)$ be arbitrary but fixed. Again we consider q as the function continued from D to $\mathbf{R}^2$ by zero. It can be shown that each map F_j is *Fréchet differentiable* at q with

$$F_j'(q)(h) = 2\,R_j(q)\,R_j(h) \quad \text{for all } h \in H_0^1(D),$$

where $R_j(q)(s) := \int_{\mathbf{R}} q(s\,\boldsymbol{\sigma}_j + r\,\boldsymbol{\sigma}_j^\perp)\,\mathrm{d}r$ for $s \in [-1,1]$. The *adjoint* of $F_j'(q)$ is given by

$$F_j'(q)^* : L^2([-1,1]) \to H_0^1(D), \, f \mapsto g(f), \tag{4}$$

whereby $g(f)$ is the solution of

$$(\text{Id} - \Delta)\, g(f) = 2R_j^{\sharp}(R_j(q)f)\,. \tag{5}$$

Here, Δ denotes the *Laplace operator* on $H_0^1(D)$ and $R_j^{\sharp}$ denotes the adjoint of R_j considered as an operator $R_j : L^2(D) \to L^2([-1,1])$, i.e.

$$R_j^{\sharp} : L^2([-1,1]) \to L^2(D),\ v \mapsto (\mathbf{x} \mapsto v(\langle \boldsymbol{\sigma}_j, \mathbf{x}\rangle))\,. \tag{6}$$

3 The Loping Landweber–Kaczmarz Method

The *Landweber–Kaczmarz method* reads as

$$q_{n+1}^{\delta} = q_n^{\delta} - \omega\, F_j'(q_n^{\delta})^*(F_j(q_n^{\delta}) - I_j^{\delta}) \tag{7}$$

with $j := n \bmod N$ and *relaxation parameter* ω satisfying $\|\omega\, F_i'(q_n)\|_{L^2} \leq 1$ for all n.

We use the following *stopping rule*: If $\|F_j(q) - I_j^{\delta}\|_{L^2} < \tau\delta_j$ then the jth update in the actual cycle is not performed. The cycle-iteration is stopped as soon as N successive updates have been omitted.

According to this stopping rule, (4) and (5), the *loping Landweber–Kaczmarz method* for *Schlieren tomography* reads as

$$q_{n+1} = q_n - 2\,\omega_n (\text{Id} - \Delta)^{-1} R_j^* \left(R_j(q_n)((R_j(q_n))^2 - I_j^{\delta})\right)\,, \tag{8}$$

with

$$\omega_n := \begin{cases} \omega & \text{if } \|(R_j(q))^2 - I_j^{\delta}\|_{L^2} > \tau\delta_j, \\ 0 & \text{otherwise} \end{cases} \tag{9}$$

4 Numerical Experiments

The numerical simulations of the Landweber–Kaczmarz method were performed without the smoothing operator $(\text{Id} - \Delta)^{-1}$, because the Landweber–Kaczmarz method itself performs smoothing. For the implementation we represented each function defined on D or $[-1,1]$ by linear splines and used $N = 250$, $\tau = 2.2$ and $q_0(\mathbf{x}) = 0.01 = \text{const.}$. The synthetic data $(I_j^{\delta})_{j \in J}$ was generated by adding 0.01% normal distributed random noise to the exact data. All simulations were performed with MATLAB.

The numerical experiment was performed for a piecewise constant pressure function. (We note that the results for the smoothed version of this pressure function was very similar.)

In Fig. 2, we compared the results from the loping Landweber–Kaczmarz method and the reconstruction via a filtered backprojection (FPB) algorithm

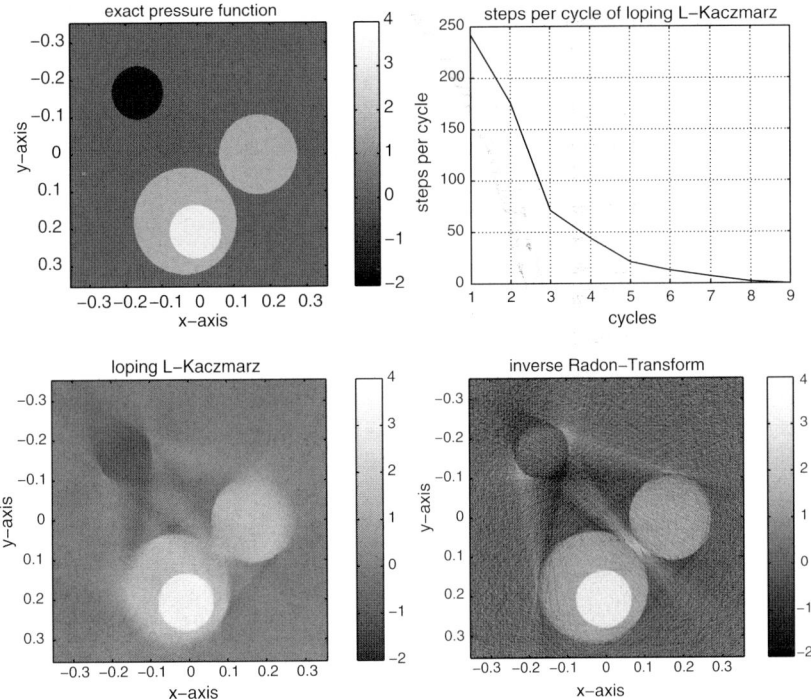

Fig. 2. The pictures in the first row show the exact pressure function q and the number of updates per cycle of the loping Landweber–Kaczmarz method. The pictures in the second row show the results obtained from the loping Landweber–Kaczmarz method and the reconstruction with a filtered backprojection algorithm for the inverse Radon transformation

for the inverse Radon transformation (with the square root of the Schlieren data as data). The reconstruction with the inverse Radon transformation was performed with the MATLAB build in function `iradon(.)`.

In contrast to the FPB algorithm the loping and Landweber–Kaczmarz methods is able to reconstruct the positive and negative part of q.

The upper right image in Fig. 2 shows that the number of updates performed in each loping Kaczmarz cycle is rapidly decreasing.

Reconstructions of a transducer pressure field from real measurement data are shown in Fig. 3.

4.1 Remark

If q is a solution of our system (3), then $-q$ is also a solution of the system. Our numerical experiments showed that a strictly positive (negative) initial guess q_0 leads to a numerical reconstruction with positive (negative) mean value.

The *local tangential cone condition* (cf. [4–6]) that guarantees stable convergence of the (loping) Landweber–Kaczmarz method is in general not

552   R. Kowar

Fig. 3. Positive and negative part of a 3-D transducer pressure field with a pulse of 30 cycles. The reconstruction is performed with the loping Landweber–Kaczmarz method and the data is provided by GE Medical Systems Kretz Ultrasound

satisfied for Schlieren tomography. According to our numerical experiments the loping Landweber–Kaczmarz method is convergent except for special cases.

References

1. Breazeale, M. A. Schlieren photography in Physics. Proc. SPIE **3581**, 41-47 (1998).
2. Charlebois, T. and Pelton, R. Quantitative 2D and 3D Schlieren imaging for acoustic power and intensity measurements. Medical Electronics, (1995), 789-792.
3. Hanafy, A. and Zanelli, C. I. Quantitative real-time pulsed Schlieren imaging of ultrasonic waves, Proc. IEEE Ultrasonics Symposium **2**, 1223-1227 (1991).
4. Kowar, R. and Scherzer, O. Convergence analysis of a Landweber-Kaczmarz method for solving nonlinear ill-posed problems. Ill posed and inverse problems (book series) **23**, 69-90 (2002).
5. Haltmeier, M. and Leitão, A. and Scherzer, O. Regularization of systems of nonlinear ill-posed equations: 1. Convergence Analysis, submitted, 2006.
6. Haltmeier, M. and Kowar, R. and Leitão, A. and Scherzer, O. Regularization of systems of nonlinear ill-posed equations: 2. Applications, submitted, 2006.
7. LeDet, E. G. and Zanelli, C. I. A Novel, Rapid Method to Measure the Effective Aperture of Array Elements. IEEE Ultrasonics Symposium, 1999.
8. Pitts, T. A. and Greenleaf, J. F. and Lu, Jian-yu and Kinnick, R. R. Tomographic Schlieren imaging for measurment of beam pressure and intensity., Proc. IEEE Ultrasonics Symposium, 1994, 1665–1668.
9. Raman, C.V. and Nath, N. S. The diffraction of light by high frequency ultrasonic waves: Part 1. Proc. Indian Acad. Sci **2**, 406-412 (1935).
10. Zanelli, C. I. and Kadri, M. M. Measurements of acoustic pressure in the nonlinear range in water using quantitative Schlieren. Proc. IEEE Ultrasonics Symposium **3**, 1765-1768 (1994).

Plant Growth Modeling

N. Morozova[1], N. Bessonov[2], and V. Volpert[3]

[1] Department of Biological Sciences, University of Illinois, Chicago, IL 60607, USA
[2] Institute of Mechanical Engineering Problems, 199178 Saint Petersburg, Russia
[3] Department of Mathematics, UMR 5208 CNRS, University Lyon 1,69622 Villeurbanne, France

1 Introduction

In this work, we use the results of plant developmental biology (such as molecular biology of pattern formation and cell cycle progression) and of plant physiology to develop a mathematical model of plant growth. Trying to describe the most essential features of growth mechanisms, we do not model some particular plant organs but the entire plant though with many simplifications, in particular, without taking into account root growth, leave and flower formation, or the biochemistry of photosynthesis.

The main processes which will be taken into account are related to the interaction of plant growth with the regulation of cell proliferation (cell cycle progression) and with fluxes of nutrients and metabolites. In plants, proliferating cells are localized in the tips of growing shoots and roots and in the cambium. Axial growth of shoots is provided by proliferation of cells in the narrow external layer called apical meristem. Outside this layer, cells differentiate, they do not divide any more and serve to transport nutrients and metabolites.

We suppose that the apical meristem, which consists only of several cell layers, is much smaller than the whole plant. Therefore, we can consider it as a mathematical surface and describe plant growth as a free boundary problem. The speed of the free boundary corresponds to proliferation rate. Proliferation of cells, in its turn, depends on available nutrients and is regulated by plant hormones, which are produced in the apical meristem or in the other growing plant tissues.

We develop in this work three different but related to each other models. These models allow us to study formation of plant organs, various growth modes, in particular, oscillating growth, apical domination, and other biological questions. On the other hand, there are many related mathematical and numerical questions including nonlinear dynamics, pattern formation, structural stability.

2 Main Principles and Assumptions (1D Model)

In the simplest 1D case without taking into account root growth, we have the following model for the axial shoot growth:

$$\frac{\partial C}{\partial t} + u\frac{\partial C}{\partial x} = d\frac{\partial^2 C}{\partial x^2} \tag{1}$$

$$h\frac{dR}{dt} = g(R)C - \sigma R, \tag{2}$$

where $0 \leq x \leq L(t)$, $L(t)$ is the shoot length, C is the concentration of nutrients coming from the root, $u = L'(t)$ is the convective speed of nutrients determined from the continuity equation for the incompressible fluid. The concentration of growth and mitosis factor (GMF) R is defined at the growing end $x = L(t)$ which corresponds to the apical meristem. Equation (1) describes diffusive and convective transport of nutrients through the plant, (2) describes the production and consumption or destruction of the GMF in the apical meristem. Growth and mitosis factor is a generic name for a number of bio-chemical products related to cell cycle. Its production is self-accelerating, which determines the specific form of the function $g(R)$. System (1), (2) should be completed by boundary conditions for C at $x = 0$ (supply of nutrients) and at $x = L(t)$ (flux of nutrients to the meristem),

$$x = 0 : C = 1, \quad x = L(t) : d\frac{\partial C}{\partial x} = -g(R)C, \tag{3}$$

and by the additional relation $L'(t) = f(R)$ which shows how the growth rate depends on the GMF. We consider f as a piece-wise constant function equal to 0 if R is less than a critical value c_f and equal some positive constant f_0 if R is greater than c_f. This means that growth begins if the concentration of the plant growth factor exceeds some critical value.

These assumptions are consistent with plant morphogenesis. It is well known, for example, that auxin, produced in the apex, stimulates mitosis and cell proliferation. Kinetin is also known to stimulate cell proliferation. Production of mitosis factors can be self-accelerating [1].

There are two directions of the development of this model: 1D model with branching and 2D model.

3 1D Model with Branching

To study branching patterns in plants, we need to specify the conditions of the appearance of new branches. We use the experimental observations on shoot and root growth from callus: if the concentrations of two hormones, auxin and cytokinin (which we denote by A and K, respectively) are in a certain proportion, then shoots will appear. For a different proportion, not shoots but roots will grow [2, 3].

Hormone A is produced in growing parts of the plant (leaves, shoots); hormone K in either roots or in growing parts. In our model, A will be produced at the moving boundary $x = L$ that corresponds to the apical meristem. The rate of its production is proportional to the growth rate. Hormone K will either be supplied solely through the stationary end of the interval $x = 0$ (the root) or will also be produced at the moving boundary.

The concentrations of nutrients C, and of hormones A and K are described by the diffusion equations with convective terms:

$$\frac{\partial C}{\partial t} + V\frac{\partial C}{\partial x} = d_C\frac{\partial^2 C}{\partial x^2} - \beta C, \tag{4}$$

$$\frac{\partial K}{\partial t} - V_K\frac{\partial K}{\partial x} = d_K\frac{\partial^2 K}{\partial x^2} - \mu K, \tag{5}$$

$$\frac{\partial A}{\partial t} - V_A\frac{\partial A}{\partial x} = d_A\frac{\partial^2 A}{\partial x^2} - \mu A. \tag{6}$$

The convective speed V in the first equation is determined as the speed of growth:

$$\frac{\mathrm{d}L}{\mathrm{d}t} = V, \quad V = f(R). \tag{7}$$

Here d_C, d_K, d_A and μ are parameters; the space variable x is defined independently for each branch. The convective speed V_A in equations (5) and (6) can be different in comparison with equation (4). It corresponds to transport in the phloem in the direction from top (meristem) to bottom (root).

The rate of production of the GMF at $x = L(t)$ is given by the equation

$$h\frac{\mathrm{d}R}{\mathrm{d}t} = F(A, K)\, g(R)\, C - \sigma R, \quad R(0) = R_0. \tag{8}$$

Here $F(A, K) = F_1(A)F_2(K)$. The form of the functions $F_1(A)$ and $F_2(K)$ is chosen in accordance with the biological observations that there are optimal concentrations of plant hormones. The model should be completed by the initial and boundary conditions.

We define next the branching conditions. A new branch appears at $x = x_0$ and $t = t_0$ if

$$A(x_0, t_0) = A_b, \quad K(x_0, t_0) = K_b, \tag{9}$$

where A_b and K_b are some given values. Appearance of a new branch means that there is an additional interval connected to the previous one at its point x_0.

A typical example of plant growth in the 1D model with branching is shown in Fig. 1. We can see that new buds appear but remain dormant while the main branch continues growing. When it stops growing, lateral buds can give new branches. This is related to apical domination.

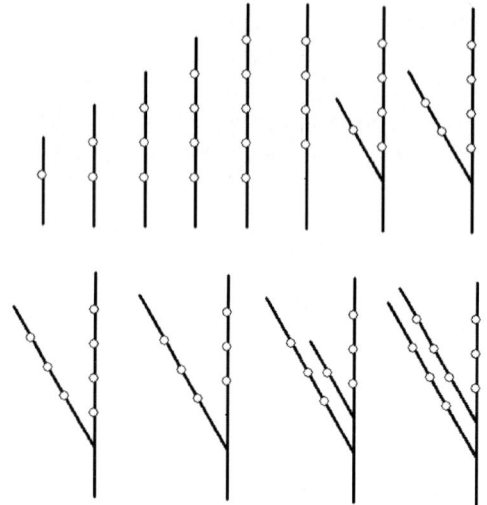

Fig. 1. Evolution of the plant structure in time, $h = 0.0005, R_0 = 0.12$

Fig. 2. 1D model with branching. Branching patterns for different values of parameters in the branching conditions

Two different final branching patterns are shown in Fig. 2. They correspond to some specific plant families (e.g., Boraginaceae, Gramineae).

4 2D Model

We formulate the 2D model of plant growth based on the same assumptions as for the 1D model presented above. Apical meristem in this case corresponds to the outer surface of the growing domain. Numerical realization of this

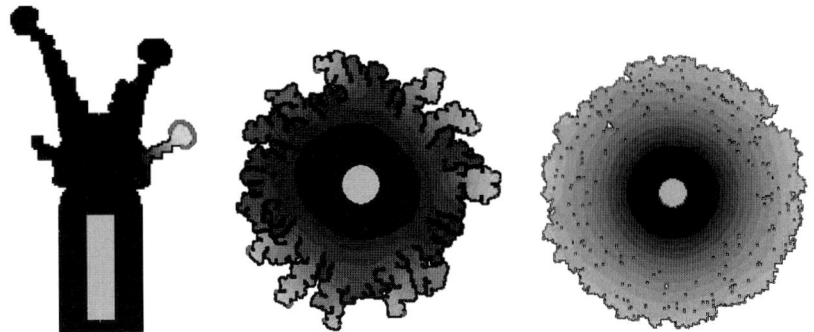

Fig. 3. 2D simulations. Nutrients are supplied through the internal rectangle (*left*) or through the internal circle: model without merging (*center*), model with merging (*right*)

model encounters essential difficulties because the free boundary can have a complex form with singularities and points of self-intersection. The growing plant fills the domain between two closed curves. For the examples in Fig. 3 (at the center and from the right) the internal curve Γ_i is a circle of a fixed radius r with the center at the origin, the external curve Γ_e represents the moving boundary Nutrients are supplied through the internal curve, while the meristem corresponds to the external boundary. Its motion describes the plant growth. The region between the two curves is considered as a porous medium.

We define now the motion of the discretized boundary $\Gamma_e(t_i)$. Let C_k be a grid cell from $\Gamma_e(t_i)$ and S_{k1}, S_{k2}, and S_{k3} its sides common with the outer grid cells. We consider a flow from the cell C_k to the neighboring outer cells through the sides S_{kj}. The flow speed is determined by the value $R_k(t_i)$ in C_k. It can be interpreted as a motion of the intervals S_{kj} in the normal direction. When an outer cell is completely filled by the flow, it becomes an inner cell. An outer cell is filled through all its boundaries common with inner cells. Thus, the motion of the discrete boundary is defined in terms of local flows. This approach is applicable for any shape of the boundary.

Figure 3 (left) presents the example of 2D modeling where nutrients are supplied through the internal rectangle. It corresponds to the process of plant embryogenesis or growth of a vegetative or of a floral bud. Figure 3 (center) presents an example of 2D modeling without merging. This means that different parts of the growing plant remain separated when they touch each other. This situation can occur in embryogenesis when different parts of the embryo form different embryogenic tissues and organs. The right picture corresponds to the specific conditions allowing merging. It can be growth of the meristematic tissue without tissue differentiation.

References

1. Hoffmann I, Clarke PR, Marcote MJ, Karsenti E, Draetta G., Phosphorylation and activation of human cdc25-C by cdc2-cyclin B and its involvement in the self-amplification of MPF at mitosis. The EMBO Journal, 1993, Vol. 12, No. 1, 53-63.
2. Skoog, F., Millar, C. O. Chemical regulation of growth and organ formation in plant tissues cultured in vitro. Symp. Soc. Exp. Biol. 1957, 11, 118-140.
3. Yamaguchi M, Kato H, Yoshida S, Yamamura S, Uchimiya H, Umeda M.Control of in vitro organogenesis by cyclin-dependent kinase activities in plants. Proc Natl Acad Sci U S A. 2003 Jun 24;100(13):8019-23.

Minisymposium "New Trends in the Analysis of Functional Genomics Data"

José María Carazo[1] and Alberto Pascual[2]

[1] Biocomputing Unit., Centro Nacional de Biotecnologia - CSIC, Madrid, Spain
[2] Universidad Complutense de Madrid, Spain

In this minisymposium, R. Armañanzas, B. Calvo, I. Inza, P. Larrañaga, I. Bernales, A. Fullaondo and A.M. Zubiaga present "Bayesian Classifiers with Consensus Gene Selection: A Case Study in the Systemic Lupus Erythematosus", A. Sánchez and J.L. Mosquera present "The Quest for Biological Significance," M. Chagoyen, H. Fernandes, J.M. Carazo and Pascual-Montano present "Functional Classification of Genes Using Non-Negative Independent Component Analysis", and D. Montaner, F. Al-Shahrour and J. Dopazo present "New Trends in the Analysis of Functional Genomic Data".

Bayesian Classifiers with Consensus Gene Selection: A Case Study in the Systemic Lupus Erythematosus

Rubén Armañanzas[1], Borja Calvo[1], Iñaki Inza[1], Pedro Larrañaga[1], Irantzu Bernales[2], Asier Fullaondo[2], and Ana M. Zubiaga[2]

[1] ISG - Department of Computer Science and Artificial Intelligence,
University of the Basque Country, P.O. Box 649 - 20080 San Sebastián, Spain
{ruben, borxa, inza, ccplamup}@si.ehu.es
[2] Department of Genetics, Physical Anthropology and Animal Physiology,
University of the Basque Country, P.O. Box 644 - 48080 Bilbao, Spain
{ggpbepui, ggpfuela, ggpzuela}@lg.ehu.es

1 Introduction

Within the wide field of classification on the Machine Learning discipline, Bayesian classifiers are very well established paradigms. They allow the user to work with probabilistic processes, as well as, with graphical representations of the relationships among the variables of a problem.

Bayesian classifiers assign the corresponding predicted class of a certain pattern as the one that has the highest a posteriori probability. This a posteriori probability is computed by means of the Bayes theorem in conjunction with assumptions about the density of the patterns conditioned to the class.

In this work three of these classification paradigms are applied to a DNA microarray database of control, systemic lupus erythematosus and antiphospholipid syndrome samples. The number of genes from which the models are induced is considerably reduced by means of a novel consensus filter gene selection technique.

Combining a nonparametric bootstrap resampling technique and the k dependence Bayesian classifier paradigm, we propose a new method to obtain gene interaction networks of high reliability. These gene networks can be seen as a tool to study the relationships among the genes of the domain. In fact, some of the previous knowledge about both pathologies is confirmed by the new approach.

2 Bayesian Classifiers

A supervised classifier is a function that assigns labels to observations,
$$\gamma : (x_1, \ldots, x_n) \rightarrow \{1, 2, \ldots, m\},$$

where $\boldsymbol{x} = (x_1, \ldots, x_n) \in \mathcal{R}^n$ conforms the observation and $\{1, 2, \ldots, m\}$ are the range of possible values for the class variable. The main assumption is the existence of an unknown underlying probability joint distribution $p(x_1, \ldots, x_n, c)$ where the observations come from:

$$p(x_1, \ldots, x_n, c) = p(c|x_1, \ldots, x_n)p(x_1, \ldots, x_n) = p(x_1, \ldots, x_n|c)p(c).$$

In practice, this joint probability distribution $p(x_1, \ldots, x_n, c)$ is estimated from a random sample, $\{(\boldsymbol{x}^{(1)}, c^{(1)}), \ldots, (\boldsymbol{x}^{(N)}, c^{(N)})\}$.

The *naïve Bayes* (NB) classifier [6] is based on two assumptions over the predictive variables and the class to predict: the class variable C can only take one of its m possible values $c_1, \ldots, c_m$; and, if this class value is known, the knowledge of some predictive variables is independent from the knowledge of the rest ones. Therefore, the search for the most probable class value, c^*, once all the variables' values are known, can be reduced to look for

$$c^* = \arg\max_c p(c) \prod_{i=1}^{n} p(x_i|c).$$

The conditional independence assumption of the naïve Bayes paradigm can be a very restrictive condition. So as to overcome this limitation, there are classification paradigms that allow conditional dependencies among the variables. One of them is the *tree augmented network* (TAN) [3], in which a tree-like classification modelization and the Bayesian classification paradigm comes together; first, a tree structure among the predictive variables is built, and then, the class node is related to all the variables.

The metric to configure edges between variables is based on the mutual information conditioned to the class variable,

$$I(X, Y|C) = \sum_{i=1}^{t} \sum_{j=1}^{w} \sum_{r=1}^{m} p(x_i, y_j, c_r) \log \frac{p(x_i, y_j|c_r)}{p(x_i|c_r)p(y_j|c_r)},$$

where X and Y are two discrete predictive variables and C is the class label. The complete learning algorithm is discussed in [3] and makes use of the Kruskall algorithm to build the maximum weight spanning tree.

In order to go trough the wide spectrum from the naïve Bayes to a complete Bayesian network, Sahami (1996) [7] presents an algorithm called *k dependence Bayesian classifier* (kDB). The algorithm has its basis on a naïve Bayes structure that allows each predictive variable to have a maximum number of parent variables. The algorithm extends the TAN algorithm allowing a variable to have a number of parents, excluding the class variable C, bounded by k. In Fig. 1 graphical examples of the three paradigms are gathered.

3 Consensus Gene Selection

Bayesian classifiers deal only with discrete data. This restraint makes it necessary to translate the microarray data from continuous to discrete value-domains. This translation can make the original data lose precision, even

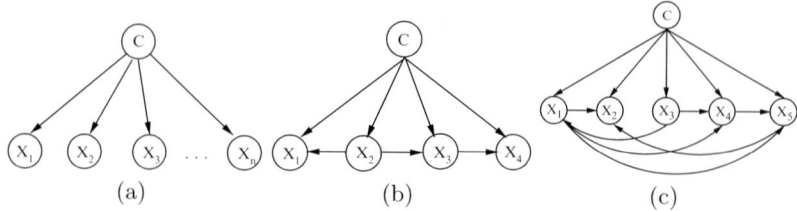

Fig. 1. Graphical structures of a naïve Bayes (**a**), tree augmented naïve Bayes (**b**) and k dependence Bayesian (**c**) classification models

degrading its original quality. Thus, if a discretization process has biased the original data, this bias affects all posterior knowledge discovery processes. Therefore, the search for a robust solution makes us rely on several rather than on a single discretization method.

Let O be the original microarray dataset with continuous features and $S_1, \ldots, S_D$ the results of D different discretizations of the O set. Using a filter subset selection method, N different feature selections are performed on the basis of the $S_1, \ldots, S_D$ discrete datasets, producing the following subsets of genes: $G_1, \ldots, G_D$. The final consensus gene subset Γ is the intersection of all of them, that is $\Gamma = \bigcap_{i=1}^{D} G_i$, with $|\Gamma| = m \leqslant \min_{i=1,\ldots,D} |G_i|$. The complete formulation of this consensus approach can be reviewed in [1].

4 Knowledge Discovery by Means of Bayesian Classifiers

For the final process of knowledge inference, the application of a suited technique that contributes certain level of reliability is crucial. For this purpose, we propose the application of a technique known as *bootstrap*, firstly presented by Efron (1979).

The bootstrap procedure allows us to compute a confidence level for each feature under study on a probabilistic graphical model. These confidence levels are calculated after repetitive runs of the induction algorithm, but, instead of inducing the models in basis of the original dataset, for each run, the original dataset is substituted by N randomly sampled instances with replacement from the original ones. The knowledge discovery process is centered in the detection of the same edges along the different induced graphical models, because this high confidence edges are expected to have a direct biological interpretation. The nonparametric bootstrap algorithm approach implemented for the present study can be found in [4].

5 Results

Both systemic lupus erythematosus (SLE) and antiphospholipid syndrome (APS) are autoimmune diseases with unknown origin. SLE is mainly an inflammatory disease with clear autoimmune features, and it can affect multiple organs and body systems. Related to the genetic basis of the disease, more than 100 genes are now thought to be involved in SLE genetic susceptibility. APS, also known as "sticky blood" syndrome, is another immunological disease characterized by the repeated appearance of thrombosis, a high number of miscarriages in the second and third gestation quarters, and thrombopenia or hemolytic anemia.

There is no clear diagnosis methodology for SLE and APS: different criteria have to be evaluated in order to assess its presence. Therefore, the study of genes that present different expression profiles among SLE, APS and control subjects is medically and biologically of great interest.

5.1 Data Preprocess

The biochip model used is the Affymetrix® *HGU133A*. The marking and hybridization processes are performed using peripheral blood obtained from 12 different Caucasian women: two with primary APS, four with SLE, and six healthy people, used as controls. Four different criteria are measured to evaluate the biochips reliability: the presence of spike control BioB, the $3'/5'$ relation of the GAPDH housekeeping control, the percentage of present probes in the array and the *dChip*[1] array outlier percentage. From the original 12 biochips, one of them does not reach a sufficient quality level in three out of the four criteria, consequently, it is removed from the dataset.

Filtering the data by the Affymetrix® *detection* algorithm, the amount of valid probes decreases from 22,067 to 8,808; these probes form our starting set of predictive variables. There are a total of 40 comparisons between the samples and the reference microarray ratios, divided in three phenotypes or classes: ten correspond to the control arrays among themselves, another ten between the five control and the two APS patient arrays, and the last 20 correspond to the control and SLE arrays.

5.2 Gene Selection Step

Three different discretization policies are used in the consensus selection: equal width, equal frequency and entropy discretization [2]. Correlation-based feature selection [5] is used as the feature selection method, and the overall process returns eight variables. These genes are considered *statistical prototypes* of gene families showing different behavior profiles over the original data. The members of each family are computed by the classical mutual information metric, obtaining a total set of 150 relevant variables.

---

[1]DNA-chip analyzer from Harvard University available in `http://biosun1.harvard.edu/complab/dchip/`

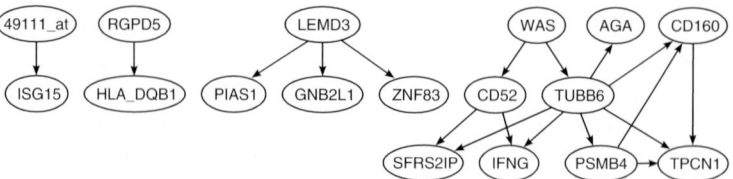

Fig. 2. Dependent structures for a 90% confidence level found by the bootstrap approach over the kDB (k=4) classification models

5.3 Classification Step

By means of the Elvira[2] platform for Bayesian networks, we induce five different models: naïve Bayes, TAN, and kDB with k values of 2, 3 and 4. Each of these models is validated using a leaving one out cross validation. The induced models by naïve Bayes, TAN, or kDB for its three k values, achieve a 100.0% classification accuracy. Due to the low number of instances in the problem, these good results in classification may come from an overfitting effect of the classifiers to the data.

5.4 Knowledge Discovery Step

Starting from the 150 relevant variables identified in the previous process and in basis of the entropy discretization dataset, we perform 1,000 loops in the bootstrap procedure. Thus, 1,000 of random samplings are performed, and 2,000 kDB models are induced (three and four are taken as values for the k parameter).

A total of three different edges are configured always (100% confidence), for both k values. When decreasing the confidence to 90%, the k=4 models configure 18 edges while the k=3 models configure only 13 of them. Notice that the edge between the class and the nodes in the graph is not taken into account. These edges allow us to construct networks of high reliability with respect to their graphical dependencies. Fig. 2 shows the structures found for a k value of four and a 90% of confidence. A deep discussion about the identified genes is collected in [1].

References

1. R. Armañanzas. Solving bioinformatics problems by means of Bayesian classifiers and feature selection. Technical Report EHU-KZAA-IK-2/06, University of the Basque Country, 2006.
2. U. M. Fayyad and K. B. Irani. Multi-interval discretization of continuous-valued attributes for classification learning. In *Proceedings of the Thirteenth International Joint Conference on Artificial Intelligence*, pages 1022–1027. Morgan Kaufmann, 1993.

---

[2]Elvira system available in `http://leo.ugr.es/elvira/`

3. N. Friedman, D. Geiger, and M. Goldszmidt. Bayesian network classifiers. *Machine Learning*, 29(2):131–164, 1997.
4. N. Friedman, M. Goldsmidt, and A. Wyner. Data analysis with Bayesian networks: A bootstrap approach. In *Proceedings of the Fifteenth Conference on Uncertainty in Artificial Intelligence*, pages 196–205, 1999.
5. M. A. Hall and L. A. Smith. Feature subset selection: A correlation based filter approach. In N. Kasabov et al., editor, *Proceedings of the Fourth International Conference on Neural Information Processing and Intelligent Information Systems*, pages 855–858, Dunedin, 1997.
6. M. Minsky. Steps toward artificial intelligence. *Transactions on Institute of Radio Engineers*, 49:8–30, 1961.
7. M. Sahami. Learning limited dependence Bayesian classifiers. In *Proceedings of the Second International Conference on Knowledge Discovery and Data Mining*, pages 335–338, 1996.

The Quest for Biological Significance

Alex Sánchez and Josep Lluis Mosquera

Departament d'Estadística, Universitat de Barcelona, Facultat de Biologia, Avda
Diagonal 645, 08028 Barcelona, Spain
asanchez@ub.edu, jlmosquera@ub.edu

1 Introduction

With the advent of genomic technologies it has become possible to perform, in
a routinely manner, new types of experiments to analyze simultaneously the
behavior of thousands of genes or proteins in different conditions. A common
trait in these type of studies is the fact that they generate huge quantities of
data what has lead to using the term "high-throughput" to describe them.
There are different types of high-throughput experiments, but we will refer
from now on to the most well known ones: microarray experiments.

A typical microarray experiment is one who looks for genes *differentially
expressed* between two or more conditions. That is, genes which behave differently in one condition, for instance healthy or untreated cells, than in another,
for instance tumor or treated cells. Such an experiment will result very often
in long lists of genes which have been selected using some criteria, such as a
t-test, to assign them *statistical significance.*

Most of the times the biological interpretation of the list is not obvious.
Sometimes the number of items selected as being statistical significant is very
high and it seems reasonable to (try to) synthesize them looking at *what the
list means from the biological point of view.* Sometimes, instead, the selected
items do not show any statistical significance, but even so, it is expected – or
it seems clear – that, biologically, they "mean something", probably related
to the process being analyzed.

In whatever of the previous situations we find, the usual way to proceed
is to shift the focus from "statistical" to "biological" significance. There is a
clear agreement about what does statistical significance mean. However there
is no consensus definition of biological significance at all. Although everyone
talks about it...

1.1 So, What Does Biological Significance Mean?

Interestingly what many authors do to define biological significance is to re-define it in terms of statistical significance. This can be clearly seen in [1] who describes it as:

> ... *to understand the biological relevance of statistical differences in gene expression data* by examining significant differences in the distribution of (GO) terms related to biological processes or molecular function.

This is not however the only possible definition. For instance `GeneSifter` (`http://www.genesifter.net/web/`) a company presenting their goals as to " ... make it easier to understand the biological significance of your microarray data" does not give any definition of the term. The nearest explanation of what they mean by this is the following:

> ... *to characterize the biology involved in a particular experiment,* and to identify particular genes of interest ... combining the identification of broad biological themes with the ability to focus on a particular gene ...

In any case, it is clear that whatever they mean by Biological Significance they do not relate this to Statistical Significance.

In short. Establishing the biological significance of high throughput experiments is an important step for their success and many efforts are addressed to this. Less efforts, it seems, than to clarifying what the term exactly means.

1.2 The Gene Ontology

Attempts to perform a biological interpretation of high throughput experiments are often based on the Gene Ontology (GO), an annotation database created and maintained by a public consortium, the Gene Ontology Consortium[1], whose main goal is, citing their mission, *to produce a controlled vocabulary that can be applied to all organisms even as knowledge of gene and protein roles in cells is accumulating and changing.* The GO is organized around three principles or basic ontologies: (1) Molecular function (MF), which describes tasks performed by individual gene products; (2) Biological process (BP), which describes broad biological goals, such as mitosis (cell division) and (3) Cellular component (CC) describing subcellular structures, locations, and macromolecular complexes such as nucleus, or other organelles. A given gene product may represent one or more molecular functions, be used in one or more biological processes and appear in one or more cellular components. Each ontology (MF, BP or CC) consists of a high number of terms or categories hierarchically related from least (top) to most (bottom) specialized

---

[1] www.geneontology.org

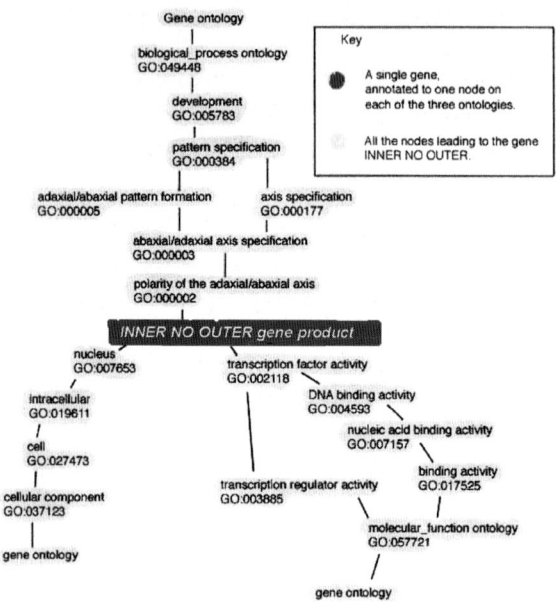

Fig. 1. A hypothetical example of GO annotations for the gene "INNER NO OUTER". Every gene is annotated in the three ontologies, MF, BP and CC

characteristics. Ontologies are indeed direct acyclic graphs (DAG) and graph theory is clearly one possible, although not yet generalized, approach for their study. Most genes are annotated in one or more categories. Annotations are made as specific as possible. As a consequence a gene is associated not only with its annotations but also with all the less specific terms associated with them. This altogether configures a network of terms for each gene integrated in the bigger network which is the GO (see Fig. 1).

2 From Biological to Statistical Significance: Gene Enrichment Analysis

In recent years there have been developed many methods intended to quantify Biological Significance (BS from now on) in terms of Statistical Significance (SS from now on). Draghici et al. ([2]) consider as many as 15 related applications which perform in different but related ways. In this chapter we will not even attempt to compare or offer a panoramic view of the existing methods, although in the appendix we describe a tool that we have developed precisely with this goal in mind. Instead we will center on what is possibly the most well-known and most used approach to obtaining BS from SS.

Assume that we have the results of a typical microarray experiment where we have selected K "interesting" genes from a wider population or Universe, of size N. Each gene is annotated to one or more GO categories so that we end up with a subset $\{A_1, A_2, ... A_G\}$ of annotated categories. Gene Enrichment Analysis (GEA) consists of performing a statistical test *separately for each category* A_i, $i = 1, ...G$ to decide if the proportion of genes in the sample which have been annotated in category A_i is the same as those in the Universe belonging to the same category. If this is so one can interpret that this category is not related to the biological phenomenon that led to select the genes in the sample. Oppositely if the proportion of genes in the sample appearing in category A_i is greater (enriched) or smaller (impoverished) that those in the Universe one can assume that this category is *Biologically Significant*. GEA can easily be formulated in terms of hypergeometric sampling allowing to use the hypergeometric distribution to compute p-values for the test having null hypothesis: H_0 *The GO category A_i is equally represented in the Universe than in the group of differentially regulated genes*. Details of this test can be found for instance in [2].

3 Discussion: Drawbacks and Limitations

Keeping in mind that, for the sake of centering on the difference between *SS* and *BS*, we have adopted a simplified view it is clear that the previous approach shows some limitations.

By one side, and this is applicable mainly to GEA, the method selects categories separately, without explicitly caring for relations between them. This, jointly with the fact that it relies on a statistical filtering criteria, suggests that is useful to highlight biologically relevant "hot spots" but it does not offer a global picture of what is happening in the biological side of the experiment.

Instead of looking at more and more methods checking their virtues and defaults (but see the appendix and [4]) it is good perhaps to remark another important flaw: When we rely in *SS* to define *BS* we depend on p-values at one or two levels, that is those p-values that have been used to select the genes, and those p-values computed to check the significance of the categories. However p-values are not free from criticisms (see [3]). They depend on underlying probability models and are often subject to misinterpretation as well as used to justify otherwise unjustifiable cutoffs. In short using p-values to define *BS* we risk to translate into it the abuse that has sometimes been observed with its use to define *SS*.

3.1 Towards a New Definition of Biological Significance

Our goal in the previous lines has been mainly to emphasize that simply relying on statistical significance to define biological significance can be as misleading as just using but not defining the term. And the interesting point is precisely this: biological significance is not an entelechy. An expert in a given biological field will often be able to distinguish between two sets of results and chose those that can be considered more relevant. The challenge for mathematicians, statisticians and other scientists working in parallel with those experts is to develop an approach which tells a story which is, at the same time, as objective as possible, but also as near to the biologist's choice as can be obtained. Probably it will requires approaches that integrate information from several sources and are able to combine weak nonsignificant evidences with more objective results into relevant conclusions that can be considered biologically significant, not because somewhere a p-value is tiny, but because they really mean something.

Appendix

In this work we have explicitly avoided making comparisons between the existing methods or tools. It is not an easy task because there exists dozens of them and they are not free of redundance at all.

This is in itself a barrier for a potential user because even if she understands clearly what she is looking for she will be faced to choose between many similar tools.

To help users in this decision process we have developed SerbGO (for **Se**arching the **b**est **GO** tool). It is a free web based tool that can be used in any two directions: One can ask for the desired functionalities and find out which are the programs that include them or one can simultaneously analyze several tools to find out which functionalities are implemented and which are missing.

SerbGO is available at http://estbioinfo.stat.ub.es/apli/serbgo/.

The program has proven useful not only to users who wish to find the tool they need or who want to compare several tools. It has helped also to classifiy the tools by their functionalities showing some interesting results such as, for example, the fact that in spite of the apparent redundance between tools most of them perform slightly different tasks, suggesting that they all may be useful, or at least that redundance is more apparent than real.

References

1. Díaz-Uriarte, R., Al-Shahrour, F. and Dopazo, J. Use of GO terms to understand the biological significance of microarray dfferential gene expression data. In K. F. Johnson and S. M. Lin, editors, *Methods of microarray data analysis* (CAMDA 2002), 2003.
2. S. Draghici, P. Khatri, R. P. Martins, G. C. Ostermeier, and S. A. Krawetz. Global functional profiling of gene expression. *Genomics* **81**(2):98-104 (2003).
3. Steven Goodman. Commentary: The P-value, devalued. *Int. J. Epidemiol.* **32**(5):699-702, (2003).
4. Mosquera, J.L. and Sánchez-Pla, A. SerbGO: Searching the best go tool. In *European Conference on Computational Biology*. ICBS, 2005.

Functional Classification of Genes Using Non-Negative Independent Component Analysis

Monica Chagoyen[1,3], Hugo Fernandes[2], Jose M. Carazo[1], and Alberto Pascual-Montano[3]

[1] Biocomputing Unit, Centro Nacional de Biotecnologia - CSIC, Madrid, Spain
[2] Integromics S.L., Granada, Spain
[3] DACYA, Universidad Complutense de Madrid, Madrid, Spain

Summary. In the last few years, several analysis methods have been proposed to assist in the functional interpretation of genome-wide data. To this aim, we explore the use of non-negative Independent Component Analysis (nnICA) for the classification of genes based on their associated functional annotations.

1 Introduction

Several statistical and machine learning techniques have been developed and applied in bioinformatics. Among them, matrix factorization techniques are used in order to reduce the dimensionality, discover patterns and aid in the interpretation of biological data. In particular, two factorization methods have been applied to the analysis of functional information in a genome-wide context: Singular Value Decomposition (SVD) and Non-negative Matrix Factorization (NMF). These factorizations have been used to create and compare gene [1, 2] and cellular process [3] profiles constructed from their associated literature, as well as to perform different analysis on gene functional annotations [4–6].

In this work we explore an alternative matrix factorization model for the classification of genes/proteins based on their functional annotations: non-negative ICA [7, 8]. Our hypothesis is that gene annotations can be represented, by means of nnICA, as a linear combination of statistically independent sources, which are a useful representation from which to classify genes. In order to assess the validity of our approach, we have analyzed the functional annotations provided by the Gene Ontology (GO) [9] in the context of a model organism (baker's yeast). Finally, we compare the results obtained by nnICA with those obtained by NMF (previously proposed for this task in [6]).

2 Methods

2.1 Data Representation

The complete functional annotations of the gene set are represented using a $n \times m$ matrix $\mathbf{X}$ similarly as in [4]. $\mathbf{X}_{ij}$ is equal to 1, if and only if the ith-gene is involved in the jth-function. In any other case, $\mathbf{X}_{ij}$ is equal to 0. In this work, functional annotations included the three broad GO categories (i.e. biological process, molecular function and cellular component), as provided by the *Saccharomyces* Genome Database (SGD) (version 10/20/2006). To construct $\mathbf{X}$, a gene is considered to be involved in the functions described by direct GO annotations, as well as all those described by their corresponding ancestor terms in the GO hierarchy.

2.2 Factorization

Non-negative ICA estimates the m-dimensional source vectors $\mathbf{S} = (s_1, \ldots, s_k)$ and the mixing matrix $\mathbf{A}$ in the linear generative model:

$$\mathbf{X} = \mathbf{AS}$$

with $\mathbf{X} = (x_1, \ldots, x_n)$ and $k \leq \min(n, m)$, where the sources are *non-negative*, i.e. $\Pr(s_i < 0) = 0$, and *independent*, i.e. $p(s_i s_j) = p(s_i)p(s_j)$ if $i \neq j$. Here $\Pr(\cdot)$ denotes the probability function and $p(\cdot)$ denotes probability density function.

We assume, as several of the algorithms for performing non-negative ICA do (see [7]), that the sources are *well-grounded*, i.e. $\Pr(s < \lambda) > 0$ for any $\lambda > 0$. Note that well-groundness is a valid assumption in our case since we expect each source not to be involved in all functions.

We use the algorithm proposed in [8]. Briefly, the first step is to reduce the dimension and whiten the data. This is done using the eigenvector–eigenvalue decomposition of $\mathbf{C_X}$, the covariance matrix of $\mathbf{X}$, to determine a $k \times m$ matrix $\mathbf{V}$ and $\mathbf{Z} = \mathbf{VX}$ such that $\mathbf{C_Z} = \mathbf{I}_k$ where $\mathbf{I}_k$ denotes the identity matrix of order k. We do not remove the mean of the data in the whitening process, since we do not want to lose information about non-negativity of the sources.

Next we need to find a $k \times k$ orthogonal matrix $\mathbf{W} \in O(k)$, i.e. $\mathbf{W}^T\mathbf{W} = \mathbf{WW}^T = \mathbf{I}_k$ for which $\mathbf{Y} = \mathbf{WZ}$ is a permutation of $\mathbf{S}$, which is equivalent to find one for which $\mathbf{Y}$ is non-negative [7]. To do this we search $\mathbf{W} \in O(k)$ that minimizes the cost function:

$$J = \frac{1}{2}\|\mathbf{Y} - \mathbf{Y}_+\|_{\mathrm{F}}^2$$

where $[\mathbf{Y}_+]_{ij} = \max\{\mathbf{Y}_{ij}, 0\}$ and $\|\cdot\|_{\mathrm{F}}$ denotes the Frobenius norm.

The matrix $\mathbf{W}$ is initialized as the identity and, at each iteration, is multiplicatively updated by an orthogonal rotation matrix $\mathbf{R} \in SO(n)$, i.e. $\mathbf{R} \in O(n)$ and $\det(\mathbf{R}) = 1$. $\mathbf{R}$ is determined by finding a Newton-like update step using a first order Fourier expansion of the cost function along the steepest-descent geodesic.

2.3 Clustering

Similarly as in [6], genes are grouped following a winner-takes-all approach: a gene is assigned to the source j for which $[\mathbf{A}_N]_{ij} = \max\{[\mathbf{A}_N]_{il}, l = 1, \ldots, k\}$ where $\mathbf{A}_N$ is the normalized mixing matrix $\mathbf{A}$. For each gene cluster, the significance of its functional annotations is tested against a reference set (in this work, the whole genome), using the hypergeometric test.

3 Results

Results for a test set containing 575 genes are reported. The set was constructed in [2], and comprises genes of eight broad categories (namely "cell cycle", "cell wall organization and biogenesis", "DNA metabolism", "lipid metabolism", "protein biosynthesis", "response to stress", "signal transduction" and "transport"). Therefore, an analysis with k=8 factors was performed.

Significant GO terms were selected for each cluster (p-value $< 10^{-6}$), and their corresponding precision (P), recall (R) and F1[1] values were calculated using the complete gene set as reference. Clusters were labeled according to the GO term with highest F1 value among the significant terms. See Table 1 for average values within each cluster. As shown, clusters reveal some of the broad categories in the set (e.g. "cell cycle", "protein biosynthesis", "transport") while hiding others (e.g. 'signal transduction', "lipid metabolism").

In order to measure the performance of nnICA, we analyzed the same gene set using NMF (as proposed in [6]). Results are shown in Table 2. Some functional annotations are revealed by both approaches (e.g. "mitotic cell cycle",

Table 1. Clustering results obtained by nnICA (ordered by decreasing F1): number of genes, average precision (P), average recall (R) and average F1

Cluster	Genes	Avg.P	Avg.R	Avg.F1
(92.0) "ATPase activity, coupled to transmembrane movement of substances"	25	69.26	56.96	48.97
(77.4) "Translation"	49	49.61	68.38	34.87
(97.3) "Protein targeting"	57	38.37	69.74	32.34
(86.8) "Transcription, DNA-dependent"	67	34.23	64.31	31.60
(69.3) "Mitotic cell cycle"	48	28.97	70.24	30.25
(81.1) "DNA repair"	95	23.92	71.28	23.64
(49.6) "Catalytic activity"	93	24.84	66.78	22.06
(57.6) "Membrane" & (57.5) 'transport'	141	13.78	79.13	17.55
Average	72	35.37	68.35	30.16

Clusters are labeled with the GO term with highest F1 value (shown in brackets)

---

[1] $F1 = \frac{2PR}{P+R}$

574   M. Chagoyen et al.

Table 2. Clustering results obtained by NMF (ordered by decreasing F1): number of genes, average precision (P), average recall (R) and average F1

Cluster	Genes	Avg.P	Avg.R	Avg.F1
(76.2) "Transcription"	75	34.27	66.66	32.13
(71.6) "Ion transporter activity"	59	29.88	78.45	31.78
(70.3) "Protein targeting"	72	26.57	65.10	26.32
(66.7) "DNA repair"	72	27.05	65.03	24.77
(67.3) "Secretory pathway"	59	24.77	68.01	24.10
(51.7) "Mitotic cell cycle"	75	19.96	75.26	23.25
(51.3) "Signal transduction"	96	18.85	71.76	21.33
(50.9) "Hydrolase activity, acting on ester bonds"	67	24.14	65.75	20.63
Average	72	25.69	69.50	25.54

Clusters are labeled with the GO term with highest F1 value (shown in brackets).

"DNA repair", "protein targeting" and "transcription"), although clusters are dissimilar in size. On average, nnICA obtained similar results in terms of recall, although precision values are higher, providing therefore higher F1. This means that nnICA produced, on average, more homogeneous clusters.

4 Discussion and Conclusions

In recent years, a significant number of methods have been proposed to perform a functional interpretation of experimental data. In this work we have explored the use of nnICA to perform a functional classification and assessment of a gene set using functional annotations. In order to demonstrate the validity of our approach, we analyzed a number of sets from a model organism, *S. cerevisiae*, and compare the results with those obtained by NMF (as in [6]).

ICA produces a linear transformation that minimizes the statistical dependence between basis vectors, producing a decomposition of data that seems to reveal truly biological functional independence. Functional classification by nnICA analysis was able to obtain similar results as NMF in terms of average cluster recall, while providing more homogeneous clusters (with higher average precision). These results encourages us to extend this study also to functional classification based on literature analysis.

Acknowledgements

This work has been partially funded by the Spanish grants CICYT BFU2004-00217/BMC, GEN2003-20235-c05-05, TIN2005-5619, PR27/05-13964-BSCH and a collaborative grant between the Spanish CSIC and the Canadian NRC (CSIC-050402040003). PCS is recipient of a grant from CAM. HF is supported by Programa Inov Contacto (ICEP-Portugal) sponsored by the Ministério da Economia e Inovação and Prime Program (EU). APM acknowledges the support of the Spanish Ramón y Cajal program.

References

1. Homayouni, R. and Heinrich, K. and Wei, L. and Berry, M. W. Gene clustering by latent semantic indexing of MEDLINE abstracts. Bioinformatics **21**, 104-15 (2005).
2. Chagoyen, M. and Carmona-Saez, P. and Shatkay, H. and Carazo, J. M. and Pascual-Montano, A. Discovering semantic features in the literature: a foundation for building functional associations. BMC Bioinformatics **7**, 41 (2006).
3. Chagoyen, M. and Carmona-Saez, P. and Gil, C. and Carazo, J. M. and Pascual-Montano, A. A literature-based similarity metric for biological processes. BMC Bioinformatics **7**, 363 (2006).
4. Khatri, P. and Done, B. and Rao, A. and Done, A. and Draghici, S. A semantic analysis of the annotations of the human genome. Bioinformatics **21**, 3416 (2005).
5. Bodenreider, O. and Aubry, M. and Burgun, A. Non-lexical approaches to identifying associative relations in the gene ontology. Pac Symp Biocomput 91-102 (2005).
6. Pehkonen, P. and Wong, G. and Toronen, P. Theme discovery from gene lists for identification and viewing of multiple functional groups. BMC Bioinformatics **6**(1), 162 (2005).
7. Plumbley, M. D. Algorithms for nonnegative independent component analysis. IEEE Transactions on Neural Networks **14**, 534-543 (2003).
8. Plumbley, M. D. Optimization using Fourier expansion over a geodesic for nonnegative ICA. Independent Component Analysis and Blind Signal Separation **3195**, 49-56 (2004).
9. Ashburner, M. et al. Gene ontology: tool for the unification of biology. The Gene Ontology Consortium. Nat Genet **25**(1), 25-9 (2000).

New Trends in the Analysis of Functional Genomic Data

David Montaner[1,2], Fatima Al-Shahrour[1], and Joaquin Dopazo[1,2]

[1] Bioinformatics Department, Centro de Investigacin Prncipe Felipe, CIPF,
Autopista del Saler 16, E-46013, Valencia, Spain
dmontaner@cipf.es
[2] Functional Genomics Node, INB, CIPF, Autopista del Saler 16, E-46013,
Valencia, Spain

1 Replications of the Same Statistical Test

Most analyses carried out using high throughput data consist of the repetition of the same statistical test for all genes in the dataset. As a result of such replicated analysis we get, for each gene, several estimates of statistical parameters: statistics, p-values or confidence intervals. Being aware that most statistical methods were developed to test for a single hypothesis, researchers will usually correct p-values for multiple testing before choosing a cut-off that will indicate the rejection of the null hypotheses, whichever it is. Once chosen the genes with alternative pattern (meaning different form the one stated in the null hypothesis) the next step is to biologically interpret such departure from hypothesis. Different repositories of functionally relevant biological information such as Gene Ontology [1], KEGG [2] or Interpro [3] are available and can be used for the functional annotation of genome-scale experiments. Thus the functional properties of the selected genes can be analysed.

The trouble of this approach is that, by discarding genes with p-values above the cut-off, we loose most of our information. Not only we loose the measurements taken over the genes but also the functional annotation that could be linked to them from repositories, making it difficult the biological interpretation of results.

2 Blocks of Functional Genes

Aiming to prevent such waste of information, some authors have recently proposed to directly analyse the behaviour of blocks of functionally related genes in a whole-genome context. The Gene Set Enrichment Analysis (GSEA) [4,5], the FatiScan [6,7] or the Global Test [8,9] constitute examples of this type of approach inspired from systems biology. This three methodologies address the issue of whether the general expression pattern of a group of genes, for example a GO term or a KEGG pathway, changes across biological conditions. Here we will discus just some particular aspects of these methods but a more general view of this and similar methods can be found in Dopazo's revision of 2006 [10].

The Global Test uses generalised linear models to study the relationship between the expression of the genes of the block of interest and a characteristic associated to each biological sample. Such characteristic may be a categorical condition, like the class of the microarray in the context of differential gene expression, or a continuous variable such as a level of a metabolite. In this approach we can see a change in the philosophy of the analysis. The unit of interest is not any more a single gene but a block of genes with a common biological meaning. This new way of looking at the data provides, among others, obvious advantages for the biological interpretation of results and for the p-value adjustment. We just need to correct by the number of blocks, usually smaller than the number of genes.

3 The Overall Approach

The block of genes is also the unit of interest of the GSEA and the FatiScan. These two methods are similar to the Global Test in that they are also used to discover groups of genes which overall expression pattern changes across biological conditions. Nevertheless, GSEA and FatiScan consider all genes in the data when analysing each of the blocks. They compare the pattern of the genes of one block with the general pattern of the genes in the whole dataset. GSEA is particularly designed for the two class comparison context while FatiScan may be applied in a wider range of studies.

The rationale underlying both methodologies is that, if a property of genes can be described using a continuous index, then the statistical distribution of such index within a functional block of genes can be compared to the general distribution of the index across all genes in the data. We can therefore asses whether the property described by the index depends on the characteristic that defines the block of genes.

As said before GSEA is developed for the two class comparison. In this methodology, a signal-to-noise ratio comparing mean expression across classes is computed for each gene in the dataset. This statistic can be seen as a continuous index that ranks the genes according to their differential expression, from those more expressed in one of the biological conditions to those more expressed the second condition, passing through those genes non differentially expressed. Then, given a block of genes, for instance a functional class that we may be interested in, we can compare the distribution of the signal-to-noise ratio of the genes in the block to the distribution of the same statistic in the remaining genes. If the values of the signal-to-noise ratio are, for instance, systematically higher in the genes of the block compared to the genes in the whole dataset, we will conclude that, as a block, the genes of the functional class of interest are overexpressed in one of the biological conditions. GSEA uses a modification of the Kolmogorov–Smirnov test to asses differences between the signal-to-noise ratio in the class of interest and in the rest of the genes. Significance of the modified Kolmogorov–Smirnov statistic is computed in GSEA using permutations of the expression data. The original expression data is permuted several times, the signal-to-noise ratios are calculated over each permuted expression dataset and the modified Kolmogorov–Smirnov statistic is computed over each new distribution of the signal-to-noise ratio. Thus GSEA can estimate the random variability of the Kolmogorov–Smirnov statistic and test its significance in the original data.

4 Detaching Concepts and Algorithms

FatiScan follows the same analytical philosophy than GSEA but with a more general and flexible approach. FatiScan implements a segmentation test which checks for asymmetrical distributions of biological labels associated to genes ranked by any index. The main difference is that FatiScan does not implement a permutation test to asses such asymmetry. Therefore, the algorithm that computes the index and the algorithm that analyses the distribution of the index are completely separated so the calculations can be done in two different steps. This means that FatiScan can be used to study the relationship between biological labels associated to genes and any type of experiment whose outcome is a sorted list of genes or a variable that can be used to rank genes according to some characteristic of interest. Block of genes sorted by differential expression between two experimental conditions can be studied as it would be done using GSEA. But with FatiScan we can also consider many other gene properties or characteristics.

We can easily explore the correlation between gene expression and a clinical continuous variable such as the level of a metabolite. First, for each gene we will compute the correlation between its expression measurements and the levels of the metabolite. Thus we can range the genes from those which expression is more positively correlated to the levels of the metabolite to those inversely correlated, passing by genes which expression does not correlate with the clinical variable. In a second step, FatiScan explores the distribution of such correlation measurements, testing whether the distribution of correlations within a block of genes is different from the overall distribution of correlation in the dataset.

We can fit a Cox proportional hazard model to each gene in our data in order to study the relationship between gene expression and survival times. The estimates of the slope coefficients may be used as an index that ranks genes from those which increased expression is associated with long time survival to those which increased expression is associated to an early death. After computing this rank-index, FatiScan will find those blocks of genes for which the distribution of the slopes differs from the global distribution of the slopes.

The complete separation of the two steps in FatiScan analysis is the key point which provides its flexibility to the method. Such flexibility makes possible to handle many different sources of information, not only microarray gene expression data. Any lists of genes ranked by any other experimental or theoretical criteria can be studied. Genes can be for example arranged by physico-chemical properties, mutability, structural parameters and so on. In order to understand whether there is some biological feature, characterised by the blocks of genes, which is related to the experimental parameter studied.

5 Coda

The three methodologies here mentioned illustrate two of the main new conceptual trends in the analysis of functional genomic data.

The first one is the change of the descriptive unit used to address biological studies, shifting from gene to functional class. Gene still remains the unit of measured information, as what we record at the end is gene expression. But the conceptual entity over which biological interpretation is done, is the functional class of genes. New analytical strategies, like those above mentioned, should consider this fact in order to use the available information in the most efficient an meaningful way.

The second one is probably more subtle but not less important. Usual genomic studies follow the classical statistical approach in which one or several hypotheses are stated, estimate statistics and p-values are computed from data and finally, hypotheses are accepted or rejected depending on such estimated values. The analytical approach explicit in FatiScan an implicit in GSEA shows how estimated values provided by one first statistical analysis are not directly interpreted in terms of acceptance or rejection of hypotheses. Instead they are treated as variables quantifying some characteristic of the genes under study. This new variables may then be analysed using statistical methodologies. Thus, statistical results of one step of the analysis become themselves a new dataset which needs to be explored in a second analytical step. As we see, modular implementations of complex data analysis strategies like FatiScan, seem to be both, conceptually useful for the analysis of biological data and computationally advantageous, calling for the development of the theoretical framework within which combinations of statistical methods can be properly done.

References

[1] Ashburner, M., Ball, C.A., Blake, J.A., Botstein, D., Butler, H., Cherry, J.M., Davis, A.P., Dolinski, K., Dwight, S.S., Eppig, J.T., et al.; Gene ontology: tool for the unification of biology. The Gene Ontology Consortium. Nature Genet., 25, 25-29 (2000)

[2]   Kanehisa, M., Goto, S., Kawashima, S., Okuno, Y., Hattori, M.; The KEGG resource for deciphering the genome. Nucleic Acids Res., 32, D277-D280 (2004)

[3]   Mulder, N.J., Apweiler, R., Attwood, T.K., Bairoch, A., Bateman, A., Binns, D., Bradley, P., Bork, P., Bucher, P., Cerutti, L., et al.; InterPro, progress and status in 2005. Nucleic Acids Res., 33, D201-D205 (2005)

[4]   Mootha, V.K., Lindgren, C.M., Eriksson, K.F., Subramanian, A., Sihag, S., Lehar, J., Puigserver, P., Carlsson, E., Ridderstrale, M., Laurila, E., et al; PGC-1alpha-responsive genes involved in oxidative phosphorylation are coordinately downregulated in human diabetes. Nature Genet., 34, 267-273 (2003)

[5]   Subramanian, A., Tamayo, P., Mootha, V.K., Mukherjee, S., Ebert, B.L., Gillette, M.A., Paulovich, A., Pomeroy, S.L., Golub, T.R., Lander, E.S., et al; Gene set enrichment analysis: a knowledge-based approach for interpreting genome-wide expression profiles. Proc. Natl. Acad. Sci. USA, 102, 15545-15550 (2005)

[6]   Al-Shahrour, F., Diaz-Uriarte, R., Dopazo, J.; Discovering molecular functions significantly related to phenotypes by combining gene expression data and biological information Bioinformatics, 21, 2988-2993 (2005)

[7]   Al-Shahrour F., Minguez P., Trraga J., Montaner D., Alloza E., Vaquerizas J.M., Conde L., Blaschke C., Vera J. and Dopazo J.; BABELOMICS: a systems biology perspective in the functional annotation of genome-scale experiments. Nucl Acids Res., 34, W472-W476 (2006)

[8]   Goeman, J.J., van de Geer, S.A., de Kort, F., van Houwelingen, H.C.; A global test for groups of genes: testing association with a clinical outcome. Bioinformatics, 20, 93-99 (2004)

[9]   Goeman J.J., Oosting J., Cleton-Jansen A.M., Anninga J.K., van Houwelingen H.C.; Testing association of a pathway with survival using gene expression data. Bioinformatics, 21, 1950-1957 (2005)

[10] Dopazo, J.; Functional Interpretation of Microarray Experiments. OMICS: A Journal of Integrative Biology, 10, 398-410 (2006)

Minisymposium: "Inverse Problems and Applications"

Oliver Dorn and Miguel Moscoso

Universidad Carlos III de Madrid, Leganés-Madrid, Spain
odorn@math.uc3m.es, moscoso@math.uc3m.es

The wide field of *Inverse Problems* plays an important role nowadays in industrial applications of applied mathematics. This minisymposium is intended to present some selected topics which have received much attention lately.

One main theme of the minisymposium is the multi-scale approach for solving inverse problems. In the chapter by D. Franceschini, M. Donelli, R. Azaro and A. Massa, the authors present and discuss an iterative multi-scaling approach with applications in inverse scattering. Furthermore, a multi-scale level set approach is presented in the chapter by I. Berre, M. Lien and T. Mannseth for the application of characterizing petroleum reservoirs.

A second main theme of the minisymposium is the investigation of fast adjoint techniques for finding descent directions (or updates) for shape-based inverse solvers in a variety of applications. Here, the chapter by N. Irishina, M. Moscoso and O. Dorn investigates a level set adjoint field approach for the early detection of breast cancer from microwave data. Microwave imaging is also the focus of the chapter by O. Dorn, M. El-Shenawee and M. Moscoso, which presents a shape-based adjoint field technique in 3D for the early detection of breast tumours. A method of moments is employed for the forward (and adjoint) modelling of the microwave fields. Another example from medical imaging is given in the contribution by A. Zacharopoulos, O. Dorn, S.R. Arridge, V. Kolehmainen and J. Sikora. It investigates an efficient adjoint-field approach for simultaneously updating shape parameters of simple geometric objects (such as ellipsoids) in the application of Diffuse Optical Tomography in 3D. The chapter by R. Villegas, O. Dorn, M. Moscoso and M. Kindelan presents a shape-based approach (with initializations obtained from geostatistics) for the history matching problem in petroleum reservoir engineering. It uses adjoint fields for calculating iterative updates for the level set function describing the unknown shapes. Finally, the chapter by N. Polydorides discusses the use of logarithmic barrier functions for regularizing high-contrast inverse problems of electrical impedance tomography with applications in biomedical imaging and industrial process tomography.

A Robustness Analysis of the Iterative Multi-Scaling Approach Integrated with Morphological Operations

D. Franceschini, M. Donelli, R. Azaro, and A. Massa

Department of Information and Communication Technology, University of Trento, Via Sommarive 14, 38050 Trento, Italy
andrea.massa@ing.unitn.it

Summary. The accuracy and the robustness of the iterative multi-resolution strategy in dealing with inverse scattering problems involving multiple objects configurations has been enhanced by introducing a suitable morphological processing. In an iterative fashion, a set of scatterers are reconstructed by progressively increasing the resolution level in the Regions-of-Interest (RoIs), where the objects are localized, identified through suitable morphological operations. Selected numerical results are presented and discussed in order to assess the accuracy and effectiveness of the morphological-based processing.

1 Introduction

In several applications (e.g., [1]), there is the need of retrieving unknown objects by processing the scattered electromagnetic radiation collected in a non-invasive fashion outside the domain under investigation.

However, although an accurate spatial resolution is required, unfortunately the information content of the data is limited [2]. In order to satisfy the accuracy requirements exploiting the limited amount of available information, different kinds of multiresolution approaches have been proposed [3, 5].

In this framework, the Iterative Multi-scaling Approach (IMSA) has shown a satisfactory robustness [5] in dealing with single-scatterer geometries. Successively, it has been extended to multiple objects [6] allowing the detection of non-connected RoIs by means of a *Clustering Procedure* [6]. However, such an implementation showed a reduced effectiveness in dealing with some complex configuration. Consequently, an improved procedure for the detection of the RoIs is needed for enhancing the reconstruction accuracy. Towards this end, a bare thresholding of the retrieved profile (with the risk of omitting the weakest scatterers of the scenario) has been substituted in this chapter by a more effective set of morphological operations.

An outline of the chapter is as follows. The key features of the morphological-based processing are presented in Sect. 2. Successively, a preliminary robustness analysis is carried out in Sect. 3 by considering a numerical benchmark. Eventually, some conclusions will be drawn in Sect. 4.

2 Mathematical Formulation

Referring to a two-dimensional geometry, a cross-section of an inhomogeneous investigation domain D invariant along the $\hat{z}$ axis is illuminated by a set of monochromatic incident electric fields TM-polarized impinging from V different directions $[E_v^{\text{inc}}(x,y)\hat{z}, v = 1, ..., V]$.

The electromagnetic interactions among the scatterers modeled through the following object function:

$$\tau(x,y) = \varepsilon_r(x,y) - 1 - j\frac{\sigma(x,y)}{2\pi f \varepsilon_o}, \tag{1}$$

where $\varepsilon_r(x,y)$ and $\sigma(x,y)$ are the relative permittivity and conductivity of the medium, and the incident fields are estimated by collecting the scattered field values $E_v^{\text{scatt}}(x_{m(v)}, y_{m(v)})$ at $m(v) = 1, ..., M(v)$, $v = 1, ..., V$ positions belonging to the measurement domain D_M located outside D. These data are related to the contrast function $\tau(x,y)$ by means of two integral equations, the so-called "*Data equation*" and "*State equation*" (for further details see [5] and the references cited therein).

Usually, the electromagnetic inversion of these integral equations is performed by looking for a finite representation of the unknowns [i.e., $\tau(x,y)$ and $E_v^{\text{tot}}(x,y)$ in D]. Because of the intrinsic bound of the information collectable from the scattering experiments, an effective choice of the basis function is mandatory for obtaining a finer resolution in the RoIs of D.

Likewise other multi-scaling approaches [5,6] and because of the non linear nature of the problem at hand and its intrinsic ill-posedness, the proposed inversion scheme (M-IMSA) recasts the reconstruction to the minimization of a suitable multi-resolution cost function [6] by means of a multi-step ($s = 1, .., S_{\text{opt}}$) procedure. At each step of the minimization procedure, the supports $\Omega_i^{(s)}$, $i = 1, ..., I(s)$ of the $I(s)$ RoIs are determined applying a morphological processing to the contrast function. This set of operations begins with a noise filtering stage where the gray-level representation of the reconstructed profile is processed in order to deblur the image from the noise thus avoiding an overestimate of the number of RoIs. Accordingly, a new distribution of the object function $\tau_{nf}^{(s)}(x_{n_{r(i)}}, y_{n_{r(i)}})$ is determined

$$\tau_{nf}^{(s)}(x_{n_{r(i)}}, y_{n_{r(i)}}) = \begin{cases} 0 & \text{if } \tau^{(s)}(x_{n_{r(i)}}, y_{n_{r(i)}}) \leq \eta_{nf} \\ \tau^{(s)}(x_{n_{r(i)}}, y_{n_{r(i)}}) & \text{if } \tau^{(s)}(x_{n_{r(i)}}, y_{n_{r(i)}}) > \eta_{nf} \end{cases} \tag{2}$$

$(x_{n_{r(i)}}, y_{n_{r(i)}})$ being the center of the nth discretization cell of the ith RoI at the rth resolution level and $\eta_{\text{cl}} = \alpha \max\{\tau^{(s)}(x_{n_{r(i)}}, y_{n_{r(i)}})\}$ where α is a threshold to be heuristically calibrated.

Successively, in order to smooth the object function and reduce the occurrence of isolated artifacts, the contrast $\tau_{nf}^{(s)}$ is further processed for defining the following distribution

$$\tau_{lpf}^{(s)}\left(x_{n_{r(i)}}, y_{n_{r(i)}}\right) = \sum_{p=-1}^{1}\sum_{t=-1}^{1} A(p,t)\tau_{nf}^{(s)}\left(x_{n_{r(i)}+p}, y_{n_{r(i)}+t}\right) \qquad (3)$$

$A(p,t)$ being a spatial averaging mask defined as

$$A(p,t) = \begin{cases} \gamma & \text{if } p = j = 0 \quad p = -1, ..., 1 \\ \delta & \text{otherwise} \qquad\quad t = -1, ..., 1 \end{cases} \qquad (4)$$

where

$$\begin{aligned} \gamma &= 1 - \frac{\beta + 20}{100} \\ \delta &= \frac{\beta + 20}{100}\frac{1}{(L_B - 1)} \end{aligned} \qquad (5)$$

β being a calibration parameter and L_B is the number of neighborhood pixels. Then, the identification of the RoIs is carried out firstly performing the thresholding of $\tau_{lpf}^{(s)}$

$$\tau_T^{(s)}\left(x_{n_{r(i)}}, y_{n_{r(i)}}\right) = \begin{cases} 0 & \text{if } \tau_{lpf}^{(s)}\left(x_{n_{r(i)}}, y_{n_{r(i)}}\right) \le \eta_T \\ 1 & \text{if } \tau_{lpf}^{(s)}\left(x_{n_{r(i)}}, y_{n_{r(i)}}\right) > \eta_T \end{cases} \qquad (6)$$

where $\eta_T = \beta \max\left\{\tau_{lpf}^{(s)}\left(x_{n_{r(i)}}, y_{n_{r(i)}}\right)\right\}$, and successively eroding $\tau_T^{(s)}$ by means of a structuring element $B_{n_{r(i)}}$ [located at $\left(x_{n_{r(i)}}, y_{n_{r(i)}}\right)$ and defined over a window W_B of $L_B = 3 \times 3$ neighborhood pixels]. These morphological operations provide the binary image $\tau_E^{(s)}$

$$\tau_E^{(s)}\left(x_{n_{r(i)}}, y_{n_{r(i)}}\right) = \begin{cases} 1 \ \ if & \begin{bmatrix}\tau_T^{(s)}\left(x_{n_{r(i)}}, y_{n_{r(i)}}\right) = 1\end{bmatrix} \text{ and} \\ & \begin{bmatrix}\sum_{p=-1}^{1}\sum_{t=-1}^{1}\tau_T^{(s)}\left(x_{n_{r(i)}+p}, y_{n_{r(i)}+t}\right) = 1\end{bmatrix} \\ 0 & \text{\textit{otherwise}} \end{cases} \qquad (7)$$

and the seeds (i.e., the isolated pixel) of the objects determine the origin of the square area (the RoI $\Omega_i^{(s)}$) whose contour has zero intersection with the object function $\tau_{lpf}^{(s)}$. Eventually, a suitable set of finer basis functions is allocated into the $I(s)$ regions, thus increasing the spatial resolution in the RoIs of D and consequently the accuracy of the reconstructions[1]. The M-IMSA terminates when the stationary condition for the reconstruction, defined in [6], is verified.

---

[1]The minimization of the multi-resolution cost function is iteratively performed with a conjugate-gradient procedure as in [6].

3 Numerical Results

The accuracy of the M-IMSA has been preliminary assessed by considering two set of experiments concerned with the scatterer configuration shown in Fig. 1a. The reference scenario is composed by three $0.60\lambda_0$-sided square homogeneous dielectric ($\tilde{\tau}_1 = \tilde{\tau}_2 = \tilde{\tau}_3 = 1.0$) objects located at ($\tilde{x}_0^{(1)} = \tilde{y}_0^{(1)} = 0.385\lambda_0$), ($\tilde{x}_0^{(2)} = -\tilde{y}_0^{(2)} = -0.385\lambda_0$), and ($\tilde{x}_0^{(3)} = 0.0\lambda_0$, $\tilde{y}_0^{(3)} = -0.385\lambda_0$) in a square investigation domain $L_D = 3.0\lambda_0$-sided and illuminated by $V = 8$ plane waves. The computational domain D has been initially partitioned into $N_{(1)} = 144$ square sub-domains and at each view ($V = 8$) $M(v) = 15$ data have been collected on a circular measurement domain of radius $\rho = 4\lambda_0$.

As far as the numerical assessment is concerned, a robustness test against noise and aspect-limited data has been carried out. In the first case, the data have been blurred with a gaussian noise characterized by a signal-to-noise ratio in the range $5\,\mathrm{dB} \le \mathrm{SNR} \le 30\,\mathrm{dB}$, while the second set refers to an aspect-limited configuration characterized by blind angle $0° \le \theta_\mathrm{b} \le 120°$. These tests are aimed at obtaining some indications on the robustness of the morphological operations[2] also in the presence of systematic errors in the scattering data or when the measurement setup is not able to collect the whole set of available information coming from scattered data.

Figure 1b shows the distribution retrieved through the M-IMSA when SNR=20 dB. As it can be observed, the RoIs are rightfully detected and a satisfactory reconstruction of scatterers is reached. On the contrary, the presence of a blind angle of $\theta_\mathrm{b} = 90°$ (Fig. 1c) notably reduces the inversion accuracy.

In order to verify such indication, the results of a wider numerical analysis has been summarized in Fig. 2 in terms of the quantitative error figures as defined in [5]. As it can be observed, the total reconstruction error is $\varepsilon_{\mathrm{tot}} < 20\%$

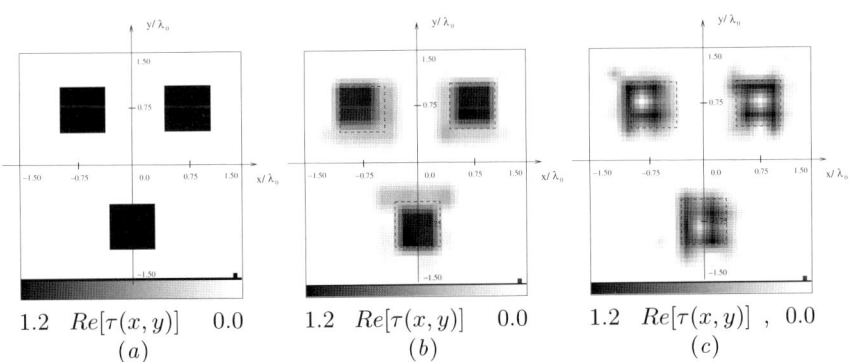

1.2 $Re[\tau(x,y)]$ 0.0   1.2 $Re[\tau(x,y)]$ 0.0   1.2 $Re[\tau(x,y)]$, 0.0
(a)   (b)   (c)

Fig. 1. Multiple scatterers [SNR = 20 dB]. (**a**) Reference distribution of the object function. Reconstructed distributions (at $S_{\mathrm{opt}} = 2$) when (**b**) $\theta_\mathrm{b} = 0°$ and (**c**) $\theta_\mathrm{b} = 90°$

---

[2]The parameters $\alpha = 15$ and $\beta = 25$ have been heuristically calibrated.

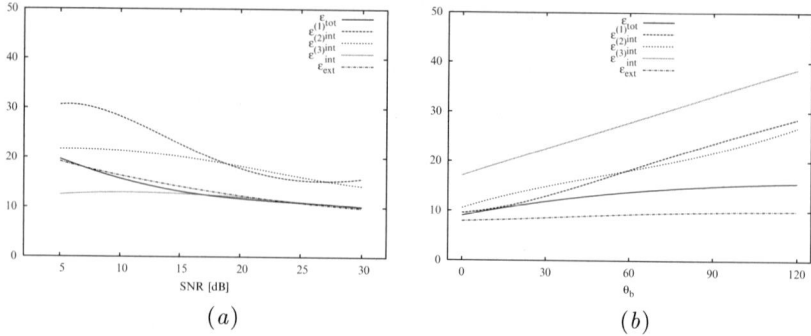

Fig. 2. Multiple scatterers. Behavior of the quantitative error figures vs. the **(a)** signal-to-noise ratio and **(b)** θ_b [SNR $= 20$ dB]

whatever the analysis. However, the sensitiveness to θ_b is greater than that to SNR, as suggested by the behaviour of the internal errors. As a matter of fact, when θ_b enlarges, the accuracy in retrieving each single scatterer decreases.

4 Conclusions

A preliminary assessment of the morphological-based IMSA for the detection of multiple and non-connected RoIs have been carried out. The results from the numerical analysis points out the robustness of the M-IMSA vs. the noise level, while the aspect limited nature of the measurement system seems to be a more critical issue to be carefully addressed for obtaining accurate reconstructions.

References

[1] Bolomey J.C.: Frontiers in Industrial Process Tomography. Eng. Foundation, London, U.K. (1995)

[2] Bucci O.M., Franceschetti G.: On the degrees of freedom of scattered fields. IEEE Trans. Antennas Propagat., **37**, 918–926 (1989)

[3] Miller L., Willsky A.S.: A multiscale, statistically based inversion scheme for linearized inverse scattering problems. IEEE Trans. Geosci. Remote Sensing, **34**, 346–357 (1996)

[4] Bucci O.M., Crocco L., Isernia T., Pascazio V.: Wavelets in nonlinear inverse scattering. Proc. Geoscience and Remote Sensing Symp., **7**, 3130–3132 (2000)

[5] Caorsi S., Donelli M., Franceschini D., Massa A.: A new methodology based on an iterative multiscaling for microwave imaging. IEEE Trans. Microwave Theory Tech., **51**, 1162–1173 (2003)

[6] Caorsi S., Donelli M., Massa A.: Detection, location, and imaging of multiple scatterers by means of the iterative multiscaling method. IEEE Trans. Microwave Theory Tech., **52**, 1217–1228, (2004)

Iterative Microwave Inversion Algorithm Based on the Adjoint-Field Method for Breast Cancer Application

Oliver Dorn[1], Magda El-Shenawee[2], and Miguel Moscoso[1]

[1] Universidad Carlos III de Madrid
   odorn@math.uc3m.es, moscoso@math.uc3m.es
[2] University of Arkansas
   magda@uark.edu

Summary. Our goal is to develop an inversion algorithm for reconstructing the shape of 3D breast tumors using electromagnetic data. The method of moments (MoM) forward solver is used to calculate the electric and magnetic equivalent surface currents at the tumor interface and consequently the scattered electromagnetic fields. Using a so-called "adjoint scheme" for gradient calculation, the mismatch between calculated and measured fields at the receivers is used as new sources at all receiver locations and is back-propagated towards the tumor. The gradient is calculated then simultaneously for all nodes of the guessed tumor surface in order to obtain a correction displacement of each individual node of the surface which points into a descent direction of a least-squares cost functional. This process is repeated iteratively until the cost has decreased satisfactorily. Numerical results in 3D are presented based on the proposed technique using multiple transmitting sources/receivers at multiple microwave frequencies.

1 Introduction

Microwave tomographic imaging is showing significant promise as a new technique for the early detection of breast cancer. Its physical basis is the high contrast between the dielectric properties of the healthy breast tissue and the malignant tumors at microwave frequencies [Gabriel et al. (1996)]. As a consequence, microwave imaging systems which aim at detecting, localizing and characterizing tumors in the breast are being developed. Among them, we mention for example confocal imaging and near-field tomographic reconstructions (see [Fear et al. (2002)] and references therein).

Mathematically, microwave medical tomography amounts to solving a nonlinear inverse problem for some form of Maxwell's equations in which a given cost functional is minimized via an iterative algorithm. Traditional iterative algorithms, well suited for nonlinear inverse problems and based on pixel reconstruction techniques suffer from several drawbacks in this application, amongst them the need of strong regularization for stabilizing the algorithms which typically is done by adding a Tikhonov–Philips term to the cost functional. This, however, has the effect of severely smoothing out interfaces between tumors and surrounding tissue. Therefore, new approaches that avoid these difficulties need to be investigated. We will present here a shape-based approach for this application which allows to reconstruct quite general shapes by moving each individual surface node until a given cost functional is minimized. For more details on shape-based reconstruction schemes in various applications see for example the discussion led in [El-Shenawee et al. (2006),2].

2 Shape Reconstruction in Microwave Imaging

Dropping out the time dependence $e^{i\omega_k t}$, we consider the system of Maxwell's equations

$$\nabla \times \mathbf{E}_{jk}(\mathbf{x}) - \alpha_k(\mathbf{x})\mathbf{H}_{jk}(\mathbf{x}) = 0 \qquad (1)$$

$$\nabla \times \mathbf{H}_{jk}(\mathbf{x}) - \beta_k(\mathbf{x})\mathbf{E}_{jk}(\mathbf{x}) = 0 \qquad (2)$$

in a domain $\Omega \subset \mathbf{R}^3$, where $\beta_k(\mathbf{x}) = \sigma(\mathbf{x}) + i\omega_k\epsilon(\mathbf{x})$ and $\alpha_k(\mathbf{x}) = -i\omega_k\mu(\mathbf{x})$ and where ω_k, $k = 1,\ldots,\underline{k}$, are the different (angular) frequencies of the applied fields. We will assume that $\alpha_k(\mathbf{x})$, $\beta_k(\mathbf{x})$ are constant outside some sufficiently large ball, with values denoted by $\alpha_{k,0}$ and $\beta_{k,0}$, respectively. With this assumption, we can apply the standard radiation condition outside this ball. The index j in (1), (2) indicates the different incoming radiation patterns (plane waves).

We will consider here the situation that the coefficient functions $\alpha_k(\mathbf{x})$ and $\beta_k(\mathbf{x})$ contain discontinuities along closed interfaces $\Gamma_m \subset \Omega$, $m = 1,\ldots,\underline{m}$, such that we add standard interface conditions to (1), (2). Given incoming plane waves corresponding to index jk, we can write the total field in the medium as

$$\mathbf{E}_{jk}^{\text{tot}} = \mathbf{E}_{jk}^{\text{inc}} + \mathbf{E}_{jk}^{\text{scat}}, \qquad \mathbf{H}_{jk}^{\text{tot}} = \mathbf{H}_{jk}^{\text{inc}} + \mathbf{H}_{jk}^{\text{scat}} \qquad (3)$$

where $\mathbf{E}_{jk}^{\text{inc}}$ and $\mathbf{H}_{jk}^{\text{inc}}$ satisfy (1), (2) with $\alpha_k = \alpha_{k,0}$ and $\beta_k = \beta_{k,0}$. Let us assume that we have $\underline{l}$ receivers available at locations $\mathbf{d}_l$, $l = 1,\ldots,\underline{l}$. At these receiver positions, we can decompose the scattered electric fields as

$$\mathbf{E}^{\text{scat}}(\mathbf{d}_l) = \mathbf{E}^r(\mathbf{d}_l)\hat{r} + \mathbf{E}^\theta(\mathbf{d}_l)\hat{\theta} + \mathbf{E}^\phi(\mathbf{d}_l)\hat{\phi}. \qquad (4)$$

Here, $\hat{r}$, $\hat{\theta}$ and $\hat{\phi}$ are the polar unit vectors at the points $\mathbf{d}_l$. With this, we can define the linear measurement operators M_{jkl} by

$$M_{jkl}\mathbf{E}_{jk} = \mathbf{E}_{jk}^\theta(\mathbf{d}_l)\hat{\theta} + \mathbf{E}_{jk}^\phi(\mathbf{d}_l)\hat{\phi} \qquad (5)$$

which measure the "plane wave components" of the scattered fields at the given receiver location. We will assume in the following that the coefficient α_k is fixed and known to be $\alpha_k = \alpha_{k,0}$. We will write then $M_{jk}\mathbf{E}_{jk}(\beta_k)$ for the vector of all measured fields which correspond to the parameter β, the frequency ω_k and the incoming plane wave with index j. Furthermore, g_{jk} will denote the corresponding physically measured ("true") data. With this notation, we can define the least squares cost functional

$$\mathbf{J}_{jk}(\beta_k) = \frac{1}{2}\|R_{jk}(\beta_k)\|^2 \tag{6}$$

where $R_{jk}(\beta_k) = M_{jk}\mathbf{E}_{jk}(\beta_k) - g_{jk}$ is the residual operator for indices (jk). In the shape inverse problem, we assume that

$$\beta_k(\mathbf{x}) = \begin{cases} \beta_i & \text{for} \quad \mathbf{x} \in D, \\ \beta_e & \text{for} \quad \mathbf{x} \in \Omega \backslash D. \end{cases} \tag{7}$$

When deforming in a given step of the iterative inversion scheme the current shape D by a vector field $\mathbf{v}$ (i.e., each point $\mathbf{x} \in D$ is displaced according to $\mathbf{x} \rightarrow \mathbf{x} + \mathbf{v}(\mathbf{x})$) then the fields and therefore also the least squares cost will change. We want to find a vector field such that $\mathbf{J}$ is reduced by the corresponding deformation. It has been shown in [2] that the deformation of the boundary $\Gamma = \partial D$ by a sufficiently small vector field $\mathbf{v}$ gives rise to a change in the cost

$$\delta\mathbf{J}_{jk} = \text{Re} \int_{\partial D} [R'_{jk}(\beta_k)^* R_{jk}(\beta_k)]\overline{\beta_i - \beta_e}\mathbf{v}(\mathbf{x}) \cdot \mathbf{n}(\mathbf{x}) \, ds(\mathbf{x}) \tag{8}$$

where Re denotes 'real part', $R'_{jk}(\beta_k)^*$ denotes the formal adjoint operator of the linearized residual operator $R'_{jk}(\beta_k)$ and $\mathbf{n}(\mathbf{x})$ is the normal direction to the boundary Γ in the point $\mathbf{x}$. Therefore, it is sufficient to find a vector field in the normal direction to the boundary $\mathbf{v}_d(\mathbf{x}) = F_d(\mathbf{x})\mathbf{n}(\mathbf{x})$ which points into a descent direction of the cost $\mathbf{J}$. Obviously, we can choose

$$F_d(\mathbf{x}) = -\gamma\text{Re}\left\{R'_{jk}(\beta_k)^* R_{jk}(\beta_k)\overline{\beta_i - \beta_e}\right\} \tag{9}$$

for a sufficiently small positive step size $\gamma > 0$. Plugging $\mathbf{v} = \mathbf{v}_d$ into (8) shows us that then the cost is reduced. The expressions $R'_{jk}(\beta_k)^* R_{jk}(\beta_k)$ are calculated by an efficient "adjoint scheme" as explained for example in [Dorn et al. (1999)]. This scheme requires us to run just one forward and one adjoint simulation for a given frequency and incoming wave in order to evalute the gradient expressions (9) at all nodes simultaneously.

3 Numerical Experiments

In our numerical experiment shown here the true object is a sphere of radius $1\,\text{cm}$ located at the center of the computational domain of $20 \times 20 \times 20\,\text{cm}^3$. Inside the object, the relative dielectric constant is $50 - j12$, and in the background it is $9 - j1.2$. The total number of transmitters (receivers) is 30 with

5 of them being located at each plane of constant azimuth angle (starting at $\theta = 0.1\pi$ and ending at $\theta = 0.9\pi$) with φ between 0 and 2π . Two frequencies are used here, namely $f = 3\,\text{GHz}$ and $f = 5\,\text{GHz}$ ($\omega = 2\pi f$). Plane waves are used to excite the object with incident polarization in the θ-direction. The results shown here are for the co-polarization case, where both the incident and scattered plane waves are in the θ-direction. Synthetic data is generated using the method of moments, where the surface of the object is discretized into surface nodes and triangular patches similar to the work of [El-Shenawee et al. (2006)]. The number of discretization points in the θ- and φ-directions are 8 and 16, respectively, for the true object, while the object generated at each inversion iteration is discretized into 10 and 16 points, respectively. The gradient-based algorithm using adjoint fields is implemented such that the location of each surface node will be corrected into the normal direction of the current boundary by an amount given in (9) for each node using a fixed small step-size factor γ (being 10^{-5} at $3\,\text{GHz}$ and 10^{-3} at $5\,\text{GHz}$). A regularization step is applied after each update which amounts to filtering neighbouring

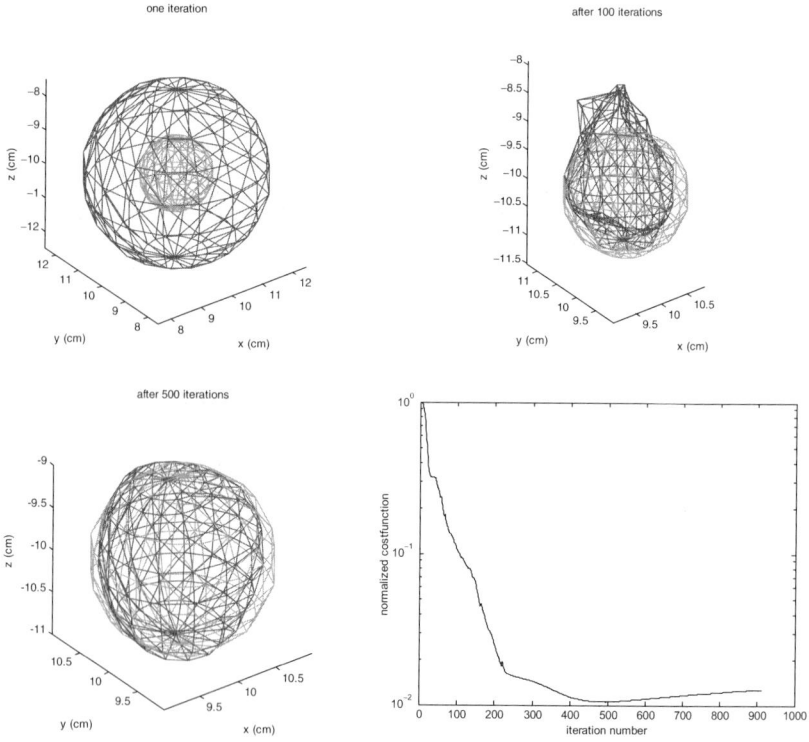

Fig. 1. Reconstruction of a small sphere. The true object is displayed in gray colour, and the reconstructed object by black colour in each iteration. *Top left*: after one iteration; *top right*: after 100 iterations; *bottom left*: after 500 iterations. The bottom right shows the evolution of the cost

nodes by an averaging filter in order to obtain a smooth surface. This smoothing operation will be discussed in more detailes in a forthcoming publication.

In this work, the main focus is on reconstructing the shape of the object assuming the knowledge of its position and electrical properties. The initial guess in this case is a sphere of radius 2 cm located at the same position as the true object. Figure 1 shows the true object (gray) and the guessed object (black) at iteration numbers 1 (top left), 100 (top right) and 500 (bottom left), where the latter one corresonds approximately to the lowest cost value which we could achieve during our reconstruction. The evolution of the total cost (summed over all indices) is displayed in the bottom right image of the figure. We mention that the reconstruction at the final iteration number 900 looks quite similar to that one at number 500. We conclude that our algorithm has converged in a stable way to the correct sphere.

Acknowledgements

This research was sponsored in part by the National Science Foundation award no. ECS-0524042, in part by Spanish Ministerio de Educacin y Ciencia (MEC), Grant n FIS2004-03767, and in part by the Women Giving Circle at the University of Arkansas grant no. WGC-22.

References

[Gabriel et al. (1996)] C. Gabriel, S. Gabriel, R.W. Lau, and E. Corthout, "The dielectric properties of biological tissues: Part I, II, and III," *Phy. Med. Biol.*, vol. 41, pp. 2231-2249. 1996.

[Fear et al. (2002)] C. E. Fear, S.C. Hagness, P.M. Meaney, M. Okoniewski, and M. A. Stuchly, "Breast tumor detection," *IEEE Microwave Mag.*, pp. 48-56, March 2002.

[Dorn et al. (2006)] Dorn, O., Lesselier, D.: Level set methods for inverse scattering. Inverse Problems, **22**, R67–R131 (2006)

[Dorn et al. (1999)] Dorn, O., Bertete-Aguirre, H., Berryman, J., Papanicolou, G.: A nonlinear inversion method for 3D electromagnetic induction tomography using adjoint fields, Inverse Problems, **15**, 1523–1558 (1999)

[El-Shenawee et al. (2006)] Magda El-Shenawee and Eric Miller, "Spherical Harmonics Microwave Algorithm for Shape and Location Reconstruction of Breast Cancer Tumor," IEEE Trans. Medical Imaging, vol. 25, no. 10, pp. 1258-1271, October 2006.

Iterative Microwave Inversion for Breast Cancer Detection Using Level Sets

Natalia Irishina, Miguel Moscoso, and Oliver Dorn

Universidad Carlos III de Madrid, Leganes-Madrid, Spain
nirishin@math.uc3m.es, moscoso@math.uc3m.es, odorn@math.uc3m.es

Summary. In this chapter we analyze the potential of a shape-based model based on a level-set technique for the early detection of breast cancer tumors from microwave data. A reconstruction using a shape-based model offers several advantages like well-defined boundaries and the incorporation of an intrinsic regularization that reduces the dimensionality of the inverse problem whereby at the same time stabilizing the reconstruction. In this chapter, we present a novel strategy that is able to detect very small tumors compared to the wavelength used for illuminating the breast.

1 Introduction

The use of microwaves for the early detection of breast tumors shows great promise as an alternative technique to the more traditional use of X-rays. The physical basis of microwave imaging in this application is the high contrast between the dielectric properties of the healthy breast tissue and the malignant tumors at microwave frequencies [Gabriel et al. (1996)]. The goal in microwave imaging is therefore to detect, localize and characterize hidden tumors in the breast.

Mathematically, microwave tomography constitutes a nonlinear inverse problem in which a given cost functional is minimized via an iterative algorithm. Traditional iterative pixel-based algorithms turn out to suffer from several drawbacks related to the smoothing property of standard regularization techniques (like Tikhonov–Philips) in this application. In particular, they do not allow for representing the sharp discontinuities that exist between tumors and the healthy tissue. Therefore, our group has developed recently a novel shape-based reconstruction technique for microwave imaging which uses a level set representation of shapes for representing the basic features in the medium, like skin, fatty tissue and tumors. See [Irishina et al. (2006), 2].

In this chapter we present an adaptation of our level set approach for the early detection of small breast tumors. Our algorithm is able to start without any pre-specified starting guess for the location, being able to create shapes at any position of the domain and thereby avoiding the often encountered problem of local minima. It does so during the early iterations taking into account the data and the sensitivity structure of the inverse problem. Once a good first approximation for the shape is found, the algorithm continues in a completely automatic way with optimizing this shape, until the data least-squares data misfit (cost functional) is sufficiently reduced.

2 Mathematical Model

In this chapter we consider a heterogeneous 2D medium Ω, embedded in a layer of skin as shown in the top left image of Fig. 1, and illuminated by TM waves. In this case, the scalar Helmholtz equation

$$\Delta u + \kappa(\mathbf{x})u = q(\mathbf{x}) \quad \text{in } \Omega \tag{1}$$

is a good approximation for describing the non-zero component of the electric field u. In (1), $\kappa(\mathbf{x}) = \omega^2 \mu_0 \epsilon_0 \left[\epsilon(\mathbf{x}) + i\frac{\sigma(\mathbf{x})}{\omega \epsilon_0}\right]$ is the complex wave number, where ϵ is the relative dielectric constant and σ is the conductivity. q is the

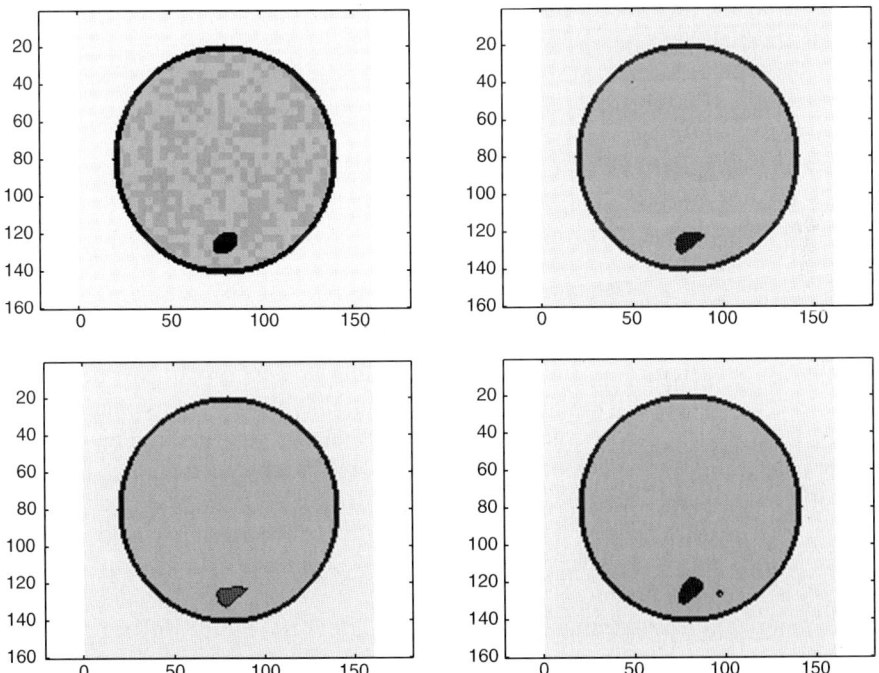

Fig. 1. *Top left*: realistic true model. *Top right*: final reconstruction $\varepsilon = 36$. *Bottom left*: final reconstruction $\varepsilon = 20$. *Bottom right*: final reconstruction $\varepsilon = 45$

source. In the shape-based approach we assume that $\kappa(\mathbf{x})$ is a piecewise constant function that can be written as

$$\kappa(\psi(\mathbf{x})) = \begin{cases} \kappa_i \text{ inside } S \text{ where } \psi(\mathbf{x}) \le 0 \\ \kappa_e \text{ outside } S \text{ where } \psi(\mathbf{x}) > 0 \end{cases} \tag{2}$$

where $\psi(\mathbf{x})$ is a sufficiently smooth level set function which represents the shape S of the tumor. The boundary of the tumor, δS, consists of all the points where $\psi(\mathbf{x}) = 0$. Here, we focus on the shape reconstruction problem assuming that the electrical properties of the tumor and the healthy tissue are given. The goal is to find an evolution law for the unknown level set function ψ which reduces, and eventually minimizes, the cost functional

$$\mathcal{J}(\psi) = \frac{1}{2} \|\mathcal{R}(\kappa(\psi))\|^2. \tag{3}$$

In (3), $\mathcal{R}(\kappa(\psi))$ denotes the mismatch between the true boundary data and those calculated by the forward model using the parameter distribution $\kappa(\psi)$. In order to obtain an evolution of the unknown function ψ, we consider the evolution law

$$\frac{\mathrm{d}\psi}{\mathrm{d}t} = f(\mathbf{x}, t) \tag{4}$$

with some forcing term $f(\mathbf{x}, t)$ which still needs to be specified. Formally differentiating the least squares cost functional $\mathcal{J}(\kappa(\psi(t)))$ with respect to the artifical time t and applying the chain rule yields

$$\frac{\mathrm{d}\mathcal{J}}{\mathrm{d}t} = \mathrm{Re} \int_\Omega \mathcal{R}_l'(\kappa)^* \mathcal{R}(\kappa)\, (\kappa_e - \kappa_i)\delta(\psi)\, f(\mathbf{x}, t)\, \mathrm{d}\mathbf{x}, \tag{5}$$

where Re indicates the real part of the corresponding quantity and $\mathcal{R}_l'(\kappa)^*$ denotes the formal adjoint of the linearized residual operator $\mathcal{R}_l'(\kappa)$. Obviously, the following choice of the forcing term

$$f(\mathbf{x}) = -\mathrm{Re}\left((\kappa_e - \kappa_i)\,\mathcal{R}_l'(\kappa)^*\mathcal{R}(\kappa)\right) \quad \text{for all } \mathbf{x} \in \Omega \tag{6}$$

defines a descent direction for the least squares cost. It can be computed efficiently by using an adjoint scheme [Irishina et al. (2006)]. We note that our search direction $f(\mathbf{x}, t)$ has the property that it can be applied even if there is no initial shape available when starting the algorithm. Therefore, it allows for the creation of objects at any point in the domain, by lowering a positive level set function until its values arrive at zero. This property is useful for avoiding certain types of local minima which often occur in level set formulations which are solely based on the propagation of an already existing shape. See the discussion in [2].

Numerically discretizing (4) by a straightforward finite difference time-discretization with time-step $\tau > 0$ and interpreting $\psi^{(n+1)} = \psi(t + \tau)$ and $\psi^{(n)} = \psi(t)$ yields the iteration rule

$$\psi^{(n+1)} = \psi^{(n)} + \tau f(\mathbf{x}), \quad \psi^{(0)} = \psi_0. \tag{7}$$

3 Numerical Experiments

Our numerical setup is a $2D$ tomographic configuration. We have a set of 40 transducers which are equidistantly located around the 12-cm-diameter breast. They illuminate the breast using different frequencies. Here we use the iterative algorithm for five frequencies, namely 500, 800, 1,000, 1,500 and 2,000 MHz. In our reconstruction algorithm we assume that the only unknown is the relative permittivity inside the tumor to which we assign different values ($\varepsilon_{in} \in \{20, 36, 45\}$) for the search of shape and location. The value of the true tumor is $\varepsilon = 36$. We use a mesh of 160×160 pixels of size 1×1 mm^2 each. In the following we briefly summarize the algorithm explained above.

Our realistic model consists of the surrounding medium with $\varepsilon_{liquid} = 2.5$, $\sigma_{liquid} = 0.04$ Siemens per meter (S m^{-1}), the breast tissue with $\varepsilon_e = 9.0$, $\sigma_e = 0.4$ S m^{-1}, the skin layer with $\varepsilon_s = 34.0$, $\sigma_s = 4.0$ S m^{-1}, and the tumor with $\varepsilon_{in} = 36.0$, $\sigma_{in} = 4.0$ S m^{-1}. To simulate the heterogeneity of human breast tissue we add random variation to the background parameter distribution of up to $\pm 5\%$, distributed over 4×4 mm^2 squares. The data in the detectors are perturbed by 5% white Gaussian noise. We suppose that all the dielectric constants of the breast and surrounding medium are known (except of the ramdom perturbations), as well as the conductivity value of the tumor. The tumor is 118 pixels size and situated at 12 pixels depth. We start with a level set function equal to one in all the domain. In the first experiment we assign $\varepsilon_{in} = 36.0$ to the reconstructed tumor. We perform a loop of 20 iterations in which we use boundary data corresponding to 5 frequencies (between 500 and 2,000 MHz) one after the other. During this loop, the level set function is iteratively updated, and the location of the tumor is finally found along with an approximation of its shape. In order to investigate the performance of our algorithm in the situation where we do not know the correct value of the permittivity of the tumor, we repeat the above experiment using instead of the true permittivity value $\varepsilon_{in} = 36.0$ two incorrect approximations during the reconstruction, namely $\varepsilon_{in} = 20.0$ and $\varepsilon_{in} = 45.0$. Figure 1 shows on the top left the true model and on the top right the final reconstruction when assuming the correct permittivity value $\varepsilon_{in} = 36.0$. On the bottom left and bottom right are shown the final reconstructions when assuming $\varepsilon_{in} = 20.0$ and $\varepsilon_{in} = 45.0$ instead, respectively. The evolution of the cost during these three reconstructions is displayed in Fig. 2. The solid line corresponds to $\varepsilon_{in} = 36.0$, the dash-dotted line to $\varepsilon_{in} = 20.0$ and the dashed line to $\varepsilon_{in} = 45.0$. We see that in all reconstructions the correct location of the tumor is recovered in an efficient and stable way even in this realistic case of noisy data, unknown fluctuations in the background and incorrectly assumed permittivity value. Certainly, some ghost objects can occur in this situation which, however, are significantly smaller than the tumor. The size of the reconstructed tumor seems slightly depending on the choice of the permittivity value which is used for the reconstruction. Avoiding the ghost objects and reconstructing

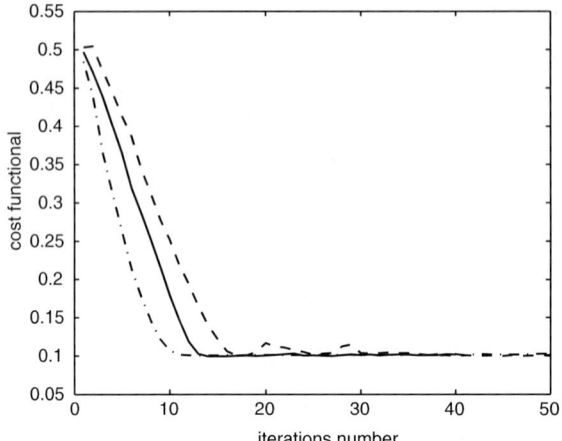

Fig. 2. Evolution of cost during shape reconstruction. *Solid*: $\varepsilon_{in} = 36.0$, *dash-dotted*: $\varepsilon_{in} = 20.0$, *dashed*: $\varepsilon_{in} = 45.0$

the correct permittivity values together with shape and location of the tumor will be the focus of our future work.

Acknowledgements

This work was financed by the Spanish Ministry of Education and Science, Grant no. FIS2004-03767.

References

[Gabriel et al. (1996)] C. Gabriel, S. Gabriel, R.W. Lau, and E. Corthout, "The dielectric properties of biological tissues: Part I, II, and III," *Phy. Med. Biol.*, vol. 41, pp. 2231-2249. 1996.

[Irishina et al. (2006)] N. Irishina, M. Moscoso, and O. Dorn,, "Detection of small tumors in microwave medical imaging using level sets and MUSIC," *Proc. PIERS, Cambridge*, pp. 43-47, 2006.

[Dorn et al. (2006)] Dorn, O., Lesselier, D.: Level set methods for inverse scattering. Inverse Problems, **22**, R67–R131 (2006)

Characterization of Reservoirs by Evolving Level Set Functions Obtained from Geostatistics

Rossmary Villegas, Oliver Dorn, Miguel Moscoso, and Manuel Kindelan

Universidad Carlos III de Madrid, Leganes-Madrid, Spain
rvillega@math.uc3m.es, odorn@math.uc3m.es, moscoso@math.uc3m.es,
kinde@ing.uc3m.es

Summary. In the chapter we discuss the use of sequential Gaussian simulations in order to create geostatistical initial guesses for an earlier introduced level set based shape reconstruction algorithm for the history matching problem in reservoir characterization. We present and discuss numerical results which compare the performance of the reconstruction algorithm for these different initial guesses.

1 Introduction

In the water-flooding process of secondary oil recovery, water is injected under high pressure into so-called injection wells in order to enhance oil production at the production wells. In order to optimize the oil production, in the history matching problem it is attempted to use the measured production data of the water-flooding process in order to estimate the physical properties (in our case permeability) inside the reservoir. The corresponding mathematical inverse problem is severely ill-posed and underdetermined such that strong regularization tools need to be employed during the inversion. Our group has recently introduced a novel shape-based reconstruction techniques which aims at reconstructing regions of different rock-types (and if necessary also smoothly varying internal profiles inside these regions) from production data. In order to represent the different geological regions, a level set technique is employed in this algorithm [Villegas et al. (2005), Villegas et al. (2006)]. In the current chapter we will concentrate on discussing the use of different initial guesses in order to start the iterative inversion algorithm for reconstructing the unknown interfaces. In particular, we will investigate the use of geostatistical techniques for constructing initial realizations of reservoirs.

Our simplified model for describing two-phase flow of oil and water in the Earth (modeled as a porous medium) is

$$-\nabla \cdot \left[T\nabla p \right] = Q \qquad \text{in} \qquad \Omega \times [0, t_{\mathrm{f}}] \tag{1}$$

$$\phi \frac{\partial S_{\mathrm{w}}}{\partial t} - \nabla \cdot [T_{\mathrm{w}} \nabla p] = Q_{\mathrm{w}} \qquad \text{in} \quad \Omega \times [0, t_{\mathrm{f}}] \tag{2}$$

for the two unknowns p (pressure) and S_{w} (water saturation). In the following, the subindices w and o will always indicate "water" and "oil", respectively. $\Omega \subset \mathbf{R}^n$ ($n = 2, 3$) is the modeling domain with boundary $\partial \Omega$, and $[0, t_{\mathrm{f}}]$ is the time interval for which production data is available. We denote by $\phi(\mathbf{x})$ the porosity, and by T_o, T_w and T the transmissibilities, which are known functions of the permeability K and the water saturation S_w:

$$T_{\mathrm{w}} = K(\mathbf{x}) \frac{K_{\mathrm{rw}}(S_{\mathrm{w}})}{\mu_{\mathrm{w}}}; \quad T_{\mathrm{o}} = K(\mathbf{x}) \frac{K_{\mathrm{ro}}(S_{\mathrm{w}})}{\mu_{\mathrm{o}}}; \quad T = T_{\mathrm{w}} + T_{\mathrm{o}}. \tag{3}$$

Here, the relative permeabilities $K_{\mathrm{rw}}(S_{\mathrm{w}})$ and $K_{\mathrm{ro}}(S_{\mathrm{w}})$ are typically available as tabulated functions, and μ_{w} and μ_{o} denote the viscosities of each phase. The quantities Q_{o}, Q_{w} and $Q = Q_{\mathrm{o}} + Q_{\mathrm{w}}$ represent the flows (oil, water, and total, resp.) at the few injection and production well locations in the reservoir. They define the measured data of our inverse problem. Equations (1)–(3) are solved with appropriate initial conditions, and a no-flux boundary condition on $\partial \Omega$.

In the shape inverse problem we assume now that the parameter K has the following specific form

$$K(\mathbf{x}) = \begin{cases} K_i(\mathbf{x}), & \text{where} \quad \psi(\mathbf{x}) \leq 0 \\ K_e(\mathbf{x}), & \text{where} \quad \psi(\mathbf{x}) > 0. \end{cases} \tag{4}$$

In this representation, $\psi(\mathbf{x})$ is the describing level set function. The two regions D (shale) and $\Omega \backslash D$ (sand) are accordingly given as $D = \{\mathbf{x} \in \Omega : \psi(\mathbf{x}) \leq 0\}$ and $\Omega \backslash D = \{\mathbf{x} \in \Omega : \psi(\mathbf{x}) > 0\}$. The boundary of D (denoted as $\Gamma = \partial D$) is defined by the zero level set of the level set function ψ, i.e., $\partial D = \{x : \psi(x) = 0\}$. For solving the inverse problem, we define an evolution of the level set function

$$\frac{\mathrm{d}\psi}{\mathrm{d}\tau} = f(\mathbf{x}, \tau) \tag{5}$$

for the level set function ψ such that upon convergence of this evolution the corresponding shape will minimize a suitably chosen cost functional (in our case the least-squares data misfit). An adjoint technique is used in order to calculate the forcing term $f(\mathbf{x}, \tau)$ in each step of this shape evolution. For more details regarding this level set based shape evolution algorithm we refer to [Villegas et al. (2005), Villegas et al. (2006)].

2 Sequential Gaussian Simulation for Constructing Initial Guesses

Sequential Gaussian Simulation (SGS) is used for estimating the reservoir characteristics (in our case permeability) in regions where these values are not available by taking into account the measurements of the values at the well positions. This procedure uses statistical assumptions on the distribution of the parameters in the reservoir which are expressed in a so-called semivariogram. Using this information, SGS creates a family of Gaussian realizations of the reservoir which are all equiprobable and honor the measured values at the well locations. By applying a threshold to these realizations, we obtain realizations of a binary reservoir (i.e. consisting of exactly two lithofacies). These realizations are used in order to calculate the corresponding level set functions as signed distance functions for these realizations, which are then used as initial level set functions for our shape evolution approach. Figure 2 shows in the upper three images of the left column three of these different (binary) realizations which have been generated by using the SGSIM program of the GSLIB Fortran library [Deutsch et al. (1997)] and thereafter applying a threshold.

In the following we briefly list the basic steps of our SGS process:

1. Generate a random path through the grid nodes of the reservoir.
2. Visit the first node along the path and use kriging to estimate a mean μ and standard deviation σ for the parameter at that node based on the available (measured or already estimated) values of this parameter at a set of close grid points.
3. Select a value at random from the corresponding (Gaussian) distribution of type (μ, σ) and put this value at the given grid node as parameter.
4. Visit each successive node in the random path and repeat the process, taking into account already estimated values inside a sufficiently small neighborhood of the current node.

For more details, see for example [Deutsch et al. (1997)].

3 Numerical Experiments and Discussion

Given the measured permeability values at the wells, we first create a semivariogram as described above and create a family of thresholded Gaussian realizations. Each of them can be used as initial guess for the level set reconstruction. In order to investigate the behavior of the cost during the reconstruction process in dependence of these initial values we have displayed in Fig. 1 the corresponding evolution of the cost for three choices of initial guesses. The selection has been made in dependence of the initial cost of these realizations when plugged into a reservoir simulator such as ECLIPSE. See the bottom left image of Fig. 1 for the initial cost values of a small selection of nine Gaussian realizations (numbers 1–9) and of our standard deterministic initial guess for the reservoir (number 10). Out of this selection, we have chosen furthermore three realizations (number 4, number 3 and number 8 representing the maximal, an intermediate and the minimal cost value, respectively) in order to run a shape reconstruction for each of them independently. In addition, we have run a reconstruction using our standard deterministic initial guess

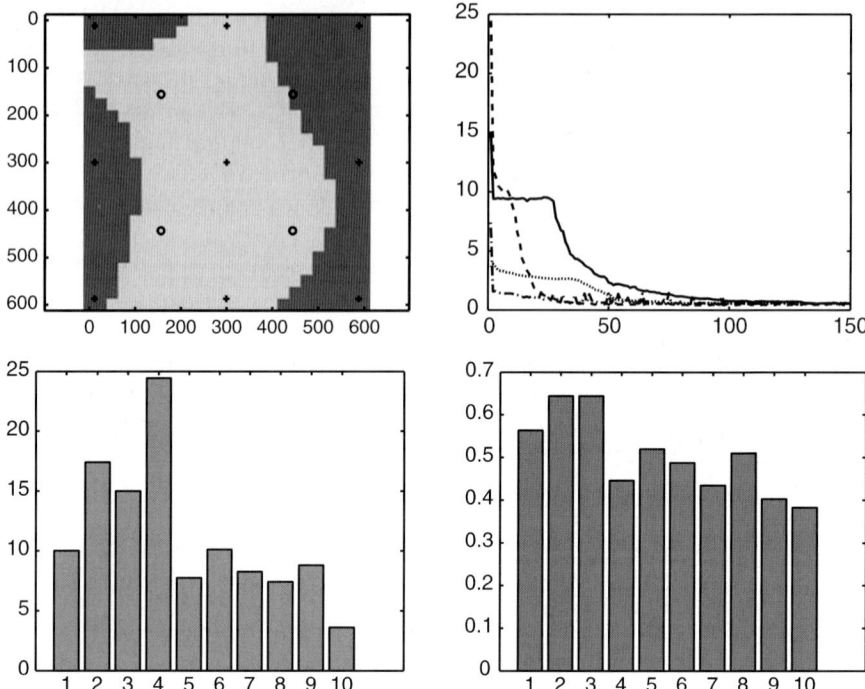

Fig. 1. *Upper left*: reference permeability distribution, shale (*dark gray*) and sand (*light gray*). *Upper right*: evolution of cost using Gaussian realizations 3 (*solid*), 4 (*dashed*), 8 (dash-dotted) and deterministic case 10 (*dotted*). *Bottom row*: Initial (left image) and final (after 150 iterations, right image) cost values for nine Gaussian realizations (columns 1–9) and one deterministic case (column 10)

based on well information. Figure 2 shows the corresponding initial realizations (left column) and the final reconstructions (right column) for these four choices.

We observe that our level set reconstruction technique yields good results for each of the Gaussian initializations. Gaussian simulations can provide us therefore with good initial guesses for our level set reconstructions. We mention that Gaussian initializations can also be used in order to design interesting alternative 'adjoint-free' statistical reconstruction techniques, which we plan to address in our future research.

Acknowledgements

Funding for this work was provided by the Dirección de Tecnología y Soporte Técnico, Repsol-YPF. R. Villegas and O. Dorn are grateful for partial support from the Institute for Mathematics and its Applications (IMA), Minneapolis.

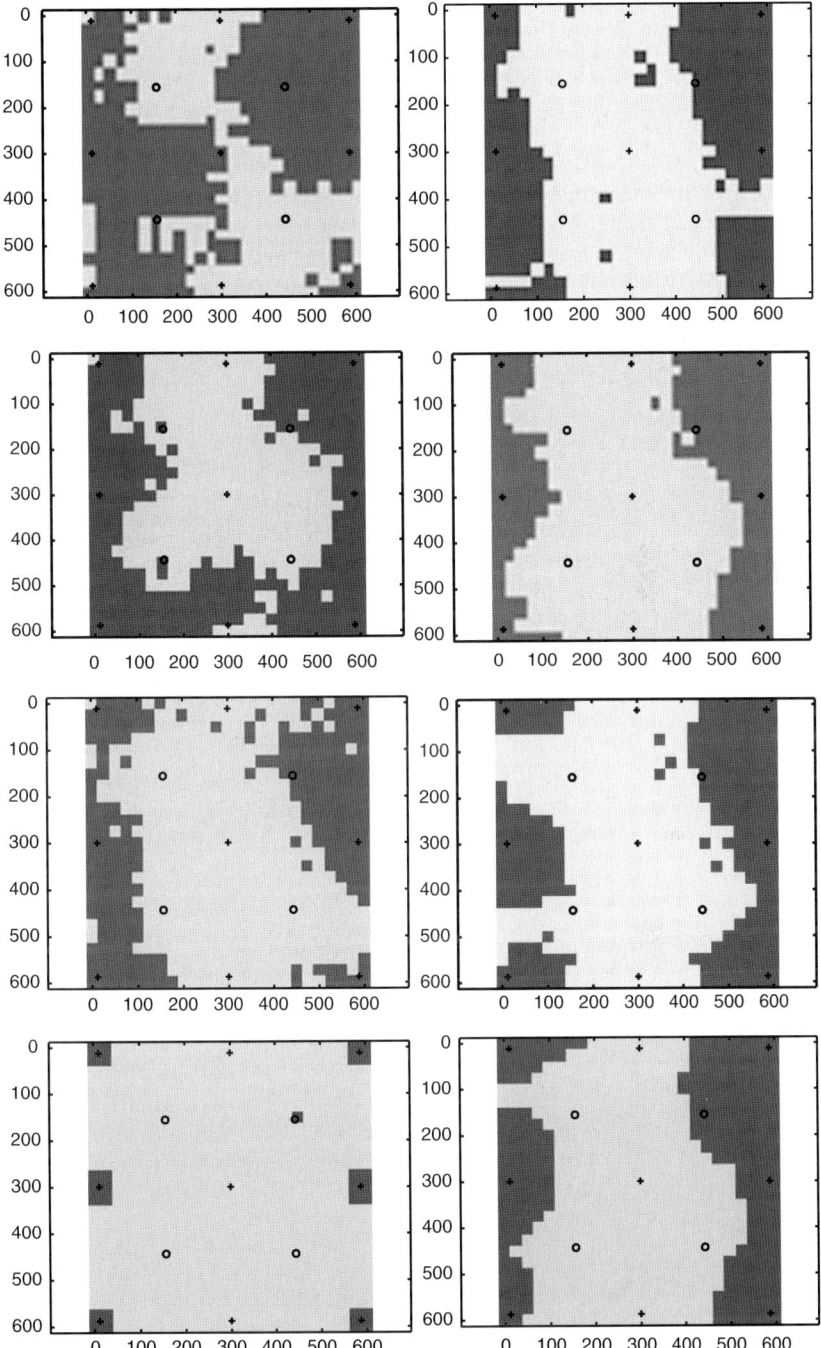

Fig. 2. *Left column*: Upper three images examples of Gaussian realizations (numbers 3, 4 and 8 of Fig. 1). *Bottom image*: deterministic initial guess (number 10). *Right column*: corresponding final reconstructions at iteration number 150

602   R. Villegas et al.

References

[Deutsch et al. (1997)] C. V. Deutsch and A. G. Journel, GSLIB: Geostatistical
    Software Library and User's Guide, Applied Geostatistics Series, Oxford Uni-
    versity Press, 1997
[Villegas et al. (2005)] R. Villegas R, O. Dorn, M. Moscoso and M. Kindelan, *Shape
    reconstruction from two-phase incompressible flow data using level sets*, Proc.
    International conference on PDE-based image processing and related inverse
    problems, CMA, Oslo: August 2005 (Springer: *to appear*) (2006).
[Villegas et al. (2006)] R. Villegas R, O. Dorn, M. Moscoso, M. Kindelan and
    F. J. Mustieles, *Simultaneous characterization of geological shapes and per-
    meability distributions in reservoirs using the level set method*, Proc. SPE
    Europec/EAGE Annual Conference and Exhibition, Vienna: June 2006 *in press*
    (2006).

Reconstruction of Simple Geometric Objects in 3D Optical Tomography Using an Adjoint Technique and a Boundary Element Method

A. Zacharopoulos[1], O. Dorn[2], S.R. Arridge[1], V. Kolehmainen[3], and J. Sikora[4]

[1] Department of Computer Science, UCL, Gower st. London, WC1E 6BT, UK
   A.Zacharopoulos@cs.ucl.ac.uk
[2] Universidad Carlos III de Madrid, Madrid, Spain
[3] University of Kuopio, Kuopio, Finland
[4] Warsaw University of Technology, Warsaw, Poland

Summary. In this paper we consider the recovery of ellipsoidal 3D shapes with piecewise constant coefficients in Diffuse Optical Tomography (DOT). We use an adjoint scheme for calculating gradients for the shape parameters defining the unknown ellipsoids, and a Newton-type optimisation process for the minimization of a least squares data misfit functional. A boundary integral formulation is used for the forward modelling. An advantage of the proposed method is the implicit regularisation effect arising from the reduced dimensionality of the inverse problem. Results of a numerical experiment in 3D are shown which demonstrate the performance of the method.

1 Introduction

In this paper, we explore an adjoint technique for the retrieval of the internal boundaries of 3D ellipsoidal regions in frequency domain Diffusive Optical Tomography (DOT), [1]. The optical parameters of interest in this application are μ_a being the absorption coefficient, μ'_s being the (reduced) scattering coefficient, and their combination $D = \frac{1}{3(\mu_a + \mu'_s)}$ being the diffusion coefficient. If the distribution of these optical parameters inside the body Ω is arranged into L disjoint regions Ω_ℓ with piecewise constant optical properties, the propagation of light can be described by a set of coupled Helmholtz equations [3]

$$-\nabla^2 \Phi_\ell + k_\ell^2 \Phi_\ell = q_j \quad \text{in } \Omega_\ell,$$

(1)

with interface conditions

$$\Phi_{\ell+1} = \Phi_\ell, \quad D_{\ell+1} \frac{\partial \Phi_{\ell+1}}{\partial \nu} = D_\ell \frac{\partial \Phi_\ell}{\partial \nu} \quad \text{on } \Gamma_\ell,$$

(2)

$$\Phi_1 + 2\alpha D_1 \frac{\partial \Phi_1}{\partial \nu} = 0 \quad \text{on } \partial\Omega.$$

(3)

Here, α models the refractive index difference at the boundary $\partial\Omega$ and the respective (complex) 'wavenumbers' are $k_\ell^2(\omega) = \frac{\mu_{a,\ell} + \frac{i\omega}{c}}{D_\ell}$. We consider in this paper the case that the domain of interest can be divided into two different zones, namely a background distribution and an embedded object whose shape can be approximately described by an ellipsoid.

2 Solution Strategy for the Inverse Problem

Discretising our forward problem (1)–(3) by the Boundary Element Method (BEM), the shapes and locations of the boundaries are described by finite sets of shape coefficients γ. In our case these shape coefficients are derived from a much smaller set of parameters $\{f_k\}$, $k = 1, \ldots, 6$, namely locations and semi-axes of the ellispoids, which are defined in the following section and which uniquely determine our ellipsoids during the reconstruction. In other words, we have $\gamma = \gamma(\{f_k\})$. Using the BEM, we construct a linear matrix equation of the form $\mathbf{T}(\gamma)\mathbf{f} = \mathbf{q}$. Introducing the measurement operator $\mathcal{M}$, we have the residual operator

$$\mathcal{R}(\{f_k\}) = \mathcal{K}(\{f_k\}) - \mathbf{g} = \mathcal{M}\mathbf{T}^{-1}(\gamma(\{f_k\}))\mathbf{q} - \mathbf{g}, \qquad (4)$$

where $\mathbf{g}$ are the physically measured data and $\mathcal{K}(\{f_k\})$ denotes the nonlinear forward operator which maps unknown ellipsoidal parameters to the corresponding measurements [1]. The least squares data misfit (cost) is given as

$$\mathcal{J}(\{f_k\}) = \frac{1}{2}\|\mathcal{R}(\{f_k\})\|^2. \qquad (5)$$

A typical way to minimise such a cost function is a Newton-type method, where we search for a minimum for $\mathcal{J}(\{f_k\})$ by iterations of local linearisation and Taylor expansion around the current estimate $\{f_k\}^{(n)}$ as

$$\{f_k\}^{(n+1)} = \{f_k\}^{(n)} + (\mathbf{J}_n^T\mathbf{J}_n + \Lambda)^{-1}\mathbf{J}_n^T(\mathbf{g} - \mathcal{K}(\{f_k\}^{(n)})). \qquad (6)$$

Here, the Jacobian $\mathbf{J}_n$ in step n of the algorithm is a discretized version of the descent directions (10), (13) derived further below (see also [3] for more details) and Λ is a Levenberg-Marquandt control term. In our implementation, we take Λ to be the identity. In addition, a quadratic fit line search method is applied in order to avoid detours in the downhill direction and speed up the optimisation.

3 Calculating Gradient Directions by an Adjoint Scheme

The boundary of an ellipsoid with center $\mathbf{w} = (w_x, w_y, w_z)$ and semi-axes $\mathbf{a} = (a_x, a_y, a_z)$ which is aligned with the cartesian axes is given by the implicit representation

$$\Phi(\mathbf{x}; \mathbf{a}, \mathbf{w}) = \frac{(x - w_x)^2}{a_x^2} + \frac{(y - w_y)^2}{a_y^2} + \frac{(z - w_z)^2}{a_z^2} - 1 = 0 \qquad (7)$$

with $\mathbf{x} = (x, y, z)$. In our inverse problem, we will transform an initial ellipsoid of this form into a final one which best fits the sought object according to some criterion. During the movement of the ellipsoid the distribution of the optical parameters $b = (\mu_a, D)^T$ changes (where T means 'transpose') inside Ω and with it the cost functional (5). The basic idea of our algorithm will be to define an evolution law (in artificial time t) for each of the six parameters $f_k \in \{w_x, w_y, w_z, a_x, a_y, a_z\}$, $k = 1, \ldots, 6$, of the form

$$\frac{\partial f_k}{\partial t} = h_k(t),$$

such that (5) is minimized upon convergence of this scheme. As shown in [2], finding such an evolution for each of the six parameters amounts to finding velocity fields $\mathbf{V}_k(x, t)$, $k = 1, \ldots, 6$, in the domain of interest which deform the given shapes (ellipsoids) in a controlled fashion, i.e., maintaining the shape of an ellipsoid. In order to find these different velocity fields for our case, we define

$$\mathbf{V}_k(\mathbf{x}, t) = F_k(t)\mathbf{e}_k \qquad (8)$$

for $k = 1, 2, 3$ where $\mathbf{e}_1 = \mathbf{e}_x$, $\mathbf{e}_2 = \mathbf{e}_y$ and $\mathbf{e}_3 = \mathbf{e}_z$ are the direction vectors of the cartesian axes x, y and z in $\mathbb{R}^3$. Here, $F_k(t)$ is a scalar function to be chosen properly. This velocity field $\mathbf{V}(\mathbf{x}, t)$ is in fact independent of the position $\mathbf{x} \in \Omega$, such that the argument $\mathbf{x}$ is not active. Motion of an object by these velocity fields corresponds to a simple translation along the three cartesian coordinate axes. Following arguments given in [2] we see that the response of the cost due to these velocity fields is described by

$$\frac{\partial \mathcal{J}(b)}{\partial t} = \left(\int_{\partial S(t)} [\mathcal{R}'(b)^* \mathcal{R}(b)] (b_i - b_e)\mathbf{e}_k \cdot \mathbf{n}(\mathbf{x}) ds \right) F_k(t), \qquad (9)$$

such that we find the steepest descent directions

$$F_k(t) = -\left(\int_{\partial S(t)} \mathcal{R}'(b)^* \mathcal{R}(b)(b_i - b_e)\mathbf{e}_k \cdot \mathbf{n}(\mathbf{x}) ds \right) \qquad (10)$$

for $k \in \{1, 2, 3\}$. Here, $\mathcal{R}'(b)^*$ is the formal adjoint to the linearized residual operator $\mathcal{R}'(b)$ which takes into account the correct interface conditions (3).

In addition to the above described translation, we can furthermore deform a given ellipsoid into another one by changing one or more of the individual semi-axes. For this purpose we define the following three velocity fields

$$\mathbf{V}_{s,l}(\mathbf{x}, t) = F_{sl}(t)\frac{l - w_l}{a_l}\mathbf{e}_l, \qquad (11)$$

with $l \in \{x, y, z\}$ corresponding to $k \in \{4, 5, 6\}$, respectively. These three velocity fields define an evolution law in the space of ellipsoids by evolving the three semi-axes a_l. In particular, we get the corresponding response of the cost due to this evolution

$$\frac{\partial \mathcal{J}(b)}{\partial t} = \int_{\partial S(t)} [\mathcal{R}'(b)^* \mathcal{R}(b)] (b_i - b_e) \frac{l - w_l}{a_l} \mathbf{e}_l \cdot \mathbf{n}(\mathbf{x}) ds\, F_{sl}(t). \quad (12)$$

Therefore, the steepest descent directions for the three semi-axes are given as

$$F_{sl}(t) = - \int_{\partial S(t)} \mathcal{R}'(b)^* \mathcal{R}(b)(b_i - b_e) \frac{l - w_l}{a_l} \mathbf{e}_l \cdot \mathbf{n}(\mathbf{x}) ds. \quad (13)$$

The extraction of the Jacobians $\mathbf{J}_n$ and of the descent directions (6) using these adjoint expressions for $F_k(t)$ and $F_{sl}(t)$ in the framework of our BEM implementation follows then the guidelines described in details in [3].

4 Results from 3D Simulations

In our experimental setup, a geometric model for an infant's head is created and treated as a homogeneous domain with an embedded inhomogeneity, which we try to recover. The optical parameters chosen for the homogeneous background are $\mu_a = 0.01\,\mathrm{cm}^{-1}$ and $\mu_s = 1\,\mathrm{cm}^{-1}$, and for the unknown embedded region (a tilted ellipsoid) Ω_2 we have $\mu_a = 0.05\,\mathrm{cm}^{-1}$ and $\mu_s = 2.0\,\mathrm{cm}^{-1}$. See Fig. 1.

Using this geometric setup, we assign 20 sources and 20 detector positions at the surface of the head. The modulation frequency of the sources is set to

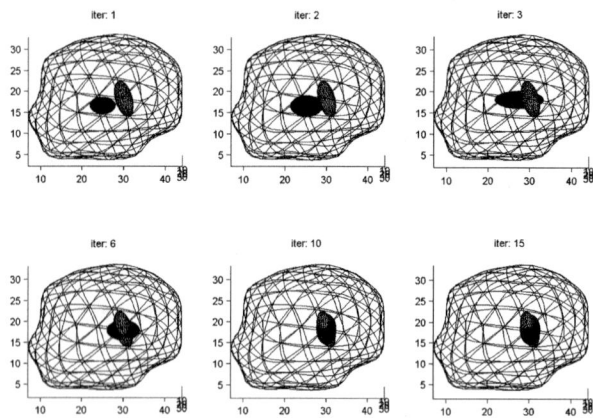

Fig. 1. Our numerical experiment for recovering an unknown object (here a tilted ellipsoid) from noisy DOT data. *Light gray:* the target inhomogeneity. *Dark gray:* the probing ellipsoid during its evolution. Upper row from *left* to *right:* iterations 1, 2, 3. Bottom row: iterations 6, 10, 15 of the Newton-type scheme evolving location and semi-axes of the probing ellipsoid which is aligned with the three cartesian axes

100 MHz. Synthetic data are then collected at the 20 detectors using our BEM forward modelling scheme. We split this data into real and imaginary parts of its logarithm to get the data vector **g**, see [3]. Gaussian random noise with a standard deviation of 1% of the measured signal is added to these data. For starting the reconstruction we select a random initial ellipsoid. The algorithm follows the residual minimization technique described above. Some steps of the evolution and the final reconstructed boundary are displayed in Fig. 1. As can be seen, the location and the approximate shape of the unknown object has been approximated with good accuracy by the reconstructed ellipsoid.

5 Conclusion

In the paper we have proposed an adjoint scheme for recovering a 3D ellipsoidal shape which best fits a given object in DOT. By using such a low-parameterized model we incorporate in our scheme a strong implicit regularisation which helps stabilizing the inverse problem. The adjoint scheme uses just one forward and one adjoint BEM simulation in each iteration for calculating gradient directions for all unknown parameters simultaneously. Even for our small set of six parameters this scheme for calculating gradient directions is significantly faster than the usually employed finite-differences scheme for parameterized representations. In our numerical experiments we have demonstrated that our scheme is able to reconstruct in a stable and efficient way a 3D ellipsoidal shape from few noisy data which approximates well an unknown object which is not representable in the space of search parameters due to the lack of rotation angles during the search.

References

1. Arridge S R 1999 Optical tomography in medical imaging *Inverse Problems* **15** 41–93
2. Dorn O and Lesselier 2006 Level set methods for inverse scattering *Topical Review, Inverse Problems* **22** R67–R131
3. Zacharopoulos A, Arridge S, Dorn O, Kolehmainen V and Sikora J 2006 Three dimensional reconstruction of shape and piecewise constant region values for Optical Tomography using spherical harmonic parameterisation and a Boundary Element Method *Inverse Problems* **22** 2175–2196

High Contrast Electrical Impedance Imaging

Nick Polydorides

Electrical and Computer Engineering, University of Cyprus
nick.polydorides@ucy.ac.cy

1 Introduction

In Electrical Impedance Imaging (EII) a finite number of electrodes are positioned at the boundaries of closed conducting domains [LPB05]. Typically some of the electrodes are used to inject low-frequency current patterns into the domain, while others sample the induced voltage potential at the boundary. In the image reconstruction problem, the interior admittivity distribution must be recovered using the acquired noise-infused boundary measurements. This nonlinear problem is ill-posed and therefore a regularization scheme is necessitated in order to yield a stable and unique solution.

2 The Forward Electrostatic Problem

In order to solve the inverse problem, one firstly attends to the forward problem of computing the boundary Dirichlet data when the domain's interior electrical properties and the excitation boundary conditions are known. This involves the solution of Maxwell's electrostatic equations according to the complete electrode model boundary conditions. In the discrete sense, one usually approximates the forward problem on finite dimensional models where the nonlinear forward operator $F := \mathcal{X} \to \mathcal{Y}$ relates the parameters of interest $\mathbf{x} \in \mathcal{X}$ to their corresponding boundary observations $\mathbf{y} \in \mathcal{Y}$ like

$$F(\mathbf{x}) = \mathbf{y}^{\text{exact}} + \epsilon = \mathbf{y} \qquad (1)$$

assuming $\mathbf{y}$ are measurements contaminated with a noise signal ϵ of level w so that $0 < \|\mathbf{y} - F(\mathbf{x}^*)\|_2 \leq w$ and $\mathbf{y} \notin R(F)$. In this work we focus attention to the case where F is twice Fréchet differentiable with non-closed range.

3 The Primal and Dual Inverse Problems

In a finite domain Ω with k degrees of freedom, given m noise contaminated boundary measurements $\mathbf{y} \in \mathbf{R}^m$, the bounds $l_j, u_j \in \mathbf{R}$ for $j = 1, \ldots, k$, with $0 < l_j \leq \mathbf{x}_j \leq u_j < +\infty$, the inverse problem is to find a solution $\mathbf{x}^* \in \mathcal{X}$ that satisfies

$$\mathbf{x}^* = \arg \min_{\mathbf{x} \in \mathcal{F}} f(\mathbf{x}) \quad \text{subject to} \quad \mathcal{F} = \{\mathbf{x} \,|\, l_j \leq \mathbf{x}_j \leq u_j, \, j = 1, \ldots, k\}, \quad (2)$$

where $f : \mathcal{X} \to \mathbf{R}$ is the twice continuously Fréchet differentiable in the interval $[l, u]$ cost function resembling the Euclidian norm of the model misfit according to the data

$$f(\mathbf{x}) = \frac{1}{2} \|F(\mathbf{x}) - \mathbf{y}\|^2. \tag{3}$$

If the prior information on the solution is cast in terms of the quadratic constraints

$$c_j(\mathbf{x}) = (u_j - \mathbf{x}_j)(\mathbf{x}_j - l_j) \exp(\theta) \geq 0, \quad j = 1, \ldots, k, \tag{4}$$

where $\theta = -2 \log\big((u_j - l_j)/2\big)$ is a scalar calibration parameter that preserves the non-negativity of the barrier, then one defines the logarithmic barrier function $\beta : \mathcal{X} \to \mathbf{R} \cup \{\infty\}$

$$\beta(\mathbf{x}, \mu) = f(\mathbf{x}) - \mu \sum_{j=1}^{k} \log c_j(\mathbf{x}), \tag{5}$$

where log is the natural logarithm. The formulation of the composite minimization functional β encompassing the misfit in the measurements according to F and the prior information constraints is differentiable for nonzero values of the barrier parameter, while for μ approaching zero the barrier term becomes singular at the bounds. The key idea behind this strategy is embodied in the interior point algorithms, where one identifies the global minimum of the regular-convex β in a Newton-Raphson methodology. This is also known as sequential unconstrained minimization process, which begins with a strictly feasible estimate on $\mathbf{x}^*$ in the interior of the feasible region $\mathcal{F}$, and then linearizes and solves a sequence of ν subproblems

$$\mathbf{x}^\nu = \arg \min_{\mathbf{x} \in \mathcal{F}} \beta(\mathbf{x}, \mu^\nu) \quad \text{for } \nu = 1, 2, \ldots \tag{6}$$

each time reducing the value of μ. This creates a sequence of unconstrained local minima which form a continuously differentiable path leading to the optimal solution $\mathbf{x}^*$ at the limit $\mu \to 0_+$. The required optimal solution will be adjacent to the boundary of $\mathcal{F}$ at a distance analogous to the level of noise in the data. Convergence criteria and properties of interior point methods can be found in the review article [FGR02]. In essence, the process implies that the optimum solution is the point where the data misfit is minimum and the constraints are just binding, which in principle is equivalent to regularization.

If the set of unconstrained minimizers of the barrier function is nonempty, then any sequence in this set has at least one convergent subsequence whose limit point is the optimal constrained solution [BER04]. If $\{\mathbf{x}^\nu\}$ is such a convergent subsequence, then the sequence of multipliers

$$\lambda_j^\nu(\mu) = \mu/c_j(\mathbf{x}^\nu) \tag{7}$$

is bounded. Moreover, dual multipliers provide estimates for the Lagrange multipliers at the optimum $\lim_{\nu \to \infty} \lambda_j^\nu(\mu) = \lambda_j^*$ where λ_j^* is the Lagrange multiplier associated with the jth constraint. Next we proceed to formulate the inverse problem treating $\mathbf{x}$ and λ as dependent primal and dual variables.

4 Primal–Dual Newton Equations

Relaxing the standard KKT conditions by μ we obtain the primal-dual Newton equations where the unique point $(\mathbf{x}^\nu, \lambda^\nu)$ that satisfies

$$\nabla f(\mathbf{x}^\nu(\mu)) - \nabla c(\mathbf{x}^\nu(\mu)) \lambda^\nu(\mu) = 0, \tag{8}$$

$$c(\mathbf{x}^\nu(\mu)) \cdot \lambda^\nu(\mu) - \mu \mathbf{1} = 0, \tag{9}$$

$$c(\mathbf{x}^\nu) > 0 \quad \text{and} \quad \lambda^\nu(\mu) > 0, \tag{10}$$

where $\mathbf{1}$ is the ones vector. In a matrix notation the above are expressed as

$$D_B(\mathbf{x}, \mu) = \begin{pmatrix} D_{B1} \\ D_{B2} \end{pmatrix} = 0, \tag{11}$$

where

$$D_{B1} = D_f - A^{\mathrm{T}} \lambda^\nu \quad \text{and} \quad D_{B2} = C\lambda^\nu - \mu \mathbf{1}. \tag{12}$$

Allowing $p, l \in \mathbf{R}^k$ to be the primal and dual Newton directions respectively, then applying Newton's method to the system (11) leads to the primal-dual formulation of the problem as $H_B \left(p \, l \right)^{\mathrm{T}} = -D_B$

$$\begin{pmatrix} H_f + I_\theta \Lambda & -A^{\mathrm{T}} \\ \Lambda A & C \end{pmatrix} \begin{pmatrix} p \\ l \end{pmatrix} = - \begin{pmatrix} D_{B1} \\ D_{B2} \end{pmatrix}, \tag{13}$$

which for a step length t yields a solution update $(\mathbf{x}^{\nu+1}, \lambda^{\nu+1}) \longleftarrow (\mathbf{x}^\nu + tp, \lambda^\nu + tl)$ where $I_\theta = 2\exp(\theta) I$, and I is the identity matrix. The Hessian is full rank and invertible only when the diagonal of the Lagrange multipliers is strictly positive, thus yielding unique and bounded updates in the primal and dual directions. We prove that following the proof of Lagrange multiplier theorem using a penalty approach. As the bound constraints are mutually independent, for ν sufficiently large the Jacobian of the constraints $A = \nabla c(x)$ has full column rank, and hence $(A^{\mathrm{T}} A)$ is invertible. At the regularized solution the dual multipliers satisfy

$$\lambda_j^* = \left(\nabla c_j(\mathbf{x}^\nu)^\mathrm{T} \nabla c_j(\mathbf{x}^\nu)\right)^{-1} \nabla c_j(\mathbf{x}^\nu)^\mathrm{T} D_f(\mathbf{x}^*), \tag{14}$$

where $D_f(\mathbf{x}^*)$ is the gradient of the objective cost term. Since $\mathbf{x}^*$ satisfies Morozov's discrepancy principle, then at the optimum constrained point

$$D_f(\mathbf{x}^*) = F'(\mathbf{x}^*)^\mathrm{T} \left(\mathbf{y}^{\mathrm{exact}} - \mathbf{y}\right) \neq 0$$

the right-hand side of (14) is bounded from below by the level of noise in the data, and thus $\lambda_j^* > 0$ for all j. Recalling that C, Λ and A are commuting diagonal matrices and iterates are strictly feasible $c(x) > 0$, the block $(2,2)$ of the Hessian in (13) can be eliminated to yield the equivalent condensed system

$$H_c(\mathbf{x}, \lambda)\, p = -\left(D_f(\mathbf{x}) - A^\mathrm{T}\lambda\right), \tag{15}$$

where the condensed primal-dual Hessian matrix at the νth iteration is expressed as

$$H_c^\nu = H_f(\mathbf{x}^\nu) - \sum_{j \in K} \frac{\mu^\nu}{c(\mathbf{x}^\nu)} H_g(\mathbf{x}^\nu) + \mu^\nu \nabla c(\mathbf{x}^\nu)^\mathrm{T} c^{-2}(\mathbf{x}^\nu) \nabla c(\mathbf{x}^\nu). \tag{16}$$

In the neighborhood of the optimum point $\nu \to \infty$ the Hessian becomes

$$H_{c,*} = H_* + \gamma A_*^\mathrm{T} A_* + M, \tag{17}$$

where $H_* = H_f(\mathbf{x}^*) - \sum \lambda^* H_g(\mathbf{x}^*)$, $\gamma > 0$, and M a positive semi-definite matrix. As the Hessian of the objective term H_f is effectively rank deficient and ill-conditioned, the Hessian of the constraints is $H_g = -2I_\theta \prec 0$, then it follows that $H_{c,*}$ is positive definite only if $H_* + \gamma A_*^\mathrm{T} A_* \succ 0$. Since H_g is essentially a constant multiple of the identity, then H_* is positive definite in a Tikhonov regularization sense. Moreover, at the regularized point $(\mathbf{x}^*, \lambda^*)$ the multipliers are strictly positive due to the impact of Morozov's principle on $D_f(\mathbf{x}^*)$. The positive definiteness of the condensed Hessian follows from Debreu's lemma for any positive coefficient γ given that the gradient of the constraints matrix A_* is full rank positive definite diagonal.

5 Numerical Results

Two inhomogeneities with magnitudes at $3\,\mathrm{S\,m}^{-1}$ and $6\,\mathrm{S\,m}^{-1}$ are immersed into an otherwise homogeneous cylindrical domain of background of $1\,\mathrm{S\,m}^{-1}$, planes of which are illustrated in left of Fig. 1. For the system a set of 464 boundary voltage measurements were simulated and subsequently infused with 0.1% Gaussian noise based on the maximum voltage reading. The contact impedance for the electrodes was fixed at $100\,\Omega\,\mathrm{m}^{-1}$, while adjacent pair drive currents were used in the excitation of the domain. Setting the lower and upper bounds on conductivities at $l = 1\,\mathrm{S\,m}^{-1}$ and $u = 6\,\mathrm{S\,m}^{-1}$ respectively, a logarithmic barrier function was constructed and calibrated using an offset $\theta = -1.8326$. The box-constrained solver was provided with an

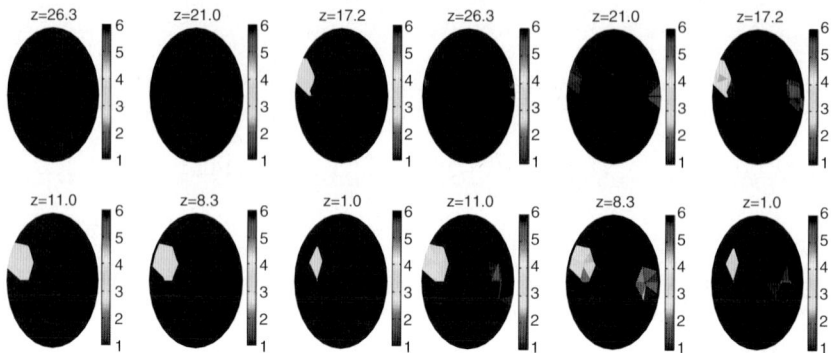

Fig. 1. Planes of the simulated (*left*) and reconstructed (*right*) conductivity distributions

initial value of the barrier parameter of $\mu^0 = 10^{-8}$ and an initial homogeneous guess on the solution of $\mathbf{x}_0 = 3.5\,\mathrm{S\,m}^{-1}$. After 325 interior point iterations, at a barrier parameter value $\mu \approx 10^{-40}$ the algorithm reached convergence at a primal solution as shown in the right of Fig. 1. At this point the upper and lower bounds in the computed primal solution are $\mathbf{x}_{\min} = 1.0013\,\mathrm{S\,m}^{-1}$ and $\mathbf{x}_{\max} = 5.9966\,\mathrm{S\,m}^{-1}$, while the dual solution is bounded within $\lambda_{\min} = 2.1669 \times 10^{-9}$ and $\lambda_{\max} = 5.5643 \times 10^{-7}$. The forward computations have been implemented numerically using the EIDORS 3D toolbox.

6 Conclusions

A regularized formulation for high-contrast electrical impedance tomography was presented where regularization is enforced by imparting upper and lower bounds on the conductivities. The inverse problem was solved using a primal-dual interior point algorithm and the numerical results presented indicate that nonlinear imaging with enhanced quantitative resolution is feasible. The technique is particularly suitable in recovering discontinuous conductivity profiles.

References

[BER04] Bertsekas D.: Nonlinear programming, Athena Scientific, Massachusets (2004).
[LPB05] Lionheart W.R.B., Polydorides N., Borsic A.: The image reconstruction problem. In: Holder D.S. (ed), Electrical Impedance Tomography: Methods, History and Applications. IoP, Bristol, (2005).
[FGR02] Forsgren A., Gill P., Wright M.: Interior methods for nonlinear optimization. SIAM Review, **44**, 525–597 (2002).

Minisymposium "Finance" (Oxford)

Sam Howison and Klaus Schmitz

Oxford Centre for Industrial and Applied Mathematics,
Mathematical Institute, Oxford University
howison@maths.ox.ac.uk, schmitz@maths.ox.ac.uk

We present three articles in this minisymposium on three areas of mathematical finance. We open up with Klaus Schmitz on simulation-based valuation where a new scheme is presented that values exotic options that possesses better convergence properties than existing schemes and is therefore more efficient. This is then followed by Helen Haworth on structural default risk modelling. A credit contagion framework is presented that captures the interdependency of firms under the consideration of credit risk and economic reality. Finally, Eric Yu demonstrates the valuation of a selection of elementary exotics with strike price resets under the same framework that keeps the valuation problem two-dimensional and is therefore competitive over existing methods.

Pricing Exotic Options with Strong Schemes: In finance, the strong convergence properties of discretisations of stochastic differential equations (SDEs) are very important for the hedging and valuation of exotic options. We show how the use of the Milstein scheme (vector case) and an orthogonal transformation can improve the convergence of the multi-level Monte Carlo method, so that the computational cost to achieve an accuracy of $O(\epsilon)$ is reduced to $O(\epsilon^{-2})$ for a Lipschitz payoff. We present examples of pricing exotic options.

Credit Contagion in a Structural Framework: We provide a multi-name structural credit model and use it to price credit products in the presence of default contagion. Based on economic fundamentals, structural models are attractive, modelling corporate default as the first time that firm value hits a lower barrier. Despite the proliferation in multi-name credit products, however, there has been little work extending the framework to multiple firms. Our results illustrate a meaningful relationship between the dependence structure and spreads.

The Valuation of Elementary Exotics with Strike Resets: We demonstrate the valuation of a selection of elementary exotic options with strike price resets in this article using a simple change of variable that keeps the valuation problem two-dimensional.

Pricing Exotic Options Using Strong Convergence Properties

Klaus Schmitz Abe and Michael Giles

Mathematical Institute, Oxford University, Oxford, U.K.
`schmitz@maths.ox.ac.uk`

Summary. In finance, the strong convergence properties of discretisations of stochastic differential equations (SDEs) are very important for the hedging and valuation of exotic options. In this paper we show how the use of the Milstein scheme can improve the convergence of the multilevel Monte Carlo method, so that the computational cost to achieve an accuracy of $O(e)$ is reduced to $O(\epsilon^{-2})$ for a Lipschitz payoff. The Milstein scheme gives first order strong convergence for all one-dimensional systems (one Wiener process). However, for processes with two or more Wiener processes, such as correlated portfolios and stochastic volatility models, there is no exact solution for the iterated integrals of second order (Lévy area) and the Milstein scheme neglecting the Lévy area gives the same order of convergence as the Euler-Maruyama scheme. The purpose of this paper is to show that if certain conditions are satisfied, we can avoid the calculation of the Lévy area and obtain first convergence order by applying an orthogonal transformation. We demonstrate when the conditions of the two-dimensional problem permit this and give an exact solution for the orthogonal transformation. We present examples of pricing exotic options to demonstrate that the use of both the orthogonal Milstein scheme and the multilevel Monte Carlo give a substantial reduction in the computation cost.

1 Introduction

We begin with a two-dimensional Itô stochastic differential equation (SDE) with a two-dimensional Wiener process

$$dx = \mu^{(x)}(x,y)\,dt + \sigma(x,y)\,d\widehat{W}_{1,t} \tag{1}$$

$$dy = \mu^{(y)}(x,y)\,dt + \xi(x,y)\,d\widehat{W}_{2,t} \qquad \rho\,dt = \left\langle d\widehat{W}_{1,t}, d\widehat{W}_{2,t} \right\rangle.$$

Alternatively, in vector form

$$dZ(t) = A_0(t,Z)\,dt + \sum_{k=1}^{2} A_k(t,Z)\,d\widehat{W}_{k,t} \qquad Z \in \mathbb{R}^2.$$

This is in fact, only a symbolic representation for the stochastic integral equation

$$Z(t) = Z(t_0) + \int_{t_0}^t A_0(s, Z)\, ds + \sum_{k=1}^2 \int_{t_0}^t A_k(s, Z)\, d\widehat{W}_{k,s}.$$

The first integral is a deterministic Riemann integral and the second is a stochastic integral [7].

Using the definition of correlation we can represent our system (1) in vector form with independent noise

$$d\begin{bmatrix} x \\ y \end{bmatrix} = \begin{bmatrix} \mu^{(x)}(x,y) \\ \mu^{(y)}(x,y) \end{bmatrix} dt + \begin{bmatrix} \sigma(x,y) \\ \rho\,\xi(x,y) \end{bmatrix} dW_{1,t} + \begin{bmatrix} 0 \\ \widehat{\rho}\,\xi(x,y) \end{bmatrix} dW_{2,t} \qquad (2)$$

$$\langle dW_{1,t}, dW_{2,t} \rangle = 0 \qquad\qquad \widehat{\rho} = \sqrt{1 - \rho^2}.$$

The Milstein approximation is

$$\begin{bmatrix} x_{t+\Delta t} \\ y_{t+\Delta t} \end{bmatrix} = \begin{bmatrix} x_t \\ y_t \end{bmatrix} + \begin{bmatrix} \mu^{(x_t)} \\ \mu^{(y_t)} \end{bmatrix} \Delta t + \begin{bmatrix} \sigma \\ \rho\xi \end{bmatrix} \Delta W_{1,t} + \begin{bmatrix} 0 \\ \widehat{\rho}\xi \end{bmatrix} \Delta W_{2,t} \qquad (3)$$

$$\frac{1}{2}\begin{bmatrix} \sigma\sigma_x + \rho\xi\sigma_y \\ \rho\sigma\xi_x + \rho^2\xi\xi_y \end{bmatrix} \left((\Delta W_{1,t})^2 - \Delta t \right) + \frac{1}{2}\begin{bmatrix} 0 \\ \widehat{\rho}^2\xi\xi_y \end{bmatrix} \left((\Delta W_{2,t})^2 - \Delta t \right)$$

$$+\frac{1}{2}\begin{bmatrix} \widehat{\rho}\xi\sigma_y \\ \widehat{\rho}\sigma\xi_x + 2\rho\widehat{\rho}\xi\xi_y \end{bmatrix} (\Delta W_{1,t}\Delta W_{2,t}) + \frac{1}{2}[A_1, A_2]\left[L_{(1,2)} \right]_t^{t+\Delta t},$$

where subscript x and y denote partial derivatives, $L_{(1,2)}$ is the Lévy area defined by

$$\left[L_{(1,2)} \right]_t^{t+\Delta t} = \int_t^{t+\Delta t}\int_t^{S+\Delta t} dW_{1,U} dW_{2,S} - \int_t^{t+\Delta t}\int_t^{S+\Delta t} dW_{2,U} dW_{1,S},$$

and $[A_1, A_2]$ is the Lie bracket defined by (∂_{A_i} is the Jacobian matrix of A_i)

$$[A_1, A_2] = (\partial_{A_2} A_1 - \partial_{A_1} A_2) = \begin{bmatrix} -\widehat{\rho}\xi\sigma_y \\ \widehat{\rho}\sigma\xi_x \end{bmatrix}.$$

The numerical difficulty is how to calculate the Lévy area $L_{(1,2)}$. The technique of Gaines and Lyons [6] can be used to sample the distribution for $L_{(1,2)}$ conditional on ΔW_1, ΔW_2. However there is no generalisation of this to higher dimensions apart from the approximation of [9], which has a significant computational cost.

2 Orthogonal Transformation

If we make an orthogonal transformation of the uncorrelated Wiener processes in (2), we do not change the distribution and we obtain

$$d\widetilde{x} = \mu^{(x)}(\widetilde{x}, \widetilde{y}) \ dt + \sigma(\widetilde{x}, \widetilde{y}) \ d\widetilde{W}_{1,t}, \tag{4}$$

$$d\widetilde{y} = \mu^{(y)}(\widetilde{x}, \widetilde{y}) \ dt + \xi(\widetilde{x}, \widetilde{y}) \ d\widetilde{W}_{2,t},$$

where

$$\begin{bmatrix} d\widetilde{W}_{1,t} \\ d\widetilde{W}_{2,t} \end{bmatrix} = \begin{bmatrix} 1 & 0 \\ \rho & \widehat{\rho} \end{bmatrix} \begin{bmatrix} \cos\theta & -\sin\theta \\ \sin\theta & \cos\theta \end{bmatrix} \begin{bmatrix} dW_{1,t} \\ dW_{2,t} \end{bmatrix}.$$

If we compute the Lie bracket for the new orthogonal process using independent Brownian paths $W_{1,t}, W_{2,t}$, we have

$$[A_1, A_2] = \begin{bmatrix} -\widehat{\rho}\xi\sigma_{\widetilde{y}} - \sigma^2\theta_{\widetilde{x}} - \rho\sigma\xi\theta_{\widetilde{y}} \\ \widehat{\rho}\sigma\xi_{\widetilde{x}} - \rho\sigma\xi\theta_{\widetilde{x}} - \xi^2\theta_{\widetilde{y}} \end{bmatrix}.$$

To avoid having to simulate the Lévy area, we need the Lie brackets to be identically zero [1], i.e., we need to impose the following conditions

$$-\widehat{\rho}\xi\sigma_{\widetilde{y}} - \sigma^2\theta_{\widetilde{x}} - \rho\sigma\xi\theta_{\widetilde{y}} = 0,$$

$$+\widehat{\rho}\sigma\xi_{\widetilde{x}} - \rho\sigma\xi\theta_{\widetilde{x}} - \xi^2\theta_{\widetilde{y}} = 0.$$

Simplifying we get

$$\Phi \doteq \frac{\partial\theta}{\partial\widetilde{x}} = \frac{-1}{\widehat{\rho}}\left(\frac{\xi\sigma_{\widetilde{y}}}{\sigma^2} + \frac{\rho\xi_{\widetilde{x}}}{\xi}\right), \tag{5}$$

$$\Psi \doteq \frac{\partial\theta}{\partial\widetilde{y}} = \frac{1}{\widehat{\rho}}\left(\frac{\rho\sigma_{\widetilde{y}}}{\sigma} + \frac{\sigma\xi_{\widetilde{x}}}{\xi^2}\right).$$

If we want to find a solution for θ, we must first determine when the system is consistent, or integrable. This requires that

$$\frac{\partial\Phi}{\partial\widetilde{y}} = \frac{\partial^2\theta}{\partial\widetilde{x}\,\partial\widetilde{y}} = \frac{\partial\Psi}{\partial\widetilde{x}} \tag{6}$$

and the solution for θ is

$$\theta(\widetilde{x}, \widetilde{y}) = \int^{(\widetilde{x}, \widetilde{y})} (\Phi \, d\widetilde{x} + \Psi \, d\widetilde{y}). \tag{7}$$

However, because not all SDEs satisfy condition (6), we also obtain the following SDE for θ

$$d\theta = \frac{\partial \theta}{\partial \widetilde{x}} \, d\widetilde{x} + \frac{\partial \theta}{\partial \widetilde{y}} \, d\widetilde{y} = \left(\Phi \mu^{(\widetilde{x})} + \Psi \mu^{(\widetilde{y})} \right) dt + \sigma \Phi d\widetilde{W}_{1,t} + \xi \Psi d\widetilde{W}_{2,t}.$$

If we choose to define θ in this way even when condition (6) is not satisfied then our system becomes a three-dimensional Itô process with two Wiener process inputs ($\theta-$ scheme)

$$\begin{bmatrix} d\widetilde{x} \\ d\widetilde{y} \\ d\theta \end{bmatrix} = \begin{bmatrix} \mu^{(\widetilde{x})} \\ \mu^{(\widetilde{y})} \\ \Phi \mu^{(\widetilde{x})} + \Psi \mu^{(\widetilde{y})} \end{bmatrix} dt + \begin{bmatrix} \sigma \\ 0 \\ \sigma \Phi \end{bmatrix} d\widetilde{W}_{1,t} + \begin{bmatrix} 0 \\ \xi \\ \xi \Psi \end{bmatrix} d\widetilde{W}_{2,t}. \qquad (8)$$

If we compute again the Lie brackets with independent noise, we obtain

$$[A_1, A_2] = \begin{bmatrix} 0 \\ 0 \\ \widehat{\rho} \sigma \xi \left(\dfrac{\partial \Psi}{\partial \widetilde{x}} - \dfrac{\partial \Phi}{\partial \widetilde{y}} \right) \end{bmatrix}. \qquad (9)$$

Note that when condition (6) is satisfied this Lie bracket (9) is identically zero. In the remainder of the paper we investigate when particular applications satisfy condition (6), in which case one can discretise either (4) or (8) and when they do not, in which case one can only discretise (8) or the original untransformed SDE (1). Our objective is to try to achieve higher order strong convergence without the simulation of the Lévy areas.

When the Lie bracket is not equal to zero, the important question to be considered is how precisely does θ need to be calculated to obtain first strong order convergence in $\widetilde{x}$ and $\widetilde{y}$? For example, does neglecting the Lie bracket affect the accuracy of θ but not $\widetilde{x}$ and $\widetilde{y}$?

3 Strong Convergence

If we apply any discrete approximation scheme to our system (1) and we want to numerically evaluate the strong convergence order of our approximations $\widehat{X}$, an exact solution is normally required. However, at present, there are no solutions available for many SDEs. Because we are only interested in the distribution of the solution, we can use the next theorems [8] to determine the order of convergence for our discrete time approximation without an exact solution.

Most models can be described through a SDE of the form

$$dx(t) = \mu(x,t) \, dt + \sigma(x,t) \, dW(t) \qquad x(0) = x_0 \qquad (10)$$

with W a M-dimensional Brownian motion, μ mapping $\mathbb{R}^N \times [0, \infty)$ into $\mathbb{R}^N$, σ mapping $\mathbb{R}^N \times [0, \infty)$ into $\mathbb{R}^{N \times M}$ and x_0 a random N-dimensional vector independent of W.

Theorem 1 (Existence and Uniqueness of Strong Solutions).
   Suppose $E\left[||x_0||^2\right]$ is finite and that there is a constant K for which for all $t \in [T_0, T]$ and all $x, y \in \mathbb{R}^d$ the following conditions are satisfied:

$$||\mu(x, t) - \mu(y, t)|| + ||\sigma(x, t) - \sigma(y, t)|| \leq K\,||x - y|| \quad \text{(Lipschitz condition)}$$

$$||\mu(x, t)|| + ||\sigma(x, t)|| \leq K(1 + ||x||) \qquad \text{(Linear growth condition)}$$

Then the SDE (10) admits a strong solution x and satisfies

$$\left(E\left[||x(t)||^2\right] < \infty\right).$$

This solution is unique in the sense that if $\widehat{x}$ is also a solution, then

$$P\left(x(t) = \widehat{x}(t).\ \forall t \in [0, T]\right) = 1.$$

Proof. Proofs and additional explanation can be found in [2] and [5].

Theorem 2 (Convergence Order without an Exact Solution).
   A) If a discrete approximation $\widehat{x}$ of (10) with time step Δt has strong convergence order η, i.e., there exist a constant C_1 such that

$$E[\ |x(T) - \widehat{x}(T, \Delta t)|\ ] \leq C_1 \Delta t^\eta \tag{11}$$

Then there exist a positive constant, C_2, such that

$$E\left[\left|\widehat{x}(T, \Delta t) - \widehat{x}\left(T, \frac{\Delta t}{2}\right)\right|\right] \leq C_2 \Delta t^\eta. \tag{12}$$

   B) Conversely, if it is known that the discretisation is strongly convergent and (12) holds for some positive constant C_2, then the strong convergence order is η.

Proof. A) If (11) is true for all Δt, then

$$E\left[\left|x(T) - \widehat{x}\left(T, \frac{\Delta t}{2}\right)\right|\right] \leq C_1 \left(\frac{\Delta t}{2}\right)^\eta. \tag{13}$$

Using the triangle law ($|A - B| \leq |A| + |B|$) and adding (11) and (13), we get

$$E\left[\left|\widehat{x}(T, \Delta t) - \widehat{x}\left(T, \frac{\Delta t}{2}\right)\right|\right] \leq C_1 \left(1 + \left(\frac{1}{2}\right)^\eta\right) \Delta t^\eta.$$

   B) Using the triangle law

$$E\left[\ |x(T) - \widehat{x}(T, \triangle t)|\ \right] \leq E\left[\ \left|x(T) - \widehat{x}\left(T, \left(\frac{1}{2}\right)^{M} \triangle t\right)\right|\ \right]$$

$$+ \sum_{m=0}^{M-1} E\left[\ \left|\widehat{x}\left(T, \left(\frac{1}{2}\right)^{m+1} \triangle t\right) - \widehat{x}\left(T, \left(\frac{1}{2}\right)^{m} \triangle t\right)\right|\ \right].$$

Due to strong convergence

$$\lim_{M \to \infty} E\left[\ \left|x(T) - \widehat{x}\left(T, \left(\frac{1}{2}\right)^{M} \triangle t\right)\right|\ \right] = 0.$$

Hence, using (12)

$$E\left[\ |x(T) - \widehat{x}(T, \triangle t)|\ \right] \leq \sum_{m=0}^{\infty} C_2 \left(\frac{1}{2}\right)^{m\ \eta} \triangle t^{\eta} = \frac{C_2}{1 - \left(\frac{1}{2}\right)^{\eta}} \triangle t^{\eta}.$$

4 Stochastic Volatility Models

In this section we consider three different stochastic volatility models. All three have the following generic form

$$dx = \mu^{(x)} dt + \alpha\, x^{\gamma_1} y^{\lambda_1} d\widehat{W}_{1,t} \qquad\qquad \rho dt = \left\langle d\widehat{W}_{1,t}, d\widehat{W}_{2,t} \right\rangle$$

$$dy = \mu^{(y)} dt + \beta\, x^{\gamma_2} y^{\lambda_2} d\widehat{W}_{2,t}.$$

The integrability condition (6) becomes

$$\lambda_C\, \lambda_1\, \beta^2\, y^{2\lambda_C} = -\gamma_C\, \gamma_2\, \alpha^2\, x^{2\gamma_C} \tag{14}$$

$$\gamma_C = \gamma_1 - \gamma_2 - 1; \qquad \lambda_C = \lambda_2 - \lambda_1 - 1$$

so then, for $\alpha, \beta, \gamma_2, \lambda_1 \neq 0$, we can conclude that θ is integrable if, and only if, $\lambda_C = \gamma_C = 0$, in which case the solution is

$$\theta = \frac{1}{\widehat{\rho}}\left(\left(\frac{-\rho\gamma_2\alpha - \lambda_1\beta}{\alpha}\right) \log x + \left(\frac{\rho\lambda_1\beta + \gamma_2\alpha}{\beta}\right) \log y\right). \tag{15}$$

4.1 Quadratic Volatility Model (Case 1)

The first case we consider is

$$dx = x\, \overline{\mu}\, dt + x\, y\, d\widehat{W}_{1,t}, \tag{16}$$

$$dy = k\,(\varpi - y)\, dt + \beta\, y^2 d\widehat{W}_{2,t},$$

with parameters $T = 1$; $\rho = 0.30$; $\overline{\mu} = 0.05$; $k = 1.8$; $\varpi = 0.26$; $\beta = 1$ and initial conditions $x(0) = 1$; $y(0) = 0.21$.

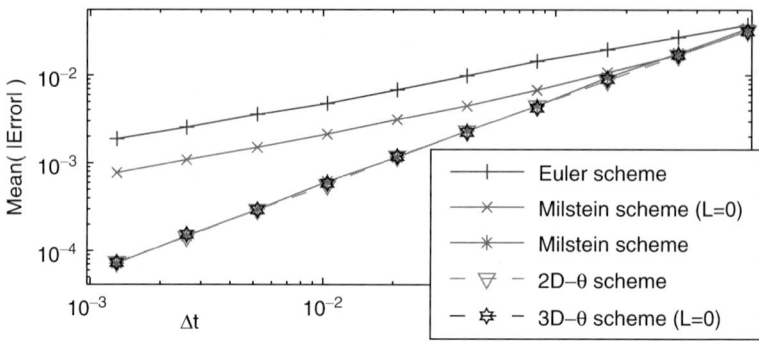

Fig. 1. Convergence test for Case 1

Table 1. Convergence orders η for all cases (N/A = not applicable)

Scheme	Description	C-1	C-2	C-3
Euler scheme	set $\triangle t = dt$, $\Delta W_i = dW_i$ in (2)	0.49	0.49	0.48
Milstein scheme ($L=0$)	Milstein–(3), set $L_{(1,2)} = 0$	0.56	0.56	0.51
Milstein scheme	Milstein–(3), simulate $L_{(1,2)}$	0.99	0.99	0.95
2D$-\theta$ scheme	Milstein–(4) with (15)	0.98	N/A	N/A
3D$-\theta$ scheme ($L=0$)	Milstein–(8), set $L_{(1,2)} = 0$	0.98	0.87	0.52
3D$-\theta$ scheme	Milstein–(8), simulate $L_{(1,2)}$	0.98	0.98	0.91

Because $\lambda_C = \gamma_C = 0$, we can use either equation (4) together with (15), or the three-dimensional θ scheme (8). Because of the orthogonal transformation, neither requires the calculation of the Lévy area. Figure 1 and Table 1 show that, as expected, the Euler scheme and the Milstein scheme with zero Lévy areas (setting $L_{(1,2)} = 0$ in (3)) give strong convergence order 0.5. On the other hand, the Milstein scheme (3) with a proper value for the distribution of the Lévy area (through simulating the Lévy area using N subintervals within each timestep) gives 1.0 order strong convergence, as do the three orthogonal θ-schemes.

Variance Model (Case 2)

The second case we consider is the following stochastic variance model.

$$dx = x\,\overline{\mu}\,dt + x\,\sqrt{y}\,d\widehat{W}_{1,t}, \tag{17}$$

$$dy = k\,(\varpi - y)\,dt + \beta\,y\,d\widehat{W}_{2,t}.$$

The parameters and initial conditions are the same except for $\beta = 0.5$; $\varpi = 0.26^2$; $y(0) = 0.21^2$, which are chosen so that x and y will have approximately the same relative volatility as in Case 1.

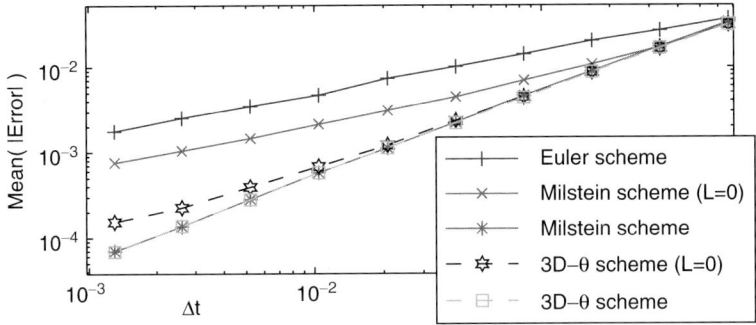

Fig. 2. Convergence test for Case 2

In this case $\lambda_C = 0.5$, and since the integrability condition is not satisfied it is not possible to use the $2D - \theta$ scheme. Figure 2 and Table 1 show that the only schemes that achieved first order convergence are the Milstein schemes which simulate the Lévy area in the simulation. However, Fig. 2 shows there is a remarkable difference between the original and the orthogonal scheme without the simulation of the Lévy area, not the improved order of convergence achieved in the first case (Table 1) but a much improved constant of proportionality.

Heston Model (Case 3)

A particularly bad case for the orthogonal transformation is the Heston model [4]

$$\mathrm{d}x = x\,\mu\,\mathrm{d}t + x\,\sqrt{y}\,\mathrm{d}\widehat{W}_{1,t}, \tag{18}$$
$$\mathrm{d}y = k\,(\varpi - y)\,\mathrm{d}t + \beta\,\sqrt{y}\,\mathrm{d}\widehat{W}_{2,t}.$$

The parameters and initial conditions are the same as in Case 2 except for $\beta = 0.25$; this is again chosen so ensure that x and y will have approximately the same relative volatility as in the first two cases.

In this case, $\lambda_c = 1$. Figure 3 and Table 1 show that neither of the Milstein schemes in which the Lévy areas are set to zero performs very well. Both have order 0.5 strong convergence, and the constant of proportionality is not much better than for the Euler scheme. When the Lévy areas are simulated correctly, the two Milstein schemes do exhibit the expected first order strong convergence. This demonstrates the importance of the Lévy areas in this case.

5 Pricing Exotic Options Using ML-MC

Usually, it is the weak convergence properties of numerical discretisations which are most important, because in financial applications one is mostly

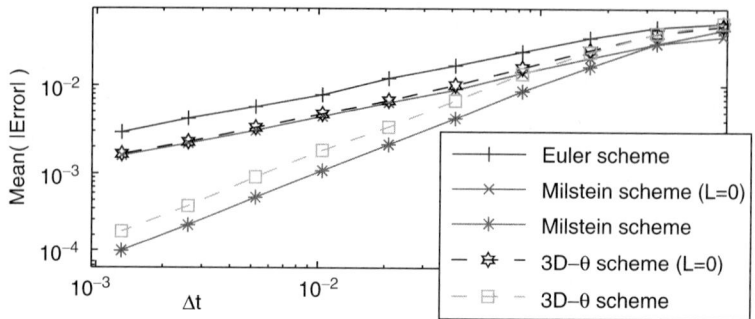

Fig. 3. Convergence test for Case 3

concerned with the accurate estimation of expected payoffs. However, in the recently developed multilevel Monte Carlo path simulation method (ML-MC [3]), the strong convergence properties play a crucial role.

The key idea in the ML-MC approach is the use of a multilevel algorithm with different timesteps Δt on each level. Suppose level l uses 2^l timesteps of size $\Delta t_l = 2^{-l} T$, and define P_l to be the numerical approximation to the payoff on this level. Let L represent the finest level, with timesteps so small that the bias due to the numerical discretisation is smaller than the accuracy ϵ which is desired. Due to the linearity of the expectation operator, we can express the expectation on the finest grid as

$$E\left[P_L\right] = E\left[P_0\right] + \sum_{l=1}^{L} E\left[P_l - P_{l-1}\right].$$

The quantity $E\left[P_l - P_{l-1}\right]$ represents the expected difference in the payoff approximation on levels l and $l-1$. This is estimated using a set of Brownian paths, with the same Brownian paths being used on both levels. This is where the strong convergence properties are crucial. The small difference between the terminal values for the paths computed on levels l and $l-1$ gives a small value for the payoff difference. Consequently, the variance

$$V_l = V[P_l - P_{l-1}]$$

decreases rapidly with level l. In particular, for a European option with a Lipschitz payoff, the order with which the variance converges to zero is double the strong order of convergence.

Using N_l independent paths to estimate $E\left[P_l - P_{l-1}\right]$, if we define the level 0 variance to be $V_0 = V[P_0]$ then the variance of the combined multilevel estimator is $\sum_{l=0}^{L} N_l^{-1} V_l$. The computational cost is proportional to the total number of timesteps: $\sum_{l=0}^{L} N_l \Delta t_l^{-1}$. Varying N_l to minimise the variance for a given computational cost gives a constrained optimisation problem whose solution is $N_l = C\sqrt{V_l \Delta t_l}$. The value for the constant of proportionality, C, is chosen to make the overall variance less than the ϵ^2, so that the r.m.s. error is less than ϵ.

The analysis in [3] shows that in the case of an Euler discretisation with a Lipschitz payoff, the computational cost of the ML-MC algorithm is $O(\epsilon^2 \log)$, which is significantly better than the $O(\epsilon^3)$ cost of the standard Monte Carlo method. Furthermore, the analysis shows that first order strong convergence should lead to an $O(\epsilon^2)$ cost for Lipschitz payoffs; this will be demonstrated in the results to come.

5.1 European Option

The first set of numerical results are for a European put option with strike K and maturity T, for which the payoff is given by

$$P = \max\left(K - x(T), 0\right).$$

Using the Case 1 volatility model (16), with the same set of parameters as before and strike $K = 1.1$, we obtain the ML-MC results in Fig. 4. The top left plot shows the weak convergence in the estimated value of the payoff as the finest grid level L is increased. All of the methods tend asymptotically to the same value. The bottom left plot shows the convergence of the quantity $V_l = V[P_l - P_{l-1}]$. The 3D $-\theta$ scheme exhibits second order convergence due to the first order strong convergence. The Milstein approximation (3) with

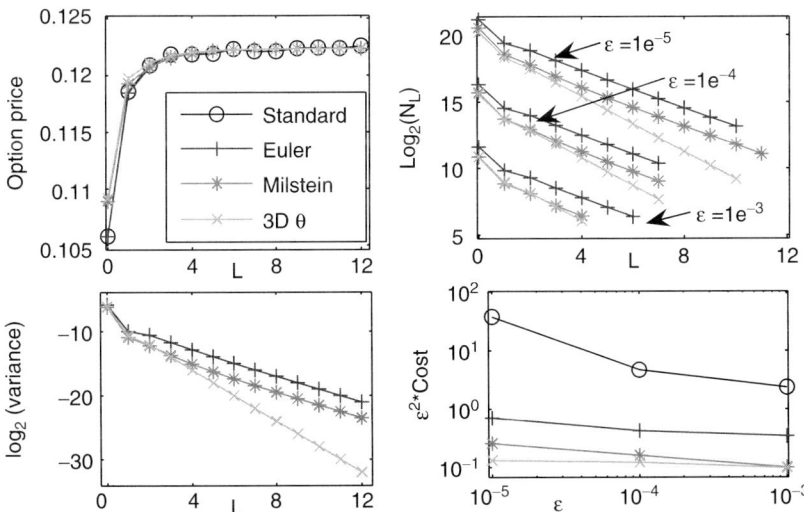

Fig. 4. European put option, Case 1. *Top left*: convergence in option value with grid level. *Bottom left*: convergence in the ML-MC variance with grid level. *Top right*: number of Monte Carlo paths N_l required on each level, depending on the desired accuracy. *Bottom right*: overall computational cost as a function of accuracy ϵ

the Lévy areas set equal to zero, and the Euler discretisation both give first order convergence, which is consistent with their 0.5 order strong convergence properties.

The top right plot shows three sets of results for different values of the desired r.m.s. accuracy ϵ. The ML-MC algorithm [3] uses the correction obtained at each level of timestep refinement to estimate the remaining bias due to the discretisation, and therefore determine the number of levels of refinement required. The results illustrate this, with the smaller values for ϵ leading to more levels of refinement. To achieve the desired accuracy, it is also necessary to reduce the variance in the combined estimator to the required level, so any more paths (roughly proportional to ϵ^{-2}) are required for smaller values of ϵ. The final point to observe in this plot is how many fewer paths are required on the fine grid levels compared to the coarsest grid level for which there is just one timestep covering the entire time interval to maturity. This is a consequence of the variance convergence in the previous plot, together with the optimal choice for N_l described earlier.

The final bottom right plot shows the overall computational cost as a function of ϵ. The cost C_ϵ is defined as the total number of timesteps, summed over all paths and all grid levels. It is expected that C_ϵ will be $O(\epsilon^{-2})$ for the best ML-MC methods, and so the quantity which is plotted is $\epsilon^2 C_\epsilon$ vs. ϵ. The results show that $\epsilon^2 C_\epsilon$ is almost perfectly independent of ϵ for the $3D - \theta$ scheme, and varies only slightly with ϵ for the Milstein scheme. The Euler ML-MC scheme shows a bit more growth as $\epsilon \to 0$, which is consistent with the analysis in [3] which predicts that $C_\epsilon = O(\epsilon^{-2}(\log \epsilon)^{-2})$. The final comparison line is the standard Monte Carlo method using the Euler discretisation, for which $C_\epsilon = O(\epsilon^{-3})$.

The use of fewer Monte Carlo paths N_L is reflected directly in the computational cost of the process. For the most accurate case, $\epsilon = 0.00001$, the Euler, Milstein and $3D - \theta$ version of the ML-MC scheme are roughly 50, 150 and 300 times more efficient than the standard Monte Carlo method using the Euler discretisation.

Figures 5 and 6 show the corresponding results for Cases 2 and 3, corresponding to the variance model (17) and the Heston model (18) respectively. For Case 2, the computational savings from using the ML-MC method are similar to Case 1, while for Case 3 the savings from the Euler, Milstein and $3D - \theta$ versions of the ML-MC scheme are roughly 20, 40 and 40 in the most accurate case.

5.2 Digital Option

The payoff for a digital option is given by

$$P = H\left(x(T) - K\right),$$

where $H(x)$ is the Heaviside function ($H(x) = 1$ if $x > 0$, else $H(x) = 0$). Figure 7 shows the results using the Case 2 variance model (17) and strike

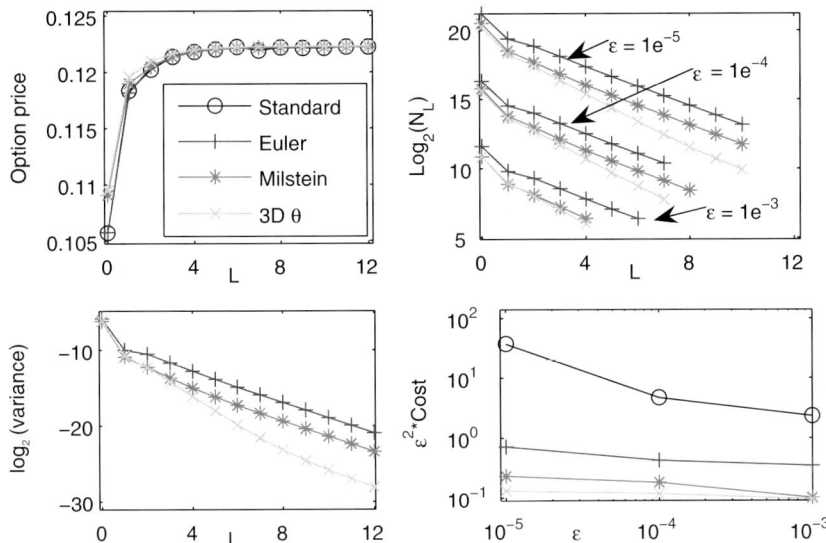

Fig. 5. European put option, Case 2. *Top left*: convergence in option value. *Bottom left*: convergence in ML-MC variance. *Top right*: number of Monte Carlo paths N_l required on each level. *Bottom right*: overall computational cost

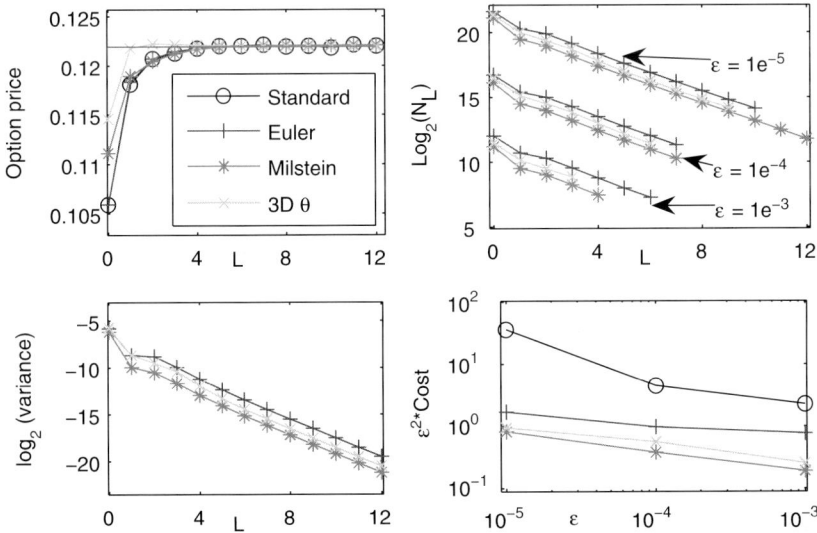

Fig. 6. European put option, Case 3. *Top left*: convergence in option value (red line is analytic value). *Bottom left*: convergence in ML-MC variance. *Top right*: number of Monte Carlo paths N_l required on each level. *Bottom right*: computational cost

$K = 1$. Because this payoff is not Lipschitz continuous, it shows the poorest benefits from the ML-MC approach. For the most accurate case, $\epsilon = 0.0001$, the Euler, Milstein and $3D - \theta$ versions of the ML-MC scheme are roughly 3, 60 and 90 times more efficient than the standard method using the Euler scheme.

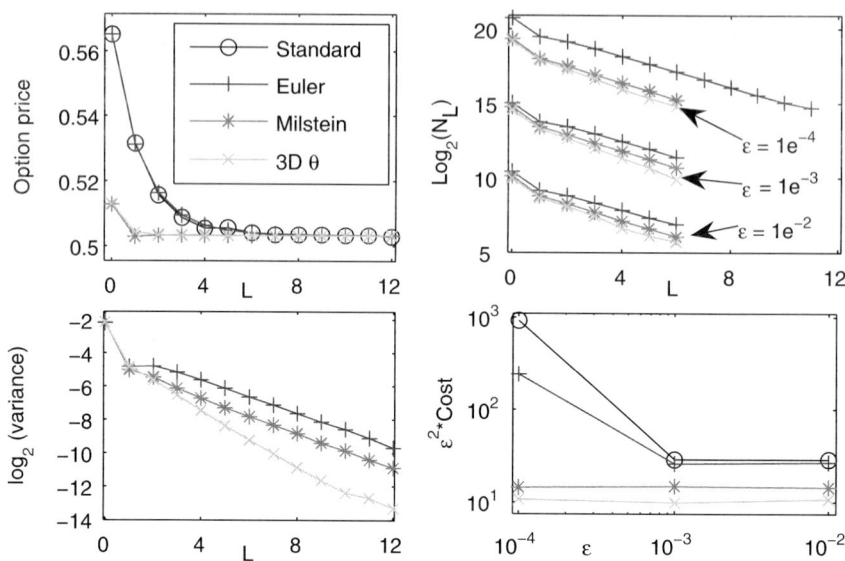

Fig. 7. Digital option, Case 2. *Top left*: convergence in option value. *Bottom left*: convergence in ML-MC variance. *Top right*: number of Monte Carlo paths N_l required on each level. *Bottom right*: overall computational cost

5.3 Asian Option

The payoff for an Asian call option is given by

$$P = \max\left(\bar{x}(T) - K,\ 0\right),$$

where $\bar{x}$ is the arithmetic average which can be approximated numerically as

$$\bar{x}(T) = \frac{1}{T}\int_0^T x(t)\mathrm{d}t \approx \frac{\Delta t}{2\,T}\sum_{n=1}^{N_{\Delta t}}\left(\widehat{x}_n + \widehat{x}_{n-1}\right).$$

Using the Case 2 variance model (17), with strike $K = 1$, Fig. 8 shows that for the most accurate case, $\epsilon = 0.00001$, the Euler, Milstein and $3\mathrm{D} - \theta$ versions of the ML-MC scheme are roughly 50, 80 and 110 times more efficient than the standard method using the Euler scheme.

5.4 Variance Swap Option

The payoff for a variance swap option is given by

$$P = N\left(\bar{y}(T) - K_{var}\right),$$

where N is the nominal price and $\bar{y}(T)$ is the average of the variance in the time interval $[0, T]$ which can be approximated numerically in the same way as $\bar{y}(T)$ in the previous example.

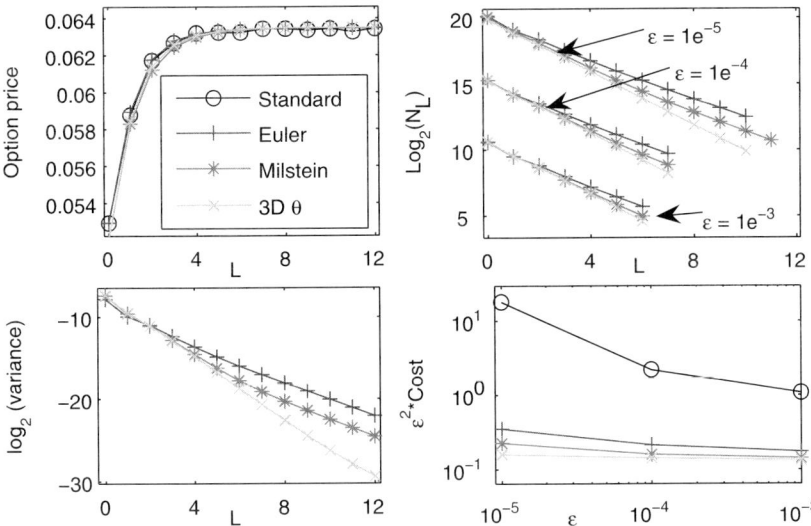

Fig. 8. Asian option, Case 2. *Top left*: convergence in option value. *Bottom left*: convergence in ML-MC variance. *Top right*: number of Monte Carlo paths N_l required on each level. *Bottom right*: overall computational cost

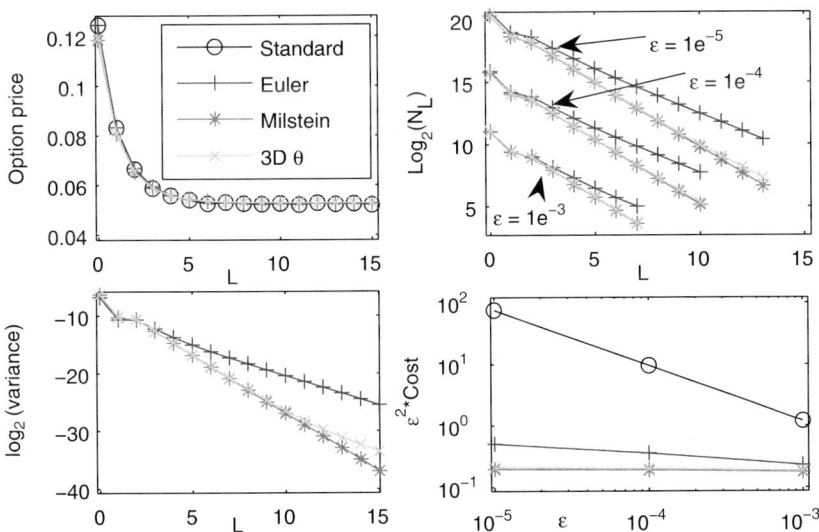

Fig. 9. Variance swap option, Case 2. *Top left*: convergence in option value. *Bottom left*: convergence in ML-MC variance. *Top right*: number of Monte Carlo paths N_l required on each level. *Bottom right*: overall computational cost

Using the Case 2 variance model (17), $K_{var} = 0.26^2$ and $N = 10$, Fig. 9 shows that for the most accurate case, $\epsilon = 0.00001$, the Euler, Milstein and $3D - \theta$ versions of the ML-MC scheme are roughly 150, 380 and 360 times more efficient than the standard method using the Euler scheme. In this case, the Milstein method gives first order strong convergence for y, whereas the three-dimensional θ scheme gives similar accuracy initially but is tailing off towards order 0.5 strong convergence on the finest grids.

6 Conclusions

In finance, stochastic variance and volatility models are very important for the valuation of exotic options. We have shown that the use of the orthogonal θ scheme can achieve the first order strong convergence properties of the Milstein numerical discretisation without the expensive simulation of Lévy areas. In combination with the recently introduced multilevel Monte Carlo method it can reduce substantially the computational cost in pricing exotic options, reducing the cost to achieve an r.m.s. error of size ϵ from $O(\epsilon^{-3})$ to $O(\epsilon^{-2})$.

The ML-MC works without any problems with all schemes and does not depend on the value of the parameters of the system. However, when a specific orthogonal transformation (θ scheme) is applied to a two-dimensional SDE it is only possible under certain conditions to avoid calculation of the Lévy area. The bias or error in the computation of the rotation angle θ that makes the Lie bracket equal to zero in the orthogonal scheme is crucial to obtain a better convergence order. When the conditions for integrability are satisfied, we can use the formula for θ to obtain the value of the rotation angle and obtain first order strong convergence. Otherwise, we have to use the three-dimensional transformation and check the magnitude of the Lie brackets to decide if it is likely to give computational savings in the solution of our system.

The numerical results demonstrate considerable computational savings when the orthogonal transformation is applied to either the quadratic volatility model (16) or the stochastic variance model (18). Unfortunately, similar savings are not achieved with the Heston model, and so the orthogonal transformation is not recommended in this case.

Acknowledgements

We are very grateful to Prof. T.J. Lyons for many very helpful discussions about this work. The first author would also like to thank CONACYT-MEXICO for their financial support.

References

1. A. B. Cruzeiro, P. Malliavin and A. Thalmaier: Geometrization of Monte-Carlo numerical analysis of an elliptic operator: strong approximation. C. R. Acad. Sci. Paris, Ser. I, **338**, 481–486, (2004).
2. P. Glasserman: Monte Carlo Methods in Financial Engineering. Springer, (2004).
3. M. Giles: Multi-level Monte Carlo path simulation. Technical Report No. **NA06/03**, Oxford University Computing Laboratory, Parks Road, Oxford, U.K, (2006).

4. S. Heston: A Closed-Form Solution for Options with Stochastic Volatility with Applications to Bond and Currency Options. The Review of Financial Studies, Vol. **6**, Issue 2, 327–343, (1993).
5. P. E. Kloeden and E. Platen: Numerical Solution of Stochastic Differential Equations. Springer, (1999).
6. J.G. Gaines and T.J. Lyons: Random Generation of Stochastic Area Integrals. SIAM Journal on Applied Mathematics, Vol. **54**, No. 4, 1132–1146, (1994).
7. P. Malliavin and A. Thalmaier: Stochastic Calculus of Variations in Mathematical Finance. Springer, (2005).
8. K. Schmitz-Abe and W. T. Shaw: Measure Order of Convergence without an Exact Solution, Euler vs Milstein Scheme. International Journal of Pure and Applied Mathematics, Vol. **24**, No 3, 365–381, (2005).
9. M. Wiktorsson: Joint characteristic function and simultaneous simulation of iterated Itô integrals for multiple independent Brownian motions. The Annals of Applied Probability, Vol. **11**, No 2, 470–487, (2001).

Credit Contagion in a Structural Framework

Helen Haworth

The Nomura Centre for Quantitative Finance, OCIAM Mathematical Institute,
Oxford University, England
haworthh@maths.ox.ac.uk

1 Introduction

Credit risk, often thought of as the risk arising from a company default, is the
risk that an obligor does not honour its obligations. It is the reason the multi-
trillion dollar credit derivatives market exists and its influence is pervasive
across global financial markets. As multi-asset credit products have increased
in popularity, the need for models incorporating a realistic dependence structure between companies has grown.

We consider the structural framework in which a firm's assets are modelled as a geometric Brownian motion with default as the first first hitting
time of an exponential default barrier. Originally proposed by [4] and [1],
these models are attractive, based as they are on economic fundamentals, and
enabling debt and equity to be valued as contingent claims on firm value.
The extension to two correlated firms is considered in [5] to calculate default
correlations. We extend this framework to incorporate default contagion and
derive results for bond yields and CDS spreads. Firm values are modelled
as correlated geometric Brownian motions, reflecting a common influence on
corporate strength whilst default contagion represents a direct link between
the fortunes of both companies. The result is a model that incorporates default causality and is asymmetric with regard to default risk, a significant
improvement on prior models. Further details of the methodology and results
are in [2].

2 Two-Firm Model

We consider two companies, firm values V_i, $i = 1, 2$. Each company issues
equity and a single homogeneous class of debt, assumed to be a zero coupon
bond, $C_i(t, T)$, par value K_i, maturity T. For each company, firm value is
assumed to follow a geometric Brownian motion, with default as the first time
that the value of the firm hits a lower default barrier $b_i(t)$. As in [4] and [1],

we assume that a firm's value can be constructed from tradable securities and so in the risk-neutral pricing measure, for $i = 1, 2$,

$$dV_i(t) = (r_f - q_i)V_i dt + \sigma_i V_i dW_i(t),$$

where the risk-free rate, r_f, dividend yields, q_i, and volatilities, σ_i, are constants, $W_i(t)$ are Brownian motions and $\mathrm{cov}(W_1(t), W_2(t)) = \rho t$ for constant correlation ρ. We assume that each company has an exponential default barrier, $b_i(t) = K_i e^{-\gamma_i(T-t)}$, reflecting the existence of debt covenants.

To value credit spreads we use the joint survival probability density function and the resultant joint survival probability, $P(t)$, as derived in [3] for the valuation of double lookbacks,

$$P(t) = \mathbb{P}(X_1(t) \geq B_1, \underline{X}_2(t) \geq B_2) \tag{1}$$

$$= \frac{2}{\beta t} e^{a_1 B_1 + a_2 B_2 + bt} \sum_{n=1}^{\infty} e^{-r_0^2/2t} \sin\left(\frac{n\pi\theta_0}{\beta}\right) \int_0^\beta \sin\left(\frac{n\pi\theta}{\beta}\right) g_n(\theta)\, d\theta,$$

where $X_i(t) = \ln\left(\frac{V_i(t)}{V_i(0)} e^{-\gamma_i t}\right)$, $B_i = \ln\left(\frac{b_i(0)}{V_i(0)}\right) \leq 0$, $\underline{X}_i(t) = \min_{0 \leq s \leq t} X_i(s)$, $\alpha_i = r_f - q_i - \gamma_i - \frac{1}{2}\sigma_i^2$, $I_{(\frac{n\pi}{\beta})}\left(\frac{rr_0}{t}\right)$ is a modified Bessel's function and

$$g_n(\theta) = \int_0^\infty r e^{-r^2/(2t)} e^{A(\theta)r} I_{(\frac{n\pi}{\beta})}\left(\frac{rr_0}{t}\right) dr$$

$$a_1 = \frac{\alpha_1\sigma_2 - \rho\alpha_2\sigma_1}{(1-\rho^2)\sigma_1^2\sigma_2}, \quad a_2 = \frac{\alpha_2\sigma_1 - \rho\alpha_1\sigma_2}{(1-\rho^2)\sigma_1\sigma_2^2}$$

$$b = -\alpha_1 a_1 - \alpha_2 a_2 + \frac{1}{2}\sigma_1^2 a_1^2 + \rho\sigma_1\sigma_2 a_1 a_2 + \frac{1}{2}\sigma_2^2 a_2^2$$

$$\tan\beta = -\frac{\sqrt{1-\rho^2}}{\rho}, \quad \beta \in [0, \pi]$$

$$r_0 = \frac{1}{\sqrt{1-\rho^2}} \left(\frac{B_1^2}{\sigma_1^2} - \frac{2\rho B_1 B_2}{\sigma_1\sigma_2} + \frac{B_2^2}{\sigma_2^2}\right)^{1/2}$$

$$\tan\theta_0 = \frac{\sigma_1 B_2\sqrt{1-\rho^2}}{\sigma_2 B_1 - \rho\sigma_1 B_2}, \quad \theta_0 \in [0, \beta]$$

$$A(\theta) = a_1\sigma_1 \sin(\beta - \theta) + a_2\sigma_2 \sin\theta.$$

2.1 Bond Yield Calculation

We calculate bond yields, $y_i(0, T) = -\ln(K_1/C_1(0, T))/T$ from the discounted expected values of the default and maturity payments,

Payment at maturity $= \min(\omega_1 V_1(T), K_1)$, provided $\tau_1 > T, \tau_2 > T$
Payment on default $= \omega_1 K_1 e^{-r_f(T-\tau)}$, where $\tau = \min\{\tau_1, \tau_2\}$,

τ_i denotes the default time of company i and $0 \leq \omega_1 \leq 1$ represents the fact that a portion of a defaulting company's value is lost to bondholders. Using (1) we obtain analytical formulae for yields as outlined in [2] which we evaluate by numerical quadrature using a sparse grid. Default contagion arises from the assumption that company one defaults on its outstanding debt the first time that the value of either company reaches its default barrier. Default can then be triggered in two ways – either as a direct result of the company's specific circumstances, or due to links with another company. An example of such a scenario would be if company two was the only supplier of a key component in company one's business, with company one unable to operate without it. Company two need not default automatically if company one does. It can continue to operate regardless of the financial viability of company one with dependence solely through the asset correlation, ρ. As a result, the model is asymmetric with respect to default risk, in contrast with the majority of previous models incorporating a credit dependence structure.

Figures 1–4 illustrate the sensitivity of yields to correlation, the shape of the default barrier and firm volatility. Initial credit quality is the initial distance of the firm from its default barrier. Since default is less likely with increasing correlation (the probability of at least one company defaulting is

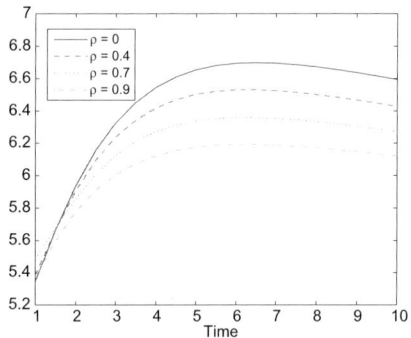

Fig. 1. Implied yield curve

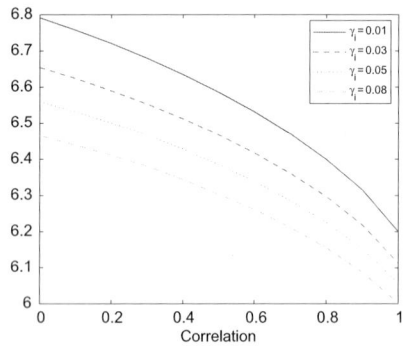

Fig. 2. Bond yield, varying γ_i

$$\sigma_i = 0.2, \ K_1 = 100, \ r_f = 0.05, \ \omega_1 = 0.7, q_i = 0,$$
initial credit quality $= 2$; $\gamma_i = 0.03$ in Fig. 1, T$=5$ in Fig. 2

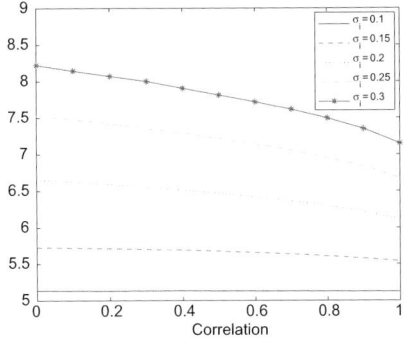

Fig. 3. Bond yield, varying σ_i

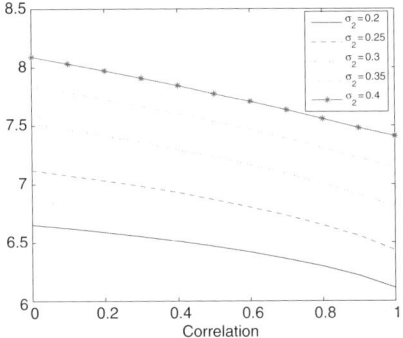

Fig. 4. Bond yield, varying σ_2

$\gamma_i = 0.3$, $K_1 = 100$, $r_f = 0.05$, $\omega_1 = 0.7$, $q_i = 0$, initial credit quality $= 2$, T=5.

higher when they are negatively correlated), yields decline as correlation increases since the bond is less risky, increasing the price and reducing the yield.

Figures 3 and 4 illustrate the impact of volatility on yields. The volatility of both firms is changed simultaneously in Fig. 3, whilst in Fig. 4, $\sigma_1 = 0.2$ is fixed and company two's volatility is increased. In both cases, higher volatility leads to a higher likelihood of default and higher yields; in the latter situation this is purely due to the correlation between the two companies and the possibility of default contagion since we are considering the yield on company one's bonds.

2.2 CDS Spread Calculations

Using a similar approach to that in Sect. 2.1, we evaluate first and second-to-default credit default swap (CDS) spreads for a two-company basket. The buyer of a kth-to-default CDS on this underlying basket pays a premium, the CDS spread, for the life of the CDS – until maturity or the kth default, whichever happens first. In the event of default by the kth underlying reference company, the buyer receives a default payment and the contract terminates. For bond recovery on default, R, and continuous spread payments, c, the kth-to-default CDS spread is

$$c_k = \frac{(1-R)\left\{1 - e^{-r_f T}\mathbb{P}(\tau_k > T) - \int_0^T r_f e^{-r_f s}\mathbb{P}(\tau_k > s)\,\mathrm{d}s\right\}}{\int_0^T e^{-r_f s}\mathbb{P}(\tau_k > s)\,\mathrm{d}s}. \tag{2}$$

Figures 5–8 show the impact of correlation, maturity and volatility on first and second-to-default CDS spreads. As correlation between the reference entities increases, first-to-default CDS spreads decrease, whilst second-to-default spreads increase since the probability of at least one company defaulting is higher for negative correlations, whilst the probability of both defaulting is greater for positive correlations. Spreads increase with maturity and volatility as expected.

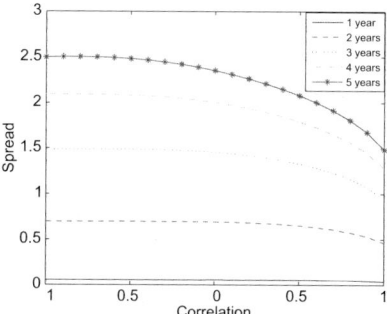

Fig. 5. First-to-default CDS

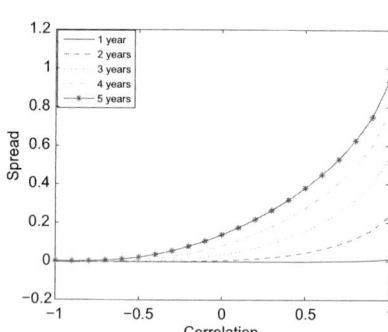

Fig. 6. Second-to-default CDS

CDS spreads for varying T
$\sigma_i = 0.2$, $r_f = 0.05$, $q_i = 0$, $\gamma_i = 0.03$, initial credit quality = 2, $R = 0.5$

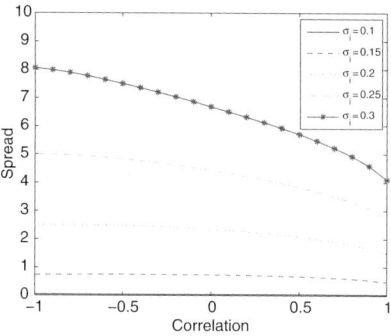

Fig. 7. First-to-default CDS

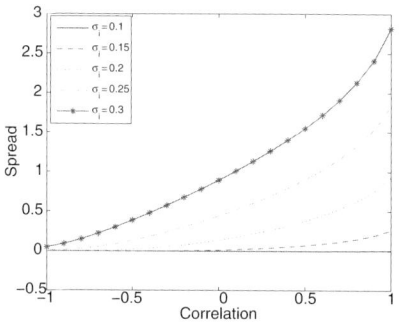

Fig. 8. Second-to-default CDS

CDS spreads for varying σ_i
$r_f = 0.05$, $q_i = 0$, $R = 0.5$, $\gamma_i = 0.03$, initial credit quality $= 2$, T $= 5$

3 Conclusion

We have incorporated default contagion within the structural framework for the first time, valuing bond yields and CDS spreads in a 2-dimensional first passage model. The result is a credit model that is asymmetric with respect to default risk and which has a dependence structure based on both long-term asset correlation and default contagion. Future work includes modelling larger baskets of companies and incorporating a more realistic specification of default contagion. A company default does not usually cause outright default at a related company, but a ripple of credit weakness at related companies.

References

1. F. Black and J. Cox. Valuing corporate securities: Some effects of bond indenture provisions. *Journal of Finance*, 31:351-367, 1976.
2. H. Haworth, C. Reisinger, and W. Shaw. Modelling bonds and credit default swaps using a structural model with contagion. *Working paper*, 2006.

3. H. Hua, W. Keirstead, and J. Rebholz. Double lookbacks. *Mathematical Finance*, 8:201-228, July 1998.
4. R. Merton. On the pricing of corporate debt: The risk structure of interest rates. *Journal of Finance*, 29:449-470, 1974.
5. C. Zhou. An analysis of default correlations and multiple defaults. *Review of Financial Studies*, 14:555-576, 2001.

The Valuation of Elementary Exotics with Strike Resets

Eric C. K. Yu

Nomura Centre for Quantitative Finance, OCIAM, Mathematical Institute,
Oxford University
yue@maths.ox.ac.uk

Generous funding from the United Kingdom Engineering and Physical Sciences
Research Council and Nomura International plc in the form of a CASE award is
gratefully acknowledged. The author would like to thank William Shaw for helpful
comments.

Summary. We demonstrate the valuation of a selection of elementary exotic options with strike price resets in this article using a simple change of variable that keeps the valuation problem two-dimensional.

1 Introduction

In the 1990s, a path-dependent feature known as resets appeared in the markets where the strike prices of call and put options may be reset at some prespecified dates with reference to the history realised by the underlying asset thus far. This feature can be especially appealing to investors if used for portfolio insurance purposes as it saves investors from having to adjust positions in the option in the event of adverse market movements, provide the reset is to the investor's interests. If the reset is against the investor's interests, the reset serves to encourage early exercise if permitted.

The literature on reset options was initiated by Gray and Whaley [GW97, GW99] who value bear warrants and put options with a single reset. A straightforward analytic solution involving bivariate integrals was derived for the European case; for the American case, solution was via a multi-layer version of the well-known binomial tree of Cox et al. [CRR79]. Subsequent contributions include analytic extensions to multiple resets of Cheng and Zhang [CZ00] and Liao and Wang [LW03], time-dependent parameters of Li and Li [LL06], and Li et al. [LLS06] where the short rate is governed by the Hull White extended Vasicek process. Numerical methods were lattice-based with an extra dimension, see Kwok and Lau [KL01] and Haug and Haug [3].

Reset options are quite similar to convertible bonds with conversion price resets, to which similar valuation methods were presented in the literature. There are lattice-based methods of Connolly [Con98, ch 9], Nelken [Nel98] and

Berger et al. [BKL00]; finite difference method of Wilmott [Wil98, p 472] based
on the framework of Dewynne and Wilmott [DW93]; numerical integration
of Shaw and Bennett [SB98] and simulation-based methods of Shaw and
Bennett [SB98] and Kimura and Shinohara [KS06].

The above mentioned methods are three-dimensional in essence, except for
Hoogland et al. [HND01] and Yu [Yu05] who achieved a similar reduction in
the valuation of reset convertible bonds by a change of variable by exploiting
first-degree homogeneity. In a subsequent contribution, Yu and Shaw [YS06]
employ the same change of variable as Yu [Yu05] to value both reset options
and reset convertible bonds that keeps the problem two-dimensional.

The resets considered in the above mentioned references, on which we will
focus, are snapshot resets where the the new strike/conversion price on reset
depends only on the share price and the prevailing strike/conversion price
just before a reset is due. In practice, the strike/conversion price is reset with
reference to some function of the share price attained during a short window
period prior to reset. These are window resets and require more effort to
model. Note that snapshot resets is a special case of window resets with the
length of the window period set to zero. For short window periods, it may
be valid to simplify the modelling by approximating window resets with their
snapshot equivalents. Refer to Yu and Shaw [YS06] for a review of window
resets in the literature.

This article is, in essence, a quick and straightforward extension of Yu and
Shaw [YS06] where we apply a simple change of variable to value a selection
of elementary exotics. The next section sets the stage by defining our prototype reset and briefly outlining the framework of Yu and Shaw [YS06]. The
third and final section demonstrates the valuation of a selection of elementary
exotics with strike resets.

2 Background

The analysis throughout is under the usual assumptions of Black and Scholes
[BS73] and Merton [Mer73]. The risk-neutral dynamics of the value of the
underlying asset, S, is governed by

$$\mathrm{d}S_t = (r - q)S_t\mathrm{d}t + \sigma S_t\mathrm{d}W_t, \tag{1}$$

where the risk-free short rate r, continuous dividend yield q and volatility of
the underlying σ are assumed to be constant and W_t is a Wiener process. We
can, of course, extend the forthcoming analysis with little additional effort if
r, q and σ are time-dependent. In particular, we can make the yield dividend
discrete by incorporating delta functions. The payout structure of the underlying is governed solely by the continuous dividend yield q and the exercise
structure is European. The value H of a derivative that depends on the value
of the underlying and time and is paid for upfront satisfies the well-known
Black–Scholes equation

$$\frac{\partial H}{\partial t} + \frac{1}{2}\sigma^2 S^2 \frac{\partial^2 H}{\partial S^2} + (r-q)S\frac{\partial H}{\partial S} - rH = 0 \tag{2}$$

for $S \in [0,\infty)$ and $t < T$ where T is the expiry date. The specification of the payoff $H(S,T)$ completes the valuation problem and uniquely identifies the derivative.

In general, valuation is to be performed numerically as analytics is relatively rare. If the finite difference method is used, upper and lower boundary conditions are to be specified along the S direction. The condition along the lower boundary is simply (2) with $S = 0$ and the upper boundary condition can be generally specified as

$$\lim_{S \to \infty} \frac{\partial^2 H}{\partial S^2} = 0$$

as the payoff is expected to be linear in S when the share price is very high [Wil98, p 624].

2.1 Snapshot Resets

A reset is given by the definition of g, it typically takes the form

$$g(K_{u-}, S_{u-}) = \begin{cases} \alpha K_{u-} & \text{if } S_{u-} > aK_{u-}, \\ \gamma S_{u-} & \text{if } bK_{u-} \le S_{u-} \le aK_{u-}, \\ \beta K_{u-} & \text{if } S_{u-} < bK_{u-}, \end{cases}$$

where α, β and γ are strictly positive and finite constants, and $0 \le b \le a$. We also insist that $a = b$ if $b = \infty$ or $a = 0$. The values of β when $b = 0$ and α when $a = \infty$ are irrelevant. If $a = b = 0$ or $a = b = \infty$, the reset would be deterministic in the sense that the reset variable would be the same and known for all values of the share price at the reset time and so is not of much interests to us.

By no arbitrage, the value of the RD is continuous across a reset date so that

$$H(S_{u-}, u^-; K_{u-}) = H(S_u, u; g(K_{u-}, S_{u-})) = H(S_u, u; K_u). \tag{3}$$

2.2 Similarity Reduction

Following the analysis of Yu and Shaw [YS06], let us define, as usual, the moneyness at time $t \le T$ as

$$M_t = \frac{S_t}{K_t},$$

where K is the strike/conversion price; it can be shown that if S satisfies (1) then so does M except at reset times when K changes. Moreover, the pricing equation (2) transforms to

$$\frac{\partial H}{\partial t} + \frac{1}{2}\sigma^2 M^2 \frac{\partial^2 H}{\partial M^2} + (r-q)M\frac{\partial H}{\partial M} - rH = 0. \tag{4}$$

Across a reset date, the moneyness changes:

$$M_u(M_{u-}) = \begin{cases} M_{u-}/\alpha \text{ if } M_{u-} > a, \\ 1/\gamma \quad\text{ if } b \le M_{u-} \le a, \\ M_{u-}/\beta \text{ if } M_{u-} < b. \end{cases} \tag{5}$$

On changing the variable from S to M, the reset g at time u can be regarded as a jump condition give by the function $f : (0,\infty) \to (0,\infty)$ with $f(M_{u-}) = M_u$ defined as per (5). The no arbitrage condition (3) becomes

$$H(M, u^-) = H(f(M), u). \tag{6}$$

A change of variable on H may be required depending on the nature of the payoff. Note that the upper and lower spatial boundary conditions are unaltered. We simply solve (4) backwards in time and apply a jump condition equivalent to (6) on reset dates. Refer to Yu and Shaw [YS06] for further details.

3 Elementary Exotics with Resets

It is well known that an European call option C with strike K can be regarded as an asset-or-nothing call A with a short position in K digital calls B with the same strike [RR91b]:

$$C(S,T) = (S-K)\theta(S-K) = S\theta(S-K) - K\theta(S-K) = A(S,T) - KB(S,T). \tag{7}$$

As we can value a reset call option with the method described in Sect. 2.2, we have reasons to believe that we can price asset-or-nothing and digital calls with strike resets.[1] If we make the change of variable

$$A(S,t) = K_t\bar{A}(M,t) \text{ and } B(S,t) = \bar{B}(M,t),$$

we see that the payoffs become

$$\bar{A}(M,T) = M\theta(M-1) \text{ and } \bar{B}(M,T) = \theta(M-1)$$

as $\theta(S-K) = \theta(M-1)$. We can therefore proceed with valuation as per Sect. 2.2. Note that asset-or-nothing and digital calls no longer synthesises call options in that (7) ceases to hold when there are strike resets.

The linearity (7) is, however, preserved in some cases. Consider a gap call option [RR91b] C_G with payoff

$$C_G(S,T) = (S - X)\theta(S - K),$$

----

[1] The value of a reset binary call, when multiplied by the accumulation factor $e^{r(T-t)}$, gives the probability of finishing in-the-money and may therefore be of interest.

where X is an absolute number that can be either side of and does not depend on the strike; the relation (7) becomes

$$C_G(S,T) = A(S,T) - XB(S,T),$$

which can be shown to hold when there is a reset on the strike K. On the other hand if we have $X = xK$ at all times, we can define $C_G(S,t) = K_t\bar{C}_G(M,t)$ so that $\bar{C}_G(M,T) = (M - x)\theta(M - 1)$ and we can proceed with valuation as before.

Our final example, similar to Haug and Haug [3], concerns standard continuously monitored barrier options [RR91a] where the barrier X is related to the strike via $X = xK$ at all times. We focus on a reverse up-and-out call option with $X > K$; its value in the classical Black–Scholes world, $C^{UO}(S,t)$, is

$$C^{UO}(S,t) = C(S,t;K) - C(S,t;X) - (X - K)B(S,t;X)$$

$$-(S/X)^k(C(X^2/S,t;K) - C(X^2/S,t;X)$$

$$-(X - K)B(X^2/S,t;X)),$$

where $k = 2(r - q - \sigma^2/2)/\sigma^2$. We can proceed as per with asset-or-nothing calls in defining

$$C^{UO}(S,t) = K_t\bar{C}^{UO}(M,t)$$

so that

$$\bar{C}^{UO}(M,t) = \bar{C}(M,t;1) - \bar{C}(M,t;x) - (x - 1)\bar{B}(M,t;x)$$

$$-(M/x)^k(\bar{C}(x^2/M,t;1) - \bar{C}(x^2/M,t;x) - (x - 1)\bar{B}(x^2/M,t;x))$$

for valuation. Likewise for the other options with "out" barriers, the incorporation of M- or t-dependent rebates is straightforward.

We have obtained satisfactory agreements amongst the results using the finite difference method as per Sect. 2.2 and numerical integration.[2] In the scenarios we calculated, we observed that a barrier option with a less generous reset may worth comparatively more over a range of share price levels. This is because although the reset may be expected to lower the moneyness, the loss in expected moneyness is more than compensated by a lower likelihood of being knocked out.

The valuation of reset options with "in" barriers requires a clear specification of the strike price in the event that the option is knocked-in after the reset time. Valuation is only slightly more complicated in that we may need to simultaneously value the contract the option can be knocked into at anytime.

---

[2]Specimen results are available from the author on request.

References

[BKL00] Berger, E., Klein, D., Levitan, B.: Modeling convertible bonds that reset. Bloomberg., **9**, 120–124, (2000)

[BS73]   Black, F., Scholes, M.: The pricing of options and corporate liabilities. Journal of Political Economy., **81**, 637–654, (1973)

[CZ00]   Cheng, W.-Y., Zhang, S.: The analytics of reset options. Journal of Derivatives., **8**, 59–71, (2000)

[Con98] Connolly, K.B.: Pricing convertible bonds. Wiley, Chichester (1998)

[CRR79] Cox, J.C., Ross, S.A., Rubinstein, M.: Option pricing: a simplified approach. Journal of Financial Economics., **7**, 229–263, (1979)

[DW93]   Dewynne, J.N., Wilmott, P.: Partial to the exotic. Risk., **6**, 38–46, (1993)

[GW97]   Gray, S.F., Whaley, R.E.: Valuing S&P 500 bear market warrants with a periodic reset. Journal of Derivatives., **5**, 99–106 (1997)

[GW99]   Gray, S.F., Whaley, R.E.: Reset put options: valuation, risk characteristics, and an application. Australian Journal of Management., **24**, 1–20, (1999)

[HH01]   Haug, E., Haug, J.: Who's on first base? Resetting strikes, barriers and time. Wilmott., (2001)

[HND01] Hoogland, J.K., Neumann, C.D.D., Bloch, D.: Converting the reset. Technical Report SEN-R0108, CWI, Amsterdam, (2001)

[KS06]   Kimura, T., Shinohara, T.: Monte Carlo analysis of convertible bonds with reset clauses. European Journal of Operational Research., **168**, 301–310, (2006)

[KL01]   Kwok, Y.K., Lau, K.W.: Pricing algorithms for options with exotic path-dependence. Journal of Derivatives., **9**, 1–11, (2001)

[LL06]   Li, S.-S., Li, S.-H.: A generalization of exotic options pricing formulae. Journal of Zhejiang University SCIENCE A., **7**, 584–590, (2006)

[LLS06] Li, S.J., Li, S.H., Sun, C.: A generalization of reset options pricing formulae with stochastic interest rates. Research in International Business and Finance., in press, (2006)

[LW03]   Liao, S.-L., Wang, C.-W.: The valuation of reset options with multiple strike resets and reset dates. Journal of Futures Markets., **23**, 87–107, (2003)

[Mer73] Merton, R.C.: Theory of rational option pricing. Bell Journal of Economics and Management Science., **4**, 141–183, (1973)

[Nel98] Nelken, I.: Reassessing the reset. AsiaRisk., 36–39, (1998)

[RR91a] Rubinstein, M., Reiner, E.: Breaking down the barriers. Risk., **4**, 28–35, (1991)

[RR91b] Rubinstein, M., Reiner, E.: Unscrambling the binary code. Risk., **4**, 75–83, (1991)

[SB98]   Shaw, W.T., Bennett, M.G.: Refixable convertible bonds '98. Research notes, Nomura Quantitative Analysis Group, London (1998)

[Wil98] Wilmott, P.: Derivatives: the theory and practice of financial engineering. Wiley, Chichester (1998)

[Yu05]   Yu, E.C.K.: Modelling convertible bonds with snapshot conversion price reset features. Proceedings of the International Conference in Economics and Finance 2005., 555–564, (2005)

[YS06]   Yu, E.C.K., Shaw, W.T.: On the valuation of derivatives with snapshot resets. Working paper, Nomura Centre for Quantitative Finance, Oxford University, (2006)

Minisymposium "On Optimal Strategies of Multivariate Passport Options"

Jörg Kampen

Weierstrass Institute, Berlin, Germany

American, and Passport Options. These options are modelled by free boundary equations and optimal stopping problems, and by HJB-equations and stochastic optimal control problems. A Bermudean Option contract allows early exercise only at discrete values of time prescribed in the contract. Bermudean Options are popular in high-dimensional fixed income markets and treated typically by Monte–Carlo simulations. At each possible date of expiration the holder of a Bermudean Option has to decide between the value of the product upon exercise and the value of the product upon non-exercise. The latter value is given in terms of conditional expectations. The approximation by conditional estimators involves Monte–Carlo errors. Christian Fries investigates the foresight bias of the Bermudean Option which he interpretes as an Option on the Monte–Carlo error of the conditional estimator. He shows how to apply an analytical correction on the foresight bias which allows for simplifications in coding and more efficient pricing. As the number of exercise dates increase and the maximal distance of two consecutive exercise times decreases, Bermudean Options approach American Options which can be exercised at any time up to expiration. The contribution of Etienne Chevalier provides a lower bound for the difference between the value function of a multivariate American Option and the payoff function. From this he obtains a convergence rate of the Bermudean exercise region to the American one. This result is important because up to now we have to rely on Monte–Carlo methods in order to price and exercise higher-dimensional American Options. A uniform approach for American options and some related early exercise problems is presented in the contribution of John Chadam. He summarizes a bunch of recent works concerning analytical and numerical treatment of American options, prepayment of mortgages, and shows that that the underlying approach can also be applied to the inverse first crossing problem of a default barrier. The approach is based on the representations of prices of early exercise options by fundamental solutions. The fundamental solution plays also a fundamental role in the contribution by Jörg Kampen. He determines the optimal strategy of a multivariate call option on a traded account where the option holder pays a premium upfront and is allowed to choose short and long positions of a portfolio within certain position limits. The so-called passport options is modelled by HJB-equations and optimal stochastic control problems.

Foresight Bias and Suboptimality Correction in Monte–Carlo Pricing of Options with Early Exercise

Christian P. Fries

www.christian-fries.de
email@christian-fries.de

Summary. We provide a definition and an analytic formula for the so called *foresight bias* that may appear in the Monte–Carlo pricing of Bermudan and compound options if the exercise criteria is calculated by the same Monte–Carlo simulation as the exercise values. The analytical correction for the foresight bias is then applied to the Monte–Carlo pricing of a Bermudan option (Bellman's principle), resulting in better prices, especially for very low number of paths.

1 Bermudan Option Pricing, Bellman's Principle

Let $\{T_i\}_{i=1,\dots,n}$ denote a set of exercise dates and $\{V_{\mathrm{underl},i}\}_{i=1,\dots,n}$ a corresponding set of underlyings. The Bermudan option is the right to receive at one and only one time T_i the corresponding underlying $V_{\mathrm{underl},i}$ (with $i = 1,\dots,n$) or receive nothing. From Bellman's principle we have that the value of the Bermudan option is given recursively by

$$V_{\mathrm{berm}}(T_i,\dots,T_n;T_i) := \max\left(V_{\mathrm{berm}}(T_{i+1},\dots,T_n;T_i)\,,\,V_{\mathrm{underl},i}(T_i)\right), \quad (1)$$

where $V_{\mathrm{berm}}(T_n;T_n) := 0$ and $V_{\mathrm{underl},i}(T_i)$ denotes the value of the underlying $V_{\mathrm{underl},i}$ at exercise date T_i.

Let $N(t)$ denote the time t value of a chosen numraire and $\mathbb{Q}^N$ the corresponding pricing measure, see [2]. From the universal pricing theorem we have that the $N(T_i)$-relative value of $V_{\mathrm{berm}}(T_{i+1},\dots,T_n;T_i)$ is given by the conditional expectation (w.r.t. the pricing measure) the $N(T_{i+1})$ relative value of $V_{\mathrm{berm}}(T_{i+1},\dots,T_n;T_{i+1})$. Defining relative prices $\tilde{V}_{\mathrm{underl},i}(T_j) := \frac{V_{\mathrm{underl},i}(T_j)}{N(T_j)}$ and $\tilde{V}_{\mathrm{berm},i}(T_j) := \frac{V_{\mathrm{berm}}(T_i,\dots,T_n;T_j)}{N(T_j)}$, we have

$$\tilde{V}_{\mathrm{berm},i}(T_i) = \max\left(\tilde{V}_{\mathrm{berm},i+1}(T_i)\,,\,\tilde{V}_{\mathrm{underl},i}(T_i)\right),$$

with $\tilde{V}_{\mathrm{berm},i+1}(T_i) = \mathrm{E}^{\mathbb{Q}^N}\left(\tilde{V}_{\mathrm{berm},i+1}(T_{i+1}) \mid \mathcal{F}_{T_i}\right)$ and $\tilde{V}_{\mathrm{berm},n} \equiv 0$, where $\{\mathcal{F}_t\}$ denotes the filtration.

1.1 Bermudan Option as Optimal Exercise Problem

The recursive Definition (1)represents the optimal exercise strategy in each exercise time. For a given path $\omega \in \Omega$ let $T(\omega) := \min\{T_i : V_{\text{berm},i+1}(T_i, \omega) < V_{\text{underl},i}(T_i, \omega)\}$. T gives a description of the exercise strategy as a stopping time. With the definition of the optimal exercise strategy T it is possible to define a random variable which allows to express the Bermudan option value as a single (unconditioned) expectation. With $\tilde{U}(T_i) := \tilde{V}_{\text{underl},i}(T_i)$ $i = 1, \ldots, n$ denoting the relative price of the i-th underlying upon its exercise date T_i we have for the Bermudan value $\tilde{V}_{\text{berm}}(T_0) = \mathrm{E}^{\mathbb{Q}}(\tilde{U}(T) \mid \mathcal{F}_{T_0})$.

The random variable $\tilde{U}(T)$ may be derived through the backward algorithm (Bellman's principle), *given* the exercise criteria (1), i.e. the conditional expectation. Induction start: $\tilde{U}_{n+1} \equiv 0$. Induction step:

$$\tilde{U}_i = \begin{cases} \tilde{U}_{i+1} & \text{if } \tilde{V}_{\text{underl},i}(T_i) < \mathrm{E}^{\mathbb{Q}}(\tilde{U}_{i+1}|\mathcal{F}_{T_i}) \\ \tilde{V}_{\text{underl},i}(T_i) & \text{else.} \end{cases} \tag{2}$$

From the tower law we have by induction $\mathrm{E}^{\mathbb{Q}}(\tilde{U}_{i+1}|\mathcal{F}_{T_i}) = \mathrm{E}^{\mathbb{Q}}(\tilde{V}_{\text{berm},i+1}(T_i)|\mathcal{F}_{T_i})$ and thus $\tilde{V}_{\text{berm}}(T_1, \ldots, T_n, T_0) = \mathrm{E}^{\mathbb{Q}}(\tilde{U}_1|\mathcal{F}_{T_0})$ and $\tilde{U}_1 = \tilde{U}(T)$. Since in the Monte–Carlo simulation the calculation of unconditional expectations is just the average value across all Monte–Carlo samples, the only missing part is the determination of the exercise criteria (2).

2 Conditional Expectation Estimators

One approach to determine the exercise criteria (2) is to give an estimator for the conditional expectation involved. If the SDE is in a Markovian form, then expectation conditional to $\mathcal{F}_{T_1}$ is a function of time T_1 state variable Z (and possibly other model parameters known in T_1). Thus we have $\mathrm{E}^{\mathbb{Q}^N}(U(T_2) \mid \mathcal{F}_{T_1}) = \mathrm{E}^{\mathbb{Q}^N}(U(T_2) \mid Z)$.

A standard method to obtain an estimate for the conditional expectation operator is to perform a regression against a suitable set of predictors $X := (X_1, \ldots, X_p)$, where the X_i's are $\mathcal{F}_{T_1}$ measurable random variable (basis functions), e.g. monomials in Z. Let $\hat{U}$, $\hat{V}$, $\hat{X}_i$ denote the n-vectors given by evaluating the random variables U, V, X_i on the Monte–Carlo sample paths $\omega_1, \ldots, \omega_n$. In case of a linear regression an estimate of $\mathrm{E}(U|X)$ is given by $\hat{X}\alpha^*$ with $\alpha^* = (\hat{X}^{\mathrm{T}}\hat{X})^{-1}\hat{X}^{\mathrm{T}}\hat{U}$. See [1, 2] an references therein.

In [1] we derive an estimate for the local Monte–Carlo error of the regression estimate $\hat{X}\alpha^*$ of the conditional expectation operator $\mathrm{E}(V \mid X)$. Let $\hat{H} := \hat{X}(\hat{X}^{\mathrm{T}}\hat{X})^{-1}\hat{X}^{\mathrm{T}}$, then the Monte–Carlo error of the regression estimate is $\sum_j \hat{H}_{i,j}^2 \mathrm{E}(\hat{\epsilon}_j^2 \mid \hat{X})$, where ϵ_j denote the residuals of the estimate, i.e. the value of $V - X\alpha^*$ on the i-th sample path. We obtain the conditional variance of the residuals by using regression once again on $\epsilon^2 = (V - \mathrm{E}(V|X))^2 \approx (V - X\alpha^*)^2$. See [1] for further details.

3 Foresight Bias: Classification, Calculation & Removal

The foresight bias is an option on the Monte Carlo error of the conditional expectation estimator. The standard deviation of the Monte Carlo error is the volatility of that option and the foresight bias is always non-negative.

Consider the optimal exercise value $\max(K, \mathrm{E}(\tilde{V} \mid Z))$ where the conditional expectation estimator has a Monte Carlo error which we denote by ϵ. Then the foresight bias is given by:

$$\mathrm{E}\big(\max(K, \mathrm{E}(\tilde{V}|Z) + \epsilon)\big|Z\big) = \max(K, \mathrm{E}(\tilde{V}|Z)) + \text{foresightbias}.$$

Here and in the following we will consider the exercise criteria $\max(K, \mathrm{E}(\tilde{V}|Z))$, i.e. with the notation used in the previous section $\tilde{V}$ stands for $\tilde{U}_{i+1}$ and K stands for $\tilde{V}_{\mathrm{underl},i}(T_i)$ for some i. The conditional expectation estimator (e.g. binning, regression) will be denoted by $\mathrm{E}^{\mathrm{est}}$ in place of E, i.e. $\mathrm{E}^{\mathrm{est}}(\tilde{V}|Z) = \mathrm{E}(\tilde{V}|Z) + \epsilon$.

3.1 Estimation of the Foresight Bias

We want to asses the foresight bias induced by a Monte–Carlo error ϵ of the conditional expectation estimator $\mathrm{E}(\tilde{V}|Z)$, i.e. we consider the optimal exercise criteria

$$\max(K, \mathrm{E}(\tilde{V}|Z) + \epsilon).$$

Conditioned on a given $Z = z^*$ we have from central limit theorem that ϵ has normal distribution with mean 0 and standard deviation σ for fixed $\mathrm{E}(\tilde{V}|Z)$. Then we have the following result for the foresight bias:

Lemma 1. *(Estimation of Foresight Bias) Given a conditional expectation estimator of $\mathrm{E}(\tilde{V}|Z)$ with (conditional) Monte–Carlo error ϵ having normal distribution with mean 0 and standard deviation σ will result in a bias of the conditional mean of $\max(K, \mathrm{E}(\tilde{V}|Z) + \epsilon)$ given by*

$$\underbrace{\underbrace{\sigma \cdot \phi(-\frac{\mu - K}{\sigma})}_{\textit{foresight bias}} + \underbrace{\underbrace{(\mu - K) \cdot \big(1 - \Phi(-\frac{\mu - K}{\sigma})\big) + K}_{\textit{smoothed payout}} - \underbrace{\max(K, \mathrm{E}(\tilde{V}|Z))}_{\textit{true payout}}}_{\textit{diffusive part, biased low}}}_{\textit{biased high}} ,$$

(3)

where $\mu := \mathrm{E}(\tilde{V}|Z)$, $\phi(x) := \frac{1}{\sqrt{2\pi}} \exp(-\frac{1}{2}x^2)$ and $\Phi(x) = \int_{-\infty}^{x} \phi(\xi)\, d\xi$.

Proof. (of Lemma 1) Let ϵ have Normal distribution with mean 0 and standard deviation σ. For $a, b \in \mathbb{R}$ we have with $\mu^* := b - a$

$$\mathrm{E}(\max(a, b + \epsilon)) = \mathrm{E}(\max(0, b - a + \epsilon)) + a = \mathrm{E}(\max(0, \mu^* + \epsilon)) + a$$

$$= \frac{1}{\sigma} \int_0^\infty x \cdot \phi\left(\frac{x - \mu^*}{\sigma}\right)\, dx + a = \frac{1}{\sigma} \int_{-\mu^*}^\infty (x + \mu^*) \cdot \phi(\frac{x}{\sigma})\, dx + a$$

$$= \int_{-\frac{\mu^*}{\sigma}}^\infty (\sigma \cdot x + \mu^*) \cdot \phi(x)\, dx + a = \sigma \cdot \phi(\frac{\mu^*}{\sigma}) + \mu^* \cdot \left(1 - \Phi\left(-\frac{\mu^*}{\sigma}\right)\right) + a.$$

The result follows with $b = \mathrm{E}(\tilde{V}|Z)$, $a := K$, i.e. $\mu^* = \mu - K$.

The bias induced by the Monte-Carlo error of the conditional expectation estimator consists of two parts: The first part in (3) consists of the systematic one sided bias resulting from the non linearity of the $\max(a, b + x)$ function. The second part is a diffusion of the original payoff function. The first part should be attributed to super-optimal exercise due to foresight, the second part to sub-optimal exercise due to Monte–Carlo uncertainty.

We define the first term in (3) as the *foresight bias correction* $\beta(\mu, \sigma) := \sigma \cdot \phi(-\frac{\mu - K}{\sigma})$ and the second term in (3) as the *suboptimal exercise correction* $\gamma(\mu, \sigma) := (\mu - K) \cdot \left(1 - \Phi(-\frac{\mu - K}{\sigma})\right) - \max(0, \mu - K)$, where $\mu := \mathrm{E}(\tilde{V}|Z)$ and σ^2 is the variance of the Monte–Carlo error ϵ of the estimator μ.

3.2 Analytical Removal of Foresight Bias

We modify the backward algorithm and correct for the foresight bias by subtracting the term $\beta^{\mathrm{est}} := \beta(\mu^{\mathrm{est}}, \sigma^{\mathrm{est}})$ and $\gamma^{\mathrm{est}} := \gamma(\mu^{\mathrm{est}}, \sigma^{\mathrm{est}})$ from the payout on each path, where $\mu^{\mathrm{est}} := \mathrm{E}^{\mathrm{est}}(\tilde{V}|Z)$ and σ^{est} is the estimator for the Monte–Carlo error ϵ.

$$\tilde{U}_i := -\beta^{\mathrm{est}} - \gamma^{\mathrm{est}} + \begin{cases} \tilde{V}_{\mathrm{underl}}(T_i) & \text{if } \tilde{V}_{\mathrm{underl}}(T_i) > \mathrm{E}^{\mathrm{est}}(\tilde{V} \mid Z) \\ \tilde{U}_{i+1} & \text{else.} \end{cases}$$

4 Numerical Results

Our Benchmark model is a simple Black–Scholes model for an asset S where S follows $dS = \mu S dt + \sigma S dW$, with $S(0) = 1.0$, $\sigma = 20\%$, and assuming the risk free asset $dB = rB dt$ with $r = 5\%$. Our benchmark product is a simple Bermudan option on S paying $N_i \cdot (S(T_i) - K_i)$ upon exercise in T_i with exercise dates $T_1 = 1.0, T_2 = 2.0, T_3 = 3.0$, notionals $N_i = 1.0$ and strikes $K_1 = 0.95, K_2 = 1.0, K_3 = 1.10$. The regression polynomial of the conditional expectation estimator is order 5 in S.

4.1 Aggregation of Monte–Carlo Prices

We setup m independent Monte–Carlo simulation with n/m paths and calculate the average price of a Bermudan option price over the set of Monte–Carlo simulations. We vary m from $m = 1$, i.e. a single Monte–Carlo simulation with a huge number of paths, to $m = 2048$, i.e. many small Monte–Carlo simulations. The aggregated prices have similar Monte–Carlo errors and for an European option the different methods should result in (almost) identical prices. For a Bermudan option the different methods of aggregation are not equivalent. The foresight bias is a systematic error being $O(\sqrt{m/n})$.

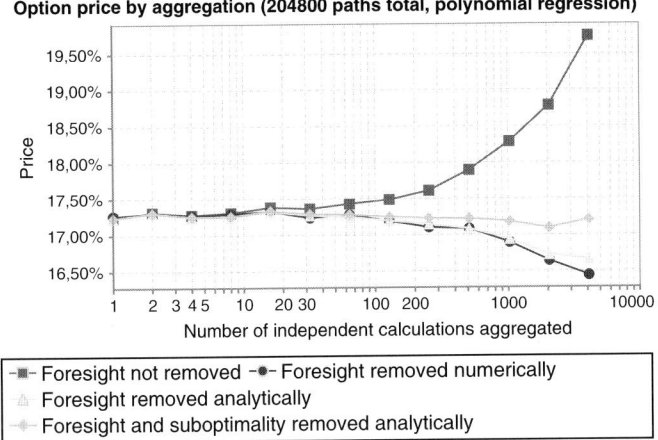

Fig. 1. Aggregation of Monte–Carlo Prices with or without Removal of Foresight

Figure 1 shows that the numerical removal of the foresight bias and our analytical removal of the foresight bias give very similar results. For m large a single Monte–Carlo simulation has a low number of paths, thus a larger foresight-bias. If foresight bias is removed the price will slowly become lower. This is due to the fact that the optimal exercise is smeared out by the diffusive term in (3).

References

1. Fries, C.P.: Foresight Bias and Suboptimality Correction in Monte-Carlo Pricing of Options with Early Exercise: Classification, Calculation and Removal (2005). http://www.christian-fries.de/finmath/foresightbias
2. Fries, C.P.: Mathematical Finance. Theory, Modeling, Implementation. Lectures Notes. Frankfurt am Main (2006). http://www.christian-fries.de/finmath/book

On the American Option Value Near its Exercise Region

Etienne Chevalier

Université d'Evry Val d'Essonne, Laboratoire d'Analyse et Probabilité,
Boulevard François Mitterrand, 91025 Evry Cedex, France
echevali@univ-evry.fr

Summary. American options valuation leads to solve an optimal stopping problem or a variational inequality. These two approaches involve the knowledge of a free boundary, boundary of the so-called exercise region. Numerical methods exist to solve this kind of problems but these methods are not very efficient in high dimension because some information on the free boundary is needed. To improve our knowledge of the value function near its exercise region, we give here a lower bound for the difference between the value function and the pay-off function near the free boundary. This result can be used, for instance, to get some estimation for the convergence rate of the Bermudean option exercise region to the American one.

1 American Options

An American option is a financial product which gives to its owner the right to earn a specific amount of money at any time he wishes between the initial date 0 and the maturity T. This amount of money, so-called the option pay-off, is very often based on the values of one or several underlying assets. The two main problems of the American option theory are then to give a price to this product and to determine the optimal strategy for the owner: the optimal time to exercise his right is assumed to be the time for which his gain is greater as possible.

The first step to solve these two linked problems is to make assumptions on the market. We will assume that the market is composed by d risky assets and denote by S_t^i their respective value at time t. We assume that $(S_t)_{0 \leq t \leq T}$ is solution of the following stochastic differential equation:

$$\mathrm{d}S_t = \mathrm{diag}(S_t)((r\mathrm{I} - \delta)\mathrm{d}t + \sigma \mathrm{d}W_t) \tag{1}$$

where $\mathrm{I} = (1)_{1 \leq i \leq d}$, $r > 0$ is the interest rate of the market, $\delta \in [0, +\infty)^d$ is such that δ_i is the dividend rate of the asset i, $\sigma \in \mathbb{R}^d \times \mathbb{R}^d$ is called the market volatility and $(W_t)_{0 \leq t \leq T}$ is a standard Brownian motion on $\mathbb{R}^d$.

Moreover, we assume that σ satisfies the following hypothesis which insures the non degeneracy for the infinitesimal generator of the diffusion S:

$$H_1 : \quad \exists M > m > 0, \; \forall x \in \mathbb{R}^d, \; m\|x\|^2 \leq x^* \sigma \sigma^* x \leq M\|x\|^2.$$

We denote by $\mathbb{F}$ the filtration associated to W and for $x \in [0, +\infty)^d$, $(S_t^x)_{0 \leq t \leq T}$ is the solution of the stochastic differential equation (1) such that $S_0^x = x$.

Our goal here is to study a specific class of options, called basket options. These options offer a pay-off which is the positive part of the difference between a positive constant (the strike price) and a linear combination of several assets. We define the pay-off function f such that:

$$\forall x \in [0, +\infty)^d, \; f(x) = (K - \langle \alpha, x \rangle)^+,$$

where $K > 0$ is the strike price, $\alpha \in \mathbb{R}^d$, and $\langle ., . \rangle$ is the usual scalary product on $\mathbb{R}^d$.

In this setting, the option theory (see [B84] and [K88]) asserts that at time $t \in [0, T]$, the price of the American option associated with the pay-off f is $P(T - t, S_t)$ where:

$$P(t, x) = \sup_{\tau \in \mathcal{T}_{0,t}} \mathbb{E}[e^{-r\tau} f(S_\tau^x)], \quad \forall x \in [0, +\infty)^d,$$

where $\mathcal{T}_{0,t}$ is the set of $\mathbb{F}$-stopping times with values in $[0, t]$.

At this point two approaches enable us to get information on the value function P. First the optimal stopping theory (see [EK81]) asserts that the supremum is attained and more precisely, we have:

$$P(t, x) = \mathbb{E}\left[e^{-r\tau^*} f(S_{\tau^*}^x) \right],$$

where $\tau^* = \inf \{ t \geq 0 : P(T - t, S_t^x) = f(S_t^x) \} \wedge T$.

A second point of view gives a variational characterisation for P. We know (see [BL82] and [JLL90]) that P is the solution of the following variational inequality

$$\begin{cases} (\mathcal{M}P - rP) \leq 0, \quad f \leq P, \quad (\mathcal{M}P - rP)(P - f) = 0 \quad \text{a.s.} \\ P(0, x) = f(x) \quad \text{on} \quad \mathbb{R}^+, \end{cases}$$

where we set:

$$\mathcal{M}h(t, x) = -\frac{\partial h}{\partial t} + \frac{1}{2} \sum_{i,j=1}^{d} (\sigma\sigma^*)_{i,j} x_i x_j \frac{\partial^2 h}{\partial x_i x_j} + \sum_{i=1}^{d} (r - \delta_i) x_i \frac{\partial h}{\partial x_i}.$$

A specific region of $(0, +\infty) \times [0, +\infty)^d$ appears in these two approaches, it is called the exercise region:

$$\mathcal{E} = \left\{ (t, x) \in]0, T) \times [0, +\infty)^d : P(T - t, x) = f(x) \right\}.$$

In fact if we know this region we would be able on one hand to compute τ^*, on the other hand to compute P as a solution of a partial differential equation.

From a financial point of view, this region is very interesting because it determines the optimal strategy the option owner has to follow. Our goal here is to give an estimation of this region and more specifically of its temporal sections:

$$\forall t \in (0, T], \quad \mathcal{E}_t = \{x \in [0, +\infty)^d : P(t, x) = f(x)\}.$$

Indeed, we are not able to get a closed formula for the price of an American option and to determine its exercise region so a lot of numerical methods have been developed to compute American options prices. The first idea is to solve the variational inequality satisfied by the value function thanks to a finite differences method. However, for problems with high dimension this approach become very difficult to implement. In this case, we solve the optimal stopping problem with Monte-Carlo methods (see [BG97], [LS01] and [BP03]). For that, we consider a Bermudean option, this is an American one which can be exercised only at a finite number of dates. From a financial point of view, it gives less rights to its owner than an American option then its price is lower than the price of the corresponding American option. However, if the number of exercise opportunities goes to infinity, it is well known that the Bermudean option price tends to the American one.

Some estimations of the convergence rate have been found (see [BP03]). Since, the Bermudean option value function is the solution of an optimal stopping problem which can be seen as a free boundary problem, our goal is then the estimation of the convergence rate of the Bermudean free boundary to the American one when the number of exercise opportunities goes to infinity.

2 Lower Bound for the Value Function Near its Exercise Region

In this section we present our main result and give the ideas of the proof.

Theorem 1. *Let $x_T^* \in \Gamma_T$ where Γ_T is the boundary of the continuation region. There exists $\varepsilon > 0$ and $C > 0$ such that*

$$\forall y \in (0, +\infty)^d - \mathcal{E}_T \text{ such that } \|y - x_T^*\| \leq \varepsilon \quad P(T, y) - f(y) \geq C\|y - x_T^*\|^2.$$

When $d = 1$, this result is quite easy to get. Indeed, thanks to the variational inequality satisfied by P in the exercise region, we can prove that there exists $C > 0$ such that

$$\frac{\partial^2 P}{\partial x^2}(t, z) \geq C,$$

for all $z \in (x_T^*, x_T^* + \varepsilon)$. We conclude the proof by integrating two times between x_T^* and y and using the continuity of $\frac{\partial P}{\partial x}$.

In the case $d > 1$, we can not get such control for the second space derivatives of P. We then follow ideas of the proof of Proposition 18.1 in [C04] and overcome this difficulty by using the following maximum principle (see [F75]).

Let D a bounded domain of $(0,T) \times \mathbb{R}^d$. We define the parabolic boundary of D by $\delta_p D = \delta D - \{(t,x) \in \delta D : t = T\}$ where δD is the boundary of D and introduce the operator $\tilde{\mathcal{M}}$ such that $\tilde{\mathcal{M}}h = \mathcal{M}h - rh$.
Let u a function defined on $[0,T] \times \mathbb{R}^d$, continous on $\bar{D}$ and such that

$$u \in \mathcal{C}^{1,2}(D), \quad \tilde{\mathcal{M}}u \geq 0 \text{ on} D \quad and \quad u \leq 0 \text{ on } \delta_p D.$$

Then we have $u \leq 0$ on D.

To prove Theorem 1, we introduce a bounded domain D included in the continuation region of the American option and such that its boundary contains x_T^*. Then we assume that for $C > 0$ small enough, there exists $y \in (0,+\infty)^d - \mathcal{E}_T$ such that $\|y - x_T^*\| \leq \varepsilon$ and $P(T,y) - f(y) \leq C\|y - x_T^*\|^2$. Under this assumption we can construct a function $\beta \in \mathcal{C}^2(D)$ such that $P - f - \beta$ satisfies the parabolic maximum principle and such that, for $(t,x) \in D$ close to the exercise region, $\beta(t,x) = 0$. That leads to a contradiction, because the maximum principle would allow us to write $0 < [P - f](t,x) \leq \beta(t,x) = 0$.

3 Application to the Bermudean Approximation of American Options

At time t, the price of a Bermudean option which offers n exercise opportunities $\{T_1, \ldots, T_n\}$, with a pay-off function f and a maturity T is $P^n(T - t, S_t)$ with:

$$P^n(t,x) = \sup_{\tau \in \mathcal{T}_{0,t}^n} \mathbb{E}[e^{-r\tau} f(S_\tau^x)],$$

where $\mathcal{T}_{0,t}^n$ is the set of stopping times with values in $\{T_1, ..., T_n\} \wedge t$

It is well known that when n goes to infinity, P^n tends to P the value function of the corresponding American option. Some estimations on the convergence rate have been found in [BP03]. More precisely, it have been proved that there exists a constant $C > 0$ such that for al $(t,x) \in (0,T) \times (0,+\infty)^d$,

$$0 \leq P(t,x) - P^n(t,x) \leq \frac{C}{n}.$$

As in the American case, we can define the exercise region of a Bermudean option by

$$\mathcal{E}^n = \{(t,x) \in [0,T] \times (0,+\infty)^d : P^n(T - t,x) = f(x)\}.$$

It is also easy to see $\mathcal{E}^n$ tends to $\mathcal{E}$ when n goes to infinity. Now we will see that Theorem 1 allows us to give an estimation for the convergence rate.

Let $u \in (0,+\infty)^d$ such that $\|u\| = 1$. As the temporal sections of $\mathcal{E}$ and $\mathcal{E}^n$ are convex and contain 0, we can introduce the following quantities:

$$s(t,u) = \inf\{\lambda \in \mathbb{R} : P(t,\lambda u) = f(\lambda u)\}$$
$$s^n(t,u) = \inf\{\lambda \in \mathbb{R} : P^n(t,\lambda u) = f(\lambda u)\}.$$

Now, we are able to define a convergence rate for the exercise region by $s^n(t, u) - s(t, u)$ because we know that $s^n(t, u) \geq s(t, u)$. We apply Theorem 1 near $s(t, u)u$. Let $\varepsilon > 0$ defined as in Theorem 1. As $s^n(l, u)$ tends to $s(t, u)$ when n goes to infinity, for n great enough we have $0 \leq s^n(t, u) - s(t, u) \leq \varepsilon$ and then we can conclude that there exists $\gamma > 0$ such that

$$\gamma \left(s^n(t, u) - s(t, u)\right)^2 \leq P(t, s^n(t, u)) - f(s^n(t, u))$$
$$= P(t, s^n(t, u)) - P^n(t, s^n(t, u)).$$

Using the estimation on the convergence rate for the value function, we obtain

$$0 \leq s^n(t, u) - s(t, u) \leq \sqrt{\frac{C}{\gamma n}}.$$

References

[BP03]   Bally V., Pagès G.: Error Analysis of the quantization algorithm for ob-stacle problems, Stochastic Processes and their Applications **106**, 1-40. (2003)

[B84]   Benssoussan A.: On the theory of option pricing, Acta Applicandae Math-ematicae **2**, 139-158. (1984)

[BL82]   Benssoussan A., Lions J.L.: Applications of variational Inequalities in Sto-chastic Control, North-Holland. (1982)

[BG97]   Broadie M., Glasserman P.: Pricing American-style securities using simu-lation, Journal of Economic Dynamics and Control **21**, 1323-1352. (1997)

[CW90]   Carverhill A. P., Webber N.: American options: theory and numerical analysis. In: Options: Recent Advances in Theory and Practice, Manchester Univ. Press. (1990)

[C04]   Chevalier E.: American options and free boundaries, University of Marne-la-Vallée (2004)

[EK81]   El Karoui N.: Les aspects probabilistes du contrôle stochastique, Lecture Notes in Mathematics **876**, 72-238. Springer-Verlag. (1981)

[F75]   Friedman A.: Stochasic differential equations and applications, vol. **1**, New York: Academic Press. (1975)

[JLL90] Jaillet P., Lamberton D., Lapeyre B.: Variational inequalities and the pric-ing of American options, Acta Applicandae Mathematicae, **21**, 263-289. (1990)

[K88]   Karatzas I.: On the pricing of American options, Applied Math. Optimiza-tion **17**, 37-60. (1988)

[L98]   Lamberton D.: Error estimates for the binomial approximation of American put options, Annals of Applied Probability **8**, 206-233. (1998)

[LS01]   Longstaff F. A., Schwartz E. S.: Valuing American Options by Simulations: a Simple Least Squares Approach, Review of Financial Studies **14**, 113-147. (2001)

[V99]   Villeneuve S.: Options américaines dans un modèle de Black-Scholes multi-dimensionnel, Thesis, Université de Marne-la-Vallée. (1999)

Free Boundary Problems in Mathematical Finance

John Chadam

Department of Mathematics, University of Pittsburgh, PA 15260

Summary. We provide a unified approach to studying a wide variety of free boundary problems that arise in modern mathematical finance. For the most part, the main ideas will be presented in the simplest case of the early exercise boundary for the American put option on a geometric Brownian motion. In addition to discussing the existence and uniqueness of the solution to the problem, and the convexity of the free boundary, we will describe several fast and accurate numerical and analytical approximations for the location of these early exercise boundaries. The same approach can be used to treat similar problems with more general underliers such as jump diffusion processes. We will also show how the techniques can be carried over to treat other classes of free boundary problems such as the inverse first crossing problem of the default barrier of a credit process as well as the pricing of mortgage prepayment options. Various parts of this work are joint efforts with Xinfu Chen (Pittsburgh) and David Saunders (Pittsburgh and Waterloo) as well as our recent Ph.D. students Lan Cheng, Ge Han and Dejun Xie.

1 American Put Option

In this section we shall outline our methods for studying free boundary problems in finance in the context of the prototypical case of the American put with the underlying asset following a geometric Brownian motion

$$\frac{\mathrm{d}S}{S} = \mu \mathrm{d}t + \sigma \mathrm{d}W(t) \tag{1}$$

The classic result of Black, Scholes and Merton risk-neutral pricing theory says that the value of the American put option satisfies the free boundary problem

$$p_t + \frac{\sigma^2}{2} p_{SS} + rSp_S - rp = 0, \ S_f(t) < S, \ 0 < t < T, \tag{2a}$$

$$p(S,t) = K - S \text{ on } S = S_f(t), \ 0 \le t < T, \tag{2b}$$

$$p_S(S,t) = -1 \text{ on } S = S_f(t), \ 0 \le t < T, \tag{2c}$$

$$p(S,t) \to 0 \text{ as } S \to \infty, \tag{2d}$$

$$p(S,T) = \max(K - S, 0), \quad S > S_f(T) = K. \tag{2e}$$

where K is the strike price and T is the expiry time of the contract and $S = S_f(t)$ is the early exercise boundary. Letting $\tau = \frac{\sigma^2}{2}(T - t)$ (the scaled time to expiry) and $x = \ell n(S/K)$, then the scaled option price price

$$p_{\text{new}} = \begin{cases} 1 - S/K & S < S_f \\ p/K & S > S_f \end{cases})$$

satisfies the transformed Black, Scholes, and Merton (BSM) problem (dropping the subscript)

$$p_\tau - \{p_{xx} + (k-1)p_x - kp\} = kH(x_f(\tau) - x) \tag{3}$$

$$p(x,0) = \max(1 - e^x, 0) \tag{4}$$

where $k = 2r/\sigma^2$, H is the Heaviside function, $x_f(\tau) = \ell n(S_f/S)$, and the coefficient k appears because the intrinsic payoff, $p_0(x) = 1 - e^x$ satisfies

$$p_{0\tau} - \{p_{0xx} + (k-1)p_{0x} - kp_0\} = k. \tag{5}$$

The solution to problem (3,4) can be written in terms of the free boundary $x_f(\tau)$ and the fundamental solution of the BSM pdo on the lhs of (3),

$$\Gamma(x,\tau) = \frac{e^{-k\tau}}{2\sqrt{\pi\tau}} e^{-(x+(k-1)\tau)^2/4\tau} \tag{6}$$

in the form

$$p(x,\tau) = \int_{-\infty}^0 p_0(y)\Gamma(x - y, \tau)dy + k\int_0^\tau \int_{-\infty}^{x_f(u)} \Gamma(x - y, \tau - u)dydu \tag{7}$$

Rather than following the usual approach in the free boundary literature of using one of the conditions

$$p(x_f(\tau), \tau) = 1 - e^{x_f(\tau)} \tag{8a}$$

$$p_x(x_f(\tau), \tau) = -e^{x_f(\tau)} \tag{8b}$$

which are the transformed versions of the usual smooth pasting conditions on the early exercise boundary, and follow from problem (3,4), we instead use a trick here, based on financial considerations, to notice that $p_\tau(x_f(\tau), \tau) = 0$. Thus, from (7)

$$p_\tau(x,\tau) = \Gamma(x,\tau) + k\int_0^\tau \Gamma(x - x_f(u), \tau - u)\,\dot{x}_f(u)du \tag{9}$$

which, upon evaluation on the early exercise boundary, provides the following non-linear integral equation for $x_f(\tau)$

$$\Gamma(x_f(\tau), \tau) = -k \int_0^\tau \Gamma(x_f(\tau) - x_f(u), \tau - u)\dot{x}_f(u)du. \qquad (10)$$

Careful estimates [1-4] show that the integral on the rhs of (10) tends to -1 as $\tau \to 0$, resulting in the near expiry estimate

$$\frac{e^{-k\tau}}{2\sqrt{\pi\tau}} e^{-(x_f(\tau)+(k-1)\tau)^2/4\tau} \cong \frac{e^{-x_f(\tau)^2/4\tau}}{2\sqrt{\pi\tau}} = k \qquad (11)$$

which leads to

$$x_f(\tau) \approx 2\sqrt{\tau}\sqrt{-\ell n(4\pi k^2\tau)^{1/2}} \text{ as } \tau \to 0. \qquad (12)$$

Writing $x_f(\tau) = -2\sqrt{\tau}\sqrt{s(\tau)}$ then the near expiry behavior (12) can be written as

$$s(\tau) \approx -\frac{1}{2} \ell n(4\pi k^2\tau) = -\frac{\xi}{2} \text{ as } \tau \to 0 \qquad (13)$$

where $\xi = \ell n(4\pi k^2\tau)$.

Using the above machinery, one can obtain [1,2] more precise analytic estimates valid for intermediate and large times. For example, using Mathematica to iterate (13) through (10), one obtains

$$s(\tau) = -\frac{\xi}{2} - \frac{1}{\xi} + \frac{1}{2\xi^2} + \frac{17}{3\xi^3} - \frac{51}{4\xi^4} - \frac{1148}{15\xi^5} + \frac{398}{\xi^6} + \cdots \qquad (14)$$

One can also imagine using (10) to express ξ as a function of s. One finds [1,2] for arbitrary, a,

$$-\frac{\xi}{2} = s + \ell n \left[1 - \frac{1/2}{s+a} - \frac{a/2}{(s+a)^2} + \frac{(1-a)^2}{2(s+a)^3} + \cdots \right]. \qquad (15)$$

or equivalently, on exponentiation,

$$\sqrt{\tau}e^s \left[1 - \frac{1}{2(s+a)} - \frac{a}{2(s+a)^2} + \frac{(1-a)^2}{2(s+a)^3} + \cdots \right) \right] = 1/\sqrt{4\pi k^2}, \qquad (16)$$

allowing for truncation, for example by taking $a = 1$. By interpolating estimates like (14), (15) above with Merton's infinite horizon solution ($S_f = Kk(k+1)^{-1}$), we obtain accurate estimates valid for all times [1-4].

We conclude this section with a discussion of how our results relate to the work of other contributors in the area and make some comments about our proofs. The first rigorous estimate for the near expiry behavior of the early exercise boundary was given by Barles et al. [5]:

$$S_f(t) \approx K \left[1 - \sigma\sqrt{(T-t)\ln(T-t)|} \right], \ t \sim T. \qquad (17)$$

where, recall, $S_f(t) = Ke^{-2\sqrt{\tau}\sqrt{s(\tau)}}$, $\tau = \frac{\sigma^2}{2}(T-t)$. Others [6,7] including the much earlier work of Barone-Adesi and Whaley [8], have provided estimates for intermediate times as well as near expiry. These estimates, near expiry, can be summarized in the present notation as

$$\sqrt{\tau}\,\sqrt{s}e^s \approx 1/\sqrt{4\pi k^2} \qquad [8] \tag{18a}$$

$$\sqrt{\tau}\,se^s \approx 1/\sqrt{9\pi k^2} \qquad [7] \tag{18b}$$

$$\sqrt{\tau}\,\sqrt{s}e^s \approx \left(\left(1 - \frac{1}{2}\left(\frac{k}{1+k}\right)^2\right)/4k^2\right)^{-1/2} \qquad [6] \tag{18c}$$

where $k = 2r/\sigma^2$. Our corresponding estimate, from (12) or (16) is

$$\sqrt{\tau}\,e^s \approx 1/\sqrt{4\pi k^2} \qquad [1,2] \tag{18d}$$

One notices that all versions of (18) agree with the Barles et al. result (17), that (18a-c) all lead to $\ell n\ \ell n$ corrections while our result (18d) through the integral equation (10) leads to corrections with inverse powers of ξ (14) and our near expiry estimate (18d) is the only one that is compatible with the implicit estimate (15) derivable from (10).

Our first rigorous derivation of the near expiry estimate (12, 18d) appeared in [3]. Several expressions, equivalent to (10), were obtained in [3], whose solution for $x_f(\tau)$ when inserted into (7) provided the existence and uniqueness of the solution to the original problem (2,3,4) for the American put. The most interesting of these was the integro-differential equation for $x_f(\tau)$

$$\dot{x}_f(\tau) = \frac{x_f(\tau)}{2k\tau}\Gamma(x_f(\tau),\tau)\Big[1 + m(\tau)\Big], \tag{19a}$$

where

$$m(\tau) = k\int_0^\tau \left[\frac{x_f(\tau) - x_f(u)}{\tau - u}\frac{2\tau}{x_f(\tau)} - 1\right]\frac{\Gamma(x_f(\tau) - x_f(u), \tau - u)}{\Gamma(x_f(\tau), \tau)}\dot{x}_f(u)du \tag{19b}$$

that is to be solved with $x_f(0) = 0$. In [3] we provided a rigorous proof of the existence and uniqueness of the solution $x_f(\tau)$ to problem (19), which, as mentioned above, when substituted in (7) solved (3, 4). We believe this to be the first proof in this integral equation formulation of the existence and uniqueness to the problem (2) for the American put. It should be noted that a probabilistic proof by Karatzas and Shreve [9] appeared much earlier and that an alternate, independent proof by Peskir [10] appeared during the revision of our manuscript [3]. It should be pointed out that our proof [3] does not rely on the convexity of the early exercise boundary and, as such, may serve as a prototype for problems for which the free boundary is not convex. The derivations in [3] are also provide an alternate proof of existence and uniqueness for (2,3) using variational methods and, in addition, we establish that the early exercise boundary is C^1 away from expiry. In [4] we provide a rigorous proof of the convexity of the early exercise boundary based on the methods of Friedman and Jensen [11]. Using the convexity we provide [4] a much simpler proof of the near expiry behavior (12, 18d) than that in [3]. We point out that during the revision of our manuscript [4] an independent proof of convexity was obtained by Ekstrom [12]. Equations (19) provide a fast and accurate iterative scheme for numerically approximating the location of the boundary. The first iteration obtained by taking $m = 0$ in (19a) can be solved numerically with Mathematica for the entire boundary, instantaneously. Each successive iteration takes approximately one minute with Mathematica and the third iterate results in a 10^{-5} relative change from the second up to one year from expiry. We find this iterative scheme based on (19) to be the most efficient and accurate of our numerical estimates.

We have also begun a program to extend these integral equation methods to jump-diffusion models. Specifically, letting $X = ln(S/K)$, we assume that the transformed asset follows the process

$$X(t) = (\mu - \sigma^2/2)t + \sigma W(t) + N(t) \tag{20}$$

where $N(t)$ is a Poisson process with rate λt and having jumps of size $\pm\epsilon$ with equal probability. In terms of the modified BSM pdo

$$\mathcal{L}p = p_\tau - \{p_{xx} + (k-1)p_x - kp\} + \lambda\{p(x+\epsilon,\tau) - 2p + p(x-\epsilon,\tau)\}, \tag{21}$$

(3, 4) becomes

$$\mathcal{L}p = \mathcal{L}(1 - e^x)H(x_f(\tau) - x), \tag{22a}$$

$$p(x,0) = \max(1 - e^x, 0). \tag{22b}$$

The analog of the integral equation (10) becomes

$$\Gamma(x_f(\tau),\tau) = -\int_0^\tau (k + \lambda\{2 - e^\epsilon - e^{-\epsilon}\}e^{x_f(u)}\Gamma(x_f(\tau) - x_f(u),\tau - u)\dot{x}_f(u)du$$

$$+\lambda\int_0^\epsilon (1 - e^{y-\epsilon})\Gamma(x_f(\tau) - y,\tau)dy. \tag{1}$$

from which we obtain the near expiry estimate

$$\sqrt{\tau}e^s \approx 1\sqrt{4\pi\tilde{k}^2}, \quad \tau \to 0 \tag{24}$$

with $\tilde{k} = k + \lambda(1 - e^{-\epsilon})$. This agrees with the results obtained by Pham [13] using other methods. With D. Saunders (Waterloo) we are developing the rest of the program outlined above in the context of the simple jump model (20), and extending it to more general cases including the degenerate case of vanishing diffusion, as well as in the context of geometric Brownian motions with stochastic volatility and variance gamma processes studied by Madan and collaborators (see, for example [14]).

Finally, we just mention that our student, Ge Han, has obtained in his doctoral dissertation [15] the $\sqrt{\tau}\sqrt{-\ell n\tau}$ behavior near expiry for the American put on the sum of geometric Brownian motions. He also established the convexity of the price as a function of the share price.

2 Credit Default

These methods can be carried over to firm value (structural) models for credit processes. Suppose $X(t)$ is a stochastic process for the default index of a company (rather than triggering changes in credit rating) satisfying an Uhlenbeck–Orstein process

$$dX(t) = a\,dt + \sigma dW(t), \quad X(t) = x_0, \tag{25}$$

(equivalently the log of such an index that originally satisfied a geometric Brownian motion). Default occurs the first time τ that $X(t)$ falls below a pre-assigned value, b. The survival pdf, $u(x,t)$, defined by

$$u(x,t)\mathrm{d}x = \Pr[x < X(t) < x + \mathrm{d}x \mid t < \tau]. \tag{26}$$

is known to satisfy the following problem for the Kolmogorov forward equation:

$$u_t = \frac{\sigma^2}{2}u_{xx} - a\,u_x, \quad b < x < \infty, \quad 0 < t < T \tag{27a}$$

$$u(x,t) = 0, \quad x = b, \quad 0 < t < T \tag{27b}$$

$$u(x,t) \to 0 \quad \text{as } x \to \infty, \quad 0 < t < T \tag{27c}$$

$$u(x,0) = \delta(x - x_0). \tag{27d}$$

The survival probability at time $t = T, p_T$, is then given by

$$p_T = \int_b^\infty u(x,T)\mathrm{d}x. \tag{28}$$

Merton [16] (see also the related work of Black and Cox [17]) posed and solved this problem (given b, find $u(x,t)$ and hence p_T) as well as the inverse problem (given p_T find $u(x,t)$ and b, such that (28) holds). In this single time horizon setting, everything follows from knowing the Greens function in the half space $b < x < \infty$.

In recent work [18] our student, Lan Cheng, solved the time dependent version of the inverse first passage problem: given $p(t), 0 < t, T$, find the time dependent absorbing boundary, $b(t)$, in (27b) such that

$$p(t) = \int_{b(t)}^\infty u(x,t),\,dx \tag{28}$$

is satisfied for all $0 < t < T$. Using viscosity solution methods she proved existence and uniqueness for this free boundary problem as well as establishing small time estimates for the location of the default barrier, $b(t)$, in terms of the default probability $q(t) = 1 - p(t)$:

$$\lim_{t \to 0} \frac{b(t)}{\sqrt{-4t\ell n(q(t))}} = -1 \tag{29}$$

She has also derived integral equations for $b(t)$ (the analog of (10)) of the form

$$\Gamma(b(t),t) = \int_0^t \Gamma(b(t) - b(\tau), t - \tau)\dot{q}(\tau)d\tau, \tag{30}$$

where $\Gamma(x,t)$ is the fundamental solution of the pdo in (27a). This can be used to provide an alternate derivation of the analytical estimate (29) as well as to develop a fast and accurate numerical scheme. Specifically, solving

$$F(x,t) = \Gamma(x,t) - \int_0^t \Gamma(x - b(\tau), t - \tau)\dot{q}(\tau)d\tau = 0 \tag{31}$$

for $x = b(t)$, using (29) as the first step in a Newton-Raphson scheme results in the iteration

$$b(t)^{\text{new}} = b(t)^{\text{old}} - \frac{F(b(t)^{\text{old}}, t)}{\dot{q}(t)/2} \tag{32}$$

where in computing F_x in the denominator of (32) we use

$$\frac{\dot{q}(t)}{2} \cong \Gamma_x(b(t)^{\text{old}}, t) - \int_0^t \Gamma_x(b(t)^{\text{old}} - b(\tau)^{\text{old}}, t - \tau)\dot{q}(\tau)d\tau, \tag{33}$$

All of our numerical simulations suggest that the boundary is concave for appropriate survival probabilities (e.g., $p(t) = t, \sqrt{t}, 1 - e^{-t}$ result in boundaries that are concave up and for $p(t) = e^{-1/2t}$ the boundary is concave down; the limiting case of a linear boundary is the only one that can be solved explicitly). It would be satisfying to prove the appropriate convexity result in this situation. Our work on this problem was motivated by that of Avellaneda and Zhu [19] and Zucca, Sacardote and Peskir [20] and our numerics have been compared with theirs in [18]. We anticipate that this fully dynamic setting ($b(t)$ depends on $p(s)$ and $b(s)$ for all $0 < s < t$) versus the single time horizon setting ($t = T$) will provide a much richer structure for the important issue of default correlations among many firms (see, for example, Schoenbucher [Chaps. 9, 10, 21] for a nice treatment of existing results).

3 Mortgage Prepayment Options

We provide one further example to show that these methods can carry over to treat free boundary problems for which the underlier is the short-term rate of interest, $r(t)$, that is assumed to follow a process of the type

$$dr = u(r,t)dt + w(r,t)dW(t) \tag{34}$$

is a risk-neutral world. The contingent claim (financial derivative) to be studied here is the American style contract providing the holder the right to prepay the outstanding balance of a mortgage

$$M(t) = \frac{m}{c}(1 - e^{c(t-T)}) \tag{35}$$

where T is the time that the mortgage is paid off (i.e., $M(T) = 0$), c is the (continuous) fixed rate of the mortgage and m is the (continuous) rate of payment of the mortgage (i.e., mdt is the premium paid in any time interval dt). Suppose the mortgage holder (borrower) would like to purchase a contract that allows for the prepayment of the current value, $M(t)$, of the mortgage at any time t up to T. Clearly the value of the contract, $V(r,t)$, depends not only on $M(t)$ but also on the rate of return, $r(t)$, that can be obtained by investing $M(t)$ in other instruments (re-mortgage, equities, bonds, etc). The BSM risk-neutral pricing of the contract $V(r,t)$ is obtained by solving the problem (the analog of (2))

$$V_t + \frac{w(r,t)^2}{2}\frac{\partial^2 V}{\partial r^2} + u(r,t)\frac{\partial V}{\partial r} + m - rV = 0, \quad R(t) < r < \infty, 0 < t < T \tag{36a}$$

$$V(r,t) = M(t), \quad r = R(t), \ 0 < t < T \tag{36b}$$

$$V_r(r,t) = 0, \quad r = R(t), 0 < t < T \tag{36c}$$

$$V(r,t) \to 0 \quad r \to \infty, 0 < t < T \tag{36d}$$

$$V(r,T) = M(T) = 0 \tag{36e}$$

The optimal strategy for the mortgage holder is to exercise the option to pay off the mortgage the first time that the rate r falls below $R(t)$ at time t.

Recent work on this problem (36) by Jiang et al. [22] for the Vasicek model for the investment rate $r(t)$ (i.e., $u(r,t) = (\eta - \theta r)$ and $w(r,t) = w$, constant in (34)) established existence and uniqueness using variational methods and provides the short-term behavior

$$R(t) = c - \sigma\bar{\kappa}\sqrt{T-t}, \ \bar{\kappa} = 0.47386\cdots \tag{37}$$

using asymptotic analysis in the framework of similarity solutions. Our student, Dejun Xie, has recast the problem in the current integral equation framework [23]. Specifically, after a sequence of changes of dependent and independent variables (too complicated for this summary), problem (36) in the Vasicek case ($u(r,t) = (\eta - \theta r), w(r,t) = w$) can be written as

$$u_s - \frac{1}{4}u_{xx} = f(x,s)H(x - x_f(s)) - \infty < x < \infty, s > 1 \tag{38a}$$

$$u(x,s) = 0, \quad x \le x_f(s), \ s > 1 \tag{38b}$$

$$u(x,s) > 0, \quad x > x_f(s), \quad s > 1 \tag{38c}$$

$$u(x,1) = 0, \quad -\infty < x < \infty.. \tag{38d}$$

where $f(x,s)$ is a specific function resulting from the transformations and $x_f(s)$ is the transformed free boundary. By analogy with (7) and (10), the solution of (38) can be written as

$$u(x,s) = \int_1^s \left[\int_{x_f(\tau)}^\infty \Gamma(x-y, s-\tau) f(y,\tau) dy \right] d\tau \qquad (39)$$

in terms of the fundamental solution $\Gamma(x,s) = \frac{1}{\sqrt{\pi s}} e^{-x^2/s}$ of the heat operator $\partial_s - \frac{1}{4}\partial_{xx}^2$, and $x_f(s)$ is determined from (38b, c) as the solution of the integral equation

$$\int_1^s d\tau \int_{x_f(\tau)}^\infty \Gamma(x_f(s) - y, s-\tau) f(y,\tau) dy = 0, \qquad (40)$$

In his Ph.D. dissertation [23] written under our supervision, Dejun Xie, has proven the existence of a unique solution to (40) which, as mentioned earlier, when substituted into (39) provides a unique solution to (38), the transformed version of (36). He also obtained from (39), along the lines of Sect. 1 (convexity was not used; in fact has not yet proven in this case) the near expiry estimate (37). He also used (40) along the lines of Sect. 2 to obtain a numerical scheme to determine $x_f(s)$ in the form

$$x_f(s)^{\text{new}} = x_f(s)^{\text{old}} + \frac{Q(x_f(s)^{\text{old}}, s)}{2f(x_f(s)^{\text{old}}, s)}. \qquad (41)$$

where $Q(x,s)$ is the rhs of (39). Finally, by careful analysis of the infinite horizon solution to (38), Dejun Xie was able to obtain precise new estimates on the behavior of $R(t)$ as $t \to -\infty$ of the form

$$R(t) \sim R^* + \rho^* e^{-c(T-t)} \text{ as } t \to -\infty. \qquad (42)$$

This was combined with the near expiry estimate (37) to give the global analytic estimate

$$R(t) \approx c - \frac{\sigma\bar{\kappa}}{\sqrt{2c}} \sqrt{2 - e^{-2c(T-t)}} + \rho^* \left[e^{-c(t-t)} - 1 \right]$$

$$+ \left[R^* - c + \frac{\sigma\bar{\kappa}}{\sqrt{2c}} + \rho^* \right] \left[1 - e^{-2c(T-t)} \right], \qquad (43)$$

It is quite surprising that this extremely simple expression (43) agrees quite well for all $0 < t < T$ with estimates obtained numerically from (41).

In conclusion, we have summarized a unified approach to studying free boundary problems in Mathematical Finance. The methods are suitable for studying basic existence and uniqueness questions, convexity and smoothness of the free boundary as well providing analytical and numerical estimates for the free boundary. The methods are applicable to a wide variety of problems with the randomness arising from the full spectrum of underliers.

References

[1] R. Stamicar, D. Sevcovic & J. Chadam, *Numerical and analytical approxima-tions of the early exercise boundary for the American put near expiry*, Can. Appl. Math. Quart. **7** (1999).

[2] X. Chen & J. Chadam, *Analytical and numerical approximations for the early exercise boundary for American put options*, Cont. Disc. and Imp. Systs., Series A: Math Anal. **10** (2003), 649-660.

[3] X. Chen & J. Chadam, *A mathematical analysis for the optimal exercise bound-ary of American put options*, preprint February 8, 2002, SIAM J. Math. Anal., accepted for publication. (*)

[4] X. Chen, J. Chadam, L. Jiang & W. Zheng, *Convexity of the exercise boundary of an American put option for a zero dividend asset*, preprint, September 2002, Mathematical Finance, accepted for publication. (*)

[5] G. Barles, J. Burdeau, M. Romano & N. Samsoen, *Critical stock price near expiration*, Mathematical Finance, **5** (1995), 77-95.

[6] D. A. Bunch & H. Johnson, *The American put option and its critical stock price*, Journal of Finance, **55** (2000).

[7] R. A. Kuske & J. B. Keller, *Optimal exercise boundary for an American put option*, Applied Mathematical Finance, **5** (1998), 107-116.

[8] G. Barone-Adesi & R. E. Whaley, *Efficient analytic approximations of Amer-ican option values*, Journal of Finance, **42**, (1987), 301-320.

[9] I. Karatzas & S. Shreve, *Methods of Mathematical Finance*, **6**, Springer-Verlag (1998).

[10] G. Peskir, *On the American option problem,* Mathematical Finance, **15** (2005).

[11] A. Friedman & R. Jensen, *Convexity of the free boundary in the Stefan problem and in the dam problem*, Arch. Rat. Mech. and Anal., **67** (1978).

[12] E. Ekstrom, *Convexity of the optimal stopping boundary for the American put option*, J. Math. Anal. Appl., **299** (2004).

[13] H. Pham, *Optimal stopping, free boundary and American option in a jump-diffusion model*, Appl. Math. Optim. **35** (1997) 145-164.

[14] D. Madan, P. Carr & E. Chang, *The variance gamma process and option pric-ing*, Eur. Fin. Rev., **2** (1998).

[15] Ge Han, *The American put on the sum of geometric Brownian motions*, Ph.D. thesis, University of Pittsburgh, (2003).

[16] R.C. Merton, *On the pricing of corporate debt: The risk structure of interest rates*, J. Fin., **29** (1974).

[17] F. Black & J.C. Cox, *Valuing corporate securities: Some effects of bond inden-ture provisions*, J. Fin., **31** (1976).

[18] L. Cheng, X. Chen, J. Chadam & D. Saunders, *An inverse first passage prob-lem from risk management*, preprint 2005, SIAM J. Math. Anal., accepted for publication. (*)

[19] M. Avellaneda & J. Zhu, *Modeling the distance-to-default of a firm*, Risk, **14** (2001).

[20] C. Zucca, L. Sacerdote & G. Peskir, *On the inverse first-passage problem for a Wiener process*, preprint.

[21] P. Schoenbucher, *Credit Derivative Pricing Models*, Wiley Finance (2003).

[22] L. Jiang, B. Bian & F. Yi, *A parabolic variational inequality arising from the valuation of fixed rate mortgages,* preprint, 2004.

[23] D. Xie, X. Chen & J. Chadam, *Optimal prepayment of mortgages,* submitted for publication. (*)

   (*) available at www.pitt.edu/~chadam

Optimal Strategies of Passport Options

Jörg Kampen

Weierstrass Institute, Mohrenstrasse 39, 10117 Berlin, Germany
kampen@wias-berlin.de

Summary. Passport options are options on traded accounts with payoff structure of a Call. Optimal strategies are naturally linked not only to hedging but also to evaluation of this type of options. We use recent results on mean stochastic comparison of [5, 6] in order to determine optimal strategies for multivariate passport options. Especially, we find that optimal strategies depend on the correlations of returns and are related to the Greeks.

1 Multivariate Passport Options

An option on a traded account consisting of a portfolio with several assets is a contract which allows the holder of the option to choose his position for each asset, subject to certain position limits, and where short and long positions are allowed. The holder of a passport option pays a premium upfront and accumulates gains and losses resulting from his trading up to expiration T where he gets the value of the traded account if it is positive and has a zero net position otherwise. Univariate passport options have been investigated intensively (cf. [1, 3, 4, 7] and the references therein). In this paper we consider optimal strategies of multivariate passport options. We use a multivariate extension of Hajek's result (cf. [2, 5, 6]). Consider n underlyings $S = (S_1, \cdots, S_n)$, where

$$\frac{dS_i}{S_i} = \mu_i dt + \sigma_i dW_i. \tag{1}$$

Here, W is an n-dimensional Brownian motion with

$$d\,[W_i, W_j] = \rho_{ij} dt, \tag{2}$$

and ρ_{ij} are constant correlations between the returns of the assets. We shall allow σ and μ to be dependent on the underlyings S. The trading account has the increment

$$d\Pi(t) = \mu_\Pi \Pi dt + \sum_i q_i (dS_i - S_i \nu_i dt)$$

$$= \left(\mu_\Pi \Pi + \sum_i q_i(\mu_i - \nu_i)S_i dt + \sum_i q_i \sigma_i S_i dZ_i\right). \tag{3}$$

Here, μ_Π is the rate of return of the traded account, which is part of the contract and does not depend on the underlying market and ν_i are the cost of carry of the ith underlying, also part of contract. We take $|q_i| \leq 1$ as position limits (which is no essential restriction). The payoff function P depends only on the portfolio value and is defined by

$$P(\Pi) = \max\{\Pi, 0\} =: \Pi^+. \tag{4}$$

In order to design a hedging portfolio we make two assumptions on the strategy $q = (q_1, \cdots, q_n)$

- $q_i(.) = q_i(S, \pi, t)$ are Markovian strategies,
- q is instantaneously constant, i.e. we have the return $q_i dS_i$ of the holder of q_i underlyings S_i.

The value function at time t will depend on $(S(t), \Pi(t)$ alone (not on the whole history) and may be denoted by $(t, S(t), \Pi(t)) \to V^q(t, S(t), \Pi(t))$ with increment

$$dV^q = \frac{\partial V^q}{\partial t} dt + \sum_i \frac{\partial V^q}{\partial S_i} dS_i + \frac{\partial V^q}{\partial \Pi} d\Pi + \frac{1}{2} \sum_{ij} \frac{\partial^2 V^q}{\partial S_i \partial S_j} [dS_i, dS_j]$$
$$+ \sum_i \frac{\partial^2 V^q}{\partial S_i \partial \Pi} [dS, d\Pi] + \frac{1}{2} \frac{\partial^2 V^q}{\partial \Pi^2} d[\Pi, \Pi]. \tag{5}$$

The hedging portfolio becomes:

$$\Pi = V^q - \sum_i \Delta_i S_i, \text{ with } \Delta_i = \frac{\partial V^q}{\partial S_i} + q_i \frac{\partial V}{\partial \Pi}. \tag{6}$$

The existence of a riskfree selffinancing duplicating strategy determines the value of V^q under the risk neutral measure Q to be

$$V^q(t, s, p) = E_Q^{(t,s,p)} \left[e^{-r(T-t)} \Pi(T)^+ \right], \tag{7}$$

assuming a flat yield curve. The value of the passport option V then becomes

$$V(t, s, p) = \max_{|q_i| \leq 1} V^q(t, s, p) = \max_{|q_i| \leq 1} E_Q^{(t,s,p)} \left[e^{-r(T-t)} \Pi(T)^+ \right], \tag{8}$$

and, by the Bellmann principle, the option value is characterized by the following Cauchy problem

$$\frac{\partial V}{\partial t} + \max_{q_i \in [-1,1]} \left\{ \sum_i \frac{1}{2} \sigma_i^2 s_i^2 \left(\frac{\partial^2 V}{\partial s_i^2} + 2q_i \frac{\partial^2 V}{\partial s_i \partial p} + q_i^2 \frac{\partial^2 V}{\partial p^2} \right) \right.$$

$$+ \sum_{i \neq j} \sigma_i \rho_{ij} \sigma_j s_i s_j \left(\frac{\partial^2 V}{\partial s_i \partial s_j} + q_i \frac{\partial^2 V}{\partial s_i \partial p} + q_j \frac{\partial^2 V}{\partial s_j \partial p} + q_i q_j \frac{\partial^2 V}{\partial p^2} \right) \tag{9}$$

$$\left. - (\mu_\Pi p + \sum_i (r - \delta_i - \nu_i) q_i S_i) \frac{\partial V}{\partial p} \right\} + \sum_i (r - \delta_i) s_i \frac{\partial V}{\partial s_i} - rV = 0,$$

with the boundary condition

$$V(T, s, p) = p^+. \tag{10}$$

In the following we shall consider the so-called symmetric case, where there is no directional bias incorporated in the contract, i.e.

- the cost of carry of the underlyings S_i are $\nu_i = r - \delta_i$,
- the rate of return of the money account equals the interest rate r, i.e. $\mu_\Pi = r$ is risk-neutral reinvestment rate.

2 Two Mean Stochastic Comparison Results

The following extension of Hajek's result to stochastic sums is used to obtain optimal strategies for multivariate passport options.

Theorem 1. *Let $n, T > 0$, $f \in C(\mathbb{R})$ be convex, and assume that f satisfies an exponential growth condition. Furthermore, let X, Y be semimartingales with $x = X(0) = Y(0) \in \mathbb{R}^n$, where*

$$X(t) = X(0) + \int_0^t \sigma(X(s))dW(s), \quad Y(t) = X(0) + \int_0^t \rho(Y(s))dW(s),$$

with $n \times n$- matrix-valued bounded continuous functions $x \to \sigma\sigma^T(x)$ and $y \to \rho\rho^T(y)$. If $\sigma\sigma^T \leq \rho\rho^T$, then for all $0 \leq t \leq T$

$$E^x\left(f\left(\sum_i X_i(t)\right)\right) \leq E^x\left(f\left(\sum_i Y_i(t)\right)\right).$$

The following theorem could be used o extend results to more general models of passport options than considered in this paper.

Theorem 2. *Let $T > 0$, $f \in C(\mathbb{R})$ nondecreasing, convex, and satisfying an exponential growth condition, X, Y semimartingales with $x = X(0) = Y(0) \in \mathbb{R}^n$, where*

$$X(t) = X(0) + \int_0^t \mu(X(s))ds + \int_0^t \sigma(X(s))dW(s),$$

$$Y(t) = X(0) + \int_0^t \nu(Y(s))ds + \int_0^t \rho(Y(s))dW(s),$$

with $n \times n$- matrix-valued functions $x \to \sigma\sigma^T(x)$ and $y \to \rho\rho^T(y)$, which have bounded continuous component functions $\sigma\sigma_{ij}^T$ and $\rho\rho_{ij}^T$ respectively. If $\mu \leq \nu$ are bounded continuous functions, and $\sigma\sigma^T \leq \rho\rho^T$, then for all $0 \leq t \leq T$

$$E^x\left(f\left(\sum_i X_i(t)\right)\right) \leq E^x\left(f\left(\sum_i Y_i(t)\right)\right).$$

For more information and proof of both theorems consider [5, 6].

3 Optimal Strategies of Multivariate Passport Options

Writing $f_1^q = \sum_i q_i S_i$ we observe that

$$Z(t) = \int_0^t 1_{\{q_1 > 0 \text{ or } q_2 > 0\}} \frac{1}{\sqrt{[f_1^q]}} df_1^q + \int_0^t 1_{\{q_i = 0\}} dB(s) \qquad (11)$$

is a Brownian motion on a suitable increased probability space (if $B(t)$ is a Brownian motion (the indicator function $1_{\{q_i = 0\}}$ equals 1 if for some i the value $q_i = 0$). Hence, the portfolio increment is

$$dΠ_q = rΠ dt + \sum_i q_i \sigma_i S_i dW_i$$

$$= rΠ_q dt + \sqrt{\sum_{ij} q_i q_j \rho_{ij} S_i S_j} dZ =: rΠ_q dt + \sigma_B(S_q) dZ, \qquad (12)$$

with the basket volatility $\sigma_B(.)$ (cf. [5, 6], and where we abbreviate $S_q = (q_1 S_1, \cdots, q_n S_n)$. Define $X(t) = e^{-rt} Π(t)$ and consider the case $n = 2$ for simplicity. Then the result of the previous section implies that the optimal strategy q^{opt} maximizes

$$q \to E_Q(X_q^+(T)). \qquad (13)$$

Hence,

$$q_1^{\mathrm{opt}} = \mathrm{sign} q_1^{\mathrm{opt}} = \mathrm{sign} q_2^{\mathrm{opt}} \text{ if } \rho \geq 0, \quad q_1^{\mathrm{opt}} = \mathrm{sign} q_1 = -\mathrm{sign} q_2 \text{ if } \rho < 0. \qquad (14)$$

We consider the case $\rho > 0$. Define $q_s = \mathrm{sign} q_1 = \mathrm{sign} q_2$. Changing to $Z_{q_s} = \frac{Π_{q_s}}{f_1}$ we find

$$dZ_{q_s} = -(Z_{q_s} - q_s) \frac{df_1}{f_1} + (Z_{q_s} - q_s) \left[\frac{df_1}{f_1} \right], \qquad (15)$$

where

$$[f_1](t) = v_{11} S_1^2 + v_{12} S_1 S_2 + v_{22} S_2^2 > 0. \qquad (16)$$

We change the measure to R, where

$$\frac{dR}{dQ}(T) = \exp \left(\int_0^T \sigma_B(S(s)) d\tilde{W}(s) - \frac{1}{2} \int_0^T \sigma_B^2(S(s)) ds \right). \qquad (17)$$

Then

$$\hat{W}(t) = \tilde{W} - \int_0^t \sigma_B(S(s)) ds \qquad (18)$$

is a Brownian motion with respect to the new measure. Since q is assumed to be a Markovian strategy

$$dZ_{q(s)} = -(Z_{q(s)} - q(s)) \sigma_B(S(s)) d\hat{W}. \qquad (19)$$

Hence, it suffices to maximize

$$q \to E_Q \left(Z_q^+ \right) = E_Q \left(Z_q(0)^+ \right) + \frac{1}{2} E \left(L_{Z_q}^0(T) \right), \qquad (20)$$

the maximizing strategy q^{opt} in case $\rho > 0$ which is

$$q_1^{\mathrm{opt}}(t) = \mathrm{sign}(q_1^{\mathrm{opt}}(t)) = \mathrm{sign}(q_2^{\mathrm{opt}}(t)) = -\mathrm{sign}(\Pi(t)). \qquad (21)$$

Note that the extensions of Hajek's result together with (9) imply in case $\rho > 0$ immediately that the mixed Gammas' (second derivatives with respect to the ith underling and the portfolio variable) have the same sign as the q_i^{opt} for each i.

References

1. DELBAEN, F., YOR, MARC *Passport Options,* Mathematical Finance, Vol. 12, pp. 299-328, 2002
2. HAJEK, B. *Mean stochastic comparison of diffusions,* Z. Wahrscheinlichkeitstheorie vrw. Geb., vol. 68, pp. 315-329, 1985.
3. HENDERSON, V., HOBSON, D. *Local time, coupling and the passport options* Finance & Stochastics 4, 69-80, 2000
4. HYER, T., LIPTON-LIFSCHITZ, A., PUGACHEVSKY, D. *Passport to success* RISK Magazine 10(9), 127-131, 1997.
5. KAMPEN, J., *The value of the American Call with dividends increases with the basket volatility,* Birkhäuser (ISBN 3-7643-7718-6), 2006, p. 260-272 (ISNM).
6. KAMPEN, J., *The WKB-Expansion of the Fundamental Solution of Linear Parabolic Equations and Its Applications* (Habilitation thesis, electronically published 2006 SSRN)
7. SHREVE, S.E., VECER, J. *Options on a traded account: Vacation calls, vacation puts and passport options* Finance & Stochastics 4., 255-274, 2000.

Minisymposium "Meshfree Methods for the Solution of PDEs"

Manuel Kindelan

Universidad Carlos III Madrid, Avda. Universidad 30, 28911 Leganes
kinde@ing.uc3m.es

During the last years, very intensive efforts have been devoted to develop meshfree methods that eliminate the need of element connectivity in the solution of PDEs. These methods are very flexible numerical tools and do not require the labor intensive step of mesh generation. At present, the fundamental theory of meshfree methods has been developed and considerable advances have been made in the implementation of the different methods which have been proposed. However, its use as a practical alternative to conventional finite element methods is still pending. In fact, many challenges still remain both in the mathematical analysis and in the practical implementation of the methods. The objective of this minisymposium is to review some of the most promising meshfree methods and analyze its application to relevant problems. In particular, the Finite Pointset Method (FPM) and the Radial Basis Function (RBF) method are reviewed and applied to relevant industrial modeling problems. Also, a new family of meshfree schemes based on local maximum-entropy approximants is proposed.

Solving One-Dimensional Moving-Boundary Problems with Meshless Method

Leopold Vrankar[1], Edward J. Kansa[2], Goran Turk[3], and Franc Runovc[4]

[1] Slovenian Nuclear Safety Administration, Železna cesta 16,
1001 Ljubljana, Slovenia
leopold.vrankar@gov.si
[2] Department of Mechanical and Aeronautical Engineering,
University of California, USA
ejkansa@ucdavis.edu
[3] University of Ljubljana, Faculty of Civil and Geodetic Engineering,
Jamova cesta 2, 1000 Ljubljana, Slovenia
gturk@fgg.uni-lj.si
[4] University of Ljubljana, Faculty of Natural Sciences and Engineering,
Aškerčeva cesta 12, 1000 Ljubljana, Slovenia
franc.runovc@ntf.uni-lj.si

1 Introduction

A large number of important physical processes involve heat conduction and materials undergoing a change of phase. Examples include nuclear reactors, casting of metals, semiconductor manufacturing, geophysics, and industrial applications involving metals, oil, and plastics. These problems are often called Stefan's or moving boundary value problems.

Several numerical methods have been developed to solve various Stefan's problems. Crank [1] provides a good introduction to the Stefan's problems and presents an elaborate collection of numerical methods for these problems. We follow front-tracking methods (moving grid method) which use an explicit representation of the interface, given by a set of points lying on the interface location, which must be updated at each time-step.

Heat treatment of metals is often used to optimize mechanical properties. During heat treatment, the metallurgical state of the alloy changes. This change can involve the phase present at a given location or the morphology of the various phases. In our case, we will study solid state phase transformation problem in binary metallic alloys.

The meshless method (e.g., radial basis functions (RBFs) – multiquadrics (MQ) approach [2]) has been widely investigated in the past and emerged as a new category of computational methods. One of its advantages is that no mesh generation is required to solve differential equations numerically.

The numerical solutions will be compared with analytical solutions. Actually, in our work we will examine usefulness of radial basis functions for one-dimensional Stefan's problems. The position of the moving boundary will be simulated by moving data centers method.

2 Radial Basis Function Methods

A radial basis function is a function $\phi_j(\mathbf{x}) = \phi(\|\mathbf{x} - \mathbf{x}_j\|)$, which depends only on the distance between $\mathbf{x} \in \mathbf{R^d}$ and a fixed point $\mathbf{x}_j \in \mathbf{R^d}$. Here, ϕ is continuous and bounded on any bounded sub-domain $\Omega \subseteq \mathbf{R^d}$. Let r denote by the Euclidean distance between any pair of points in the domain Ω. The commonly used RBFs are linear, cubic, thin-plate spline, Gaussian, multiquadric and inverse multiquadric.

To introduce RBF collocation methods, we consider a PDE in the form of

$$L\,u = f(\mathbf{x}) \text{ in } \Omega \subset \mathbf{R^d}, \tag{1}$$

$$B\,u = g(\mathbf{x}) \text{ on } \partial\Omega, \tag{2}$$

where u is concentration, $\mathbf{d}$ is the dimension, $\partial\Omega$ denotes the boundary of the domain Ω, L is the differential operator on the interior, and B is an operator that specifies the boundary conditions of the Dirichlet, Neumann or mixed type. Both, f and g, are given functions mapping $\mathbf{R^d} \to \mathbf{R}$.

The solution, u, to the PDE is approximated by linear combination of RBFs and polynomials

$$u \approx U(\mathbf{x}) = \sum_{j=1}^{N} \alpha_j \phi_j(\mathbf{x}) + \sum_{l=1}^{M} \gamma_l v_l(\mathbf{x}), \tag{3}$$

where $\phi_j(\mathbf{x}) = \phi(\|\mathbf{x} - \mathbf{x}_j\|)$, and ϕ can be any radial basis function from the list, $v_1, \ldots, v_M \in \Pi_m^d$ is a polynomial of degree m or less, $M := \binom{m-1+d}{d}$ [3] and $\|\cdot\|$ indicates the Euclidean norm. Let $\{(\mathbf{x}_j)\}_{j=1}^{N}$ be the $N = N_I + N_B$ collocation points in $\Omega \cup \partial\Omega$. We assume the collocation points are arranged in such a way that the first N_I points are in Ω, whereas the last N_B points are on $\partial\Omega$. To solve for the $N + M$ unknown coefficients, $N + M$ linearly independent equations are needed. By choosing N distinct collocation points $X_I = \{\mathbf{x}_1, \ldots, \mathbf{x}_{N_I}\} \subset \Omega$ and $X_B = \{\mathbf{x}_{N_I+1}, \ldots, \mathbf{x}_N\} \subset \partial\Omega$ and ensuring that $U(\mathbf{x})$ satisfies (1) and (2) at the collocation points results in a good approximation of the solution u. The first N equations are given by

$$\sum_{j=1}^{N} \alpha_j\, L\, \phi_j(\mathbf{x}_i) + \sum_{l=1}^{M} \gamma_l\, L\, v_l(\mathbf{x}_i) = f(\mathbf{x}_i) \quad \text{for} \quad i = 1, \ldots, N_I,$$

$$\sum_{j=1}^{N} \alpha_j\, B\, \phi_j(\mathbf{x}_i) + \sum_{l=1}^{M} \gamma_l\, B\, v_l(\mathbf{x}_i) = g(\mathbf{x}_i) \quad \text{for} \quad i = N_I+1, \ldots, N. \tag{4}$$

The last M equations could be obtained by imposing some extra condition on $v(\cdot)$

$$\sum_{j=1}^{N} \alpha_j v_k(\mathbf{x}_j) = 0, \quad k = 1, \ldots, M. \qquad (5)$$

In our study, we have used a general multiquadric MQ RBF. The generalized form of the MQ basis function is $\phi_j(\mathbf{x}) = [(\mathbf{x} - \mathbf{x}_i)^2 + c_i^2]^\beta$, where $\mathbf{x}, \mathbf{x}_i \in \mathbf{R}^d$, and β is a noninteger $\geq -1/2$.

The choice of basis function is another flexible feature of RBF methods. RBFs can be globally supported, infinitely differentiable, and contain a free parameter, c, called the *shape parameter*. This leads to a full coefficient matrix or a dense interpolation matrix. The shape parameter affects both the accuracy of the approximation and the conditioning of the interpolation matrix. The optimal shape parameter c is still an open question. In our case, we used an iterative mode by monitoring the spatial distribution of the residual errors in Ω and $\partial\Omega$ as a function of c. The iterations are terminated when errors are smaller than a specified value. This map is then used to guide the search of the optimal shape parameter c which gives the best approximate solution.

3 The Problem

3.1 The Mathematical Model

We consider the domain Ω to be the union of Ω_{dp} where dp represents the diffusion phase and Ω_{cc} where cc represents constant composition. The particle dissolves due to Fickian diffusion in the diffusive phase. The governing equations and boundary conditions of this problem are

$$\frac{\partial u}{\partial t}(\mathbf{x}, t) = D \Delta u(\mathbf{x}, t), \quad \mathbf{x} \in \Omega_{\mathrm{dp}}(t), \, t > 0, \qquad (6)$$

$$u(\mathbf{x}, t) = u^{\mathrm{cc}}, \quad \mathbf{x} \in \Omega_{\mathrm{cc}}(t), \, t \geq 0, \qquad (7)$$

$$u(\mathbf{x}, t) = u^{\mathrm{sol}}, \quad \mathbf{x} \in \Gamma(t), \, t \geq 0, \qquad (8)$$

$$(u^{\mathrm{cc}} - u^{\mathrm{sol}})v_n(\mathbf{x}, t) = D \frac{\partial u}{\partial \mathbf{n}}(\mathbf{x}, t), \quad \mathbf{x} \in \Gamma(t), \, t > 0, \qquad (9)$$

where $\mathbf{x}$ is coordinate vector of a point in Ω, D means the diffusivity constant, $\mathbf{n}$ is the unit normal vector on the interface pointing outward with respect to $\Omega_{\mathrm{cc}}(t)$, u^{sol} is the interface concentration and v_n is the normal component of the velocity of the interface. The initial concentration $u(\mathbf{x}, 0)$ inside the diffusive phase is given. We assume no flux of the concentration through the boundary

$$\frac{\partial u}{\partial \mathbf{n}}(\mathbf{x}, t) = 0, \quad \mathbf{x} \in \partial\Omega_{\mathrm{dp}}(t)\backslash\Gamma(t), \, t > 0, \qquad (10)$$

hence mass is conserved.

4 The Numerical Solution Methods

In our model the motion of the interface is determined by the gradient of concentration, which can be computed from the solution of the diffusion equation. Here we present an interpolative moving data center method, in which the data centers (make a global substitution) are computed for each time-step and the solution is interpolated from the old data centers to the new. The equations are solved with collocation methods using MQ RBF. The position of the points depend on time. An outline of the algorithm is

- Compute the concentrations profiles solving (6)–(8) and (10)
- Predict the position of boundary s_1 at the new time-step: $s_1(t + \Delta t)$ using boundary condition (9)
- Once the boundary is moved, the concentration u can be computed in the new region using (6). The solution is interpolated from the old grid to the new

5 Numerical Example

In numerical experiments, we will compare our numerical solutions with the analytical solutions (see [4]) which exist for the problem presented in Chap. 3.

For the simulations, we used data from [4]: the concentration inside the part where the material characteristics remain constant $u^{cc} = 0.53$, the concentration on the interface $u^{sol} = 0$, the initial concentration of the diffusive phase $u^0 = 0.1$, the diffusivity constant $D = 1$, the domain length $l = 1$ and the initial position of the interface $s_0 = 0.2$. In numerical experiments we will include MQ exponent, β as additional parameter needs to be optimized.

On the left side of the Fig. 1 we show the results obtained by shape parameter c based on residual error calculated at collocation points, and on the

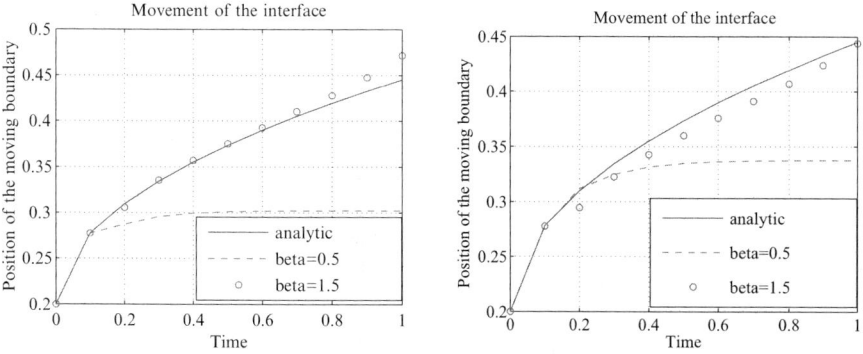

Fig. 1. Interface position versus time simulated with MQ

right side of the Fig. 1 we show the results obtained by shape parameter c based on residual error calculated at other points (those which lie between collocation points). During the time simulation steps the shape parameter c had values between 0.01 and 0.09. In the case of the collocation points the tolerance for residual error was $\epsilon = 10^{-8}$. On the other hand, in the case of other points the tolerance for residual error was $\epsilon = 0.4$.

6 Discussion and Conclusions

Comparison of positions of the moving boundary calculated with MQ ($\beta = 0.5$) and MQ ($\beta = 1.5$) (Fig. 1) shows that MQ ($\beta = 1.5$) determines the position of the interfaces much more accurately than MQ ($\beta = 0.5$). The simulations have also show that the value of the shape parameter c which was computed by residual error procedure was in range between 0.01 and 0.09. This confirms the fact that for a fixed number of centers N, smaller shape parameters produce the more accurate approximations. The results have shown that β should be greater than 0.5 if we want to get reasonable results. Probably reasons for bad results in Fig. 1 (right side) could be found in the facts that some centers were clustered (too close to each other).

This study presents modeling of moving boundary value problems using a MQ RBF. Simulations show that MQ ($\beta = 1.5$) scheme give good results. In this case the method of evaluation was verified by comparing results with the analytical solutions. We explore the residual error from the equation as an indicator which provides a road map to the optimal selection of the shape parameter value c. In our future work, we will employ a Lagrangian approach and level set methods to track-capture the movement of the moving boundary and transform the problem to a time-independent domain.

References

[1] Crank, J.: Free and Moving Boundary Problems. Clarendon Press, Oxford (1984)
[2] Kansa, E. J.: Multiquadrics–A scattered data approximation scheme with applications to computational fluid dynamics-II. Solutions to parabolic, hyperbolic and elliptic partial differential equations, Computers Math. Applic., Vol. 19, No. 8–9, pp. 147–161, (1990)
[3] Iske, A.: Charakterisierung bedingt positiv definiter Funktionen für multivariate Interpolationsmethoden mit radialen Basisfunktionen. PhD Thesis, University Göttingen (1994)
[4] Javierre, E., Vuik, C., Vermolen, F. J. and van der Zwaag, S.: A comparison of numerical models for one–dimensional Stefan problems, Reports of the Delft Institute of Applied Mathematics, Netherlands (2005)

Meshless Simulation of Hele-Shaw Flow

Francisco Bernal and Manuel Kindelan

Universidad Carlos III de Madrid, Avda. Universidad 30, 28911 Leganés
fcoberna@math.uc3m.es, kinde@ing.uc3m.es

Summary. In a previous chapter [F. Bernal and M. Kindelan An RBF Meshless Method for Injection Molding Modelling, Lecture Notes in Computational Science and Engineering, Springer (2006)], a novel meshless approach was proposed for solving the Hele-Shaw flow which models plastic injection molding, in the case of a Newtonian fluid. Here, we have extended this idea to non-Newtonian Hele-Shaw flow via a Newton algorithm for the resulting nonlinear PDE.

1 Introduction

In this chapter, we address the simulation of injection molding, a process of industrial relevance whereby molten polymer is driven into a cavity (the mold) in order to manufacture small plastic parts. If the polymer viscosity obeys a power law and the mold is thin compared to its planar dimensions, the classical mathematical model of injection molding is the Hele-Shaw approximation [4]. In the remainder of this chapter, we will restrict ourselves to isothermal Hele-Shaw flows, which physically arise whenever the fluid viscosity does not depend on temperature. In this case it suffices to solve the following 2D, nonlinear, elliptic equation

$$\operatorname{div}(\,|\nabla p|^{\gamma}\,\nabla p) \;=\; 0, \tag{1}$$

whose solution yields the pressure distribution $p(x, y)$ in the filled region of the mold. Exponent γ completely characterizes the polymer rheology, and is typically about 0.5. If the pressure profile is set (p_{IN}) along the injection gates by the injection machine, the boundary conditions are

$$p = p_{\text{IN}} \text{ (injection gates)} \qquad \partial p/\partial n = 0 \text{ (walls)} \qquad p = 0 \text{ (front)}. \tag{2}$$

From this pressure field, the average planar velocity can be computed and the location of the advancing front can be updated. In dimensionless units

$$< \mathbf{v} > = -\,|\nabla p|^{\gamma}\,\nabla p. \tag{3}$$

Therefore, the numerical simulation of the Hele-Shaw flow requires coupling (a) some method for solving (1) at every time-step with (b) some technique to advance the front to its new position, until the mold domain has been completely filled.

In the state-of-the-art approach (a) is accomplished through finite elements (FEM), whereas for (b) the volume-of-flow method is used, or the nodes along the front are tracked to their new positions. The latter option entails remeshing around the front at every time-step, while the former avoids it at the price of forgoing a sharp frontline. In [1], an alternative, meshless framework was introduced for solving this problem in the linear (Newtonian) case, namely combining the method of asymmetric Radial Basis Function (RBF) collocation for pressure with Level Sets for capturing the front motion. Although such approach is still under research, it has the potential to overcome some difficulties inherent to finite elements. In this chapter, we have extended these ideas to the non-Newtonian flow. However, and due to space limitations, we cannot treat Level Sets here. Instead, we refer the reader to [7].

2 Asymmetric RBF Collocation

2.1 Kansa's Method

The idea of using RBFs to solve PDEs was first introduced by Kansa [5, 6]. Consider the BVP $L(u) = f(\mathbf{x})$ in domain Ω with boundary conditions along $\partial\Omega$ given by $B(u) = g(\mathbf{x})$, where L and B are linear operators. Ω is discretized into a set of $N = N_I + N_B$ scattered nodes $\chi = \{\mathbf{x}_i \in \Omega,\ i = 1...N_I\} \cup \{\mathbf{x}_j \in \partial\Omega,\ j = N_I + 1,\ldots, N_I + N_B\}$ (called *centers*) and an approximate solution to the PDE is sought in the form of a linear combination of RBFs $\{\phi_k(\mathbf{x}),\ k = 1...N\}$ centered at each of them,

$$u(\mathbf{x}) = \sum_{k=1}^{N} \alpha_k\, \phi_k(\mathbf{x}), \qquad \phi_k(\mathbf{x}) \equiv \phi(\| \mathbf{x} - \mathbf{x}_k \|). \tag{4}$$

Having L and B operate on the RBF, the unknown coefficients α_k are determined by appropriate collocation of either the PDE or the BC on N points, which usually – but not necessarily – are the same set of centers

$$\sum_{k=1}^{N} \alpha_k\, L\phi_k(\mathbf{x}_i) = f(\mathbf{x}_i), \qquad i = 1,\ldots, N_I, \tag{5}$$

$$\sum_{k=1}^{N} \alpha_k\, B\phi_k(\mathbf{x}_j) = g(\mathbf{x}_j), \qquad j = N_I + 1,\ldots, N_I + N_B. \tag{6}$$

Among the many RBFs available, we have chosen the multiquadric (MQ), which has been extensively used both for interpolation and for the solution of PDEs. The MQ–RBF depends on a tunable parameter c^2 (the *shape parameter*) which can be the same for each RBF in the set or vary among them.

2.2 PDE Collocation on Boundary (PDEBC)

Accuracy can be greatly improved by enforcing the PDE on the boundary nodes also [3]. In this case, expansion (4) must be supplemented with N_B extra RBF centers $\{\mathbf{x_m}, m = N+1, \ldots, N+N_B\}$ in order to match the N_B new collocation equations. Since these centers are not to be collocated on, they may lie outside the PDE domain. With them, the RBF interpolant takes on the form

$$u(\mathbf{x}) = \sum_{k=1}^{N} \alpha_k\, \phi_k(\mathbf{x}) + \sum_{m=N+1}^{N+N_B} \alpha_m\, \phi_m(\mathbf{x}). \tag{7}$$

2.3 Operator-Newton Scheme

Although Kansa's method is intended for linear operators, it can be used to solve nonlinear PDEs through iteration or continuation [3]. Instead, we have adapted an operator-Newton algorithm with MQs introduced by Fasshauer (see [2] for details).

- Let $Hu = 0$ be the nonlinear PDE in Ω and L a linearization of it
- Pick an initial guess u_0 of solution. We seek w such that $H(u + w) = 0$
- *For $k = 1, 2\ldots$ until convergence*
  - Compute residual $R_k = -Hu_{k-1}$
  - Solve $L_k w_k = R_k$ by Kansa's method, where $L_k = L(u_{k-1})$
  - Update the previous iterate, $u_k = u_{k-1} + w_k$

3 Non-Newtonian Flow

In this section, we will show our preliminary results concerning the solution of non-Newtonian Hele-Shaw flow with MQs. There are a number of qualitative differences between this problem and the test problem analyzed in [2]. First, the nonlinearity is a *differential operator* rather than a *function* of the solution. Secondly, both Dirichlet *and* Neumann BCs must be enforced, instead of only Dirichlet. Finally, the highest gradients take place along the boundary. In order to meet the latter two features, we have slightly modified Fasshauer's algorithm to incorporate PDEBC, seeking better performance along the boundary.

Let us define *flow fluidity* as $S(p) = |\nabla p|^\gamma$ so that (1) may be rewritten as $H(p) := \operatorname{div}(S(p)\nabla p) = 0$. In order to linearize this equation, we observe that, to first order in $|\nabla w| / |\nabla p|$, $S(p+w) \approx S(p) + \gamma\, |\nabla p|^{\gamma-2}\, (\nabla p \cdot \nabla w)$. Now define

$$\mathbf{K}(p) = \left(K^{(x)}(p), K^{(y)}(p)\right) = \gamma\,|\nabla p|^{\gamma-2}\,\nabla p, \qquad (8)$$

$$\mathbf{J}(p) = \left(\nabla p \cdot \nabla K^{(x)}(p),\, \nabla p \cdot \nabla K^{(y)}(p)\right), \qquad (9)$$

$$Q(p)w = \left(\frac{\partial p}{\partial x}\mathbf{K}(p)\right) \cdot \frac{\partial \nabla w}{\partial x} + \left(\frac{\partial p}{\partial y}\mathbf{K}(p)\right) \cdot \frac{\partial \nabla w}{\partial y}. \qquad (10)$$

After some calculus and keeping only terms to order $|\nabla w|\,/\,|\nabla p|$:

$$H(p+w) = \mathrm{div}\!\left(S(p+w)\nabla(p+w)\right)$$

$$\approx H(p) + S(p)\nabla^2 w + Q(p)w + \left[\nabla S(p) + (\nabla^2 p)\mathbf{K}(p) + \mathbf{J}(p)\right] \cdot \nabla w. \qquad (11)$$

In this work we are primarily interested in the performance of Fasshauer's algorithm, rather than in front motion, and have therefore restricted ourselves to solving (1) in a square $[0,\,1]\times[-1/2,\,1/2]$. The front is the side $x = 1$ and the injection segment is $x = 0, |y| < E$, $E{=}0.25$, while the remainder of the boundary are walls. This domain has been modeled by 190 scattered collocation nodes, along with 45 extra RBF centers (needed for PDEBC). Such extra centers are placed at a distance $\lambda = 0.05$ of the boundary along the outward normal (see Fig. 1). Moreover, we have set $\gamma = 0.6$, and $c = k/\sqrt{N}$ for the MQ–RBF, where $k = 5$ and $N = 135$. For monitoring purposes, (1) has been solved on a FEM mesh of 2,715 vertices with the following BCs (which give rise to a smooth pressure field)

$$\partial p/\partial n = 10(1 - (y/E)^2)\,(\text{injection}) \qquad \partial p/\partial n = 0\,(\text{walls}) \qquad p = 0\,(\text{front}).$$
$$(12)$$

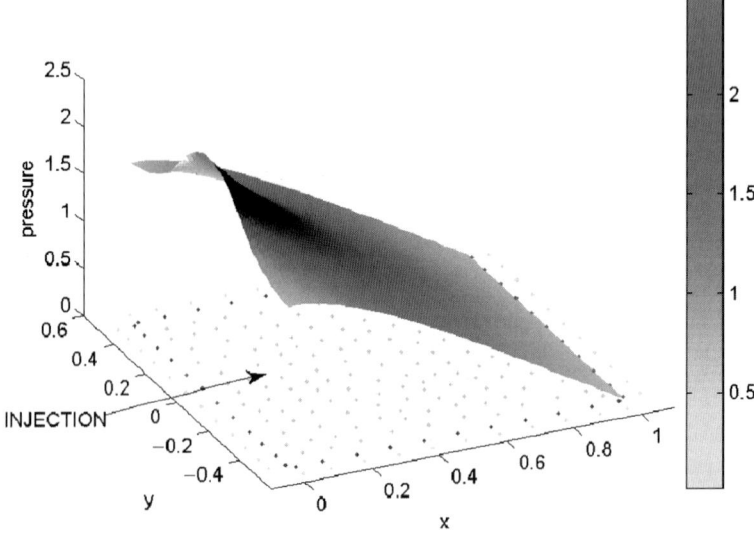

Fig. 1. RBF solution of the test case. Also shown is the pointset used to compute the solution.

Table 1. Convergence of Newton iterations

Iteration	RMS(CR)	RMS(IR)	RMS(CE)	RMS(IE)
0	0.8063	0.2308	0.0051	0.0013
1	0.1372	0.1024	0.0003	9.5×10^{-5}
2	0.0062	0.0621	0.0002	6.5×10^{-5}
3	6.9×10^{-5}	0.0603	0.0002	6.4×10^{-5}
4	1.2×10^{-8}	0.0603	0.0002	6.4×10^{-5}
5	7.9×10^{-16}	0.0603	0.0002	6.4×10^{-5}

From the resulting FEM approximation we interpolate the values of p along the injection segment, $p_{\text{IN}} = p_{\text{FEM}}(0, |y| < E)$. Now we have solved a Laplace equation ($\gamma = 0$) with BCs like in (2) as the initial guess p_0 to trigger Newton iterations. At every Newton step we have solved the linearized PDE for w with homogeneous BCs and computed the RBF approximation on the collocation nodes and the root mean square (RMS) values of the residual to (1) and the error (to FEM solution). They are denoted as CR and CE (collocation residual/error). Since the PDE has been collocated on such nodes, residuals on them are lower than anywhere else. A far better estimation of the goodness of the approximation is got if the RBF interpolant is evaluated on non-collocation points. We have also done this on the vertices of the FEM mesh, and called them IR/IE (interpolation residual/error). These values are listed in Table 1.

From Table 1, it can be seen that the operator-Newton idea basically works. Residual on collocation nodes drops within very few iterations below machine error, which amounts to solving the nonlinear PDE. On the other hand, error (both CE and IE) and interpolation residual drop fast at the beginning but soon level off even though CR continues to decrease. This threshold is the bottomline of the numerical approximation, which depends on the size and distribution of the support, and the shape parameter of the MQ (which has not been optimized). For comparison purposes, it is interesting to notice that, if the FEM solution is RBF-interpolated with the same parameters, the error RMS as computed on the FEM mesh is 1.12×10^{-5}.

4 Conclusions

A operator-Newton algorithm introduced by Fasshauer has been used to solve the isothermal nonlinear Hele-Shaw equation. Certain modifications of the algorithm have been carried out in order to meet the problem features. The method has then been tested against a FEM approximation. Results are preliminary yet promising, showing fast convergence.

References

1. F. Bernal and M. Kindelan *An RBF Meshless Method for Injection Molding Modelling*, Lecture Notes in Computational Science and Engineering, Springer, to appear (2006).
2. G. E. Fasshauer, *Newton iteration with multiquadrics for the solution of nonlinear PDEs*, Comput. Math. Appls., **43**, 423-438 (2002).
3. A. I. Fedoseyev, M. J. Friedman, and E. J. Kansa, *Improved multiquadric method for elliptic partial differential equations via PDE collocation on the Boundary*, Comput. Math. Appl., **43**, 439-455 (2002).
4. C. A. Hieber and S. F. Shen, *A finite-element/finite difference simulation of the injection-molding filling process*, Journal of Non-Newtonian Fluid Mechanics, **7**, 1-32 (1979).
5. E. J. Kansa, *Multiquadrics - a scattered data approximation scheme with applications to computational fluid-dynamics. I. Surface approximations and partial derivative estimates*, Comput. Math. Appls., **19**, 127-145 (1990).
6. E. J. Kansa, *Multiquadrics - a scattered data approximation scheme with applications to computational fluid-dynamics. II. Solutions to parabolic, hyperbolic and elliptic partial differential equations*, Comput. Math. Appls., **19**, 147-161 (1990).
7. J. A. Sethian, *Level set methods and fast marching methods. Evolving interfaces in computational geometry, fluid mechanics, computer vision and materials science*, Cambridge University Press (1999).

Minisymposium: "Mathematical Models for the Textile Industry"

T. Götz

TU Kaiserslautern, Germany

Nowadays artificial fibres made of polymers or glass are of increasing industrial importance. Worldwide a total amount of 37.9 million tons of chemical fibres was produced in 2004 and the production still increases by around 5% annually.

The textile industry and especially the production of artificial fibres involves a vast number of different processes which present modelling challenges to mathematicians. Aiming to a mathematical description of the complete production cycle for artificial fibres, different subprocesses can be identified:

- Extrusion of polymer melt through a nozzle
- Fibre formation during the cooling of the polymer melt in ambient air, solidification and crystallisation
- Entanglement of fibres in turbulent air streams and lay-down on a conveyor belt

Further problems arise in the use of fibres (both natural and man made) and there has been a recent surge in the use of fibre assemblies for industrial purposes. Fibre assemblies represent a multiscale problem and several different models have been proposed. One is to treat the structure as a hyperelastic membrane at the macroscale and as three-elastica with frictional interactions at the micro- and mesoscale. Continuum models have also been derived to deal with dynamic deformations of fibre assemblies which in tension behave somewhat like a anisotropic viscoelastic material.

Another approach is to geometrically and statistically model the woven or non-woven textile and to compute properties such as flow resistivity or filter efficiency based on a three-dimensional realisation of the model.

Polymer fibres like Nylon are produced in a melt spinning process. Thereby, hot, molten polymer is pressed through narrow nozzles and solidifies afterwards while being cooled by air. A mathematical description of this fibre formation process typically starts from a full three-dimensional model for the flow of the visco-elastic polymer including the energy exchange with the ambient air. Based on the – typically small – slenderness ratio of the fibre, asymptotic expansions allow the derivation of a simplified one-dimensional model of balance equations. Including phenomena like crystallisation yields a highly non-linear system of differential equations which has to be treated numerically.

When the fibre is completely solidified, further production steps can be taken into account. For the production of non-wovens a turbulent air stream is blown against bundles of fibres. Due to aerodynamic forces the fibres entangle and form a fleece laid down on a conveyor belt.

Based on a mathematical model reflecting all the above mentioned processes involved in the fibre production, optimisation methods can be used to maximise the output of a production plant or to design the process parameters resulting in the production of fibres with specified properties.

The minisymposium contains presentations by

1. A. Wiegmann (Fraunhofer ITWM, Kaiserslautern, Germany)
2. N. Marheineke (TU Kaiserslautern, Germany)
3. T. Götz (TU Kaiserslautern, Germany)
4. M. Günther (Fraunhofer ITWM, Kaiserslautern, Germany)
5. H. Ockendon (OCIAM Oxford, UK)
6. P. Potluri (Manchester, UK)

and gives an overview of the research recently carried out on this non-standard application of industrial mathematics.

Dynamics of Curved Viscous Fibers

Satyananda Panda[1], Nicole Marheineke[2], and Raimund Wegener[1]

[1] Fraunhofer-Institute for Industrial Mathematics (ITWM), Fraunhofer-Platz 1,
   D-67663 Kaiserslautern, Germany
   {panda, wegener}@itwm.fhg.de
[2] Technical University Kaiserslautern, Department of Mathematics,
   P.O. Box 3049, D-67653 Kaiserslautern, Germany
   nicole@mathematik.uni-kl.de

Summary. This work deals with the modeling and simulation of the dynamics of a curved inertial viscous Newtonian fiber. Neglecting surface tension and temperature dependence, the fiber flow is modeled as a 3D free boundary value problem (BVP) via instationary incompressible Navier–Stokes equations (NSE). From regular asymptotic expansions in powers of the slenderness parameter leading-order balance laws for mass and momentum are derived that combine the unrestricted motion of the fiber center-line with the inner viscous transport. The form of the 1D fiber model results from the introduction of the intrinsic velocity characterizing the convective terms. For the numerical simulations of the fiber evolution a finite volume approach is applied.

1 Motivation

In the glass wool production, hot molten glass is pressed through narrow nozzles of a rotating cylindrical drum by the acting centrifugal forces. Thereby, thin fibers are formed that break into filaments due to the surrounding air flow and fall down onto a conveyor belt, see Fig. 1.

Focusing on the spinning process, we consider a single slender curved viscous fiber in motion. Neglecting temperature dependence and surface tension, the fiber medium is modeled as an incompressible Newtonian fluid and the fiber forming as 3D free BVP in terms of incompressible NSE with inflow boundary and stress-free surface conditions. The slender fiber geometry enables its asymptotic reduction to an 1D fiber model for mass and momentum. The demand on this work is the systematic derivation analogously to [4] without any restrictions on the center-line shape and motion nor on the inner viscous transport, which is an extension to [1, 3, 4, 6].

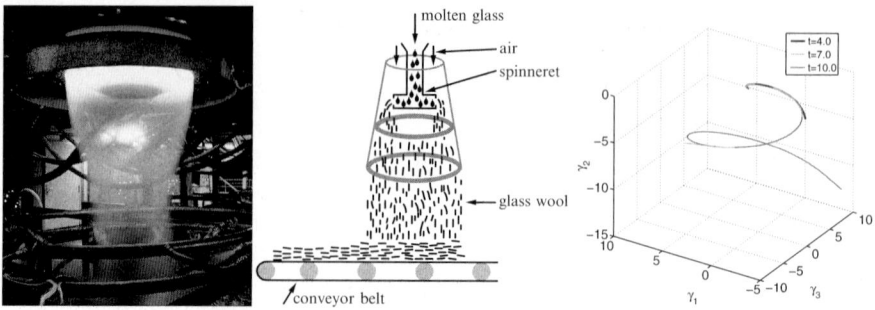

Fig. 1. Glass wool production: plant, sketch, simulated fiber motion

2 Systematic Asymptotic Derivation of the Model

2.1 Free Boundary Value Problem

Let the flow domain at time $t \in \mathbb{R}^+$ be $\Omega(t) \subset \mathbb{R}^3$ and its boundary $\partial\Omega(t) = \Gamma_{\text{fr}}(t) \cup \Gamma_{\text{in}}$ with $\Gamma_{\text{fr}}(t) \cap \Gamma_{\text{in}} = \emptyset$. Here, $\Gamma_{\text{fr}}(t)$ and Γ_{in} prescribe the time-dependent free surface and the time-independent planar inflow boundary (nozzle), respectively. The nondimensionalized model for the BVP reads

$$\nabla_{\mathbf{r}} \cdot \mathbf{v}(\mathbf{r}, t) = 0$$

$$\partial_t \mathbf{v}(\mathbf{r}, t) + \nabla_{\mathbf{r}} \cdot (\mathbf{v} \otimes \mathbf{v})(\mathbf{r}, t) = \nabla_{\mathbf{r}} \cdot \mathbf{S}^{\text{T}}(\mathbf{r}, t) + \mathbf{f}(\mathbf{r}, t) \qquad \mathbf{r} \in \Omega(t)$$

$$\mathbf{S} = -p\mathbf{I} + \frac{1}{Re}(\nabla_{\mathbf{r}}\mathbf{v} + (\nabla_{\mathbf{r}}\mathbf{v})^{\text{T}})$$

$$(\mathbf{v} \cdot \mathbf{n})(\mathbf{r}, t) = w(\mathbf{r}, t), \qquad (\mathbf{S} \cdot \mathbf{n})(\mathbf{r}, t) = \mathbf{0} \qquad \mathbf{r} \in \Gamma_{\text{fr}}(t)$$

$$\mathbf{v}(\mathbf{r}, t) = \mathbf{v}_{\text{in}}(\mathbf{r}) \qquad \mathbf{r} \in \Gamma_{\text{in}}$$

with Reynolds number Re. Apart from the unknown field variables for velocity $\mathbf{v}$ and hydrodynamic pressure p, the BVP determines the geometry $\Omega(t)$ specified by the outer normal vectors $\mathbf{n}$ and the scalar speed w of $\Gamma_{\text{fr}}(t)$. By choosing homogeneous dynamic boundary conditions for the stress tensor $\mathbf{S}$ the effects of surface tension are neglected. Body forces $\mathbf{f}$ complete the BVP.

The radius of the nozzle is comparably small to the typical length of the spun fiber, thus we introduce a slenderness parameter ϵ. Due to the scaling, the dimensionless inflow velocity profile $\mathbf{v}_{\text{in}}$ at the nozzle satisfies

$$|\Gamma_{\text{in}}|^{1/2} = \epsilon \ll 1, \qquad \int_{\Gamma_{\text{in}}} \mathbf{v}_{\text{in}} \cdot \boldsymbol{\tau_0} \, d\mathcal{A} = \int_{\Gamma_{\text{in}}} d\mathcal{A} = \epsilon^2,$$

where $|\Gamma_{\text{in}}| = \int_{\Gamma_{\text{in}}} d\mathcal{A}$. and $\boldsymbol{\tau_0}$ inner normal vector of Γ_{in}.

2.2 Coordinate Transformation

For the asymptotic reduction, we transform the free BVP into general coordinates being specified as scaled curvilinear. These coordinates can be understood as generalization of cylindrical ones along an arbitrary curve for which

the fiber center-line is taken. Scaling leads to inflow conditions independent of
the slenderness parameter ϵ, instead ϵ occurs explicitely in the balance laws.
So, the free BVP is embedded into a family of self-similar problems with fixed
inflow domain and fixed inflow velocity.

Definition 1 (General Coordinate Transformation). *A function $\check{\mathbf{r}}$ de-
fined by $\check{\mathbf{r}}(\cdot,t)\colon \hat{\Omega}(t) \subset \mathbb{R}^3 \mapsto \Omega(t) \subset \mathbb{R}^3$ for $t \in \mathbb{R}^+$ is called time-dependent
general coordinate transformation if $\check{\mathbf{r}} \in \mathcal{C}^2$ and if $\check{\mathbf{r}}(\cdot,t)$ are bijective.*

Related to $\check{\mathbf{r}}$, we introduce the following characteristic quantities: coordinate
transformation matrix $\mathbf{F} = \nabla_{\mathbf{x}}\check{\mathbf{r}}$, functional determinant $J = \det(\mathbf{F})$, inverse
matrix $\mathbf{G} = \mathbf{F}^{-1}$ and coordinate velocity $\mathbf{q} = \partial_t\check{\mathbf{r}}$. Then, the governing equations of the free BVP in general coordinates $\mathbf{x} \in \hat{\Omega}(t)$ read

$$\partial_t J(\mathbf{x},t) + \nabla_{\mathbf{x}} \cdot (J\mathbf{u})(\mathbf{x},t) = 0,$$
$$\partial_t(J\mathbf{v})(\mathbf{x},t) + \nabla_{\mathbf{x}} \cdot (\mathbf{u} \otimes J\mathbf{v})(\mathbf{x},t) = \nabla_{\mathbf{x}} \cdot \mathbf{T}^{\mathrm{T}}(\mathbf{x},t) + (J\mathbf{f})(\mathbf{x},t)$$
$$\mathbf{u} = (\mathbf{v} - \mathbf{q}) \cdot \mathbf{G}$$
$$\mathbf{T} = J\mathbf{S} \cdot \mathbf{G}.$$

The physical and geometrical properties of the observables are kept under
the transformation [7]. The intrinsic velocity $\mathbf{u}$ describes the transport of the
unknowns, whereas the velocity $\mathbf{v}$ is associated with the original momentum,
i.e., transported quantity. Their relation is expressed in the coupling condition.

Definition 2 (Scaled Curvilinear Coordinate Transformation). *For a
given smooth, arc-length parameterized, time-dependent curve $\boldsymbol{\gamma}$ in $\Omega(t) \subset \mathbb{R}^3$
and a fixed parameter $\epsilon \in \mathbb{R}^+$ the special choice of a time-dependent general
coordinate transformation*

$$\check{\mathbf{r}}(\mathbf{x},t) = \boldsymbol{\gamma}(s,t) + \epsilon x_1\boldsymbol{\eta}_1(s,t) + \epsilon x_2\boldsymbol{\eta}_2(s,t) \quad with \quad s = x_3$$

*is called scaled curvilinear coordinate transformation, if $\check{\mathbf{r}}(\cdot,t)$ is bijective for
$t \in \mathbb{R}^+$. The normal vectors $\boldsymbol{\eta}_1$, $\boldsymbol{\eta}_2$ form an orthonormal basis with $\boldsymbol{\tau} = \partial_s\boldsymbol{\gamma}$.*

The curve $\boldsymbol{\gamma}$ and the scaling factor ϵ of Definition 2 are specified as fiber
center-line and slenderness parameter, Fig. 2. This choice has some crucial
consequences for the logical structure of our problem. As the dynamics
of the center-line depends on the solution of the free BVP, the coordinate transformation becomes part of the problem. We assume that $\hat{\Omega}(t)$ is
given by the fiber length $L(t)$ and the smooth, 2π-periodic radius function
$R(\cdot,t) : [0,2\pi) \times [0,L(t)) \mapsto \mathbb{R}^+$ in such a way that $\hat{\Omega}(t) = \{\,\mathbf{x} = (x_1,x_2,s) \in
\mathbb{R}^3 \,|\, (x_1,x_2) \in \mathcal{A}(s,t), s \in [0,L(t))\}$ with cross-sections $\mathcal{A}(s,t) = \{(x_1,x_2) \in
\mathbb{R}^2 \,|\, x_1 = \varrho\cos(\psi), x_2 = \varrho\sin(\psi), \varrho \in [0,R(\psi,s,t)], \; \psi \in [0,2\pi)\}$. Then the
fiber domain is described by L, $\boldsymbol{\gamma}$, R.

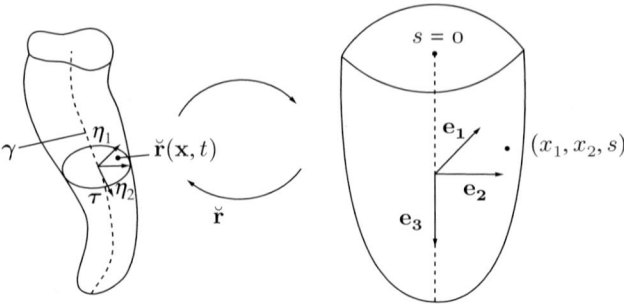

Fig. 2. Scaled curvilinear coordinate transformation

2.3 Asymptotic Analysis

The derivation of the 1D model from the 3D free BVP is based on the cross-sectional averaging of the balance laws. Thereby, the regular power expansions, e.g., $\mathbf{v}_\epsilon = \mathbf{v}^{(0)} + \epsilon \mathbf{v}^{(1)} + \mathcal{O}(\epsilon^2)$, in zeroth and first order yield the necessary cross-sectional profile properties of the unknowns.

Theorem 1 (Cross-Sectional Averaged Balance Laws). *Let a solution of the fiber family BVP exist. Denote* $\langle f \rangle_{\mathcal{A}_\epsilon(s,t)} = \int_{\mathcal{A}_\epsilon(s,t)} f(x_1, x_2, s, t) \mathrm{d}x_1 \mathrm{d}x_2$. *Then the following cross-sectional integral relations hold [4]*

$$\partial_t \langle J_\epsilon \rangle_{\mathcal{A}_\epsilon(s,t)} + \partial_s \langle J_\epsilon (\mathbf{u}_\epsilon \cdot \mathbf{e_3}) \rangle_{\mathcal{A}_\epsilon(s,t)} = 0,$$

$$\partial_t \langle J_\epsilon \mathbf{v}_\epsilon \rangle_{\mathcal{A}_\epsilon(s,t)} + \partial_s \langle J_\epsilon (\mathbf{u}_\epsilon \cdot \mathbf{e_3}) \mathbf{v}_\epsilon \rangle_{\mathcal{A}_\epsilon(s,t)} = \partial_s \langle \mathbf{T}_\epsilon \cdot \mathbf{e_3} \rangle_{\mathcal{A}_\epsilon(s,t)} + \langle J_\epsilon \mathbf{f}_\epsilon \rangle_{\mathcal{A}_\epsilon(s,t)}.$$

From the full model we derive in zeroth and first order [7]

$$\mathbf{u}^{(-1)} = \mathbf{0}, \quad u_3^{(0)} = u_3^{(0)}(s,t), \quad \mathbf{v}^{(0)} = u_3^{(0)} \partial_s \boldsymbol{\gamma}^{(0)} + \partial_t \boldsymbol{\gamma}^{(0)},$$

$$\mathbf{T}^{(0)} = \mathbf{T}^{(1)} = \mathbf{0}, \qquad \mathbf{T}^{(2)} \cdot \mathbf{e_3} = \frac{3}{Re} \partial_s u_3^{(0)} \partial_s \boldsymbol{\gamma}^{(0)}.$$

Abbreviating $u = u_3^{(0)}$, $A = |\mathcal{A}_0|$ and dropping the superscripts of leading order, we hence obtain the following result.

Theorem 2 (Asymptotic Fiber Model). *The spinning of a slender curved inertial viscous Newtonian fiber is modeled by*

$$\partial_t A + \partial_s (uA) = 0,$$

$$\partial_t (\mathbf{v}A) + \partial_s (u\mathbf{v}A) = \frac{3}{Re} \partial_s (\partial_s u \partial_s \boldsymbol{\gamma} A) + \mathbf{f}A$$

$$\mathbf{v} = u \partial_s \boldsymbol{\gamma} + \partial_t \boldsymbol{\gamma},$$

$$\frac{\mathrm{d}L(t)}{\mathrm{d}t} = u(L(t), t), \quad L(0) = 0, \qquad \partial_s u(L(t), t) = 0,$$

$$|\mathcal{A}(0, t)| = 1, \quad u(0, t) = 1, \quad \boldsymbol{\gamma}(0, t) = \boldsymbol{\gamma}_0, \quad \partial_s \boldsymbol{\gamma}(0, t) = \boldsymbol{\tau}_0, \quad \|\partial_s \boldsymbol{\gamma}\| = 1.$$

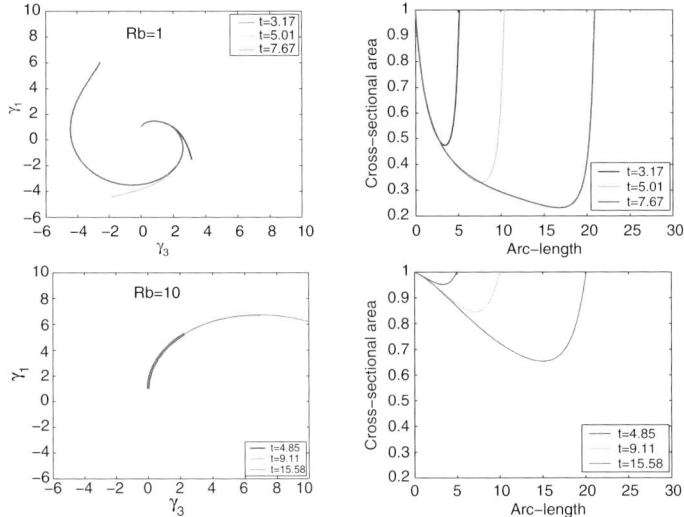

Fig. 3. Evolution of fiber, cross-sectional area, $Re = 4$, $Fr = 2$. Top: $Rb = 1$; Bottom: $Rb = 10$

The model coincides with the experiments by Trouton [8] in the factor 3 of the Trouton viscosity. In the rotational spinning process the body densities **f** stem from gravitation and rotation with Froude Fr and Rossby number Rb

$$\mathbf{f} = Fr^{-2}\mathbf{e_g} - 2Rb^{-1}(\mathbf{e}_\omega \times \mathbf{v}) - Rb^{-2}(\mathbf{e}_\omega \times (\mathbf{e}_\omega \times \boldsymbol{\gamma})).$$

3 Numerical Results

The numerical results for the evolution of the fiber in Fig. 3 show an artificial boundary layer in the cross-sections that comes from the momentary neglect of surface tension $\partial_s u(L(t), t) = 0$. This leads with $dL(t)/dt = u(L(t), t)$ and the conservation of mass to a constant bottom surface since $dA(L(t), t)/dt = 0$. In particular, together with the chosen initial condition it gives $A(L(t), t) = 1$.

The planned incorporation of surface tension in this framework will cause a source term in the averaged momentum equations and a reasonable change in the boundary conditions at the free fiber end.

Acknowledgements

Work supported by the Kaiserslautern Excellence Cluster *Dependable Adaptive Systems and Mathematical Modeling*.

References

1. Cummings L. J. and Howell P. D., On the evolution of non-axissymetric viscous fibres with surface tension inertia and gravity, *JFM*, 389: 361–389, 1999

2. Decent S. P., Simmons M., Parau E., Wong D., King A. and Partridge L., Liquid jets from a rotating orifice, In *Proceedings of the 5th Int. Conf. on Multiphase Flow, ICMF' 04*, Yokohama, Japan, 2004
3. Dewynne J. N., Howell P. D. and Wilmott P., Slender viscous fibers with inertia and gravity, *Quart J Mech Appl Math*, 47: 541–555, 1994
4. Dewynne J. N., Ockendon J. R. and Wilmott P., A systematic derivation of the leading-order equations for extensional flows in slender geometries, *JFM*, 244: 323–338, 1992
5. Entov V. M. and Yarin A. L., The dynamics of thin liquid jets in air, *JFM*, 140: 91–111, 1984
6. Howell P. D., Extensional thin layer flows, PhD thesis, St. Catherine's College, Oxford, 1994
7. Panda S., Marheineke N. and Wegener R., Systematic Derivation of an asymptotic model for the dynamics of curved viscous fibers, *M2AS*, 2006, submitted
8. Trouton F. R. S., On the coefficient of viscous traction and its relation to that of viscosity, *PRS London*, A 77: 426–440, 1906

Modeling and Simulation of Non-Woven Processes

Marco Günther[1], Raimund Wegener[2], and Ferdinand Olawsky[3]

[1] Fraunhofer Institut für Techno- und Wirtschaftsmathematik (ITWM),
    Fraunhofer-Platz 1, 67665 Kaiserslautern, Germany
    marco.guenther@itwm.fraunhofer.de
[2] raimund.wegener@itwm.fraunhofer.de
[3] ferdinand.olawsky@itwm.fraunhofer.de

1 Introduction

In our life-world non-woven products play an important role. Many products such as wipes, filters (air conditioning, cleaner, oil, ...), hygiene products, floor-covering, etc., are made out of non-woven. New needs require improvement and development of new products which constitutes a permanent task. Modeling and numerical simulations can support the development.

There are several production processes to manufacture non-woven. In melt-spinning processes granular material (e.g., polypropylen or polyethylen) is melted and pressed through nozzles in a spinneret (Fig. 1). The resulting filaments solidifies and are extruded (close to the nozzles or in some distance) by high speed air flow along the fiber. Finally, the filaments are deposited with support of a suction unit onto a conveyor belt such that a continuous non-woven is generated and removed. There exists usually a lay-down unit that gives a better control on the deposition of the filaments (Fig. 2).

One manufacture method of fibers for glass wool is based on a rotational principle (Fig. 3). A liquid glass beam falls into a rotating disk. Due to centrifugal forces the glass flows outwards and is pressed through the outer wall containing many holes. The filaments formed outside the disk are extruded by air drag forces caused by a hot gas stream and rotation. The glass filaments cool down and break at a certain distance from the disk. The hot gas stream is also used to heat up the rotating disk.

Simulations of the fiber dynamics can give new insights into the processes and can help to improve the processes and products. The modeling of the fiber dynamics in the following is based on an asymptotic approach [Pan06] where the fiber is regarded as a one-dimensional curve. It considers the balancing of effecting forces along the fiber. The effect of turbulent flows is considered by stochastical forces [Mar05].

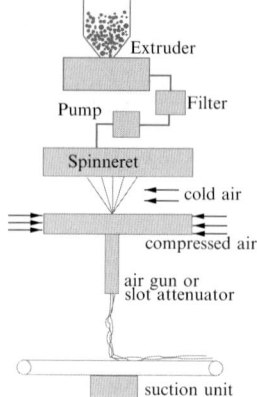

Fig. 1. Scheme of a melt-spinning process

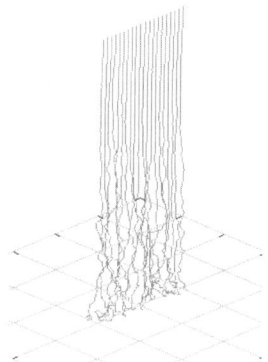

Fig. 2. Result of fiber dynamics simulation for the lay-down process

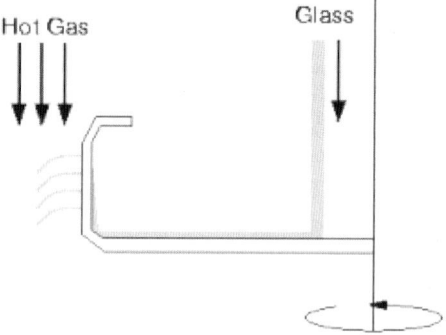

Fig. 3. Scheme of glass fiber spinning

The fiber dynamics and material laws can be simplier described in Euler than in Lagrange form. Unfortunately, numerical simulations basing on the Euler description show problems under industrial conditions. Therefore, in Sect. 2 we show the derivation of the Lagrange description from the Euler form. The resulting model is applied on the glass fiber spinning.

Usually, the effect of the fibers onto the fluid can be neglected. In the case of many fibers the complete interaction of fibers and fluid flow has to be considered which we demonstrate in the last section.

2 Fiber Modeling in Euler and Lagrange Description

A fiber will be considered as a time-dependant sufficient smooth curve in $\mathbb{R}^3$. There are mainly two parameter representations which are called Eulerian and Lagrangian representations.

In the Euler description $X(s,t) \in \mathbb{R}^3$ is a fiber point at time $t \in \mathbb{R}_0^+$ and arc length $s \in \mathbb{R}$, i.e., we have $\|X'\| = 1$. In the Lagrange description $Y(l,t) \in \mathbb{R}^3$ is a fiber point at time t with parameter $l \in \mathbb{R}$, which is assigned to that material point for all times. The transformation between both descriptions is realized by a function $S(l,t)$ with

$$Y(l,t) = X(S(l,t),t).$$

Additionally, we choose $S(l,t)$ such that $S(l,0) = l$, i.e., parameter l and arc length s are identical at time $t = 0$. The function S arises from the solution of the differential equation

$$\dot{S}(l,t) = u(S(l,t),t), \quad \text{with } S(l,0) = l.$$

Here, we denote the function $u(s,t) \in \mathbb{R}$ as transport velocity inside the fiber geometry. It will be determined later by a coupling condition.

The fiber dynamics is modeled by balancing of mass and momentum. The balancing of energy is neglected. For the Euler description we obtain

$$\dot{\sigma}_E + (\sigma_E\, u)' = 0, \tag{1}$$
$$(\sigma_E\, v_E)^{\cdot} + (\sigma_E\, v_E\, u)' = Q'_E + f_E, \tag{2}$$

where $\sigma_E(s,t)$ denotes the line density, $v_E(s,t) \in \mathbb{R}^3$ is the Eulerian fiber velocity, $Q_E \in \mathbb{R}^3$ represents the stresses and $f_E \in \mathbb{R}^3$ are the outer forces.

The Lagrangian quantities are defined as follows:

$$\sigma_L(l,t) = S'(l,t)\,\sigma_E(S(l,t),t) \quad v_L(l,t) = v_E(S(l,t),t), \tag{3}$$
$$Q_L(l,t) = Q_E(S(l,t),t) \qquad f_L(l,t) = S'(l,t)\,f_E(S(l,t),t). \tag{4}$$

The mass balance equation is obtained by $\dot{\sigma}_L(l,t) = \frac{d}{dt}[S'(l,t)\,\sigma_E(S(l,t),t)]$. Simple computation shows $\dot{\sigma}_L(l,t) = 0$ or

$$\sigma_L(l,t) = \sigma_L(l,0).$$

In order to receive the Lagrange representation for the momentum we regard $\dot{v}_L(l,t) = \dot{v}_E(S(l,t),t) + u\,v'_E(S(l,t),t)$. By (2)–(4) this gives

$$\sigma_L \dot{v}_L = Q'_L + f_L. \tag{5}$$

Up to now, the geometric description of the fiber in $\mathbb{R}^3$ is not coupled on the balance equations, i.e., we have no definition for u. In the sense of a material description this is simply given by $v_L = \dot{Y}$. Transforming this equation into the Euler description, we obtain

$$v_E = u\,X' + \dot{X} \quad \text{or} \quad u = (v_E - \dot{X}) \cdot X'$$

by using $\|X'\| = 1$.

For the constitutive equations we consider visco-elastic material laws and outer forces of the form

$$Q_E = N(u', \sigma_E, \sigma_E(\cdot, 0))X' \qquad f_E = f(X, X', v_E, \sigma_E, \sigma_L).$$

Using the transformations between Euler and Lagrange description we receive

$$Q_L = N\left(\frac{\|Y'\|^\bullet}{\|Y'\|}, \frac{\sigma_L}{\|Y'\|}, \sigma_L\right)\frac{Y'}{\|Y'\|} \qquad f_L = f\left(Y, \frac{\|Y'\|}{\|Y'\|}, \dot{Y}, \frac{\sigma_L}{\|Y'\|}, \sigma_L\right)\|Y'\|. \tag{6}$$

3 Interaction of Fibers and Fluid Flow

The interaction of fibers and fluid flow consists in an exchange of momentum and heat. As long as there are only a few fibers with moderate velocity the effect of them onto the flow can be neglected. This assumption fails in situations with many fibers where we have some influence onto the flow.

Due to the amount, diameter and dynamics of fibers it is not possible to compute numerically the fluid flow around the fibers. Therefore, an alternative modeling of the interaction of fluid and fibers has to be found.

A possibility is given by an iteration procedure of computing the fluid flow and the fiber dynamics. First, a numerical simulation of the fluid flow without the presence of fibers is performed. Using the obtained velocity and temperature distribution of the fluid the outer forces and heat transfer for the fibers can be determined and the fiber dynamics computation can be performed. The effect of the fibers onto the flow can be determined by a homogenization strategy. Considering a representative volume V of the flow domain (e.g., a grid cell from the fluid flow computation) the momentum m_i and heat transfer rate h_i of all fibers i covered by the volume can be averaged,

$$\bar{m} = \sum_i \frac{1}{V} \int_V m_i(x)\,\mathrm{d}x \qquad \bar{h} = \sum_i \frac{1}{V} \int_V h_i(x)\,\mathrm{d}x. \tag{7}$$

Then, the resulting momentum $\bar{m}$ and heat transfer rate $\bar{h}$ can be used as additional sources in the momentum and heat equation of the fluid flow. Finally, a fluid flow simulation can be done and the procedure can be repeated.

4 Example: Spinning of Glass Wool Fibers

In this section, the simulation of fiber dynamics using the Lagrange description and the interaction of fluid flow and fibers is shown by the spinning process of glass wool production.

4.1 Steady-State of Viscous Fibers

Outside the rotating disk there is an interaction between the gas flow and fiber dynamics. The flow extrudes the filaments and forces them to move downwards (Fig. 1). Since many thousands filaments are created they have also an effect onto the flow. The large number of filaments generates a flow resistance and deflects the gas flow. Additionally, there exists a heat exchange between flow and filaments.

Changing the reference frame to the rotating system we receive a steady-state situation because the boundary conditions are constant in time, i.e., we have $u = u(s)$ and $X = X(s)$. Using the parametrization $\dot{S}(t) = u(S(t))$ with $S(t) = 0$ the fiber dynamics is described by

$$\Omega \ddot{Y} = \dot{Q}_L + f_L, \tag{8}$$

where $Y(t) = X(S(t))$ and mass flow $\Omega = \sigma_E(0)u(0)$.

In the spinning process the filaments are liquid and can be extruded. The viscous tension in (8) is given by $Q_L = 3\nu\,\Omega\,\frac{\dot{Y}\cdot\ddot{Y}}{\|\dot{Y}\|^4}\,\dot{Y}$ [Pan06] with fluid viscosity ν. Additionally, surface tension, gravitation, air drag, coriolis and centrifugal forces effect the fiber dynamics.

4.2 Simulation and Numerical Results

The simulation of the fiber dynamics is done by the software tool FIDYST developed at Fraunhofer-ITWM. For the fluid flow simulation the commercial software tool FLUENT is used.

The results of the iterated simulation of the fluid flow and fiber dynamics can be seen in Fig. 4. The results show the temperature distribution at different iteration steps in front of the rotating disk. First, the temperature without considering filaments is shown. Then, the temperature of the iteration steps can be seen. It is obvious that there is a fast convergence. Already after the third iteration the simulation becomes nearly stationary.

The result for the temperature without filaments shows a hot gas stream close to the side wall of the rotating disk. The gas flow is not deflected by fibers and there is a good heat exchange between flow and disk. The final result shows some deflection and the hot gas does not anymore flow closely to the side wall. The filaments cool down the gas which can be seen by the reduced dark colored domain in front of the disk.

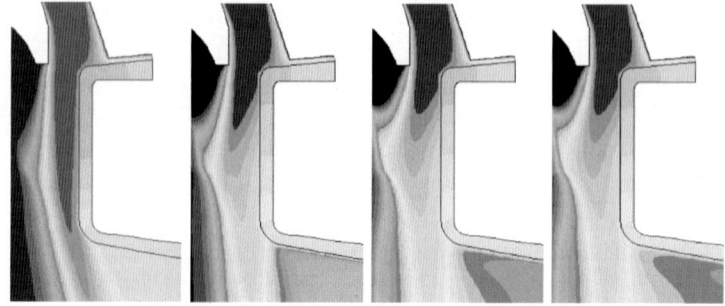

Fig. 4. Iteraction between flow and fiber movement with respect to the temperature distribution: convergence over the number of iteration steps, beginning with a flow without fibers

The numerical simulation shows the influence of the filaments onto the flow. Considering the interaction, this enables to obtain the correct understanding of the extruding process. It allows to explore the process by changing process parameters and to find an improved production.

References

[Mar05]   Marheineke, N.: Turbulent Fibers - On the Motion of Long, Flexible Fibers in Turbulent Flows. PhD Thesis, Technical University, Kaiserslautern (2005)

[Pan06]   Panda, S.: The Dynamics of Viscous Fibers, PhD Thesis, Technical University, Kaiserslautern (2006)

Asymptotics of Fiber Spinning Equations

Thomas Götz[1], Axel Klar[1], and Andreas Unterreiter[2]

[1] Dept. of Mathematics, TU Kaiserslautern, Germany
   `goetz, klar@mathematik.uni-kl.de`
[2] Dept. of Mathematics, TU Berlin, Germany
   `unterreiter@math.tu-berlin.de`

Summary. We consider a model for the production of glass fibers in a rotating spinning device. The model depends on two small parameters, namely the Reynolds-number δ and the Rossby-number ε. For small Rossby-numbers, i.e., a fast rotating spinning drum, numerical difficulties arise. To overcome this problems, asymptotic expansions are carried out for the stationary, inviscid case $\delta = 0$ as well as for the instationary, viscous case $\delta = \mathcal{O}(1)$.

1 The Problem

The following model for the motion of viscous, isothermal glass fibers in an rotation spinning process was derived by S. Panda, see [Pan06]. The model includes the conservation of mass and momentum, where inertial, viscous, Coriolis and centrifugal forces are taken into account.

$$\dot{A} + (Au)' = 0 \qquad\qquad A(t,0) = 1, \qquad (1a)$$

$$\dot{v}_\tau + uv_\tau' - v_n(v_n' + v_\tau \alpha') = \delta\frac{(Au')'}{A} + \frac{2}{\varepsilon}v_n + \frac{1}{\varepsilon^2}\gamma\cdot\tau \qquad \begin{aligned} u(t,0) &= 1 \\ u'(t,L) &= 0, \end{aligned} \qquad (1b)$$

$$\dot{v}_n + uv_n' + v_\tau(v_n' + v_\tau\alpha') = \delta u'\alpha' - \frac{2}{\varepsilon}v_\tau + \frac{1}{\varepsilon^2}\gamma\cdot n \qquad v_n(t,0) = 0, \qquad (1c)$$

$$\dot{\alpha} + u\alpha' - (v_n' + v_\tau\alpha') = 0 \qquad\qquad \alpha(t,0) = 0, \qquad (1d)$$

$$u' - (v_\tau' - v_n\alpha') = 0 \qquad\qquad v_\tau(t,0) = 1, \qquad (1e)$$

$$\gamma' = (\cos\alpha, \sin\alpha) \qquad\qquad \gamma(t,0) = (1,0). \qquad (1f)$$

Here, A denotes the cross-sectional area of the fiber, u the intrinsic velocity of the molten glass, v_τ and v_n the tangential and the normal component of the centerline velocity and γ the position of the fiber's centerline. By α we denote the angle between the tangent of the centerline and the x-axis. Furthermore,

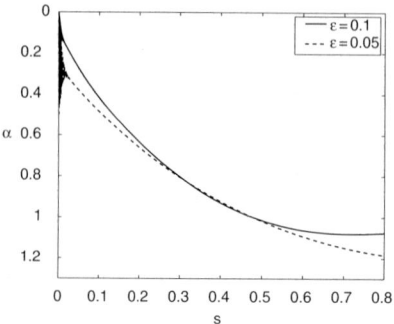

Fig. 1. Numerical solution for the angle α for different Rossby-numbers. Note the instability close to $s = 0$

$\dot{f} = \partial_t f$ denotes the time derivative and $f' = \partial_s f$ the derivative with respect to the arc-length parameter s. The system is given in an Eulerian frame fixed at the rotating spin drum. Lateron in the analysis we will also make use of a Lagrangian description.

The parameters δ and ε are related to the Reynolds and Rossby number. Both, the Reynolds-number $Re = 3/\delta$ and the Rossby-number ε can be small in application relevant cases. In his work, Panda faced severe numerical difficulties when solving the above system for small Rossby-numbers, e.g., $\varepsilon \sim 10^{-3}$. An example, where oscillations due to numerical instabilities arise, is given in Fig. 1. The purpose of this work is to investigate the limiting behavior of the above system for small ε, both in the inviscid case $\delta = 0$ as well as in the viscous case $\delta > 0$.

In Sect. 2, we will analyze the limit $\varepsilon \to 0$ for the stationary, inviscid case $\delta = 0$ and Sect. 3 is devoted to the instationary, viscous case $\delta > 0$.

2 The Stationary, Inviscid Case

In the stationary, inviscid case, the continuity equation $(Au)' = 0$ separates and we immediately obtain $A = 1/u$. For the centerline velocities v_τ and v_n, we get in the stationary case the solutions $v_n \equiv 0$ and $v_\tau \equiv u$. Hence the remaining system reads as

$$uu' = \frac{1}{\varepsilon^2}\gamma \cdot \gamma' \qquad\qquad u(0) = 1, \qquad\qquad (2a)$$

$$u^2\alpha' = -\frac{2}{\varepsilon}u + \frac{1}{\varepsilon^2}\gamma \cdot n \qquad\qquad \alpha(0) = 0, \qquad\qquad (2b)$$

$$\gamma' = (\cos\alpha, \sin\alpha) \qquad\qquad \gamma(0) = (1,0). \qquad\qquad (2c)$$

Next, we introduce polar coordinates (r, ϕ) for the centerline $\gamma = r(\cos\phi, \sin\phi)$ and the angle $\beta = \alpha - \phi$ between the centerline and the tangent. Solving the

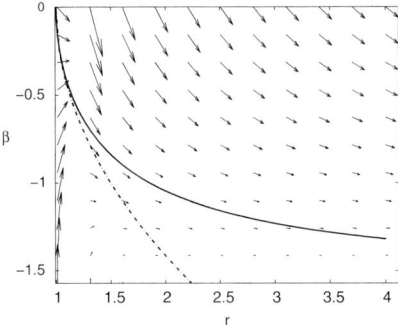

Fig. 2. Phaseportrait of (3). The solution is given by the solid line and the dash dotted line is the asymptotic expansion for $\beta \ll 1$

velocity (2a) explicitely, i.e., $u^2 = C + (r/\varepsilon)^2$ with $C = 1 - 1/\varepsilon^2$, we end up with a reduced system which has a regular limit for $\varepsilon \to 0$.

$$r' = \cos \beta \qquad\qquad\qquad\qquad r(0) = 1, \qquad (3a)$$

$$\beta' = -\frac{2}{\sqrt{r^2 - 1}} - \frac{2r^2 - 1}{r^2 - 1}\frac{\sin \beta}{r} \qquad \beta(0) = 0. \qquad (3b)$$

We consider this equation in the phase-space $(r, \beta) \in D = [1, \infty) \times [-\pi/2, 0]$, see Figure 2. Note, that r is increasing and β is decreasing in D. For the parameter $s \to \infty$, the solution tends to $(r, \beta) = (\infty, -\pi/2)$.

The solution to the above system (3) is given by

$$r(s) = \sqrt{2s + 1} \qquad \beta(s) = -\sqrt{2s} - \arctan \frac{\sqrt{2s} \cos \sqrt{2s} - \sin \sqrt{2s}}{\cos \sqrt{2s} + \sqrt{2s} \sin \sqrt{2s}}. \qquad (4)$$

This solution has a singular slope at $s = 0$. To resolve this singularity, we carry out an asymptotic expansion for $\beta \ll 1$ and get $\beta \simeq -\sqrt{2s}$.

3 The Instationary, Viscous Case

In this section, we will analyze the instationary, viscous case of the glass fiber spinning equations in the limit of vanishing Rossby-number $\varepsilon \to 0$ and fixed Reynolds-number $\delta > 0$. Contrary to the stationary, inviscid case a treatment in Eulerian coordinates seems to be quite difficult. Hence, we transform the equations into a Lagrangian frame moving with the fiber. The arc-length parameter S can now be obtained as the solution of the ODE

$$\partial_t S(t, \lambda) = u(t, S(t, \lambda)), \qquad S(t, t) = 0, \qquad (5)$$

where the Lagrangian parameter $\lambda \in [0, t]$ denotes the time, when the fluid particle leaves the nozzle. The centerline of the fiber, i.e., the trajectory of the fluid particles, is denoted by $\xi(t, \lambda) = \gamma(t, S(t, \lambda))$. We rescale $t^* = t/\varepsilon$, $\lambda^* = \lambda/\varepsilon$ and $\xi^* = \xi/\sqrt{\varepsilon}$ and seek similarity solutions of the form $w(s) = \xi(t-\lambda, \lambda)$ for $s = t - \lambda$ being the time since a fluid particle has left the nozzle. Now w solves

$$w'' = \delta \left(\frac{w' \cdot w''}{|w'|^4} w' \right)' + 2(w')^{\perp} + w \tag{6}$$

subject to $w(0) = (1/\sqrt{\varepsilon}, 0)$, $w'(0) = (\sqrt{\varepsilon}, 0)$ and $w'(0) \cdot w''(0) = \sqrt{\varepsilon^3} z_0$.

Introducing polar coordinates $w' = x(s) \, (\cos\phi(s), \sin\phi(s))$ and accordingly $w = r(s) \, (\cos\phi(s), \sin\phi(s)) + q(s) \, (-\sin\phi(s), \cos\phi(s))$ yields the system

$$\phi' = -x^2 \frac{2x - q}{x^3 - \delta z} \qquad\qquad \phi(0) = 0, \tag{7a}$$

$$r' = x - qx^2 \frac{2x - q}{x^3 - \delta z} \qquad\qquad r(0) = \frac{1}{\sqrt{\varepsilon}}, \tag{7b}$$

$$q' = rx^2 \frac{2x - q}{x^3 - \delta z} \qquad\qquad q(0) = 0, \tag{7c}$$

$$x' = z \qquad\qquad x(0) = \sqrt{\varepsilon}, \tag{7d}$$

$$\delta z' = x^2 z + 2\delta \frac{z^2}{x} - rx^2 \qquad\qquad z(0) = \sqrt{\varepsilon^3} z_0. \tag{7e}$$

The artificial initial condition $z(0) = \sqrt{\varepsilon^3} z_0 \in [0, 1/\delta)$ leads to $\phi'(0) \leq 0$. Numerical simulations show, that the solution is almost independent of the initial value z_0.

The numerical simulations also indicate, that there exists a boundary layer at $s = 0$ (see Fig. 3) of thickness $s = \mathcal{O}(\varepsilon)$ for ϕ, q and z, while q and x are $\mathcal{O}(\sqrt{\varepsilon})$ uniformly in s, see Fig. 3. Inside the boundary layer, we find $\phi = \mathcal{O}(\varepsilon)$, $r = 1/\sqrt{\varepsilon} + \mathcal{O}(\sqrt{\varepsilon^3})$ and $z = \mathcal{O}(\sqrt{\varepsilon^3})$. Outside the boundary layer, we have $\phi = \mathcal{O}(\varepsilon)$, $r = 1/\sqrt{\varepsilon} + \mathcal{O}(\sqrt{\varepsilon})$ and $z = \mathcal{O}(\sqrt{\varepsilon})$.

To resolve the boundary layer, we rescale $\tau = s/\varepsilon$, $\phi = \varepsilon\Phi$, $q = \sqrt{\varepsilon}\beta$, $x = \sqrt{\varepsilon}X$, $z = \sqrt{\varepsilon^3}Z$ and $r = 1/\sqrt{\varepsilon} + \sqrt{\varepsilon^3}\rho$. In this inner scaling the leading order of the solution is given by

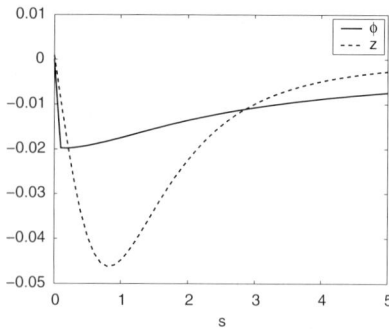

Fig. 3. Numerical simulation for ϕ and z with boundary layers at $s = 0$

$$\Phi_0(\tau) = -\frac{2\tau}{1 - \delta z_0 + \tau}, \tag{8a}$$

$$\rho_0(\tau) = \tau - 3\left(\frac{\tau}{2 - \delta z_0 + \tau}\right)^2, \tag{8b}$$

$$\beta_0(\tau) = \frac{2\tau}{1 - \delta z_0 + \tau}, \tag{8c}$$

$$X_0(\tau) = 1, \tag{8d}$$

$$Z_0(\tau) = z_0 - \frac{\tau}{\delta}. \tag{8e}$$

To determine the conditions needed lateron for matching the inner and outer expansion, we undo the inner scaling and obtain for $s \ll 1$

$$\phi \sim -2\varepsilon, \quad q \sim 2\sqrt{\varepsilon}, \quad x \sim \sqrt{\varepsilon}, \quad r \sim \frac{1}{\sqrt{\varepsilon}} + \sqrt{\varepsilon}s, \quad z \sim \sqrt{\varepsilon}\frac{s}{\delta} + \sqrt{\varepsilon^3} \tag{9}$$

For the outer expansion, we use the scalings $\phi = \varepsilon\Phi_a$, $q = \sqrt{\varepsilon}\beta_a$, $x = \sqrt{\varepsilon}X_a$, $z = \sqrt{\varepsilon}Z_a$ and $r = 1/\sqrt{\varepsilon} + \sqrt{\varepsilon}\rho_a$. In this new scaling, the leading order solution reads as

$$\Phi_a(s) = -2\frac{\sqrt{2\delta}\arctan\frac{s}{\sqrt{2\delta}}}{s}, \tag{10a}$$

$$\rho_a(s) = \sqrt{2\delta}\arctan\frac{s}{\sqrt{2\delta}}, \tag{10b}$$

$$\beta_a(s) = 2\frac{\sqrt{2\delta}\arctan\frac{s}{\sqrt{2\delta}}}{s}, \tag{10c}$$

$$X_a(s) = \frac{2\delta}{s^2 + 2\delta}, \tag{10d}$$

$$Z_a(s) = -\frac{4\delta s}{(s^2 + 2\delta)^2}. \tag{10e}$$

Figures 4 and 5 show an excellent agreement of the inner and outer expansion with the numerical solution of (7).

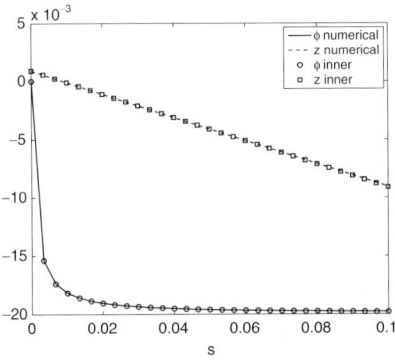

Fig. 4. Comparison of the numerical solution and inner expansion for ϕ and z

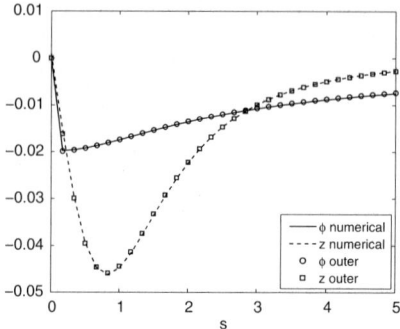

Fig. 5. Comparison of the numerical solution and outer expansion for ϕ and z

References

[Pan06] Panda, S.: The dynamics of viscous fibers. PhD–Thesis, University of Kaiserslautern, Germany (2006)

Three-Dimensional Elastica for Modelling Fibre Assemblies

R.B. Ramgulam and P. Potluri

University of Manchester, P.O. Box 88, Manchester M60 1QD
r.ramgulam@manchester.ac.uk, prasad.potluri@manchester.ac.uk

1 Introduction

In knitted and woven fabrics, inter-yarn forces at the crossover regions tend to compress the yarns thus affecting their mechanical properties. Hence the need to relate lateral forces and yarn cross-sectional deformation. Some theoretical analysis [1, 2] assumed filaments are elasticas and forces are applied at discrete points [1, 2] or uniformly distributed [3] on the filaments. Harwood et al. [4] have proposed a model of yarn compressibility by extending the work of van Wyk [5] to oriented fibres. The present chapter is concerned with compression of continuous filament yarns. The filaments are assumed to be elasticas and follow helical paths in the yarn. The theory of three-dimensional elastica is developed from differential geometry of curves and is applied to lateral deformation of a single helix. Finally, a yarn geometrical model and the algorithm for yarn compression are described.

2 Differential Geometry of Centreline of Elastica

The objective is to describe the shape of the centreline of an elastica, generally a space curve, subjected to external forces and moments. The obvious way of doing so would be in terms of a moving trihedron $(\mathbf{t}, \mathbf{n}, \mathbf{b})$, where $\mathbf{t}$ is the tangent to the curve, $\mathbf{n}$ the principal normal and $\mathbf{b}$ the binormal vector. The derivatives of the moving trihedron with respect to arc-length s are the Serret–Frenet equations [6] which define the curvature, k, and torsion τ of a curve. However these parameters are purely geometrical quantities. For the study of an elastica we require measures which relate to both geometry and mechanical properties. A new moving trihedron $(\mathbf{u}, \mathbf{v}, \mathbf{w})$, with $\mathbf{n.u} = \mathbf{b.v} = \cos \lambda$ and $\mathbf{w} = \mathbf{t}$ is defined [4]. In practical situation, the vectors $\mathbf{u}$ and $\mathbf{v}$ represent the principal directions of the elastica cross-section and vector $\mathbf{w}$ is the tangent of the centreline. With this change of local coordinates the following differential equations can be derived to express the geometry of a curved elastica in three dimensions in terms of cross-section parameters and twist.

$$\frac{d\mathbf{u}}{ds} = \sigma\mathbf{n} - q\mathbf{w}, \tag{1}$$

$$\frac{d\mathbf{v}}{ds} = p\mathbf{w} - \sigma\mathbf{u}, \tag{2}$$

$$\frac{d\mathbf{w}}{ds} = q\mathbf{u} - p\mathbf{v}. \tag{3}$$

where p and q are components of curvature in the $\mathbf{u}$ and $\mathbf{v}$ directions and σ is the twist, the relative displacement of two neighbouring cross-sections.

3 Constitutive Equations

The elastica is assumed to be inextensible, of constant cross-section and unaffected by shear. In that case the elastica shows resistance to bending and torsion according to Love's 'Ordinary Approximate Theory': $M_u = A\,(p - p_o)$; $M_v = B\,(q - q_o)$ and $M_w = C\,(\sigma - \sigma_o)$. The parameters M_u and M_v are the components of internal moments and constants A and B are the bending rigidities while M_w is the internal torque and C is the torsional rigidity. The curvatures p_o, q_o and twist σ_o refers to the undeformed elastica.

4 Equilibrium Equations

The vector form of equations of equilibrium of forces and moments on the small element, Fig. 1, within a moving basis, such as the $(\mathbf{u},\ \mathbf{v},\ \mathbf{w})$ are as follows:

$$\frac{d\mathbf{M}}{ds} + \mathbf{M} \times \psi + \mathbf{w} \times \mathbf{F} + \mathbf{m} = 0, \tag{4}$$

$$\frac{d\mathbf{F}}{ds} + \mathbf{F} \times \psi + \mathbf{f} = 0. \tag{5}$$

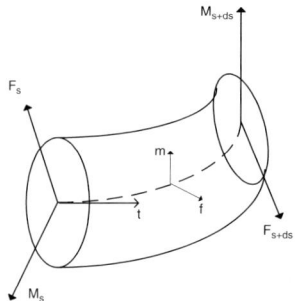

Fig. 1. Forces and moments on a small element of elastica

M and **m** are concentrated and distributed moments. **F** and **f** are concentrated and distributed forces. Vector $\boldsymbol{\psi} = p\mathbf{u} + q\mathbf{v} + \sigma\mathbf{w}$. Combining (4) and (5) with the constitutive equations relate the geometric parameters, curvature and twist, to external loads and moments. Hence the complete mathematical model of three-dimensional elastica consists of (1)–(3), (4) and (5) modified by the constitutive model and the orientation-coordinate equation $\mathbf{w} = \frac{d\mathbf{r}}{ds}$. In scalar form the model consists of 18 first-order linear differential equations.

5 Single Helix Compression Model

The model for three-dimensional elastica can be solved by numerical methods such as Runge–Kutta. Nine scalar equations, (1)–(3) in vector form, relate to direction cosines. These equations are not strictly independent and judicious choice of only three initial values lead to specification of the other six. In the case of the helix the following initial direction cosines are defined: $u_x(0) = -\cos\theta$; $w_x(0) = -\sin\theta$; $w_y(0) = 0$ where θ is the helix angle. If radial deformation of the helix is specified then the objective is to solve a boundary value problem with four unknown initial values: p, q, σ and lateral force F_y. The methodology adopted is illustrated in Fig. 2. The length of the elastica over which the integration is performed is half the helix length L. The four boundary values are: $y(L/2) = \delta$; $y(L/4) = 0$; $x(L/2) = 0$; $z(L/2) = 0.5L\cos\theta$ where δ is the yarn radial deformation and (x, y, z) are the fixed Cartesian coordinates.

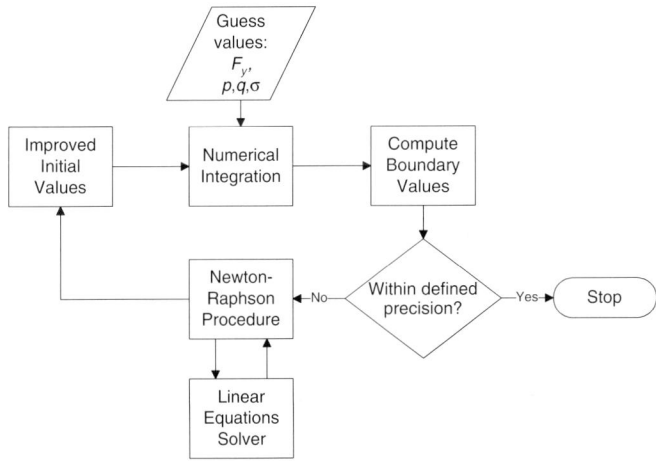

Fig. 2. Flowchart of boundary value solution procedure

6 Yarn Compression Model

The model is based on the deformation of helices under forces applied perpendicular to the helix axis. In addition a model that describes the distribution of filaments in the yarn is required.

6.1 Geometrical Model of Filament Yarn

The filaments are arranged in concentric layers as illustrated in Fig. 3. Each filament follow a helical path along the yarn length. Layer $n = 1$ surrounds the core. The helix radius, r_n of a filament in layer n is $2rn$. Helix angle $\theta = \arctan\left(2\pi \times r_n T\right)$ and length $L = \frac{2\pi \times r_n}{\sin\theta}$. The twist per unit length, T, is assumed to be the same for all the filaments. The deformation of each filament is assumed to vary linearly with distance from yarn centre. If a deformation δ is imposed on the yarn then a given filament within the yarn will be deformed by $\zeta = \delta \frac{r_n}{R-r}$.

6.2 Algorithm for Yarn Compression Simulation

The following procedure is used to obtain the force-compression behaviour of filament type yarns.

1. Apply defined deformation, δ perpendicular to yarn axis.
2. Calculate filament deformation, ζ for each filament.
3. Use single helix compression model to compute energy required.
4. Compute total energy required to deform all the filaments in the yarn.
5. Repeat steps 1–4 for different yarn deformation values.
6. From total energy versus yarn deformation relationship, the force-deformation characteristic of the yarn is simulated.

The force in step 6 is the force acting over a length $z\,(L/2) = 0.5L\cos\theta$. Fig. 4 indicates that the simulation results compares favourably with experimental data.

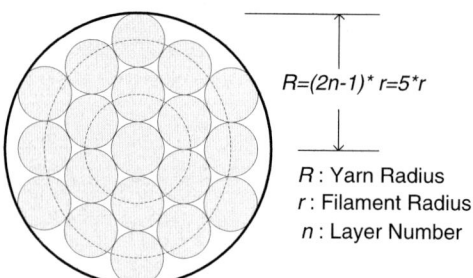

$R=(2n-1)^{*}\,r=5^{*}r$

R : Yarn Radius
r : Filament Radius
n : Layer Number

Fig. 3. Yarn cross-section

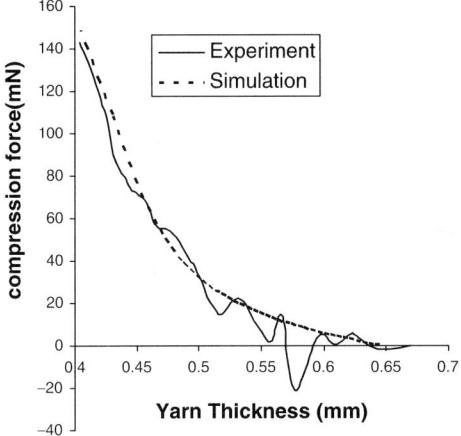

Fig. 4. Comparison of simulation with experimental data

7 Conclusions

A mathematical model for three-dimensional elastica has been presented and its application to the compression of yarns discussed. Though the model makes a number of simplifying assumptions the results are not too far from experimental data. To improve the usefulness of the model for practical applications, friction between filaments has to included as well. This part of the research is still ongoing.

References

1. Leaf G.A.V, Oxenham W, The compression of yarns, I. The Compression-energy function, J.Text.Inst, 4, T168, 1981.
2. Leaf G.A.V, Oxenham W, The compression of yarns, II. The load-compression relation, J.Text.Inst, 4, T176, 1981.
3. Harwood R.J, Grishanov S.A, Lomov S.V, Cassidy T, Modelling of two-component yarns part 1: the compressibility of yarns, J.Text.Inst, 88, Part 1, No.4, 1997.
4. van Wyk C.M, Note on the compressibility of wool, J.Text.Inst, 37, T285-T292, 1946.
5. M. Konopasek, Improved Procedures for calculating the mechanical properties of textile structures, PhD Thesis, UMIST, 1970.
6. Oprea J, Differential Geometry ad its Applications, Prentice-Hall,NJ, 1997, pp 18-27.

Effective Properties of Nonwoven Textiles from Microstructure Simulations

Andreas Wiegmann

Fraunhofer Institut für Techno- und Wirschaftsmathematik
Fraunhofer Platz 1, 67663 Kaiserslautern, Germany
`wiegmann@itwm.fhg.de`

Summary. Nonwoven are technical textiles that may be described as random collections of straight fibers. Infinitely long fibers have the disadvantage that periodicity can not be ensured. We give a short fiber model with two anisotropy parameters that allows to account for periodicity and compute the nonwoven's permeability for some such models.

1 Introduction

Spun-bond nonwoven are textiles of significant commercial interest. They are produced by spinning fibers directly onto a moving conveyor belt. This results in a strongly anisotropic distribution of random fiber directions. First, fibers tend to lie parallel with the transporting belt. Second, the motion of the belt induces different probabilities of the fiber directions in this plane. Schladitz [3] introduced an infinite fiber model using only the porosity, fiber diameter, and fiber direction distribution. Here, we consider a more general anisotropy and introduce short fibers mostly in order to be able to generate periodic nonwoven representations. Such periodic representations are helpful for the computation of effective material properties from microstructure simulations [2,4], because the (Navier-) Stokes solver codes may use periodic boundary conditions as assumed in homogenization theory for the computation of effective properties [1]. Finally, the influence of the anisotropy, fiber diameter, and porosity on the permeability tensor of the nonwoven are studied.

2 Nonwoven Model

The nonwoven model consists of two components, Sect. 1 dealing with individual fibers and Sect. 2 dealing with the discrete representative elementary volume, or **discrete REV**.

2.1 Fiber Diameter, Length, Position, and Direction

Fibers considered here are idealized to cylinders with flat end or capped by a half-sphere. Fiber crimp is not considered, as it often can be neglected as occurring on the scale of millimeter to centimeter, while the fiber diameter is on the order of micrometer. A fiber is a 9-tuple

$$F : (x, y, z, l, r, n_x, n_y, n_z, b),$$

where (x, y, z) is the center of gravity of the fiber, l and r are half the length of the cylinder and half the diameter of the fiber, $\mathbf{n} = (n_x, n_y, n_z)$ with $\|\mathbf{n}\| = 1$ is the direction of the fiber and $b \in \{0, 1\}$ indicates a half-sphere (1) or flat cap (0). Note that the representation is not unique because $\mathbf{n}$ and $-\mathbf{n}$ yield the same fiber, and that this representation also includes spheres.

The length $2l$, diameter $2r$, and end shape b are usually specified by the manufacturer, and the position (x, y, z) is usually uniformly distributed in either the original or an enlarged image, so the most interesting parameter from a mathematical point of view is the distribution of fiber directions $\mathbf{n}$. Due to the manufacturing process, not all directions occur with the same probability. In [3], a one-parameter model was used that accounts for the fiber spinning process by orienting fibers "almost parallel to an axis" or "almost parallel to a plane."

Algorithm 1:
$$\Phi := rand([0, 2\pi])$$
$$z := rand([-1, 1])$$
$$q := \sqrt{z^2 + \beta^2 - z^2\beta^2}$$
$$n_x := \sqrt{\beta^2 - z^2\beta^2}/q \, \sin\Phi$$
$$n_y := \sqrt{\beta^2 - z^2\beta^2}/q \, \cos\Phi$$
$$n_z := z/q$$

The probability density for (n_x, n_y, n_z) in this case is

$$\frac{\beta \sin\theta}{2(1 + (\beta^2 - 1)\cos^2\theta)^{3/2}},$$

where $\beta \in (0, \infty)$ governs the compression or stretching perpendicular to the z-direction and (Φ, θ) are the angles in standard spherical coordinates.

Here, we add the additional parameter $\alpha \in (0, \infty)$ to govern an additional compression or stretching only in the y-direction.

Algorithm 2:
$$\Phi := rand([0, 2\pi])$$
$$z := rand([-1, 1])$$
$$x := \beta\sqrt{1 - z^2} \, \sin\Phi$$
$$y := \beta\alpha\sqrt{1 - z^2} \, \cos\Phi$$
$$\mathbf{n} := \frac{(x, y, z)}{\|(x, y, z)\|_2}$$

Lemma 1. *If $\alpha = 1$ and Φ and z agree with those from Algorithm 1 then the direction $\mathbf{n}$ also agrees with the one from Algorithm 1.*

Proof. First, $q = ||(x, y, z)||_2$ since

$$\beta^2(1 - z^2) \sin^2 \Phi + \beta^2(1 - z^2) \cos^2 \Phi + z^2 = \beta^2 - \beta^2 z^2 + z^2.$$

This also implies that z has the same meaning in both algorithms. Second, $x = \sqrt{\beta^2 - z^2 \beta^2} \sin \Phi$ since

$$\beta^2(1 - z^2) \sin^2 \Phi + \beta^2(1 - z^2) \cos^2 \Phi = \beta^2 - \beta^2 z^2.$$

Similarly, $y = \sqrt{\beta^2 - z^2 \beta^2} \cos \Phi$.

2.2 Domain Periodicity, Overlap, and Porosity

For sake of simplicity and efficiency, the fibers are discretized into three-dimensional *images*, or parallelepipeds with some $N_x \times N_y \times N_z$ cube-shaped volume elements or *voxels*. A voxel is set to 1 to represent volume occupied by a fiber and set to 0 to represent an empty volume.

Due to the typical scales of the fibers of about $10\,\mu$m, the 3D model can only represent about $1\,\text{mm}^3$ at a resolution of $2\,\mu$m, the minimum required to get somewhat round fibers. This yields already 500^3 voxels or grid points, a formidable challenge beyond the capabilities of most flow solvers.

To have some flexibility when discretizing given objects from, for example, CAD systems, a discrete REV is defined by an 11-tuple

$$D : (x_0, y_0, z_0, h, i_{\min}, j_{\min}, k_{\min}, i_{\max}, j_{\max}, k_{\max}, I),$$

where (x_0, y_0, z_0) is a reference point in space, h is the length of a voxel, $(i_{\min}, j_{\min}, k_{\min})$ indicates the left front bottom voxel center with coordinates $(x_0 + i_{\min}h - h/2, y_0 + j_{\min}h - h/2, z_0 + k_{\min}h - h/2)$ and $(i_{\max}, j_{\max}, k_{\max})$ indicates the right back top voxel center with coordinates $(x_0 + i_{\max}h - h/2, y_0 + j_{\max}h - h/2, z_0 + k_{\max}h - h/2)$. I is the $N_x \times N_y \times N_z$ dimensional binary image with $N_x = i_{\max} - i_{\min} + 1$, $N_y = j_{\max} - j_{\min} + 1$, and $N_z = k_{\max} - k_{\min} + 1$. This choice ensures that the simplified discrete REV

$$D : (h, N_x, N_y, N_z, I) = (0, 0, 0, h, 1, 1, 1, N_x, N_y, N_z, I)$$

is a discretization of the parallelepiped $[0, N_x h] \times [0, N_y h] \times [0, N_z h]$ into $N_x \times N_y \times N_z$ voxels. Unless noted otherwise, we usually consider this simpler type of REV and use $L_x = hN_x$, $L_y = hN_y$, and $L_z = hN_z$.

Periodicity. In the periodic case, the center of gravity of a fiber is generated as three uniformly distributed random numbers in $[0, L_x] \times [0, L_y] \times [0, L_z]$. Then, a fiber is viewed as the 27 copies with centers of gravities $(x + dx, y + dy, z + dz)$ that result from $dx \in \{-L_x, 0, L_x\}$, $dy \in \{-L_y, 0, L_y\}$, and $dz \in \{-L_z, 0, L_z\}$.

In the non-periodic case, the center of gravity of a fiber is generated as three uniformly distributed random numbers in $[-l, L_x + l] \times [-l, L_y + l] \times [-l, L_z + l]$ to avoid boundary artefacts in the local porosity.

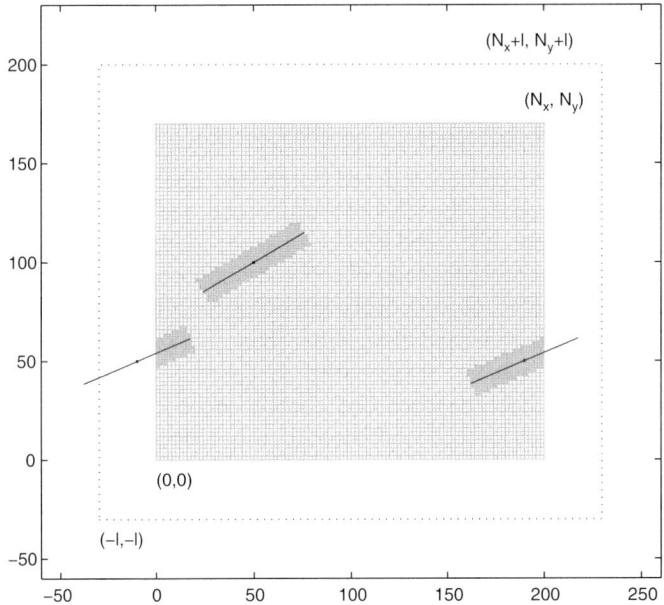

Fig. 1. Discrete REV with periodic copies of fibers. A fiber exits the REV on the right, so the copy shifted to the left enters into the discretization. The dotted lines illustrate the area where centers of gravity may be placed in the non-periodic case.

Fiber discretization. A voxel (i, j, k) in the image I is considered to belong to the fiber if the voxel center $(x_i, y_i, z_i) = ((i-\frac{1}{2})h, (j-\frac{1}{2})h, (k-\frac{1}{2})h)$ lies within the radius r of the center line of the fiber axis given by $(x, y, z) + t\mathbf{n}$, and if the projection point on this line has a line coordinate t_p with $t_p \in [-l, l]$.

For semispherical cap on the fiber ends, also voxels with center coordinates within the radius r of the fiber end points are considered to lie in the fiber.

Nonwoven porosity. The porosity p of an image I is $\sum_{i=1,j=1,k=1}^{i=N_x,j=N_y,k=N_z} I(i, j, k)/ (N_x N_y N_z)$. p is evaluated after a fiber is discretized, and if the desired porosity is not reached, another fiber is randomly generated and discretized. p can attain only a discrete set of values, and so an interval of desired porosity must be specified. If the porosity reaches a value below the lower end of this interval, the process must be started again with an empty discrete REV (Fig. 1). After a fixed number of tries the algorithm decides that the porosity requirement cannot be achieved.

Fiber overlap. To achieve realistic, nonoverlapping fibers, the 27 copies of a fiber are not discretized into the original image right away. Instead, all voxels that would be occupied by a newly entered fiber are checked for availability, and the fiber is discretized into an auxiliary image. By testing voxels also with $r_1 > r$, but only entering those for r, it is also possible to achieve a minimum distance between fibers.

Table 1. Anisotropy, porosity, fiber radius, and computed permeabilities

Parameters	κ_{xx} (μm^2)	κ_{yy} (μm^2)	κ_{zz} (μm^2)
$\beta = 10$, $\alpha = 3$, $p = 0.95$, $r = 4\,\mu m$	9.3	12.0	8.0
$\beta = 0.1$, $\alpha = 1$, $p = 0.95$, $r = 4\,\mu m$	8.2	7.9	12.4
$\beta = 10$, $\alpha = 1$, $p = 0.95$, $r = 4\,\mu m$	10.3	10.8	8.0
$\beta = 10$, $\alpha = 1$, $p = 0.93$, $r = 4\,\mu m$	4.0	4.0	6.7
$\beta = 10$, $\alpha = 1$, $p = 0.91$, $r = 4\,\mu m$	2.8	2.9	4.8
$\beta = 10$, $\alpha = 1$, $p = 0.95$, $r = 3\,\mu m$	5.5	4.8	8.7
$\beta = 10$, $\alpha = 1$, $p = 0.95$, $r = 5\,\mu m$	10.4	12.1	19.5

3 Computed Permeability

To compute the permeability of the nonwoven, the Stokes problem with periodic boundary conditions on the image boundaries and nonslip boundary conditions on the fiber surfaces is solved for a given pressure drop and fluid viscosity 1. Then, material permeability in the direction of the pressure drop is the quotient of the average velocity over the pressure drop [2, 5]. Table 1 illustrates nicely that the resulting permeabilities depend in an intuitive way on the selected nonwoven parameters. By judicious choice of the nonwoven parameters, good agreement with measurements has been observed with measurements on real nonwoven in several industrial projects at Fraunhofer ITWM.

References

1. U. Hornung. Introduction. In U. Hornung, editor, *Homogenization and Porous Media*, pages 1—25. Springer, 1997.
2. D. Kehrwald. Parallel lattice Boltzmann simulation of complex flows. In *NAFEMS Seminar: Simulation of Complex Flows (CFD) - Application and Trends*, May 2004.
3. K. Schladitz, S. Peters, D. Reinel-Bitzer, A. Wiegmann, and J. Ohser. Design of acoustic trim based on geometric modeling and flow simulation for non-woven. Technical Report 72, Fraunhofer ITWM Kaiserslautern, 2005.
4. V. Schulz, D. Kehrwald, A. Wiegmann, and K. Steiner. Flow, heat conductivity, and gas diffusion in partly saturated microstructures. In *NAFEMS Seminar: Simulation of Complex Flows (CFD)*, April 2005.
5. A. Wiegmann. FFF-Stokes: A fast fictitious force 3d Stokes solver. In preparation, 2006.

Minisymposium "Approximate Algebraic Techniques for Curves and Surfaces"

Bert Jüttler

Johannes Kepler University Linz, Austria
bert.juettler@jku.at

The two fields of computer aided design (CAD) and algebraic geometry deal with curves and surfaces defined by algebraic equations, but the objects are studied using different approaches. On the one hand, algebraic geometry has developed impressive results for understanding the theoretical nature of these objects. On the other hand, the CAD community focuses on practical applications of virtual shapes defined by algebraic equations, and the curves and surfaces are typically represented with limited precision, using floating point numbers.

The two fields can benefit from mutual interaction, e.g., the need for analyzing singularities occurs frequently when dealing with results of *offsetting*, which is one of the fundamental operations in CAD. Also, the applicability of results from algebraic geometry can be enlarged by using the novel technique of approximate implicitization [Dok2]. These and similar interactions were explored in the European project IST 2001-35512[1] GAIA II.

The minisymposium on Approximate Algebraic Techniques for Curves and Surfaces at ECMI 2006 presented results related to this project, by discussing topics such as offset curves and surfaces, symbolic-numerical algorithms, and approximate implicitization. More information about the results of the GAIA II project, which include scientific publications and software prototypes, are described in [Dok1].

References

[Dok1] T. Dokken, The GAIA Project on Intersection and Implicitization, in R. Piene and B. Jüttler (eds.), Computational methods for Algebraic Spline Surfaces II, Springer, to appear.

[Dok2] T. Dokken and J. B. Thomassen. Overview of approximate implicitization. In *Topics in algebraic geometry and geometric modeling*, AMS, Providence, RI, 2003, 169–184.

---

[1]see project Web Site at www.sintef.no/IST_GAIA.

Computing the Intersection Curve Between a Plane and the Offset of a Parametric Surface

Fernando Carreras[1], Laureano Gonzalez–Vega[1], and Jaime Puig-Pey[2]

[1] Departamento de Matematicas, Estadística y Computacion
[2] Departamento de Matematica Aplicada y Ciencias de la Computacion
    Universidad de Cantabria, Santander, Spain
    carreras.fernando@gmail.com, gonzalezl@unican.es, puigpeyj@unican.es

Summary. A new seminumerical algorithm for computing the intersection curve between a plane and the offset of a parametric surface is presented. The corresponding implementation and the performed experimentation are also reported.

1 Introduction

Given a rational surface $\mathcal{S}$ in $\mathbb{R}^3$ presented by their parametric equations,

$$x = f(s,t), y = g(s,t), z = h(s,t)$$

a point $(x, y, z) \in \mathbb{R}^3$ is in the offset surface to $\mathcal{S}$ at a distance $d > 0$ (see for example [Ho90]) if there exists $(s, t) \in \mathbb{R}^2$ such that

$$(x - f(s,t))^2 + (y - g(s,t))^2 + (z - h(s,t))^2 = d^2$$
$$f_s(s,t)(x - f(s,t)) + g_s(s,t)(y - g(s,t)) + h_s(s,t)(z - h(s,t)) = 0$$
$$f_t(s,t)(x - f(s,t)) + g_t(s,t)(y - g(s,t)) + h_t(s,t)(z - h(s,t)) = 0$$

where the subscripts denote partial differentiation.

The purpose of this note is to introduce a new method to compute the intersection curve $\mathcal{C}$ between a plane Π and the offset to distance $d > 0$ of a surface $\mathcal{S}$ presented parametrically (by polynomial functions). This is a critical problem in computer aided design arising in many practical situations such as tool path generation, 3D NC machining, etc. (see, for example, [Ma99,Stu76]).

The algorithm here presented includes the use of several symbolic tools (like polynomial manipulations) together with several seminumerical techniques such as the determination of the topology of a real algebraic plane curve presented implicitly.

The generation of the points in the searched intersection curve is to be performed numerically by applying a Runge–Kutta scheme on a very controlled

way (in the sense that the shape of the final result is known in advance due
to the previous topology determination).

2 The Algorithm

The points in the intersection curve between a plane Π and the offset to
distance $d > 0$ of a surface $\mathcal{S}$ presented parametrically can be characterized
in the following way: if Π is presented by

$$X(u, v) = a_1 u + a_2 v + a_3$$
$$Y(u, v) = b_1 u + b_2 v + b_3$$
$$Z(u, v) = c_1 u + c_2 v + c_3$$

$(a_i, b_i, c_i \in \mathbb{R})$ then the point $(X(u, v), Y(u, v), Z(u, v))$ is in the searched
intersection curve if there exists $(s, t) \in \mathbb{R}^2$ such that

$$(X(u, v) - f(s, t))^2 + (Y(u, v) - g(s, t))^2$$
$$+ (Z(u, v) - h(s, t))^2 - d^2 = 0 \qquad (1)$$
$$f_s(s, t)(X(u, v) - f(s, t)) + g_s(s, t)(Y(u, v) - g(s, t))$$
$$+ h_s(s, t)(Z(u, v) - h(s, t)) = 0 \qquad (2)$$
$$f_t(s, t)(X(u, v) - f(s, t)) + g_t(s, t)(Y(u, v) - g(s, t))$$
$$+ h_t(s, t)(Z(u, v) - h(s, t)) = 0 \qquad (3)$$

Note that the considered points in the offset of $\mathcal{S}$ to distance d belong to
one of the two components of the offset defined depending on which normal
direction to $\mathcal{S}$ is taken.

The linearity of u and v in (2) and (3),

$$f_s(s, t)(a_1 u + a_2 v + a_3 - f(s, t)) + g_s(s, t)(b_1 u + b_2 v + b_3 - g(s, t))$$
$$+ h_s(s, t)(c_1 u + c_2 v + c_3 - h(s, t)) = 0$$
$$f_t(s, t)(a_1 u + a_2 v + a_3 - f(s, t)) + g_t(s, t)(b_1 u + b_2 v + b_3 - g(s, t))$$
$$+ h_t(s, t)(c_1 u + c_2 v + c_3 - h(s, t)) = 0$$

allows to describe them in terms of s and t:

$$u = U(s, t) \quad v = V(s, t).$$

Replacing these expressions into (1) the following equation is obtained:

$$W(s, t) = (a_1 U(s, t) + a_2 V(s, t) + a_3 - f(s, t))^2$$
$$+ (b_1 U(s, t) + b_2 V(s, t) + b_3 - g(s, t))^2 \qquad (4)$$
$$+ (c_1 U(s, t) + c_2 V(s, t) + c_3 - h(s, t))^2 - d^2 = 0$$

representing the set of points in the st–domain whose lifting to $\mathcal{S}$ provides
a curve $\mathcal{C}$ providing the intersection curve between Π and the offset of $\mathcal{S}$
to distance d. This is done by merely evaluating these points into $U(s, t)$
and $V(s, t)$, providing the plane parameters whose evaluation at the plane
parameterization $(X(u, v), Y(u, v), Z(u, v))$ produces the points on the plane

corresponding to the searched intersection curve. Note that in this way the intersection curve between the offset and the plane is determined without computing any explicit description or approximation of the offset surface.

If the normal vector to the surface is contained in the sectioning plane, then we cannot describe u and v as an expression of s and t. But, in this case the searched intersection curve between the offset to the surface and the plane can be determined by computing the offset to distance d of the intersection curve between the surface and the plane, which is an easier problem.

Difficulties can arise when analyzing the curve defined by $W(s,t) = 0$. An exact computation requires a big amount of time due the way singularities like critical or isolated points are handled. The algorithm introduced in [GN02] to compute in a very efficient way the topology of $W(s,t) = 0$ will be used for determining the points of the curve $W(s,t) = 0$. This will be the main tool to get in an easy and fast way the topology or shape of the curve $W(s,t) = 0$ in the st domain, providing the searched intersection curve between Π and the offset of S to distance d.

3 Implementation and Experimentation

The first step of the algorithm manipulates equations (1), (2), and (3) in order to get the resulting expression (4). This is a completely symbolic task. Second step computes the topological graph of $W(s,t) = 0$. Itself constitutes a seminumerical procedure as described in [GN02]. There is described that if it is required, coordinate changes are carried out on $W(s,t) = 0$ until obtain the curve in generic position.

Third step concerns with the numerical integration from the outgoing information from the topological graph $W(s,t) = 0$, which is done by using a Runge–Kutta like method on

$$\frac{W_s(s,t)}{W_t(s,t)} + \frac{dt(s)}{ds} = 0$$

This is a very fast and simple procedure because all the points inside of the integration interval are free of singularities of any kind or vertical tangent points (thanks to the performed topological analysis). In the forth step, the obtained points (note that if coordinate changes were done at the second step, they must me reversed here) are substituted into the expressions of $u = U(s,t)$ and $v = V(s,t)$ and the outgoing information is used to find the points $(X(u,v), Y(u,v), Z(u,v)$, that is, the points of the searched intersection curve.

The above sketched algorithm has already been implemented in the computer algebra system Maple. Next, one example is presented and Table 1 shows the computing time and how it is distributed between the topological graph computation, the integration process, and the required manipulations needed to get the offset intersection curve.

Table 1. Computing time (in seconds)

Topology of $W(s,t) = 0$	Integration	Polynomial and point manipulations
2.715	4.993	2.425

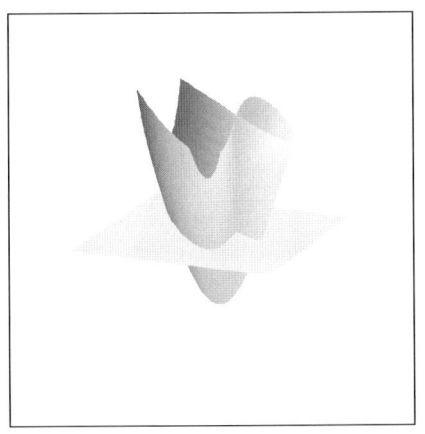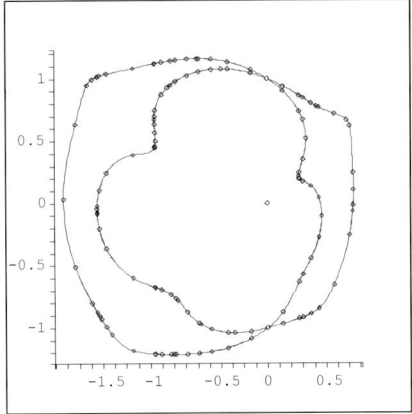

Fig. 1. The surface $\mathcal{S}$ and the plane Π (*left*). The curve $W(s,t) = 0$ (*right*)

Example 1. Let $\mathcal{S}$ be the surface defined by the parameterization

$$x = st + 1, \ y = st + s, \ z = s^2 + t^2 + s,$$

Π the plane with equation $z = 1$ and $d = 1$. For this case $W(s,t)$ is (see (4)):

$$\begin{aligned}
W(s,t) = {} & 20\,s^3t + 19\,s^2t^4 + 38\,s^3t^2 - 10\,s^2t^2 - 8\,st^3 + 8\,st^4 + 8\,s^2t^5 - 8\,s^3t^3 \\
& - 20\,s^5t - 8\,s^6t + 10\,s^4t^2 - 24\,s^3t^4 - 8\,s^4t^3 - 16\,s^4t^4 - 8\,s^5t^2 \\
& + 12\,st^5 + 8\,st^6 + 2\,s^3 - 16\,st^2 + 8\,t^3 + 8\,t^7 + 8\,t^8 + 24\,s^7 + 8\,s^8 \\
& - 16\,t^5 - 12\,t^6 - 18\,s^5 + 11\,s^6 + 2\,s^2 + 4\,t^2 - 4\,st - 11\,s^4.
\end{aligned}$$

Figure 1 shows the surface $\mathcal{S}$ and the plane Π together with the curve $W(s,t) = 0$.

The curve $W(s,t) = 0$ contains two self-intersection points and one isolated point. This isolated point, $(0,0)$ in the (s,t) domain, is an extraneous point due to its singularity since the considered surface has no normal vector at the corresponding point: $(1,0,0)$.

Figure 2 shows the topology of $W(s,t) = 0$ together with the inner and outer offset sections. The different segments in the graph providing $W(s,t) = 0$ are those guiding the integration step since they are free of singularities of any kind. Note that the self-intersection points of $W(s,t) = 0$ correspond necessarily to self-intersection points of the intersection curve but they can come from the intersection between the inner and outer offset sections.

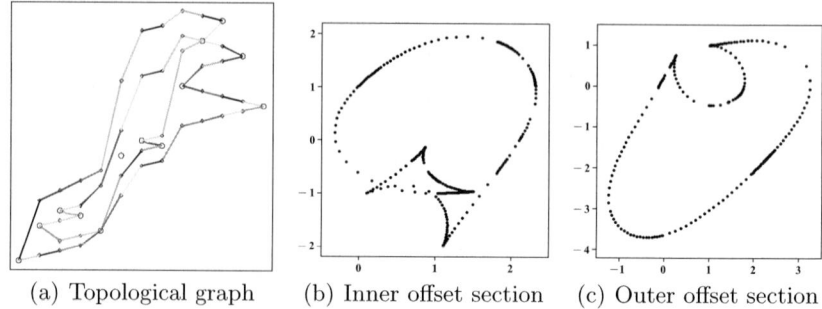

(a) Topological graph   (b) Inner offset section   (c) Outer offset section

Fig. 2. Topology of $W(s, t) = 0$ and inner and outer offset sections

4 Conclusions and Further Work

The introduced algorithm to compute the intersection curve between a plane and the offset of a parametric surface to distance d mixes symbolic and numerical techniques in order to guarantee that the final result is topologically reliable and that no components are missed.

The consideration of implicit surfaces, instead of parametric ones, is in progress and the intersection of offsets with other simple surfaces, instead of planes, is currently being analyzed by using a similar approach to the one introduced here.

Acknowledgements

Partially supported by the spanish Ministerio de Educacion y Ciencia grant MTM2005-08690-C02-02.

References

[GN02]   L. Gonzalez-Vega and I. Necula: *Efficient topology determination of implicitly defined algebraic plane curves.* Computer Aided Geometric Design 19, 719–743, 2002.

[Ho90]   C. M. Hoffmann: *Algebraic and numerical techniques for offsets and blends.* In *Computation of curves and surfaces*, 499–528, Kluwer Acad. Publ., 1990.

[Ma99]   T. Maekawa: *An overview to offset curves and surfaces.* Computer–Aided Design 31, 165–173, 1999.

[Sa02]   M. Sabin: *Interrogation of subdivision surfaces.* In *The Handbook of Computer Aided Geometric Design* (G. Farin, J. Hoschek eds.), North-Holland, 327–342, 2002.

Approximating Offsets of Surfaces by using the Support Function Representation

Jens Gravesen[1], Bert Jüttler[2], and Zbyněk Šír[2]

[1] Technical University of Denmark
   j.gravesen@mat.dtu.dk
[2] Kepler University Linz, Austria
   {zbynek.sir|bert.juettler}@jku.at

Summary. The support function (SF) representation of surfaces is useful for analyzing curvatures and for representing offset surfaces. After reviewing basic properties of the SF representation, we discuss several techniques for approximating the SF of a given surface.

1 Introduction

Robust and efficient methods for dealing with offset curves and surfaces are one of the major challenges in computer aided design. Offset to (piecewise) rational curves and surfaces (i.e., NURBS) are not rational and need to be approximated. Also, singularities and self-intersections can easily be generated and have to be dealt with [Mae].

Certain subsets of the set of rational curves and surfaces are closed under offsetting, or even under the (more general) convolution operator [PP]. In particular, such subsets can be obtained by using the support function (SF) representation, where the support functions vary in the space of polynomials [SGJ1]. The SF representation is one of the classical tools in the field of convex geometry, see e.g., [Gro]. Its application to problems in computer aided design can be traced back to a classical paper of Sabin [Sab]. It does not only provide computational advantages for dealing with offsets, but also leads to particularly simple expressions for quantities and mappings governing the differential geometry of surfaces.

2 Support Function Representation of Surfaces

For any smooth surface Σ in three-dimensional space, the so-called *Gauss map* $\gamma : \Sigma \to \mathbb{S}^2$ assigns to each point $\mathbf{x} \in \Sigma$ the associated unit normal $\mathbf{n}(\mathbf{x})$, which is identified with a point on the unit sphere, cf. Fig. 1. It can be

720   J. Gravesen et al.

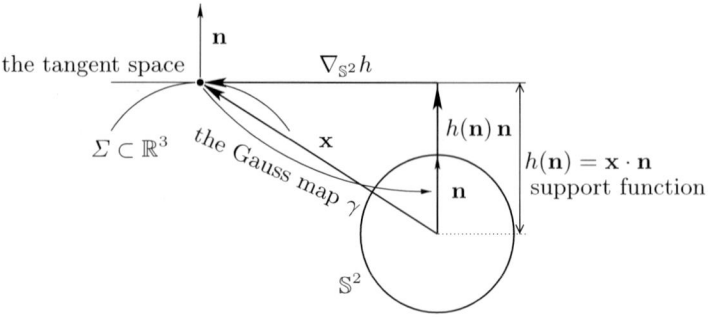

Fig. 1. The Gauss map and the SF of a surface

used to analyze the curvature of the surface. In particular, the Weingarten map equals $-d\gamma$ and the principal curvatures and principal directions are its eigenvalues and eigenvectors, respectively. The Gaussian curvature is the product of the principal curvatures, i.e., the determinant of the Weingarten map. So if the Gaussian curvature does not vanish, then the Weingarten map is invertible and Gauss map is locally invertible.

Consequently, any surface with nonvanishing Gaussian curvature can locally be described by its inverse Gauss map. Since the Gauss map is geometrically significant, many geometric constructions simplify if its inverse is explicitly known. The function

$$h_0 : \Sigma \to \mathbb{R} : \mathbf{x} \mapsto \mathbf{x} \cdot \mathbf{n}(\mathbf{x}) \tag{1}$$

associates with each point the distance of its tangent plane to the origin. The *support function* (SF) $h : \mathbb{S}^2 \to \mathbb{R}$ is then obtained by composing this function with the inverse Gauss map, $h = \gamma^{-1} \circ h_0$. Under certain technical assumptions, the surface can be reconstructed from its SF (cf. [Gra, SGJ1]):

Theorem 1. *Let U be an open subset of the unit sphere and $h \in C^k(U, \mathbb{R})$, where $k > 2$. Define $\mathbf{x}_h \in C^{k-1}(U, \mathbb{R}^3)$ by*

$$\mathbf{x}_h(\mathbf{n}) = h(\mathbf{n})\mathbf{n} + \nabla_{\mathbb{S}^2} h|_{\mathbf{n}} , \tag{2}$$

where $\nabla_{\mathbb{S}^2}$ denotes the intrinsic gradient. If $\det(\mathrm{Hess}_{\mathbb{S}^2}(h) + h\,\mathrm{id})$ does not vanish in U, where $\mathrm{Hess}_{\mathbb{S}^2}(h)$ denotes the intrinsic Hessian of h, then

1. *The image $\mathbf{x}_h(U)$ is a C^k-surface and its SF is h.*
2. *The Weingarten map of the surface is $-(\mathrm{Hess}_{\mathbb{S}^2}(h) + h\,\mathrm{id})^{-1}$.*
3. *If λ is an eigenvalue of $\mathrm{Hess}_{\mathbb{S}^2}(h)$ and $\mathbf{e}$ the associated eigenvector, then $-1/(h+\lambda)$ is a principal curvature and $\mathbf{e}$ is a principal curvature direction.*
4. *The Gaussian and the mean curvatures are*

$$K = \frac{1}{\det(\mathrm{Hess}_{\mathbb{S}^2}(h) + h\,\mathrm{id})}, \quad M = \frac{-\operatorname{tr}(\mathrm{Hess}_{\mathbb{S}^2}(h) + h\,\mathrm{id})}{2\det(\mathrm{Hess}_{\mathbb{S}^2}(h) + h\,\mathrm{id})} \tag{3}$$

5. *Point-wise the absolute value of h and the norms of its gradient and $\mathbf{x}_h$ are related by*

$$\|\mathbf{x}_h(\mathbf{n})\|^2 = h(\mathbf{n})^2 + \|\nabla_{S^2} h(\mathbf{n})\|^2, \tag{4}$$

6. *The L^2 norms of h and $\mathbf{x}_h$ are related by $\|\mathbf{x}_h\|_2^2 = \|h\|_2^2 + \|\nabla_{S^2} h\|_2^2$.*
7. *The maximum norms satisfy $\|\mathbf{x}_h\|_\infty^2 \le \|h\|_\infty^2 + \|\nabla_{S^2} h\|_\infty^2$. In particular, if $U = S^2$ and the surface $\mathbf{x}_h$ is regular everywhere, then this inequality becomes an equation.*

The SF of a surface behaves nicely under geometrical transformations. Translation and offsetting correspond to adding linear and constant functions, respectively, while rotations have to be composed with h. Consequently, the maximum allowed offsetting distance that does not introduce self-intersections or singularities can be computed by analyzing the eigenvalues of the Hessian.

Note that the mapping $h \to \mathbf{x}_h$ is linear; it introduces an isomorphism between the linear spaces $C^k(U, \mathbb{R})$ and its images, where the addition in the image spaces is given by the so-called convolution (in the sense of [SPJ]) of surfaces, see [SGJ2].

The linearity implies in particular that the norm estimates above are invariant under offsetting.

If $k = 1$, then the Hessian cannot be used to analyze the regularity. However, if h is globally C^1 and piecewise C^2 and the sign of $\det(\mathrm{Hess}_{S^2}(h) + h\,\mathrm{id})$ is the same on each patch, then the surface is of class C^1, see [Gra, SGJ1].

3 Approximation of Surfaces

According to results 6 and 7 of the theorem, we can translate questions concerning approximation of surfaces with nonvanishing Gaussian curvature to questions concerning the approximation of scalar fields on S^2, cf. [ANS].

Approximation by Harmonic Expansions. If we consider a surface whose support is either defined or can smoothly be extended to S^2, then it is possible to apply the tools from harmonic analysis. Note that the harmonic expansion leads to rational surfaces with rational offsets. Indeed, by composing the harmonic expansion with a rational parameterization of the sphere, (2) gives a rational parametric representation, which complies with the CAD standard.

This applies immediately to closed convex surfaces, which are studied in convex geometry (see Example 22 of [SGJ1]). Here we present a nonconvex one. We consider a one-sheeted hyperboloid of revolution with the support function $h_0 = \sqrt{x^2 + y^2 - z^2}$. In order to approximate this surface and its offsets, we restrict h_0 to the sphere zone $|z| \le \frac{1}{2}\sqrt{2} - \epsilon$, where ϵ is a small constant, and extend the restriction to a function $h^* \in C^3(S^2, \mathbb{R})$. The results are shown in Fig. 2.

Approximation by piecewise linear functions. Another very interesting way to approximate the SF h is by using a piecewise linear function $\overline{h}$ defined over a triangulation of (a part of) the unit sphere. Each vertex $\mathbf{n}_i$ defines the plane

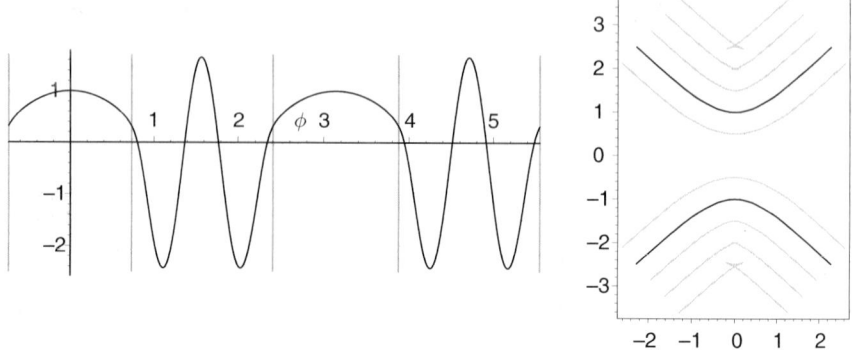

Fig. 2. Support function of a nonconvex surface of revolution (the concave region between the vertical grey bars) and its C^3 smooth extension (*left*). Approximation of the surface and of its offsets (*right*). In both cases, only the intersections with the plane $y = 0$ are shown, and the support function is parameterized by the angle

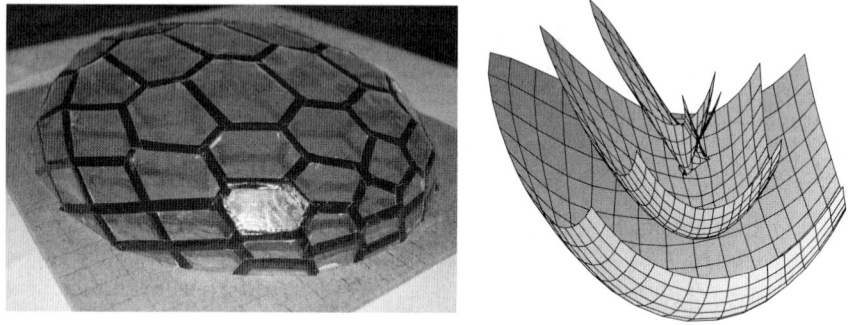

Fig. 3. Approximations constructed via the SF

$\mathbf{x} \cdot \mathbf{n}_i = h_i = h(\mathbf{n_i})$ in $\mathbb{R}^3$. Each triangle defines a point $\mathbf{v} \in \mathbb{R}^3$ where the linear function $\mathbf{v} \cdot \mathbf{n}$ interpolates the values of the SF in the corners of the triangle. Clearly $\mathbf{v}$ is the point of intersection between the three planes defined by the corners of the triangle.

The triangles around a vertex define a polygon in the plane defined by the vertex. We obtain a graph embedded in $\mathbb{R}^3$ with planar faces, which is the dual to the triangulation. Figure 3 left shows a photograph of a physical model of a surface with planar faces approximating half of an ellipsoid. The technique can be applied to nonconvex surfaces too, see [SGJ1].

Note that the planar faces may have self-intersections ("swallowtails"). In order to avoid these problems, the spherical triangulation may have to be modified by "edge flipping."

Least-squares fitting. In many cases the SF is not (explicitly) available and only a surface patch or point cloud may be given. For these cases we propose the following approximation scheme, which is to be applied to a given surface represented by sample points $\mathbf{X}_i$, possibly with associated normals $\mathbf{n}_i$.

1. Sample points $\mathbf{X}_i$ and associated unit normals $\mathbf{n}_i$ from the patch. If the points $\mathbf{X}_i$ are the input, then estimate the normal $\mathbf{n}_i$ (e.g., based on local planes of regression).
2. Consider a suitable[1] finite-dimensional space $\mathcal{H}$ of support functions with basis h_j.
3. Find the SF $h = \sum_j \alpha_j h_j$ such that the associated surface $\mathbf{x}_h$ approximates the data in the least-squares sense, by minimizing the objective function

$$\sum_{i=1}^{N} \left(\left(\mathbf{X}_i \cdot \mathbf{n}_i - \sum_j \alpha_j h_j(\mathbf{n}_i) \right)^2 + \left\| \mathbf{X}_i - (\mathbf{X}_i \cdot \mathbf{n}_i)\mathbf{n}_i - \sum_j \alpha_j \nabla_{\mathbb{S}^2} h_j \mid_{\mathbf{n}_i} \right\|^2 \right).$$

As an example, we approximated the support function of a biquadratic patch by a support function of degree 9, see Fig. 3, right. In this case, 256 sample points were used in order to define the objective function. In the same picture two offsets are also depicted and it is an important fact that they are approximated by exactly the same precision as the surface itself.

References

[ANS] Alfeld, P., Neamtu, M., Schumaker, L. L., Fitting scattered data on sphere-like surfaces using spherical splines, *J. Comput. Appl. Math.*, 73, 5–43 (1996).

[Gra] Gravesen, J.: Surfaces Parametrised by the Normals. Computing, to appear.

[Gro] Groemer, H. *Geometric Applications of Fourier Series and Spherical Harmonics* Cambridge University Press, Cambridge, 1996.

[Mae] Maekawa, T., An overview of offset curves and surfaces, Comput.-Aided Des. **31**, 165-173 (1999).

[PP] Peternell, M., and Pottmann, H., A Laguerre geometric approach to rational offsets, *Computer Aided Geometric Design* **15**, 223–249 (1998).

[Sab] Sabin, M.: A Class of Surfaces Closed under Five Important Geometric Operations, Technical Report, British Aircraft Corporation Ltd. (1974)

[SPJ] Sampoli, M.L., Peternell, M., Jüttler, B.: Rational surfaces with linear normals and their convolutions with rational surfaces, Comput. Aided Geom. Design **23**, 179–192 (2006)

[SGJ1] Šír, Z., Gravesen, J., Jüttler, B.: Curves and surfaces represented by polynomial support functions, submitted. Available as SFB report at www.sfb013.uni-linz.ac.at.

[SGJ2] Šír, Z., Gravesen, J., Jüttler, B.: Computing Minkowski sums via Support Function Representation, submitted, available as FSP report at www.ig.jku.at

---

[1] In order to be invariant with respect to translations and offsetting this space should contain all polynomials of degree 1.

Semantic Modelling for Styling and Design

C.E. Catalano[1], V. Cheutet[2], F. Giannini[1], B. Falcidieno[1], and J.C. Leon[2]

[1] Istituto per la Matematica e Tecnologie Informatiche, CNR IMATI-Ge
{catalano, giannini, falcidieno}@ge.imati.cnr.it
[2] Laboratory Soils, Solids and Structures, Integrated Design Project,
Grenoble, France
{vincent.cheutet, jean-claude.leon}@hmg.inpg.fr

Summary. Starting from the modelling requirements of the early design phase of the product development, the paper will show a possible strategy to overcome some limitations of current CAS/CAD systems. In fact, the styling stage involves both technical knowledge and fuzzy and dynamic aspects, which have to be taken into account for a proper management. The paper focuses on high-level modelling tools developed to deform surfaces with semantic (aesthetic) constraints, i.e. the crucial design elements for the stylist. Furthermore, the communication among the other actors of the design process is consequently facilitated.

1 Introduction

The styling stage of the product development process differs from the downstream product design activities since fixing the aesthetics of an object involves not only a technical knowledge but also fuzzy and dynamic aspects, such as creativity and subjectivity. Unfortunately, currently available CAS/CAD systems address mainly the geometry of the product shape, while do not support the direct management of the aesthetic knowledge [Pie05].

Our research activity aims at developing modelling tools able to act directly on the design elements relevant to the perception of the object, the ones stylists have in mind when creating a new product. In this way, the modelling phase takes directly into account the semantics of the context, thus facilitating also the communication among the different actors of the design process.

To enrich geometric models with semantics, the strategy adopted in mechanical CAD has been designed by *features*, where features are groups of geometric entities treated as a single unit, significant in the application context [SM95]. Due to the nature of styling products, an analogous concept to form features should be related to the shape of the object, where the shape features influencing the customer's eye have to be characterised. The identification of such entities and their incorporation into a high-level semantic modelling process have been the goal of the European projects FIORES and FIORES II [FIOR].

From the projects, we learnt that there are two main aspects to take into account when proposing innovative CAS/CAD tools for the conceptual design phase. The first one is that stylists generally use 2D curves in their sketch to give a certain character to the product designed. Moreover, also when the first digital model is created directly in 3D, curves have a leading role during the modelling phase. This induces to consider a curve-driven methodology as the most appropriate for shaping an object. Product style semantics is expressed through a special use of such characteristic curves, which we called *aesthetic key lines*. In fact, from the geometrical point of view, such curves will be the complex entities constraining the deformation process together with few side numerical parameters.

The second aspect to consider is that the early phase of the design is dominated by uncertainty. A modelling tool supporting sketching should incorporate the possibility not to constrain the shape univocally, possibly producing undesired effects, but giving some freedom in the areas with less visual impact, e.g. at the boundary of a local modification. In this context the most central issue is the visual perception of the object as a whole rather than its precise geometry, which is instead needed in the subsequent phases of the development process.

In the paper, ad-hoc feature-based modelling approaches have been outlined, working on NURBS and subdivision surfaces, respectively. In fact, while NURBS are the standard in CAS/CAD modelling, subdivision surfaces can be considered as a valid alternative geometric representation. Therefore, two different methods have been developed starting from the same feature conceptualisation. In Sect. 2 the *Fully Freeform Deformation Features* ($\delta - F^4$) applied to NURBS will be described, while in Sect. 3 *Sweep-like Features* acting on subdivision surfaces will be treated. Finally, Sect. 4 outlines the semantic environment we have been developing more recently and concludes the paper.

2 Fully Freeform Deformation Features

Since in the styling activity shapes are generally arbitrary and different alternatives are often needed, the concept of $\delta - F^4$, *Fully Free-Form Deformation Features* has been detailed [PF*05], by defining a styling feature as a subset of the shape having an aesthetic meaning and obtained through a set of surface deformation operations guided by the aesthetic key lines drawn by the designer [FGM00].

Implemented through a deformation technique applicable not only to the standard NURBS representations but also to tessellated representations, such styling features establish a link between the geometric and semantic level, thus making the integration of the stylist's intent during the whole design process easier. The adopted deformation approach is based on the force density method and applied to a bar network coupled with the control network of the NURBS surface. The bar network used is made by nodes (coincident with

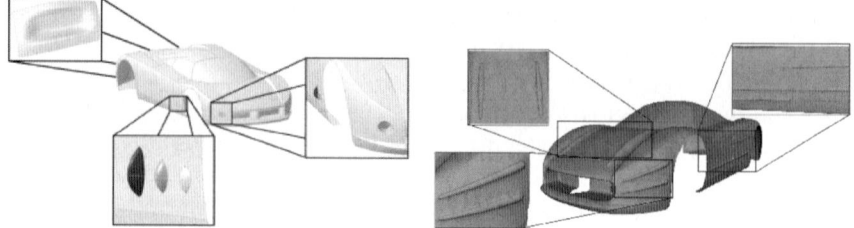

Fig. 1. (*Left*) Fully freeform deformation features, and (*Right*) Sweep-like features inserted on a car model

the surface control points) and bars joining the nodes to which a force is applied, determining the equilibrium status. Due to the association between the mechanical model and the surface, finding the correct position of the surface control points coincides with finding the new equilibrium configuration of the bar network depending on the given curve constraints.

To be suited for a sketching environment, the shape features generated by the designer can be constrained also with numerical parameters, e.g. dimension, relative positions, as well as tangency conditions if higher-order accuracy is needed. On the other hand, the designer's input may also introduce some inaccuracies and inconstancies because the mental shape perception is not always consistent with the input parameters monitoring the shape feature generation. Therefore, the capability of tuning the shape in accordance with the user's intent through a progressive shape refinement process becomes very helpful.

Moreover, users can also prescribe a predefined behavior corresponding to a planar area and introduce sharp edges. Finally, they can indicate a tendency for the surface, thanks to different minimisation criteria for the geometric and mechanical parameters used [CC*05]. In Fig. 1 (left) the insertion of a pattern of $\delta - F^4$ and planar areas are shown on a car model.

3 Sweep-Like Features

The second application of the concept of aesthetic feature to 3D product models has been provided for subdivision surfaces. It is clear that the integration among the different phases is a key issue for the optimisation of the whole development process, and subdivision surfaces have been gaining attention as an alternative geometric representation moving to this direction [CS*02,BRB05]. Moreover, several activities – such as reverse engineering, rapid prototyping and tooling, FEM analysis, virtual inspection and navigation – require a discrete model. Certainly, suitable and effective modelling tools are needed in all the phases where a continuous model is adopted such as styling.

Among the *detail features* defined in [FGM00], *Sweep-like features* have been implemented [Cat05]. Such a class of features can be obtained by performing a sweep operation of a given profile *s* (*section*), possibly varying in the size, along a specific curve *d* (*directrix*).

The semantic parameters of this class of features are the two driving curves *s* and *d*; according to the properties and the position of the parameters with respect to the surface, a classification has been formalised. Moreover, the possibility of scaling the section arbitrarily along the directrix have been given to users to support free shaping, while a friendly interface assists the intuitive insertion of the features.

The algorithm elaborates the parameters of the feature in order to create the feature surface and adjusts the shape according to the evolution of the section along the directrix and the behaviour of the surface in the area affected by the feature. Then, a local remeshing is performed to merge the feature surface with the reference model. The Catmull–Clark scheme has been adopted in the insertion algorithm of the sweep-like features, since it is an extension of cubic B-Splines: the initial tessellation is quadrangular almost everywhere, similarly to a NURBS control polyhedron, and it converges to a bi-cubic at the limit. In this way, the new geometry can be understood and manipulated by designers in an easier way.

The algorithm does not work directly on the dense final mesh, but on a coarser version, so that the computation results much faster. After the insertion, the refinement rules lead to a very smooth mesh together with the new feature added. Also in this case, the visual effect of the product aesthetics is more important than the precision of the geometry. In Fig. 1 (right) different types of sweep-like features and patterns of sweep-like features have been inserted in the same car model as in Fig. 1 (left), after a conversion from a trimmed NURBS surface to a subdivision one.

4 Conclusions

In this paper, the authors summarised part of their research activity on semantic modelling tools for industrial design, particularly the ones related to the aesthetic features.

On the other hand, the increasing demand for accessing and sharing digital shapes enhances the needs of structuring the shape and design knowledge at any step of the design workflow, thus making a mapping process among the various stages also possible. The most recent activity goes in this direction [CC*06], devising an ontology to formalise the knowledge embedded in car styling. It also provides the basic framework of a design environment for 2D digital sketches and 3D digital shapes in which the traditional modelling systems may be completed by semantic-based and context-aware tools; in this way, stylists and engineers are allowed to create and manipulate shapes more intuitively. As a natural future activity, a semantic modeller handling the aesthetic key lines and features will be coupled with this aesthetic environment, integrating the geometric representation inside the ontology.

Acknowledgements

The work is being carried out in the scope of the activity of AIM@SHAPE Network of Excellence supported by the European Commission, VI Framework, IST Contract N. 506766.

References

[SM95]   Shah, J.J., Mantyla M.: Parametric and feature-based CAD/CAM. Wiley-Interscience Publication, John Wiley & Sons, Inc. (1995)

[CC*06]   Cheutet, V., Catalano C.E., Giannini, F., Monti, M., Falcidieno, B., Leon, J.C.: Semantic-based environment for aesthetic design. In: Proceedings of TMCE06, Ljubljana, 185–196 (2006)

[CC*05]   Cheutet, V., Catalano, C.E., Pernot, J.P., Falcidieno, B., Giannini, F., Leon, J.C.: 3D Sketching for Aesthetic Design using Fully Free Form Deformation Features. In: Computers & Graphics, Elsevier, **29-6**, 916–930 (2005)

[Pie05]   Piegl, L.A.: Ten challenges in computer-aided design. In: Computer Aided Design Journal, **37-4**, 461–470 (2005)

[Cat05]   Catalano, C.E.: Introducing design intent in discrete surface modelling. In: International Journal of Computer Applications in Technology (IJCAT). Interscience Publishers, **23-2/3/4**, 108–119 (2005)

[PF*05]   Pernot, J.P., Falcidieno, B., Giannini, F., Leon, J.C.: Fully Free Form Deformation Features for aesthetic shapes design. In: Journal of Engineering Design, **16-2**, 115–133 (2005)

[FGM00]   Fontana, M., Giannini, F., Meirana, M.: Free Form Features for Aesthetic Design. In: Int. Journal of Shape Modeling, **6-2**, 273–302 (2000)

[CS*02]   Cirak, F., Scott, M.J., Antonsson, E.K., Ortiz, M., Schröder, P.: Integrated Modelling, finite-element analysis, and engineering design for thin-shell structures using subdivision. In: Computer-Aided Design, **34**, 137–148 (2002)

[BRB05]   Boier-Martin, I., Ronfard, R., Bernardini, F.: Detail-Preserving Variational Surface Design with Multiresolution Constraints. In: JCISE, **5-2**, 104–110 (2005)

[FIOR]   http://www.fiores.com/home.html

Minisymposium "Web-based Learning Environments in Applied Mathematics"

Matti Heiliö

Lappeenranta University of Technology
matti.heilio@lut.fi

Modeling and mathematical technologies are a vital resource for R&D and innovation in Europe. Ingenious exploitation of mathematics means opportunity to achieve competitive edge in effective design process, accelerate test cycles, support systems integration schemes, redesign production models.

Virtual technologies and digital educational environments are a viable media to facilitate innovative intellectual processes and organize knowledge transfer. It enables learning, brings solutions to training and educational processes, helps to facilitate distributed and concurrent planning and development processes, share software, and provide remote access to software libraries.

An evolution of educational methods, materials, and means of delivery is taking place. Traditional textbook will in some cases be replaced by an interactive cross-media environment. Advantages are easy access and portability, flexible updates, dynamic edition and the benefits of media technology, hypertext properties, links and navigation, animations, interactive exercises. This creates also challenge for educational research on learning processes.

In this Minisymposium we discussed the challenge of web based solutions in organizing education in modeling and applied mathematics. We shared experience, described examples of virtual courses and technologies used for web based delivery and publication of interactive documents.

Knowledge in industrial mathematics is dispersed in small nodes. There is an obvious need for collaboration, knowledge sharing, and retrieval from the scattered pockets of expertise in Europe. The long term goal would be a service portal containing an integrated network of software libraries, menu of up-to-date educational products, training environments, and flexible network services.

A vision of the future is a Netcampus of educational services, a menu of profession level courses built on the concepts of current eLearning technologies. Flexible access to an integrated library of contemporary and proto-level scientific software in modeling, numerical methods, and scientific computing.

Many interesting pathfinder projects are underway in the community of applied mathematics in Europe. Some of those projects are presented in this section. The time seems to be mature for launching a more systematic European collaboration. We aim at a collaborative project involving several universities who will develop forefront technology in building digital environment for applied mathematics. Based on an up-to-date assessment of user needs in universities and industry such a consortium could produce important added value for the academic and industrial community.

Included are the following contributions. Simona Runci described an integrated framework for production of interactive documents and distant education materials. Giuseppe Ali described "An e-learning system for applications of mathematics to microelectronic industry." Matti Heilio presented a "Web-based system for graduate studies – optimization, games, and markets," describing a virtual learning environment developed by professor Mockus, consisting of several course modules, interactive features, case examples, etc. Finally a web based course and learning module "Web-tool on Differential Equations" by Peep Miidla was presented.

An Industrial Application of an Integrated Framework for Production of Interactive Documents

G.M. Grasso[1], C.L.R. Milazzo[2], and S. Runci[3]

[1] Department of Physics, University of Messina
`ggrasso@informatica.unime.it`
[2] Department of Mathematics and Computer Science, University of Catania
`cmilazzo@dmi.unict.it`
[3] Catania Research Consortium
`simona.runci@yahoo.it`

Summary. In this paper we will show an industrial application of a new framework, called LaTeX2WEB, which translates LaTeX material into an interactive Web-based document. The more important characteristic of LaTeX2WEB is the possibility of integrating, in the Web-based document, external programs produced in every languages. We exploited LaTeX2WEB to create an interactive Web-based manual, which illustrates a new software for the multiobjective optimization applied to the parameter extraction in circuit design. Thanks to LaTeX2WEB it was possible to simulate the algorithms written in C, C++, and FORTRAN, used in the multiobjective optimization software.

1 Introduction

Today, most universities and an increasing number of companies and industries all over the world feel the need to offer virtual Web-based courses to educate their students or employees. The main reason of this trend is the suitability of Web for publishing material of educational nature at very low cost and with a high degree of reachability. Due to the fact that setting up virtual Web-based courses has become very popular among teachers and educators, many of them are in search of a complete instrument to create virtual and interactive documents for the Web from text based scripts and books, used for traditional lectures.

Hyper Text Markup Language (HTML) is a predominant markup language for the creation of web pages. It provides a means to describe the structure of text-based information in a document and to supplement texts with interactive forms, embedded images, and other objects. HTML can also describe, to some degree, the appearance and semantics of a document, and can provide additional cues, such as embedded scripting language code, that can affect the behavior of web browsers and other HTML processors. HTML's exact rendering is not specified by the document that is published but is, to some degree, left to the discretion of the browser. It recognizes that the window size, resolution, or shape on which a document is viewed will vary from reader to reader, and that therefore layout, font size, and other choices for good readability should be at least partly up to the reader, not the author. The result is that well designed HTML is excellent for browsing and for this reason it is well suited for virtual courses.

LaTeX is a typesetting system which is very popular with computer scientists, engineers, mathematicians, physicists etc. It is especially good for mathematical work, but is also used by many nonscientists. It is well suited for producing electronically publishable documents, and it is capable of extremely detailed page layout, specifying precisely where on the page symbols go. Most scientists utilize LaTeX to produce their research papers and all their teaching materials, but often they are not very familiar with the implementation method of material on the Web. Therefore, they need a simple method to convert their LaTeX documents into HTML, in order to publish it on the Web. Softwares that allow the translation of a LaTeX source document into HTML already exist, and some of them are Latex2Html, HEVEA, and TTH.

Latex2Html is a Perl program that translates LaTeX source code into HTML source code. Latex2Html extends LaTeX by supporting arbitrary hypertext links and symbolic cross-references between evolving remote documents. Latex2Html replicates the basic structure of a LaTeX document as a set of interconnected HTML files which can be traversed like any hypertext document. All of the parts of typical LaTeX documents are translated into their hypertext equivalent, including chapters, sections, formulas, pictures, etc.

TTH translates TeX into HTML. Document structure, using either the Plain or LaTeX macro packages, is also translated and incorporated in the form of hyperlinks. TTH produces more compact, faster viewing, web documents than other converters, because it really translates the equations, instead of converting them to images. The disadvantages of this choice for representing equations are that it depends on having the symbol font accessible on the browser, and that the equation layout is not as compact or elegant as LaTeX's.

HEVEA is a translator whose input language is a fairly complete subset of LaTeX and the output language is HTML. HEVEA translates various math symbols used in LaTeX, almost the entire set of math symbols, including the amssymb ones, are correctly rendered.

A great drawback of all of this LaTeX translators is the fact that they do not include interactivity, which would be a desirable feature for a document available through the Web. In fact, without interactivity the consultation of an HTML document is equivalent to turning over the pages of a book, except it appears on a computer screen.

Recently, a new software called LaTeX2WEB has been realized from our research group. It does not only allows the translation of a document from LaTeX into HTML, but also makes it interactive. This new framework is capable of creating a Web-based document which includes the links to external programs, written in C, C++, MatLAB, Fortran, etc. As a result it is possible to execute them directly on the Web server. It makes use of LaTeX2Html and of the dynamic PHP language, which is a software package installed on web servers to provide scripting capabilities. In the following sections is illustrated this new framework LaTeX2WEB and is shown an industrial application of it.

2 LaTeX2WEB

The LaTeX2WEB has one important charateristic: in a single LaTeX source there are both the source of the traditional printable version of the document and the tags for the automatic generation of the navigation tools and links that produce interactivity. By introducing a few command lines into the LaTeX document it is also possible to create a Web-based document with different learning paths. In this way the students can decide among these different paths, and therefore they can follow a very personalized mode of learning, approach a new subject at the elementary level, and gradually add details, applications and exercises. For this purpose the LaTeX2WEB has an embedded mechanism to define different paths through the same document. To realize this it is sufficient, as previously mentioned, to insert in the LaTeX document, for each section and subsection, tags which define a pointer to the next relevant section. To define paths in the TeX document it is sufficient to insert a tag such as the following:

```
\nextnode{path-name}{next-section-name}\vspace*{-2pt}
```

With this new command, LaTeX2WEB adds a new path to the HTML document every time it encounters a new path-name definition and appends nodes to the same path. Also the introduction of interactivity can be produced with few command lines in the LaTeX source. Reading these commands, LaTeX2WEB provides an integrated way to add links to external programs that can be written in C, C++, MatLAB, Fortran, and any other language, including shell script. To add a link to an external program it is sufficient to write in the LaTeX document some command lines such as the following example:

```
\proglink{esegui_paes}{\progarg{depth}{depth}{6}
\progarg{geni}{genes}{20}\progarg{archivio}{archive}{440}
\progarg{iterazioni}{iterations}{50000}\progarg{pm}{pm}{0.03}
\progarg{seme}{seed}{42344}}{gif}
```

The above example automatically generate the Web form.

This link returns a gif picture as specified in the corresponding proglink tag, see Figure 1. The arguments can be changed by the user interactively, and when the RUN button is pressed the computed results are shown, as can be seen in Fig. 2.

The translation of LaTeX documents into HTML is done by means of the Latex2Html, which is publically available software. This software has many characteristics of LaTeX but moreover it is capable of creating a Web-based

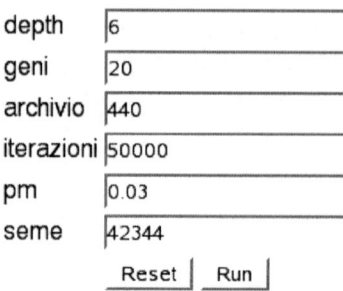

depth	6
geni	20
archivio	440
iterazioni	50000
pm	0.03
seme	42344

Reset | Run

Fig. 1. Web form generated by LaTeX2WEB

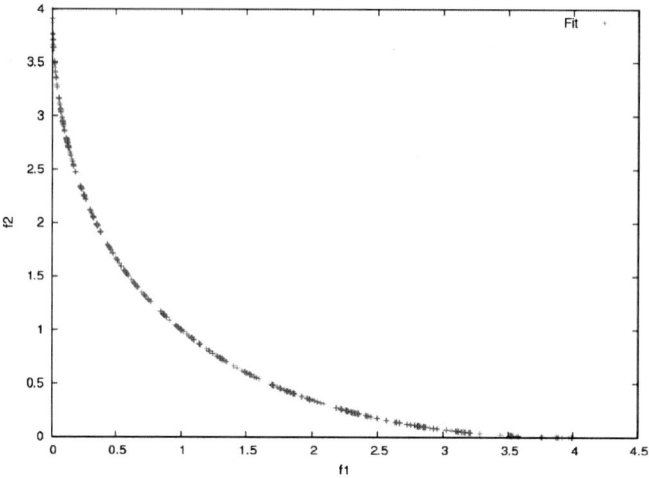

Fig. 2. The graphical plot of the Pareto front of a multiobjective optimization obtained by using a genetic algorithm

document with the same structure as the LaTeX document. All the navigation links are generated automatically on the basis of the initial contents of the LaTeX source. In this way LaTeX2WEB creates not only the links that permit going forwards or backwards, to the beginning or to the end of the document, but also an applet Java which produces a tree of the entire document, and all the links to external programs. Only LaTeX documents and executable files are necessary for the Web-maker. For this reason the usage of LaTeX2WEB is very simple.

3 An Industrial Application of LaTeX2WEB

LaTeX2WEB has been exploited in order to create an interactive manual, which illustrates new software for the parameter extraction in integrated circuit design.

This software is the result of work carried out by the Applied Mathematics research group at the Department of Mathematics and Computer Science at University of Catania. The software is aimed at industrial use. This project has been realized in cooperation with ST-Microelectronics Catania and Consorzio Catania Ricerche.

The software realized by the applied mathematics research group includes a software library of programs for the global multiobjective optimization. The optimization algorithms included in this software library will be used, as previously mentioned, as a support for the parameter extraction for integrated circuits design.

Since this optimization software will be employed by analysts and engineers, the creation of an instruction manual is crucial.

The manual that has been realized with the aid of LATEX2WEB contains the entire documentation related to the structure as well as the functioning of the software. A user who wishes to know how to make use of the optimization software not only has the opportunity to read the manual as a simple book, but also he can try the optimization algorithms on the remote server, using the default parameters or changing them by the special forms produced by LATEX2WEB; in this way any user can choose the algorithm they think is best for their personal optimization problem, with the optimal parameters, and they can do it without interfering with the performance of their computer.

References

[Anile, Grasso] Grasso G. and Anile M. (2001). An Integrated Framework for Web Publication of Interactive Documents. in Proc. World Multiconference on Systemics, Cybernetics and Informatics (SCI2001). Orlando, USA, pp 326-329

An e-Learning Platform for Applications of Mathematics to Microelectronic Industry

G. Alì[1], E. Bilotta[2], L. Gabriele[2], P. Pantano[3], and R. Servidio[2]

[1] IAC, National Research Council, and INFN-Gruppo c. Cosenza
    g.ali@iac.cnr.it
[2] Department of Linguistics, University of Calabria
    bilotta@unical.it, lgabriele@unical.it, servidio@unical.it
[3] Department of Mathematics, University of Calabria
    piepa@unical.it

1 Introduction

The European project CoMSON (Coupled Multiscale Simulation and Optimization in Nanoelectronics) is an FP6 Marie Curie RTN (Research and Training Network) action. This project involves five universities ("Bergische" University of Wuppertal, "Politehnica" University of Bucharest, University of Calabria, University of Catania, TU Eindhoven) and three microelectronics companies, (NXP-Philips, Qimonda, STMicroelectronics).

The key objective of the CoMSON project is to realize an experimental demonstrator platform (DP) in software code (see Fig. 1), which comprises coupled simulation of devices, interconnects, circuits, EM fields, and thermal effects in one single framework [ALI06]. The basis is the development and validation of appropriate mathematical models to describe the coupling of different physical effects, their analysis (well-posedness), and related numerical schemes. The DP will be interfaced with an e-Learning platform (e-LP) for micro- and nanoelectronics. The aim is the education and training of young researchers in mathematics applied to technology, both from a theoretical and a practical viewpoint.

In this paper we describe the main platforms foreseen by the project, concentrating on the concept design of the CoMSON e-LP.

2 The Demonstrator Platform and the e-Learning Platform

The main components of DP (see Fig. 1; for more details, see [DEF06]) are (1) A library of test examples and experimental measurements to be used as benchmarks for any new method. (2) A set of modules consisting each of a collection of functions providing the basic functionality of the single domain

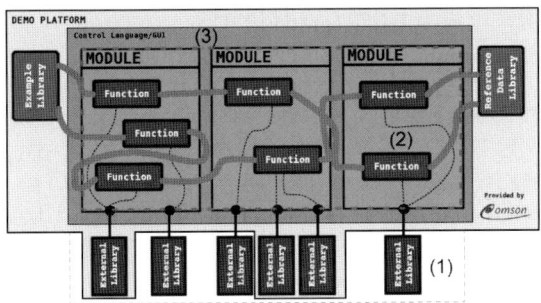

Fig. 1. The demonstrator platform architecture

simulators. (3) A control programming language that enables to connect the aforementioned functions and form simulation algorithms.

From the CoMSON web server (`http://www.comson.org`), maintained at the University of Calabria, is possible to access both to the source code of DP components, which is developed in a CVS (Concurrent Versions System), and to the main development and testing system. The DP documentation is, in part, inside the source code. The rest of the documentation is maintained in the form of LaTeX documents and is shared through the CVS repository. In addition to the CVS server,the DP will be also distributed via live CD/DVD, containing DP and all the libraries it depends on.

DP provides a natural test bench with state-of-the art models and parameters from different domains rather than academic simplifications, which will prove very precious for educating and training young researchers by hands-on experience. For this purpose, this experimental platform will be interfaced with an e-LP, by means of a visual programming language (VPL).

The e-LP will include the principal aspects of the learning-by-doing theory, which emphasizes the active role of the student in building his/her knowledge. The active dimension of learning is realized by means of virtual laboratories, which allow to visualizing (with animation) and manipulating interactively, step by step, metaphoric representations of the functions, modules, and coupling paradigms of DP, for a deeper understanding of them. In particular, the e-learning environment foresees the development of a new generation of educational tools (e.g., 3D visualization, intelligent agents, etc.). This approach is supported by a full integration between virtual tools and remote simulation by DP technology.

The learning materials [MOR04] such as web pages, Acrobat documents, Video, animation, and Java file source code will be included within a repository denominated Learning Content Management System (LCMS) and can be used freely by the students enrolled in the course. The system will allow the teachers to create, register, store, assemble, reuse, and publish digital learning material for delivering by web. Also, the environment will offer different strategies of interaction, such as video-lesson, presentation of contents, practice, feedback, and self-evaluation methodology.

3 Innovative e-Learning Methods

The e-LP environment will allow students to interact by using information and communication technologies. This environment is based on the constructivist paradigm [PAP80, Tuy83], which asserts that learning environments should support multiple perspectives or interpretations of reality, knowledge construction, context-rich, experience-based activities [SAA01]. The constructivist paradigm guides learners to conduct and manage their personalized learning activities, and encourage collaborative and cooperative learning for critical thinking and problem-solving. The knowledge is constructed through interaction with the environment in which a process of personal interpretation of the perceived world and the negotiation of meaning from multiple perspectives takes place.

In our system, learners and teacher can interact with different technologies, which support the students in the acquisition of skill on specific topics. The integration of different tools allows to applying innovative e-learning methods and technologies based on the following aspects:

- Definition and development of educational plans for all researchers, including internal training: using the "CoMSON Virtual Working Place," a web-supported documentation and transfer-of-knowledge system
- Adaptation of the DP to training and educational needs: developing suitable graphical interfaces which highlight coupling paradigms, important modeling issues, algorithmic issues, and all other issues analyzed in the training and educational plans
- Creation of a virtual educational system, which transfers traditional classrooms on electronic environment based on remote access for all system users, direct interaction between students and lecturers/tutors, support to communication among students
- A continuing education environment supplying information about the materials and some general documentation of the platform: annual progress reports on the project, software, on-line lectures, communication tools

These technologies allow to create and to manage courses, and support the students in the learning process. The teacher can design many events, and can be asked for additional support.

4 Design of the e-Learning Platform

In order to realize the e-LP, we have adopted a methodology based on a user-centered approach, starting by the following:

- Identifying the potential users and educational goals of the e-LP
- Designing of the authoring

– Identifying the principal aims of the system (collaborative and communication tools) and possible integration with specific software (tools for the simulation of electronic circuit, etc.)
– Designing preliminary architecture of the e-LP (conceptual model)

To this aim, we have delivered a questionnaire to each responsible of a CoMSON node, to gather information on the required e-LP specifications. The questionnaire consists of six sections: (1) Users of the e-LP; (2) Authoring; (3) Educational aims of the e-LP; (4) General architecture of the e-LP; (5) General characteristics of the e-LP; and (6) Standards. Subsequently, the results of the questionnaire, with the exception of sections (5) and (6), have been discussed jointly by all partners.

From analysis of the answers to the questionnaire, we have identified the main components of the e-LP. The functional specification of the e-LP, which describes the educational contents and the architecture of the environment, are the following ones:

Users of the e-LP. The final users of the e-LP will be "students" in microelectronics, but the system will be usable by microelectronics companies for training employees. At this stage of the project, e-LP users are CoMSON researchers, ERs (Experienced Researchers), and ESRs (Early-Stage Researchers) as "testing people," before making the e-LP available to the more general audience described above.

Authoring. The underlying problems are production of didactic materials; collection of existing materials; standardization of the material (post-production); definition of standards. All CoMSON partners agree on the following points: Each contributing professor can decide whether to take, or not, authoring responsibility. If some contributing professor does not want to take the authoring responsibility, that is fine, but he/she should provide the contributed material in the correct format. The professors will have the responsibility of the written contents (even if researchers will collaborate to write them). The writers will own the copyright of the written documents. CoMSON has to certificate the quality of the contents of the Learning Units, by university standards (certification of quality).

Educational aims of the e-LP. The main aims of the e-LP are Fostering research in Mathematics dedicated to industrial needs; Training to use the main simulation tools in micro- and nanoelectronics; Design Flow. The users' future professional career will be advanced modelling and simulation expert and Designer.

General architecture of the e-LP. The e-LP will be a multiplatform, independent of the operative system of the final user. The operative system will be optimized for RedHat/Linux and Windows XP. The system will offer a friendly graphical user interface (GUI) and will be based on Java language. The didactic contents should be importable by the main e-learning platform used by microelectronic companies, according to the standards of IEEE P1484 and SCORM 1.2 [CHU04]. No specific software is required to be known by the user in advance. The e-LP should provide for tutorials on simulation steps (process, device, circuit, EM, optimization), including related software packages as examples. In general, no prerequisite topics are required to be known by the user, but each learning unit has its own prerequisites. The full list of prerequisite topics is modelling of semiconductor devices; introduction to electrical circuits; electromagnetism; interconnects; Basic numerical analysis; Numerical methods for DAEs. The educational contents have been split in two categories, Basic and Advanced contents. Each content will consist of a minimal number of learning units (modules). Each node will provide the modules on specific topics.

5 Conclusions

In this paper we have presented an e-LP for the European CoMSON project, which includes a collaborative virtual environment to deliver educational content. Also, we have described the methodology used to specify users, authoring, educational aims, and general architecture of e-L. We are currently implementing a prototype of the e-LP, which includes different technologies in order to satisfy the needs of different user groups (University and Industry).

References

[ALI06] Alì, G., Bilotta, E., Gabriele, L., Pantano, P.: An E-Learning Platform for Academy and Industry Networks. In: Proceedings of the Fourth Annual IEEE International Conference on Pervasive Computing and Communications Workshops (PERCOMW'06), IEEE Computer Society, 231–234.(2006)

[CHU04] Chu, C.P., Chang, C.P., Yeh, C.W., Yeh, Y.F.: A Web-service oriented framework for building SCORM compatible learning management systems. In: Proceedings of the International Conference on Information Technology: Coding and Computing (ITCC), 156–161 (2004)

[DEF06] De Falco, C., Denk, G., Schultz, R.: A Demonstrator Platform for Coupled Multiscale Simulation. In: Proceedings of the SCEE 2006 International Conference, Sinaia, Romania (Submitted)

[JON91] Jonassen, D.H.: Evaluating constructivistic learning. Educational Technology, **31**, 28–33 (1991)

[MOR04] Morrison, G.R., Ross, S.M., Kemp, S.E.: Designing effective instruction. Wiley, New Jersey (2004)

[PAP80] Papert, S.: Mindstorms. Children, Computers and Powerful Ideas. Basic books, New York (1980)

[SAA01] Saarikoski, L., Salojrvi, S., Del Corso, D., Ovcin, E.: The 3DE: An Environment for the Development of Learner-Oriented Customised Educational Packages. In: Proceedings of the International Conference on Information Technology Based Higher Education and Training, (ITHET2001), (2001)

Web Based System for Graduate Studies: Optimization, Games, and Markets

Matti Heiliö[1] and Jonas Mockus[2]

[1] Lappeenranta University of Technology, P.O. Box 20 SF-53851 Lappeenranta
   Finland
   `matti.heilio@lut.fi`
[2] Institute of Mathematics and Informatics, Akademijos 4, LT-2600 Vilnius,
   Lithuania
   `jmockus@gmail.com`

1 Introduction

The well known results of algorithm complexity show the limitations of exact analysis. That explains popularity of heuristic algorithms. It is well known that efficiency of heuristics depends on the parameters. Thus we need some automatic procedures for tuning the parameters of heuristics. That helps to compare results of different heuristics. This enhance their efficiency, too.

The paper shows how optimization models can be implemented and updated by graduate students themselves. That reflects the usual procedures of the open source development. This way students not just learn the underlying model but obtain the experience in the development of open source software. The step-by-step improvement of the model and software is at least as important as the final result.

Doing this we accumulate some experience in the completely new field of education when all the information can be easily obtained by internet. The internet users are filtering and transforming the information to meet their own objectives, to build their own models. Here creative approach is needed. No well defined patterns and no well tested models exist yet. The natural way of research is by computer experimentation. This approach is natural and convenient for scientific collaboration, too. To simplify the task all the algorithms are implemented as platform independent Java applets or servlets. Readers can easily verify and apply the results for studies and for real life optimization models. All the examples of economic, social, and engineering models are regarded as optimization problems. Simplified versions of the models are presented for better understanding.

No "perfect" examples are presented in these websites. Improvement of "nonperfect" models is useful both for students and for colleagues. The main objective of this paper is to help establish scientific collaboration in the

Internet environment with distant colleagues and students by creating an environment of e-education and scientific collaboration in the fields related to optimization.

2 Heuristics

In the Internet environment computer simulation is the main tool of experimental research. The well known results of algorithm complexity show the limitations of exact analysis. That explains popularity of heuristic algorithms.

Investigating heuristic algorithms subjective factors are important. It is well known that efficiency of heuristics depends on some parameters. Thus the published results reflects not just the quality of proposed heuristic method but authors experience, too. Thus we need some automatic procedures for tuning the parameters of heuristics. That helps comparison of different heuristics. This enhance their efficiency, too. It is difficult to regard the problem in general. Therefore, we investigate a set of relevant examples.

To make this a part of more general e-education environment we need a theoretical background and some basic software tools first. All the examples should be united by some general concept. In this paper that is Bayesian heuristic approach. Therefore, we shall discuss the theory and applications of the Bayesian heuristic approach. Examples of Bayesian approach to automated tuning of heuristic parameters will be regarded.

3 Bayesian Heuristic Approach

An initial presentation of the basic ideas is in [1]. Preliminary results of distance graduate studies are in [2,3]. We regard various examples as optimization models. That is a general concept. First we investigate heuristic algorithms that reflects real life conditions. Comparing various heuristics and improving the efficiency we need specific optimization methods. A convenient theoretical concept is the Bayesian approach. We apply this approach for automatic tuning of heuristic parameters and for search of optimal mixtures of heuristics.

The traditional numerical analysis considers optimization algorithms that guarantee some accuracy for all functions to be optimized. Limiting the maximal error requires a computational effort that often increases exponentially with the size of the problem [4]. An alternative is the average analysis where the expected error is made as small as possible [5]. The average is taken over a set of functions to be optimized. The average analysis is called the Bayesian approach (BA) [6]. Application of BA to optimization of heuristics is called the Bayesian heuristic approach (BHA) [1].

Possibilities of application are illustrated by several examples designed for distance graduate studies in the Internet environment. All the algorithms are implemented as platform independent Java applets or servlets; therefore, readers can easily verify and apply the results for studies and for real life heuristic optimization problems.

4 Improving Expert Heuristics

The main objective of BHA is to improve any given heuristic by defining the best parameters and the best "mixtures" of different heuristics. The examples indicate that heuristic decision rules mixed and adapted by BHA often outperform the best individual heuristics. In addition, BHA provides almost sure convergence. However, the final results of BHA depend on the quality of the specific heuristics, including the expert knowledge. Therefore, the BHA should be regarded as a tool for enhancing the heuristics but not for replacing them.

Many well known optimization algorithms, such as genetic algorithms [7] and Tabu search [8], may be regarded as metaheuristics that can be improved using BHA.

In optimization problems, theory and software are interconnected. The final results depend on the mathematical theory of optimization and the software implementation. Thus we have to regard them both.

Representing different examples as a part of some general set-up we need basic theoretical and software tools. The examples should be united by some common framework. We call that GMJ (Global Minimizer by Java). The Bayesian heuristic approach is a proper theoretical concept. We apply this approach both for automatic tuning of heuristic parameters and for search of optimal mixtures of heuristics.

That is just an initial part of the GMJ. Important is to make GMJ open for development by users. Users contribute their own optimization methods in addition to the Bayesian ones. User optimization models are included as GMJ tasks. The results of optimization are represented by GMJ analysis objects. A minimal set of methods, tasks, and analysis objects is implemented by default. The rest depends on users.

5 Distance Studies

The video-conferencing is regular: each Friday from 8:00 until 9:30 EET (EEST). Broadcasting is in Lithuanian if no foreign students are connected. However, essential part of the web-site is in English so the English broadcasts are used, too.

The example is a joint Lithuanian-Finnish video-conference:
http://distance.ktu.lt/vips/join.php?sr=242   2006-05-05   record.   Figure 1 shows a snapshot of the Finish view. Figure 2 shows a snapshot of the Lithuanian students.

There is the main web-site: *http://pilis.if.ktlu.lt/jmockus* and five mirror-sites. The theoretical background and the complete description of the software is in the section "General Description" of the websites. Software tools are in the section "Software Systems" Examples of continuous global optimization are in "Global Optimization." The section "Discrete Optimization" is for examples of discrete optimization and linear and dynamic programming.

Fig. 1. Distance studies, Finnish view

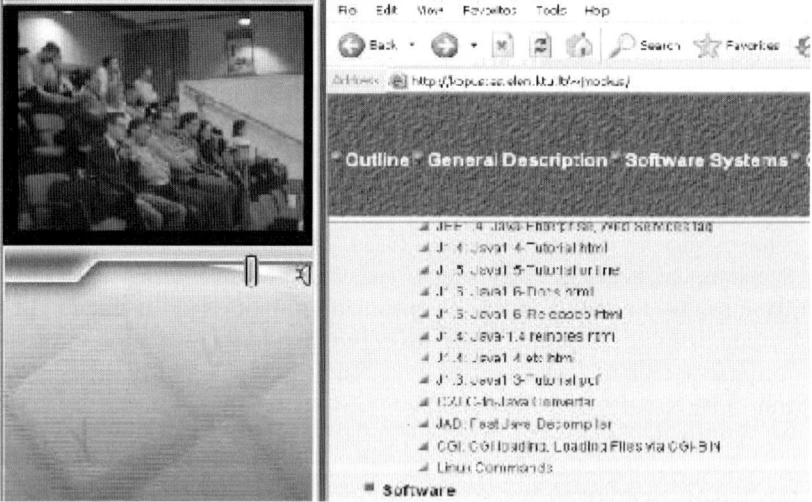

Fig. 2. Distance studies, Lithuanian students

5.1 Suggestions by Viewers

The significance of the voice channel in theoretically intense presentation is crucial. One blurred word or few missing syllables will destroy a sentence and then you loose the story line. A crude rule of thumb estimate is that the written material/slides will carry 40% of the message, voice track another 40% and the live picture will carry 20% and support the attention, give socially amicable atmosphere, etc.

More effort is needed to improve the sound transfer. Perhaps one should use gadgets familiar from TV-studios, where the speakers have personal microphone attached to clothing or a headset. Also in the receiving side the sound managements and proper loudspeakers are important.

Regarding the realtime demos with the software some added illustrative effects would be useful. The computational performance happens mostly in the background. When you see (1) the initial values given to the parameters and then seconds later (2) the output numbers appear, then one would hope to see some sort of visualization, sequence of intermediate steps or something else to get a grasp of the computational journey.

6 Conclusions

1. An objective of the paper is to start the scientific collaboration with colleagues on similar lines.
2. The growing power of internet presents new problems and opens new possibilities for distant scientific collaboration and graduate studies. Therefore, some nontraditional ways for presentation of scientific results should be defined.
3. The results of optimization show the possibilities of some nontraditional ways of graduate studies and scientific collaboration by creating and using a specific environment for e-education.
4. Examples of applications of the Bayesian heuristic approach show the efficiency of automated tuning of heuristics.

References

1. Mockus, J.: A Set of Examples of Global and Discrete Optimization: Application of Bayesian Heuristic Approach. Kluwer Academic Publishers (2000) ISBN 0-7923-6359-0.
2. Mockus, J.: A System for Distance Studies and Applications of Metaheuristics. Journal of Global Optimization 35 (2006) 637-665.
3. Mockus, J.: Stock exchange game model as an example for graduate level distance studies. Computer Applications in Engineering Education 10 (2002) 229-237 ISBN 1061-3773.
4. Horst, R., Pardalos, P.: Handbook of Global Optimization. Kluwer Academic Publishers, Dordrecht/Boston/London (1995).
5. Calvin, J., Zilinskas, A.: One-Dimensional P-Algorithm with Convergence Rate O (n 3C) for Smooth Functions. JOTA Journal pf Optimization Theory and Applications 106 (2000) 297-307.
6. Diaconis, P.: Bayesian Numerical Analysis. In: Statistical Decision Theory and Related Topics. Springer Verlag (1998) 163-175.
7. Goldberg, D.E.: Genetic Algorithms in Search, Optimization, and Machine Learning. Addison-Wesley, Reading, MA, (1989).
8. Glover, F.: Tabu search: improved solution alternatives. In: Mathematical Programming. State of the Art 1994. University of Michigan (1994) 64-92.

Web-Tool on Differential Equations

Peep Miidla

Institute of Applied Mathematics, University of Tartu, Liivi Str. 2,
50409 Tartu, Estonia
`peep.miidla@ut.ee`

1 Introduction

This paper introduces the principles of creating the web-tool on differential equations. It can be used to support European Master Program for Mathematics in Industry. Such a Program is working already on the leading partner universities of ECMI and now the use of e-study as an innovational step is being discussed.

Differential equations are very important tools of continuous mathematical modeling and so they are in great importance also in postgraduate program on Industrial Mathematics of ECMI. Web-based learning is one of the tools of the broader term "e-learning" with which education might to be delivered. In the educational system of ECMI this form of learning would have an important place because of the need of collaboration of academic institutions in this educational system.

Beside of e-learning the second part of e-study, e-teaching and with those two parts also the communication between professor and student must be under consideration. For a profound overview of e-study (e-learning and teaching) see [2].

2 Strategy of e-Study

Many Universities have approved already some strategy of e-study and fixed the general aims of this. Following ideas are taken from the document [6] and those are appropriate to be in the basis also for ECMI e-study system. In 2010, e-study must be natural component of education in the University of Tartu and why not also in the ECMI educational system.

The general aim of the developing of e-study is to create a modern, flexible, and internationally open educational process supportive of efficient and independent learning centered on student. E-study has to become a natural part of the learning process, ensuring the quality and flexibility of studies, supporting involvement of new target groups, and internationalization of the education. This purpose contributes to the planning of the development of e-study and international cooperation and will be used to assess todays achievements. The goals will be implemented in cooperation with all member-Universities of ECMI educational system.

In order to achieve the established objective, the following strategic tasks in developing of e-study are useful.

A. Support high-quality studies of high levels of interactivity centered on the student and involvement of new target groups.

This means development of efficient and high-quality combined models of e-learning and traditional learning (so called blended learning, see [4]) and implementation of these models in the studies in cooperation with ECMI Universities through involvement of lecturers and joint courses. The demand of high quality brings along the need of a system for ensuring and auditing the quality of e-learning courses, including quality criteria for e-study, continuous internal evaluation of the courses, quality signs, etc. Technically new will be the creating of multimedia objects for independent learning and ensure their realization with modern solutions of Information and Communications Technology (ICT). To broaden the circle of users it is useful from the very beginning to develop and update web-based in-service training courses besides of students also for various other target groups (specialists working in industry, teachers, etc.) and ensure active marketing of these courses.

B. Increase the e-teaching competence of the teaching staff, students, and assistance personnel and develop cooperation models for e-teaching (education of educators).

This means to introduce to the teaching staff the usage opportunities of e-learning and e-teaching in the study process. This means also development of the ICT skills of professors, teachers, tutors, and students to enable their effective participation in e-study, as well as development of their teaching and learning skills through training, guidelines, and counseling. Promote acquisition and use of modern teaching skills and innovation in the study process. Offer to the teaching staff methodological and technological support in conducting e-study. Develop and implement a cooperation model for the teaching staff, education technologists, tutors, and programme managers in the development and guide of e-study.

C. Ensure high level of infrastructure and support services for e-learning.

To ensure the perfect technical conditions for e-study is the care of each single university but there are some common recommendations to the participating academic organizations. It is desirable to install wireless Internet connections in each study building and dormitory of the universities. Also it is reasonable to develop the library as the central e-learning environment with integrated traditional and e-learning both for individuals and group work. Ensure the use of functional e-learning environments that offer optimal tools and support international standards (Instructional Management Systems IMS [9], Sharable Courseware Object Reference Model SCORM [1, 10]).

Develop in cooperation with other ECMI universities a virtual learning portal to support the magister prorgam on Industrial Mathematics, which would collect all necessary course databases, tutor databases, and e-portfolios of universities. Develop in cooperation with all ECMI universities a collection of electronic learning objects and principles for its usage (including copyright issues, see [7]). Ensure required exchange of data between different information systems (course database, e-learning environment, database of learning objects, tutor database) and develop global authentication system (university computer network, e learning environment). Ensure the possibility for storing large video and audio files in a special media server.

Web-based learning systems are generally housed administratively in a special "distance education" department alongside other at-distance delivery methods. All such tools seek to serve learners at some distance from their learning facilitator, these attempt to serve learners interacting with the learning source at different chronological times. E-education is often referred to as those delivery modalities that seek to reduce the barriers of time and space to learning, thus the frequently used phrase "anytime, anywhere learning".

This might to be the background for creating all web-tools in the framework of introduced by ECMI the postgraduate program in Industrial Mathematics, among this the one on Differential Equations.

3 Design of the Tool on Differential Equations

The final fixed program of the course on differential equations is a topic to consultations between partner universities and is not considered here. Instead some more general aspects are discussed. The first thing to do is the choice of e-study environment. Today many appropriate frameworks are available. Some of them are commercial, but there are also free learning environments. Maybe the most famous is Blackboard/WebCT (see [3]), which is a commercial software, but is highly evaluated and so appropriate for development of web-tool on Differential Equations. Let us mention also Moodle, a good example of freeware. It would be reasonable to develop all e-courses for ECMI postgraduate program in the same environment.

There are several types of e-learning tools: discussion, portfolio, group-work, learning material, testing, management, community building, etc. For presentation of differential equations it seems reasonable to use the ideas of blended learning, which is designed to offer a number of tested ways that integrate traditional learning methods with methods offered by new technology (see [4]). Blended learning allow to benefit from good sides of both traditional and new ways of learning, make innovation in otherwise traditional university teaching easier and acceptable. Integrating research and practical examples offers a good bases for initiating change in universities that by definition are based on research.

The design of an e-course means in addition to establishing the environment of the course, adding neccessary tools, objects, and instruments to this environment and establishing the settings of the course. Naturally the designer must be acquainted with the environment of e-study and aware about the possibilities of this. In the case of WebCT the course will be introduced by the Homepage of this, where one can find all neccessary links to the parts of the e-course.

The structural components of the web-tool on differential equations are following.

Official documents. ECMI centres are universities that have joined the network and fulfilled certain criteria in their educational programme. Orientation towards real life applications and industrial problems must be visible in educational style and contents. The educational principles of ECMI during some last years allow the partner Universities to keep their specifity and the Educational Committee does not intervene in the local educational habits. Various implementations of industrial mathematics programmes are available at the ECMI partner universities. Although the course on differential equations is classical in some sense, the corresponding syllabuses are quite similar in Universities. The students can chose between two lines of study, the first one is technomathematics and the second one is economathematics (see [5]). The course of ordinary and partial differential equations belong to the technomathematical branch (Syllabus, Program, schedule, etc.). The registration system and rules, individualization of access and assessment, management of user names and passwords are also the parts of this section of web-tool.

Lecture notes and tutorial texts. Open CourseWare system allows use of existing and available Web files, but these are usable as additional material. For ECMI purposes we need to rewrite all lectures to obtain unified notations and style. The language is English.

Exercises and examples. There exists a Web-tool in Estonian [8]. This is a good starting point to complete the set of exercises in English, also corresponding tutorial texts and examples for ECMI purposes. However, this tool must be extended and revised.

Self-training. Very important part of the web-tool because it allows learning in the suitable time and place. Self-training learning objects are texts, exercises, audio and video files, examples and case studies, and also the instruments designed for self-control. Here it is observed the possibility for interactive communication with server and with corresponding part of evaluation of the results obtained by student after solving training exercises.

Consultation. Here the interaction between student and professor or tutor is needed. The questions and answers may be transferred in real time conditions, but also via protocols like e-mail and others. In the first case the videoconference is a good possibility of realization.

Demos, simulations. There are many possibilities to illustrate the course on differential equations. Various software solutions, demo programs, and examples are available in the Internet.

Testing of students. The most responsible part of the tool. Final testing and evaluation of students' knowledge after passing the course must be in maximum objectivity and neutrality. Here unification of requirements through ECMI is needed.

References

[1]   Advanced Distributed Learning Homepage. `http://www.adlnet.gov/index.cfm`. (Last visited November 1, 2006).

[2]   Anderson, T., Elloumi, F. (editors). Theory and Practice of Online Learning. Printed at Athabasca University, 2004. `http://cde.athabascau.ca/online_book`. (Last visited November 1, 2006).

[3]   Blackboard & WebCT. `http://www.webct.com/`. (Last visited October 30, 2006).

[4]   B-learn Assisting teachers of traditional universities in design blended learning. `http://www.ut.ee/blearn`. (Last visited November 1, 2006).

[5]   ECMI Postgraduate Programmes. `http://www.ecmi-indmath.org/edu/index.php`. (Last visited November 1, 2006).

[6]   E-Learning Strategy of the University of Tartu 2006-2010. `http://www.ut.ee/154573` (in Estonian). (Last visited November 1, 2006).

[7]   Enabling the legal sharing and reuse of cultural,educational and scientific works. `http://creativecommons.org/`. (Last visited November 1, 2006).

[8]   Exercises on Ordinary Differential Equations (in Estonian). `http://math.ut.ee/peepm/dv/index.html`. (Last visited November 1, 2006).

[9]   IMS Global Learning Consortium. `http://www.imsglobal.org/`. (Last visited November 1, 2006).

[10]   Sharable Courseware Object Reference Model. `http://www.rhassociates.com/scorm.htm`. (Last visited November 1, 2006).

Part III

Contributed Papers

Model and Method to Increase the Thermal Efficiency of Micro-Heat Exchangers for Aerospace Applications

A. Velazquez[1], J.R. Arias[2], and B. Mendez[3]

[1] Micronics Thermal Microsystems S.L. Cardenal Cisneros 25, 2B,
28010 Madrid, Spain
angel.velazquez@micronics.es
[2] Propulsion and Fluid Mechanics Department, School of Aeronautics, Universidad
Politécnica de Madrid, Plaza Cardenal Cisneros 3, 28040 Madrid, Spain
[3] Fluid Mechanics Area, Universidad Carlos III de Madrid, Avenida de la
Universidad 30, 28911 Leganés, Madrid, Spain

Summary. The aim of this study is to test numerically the influence that incompressible flow pulsation has on heat transfer in configurations, such as the backward facing step, that appear in micro-electro mechanical systems (MEMS) and that are not very efficient from the thermal point of view. Two control parameters have been used to increase heat transfer: velocity pulsation frequency and pressure gradient amplitude at the inlet section. The working fluid is water with temperature dependent viscosity and thermal conductivity. The results obtained show that the time averaged Nusselt number grows when using appropriate flow pulsations.

1 Introduction

The objective of this study is to assess numerically the influence that flow pulsation has on heat transfer behind a backward facing step. This configuration appears in practical applications of micro-industrial products such as micro-motors, micro-cooling devices and power MEMS.

This work deals with the low Reynolds number, unsteady, 2D, incompressible flow regime. Two-dimensional flow is ensured because the onset of three-dimensional effects appears at higher Reynolds numbers (of the order of 800) in this type of configuration, Armaly et al. [ADP83], Durst and Pereira [DP88], Kaiktsis et al. [KKO91], Barkley et al. [BGH02]. Temperature dependent fluid properties (viscosity and thermal conductivity) have been used in this analysis so as to simulate realistic configurations. It is to be noted that water viscosity changes by a factor of 3 in the range from 20 to 80°C that is typical of cooling devices.

The use of pulsating flows to enhance heat transfer in several types of configurations has been addressed previously, see, for instance Yu et al. [YLZ04]. The outcome of these previous studies has been somewhat controversial

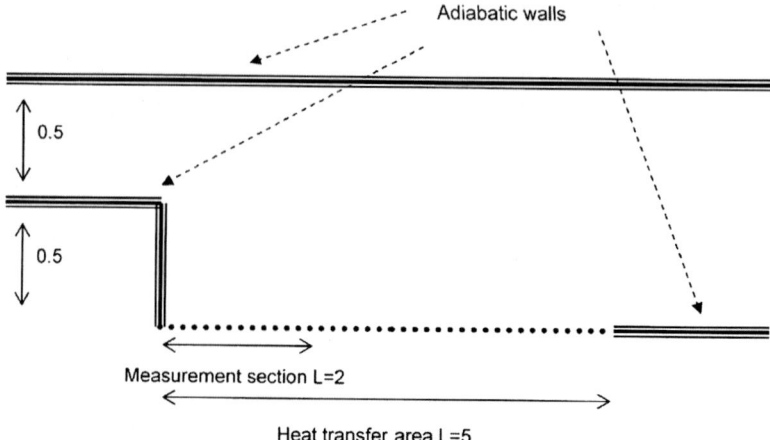

Fig. 1. Geometry of the problem

because some authors report heat transfer increase while others report no enhancement at all. Nevertheless, to our knowledge, there are no previous studies dealing with the geometry, boundary conditions, and flow regime that is addressed is this paper.

2 Problem Description

This study focuses on 2D, laminar, incompressible, and unsteady flows. The cooling fluid (water in this study) enters the channel at 20°C. Dimensionless step height is taken to be 0.5. All walls are adiabatic except a portion of length equal to 5 just downstream of the channel where temperature is prescribed to be 80°C. Distances are made dimensionless using the hydraulic diameter of the inlet channel. Nusselt number is measured in a smaller region of length 2 (see Fig. 1). This length is selected to accommodate the size of the expected unsteady recirculation region.

Pulsating flow moves from left to right in the computational domain and the presence of the step induces unsteady recirculation regions that appear and disappear periodically.

3 Governing Equations and Boundary Conditions

Dimensionless equations that govern the problem under consideration are

$$\frac{\partial u}{\partial x} + \frac{\partial v}{\partial y} = 0 \tag{1}$$

$$\frac{\partial u}{\partial t} + u\frac{\partial u}{\partial x} + v\frac{\partial u}{\partial y} = -\frac{\partial P}{\partial x} + \frac{1}{Re}\left[\mu\left(\frac{\partial^2 u}{\partial x^2} + \frac{\partial^2 v}{\partial y^2}\right) + 2\frac{\partial \mu}{\partial x}\frac{\partial u}{\partial x} + \frac{\partial \mu}{\partial y}\left(\frac{\partial u}{\partial y} + \frac{\partial v}{\partial x}\right)\right] \tag{2}$$

$$\frac{\partial v}{\partial t}+u\frac{\partial v}{\partial x}+v\frac{\partial v}{\partial y}=-\frac{\partial P}{\partial y}+\frac{1}{Re}\left[\mu\left(\frac{\partial^2 v}{\partial x^2}+\frac{\partial^2 v}{\partial y^2}\right)+2\frac{\partial \mu}{\partial y}\frac{\partial v}{\partial y}+\frac{\partial \mu}{\partial x}\left(\frac{\partial v}{\partial x}+\frac{\partial u}{\partial y}\right)\right] \tag{3}$$

$$\frac{\partial T}{\partial t}+u\frac{\partial T}{\partial x}+v\frac{\partial T}{\partial y}=\frac{1}{Re\,Pr}\left[k\left(\frac{\partial^2 T}{\partial x^2}+\frac{\partial^2 T}{\partial y^2}\right)+\frac{\partial k}{\partial x}\frac{\partial T}{\partial x}+\frac{\partial k}{\partial y}\frac{\partial T}{\partial y}\right] \tag{4}$$

In which variables are made dimensionless by using their upstream values.

The dimensionless viscosity and thermal conductivity are temperature dependent and follow experimental laws:

$$\mu=\frac{\mu'}{\mu_{293K}}=1-5.646(T-1)+12.259(T-1)^2 \tag{5}$$

$$k=\frac{k'}{k_{293K}}=1+0.786(T-1)+1.176(T-1)^2 \tag{6}$$

Boundary conditions are of the unsteady Poiseuille type:

Inlet section:

$$u(y,t)=u_1(y)+u_2(y,t) \tag{7}$$

where

$$u_1(y)=-24\left(y^2-\frac{y}{2}\right) \tag{8}$$

$$u_2(y,t)=\mathrm{Real}\left[\frac{ia_1a_2e^{i2\pi\omega t}}{2\pi\omega}\left(\frac{1}{\alpha+1}e^{\psi y}+\frac{\alpha}{\alpha+1}e^{-\psi y}-1\right)\right] \tag{9}$$

$$a_1=\frac{48}{Re}\quad \psi=(i2\pi\omega Re)^{1/2}\quad \alpha=e^{\psi/2} \tag{10}$$

$$\frac{\partial P}{\partial x}=-a_1+a_1a_2\,\cos(2\pi\omega t) \tag{11}$$

$$T=1\quad v=0 \tag{12}$$

Outlet section:

$$\frac{\partial^2 u}{\partial x^2}=\frac{\partial^2 v}{\partial x^2}=\frac{\partial^2 P}{\partial x^2}=\frac{\partial^2 T}{\partial x^2}=0 \tag{13}$$

Walls:

$$u=v=0,\quad T=T_{wall}\text{ for }5\le x\le 10,\quad \frac{\partial T}{\partial n}=0\text{ for any other wall} \tag{14}$$

Concerning pressure, momentum equations are solved at the wall with one sided derivatives.

The spatial discretization of the equation has been carried out by using the finite point formulation developed by the authors of this paper in previous works (Méndez and Velázquez [MV04]). Time integration used in this work is the standard semi-implicit pseudo-compressibility approach described by Tannehil [TAP97]. The cartesian grid used for this numerical study contains 32,051 points with spatial discretization of $\Delta x=\Delta y=0.2$. The dimensionless time step used in the temporal integration is 2.5e-4.

4 Results

The local time-dependent Nusselt number is defined as follows:

$$Nu_x(t) = \frac{h_x(t)D_h}{k_{wall}} = \frac{D_h}{\Delta x'}\frac{T'_{wall} - T'_{wall+1}(t')}{T'_{wall} - T'_\infty} = \frac{1}{\Delta x}\frac{T'_{wall} - T'_{wall+1}(t)}{T_{wall} - 1} \tag{15}$$

Where $h_x(t)$ is local the convection coefficient. The time-averaged Nusselt number is defined as

$$Nu_{average} = \int_{t=0}^{t=tc}\left(\frac{1}{2}\int_{x=5}^{x=7} Nu_x \, dx\right) dt \tag{16}$$

Three different cases have been computed and results are presented in the following table. For all the cases Prandtl and average Reynolds number, based on the inlet channel hydraulic diameter, at the inlet section of the channel are 6.62 and 100, respectively.

The Nusselt number in the steady case is 5.83, so inlet pulsation (Case A) can increase this figure by a factor of 44%. Figure 2 shows velocity contours and streamlines (Case A) for several equally spaced instants along the pulsating cycle.

Table 1. Cases studied

	ω	a_2	Nusselt$_{average}$
Case A	0.15	1.50	8.41
Case B	0.15	0.75	7.22
Case C	0.15	0.25	6.05

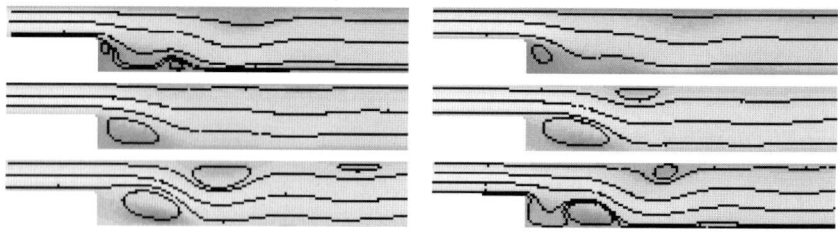

Fig. 2. Velocity contours and streamlines for Case A

5 Conclusions

Forcing flow pulsation at the inlet section of a channel containing a backward facing step produces a heat transfer enhancement. This increase of the rate of heat exchanged depends on the value of the two control parameters that are the frequency of the velocity pulsation and amplitude of the oscillating pressure gradient at the inlet section. For the two cases presented in this work the maximum Nusselt number obtained is for the bigger amplitude and is 44% higher that the one obtained in the steady case.

In addition, flow pulsation has a large impact on the flow topology behind the backward facing step. Especially, it is worth mentioning the appearance and disappearance of several recirculation regions downstream of the step that is strongly related to the increase of heat transfer. Remember that in the steady case a single recirculation region appears after the step and its size depends exclusively on the Reynolds number. This fact suggests the existence of a strong coupling between thermal effects and fluid dynamics parameters.

References

[ADP83] B. F. Armaly, F. Durst and J. Pereira. Experimental and theoretical investigation of backwards facing step flow. J. Fluid. Mech. **127** (1983), 473-496.

[DP88] F. Durst and J. C. F. Pereira. Time dependent laminar backward facing step flow in a two dimensional duct. J. Fluids. Eng. **110** (1988). 289-296.

[KKO91] L. Kaiktsis, G. E. Kairnaidakis and S. A. Orszag. Onset of three dimensionality, equilibria, and early transition in flow over a backward facing step. J. Fluid. Mech. **231** (1991), 501-558.

[BGH02] D. Barkley, M. G. Gomes, and R. Henderson. Three dimensional instability in flow over a backward facing step. J. Fluid. Mech. **473** (2002) 167-190.

[YLZ04] J. C. Yu, Z. X. Li, and T. S. Zhao. An analytical study of pulsating laminar heat convection in a circular tube with constant heat flux. Int. J. Heat Mass Transfer. **47** (2004) 5297-5301.

[MV04] B. Mendez and A. Velazquez. Finite Point Solver for the simulation of 2-D laminar incompressible unsteady flows. Comput. Methods Appl. Mech. Engrg. **193** (2004) 825-848.

[TAP97] J. C. Tannehill, D. A. Anderson and R. H. Pletcher. Computational fluid mechanics and heat transfer. Taylor and Francis. Philadelphia, 1997.

Influence of Trailing Jet Instability on the Dynamics of Starting Jets

Carolina Marugan-Cruz[1], Marcos Vera[1], Carlos Martinez-Bazan[2], and Geno Pawlak[3]

[1] Universidad Carlos III de Madrid
   cmarugan@ing.uc3m.es, marcos.vera@uc3m.es
[2] Universidad de Jaen
   cmbazan@ujaen.es
[3] University of Hawaii at Manoa

Summary. The starting jet produced by the discharge of a submerged fluid stream through a circular orifice is investigated both numerically and experimentally for moderately large values of the jet Reynolds number, Re. Low-amplitude sinusoidal perturbations were superimposed to the jet exit velocity to reproduce the effect of flow perturbations on the trailing jet and leading vortex ring dynamics. While the trailing jet is strongly modified by flow perturbations, the evolution of the total circulation, as well as the leading vortex dynamics, remain relatively unaffected, and thus can be considered as more robust indicators of the dynamics of starting jets.

1 Introduction

The initial development of a jet produced by the discharge of fluid into a quiescent atmosphere involves the roll-up of the shear layer into the leading vortex ring followed by a column of fluid subject to shear instabilities [3]. Starting jets can be found in several industrial applications: intake of reactants into combustion chambers, gas flows produced by sprays in diesel engines, or cracks in pressurized vessels. Moreover, starting jets also appear in a large variety of natural flows: tidal jets, animal propulsion (e.g., squids), or the blood entering the heart. In addition, in laboratory scale, starting jets are used to generate vortex rings.

Previous studies have addressed the study of transient jets. For example, Gharib et al. [1] tested a range of piston stroke to diameter ratios (L/D), and velocity programmes and demonstrated the existence of a ratio (about 4) after which the vortex ring pinches off from the trailing jet. Rossenfeld et al. [4] and Zhao et al. [5] found a strong dependence on the velocity profile and a weak but significant dependence on the velocity program. Iglesias et al. [2] studied the effect of temperature ratio (T_j/T_o) in the dynamics of the jet. In the present paper we analyze the effect of the perturbations generated in

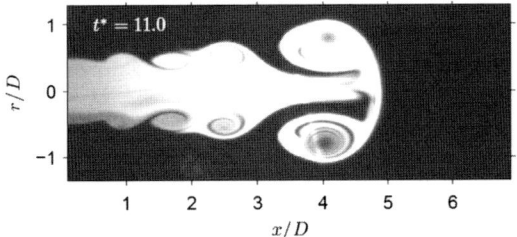

Fig. 1. Flow visualization at $t^* = 11$

the trailing jet on the global characteristic of the leading vortex from the experimental and numerical point of view.

2 Experimental Set-Up

To perform the experiments, a transient water jet was generated using a piston–cylinder mechanism [3]. The piston pushed the water inside the cylinder through a circular nozzle of exit diameter $D=2$ cm into a quiescent water pool. The motion of the piston was controlled by a computer, which allowed us to test at various jet velocities ($U_j=$ 4.2, 6.3, 8.3, 10.4, and 12.5 cm s^{-1}). The walls of the tank were made of plexiglas, enabling measurements with nonintrusive techniques such as digital particle image velocimetry (DPIV) and laser induced fluorescence (LIF). Figure 1 shows a LIF visualization of a well developed leading vortex followed by a trailing jet with secondary instabilities at nondimensional time $t^* = t\,U_j/D = 11$. These secondary vortices develop a rich dynamics that include vortex pairing and ingestion by the leading vortex.

3 Numerical Simulation

The experimental results were compared with numerical simulations carried out with a commercial CFD code. The flow was simulated with a finite-volume discretization of the incompressible axisymmetric Navier–Stokes equations, which were solved with a segregated solver using a second-order up-wind scheme to discretize the convective term and a central-differenced second-order scheme for the diffusion term. A second-order implicit temporal discretization scheme was used with a staggered control volume method. At time $t = 0$ the flow inside the computational domain was assumed to be at rest. A nonslip condition was imposed at the wall boundary, while the pressure was specified at the lateral and downstream boundaries. The size of the computational domain was $r_{max} = 20D$ and $x_{max} = 40D$. An adaptive grid was used to follow the jet evolution: the initial size of the grid was around 30,000 cells. In addition a grid sensitivity analysis was performed to ensure that the grid size was appropriate for this problem. Three different velocity programes were specified at the inlet to show their effect on the trailing jet and vortex head. First, the piston was programmed to give an (uniform) impulsive velocity for $t > 0$ as given by (1). Second, an accelerated model, given by (2), was proposed with an initial moderate acceleration, until it reached constant speed at $t = T$. Finally, we superimposed an impulsive motion and an oscillatory perturbation of small amplitude (3), only 1% of the maximum velocity and a dimensionless frequency of 0.5.

$$u(t) = U_\mathrm{j} \tag{1}$$

$$u(t) = \begin{cases} U_\mathrm{j}t/T & t < T \\ U_\mathrm{j} & t > T \end{cases} \tag{2}$$

$$u(t) = U_\mathrm{j}(1 + \delta \cos \omega t) \tag{3}$$

4 Analysis

The most relevant features of the transient jet obtained experimentally (such as the location of the vortex core, the stagnation point, or the total circulation) were compared with the results of the numerical simulations. Figure 2 shows the vorticity contours at $t^* = 11$ for the accelerated and the forced case. Comparing the numerical results with the flow visualization shown in Fig. 1 it can be observed that in the accelerated case the vortex front location compares well with the experimental value. However, only the forced case is able to reproduce appropriately the trailing secondary vortices.

To describe the time evolution of the leading stagnation point (or vortex front), its axial position, x_f, was optically determined from the experiments, while it was defined as the point of maximum pressure along the axis in the numerical simulations. Figure 3 shows the comparisons between the three numerical models and the experiments. The accelerated case compares the best since the piston started with an initial acceleration as well. The position of the stagnation point given by the impulsive and forced cases are nearly the same during the initial instants. However, for longer times the stagnation point obtained for the forced case is ahead of that given by the impulsive one due to the pairing between the leading vortex and the trailing vortices.

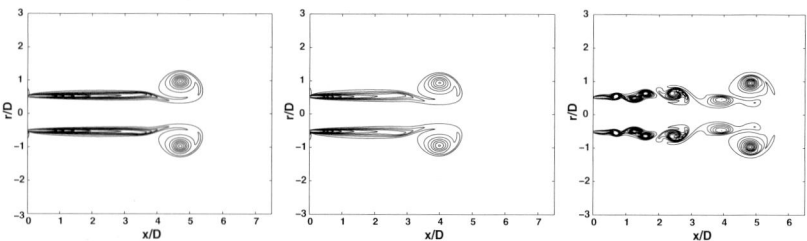

Fig. 2. Vorticity contours at $t^* = 11$ for the impulsive (*left*), accelerated (*center*), and pulsed (*right*) cases

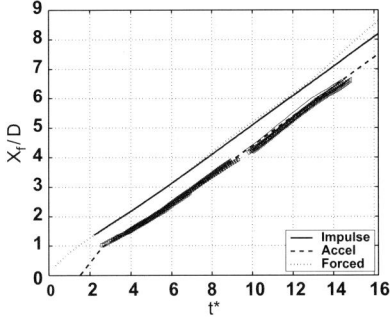

Fig. 3. Nondimensional position of the stagnation point as a function of time. Solid line indicates the impulsive case, dotted line represents the forced case, and dashed line corresponds to the accelerated case. The experiments are shown with symbols

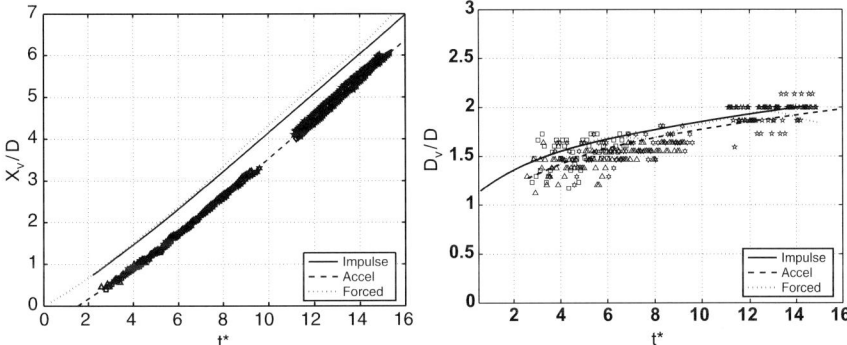

Fig. 4. Axial position of vortex head (*left*) and vortex diameter (*right*). Legend as in Fig. 3

Other important features of the experimentally generated transient jets were also identified, i.e., the position of the leading vortex. For example, the axial position was calculated using two integral magnitudes, $Q_r = \int_0^{r_{\max}} u_r\, r\, dr$ and $J_x = \int_0^{r_{\max}} u_x^2\, r\, dr$, that are zero and maximum at the vortex location, respectively and where u_r and u_x are the radial and axial velocities. Similarly, the radial location was calculated with two additional magnitudes that are both zero at the radial vortex center, $Q_r^{\mathrm{R}} = \int_{x_v+L_i}^{x_v} u_r\, dx$ and $J_x^{\mathrm{R}} = \int_{x_v+L_i}^{x_v-L_i} u_x^2\, dx$, where x_v is the axial vortex position and $L_i = 0.75D$. The vortex position was calculated numerically as the point of minimum pressure.

Figure 4 shows the axial vortex position, x_v, and the vortex diameter, D_v, as a function of time. It can be observed that, as in the front location, the accelerated case provides the best comparison with the experiments in terms of the axial position. However, the vortex diameter of the three numerical simulations compares fairly well with the experimental data. The impulsive and accelerated models show a monotonic increase of the vortex diameter,

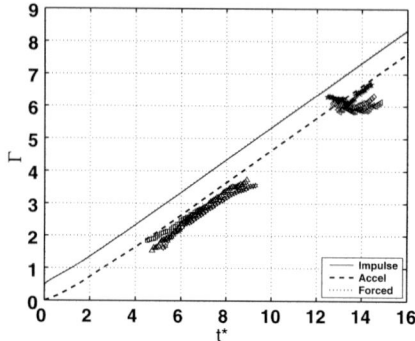

Fig. 5. Temporal evolution of the total circulation. Legend as in Fig. 3

while the vortex diameter given by the forced case starts increasing with time until the vortex pairing between the leading vortex and a pair of vortices from the trailing jet. The total circulation of the starting jet is calculated through the integral of the vorticity in the whole domain, $\Gamma = \int_{\Sigma} \nabla \wedge u(r,x)\,\mathrm{d}\sigma$, see Figure 5. The values for the forced and impulsive simulations are coincident and again the accelerated model compares the best with the experiments.

5 Discussion

The computations for the impulsive velocity program show a time lag between the experimental values and the numerical simulation, and no vortices in the trailing jet appear. The effect of the secondary instabilities on the axial vortex position, stagnation point, or total circulation lacks of importance compared to the effect of the initial acceleration; however, the radial vortex position seems to be affected by these vortices specially after a pairing.

References

1. Gharib, M., Rambod, E. and Shariff, K. A universal time scale for vortex ring formation. J. Fluid Mech. **360**, (1998) 121-140.
2. Iglesias, I. Vera, M., Sánchez, A. L. and Liñán, A. Simulations of starting jets at low mach numbers. Phys. Fluids **17**, (2005) 38105.
3. Pawlak, G., Marugán-Cruz, C., Martínez-Bazán, C. and García-Hrdy, P. Experimental Characterization of Starting Jet Dynamics, Fluid Dynamics Research, submitted, 2006.
4. Rosenfeld, M., Rambod, E. and Gharib, M. Circulation and formation number of laminar vortex rings. J. Fluid Mech. **376**, (1998) 297-318.
5. Zhao, W., Frankel, S. H. and Mongeau, L. G. Effects of trailing jet instability on vortex ring formation. Phys. Fluids **12**, (2000) 589-596.

Modelling and Computational Analysis of the Dynamic Crash Behaviour of Fabric Reinforced Composite Automotive Structures

E.V. Morozov[1] and V.A. Thomson[2]

[1] University of New South Wales at ADFA, Canberra, Australia 2600
   e.morozov@unswasia.edu.sg
[2] School of Mechanical Engineering, Howard College,
   University of KwaZulu-Natal, Durban 4041, South Africa

Summary. This chapter is concerned with the modelling and computer simulation of a dynamic failure development in the thin-walled automotive structural components made from the laminated polymer composite materials reinforced with fabric layers. The scope of the work includes geometrically non-linear numerical structural analysis coupled with the progressive damage material modelling and applicable to structures with complex geometries. Impact crash simulation of a scaled-down automotive composite spare-wheel compartment has been performed using the explicit finite element code (PAM-CRASH). Simulation results are compared to experimentally recorded data, and the predicted deformation states and failure patterns show good agreement with the experimental data.

1 Introduction

The progressive damage modelling is required to predict the material and structural response up to the point of ultimate failure. The non-linear explicit finite element analysis code, PAM-CRASH, is used in this work for simulation of the dynamic structural response of thin-walled composite components to a crushing load. Progressive damage modelling of fibre composite materials is integrated into mathematical models using continuum damage mechanics. The present work aims to model, simulate and predict the dynamic response of the composite structural components with complex shape and geometry to the crushing loads.

2 Progressive Damage Modelling

The bi-phase model is a heterogeneous material model adapted to unidirectional continuous fibre reinforced composites or composite fabrics. Damage law for the bi-phase model is implemented by a reduction in stiffness $C(d) = C_0 \times (1 - d)$, where C is the instantaneous stiffness matrix, C_0 is the initial undamaged stiffness matrix and d is a dimensionless scalar damage parameter, which is a function of strain. Damage function is separated into volumetric and shear components $d(\epsilon) = d_v(\epsilon_v) + d_s(\epsilon_s)$, where d_v is the volumetric damage as a function of volumetric equivalent strain and d_s is a shear induced damage as a function of equivalent shear strain [2]. The damage functions are determined by choosing the critical damage points from the relevant stress–strain diagram [2].

3 Structural Prototype and Model Development

The geometry of the structural prototype is selected to be representative of an automotive compartment typically used to house a spare wheel 1. The laminated demonstrator is constructed using 8 layers of $290\,\mathrm{g\,m^{-2}}$ twill weave glass fabric in an Ampreg 20 epoxy resin system. With each fabric layer represented according to its warp direction, the following stacking sequences were obtained during prototype manufacture (orientation code as per ASTM D6507):

$$\text{FLATS}(\phi_1) : [0]_8 \quad \text{SIDEWALL}(\phi_2) : [045]_{2S}$$

The prototypes were manufactured using a hot vacuum bagging process together with hand layup of resin and fabric patterns [3]. A four node quadrilateral shell element has been employed for modelling of the prototype geometry. Application of the bi-phase model for shell elements in the PAM-CRASH code corresponds to Material Type 130, which is a multi-layered material tailored for the description of orthotropic laminates. In order to represent the experiment, the velocity in the x-direction (see Fig. 1) of the moving edge is set to equal to the velocity of loading for the experimental case concerned.

4 Prototype Testing and Model Verification

Testing of the laminated prototype was completed using an MTS servo hydraulic dynamic testing rig. A comparison of the real, schematic and simulated failure patterns is provided in Fig. 2. The real failure pattern (Fig. 2a) is a

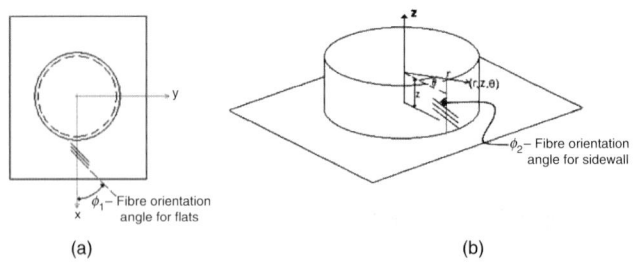

(a)   (b)

Fig. 1. Definitions of fibre orientation for flats (**a**) and sidewall (**b**)

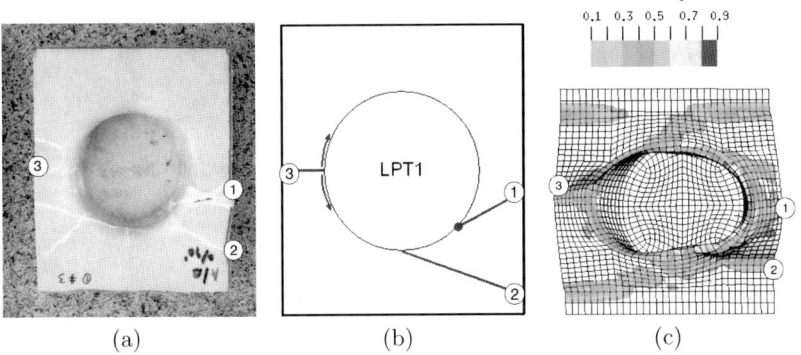

Fig. 2. Comparison of the prototype failure patterns: (**a**) real, (**b**) schematic, (**c**) simulated

photographic record of the component after testing and shows the location of the major cracks and fold lines that develop during the crushing process. The schematic failure pattern (Fig. 2b) is a representation of the real failure pattern, where the three major cracks (numbered 1–3) that have been observed during the crushing experiment are indicated. The areas of damage reported from the simulation that correspond to the major failures observed from testing are numbered accordingly (Fig. 2c). The comparison of the real, schematic and simulated failure patterns shows the areas of damage reported from simulation that correspond to failure in the real component.

5 Simulation of Impact Loading

Simulation of the laminated prototype's response to a case of impact loading has been undertaken in order to demonstrate the capability of the model to simulate the crash response of laminated structures. For the crash simulation, an added mass of 1,000 kg is applied and distributed over the nodes that make up the moving end of the component. The nodes with added mass are then assigned an initial velocity in the x-axis direction (see Fig. 1) and the two cases investigated: for an initial velocity of $1\,\mathrm{m\,s}^{-1}$ and an initial velocity of $5\,\mathrm{m\,s}^{-1}$, respectively. For implementation of the prototype crash models, alternative boundary conditions are applied to the FE model previously developed for simulation of the laminated component's response to a constant velocity crushing load. One edge of the structure is fixed and has all nodal degrees of freedom constrained and the row of nodes at the inside edge of the fixed end are defined as a section for reporting the reaction force. The moving end is allowed to translate in the x-direction and has all other degrees of freedom fixed. Loading applied to the moving end consists of an added mass of 1,000 kg, distributed over the constrained nodes, and an initial

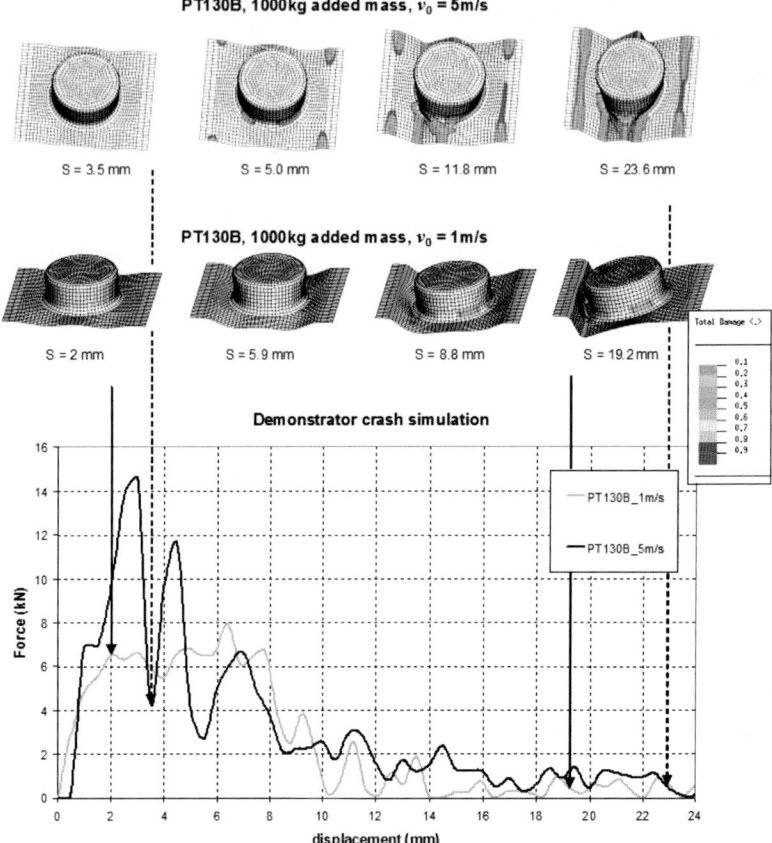

Fig. 3. Simulated crash response for the laminated component (1,000 kg added mass impact)

velocity of either 1 or $5\,\mathrm{m\,s^{-1}}$, depending on the case being investigated, is applied to the same nodes. The results obtained from the crash simulation in the form of a force–displacement plot (Fig. 3) show how an increase in the initial velocity results in an increase in the peak load reached, as well as more noticeable oscillations in the force displacement response. New deformation modes are predicted for the crash simulation of the laminated prototype (see Fig. 3) when compared to the simulated deformation obtained for lower velocity loading (Fig. 4), accounting for the significantly varied load–displacement behaviour. For the crash simulation at $1\,\mathrm{m\,s^{-1}}$, complete folding of the base plate occurs, which accounts for the abrupt drop in load carrying capability between 8 and 10 mm of displacement. For the $5\,\mathrm{m\,s^{-1}}$ crash, a high peak load is predicted (just over 14 kN) and folding at the sides of the cylindrical shell, as well as large oscillations in the load transmitted to the fixed support, are predicted.

PT130B, constant velocity loading v = 0.15m/s

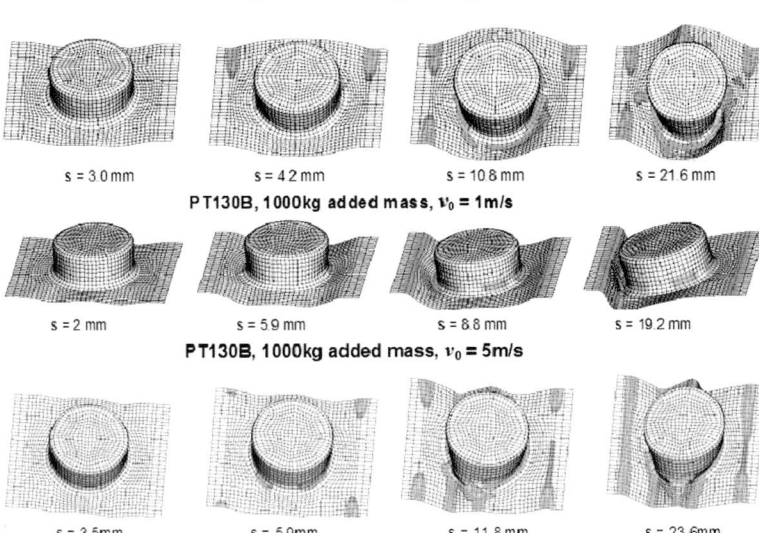

s = 3.0mm   s = 4.2mm   s = 10.8mm   s = 21.6mm

PT130B, 1000kg added mass, v_0 = 1m/s

s = 2 mm   s = 5.9mm   s = 8.8mm   s = 19.2 mm

PT130B, 1000kg added mass, v_0 = 5m/s

s = 3.5mm   s = 5.0mm   s = 11.8mm   s = 23.6mm

Fig. 4. Comparison of deformation modes with increased loading rate

6 Conclusions

Modelling the crush response of a laminated structural component with complex geometry has been undertaken using the bi-phase material model included with the PAM-CRASH analysis tool. Development of the modelling methodology has been completed to a level that results in simulated deformation states showing good resemblance to the real deformation states recorded during the crushing experiment. Critical regions of failure have been predicted effectively as indicated by the onset and further development of damage in those elements of the discretised structure where failure occurs in the real component. Simulation of the demonstrator's response to the impact loading indicates alternative deformation modes and failure patterns when compared to deformation modes and failure patterns observed through simulation of the demonstrator's response to lower velocity loading.

References

1. D.W. Schmueser and L.E. Wickliffe: Impact energy absorption of continuous fiber composite tubes. J. Eng. Mater. Technol., 1987, 109, pp. 72-77.
2. E.V. Morozov and V.A. Thomson: Simulating the progressive crushing of fabric reinforced composite structures. Composite Structures, 2006, 76, pp. 130–137.
3. E.V. Morozov and V.A. Thomson: Processing effects and structural integrity of fabric reinforced thin-walled composite components. In: Structural Health Monitoring. Proc. of the Third European Workshop. (A. Guemes, Ed.). DEStech Publications, Inc., Lancaster, PA, USA, 2006, pp. 872-879.

Theoretical Modeling of Flame–Acoustic Interaction

M.L. Bondar, J.H.M. ten Thije Boonkkamp, and R.M.M. Mattheij

Eindhoven University of Technology, Centre for Analysis, Scientific Computing and Applications, P.O. Box 513, 5600 MB, Eindhoven, The Netherlands
`mbondar@win.tue.nl`

1 Introduction

The interaction between premixed flames and acoustic perturbations of the gas velocity leads to combustion noise. In order to design noise-free combustion devices, one needs to understand the detailed mechanism by which the combustion noise is produced. The conical Bunsen flame is an excellent model for theoretical studies of the combustion noise.

The response of a Bunsen flame to velocity perturbations is evaluated in terms of the flame transfer function (TF), which is the ratio of the heat release rate perturbation to velocity perturbation in the frequency domain. The main features of the flame dynamics can be described with the G-equation model. In this model, the flame is treated as a surface (flame front) that separates the burnt from the unburnt gas. The dynamics of the flame front is then described by the G-equation, whose solution gives the instantaneous position of the flame front. The perturbation of the heat release rate can be considered proportional with the area of the flame. As a consequence, the first step in evaluating the flame TF involves computing the area of the flame. This requires the solution of the G-equation. To overcome the difficulties arising from the G-equation being nonlinear, a set of constraints can be used to derive a linear form of the G-equation. For example, linear G-equations were obtained by assuming very long flames parallel with the stream lines [Fleifil96], or that the laminar burning velocity has a constant direction, normal to a stationary position of the flame [DDC00]. These linear models (Fleifil96, DDC00) predict the correct behavior of the magnitude of the flame TF (low pass filter), but fail in describing the phase of the TF. Unlike the phase of the measured TF which increases linearly with the increase in the excitation frequency, the phase of the theoretical TF saturates to a level of $\pi/2$ (DDC00).

To understand the origin of the discrepancy between experiments and theory in describing the behavior of the phase TF, here we extend the kinematic models proposed previously in [Fleifil96,DDC00]. Here we consider flames that have an arbitrary cone angle and a burning velocity whose direction relative to the stationary flame front position is allowed to change.

2 Flame Model

The flame front is described by the G_0 level set of some combustion variable, i.e., $G(r, z, t) = G_0$, where r and z are the axial and the radial coordinates, respectively, and t is time. The flame front moves under the action of the perturbed gas velocity v and of the laminar burning velocity S_L, which is assumed to be constant. Assuming that the flame front is not locally vertical and that the flame and the flame front oscillations are axisymmetric implies that the flame axial location above the burner rim, z, is a single-valued function of the time and radial coordinate, i.e., $z = \zeta(r, t)$. The movement of the flame front is given by the following kinematic relation:

$$\frac{\partial \zeta(r, t)}{\partial t} + u \frac{\partial \zeta(r, t)}{\partial r} - v + S_L \sqrt{1 + \left(\frac{\partial \zeta(r, t)}{\partial r} \right)^2} = 0, \tag{1}$$

where $u(r, z, t)$ and $v(r, z, t)$ are the radial and the axial components of the gas velocity, respectively. For the mean gas velocity we assume as in [Fleifil96] a Poiseuille profile, i.e.,

$$\bar{u}(r) = 0, \qquad \bar{v}(r) = v_0 \left(1 - \left(\frac{r}{R} \right)^2 \right), \tag{2}$$

where $v_0 > 0$ and R are the maximum velocity at the centerline and the radius of the duct, respectively. The perturbation of the gas velocity is modeled by

$$u'(r, t) = 0, \qquad v'(r, t) = \varepsilon v_0 \sin(\omega t), \tag{3}$$

where ω and ε are the frequency and the amplitude of the velocity oscillation, respectively. Here we consider as in [DDC00, Fleifil96] that the amplitude ε is small. To find the response of the flame to velocity perturbations, (1) needs to be solved. Equation (1) cannot be solved analytically for $\varepsilon \neq 0$. Instead, we derive a system of linear advection equations which can be easily solved numerically by applying the following asymptotic expansion for ζ:

$$\zeta(r, t) \approx \zeta_0(r, t) + \varepsilon \zeta_1(r, t) + \varepsilon^2 \zeta_2(r, t) + \cdots . \tag{4}$$

The G-equation model assumes that the flame is attached to the burner rim ($r = R$), i.e., $\zeta(R, t) = 0$. In reality the flame is not attached, but there is a stand off distance between the flame and the burner rim. The attachment of the flame implies that at the point $r = R$ the slope at the flame front is given by

$$S_L \frac{\partial \zeta}{\partial r}(R, t) = -\sqrt{v(R, t)^2 - S_L^2}. \tag{5}$$

Then, the attachment of the flame at the burner rim $r = R$ can be imposed only if the square of the gas velocity is larger than the square of the burning velocity, requiring the fulfillment of the condition $v(R,t)^2 - S_L^2 \geq 0$. In our case this latter condition implies

$$(v_0\varepsilon \sin(\omega t) - S_L)(v_0\varepsilon \sin(\omega t) + S_L) \geq 0. \tag{6}$$

Condition (6) cannot be satisfied for an ε independent of t. Thus, we compute the maximum interval $[0, r_\varepsilon(t)]$ on which condition $v(r,t)^2 - S_L^2 \geq 0$ is fulfilled and thus the slope at the flame front is well defined. To avoid the difficulties of working on the time dependent interval $[0, r_\varepsilon(t)]$ we restrict the domain to $[0, \delta_\varepsilon]$ where,

$$\delta_\varepsilon = R\sqrt{1 - \varepsilon - \frac{S_L}{v_0}}, \tag{7}$$

under the condition $\varepsilon < (v_0 - S_L)/v_0$. In the following we will derive the analytical solution of (1) by imposing $\zeta(\delta_\varepsilon, t) = 0$.

3 Location of the Flame Front

The following dimensionless variables are introduced, $r^* := r/R$, $t^* := t/\tau$, $z^* := z/R$, $\tau := R/S_L$, $\omega^* = \omega R/S_L$, and $\hat{v} := v_0/S_L$. Substituting expression (4) into (1) and collecting terms of the same order leads to a system of equations. The small amplitude assumption allows us to take into consideration only the leading and the first order equations resulting in the following system (we omitted the $*$)

$$\frac{\partial \zeta_0}{\partial t} - \hat{v}(1 - r^2) + \sqrt{\left(\frac{\partial \zeta_0}{\partial r}\right)^2 + 1} = 0, \tag{8}$$

$$\frac{\partial \zeta_1}{\partial t} - \frac{\sqrt{\hat{v}^2(1 - r^2)^2 - 1}}{\hat{v}(1 - r^2)}\frac{\partial \zeta_1}{\partial r} = \hat{v}\sin(\omega t). \tag{9}$$

To simplify the calculations, we assume that the perturbation in the gas velocity is introduced after the stabilization of the flame above the burner rim, so that the leading order term in (4) can be replaced by the steady solution of (8). The steady solution ζ_0 can be expressed in terms of elliptic integrals [BMTB05]. The expression for ζ_0 is rather lengthy and will be omitted here. Equation (9) along with the boundary condition, $\zeta_1(\delta_\varepsilon) = 0$, and the initial condition, $\zeta_1(r,0) = 0$, $0 \leq r \leq \delta_\varepsilon$, is integrated using the Laplace transform to yield

$$\zeta_1(r,t) = \frac{\hat{v}}{\omega}\Big(\cos(\omega(-B(\delta_\varepsilon) + B(r) + t)) - \cos(\omega t)\Big), \tag{10}$$

where

$$B(r) := \frac{1}{\sqrt{\hat{v}}\sqrt{\hat{v}+1}} \left((\hat{v}+1) \, E\left(\theta(r),\chi\right) - F\left(\theta(r),\chi\right) \right). \tag{11}$$

Here

$$\theta(r) := \arcsin\left(\frac{r\sqrt{\hat{v}}}{\sqrt{\hat{v}-1}}\right), \quad \chi := \sqrt{\frac{\hat{v}-1}{\hat{v}+1}}, \tag{12}$$

and E and F are the elliptic integrals of first and second kind, respectively (see [BF71]). The first order solution ζ_1 is similar to the first order solution ζ^f derived in [Fleifil96] after transforming (1) into a linear equation by approximating the square root term with $-\partial\zeta/\partial r$. This leads to (in our notation)

$$\zeta_1^f = \frac{\hat{v}}{\omega}\left(\cos(\omega(-B^f(1) + B^f(r) + t)) - \cos(\omega t) \right), \tag{13}$$

where $B^f(r) = r$.

The difference between the functions B and B^f above leads to differences between the TF obtained with our model as compared to the simpler models from [Fleifil96,DDC00]. B is a nonlinear function that increases almost linearly with increasing r, and a simple analysis show that

$$\max_{r\in[0,\delta_\varepsilon]} |B(r) - B^f(r)| < |B(\delta_\varepsilon) - \delta_\varepsilon| < |B(\delta) - \delta|, \tag{14}$$

where $\delta = \sqrt{1 - 1/\hat{v}}$. Using Lemma 6.1 from [BMTB05] we can prove that $B(\delta) \in [1, 1.1107]$. The parameter $\hat{v}$ is typically in the interval $\hat{v} \in [2, 10]$, which implies that the difference $|B(r) - B^f(r)|$ is small, and hence a small improvement in the flame TF is to be expected. The difference $|B(r) - B^f(r)|$ reaches its maximum at the boundary where the flame is attached.

4 Transfer Function

Assuming that the heat release rate is proportional with the area of the flame the transfer function of the flame $H(\omega)$ is given by

$$H(\omega) = \frac{A'}{A_0} \frac{v_0}{v'}(\omega), \tag{15}$$

where A_0 and A' are the mean value of the flame area and its variation, respectively. The area is computed by evaluating numerically with the composite trapezoid rule the following expression:

$$A(t) = 2\pi \int_0^{\delta_\varepsilon} r \sqrt{\left(\frac{\partial\zeta}{\partial r}\right)^2 + 1} \, \mathrm{d}r. \tag{16}$$

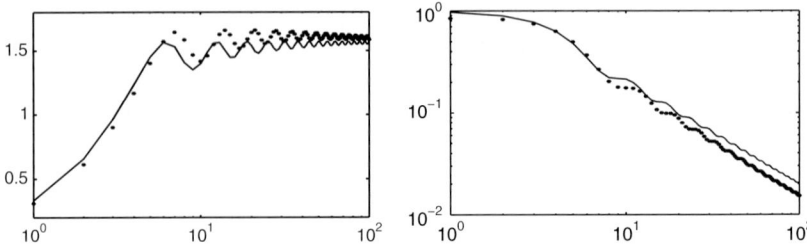

Fig. 1. The phase difference (*left*) and the magnitude (*right*) of the transfer function as function of dimensionless frequency. ∗ our model; *solidlines* model from [Fleifil96]. The following parameters were used: $\hat{v} = 8$, $\varepsilon = 0.1$, $\delta_\varepsilon = 0.8803$

The TF derived with our model is depicted in Fig. 1 along with the TF derived in [Fleifil96]. In agreement with experiments, in both models the magnitude of the TF has a low pass filter behavior. The phase of both models agrees well with the experiments up to the dimensionless frequency $\omega = 6$, [DDC00], but the linear increase of the phase with the increase of the excitation frequency is not captured. Nevertheless, due to the difference $|B(r) - B^f(r)|$ at the boundary, our model gives a value for the saturation of the flame TF that is slightly larger than $\pi/2$. The better description of the flame TF provided by our model suggests that the discrepancy between the theoretical models and the experiments might reside in the boundary conditions that force the flame attachment.

5 Conclusion

Here we extended the previous flame kinematic models ([Fleifil96, DDC00]) by addressing flames with arbitrary cone angles, and a burning velocity with variable direction relative to the stationary flame front. An analytic expression of the perturbed flame front position was derived in terms of elliptic integrals. Compared to the previous models [Fleifil96], the model proposed here improves the description of the front close to the boundary, and consequently, of the phase behavior. This suggests that better flame models should account for more realistic boundary conditions.

References

[BMTB05]   Bondar, M. L., Mattheij, R.M.M., ten Thije Boonkkamp, J.H.M.: Investigation of Bunsen flame dynamics by the method of characteristics. CASA-Report, 06–05 (2005)
[BF71]   Byrd, P.F., Friedman, M.D.: Handbook of elliptic integrals for engineers and scientists.Springer-Verlag (1971)

[DDC00]   Ducruix, S., Durox., D., Candel, S.: Theoretical and experimental determination of the transfer function of a laminar premixed flame. Proc. of Combust. Inst., **28**, 765–773 (2000)

[Fleifil96]   Fleifil, M., Annaswamy, A. M., Ghoneim, Z. A., Gnoniem, A. F.: Response of a laminar premixed flame to flow oscillations: A kinematic model and thermoacoustic instability results. Comb. and Flame, **106**, 487–510 (1996)

Air-Blown Rivulet Flow of a Perfectly Wetting Fluid on an Inclined Substrate

Julie M. Sullivan, Stephen K. Wilson, and Brian R. Duffy

Department of Mathematics, University of Strathclyde, Livingstone Tower, 26 Richmond Street, Glasgow G1 1XH, United Kingdom

1 Introduction

Thin-film flows occur in a variety of physical contexts including, for example, industry, biology and nature, and have been the subject of considerable theoretical research. (See, for example, the review by Oron, Davis and Bankoff [4].) In particular, there are several practically important situations in which an external airflow has a significant effect on the behaviour of a film of fluid, and consequently there has been considerable theoretical and numerical work done to try to understand better the various flows that can occur. (See, for example, the studies by King and Tuck [2] and Villegas-Díaz, Power and Riley [6].) The flow of a rivulet on a planar substrate subject to a shear stress at its free surface has been investigated by several authors, notably Myers, Liang and Wetton [3], Saber and El-Genk [5], and Wilson and Duffy [9]. All of these works concern a non-perfectly wetting fluid; the flow of a rivulet of a perfectly wetting fluid in the absence of a shear stress at its free surface has been treated by Alekseenko, Geshev and Kuibin [1], and by Wilson and Duffy [7,8]. In the present short paper we use the lubrication approximation to obtain a complete description of the steady unidirectional flow of a thin rivulet of a perfectly wetting fluid on an inclined substrate subject to a prescribed uniform longitudinal shear stress at its free surface.

2 Problem Formulation

Consider the steady unidirectional flow of a thin rivulet with constant semi-width a and constant volume flux Q of a perfectly wetting fluid subject to a prescribed uniform longitudinal shear stress τ at its free surface on a planar substrate inclined at an angle α to the horizontal. Cartesian axes $Oxyz$ are chosen with the x-axis down the slope, the y-axis parallel to the substrate $z = 0$, and the z-axis normal to the substrate. The fluid is assumed to be Newtonian with constant density ρ, viscosity μ, and surface tension γ. The velocity $\mathbf{u} = u(y, z)\mathbf{i}$ and pressure $p = p(x, y, z)$ of the fluid are governed by the familiar mass-conservation and Navier–Stokes equations subject to the usual normal and tangential stress balances and the kinematic condition at the free surface $z = h(y)$, and no slip at the substrate $z = 0$. Since the fluid is perfectly wetting the contact angle is zero at the contact lines $y = \pm a$ (where, by definition, the rivulet has zero thickness).

We consider a thin rivulet with a small transverse aspect ratio $\epsilon \ll 1$; in this case it is appropriate to non-dimensionalise y and a with l, z and h with ϵl, u with $U = \rho g \epsilon^2 l^2 / \mu$, Q with $\epsilon l^2 U = \rho g \epsilon^3 l^4 / \mu$, $p - p_\infty$ and τ with $\rho g \epsilon l$, where $l = (\gamma / \rho g)^{1/2}$ is the capillary length, g is acceleration due to gravity, and p_∞ is the uniform atmospheric pressure.

Since the flow is unidirectional, the mass-conservation equation and kinematic boundary condition are satisfied identically, and at leading order in ϵ the Navier–Stokes equation reduces to

$$0 = \sin \alpha + u_{zz}, \quad 0 = -p_y, \quad 0 = -p_z - \cos \alpha, \tag{1}$$

which can readily be solved subject to boundary conditions of no slip at the substrate, $u = 0$ on $z = 0$, balances of normal and tangential stress at the free surface, $p = -h''$ and $u_z = \tau$ on $z = h$, and appropriate conditions at the contact lines, $h = 0$ and $h' = 0$ at $y = \pm a$, where the prime denotes differentiation with respect to argument, to give the solution

$$u = \frac{\sin \alpha}{2}(2h - z)z + \tau z, \quad p = (h - z) \cos \alpha - h''. \tag{2}$$

Substituting the solution for p into the second equation in (1) yields a third-order ordinary differential equation for the free surface profile h, namely $(h'' - \cos \alpha \, h)' = 0$ to be solved subject to $h = h' = 0$ at $y = \pm a$. This elementary problem was solved by Wilson and Duffy [7] who showed that there is no solution for h when $0 \le \alpha \le \pi/2$ (i.e. no solution corresponding to a sessile rivulet or a rivulet on a vertical substrate), but that there is a solution when $\pi/2 < \alpha \le \pi$ (corresponding to a pendent rivulet), namely

$$a = \frac{\pi}{m}, \quad h = \frac{h_m}{2}(1 + \cos my), \tag{3}$$

where $m = \sqrt{|\cos \alpha|}$ and $h_m = h(0)$ is the maximum height of the rivulet. The volume flux down the rivulet Q is given by

$$Q = \int_{-a}^{+a} \int_0^h u \, dz \, dy = \frac{\pi}{24m}(5 \sin \alpha \, h_m + 9\tau) h_m^2. \tag{4}$$

If the flux takes the prescribed value $Q = \bar{Q}$, then (4) determines the appropriate value(s) of h_m. Once h_m is known the rivulet solution given by (2) and (3) is completely determined. In the special case of no prescribed shear stress, $\tau = 0$, and in the limit of large prescribed shear stress, $|\tau| \to \infty$, we obtain the simple explicit solutions

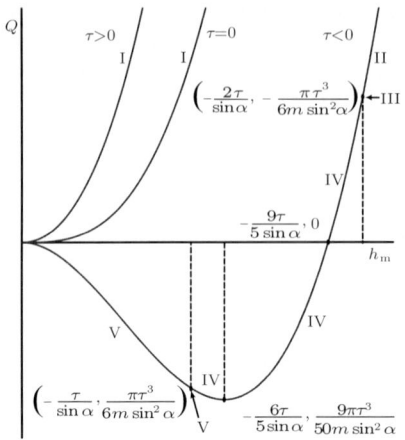

Fig. 1. Sketch of Q as a function of h_m for $\tau > 0$, $\tau = 0$ and $\tau < 0$, showing when the different types of flow pattern occur

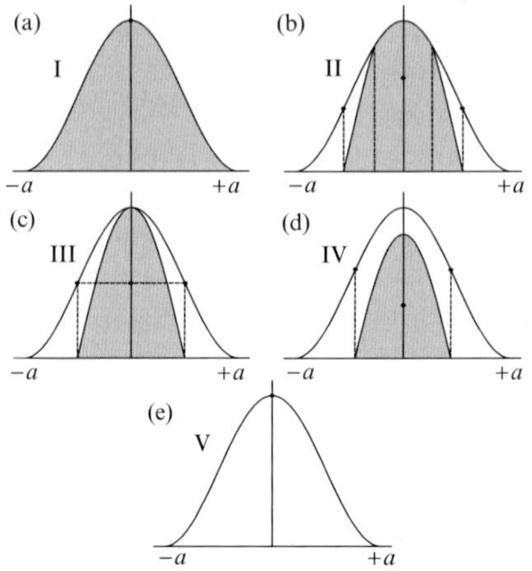

Fig. 2. Sketch of the five different types of flow pattern. Regions of downward flow are *shaded* and regions of upward flow are *unshaded*

$$h_m = \left(\frac{24m\bar{Q}}{5\pi \sin \alpha}\right)^{1/3} \quad \text{and} \quad h_m = \left(\frac{8m\bar{Q}}{3\pi\tau}\right)^{1/2}, \tag{5}$$

respectively.

3 Rivulet Solutions

Figure 1 shows a sketch of Q given by (4) as a function of h_m for $\tau > 0$, $\tau = 0$ and $\tau < 0$. For $\tau \geq 0$, Q is a monotonically increasing function of h_m tending to infinity as $h_m \to \infty$. In contrast, for $\tau < 0$, Q initially decreases monotonically to a minimum value $Q = Q_{\min}$, where $Q_{\min} = 9\pi\tau^3/50m \sin^2 \alpha \ (< 0)$, at $h_m = h_{\min} = -6\tau/5 \sin \alpha$, before increasing monotonically through the value $Q = 0$ at $h_m = h_{m0}$, where $h_{m0} = -9\tau/5 \sin \alpha$, and eventually tending to infinity as $h_m \to \infty$. The number of solutions for h_m thus depends on the sign of τ and the value of $\bar{Q}$. When $\tau \geq 0$, there is one solution when $\bar{Q} > 0$, but there are no solutions when $\bar{Q} \leq 0$. When $\tau < 0$, there is one solution when $\bar{Q} \geq 0$ with $h_m \geq h_{m0}$ and there are two solutions when $Q_{\min} < \bar{Q} < 0$, a "thin" solution with $0 < h_m < h_{\min}$ and a "thick" solution with $h_{\min} < h_m < h_{m0}$; when $\bar{Q} = Q_{\min}$ there is a single solution $h_m = h_{\min}$, and when $\bar{Q} < Q_{\min}$ there are no solutions.

4 Classification of Flow Patterns

Figure 2 shows a sketch of the five different types of cross-sectional flow patterns that can occur; regions of downward flow (i.e. $u > 0$) are shaded and regions of upward flow (i.e. $u < 0$) are unshaded. When $\tau > 0$ the prescribed shear stress acts down the substrate in cooperation with the effect of gravity. As a result, the flow is downward throughout the rivulet (we refer to this flow pattern as type I; see Fig. 2a). When $\tau < 0$, the prescribed shear stress acts up the substrate in opposition to the effect of gravity, which leads to more interesting behaviour than in the case $\tau \geq 0$. In particular, we find that although the velocity can be downward within the rivulet, it is always upward near the contact lines. When $h_{\mathrm{m}} > h_{\mathrm{III}} = -2\tau/\sin\alpha$ there is both upward and downward flow on the free surface (type II, Fig. 2b). When $h_{\mathrm{m}} = h_{\mathrm{III}}$ the flow is upward on the free surface except at $y = 0$ and $z = h_{\mathrm{m}}$, where the velocity is zero (type III, Fig. 2c). When $h_{\mathrm{V}} < h_{\mathrm{m}} < h_{\mathrm{III}}$, where $h_{\mathrm{V}} = -\tau/\sin\alpha$, the flow is always upward on the free surface (type IV, Fig. 2d) but is downward within part of the rivulet. Finally, when $h_{\mathrm{m}} \leq h_{\mathrm{V}}$ the effect of the prescribed shear stress dominates that of gravity and the flow is upward throughout the rivulet (type V, Fig. 2e). Figure 1 summarises when the different types of flow pattern occur.

5 Solutions for Prescribed τ and Varying α

Figure 3 shows a plot of h_{m} as a function of α/π when $\tau = 1$ and is typical of all such plots for $\tau > 0$. When $\bar{Q} > 0$ there is a single solution for h_{m} for all $\pi/2 < \alpha \leq \pi$ and all solutions are of type I. Figure 4 shows a plot of h_{m} as a function of α/π when $\tau = -1$ and is typical of all such plots for $\tau < 0$. When $\bar{Q} \geq 0$ there is a single solution for h_{m} $(\geq h_{\mathrm{m0}})$ for all $\pi/2 < \alpha \leq \pi$. However,

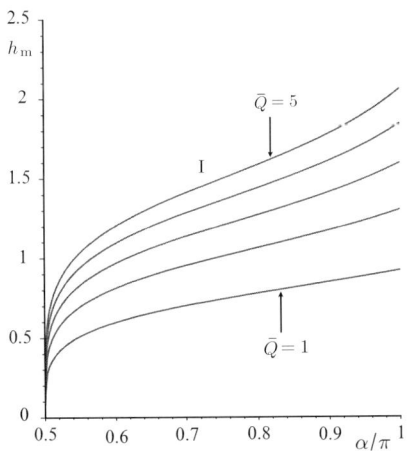

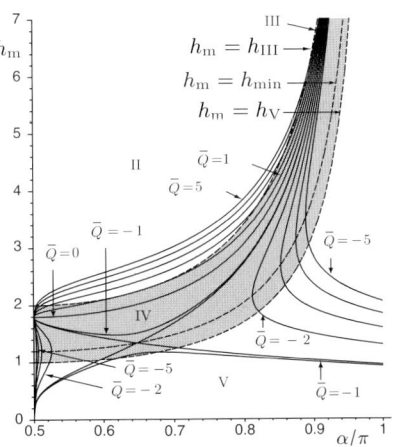

Fig. 3. Plot of h_{m} as a function of α/π when $\tau = 1$ for $\bar{Q} = 1, \ldots, 5$. Note that all solutions are of type I

Fig. 4. Plot of h_{m} as a function of α/π when $\tau = -1$ for $\bar{Q} = -5, \ldots, 5$ showing when the different types of flow pattern occur

when $\bar{Q} < 0$ there can be no, one or two solution(s) for h_{m} ($< h_{\mathrm{m0}}$). When $Q_{\mathrm{c}} < \bar{Q} < 0$, where $Q_{\mathrm{c}} = 9\pi 5^{1/4}\tau^3/40 \simeq 1.0570\tau^3$ (< 0), there is a thick and a thin solution for all $\pi/2 < \alpha \leq \pi$. When $\bar{Q} = Q_{\mathrm{c}}$ these solutions coincide at $\alpha = \alpha_{\mathrm{c}} = \pi - \tan^{-1} 2 \simeq 0.6476\pi$ and $h_{\mathrm{m}} = h_{\mathrm{mc}} = h_{\min}(\alpha_{\mathrm{c}}) = -3\tau/\sqrt{5} \simeq -1.3416\tau$, while for $\bar{Q} < Q_{\mathrm{c}}$ there are two disconnected branches of solutions, each consisting of a thick and a thin solution which coincide on the curve $h_{\mathrm{m}} = h_{\min}$. Figure 4 also shows how the curves $h_{\mathrm{m}} = h_{\mathrm{III}}$ and $h_{\mathrm{m}} = h_{\mathrm{V}}$ divide the α–h_{m} plane into regions in which different types of flow pattern occur.

6 Conclusions

We have obtained a complete description of the steady unidirectional flow of a thin rivulet of a perfectly wetting fluid on an inclined substrate subject to a prescribed uniform longitudinal shear stress at its free surface. In ongoing work we are analysing the stability of such a rivulet to small perturbations and investigating when it is energetically favourable for it to split into sub-rivulets.

References

1. S. V. Alekseenko, P. I. Geshev, and P. A. Kuibin "Free-boundary fluid flow on an inclined cylinder," Soviet Phys. Dokl. **42**, 269–272 (1997).
2. A. C. King and E. O. Tuck, "Thin liquid layers supported by steady air-flow surface traction," J. Fluid Mech. **251**, 709–718 (1993).
3. T. G. Myers, H. X. Liang, and B. Wetton, "The stability and flow of a rivulet driven by interfacial shear and gravity," Int. J. Non-Linear Mech. **39**, 1239–1249 (2004).
4. A. Oron, S. H. Davis, and S. G. Bankoff, "Long-scale evolution of thin liquid films," Rev. Mod. Phys. **69**, 931–980 (1997).
5. H. H. Saber and M. S. El-Genk, "On the breakup of a thin liquid film subject to interfacial shear," J. Fluid Mech. **500**, 113–133 (2004).
6. M. Villegas-Díaz, H. Power and D. S. Riley, "On the stability of rimming flows to two-dimensional disturbances," Fluid Dyn. Res. **33**, 141–172 (2003).
7. S. K. Wilson and B. R. Duffy, "A rivulet of perfectly wetting fluid draining steadily down a slowly varying substrate," IMA J. Appl. Math **70**, 293–322 (2005).
8. S. K. Wilson and B. R. Duffy, "When is it energetically favorable for a rivulet of perfectly wetting fluid to split?" Phys. Fluids **17**, 078104-1–07801-3 (2005).
9. S. K. Wilson and B. R. Duffy, "Unidirectional flow of a thin rivulet on a vertical substrate subject to a prescribed uniform shear stress at its free surface," Phys. Fluids **17**, 108105-1–108105-4 (2005).

The Effect of the Thermal Conductivity of the Substrate on Droplet Evaporation

Gavin J. Dunn[1], Stephen K. Wilson[1], Brian R. Duffy[1], Samuel David[2], and Khellil Sefiane[2]

[1] Department of Mathematics, University of Strathclyde, Livingstone Tower, 26 Richmond Street, Glasgow G1 1XH, UK
[2] School of Engineering and Electronics, University of Edinburgh, Sanderson Building, The King's Buildings, Mayfield Road, Edinburgh EH9 3JL, UK

1 Introduction

The evaporation of liquid droplets is of fundamental importance to industry, with a vast number of applications including ink-jet printing, spray cooling and DNA mapping, and has been the subject of considerable theoretical and experimental research in recent years. Significant recent papers include those by Deegan [1], Deegan et al. [2], Hu and Larson [3], Poulard et al. [4], Sultan et al. [5], and Shahidzadeh-Bonn et al. [6].

New experiments that we have conducted recently using a variety of liquids and substrates show that the thermal conductivity of the substrate can have a significant effect on the total evaporation rate, behaviour not captured by the widely used theoretical model of Deegan et al. [2] (hereafter referred to simply as "the Deegan model" for brevity). In this short paper a mathematical model for the quasi-steady diffusion-limited evaporation of a thin axisymmetric sessile droplet of liquid with a pinned contact line is formulated and solved. This model generalises the Deegan model to include the effect of evaporative cooling on the concentration of vapour at the free surface of the droplet. The results presented here show that the predictions of this new model are in good qualitative, and in some cases also quantitative, agreement with the new experimental results.

2 Mathematical Model

Consider the quasi-steady diffusion-limited evaporation of a thin axisymmetric sessile droplet of Newtonian fluid with constant viscosity, density ρ, surface tension, and thermal conductivity k on a thin horizontal substrate of constant thickness h^s with constant thermal conductivity k^s. Referred to polar coordinates (r, ϕ, z) with origin on the substrate at the centre of the droplet with

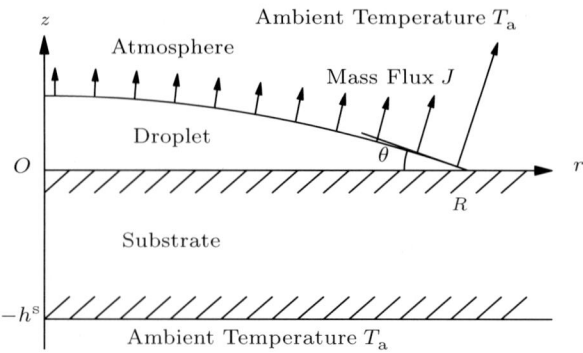

Fig. 1. Geometry of the problem

the z axis vertically upwards, the shape of the free surface of the droplet is denoted by $z = h(r, t)$, the upper surface of the substrate by $z = 0$, and the lower surface of the substrate by $z = -h^{\mathrm{s}}$, as shown in Fig. 1.

Motivated by the experimental results we assume that the contact line of the droplet is pinned by surface roughness (or other) effects so that the droplet radius R remains constant, and that the droplet is sufficiently small that surface tension effects dominate gravitational effects and hence that the droplet has the simple quasi-static shape $h = \theta(R^2 - r^2)/2R$ with volume $V = V(t)$ and contact angle $\theta = \theta(t)$ related by $\theta = 4V/\pi R^3$. The total evaporation rate is given by

$$\rho \frac{\mathrm{d}V}{\mathrm{d}t} = -2\pi \int_0^R J(r, t)\, r\, \mathrm{d}r, \tag{1}$$

where $J = J(r, t)$ (>0) is the local evaporative mass flux from the droplet.

The atmosphere surrounding the droplet and the substrate is assumed to be at a constant ambient temperature T_{a}. Since both the droplet and the substrate are thin, their temperatures, denoted by $T = T(r, z, t)$ and $T^{\mathrm{s}} = T^{\mathrm{s}}(r, z, t)$, satisfy $\partial^2 T/\partial z^2 = 0$ and $\partial^2 T^{\mathrm{s}}/\partial z^2 = 0$, respectively. The mass flux from the droplet satisfies the local energy balance $LJ = -k\partial T/\partial z$ on $z = h$ for $z < R$, where L is the latent heat of vaporisation. Hence, assuming that both the temperature and the heat flux are continuous between the droplet and the wetted part of the substrate, and that the lower surface of the substrate is at the ambient temperature T_{a}, we have

$$T = T_{\mathrm{a}} - LJ \left(\frac{z}{k} + \frac{h^{\mathrm{s}}}{k^{\mathrm{s}}} \right), \quad T^{\mathrm{s}} = T_{\mathrm{a}} - \frac{LJ}{k^{\mathrm{s}}}(z + h^{\mathrm{s}}), \tag{2}$$

showing clearly the evaporative cooling of both the droplet and the substrate.

Assuming that transport of vapour in the atmosphere is solely by diffusion, the concentration of vapour, denoted by $c = c(r, z, t)$, satisfies Laplace's equation, $\nabla^2 c = 0$. At the free surface of the droplet we assume that the atmosphere is saturated with vapour and hence, since the droplet is thin, that

$c = c_{\text{sat}}(T)$ on $z = 0$ for $r < R$, where the saturation value of the concentration $c_{\text{sat}} = c_{\text{sat}}(T)$ is assumed to be a linearly increasing function of temperature given by

$$c_{\text{sat}}(T) = c_{\text{sat}}(T_a) + \left.\frac{dc_{\text{sat}}}{dT}\right|_{T=T_a} (T - T_a). \tag{3}$$

On the dry part of the substrate there is no mass flux, i.e. $\partial c/\partial z = 0$ on $z = 0$ for $r > R$, and far from the droplet the concentration of vapour approaches its ambient value, i.e. for acetone and methanol $c \to 0$ while for water $c \to Hc_{\text{sat}}(T_a)$, where H is the relative humidity of the atmosphere, as $(r^2 + z^2)^{1/2} \to \infty$. Once c is known the mass flux from the droplet is given by $J = -D\partial c/\partial z$ on $z = 0$ for $r < R$, where D is the diffusion coefficient of vapour in the atmosphere.

In the special case when $dc_{\text{sat}}/dT|_{T=T_a}$ is negligible, the saturation concentration constant and the Deegan model is recovered. In this case the problem for c is independent of the temperature, and the solution for c (not repeated here for brevity) is well known and yields

$$J = \frac{2D(1-H)c_{\text{sat}}(T_a)}{\pi\sqrt{R^2 - r^2}} \quad \text{for} \quad r < R \tag{4}$$

and hence

$$\rho\frac{dV}{dt} = -4RD(1-H)c_{\text{sat}}(T_a), \tag{5}$$

and so, in particular, the evaporation rate is proportional to the radius of the droplet but independent of the thermal conductivity of both the liquid and the substrate. In general, the problem for c depends on the temperature and has to be solved numerically. This was done using the MATLAB-based numerical analysis package COMSOL (formerly FEMLAB).

3 Comparison with Experiments

Physical experiments were undertaken to investigate the effect of the thermal conductivity of the substrate on the evaporation of small droplets of three different liquids (specifically, acetone, methanol and water) on thin substrates of thickness 1 mm made of two materials with very different thermal conductivities, namely aluminium (Al) ($k^s = 237\,\text{W m}^{-1}\,\text{K}^{-1}$) and PTFE ($k^s = 0.25\,\text{W m}^{-1}\,\text{K}^{-1}$). The values of the relevant physical parameters for the three liquids used are listed in Table 1. Both substrates were coated with a very thin (3 μm) layer of Al to ensure that they had the same surface energy and roughness properties. The experiments were conducted in a controlled atmosphere with fixed temperature $T_a = 295\,\text{K}$, pressure 99.8 kPa, and relative humidity $H = 0.4$. Droplets with various volumes ranging from 0.5 to 8 μl were deposited on the substrates and left to evaporate spontaneously. The evaporation rates were measured using a KRUSS DSA 100 contact-angle

Table 1. Values of the relevant physical parameters for the three liquids used at temperature $T_a = 295\,\mathrm{K}$ and pressure $99.8\,\mathrm{kPa}$

Parameter	Units	Acetone	Methanol	Water
ρ	$\mathrm{kg\,m^{-3}}$	788	790	998
L	$\mathrm{J\,kg^{-1}}$	5.49×10^5	1.20×10^6	2.45×10^6
k	$\mathrm{W\,m^{-1}\,K^{-1}}$	0.161	0.203	0.604
c_{sat}	$\mathrm{kg\,m^{-3}}$	0.637	0.186	1.94×10^{-2}
$\mathrm{d}c_{\mathrm{sat}}/\mathrm{d}T$	$\mathrm{kg\,m^{-3}\,K^{-1}}$	2.84×10^{-2}	9.47×10^{-3}	1.11×10^{-3}
D	$\mathrm{m^2\,s^{-1}}$	1.06×10^{-5}	1.50×10^{-5}	2.46×10^{-5}

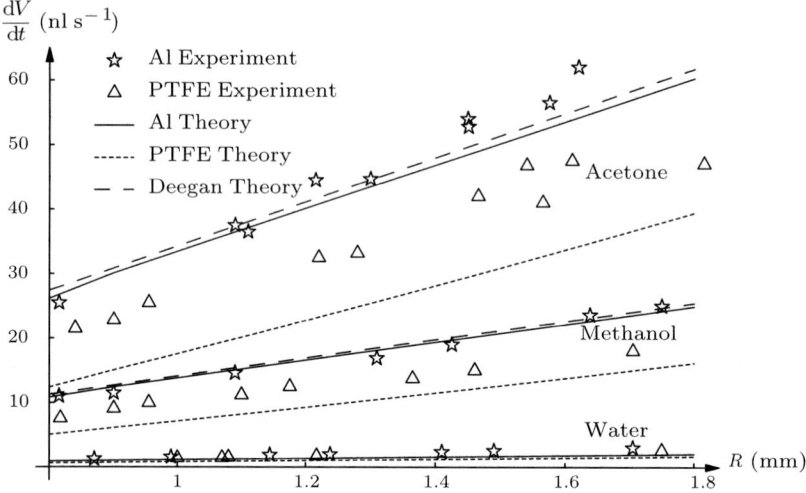

Fig. 2. Comparison between the experimentally measured values of the average evaporation rate for droplets of various radii and the corresponding theoretical predictions of the present model and the Deegan model given by (5)

analyser, and the accuracy of the results obtained was confirmed by using a micro-balance technique.

Figure 2 shows the comparison between the experimentally measured values of the average evaporation rate for droplets of various radii and the corresponding theoretical predictions of the present model and the Deegan model given by (5). Figure 2 shows that there is good qualitative agreement between the experimental and theoretical results. For acetone and methanol there is good quantitative agreement for the Al substrate, but the theory under-estimates the evaporation rate for the PTFE substrate. For water the theory under-estimates the evaporation rate for both substrates. Nevertheless, in view of the many assumptions made in deriving the model, the agreement is remarkably good, especially as there are no "fitting" parameters in the theory, and no "tuning" of the values of the physical parameters has taken place.

Perhaps the most satisfying aspect of the agreement shown in Fig. 2 is the manner in which the present model reproduces the significant difference in evaporation rate between droplets of the same liquid on different substrates. Figure 2 also shows that the predictions of the Deegan model are close to those of the present model for the Al but not the PTFE substrate. This is because Al is a much better conductor than PTFE and hence the evaporative cooling of a droplet on Al is much less than that of a droplet on PTFE, and hence the saturation concentration of vapour at the free surface is much closer to the constant value of $c_{\mathrm{sat}}(T_{\mathrm{a}})$ assumed in the Deegan model.

4 Further Work

The present work is restricted to the special case of a thin droplet on a thin substrate. In ongoing work we are currently extending the model to the general case of a non-thin droplet on a non-thin substrate, and preliminary results indicate that this generalisation significantly improves the quantitative agreement between the theoretical predictions and experimental results for acetone and methanol, but still leads to an under-estimation of the evaporation rate for water. A possible explanation for this latter under-estimation may be, as suggested recently by Shahidzadeh-Bonn et al. [6], that because water vapour (unlike acetone or methanol vapour) is lighter than air, buoyancy effects may play a significant role in enhancing the diffusion of vapour away from the droplet in this case.

Acknowledgements

This work is supported by the United Kingdom Engineering and Physical Sciences Research Council via grants GR/S59444 and GR/S59451.

References

1. R. D. Deegan, "Pattern formation in drying drops," Phys. Rev. E **61**, 475–485 (2000).
2. R. D. Deegan, O. Bakajin, T. F. Dupont, G. Huber, S. R. Nagel and T. A. Witten, "Contact line deposits in an evaporating drop," Phys. Rev. E **62**, 756–765 (2000).
3. H. Hu and R. G. Larson, "Evaporation of a sessile droplet on a substrate," J. Phys. Chem. B **106**, 1334–1344 (2002).
4. C. Poulard, G. Guéna and A. M. Cazabat, "Diffusion-driven evaporation of sessile drops," J. Phys.: Condens. Matter **17**, S4213–S4227 (2005).
5. E. Sultan, A. Boudaoud and M. Ben Amar, "Evaporation of a thin film: diffusion of the vapour and Marangoni instabilities," J. Fluid Mech. **543**, 183–202 (2005).
6. N. Shahidzadeh-Bonn, S. Rafaï, A. Azouni and D. Bonn, "Evaporating droplets," J. Fluid Mech. **549**, 307–313 (2006).

The Effect of Particles on Linear and Weakly Nonlinear Instability of a Two-Phase Shallow Flows

Andrei Kolyshkin and Sergejs Nazarovs

Riga Technical University, Riga Latvia
akoliskins@rbs.lv, nazarow@yahoo.com

1 Introduction

Shallow flows are widespread in nature and engineering. Examples include shallow wakes (flows behind obstacles such as islands), shallow mixing layers (flows at river junctions) and shallow jets. Shallow flows, where the transverse length scale of the flow, d, is much larger than water depth, h, i.e., $d/h \gg 1$, are very different from deep water flows. This difference is associated with the fact that bottom friction plays an important role in suppressing flow instability. In addition, limited water depth prevents the development of three-dimensional instabilities.

Different methods of analysis of two-dimensional structures in shallow water flows are considered in a recent review [1]. Methods of linear stability theory are widely used in the analysis of shallow flows. Different aspects of the linear stability of shallow flows are analyzed in [2–6]. Experiments in [2] showed that the following three flow regimes can be identified in shallow wake flows: steady bubble, unsteady bubble and vortex street. Theoretical studies in [4–6] demonstrated that these regimes are related to convective/absolute instabilities in the shallow water layer.

Two-phase shear flows can be found in many engineering applications. Examples include gas–solid particle flows, gas–droplet flows and liquid–gas bubble flows. Several simplifying assumptions are used in stability analyses of two-phase flows (see [7] and [8]). These assumptions are as follows. First, the particle distribution is assumed to be uniform. Second, it is assumed that small perturbations imposed on the flow have no effect on the particles during the initial moment. The case of a dynamic interaction of particles with the base flow is considered in [9].

Stability of shallow water flows is usually analyzed under the assumption that the base flow is parallel. It is well-known, however, that the width of shallow mixing layer and the width of shallow wake are not constant and are slowly changing with respect to the longitudinal coordinate (see, for example, [10]). The goal of the present paper is to take into account slow variation of the flow in the longitudinal direction and construct an asymptotic scheme where the local parallel stability results appear as the leading-order approximation. The analysis is performed for two-phase flows under some simplifying assumptions (see [8]). Such schemes have been already applied to spatial stability studies of slowly diverging single-phase shear flows in deep water (see [11]) and are based on the WKB approach.

2 Weakly Nonlinear Spatial Instability of Two-Phase Flows

Consider the two-dimensional shallow water equations of the form

$$\frac{\partial u}{\partial x} + \frac{\partial v}{\partial y} = 0, \tag{1}$$

$$\frac{\partial u}{\partial t} + u\frac{\partial u}{\partial x} + v\frac{\partial u}{\partial y} + \frac{\partial p}{\partial x} + \frac{c_f}{2h}u\sqrt{u^2 + v^2} - B(u^p - u) = 0, \tag{2}$$

$$\frac{\partial v}{\partial t} + u\frac{\partial v}{\partial x} + v\frac{\partial v}{\partial y} + \frac{\partial p}{\partial y} + \frac{c_f}{2h}v\sqrt{u^2 + v^2} - B(v^p - v) = 0, \tag{3}$$

where u and v are the depth-averaged velocity components in the x and y-directions, respectively, u^p and v^p are particle velocities, h is water depth, c_f is the friction coefficient, p is the pressure and B is the particle loading parameter (see [7,8]). Introducing the stream function $\psi(x, y, t)$ by the relations

$$u = \psi_y, \quad v = -\psi_x \tag{4}$$

and using the simplifying assumptions [7,8] we rewrite the system (1)–(3) in the form

$$(\Delta\psi)_t + \psi_y(\Delta\psi)_x - \psi_x(\Delta\psi)_y + \frac{c_f}{2h}\Delta\psi\sqrt{\psi_x^2 + \psi_y^2}$$

$$+ \frac{c_f}{2h\sqrt{\psi_x^2 + \psi_y^2}}[\psi_y^2\psi_{yy} + 2\psi_x\psi_y\psi_{xy} + \psi_x^2\psi_{xx}] + B\Delta\psi = 0. \tag{5}$$

Assuming that the normal velocity component is small in comparison with the streamwise component and that the base flow quantities are weakly varying functions of the streamwise coordinate, we introduce a slow streamwise coordinate $X = \varepsilon x$, where the small parameter $\varepsilon = \lambda/L \ll 1$ represents the degree of non-parallelism of the flow. Here λ is the instability wavelength and L is the length scale which is associated with streamwise inhomogeneities of the base flow.

The stream function of the flow is represented in the form

$$\psi(x, y, t) = \psi_0(y, X) + \psi_f(x, y, t), \tag{6}$$

where the first and the second term on the right-hand side of (6) represent the base flow and the perturbed flow, respectively. In this case $\psi_f(x, y, t)$ in (6), in general, is not small. Linearizing (5) in the neighborhood of the base flow, dropping the subscript "f" and retaining only the terms of order ε we obtain

$$
\frac{\partial}{\partial t}\left(\frac{\partial^2 \psi}{\partial x^2} + \frac{\partial^2 \psi}{\partial y^2}\right) + U\frac{\partial}{\partial x}\left(\frac{\partial^2 \psi}{\partial x^2} + \frac{\partial^2 \psi}{\partial y^2}\right) - \frac{\partial \psi}{\partial x}\frac{\partial^2 U}{\partial y^2} + B\left(\frac{\partial^2 \psi}{\partial x^2} + \frac{\partial^2 \psi}{\partial y^2}\right)
$$
$$
+ \frac{c_f}{2h}\left[U\left(\frac{\partial^2 \psi}{\partial x^2} + 2\frac{\partial^2 \psi}{\partial y^2}\right) + 2\frac{\partial U}{\partial y}\frac{\partial \psi}{\partial y}\right] + \varepsilon\left[\frac{\partial^2 U}{\partial y \partial X}\frac{\partial \psi}{\partial y} + V\frac{\partial}{\partial y}\left(\frac{\partial^2 \psi}{\partial x^2} + \frac{\partial^2 \psi}{\partial y^2}\right)\right.
$$
$$
\left. + \frac{c_f}{2h}\left(2\frac{\partial U}{\partial X}\frac{\partial \psi}{\partial x} - 2V\frac{\partial^2 \psi}{\partial y \partial X} + \frac{\partial U}{\partial y}\frac{V}{U}\frac{\partial \psi}{\partial x}\right)\right] = 0, \tag{7}
$$

where $U = \psi_{0y}$ and $V = -\psi_{0X}$.

We decompose the perturbation stream function $\psi(x, y, t)$ into a slowly varying amplitude function $\varphi(y, X, \omega)$ and a fast varying phase function $\theta(X, \omega)/\varepsilon$:

$$
\psi(x, y, \omega, t) = \varphi(y, X, \omega)\exp\left[i\left(\frac{\theta(X, \omega)}{\varepsilon} - \omega t\right)\right]. \tag{8}
$$

Next, the function $\varphi(y, X, \omega)$ is expanded into a power series in ε in the form

$$
\varphi(y, X, \omega) = \varphi_1(y, X, \omega) + \varepsilon\varphi_2(y, X, \omega) + \cdots \tag{9}
$$

Substituting (8) and (9) into (7) and collecting the terms that do not contain ε we obtain

$$
\mathcal{L}\varphi_1 = 0, \tag{10}
$$

where

$$
\mathcal{L}\varphi_1 = \varphi_1'' - k^2\varphi_1 - \frac{U''}{U - c}\varphi_1 + \frac{c_f i}{2hk(U - c)}\left(-k^2 U\varphi_1\right.
$$
$$
\left. + 2U\varphi_1'' + 2U'\varphi_1'\right) + B\frac{ik}{U - c}\varphi_1 = 0, \tag{11}
$$

and $c = \omega/k$. The primes in (11) denote the derivatives with respect to y and $k = k(X, \omega) = \theta_X$. Equation (11) is the modified Rayleigh equation which is obtained in [12] under parallel flow assumption. Equation (11) together with zero boundary conditions forms an eigenvalue problem (where the eigenvalues are $k = k(X, \omega)$). Temporal linear stability calculations for several wake profiles in shallow water are performed in [12]. It is shown that growth rates of the most unstable mode decrease as the particle loading parameter B increases. Calculations show that the critical values of the parameter characterizing bottom friction decrease almost linearly as the parameter B increases. The eigenvalues $k = k(X, \omega) = \theta_X$ can be also obtained as a result of the numerical solution of the spatial stability problem. Note that the coordinate X appears in (11) as a parameter.

In order to derive the amplitude evolution equation under the assumption of weak non-parallelism of the flow we assume that

$$\varphi_1(y, X, \omega) = A(X, \omega)\Phi(y, X, \omega), \tag{12}$$

where $A(X, \omega)$ is an unknown complex amplitude and $\Phi(y, X, \omega)$ is a normalized eigenfunction of the linear stability problem.

Substituting (8), (9) and (12) into (7) and collecting the terms containing ε we obtain

$$\mathcal{L}\varphi_2 = g, \tag{13}$$

where

$$
\begin{aligned}
g = {} & \frac{\mathrm{i}}{kU - \omega}\frac{\mathrm{d}A}{\mathrm{d}X}\left(2\omega k\Phi - 3Uk^2\Phi + U\Phi'' - \Phi U'' + \frac{c_f}{2h}2\mathrm{i}kU\Phi + 2\mathrm{i}kB\Phi\right) \\
& + \frac{\mathrm{i}}{kU - \omega}A\left[2\omega k\frac{\partial\Phi}{\partial X} + \omega\Phi\frac{\mathrm{d}k}{\mathrm{d}X} - 3Uk^2\frac{\partial\Phi}{\partial X} - 3Uk\Phi\frac{\mathrm{d}k}{\mathrm{d}X} + U\frac{\partial\Phi''}{\partial X}\right. \\
& - U''\frac{\partial\Phi}{\partial X} + \frac{c_f}{2h}\left(2\mathrm{i}kU\frac{\partial\Phi}{\partial X} + \mathrm{i}U\Phi\frac{\mathrm{d}k}{\mathrm{d}X} + 2\mathrm{i}k\Phi\frac{\partial U}{\partial X} - 2\mathrm{i}kV\Phi'\right) \\
& \left. + \frac{\partial^2 U}{\partial y\partial X}\Phi' - Vk^2\Phi' + V\Phi''' + B\left(2\mathrm{i}k\frac{\partial\Phi}{\partial X} + \mathrm{i}\Phi\frac{\mathrm{d}k}{\mathrm{d}X}\right)\right].
\end{aligned}
\tag{14}
$$

In accordance with the Fredholm's alternative, (13) has a solution if and only if the function g in (14) is orthogonal to all eigenfunctions $\tilde{\Phi}$ of the corresponding adjoint problem. Thus,

$$\int_{-\infty}^{\infty} g\tilde{\Phi}\,\mathrm{d}y = 0. \tag{15}$$

The equation for the function $A(X, \omega)$ is obtained from (15) in the form

$$M(X, \omega)\frac{\mathrm{d}A}{\mathrm{d}X} + N(X, \omega)A = 0, \tag{16}$$

where

$$
\begin{aligned}
M(X, \omega) = \mathrm{i}\int_{-\infty}^{\infty}\frac{1}{kU - \omega}\left(2\omega k\Phi - 3Uk^2\Phi + U\Phi''\right. \\
\left. - \Phi U'' + \frac{c_f}{2h}2\mathrm{i}kU\Phi + 2\mathrm{i}kB\Phi\right)\tilde{\Phi}\,\mathrm{d}y
\end{aligned}
\tag{17}
$$

$$
\begin{aligned}
N(X,\omega) = \mathrm{i} \int_{-\infty}^{\infty} \frac{1}{kU-\omega} &\left[2\omega k \frac{\partial \Phi}{\partial X} + \omega \frac{\mathrm{d}k}{\mathrm{d}X} \Phi - 3Uk^2 \frac{\partial \Phi}{\partial X} \right.\\
&- 3Uk\Phi \frac{\mathrm{d}k}{\mathrm{d}X} + U \frac{\partial \Phi''}{\partial X} - U'' \frac{\partial \Phi}{\partial X} + \frac{c_f}{2h}\left(2\mathrm{i}kU \frac{\partial \Phi}{\partial X} + \mathrm{i}U\Phi \frac{\mathrm{d}k}{\mathrm{d}X} \right.\\
&\left. + 2\mathrm{i}k\Phi \frac{\partial U}{\partial X} - 2\mathrm{i}kV\Phi' \right) + \frac{\partial^2 U}{\partial y \partial X} \Phi' - Vk^2\Phi' + V\Phi'''\\
&\left. + B\left(2\mathrm{i}k \frac{\partial \Phi}{\partial X} + \mathrm{i}\frac{\mathrm{d}k}{\mathrm{d}X} \Phi \right) \right] \tilde{\Phi} \, \mathrm{d}y.
\end{aligned} \tag{18}
$$

Thus, the leading order approximation of the stream function $\psi(x,y,\omega,t)$ is

$$
\psi(x,y,\omega,t) \sim A(X,\omega)\Phi(y,X,\omega) \exp\left[\mathrm{i}\left(\frac{1}{\varepsilon} \int_0^X k(X,\omega) \, \mathrm{d}X - \omega t \right) \right]. \tag{19}
$$

Formula (19) provides the connection between local parallel flow approximations and takes into account slow streamwise variation of the base flow. It follows from (19) that each of the three terms on the right-hand side of (19) contain information related to the amplitude and phase of perturbations (see [13]). It can also be shown from (19) that the growth rate and phase speed of the perturbation at any given station x downstream are different for different choices of the perturbed quantities and even depend on cross-stream location at which they are evaluated. Following [13] we define a local wavenumber $\bar{k}$ for any flow variable Q as follows

$$
\bar{k}(x,y|Q) = -\mathrm{i}\frac{\partial}{\partial x} \ln Q(x,y). \tag{20}
$$

The real and imaginary parts of $\bar{k}$ can be interpreted as the local phase speed and the local spatial growth rate. In order to make a meaningful comparison between experimental data and theory one needs to select a particular flow quantity Q (for example, pressure, velocity, etc.), measure it at a particular point (x,y) and evaluate the right-hand side of (20) at the same point. Finally, using the weakly nonlinear model (19) one can calculate the right-hand side of (20) and compare the results with experimental data. This means that in order to test the validity of the proposed model either detailed experimental data or numerical solution of full nonlinear two-dimensional shallow water equations are needed.

References

1. Jirka, G.H.: Large scale flow structures and mixing processes in shallow flows. J. Hydr. Res., **39**, 567–573 (2001)
2. Chen D., Jirka G.H.: Experimental study of plane turbulent wake in a shallow water layer. Fluid Dyn. Res., **16**, 11–41 (1995)

3. Chu V.H., Wu J.H., Khayat R.E. : Stability of transverse shear flows in shallow open channel. J. Hydr. Eng., **117**, 1370–1388 (1991)
4. Chen D., Jirka G.H.: Absolute and convective instabilities of plane turbulent wakes in a shallow water layer. J. Fluid Mech., **338**, 157–172 (1997)
5. Ghidaoui M.S., Kolyshkin A.A.: Stability analysis of shallow wake flows. J. Fluid Mech., **494**, 355–377 (2003)
6. Ghidaoui, M.S., Kolyshkin, A.A., Liang, J.H., Chan, F.C., Li, Q., Xu, K.: Linear and nonlinear analysis of shallow wakes. J. Fluid Mech., **548**, 309–340 (2006)
7. Yang, Y., Chung, J.N., Troutt, T.R., Crowe, C.T.: The influence of particles on the spatial stability of two-phase shallow mixing layers. Phys. Fluids, **A2(10)**, 1839–1845 (1990)
8. Yang, Y., Chung, J.N., Troutt, T.R., Crowe, C.T.: The effect of particles on the stability of a two-phase wake flow. Int. J. Multiphase Flow, **19**, 137–149 (1993)
9. Dimas, A.A., Kiger, K.T: Linear instability of a particle-laden mixing layer with a dynamic dispersed phase. Phys. Fluids, **10**, 2539–2557 (1998)
10. van Prooijen, B.C., Uijttewaal, W.S.J.: A linear approach for the evolution of coherent structures in shallow mixing layer. Phys. Fluids, **14**, 4105–4114 (2002)
11. Huerre P., Rossi M.: Hydrodynamic instabilities in open flows. In: Godreche C., Manneville P., (eds) Hydrodynamics and Nonlinear Instabilities. Cambridge University Press, New York (1998)
12. Kolyshkin, A.A., Nazarovs, S.: Linear and weakly nonlinear analysis of two-phase shallow wake flows. WSEAS Transactions on Mathematics, **6**, No.1, 1–8 (2006)
13. Crighton D.G., Gaster M.: Stability of slowly diverging jet flow. J. Fluid Mech., **77**, part 2, 397–413 (1976)

Water Quality Simulation of a Future Pit Lake

A. Bermúdez[1], L.M. García García[1], P. Quintela[1], and J.L. Delgado[2]

[1] Universidade de Santiago de Compostela
   mabermud@usc.es, luzgar@usc.es, mapere@usc.es
[2] Lignitos de Meirama, S.A
   jldelgado@limeisa.es

1 Introduction

By the end of 2007, the "Lignitos de Meirama" open pit coal mine will cease its extraction activities definitively and a lake is expected to develop due to the hydrological conditions of the region. Its possible connection to a reservoir which supplies water to the city of A Coruña (NW Spain) enforces the mining company to fulfill the Spanish water quality standards. In this frame, a numerical model to predict the future lake water quality has been developed. The water quality of a lake generated by filling a former open pit coal mine depends on several factors, such as the presence of iron sulfides at the pit walls and the environmentally hazardous consequences of their oxidation (heavy metals release and water acidification), the establishment of a flow regime on the future lake as a result of the water discharges on the pit and the possible stratification of the water column due to seasonal changes of the solar radiation [2].

The concurrence of all these phenomena makes the water quality modelling a difficult task in which hydrodynamic, thermal and geochemical considerations have to be studied in a coupled way. For this reason Mike 3 Flow Model FM software package from DHI Water and Environment (Denmark) was selected. Mike 3 comprises three modules, two of which were used in this work: the *hydrodynamic module* (HD), based on the solution of the three-dimensional shallow water equations considering the temperature evolution, and the *Ecolab module*, that allows the implementation of any environmental problem in which the evolution of the state variables can be described by means of a system of ordinary differential equations (ODEs, from now on).

2 Model Development

As it was stated before, Mike 3 is a 3D model in which the evolution of the concentration of a chemical species can be written as

$$\frac{\partial y_i}{\partial t} + \frac{\partial u y_i}{\partial x} + \frac{\partial v y_i}{\partial y} + \frac{\partial w y_i}{\partial z} = F_{y_i} + \frac{\partial}{\partial z}\left(D_v \frac{\partial y_i}{\partial z}\right) + y_{i_s}S + \phi_i, \qquad (1)$$

where y_i represents the concentration of species i, (u, v, w) are the flow velocities in the (x, y, z) directions respectively, F_{y_i} is the horizontal diffusion term, D_v is the vertical diffusion coefficient, S is the rate flow of the water sources, y_{i_s} is the concentration of the i-th species in the water sources and ϕ_i is a term representing the rate of change in concentration due to production/consumption mechanisms which, in this case, are related to (bio) geochemical phenomena [Mike 3 HD (2005)].

The current velocities and temperature distribution of the future lake, which are needed to solve (1), were estimated using Mike 3 HD.

This paper will be focused on the calculation of ϕ_i. In other words, for the sake of simplicity in the exposition, the lake will be treated as if it were a stirred tank.

2.1 Geochemical Model

This part of the work started with a literature review regarding all the environmentally relevant chemical reactions related to the existence of iron sulfides and silicates at the pit walls (Table 1).

On this basis, sixteen chemical species (twelve of them dissolved species and four precipitates) were chosen to define the future lake water quality. Table 2 collects all these species together with the adopted notation for each one. Iron, protons, aluminium and manganese represent the most important ones in terms of environmental hazard.

Let us represent the L chemical reactions involving our dissolved chemical species E_i by

$$\xi_1^l E_1 + \cdots + \xi_{12}^l E_{12} \longrightarrow \zeta_1^l E_1 + \cdots + \zeta_{12}^l E_{12}, \qquad l = 1, \ldots, L \qquad (2)$$

ξ_i and ζ_i being the stoichiometric coefficients. The evolution of the concentration of these chemical species due to geochemical processes (term ϕ_i in (1)), would be represented by a system of ODEs:

$$\frac{dy_i}{dt} = \sum_{l=1}^{L}(\zeta_i^l - \xi_i^l)v_{r_l}, \quad i = 1, \ldots, 12, \qquad (3)$$

v_{r_l} being the velocity of the l-th chemical reaction, which is generally expressed by an empirical law

$$v_{r_l} = h_l(y_1, \ldots, y_{12}) \qquad (4)$$

for some functions h_l, $l = 1, \ldots, 12$.

Table 1. Most relevant chemical reactions at Lignitos de Meirama coal mine

---

1. Pyrite oxidation via O_2: $FeS_2(s) + 7/2O_2(aq) + H_2O \longrightarrow Fe^{2+}(aq) + 2SO_4^{2-}(aq) + 2H^+(aq)$

2. Pyrite oxidation via Fe^{3+}: $FeS_2(s) + 14Fe^{3+}(aq) + 8H_2O \longrightarrow$
$$15Fe^{2+}(aq) + 2SO_4^{2-}(aq) + 16H^+(aq)$$

3. Chalcopyrite oxidation via O_2: $CuFeS_2(s) + 4O_2(aq) \longrightarrow Fe^{2+}(aq) + Cu^{2+}(aq) + 2SO_4^{2-}(aq)$

4. Chalcopyrite oxidation via Fe^{3+}:
$CuFeS_2(s) + 16Fe^{3+}(aq) + 8H_2O \longrightarrow 17Fe^{2+}(aq) + Cu^{2+}(aq) + 2SO_4^{2-}(aq) + 16H^+$

5. Ferrous ion oxidation: $Fe^{2+}(aq) + 1/4O_2(aq) + H^+(aq) \longrightarrow Fe^{3+}(aq) + 1/2H_2O(aq)$

6. Chlorite weathering*:
$(Mg_aFe_b{}^{2+}Fe_c{}^{3+}Mn_d)Al_2Si_3O_{10}(OH)_8(s) + 16H^+ \longrightarrow$
$$aMg^{2+} + bFe^{2+} + cFe^3 + dMn^{2+} + 2Al^{3+} + 3SiO_2(s) + 12H_2O$$

7. Muscovite weathering: $KAl_3Si_3O_{10}(OH)_2(s) + 10H^+ \longrightarrow K^+ + 3Al^{3+} + 3SiO_2(s) + 6H_2O$

8. Plagioclase weathering: $Na_eCa_fAlSi_3O_8 + 4H^+ \longrightarrow eNa^+ + fCa^{2+} + Al^{3+} + 3SiO_2(s) + 2H_2O$

9. Microcline weathering: $KAlSi_3O_8 + 4H^+ \longrightarrow K^+ + Al^{3+} + 3SiO_2 + 2H_2O$

10. Biotite weathering :
$K(Fe_g{}^{2+}, Mg_h)_3(Al_l, Fe_m{}^{3+})Si_3O_{10}(OH_q, F_r)_2 + 2(2q + 10)H^+ \longrightarrow$
$$K^+ + 3gFe^{2+} + 3hMg^{2+} + 1Al^{3+} + mFe^{3+} + 2rF^- + 3SiO_2 + (2q + 10)H_2O$$

11. Kaolinite weathering: $Al_2Si_2O_5(OH)_4 + 6H^+ \longrightarrow H_2O + 2H_4SiO_4 + 2Al^{3+}$

---

*The exact composition of Chlorite, Plagioclase and Biotite is not known, so subindexes a, b, c, d, e, f, g, h, l, m and q have been used to express the composition with respect to certain species.

Table 2. Studied chemical species and notation

Not.	Chem.sp	Not.	Chem.sp	Not.	Chem.sp
y_1	$[Fe^{2+}]$	y_7	$[Ca^{2+}]$	y_{13}	$[Fe(OH)_3]$
y_2	$[Fe^{3+}]$	y_8	$[Mn^{2+}]$	y_{14}	$[Al(OH)_3]$
y_3	$[O_2]$	y_9	$[Na^+]$	y_{15}	*gypsum*
y_4	$[H^+]$	y_{10}	$[K^+]$	y_{16}	*manganite*
y_5	$[Cu^{2+}]$	y_{11}	$[Mg^{2+}]$	y_{17}	$[SO_4]_L$
y_6	$[Al^{3+}]$	y_{12}	$[SO_4]$	$\mathbf{y}$	$(y_1,\ldots,y_{17})$

In addition to reactions on Table 1, some of the selected dissolved species are subjected to restrictions derived from the solubility equilibria they establish with secondary minerals. This means that, if one of these species overcomes a certain threshold concentration, a precipitate will form. For example, if manganese concentration is higher than a limit (the *saturation concentration*), then manganite will precipitate. Mathematically, this implies that certain constraints should be satisfied:

$$g_r(\mathbf{y}) \leq 0; \quad r = 1,\ldots,4. \tag{5}$$

Following the example above, the restriction function for manganese (y_8, Table 2) is

$$g_4(\mathbf{y}) = y_8 - y_{8_{sat}} \leq 0, \tag{6}$$

where $y_{8_{sat}}$ is the manganese saturation concentration.

Taking into account all the considerations above and following notation on Table 2, the evolution of manganese concentration could be written as

$$\frac{dy_8}{dt} = \frac{\Gamma_{chl}\,\mathcal{S}}{V} d(k_{clo1}y_4^{0.5} + k_{clo2}) - k_{adsmn}\frac{y_8}{\alpha + K_{eq}\,y_{17}}\,[\text{hydroxide}] - \text{pmn}, \quad (7)$$

where the first term on the right is the rate of manganese production due to reaction 6 in Table 1 (Γ_{chl} is the fraction of chlorite at the rock surface ($\mathcal{S}$), V is the lake volume, d is the stoichiometric coefficient of the reaction and k_{clo1} and k_{clo2} are the rate constants). The second term represents the rate of manganese adsorption onto hydroxides (k_{adsmn} is the rate constant for this reaction, α is a function of y_4 and K_{eq} is an equilibrium constant) and the last term accounts for the rate of manganese consumption due to manganite precipitation. Equations like (7) were obtained for the rest of the species. Summarizing, the future lake water quality is determined by

$$(\mathcal{P})\quad\begin{cases}\dfrac{dy_i}{dt} = f_i(\mathbf{y}) - \displaystyle\sum_{r=1}^{4} p_r\dfrac{\partial g_r(\mathbf{y})}{\partial y_i} + y_{i_s}\mathcal{S}, & 1 \le i \le 12, \quad (8)\\[3mm] \dfrac{dy_{j_r}}{dt} = p_r\displaystyle\sum_{i=1}^{12}\dfrac{\partial g_r(\mathbf{y})}{\partial y_i} + y_{j_{r_s}}\mathcal{S}, & r = 1,\ldots,4, \quad (9)\\[3mm] F(\mathbf{y}) = 0, & (10)\\[2mm] p_r = \max\{0, p_r + \dfrac{1}{\lambda}g_r(\mathbf{y})\} \quad \forall\lambda > 0, & r = 1,\ldots,4, \quad (11)\\[2mm] \mathbf{y}(0) = \mathbf{y}_0. & (12)\end{cases}$$

Equation (8) represents the evolution of the dissolved species, with $p_r(t)$ the Lagrange multiplier subjected to restriction $g_r(\mathbf{y})$ (see (11) obtained from [1]). Term $f_i(\mathbf{y})$ is equal to the right-hand side of (3). Equation (9) accounts for the evolution of the precipitates ($j_r = 12 + r$); (10) is an algebraic equation for the calculation of y_{17}, and finally, the initial conditions of the problem are included in (12).

Reminding that Mike 3 EcoLab (cf. [Ecolab DHI (2005)]) only allows the implementation of ODEs to describe an environmental problem, further work was required in order to adapt problem $\mathcal{P}$ to the software constraints. A more precise description of the model can be found in [Garcia (2005)].

3 Model Results

Results for some hydrodynamic and geochemical parameters are shown in Fig. 1. Figure 1b displays a typical summer vertical density profile (arrows representing current velocities) corresponding to the section depicted onto the lake bathymetry (Fig. 1a). As it can be seen, the densities at the surface are lower than at the bottom, as it is expected for this time of the year. Figures 1c

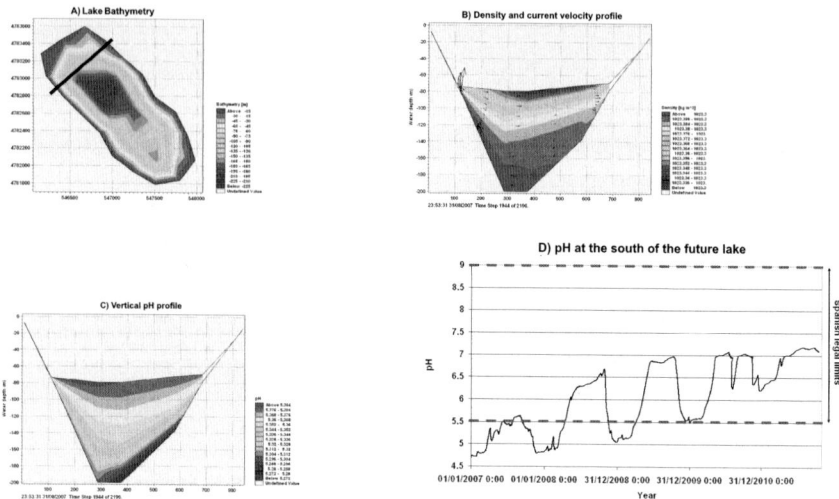

Fig. 1. Model results

and d show results for one of the most important environmental parameters, the pH (a measure of the water acidity). Figure 1c represents a vertical pH profile in which we can observe that the pH is higher at the surface than at the bottom. These results are consequence of sulfide oxidation at the bottom and the entrance of polluted sources. Focusing on a point located at the south of the lake (very close to the place where the water would be diverted to the water reservoir) and considering the complete filling period (approximately five years), we see in Fig. 1d that, after two years of filling, surface water quality would be good enough to satisfy the Spanish water quality standards.

References

[Bermudez et al. (1994)] Bermúdez, A. and Moreno, C.: Duality methods for solving variational inequalities. *Comput. Math. Appl.*, 7:43-58 (1994)
[Ecolab DHI (2005)] Mike 3 Flow Model FM. EcoLab Module. User guide. DHI Software (2005)
[Mike 3 HD (2005)] Mike 21/3 Flow Model FM. Hydrodynamic and Transport Module (Scientific Documentation). DHI Water and Environment (2005)
[Garcia (2005)] García García, L.M.: Numerical simulation of the water quality of a lake. (In spanish). DEA Research work. Dpto. Matemática Aplicada. Universidad de Santiago de Compostela (2005)
[Davis et al. (1996)] Davis, A., Lyons, W.B. and Miller, G.C.: Understanding the water quality of pit lakes. *Environ. Sci. Technol.*, 30:118-123 (1996)

Optimal Management and Design
of a Wastewater Purification System

Lino J. Alvarez-Vázquez[1], Eva Balsa-Canto[2], and Aurea Martínez[1]

[1] Departamento de Matemática Aplicada II. E.T.S.I. Telecomunicación,
  Universidad de Vigo, 36310 Vigo, Spain
  lino@dma.uvigo.es,aurea@dma.uvigo.es
[2] Instituto de Investigaciones Marinas, C.S.I.C., Vigo, Spain
  ebalsa@iim.csic.es

1 Introduction

Coastal areas are continually exposed to land-based sources of pollution resulting from domestic and industrial activities including oil spills, discharge of sewage and industrial effluents, among others. These contaminants arrive in the sea through wastewater discharges from sewage farms where contaminant concentrations are reduced by means of biological or chemical processes.

Biological processes are commonly used to treat domestic or combined domestic and some specific industrial wastewater. The basic idea is to reproduce the same processes that would occur naturally in the receiving water (river, estuary, etc.), but under controlled conditions, so that the cleansing reactions are completed before the water is discharged into the environment. The objective is, therefore, to provide an optimum environment for the microbial population to decompose the organic matter.

Microorganisms utilize the organic matter for the production of energy by cellular respiration and the manufacture of new cells. As the pollution increases, the *Biochemical Oxygen Demand* (BOD) also increases and, as a result, the *Dissolved Oxygen* (DO) decreases. This damages the marine life and causes the organic matter decomposition by means of anaerobic processes, which do not use oxygen but produce sulphide of hydrogen and methane, both having a nauseating smell. To avoid this problem, we have to guarantee a minimum level of DO and a maximum level of BOD in each region to be protected.

Is, therefore, of the highest interest to implement a wastewater treatment system that is able to assure water quality standards, corresponding to pollution concentrations lower than a certain allowed value imposed by the regional legislation. Note, however, that the economic cost of the process may be excessively large, depending on both the purification cost at each plant and the distance from the plant to the outfall (design cost).

The authors have recently considered, both from the theoretical and the numerical point of view, two related optimization problems. The first problem was devoted to determine the optimal level of the oxygen discharges in order to minimize the global purification cost (see Martínez et al. [5, 6]), and the second was aimed to obtain the optimal locations for the wastewater outfalls, assuming constant oxygen discharges (see Alvarez-Vázquez et al. [1, 2]) while keeping, in both cases, the constraints on the water quality.

This work addresses, for the first time to the authors knowledge, the combined design and operation optimization problem. This problem can be formulated as finding the optimal design (outfall locations) and the optimal operation conditions (that is, the optimal oxygen discharge levels) which minimize the total economic cost of the system while guarantying the above mentioned constraints on the water quality.

Note that this is a two-objective problem: in one hand, one would be interested in minimizing the initial cost due to the installation of the outfalls and in other hand it is also desirable to minimize the expenses of the purification process which will be repeated along time. The simultaneous optimization of multiple, usually competing, objectives, deviates from the single objective case in that it does not admit an unique optimal solution. Instead, a number of solutions, the so-called *Pareto-optimal* solutions, may be found that must be considered equivalent in the absence of information concerning the relative importance of the different objectives. In contrast to single-objective optimization, multi-objective optimization problems require the involvement of a decision maker who has to select one Pareto solution from the set.

This work proposes the use of a classical approach, the weighted sum method, to combine the two objectives in a single-objective with the aim of, from the theoretical point of view, obtaining the optimality conditions and demonstrating the existence of a solution, for any given set of weights. From the numerical point of view, a Pareto front is obtained for the two-objective case, using a control vector parametrization approach to approximate the control variables and an evolutionary algorithm to deal with the non-convex character of the resultant nonlinear programming problem.

2 Optimal Operation and Design: Problem Formulation

A domain $\Omega \subset R^2$ occupied by shallow waters, for instance an estuary, is considered. The sewage is dumped into the domain Ω through N submarine outfalls, each of them located at a point $b_j \in \Omega$ (that must be determined), and connected to a purification plant, located at a point a_j, which discharges an amount $m_j(t)$ (also to be determined). Moreover, there are M areas in Ω, denoted by A_i, for example beaches or fish nurseries, that must be protected guarantying that the levels of pollution are bellow previously fixed thresholds.

As it was stated in the introduction two of the most important parameters used to control pollution levels are the *dissolved oxygen* (DO) and the *biochemical oxygen demand* (BOD). As the pollution increases, the oxygen demand also increases and, as a result, the dissolved oxygen decreases with undesirable environmental consequences. To avoid this problem, we have to guarantee a minimum level ζ_i of DO and a maximum level σ_i of BOD in each region $\bar{A}_i$ to be protected. The evolution of the concentrations of BOD $(\rho_1(x,t))$ and DO $(\rho_2(x,t))$ in the domain Ω along the time interval $[0,T]$ is governed by a complex system of partial differential equations coupled with the shallow water equations (see, for instance, Martínez et al. [5]). The pair (m,b) formed by the set of optimal discharges $m = (m_j)_{j=1}^N$ and the set optimal locations $b = (b_j)_{j=1}^N$ are the control variables of the problem.

Now, we assume that inside the domain Ω there are M protected zones $\bar{A}_i$ where a maximum level of BOD and a minimum level of DO must be ensured, that is,

$$\rho_1|_{\bar{A}_i \times [0,T]} \le \sigma_i, \qquad \rho_2|_{\bar{A}_i \times [0,T]} \ge \zeta_i, \qquad \forall i = 1, \ldots, M. \tag{1}$$

Moreover, taking into account technological limitations, the j-th discharge must verify that $\underline{m}_j \le m_j(t) \le \overline{m}_j$, $\forall t \in (0,T)$, and the j-th outfall must be placed in a suitable region U_j, where $U_j \subset \Omega \backslash \cup_{i=1}^M \bar{A}_i$ is a compact, convex, polyhedral set representing all the admissible points to locate the outfalls. Thus, the optimal pair (m,b) must verify $\underline{m}_j \le m_j(t) \le \overline{m}_j$, $b_j \in U_j$, $\forall j = 1, \ldots, N$. If we define $U_{ad} = \{m \in [L^\infty(0,T)]^N : \underline{m}_j \le m_j(t) \le \overline{m}_j, \forall t \in (0,T) \quad \forall j = 1, \ldots, N\} \times \Pi_{j=1}^N U_j$, technological constraints can be written in the simpler way:

$$(m,b) \in U_{ad}. \tag{2}$$

Consider now that the purification cost in each plant is given by $J_1(m,b) = \sum_{j=1}^N \int_0^T f_j(m_j(t)) \, \mathrm{d}t$, where function f_j represents the cost of the purification in the j-th plant; and that the design cost depends on the distance from the farm to the outfall in the following manner $J_2(m,b) = \frac{1}{2} \sum_{j=1}^N \|b_j - a_j\|^2$.

In contrast to the single objective cases where an unique objective was pursued, the aim in this contribution is to solve the two-objective case, finding good compromises between the two different objectives. The notion of optimal solution is substituted by the notion of Pareto-optimal solution: a solution is said to be Pareto-optimal if there exists no feasible solution which would decrease one objective without causing a simultaneous increase in the other.

Under certain conditions the *weighted sum method* can be used to obtain the set of Pareto-optimal solutions. This method transforms the original multi-objective case into a single-objective optimization problem in which the objective function is a weighted sum of the original objectives. In practice, the objectives are usually scaled and combined to form a composite objective:

$$J(m,b) = \sum_{j=1}^N \int_0^T f_j(m_j(t)) \, \mathrm{d}t + \frac{\alpha}{2} \sum_{j=1}^N \|b_j - a_j\|^2, \tag{3}$$

where $\alpha > 0$ is a weight parameter.

Therefore the problem, denoted by $(\mathcal{P})$, consists of finding the time varying discharges $m_j(t)$, $j = 1, \ldots, N$, and the points b_j, $j = 1, \ldots, N$, to minimize the cost function J subject to the system dynamics, the state constraints (1) and the control constraints (2). This is a parabolic optimal control problem with non-convex pointwise state constraints, which makes difficult its analysis and resolution.

By using minimizing sequences we can prove that the optimal control problem has, at least, one solution. We are also able to obtain, introducing the adjoint state, a first order optimality system satisfied by the solutions of the optimal control problem (see the details in Alvarez-Vázquez et al. [3]).

3 Numerical Solution

The usual approach to solve optimal control problems is to transform the original infinite dimension problem into a finite dimension nonlinear programming (NLP) problem. With this aim, the control vector parametrization approach proceeds by dividing the duration of the process $[0, T]$ into a reduced number of non-equidistant intervals and approximating the control variables (m_j) using low order Lagrange polynomials within each interval. As a result, a NLP problem is obtained where the vector of decision variables includes the coefficients in the polynomials, the switching times, and the time independent parameters (outfall locations, in our case). We must remark that the calculation of the objective function requires the solution of the state system, that in this work is approached using a characteristics-finite element method.

As it was stated in previous sections the weighted sum method transforms the original two-objective case into a single-objective optimization problem in which the objective function is a weighted sum of the original objectives. The Pareto front can then be generated by varying the weight α in the objective expression (3) and solving the corresponding nonlinear programming problems with a suitable technique.

The optimization literature offers a large number of methods to solve nonlinear programming problems: local and global strategies, deterministic and stochastic methods... (cf. [4] and the references therein). In this work, a evolutionary global optimization method, Differential Evolution (DE) [7], will be used due mainly to its efficiency in solving real valued multimodal objective functions. DE is basically a parallel, population-based, direct search algorithm. In addition to its good convergence properties some of its main advantages are its conceptual simplicity and ease of use.

Numerical results (Pareto front and optimal control profiles for a particular Pareto solution) for a realistic problem posed in the *ría* of Vigo (Spain) are shown, respectively, in Figs. 1 and 2. More details can be found in [3].

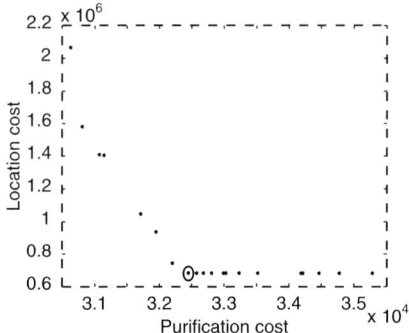

Fig. 1. Front of Pareto solutions

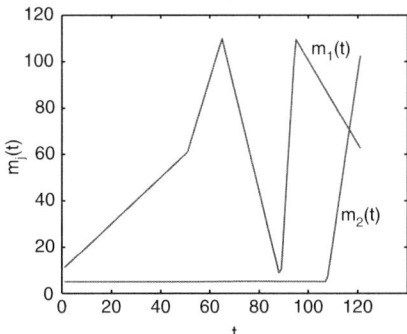

Fig. 2. Optimal control profiles for previous circled Pareto solution

Acknowledgements

The research contained in this work was partially supported by Project MTM2006-01177 of Ministerio de Educación y Ciencia (Spain). The authors also thank the help of C. Rodríguez and M.E. Vázquez-Méndez.

References

1. Alvarez-Vázquez, L.J., Martínez, A., Rodríguez, C., Vázquez-Méndez, M.E.: Mathematical Analysis of the optimal location of wastewater outfalls. IMA J. Appl. Math., **67**, 23–39 (2002).
2. Alvarez-Vázquez, L.J., Martínez, A., Rodríguez, C., Vázquez-Méndez, M.E.: Numerical optimization for the location of wastewater outfalls. Comput. Optim. Appl., **22**, 399–417 (2002).
3. Alvarez-Vázquez, L.J., Balsa-Canto, E., Martínez, A.: Optimal design and operation of a wastewater purification system. Submitted (2006).
4. Balsa-Canto, E., Vassiliadis. V.S., Banga. J.R.: Dynamic Optimization of Single- and Multi-Stage Systems using a Hybrid Stochastic-Deterministic Method. Ind. Eng. Chem. Res., **44**, 1514–1523 (2005).

5. Martínez, A., Rodríguez. C., Vázquez-Méndez. M.E.: Theoretical and numerical analysis of an optimal control problem related to wastewater treatment. SIAM J. Control Optim., **38**, 1534–1553 (2000).
6. Martínez, A., Rodríguez. C., Vázquez-Méndez. M.E.: A control problem arising in the process of waste water purification. J. Comp. Appl. Math., **114**, 67–79 (2000).
7. Storn, R., Price, K.: Differential Evolution - a Simple and Efficient Heuristic for Global Optimization over Continuous Spaces. J. Global Optim., **11**, 341–359 (1997).

Estimation of Fuzzy Anomalies in Water Distribution Systems

J. Izquierdo[1], M.M. Tung[2], R. Peréz[1], and F.J. Martínez[1]

[1] Centro Multidisciplinar de Modelación de Fluidos
   jizquier@gmmf.upv.es
[2] Instituto de Matemática Multidisciplinar, Universidad Politécnica
   de Valencia, Spain
   mtung@imm.upv.es

Summary. State estimation is necessary in diagnosing anomalies in Water Demand Systems (WDS). In this paper we present a neural network performing such a task. State estimation is performed by using optimization, which tries to reconcile all the available information. Quantification of the uncertainty of the input data (telemetry measures and demand predictions) can be achieved by means of robust estate estimation. Using a mathematical model of the network, fuzzy estimated states for anomalous states of the network can be obtained. They are used to train a neural network capable of assessing WDS anomalies associated with particular sets of measurements.

1 Introduction

Water companies use telemetry systems for control and operation purposes. By considering the data provided by telemetry, the engineer on duty makes operation decisions trying to optimize the system utilization. Nevertheless, the system complexity does not permit but to take a few real-time measures, which only incompletely represent the network state. They give indication of only certain aspects of the system, leaving out other more specific or "less relevant" ones. Thus, suitable techniques that allow for more accurate network health estimation are necessary so that anomalies can be detected more rapidly, and light anomalies, which develop progressively and insidiously, can be identified. This will enable to control their consequences in earlier stages, thus avoiding, among other things, losses of water, which can be of great importance.

The state of a WDS is obtained by interrelating different measures within a mathematical model of the network, [Mar95]. Different tools to analyze water networks have been developed in the last years, SARA [GMF98], and EPANET [Ros97], among others. But state estimation cannot be accurately performed if there are missing or uncertain data. Thus, system operators need error limits for the state variables. Yet, data are abundant since they are permanently received. Therefore, operators cannot evaluate errors easily or in real time. It is expected that suitable techniques borrowed from Artificial Intelligence (AI) could encapsulate the necessary knowledge to assess the network state.

In this paper, we present an approach for the diagnosis and decision making process which is necessary on a neural network for clustering and pattern classification. First, the mathematical model, a state estimation procedure and a mechanism for treating uncertainties, already presented in [Izq04] and [Izq05], are briefly presented. The state estimator, together with the error limits will be used as a surrogate of the real WDS to generate data to train and check the neural network (NN). Then, the inherent procedures to neural techniques will be described. Specifically, the NN architecture, the classification and clustering mechanisms of both, crisp and fuzzy, patterns and the training technique will be presented.

2 Mathematical Model and State Estimation

Analyzing pressurized water systems is a complex task, especially for big systems. But even for moderately sized cities, it involves solving a big number of non-linear simultaneous equations. The complete set of equations may be written by using block-matrix notation,

$$\begin{pmatrix} A_{11}(q) & A_{12} \\ A_{12}^t & 0 \end{pmatrix} \begin{pmatrix} q \\ H \end{pmatrix} = \begin{pmatrix} -A_{10}H_f \\ Q \end{pmatrix}, \tag{1}$$

where A_{12} is the so-called connectivity matrix describing the way demand nodes are connected through the lines. Its size is $L \times N_p$, N_p being the number of demand nodes and L the number of lines; q is the vector of the flow rates through the lines, H the vector of unknown heads at demand nodes; A_{10} is an $L \times N_f$ matrix, N_f being the number of fixed-head nodes with known head H_f, and Q is the N_p-dimensional vector of demands. Finally, $A_{11}(q)$ is an $L \times L$ diagonal matrix. System (1) is a non-linear problem whose solution is the state vector $x = (q, H)^t$ of the system.

The non-linear relations describing the system balances are complemented by the specific telemetry measurements. These measurements are integrated into the model by expanding system (1) to a new system, typically overdetermined:

$$\begin{pmatrix} A_{11}(q) & A_{12} \\ A_{12}^t & 0 \\ A_{31} & A_{32} \end{pmatrix} \begin{pmatrix} q \\ H \end{pmatrix} = \begin{pmatrix} -A_{10}H_f \\ Q \\ M_t \end{pmatrix}. \tag{2}$$

The components A_{31} and A_{32} in system (2) were introduced to account for additional telemetry measurements M_t with uncertainties in the demand predictions. System (2) is usually solved using least-square methods for a state estimation by an over-relaxation iterative process applied to a linearized version of (2):

$$
\begin{pmatrix} A'_{11}(q^{(k)}) & A_{12} \\ A^t_{12} & 0 \\ A_{31} & A_{32} \end{pmatrix} \begin{pmatrix} \Delta q \\ \Delta H \end{pmatrix} = \begin{pmatrix} -A_{10}H_f - A_{11}(q^{(k)})q^{(k)} - A_{12}H^{(k)} \\ Q - A_{21}q^{(k)} \\ M_t - A_{31}q^{(k)} - A_{32}H^{(k)} \end{pmatrix}, \quad (3)
$$

where A'_{11} is the Jacobian matrix corresponding to A_{11}.

3 Error Limit Analysis

Error limit analysis is a process to determine uncertainty bounds for the state estimation originated by the lack of precision of measurements and data. To put it in a nutshell, the question is what is the reliability of the estimated state x^*, if measurement vectors y are not crisp but may vary in some region, $[y - \delta y, y + \delta y]$?

Different techniques may be used to estimate this unknown but bounded error, [Mil96], [Nor86], [Kur97]. We use a variant of the so-called sensitivity matrix analysis, [Bar03], which uses the state estimator presented above.

In [Izq05], it is proved that a component by component bound, e^*, for δx^* can be obtained by means of

$$
e^* = \left| \left(A_k^{*t} W A_k^* \right)^{-1} A_k^{*t} W \right| |\delta y|, \quad (4)
$$

where W is a diagonal matrix that weights the equations according to the nature of the right-hand sides, and the vertical bars indicate absolute values of all matrix and vector entries. Because of linearity, the bounds calculated by (4) are symmetrical and the error limit may be expressed as a multidimensional interval (see cell definition in next section) $[x^*]$ in the state space

$$
[x^*] = \left[x^*_{\inf}, x^*_{\sup} \right] = \left[x^* - e^*, x^* + e^* \right]. \quad (5)
$$

4 The Neural Network

A neural network for clustering and classification is a mechanism for pattern recognition. Here, we use multidimensional cells, [Sim92], [Lik94]. Voronoi diagrams are used in [Ble97].

A cell C is a region of the pattern space of n-dimensional vectors obtained as the intersection of n pairs of half-spaces of the form $m_i \leq x_i \leq M_i$, for $i = 1, 2, \ldots, n$, where m_i and M_i are real numbers. Vectors $m = (m_i, i = 1, \ldots, n)$ and $M = (M_i, i = 1, \ldots, n)$ are called min and max points of C and completely determine C. Membership of patterns to a cell is defined from fuzzy grounds. For fuzzy patterns, $P = \left[P^{\inf}, P^{\sup} \right]$, like the ones obtained in (5), membership values are given by the membership function

$$c(P) = \max_{i=1,\ldots,n} \left\{ \max \left\{ \varphi_i \left(P_i^{\text{sup}} - M_i \right), \varphi_i \left(m_i - P_i^{\text{inf}} \right) \right\} \right\}, \qquad (6)$$

where each $\varphi_i(x)$ controls the cell fuzziness.

Values taken by membership function (6) are used during the operation phase to decide the membership degree to the class associated with a cell exhibited by certain pattern presented to it and, as a consequence, to recognize the potential anomalous state of the water distribution system corresponding to the associated label of each class. Patterns presented to the network during the training phase are ordered pairs (P, l), where l is a label associated to pattern P describing the type of anomaly it represents.

The NN implementing the classification process is a three-layer network that grows adapting itself to the problem characteristics. The input layer has $2n$ neurons, two for any of the dimensions of the patterns $P = \left[P^{\text{inf}}, P^{\text{sup}} \right]$. When a new pattern is presented to the network through the input layer, the components of vectors P^{inf} and P^{sup} are compared, respectively, with those of the minimum point, m, and the maximum point, M, of the J existing cells. Specifically, numbers in the inner brackets of (6) are calculated.

This way, each neuron on the hidden layer has two n-dimensional vectors φ^{inf} and φ^{sup} as its input, formed by numbers between 0 and 1, ready to be processed, first component by component with the max operator, and then with the max operator, but now through all the components. Specifically,

$$c(P) = 1 - \max_{i=1,\ldots,n} \left\{ \max \left\{ \varphi_i^{\text{sup}}, \varphi_i^{\text{inf}} \right\} \right\}$$

is calculated for each cell. This process gives the membership degree of P to every one of the cells. Thus, membership functions may be considered as the transfer or activation functions for all the J existing hidden neurons. And the values of the minimum and maximum points of those existing cells, which will be adjusted during the training phase, must be regarded precisely as the synaptic weights between the input and the hidden layer.

The values produced by the membership functions of the existing cells constitute the outputs of the hidden layer. These values must be operated with the weights between the hidden and the output layers. This process will produce a class, a diagnosis of the hydraulic system represented by pattern P. This procedure facilitates the decisions to be made by the system managers.

5 Conclusions

The described neural procedure does not fit into any standard paradigm, since it is made of several sub-nets that evolve by accumulating experience as new loads (peak, valley, seasonal-dependent, etc.) are observed, which mimics human knowledge acquisition. From the reduced number of tests performed we conclude that the classification ability of the NN is excellent. Since the response given by the NN is graded, as a consequence of its fuzziness, the information it provides is not only qualitative (pointing out an anomaly) but also quantitative (weighting the distributed importance of the problem). The tool presented here, once completed, calibrated and implemented, will provide WDS managers with a decision support mechanism allowing early identification of anomalies and, as a consequence, better Integrated Water Management.

Acknowledgements

This work is been performed under the support of the projects Investigación Interdisciplinar nº 5706 (UPV), **DPI2004-04430** of the Dirección General de Investigación del Ministerio de Educación y Ciencia (Spain) and FEDER funds.

References

[Bar03] Bargiela, A., Pedrycz, W.: Granular Computing. Kluwer Academia Press, Boston, Dordrecht, London (2003).

[Ble97] Blekas, K., Likas, A., Safylopatis, A.: A Fuzzy Neural Network Approach to Classification Based on Proximity Characteristics Patterns. Proc. 9th IEEE Int. Conference on tools with Artificial Intelligence, Newport Beach, CA (1997).

[GMF98] GMF (Grupo Mecánica de Fluidos): SARA, Software de Análisis de Redes de Agua, Manual de Usuario. Ed. Grupo Mecánica de Fluidos, UPV (1998).

[Izq04] Izquierdo, J., Iglesias, P.L., Díaz, J.L., Pérez, R.: Identificación de estados en sistemas de distribución de agua mediante técnicas de optimización, IV SEREA Seminario hispano-brasileño: Planificación, proyecto y operación de redes de abastecimiento de agua, CD-ROM, Valencia (2004).

[Izq05] Izquierdo, J., López, P.A., Pérez, R., Ledesma, B.: Intervalos de confianza en la estimación de estados en sistemas de distribución de agua. V SEREA Seminario hispano-brasileño: Planificación, proyecto y operación de redes de abastecimiento de agua, CD-ROM, Valencia (2005).

[Kur97] Kurzhanski, A.B., Valyi, I.: Ellipsoidal Calculus for Estimation and Control. Birkhäuser (1997).

[Lik94] Likas, A., Blekas, K., Safylopatis, A.: Application of the Fuzzy Min-Max Neural Network Classifier to Problems with Continuous and Discrete Attributes. Proc. of IEEE Workshop on Neural Networks for Signal Processing (NNSP'94), 163–170 (1994).

[Mar95] Martínez, F., R. Pérez and J. Izquierdo: Optimum Design and Reliability in Water Distribution Systems, in Improving efficiency and reliability in water distribution systems. Kluwer Academic Pub., Dordrecht, Boston, London (1995).

[Mil96] Milanese, M. et al. (Eds.): Bounding Approaches to system identification, Milanese, M. et al. (Eds.), Plenum Press, NY (1996).

[Nor86] Norton, J.P.: An Introduction to Identification. Academic Press (1986).

[Ros97] Rossman, L.A.: Manual de usuario de EPANET. Drinking Water Research Group. Risk Reduction Engineering Laboratory. US EPA. Translated by GM Fluidos, UPV (1997).

[Sim92] Simpson, P.: Fuzzy Min-Max Neural Networks-Part 1: Classification. IEEE Trans. On Neural Networks **3**(5), 776–786 (1992).

Investigation of the Evolution and Breakup of Electrically Charged Drops

S.I. Betel[1], M.A. Fontelos[2], U. Kindelán[3], and O. Vantzos[1]

[1] Dpt. of Mathematics, University of North Texas, P.O. Box 311430,
    Denton, TX 76203-1430
    betelu@unt.edu
[2] Instituto de Matemáticas y Física Fundamental, C.S.I.C., C/Serrano 123,
    28006 Madrid, Spain
    marco.fontelos@uam.es
[3] Dpt. de Matemática Aplicada y Met. Inf., Universidad Politécnica de Madrid,
    Alenza 4, 28003 Madrid, Spain
    ultano.kindelan@upm.es

Summary. We study the evolution of charged droplets of a conducting viscous liquid. The flow is driven by electrostatic repulsion and capillarity and may lead to the breakup of the droplets. These droplets are known to be linearly unstable when the electric charge is above the Rayleigh critical value. Here we investigate the nonlinear evolution that develops after the linear regime.

1 Introduction

The formation of singularities on charged masses of fluid is relevant in a variety of physical and technological situations, such as the breakup of water droplets in thunderstorms, electrospraying, electrospinning, an electropainting. The interest in the shape of electrified drops dates back to Lord Rayleigh [Ra], who showed that if the electric charge is larger than some critical value, the spherical drop becomes unstable. For a drop with total charge Q, surface tension coefficient γ, and radius R suspended in a medium of dielectric constant ϵ_0, this critical value is $Q_c = \sqrt{32\gamma\pi^2\epsilon_0 R^3}$. After de drop becomes unstable, it disintegrates into droplets of smaller size. However, in recent experiments (see [Duf]) it has been noticed that, previous to drop disintegration, the drop evolves into a prolate spheroid which, after a finite time, develops conical tips from which thin jets emerge.

Here we describe the numerical method used to solve the PDE system that models the drop behavior. We implement boundary element methods to compute: 1) the velocity field inside the drop and 2) the electrostatic potential anywhere in the space as well as the surface charge density. The methods are complemented with suitable iterative procedures to accurately compute the mean curvature of the drop surface.

2 The Model Equations

We assume that the drop occupies a region $\Omega(t)$ and the liquid of the drop is a perfect conductor with infinite conductivity. Hence the electric potential V is constant inside and at the drop surface, and all the electric charge will be located at the boundary $\partial\Omega$, and since the surrounding medium is a dielectric, the total charge Q remains constant. The electric field $\mathbf{E}$ outside the drop is given by $\mathbf{E} = -\nabla V$ where $\Delta V = 0$ in $R^3\backslash\Omega$, $V = C$ on $\partial\Omega$ and V decays at infinity. At the surface of a conductor, the surface charge density σ is given by the normal derivative of the potential, $\sigma = -\varepsilon_0 \frac{\partial V}{\partial n}$, so that the repulsive electrostatic force per unit area is $\mathbf{F_e} = \frac{\mathbf{E}\sigma}{2} = \frac{\varepsilon_0}{2}\left(\frac{\partial V}{\partial n}\right)^2 \mathbf{n} = \frac{\sigma^2}{2\varepsilon_0}\mathbf{n}$, where $\mathbf{n}$ is the outward normal to the surface.

The fluid velocity $\mathbf{u}$ and the fluid pressure p inside the drop satisfy the Stokes equations $-\nabla p + \mu_1 \Delta\mathbf{u} = 0$ in $\Omega(t)$ and $\nabla \cdot \mathbf{u} = 0$ in $\Omega(t)$; where μ_1 is the viscosity of the liquid inside the drop. Similar equations must be satisfied by the velocity and the pressure outside of the drop, $\mathbf{R}^3\backslash\Omega(t)$, with μ_1 replaced by μ_2, the viscosity of the surrounding liquid.

The boundary condition for the stress is $(T^{(2)} - T^{(1)})\mathbf{n} = \left(\gamma\kappa - \frac{\sigma^2}{2\varepsilon_0}\right)\mathbf{n}$ on $\partial\Omega(t)$, where κ is the mean curvature of the surface and $T^{(k)}$ is the stress tensor inside ($k = 1$) or outside ($k = 2$) the drop, given by $T_{ij}^{(k)} = -p\delta_{ij} + \mu_k\left(\frac{\partial u_i}{\partial x_j} + \frac{\partial u_j}{\partial x_i}\right)$, $k = 1, 2$. The boundary condition expresses the balance between viscous stress, capillary forces and electrostatic repulsion.

3 The Numerical Method

Our numerical method to compute the evolution of the drop is based on the boundary integral method for the Stokes system (see [Poz] for a comprehensive explanation). In this method, the equation for the velocity at $\partial\Omega(t)$ is written in integral form as

$$
u_j(\mathbf{r}_p) = -\frac{1}{4\pi}\frac{1}{\mu_1 + \mu_2}\int_{\partial\Omega(t)} f_i(\mathbf{r})G_{ij}(\mathbf{r}, \mathbf{r}_p)\mathrm{d}S(\mathbf{r})
$$
$$
-\frac{1}{4\pi}\frac{\mu_2 - \mu_1}{\mu_2 + \mu_1}\int_{\partial\Omega(t)} u_i(\mathbf{r})T_{ijk}(\mathbf{r}, \mathbf{r}_p)n_k(\mathbf{r})\mathrm{d}S(\mathbf{r}) , \tag{1}
$$

where $G_{ij}(\mathbf{r}, \mathbf{r}_p) = \frac{\delta_{ij}}{|\hat{\mathbf{r}}|} + \frac{(\hat{r}_i)(\hat{r}_j)}{|\hat{\mathbf{r}}|^3}$, $T_{ijk}(\mathbf{r}, \mathbf{r}_p) = -6\frac{(\hat{r}_i)(\hat{r}_j)(\hat{r}_k)}{|\hat{\mathbf{r}}|^5}$ and $f_i(\mathbf{r}) = [\gamma\kappa(\mathbf{r}) - \frac{\varepsilon_0}{2}(\frac{\partial V}{\partial n})^2(\mathbf{r})]n_i(\mathbf{r})$ with $\hat{\mathbf{r}} = \mathbf{r} - \mathbf{r}_p$.

The equation for the charge density is

$$
V(\mathbf{r}_p) = \frac{1}{4\pi\epsilon_0}\int_{\partial\Omega(t)} \frac{\sigma(\mathbf{r})}{|\mathbf{r} - \mathbf{r}_0|}\mathrm{d}S(\mathbf{r}). \tag{2}
$$

This integral equation must be inverted numerically to obtain the charge density. $V(\mathbf{r}_p)$ is a constant along the surface, and it is determined by the condition

$$Q = \int_{\partial\Omega(t)} \sigma(\mathbf{r})\mathrm{d}S(\mathbf{r}). \tag{3}$$

Both integral equations (1) and (2) are coupled through the charge density σ. First we invert numerically the potential equation for the surface charge density. Once σ is known we calculate the mean curvature κ at the nodes. When σ and κ are known we can obtain the balance force term $\mathbf{f}$, replace it in (1) and solve the equation in order to get the velocity $\mathbf{u}$. Given the velocity $\mathbf{u}$, we move the points of the surface using an Euler explicit scheme. In the following subsections we explain with detail this procedure.

3.1 Charge Density

At any given time $t > 0$, we approximate the free boundary $\partial\Omega$ with a triangular mesh. The mesh is made up of N vertices and M (triangular) faces. On each face, we approximate the various physical quantities that are defined in the surface (curvature, surface charge density, velocity) with elementwise constant functions over a "virtual" element centered in each node with an area equal to $1/3$ of the total area of the elements that share the node (see [Zin]). We obtain the charge density from equations (2) and (3). From (2) and taking into account that the potential is constant on the surface of the drop we get:

$$4\pi\varepsilon_0 V(\mathbf{r}_i) = C_1 = \int_{\partial\Omega(t)} \sigma(\mathbf{r}) \frac{1}{|\mathbf{r} - \mathbf{r}_i|}\mathrm{d}s(\mathbf{r}) \qquad i = 1, \ldots, M, \tag{4}$$

where $\mathbf{r}_i$ is the barycenter of the mesh element i and M is the number of mesh elements. We approximate the integral that appears in (4) as follows: $\int_{\partial\Omega(t)} \sigma(\mathbf{r}) \frac{1}{|\mathbf{r}-\mathbf{r}_i|}\mathrm{d}s(\mathbf{r}) \approx \sum_{j=1}^{N_e} \lambda_{ij}\sigma_j$, with $\lambda_{ij} = \int_{T_j} \frac{1}{|\mathbf{r}-\mathbf{r}_i|}\mathrm{d}s(\mathbf{r})$ and $\sigma_j = \sigma(\mathbf{r}_j)$. Two cases are considered when calculating λ_{ij}:

Potential created by one element onto himself, $i = j$: λ_{ii} can be calculated exactly dividing the element T_i in six subtriangles joining $\mathbf{r}_i$ with the three vertices of T_i and splitting these three triangles with the lines that join $\mathbf{r}_i$ with its projection in each edge of T_i,

$$\lambda_{ii} = \int_{T_i} \frac{\mathrm{d}s(\mathbf{r})}{|\mathbf{r} - \mathbf{r}_i|} = \int\int_{T_i} \frac{1}{\rho}\rho\,\mathrm{d}\rho\,\mathrm{d}\theta = \sum_{k=1}^{6} \int\int_{T_{ik}} \mathrm{d}\rho\,\mathrm{d}\sigma$$

$$= \sum_{k=1}^{6} a_k \ln(\sec(\alpha_k) + \tan(\alpha_k)),$$

where α_k are the angles with vertex $\mathbf{r}_i$ and a_k are the lengths of the lines that join $\mathbf{r}_i$ with its projection ($a_1 = a_2$, $a_3 = a_4$, $a_5 = a_6$).

Potential created by element j onto element i, $i \neq j$: in this case we subdivide the element T_j in a variable number of subelements. If we suppose that all the charge of an element is concentrated in its center of gravity: $\lambda_{ij} = \sum_{k=1}^{N_s} \lambda_{ij,k}$ with $\lambda_{ij,k} = \frac{A_{T_{jk}}}{|\mathbf{b}_i - \mathbf{b}_{jk}|}$, where $A_{T_{jk}}$ = area of subelement T_{jk}, $\mathbf{b}_{jk}$ = barycenter of subelement T_{jk}, $\mathbf{b}_i$ = barycenter of element i and N_s = total number of subelements T_{jk} of T_j.

Once all the coefficients λ_{ij} are known, we calculate a fictitious charge density proportional to the actual charge density solving the system

$$\sum_{j=1}^{N_e} \lambda_{ij} \overline{\sigma}_j = C_1 \qquad i = 1, \ldots, M, \tag{5}$$

where an arbitrary value is given to C_1. From (3) we get $\overline{Q} = \sum_{i=1}^{N_e} \overline{\sigma}_i A_i$ and $Q = \sum_{i=1}^{N_e} \sigma_i A_i$. We can use these two identities to scale the fictitious charge density $\overline{\sigma}$, obtaining the actual charge density σ corresponding to a total charge Q: $\sigma_i = \frac{Q}{\overline{Q}} \overline{\sigma}_i$ $i = 1, \ldots, M$. Finally we get the charge density in each node of the mesh as an average of the charge densities of all the elements that share that node.

3.2 Curvature

We calculate the mean curvature in each node $\mathbf{p}$ of the mesh following a method proposed in [Zin]. The algorithm is based on the following idea. If the z' axis of the local cartesian coordinates $(\mathbf{p}, x', y', z')$ was directed along the normal vector $\mathbf{n}(\mathbf{p})$, then z' as a quadratic function of x' and y' would be a good local representation of the surface $\partial\Omega$. This quadratic function can be obtained finding a paraboloid which passes through $\mathbf{p}$, has its axis parallel to z' and best fits its neighbors by the least-squares method. However $\mathbf{n}(\mathbf{p})$ is not known *a priori*, and so the method is iterative.

Also we will use this algorithm to calculate the normal to the surface $(\mathbf{n}(\mathbf{p}))$ in each node of the mesh.

3.3 Velocity

Once the curvature and the surface charge density are known in each node of the mesh, we are able to evaluate the balance force term $\mathbf{f}$ and thus, replacing $\mathbf{f}$ in equation (1), calculate $\mathbf{u}$ in each node of the mesh.

Both integrals in (1) are singular in $\mathbf{r} = \mathbf{r}_p$. We will remove both singularities with a well known technique proposed, for example, in [Poz], which is based in the fact that

$$\int_{\partial\Omega(t)} G_{ij}(\mathbf{r}, \mathbf{r}_p) n_i(\mathbf{r}) \mathrm{d}S(\mathbf{r}) = 0 \quad \text{and} \quad \int_{\partial\Omega(t)} T_{ijk}(\mathbf{r}, \mathbf{r}_p) n_k(\mathbf{r}) \mathrm{d}S(\mathbf{r}) = -4\pi\delta_{ij}.$$
$$\tag{6}$$

Using (6) we get an equivalent equation to (1):

$$u_j(\mathbf{r}_p) = -\lambda_S \int_{\partial\Omega(t)} (b(\mathbf{r}) - b(\mathbf{r}_p)) \, n_i(\mathbf{r}) G_{ij}(\mathbf{r}, \mathbf{r}_p) \mathrm{d}S(\mathbf{r})$$

$$-\lambda_D \int_{\partial\Omega(t)} (u_i(\mathbf{r}) - u_i(\mathbf{r}_p)) \, T_{ijk}(\mathbf{r}, \mathbf{r}_p) n_k(\mathbf{r}) \mathrm{d}S(\mathbf{r}) + 4\lambda_D \pi u_i(\mathbf{r}_p)\delta_{ij}, \quad (7)$$

where $\lambda_S = \frac{1}{4\pi} \frac{1}{\mu_1 + \mu_2}$, $\lambda_D = \frac{1}{4\pi} \frac{\mu_2 - \mu_1}{\mu_2 + \mu_1}$ and $\mathbf{f}(\mathbf{r}) = b(\mathbf{r})\mathbf{n}(\mathbf{r})$.

The integrals in (7) are approximated by a trapezoidal rule that requires the integrands only at nodes of the mesh: $\int_{\partial\Omega} \phi(\mathbf{x})ds(\mathbf{x}) = \sum_{i=1}^{N} \phi(\mathbf{x}_i)S_i$, where S_i is the area of the virtual element associate to node i. Applying this rule to the discretization of (7) we get:

$$(1 - 4\pi\lambda_D)u_j(\mathbf{r}_p) + \lambda_D \sum_{\substack{l=1 \\ l \neq p}}^{N} (u_i(\mathbf{r}_l) - u_i(\mathbf{r}_p)) T_{ijk}(\mathbf{r}_l, \mathbf{r}_p) n_k(\mathbf{r}_l) S_l$$

$$= -\lambda_S \sum_{\substack{l=1 \\ l \neq p}}^{N} (b(\mathbf{r}_l) - b(\mathbf{r}_p)) \, n_i(\mathbf{r}_l) G_{ij}(\mathbf{r}_l, \mathbf{r}_p) S_l. \quad (8)$$

The equation (8), for all values of $p \in \{1, \ldots, N\}$ and $j \in \{1, 2, 3\}$, forms a linear system of $3N$ equations in the $3N$ unknown velocity components $u_j(\mathbf{r}_p)$ at the nodes. After solving the system, we move the nodes of the surface with an euler explicit scheme: $\mathbf{r}_i(t_{n+1}) = \mathbf{r}_i(t_n) + \mathbf{u}_i(t_{n+1})\triangle t$, $i \in \{1, \ldots, p\}$.

4 Results and Conclusions

We have implemented the numerical method explained in Sect. 3 in a computer code written in Matlab. It is important to note that this code is a fully 3D code, i.e. it can handle non-axisymmetric initial configurations. Nevertheless we are going to show two results with initial axisymmetric configurations in order to compare them with the results obtained with a different 1D program that solves the axisymmetric formulation of the problem (see [BFKV]). We will simulate the time evolution of a drop that initially is a prolate spheroid ($a = c = 0.8, b = 1$). The inner fluid of the fluid has a viscosity of $\mu_1 = 0.4$ and the outer fluid a viscosity of $\mu_2 = 1$. We will consider two cases, the first one with total charge $Q = 1.27Q_c$ and the second one with total charge $Q = 2Q_c$. The results are depicted in Fig. 1.

Figure 1 provides evidence of formation of finite time singularities in both cases but with different shapes. In the case $Q = 1.27Q_c$ there is a cone-like singularity that appears in both tips of the drop. This result matches the experimental result obtained in [Duf] just before the drop emits a thin jet from both tips. In the second case a "neck" singularity appears in both

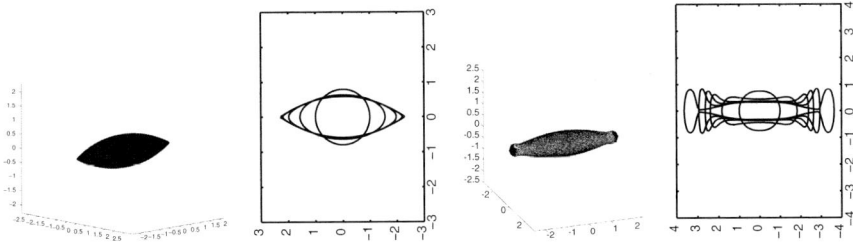

Fig. 1. On the *left*: shape of the drop just before breakup with $Q = 1.27Q_c$ On the *right*: shape of the drop just before breakup with $Q = 2Q_c$

sides of the drop, but not in the tips, It seems that the drop will break into three smaller drops. Another important conclusion is the agreement with the axisymmetric results which validates the axisymmetric program and supports the hypothesis of the axisymmetry of the singularities.

References

[BFKV]   Betel, S.I., Fontelos, M.A., Kindelán, U., Vantzos, O.: Singularities on charged viscous droplets. Physics of Fluids., **18**, 051706 (2006)

[Duf]   Duft, D., Achtzehn, T., Müller, R., Huber, B. A., Leisner, T.: Rayleigh jets from levitated microdroplets. Nature, vol. **421**, 9 Jan., pg. 128 (2003)

[Poz]   Pozrikidis, C.: Boundary integral methods for linearized viscous flow. Cambridge texts in App. Math., Cambridge University Press, (1992)

[Ra]   Lord Rayleigh, On the equilibrium of liquid conducting masses charged with electricity. Phil. Mag., **14**, 184-186 (1882)

[Zin]   Zinchenko, A. Z., Rother, M. A., Davis, R. H.: A novel boundary-integral algorithm for viscous interaction of deformable drops. Phys. Fluids, **9**, No. 6 (1997)

Homogeneous Nucleation of Dipole Domains and Current Self-Oscillations in Photoexcited Semiconductor Superlattices

J.I. Arana and L.L. Bonilla

Modeling, Simulation and Industrial Mathematics, Universidad Carlos III de Madrid, 28911 Leganés, Spain

Summary. A model for charge transport in photoexcited undoped semiconductor superlattices is proposed and analyzed. Under dc voltage bias, self-sustained oscillations of the current due to repeated homogeneous nucleation of pairs of charge dipole waves inside the sample, followed by wave splitting and motion in opposite directions are among the numerical solutions of the model.

1 Introduction

Semiconductor superlattices (SL) are used as fast-oscillator nanodevices in communications and are the basis of quantum cascade lasers [1]. The latter cover all the mid-infrared spectrum with the same material and are used in industrial applications such as environmental sensing and pollution monitoring, combustion control and catalytic converter diagnostics in the automotive industry, as breath analyzers in medical applications, etc. Nonlinear electronic transport in weakly coupled undoped photoexcited type-I SL is well described by spatially discrete drift-diffusion equations [2]. Nonlinear behavior at high fields include formation and dynamics of electric field domains, self-sustained oscillations of the current through voltage biased SL, chaos, etc [1].

In this work, we show that the electron-hole recombination coefficient decreases with increasing electric field by using a simple model that takes the overlap integral between electron and hole wave functions into account. The consequences for the nonlinear dynamics of electric field domains appearing inside the SL may be striking. With field-independent recombination, there are two values of the electric field that can be used to form profiles with coexistence of two domains, a low-field domain and a high-field domain. Under a dc voltage bias and depending on the laser intensity (which excites electron-hole pairs), these profiles can be stationary and stable or they can become unstable, and self-sustained oscillations of the current may appear as a consequence of domain wall dynamics [2]. With field-dependent recombination and high laser intensity, it is possible to find only one stable electric field domain.

Under dc voltage bias and for high conductivity contacts, the resulting field profiles giving rise to self-oscillations of the current comprise dipole waves resembling the action potential in nerve impulses [3]. Numerical simulations of the model show homogeneous nucleation of two of these dipole waves inside the SL. They then split and one resulting dipole wave moves toward the injector and the other toward the collector contact.

2 Model Equations

The equations governing nonlinear charge transport in a weakly coupled photoexcited undoped SL are [2]

$$\varepsilon \left(F_i - F_{i-1} \right) = e \left(n_i - p_i \right), \tag{1}$$

$$\varepsilon \frac{dF_i}{dt} + J_{i \to i+1} = J(t), \tag{2}$$

$$\frac{dp_i}{dt} = \gamma(I) - r(F_i, I) n_i p_i, \tag{3}$$

$$\frac{1}{N+1} \sum_{i=0}^{N} F_i = \phi \equiv \frac{V}{l\,(N+1)}, \tag{4}$$

$$J_{i \to i+1} = \frac{e n_i v(F_i)}{l} - e D_i(F_i) \frac{n_{i+1} - n_i}{l^2}, \tag{5}$$

$$J_{0 \to 1} = \sigma F_0, \qquad J_{N \to N+1} = \frac{n_N}{N_D} \sigma F_N. \tag{6}$$

Here $-F_i$, n_i, p_i are the average electric field, electron and hole surface densities at the ith period of the SL. The periods $i = 0$ and $i = N+1$ represent the injecting and collecting contacts of the SL. Equation (1) is the averaged Poisson equation, in which $-e < 0$, ε and l are the electron charge, the average permittivity and the period of the SL, respectively. Equation(2) is a form of Àmpere's law, yielding the total current density $J(t)$ as sum of the displacement current and the tunneling current from period i to period $(i+1)$, $J_{i \to i+1}$. We are assuming that only the electrons may tunnel across barriers. Then, at high temperature, the tunneling current is a function of F_i, n_i and n_{i+1} given by (5), which has the form of a discrete drift-diffusion equation [1]. Equation(4) establishes that the change in the hole density per unit time is due to photogeneration of electron-hole pairs (at a rate γ which depends on the laser intensity I) and to recombination thereof (with coefficient r). The electron drift and diffusion and the recombination coefficient are functions of the electric field illustrated in Fig. 1. The expressions for $v(F)$ and $D(F)$ can be found in [1], whereas we explain how to calculate $r(F, I)$ below. Equation(4) is the voltage bias condition, indicating that the voltage drop between injecting and collecting contacts of the SL is V. The tunneling current at the contacts $J_{0 \to 1}$ and $J_{N \to N+1}$ are approximated by Ohm's law (with a

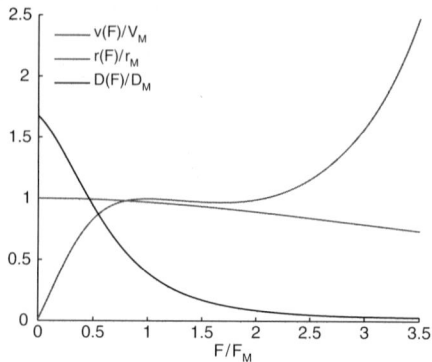

Fig. 1. Drift velocity, diffusion and recombination coefficients as functions of field for a 4 nm $Al_{0.3}Ga_{0.7}As$/10 nmGaAs SL under a laser intensity of 60 mW at a frequency 4.56×10^{15} Hz

contact conductivity σ) and by a linear function of the field, respectively [1]. N_D is the doping at the collecting contact. In (1) and (3), i goes from 1 to N, whereas i goes from 0 to N in (2). Then (1) to (3) are 3N+1 equations for the 3N+2 unknowns n_i, p_i ($i = 1, \ldots N$), F_i (i=0,1,...N) and J, provided the expressions (5) and (6) are taken into consideration. The bias condition (4) yields the missing equation. To solve (1) to (6), initial conditions for F_i and p_i should also be given.

Note that time differencing (2) and using the Poisson equation (1), we obtain the equation of charge continuity for the charge density $e\,(n_i - p_i)$. The coefficients γ and $r(F, I)$ require additional modeling. $\gamma = I\alpha_{3D}w/(\hbar\omega_o)$, where α_{3D}, w and ω_o are the 3D absorption coefficient, the width of the GaAs layer (quantum well) and the laser frequency, respectively [4]. $r(F, I)$ is

$$r(F, I) = \frac{n_r^2}{n_{in}^2 \pi^2 c^2} \int_0^\infty \frac{\alpha_{2D}(\omega, F)\omega^2}{\exp\left(\frac{\hbar\omega}{k_B T}\right) - 1}\, d\omega. \tag{7}$$

Here n_r is the SL refraction index, n_{in} is the density of intrinsic charge carriers in a quantum well. For large I, $n_{in} \approx \gamma n_r L_y/c$, where c and L_y are the velocity of light and the SL lateral extension, respectively. The 2D absorption coefficient α_{2D} in (7) is proportional to the modulus square of the overlap integral $\int_{-l/2}^{l/2} \Psi_e \Psi_h dx$ [4], in which Ψ_e and Ψ_h solve the stationary Schrödinger equation inside one SL period $-l/2 < x < l/2$ for the electrons and holes, respectively. In this equation, the electric field F is considered constant, and $\Psi(\pm(w+l_p)/2) = 0$. l_p is a penetration length inside the AlAs layer (quantum barrier), which depends self-consistently on the energy eigenvalue [5]. The resulting recombination coefficient is depicted in Fig. 1.

3 Numerical Results

In the phase plane of hole density and electric field (for space independent solutions), the curve $\mathrm{d}F_i/\mathrm{d}t = 0$ has three branches whereas the curve $\mathrm{d}p_i/\mathrm{d}t = 0$ has only one. With the parameters we are using (high laser intensity), we find that only one critical point in which both curves intersect is outside the middle branch of the curve $\mathrm{d}F_i/\mathrm{d}t = 0$; cf. Fig. 2. The resulting dipole waves depicted in Fig. 3 for constant J resemble the action potential waves in nerve impulses [3]: it comprises regions of slow variation of the electric field bound by moving wave fronts in which the electric field varies rapidly. In the wave fronts, the hole density is constant whereas the field jumps abruptly from one stable branch to the other one in the curve $\mathrm{d}F_i/\mathrm{d}t = 0$.

Under dc voltage bias and for high conductivity contacts, there are self-sustained oscillations of the current through the SL triggered inside the sample, as in Fig. 4. At an interior point, two dipole waves are created, split and each move to one of the contacts. When they reach the contacts, a new pair of

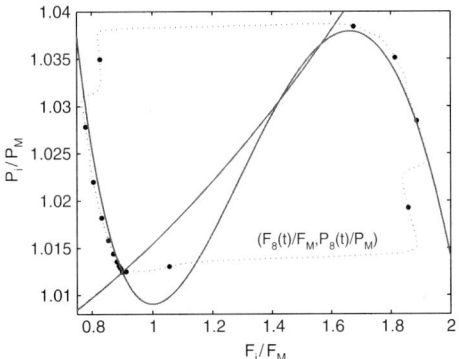

Fig. 2. Nullclines $\mathrm{d}F_i/\mathrm{d}t = 0$ and $\mathrm{d}p_i/\mathrm{d}t = 0$ for the parameter values in Fig. 1

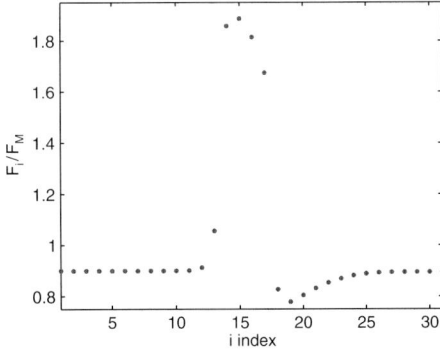

Fig. 3. Electric field profile of the pulse corresponding to the phase plane in Fig. 2 for $J = 132\ \mathrm{kA\,cm}^{-2}$

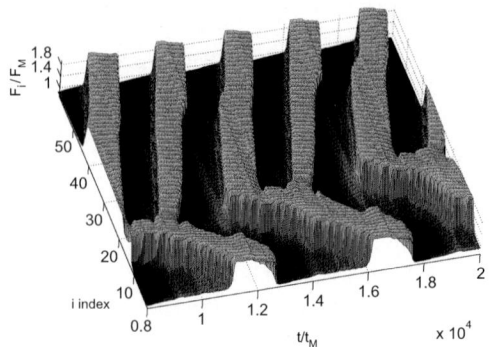

Fig. 4. Electric field profiles during self-oscillations in a dc voltage biased 61-period undoped SL having a configuration as in Fig. 1. The voltage between the two ends of the SL is 1.765 V and the contact resistivities are 9.07 ($i = 0$) and 8.87 Ωcm ($i = N$)

dipole waves is created inside the sample and the process repeats itself. Other unusual spatio-temporal patterns are observed [5]. At low laser intensity, current self-oscillations involve the dynamics of two electric field domains, which resembles more the case of doped SL [1]. An interpretation of the numerical results can be given using the construction of dipole domains under current bias and the difference in time scales as building blocks of the theory [5].

Support from the MEC grant MAT2005-05730-C02-01 and the Universidad Carlos III Foundation (JIA) are gratefully acknowledged.

References

1. L. L. Bonilla and H. T. Grahn, Nonlinear dynamics of semiconductor superlattices. Rep. Prog. Phys. **68**, 577-683 (2005).
2. L.L.Bonilla, J. Galán, J. A. Cuesta, F. C. Martínez and J. M. Molera, Dynamics of electric field domains and oscillations of the photocurrent in a simple superlattice model. Phys. Rev. B **50**, 8644-8657 (1994).
3. J.P. Keener and J. Sneyd, Mathematical Physiology. Springer, NY 1998.
4. H.T. Grahn, Introduction to Semiconductor Physics. World Sci., Singapore 1999.
5. J.I. Arana and L.L. Bonilla, preprint, 2007.

Numerical Analysis of a Nickel-Iron Electrodeposition Process

N. Alaa[1], M. Iguernane[2], and J.R. Roche[2]

[1] Faculté des Sciences et Téchniques Gueliz, Département de Mathématiques et Informatique, B.P. 549 Marrakech-Morocco
   alaa@fstg-marrakech.ac.ma
[2] I.E.C.N., Université Henri Poincaré, B.P. 239, 54506 Vandoeuvre-lès-Nancy, France
   iguernan@iecn.u-nancy.fr, roche@iecn.u-nancy.fr

Summary. This paper deals with a coupled system of non-linear elliptic differential equations arising in electrodeposition modelling process. We show the existence and uniqueness of the solution. A numerical algorithm to compute an approximation of the weak solution is described. We introduce a domain decomposition method to take in account the anisotropy of the solution. We show the domain decomposition method convergence. A numerical example is presented and commented.

1 Introduction

Electrodeposition of alloys based on the iron group of metals is one of the most important recent developments in the field of alloy deposition. In [Nat86], [Tuy83] Pritzker et al have proposed a model which involves the one-dimensional steady-state transport of the various species with simultaneous homogeneous reactions. The concentration of different species that are involved satisfies a system of non-linear differential equations. In this paper we are concerned with a reduced problem arising in one step of an iterative method solving the whole system. More precisely we consider the following system:

$$
\begin{cases}
-dv'' + b(x)v' - m(v\Phi')' = f & \text{in } (0, \delta) \\
v(\delta) = v^* \\
-dv'(0) - m\, v(0)\Phi'(0) = -\gamma\, v(0) \\
-[p(v)\Phi']' = q(v) & \text{in } (0, \delta) \\
\Phi(0) = V_0, \Phi(\delta) = 0,
\end{cases}
\tag{1}
$$

where v is the concentration, Φ is the potential, f denotes the production rate, d is the diffusion coefficient, m is the electrical mobility, δ is a fixed nonnegative real, v^*, V_0, γ are constants, p, q are nonnegative functions and $b(x) = -ax^2$ is the fluid velocity vector, with a a nonnegative constant.

In Sect. 2 we give a proof of existence and uniqueness of the solution (v, Φ) of system (1) in $C^2([0, \delta]) \times C^2([0, \delta])$.

The numerical solution of the system considered in the electrodeposition are characterized by stiff variations near the boundary $x = 0$. In order to take account of the anisotropy of the solution we introduce in Sect. 3 a generalized version of the two domain decomposition method due to F. Gastaldi, L. Gastaldi and A. Quarteroni (see [GGQ96]). We give a sketch of the proof for the convergence in the new case of non constant coefficients and Robin boundary conditions in $x = 0$. In Sect. 4 we present and discuss the result of a numerical example.

2 Existence and Uniqueness Result

Let $\varepsilon > 0$. We introduce the following assumptions:

H01) $p \in C^1(\mathbb{R})$ and there exist nonnegative constants η_0 and η_1 such that: $\eta_0 \le p \le \eta_1$.

Let $k_1 > 0$ such that $\mid p(x) - p(y) \mid \le k_1 \mid x - y \mid \quad \forall x, y \in [0, v^* + \varepsilon]$.

H02) There exist two nonnegative constants k_2 and η_2 such that $-\eta_2 \le q \le \eta_2$ and $\mid q(x) - q(y) \mid \le k_2 \mid x - y \mid \quad \forall x, y \in [0, v^* + \varepsilon]$.

H03) The constant d is such that:

1. $d > \gamma\delta + \dfrac{2a\delta^3}{3} + \dfrac{m(V_0 + 2\eta_2\delta^2)((v^* + \varepsilon)k_1 + \eta_1)}{\eta_0^2} + \dfrac{2m(v^* + \varepsilon)k_2\delta^2}{\eta_0}$.

2. $d \ge \dfrac{1}{\min(v^*, \varepsilon)} \left\{ \| f \| \delta^2 + \left(\gamma\delta + \dfrac{2a\delta^3}{3} + \dfrac{mV_0}{\eta_0} + \dfrac{2m\eta_2\delta^2}{\eta_0} \right)(v^* + \varepsilon) \right\}$.

Theorem 1. *Under assumptions H01–H03 the system (1) has a unique solution $(v, \Phi) \in C^2([0, \delta]) \times C^2([0, \delta])$.*

Proof. Let Π the map defined from $C([0, \delta])$ to $C([0, \delta])$ by $\Pi v = u$, where for $x \in [0, \delta]$

$$u(x) = v^* + \frac{\gamma}{d}(x - \delta)v(0) + \frac{1}{d} \int_\delta^x \left[(bv)(y) - \int_0^y (b'v)(t)dt - \int_0^y f(t)dt \right] dy$$

$$- \frac{m}{d} \int_\delta^x \left\{ \frac{v(y)}{p(v)(y)} \left[-\frac{V_0}{\delta} + \frac{1}{\delta} \int_0^\delta \int_0^t q(v)(s)dsdt - \int_0^y q(v)(t)dt \right] \right\} dy. \quad (2)$$

By integration of (1) it follows that a solution of the system is a fixed point of application Π. We set $D = \{u \in C([0, \delta]), 0 \le v \le v^* + \varepsilon\}$ equipped with the uniform norm. Using hypotheses H01-H03 we prove that the map Π is a contraction from D into itself. By Banach fixed point theorem it comes that Π has a unique fixed point $v \in D$ and by (2) $v \in C^2([0, \delta])$. Then (1) has a unique solution $(v, \Phi) \in C^2([0, \delta]) \times C^2([0, \delta])$. With

$$\Phi(x) = - \int_\delta^x \left\{ \frac{1}{p(v)(s)} \left(\frac{V_0}{\delta} + \frac{1}{\delta} \int_0^\delta \int_0^y q(v)(t)dt\, dy - \int_0^s q(v)(y)dy \right) \right\} ds. \quad (3)$$

3 Numerical Methods

For convenience we introduce the following new unknowns:

$$\psi(x) = \Phi(x) - \frac{V_0}{\delta}(\delta - x) \text{ and } w(x) = v(x) - v^* \text{ for all } x \in [0, \delta]. \quad (4)$$

System (1) is then equivalent to the following systems:

$$\begin{cases} L_1 w = F(w, \psi) & \text{in } (0, \delta), \\ w(\delta) = 0, \ -dw'(0) = G(w, \psi)(0). \end{cases} \quad (5)$$

and

$$\begin{cases} -[p(w + v^*)\psi']' = q(w + v^*) \text{ in } (0, \delta) \\ \psi(0) = 0, \ \psi(\delta) = 0, \end{cases} \quad (6)$$

where:

$$\begin{cases} L_1 w = -dw'^{f'} + B_0(x)w', \ B_0(x) = b(x) + m\dfrac{V_0}{\delta} \quad x \in (0, \delta). \\ F(w, \psi) = m[(w + v^*)\psi']' + f \quad x \in (0, \delta). \\ G(w, \psi) = \left[m\left(\psi'(0) - \dfrac{V_0}{\delta} \right) - \gamma \right](w(0) + v^*) \quad x \in (0, \delta). \end{cases} \quad (7)$$

The iterative method considered to solve this coupled problem first solves the equation (5) for a given potential ψ_n and then using the same algorithm solves equation (6) for a given concentration w_n. Let w_0 be the solution of (5) with $F = 0$ and then for any $n \in N$, w_{n+1} is the solution of the linear system:

$$\begin{cases} L_1 w = F(w_n, \psi) & \text{in } (0, \delta), \\ w(\delta) = 0, \ -dw'(0) = G(w_n, \psi)(0). \end{cases} \quad (8)$$

The existence and uniqueness of a solution of problem (8) is trivial in $C^2([0, \delta])$.

3.1 Iterative Method to Solve the Equation (8)

Let $c \in (0, \delta)$ be fixed. To solve equation (8) using the iterative domain decomposition method we decompose the set $(0, \delta)$ in two non-overlapping subdomains, $\Omega_1 = (0, c)$ and $\Omega_2 = (c, \delta)$. In the subdomain Ω_1 we consider a finer mesh structure than in Ω_2.

Let $n \in N$, A and B two reals parameters such that $AB \leq 0$, $A \neq B$.

Given $w_{1,0} = w_{2,0} = w_n$ and $\lambda^0 = d(w_{2,0})'(c) - \left(\dfrac{1}{2} B_0(c) + A \right) w_{2,0}(c)$, for each $k \geq 0$ we have to solve

$$\begin{cases} L_1 w_{1,k+1} = F(w_n, \psi) \text{ in } H^1(0, c), \\ -d(w_{1,k+1})'(0) = G(w_n, \psi)(0), \\ d(w_{1,k+1})'(c) - \left(\dfrac{1}{2} B_0(c) + A \right) w_{1,k+1}(c) = \lambda^k, \end{cases} \quad (9)$$

and then

$$
\begin{cases}
L_1 w_{2,k+1} = F(w_n, \psi) \text{ in } H^1(c, \delta), \\
w_{2,k+1}(\delta) = 0, \\
d(w_{2,k+1})'(c) - \left(\dfrac{1}{2} B_0(c) + B \right) w_{2,k+1}(c) \\
= d(w_{1,k+1})'(c) - \left(\dfrac{1}{2} B_0(c) + B \right) w_{1,k+1}(c),
\end{cases}
\tag{10}
$$

with

$$
\lambda^{k+1} = d(w_{2,k+1})'(c) - \left(\frac{1}{2} B_0(c) + A \right) w_{2,k+1}(c).
\tag{11}
$$

Thanks to the Lax-Milgram Theorem we can able to prove the:

Proposition 1. *If $A \leq 0$, then the problem (9) has a unique solution $w_{1,k+1} \in C^2([0, c])$ and if $B \geq 0$, then the problem (10) has a unique solution $w_{2,k+1} \in C^2([c, \delta])$.*

We will now give a sketch of the proof of convergence of the subdomain decomposition algorithm (9) and (10) applied to the solution of the linear problem (8) taking in account an anisotropic advective field and non constant absorption terms.

Proposition 2. *Let $c \in (0, \delta)$ such that $2d > | B_0(c) + A + B |$. Then the sequence $(w_{1,k}, w_{2,k})$ converge to (v, v) in $C(0, c) \times C(0, c)$.*

Proof. Let us define the errors $e_{j,k} = v - w_{j,k}$; $j = 1, 2$, and study their behavior as k grows. We can prove the following inequality:

$$
\| e_{1,k+1} \|_\infty \leq \gamma_0 \| e_{1,k} \|_\infty \text{ and } \| e_{2,k+1} \|_\infty \leq \gamma_0 \| e_{2,k} \|_\infty,
\tag{12}
$$

with $\gamma_0 > 0$. Conditions $A < B$ and $2d > | B_0(c) + A + B |$ imply that $\gamma_0^2 < 1$ which finish the proof.

4 Numerical Result

The algorithm introduced in the previous section has been implemented numerically for one example of problem (1) with $\delta = 11341 * 10^{-9}$, $c = \delta/10$, $m = 52133 * 10^{-12}$, $d = 68 * 10^{-11}$, $\gamma = 0.05$, $v^* = 1$, $V_0 = -0.85$ and $a = 660.45$, $p = 1 + \dfrac{1}{x^2 + x + 1}$, $q = \dfrac{1}{|x|+1}$ and $\dfrac{10\delta}{m+x}$. This is a nonlinear system with nondifferentiable second member. The numerical concentration was plotted in Fig. 1.

We remark that the variation rate of v is very strong near the boundary 0. This property justifies the use of the domain decomposition method and the choice of the fictitious boundary c near 0.

The algorithm (9)–(10) converges with $N_1 = 70$ finite element at the subdomain $[0, c]$ and $N_2 = 50$ finite element at the sub-domain $[c, \delta]$. We stop when the error is of order 10^{-19}.

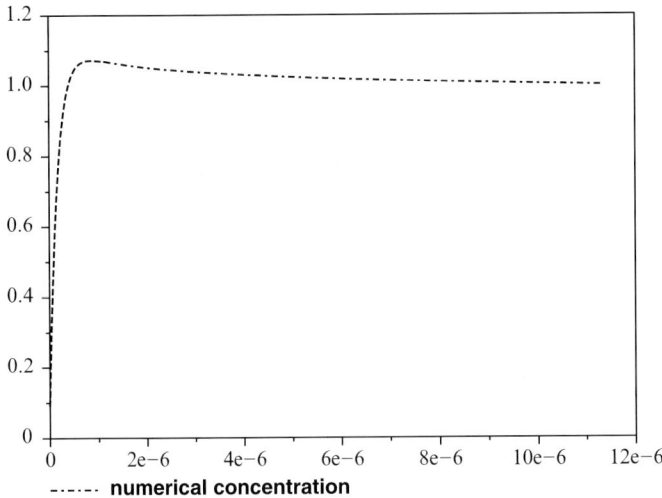

Fig. 1. Numerical concentration solution for $f = 10 * \delta/(m + x)$

References

[KAP97] T. Krause, L. Arulnayagam, and M. Pritzker, Model for Nickel-Iron Alloy Electrodeposition on a Rotating Disk Electrode, J. Electrochem. Soc. 144, 960-969, (1997).

[GGQ96] F. Gastaldi, L. Gastaldi and A. Quarteroni, Adaptive domain decomposition methods for advection dominated equation, East-West J. Numer. Math, 4, 165–206. (1996).

[SP98] H. Schultz, and M. Pritzker, Modeling the Galvanostatic Pulse and Pulse Reverse Plating of Nickel-Iron Alloys on a Rotating Disk Electrode, J. Electrochem. Soc. 145, 2033-2042, (1998).

A Simplified Finite Element Formulation for Spray Transfer GMA Weld Pools

Marcus Edstorp

University West, Department of Technology, Mathematics & Computer Science,
P.O. Box 957, SE-461 29, Trollhättan
marcus.edstorp@hv.se

1 Introduction

This chapter is concerned with the matter of mathematically modelling and computationally simulating the thermo and fluid dynamical phenomena occuring in the workpiece during a gas metal arc welding (GMAW) process, and does so by employing a continuum mechanical approach and a finite element formulation for approximating the solution of equations expressing the continuity of mass, the balance of linear momentum, the conservation of energy and the motion of the weld pool surface. GMAW is an electrode arc fusion welding process. The designation *arc fusion* signifies that an electric arc is struck between the welding electrode and the workpiece, and this causes the base material to melt on either side of the joint. During the subsequent solidification this will cause fusion between the workpiece parts. The electrode consist in a filler metal, and it is hence consumed during the process and molten droplets are, under the influence of electromagnetical and gravitational forces, transferred to the liquid weld pool. Mass is thus added to the workpiece and this causes a reinforcement of the joint.

Weld pool simulations may be used for predicting the interior geometry and reinforcement geometry of the resulting weld. The aim of this study is to provide a tool for predicting such quantities as weld penetration depth and weld toe radii. The weld toe radius is determined by the outer geometry of the weldment, and this is why we are required to model the motion of the freely moving surface of the weld pool. The simulation will also output the thermal history of the workpiece, which can be input to further simulations concerning the microstructure of the workpiece.

The arc affects the workpiece not only by generating heat at the weld surface, but also by exerting forces on the pool surface. In this study the action of these forces, as well as the influence of the energy source, are hypothesized and incorporated via boundary conditions for the governing equations.

2 Mathematical Modelling

We begin by deriving a generic semi-weak balance equation which is then instantiated for the mass density ρ^t, the fluid momentum density $\rho^t \mathbf{u}^t$, and the energy density H^t. We also propose an equation for the workpiece motion $\{\omega^t\}_t$, which is based on physical modelling.

The workpiece will at every time instant t be identified with a subset $\Omega(t)$ of $\Re^3$. The electrode tip travels at the constant velocity $\mathbf{v}$ and at constant height over the base plate. The generic equation for analyzing the weld pool behaviour was derived, developed and applied in [1] and [2], and we now present it in its entirety. We consider an arbitrary physical quantity $\mathcal{X}$, the density distribution of which at time t is represented spatially by the function $\mathcal{X}^t : \Omega(t) \to \Re$. We let the total flux $\mathbf{q}^t_{tot}$ and the source density F^t be functions such that, for every $t \in [0, t_{end}]$, the conservation of $\mathcal{X}$ in any open control volume W fixed in the interior of $\Omega(t)$ is expressed by the following relation:

$$\frac{\mathrm{d}}{\mathrm{d}t} \int_W \mathcal{X}^t = - \int_{\partial W} \mathbf{q}^t_{tot} \cdot \mathbf{n}^t + \int_W F^t$$

By assuming that for every physical quantity $\mathcal{X}$, including the flux, it is valid that $\mathcal{X}^t(x, y, z) = \mathcal{X}^{t - \frac{z - Z^\star}{v}}(x, y, Z^\star)$ for any fixed $Z^\star \in \Re$, where $v = |\mathbf{v}|$, and considering space-time material motions on the form $\omega^\star(X, Y, Z, \hat{t}) = [\omega_1(X, Y, \hat{t}), \omega_2(X, Y, \hat{t}), Z, \hat{t} + \frac{Z - Z^\star}{v}]^T$, where $[X, Y, Z]^T$ are the coordinates of an arbitrary point in the reference domain Ω^0, we find that all material representations (hatted) are independent of Z. Hence it suffices to solve only for the plane $Z = Z^\star$ where transformed and untransformed time coincide. By applying a change of coordinates in the integrals, and transforming the derivatives accordingly, we arrive at an equation posed on the two dimensional computational domain $\Omega_{Z^\star} = \{[X, Y] : \mathbf{X} \in \Omega^0, Z = Z^\star\}$, in which the influence of $\omega^\star$ is via the motion $\omega(X, Y; t) = [\omega_1(X, Y, t), \omega_2(X, Y, t)]^T$ of $\Omega_{Z^\star}$, and via its 2×2 Jacobian for fixed t;

$$\int_{\Omega_{Z^\star}} \frac{\partial}{\partial \hat{t}} \left[\hat{\mathcal{X}} - \frac{\hat{q}_3 + \hat{\mathcal{X}}\hat{u}_3}{v} \right] \hat{\phi} |\mathcal{J}_\omega| + \int_{\Omega_{Z^\star}} \nabla_{(X,Y)} \hat{\mathcal{X}} \mathcal{J}_\omega^{-1} \left(\begin{bmatrix} \hat{u}_1 \\ \hat{u}_2 \end{bmatrix} - \frac{\partial \omega}{\partial \hat{t}} \right) \hat{\phi} |\mathcal{J}_\omega|$$

$$+ \int_{\Omega_{Z^\star}} \frac{\partial \omega}{\partial \hat{t}} \cdot \nabla_{(X,Y)} \frac{\hat{q}_3 + \hat{\mathcal{X}}\hat{u}_3}{v} \mathcal{J}_\omega^{-1} \hat{\phi} |\mathcal{J}_\omega| = - \int_{\Omega_{Z^\star}} \hat{\mathcal{X}} \nabla_{(X,Y)} \begin{bmatrix} \hat{u}_1 \\ \hat{u}_2 \end{bmatrix} : \mathcal{J}_\omega^{-T} \hat{\phi} |\mathcal{J}_\omega|$$

$$+ \int_{\Omega_{Z^\star}} [\hat{q}_1, \hat{q}_2] \cdot \nabla_{(X,Y)} \hat{\phi} \mathcal{J}_\omega^{-1} |\mathcal{J}_\omega| - \int_{\partial \Omega_{Z^\star}} [-\hat{q}_2, \hat{q}_1] \mathcal{J}_\omega \hat{\mathbf{t}} \hat{\phi} + \int_{\Omega_{Z^\star}} \hat{F} \hat{\phi} |\mathcal{J}_\omega|, \forall \phi^t \in V(\mathcal{X}^t),$$

where $\hat{\mathbf{t}}$ is the positively oriented unit tangent to $\partial \Omega_{Z^\star}$, $\mathbf{q}^t_{\text{tot}} = \mathbf{q}^t + \mathcal{X}^t \mathbf{u}^t$ and $V(\mathcal{X}^t)$ is the weighting space for $\mathcal{X}^t$. When disregarding $\hat{q}_3$, and incorporating the pointwise incompressibility constraint, we arrive at

$$\int_{\Omega_{Z^\star}} \frac{\partial}{\partial \hat{t}} \hat{\mathcal{X}} \hat{\phi} |\mathcal{J}_\omega| + \int_{\Omega_{Z^\star}} \nabla_{(X,Y)} \hat{\mathcal{X}} \mathcal{J}_\omega^{-1} \left(\begin{bmatrix} \hat{u}_1 \\ \hat{u}_2 \end{bmatrix} - \frac{\partial \omega}{\partial \hat{t}} \right) \hat{\phi} |\mathcal{J}_\omega|$$

$$= + \int_{\Omega_{Z^\star}} [\hat{q}_1, \hat{q}_2] \cdot \nabla_{(X,Y)} \hat{\phi} \mathcal{J}_\omega^{-1} |\mathcal{J}_\omega| - \int_{\partial \Omega_{Z^\star}} [-\hat{q}_2, \hat{q}_1] \mathcal{J}_\omega \hat{\mathbf{t}} \hat{\phi} + \int_{\Omega_{Z^\star}} \hat{F} \hat{\phi} |\mathcal{J}_\omega| \quad (1)$$

We now instantiate (1) for the mass density, i.e. we take $\mathcal{X}^t = \rho^t$. No interior sources of mass are present, however by assuming that the amperage of the welding current is high enough (c.f. [3]) for the process to operate in spray transfer mode, that is, the metal droplets transferred to the weld pool are quite small and frequent, we may specify the flux across the workpiece boundaries such that it models the addition of filler metal. By incorporating the Boussinesq approximation and relaxation terms (c.f. [2]), we obtain

$$0 = -\rho_{ref} \int_{\Omega_0} \nabla_{(X,Y)} \hat{\mathbf{u}} : \mathcal{J}_\omega^{-T} \hat{\bar{p}} |\mathcal{J}_\omega| - \epsilon \int_{\Omega_0} \nabla_{(X,Y)} \hat{p} \mathcal{J}_\omega^{-1} \mathcal{J}_\omega^{-T} \nabla_{(X,Y)}^T \hat{\bar{p}} |\mathcal{J}_\omega|$$
$$- \epsilon^2 \int_{\Omega_0} \hat{p} \hat{\bar{p}} |\mathcal{J}_\omega| + \hat{M}^t \left[\hat{\bar{p}} \right], \forall \bar{p} \in V(p^t)$$

where $M^t : V(p^t) \longrightarrow \Re$ is a boundary source of mass density such that $M[1] =: M$ is an approximation of the deposition rate, i.e. the mass added to the workpiece per unit time. By taking the computational domain Ω_0 as a rectangle, the part Γ_0^w of the boundary $\partial \Omega_0$ that is mapped on a subset of the welding surface is a subset of the $Y = 0$ line, and we may write

$$\hat{M}^t \left[\bar{p} \right] = \frac{M}{\pi b^2} \int_{\Gamma_0^w} f \circ \hat{T}(X, 0, t) \, exp \left(\frac{-||(\omega_1(X, 0, t), P_3(t))||^2}{b^2} \right) \frac{\partial \omega_1}{\partial X}(X, 0, t) \hat{\bar{p}} \, \mathrm{d}X$$

where b is the spot radius of the spray, $P_3(t)$ is the coordinate of the electrode tip in the welding direction, and the local liquid fraction f has been incorporated in order to avoid adding mass to the solid workpiece. If the weld pool is wider than $b\sqrt{-\ln(0.01)}$, the theoretical deposition rate does not deviate more than 1% from M.

The instantiation of equation (1) for $\mathcal{X}^t = \rho^t \mathbf{u}^t$ and $\mathcal{X}^t = H^t$ is as in [2].

The equation for the workpiece motion is defined in such a fashion that it mimics the motion of an inertia-less fluid constrained by the kinematic condition. By taking the viscosity μ_{mesh} of this pseudo-fluid to depend upon the element domain size h, small elements are less distorted, and convergence is speed up. We have experienced that $\mu_{mesh}(h) = 1 - h$ is a good choice in our applications. We thus require that the relation

$$0 = - \int_{\Omega^0} \mu_{mesh} \left[\nabla_{\mathbf{x}} \frac{\partial \omega}{\partial t} \mathcal{J}_\omega^{-1} \right] : \left[\nabla_{\mathbf{x}} \bar{\omega} \mathcal{J}_\omega^{-1} \right] |\mathcal{J}_\omega| + \hat{\Gamma}_\omega \left[\bar{\omega} \right],$$

holds for every $\bar{\omega}$ in $V(\frac{\partial \omega}{\partial t})$, where $\hat{\Gamma}_\omega : V(\frac{\partial \omega}{\partial t}) \to \Re$ is the material formulation of the total force acting on the surface of the pseudo-fluid. This force is determined via a Lagrange multiplier for the kinematic constraint. As an alternative to having the surface move as a result of adding mass in the continuity equation, one may modify the velocity of the boundary in the expression for $\hat{\Gamma}_\omega$. This way, the motion of the free surface is not a result of simulating the response of an incompressible material to a mass source, but the the motion of the "freely" moving surface is somewhat forced. The two approaches produce different outputs. We have experienced that the approach we have described above computationally outperforms the approach employing a modified $\hat{\Gamma}_\omega$. As an example, we notice that the alternative approach does not converge for the case depicted by the rightmost circle in Fig. 2. What more is, we think that the current approach is more physically reasonable.

3 Computational and Numerical Modelling

We discretize the governing equations in space using a finite element method employing Taylor–Hood triangles for the Navier–Stokes equations, and second order Lagrange triangles for the other unknowns except for the multplier for the kinematic constraint, for which first order one-dimensional elements are used. One may expect that a first order geometrically continuous representation of the surface shape would improve the convergence with respect to the surface tension forces. Since the surface is represented by a parametrization determined by the FE approximation, triangular Hermite elements would achieve this due to the continuity of the derivatives of the approximation at the nodes. However, using Hermite elements for the motion makes the computations perform much worse, possibly due to a poor combination of element types.

A numerical method was implemented using the software *Comsol Script*, which applies the DASPK algorithm [4,5] for solving the differential-algebraic system. The reason we employ the formulation based on space-time material motions instead of the equivalent three dimensional steady state formulation is that it is generally easier to obtain convergence when applying the DASPK algorithm for time-dependant problems, than for the corresponding static problems.

4 Qualitative Behaviour of the Model

Investigations into the qualitative behaviour of the model were performed. Even though this includes quantitative predictions of pool temperatures and melting efficiency, validations of these results are not included in this chapter.

A study with respect to the assumption that the total flux vanishes in the welding direction was performed, and the important features were reported in [2]. We now add to this study a parameter study of the effect of the electrode tip travel speed on the maximum pool temperature predicted by the simplified formulation, measured as a percentage of the maximum temperature predicted by a three dimensional static simulation. Figure 1 shows that the maximum temperature is predicted quite well even for low values for v.

In order to quantify the influence of the heat input to the workpiece that is due to the addition of filler metal, and compare it to the significance of the boundary source term in the energy equation, we study the melting efficiency predicted by the simulation. We define the melting efficiency of a steady-state linear welding process as the ratio of the theoretical minimum amount of energy required to establish the weld pool and mushy zone, to the total amount of thermal energy contained in the workpiece. This ratio was calculated for a number of typical linear bead-on-plate welding processes (on steel 316 plates), the parameter settings of which were the same, except that we beteween the runs modified the deposition rate. This way we may study the dependence of the melting efficiency on the reinforcement cross-sectional area. We obtained nine data points in the range 0–4 mm^2, see Fig. 2. Our model behaves qualitatively correct in that the melting efficiency increases with the

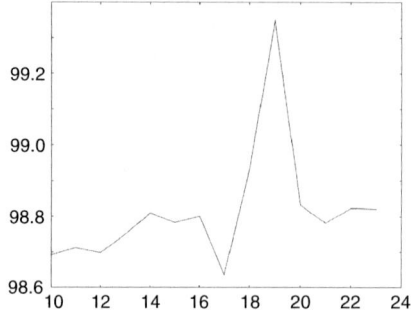

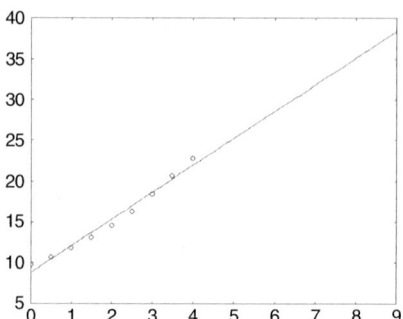

Fig. 1. Maximum pool temperatures (*percent*) vs. arc speed (*mm s⁻¹*).

Fig. 2. Melting efficiencies (*percent*) vs. reinforcement cross-sectional area (*mm s⁻²*)

actual heat input to the workpiece (c.f. [6]) which is strongly affected by the deposition rate. By extrapolating a quadratic fit, we find that the performed parameter study predicts that, for greater M, our model would simulate a melting efficiency the size of which agrees with what is usually experienced during non-autogenous arc welding (40-50%), see for example [6].

5 Conclusions and Future Work

We have defined a tool for simulating the dynamical behaviour of an arc fusion weld pool with respect to fluid and thermo dynamical phenomena. Parameter studies were performed, and the tool behaves qualitatively according to physical expectations.

The author would like to thank the following companies; Volvo Construction Equipment, ESAB AB, SSAB Tunnplåt AB, Volvo Aero Corporation.

References

1. Edstorp, M., Eriksson, K.: Modelling and Simulation of Moving Boundaries and Convective Heat Transfer in Non-Autogenous Fusion Weld Pools using FEMLAB 3.1. In: Yström, J. (ed) Proceedings of the FEMLAB Conference 2005, 85–90, (2005)
2. Edstorp, M.: A Comparison Between Moving Mesh Implementations for Metal Deposition Simulations. In: Gregersen, L. (ed) Proceedings of the Nordic Comsol Conference 2006, 107–110, (2006)
3. Kou, S.: Welding Metallurgy, second edition. Wiley-Interscience (2003)
4. Brenan, K.E., Campbell, S.L., Petzold, L.R.: Numerical Solution of Initial-Value Problems in Differential-Algebraic Equations. Society for Industrial and Applied Mathematics (1996)
5. Brown, P.N., Hindmarsh, A.C., Petzold, L.R.: Using Krylov Methods in the Solution of Large-Scale Differential-Algebraic Systems. Journal of Scientific Computing., **15:6**, 1467–1488, (1994)
6. DuPont, J.N., Marder, A.R.: Thermal Efficiency of Arc Welding Processes. Welding Journal, **74:12**, 406–416, (1995)

Numerical Solution of a Non-Local Elliptic Problem Modeling a Thermistor with a Finite Element and a Finite Volume Method

C.V. Nikolopoulos and G.E. Zouraris

Department of Mathematics, University of the Aegean, 83200 Samos, Greece
cnikolo@aegean.gr, zouraris@aegean.gr

1 Introduction

Let $D := (-1, 1)$, and the following non-local elliptic boundary value problem:

$$w''(x) + \lambda \frac{f(w(x))}{\left(\int_D f(w) dx \right)^2} = 0 \quad \forall\, x \in D, \tag{1}$$

$$w'(1) + a\, w(1) = 0, \qquad w'(-1) - a\, w(-1) = 0, \tag{2}$$

where $w = w(x; \lambda)$ and λ is a dimensionless parameter(see, e.g., [3,6,7]). The problem (1) and (2) models the steady state temperature profile of the thermistor device (see [2,6]). We will focus in the case of the Negative Temperature Coefficient thermistor (NTC-thermistor), where the electrical resistivity decreases with temperature, e.g. $f(s) = e^{-s}$ or $f(s) = (1 + s)^{-p}$. It is has been proved that if $f(s) > 0$, $f'(s) < 0$, $f''(s) > 0$ for $s > 0$, and $\int_0^\infty f(s)\, ds < \infty$, then the problem (1) and (2), has at least one classical (regular) solution for the critical value of the parameter λ, say λ^*, has no solution for $\lambda > \lambda^*$, and for $\lambda < \lambda^*$ attains at least two regular solutions $(\overline{w}, \underline{w})$ where $\underline{w}(x) < \overline{w}(x)$ for $x \in D$, $\underline{w}$ is stable and $\overline{w}$ is unstable for λ close to λ^*. In addition we may scale f so that $\int_0^\infty f(s)\, ds = 1$, and then $\lambda^* < 8$. For the (1) but with Dirichlet b.c.s, $w(\pm 1) = 0$, we have that $\lambda^* = 8$ (if $\int_0^\infty f(s) ds = 1$). For $\lambda < \lambda^*$ we have a unique stable solution w, while for $\lambda \geq \lambda^*$ we have no solution. For $f(s) = e^{-s}$ the analytical solution for problem (1) and (2) is known (see [4]). It holds that $w(x) = \frac{2\gamma}{\alpha} \tan(\gamma) + 2\ln(\frac{\cos(\gamma x)}{\cos(\gamma)})$ where γ solves the equation

$$\lambda = 8 \sin^2(\gamma) \exp\left(-\frac{2\gamma}{\alpha} \tan(\gamma) \right)$$

for λ, a known. Also for $\alpha = 1$, λ^* can be computed and is found to be $\lambda^* \simeq 1.1239$. When the Dirichlet b.c.s, $w(\pm 1) = 0$, imposed to (1) with $f(s) = e^{-s}$, then $w(x) = 2\ln\left[\frac{\cos(\gamma x)}{\cos(\gamma)} \right]$ for

$$\gamma = \sin^{-1}\left(\frac{\lambda}{8}\right)^{1/2}.$$

These analytical solutions can be used for the comparison with the numerical results presented in this work. The accurate knowledge for the steady solution is needed, in order to obtain estimates for the evolutionary problem and for the practical point of view of applications (cf. [2, 4, 5]).

In the chapter at hand, in order to approximate the solution of (1) with Robin or Dirichlet b.c.s, we construct a finite element and a finite volume method based on piecewise continuous piecewise quadratic functions. In particular, the proposed finite volume method extends a new finite volume method derived recently in [8] for general linear two-point boundary value problems. Both methods leads to a nonlinear system of algebraic equations that we solve by applying an iterative method. In the case of the Robin boundary conditions (2), when we start the iterative method below this solution, e.g., from zero, it is expected that we approximate the solution belonging to the stable branch of the response diagram (minimal solution $\underline{w}$) which is the situation of interest regarding the application of the model (see, e.g., [1]). Apart from this it is useful to compare the finite element method and the finite volume method for a nonlinear elliptic problem, since the general theory for finite volume methods is not as extensive as for the finite element methods.

2 Formulation of the Numerical Methods

We consider a partition of D with $J+1$ nodes $\{x_j\}_{j=0}^{J}$ where $J \geq 3$, $x_0 = -1$, $x_J = 1$ and $x_j < x_{j+1}$ for $j = 0, \ldots, J-1$. Then, set $I_j := (x_{j-1}, x_j)$ for $j = 1, \ldots, J$, $x_{j+z} := x_j + z(x_{j+1} - x_j)$ for $j = 0, \ldots, J-1$ and $z \in [0,1]$, and $\xi_j : I_j \to [0,1]$ by $\xi_j(x) := \frac{x - x_{j-1}}{x_j - x_{j-1}}$ for $x \in I_j$ and $j = 1, \ldots, J$. Let $I = (y_L, y_R)$. Then, we denote by $\mathcal{X}_I$ the characteristic function of the interval I, and we write $[\![v]\!]_{\partial I} = v(y_R^-) - v(y_L^+)$, where $v(x^{\pm}) := \lim_{\epsilon \to 0^+} v(x \pm \epsilon)$.

The methods we propose construct an approximation of the solution of problem (1) and (2) from the space $S_{\mathcal{R}}^2$ consisting of functions which are continuous on $[-1, 1]$ and reduce to polynomials of degree less than or equal to 2 on each $I \in \{I_j\}_{j=1}^{J}$. When we consider the (1) with Dirichlet boundary conditions the methods construct an approximations of the solution from the space $S_{\mathcal{D}}^2 := \{\phi \in S_{\mathcal{R}}^2 : \phi(\pm 1) = 0\}$. We note that $\dim(S_{\mathcal{R}}^2) = 2J + 1$ and $\dim(S_{\mathcal{D}}^2) = 2J - 1$.

2.1 The Finite Volume Method

Let $\rho \in (0, 1)$ be a real parameter, and $\{\Delta_j\}_{j=1}^{2J+1}$ be control volumes given by $\Delta_{2\ell} := (x_{\ell-1}, x_\ell)$ for $j = 1, \ldots, J$, $\Delta_{2\ell+1} := (x_{\ell-1+\rho}, x_{\ell+\rho})$ for $j = 1, \ldots, J-1$, $\Delta_1 := (x_0, x_{0+\rho})$ and $\Delta_{2J+1} := (x_{J-1+\rho}, x_J)$. The proposed finite volume method (FVM) (cf. Proposition 3.7 in [8]) is formulated as follows: find $w_h \in S_{\mathcal{R}}^2$ such that

$$
\begin{aligned}
&- \left[w_h'(x_{0+\rho}) - a\, w_h(x_0) \right] = V(w_h; \Delta_1) \\
&- \left[w_h'(x_1) - a\, w_h(x_0) \right] = V(w_h; \Delta_2), \\
&- [\![w_h']\!]_{\partial \Delta_j} = V(w_h; \Delta_j), \qquad j = 3, \ldots, 2J-1, \\
&- \left[-a\, w_h(x_J) - w_h'(x_{J-1}) \right] = V(w_h; \Delta_{2J}), \\
&- \left[-a\, w_h(x_J) - w_h'(x_{J-1+\rho}) \right] = V(w_h; \Delta_{2J+1})
\end{aligned}
\tag{3}
$$

where $V(w_h; \Delta) := \lambda \dfrac{\int_\Delta f(w_h(x))\,\mathrm{d}x}{\left(\int_D f(w_h(x))\,\mathrm{d}x\right)^2}$. Using the auxiliary functions $\widehat{\phi}_0(x) :=$ $\frac{2-3\rho}{1-\rho}(1-x^2) + \frac{2\rho-1}{1-\rho}(1-x)$, $\widehat{\phi}_{\frac12}(x) := 6x(1-x)$, $\widehat{\phi}_1(x) := \frac{3\rho-1}{\rho}x^2 + \frac{1-2\rho}{\rho}x$ (see [8]), we construct a basis $\mathcal{B}_{\mathcal{R}}^{FV} = \{\varphi_j\}_{j=1}^{J+1} \cup \{\varphi_{j-\frac12}\}_{j=1}^{J}$ of $S_{\mathcal{R}}^2$ by $\varphi_1(x) = \widehat{\phi}_0(\xi_1(x))\,\mathcal{X}_{I_1}(x)$, $\varphi_{j-\frac12}(x) = \widehat{\phi}_{\frac12}(\xi_j(x))\,\mathcal{X}_{I_j}$ for $j = 1, \ldots, J$, $\varphi_j(x) = \widehat{\phi}_1(\xi_{j-1}(x))\,\mathcal{X}_{I_{j-1}}(x) + \widehat{\phi}_0(\xi_j(x))\,\mathcal{X}_{I_j}(x)$ for $j = 2, \ldots, J$, and $\varphi_{J+1}(x) = \widehat{\phi}_1(\xi_J(x))\,\mathcal{X}_{I_J}(x)$. Hence, $w_h = \sum_{i=1}^{J+1} \beta_i^{FV}\varphi_i + \sum_{i=1}^{J} \beta_{J+1+i}^{FV}\varphi_{i-\frac12}$, where $\beta = \{\beta_i^{FV}\}_{i=1}^{2J+1}$ is the coefficients vector to be determined. Then, (3) is equivalent to a nonlinear system of algebraic equations of the form $A^{FV}\beta^{FV} = F^{FV}(\beta^{FV})$, where $A^{FV} \in \mathbf{R}^{(2J+1)\times(2J+1)}$ is a matrix and $F^{FV} : \mathbf{R}^{2J+1} \to \mathbf{R}^{2J+1}$ is a nonlinear map defined by $(F^{FV}(y))_i := V\left(\sum_{j=1}^{J+1} y_j\varphi_j + \sum_{j=1}^{J} y_{J+1+j}\,\varphi_{j-\frac12}; \Delta_i \right)$ for $y \in \mathbf{R}^{2J+1}$ and $i = 1, \ldots, 2J+1$. We solve the obtained nonlinear system by an iterative process based on Broyden's method with initial approximation $\beta_{(1)}^{FV} = 0 \in \mathbf{R}^{2J+1}$. When, we consider the (1) with Dirichlet boundary conditions, the finite volume method is formulated as follows: find $w_h \in S_D^2$ such that:

$$
-[\![w_h']\!]_{\partial \Delta_j} = V(w_h; \Delta_j), \quad j = 2, \ldots, 2J.
$$

To formulate the analogous nonlinear system of algebraic equations, we choose the basis $\mathcal{B}_D^{FV} = \{\varphi_j\}_{j=2}^{J} \cup \{\varphi_{j-\frac12}\}_{j=1}^{J}$ of S_D^2. In the numerical experiments, we choose $\rho = \frac12 - \frac{\sqrt3}{6}$ because, according to the error analysis in [8], this is one of the values which, in the linear case, ensure an optimal order of convergence in the L^2, H^1 and L^∞ norms.

2.2 The Finite Element Method

The finite element method (FEM) for problem (1) and (2) is formulated as follows: find $w_h \in S_{\mathcal{R}}^2$ such that

$$
a\left[w_h(1)\,\phi(1) + w_h(-1)\,\phi(-1) \right] + (w_h', \phi')_{0,D} = W(w_h, \phi) \quad \forall \phi \in S_{\mathcal{R}}^2, \tag{4}
$$

where $W(w_h, \phi) := \lambda \dfrac{(f(w_h), \phi)_{0,D}}{\left(\int_D f(w_h)\,\mathrm{d}x\right)^2}$. In the numerical experiments we use a basis $\mathcal{B}_{\mathcal{R}}^{FE} = \{\phi_j\}_{j=1}^{2J+1}$ of $S_{\mathcal{R}}^2$ determined by $\phi_{J+1+j}(x) = \widehat{\phi}_{\frac12}(\xi_j(x))\mathcal{X}_{I_j}(x)$

for $j = 1, \ldots, J$, $\phi_1(x) = \widehat{\phi}_0(\xi_1(x)) \mathcal{X}_{I_1}(x)$, $\phi_j(x) = \widehat{\phi}_1(\xi_{j-1}(x)) \mathcal{X}_{I_{j-1}}(x) + \widehat{\phi}_0(\xi_j(x)) \mathcal{X}_{I_j}$ for $j = 2, \ldots, J$, and $\phi_{J+1} = \widehat{\phi}_1(\xi_J(x)) \mathcal{X}_{I_J}$, where $\widehat{\phi}_0(x) := 1 - x$, $\widehat{\phi}_{\frac{1}{2}}(x) := 4x(1-x)$ and $\widehat{\phi}_1(x) := x$. Thus $w_h = \sum_{i=1}^{2J+1} \beta_i^{\mathrm{FE}} \phi_i$, where $\beta^{\mathrm{FE}} = \{\beta_i^{\mathrm{FE}}\}_{i=1}^{2J+1}$ is the coefficients vector to be specified. It is easily seen that (4) is equivalent to a nonlinear system of algebraic equations of the form $A^{\mathrm{FE}} \beta^{\mathrm{FE}} = F^{\mathrm{FE}}(\beta^{\mathrm{FE}})$, where $A^{\mathrm{FE}} \in \mathbf{R}^{(2J+1) \times (2J+1)}$ is a matrix and $F^{\mathrm{FE}} : \mathbf{R}^{2J+1} \to \mathbf{R}^{2J+1}$ is a nonlinear map defined by $(F^{\mathrm{FE}}(y))_i := W(\sum_{j=1}^{2J+1} y_j \phi_j, \phi_i)$ for $y \in \mathbf{R}^{2J+1}$ and $i = 1, \ldots, 2J+1$. As in the FVM the resulting nonlinear system is solved by an iterative process based on Broyden's method. When we consider the (1) with Dirichlet boundary conditions, the FEM is formulated as follows: find $w_h \in S_{\mathcal{D}}^2$ such that:

$$(w_h', \phi')_{0, \mathcal{D}} = W(w_h, \phi) \quad \forall \phi \in S_{\mathcal{D}}^2.$$

The corresponding nonlinear system of algebraic equations, is obtained choosing the basis $\mathcal{B}_{\mathcal{D}}^{\mathrm{FE}} = \{\phi_j\}_{j=2}^{J} \cup \{\phi_{j-\frac{1}{2}}\}_{j=1}^{J}$ of $S_{\mathcal{D}}^2$.

3 Numerical Results and Comparison

In this section we present results of numerical experiments performed with the numerical methods presented in Sect. 2. All numerical schemes were implemented in a MATLAB program. The problem was solved numerically on a uniform grid with $J + 1$, and using tolerance TOL $= 10^{-10}$ in the Newton-type method. Also, we choose $a = 1$, $\lambda = 1$ and $f(s) = e^{-s}$, i.e., λ is chosen so that $\lambda < \lambda^*$. The L^2 and H^1 norms of the error $w - w_h$ were computed using Simpson's rule, and the L^∞ norm of the error was estimated by a finite sampling at the abscissae of the aforementioned quadrature rule. The results are summarized in Tables 1 and 2 indicating that the computational order of convergence agrees with the order of convergence in the linear case (see [8]), which is equal to 3 in the L^2 and L^∞ norms, and 2 in the H^1 norm.

In Tables 1 and 2 are shown that using the same uniform partition, the FEM for (1) with Robin b.c.'s is more accurate and faster than the FVM.

In Tables 3 and 4 we see that, in the case of Dirichlet b.c.'s, the methods are similarly accurate with error of the same order, however the error of the FVM is slightly bigger. This difference with the corresponding results for the

Table 1. Rates of convergence of the FEM for (1) with Robin b.c.'s

$J+1$	$\|w - w_h\|_{0, \mathcal{D}}$	Rate	$\|w - w_h\|_{1, \mathcal{D}}$	Rate	$\|w - w_h\|_{\infty, \mathcal{D}}$	Rate
20	$1.513(-6)$		$6.847(-5)$		$2.585(-6)$	
40	$1.595(-7)$	3.13	$1.647(-5)$	1.98	$2.953(-7)$	3.02
80	$1.933(-8)$	2.99	$4.042(-6)$	1.99	$3.626(-8)$	2.97
160	$2.265(-9)$	3.07	$1.001(-6)$	2.00	$4.360(-9)$	3.02

Table 2. Rates of convergence of the FVM for (1) with Robin b.c.'s

$J+1$	$\|w-w_h\|_{0,D}$	Rate	$\|w-w_h\|_{1,D}$	Rate	$\|w-w_h\|_{\infty,D}$	Rate
20	$3.648(-5)$		$5.743(-4)$		$3.169(-5)$	
40	$4.659(-6)$	2.86	$1.428(-4)$	1.94	$3.975(-6)$	2.89
80	$5.546(-7)$	3.02	$3.643(-5)$	1.94	$4.765(-7)$	3.00
160	$6.063(-8)$	2.93	$9.664(-6)$	1.90	$6.063(-8)$	2.95

Table 3. Rates of convergence of the FEM for (1) with Dirichlet b.c.'s

$J+1$	$\|w-w_h\|_{0,D}$	Rate	$\|w-w_h\|_{1,D}$	Rate	$\|w-w_h\|_{\infty,D}$	Rate
20	$4.171(-7)$		$2.427(-5)$		$7.112(-7)$	
40	$4.831(-8)$	2.99	$5.846(-6)$	1.98	$8.514(-8)$	2.95
80	$5.815(-9)$	2.99	$1.435(-6)$	1.99	$1.041(-8)$	2.98
160	$7.133(-10)$	3.00	$3.556(-7)$	1.99	$1.288(-9)$	2.99

Table 4. Rates of convergence of the FVM for (1) with Dirichlet b.c.'s

M	$\|w-w_h\|_{0,D}$	Rate	$\|w-w_h\|_{1,D}$	Rate	$\|w-w_h\|_{\infty,D}$	Rate
20	$4.206(-7)$		$2.431(-5)$		$7.403(-7)$	
40	$4.866(-8)$	2.99	$5.848(-6)$	1.98	$8.691(-8)$	2.98
80	$5.855(-9)$	2.99	$1.435(-6)$	1.99	$1.052(-8)$	2.99
160	$7.182(-10)$	3.00	$3.556(-7)$	1.99	$1.295(-9)$	3.00

problem with Robin b.c.'s, may be, is related with an error increased in the boundary volumes when the FVM is used with Robin b.c.'s.

Acknowledgements

Work supported by The University of the Aegean under the Research Project no.1356/EPEAEK II-PITHAGORAS-TDY12.

References

[1] M. Crouzeix, J. Rappaz: On numerical approximation in bifurcation theory, RMA, Springer Verlag, 1990.

[2] A.C. Fowler, I. Frigaard and S. D. Howison: Temperature surges in current–limiting circuit devices. SIAM J. Appl. Math. **52**, pp. 998–1011, 1992.

[3] P. Freitas and M. Grinfeld : Stationary solutions of an equation modelling Ohmic heating. Appl. Math. Lett. **7**, pp. 1–6, 1994.

[4] N.I. Kavallaris, C.V. Nikolopoulos, D.E. Tzanetis: Estimates of blow-up time for a non-local problem modelling an Ohmic heating process. European J. Appl. Maths. **13**, pp. 337–351, 2002.

[5] S. Kutluay and A.S. Wood: Numerical solutions of the thermistor problem with a ramp electrical conductivity, Applied Mathematics and Computation **148**, pp. 145–162, 2004.

[6] A.A. Lacey: Thermal runaway in a non-local problem modelling Ohmic heating. Part I : Model derivation and some special cases, European J. Appl. Maths. **6**, pp. 127–144, 1995.

[7] A.A. Lacey: Thermal runaway in a non-local problem modelling Ohmic heating. Part II : General proof of blow-up and asymptotics of runaway. Euro. Jl. Appl. Maths. **6**, pp. 201–224, 1995.

[8] M. Plexousakis and G.E. Zouraris: On the construction and analysis of high order locally conservative finite volume-type methods for one dimensional elliptic problems. SIAM Journal of Numerical Analysis **42**, pp. 1226–1260, 2004.

Numerical Solution of 3D Magnetostatic Problems in Terms of Scalar Potentials

A. Bermúdez[1], R. Rodríguez[2], and P. Salgado[1]

[1] Depto. de Matemática Aplicada, Universidade de Santiago de Compostela, 15706, Santiago de Compostela, Spain
mabermud@usc.es, mpilar@usc.es
[2] GI²MA, Depto. de Ingeniería Matemática, Universidad de Concepción, Casilla 160-C, Concepción, Chile
rodolfo@ing-mat.udec.cl

1 Introduction

The goal of this chapter is to analyze a finite element method to solve the magnetostatic problem in terms of scalar potentials. Several FEM have been developed to solve the magnetostatic problem in the last decades, because of its application in engineering; see, for instance, [BAL96, MSP98, PAL91, ST79, ST80]. The main difference in the numerical methods lies in the choice of the primary unknowns (vector potential, magnetic field or scalar potentials). The published numerical results ([MSP98, PAL91]) show that the combination of two different potentials, the so called *reduced scalar potential* and *total scalar potential*, seems to be the most effective in terms of accuracy and computer cost. This formulation, was introduced by Simkin and Trowbridge in [ST79] and is very well known in the engineering literature; however, to the best of the author's knowledge, the approximation of this formulation in bounded domains by standard finite elements has not been analyzed from a mathematical point of view. This is the aim of the present chapter in the context of three-dimensional domains.

2 Scalar Formulation of the Magnetostatic Problem

The classical magnetostatic model is obtained by neglecting the time derivatives in Maxwell equations. Thus, given a divergence-free stationary source current density $\boldsymbol{J}$, the magnetic field $\boldsymbol{H}$ satisfies the following equations:

$$\operatorname{curl} \boldsymbol{H} = \boldsymbol{J}, \tag{1}$$

$$\operatorname{div}(\mu \boldsymbol{H}) = 0, \tag{2}$$

where μ is the magnetic permeability which satisfies $0 < \mu_{\min} \leq \mu \leq \mu_{\max}$.

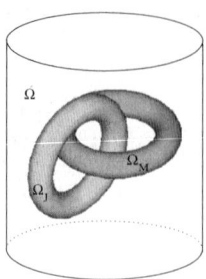
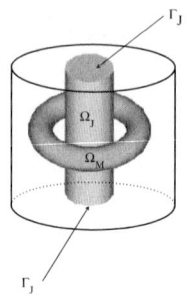

Fig. 1. Sketch of the domains. Case $\overline{\Omega}_J \subset \Omega$.

Fig. 2. Sketch of the domains. Case $\Gamma_J \neq \emptyset$

The typical magnetostatic problem involves magnetic materials with permeability $\mu \neq \mu_0$ ($\mu_0 > 0$ being the magnetic permeability of vacuum), current sources, and eventually magnets. In particular, we are interested in solving the problem (1) and (2) in a bounded 3D domain Ω containing prescribed currents and magnetic materials. This domain is assumed to be simply connected with a Lipschitz-continuous connected boundary Γ.

We denote by Ω_M an open subset of Ω containing all the magnetic materials and by Ω_J another open subset of Ω such that $\overline{\Omega}_J$ contains the support of the current source $\boldsymbol{J}$ in $\overline{\Omega}$. We assume that $\Omega_M \cap \Omega_J = \emptyset$ and the set Ω_M is assumed to be connected although in general not simply connected. We also assume that the boundary of Ω_M is connected and that $\overline{\Omega}_M \subset \Omega$. However, our analysis covers problems in which the domain Ω contains all the source currents (closed circuits), as that shown in Fig. 1, and also problems in which there is a current flow through a part $\Gamma_J := \partial\Omega_J \cap \Gamma$ of the boundary (open circuits), as that shown in Fig. 2.

To pose the magnetostatic problem in the bounded domain Ω, we add to (1) and (2), the following boundary condition:

$$\mu \boldsymbol{H} \cdot \boldsymbol{n} = g \qquad \text{on } \Gamma,$$

where g is a given data function and $\boldsymbol{n}$ the outward unit normal vector to Γ.

Simkin and Trowbridge have introduced in [ST79, ST80] a formulation of this problem based on two scalar potentials defined in different regions of the domain. The aim of this chapter is to analyze this formulation and a finite element method to compute its solution in the three-dimensional case.

To introduce the scalar potentials, we start noticing that the Biot-Savart law allows us to compute a vector field $\boldsymbol{T}$ such that $\mathbf{curl}\,\boldsymbol{T} = \boldsymbol{J}$ and $\mathrm{div}\,\boldsymbol{T} = 0$ in Ω (see [BRS06] for further details). Thus, since Ω is simply connected, there exists a scalar potential ϕ^R such that

$$\boldsymbol{H} = \boldsymbol{T} - \mathbf{grad}\,\phi^R \qquad \text{in } \Omega.$$

The scalar field ϕ^R is known as the *reduced scalar potential*.

In the domain Ω_{M} we are going to introduce another scalar potential. To do this, we denote $\Omega_{\mathrm{R}} := \Omega \setminus \overline{\Omega}_{\mathrm{M}}$, $\Gamma_{\mathrm{I}} := \partial\Omega_{\mathrm{R}} \cap \partial\Omega_{\mathrm{M}}$ and $\boldsymbol{\nu}$ the unit normal vector to Γ_{I} pointing outwards Ω_{M}. The domain Ω_{M} is in general not simply connected, but we assume that there exists a finite number of open connected surfaces Σ_j, $j = 1, \ldots, J$, such that $\Sigma_j \subset \Omega_{\mathrm{M}}$, $\partial\Sigma_j \subset \partial\Omega_{\mathrm{M}}$, $\overline{\Sigma}_j \cap \overline{\Sigma}_k = \emptyset$, for $j \neq k$, and the open set $\widetilde{\Omega}_{\mathrm{M}} := \Omega_{\mathrm{M}} \setminus \bigcup_{j=1}^{j=J} \Sigma_j$ is simply connected ([AAL98]).

For any function $\widetilde{\psi} \in \mathrm{H}^1(\widetilde{\Omega}_{\mathrm{M}})$, we denote by $[\![\widetilde{\psi}]\!]_{\Sigma_j}$ the jump of $\widetilde{\psi}$ through Σ_j. The gradient of $\widetilde{\psi}$ in $\mathcal{D}'(\widetilde{\Omega}_{\mathrm{M}})$ can be extended to $\mathrm{L}^2(\Omega_{\mathrm{M}})^3$ and will be denoted by $\widetilde{\mathbf{grad}}\,\widetilde{\psi}$. Let Θ be the subspace of $\mathrm{H}^1(\widetilde{\Omega}_{\mathrm{M}})$ defined by

$$\Theta = \left\{ \widetilde{\psi} \in \mathrm{H}^1(\widetilde{\Omega}_{\mathrm{M}}) : \ [\![\widetilde{\psi}]\!]_{\Sigma_j} = \text{constant}, \ j = 1, \ldots, J \right\}.$$

For all function $\boldsymbol{G} \in \mathrm{H}(\mathbf{curl}, \Omega_{\mathrm{M}})$ such that $\mathbf{curl}\,\boldsymbol{G} = \boldsymbol{0}$ in Ω_{M}, there exist $\widetilde{\psi} \in \Theta$ such that $\boldsymbol{G}|_{\Omega_{\mathrm{M}}} = -\widetilde{\mathbf{grad}}\,\widetilde{\psi}$ (see again [AAL98]). Then, since $\boldsymbol{J}|_{\Omega_{\mathrm{M}}} = \boldsymbol{0}$, we can write $\boldsymbol{H}|_{\Omega_{\mathrm{M}}} = -\widetilde{\mathbf{grad}}\,\widetilde{\phi}$ with $\widetilde{\phi} \in \Theta$ and the scalar multivalued function $\widetilde{\phi}$ is known as the *total scalar potential*. Then, we consider:

$$\boldsymbol{H} = \begin{cases} -\widetilde{\mathbf{grad}}\,\widetilde{\phi}, & \text{in } \Omega_{\mathrm{M}}, \\ \boldsymbol{T} - \mathbf{grad}\,\phi^{\mathrm{R}}, & \text{in } \Omega_{\mathrm{R}}. \end{cases}$$

We introduce the space $\mathcal{X} := \Theta/\mathbb{R} \times \mathrm{H}^1(\Omega_{\mathrm{R}})/\mathbb{R}$, endowed with the norm $\left\|(\widetilde{\psi}, \psi^{\mathrm{R}})\right\|_{\mathcal{X}} := \left(\left\|\widetilde{\mathbf{grad}}\,\widetilde{\psi}\right\|_{0,\Omega_{\mathrm{M}}}^2 + \left\|\mathbf{grad}\,\psi^{\mathrm{R}}\right\|_{0,\Omega_{\mathrm{R}}}^2 \right)^{1/2}$. Taking into account that $\boldsymbol{H} \times \boldsymbol{n}$ and $\mu\boldsymbol{H} \cdot \boldsymbol{n}$ does not have jumps across Γ_{I}, we introduce the closed linear manifold of $\mathcal{X}$

$$\mathcal{V}(\boldsymbol{T}) := \left\{ (\widetilde{\psi}, \psi^{\mathrm{R}}) \in \mathcal{X} : \ \mathbf{grad}\,\psi^{\mathrm{R}} \times \boldsymbol{\nu} - \widetilde{\mathbf{grad}}\,\widetilde{\psi} \times \boldsymbol{\nu} = \boldsymbol{T} \times \boldsymbol{\nu} \text{ on } \Gamma_{\mathrm{I}} \right\},$$

and the weak problem in terms of scalar potentials reads as follows:

PROBLEM **P**. *Find* $(\widetilde{\phi}, \phi^{\mathrm{R}}) \in \mathcal{V}(\boldsymbol{T})$, *such that*

$$\int_{\Omega_{\mathrm{M}}} \mu\,\widetilde{\mathbf{grad}}\,\widetilde{\phi} \cdot \widetilde{\mathbf{grad}}\,\widetilde{\psi} + \int_{\Omega_{\mathrm{R}}} \mu_0\,\mathbf{grad}\,\phi^{\mathrm{R}} \cdot \mathbf{grad}\,\psi^{\mathrm{R}}$$

$$= \int_{\Gamma} \mu_0 \boldsymbol{T} \cdot \boldsymbol{n}\,\psi^{\mathrm{R}} - \int_{\Gamma_{\mathrm{I}}} \mu_0 \boldsymbol{T} \cdot \boldsymbol{\nu}\,\psi^{\mathrm{R}} - \int_{\Gamma} g\psi^{\mathrm{R}} \qquad \forall(\widetilde{\psi}, \psi^{\mathrm{R}}) \in \mathcal{V}(\boldsymbol{0}).$$

Theorem 1. *Problem* **P** *has a unique solution.*

3 Finite Element Discretization and Numerical Results

In this section we introduce a discretization of Problem **P**, state an error estimate and show some numerical results. We assume that Ω, Ω_{M} and Ω_{R} are Lipschitz polyhedra and consider a family $\{\mathcal{T}_h\}$ of regular tetrahedral meshes of Ω, such that each element $K \in \mathcal{T}_h$ is contained either in $\overline{\Omega}_{\mathrm{M}}$ or in $\overline{\Omega}_{\mathrm{R}}$. We define $\mathcal{T}_h^{\Omega_{\mathrm{R}}} := \{K \in \mathcal{T}_h : \ K \subset \overline{\Omega}_{\mathrm{R}}\}$ and $\mathcal{T}_h^{\Omega_{\mathrm{M}}} := \{K \in \mathcal{T}_h : \ K \subset \overline{\Omega}_{\mathrm{M}}\}$;

each $\overline{\Sigma}_j$ is assumed to be a union of faces of tetrahedra for each mesh $\mathcal{T}_h$. We introduce the following finite element spaces:

$$\mathcal{L}_h(\widetilde{\Omega}_{\mathrm{M}}) := \left\{ \widetilde{\psi}_h \in \mathrm{H}^1(\widetilde{\Omega}_{\mathrm{M}}) : \ \widetilde{\psi}_h|_K \in \mathcal{P}_1(K) \ \forall K \in \mathcal{T}_h^{\Omega_{\mathrm{M}}} \right\},$$

$$\Theta_h := \left\{ \widetilde{\psi}_h \in \mathcal{L}_h(\widetilde{\Omega}_{\mathrm{M}}) : \ [\![\widetilde{\psi}_h]\!]_{\Sigma_j} = \text{constant}, \ j = 1, \dots, J \right\},$$

$$\mathcal{L}_h(\Omega_{\mathrm{R}}) := \left\{ \psi_h^{\mathrm{R}} \in \mathrm{H}^1(\Omega_{\mathrm{R}}) : \ \psi_h^{\mathrm{R}}|_K \in \mathcal{P}_1(K) \ \forall K \in \mathcal{T}_h^{\Omega_{\mathrm{R}}} \right\},$$

$$\mathcal{X}_h := \Theta_h/\mathbb{R} \times \mathcal{L}_h(\Omega_{\mathrm{R}})/\mathbb{R}.$$

We define the finite-dimensional approximation of $\mathcal{V}(T)$ as follows:

$$\mathcal{V}_h(T) := \left\{ (\widetilde{\psi}_h, \psi_h^{\mathrm{R}}) \in \mathcal{X}_h : \ \mathbf{grad}\,\psi_h^{\mathrm{R}} \times \boldsymbol{\nu} - \widetilde{\mathbf{grad}}\,\widetilde{\psi}_h \times \boldsymbol{\nu} = T_{\mathrm{N}} \times \boldsymbol{\nu} \ \text{on} \ \Gamma_{\mathrm{I}} \right\},$$

where T_{N} denotes the Nédélec interpolant of T; to approximate the normal component of T we will use its Raviart–Thomas interpolant, denoted by T_{RT} (see [Mon03] for further details about these interpolants). We obtain the following discrete version of Problem **P**:

PROBLEM **DP**. *Find* $(\widetilde{\phi}_h, \phi_h^{\mathrm{R}}) \in \mathcal{V}_h(T)$ *such that*

$$\int_{\Omega_{\mathrm{M}}} \mu\,\widetilde{\mathbf{grad}}\,\widetilde{\phi}_h \cdot \widetilde{\mathbf{grad}}\,\widetilde{\psi}_h + \int_{\Omega_{\mathrm{R}}} \mu_0\,\mathbf{grad}\,\phi_h^{\mathrm{R}} \cdot \mathbf{grad}\,\psi_h^{\mathrm{R}}$$

$$= \int_{\Gamma} \mu_0 T_{\mathrm{RT}} \cdot \boldsymbol{n}\,\psi_h^{\mathrm{R}} - \int_{\Gamma_{\mathrm{I}}} \mu_0 T_{\mathrm{RT}} \cdot \boldsymbol{\nu}\,\psi_h^{\mathrm{R}} - \int_{\Gamma} g\psi_h^{\mathrm{R}} \quad \forall (\widetilde{\psi}_h, \psi_h^{\mathrm{R}}) \in \mathcal{V}_h(\mathbf{0}).$$

Theorem 2. *Problem* **DP** *has a unique solution* $(\widetilde{\phi}_h, \phi_h^{\mathrm{R}})$. *If the solution of Problem* **P** *is such that* $H|_{\Omega_{\mathrm{M}}} = -\widetilde{\mathbf{grad}}\,\widetilde{\phi} \in \mathrm{H}^r(\Omega_{\mathrm{M}})^3$ *and* $H|_{\Omega_{\mathrm{R}}} = T|_{\Omega_{\mathrm{R}}} - \mathbf{grad}\,\phi^{\mathrm{R}} \in \mathrm{H}^r(\Omega_{\mathrm{R}})^3$, *with* $0 < r \leq 1$, *then there exists* $C > 0$ *such that,*

$$\left\| (\widetilde{\phi} - \widetilde{\phi}_h, \phi^{\mathrm{R}} - \phi_h^{\mathrm{R}}) \right\|_{\mathcal{X}} \leq Ch^r \left(\|H\|_{r,\Omega_{\mathrm{M}}} + \|H\|_{r,\Omega_{\mathrm{R}}} + \|J\|_{0,\Omega} \right).$$

Notice that the numerical solution of Problem **DP** requires to impose somehow the constraint on Γ_{I} appearing in the definition of $\mathcal{V}_h(T)$. We have imposed this condition in a weak sense, by means of a Lagrange multiplier defined on Γ_{I}; in this way, we increase the number of unknowns but with the advantage that the computer implementation is quite straightforward.

We have developed a MATLAB code which implements the method described above. Next, we show the numerical results obtained in the simulation of a typical axisymmetric electromagnet: a ferromagnetic cylindrical core (iron), surrounded by a toroidal coil (copper) with a rectangular cross section (see Fig. 3). A stationary uniform current with intensity $I = 1$ A flows through the coil. The magnetic permeabilities are μ_0 for the air and the copper coil (Ω_{R}) and $10^4\mu_0$ for the iron core (Ω_{M}). Figure 4 shows the intensity of the magnetic induction field, $|B| = |\mu H|$, on a plane containing the symmetry axis.

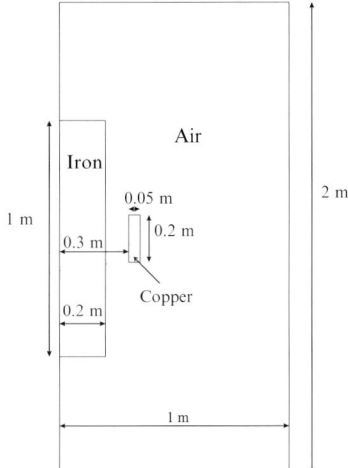

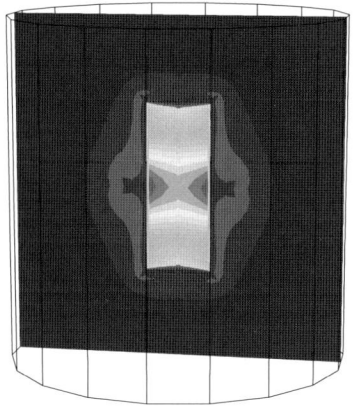

Fig. 3. Electromagnet. Axial section of the domain

Fig. 4. Electromagnet. Intensity of the magnetic induction field, $|\boldsymbol{B}|$ in Ω

References

[AAL98] Amrouche, C., Bernardi, C., Dauge, M., Girault, V.: Vector potentials in three-dimensional non-smooth domains, Math. Meth. Appl. Sci., **21**, 823–864 (1998).

[BRS06] Bermúdez, A., Rodríguez, R., Salgado, P.: A finite element method for the magnetostatic problem in terms of scalar potentials. Preprint DIM 2006-27, Universidad de Concepción, Concepción (2006).

[BAL96] Biro, O., Preis, K., Richter, K. R.: On the use of the magnetic vector potential in the nodal and edge finite element analysis of 3D magnetostatic problems, IEEE Trans. Magnetics, **32**, 651–653 (1996).

[MSP98] Magele, Ch., Stögner, H., Preis, K.: Comparison of different finite element formulations for 3D magnetostatic problems, IEEE Trans. Magnetics, **24**, 31–34 (1988).

[Mon03] Monk, P.: Finite element methods for Maxwell's equations, Oxford University Press, New York (2003).

[PAL91] Preis, K., Bardi, I., Biro, O., Magele, C., Rhenhart, W., Ritcher, K. R., Vrisk, G.: Numerical analysis of 3D magnetostatic fields, IEEE Trans. Magnetics, **27**, 3798–3803 (1991).

[ST79] Simkin, J., Trowbridge, C. W.: Use of the total scalar potential in the numerical solution of field problems in electromagnetics, Internat. J. Numer. Methods Engrg., **14**, 423–440 (1979).

[ST80] Simkin, J., Trowbridge, C. W.: Three-dimensional nonlinear electromagnetic field computations, using scalar potentials, IEE Proc.-B, **127**, 368–374 (1980).

Optimization Methods for a Wifi Location System

A. Martínez[1], L.J. Alvarez-Vázquez[1], F. Aguado-Agelet[1],
and E. Balsa-Canto[2]

[1] Departamento de Matemática Aplicada II. E.T.S.I. Telecomunicación,
Universidad de Vigo, 36310 Vigo, Spain
aurea@dma.uvigo.es, lino@dma.uvigo.es, faguado@tsc.uvigo.es
[2] Instituto de Investigaciones Marinas, C.S.I.C., Vigo, Spain
ebalsa@iim.csic.es

Summary. Indoor location systems using 802.11 standard, based on the comparison of the received and predicted levels of the received signals from the Access Points, are a very interesting research area. The location information is computed by searching the nearest neighbour of the measured signal strength within the radio map. In this chapter, we apply a global optimization algorithm to obtain the Access Points location distribution that yields the best performance of these location systems.

1 Introduction

The wireless networks based on 802.11 standard, which operate in ISM band (destined for industrial, scientific and medical issues), have been greatly developed in the last years. They are designed for the deployment of small local area network inside buildings.

User location service for context-aware applications built on a general purpose 802.11 data network is a very interesting research area. In RADAR like algorithm [1, 2], the signal strength is measured when transmitting beacon packet between the mobile host and the transmitter antennas, located at Access Points (APs). Prior to the real-time localization, RADAR-like algorithm needs to build up a radio map for the area interested by doing random or uniform sampling in that area. After that the location information is computed by searching the nearest neighbour of the measured signal strength within the radio map. Usually, at least three APs are used to carry out the communication task with the mobile host and at the same time they act as the fixed location reference points.

The accuracy of these location systems depends on many factors, such as: accuracy of the propagation model, measurement system and geometrical location of the APs.

In this chapter, we present a global optimization method based on a Differential Evolution algorithm to obtain the optimization of the geometrical locations of the transmitter antennas (APs). This kind of optimization method has been successfully used in other related wireless problems [3–5].

The developed optimization software takes into account that the propagation inside buildings is influenced by building's characteristics and, consequently, the propagation models [6] must include these characteristics, like building's layout (number of walls and their locations), the materials that have been used on its construction (they own different dielectric's characteristics, attenuations and reflection coefficients) and the building type. Any propagation model could be used to compute the cost function in the optimization procedure. In this chapter, the simulation results have been done for a simple model [6] which computes the propagation losses with the distance d between the transmitter and receiver position using an average loss parameter γ obtained in different measurement campaigns and the number of intermediate walls that the ray, starting in the transmitter and ending in the receiver, crosses. The following equation summarizes the received potency:

$$P(d)[\text{dBm}] = P(d_0)[\text{dBm}] - 10\gamma \log\left(\frac{d}{d_0}\right) - n_w \Delta$$

where γ is the average loss factor (it ranges from 0.9 to 1.3), d is the distance between receiver and transmitter position, n_w is the number of walls and Δ is the attenuation in dB for each wall transmission (the recommended value for Δ is between 2 and 3 dB).

2 The Optimization Method

We have to solve a nonlinear programming problem corresponding to the maximization of a cost function measuring the good design of the wireless system (in our case, the design variables are the locations a_1, a_2, a_3 of the three APs). The main factor in the good performance of this indoor strategy consist of assuring that the received potencies in all the reference points $\{x_i, \ i = 1, \ldots, N\}$ be as different as possible, in order to avoid identification problems. This fact leads us to consider as the objective function a linear combination (with a weight parameter $\alpha > 0$) of the two objective functions f_1 and f_2, given by:

$$f_1(a_1, a_2, a_3) = \frac{2}{N(N+1)} \sum_{k=1}^{3} \sum_{i<j=1}^{N} |p_k(x_i) - p_k(x_j)|$$

the mean of all the differences between the received potencies, and f_2 the minimum of them:

$$f_2(a_1, a_2, a_3) = \min_{i<j=1...N} \sum_{k=1}^{3} |p_k(x_i) - p_k(x_j)|$$

where the function $p_k(x_i)$ represents the potency received at reference point x_i corresponding to transmitter location a_k.

Thus, the problem can be formulated as the maximization of function $f = f_1 + \alpha f_2$, where α is a weight parameter. The problem under consideration is a nonlinear programming problem whose solution corresponds to the location of the three APs which maximize an objective function related to the received potency. Unfortunately, this objective function is nonsmooth and multimodal, therefore a local optimization method may get trapped into a suboptimal solution (suboptimal design). The use of a global optimization method may help to surmount this difficulty.

Regarding global optimization methods for nonsmooth functions, several alternatives are available in the operations research literature. However, population based strategies, which generate several solutions in each iteration, are becoming more and more popular due to their ability to build up an overall picture of the search space and to locate the vicinity of the global solution with reasonable effort.

In this work an evolutionary strategy, Differential Evolution [7], will be used mainly due to its efficiency in solving real valued multimodal objective functions (codes and reports on results might be found in `http://www.icsi.berkeley.edu/~storn`). In addition to its good convergence properties some of its main advantages are its conceptual simplicity and ease of use.

Differential evolution is a direct search method which utilizes a population for each generation, that is, for each iteration a number of vectors of decision variables are explored and the corresponding objective value calculated. The initial population is chosen randomly and should try to cover the entire search space uniformly. Basically, the method generates a perturbed vector by adding the weighted difference between two population vectors to a third vector. If the resulting vector yields a better objective value than a predetermined population member, the newly generated vector replaces the vector with which it was compared, in the next generation; otherwise, the old vector is retained.

3 Numerical Results

In our computational experiences, which we have completely developed in MATLAB, we have obtained a good performance that we assure by comparing the initial results for the random data with the highly improved ones for the optimal locations achieved by the Differential Evolution algorithm. We have used a plant of 80×40 m (see Fig. 1). For the reference points we have tried two different grids: a triangular grid (Fig. 1) and a rectangular grid (Fig. 2). For both grids the results have been similar, but slightly better for the latter. We present the results for both grids.

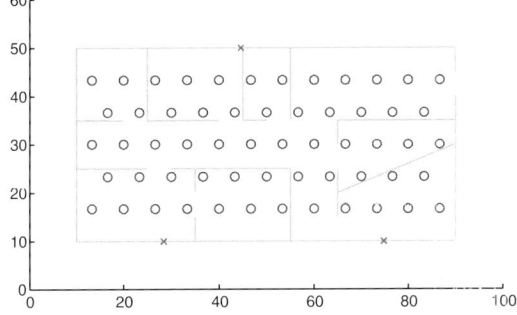

Fig. 1. Optimal locations (*cross*) for a plant with triangular grid of reference points (*circles*)

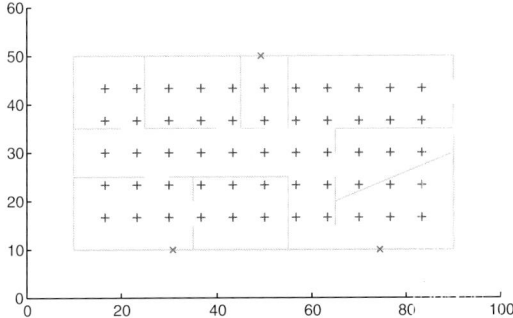

Fig. 2. Optimal locations (*cross*) for a pslant with rectangular grid of reference points (*plus*)

Applying our algorithm to the cost function f for parameter value $\alpha = 100$ (we emphasize the importance of the minimum over the mean value), in the case of a triangular grid of $N = 58$ reference points, we pass, after 273 iterations, from a random initial low cost $f_1 = 14.978$ and $f_2 = 0.617$ to the maximum cost $f_1 = 18.380$ and $f_2 = 3.172$, corresponding to the optimal locations $a_1 = (28.46, 10)$, $a_2 = (44.60, 50)$, $a_3 = (74.83, 10)$, as can be seen in Fig. 1.

In the case of a rectangular grid of $N = 55$ reference points, we reach, after 499 iterations, the slightly better maximum cost $f_1 = 17.810$ and $f_2 = 3.409$, corresponding to the optimal locations $a_1 = (30.78, 10)$, $a_2 = (49.25, 50)$, $a_3 = (74.46, 10)$, as shown in Fig. 2.

4 Conclusions

The Differential Evolution algorithm, in addition to its easier implementation, appears to be the most robust method with respect to the optimality of the achieved final solution. Moreover, the method improves the results previously obtained by the authors [8] by other direct search methods (Nelder–Mead, Hooke–Jeeves...) which are basically local maximization methods and, consequently, not powerful enough for global maximization tasks.

References

1. Bahl, P., Padmanabhan, V.: RADAR: An in-building RF-based user location and tracking system. In Proceedings of IEEE INFOCOM, Vol. 2, 775–784 (2000).
2. Bahl, P., Padmanabhan, V.: Enhancement to the RADAR User Location and Tracking System. Technical Report MSR-TR-2000-12, Microsoft Research (2000).
3. Martínez, A., Aguado-Agelet, F., Alvarez-Vázquez, L.J., Hernando, J.M., Mosteiro, D.: Optimal tranmsmitter location in an indoor wireless system by Nelder-Mead method. Microwave Opt. Technol. Lett., 27, 146–148 (2000).
4. Aguado-Agelet, F., Martínez, A., Hernando, J.M., Isasi, F., Mosteiro, D.: Bundle methods for optimal transmitter location in indoor wireless system. Electronics Letters, 36, 573–574 (2000).
5. Aguado-Agelet, F., Martínez, A., Alvarez-Vázquez, L.J., Hernando, J.M., Formella, A.: Optimization methods for optimal transmitter locations in a mobile wireless system. IEEE Trans. Veh. Technol., 51, 1316–1321 (2002).
6. Seidel, S.Y., Rappaport, T.S.: 914 MHz path loss prediction models for indoor wireless communication in multi-floored buildings. IEEE Trans. Antennas and Propagation, 40, 207–217 (1984).
7. Storn, R., Price, K.: Differential Evolution - a Simple and Efficient Heuristic for Global Optimization over Continuous Spaces. J. Global Optimization, 11, 341–359 (1997).
8. Aguado-Agelet, F., Alvarez-Vázquez, L.J., Martínez, A.: Nelder-Mead optimization algorithm applied to a wifi 802.11x location system. Submitted (2006).

Flow in the Canal of Schlemm and its Influence on Primary Open Angle Glaucoma

A.D. Fitt

School of Mathematics, University of Southampton, Southampton SO17 1BJ, UK
adf@maths.soton.ac.uk

1 POAG in Human Eyes

Primary Open Angle Glaucoma (POAG) is a major cause of blindness, affecting 65-70 million sufferers worldwide ([ERC04]). The eye produces aqueous humour (AH: a water-like substance secreted by the ciliary body) which flows behind the iris, through the pupil aperture, out into the anterior chamber (AC) and drains from the eye via the drainage angle. From the drainage angle the AH passes through a biological filter (the trabecular meshwork or TM) into the canal of Schlemm (SC), the main drainage route from the eye, and finally exhausts into "collector channels". POAG occurs when this drainage mechanism is somehow compromised [FW92]. Essentially the AH cannot be removed quickly enough and as a result the intraocular pressure (IOP) increases in the eye. Contrary to popular belief, glaucoma and elevated IOP are not synonymous. Though very often associated with elevated IOP, glaucoma is, in reality, an optic nerve neuropathy. Notwithstanding this, elevated IOP is always regarded as potentially harmful to the eye. In the current study we therefore seek to model the flow of AH from the AC through the TM and into the SC and to couple this flow to predictions of changes in IOP.

2 Governing Equations

2.1 Fluid Modelling

The flow of AH through the TM, into the SC and out into the collector channels was studied in [JK83], [TA89] and [AS06]. None of these studies appeared to realise that the flows involved may be thought of as lubrication theory flows. Though in each case the final equations were very similar to the equations derived in Sect. 2.3 below, none of these studies attempted to couple the SC flow to a model for the evolution of overall IOP changes, which is the main aim of the current study.

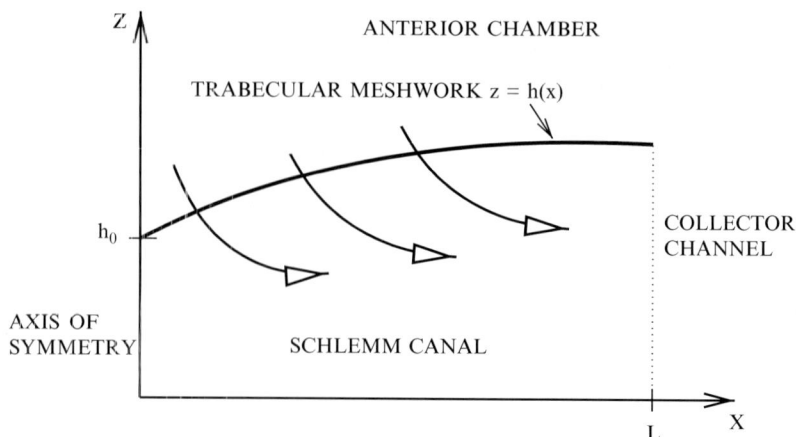

Fig. 1. Schematic diagram of flow through the TM into the SC

The SC typically has half-length (i.e. length between a symmetry axis and a collector channel) $L = 600\,\mu$m, ([JK83]) undeformed depth $h_0 = 25\,\mu$m ([JK83]) and breadth $B = 300\,\mu$m ([JK83]). The aspect ratio $\epsilon = h_0/L$ is thus about 0.04. Using the values $\mu = 0.75 \times 10^{-3}\,$Pa s ([JK83]) and $\rho = 1,003\,$kg m^{-3} ([FW92]) for the density and dynamic viscosity respectively, the Reynolds number is $Re = LU/\nu \sim 4$ and the reduced Reynolds number $\epsilon^2 Re \sim 0.004$. The lubrication theory equations may therefore indeed be used (see, for example [Ock95]). The BVP to be studied (see Fig. 1 for nomenclature) is therefore

$$p_x = \mu u_{zz}, \quad p_z = 0, \quad u_x + w_z = 0 \quad (x \in [0, L], 0 \le z \le h(x)) \qquad (1)$$

with boundary conditions

$$u(x,0) = w(x,0) = 0, \quad u(x,h(x)) = 0, \quad w(x,h(x)) = w_h(x),$$

$$p_x(0,z) = 0, \quad p(L,z) = p_{\text{out}}, \qquad (2)$$

Here p denotes pressure, $\mathbf{q} = (u(x,z), w(x,z))$ fluid velocity, subscripts denote derivatives, w_h is the flow speed through the TM and $p_{\text{out}} \sim 9\,$mmHg ([JK83]) is the IOP at a collector channel.

2.2 Friedenwald's Law

To close the model we must relate the eye's AH production and removal to the IOP. It has long been accepted that the volume and IOP of a human eye are related by Friedenwald's law [F37]. This states that two IOPs p_1 and p_2 are related to respective ocular volumes V_1 and V_2 (measured in $\mu\ell$) via

$$K(V_1 - V_2) = \log_{10} p_1 - \log_{10} p_2$$

where $K \sim 0.025/\mu\ell$ [FW92] is a known constant. Denoting normal conditions using a subscript n ($p_n \sim 14\,\text{mmHg} \sim 1,867\,\text{Pa}$ ([BRM05])) and altered conditions using a subscript i, we therefore find that $p_i = p_n \exp(\tilde{K}(V_i - V_n))$ where $\tilde{K} \sim 5.75646 \times 10^7/\text{m}^3$. Differentiation now shows that

$$\frac{dp_i}{dt} = \tilde{K}p_i(\dot{V}_{\text{in}} - \dot{V}_{\text{out}}) \tag{3}$$

where $\dot{V}_{\text{in}}$ ($\sim 2\mu\ell/\text{min}$ ([BRM05])) and $\dot{V}_{\text{out}}$ (m^3/sec) denote the respective total amounts of fluid flowing in and out of the eye.

2.3 Fluid Flow/IOP Evolution Equations

We assume that the temporal changes in the IOP take place on a much longer time scale than that associated with the passage of an individual fluid particle from the AC to a collector channel, so that the flow may be treated as quasi-steady. The flow problem (1) and (2) may now be solved to yield

$$u = \frac{p_x}{2\mu}(z^2 - hz), \quad w = \frac{p_{xx}}{2\mu}\left(\frac{hz^2}{2} - \frac{z^3}{3}\right) + \frac{p_x h_x z^2}{4\mu},$$

where the flow pressure $p(x)$ satisfies

$$\left(\frac{p_x h^3}{12\mu}\right)_x = w_h(x) \quad (p(L) = p_{\text{out}}, \quad p_x(0) = 0).$$

In general both $w_h(x)$ and $h(x)$ are unknown and must be determined. The outflow $\dot{V}_C$ (m^3/s) from a single collector channel is therefore

$$\dot{V}_C = \int_0^{h(L)} Bu \mid_{x=L} \, dz = -\frac{B}{12\mu}(h^3 p_x) \mid_{x=L}$$

so that $\dot{V}_{\text{out}} = N\dot{V}_C$ where N is the total number of collector channels ($N \sim 30$ for a human eye ([ERC04])) and the IOP $p_i(t)$ is determined by (3) with $p_i(0) = p_{io}$.

3 Results

We now examine a number of different cases, relating aqueous outflow to changes in IOP for various structure submodels.

3.1 Simple Modelling Cases

First we consider the (unrealistic) case where $h(x) \equiv h_0$ and $w_h(x) \equiv \alpha < 0$ are both constant. We find that

$$p = p_{\text{out}} - \frac{6\alpha\mu}{h_0^3}(L^2 - x^2), \quad \dot{V}_{\text{out}} = -NB\alpha L$$

and thus

$$\frac{\mathrm{d}p_i}{\mathrm{d}t} = \tilde{K}p_i(\dot{V}_{\text{in}} + NB\alpha L) \quad (p_i(0) = p_{io}).$$

Thus $p_i = p_{io}\exp(\beta t)$ where $\beta = \tilde{K}(\dot{V}_{\text{in}} + NB\alpha L)$. The IOP thus increases/decreases exponentially depending on whether the quantity $-\alpha$ is less than/greater than $\dot{V}_{\text{in}}/NBL$. A "worst-case" scenario arises if all aqueous outflow ceases so that $\dot{V}_{\text{out}}$ suddenly becomes zero. The IOP rises exponentially, on a timescale $(\tilde{K}\dot{V}_{\text{in}})^{-1} \sim 520\text{s}$. Starting from a normal IOP of $14\,\text{mmHg}$, the IOP rises to a dangerous value of $30\,\text{mmHg}$ in just under seven minutes.

3.2 Flow Through TM Determined by Darcy's Law

The previous case is unrealistic: the TM is acts as a porous filter, so that the speed $w_h(x)$ of the flow into the SC is determined by both the IOP in the AC and the flow pressure. Assuming that $h(x)$ is still given by the constant h_0, we therefore now consider the consequences of using Darcy's law $\mathbf{q} \propto \nabla p$ to model the flow through the TM by setting $w_h = -\frac{k}{d\mu}(p_i - p)$. Here d is the width of the TM and the (constant) permeability k (dimensions m^2) has been measured for the TM in the form of a "TM resistance" $R_T = \mu d/(kBL)$ (dimensions $\text{kg}\,\text{s}^{-1}\,\text{m}^{-4}$) where d is the width of the TM. Thence

$$p_{xx} - \beta^2 p = -\beta^2 p_i \quad \left(\beta^2 = \frac{12k}{dh_0^3}\right),$$

the total outflow is given by

$$\dot{V}_{\text{out}} = N\frac{B\beta h_0^3}{12\mu}(p_i - p_{\text{out}})\tanh\beta L,$$

and the IOP is therefore governed by

$$\frac{\mathrm{d}p_i}{\mathrm{d}t} = p_i(A + Bp_i) \quad (p_i(0) = p_{i0}), \tag{4}$$

where

$$A = \tilde{K}\left(\dot{V}_{\text{in}} + \frac{N\tanh(\beta L)p_{\text{out}}}{R_T\beta L}\right), \quad B = -\frac{\tilde{K}N\tanh(\beta L)}{R_T\beta L}.$$

Equation (4) has two steady states: an unphysical one at $p_i = 0$ and another (which may easily be shown to be stable) at $p_i = -A/B$. Normally the IOP remains constant at $p_i = p_n = 14\,\text{mmHg} \sim 1{,}867\,\text{Pa}$ say. Thence

$$p_n = -\frac{A}{B} = p_{\text{out}} + \frac{\dot{V}_{\text{in}}R_T\beta L}{N}\coth(\beta L).$$

We can now "back out" a value for R_T. Using the parameter values previously considered, we find that $R_T \sim 1.96 \times 10^{13}$ kg s^{-1} m^{-4}, in very close agreement with measured values ([JK83], [ERC04]). Equation (4) has exact solution

$$p_i(t) = \frac{Ap_{i0}}{(A + Bp_{i0})e^{-At} - Bp_{i0}}.$$

A TM blockage thus causes the IOP to rise on a timescale $1/A$ to its new elevated value. It is now easy to calculate the IOP rise that would occur if collector channels become blocked or the TM resistance should increase for some reason (such as blockage by particles).

4 Conclusions and Further Work

We have neglected both the effects of uveoscleral outflow which is another (much weaker) AH drainage mechanism, and "pseudofacility", whereby the production of AH by the ciliary body is suppressed at elevated IOP. Both could be included if desired. Further study will include cases where (1) the TM is deformable and (2) The permeability K in Darcy's law is not constant (the pores in the TM close as the pressure difference across it increase).

References

[AS06]   Avtar, R., Srivastava, R.: Aqueous outflow in Schlemm's canal. Appl. Math. Comp., **174**, 316–328 (2006)

[BRM05] Broquet, J., Roy, S., Mermoud, A.: Visualization of the aqueous outflow pathway in the glaucoma surgery by finite element model of the eye. Comp. Meth. Biomech. Biomed. Eng., **1**, 43–44 (2005)

[ERC04] Ethier, C.R., Read, T.A., Chan, D.: Biomechanics of Schlemm's canal endothelial cells: influence on F-Actin architecture. Biophys. J., **87**, 2828–2837 (2004)

[FW92]   Fatt, I., Weissman, B.A.: Physiology of the eye, 2nd Edition. Butterworth-Heinemann, Boston (1992)

[F37]   Friedenwald, J.S.: Contribution to the theory and practice of tonometry. Am. J. Ophthalmol., **20**, 985–1024 (1937)

[JK83]   Johnson, M.C., Kamm, R.D.: The role of Schlemm's canal in aqueous outflow from the human eye. Inv. Ophth. Vis. Sci., **24**, 320–325 (1983)

[Ock95] Ockendon, H., Ockendon, J.R.: Viscous Flow. Cambridge University Press, Cambridge (1995)

[TA89]   Tandon, P.N., Autar, R.: Flow of aqueous humour in the canal of Schlemm. Math. Biosciences, **93**, 53–78 (1989)

A One-Phase Model for Air-Breathing DMFC Cells with Non-Tafel Kinetics

Marcos Vera and Francisco J. Sánchez-Cabo

Universidad Carlos III de Madrid
marcos.vera@uc3m.es

Summary. An isothermal single-phase 3D/1D model for liquid-feed direct methanol fuel cells (DMFC) is presented and validated against experimental results. 3D mass, momentum and species transport in the anode channel and gas diffusion layer is modelled using a commercial CFD code complemented with user supplied subroutines. The 3D model is locally coupled to a 1D model that imposes a physically sound boundary condition for the velocity and the methanol concentration field at the anode gas-diffusion-layer/catalyst-layer interface. The 1D model assumes non-Tafel kinetics to account for the complex kinetics of the (multi-step) methanol oxidation reaction at the anode, and includes the mixed potential induced by methanol crossover due to diffusion and electro-osmotic drag. Polarization curves obtained for various methanol feed concentrations, temperatures, and methanol feed velocities show good agreement with recent experimental results.

1 Introduction

Direct methanol fuel cells (DMFCs) are electrochemical devices that convert the chemical energy of methanol directly into electricity, according to reactions

$$\text{Anode} \qquad : \qquad \text{CH}_3\text{OH} + \text{H}_2\text{O} \rightarrow \text{CO}_2 + 6\text{H}^+ + 6\text{e}^- \qquad (1)$$

$$\text{Cathode} \qquad : \qquad (3/2)\text{O}_2 + 6\text{H}^+ + 6\text{e}^- \rightarrow 3\text{H}_2\text{O} \qquad (2)$$

$$\text{Overall (Cell)} : \qquad \text{CH}_3\text{OH} + (3/2)\text{O}_2 \rightarrow \text{CO}_2 + 2\text{H}_2\text{O} \qquad (3)$$

with standard reduction potentials $E_a^0 = 0.02\,\text{V}$, $E_c^0 = 1.23\,\text{V}$, and $E_{\text{cell}}^0 = 1.21\,\text{V}$ vs. SHE at 298 K, respectively.

Liquid-feed DMFCs use liquid methanol as energy carrier, which makes them good candidates as small autonomous power sources due to the high energy density of methanol. However, DMFCs suffer from two fundamental problems: the slow kinetics of the methanol electro-oxidation reaction, and the ability of methanol to permeate through the polymeric membrane from the anode to the cathode (methanol crossover). The above difficulties, together with additional technological problems concerning auxiliar devices, such as pumps, fuel storage tanks, power conditioning devices, etc. have motivated a large amount of work on this field during the last decade, combining mathematical and numerical modeling with detailed experimental research. In particular, the progress in DMFC modeling has been significant. An extensive review of this work can be found elsewhere [1, 2] and will not be repeated here for brevity.

2 Mathematical Model

2.1 Cell Geometry

For simplicity we shall assume a parallel channel geometry for the anode current collector. Therefore, when describing the flow in a single channel we shall use periodic boundary conditions at the channel/rib mid-planes to reduce the computational cost. Figure 1 shows a sketch of the physical domain under consideration. The cell is divided into seven regions: anode channel (ac), anode gas diffusion layer (agdl), anode catalyst layer (acl), polymeric membrane (mem), cathode catalyst layer (ccl), cathode gas diffusion layer (cgdl), and cathode channel (cc). Since we are not solving neither the electric field nor the temperature field, we omit from the computations both the anode current colector (acc) and the cathode current colector (ccc).

2.2 Model Assumptions

- The flow is laminar, steady, isothermal, and monophasic.
- The reactant concentrations are constant across the catalyst layers.
- The concentration of methanol is sufficiently small in the anode to consider the liquid phase to be a diluted methanol aqueous solution.

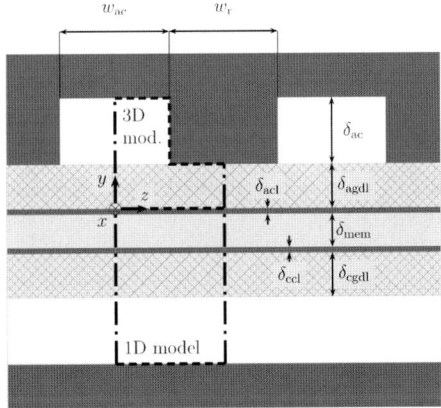

Fig. 1. Cross-section of the cell geometry

- The methanol that crosses-over from the anode to the cathode is completely oxidized at the cathode catalyst layer.
- The membrane is fully hydrated and is impermeable to gases.
- The pressure gradient across the different cell layers is neglected.
- Negligible ohmic losses in gas diffusion layers, channels and bipolar plates.

2.3 3D Model

The complete Navier–Stokes equations, complemented with the conservation equation for methanol, are solved in a three-dimensional (3D) domain with a control volume based discretization to obtain the velocity and pressure distributions, as well as the concentration of methanol, in the anode channels and gas diffusion layer. The governing equations are

$$\nabla \cdot \mathbf{u} = 0, \quad \frac{\rho}{\epsilon^2}(\mathbf{u} \cdot \nabla)\mathbf{u} = -\nabla p + \frac{\mu}{\epsilon}\nabla^2\mathbf{u} + \mathbf{S}_u \tag{4}$$

$$\nabla \cdot (\mathbf{u}C_m) = D_m^{\text{eff}}\nabla^2 C_m, \tag{5}$$

where $\mathbf{u}$ is the (superficial) velocity, ϵ is the porosity of the porous matrix, C_m is the methanol concentration field, D_m^{eff} is the effective diffusivity of methanol, ρ and μ are the density and viscosity of water at the operating temperature, T, and ϵ is the porosity of the medium (1 in the ac and 0.6 in the agdl). The momentum source term appearing in (4) is $\mathbf{S}_u = (\mu/K)\mathbf{u}$ in the agdl and 0 in the ac, with $K = 10^{-12}\,\text{m}^{-2}$ being the hydraulic permeability of the porous layer.

In integrating (4) and (5), the only non-trivial boundary condition is that imposed at the gas-diffusion-layer/catalyst interface, where we prescribe the normal velocity and the molar flux of methanol

$$\mathbf{u}\big|_{y=0} \cdot \mathbf{n} = \left(\frac{1}{6} + n_d^w\right)\frac{i}{F}\frac{W_w}{\rho} \tag{6a}$$

$$\left(\mathbf{u}\,C_m - D_m^{\text{eff}}\nabla C_m\right)\big|_{y=0} \cdot \mathbf{n} = N_m \tag{6b}$$

Here F is Faraday's constant, n_d^w is the electroosmotic drag coefficient of water, W_w is the molecular weight of water, and $\mathbf{n}$ is the unit normal vector pointing outward from the computational domain. Once we know the local methanol concentration at a given point of the acl, $C_{m,\text{acl}} = C_m|_{y=0}$, and the overall cell voltage, V, the 1D model presented below provides the local current density, i, and the local molar flux of methanol reaching the catalyst layer from the acl, N_m, thus closing the mathematical problem through (6).

2.4 1D Model

The 1D model is composed by the following equations

$$N_{O_2} = \alpha_2(C_{O_2,\text{amb}} - C_{O_2,\text{ccl}}) \tag{7}$$

$$i + i_p = (a_c i_{0,c})\delta_{\text{ccl}} \frac{C_{O_2,\text{ccl}}}{C_{O_2,\text{ref}}} \exp\left(\frac{\alpha_c F}{RT}\eta_c\right) \tag{8}$$

$$i_a = j\,\delta_{\text{acl}} = (a_a i_{0,a})\delta_{\text{acl}} \frac{\kappa\, C_{\text{m,acl}} \exp\left(\dfrac{\alpha_a F}{RT}\eta_a\right)}{C_{\text{m,acl}} + \lambda \exp\left(\dfrac{\alpha_a F}{RT}\eta_a\right)} \tag{9}$$

$$i_p = 6F N_{\text{cross}}, \quad N_{\text{cross}} = n_{\text{d}}^{\text{m}}\frac{i}{F} - D_{\text{m,mem}}^{\text{eff}} \left.\frac{dC_{\text{m}}}{dy}\right|_{\text{mem}} \tag{10}$$

$$N_{O_2} = \frac{i}{4F} + \frac{3}{2}N_{\text{cross}}, \quad N_{\text{m}} = \frac{i}{6F} + N_{\text{cross}} \tag{11}$$

where i_0 is the exchange current density (different for anode and cathode), κ and λ are two experimentally fitted constants [3], and n_{d}^{m} is the electroosmotic drag coefficient of methanol. Equation (7) models the convective-diffusive transport of oxygen from the ambient air (amb) to the ccl through an overall mass transfer coefficient α_2, (8) and (9) model the electrochemical kinetics of the cathodic and anodic reactions [3, 4], (10) expresses the molar flux of methanol that crosses-over the membrane as sum of that due to electroosmotic drag and that induced by molecular diffusion, and (11) are the oxygen and methanol mass balances at the catalyst layers.

Appropriate manipulation of (7)–(11) yields analytic expressions for the unknowns i, η_c, η_a, N_{cross}, i_p, N_{O_2}, $C_{O_2,\text{ccl}}$ as a function of N_{m} and $C_{\text{m,acl}}$. Substituting these expressions in the equation for the cell voltage provides the following non-lineal relation between N_{m}, $C_{\text{m,acl}}$, and V:

$$f(N_{\text{m}}, C_{\text{m,acl}}) \equiv E_{\text{cell}} - V - \eta_a(N_{\text{m}}, C_{\text{m,acl}})$$
$$- \eta_c(N_{\text{m}}, C_{\text{m,acl}}) - i(N_{\text{m}}, C_{\text{m,acl}})\frac{\delta_{\text{mem}}}{\sigma_{\text{mem}}} = 0 \tag{12}$$

which can be readily solved for N_{m} using a Newton–Raphson method for given values of $C_{\text{m,acl}}$ and V. After solving for N_{m}, the remaining unknowns can be obtained using the analytic formulae.

3 Results and Discussions

Figure 2 shows the effect of inlet methanol concentration $C_{\text{m},in}$ in the polarization curve of a DMFC with parallel channels. The figure also shows the experimental results of Sundmacher et al. [5] under similar operating conditions. As can be observed, after a careful selection of the electrochemical parameters, the mathematical model was able to reproduce the experimentally observed trends published in the literature.

852   M. Vera and F.J. Sánchez-Cabo

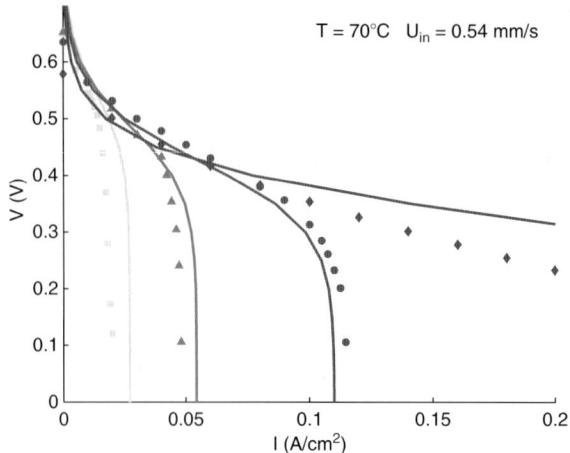

Fig. 2. Comparison of the experimental results of [5] (*symbols*) and the results provided by the mathematical model (*solid lines*) under similar operating conditions and different methanol feed concentrations. □, 0.125 M; △, 0.25 M; ○, 0.5 M; ◇, 2 M

4 Conclusions

A new hybrid 3D/1D isothermal, single-phase, mathematical model has been developed for liquid-feed direct methanol fuel cells. The model was integrated using a commercial CFD software package, yielding polarization curves for different methanol feed concentrations, temperatures, and volumetric methanol flow rates that are in qualitative agreement with the experimental results found in the literature.

Acknowledgements

This research was supported by the Spanish Comunidad de Madrid under Project# S-505/ENE/0229.

References

1. G. Lu, C.Y. Wang: Two-phase microfluidics, heat and mass transport in direct methanol fuel cells. In: B. Sundén, M. Faghri (eds) Transport Phenomena in Fuel Cells. WIT Press, Southampton, pp. 317–358 (2005)
2. Z.H. Wang, C.Y. Wang: Mathematical modeling of liquid-feed direct methanol fuel cells. J. Electrochem. Soc., **150**, A508–A519 (2003)
3. B.L. García, V.A. Sethuraman, J.W. Weidner, R.E. White: Mathematical model of a DMFC. J. Fuel Cell Sci. Tech., **1**, 43–48 (2004)
4. J. P. Meyers, J. Newman: Simulation of the direct methanol fuel cell II. Modeling and data analysis of transport and kinetic phenomena. J. Electrochem. Soc. 149 (2002) A718-A728.
5. K. Sundmacher, T. Schultz, S. Zhou, K. Scott, M. Ginkel, E.D. Gilles: Dynamics of the Direct Methanol Fuel Cell (DMFC): experiments and model-based analysis. Chem. Eng. Sci. 56 (2001) 333.

Optimising Design Parameters of Enzyme-Channelling Biosensors

D. Mackey[1] and A.J. Killard[2]

[1] School of Mathematical Sciences, Dublin Institute of Technology, Ireland
Dana.Mackey@dit.ie
[2] National Centre for Sensor Research, Dublin City University, Ireland
Tony.Killard@dcu.ie

Summary. Two mathematical models for an electrochemical biosensor are proposed and compared with a view to determining the ratio of two immobilized enzymes which maximizes the amperometric signal amplitude.

1 Introduction

Biosensors are devices in constant development due to their wide use in bio-medical diagnostics and environmental monitoring. Of particular interest to developing electrochemical immunosensors are enzyme channelling systems, where two enzymes are brought in close proximity to an electrode surface thus facilitating the fast conversion of initial substrate to final product. Moreover, cascade schemes, where an enzyme is catalytically linked to another can produce signal amplification and therefore increase the device sensitivity.

This work investigates a biosensor employing the enzymes glucose oxidase (GOX) and horseradish peroxidase (HRP), immobilized on an electrode modified with a conducting polymer. After the immobilisation, the electrode is inserted in a flow-cell for an amperometric flow-injection analysis where glucose solutions at different concentrations are passed over the electrode and the signals recorded (see [KZZ99]). A mathematical model was proposed in [MKA07] and numerical simulations were carried out in order to determine the ratio of the two enzymes which maximizes current amplitude. In this chapter, the optimal ratio of HRP and GOX, $\xi_{\max}$, is further investigated as a function of the kinetic rate constants and two different parameter regimes are identified, characterized by different qualitative behaviour of $\xi_{\max}$. A simplified model is also introduced, which allows for an explicit formula for $\xi_{\max}$ to be derived and compared with the numerical simulations of the previous model.

2 Spatially Extended Model

The flow effects are not explicitly modelled and the existence of the convective zone (where the glucose concentration is constant) is only reflected in the boundary conditions imposed at the top of the diffusion layer. The immobilized enzymes form a monolayer so all reactions can be assumed to take place at the lower boundary of the diffusion domain. The equations are one-dimensional, where the variable x measures the distance from the electrode.

A cascade reaction takes place at the electrode. Glucose oxidase catalyses the oxidation of glucose to gluconic acid, with production of hydrogen peroxide (H_2O_2). HRP is oxidised by H_2O_2 and then subsequently reduced by electrons provided by the electrode. We model these reactions by a standard Michaelis–Menten kinetics scheme, (1), and we use the notation $E_1(t) =$ first enzyme (GOX) concentration, $E_2(t) =$ second enzyme (HRP) concentration, $S_1(x,t) =$ first substrate (glucose), $S_2(x,t) =$ second substrate (H_2O_2), $C_1(t) =$ first complex, $C_2(t) =$ second complex, $P(x,t) =$ final product,

$$E_1 + S_1 \underset{k_{-1}}{\overset{k_1}{\rightleftharpoons}} C_1 \overset{k_2}{\longrightarrow} E_1 + S_2, \qquad E_2 + S_2 \underset{k_{-3}}{\overset{k_3}{\rightleftharpoons}} C_2 \overset{k_4}{\longrightarrow} E_2 + P. \qquad (1)$$

We now write down the differential equations governing the behaviour of the relevant chemical species. The two substrates, glucose and hydrogen peroxide are free to diffuse throughout the domain, hence

$$\frac{\partial S_i}{\partial t} = D_i \frac{\partial^2 S_i}{\partial x^2}, \qquad i = 1, 2, \qquad 0 \le x \le L, \quad t \ge 0 \qquad (2)$$

$$S_1(L,t) = S_0; \qquad S_2(L,t) = 0, \qquad (3)$$

and the following boundary conditions hold on $x = 0$

$$D_1 \frac{\partial S_1}{\partial x}(0,t) = k_1 E_1 S_1 - k_{-1} C_1, \qquad (4)$$

$$D_2 \frac{\partial S_2}{\partial x}(0,t) = k_3 E_2 S_2 - (k_2 + k_{-3}) C_1, \qquad (5)$$

$$\frac{dE_1}{dt} = -k_1 E_1 S_1 + (k_{-1} + k_2) C_1, \qquad \frac{dE_2}{dt} = -k_3 E_2 S_2 + (k_4 + k_{-3}) C_2, \quad (6)$$

$$\frac{dC_1}{dt} = k_1 E_1 S_1 - (k_2 + k_{-1}) C_1, \qquad \frac{dC_2}{dt} = k_3 E_2 S_2 - (k_4 + k_{-3}) C_2, \quad (7)$$

$$\frac{dP}{dt} = k_4 C_2. \qquad (8)$$

The initial conditions are

$$S_1(x,0) = S_0(x), \qquad S_2(x,0) = 0, \qquad P(x,0) = 0,$$

$$E_1(0) = \frac{\xi e}{1+\xi}, \qquad E_2(0) = \frac{e}{1+\xi}, \qquad C_1(0) = 0, \qquad C_2(0) = 0, \qquad (9)$$

where e is the total amount of enzyme present on the electrode and $S_0(x) = S_0$ if $x = L$ and 0 otherwise. The purpose of this study is to determine ξ, the ratio of GOX to HRP on the electrode, which maximizes the signal at the electrode. The measured current is given by the electron transfer rate, which can be assumed proportional to the rate of formation of product P from (8).

3 Simplified Model

In order to obtain an analytical expression for the dependence of the optimal enzyme ratio on other system parameters, we consider a simplified model which focuses on the kinetic surface processes, while neglecting transport of chemical species to and from the electrode. The main limitation in this case is failing to model the possibility of H_2O_2 to diffuse away from the electrode therefore assuming that all the product from the first reaction is readily available for the second.

With the assumption that the concentration of glucose is maintained constant at the reaction point, $S_1(t) = S_0$ for all $t \geq 0$, the model in the previous section now reduces to the following set of ordinary differential equations

$$\frac{dC_1}{dt} = -\left(k_1 S_0 + k_{-1} + k_2\right) C_1 + \frac{\xi e k_1}{1 + \xi} S_0 \tag{10}$$

$$\frac{dC_2}{dt} = \frac{e k_3}{1 + \xi} S_2 - k_3 S_2 C_2 - (k_{-3} + k_4) C_2 \tag{11}$$

$$\frac{dS_2}{dt} = k_2 C_1 + k_{-3} C_2 - \frac{e k_3}{1 + \xi} S_2 + k_3 S_2 C_2, \tag{12}$$

with the initial conditions $C_1(0) = C_2(0) = S_2(0) = 0$. There is a unique equilibrium point at (C_1^*, C_2^*, S_2^*), where

$$C_1^* = \frac{\xi e}{1 + \xi} \frac{k_1 S_0}{k_1 S_0 + k_{-1} + k_2}, \qquad C_2^* = \frac{\xi e}{1 + \xi} \frac{k_1 k_2 S_0}{k_4 \left(k_1 S_0 + k_{-1} + k_2\right)},$$

$$S_2^* = \frac{\xi k_1 k_2 (k_{-3} + k_4) S_0}{(k_1 k_4 S_0 + k_{-1} k_4 + k_2 k_4 - \xi k_1 k_2 S_0) k_3}.$$

The necessary condition for this equilibrium to be stable and positive is

$$\xi k_1 k_2 S_0 < k_4 \left(k_1 S_0 + k_{-1} + k_2\right)$$

and so, the value of ξ which yields the highest C_2^* value is

$$\xi_{\max} = \frac{k_4}{k_2} \left(1 + \frac{K_M^1}{S_0}\right), \tag{13}$$

where $K_M^1 = (k_{-1} + k_2)/k_1$ is the Michaelis constant of the first reaction. Hence, the simplified model shows that the optimal GOX:HRP ratio depends on the ratio of the turnover rates for the two consecutive reactions, as well as the number $1 + K_M^1/S_0$ (the factor by which the velocity of the first reaction differs from its maximal velocity).

4 Numerical Simulations and Discussions

Numerical simulations were carried out to establish how different kinetic parameters affect the current magnitude and optimal enzyme ratio. The system
of equations (2)–(9) were integrated numerically using a standard finite difference method. In Fig. 1 we plot the steady state values of $\frac{dP}{dt}$, a measure
of the amperometric signal, for different values of the molar ratio ξ, reaction
speed ratio k_4/k_2 and various orders of magnitude of K_M^1/S_0. The optimal
GOX:HRP ratio, $\xi_{\max}$, is then plotted in Fig. 2 as a function of k_4/k_2.

When $K_M^1/S_0 \ll 1$ (see Fig. 1a), the signal amplitude will increase with ξ,
reach a maximum and then decrease. This is due to the fact that when the
concentration of GOX increases, more H_2O_2 is produced. However, with increased production of hydrogen peroxide (as well as the reduced amount of
HRP at higher values of ξ), there is a point where diffusion effects will dominate the second reaction and the resulting signal will decrease. This decay
is faster when the second reaction is slow (small k_4) which explains why the
lower curves in Fig. 1a are steeper than the higher ones. As a consequence, for

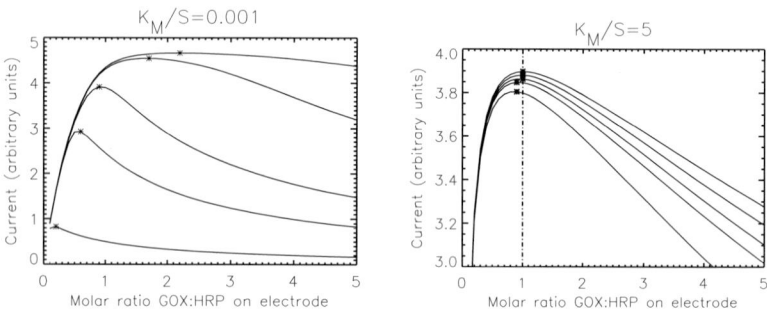

Fig. 1. Dependence of current on ξ (electrode GOX:HRP ratio). From bottom to
top, the curves correspond to $k_4/k_2 = 0.1, 0.5, 0.9, 2.1, 4.1$ for $K_M^1/S_0 = 0.001$ and
$k_4/k_2 = 0.1, 0.18, 0.26, 0.5, 4$ for $K_M^1/S_0 = 5$

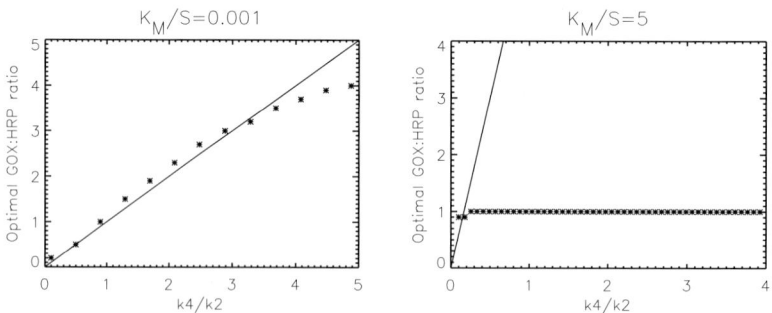

Fig. 2. Optimal GOX:HRP ratio as a function of k_4/k_2. Comparison of simplified
model (*straight line*) and spatially extended model (*curve*)

high k_4, the optimal enzyme ratio is less relevant and the signal is not very sensitive to the concentrations of immobilized enzymes. For large K_M^1/S_0 in Fig. 1b the optimal ξ value seems to be the same for most values of k_4. This can be explained by noting that since GOX is already idle, due to a relatively low amount of substrate, increasing its concentration will not result in increased production of hydrogen peroxide and will therefore not improve the efficiency of the system, regardless of how fast the second reaction is.

The main conclusion of these numerical simulations is the existence of two parameter regimes, $K_M^1/S_0 \ll 1$ and $K_M^1/S_0 = O(1)$, characterized by different qualitative behaviour of the current amplitude and optimal enzyme ratio, $\xi_{\max}$, as functions of the kinetic system variables. Moreover, for $K_M^1/S_0 \ll 1$, there is good agreement between the one-point model and the spatially extended model, as the optimal ratio increases almost linearly with k_4/k_2. The slight divergence of results observed in Fig. 2a is only apparent since, for high values of k_4/k_2 it is more appropriate to speak of optimal ξ intervals, rather than values. A rigorous asymptotic study of these parameter regions will be published separately.

Mathematical modelling is an excellent tool for biosensor design as it provides a theoretical framework through which to explore all the variables of a system without immediate recourse to experiment. For example, the high substrate concentration regime, $K_M^1/S_0 \ll 1$, is more difficult to achieve experimentally due to limitations imposed by the physical solubility of glucose. In addition, our model can also assist in establishing values for constants that are difficult to establish experimentally. For instance, although an enzyme may possess a known ideal kinetic rate constant from solution-phase studies, this may change significantly when the enzyme is deposited on a surface.

References

[KZZ99] A.J. Killard, S. Zhang, H. Zhao, R. John, E.I. Iwuoha, and M.R. Smyth. Development of an electrochemical flow-injection immunoassay (FIIA) for the real-time monitoring of biospecific interactions. Anal Chim Acta, **400**, 109–119 (1999).

[MKA07] D. Mackey, A.J. Killard, A. Ambrosi, and M.R. Smyth. Optimizing the ratio of horseradish peroxidase and glucose oxidase on a bienzyme electrode: comparison of a theoretical and experimental approach. Sensors and Actuators B: Chemical, **122**, 395-402 (2007).

Breast Nodule Ultrasound Segmentation Through Texture-Based Active Contours

Miguel Alemán-Flores[1], Luis Álvarez[1], and Vicent Caselles[2]

[1] Departamento de Informática y Sistemas
Universidad de Las Palmas de Gran Canaria, 35017, Las Palmas, Spain
maleman@dis.ulpgc.es, lalvarez@dis.ulpgc.es
[2] Departament de Tecnologia
Universitat Pompeu Fabra, 08003, Barcelona, Spain
vicent.caselles@upf.edu

Summary. This work presents a method for the segmentation of breast nodules in ultrasonography. Speckle noise is reduced using an anisotropic filter for which texture is described using Gabor filters. Afterwards, an initial segmentation is extracted using a front propagation scheme. Finally, the initial segmentation is refined using active contours. In order to delimit the nodules not only by means of image intensity, but also by texture pattern, we introduce certain terms in the classical active contours equations.

1 Introduction

Breast cancer early diagnosis is based on clinical examination, medical imaging and biopsy. Although the latter is the most reliable test, it is time-consuming, very invasive and expensive. Consequently, different medical imaging techniques are used to reduce as much as possible the number of biopsies which are carried out. These techniques include mammography, ultrasonography and magnetic resonance imaging. This work focusses on ultrasonography, for which a series of criteria have been described to help distinguish benign from malignant lesions. These factors are related to the shape of the nodule, the regularity of its contour and the contrast between certain areas [1].

In order to improve the analysis of the ultrasound images, we propose a semi-automatic segmentation of the nodules, in such a way that the physicians can determine the presence, absence or coincidence of the different criteria in a more robust, reproducible and reliable way. Furthermore, this will simplify an automatic extraction of the different measurements. To this aim, the first phase consists in the reduction of the characteristic speckle noise of ultrasound images, which makes it very difficult to apply directly an automatic segmentation method. Afterwards, an initial pre-segmentation is obtained by means of a front propagation scheme. Finally, this segmentation is improved using active contours. Since the difference between two adjacent regions, e.g. a nodule and the surrounding tissues, may be based not only on intensity, but also in texture, we include a set of texture descriptors to improve the identification of the borders. Furthermore, as the dissimilarities may be different along the contour, we transform the global descriptors into local ones, so that it is easier to find contours which are not uniform.

2 Image Filtering

In order to reduce speckle noise, we apply an anisotropic filtering. A typical anisotropic filter applied to an image I_0 is given by the solution $I(t, x, y)$, $t > 0$, $(x, y) \in \Omega$, of Perona–Malik equation [2].

$$\frac{\partial I}{\partial t} = \text{div}\left(c(\|\nabla I\|)\nabla I\right),$$

where $c(r)$ is a monotonic decreasing function of $r > 0$, such as, for example $c(r) = e^{-\left(\frac{r}{k}\right)^2}$, and k is a constant which determines the contrast of the edges to preserve. In this scheme, it is the magnitude of intensity gradient that determines how strongly the diffusion is performed and what region boundaries must be preserved. Since there may also be differences in the texture, we introduce Gabor descriptors to measure the gradient between the regions, so that the resulting scheme is as follows: let $\mathbf{R}(x, y)$ be the vector formed by the responses of a family of Gabor filters applied to the image I_0 at point (x, y), we use the following anisotropic diffusion equation:

$$\frac{\partial I}{\partial t} = \text{div}\left(c(\|\nabla \mathbf{R}\|)\nabla I\right) \tag{1}$$

where $\|\nabla \mathbf{R}\|^2 = \text{trace}((\nabla \mathbf{R})^t \nabla \mathbf{R})$ (if A is a matrix, A^t denotes the transpose matrix of A). Note that diffusion is inhibited at large values of $\|\nabla \mathbf{R}\|$, i.e. at points where there is a rapid transition of the texture characteristics of the image.

3 Front Propagation

Once the image has been filtered, it is necessary to determine the boundaries of the nodule. To avoid the tedious and variable task of a manual delineation, we ask the physicians to select a single inner point of the nodule. From this point, we apply a front propagation approach to reach an initial segmentation. We introduce an evolving function $u(t, x, y)$ whose zero level set is the curve $C(t)$ (positive inside and negative outside $C(t)$) and we write the geometric evolution for C in terms of u as

$$\frac{\partial u}{\partial t} = F \|\nabla u\|. \tag{2}$$

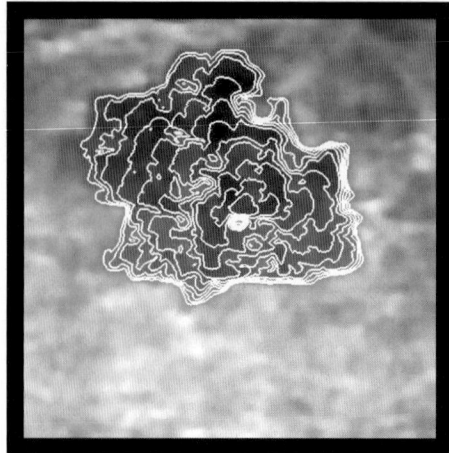

Fig. 1. Example of the evolution of the front propagation using the gradient-guided scheme

To discretize it, we have used the following numerical scheme:

$$\frac{u_{i,j}^{n+1} - u_{i,j}^{n}}{\tau} = F_{i,j} \left\| \nabla u_{ij} \right\|, \tag{3}$$

where $F_{i,j}$ is the discretization of $F(x,y)$, and $\left\| \nabla u_{ij} \right\|$ is calculated as in [3].

We intend to make the front evolve faster where low gradients are found, and reduce its speed where high gradients are present. For this reason we make $F(x,y)$ depend on the inverse of the gradient, which is calculated in a robust way [4].

In these equation, τ represents the time discretization step and h_1, h_2 the pixel dimensions ($\tau, h_1, h_2 > 0$; in our experiments, we consider $h_1, h_2 = 1$). Figure 1 displays an example of the evolution of the front for a breast nodule. Once a whole contour with high enough gradients is reached or the maximum number of iterations have been performed, the process is stopped and the final front is used as initial approximation for the active contours technique.

4 Active Contours

The initial segmentation obtained by means of the front propagation described above is not always as accurate as desired. Thus, we use active contours to improve our results [5]. If we consider image intensity to separate two regions, we can use the basic active contours equation, given by:

$$\frac{\partial u}{\partial t} = g_\sigma \left(I \right) \left\| \nabla u \right\| \operatorname{div} \left(\frac{\nabla u}{\left\| \nabla u \right\|} \right) + \nabla u \nabla g_\sigma \left(I \right), \tag{4}$$

where $g_\sigma(I)$ is a stopping function, such as:

$$g_\sigma(I) = \frac{1}{\sqrt{1 + \alpha \|\nabla I_\sigma\|^2}} \qquad (5)$$

and I_σ is a smoothed version of the image. In (4), the first term aims at obtaining a rounded contour, while the second one attracts the contour to the higher gradients. However, as in the case of image filtering, texture can be helpful to find more representative boundaries and a vector descriptor can represent the regions in a better way [6], [7]. Consequently, we have included an additional term which allows considering the information provided by Gabor descriptors in order to obtain more accurate limits. Thus, we obtain the PDE:

$$\frac{\partial u}{\partial t} = \|\nabla u\| \, \mathrm{div}\left(g_\sigma(I)\frac{\nabla u}{\|\nabla u\|}\right) + \|\nabla u\| \, M_G(I, u) \qquad (6)$$

in which we use a separation term given by:

$$M_G(I, u)(x, y) = \sum_{1 \leq i \leq m} \lambda_-^i \int_{\Omega \cap B(x,y)} \left| I^i(x, y) - c_-^i(x', y') \right| \, \mathrm{d}x' \mathrm{d}y'$$
$$- \sum_{1 \leq i \leq m} \lambda_+^i \int_{\Omega \cap B(x,y)} \left| I^i(x, y) - c_+^i(x', y') \right| \, \mathrm{d}x' \mathrm{d}y',$$

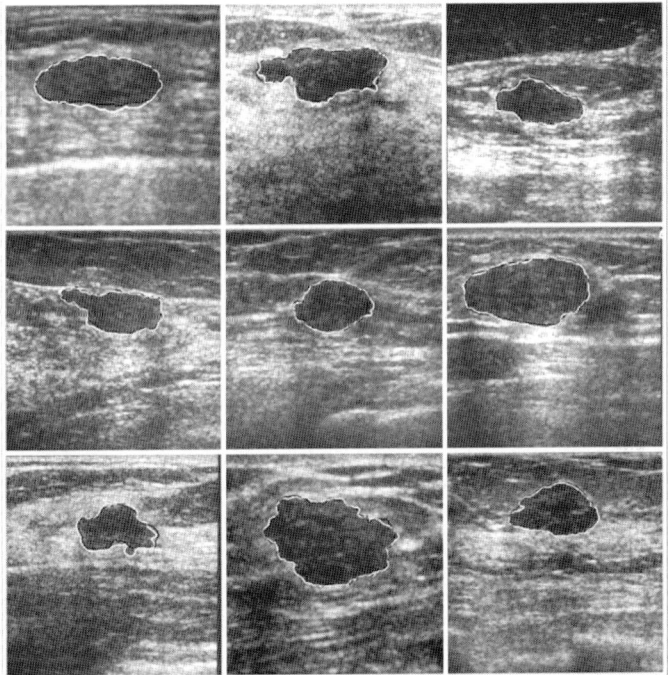

Fig. 2. Manual segmentations (*black*) and semi-automatic segmentations (*white*) for a sample set of nodules

where $I^i(x,y)$ is the response of the ith filter at point (x,y) and $c_+^i(x,y)$, $c_-^i(x,y)$ are the median values of the Gabor channel I^i around (x,y) inside and outside the curve C, respectively. This local formulation can be used because we have previously calculated an initial approximation, and we assume we are close to the final contour. Figure 2 shows some examples of the results of the automatic segmentations and those performed by the radiologists.

5 Conclusion

We have presented a combined method to extract the contours of breast nodules in ultrasound images. The process we have described consists of a filtering phase to reduce speckle noise, a front propagation scheme which allows extracting an initial rough segmentation, and an adaptation of the active contours technique to improve the results when dealing with this kind of images. This combination provides quite satisfactory results, and these are even better when texture information is included. Furthermore, if the information is local, we can tackle the problem of non-uniform or diffuse contours.

The results have been tested using manual segmentations performed by different specialists, and the numerical comparisons show the accuracy of our method. Furthermore, the contours which are obtained are suitable for a further examination of the shape of the nodule which can help in the computer-aided diagnosis of breast cancer.

References

1. Stavros, A.T., Thickman, D., Rapp, C.L., Dennis, M.A., Parker, S.H., Sisney, G.A.: Solid breast nodules: Use of sonography to distinguish between benign and malignant lesions. Radiology, **196(1)**, 123-134 (1995)
2. Perona, P., Malik, J.: Scale space and edge detection using anisotropic diffusion. IEEE Transactions on Pattern Analysis and Machine Intelligence, **12**, 629-639 (1990)
3. Osher, S., Sethian, J.: Fronts propagating with curvature dependent speed: algorithms based on the Hamilton-Jacobi formulation. Journal of Computational Physics, **79**, 12-49 (1988)
4. Brox, T., Weickert, J., Burgeth, B., Mrázek, P.: Nonlinear structure tensors. Image and Vision Computing, **24(1)**, 41-55 (2006)
5. Caselles, V., Kimmel, R., Sapiro, G.: Geodesic active contours. International Journal of Computer Vision, **22(1)**, 61-79 (1997)
6. Chan, T.F., Sandberg, B.Y., Vese, L.A.: Active contours without edges for vector-valued images. J. Visual Communication and Image Representation, **11(2)**, 130-141 (2000)
7. Paragios, N., Deriche, R.: Geodesic active regions for supervised texture segmentation. In: Proc. of International Conference on Computer Vision (1999)

A Contrast Invariant Approach to Motion Estimation: Validation and Application to Motion Estimation Improvement

Vicent Caselles, Luis Garrido, and Laura Igual

Universitat Pompeu Fabra, P. Circumval.lació, 8, Barcelona, Spain
{vicent.caselles, luis.garrido, laura.igual}@upf.edu

Summary. We consider a contrast invariant approach to motion estimation which uses the direction of the gradient fields. The approach is region-based and assumes an affine motion model for each region. We propose to check if the estimated motion parameters fit properly the apparent motion of the region by a motion significance analysis. Moreover, we propose a motion field improvement which consider those regions that are not properly estimated according to the significance analysis and reassign them a motion model of a properly estimated neighboring region.

1 Introduction

Most known motion estimation methods employ the *intensity constancy assumption*, however, global or local illumination changes may violate this assumption and prevent the correct motion to be estimated. In [1] a constraint based on *spatial gradient's constancy* is proposed [2]. The *direction of the spatial gradient* is invariant with respect to global light changes and is insensitive to changes in illumination direction [3]. The work presented in [4] is based on the last property. The contrast invariance is incorporated in our approach [5] by the assumption that the shapes of the image move along the sequence.

In this work we present a hypothesis testing analysis approach that allows to measure how well the motion has been correctly estimated. This measure is used to validate the estimated motion parameters. Moreover, the validation output is used to reassign to the not properly estimated regions a motion model of a neighboring region which has been properly estimated.

The chapter is organized as follows: Sect. 2 summarizes our motion estimation approach and presents the motion validation approach, Sect. 3 introduces the motion model reassignment approach, Sect. 4 presents some results and Sect. 5 ends up with the conclusions and future research work.

2 Motion Estimation and Significance Analysis

2.1 Region-Based Contrast Invariant Motion Estimation

Let $I: \Omega \to R$ be a given image, where Ω is the image domain. The shapes of the image are identified with the family of its level lines which is a contrast invariant geometric description of the image [6]. The main assumption is they move along the image sequence (with possible deformation).

Motion is estimated between two frames of the sequence, denoted by I_0 and I_1, and $\phi(\mathbf{x})$ denotes the coordinates of the point at image I_1 whose coordinates are $\mathbf{x}$ at image I_0. Using the unit normals to describe the level lines, we propose to compute the optical flow ϕ by aligning the unit normal vector field $Z^1(\mathbf{x})$ to the level lines of I_1 with the transformed vector field of $Z(\mathbf{x})$ by the map ϕ, denoted by $\overline{Z}_\phi = (D\phi)^\dagger Z / \|(D\phi)^\dagger Z\|$ if $(D\phi)^\dagger Z \neq 0$ and 0 otherwise. $(D\phi)^\dagger$ denotes the cofactor matrix associated to $D\phi$, see [5].

Moreover we follow a region-based strategy, assuming that the motion fields can be expressed locally by a six parameter affine model. Let $\mathcal{R}$ be a partition into disjoint connected regions of the image I_0 bounded by level lines. The partition may be computed for instance with a segmentation algorithm like the Mumford-Shah functional subordinated to the topographic map [7].

Motion is estimated by minimizing the energy functional:

$$E_{\mathcal{R}}(\phi) := \frac{1}{\mathcal{N}_P} \sum_{j=1}^{\mathcal{N}_P} E_{R_j}(\phi) := \frac{1}{\mathcal{N}_P} \sum_{j=1}^{\mathcal{N}_P} \frac{1}{N_{R_j}} \sum_{\mathbf{x} \in R_j} \Psi\left(\frac{1}{4} \| Z^1(\phi(\mathbf{x})) - \overline{Z}_\phi(\mathbf{x}) \|^2\right),$$

$$(1)$$

where $\Psi(.)$ may represent a robust function, the factor $1/4$ is used to normalize the cost term to the range $[0,1]$, $\mathcal{N}_P = \text{card}(\mathcal{R})$, and $N_{R_j} = \text{card}(\{x \in R_j\})$. Motion is estimated using a gradient descent technique applied over E_R for each region R. For more details on this issue we refer to [5].

2.2 Motion Significance Analysis

The minimization of E_R to estimate the motion parameters ϕ_R for any particular region R will always find a certain minimum, be it local or global. We cannot ensure that such minimum corresponds to the correct motion. Our purpose is to give a measure of the degree in which the motion has been correctly estimated. Let $R \in \mathcal{R}$ be a region of the image I_0. We consider the following hypotheses (which will be interpreted below): H_0: "the motion field of R is correct" and H_1: "the motion field of R is not correct".

2.3 Hypothesis Testing

Given a statistical model of the population, the observed sample is analyzed in order to see if it can be explained by it. If the observation diverges too much from the statistical model, the observation is rejected as belonging to the population.

For each region R, let ϕ_R the estimated motion parameters, and $\{\mathbf{x}_i\}_R = \{\mathbf{x}_i \in R \, / \, \|\nabla I_1(\phi(\mathbf{x}_i))\| > \gamma\}$, and $L = \mathrm{card}(\{\mathbf{x}_i\}_R)$. The threshold γ is used to ensure that the gradient orientations are not much affected by the presence of noise. We assume that the points in $\{\mathbf{x}_i\}_R$ are "independent" [8]. For each $\mathbf{x}_i \in \{\mathbf{x}_i\}_R$ we consider the unitary vectors $\mathcal{Z}^1(\phi(\mathbf{x_i}))$ as a random variable and thus $\mathcal{Y}^i(\phi) = \Psi(1/4 \, \| \mathcal{Z}^1(\phi(\mathbf{x_i})) - \overline{\mathcal{Z}}_\phi(\mathbf{x_i}) \|^2)$ may be interpreted as a random variable measuring the alignment of the two normal vectors. As in [8], we may consider that the vectors $\mathcal{Z}^1(\phi(\mathbf{x_i}))$ and $\overline{\mathcal{Z}}_\phi(\mathbf{x_i})$ are not aligned if they form an angle larger than a given threshold. We define the random variable $\mathcal{E}_R = \frac{1}{L} \sum_{i=1}^{L} \rho\left(\mathcal{Y}^i(\phi)\right)$, where $\rho : [0, +\infty[\rightarrow [0, +\infty[$ is an increasing function. Since $\mathcal{Y}^i(\phi)$ is directly related to the angle forming the two unitary vectors, its non-alignment can be subsumed into $\mathcal{E}_R$ by taking $\rho(x) := \rho_\beta(x) - 1$ if $x > \beta$, and 0 otherwise. In that case, $\mathcal{E}_R$ is a measure of the number of non-alignments for a given region R. We denote $E_R(\phi_R)$ the observed value of $\mathcal{E}_R$ corresponding to the data. The motion field ϕ_R assigned to R is correct if the error $E_R(\phi_R)$ is "sufficiently" small. Our purpose in this work is to define the region of rejection or acceptance using probability theory. If we assume that H_0 is true, the rejection region is of the form $[\mathcal{E}_R \geq \delta]$, $\delta > 0$ [9], but instead of computing the value of δ for a given level of significance as is usually done in hypothesis testing, we compute the probability $P[\mathcal{E}_R \geq E_R(\phi_R)|H_0]$ which corresponds to the probability of *miss-detection* or error of type I (to reject H_0 erroneously) for the observed value $\delta = E_R(\phi_R)$.

The probability that at least k_0 non-alignments occur is given by the binomial tail: $P[\mathcal{E}_R \geq k_0/L] = \mathcal{B}(p_0, k_0, L)$, where the probability of non-alignment p_0 is computed from the empirical data. Thus, the validation can be based in the expected number of miss-detections which is defined as follow.

The number of miss-detections (NMD) of a region is defined as NMD $(R, \phi) = \mathcal{N}_P \cdot P[\mathcal{E}_R \geq E_R(\phi_R)|H_0]$, where $\mathcal{N}_P$ is the number of tested regions.

For a given region R, we reject H_0 if $\mathrm{NMD}(R, \phi) < \epsilon_0$. In that case we say that the motion of R is not properly estimated. Assuming that the motion model has been correctly estimated, the differences $\| \mathcal{Z}^1(\phi(\mathbf{x_i})) - \overline{\mathcal{Z}}_\phi(\mathbf{x_i}) \|$ (and therefore also $\mathcal{Y}^i(\phi)$) should be interpreted as noise. The probability p_0 is computed by $p_0 \approx \sum_{j=1}^{\mathcal{N}_P} \sum_{i=1}^{L_j} \rho_\beta(Y^i(\phi_{R_j})) / \sum_{j=1}^{\mathcal{N}_P} L_j$, where $L_j = \mathrm{card}(\{\mathbf{x}_i\}_{R_j})$ and $Y^i = \Psi(1/4 \, \| Z^1(\phi(\mathbf{x_i})) - \overline{Z}_\phi(\mathbf{x_i}) \|^2)$.

2.4 A Contrario Model

The contratio models were introduced in [8] as a tool for Gestalt's detection. In this context, as it is usual in this type of approach, we check that our observations cannot be explained by random selection of the motion model, that is, the number of coincidences between both vector fields $\mathcal{Z}^1(\phi(x))$ and $\overline{\mathcal{Z}}_\phi(x)$ is too large to be explained by a fortuite coincidence. In this case, we reject H_1 and we consider the motion model to be validated. For the precise details of this approach, we refer to [10].

3 Application: Motion Reassignment

An interesting application of the previous validation analysis is the enhancement of the overall motion estimation field. We propose to reassign to the not
properly estimated regions the motion models of neighboring regions which
have been correctly estimated. Neighboring regions may belong to the same
moving object and thus they may have similar motions. The $NMD(R, \phi)$ may
be used to compare the different motion models that can be assigned to R.
The higher the NMD is the better does the motion model explain the region
movement. For each not properly estimated region the neighboring motion
model leading to the highest NMD is assigned.

4 Results

In all experiments below $\Psi(r) = \sqrt{r^2}$, two vectors are aligned if they form an
angle below 16° and $\epsilon_0 = 0.1$ for the statistical models.

Figure 1 (top) shows frames #9 and #10 of the *vectra* sequence. The apparent motion of the car is zero while the background translates from right to left.
The partition and the resulting motion vector field are shown in Fig. 1 (bottom). Some regions in the boundaries of the image and others near the car
have an incorrect estimated motion field. A validity process becomes necessary to detect these errors. Figure 2 displays the outcome of each of the
strategies presented in Sect. 2. Regions that are found as wrongly (resp. well)
estimated are grey-shaded (resp. white). We have set the modulus gradient
threshold γ to 7. Note that the validation strategies have been able to detect
most of the wrongly estimated regions. It can be seen that the contrario model
is more demanding than the hypothesis testing one. The contrario model can
be considered a validation method, whereas the hypothesis testing model only

Fig. 1. Region-based motion example. From left to right and top to bottom. Original
frame I_0, original frame I_1, partition of original frame I_0, recovered motion field

Fig. 2. Validation and reassignment example. From left to right. Validation using hypothesis testing and the contrario model. Motion reassignment for the Vectra sequence

performs an error control. The obtained motion field after the reassignment can be seen in Fig. 2 (right). We can identify different coherent moving regions of the scenes in these motion field.

5 Conclusions

Following [5], we interpret the image sequence as a set of moving level lines and we compute the optical flow as generated by a deformation between the level lines of two consecutive frames. We have introduced a motion significance measure based on hypothesis testing. These measures are useful both to detect the possible motion estimation errors and as a basic criterion for motion reassignment. Some issues have to be further developed in the future: automatic selection of the modulus gradient threshold γ, and the sensibility of the validation analysis to motion bias with respect the correct motion.

References

1. Bertero, M., Poggio, T.A., Torre, V.: Ill-posed problems in early vision. Procedings of the IEEE **76** (1988) 869889.
2. Papenberg, N., Bruhn, A., Brox, T., Didas, S., Weickert, J.: Highly accurate optic flow computation with theoretically justified warping. IJCV (2005).
3. Chen, H., Belhumeur, P., Jacobs, D.: In search of illumination invariants. In: ICCVPR. (2000) 254261.
4. Burgi, P.Y.: Motion estimation based on the direction of intensity gradient. IVC **22** (2004) 637653.
5. Caselles, V., Garrido, L., Igual, L.: A Contrast Invariant Approach to Motion Estimation. In: Int. Conf. on Scale Space, Springer Verlag (2005)
6. Caselles, V., Coll, B., Morel, J.M.: Topographic maps and local contrast changes in natural images. IJCV 33 (1999) 527.
7. Ballester, C., Caselles, V., Igual, L.: Minimal morphological shape selection for segmentation and encoding. Preprint (2005).
8. Desolneux, A., Moisan, L., Morel, J.M.: Maximal Meaningful Events and Applications to Image Analysis. The Annals of Statistics **31** (2003) 18221851.
9. Rohatgi, V.K., Saleh, M.E.: An Introduction to Probability and Statistics. Wiley-Interscience (2001).
10. Igual, L.: Image Segmentation and Compression using The Tree of Shapes of an Image. Motion Estimation. PhD thesis, Universitat Pompeu Fabra (2005).

A Mathematical Model for Prediction of Recurrence in Bladder Cancer Patients

Cristina Santamaría,[1] María Belén García-Mora,[1] Gregorio Rubio,[1] and Jose Luis Pontones[2]

[1] Instituto de Matemática Multidisciplinar, Universidad Politécnica, Valencia, Spain
   crisanna@imm.upv.es, magarmo5@imm.upv.es, grubio@imm.upv.es
[2] Hospital Universitario La Fe, Valencia, Spain
   pontones_jos@gva.es

Summary. Multiple sequential recurrences are one of the most important characteristics of superficial transitional cell carcinoma (TCC) of the bladder. Many investigations have been done to identify predictive factors for the first recurrence, but very few studies have investigated multiple recurrences of this cancer and its clinicopathologic factors associated. We consider counting process methods for analysing time-to-event data with recurrent outcomes using the models developed by Andersen and Gill and, Prentice, Williams, and Peterson. A postoperative nomogram is developed to predict recurrences based on those predictive factors.

1 Introduction

Bladder carcinoma is the fourth most frequent solid tumor among men and the seventh most frequent solid tumor among women, with more than 350,000 new cases diagnosed annually worldwide [1]. Approximately 80% of patients with newly diagnosed bladder carcinoma have *superficial* transitional cell carcinoma (TCC), which can be managed with transurethral resection (TUR), surgical endoscopic technique used to remove the macroscopic tumor from the inner of the bladder. However, more than 50% of the patients will have *recurrences* (reappearance of a new tumor) [2].

Many investigations have been done to identify prognostic factors for the *first recurrence* of *superficial* TCC of the bladder after initial treatment [3,4], very few studies have investigated *multiple recurrences* of this cancer. An analysis of the patterns of multiple recurrences of bladder carcinoma and its clinicopathologic factors associated could help us better to understand the disease course and to select patients into risk groups of good, intermediate, and poor prognosis for a right treatment.

The current article describes multiple sequential recurrences patterns among *superficial* bladder carcinoma patients and identifies clinicopathologic factors associated. Finally, a postoperative nomogram is proposed as a method

that avoids the arbitrary division of patients into risk groups and can be used
to predict 1, 3, and 5-year disease recurrence probability.

2 Material and Methods

2.1 Patient Population

La Fe University Hospital of Valencia (Spain) provided us with the TUR
database containing detailed information on *superficial* TCC of the bladder
(patient characteristics and pathologic details). Clinicopathologic data were
collected at the time of TUR for all 380 patients, between January 1995
and January 2006, and included stage and grade tumor, size, and number
of tumors. The treatment consisted of a randomized trial of three groups: no
treatment, a single dose of adriamicine or mitomicine 48 h post TUR and a
chemoteraphy (multiple instillations starting at 15 days post TUR, weekly for
4 weeks and monthly for 1 year).

The TNM system (classification of 1997) is generally used to establish the
stage of the bladder tumors [5]: stages Ta and T1 (*superficial* tumors limited to
the mucosa with tendency to produce recurrences of similar stage) and, stages
Tis and T2–T4 (tumors that invade the bladder muscle, highly aggressive, and
with a strong potential to metastatize). The histologic *grade* identifies three
cases according to the World Health Organization 1999 classification [5] of low
aggressiveness to high aggressiveness: G1, G2 and G3 (grade I, II, and III).

All tumors were graded in G1 and G2. We have eliminated the combination T1 G3 because it is a very aggressive tumor with a particular and specific treatment. Indicator or dummies variables are generated for the analysis.
For treatment (three categories) two dummies are defined: single dose and
chemotherapy. Sex, grade (G1 and G2), number (one and two or more), size
(≤ 3 cm and >3 cm) and stage (Ta and T1) are dichotomic variables. The reference individual is a man with a mean (SD) age of 64.63 (11.99) years, with
Ta stage and G1 grade, with one tumor of size minor or equal than 3 cm. He
was assigned to the no treatment group after TUR.

2.2 Statistical Analysis

For multiple sequential recurrences of bladder carcinoma we suggest two models: Andersen and Gill (AG) model [6] and Prentice, Williams, and Peterson
(PWP) model [7]. Both methods model each of the sequential recurrences by
a Cox proportional hazard model [8], taking into consideration correlation
among patient multiple recurrence times. The effect of an individual factor in
the presence of other factors is determined by the statistic -2 log likelihood.
All statistical analysis were performed using S-Plus software (PC Version 2000
Professional; Insightful Corp, Redmond, WA) with additional functions (called
Design) added [9].

If we assume that the risk of recurrence remains constant regardless of the previous number of recurrences, we consider the AG model, but if we assume that after experiencing the first recurrence, the risk of the next recurrence may increase, this suggests the use of the stratified model PWP or AG method with an appropriate time-dependent covariate (*previous number* of recurrences) to capture the dependence between recurrences of the same patient. We use the counting process style of data input for both models, where each subject is represented as observations with time intervals of (entry time (TUR), first recurrence], (first recurrence, second recurrence], ..., (mth recurrence, last followup]. For the AG model, the *risk set* at the time of the kth recurrence would be all patients who are under observation and, for the PWP model all patients who are under observation and have had *k-1* recurrences.

3 Results

Age, grade, number, size, and treatment were significant at the 5% level (Table 1). Including *previous number* of recurrences (up to 5), the value of the statistic $-2 \log \hat{L}$ is reduced in 79.002 on 1 d.f., statistically significant at the 1% level (AG model). We conclude that patients with previous number of recurrences presents an increasing risk (47.3%) of experiencing the next recurrence.

Patients with grade G2 have a risk of recurrence 51.7% with respect to patients with grade G1 (PWP model), and patients with more than one tumor have a risk 76.8% bigger than patients with only one tumor. Individuals with tumor size larger than 3 cm were 1.64 times likelier to have recurrences. The hazard ratio for a patient on chemoteraphy group is reduced about 28% compared to patients with no treatment.

Table 1. Relative risk and 95% CI of tumor recurrences from the AG and PWP

Variable	$\hat{\beta}$	se($\hat{\beta}$)	Robust se	z	*p_value*	RR	Lower.95	Upper.95
AG model with previous number								
Age	0.012	0.005	0.005	2.567	0.010	1.012	1.003	1.022
Grade	0.339	0.115	0.111	3.065	0.002	1.404	1.130	1.744
Number	0.602	0.159	0.149	4.044	0.000	1.825	1.364	2.444
Size	0.441	0.162	0.166	2.662	0.008	1.554	1.123	2.150
Single dose	0.026	0.164	0.157	0.165	0.870	1.026	0.754	1.396
Chemotherapy	−0.498	0.123	0.115	−4.317	0.000	0.608	0.485	0.762
Prev. num	0.387	0.041	0.038	10.116	0.000	1.473	1.367	1.588
PWP model								
age	0.010	0.005	0.005	2.24	0.025	1.010	1.001	1.019
grade	0.417	0.119	0.114	3.65	0.000	1.517	1.212	1.897
number	0.569	0.163	0.151	3.78	0.000	1.768	1.316	2.375
size	0.498	0.166	0.163	3.05	0.002	1.645	1.195	2.265
Single dose	0.079	0.168	0.170	0.47	0.640	1.082	0.776	1.510
Chemotherapy	−0.321	0.135	0.119	−2.69	0.007	0.726	0.574	0.917

CI: confidence interval; RR: relative risk

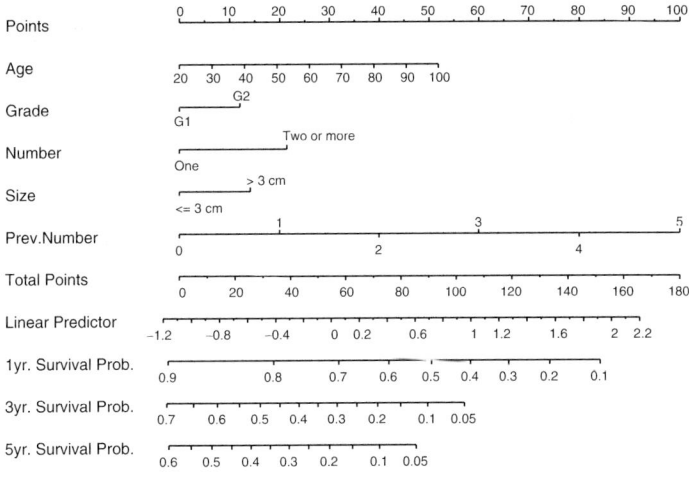

Fig. 1. Nomogram for *superficial* bladder carcinoma recurrences (*AG model*)

We construct a postoperative nomogram from the AG model. We do not include treatment because the aim is providing a graphical and simple tool for the physician and patient to decide the protocol of follow-up and the more adapted treatment to the situation of his (her) disease. The nomogram is used by first locating a patient's position on each predictor variable scale (see Fig. 1). Each scale position has corresponding points (*Points*). A patient with a G2 tumor contributes 13 points approximately, determined by comparing the location of G2 on the grade axis to the points scale above and drawing a vertical line between the two axes. The point values for the other predictor variables are determined in a similar manner. They are summed to obtain a total points value that is located on the *Total Points* axis. A vertical line drawn from the total points axis straight down to find the patient's probabilities of recurrence at 1, 3, and 5 years. So, a 50 years old patient with G2 grade, size minor than 3 cm, with only one tumor and two previous recurrences has $20 + 13 + 0 + 0 + 40 = 73$ total points, which implies about 63% of disease free probability at 1 year, 20% at 3 years and 11% at 5 years.

To assess model accuracy at 1, 3 and 5 years, Harrell's bias corrected concordance index c was calculated [9]. This index is the percentage of patients pairs in which the predicted and observed outcomes are in agreement; i.e., the probability that for two patients chosen at random, the patient who had the event first had a higher probability of having the event according to the model. $c = 0.5$ represents agreement by chance; $c = 1$ represents perfect discrimination [10]. In our analysis the concordance index was 0.67.

4 Conclusions

We have used two counting process methods (AG and PWP) to analyse the multiple sequential recurrences, that although is one of the most important characteristic of superficial TCC of the bladder, there are a few studies about this. Both models give the same predictor variables and similar relative recurrence risks. As in the clinic practice it arises the need for getting together every crumb of information in a graphic tool we have constructed a *nomogram* on the basis of the AG model because it is the simplest method to visualize since PWP is a stratified model.

References

1. Greenlee, R.T., Hill-Harmon, M.B., Murray, T., Thun, M.: Cancer statistics. CA Cancer J Clin, **51**, 15–36 (2001)
2. Hlmang, S., Hedelin, H., Anderstrm, C., Johansson, S.L.: The relationship among multiple recurrences, progression and prognosis of patients with Stage Ta and T1 transitional cell cancer of the bladder followed for at least 20 years. J Urol, **153**, 1823–1827 (1995)
3. García, B., Rubio, G, Santamaría, C., Pontones, J.L., Vera, C.D., Jiménez, J.F.: A predictive mathematical model in the recurrence of bladder cancer. Math Comp Model, **42**, 621–634 (2005)
4. Jaemal, A., Murray, T., Samuels, A., Ghafoor, A., Ward, E., Thun, M.: Cancer statistics 2002. CA Cancer J Clin, **53**, 5–26 (2003)
5. Hermanek, P., Sobin, L.H.: TNM Classification of malignant tumours 4th ed. Springer–Verlag, Berlin (1998)
6. Andersen, P. K., Gill, R. D.: Cox's regression model for counting processes: a large sample study. Ann Statis, **10**, 1100-1120 (1982)
7. Prentice, R. L., Williams, B. J., Peterson, A. V.: On the regression analysis of multivariate failure time data. Biometrika, **68**, 373-389 (1989)
8. Cox, D. R.: Partial likelihood. Biometrika, **62**, 269-276 (1975)
9. Harrell, F.E.: Regression Modeling Strategies with Aplications to Linear Models, Logistic Regression and Survival Analysis. Springer, N.Y. (2002)
10. Harrell F. E., Lee, K.L., Mark, D. B.: Multivariate prognostic models: issues in developing models, evaluating assumptions and adequacy, and measuring and reducing errors. Statis Med, **15**, 361–87 (1996)

Use of the Fourier Transform in the Distributions Sense for Creation Numerical Algorithms for Cone-Beam Tomography

O.E. Trofimov

Institute of Automation and Electrometry, Siberian Branch of Russian Academy of Sciences
trofimov@iae.nsk.su

The research was supported by Russian Foundation for Basic Research (project 06-01-81000).

Summary. Let the homogeneous of L degree function $g(x)$ be defined in N-dimensional space, and let the function $G(y)$ be its Fourier transform in the distribution sense. The theorem that allows to present the function $G(y)$ using only the values of function $g(x)$ on the unit sphere is proved in the chapter for the case $L > -N$. The case $N=3$ and $L = -1$ corresponds to the properties of beam transform in 3D space. In the chapter it is shown how the theorem may be used for creation of numerical algorithms for cone-beam tomography.

Let the homogeneous of L degree function $g(x)$ be defined in N-dimensional space, and let the function $G(y)$ be its Fourier transform. In view of homogeneity, the function $g(x)$ and its Fourier transform in sense of distributions are defined by their values on the unit sphere [GS00]. We will prove the theorem that allows to present the function $G(y)$ using only the value of function $g(x)$ on the unit sphere for the case $L > -N$.

Theorem 1. *Let the function $G(\xi)$ be the Fourier transform in sense of distributions of homogeneous function $g(\alpha)$ of λ degree, where $\lambda > -N$ and $\lambda \neq 0, \pm 1, \pm 2, \ldots, \pm(N-1), N, (N+1), \ldots$, then*

$$G(\xi) = i\Gamma(\lambda + N) \times [\exp(i(\lambda + N)(\pi/2)) \int_{|\beta|=1} g(\beta)(\beta, \xi)_+^{-(\lambda+N)} d^{(N-1)}\beta$$
$$- \exp(-i(\lambda + N)(\pi/2)) \int_{|\beta|=1} g(\beta)(\beta, \xi)_-^{-(\lambda+N)} d^{(N-1)}\beta] \qquad (1)$$

If $\lambda = 0, \pm 1, \pm 2, \ldots, \pm(N-1), N, (N+1), \ldots$, then

$$G(\xi) = i^{(n+N)}[(n + N - 1)! \int_{|\beta|=1} g(\beta)(\beta, \xi)^{-(n+N)} d^{(N-1)}\beta$$
$$+ (-1)^{n+N} i\pi \int_{|\beta|=1} g(\beta)\delta^{(n+N-1)}((\beta, \xi)) d^{(N-1)}\beta]. \qquad (2)$$

Here β is the point on the $N-1$ dimensional unit sphere, and symbol $\mathrm{d}^{N-1}\beta$ means integration on this sphere.

For rigorous proof, it is appropriate to use the functions $g_\tau(\alpha) = g(\alpha)e^{-\tau|\alpha|}$,

$$G_\tau(\xi) = \int g(\alpha)e^{-\tau|\alpha|}\exp(-2i\pi(\alpha,\xi))\mathrm{d}\alpha,$$

and formulas for the Fourier transforms of distributions [GS00].

The case $N=3$ and $\lambda=-1$ corresponds to properties of beam transform in 3D-spase. For this case the theorem was proved in [Tr04].

Let the function $f(x) = f(x_1, x_2, x_3)$, the point $S = (s_1, s_2, s_3)$ and the vector $\alpha = (\alpha_1, \alpha_2, \alpha_3)$ be given. The beam transform of function $f(x)$ is the function

$$(R_1^+ f)(\alpha, S) = \int_0^\infty f(t\alpha + S)\mathrm{d}t.$$

The function $(R_1^+ f)(\alpha, S)$ is the integral of function $f(x)$ along the beam that comes out from the point S in the direction of the vector α.

From mathematical point of view, the task of cone-beam tomography is a determination of the function $f(x)$, if the function $(R_1^+ f)(\alpha, S)$ is known. The set of points for which the beam-transform is known, usually is the set of points of some curve, which is the trajectory of X-ray source.

Let the curve $\Phi(\lambda) = (\Phi_1(\lambda), \Phi_2(\lambda), \Phi_3(\lambda))$ be given, $\lambda \in \Lambda$, Λ is some interval in R^1. For $\alpha = (\alpha_1, \alpha_2, \alpha_3)$ and $\lambda \in \Lambda$ we have the function

$$g^+(\alpha, \lambda) = (R_1^+ f)(\alpha, \Phi(\lambda)) = \int_0^\infty f(t\alpha + \Phi(\lambda))\mathrm{d}t.$$

The function $g(\alpha, \lambda)$ is the integral of function $f(x)$ along the beam that comes out from the point $\Phi(\lambda)$ in the direction of the vector α. If λ is fixed, the function $g^+(\alpha, \lambda)$ is homogeneous function with respect to α with homogeneous degree -1:

$$g^+(\alpha, \lambda) = \frac{1}{|\alpha|}\int_0^\infty f\left(\Phi(\lambda) + \frac{\alpha}{|\alpha|}\right)\mathrm{d}t \tag{3}$$

The following inverse formula for beam transform was presented in [Tuy83]

$$f(x) = \int_0^{2\pi}\int_{-\frac{\pi}{2}}^{\frac{\pi}{2}} \cos\phi \frac{1}{2i\pi(\Phi'(\lambda), \beta)}\frac{\partial G^+(\beta, \lambda)}{\partial \lambda}\mathrm{d}\phi\,\mathrm{d}\theta \tag{4}$$

If the parameter λ is fixed, the function $G^+(\beta, \lambda)$ is the Fourier transform of function $g^+(\alpha, \lambda)$ with respect to variable α. The vector β is $\beta(\phi, \theta) = (\cos\theta\cos\phi, \sin\theta\cos\phi, \sin\varphi)$. The parameter λ in formula (8) depends on x, and β and satisfies to the conditions: the scalar product (β, x) is equal to $(\Phi(\lambda), \beta)$, but $(\Phi'(\lambda), \beta) \neq 0$. The function $f(x)$ may be determined in the point x if such λ exists for any β(Kirillov–Tuy's conditions). From geometrical point of view, it means that any plane that intersects the point x, intersects also the curve $\Phi(\lambda)$ transversally (so that denominator in (4) does not equal to zero).

If the support of function $f(x)$ belongs to the unit ball, the union of two unit circles having the centers in the point $(0,0,0)$ and belonging to planes $Z = 0$ and $Y = 0$ satisfies to Kirillov–Tuy's conditions.

Creating of numerical algorithms on the base of formula (4) directly is difficult [Nat86].

The point is that the Fourier transform in the distributions sense is used in (4), and the Fourier transform of function $g^+(\alpha, \lambda)$, in classical sense

$$G^+(\xi, \lambda) = \int_{-\infty}^{\infty} g^+(\alpha, \lambda) \exp(2i\pi(\alpha, \xi)) d\alpha,$$

diverges (i.e. does not exist), because the function $g^+(\alpha, \lambda)$ is homogeneous with respect to α and therefore has the order $1/|\alpha|$ on infinity.

To use the formula (4) for the numerical algorithms, it is necessary to show that the function $G^+(\xi, \lambda)$ may be determined as a regular function, and to have formulas that connect the functions $g^+(\alpha, \lambda)$ and $G^+(\xi, \lambda)$.

Now we will get over to the receipt of corresponding formulas.

From the theorem proved above, we have the following

Corollary. If $G(\xi)$ is the Fourier transform in the distributions sense of homogeneous function $g(\alpha)$ in 3D space, then

$$G(\xi) = (-1/4\pi^2) \int_{|\beta|=1} g(\beta)[(\beta, \xi)^{-2} - i\pi\delta'((\beta, \xi))] d^2\beta. \qquad (5)$$

In tomography tasks the functions $f(x)$ are real ones; only imaginary part of $G(\xi)$ is used in formula (4):

$$\operatorname{Im} G(\xi) = (i/4\pi) \int_{|\beta|=1} g(\beta)\delta'(\beta, \xi) d^2\beta. \qquad (6)$$

Here $d^2\beta$ means the integration along the circle that is the intersection of unit sphere and the plane $(\beta, \xi) = 0$.

Using the formula (6) and δ-sequences of regular functions, it is possible to create the numerical algorithms on the base of formula (4).

If the function $g(\alpha)$ is smooth, then it is possible to derive the inverse formula for beam transform that uses only standard operations of differentiation and integration of measured functions. The following formula was presented in [Tr04]:

$$\operatorname{Im} G(\xi) = (i/4\pi) \int_{S(\xi)} L(\xi, D) g(\alpha) \Omega(\alpha).$$

Here $S(\xi) = (\alpha \in S^2 | (\xi, \alpha) = 0)$, $L(\xi, D) = \sum_{k=1}^{3} \xi_k \frac{\partial}{\partial y} k$ is the derivative in the direction ξ. Symbol $\Omega(\alpha)$ means integration along the circle $S(\xi)$.

Substituting the function $G^+(\beta, \lambda)$ in formula (8), we obtain:

$$f(x) = f(x_1, x_2, x_3)$$

$$= \frac{-1}{8\pi^2} \int_0^{2\pi} \int_{-\pi/2}^{\pi/2} \cos\phi \frac{1}{(\Phi', \beta)} \frac{\partial}{\partial\lambda} \left[\int_{S(\beta)} L(\beta, D) g^+(\alpha, \lambda) \Omega(\alpha) \right] d\phi d\theta. \qquad (7)$$

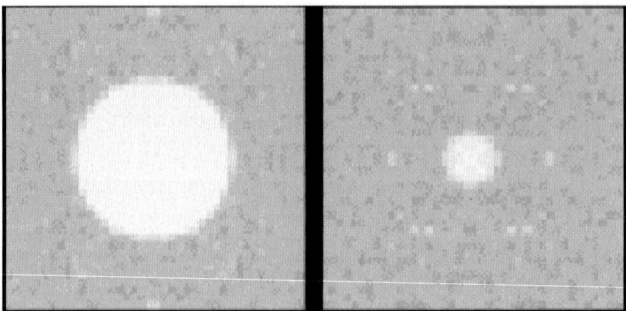

Fig. 1. Numerical simulation

Here $S(\beta)$ is the circle that is the intersection of the unit sphere and the plane $P(\beta)$. The plane $P(\beta)$ intersects the point $(0,0,0)$ and is orthogonal to vector β. As we said above, the symbol $\Omega(\alpha)$ means integration along the circle $S(\beta)$, operator $L(\beta, D)$ means the differentiation of function $g(\alpha)$ in the direction of vector β:

$$L(\beta, D)g^+(\xi, \lambda) = \beta_1 \frac{\partial}{\partial \xi_1} g^+(\xi, \lambda) + \beta_2 \frac{\partial}{\partial \xi_2} g^+(\xi, \lambda) + \beta_3 \frac{\partial}{\partial \xi_3} g^+(\xi, \lambda),$$

parameter λ depends on β and x, and is fixed, the vector $\beta = \beta(\theta, \varphi) = (\cos\theta\cos\phi, \cos\theta\sin\phi, \sin\theta)$. The parameter $\lambda = \lambda(\theta, \phi) = \lambda(x, \beta)$ is found from the conditions: $(\beta, x) = (\beta, \Phi(\lambda))$ and $(\Phi'(\lambda), \beta) \neq 0$, here $(.,.)$ means the scalar product. Only the regular functions are used in (7). This formula can be used in numerical algorithms for full 3D reconstruction.

The results of three-dimensional tomography reconstruction on base of formula (7) are presented on Figure 1. The model object is the sphere with a cavity. Different sections are presented ($z=0$, and $z=0.096$).

References

[GS00]   Gelfand, I.M. and Shilov, G.E. Distributions, v.1, Distributions and Actions over them. Moscow.2000.
[LZT01]   Lavrentiev M.M., Zerkal S.M., Trofimov O.E. Computer Modelling in Tomography and Ill-Posed Problems, VSP (The Netherlands), 2001, 128 pp.
[Nat86]   F. Natterer, The Mathematics of Computerized Tomography. B.G. Teubner, Stutgart, and John Willey & Sons Ltd, 1986.
[Tr04]   Trofimov O.E. On Fourier Transform of a Class of Generalized Homogeneous Functions. Siberian Journal of Industrial Mathematics, 2004. Vol. 7, No. 1. pp. 30-34. (in Russian)
[Tuy83]   Tuy H. K., An Inversion Formula for Cone-beam Reconstruction. SIAM. J. APPL. MATH. 1983, vol. 43, N 3, PP. 546-552.

Shapley Value vs. Proportional Rule in Cooperative Affairs

Rafel Amer, Francesc Carreras, and Antonio Magaña

Technical University of Catalonia, ETSEIAT, P.O. Box 577, E-08220 Terrassa, Spain
francesc.carreras@upc.edu

1 Introduction

The proportional rule has a long tradition in collective problems where some kind of utility (costs, profits, savings) is to be shared among the agents. However, while its (apparent) simplicity might be a reason for applying it in pure bargaining affairs, where only the whole and the individual utilities matter, it seems much more questionable in the case of general cooperative problems, where all marginal contributions should be taken into account. We will contrast the proportional rule with the Shapley value in this kind of problems.

2 Cooperative Affairs

Let $N = \{1, 2, \ldots, n\}$ be a set of agents (*players*). Each subset $S \subseteq N$ is a *coalition*. A *cooperative game* on N (a *game*, for short, in the sequel) is a function u that assigns to each coalition S a real number $u(S)$, which is interpreted as the *worth* or utility that coalition S is able to obtain independently of the behavior of the outside players (the members of $N \backslash S$). The only condition imposed to u is that $u(\emptyset) = 0$, where $\emptyset$ denotes the empty coalition. A game u is (a) *additive* if $u(S \cup T) = u(S) + u(T)$ whenever $S \cap T = \emptyset$ and (b) *symmetric* if $u(S)$ depends only on the cardinality of S for all $S \subseteq N$.

Usually, u represents profits or savings arising from cooperation between the members of any coalition. In other cases, u may well represent costs. Rather than on the individual strategic possibilities of the agents, the cooperative theory is merely based on the amounts of utility that coalitions can get. The main question related to the problems represented by cooperative games is the way to share among the players, in a rational way, the total utility of the grand coalition, that is, $u(N)$, in such a manner that all players agree and become satisfied with the outcome. In this respect, an essential notion, of great relevance in economics, should be the following: the *marginal contribution* of a player $i \in N$ in a game u to a coalition S containing i is $u(S) - u(S \backslash \{i\})$.

It will also be of interest to have in mind that $i \in N$ is a *null* player in a game u on N if all his marginal contributions vanish, i.e., if $u(S \cup \{i\}) = u(S)$ for every $S \subseteq N \backslash \{i\}$, and $i, j \in N$ are *equivalent* players in a game u on N if their marginal contributions to each coalition coincide, i.e., if $u(S \cup \{i\}) = u(S \cup \{j\})$ for every $S \subseteq N \backslash \{i, j\}$. The set $\mathcal{G}_N$ of all cooperative games on a given player set N becomes a vector space of dimension $2^n - 1$ endowed with the natural linear operations for real-valued functions defined by $(u + v)(S) = u(S) + v(S)$ and $(\lambda u)(S) = \lambda u(S)$ for all $S \subseteq N$, $u, v \in \mathcal{G}_N$ and $\lambda \in R$ (set of real numbers). We will pay special attention to the sum of games.

3 The Proportional Rule

A well-known method to share $u(N)$ is given by the so-called *proportional rule*. The amount that this rule allocates to each agent $i \in N$ in game u is

$$\pi_i[u] = \frac{u(\{i\})}{u(\{1\}) + u(\{2\}) + \cdots + u(\{n\})} u(N). \tag{1}$$

Although an obvious "advantage" of the proportional rule is the easiness of its calculation, a first problem arises since the rule can be applied only if

$$u(\{1\}) + u(\{2\}) + \cdots + u(\{n\}) \neq 0, \tag{2}$$

so that its domain is limited to the class of cooperative games on N that satisfy this condition. This is by no means a trivial restriction.

4 An Axiomatic Approach to a Value Notion

In a seminal work [Sha53] (also in [Rot88]), Shapley addressed the sharing problem in completely general terms and introduced the axiomatic method in game theory. Indeed, he stated the problem of finding a *value*, that is, a map

$$f : \mathcal{G}_N \longrightarrow R^N$$

which would assign a payoff vector $f[u] = (f_1[u], f_2[u], \ldots, f_n[u])$ to every game u on N. This payoff vector should be viewed as the solution to the problem represented by game u.

Of course, there are infinitely many ways to define such a map. Shapley stated four appealing conditions that a value f should satisfy. They are:

1. *Efficiency.* $\sum_{i \in N} f_i[u] = u(N)$ for every $u \in \mathcal{G}_N$.
2. *Null player property.* If $i \in N$ is a null player in a game $u \in \mathcal{G}_N$ then $f_i[u] = 0$.

3. *Symmetry.* If $i, j \in N$ are equivalent players in a game $u \in \mathcal{G}_N$ then $f_i[u] = f_j[u]$.
4. *Additivity.* For all $u, v \in \mathcal{G}_N$, $f[u + v] = f[u] + f[v]$.

These properties deserve to be called "axioms" because of their elegant simplicity. It is hard to claim that they are not compelling... Efficiency means that the players are going to share the total amount available to them. The null player property states that if a player does not contribute anything to all coalitions to which belongs then this player must get 0. Symmetry establishes that two players that are equally interesting as coalition partners should receive the same payoff. Finally, additivity implies that the allocation in the sum game has to coincide with the sum of allocations in each game. Maybe, in spite of its simplicity and mathematical tradition, this latter property is, in principle, the least clear one: the reason is that one does not capture easily the meaning of the sum game in practice.

In his chapter, Shapley proved that there is one and only one function f that satisfies these four axioms (a logically independent system). He denoted it by φ and found that its expression is, for arbitrary $i \in N$ and $u \in \mathcal{G}_N$,

$$\varphi_i[u] = \sum_{S \subseteq N:\, i \in S} \frac{(s-1)!(n-s)!}{n!}[u(S) - u(S \backslash \{i\})], \qquad (3)$$

where $s = |S|$ for each $S \subseteq N$.

Of course, the subsequent literature has been referring to φ as the *Shapley value*. Notice that, in fact, the Shapley value becomes a linear map from $\mathcal{G}_N$ to R^N. Also notice that all marginal contributions of a player matter when computing the Shapley value for this player. This implies a great deal of strategic sensibility on the side of the Shapley value that should be highly appreciated by practitioners.

5 Checking the Proportional Rule

The framework established by Shapley allows us to evaluate any allocation rule and, in particular, to compare the proportional rule and the Shapley value. The relevant points are the following:

1. As was pointed out before, a first essential failure of the proportional rule is its restricted domain, defined by condition (2). Instead, the Shapley value applies without any restriction to all cooperative games.
2. A second important problem is that the proportional rule does not take into account most of the marginal contributions; this becomes more and more critical as the number of players increases and results in the very low sensibility already mentioned. On the contrary, the Shapley value is always concerned with all marginal contributions without exception and enjoys therefore a nice sensibility with regard to the problem data.

3. It is instructive to put together formulas (1) and (3) for two-player games. If i and j are these players (so that $\{i, j\} = \{1, 2\}$) then we have

$$\pi_i[u] = u(\{i\}) + \frac{u(\{i\})}{u(\{i\}) + u(\{j\})}[u(\{i, j\}) - u(\{i\}) - u(\{j\})] \quad (4)$$

and

$$\varphi_i[u] = u(\{i\}) + \frac{1}{2}[u(\{i, j\}) - u(\{i\}) - u(\{j\})]. \quad (5)$$

The procedures look partially similar: first, each player is allocated his claim (say, $u(\{i\})$ in case of player i); then, the remaining worth is shared among both. However, it is worthy of mention that the Shapley value simply shares the residual utility equitably, whereas the proportional rule shares it proportionally to the individual utilities. This means that the proportional rule is, conceptually, more complicated than the Shapley value and includes a hardly justified double discriminatory level that rewards twice the player that individually can get the highest utility by his own. The phenomenon extends to more than two players.

4. In which cases do both allocation rules coincide? It is not difficult to see that, in the two-player case, the Shapley value and the proportional rule coincide on a game u (such that $u(\{1\}) + u(\{2\}) \neq 0$, of course) if and only if this game satisfies some of the following conditions:
   (i) $u(\{1\}) = u(\{2\})$ (symmetric game).
   (ii) $u(\{1, 2\}) = u(\{1\}) + u(\{2\})$ (additive game).
   In general (arbitrary n), for any additive game u satisfying (2) we have $\varphi[u] = \pi[u]$ (in fact, the i–component of both values coincides with $u(\{i\})$ for each $i \in N$). Also for any symmetric game u satisfying (2) we get $\varphi_i[u] = \pi_i[u] = \frac{u(N)}{n}$ for all $i \in N$.

5. As to Shapley's axioms, it is easy to check that, in its restricted domain defined by inequality (2), the proportional rule satisfies the axioms of efficiency, null player and symmetry.

6. This leaves us with the lack of additivity for this rule (otherwise, it would coincide with the Shapley value by the uniqueness of the value). Then, although the proportional rule satisfies $\pi[\lambda u] = \lambda \pi[u]$ for all game u in its domain and all $\lambda \in R$, the proportional rule is not linear. Let us raise the following question: is this failure important or, on the contrary, additivity is simply a standard mathematical property, just of a technical nature, without special relevance for practitioners? The answer is quite surprising. From the lack of additivity, serious inconsistencies of the proportional rule derive when applying it to certain problems.

6 Conclusions

So far we have analyzed the proportional rule, from the axiomatic viewpoint established by Shapley when defining the value notion for cooperative games but also from a practical viewpoint. Several properties and failures of the proportional rule have been remarked (items 1–6 in Sect. 5) and, especially, practical implications of the non-additivity of this rule might be evidenced that result in a serious inconsistency when dealing with costs-savings related problems and added costs problems. (For space reasons, we have not included any example.)

Summing up, it might be said that, in spite of its greater difficulty of calculus (easily solved by using computer programs or approximative methods for a high number of players), the Shapley value should replace in practice the proportional rule in cooperative affairs where coalitions of intermediate size $(1 < |S| < n)$ matter, but also in pure bargaining problems.

We would like to end the chapter by mentioning several references. First, the material included here might be completed with references [ACM01] and [ACM06], where additional information is provided. Applications of cooperative games to economic problems may be found in [BF99] and [Raf99], and even in (the chapters on cooperative games of) [Owe95], a great classical book on game theory. For an attempt to give to the proportional rule a greater relevance in cooperative games, see [Ort00].

Acknowledgements

Research partially supported by Grants BFM 2003–01314 of the Science and Technology Spanish Ministry and the European Regional Development Fund and SGR 2005–00651 of the Catalonia Government.

References

[ACM01] Amer, R., Carreras, F., Magaña, A.: Game Theory. Edicions UPC, Barcelona (2001)

[ACM06] Amer, R., Carreras, F., Magaña, A.: Cooperative games and Shapley value. Alta Dirección (2006), forthcoming

[BF99] Bilbao, J.M., Fernández, F.R. (eds.): Advances in Game Theory with Economic and Social Applications. Publication Secretariat of the Sevilla University (1999)

[Ort00] Ortmann, K.M.: The proportional value for positive cooperative games. Mathematical Methods of Operations Research **51**, 235–248 (2000)

[Owe95] Owen, G.: Game Theory. Academic Press (1995) (3rd edition)

[Raf99] Rafels, C. (coord.): Cooperative Games and Economic Applications. University of Barcelona Editions (1999)

[Rot88] Roth, A.E. (ed.): The Shapley Value: Essays in Honor of Lloyd S. Shapley. Cambridge University Press, Cambridge (1988)

[Sha53] Shapley, L.S.: A value for n–person games. Annals of Mathematical Studies **28**, 307–317 (1953)

A Wide Family of Solutions Based on Marginal Contributions for Situations of Competence–Cooperation with Structure of a Priori Coalition Blocks

José Miguel Giménez

Department of Applied Mathematics 3, Engineering School of Manresa, Technical University of Catalonia, Avda. Bases de Manresa 61, E-08242 Manresa, Spain
jose.miguel.gimenez@upc.edu

Summary. Game Theory provides suitable tools to share the total utility among the economic agents or players when the possibilities of cooperation enable obtaining the utility of each group or coalition. The semivalues are solution concepts for situations of competence–cooperation that assign to each player a weighted sum of his/her marginal contributions to the coalitions, where the weights only depend on the coalition size. The Shapley value and the Banzhaf value are semivalues. The solutions introduced here are modifications of the semivalues when we consider a priori coalition blocks in the player set. A computation procedure is also offered.

1 Introduction and Preliminaries

A *cooperative game* with transferable utility is a pair (N, v), where N is a finite set of *players* and $v : 2^N \to \mathbb{R}$ is the so-called *characteristic function*, which assigns to every *coalition* $S \subseteq N$ a real number $v(S)$, the *utility* or *worth* that the coalition S can obtain in the situation described by the game, independently of the remaining players. The function v should satisfy the natural condition $v(\emptyset) = 0$. From now on we suppose $N = \{1, \dots, n\}$ and we denote with G_N the set of all cooperative games on N. Given N, we identify each game (N, v) with its characteristic function v.

A central problem of Game Theory consists of distributing the total utility by using acceptable allocation rules. Ideas as fairness, equity or equal treatment appear in this context. Several solutions can be proposed according to these ideas are expressed by means of mathematical formulations.

A *solution* on the set of cooperative games G_N is an allocation rule that assigns a payoff to each game player, i.e., a function $\Psi : G_N \to \mathbb{R}^N$, where $\Psi[v] = (\Psi_1[v], \dots, \Psi_n[v])$. It represents a method to measure the negotiation strength of the players in the game. In order to calibrate the importance of each player i in the different coalitions S, we can look at his/her *marginal* contribution $v(S) - v(S \setminus \{i\})$. If these marginal contributions are weighted by means of coefficients depending only on the coalition size, we arrive at the solution concept known as semivalue.

The semivalues (Dubey et al., 1981 [DNW81]) are characterized by means of four axioms:

A1. *Additivity.* $\Psi[u + v] = \Psi[u] + \Psi[v] \ \forall u, v \in G_N$.

A2. *Symmetry.* $\Psi_{\pi i}[\pi v] = \Psi_i[v] \ \forall v \in G_N, \ \forall i \in N, \ \forall \pi$ permutation of N, where game πv is defined by $(\pi v)(\pi S) = v(S) \ \forall S \subseteq N$.

A3. *Positivity.* Game v monotonic $[S \subseteq T \subseteq N \Rightarrow v(S) \leq v(T)]$ implies $\Psi_i[v] \geq 0 \ \forall i \in N$.

A4. *Projection.* $\Psi_i[v] = v(\{i\}) \ \forall v \in A_N$, where A_N denotes the set of additive games in G_N: games v such that $v(S \cup T) = v(S) + v(T)$ if $S \cap T = \emptyset$ and $S, T \subseteq N$.

Theorem 1 (Dubey et al., 1981 [DNW81]). *(a) Every weighting vector* $(p_1, p_2, \ldots, p_n)$ *verifying conditions*

$$\sum_{s=1}^{n} \binom{n-1}{s-1} p_s = 1 \quad and \quad p_s \geq 0 \ for \ 1 \leq s \leq n \qquad (1)$$

defines a semivalue $\psi : G_N \to \mathbb{R}^N$ *whose allocations are given by*

$$\psi_i[v] = \sum_{S \subseteq N: \ i \in S} p_s[v(S) - v(S \setminus \{i\})] \quad \forall i \in N \ (where \ s = |S|).$$

(b) Conversely, every semivalue defined on G_N is of this form, so that, there exists a one-to-one map between the semivalues on G_N and the vectors $(p_s)_{s=1}^{n}$ that verify conditions (1).

Definition 1. *A semivalue on G_N is called binomial semivalue if its weighting coefficients are in geometric progression.*

The binomial semivalues are related with the numbers $\alpha \in (0, 1)$. Given a number $\alpha \in \mathbb{R}$, $0 < \alpha < 1$, we call α-binomial semivalue ψ_α to the semivalue on G_N whose weighting coefficients are $p_{\alpha,s} = \alpha^{s-1}(1-\alpha)^{n-s}$ for $1 \leq s \leq n$.

The extreme cases of binomial semivalues correspond to values $\alpha = 0$ and $\alpha = 1$. For $\alpha = 0$ we obtain the dictatorial index ψ_0, with coefficients $(1, 0, ..., 0)$, whereas for $\alpha = 1$ we obtain the marginal index ψ_1, with coefficients $(0, ..., 0, 1)$. The Banzhaf value is the binomial semivalue for $\alpha = 1/2$.

Definition 2. *The family of semivalues on G_N $\{\psi_j\}_{j=1}^{n}$, with respective weighting coefficients $(p_{j,s})_{s=1}^{n}$, $1 \leq j \leq n$, forms a reference system of $Sem(G_N)$ if, and only if, the family of points $\{P_j \ (p_{j,s})_{s=1}^{n}\}_{j=1}^{n}$ forms a reference system of the hyperplane of $\mathbb{R}^n$ with equation $\sum_{s=1}^{n} \binom{n-1}{s-1} p_s = 1$.*

Theorem 2 ([Gim01], [AG03]). *For $n > 1$, given n real numbers $\alpha_j \in [0, 1]$, such that $\alpha_j \neq \alpha_k$ if $j \neq k$, the family of binomial semivalues $\{\psi_{\alpha_j}\}_{j=1}^{n}$ forms a reference system of $Sem(G_N)$ $(n = |N|)$.*

Fixed a reference system of binomial semivalues $\{\psi_{\alpha_j}\}_{j=1}^n$ in $Sem(G_N)$, for each semivalue ψ defined on G_N there exists a unique family of real numbers λ_j, $1 \le j \le n$, such that

$$\psi = \sum_{j=1}^n \lambda_j \psi_{\alpha_j} \quad \text{with} \quad \sum_{j=1}^n \lambda_j = 1.$$

The components of semivalue ψ in the reference system $\{\psi_{\alpha_j}\}_{j=1}^n$ are grouped according to the following notation:

$$\Lambda^t = (\lambda_1 \ \lambda_2 \ \cdots \ \lambda_n). \tag{2}$$

Definition 3 (Owen, 1972 [Owe72]). *The multilinear extension (MLE, in the sequel) of a game $v \in G_N$ is the function $f_v : [0,1]^n \longrightarrow \mathbb{R}$ defined by*

$$f_v(x_1, x_2, \dots, x_n) = \sum_{S \subseteq N} \prod_{i \in S} x_i \prod_{j \in N \setminus S} (1 - x_j) v(S).$$

Theorem 3 ([Gim01], [AG03]). *Let $f_v = f_v(x_1, x_2, \dots, x_n)$ be the MLE of game $v \in G_N$.*

(a) The payoff vector that the binomial semivalue ψ_α assigns to the players of game $v \in G_N$ is

$$\psi_\alpha[v] = \nabla f_v(\overline{\alpha}) \quad \forall \alpha \in [0,1] \quad \text{where } \overline{\alpha} = (\alpha, \dots, \alpha).$$

(b) The payoff vector that every semivalue ψ assigns to the players of game $v \in G_N$ is

$$\psi[v] = B \, \Lambda,$$

where the matrix $B = (\, b_{ij} \,)$ depends on each reference system of semivalues $\{\psi_{\alpha_j}\}_{j=1}^n$,

$$b_{ij} = (\psi_{\alpha_j})_i[v] = \frac{\partial f_v}{\partial x_i}(\overline{\alpha}_j), \quad 1 \le i, j \le n.$$

Λ is the matrix of the components of ψ in the reference system (as in (2)).

2 Mixed Modified Semivalues

Definition 4. *Let ψ^n be a semivalue on G_N with weighting coefficients $(p_s^n)_{s=1}^n$. The family of induced semivalues by ψ^n on sets of games with less than n players is*

$$\{\psi^m \in Sem(G_M) \text{ with } 1 \le m \le n \text{ and } m = |M|\,\},$$

where the respective weighting coefficients are recursively obtained according to the Pascal triangle (inverse) formula

$$p_s^m = p_s^{m+1} + p_{s+1}^{m+1} \quad 1 \le s \le m < n.$$

We can find the above definition in [Dra99]. By convenience, we have included the initial semivalue in its induced family. It is not difficult to see that the induced semivalues of the Shapley value, the Banzhaf value or, in general, the α-binomial semivalues are of the same initial types.

Definition 5. *Let us consider cooperative games v defined on a given finite set of players N. A structure of coalition blocks in the player set is a partition of N, $B = \{B_1, \ldots, B_m\}$. With $\mathcal{B}_N$ we denote the set of all coalition structures defined in N. A solution for cooperative games with coalition structure is a function $\Psi : G_N \times \mathcal{B}_N \rightarrow \mathbb{R}^N$ that assigns a payoff to each player, $\Psi[v; B] = (\Psi_1[v; B], \ldots, \Psi_n[v; B])$.*

Let us suppose that two semivalues ψ^n and φ^n are defined on games with n players (eventually $\varphi^n = \psi^n$). The consideration of induced semivalues allows us to define a concept of *mixed modified semivalue for games with coalition structure* following a similar process to which Owen uses to derive the coalition value [Owe77] from the Shapley value [Sha53] or the modified Banzhaf value for games with coalition structure [Owe81] from the Banzhaf value [Ban65, Owe75].

Theorem 4. *Let v be a game on N and let ψ^n, φ^n be two semivalues defined on G_N with respective weighting coefficients $(p_s^n)_{s=1}^n$ and $(q_s^n)_{s=1}^n$. Given a coalition structure $B = \{B_1, B_2, \ldots, B_m\}$, the payoff to any player i in a coalition block $B_j \in B$ according to the mixed semivalue ψ^n / φ^n modified by B is*

$$\left(\psi^n / \varphi^n\right)_i[v; B] = \sum_{S \subseteq B_j \setminus \{i\}} \sum_{T \subseteq M \setminus \{j\}} q_{s+1}^{b_j} p_{t+1}^m \left[v\left(\bigcup_{t' \in T} B_{t'} \cup S \cup \{i\} \right) - v\left(\bigcup_{t' \in T} B_{t'} \cup S \right) \right],$$

where $p_{t+1}^m = \sum_{h=0}^{n-m} \binom{n-m}{h} p_{t+1+h}^n$ $(t = |T|)$ and $q_{s+1}^{b_j} = \sum_{h=0}^{n-b_j} \binom{n-b_j}{h} q_{s+1+h}^n$ $(s = |S|)$.

3 Computation Procedure of Mixed Modified Semivalues

Definition 6. *Let $B = \{B_1, \ldots, B_m\}$ be a coalition structure in N and let $M = \{1, \ldots, m\}$ be the set of classes in N according to the coalition structure B. From the MLE f_v of game v, a modified multilinear extension for each coalition block $B_j \in B$ can be obtained by means of the following rules:*

(1) For each $t \in M$, $t \neq j$, and each $u \in B_t$ replace in f_v the variable x_u with y_t.

(2) In the above function, reduce all exponents that appear in y_t to 1, i.e., replace y_t^r $(r > 1)$ with y_t, obtaining another MLE

$$f_{v,j}(x_k, y_t) \quad k \in B_j \text{ and } t \in M \setminus \{j\}.$$

Theorem 5. *Let us assume that $\{\psi^n_{\alpha_k}\}^n_{k=1}$ is a reference system of binomial semivalues in $Sem(G_N)$. If v is a cooperative game on N, ψ^n and φ^n are two semivalues on G_N with respective expressions $\psi^n = \sum_{k=1}^n \lambda_k \psi^n_{\alpha_k}$, $\varphi^n = \sum_{l=1}^n \widetilde{\lambda}_l \psi^n_{\alpha_l}$ and $B = \{B_1, \ldots, B_m\}$ is a structure of coalition blocks defined in N, then the allocation to each player i in block B_j according to the modified solution ψ^n/φ^n can be computed by means of the following expression*

$$(\psi^n/\varphi^n)_i[v; B] = \Lambda^t\, A(i)\, \widetilde{\Lambda},$$

where the matrix $A(i) = (a_{kl}(i))$ depends on the reference system of $Sem(G_N)$ and can be obtained from the MLE of block B_j,

$$a_{kl}(i) = (\psi^n_{\alpha_k}/\psi^n_{\alpha_l})_i[v; B] = \frac{\partial f_{v,j}}{\partial x_i}(\overline{\alpha}_l, \overline{\alpha}_k), \quad 1 \le k, l \le n.$$

Λ and $\widetilde{\Lambda}$ are, respectively, the matrices of the components of semivalues ψ^n and φ^n in the reference system $\{\psi^n_{\alpha_k}\}^n_{k=1}$: $\Lambda^t = (\lambda_1 \; \cdots \; \lambda_n)$, $\widetilde{\Lambda}^t = (\widetilde{\lambda}_1 \; \cdots \; \widetilde{\lambda}_n)$.

Acknowledgements

Research partially supported by Grant MTM 2006-06064 of the Education and Science Spanish Ministry and the European Regional Development Fund and Grant SGR 2005-00651 of the Catalonia Government (*Generalitat de Catalunya*).

References

[AG03]   Amer, R., Giménez, J.M.: Modification of semivalues for games with coalition structures. Theory and Decision **54**, 185–205 (2003).

[Ban65]   Banzhaf, J.F.: Weighted voting doesn't work: A mathematical analysis. Rutgers Law Review **19**, 317–343 (1965).

[Dra99]   Dragan, I.: Potential and consistency for semivalues of finite cooperative TU games. International Journal of Mathematics, Game Theory and Algebra **9**, 85–97 (1999).

[DNW81]   Dubey, P., Neyman, A., Weber, R.J.: Value theory without efficiency. Mathematics of Operations Research **6**, 122–128 (1981).

[Gim01]   Giménez, J.M.: Contributions to the study of solutions for cooperative games. Ph.D. Thesis. Technical University of Catalonia, Spain (2001).

[Owe72]   Owen, G.: Multilinear extensions of games. Management Science **18**, 64–79 (1972).

[Owe75]   Owen, G.: Multilinear extensions and the Banzhaf value. Naval Research Logistics Quarterly **22**, 741–750 (1975).

[Owe77]   Owen, G.: Values of games with a priori unions. In: Henn, R., Moeschlin, O. (eds) Essays in Mathematical Economics and Game Theory. Springer-Verlag, pp. 76–88 (1977).

[Owe81]   Owen, G.: Modification of the Banzhaf-Coleman index for games with a priori unions. In: Holler, M.J. (ed) Power, Voting and Voting Power. Physica-Verlag, pp. 232–238 (1981).

[Sha53]   Shapley, L.S.: A value for n-person games. In: Kuhn, H.W., Tucker, A.W. (eds) Contributions to the Theory of Games II. Princeton University Press, pp. 307–317 (1953).

Time-Varying Grids for Gas Dynamics

F. Coquel[1], Q.L. Nguyen[2], M. Postel[1], and Q.H. Tran[2]

[1] Université Pierre et Marie Curie-Paris6, UMR 7598 LJLL, F-75005 France;
CNRS, UMR 7598 LJLL, Paris, F-75005 France
coquel@ann.jussieu.fr, postel@ann.jussieu.fr

[2] Département Mathématiques Appliquées, Institut Français du Pétrole, 1 et 4
avenue de Bois-Préau, 92852 Rueil-Malmaison Cedex, France
q-long.nguyen@ifp.fr, q-huy.tran@ifp.fr

Summary. In the context of offshore oil production, we are interested in accurate and fast computation of two-phase flows in pipelines. A one-dimensional model of hyperbolic equations is solved numerically by an explicit Lagrange-projection method. This chapter shows that adaptive multiresolution techniques can speed up the computation significantly. Even more so when local time-stepping enhancement is used.

1 Modeling of the Physical Problem

In this short chapter we restrict ourselves to a homogeneous model for two-phase flows. The density ρ, velocity u and the gas mass fraction Y of the mixture of oil and gas are related through a PDE system

$$
\begin{cases}
\partial_t(\rho) \ + \partial_x(\rho u) \ = 0, \\
\partial_t(\rho Y) + \partial_x(\rho Y u) \ = 0, \\
\partial_t(\rho u) \ + \partial_x(\rho u^2 + P) = 0
\end{cases}
\tag{1}
$$

where the thermodynamical closure law $P(\rho, Y)$ can be in practice very costly to evaluate. This system is hyperbolic with three distinct eigenvalues $u - c < u < u+c$. The intermediate eigenvalue corresponds to the slow transport wave and will linearly degenerate, the two others are much bigger and correspond to genuinely nonlinear acoustic waves.

Improving the previous numerical treatments of the system (1) (see [1,5]), the scheme we present here enables the obtention of a maximum principle on the density and gas mass fraction and it can be expressed as a flux scheme, which makes local time-stepping techniques applicable. The main idea consists in decomposing the flux in an acoustic part – associated with the genuinely nonlinear waves and a transport part associated to the linearly degenerated waves. This is introduced for instance in [6] using Lagrangian coordinates for gas dynamics and detailed in [4] in the case of our complex system of equations. Eventually the Lagrangian step where we deal with the acoustic part of the flux will be treated implicitly, which will enable us to use a larger time step. The slow transport phenomenon will still be treated explicitly for better accuracy. In this chapter, we concentrate on the adaptive grid and local time-stepping extensions and restrict ourselves to the explicit version of the scheme.

2 Numerical Scheme

The Lagrange-projection (LP) splitting is performed at the numerical level in a two-step scheme that we briefly present in this section referring to [4] for the details. In the framework of adaptive schemes on nonuniform grids we discretize the domain in N cells $\Omega_i = [x_{i-1/2}, x_{i+1/2}]$ of size Δx_i such that $\sum_{i=0}^{N-1} \Delta x_i = L$. We denote by $\mathbb{U}_i^n = (\tau_i^n, Y_i^n, u_i^n)$ the solution on cell Ω_i at time n, and by $\mathbb{U}_i^{n\sharp} = (\tau_i^{n\sharp}, Y_i^{n\sharp}, u_i^{n\sharp})$ the solution at the end of the Lagrangian step.

Explicit scheme for the Lagrangian step gives

$$\tau_i^{n\sharp} = \tau_i^n + \Delta t \frac{\tilde{u}_{i+\frac{1}{2}}^n - \tilde{u}_{i-\frac{1}{2}}^n}{\rho_i^n \Delta x_i}, \quad Y_i^{n\sharp} = Y_i^n, \quad u_i^{n\sharp} = u_i - \Delta t \frac{\tilde{P}_{i+\frac{1}{2}}^n - \tilde{P}_{i-\frac{1}{2}}^n}{\rho_i^n \Delta x_i},$$

where $\tilde{u}_{i-\frac{1}{2}}^n$ and $\tilde{P}_{i-\frac{1}{2}}^n$ are built from the solution of the Riemann problem between states $\mathbb{U}_{i-1}^n$ and $\mathbb{U}_i^n$:

$$\tilde{P}_{i-\frac{1}{2}}^n = \frac{P_{i-1}^n + P_i^{\ n}}{2} - a \frac{u_i^n - u_{i-1}^n}{2}, \qquad \tilde{u}_{i-\frac{1}{2}}^n = \frac{u_{i-1}^n + u_i^n}{2} - \frac{P_i^n - P_{i-1}^n}{2a}. \quad (2)$$

In (2), a is a stabilizing coefficient coming from the relaxation formulation of problem (1), as described in [2].

In the Euler projection step we advect the intermediate conservative state $W^{n\sharp} = (\rho^{n\sharp}, (\rho Y)^{n\sharp}, (\rho u)^{n\sharp})^{\mathrm{T}}$ with the edge velocities $\tilde{u}_{i\pm\frac{1}{2}}^n$. Combining the two steps together provides the locally conservative flux formulation

$$W_i^{n+1} = W_i^n - \frac{\Delta t}{\Delta x_i} \left(F_{i+\frac{1}{2}}^{n\sharp} - F_{i-\frac{1}{2}}^{n\sharp} \right), \quad (3)$$

where

$$F_{i-\frac{1}{2}}^{n\sharp} = (0, 0, \tilde{P}_{i-\frac{1}{2}}^n)^{\mathrm{T}} + (\tilde{u}_{i-\frac{1}{2}}^n)^+ W_{i-1}^{n\sharp} + (\tilde{u}_{i-\frac{1}{2}}^n)^- W_i^{n\sharp}. \quad (4)$$

In the above definition of the fluxes, the superscript $(.)^+$ (respectively $(.)^-$) denotes the positive (respectively negative) part. Imposing maximum principle on ρ and Y leads to the following stability criterion

$$\Delta t \leq \min_i \frac{\min(\Delta x_i, \Delta x_{i-1})}{\left| \tilde{u}_{i-\frac{1}{2}}^n \right|}. \quad (5)$$

3 Adaptive Multiresolution

In the context of hyperbolic equations, where the solutions exhibit localized singularities, it is of course natural to discretize the solution finely in the region of these singularities and more coarsely elsewhere where it is smooth. In answer to this observation, we have adapted the multiresolution techniques established in [3] and based on ideas introduced in the context of systems of conservation laws by Harten at the beginning of the nineties. The multiscale analysis of the solution is used to design an adaptive grid by selecting the correct level out of a hierarchy of nested grids according to the local smoothness of the solution. This non-uniform grid evolves with time, with a strategy based on the prediction of the displacement and formation of the singularities in the solution. The wavelet basis used to perform the multiscale analysis enables to reconstruct the solution at any time back to the finest level of discretization, within an error tolerance controlled by a threshold parameter.

The coupling of multiresolution with a semi-implicit Euler relaxation scheme is detailed in [5] for the non-drift model and in [1] for the complete model with drift and friction. The adaptation of the method to our new Lagrange-projection scheme is straightforward, at least in the explicit case, and we present here some simulation, referring to [4] for the details of the implementation.

The test case consists of a Riemann problem set in a 16 km-long pipeline. At the initial time the density of the mixture is $400 \, \mathrm{kg \, m^{-3}}$ until $x{=}8 \, \mathrm{km}$ and $500 \, \mathrm{kg \, m^{-3}}$ beyond. The gas mass fraction is respectively 0.4 and 0.2 and the speed is respectively -10 and $+10 \, \mathrm{m \, s^{-1}}$. The contact discontinuity moves slowly towards the right at a speed $20 \, \mathrm{m \, s^{-1}}$ given by Rankine Hugoniot while two acoustic waves are visible on the density and speed components moving in opposite directions at roughly $\pm 254 \, \mathrm{m \, s^{-1}}$. Figure 1 displays the density, gas mass fraction and velocity fields obtained using the Lagrange-projection method on a uniform grid of 8,000 cells, along with the multiresolution solution, obtained using a hierarchy of 7 levels. On each graph the straight line indicates the initial solution, the dash line the uniform grid solution at time $t = 15\mathrm{s}$, and the dotted line the multiresolution solution at time $t = 15\mathrm{s}$. The crosses $\times$ indicate the level of resolution used locally at this final time. It is numbered from 1 to 7 on the right axis. The uniform and multiresolution solutions are basically undistinguishable even on the gas mass fraction graph which is zoomed-in in the unique region of variation. This computation is done by throwing away all details in the multiscale basis that fall below 1‰. For this threshold ratio, we gain a factor 8 in CPU between the uniform and adaptive computations, and the number of calls to the state law (in our case $P(\rho, \rho Y) = a_g^2(\rho_l \rho Y / \rho_l - \rho(1 - Y)))$ is 27 times less in the adaptive case. This means that when we use the scheme for realistic test cases with expensive state laws, the multiresolution will be more advantageous yet.

The fourth graph labelled (d) on the figure displays the error with respect to the uniform solution and also a reference solution computed on a four times

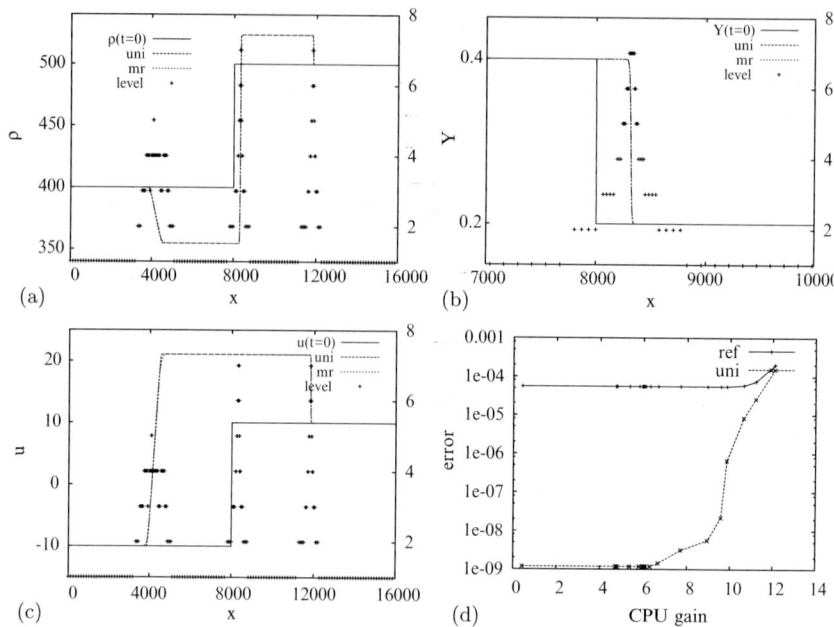

Fig. 1. Multiresolution for the Lagrange-projection scheme (**a**) $\rho(x, t = 15)$ (**b**) $Y(t = 15)$ (**c**) $u(x, t = 15)$ (**d**) error

finer uniform grid, as a function of the CPU gain. Each point on the two curves corresponds to different threshold ratio in the multiscale smoothness analysis. This parameter study shows that the error introduced by the multiresolution is negligible compared to the discretization error for CPU gains up to 10.

In the previous simulation the time step is dictated by the size of the smallest cell in the adaptive grid which enters into the stability condition (5). However, an important advantage of the Lagrange-projection scheme is that it is locally conservative and can therefore be used in the framework of a local time-stepping multiresolution method such as the one proposed in [7]. It is impossible in the allotted space to describe the algorithm; the general idea is to define now a macro time step which will be used on the cells of the adaptive grid belonging to the coarsest level. The solution on the cells belonging to the finer resolution levels is advanced using intermediate time steps, in order to synchronize the solution at the end of each macro time step.

We present in Fig. 2 the density fields and adaptive grids of a very promising simulation. Using first-order fluxes and constant global time step throughout the simulation we get a factor of 5 in CPU gain with the standard multiresolution and a factor of 15 using local time-stepping.

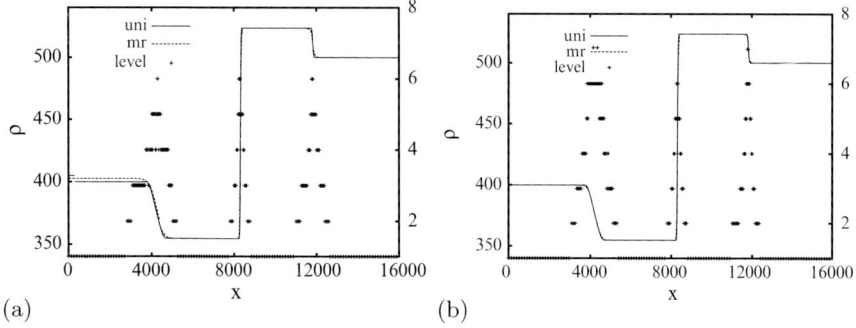

Fig. 2. Density field and adaptive grid using (**a**) multiresolution (**b**) local time-stepping enhancement

Acknowledgements

This work was supported by the Ministère de la Recherche under grant ERT-20052274: *Simulation avancée du transport des hydrocarbures* and by the Institut Français du Pétrole.

References

1. N. Andrianov, F. Coquel, M. Postel, and Q. H. Tran. A relaxation multiresolution scheme for accelerating realistic two-phase flows calculations in pipelines. *Int. J. for Num. Meth. in Fluids.* published online Dec 2006.
2. M . Baudin, C. Berthon, F. Coquel, R. Masson, and Q. H. Tran. A relaxation method for two-phase flow models with hydrodynamic closure law. *Numer. Math.*, 99:411–440, 2005.
3. A. Cohen, S. M. Kaber, S. Müller, and M. Postel. Fully adaptive multiresolution finite volume schemes for conservation laws. *Math. Comp.*, 72(241):183–225 (electronic), 2003.
4. F. Coquel, Q. L. Nguyen, M. Postel, and Q. H. Tran. Adaptive multiresolution and local time stepping for implicit-explicit Lagrange-projection schemes. Work in progress.
5. F. Coquel, M. Postel, N. Poussineau, and Q. H. Tran. Multiresolution technique and explicit-implicit scheme for multicomponent flows. *J. of Num. Math.*, Vol.14, No.3, pp. 187-216, 2006.
6. E. Godlewski and P.A. Raviart. *Numerical approximation of hyperbolic systems of conservation laws.* Springer Verlag, 1996.
7. S. Müller and Y. Stiriba. Fully adaptive multiscale schemes for conservation laws employing locally varying time stepping. *Journal for Scientific Computing*, pp. 493-531, Volume 30, Number 3, 2007. Report No. 238, IGPM, RWTH Aachen.

Meshless Poisson Problems in the Finite Pointset Method: Positive Stencils and Multigrid

Benjamin Seibold

University of Kaiserslautern, Germany
`seibold@mathematik.uni-kl.de`

Summary. The finite pointset method is a meshless Lagrangian particle method. In the application to incompressible viscous fluid flow the solution of Poisson problems on the cloud of particles is a fundamental subproblem. A valuable property of finite difference approximations to the Laplace operator is the positivity of stencils, i.e., all weights of neighboring points are positive. Classical least squares approaches do not guarantee positive stencils. We present a new approach, based on linear minimization, which enforces positivity of stencils and additionally yields a minimal number of nonzero stencil entries. The resulting system matrices are M-matrices, which is of particular interest with respect to multigrid solvers.

1 Introduction

The *finite pointset method* (FPM) was introduced [3] as a generalization of *smoothed particle hydrodynamics* (SPH) [6]. It is a meshless Lagrangian particle method, i.e., all computations are performed on a point cloud which is moving with the flow. Incompressible viscous flows can be computed with the FPM [4], using a projection method [1]. This requires the solution of at least one Poisson problem in each time step. Since the particles move with the flow, the Poisson equation has to be solved on the point cloud. In Sect. 2 we outline the general meshless finite difference approach. Classical least squares methods are presented in Sect. 3. In Sect. 4 we present a new approach, which yields minimal positive Laplace stencils. We present in Sect. 5 that the resulting system matrices are M-matrices, which is of advantage for multigrid solvers. Section 6 shows numerical results.

2 Meshless Finite Differences for Poisson Equation

Consider a smooth function u and a point cloud on a domain $\Omega \subset \mathbb{R}^d$. To an interior point $\mathbf{x}_0$ consider all points $\mathbf{x}_1, \ldots, \mathbf{x}_m$ in a circular neighborhood, i.e., $\|\mathbf{x}_i-\mathbf{x}_0\|_2 < r$. Define $\bar{\mathbf{x}}_i = \mathbf{x}_i-\mathbf{x}_0$. The function value at each neighboring point can be expressed by a Taylor expansion

$$u(\mathbf{x}_i) = u(\mathbf{x}_0) + \nabla u(\mathbf{x}_0) \cdot \bar{\mathbf{x}}_i + \tfrac{1}{2}\nabla^2 u(\mathbf{x}_0) : \left(\bar{\mathbf{x}}_i \cdot \bar{\mathbf{x}}_i^{\mathsf{T}}\right) + e_i. \tag{1}$$

The colon denotes the matrix scalar product $A : B = \sum_{i,j} A_{ij}B_{ij}$, and e_i is the error in the expansion, which is of order $e_i = O(r^3)$. A linear combination with coefficients $(\mathfrak{s}_0,\ldots,\mathfrak{s}_m)$ equals

$$\sum_{i=0}^{m} \mathfrak{s}_i u(\mathbf{x}_i) = u(\mathbf{x}_0) \underbrace{\left(\sum_{i=0}^{m} \mathfrak{s}_i\right)}_{=r_0} + \nabla u(\mathbf{x}_0) \cdot \underbrace{\left(\sum_{i=1}^{m} \mathfrak{s}_i \bar{\mathbf{x}}_i\right)}_{=\mathbf{r}_1} \tag{2}$$

$$+ \nabla^2 u(\mathbf{x}_0) : \underbrace{\left(\frac{1}{2}\sum_{i=1}^{m} \mathfrak{s}_i \left(\bar{\mathbf{x}}_i \cdot \bar{\mathbf{x}}_i^{T}\right)\right)}_{=R_2} + \underbrace{\left(\sum_{i=1}^{m} \mathfrak{s}_i e_i\right)}_{=e}$$

For expression (2) to approximate the Laplacian, the above terms must satisfy $r_0 = 0$, $\mathbf{r}_1 = 0$ and $R_2 = 2I$. Such a stencil $\mathfrak{s}$ in denoted *consistent*. The total error e shall be small. Due to the constant constraint $r_0 = 0$, the central entry $\mathfrak{s}_0$ equals minus the sum of all other entries. The linear and quadratic constraints can be written as a linear system $V \cdot \mathfrak{s} = \mathbf{b}$, where $V \in \mathbb{R}^{k\times m}$ is the Vandermonde matrix, and $\mathfrak{s} \in \mathbb{R}^m$ is the sought stencil. In 2d the system reads as

$$V = \begin{pmatrix} \bar{x}_1 & \cdots & \bar{x}_m \\ \bar{y}_1 & \cdots & \bar{y}_m \\ \bar{x}_1\bar{y}_1 & \cdots & \bar{x}_m\bar{y}_m \\ \bar{x}_1^2 & \cdots & \bar{x}_m^2 \\ \bar{y}_1^2 & \cdots & \bar{y}_m^2 \end{pmatrix}, \quad \mathbf{b} = \begin{pmatrix} 0 \\ 0 \\ 0 \\ 2 \\ 2 \end{pmatrix} \tag{3}$$

One has $k = \frac{d(d+3)}{2}$ constraints. In general, r is chosen large enough such that $m > k$. A minimization problem is formulated to single out a unique stencil.

3 Least Squares Methods

Classical approaches formulate a quadratic minimization problem to select a unique consistent stencil. Examples are the local and the moving least squares method. A comparison of such methods for the meshless approximation of the Poisson equation is given in [7, Chap. 5]. The local least squares method, for instance, yields the following quadratic minimization (QM) formulation

$$\min \sum_{i=1}^{m} \frac{\mathfrak{s}_i^2}{w_i}, \text{ s.t. } V \cdot \mathfrak{s} = \mathbf{b}. \tag{4}$$

The weights decrease with the distance to the central point, e.g., $w_i = \|\bar{\mathbf{x}}_i\|_2^{-4}$. The solution can be given explicitly in terms of Lagrange multipliers ($\mathfrak{s} = WV^T \left(VWV^T\right)^{-1} \mathbf{b}$), where W is a diagonal matrix containing the w_i. Approximating the Poisson equation by QM requires for each interior point a small (9×9 in 3d) linear system to be set up and solved.

Of particular interest for the Laplace operator are positive stencils, i.e., all non-central entries are non-negative. QM formulations do in general not yield positive stencils: Consider a central point $\mathbf{x}_0 = (0,0)$, and six neighboring points $\mathbf{x}_i = (\cos(\frac{\pi}{2}t_i), \sin(\frac{\pi}{2}t_i))$, where $(t_1, \ldots, t_6) = (0, 1, 2, 3, 0.1, 0.2)$. As the points lie of the unit circle, the weights w_i do not influence the result. QM yields the solution $\mathfrak{s} = (0.846, 1.005, 0.998, 1.003, 0.312, -0.164)$, the stencil is not positive. However, there is a positive stencil solution, namely the regular five-point stencil $\mathfrak{s} = (1, 1, 1, 1, 0, 0)$.

4 Minimal Positive Stencils

As a new approach we enforce positive stencils, i.e., we search for solutions contained in the polyhedron $P = \{\mathfrak{s} \in \mathbb{R}^m : V \cdot \mathfrak{s} = \mathbf{b}, \; \mathfrak{s} \geq 0\}$. Point clouds exist which do not admit a positive stencil: Consider a central point in the origin, and all other points in the right half plane ($x_i > 0$). The first row of system (3) cannot be satisfied with all $\mathfrak{s}_i \geq 0$ in any feasible way. In [7] conditions on the geometry are provided under which positive stencils exist: If the maximum hole size in the point cloud is d, then considering all points in a radius $r = \gamma d$ in the LM problem (5) guarantees the existence of a positive solution ($\gamma = \sqrt{4 + 2\sqrt{2}} \approx 2.6$ in 2d, $\gamma = \sqrt{7 + 2\sqrt{6}} \approx 3.4$ in 3d). In simulations, the maximum hole size is controlled by particle management. Points close to the domain boundary require special treatment. Now we select a unique stencil by formulating a linear minimization (LM) problem

$$\min \sum_{i=1}^{m} \frac{\mathfrak{s}_i}{w_i}, \; \text{s.t. } V \cdot \mathfrak{s} = \mathbf{b}, \; \mathfrak{s} \geq 0 \, . \tag{5}$$

As for QM, the weights are chosen to decay with distance to the central point. The sign constraints impose bounds on the stencil entries, thus making the LM formulation feasible. The following facts support the LM approach:

- *L-matrix property*: If a positive stencil $\mathfrak{s} \geq 0$ can be obtained for every point, then the resulting system matrix is an L-matrix.
- *Error in Taylor expansion*: The error in approximation (2) is $\sum_{i=1}^{m} \mathfrak{s}_i e_i$, where e_i are the local errors of the Taylor expansion. In the worst case, when the local errors accumulate, minimizing the total error equals the LM formulation (5) with distance weight function $w_i = \|\bar{\mathbf{x}}_i\|^{-3}$.
- *Minimal stencil*: Problem (5) is a linear program in standard form. If the constraints admit a solution, then due to the *fundamental theorem of linear*

programming [2] there is a basic solution, in which only k of the m stencil entries $\mathfrak{s}_i$ are different from zero, where k is the number of constraints.

The resulting approach is denoted *minimal positive stencil* (MPS) method.

5 M-Matrices and Multigrid Solvers

The MPS guarantees a positive stencil for every approximation point, thus yielding an L-matrix structure. Under relaxed conditions on the matrix graph the system matrices are M-matrices. Note that the finite difference matrices are in general not symmetric, thus the M-matrix property does not imply positive definiteness. Still, it has various beneficial implications. M-matrices are inverse-positive, i.e., they satisfy a discrete maximum principle. A Gauß-Seidel iteration is guaranteed to converge, which is of interest for multigrid solvers. In [5] it is shown that the M-matrix structure is sufficient for the convergence of a two-grid AMLI method. The main advantage of the MPS matrices, however, is their sparsity. In 3d problems, least squares approaches yield about 50 nonzero entries per row, while the LM approach yields merely 10 entries. Since linear solvers rely on applying the sparse matrix to a vector, computation speed increases significantly. In addition, less memory is consumed.

6 Numerical Results

Figure 1 shows the run times of generating the Poisson system matrices in dependence on the number of unknowns, on the one hand with a least squares approach (solid), on the other hand with the MPS method (dashed). The latter is slightly more expensive. However, this higher expense is more than made up for with the speedup in solving the linear system. Figure 2 shows the run times for solving the arising systems with a BiCGstab method. Figure 3 shows the corresponding run times when solved with SAMG, an algebraic multigrid solver developed by the *Fraunhofer Institut für Algorithmen und Wissenschaftliches Rechnen*. The dash-dotted line shows the computation times for MPS matrices without zeros from non-basis entries removed. It can

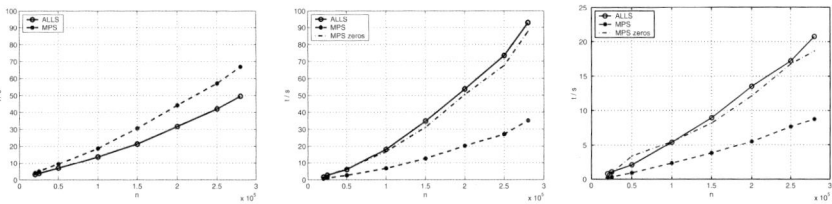

Fig. 1. Matrix generation   **Fig. 2.** Solve with BiCG   **Fig. 3.** Solve with SAMG

be observed that the M-matrix structure itself does not result in a higher convergence rate. The sparsity of the MPS matrices, however, yields a significant speedup compared to least squares approaches. One can further observe that multigrid solvers run fast and efficient on the arising MPS matrices.

7 Conclusions and Outlook

Classical least squares finite difference approaches do not guarantee positive stencils when approximating the Laplace operator. The presented MPS approach does. Additionally, it yields minimal stencils. The approach can be interpreted as a method to select a suitable minimal number of neighboring points. The resulting matrices are M-matrices, which can be efficiently solved with algebraic multigrid approaches. On the other hand, the MPS method has various drawbacks: The resulting matrices are not reciprocal. Also, the selected stencil does not depend continuously on the point positions. This aspect, however, could not be observed to have a significant negative impact. Numerical experiments indicate that least squares matrices are, although not M-matrices, often times inverse positive, which can be an explanation for their good convergence rate. The speedup due to the sparsity of the MPS matrices, however, cannot be beaten by classical least squares approaches. A deeper discussion of these aspects is provided in [7].

Acknowledgements

We would like to thank Axel Klar, Jörg Kuhnert, Christian Mense, Helmut Neunzert, Klaus Stüben, and Sudarshan Tiwari for helpful discussions and comments. This work was supported by the DFG Graduiertenkolleg "Mathematik und Praxis".

References

1. A.J. Chorin. Numerical solution of the Navier-Stokes equations. *Math. Comput.*, 22:745–762, 1968.
2. V. Chvátal. *Linear programming*. W. H. Freeman and Company, 1983.
3. J. Kuhnert. *General Smoothed Particle Hydrodynamics*. Dissertation, Department of Mathematics, University of Kaiserslautern. Shaker, 1999.
4. S. Manservisi and S. Tiwari. Modeling incompressible flows by least-squares approximation. *The Nepali Math. Sc. Report*, 20, 1&2:1–23, 2002.
5. C. Mense and R. Nabben. On algebraic multilevel methods for nonsymmetric system – Convergence results. *SIAM J. Numer. Anal*, to appear, 2006.
6. J.J. Monaghan. An Introduction to SPH. *Comput. Phys. Commun.*, 48:89–96, 1988.
7. B. Seibold. *M-matrices in meshless finite difference methods*. Dissertation, Department of Mathematics, University of Kaiserslautern. Shaker, 2006.

Basics of a Differential-Geometric Approach to Diffusion: Uniting Lagrangian and Eulerian Models on a Manifold

Michael M. Tung

Instituto de Matemática Multidisciplinar, Universidad Politécnica de Valencia, Spain
mtung@imm.upv.es

Summary. We combine the Lagrangian and Eulerian models of linear diffusion phenomena in a coherent differential-geometric framework. This approach is applied to the diffusion–advection equation and its implications are discussed.

1 Introduction

Fluid dynamics and diffusion phenomena govern the behavior of liquids and gases alike, and are a field of broad impact in physics, engineering, and meteorology. For their full understanding and conceptual development adequate mathematical tools and techniques are essential. Differential-geometric methods are particularly well-suited for this task.

In this work, we deal with one interesting aspect of fluid behavior, focussing on an inherently covariant treatment of the diffusion–advection equation on a smooth manifold. For this purpose, we introduce a general Lagrangian density $\mathcal{L}$ for a dissipative system describing the diffusion process over a properly defined configuration bundle. This will make it possible to derive the corresponding diffusion equations (equations of motion) on the manifold via a simple variational principle. Further, we can show that underlying symmetries of $\mathcal{L}$ directly relate to certain symmetry properties of the associated diffusion tensor.

The discussion continues with various other implications, dealing for example with energy and mass conservation, and the static-metric assumption in connection with the divergence of the fluid on the manifold. This leads to Eulerian and Lagrangian dispersion models, which both solve the same diffusion–advection equation with identical solutions, but in different coordinate representations. We demonstrate how both of these models are combined in a coherent differential-geometric framework, and finally conclude with an outlook on practical applications of this approach.

2 The Geometry of Diffusion

Diffusion is the spontaneous intermingling of particles or fluid elements of two or more distinguishable substances as a result of random motion. Advection is usually defined as a horizontal movement of the diffusing substance due to local changes in the properties of the diffusing medium.

The diffusion–advection equation which governs these phenomena is a non-homogeneous parabolic partial differential equation:

$$\frac{\partial C}{\partial t} = \nabla \cdot \left(D \, \nabla C \right) - \mathbf{u} \cdot \nabla C + S, \tag{1}$$

where $C : \mathbb{R}^3 \times \mathbb{R} \to \mathbb{R}$ is the concentration of the diffusing substance and $D : \mathbb{R}^3 \times \mathbb{R} \to M_{3\times3}(\mathbb{R})$ a general diffusion matrix. In this description, we also include a velocity field $\mathbf{u} : \mathbb{R}^3 \times \mathbb{R} \to \mathbb{R}^3$ and the possibility for creation or destruction of the diffusing substance by a source density $S : \mathbb{R}^3 \to \mathbb{R}$. Note that the physical condition of energy conservation imposes S to be time-independent.

Many analytical or even numerical models dealing with the diffusion–advection (1) restrict themselves to the case where $\mathbf{u}$ is constant. In more realistic scenarios, however, the methods of differential geometry may help to tackle the general case considering a variable velocity field $\mathbf{u} = \mathbf{u}(\mathbf{x}, t)$.

In the Eulerian coordinate representation, one obtains the solution of the diffusion–advection equation with respect to a fixed grid. In the Lagrangian coordinate representation, on the other hand, the local coordinates move with the elements in the medium, so in principal it can be chosen such that the transport term $\mathbf{u}$ vanishes locally.

For this purpose, we need to set up a general differential-geometric framework with a smooth manifold M and a local vector field $u = X(p) \in T_p(M)$ (i.e. element of the associated tangent space) in every point $p \in M$.

In the following, it will be necessary to "geometrize" the diffusion equation, neglecting advection for the time being, from its Euclidean form

$$\frac{\partial C}{\partial t} = \nabla \cdot \left(D \, \nabla C \right) + S \tag{2}$$

to its curvilinear equivalent allowing for an underlying curved geometry

$$\dot{C} = \left(D^{ij} C_{;i} \right)_{;j} + S, \tag{3}$$

where $D : (T_p^* M)^2 \to \mathbb{R}$ is the molecular diffusion tensor and the semicolon denotes as usual the covariant derivative. Here, being (M, g) a Riemannian manifold with metric g, $T_p^* M$ is the associated dual tangent space in $p \in M$. Then, in local coordinates $x^i(p), i = 1, 2, 3$ (for the three-dimensional case), one has

$$D = D^{ij} \frac{\partial}{\partial x^i} \otimes \frac{\partial}{\partial x^j}. \tag{4}$$

The corresponding base space is $B = M \times \mathbb{R}_+$, representing space–time with coordinates (x^1, x^2, x^3, t). Then, concentration and source density are in general functions of type $C : B \to \mathbb{R}$ and $S : M \to \mathbb{R}$.

3 Lagrangian Formalism

For dissipative systems with diffusing fluids one requires an additional "mirror-image" concentration to compensate for the energy which would otherwise be removed [Mor53]. Hence, we introduce a mirror concentration $C^* : B \to \mathbb{R}$. The ambient space P is then described in terms of the parameters (C, C^*) (see e.g., [Lew03]). Then, the configuration space is defined as the product space $N = B \times P$ with coordinates $(x^1, x^2, x^3, t, C, C^*)$, representing all physical observables.

In classical mechanics, for the construction of the Lagrangian function $\mathcal{L}$, which describes a deterministic system completely, it suffices to define $\mathcal{L}$ on a tangent bundle TM (identical with the corresponding phase space). Figure 1 represents the tangent bundle TM consisting of manifold M and its tangent spaces T_pM for all $p \in M$.

For the diffusion case, we also require the partial derivatives of a configuration with respect to all space–time coordinates. This leads to a jet bundle J^1N with coordinates $(C, C^*, \dot{C}, \dot{C}^*, C_{;i}, C^*_{;i}) \simeq \mathbb{R}^{10}$ as natural extension of the tangent bundle.

The Lagrangian function for diffusion will then in general be a mapping

$$\mathcal{L} : J^1N \to \mathbb{R}, \tag{5}$$

and specifically we *define*

$$\mathcal{L} = -D^{ij}C_{;i}\,C^*_{;j} - \tfrac{1}{2}\big(\dot{C}C^* - C\dot{C}^*\big) + S\big(C + C^*\big). \tag{6}$$

With the additional constraints $D^* = D$ and $S^* = S$, the Lagrangian will also possess mirror symmetry, i.e., s $\mathcal{L}^* = \mathcal{L}$. It is not difficult to see that mirror

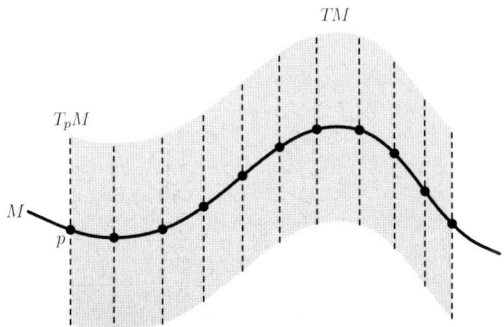

Fig. 1. Usual tangent bundle for geometric mechanics

symmetry relates to time reversal $t \leftrightarrow -t$, so that the mirror symmetry of $\mathcal{L}$ itself implies temporal forward–backward invariance of $\mathcal{L}$, i.e., C and C^* will only change their respective rôles.

The equations of motion are readily derived from the action integral over a bounded and closed set $V \subset M$

$$L = \int_V \mathcal{L} \sqrt{g}\, d\tau, \tag{7}$$

where $\sqrt{g}\, d\tau$ is the invariant volume element. After a lengthy but straightforward calculation, the variations $\delta L/\delta C = \delta L/\delta C^* = 0$ yield:

$$\dot{C} = \left(D^{ij} C_{;i} \right)_{;j} + S, \qquad \dot{C}^* = -\left(D^{ij} C^*_{;j} \right)_{;i} - S, \tag{8}$$

which is exactly (3) and its time-reversed counterpart. This shows that (6) indeed represents the Lagrangian function which fully describes diffusion phenomena.

4 Energy and Mass Conservation

For the Lagrangian (6) to represent a viable physical model, several requirements have to be fulfilled. Prime requisites are energy and mass conservation, which we show can in fact be derived from it.

Knowing the Lagrangian, a standard procedure of geometric mechanics is to apply Hamilton's formalism (see e.g. [Cal05]) and directly work with the Hamiltonian H, which represents the total energy of the system. Hence, energy conservation is expressed as $\dot{H} \equiv 0$. However, Noether's Theorem allows for a more elegant description: Symmetry properties of $\mathcal{L}$ manifest themselves as conservation laws. Here, the time invariance of $\mathcal{L}$ corresponds to energy conservation. In fact, assuming $D^{ij} = D^{ji}$ (isotropy) and adding (8) yields $\dot{C} + \dot{C}^* = 0$, and for an incompressible medium with $\partial \sqrt{g}/\partial t = 0$ it follows

$$\dot{L} = \frac{d}{dt} \int_V \mathcal{L} \sqrt{g}\, d\tau = \int_V \dot{\mathcal{L}} \sqrt{g}\, d\tau. \tag{9}$$

After substituting (8) into integral (9) and using vanishing boundary integrals over ∂V, one finds

$$\dot{L} = 0 \quad \text{and} \quad H = \int_V \left[D^{ij} C_{;i}\, C^*_{;j} - S(C + C^*) \right] \sqrt{g}\, d\tau, \tag{10}$$

where the H is readily obtained from a Legendre transformation and represents the constant total energy of the system. Note that the integrand of (10) contains a kinetic and potential energy contribution, displaying a quadratic term in the concentration gradients and a linear term in the concentrations, respectively.

Mass conservation can be demonstrated in a similar fashion. The first (physical) equation of motion (8) serves as a starting point. Taking its integral over the bounded and closed set $V \subset M$ gives

$$\int_V \left[\dot{C} - \left(D^{ij}C_{;i} \right)_{;j} - S \right] \sqrt{g} \, d\tau = 0, \tag{11}$$

which can be further manipulated by integration by parts and Gauss' Theorem with vanishing boundary conditions $(Cu)|_{\partial V} = 0$, meaning that either concentrations and/or velocities on the contours are negligible. Here, the vector field u arises by transforming Euclidean coordinates $x^i(p)$ at the point $p \in M$ to a new set of coordinates $x^i \mapsto \bar{x}^i$. Then, with the shorthand $\bar{u}^i = \dot{\bar{x}}^i$, it can be shown that

$$\partial \sqrt{g} / \partial t = \sqrt{g} \, \bar{u}^i_{;i} \tag{12}$$

which is the covariant equivalent of the Convection Theorem [Mey71].

A straightforward but tedious calculation gives for the total mass rate

$$\dot{M} = \frac{d}{dt} \int_V C \sqrt{g} \, d\tau - \oint_{\partial V} D^{ij}C_{;i} \sqrt{g} \, d\sigma_j + \int_V u^i C_{;i} \sqrt{g} \, d\tau - \int_V S \sqrt{g} \, d\tau = 0.$$

Each term has an immediate physical interpretation. The term $u^i C_{;i}$ accounts for the translation of the material due to a moving coordinate frame with local velocity u. This advection term naturally enters in the calculation by using (12). In the Lagrangian system, where we do not observe any relative motion of the medium, it is $u = 0$. On the other hand, for an observer in a particular Eulerian system, it is $u \neq 0$.

5 Conclusion and Outlook

We have presented a basic differential-geometric framework for the diffusion process of fluids on a manifold. This general coordinate-free formulation in terms of a fundamental Lagrangian on arbitrary geometries combines effectively Eulerian and Lagrangian models. Energy and mass are shown to be conserved. In the future, this framework could serve as an interesting starting point for numerical models of diffusion on curved surfaces or volumes.

References

[Cal05] Calin, O., Chang, D.-C.: Geometric Mechanics on Riemannian Manifolds, Birkäuser & Springer Science, Boston (2005).

[Lew03] Lew, A., Marsden, J.E., Ortiz, M., West, M.: Asynchronous variational integrators, *Arch. Ration. Mech. Anal.* **167**, 85–146 (2003).

[Mey71] Meyer, R.E.: Introduction to Mathematical Fluid Dynamics, John Wiley & Sons, New York (1971).

[Mor53] Morse, P.M., Feshbach, H.: Methods of Theoretical Physics, McGraw-Hill Publishing Company, New York (1953).

Diagnostic Modelling of Digital Systems with Binary and High-Level Decision Diagrams

Raimund Ubar, Jaan Raik, Helena Kruus, Harri Lensen, and Teet Evartson

Tallinn University of Technology, Department of Computer Engineering, Raja 15, 12618 Tallinn, Estonia
raiub@pld.ttu.ee

Summary. A novel hierarchical approach based on decision diagrams (DD) to modelling digital systems is introduced. Two new contributions are proposed: a new class of structurally synthesized binary DDs for modelling structural aspects of digital circuits, and DDs for high-level modelling of systems. Combination of both types of graphs allows to implement uniform formal approach to low- and high-level diagnostic modelling with increased efficiency of fault simulation and test generation for digital systems.

1 Introduction

The drawback of traditional multi-level and hierarchical approaches to diagnostic modelling of digital systems lies in the need of different languages and models for different levels. Most frequent examples are logic expressions for combinational circuits, state transition diagrams for finite state machines (FSM), abstract execution graphs, system graphs, instruction set architecture (ISA) descriptions, flow-charts, hardware description languages (HDL, VHDL, Verilog, etc.), Petri nets for system level description, etc. All these models need different manipulation algorithms and fault models which are difficult to merge in hierarchical test methods. Better opportunities for hierarchical diagnostic modelling of digital systems provide decision diagrams (DD) [1,2]. In this chapter, a multi-level method for diagnostic modelling digital systems with DDs is used. DDs serve as a basis for a general theory of test design for mixed-level representations of systems. The fault model defined on DDs represents a generalization of the classical stuck-at fault model [2].

2 Modelling Digital Systems by Binary Decision Diagrams

DDs can serve as a basis for a uniform approach to diagnostic modelling and test generation for mixed-level representations of systems, similarly as we use the Boolean algebra for the plain logic level.

In [2] structurally synthesized BDDs (SSBDD) as a special class of BDDs [1] was introduced to represent the topology of gate-level circuits in terms of signal paths. Unlike the traditional BDDs [1], SSBDDs directly support test generation for gate-level structural faults without representing these faults explicitly. The advantage of the SSBDD based approach is that the library of components is not needed for structural path activation. This is the reason why SSBDD based test generation do not depend on whether the circuit is represented on the gate level or on the macro-level whereas the macro means an arbitrary single-output gate-level subcircuit of the network. Moreover, the test generation procedures for SSBDDs can be easily generalized for higher level DDs to handle digital systems represented at higher levels [2].

In the following, we use graph-theoretical definitions instead of tradition-alite expressions [1] because all the procedures defined further for both types of DDs are based on the topological reasoning rather than on graph symbolic manipulations as traditionally in the case of BDDs.

Definition 1. A BDD that represents a Boolean function $y = f(Z), Z = (z_1, z_2, \ldots, z_n)$, is a directed acyclic graph $G_y = (M, \Gamma, Z)$ with a set of nodes M and a mapping Γ from M to M. A terminal node $m^T \in M^T = \{m^{T,0}, m^{T,1}\}$ is labelled by a constant $e \in \{0,1\}$, while nonterminal nodes $m \in M^N$ are labelled by variables $z \in Z$, and have exactly two successors. Denote by m^0 the successor of m for $z(m) = 0$ and m^1 is the successor of m for $z(m) = 1$. For $z(m) = e, e \in \{0,1\}$, we say the edge between nodes $m \in M$ and $m^e \in M$ is activated. Let all $z \in Z$ are assigned by a Boolean vector $X^t \in \{0,1\}^n$ to some value. The edges activated by X^t form an activated path $l(m_0, m^T) \subseteq M$ from the root node m_0 to one of the terminal nodes $m^T \in M^T$. We say that a BDD $G_y = (M, \Gamma, Z)$ represents a Boolean function $y = f(Z)$, iff for all the possible vectors $X^t \in \{0,1\}^n$ a path $l(m_0, m^T) \subseteq M$ is activated so that $y = f(X^t) = z(m^T)$.

Definition 2. Consider a BDD $G_y = (M, \Gamma, Z)$, where Z is the vector of literals of a function $y = P(Z)$ represented in the equivalent parenthesis form, the nodes $m \in M^N$ are labelled by $z(m)$, where $z \in Z$ and $| M | = | Z |$. The BDD is called a structurally synthesized BDD (SSBDD) iff there exists one-to-one correspondence between literals $z \in Z$ and nodes $m \in M^N$ given by the set of labels $\{z(m) \mid z \in Z, m \in M^N\}$, and iff for all the possible vectors $X^t \in \{0,1\}^n$ a path $l(m_0, m^T)$ is activated, so that $y = f(X^t) = z(m^T)$.

For synthesis of SSBDDs for a given gate network, the graph superposition procedure can be used [2]. Unlike the traditional BDDs [1], SSBDDs [1,6] support structural representation of gate-level circuits in terms of signal paths. By superposition of DDs [2,5,7], we can create SSBDDs with one-to-one correspondence between graph nodes and signal paths in the circuit. The one-to-one correspondence between nodes m in a SSBDD and paths $l(m)$ in the corresponding gate-level circuit is the direct result of the synthesis proce-dure of SSBDDs. Using SSBDDs, it is possible to rise from the gate-level to a higher macro level without loosing accuracy of representing gate-level signal paths.

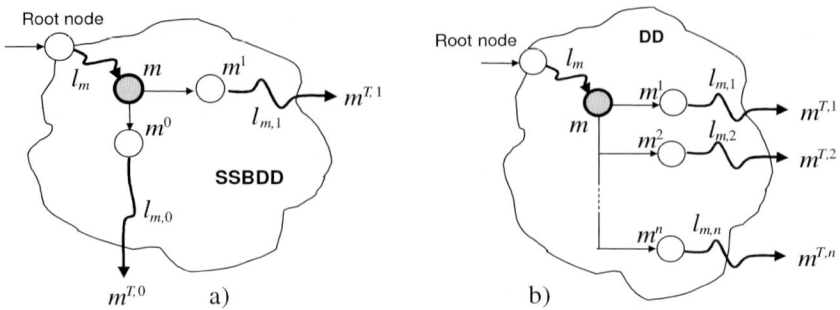

Fig. 1. Test generation for the node m with SSBDD

Test generation with SSBDDs. Consider a combinational circuit as a network of gates, which is partitioned into interconnected tree-like subcircuits (macros). Each macro is represented by a SSBDD where each node corresponds to a signal path from an input of the macro to its output.

Test generation for a node m in SSBDD G_y, which represents a function $y = f(Z)$ of a tree-like subcircuit (macro), is presented in Fig. 1a and is carried out by *Algorithm 1*:

1. Activate a path l_m from the root node of SSBDD to the node m.
2. Activate two paths $l_{m,e}$ consistent with l_m, where $e \in \{0, 1\}$, from m^e of m to the terminal nodes $m^{T,e}$.

All the values assigned to node variables (to variables of Z) build the local test pattern $T(Z, y)$ (input pattern of the macro) for testing the node m in G_y (for testing the corresponding path $l(m)$ on the output y of the given tree-like circuit).

3 Modelling Systems by a Single DD on Higher Levels

Test generation methods developed for SSBDDs have an advantage compared to other logic level methods. Namely, that they can be easily generalized to handle test generation at higher system levels. Consider a digital system $S = (Z, F)$ as a network of components where Z is the set of variables (Boolean, Boolean vectors or integers), which represent connections between components, inputs and outputs of the network. Denote by $X \subset Z$ and $Y \subset Z$, correspondingly, the subsets of input and output variables. $V(z)$ denotes the set of possible values for $z \in Z$, which are finite. Let F be the set of digital functions on Z : $z_k = f_k(z_{k,1}, z_{k,2}, \ldots, z_{k,p}) = f_k(Z_k)$ where $z_k \in Z, f_k \in F$ and $Z_k \subset Z$. Some of the functions $f_k \in F$, for the state variables $z \in Z_S TATE \subset Z$, are the next state functions.

Definition 3. A DD is a directed acyclic graph $G = (M, \Gamma, Z)$ where M is a set of nodes, Γ is a relation in M, and $\Gamma(m) \in M$ denotes the set of successor

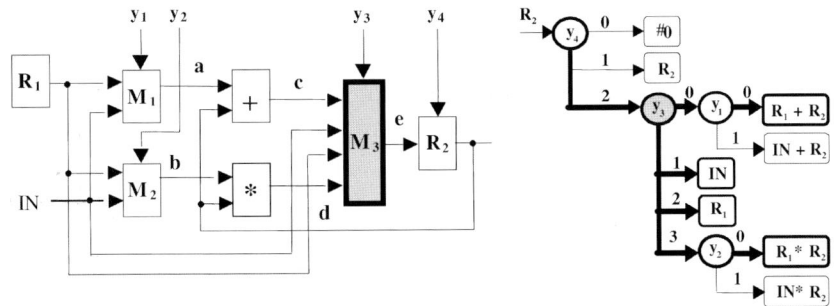

Fig. 2. Test generation for the node m with SSBDD

nodes of $m \in M$. The nodes m are labelled by $z(m)$. The labels can be either variables $z \in Z$, or algebraic expressions $f_m(Z(m)), Z(m) \subseteq Z$, or constants. For non-terminal nodes $m \in M^N$, an onto function exists between the values of $z(m)$ and the successors $m^e \in \Gamma(m)$ of m. The edge (m, m^e) which connects nodes m and m^e is called *activated* iff there exists an assignment $z(m) = e$. Activated edges, which connect m_i and m_j make up an *activated path* $l(m_i, m_j) \subseteq M$. A DD $G_{z,k}$ represents a function $z_k = f_k(z_{k,1}, z_{k,2}, \ldots, z_{k,p}) = f_k(Z_k), z_k \in Z$ iff for each value $v(Z_k) = v(z_{k,1}) \times v(z_{k,2}) \times \cdots \times v(z_{k,p})$, a full path in $G_{z,k}$ to $m^T \in M^T$ is activated, so that $z_k = z(m^T)$ is valid.

In Fig. 2, an RTL data-path and its DD is presented. Variables R_1, R_2 and R_3 represent registers, IN represents the input bus, integer variables y_1, y_2, y_3, y_4 represent control signals, M_1, M_2, M_3 are multiplexers, and the functions $R1 + R2$ and $R1 \times R2$ represent an adder and a multiplier, respectively. Each node in DD represents a subcircuit of the system (e.g. the nodes y_1, y_2, y_3, y_4 represent multiplexers and decoders). The DD describes the behaviour of the input logic of the register R_2.

In test pattern simulation, a path is traced in the graph, guided by the values of node variables until a terminal node is reached, similarly as in the case of SSBDDs. In Fig. 2 the result of simulating the vector $(y_1, y_2, y_3, y_4, R_1, R_2, IN) = (0, 0, 3, 2, 10, 6, 12)$ is $R_2 = R_1 \times R_2 = 60$ (bold arrows mark the activated path). Instead of simulating by a traditional approach all the components in the circuit, in the DD only three control variables are visited during simulation (y_4, y_3, y_2), and only a single data manipulation $R_2 = R_1 \times R_2$ is carried out.

A test for such a RT level data path represented by a single DD as shown in Fig. 1b can be created in two parts [2], as a conformity test (test for non-terminal nodes) which makes sure that the different working modes chosen by control signals are properly functioning, and as a scanning test (test for terminal nodes), which makes sure that the different functional blocks are working correctly.

Conformity test. Consider a nonterminal node m labelled by a control variable $z(m)$ in DD Gz, k, representing a digital system with a function $z_k = f_k(Z_k)$, Let $Z = (Z_C, Z_D)$ where Z_C is the vector of control variables and Z_D is the vector of data variables. To generate a test for m means to generate a test for the control variable $z(m) \in Z_C$. Suppose that $z(m)$ may have $n = | z(m) |$ different values. For testing $z(m)$, we have to activate and exercise all the proper working modes controlled at least once by each value of $z(m)$. At the same time, for each of such a working mode a current state of the system should be generated, so that every possible faulty change of $z(m)$ should produce a faulty next state which is different compared to the expected next state for the given working mode.

Denote by me the successor node of the node m for the value $z(m) = e$, where $e = 1, 2, \ldots, n$. For generating a test for m we use *Algorithm 2*:

1. Activate a path $l(m_0, m) \setminus m \subseteq M$ by assigning proper values $z(m)*$ at nodes $m \in l(m_0, m) \setminus m$;
2. Activate for all neighbours m^e of m nonoverlapping paths $l(m^e, m^{e,T})$ by assigning values $z(m)*$ at $m \in l(m^e, m^{e,T})$;
3. Find the proper set of data, by solving the inequality $z(m^{T,1}) \neq z(m^{T,2}) \neq \ldots \neq z(m^{T,n})$, where $n = | v(z(m)) |$.

The test of terminal nodes can be considered as a scanning test [2]. In terms of DDs the scanning test can be regarded as a special case of conformity test.

Example. Generate a test program for testing the node m labelled by y_3 in Fig. 2. First, we activate the path $l(m_0, m) \setminus m$, which results in $y_3 = 2$. Then we activate four paths $l(m, m^{e,T})$ for each value $e = 1, 2, 3, 4$ of y_3, which results in $y_1 = 0$ and $y_2 = 0$. Two of the four paths for values $y_3 = 1$ and $y_3 = 2$ are automatically activated since the successors of the node y_3 for these values are terminal nodes. The test data $R_1 = D_1, R_2 = D_2, IN = D_3$ are found by solving the inequality: $R_1 + R_2 \neq IN \neq R1 \neq R_1 \times R_2$.

4 Experimental Results and Conclusions

The feasibility and advantages of using DDs in diagnostic modelling of digital systems was demonstrated by using the test generator DECIDER [3]. The results (Table 1) were compared with other known test generators HITEC

Table 1. Comparison of test generators

Circuit	No. of faults	HITEC		GATEST		DECIDER	
		Fault cover (%)	Time (s)	Fault cover (%)	Time (s)	Fault cover (%)	Time (s)
gcd	454	81.1	170	91.0	75	89.9	14
sosq	1,938	77.3	728	79.9	739	80.0	79
mult	2,036	65.9	1,243	69.2	822	74.1	50
ellipf	5,388	87.9	2,090	94.7	6,229	95.0	1,198
risc	6,434	52.8	49,020	96.0	2,459	96.5	151
diffeq	10,008	96.2	13,320	96.4	3,000	96.5	296

and GATEST from the University of Illinois. The experiments were run on a 366 MHz SUN UltraSPARC 60 server with 512 MB RAM under SOLARIS 2.8 operating system. The experimental results show the high speed of the DECIDER.

Current chapter describes a novel multi-level diagnostic modelling approach based on using DD. Differently from known methods, both, higher and lower design abstraction levels are handled in a uniform topological manner. Joint formal basis for gate- and higher level descriptions allowed the first time to adopt and generalize gate-level methods to high-level ones. The feasibility and advantages of the model for using it in hierarchical test generation were demonstrated by experimental research. It was shown that high fault coverages of generated tests can be very quickly achieved compared to other methods.

Acknowledgements

The work was supported by EC VI Framework project "Vertigo", by Enterprise Estonia and by Estonian Science Foundation grants 5649, 5910.

References

[1] R.E.Bryant. Graph-based algorithms for Boolean function manipulation. IEEE Trans. on Computers, Vol.C-35, No8, 1986, pp.667-690.
[2] R.Ubar. Test Synthesis with Alternative Graphs. IEEE Design and Test of Computers, Spring 1996, pp.48-57.
[3] J.Raik, R.Ubar. Fast test pattern generation for sequential circuits using DD representations. JETTA, Kluwer, Vol. 16, No. 3, pp. 213-226, 2000.

Numerical Integration in Bayesian Positioning

Henri Pesonen and Robert Piché

Institute of Mathematics, Tampere University of Technology, POB 553, FIN-33101 Tampere, Finland
`henri.pesonen@tut.fi`, `robert.piche@tut.fi`

Summary. Multidimensional integrals arise in the bayesian approach to positioning using measurements from satellites, mobile phone networks, wireless data networks, etc. Measurement geometries and nongaussian measurement errors produce distinctive features such as multiple peaks and curved ridges. In this chapter compare several subregion adaptive simplicial cubature methods and a Monte Carlo method for typical positioning situations. We find that subregion adaptive methods give the best accuracy for the same number of samples in many two- and three-dimensional problems but that in four dimensions the dimensionality effect favors the Monte Carlo method.

1 Bayesian Positioning

Positioning and tracking are interesting scientific problems that have many commercial and industrial applications. The requirements of positioning methods are problem specific but often the computing has to be done online in limited computing environment. One of these applications is mobile phone positioning which is a widely studied problem [1]. The positioning can be done using signals from various sources such as satellites, mobile phone base stations and wireless local area networks. It is important that we use all the available information as efficiently as possible because often we have barely sufficient number of measurements. If we are to use a wide range of measurements, we have to take into account that the measurements can be strongly corrupted by noise. This is why we need to represent uncertainties with statistical models.

In the bayesian approach [2] we represent the state estimate of the object to be located, referred to as the mobile station (MS), with a posterior probability density function. It is the normalized product of the measurement likelihood function and prior probability density function. In a typical positioning model the state includes position coordinates and possibly some other data such as velocity. Measurement likelihood function includes the information of the obtained measurement and the prior represents our subjective knowledge of the state.

Let
$$y_i = h_i(x) + \mathbf{v_i} \tag{1}$$
be the measurement model that links the state x to the observed measurement y_i. The measurement function $h_i(\cdot)$ can be for example a range measurement or a restrictive measurement that bounds the state to a certain part of the state space.

If the random measurement error $\mathbf{v_i}$ has a probability density $p_{\mathbf{v_i}}$ then the probability density of an event that we have observed y_i while we are in the state x is
$$p(y_i|x) = p_{\mathbf{v_i}}(y_i - h_i(x)). \tag{2}$$
If we have obtained multiple measurements $y_i, i = 1, \ldots, n$ and take the errors in measurements to be independent, we can combine these measurements into $y = [y_1, \ldots, y_n]^\top$ as
$$p(y|x) = \prod_{i=1}^{n} p(y_i|x). \tag{3}$$
Considering (3) as a function of x, it gives the chances that MS is at x when we have observed y. This is the likelihood function of the model. Given our beliefs of the state in the form of a prior distribution $p(x)$ we can form the posterior density of the state
$$p(x|y) = \frac{p(y|x)p(x)}{\int p(y|x)p(x)\mathrm{d}x} \tag{4}$$

To represent the position with a single point, we can use for example the maximum or the mean of the posterior. Finding the maximum point of the posterior when it has multiple peaks can be a difficult task. If we use the mean
$$\widehat{x} = \int xp(x|y)dx, \tag{5}$$

we have to solve an integral that has as many dimensions as the state vector x. This integral is frequently analytically intractable and this is why we have to consider numerical methods to approximately solve it.

2 Integration Methods

To solve integrals of the form (5), different approximate methods have been considered. In this work we consider only numerical approximations although analytical approximations could also be applied. Multidimensionality of the problem makes Monte Carlo methods an attractive choice because they are not affected by the dimensionality effect. In addition, we consider different subregion adaptive methods. All methods considered here approximate integrals using weighted sums of the integrand evaluations
$$\int f(x)dx \approx Q_n f = \sum_{k=1}^{n} c_k f(x_k), \tag{6}$$

where x_k are chosen using different criteria. Monte Carlo methods use random samples drawn from certain importance distributions. Deterministic numerical integration rules are often designed to be exact for N-variate polynomials of degree at most d integrated over some standard region.

Subregion adaptive methods first divide the integration region into subregions with disjoint interiors dynamically according to the behaviour of the integrand. In practice this can be done for example by estimating the error in each subregion and then subdividing the region with largest error. The integral is computed in each subregion separately using a quadrature rule of low order accuracy. In this chapter integration region was taken to be polygonal so we used a simplex as a basic block into which the integration region was divided.

3 Numerical Results

We tested the integration methods with a positioning scenarios similar to one arising in a mobile phone positioning. The tested integration methods were plain Monte Carlo (mc) and subregion adaptive methods with accuracy of degree $d = 1, 2, 7$ ($Q_{n_1}, Q_{n_2}, Q_{n_7}$). Degree 1 rule uses the vertices of simplex as nodes, 2nd degree rule the vertices and midpoints of edges and degree 7 rule is Grundmann–Möller cubature rule [3]. We simulated range measurements with restrictive information in form of sector information as illustrated in Fig. 1. The gaussian noise in the measurements were taken to have a randomly generated variance as given in Table 1. We generated 200 test cases with uniform prior distribution in closed region and 200 cases with gaussian prior distribution. No clear difference between the different priors was found so they are combined in the results. As a reference result we computed the integral using adaptive method with very high number of function evaluations.

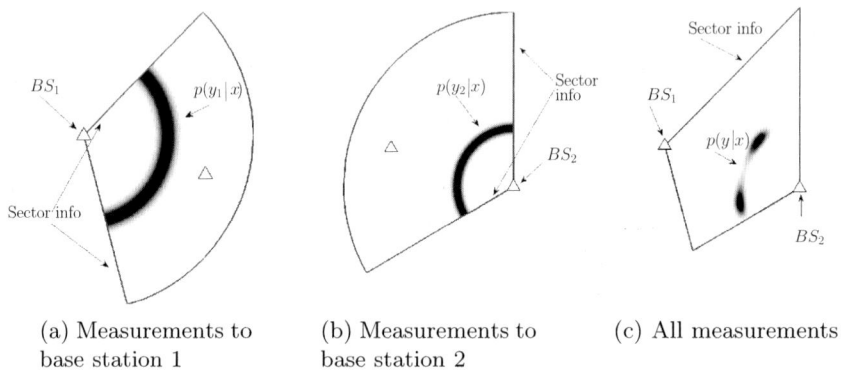

(a) Measurements to base station 1

(b) Measurements to base station 2

(c) All measurements

Fig. 1. Combination of all available measurements

Table 1. Test cases

	Case 1	Case 2
$\sigma_1, \sigma_2 \in$	$[50, 100]$	$[100, 150]$
3D: $\sigma_3 \in$	$[25, 35]$	$[75, 85]$
4D: $\sigma_3, \sigma_4 \in$	$[15, 25]$	$[45, 55]$

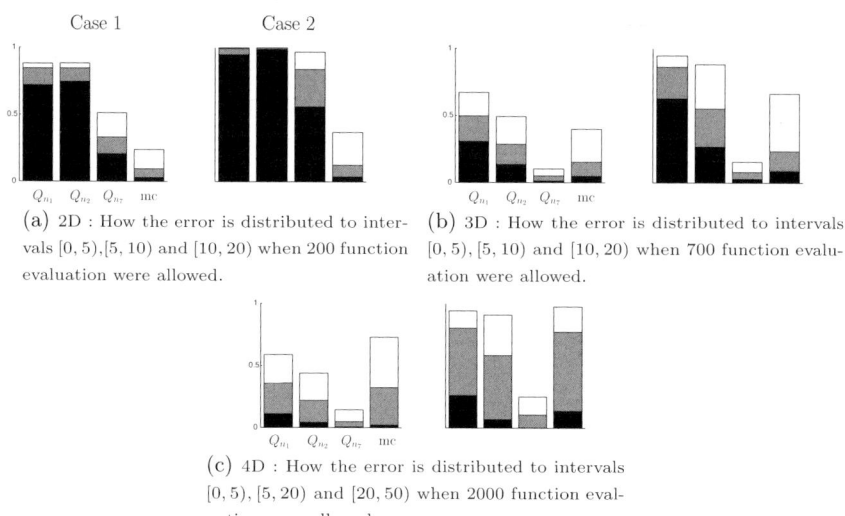

(a) 2D : How the error is distributed to intervals $[0, 5), [5, 10)$ and $[10, 20)$ when 200 function evaluation were allowed.

(b) 3D : How the error is distributed to intervals $[0, 5), [5, 10)$ and $[10, 20)$ when 700 function evaluation were allowed.

(c) 4D : How the error is distributed to intervals $[0, 5), [5, 20)$ and $[20, 50)$ when 2000 function evaluation were allowed.

Fig. 2. The distribution of errors to fixed intervals when fixed amount of function evaluation is allowed in integration formula

Results show that subregion adaptive methods perform much better than the basic Monte Carlo integration in this kind of scenario. Adaptive methods almost always give accuracy better than 5 m with as low as 200 abscissas as seen from Fig. 2a.

In three-dimensional tests we added an altitude as a third dimension to two-dimensional tests. We also added a new measurement to the model representing an altitude measurement. We restricted the integration region with an altitude bound $|x_3| \leq 100$. Figure 2b shows that the dimensionality effect is beginning to show as we have to increase the number of abscissas to obtain good accuracies.

For four-dimensional tests we added two velocity dimensions to the model. We also simulated two velocity measurements in each velocity dimension. We bounded velocity dimensions by taking $\| [x_3, x_4]^\top \|_\infty \leq 100$. Maximum-norm was taken to simplify the model, although velocity bounded by euclidian norm would be more realistic choice. The dimensionality effect is clearly shown in Fig. 2c.

4 Conclusions

Four different numerical integration formulas were compared in different positioning scenarios. The results show that adaptive methods are much more efficient than the plain Monte Carlo method in two- and three-dimensional tests as expected as seen from Fig. 2 and Tables 2 and 3. In four-dimensional tests the dimensionality effect has clearly an influence on the results. This can be seen from Fig. 2 and Table 4. Even though the best adaptive method still

Table 2. 2D: How many times (%) each method gave the best answer

	Case 1				Case 2			
pts	Q_{n_1}	Q_{n_2}	Q_{n_7}	mc	Q_{n_1}	Q_{n_2}	Q_{n_7}	mc
200	33	53	7	7	32	60	8	0
400	23	46	32	0	9	32	60	0
600	16	45	39	0	4	23	74	0
800	9	34	57	0	2	16	82	0
1000	6	29	65	0	0	13	87	0
1200	4	20	76	0	1	10	90	0

Table 3. 3D: How many times (%) each method gave the best answer

	Case 1				Case 2			
pts	Q_{n_1}	Q_{n_2}	Q_{n_7}	mc	Q_{n_1}	Q_{n_2}	Q_{n_7}	mc
300	40	12	7	42	58	16	5	21
700	51	21	3	25	70	19	4	7
1100	55	23	4	19	56	34	5	6
1500	46	30	8	17	47	39	10	5
1900	41	42	6	12	34	56	7	3
2300	36	46	6	12	27	65	6	2

Table 4. 4D: How many times (%) each method gave the best answer

	Case 1				Case 2			
pts	Q_{n_1}	Q_{n_2}	Q_{n_7}	mc	Q_{n_1}	Q_{n_2}	Q_{n_7}	mc
1000	33	8	0	59	37	13	0	50
2000	35	15	4	46	46	14	4	36
3000	36	21	4	40	56	10	2	32
4000	41	19	5	35	56	14	2	28
5000	37	25	3	35	58	14	3	26
6000	36	30	3	31	59	14	3	24

gives slightly more accurate results, the Monte Carlo method would be the method of choice in four dimensions because of its greater simplicity compared to adaptive methods.

References

1. Christopher Drane and Malcolm Macnaughtan and Craig Scott, IEEE Communications Magazine, April, 46-59, Positioning GSM Telephones, 1998.
2. Dieter Fox, Jeffrey Hightower, Lin Liao, Dirk Schulz and Gaetano Borriello, IEEE Pervasive Computation, September, 24-33, 2003.
3. Axel Grundmann and H. M. Möller, SIAM Journal on Numerical Analysis **15**(2), 1978.
4. Henri Pesonen, Numerical Integration in Bayesian Positioning, MA Thesis, Tampere University of Technology, Finland (2006).

Singular Problems With Quadratic Gradient Term

Antonio Vitolo

Department of Mathematics and Informatics, University of Salerno, Grahamstown, 84084 Fisciano, Salerno, Italy
`vitolo@unisa.it`

1 Introduction

We are concerned with the following boundary value problem (BVP):

$$\Delta u + g(x, u)|Du|^2 + f(x, u) = 0 \quad in \quad D, \tag{1}$$

$$u > 0 \quad in \quad D, \quad u = 0 \quad on \quad \partial D. \tag{2}$$

We will look for classical solutions $u \in C^2(D) \cap C(\bar{D})$, supposing that D is a bounded regular domain of $\mathbf{R}^n$, while $f(x, t)$ and $g(x, t)$ are smooth functions in $D \times (0, +\infty)$. Moreover, f will be positive function with a singularity at the origin and g may have or not such a singularity.

Problems of this kind arise, for instance, in the theory of non-Newtonian fluids, see Nachman and Callegari [NC80], and also of heat conduction in electrically conducting materials, see Cohen and Keller [CK67].

A long series of papers have been devoted to positive solution of the Dirichlet problem for the semi-linear equation

$$Lu + f(x, u) = 0 \tag{3}$$

for a linear second-order elliptic differential operator L. See for instance Fulks and Maybee [FM60], Stuart [Stu76], Lazer and McKenna [LMK91], Crandall, Rabinowitz and Tartar [CRT97]. A nonlinear convection term is considered by Zhang and Yu [ZY00] and by Giarrusso and Porru [GP06].

In the last chapter existence and uniqueness results together with boundary estimates have been shown for the problem

$$\Delta u + g(u)|Du|^q + f(u) = 0 \quad in \quad D, \quad u > 0 \quad in \quad D, \quad u = 0 \quad on \quad \partial D, \tag{4}$$

with an almost-quadratic gradient term $(0 \le q < 2)$.

Actually, our attention is concentrated on the limit case $q = 2$ exploring in depth the influence of the convection term. This is the subject of a research line in progress in collaboration with Giovanni Porru.

2 Radial Solutions

We start considering the radial symmetric boundary value problem (RBVP)

$$\Delta u + g(u)|Du|^2 + f(u) = 0 \ \text{ in } \ B_R, \tag{5}$$

$$u > 0 \ \text{ in } \ B_R, \quad u = 0 \ \text{ on } \ \partial B_R, \tag{6}$$

where $B_R = B_R(x_0)$ is a ball of radius R, centered at x_0, while $f(t) > 0$ and $g(t)$ are continuous non-increasing functions for $t \in (0, +\infty)$. Typical singularity for the function f will be $f(t) = 1/t^\gamma$ with $\gamma \geq 1$. More generally we will assume

$$\int_0^\delta f(t)\mathrm{d}t = +\infty, \tag{7}$$

for any $\delta > 0$.

We observe that (RBVP) can be solved with the aid of the following Cauchy Problem (CP) for ordinary differential equations:

$$v'' + \frac{n-1}{r}v' + g(v)|v'|^2 + f(v) = 0, \tag{8}$$

$$v(0) = v_0, \quad v'(0) = 0, \tag{9}$$

for a positive number v_0.

In fact ODE (8) is precisely the ordinary differential equation arising from RPDE (5) when searching for radial solutions.

Lemma 1. *Let $v \in C^2([0, R))$ be the maximal positive solution of (CP), then v is a decreasing and concave function such that*

$$\lim_{r \to R^-} v(r) = 0 \tag{10}$$

and

$$\lim_{r \to R^-} v'(r) = -\infty. \tag{11}$$

From the above Lemma we deduce that $u(x) = v(|x - x_0|)$ is a radial solution of (RBVP) for maximal intervals of (CP) corresponding to any initial value $v_0 > 0$. The uniqueness and the monotonicity result contained in the next Lemma follow from comparison results for quasilinear equations of Gilbarg and Trudinger [GT83].

Lemma 2. *Let $R = R(v_0)$ be the length of the maximal interval $[0, R)$ for positive solutions of (CP), then $R(v_0)$ is an increasing function such that*

$$\lim_{v_0 \to 0^+} R(v_0) = 0. \tag{12}$$

Indeed, as a consequence of the results of the next section, the function $R = R(v_0)$ turns out to be continuous, too. In fact, under the above assumptions, existence and uniqueness of (RBVP) can be obtained in every ball $B_R(x_0)$ with $R < \sup_{v_0 > 0} R(v_0)$.

3 Existence Results

Throughout this section, for a general smooth bounded domain D and functions $f = f(x,t)$ and $g = g(x,t)$, we will make the following assumptions:

(0_f) $f(x,t) \geq \underline{f}(t)$ as $t \to 0^+$; (∞_f) $f(x,t) \leq M$ as $t \to +\infty$;
(0_g) $g(x,t) \geq \underline{g}(t)$ as $t \to 0^+$; (∞_g) $g(x,t) \leq N$ as $t \to +\infty$.

Here M and N are positive constants, while $\underline{f}$ and $\underline{g}$ are smooth non-increasing functions. Moreover,

$$\underline{f} > 0, \quad \int_0^\delta \underline{f}(t)\mathrm{d}t = +\infty. \tag{13}$$

For $\varepsilon \geq 0$ we consider the approximating boundary value problems $(\mathrm{BVP}_\varepsilon)$,

$$\Delta u + g(x,u)|Du|^2 + f(x,u) = 0 \quad in \quad D, \tag{14}$$

$$u \geq \varepsilon \quad in \quad D, \quad u = \varepsilon \quad on \quad \partial D. \tag{15}$$

To find a solution we employ the monotone method of Kazdan and Kramer [KK78] and the classical regularity theory [LU68], [GT83]. We also need a cross condition between the diameter of D and the product of upper bounds at infinity for f and g, namely we suppose that
(d) $D \subset B_d$, $MNd^2 < \frac{\pi^2 N}{4}$.
It is worth to observe that condition (d)s satisfied in all bounded domains, no matter how large is the diameter, in one of the following cases:
(i) $\limsup_{t \to +\infty} \sup_{x \in D} f(x,t) = 0$;
(ii) $\limsup_{t \to +\infty} \sup_{x \in D} g(x,t) \leq 0$.

Lemma 3. *Assuming (∞_f), (∞_g) and (d), let $a \geq 1$ be such that*

$$f(x,t) \leq M, \quad g(x,t) \leq N, \quad t \geq a. \tag{16}$$

Then for every $0 < \varepsilon < 1$ there exists a solution $u = u_\varepsilon$ of (BVP_ε) such that

$$\varepsilon \leq u_\varepsilon(x) \leq a + \frac{\log \sec(\alpha d)}{N}, \tag{17}$$

where $\alpha = \sqrt{MN/n}$.

Under the additional assumptions at the origin for f ang g, using radial solutions of the previous section in small balls, comparison principles, interior gradient estimates and Schauder estimates, we can show that the solutions of the approximating problem $(\mathrm{BVP}_\varepsilon)$ in fact converge to a solution of (BVP).

Theorem 1. *Suppose that (∞_f), (∞_g), (d), (0_f) and (0_g) hold. Then (BVP) has a classical solution. If we also suppose $f(x,t)$ and $g(x,t)$ to be non-increasing in the t-variable for all $x \in D$, such a solution is unique.*

We observe that condition (0_g) implies that $g(t)$ is bounded from below as $t \to 0^+$. Thus we are not able to treat, with Theorem 1, the cases in which $\liminf_{t\to 0^+} g(t) = -\infty$.

How get existence and uniqueness results even in this case is under investigation.

4 Boundary Estimates

For sake of brevity we study the asymptotic behaviour of the solutions of (BVP) when the functions f and g are independent on x. For the same reason, we consider the model problem in which the smooth functions $f > 0$ and g are non-increasing, as it will be supposed throughout this Section. We also consider a primitive F of the function f, i.e. $F'(t) = f(t)$. According to (7), $\lim_{t\to 0^+} F(t) = +\infty$. An important role will be played by the functions

$$\psi(r) = \int_0^r \frac{ds}{\sqrt{2F(s)}}, \tag{18}$$

which turns out to be increasing, and its inverse φ, that satisfies

$$\varphi'' + f(\varphi) = 0, \quad \varphi(0) = 0. \tag{19}$$

For instance, if $f(t) = 1/t^\gamma$, $\gamma > 1$, we get

$$\psi(r) = A_\gamma r^{(\gamma+1)/2}, \tag{20}$$

and therefore

$$\varphi(s) = B_\gamma s^{2/(\gamma+1)} \tag{21}$$

for positive constants A_γ and B_γ.

We also suppose (∞_f), (∞_g) and the cross condition (d), as in the previous section, but we need an additional hypothesis on g, namely $g \in L^1(I)$ for all bounded intervals $I \subset (0, +\infty)$.

Finally, we make a technical assumption: there exists a positive constant ρ such that

$$f(t) \le \rho F(2t), \quad t \to 0^+. \tag{22}$$

In the case of positive radial solutions $u(x)$ of (BVP) in the ball $B_R(0)$ we have, for all $\varepsilon > 0$,

$$u(x) \ge \frac{\varphi(R - |x|)}{\sqrt{1+\varepsilon}} - \beta(R - |x|), \tag{23}$$

for some positive constant β, provided that $R - |x| \le \delta_\varepsilon$.

Similarly, for solutions $u(x)$ in an annular domain $B_{R'}(0)\backslash B_R(0)$,

$$u(x) \le \frac{\varphi(|x| - R)}{\sqrt{1-\varepsilon}} + \beta(|x| - R), \tag{24}$$

provided that $|x| - R \le \delta_\varepsilon$, i.e. approaching the internal boundary.

Gathering the above estimates, a typical boundary asymptotic result in a sufficiently smooth domain is the following.

Theorem 2. *Suppose that D is a smooth domain satisfying both internal and external uniform sphere conditions. Under above assumptions, let u be the classical positive solution of (BVP). Then for all $\varepsilon > 0$ there exists $\delta_\varepsilon > 0$ such that*

$$\frac{\varphi(\delta(x))}{\sqrt{1+\varepsilon}} - \beta\delta(x) \le u(x) \le \frac{\varphi(\delta(x))}{\sqrt{1-\varepsilon}} + \beta\delta(x) \qquad (25)$$

for $x \in D$ such that $\delta(x) := dist(x, \partial D) \le \delta_\varepsilon$ with some positive constant β

In the previous example, where $f(t) = 1/t^\gamma$, the above Theorem yields $u(x)/\delta(x)^{2/(\gamma+1)} \to B_\gamma$ as $x \to \partial D$.

Note that the presence of gradient term does not affects the first-order approximation of the solution near the boundary, which is also independent, as usual in this kind of problems, of the shape of the domain.

References

[CK67]   Cohen, D.S., Keller, H.B.: Some positive problems suggested by nonlinear heat generators. J. Math. Mech., **16**, 1361–1376 (1967)

[CRT97] Crandall, M.G., Rabinowitz, P.H., Tartar, L.: On a Dirichlet problem with a singular nonlinearity. Comm. Part. Diff. Eq., **2**, 193–222 (1997)

[FM60]   Fulks, W., Maybee, J.S.: A singular nonlinear equation. Osaka Math. J., **12**, 1–19 (1960)

[GP06]   Giarrusso, E., Porru, G.: Problems for elliptic singular equations with a gradient term. Nonlinear Analysis, **65**, 107–128 (2006)

[GT83]   Gilbarg, D., Trudinger, N.S.: Elliptic Partial Differential Equations of Second Order. 2nd Ed. Springer Verlag, Berlin (1983)

[KK78]   Kazdan, J.L., Kramer, R.J.: Invariant criteria for existence of solutions to quasilinear elliptic equations. Commun. Pure Applied Math., **31**, 619–645 (1978)

[LU68]   Ladyzhenskaya O.A., Ural'tseva, N.N.: Linear and Quasilinear Elliptic Equations. Academic Press, New York (1968)

[LMK91] Lazer, A.C., McKenna, P.J.: On a singular nonlinear elliptic boundary value problem. Proc. American. Math. Soc., **111**, 721–730 (1991)

[NC80]   Nachman, A., Callegari, A.: A nonlinear singular boundary value problem in the theory of pseudoplastic fluids. SIAM J. Appl. Math., **38**, 275–281 (1980)

[Stu76] Stuart, C.A.: Existence and approximation of solutions of nonlinear elliptic equations. Math. Z., **147**, 53–63 (1976)

[ZY00]   Zhang Z., Yu J.: On a singular nonlinear Dirichlet problem with a convection term. SIAM J. Math. Anal., **32**, 916–927 (2000)

Pattern Matching for Control Chart Monitoring

Domenico Cantone and Simone Faro

Università di Catania, Dipartimento di Matematica e Informatica
Viale Andrea Doria 6, I-95125 Catania, Italy
{cantone | faro}@dmi.unict.it

Summary. Recognition of control chart patterns (CCPs) is one of the most important technique for monitoring and achieving appropriate control of process environments to raise production quality. In the last 10 years several approaches have been proposed for precise and fast CCP recognition, including rule-based and expert systems, or artificial neural networks, and many efforts have been focused on comparative studies of approximate training algorithms.

This chapter presents a new approach for the identification of control chart patterns by using features dynamically extracted from raw data. Our strategy has the further advantage of avoiding the use of complex data structures and training processes.

1 Introduction

Statistical process control (SPC) is a method for achieving quality control in manufacturing processes. It comprises a collection of techniques, based on the analysis of statistical quantities such as mean, variance, and others, to detect at an early stage whether significant deviations of a manufacturing process from its normal behaviour are taking place.

Control chart patterns (CCPs) are used in SPC to provide information on the state of a process. Their identification is an important issue in SPC, as abnormal CCPs can be associated with specific assignable causes which affect the normal process execution.

Several approaches have been proposed for CCPs recognition, including rule-based methods [PW97], expert systems [PO92], and, in particular, artificial neural networks, divided in supervised [PS00], unsupervised [PC98], and self-organizing with decision tree learning [GS05]. While such approaches rely on raw data as input vector representation, other possible techniques are based on enriched data representations by means of features extracted from raw data [HSSJ03].

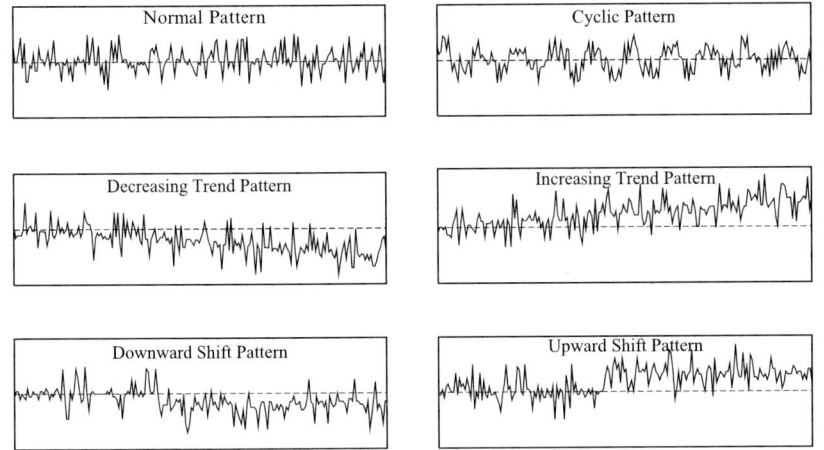

Fig. 1. The six main types of CPPs

A control chart is a run chart of a sequence of data points with five horizontal lines: a *mean line*, drawn at the process mean η, an upper and a lower *warning limit* drawn at $\pm 2\sigma$, an upper and lower *control-limit* drawn at $\pm 3\sigma$. CCPs are used to identify possible causes behind observations which fall outside control-limits. In general, CCPs can be divided into six types of patterns: normal, cyclic, increasing/decreasing trend, and upward/downward shift (see Fig. 1). With the obvious exception of normal patterns, the remaining patterns indicate that the process being monitored is not functioning correctly and requires some adjustments. These six types of patterns can be described, respectively, by the following functions:

(1) Normal pattern: $\qquad\qquad\qquad\qquad\qquad p(t) = \eta + r(t) \cdot \sigma$

(2) Cyclic pattern: $\qquad\qquad\qquad\qquad\qquad p(t) = \eta + r(t) \cdot \sigma + a \cdot \sin(2\pi t / T)$

(3) Increasing/decreasing trend pattern: $\quad p(t) = \eta + r(t) \cdot \sigma + g \cdot (t - t_0)$

(4) Upward/downward shift pattern: $\qquad p(t) = \eta + r(t) \cdot \sigma + s \cdot step(t),$

where η and σ are, respectively, the nominal mean value and the standard deviation of the process variable under observation, t is the discrete time at which the monitored process variable is sampled, $p(t)$ is the value of the sampled data point at time t, the function $r(t)$ generates random numbers normally distributed between -3 and 3, and $step(t)$ is a 0/1-step function.

One of the major difficulties in CCPs recognition lies in detecting increasing or decreasing patterns [LP05], especially when slopes are small.

In this chapter we briefly present a new simpler approach for the identification of increasing or decreasing trend patterns in CCPs, based on dynamically computed raw data features such as nominal mean and standard deviation.

2 Identifying Normal Patterns

A normal pattern, which identifies a controlled process, will exhibit only random variations. Thus, for instance, a pattern in which the items at positions $k \cdot n$, for $k = 1, 2, \ldots$, are equal is non-random. However, to identify such patterns is often quite subtle and difficult, and is beyond the scope of this chapter.

We will make the simplifying hypothesis that *random* patterns can be described by equations of type (1) above, namely equations of the form $p(t) = \eta + r(t)\sigma$, and therefore they are characterized by a nominal mean $\hat{\eta} \approx \eta$ and a standard deviation $\hat{\sigma} \le \sigma$.

Given a sample dimension n, we define the CCP at time $t = s + n - 1$ as the sequence of sample data points $\bar{P} = \langle p(s), p(s+1), \ldots, p(s+n-1) \rangle$ of length n, with nominal mean $\bar{\eta}$ and standard deviation $\bar{\sigma}$.

Associated to the pattern $\bar{P}$ we define the quantities

$$\Gamma = \sum_{t=s}^{s+n-1} p(t) \quad \text{and} \quad \Delta = \sum_{t=s}^{s+n-1} p(t)(t-s+1).$$

The dynamic nature of SPC requires that information on monitored sample patterns are quickly computed. Thus, at time $t = s + n$ the sampled pattern is updated with $\bar{P}' = \langle p(s+1), p(s+2), \ldots, p(s+n) \rangle$, having nominal mean $\bar{\eta}'$ and standard deviation $\bar{\sigma}'$. The values Γ and Δ can be dynamically updated in time $\mathcal{O}(1)$ for pattern $\bar{P}'$, while processing the raw data, by simply applying the following rules:

$$\Gamma' = \Gamma - p(s) + p(s+n), \qquad \Delta' = \Delta - p(s) + (n+1)p(s+n) - \Gamma'.$$

Thus, the values of the mean, $\bar{\eta}'$, and standard deviation, $\bar{\sigma}'$, of pattern $\bar{P}'$ can be updated in time $\mathcal{O}(1)$ using the following rules, where $\rho = \bar{\eta} - \bar{\eta}'$:

$$\bar{\eta}' = \Gamma'/n;$$

$$\bar{\sigma}' = \left(\frac{1}{n} \sum_{t=s+1}^{s+n} \left(p(t) - \bar{\eta}' \right)^2 \right)^{\frac{1}{2}} = \left(\frac{1}{n} \sum_{t=s+1}^{s+n} \left(p(t) - \bar{\eta} + \rho \right)^2 \right)^{\frac{1}{2}}$$

$$= \left(\frac{1}{n} \left(n\bar{\sigma}^2 + n\rho^2 - 2n\rho\bar{\eta} + 2\rho\Gamma' + (p(s+n) - \bar{\eta})^2 - (p(s) - \bar{\eta})^2 \right) \right)^{\frac{1}{2}}.$$

Such rules allow the automatic identification in constant time of a normal pattern, as the process is taking place.

3 Identifying Decreasing and Increasing Trend Patterns

A decreasing or increasing trend pattern, as described by (3), is a sequence of data points obeying an equation of the form $p(t) = \eta + r(t)\sigma + gt$, where g is the slope. This type of pattern indicates a drift in the process average. Often, such drift can be the result of tool wear, deteriorating maintenance, or even skill improvement.

Suppose to observe, at a given time $t = s+n-1$ of the production process, a drift of the mean value $\bar{\eta}$, such that $|\bar{\eta} - \eta| \geq \delta$, for same threshold $\delta > 0$. If the observed pattern P is of the form given by equation (3) and the change in trend started at time $t_0 = s - 1$, we can rewrite its nominal mean $\bar{\eta}$ as

$$\bar{\eta} = \frac{1}{n}\sum_{t=s}^{s+n-1}(\eta + r(t)\sigma + g\cdot(t-s+1)) \approx \eta + \frac{1}{n}g\sum_{t=1}^{n}t = \eta + \frac{1}{2}g(n+1)$$

and hence we can estimate the value of the slope by $g \approx 2(\bar{\eta} - \eta)/(n+1)$. Observe that if the sampled pattern P is a decreasing or increasing trend pattern, with slope g, then the pattern $\hat{P} = \langle p(s) - g, p(s+1) - 2g, ..., p(s) - ng \rangle$ is a normal pattern. Figure 2 shows a decreasing trend pattern P and the pattern $\hat{P}$ obtained by removing from P the effects of the gradient g. The standard variation $\hat{\sigma}$ of pattern $\hat{P}$ can be computed in time $\mathcal{O}(1)$ using the following formula

$$\hat{\sigma} = \left(\frac{1}{n}\sum_{t=s}^{s+n-1}(p(t) - (t-s+1)g - \eta)^2\right)^{\frac{1}{2}}$$
$$= \left(\bar{\sigma}^2 + g\eta(n+1) + (\bar{\eta} - \eta)^2 - \frac{2}{n}g\Delta + \frac{2}{n}(\bar{\eta} - \eta)\Gamma + \frac{1}{6}g^2(n+1)(2n+1)\right)^{\frac{1}{2}}.$$

We have carried out four sets of experiments to test the performance of our proposed algorithm, TEST-SLOPE, shown below. For each test, 200 patterns have been generated, using (3), with $\eta = 10$, $\sigma = 4.5$, and where $n \in \{60, 120\}$ and the slope g is randomly selected in the interval $[-\alpha, \alpha]$, for $\alpha \in \{0.1, 0.2\}$.

In our tests, we have assumed that a decreasing or increasing pattern with a very small slope g such that $|g| \leq 0.03$ has to be considered as a normal pattern. The following table shows the percentage errors in our tests. It turns out from our experimental results that the percentage of error is very low and decreases for increasing values of the dimension n and of the slope g.

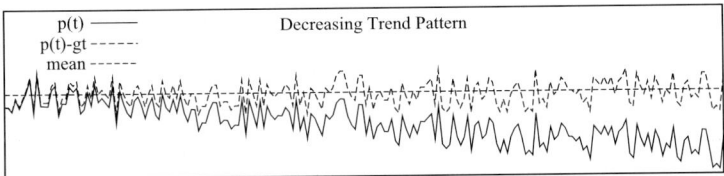

Fig. 2. A decreasing trend pattern and the pattern obtained by deleting the effects of the slope g

TEST-SLOPE(P, η, σ, n)	Experimental results		
compute $\bar{\eta}$ and $\bar{\sigma}$		n	α	Error (%)
$g = 2(\eta - \bar{\eta})/(n+1)$	test n.1	60	0.2	7
compute $\hat{\sigma}$	test n.2	60	0.1	11.5
if $|g| > 0.03$ then return TRUE	test n.3	120	0.2	2
else return FALSE	test n.4	120	0.1	4

4 Conclusions and Future Works

In this chapter we have presented a simple method for the identification of increasing and decreasing trend patterns in control charts, by using features dynamically extracted from raw data. Such patterns are among the most difficult ones to be detected, especially for small values of the slope. Experimental results show that our proposed approach turns out to be precise, simple, and flexible.

Future works will be directed to apply similar techniques for the identification of other classes of CCPs. In addition, we plan to generalize our technique also to cases in which standard deviation is allowed to change over time.

References

[GS05]   Ruey-Shiang Guh, Yeou-Ren Shiue: On-line identification of control chart patterns using self-organizing approaches, International Journal of Production Research, Taylor and Francis, vol. 43, n.6, pp. 1225-1254 (2005)

[HSSJ03] Hassan, A., Shariff Nabi Baksh, M., Shaharoun, A.M., Jamaluddin, H.: Improved SPC chart pattern recognition using statistical features, International journal of production research, Taylor and Francis, London, vol. 41, n.7, pp. 1587-1603 (2003)

[LP05]   Lavangnananda, K., Piyatumrong, A.: Image processing approach to features extraction in classification of control chart patterns, Proceedings of the 2005 IEEE Mid-Summer Workshop on Soft Computing in Industrial Applications, pp.85-90 (2005)

[PC98]   Pham, D.T., Chan, A.B.: A novel self-organising neural network for control chart pattern recognition, in Soft Computing in Engineering Design and Manufacturing, P. K. Chawdhry, R. Roy and R. K. Pant (Eds.), Springer-Verlag, London, pp.381-390, (1998)

[PO92]   Pham, D.T., Oztemel, E.: XPC: an on-line expert system for statistical process control, Int. J. Production Research, 30(12), pp.2857-2872, (1992)

[PS00]   Pham, D.T., Sagiroglu, S.: Neural network classification for defects in veneer boards, Proc. IMechE, Part E, Journal of Process Mechanical Engineering, Vol.214, Part B, (2000)

[PW97]   Pham, D.T., Wani, A.: Feature-based control chart pattern recognition, Int. J. Production Research, 35(7), pp.1875-1890, (1997)

Index Characterization in DAE Circuit Models Without Passivity Assumptions

Alfonso J. Encinas and Ricardo Riaza

Departamento de Matemática Aplicada a las Tecnologías de la Información
ETSI Telecomunicación, Universidad Politécnica de Madrid
Ciudad Universitaria s/n - 28040 Madrid, Spain
`ajencinas@ieee.org`, `rrr@mat.upm.es`

Summary. We present in this communication an index-1 characterization for differential-algebraic circuit models without passivity assumptions. The use of tree-based methods together with the Cauchy–Binet formula makes it possible to generalize previous results in the literature. The approach can be extended to modified node analysis (MNA) models and higher index configurations.

1 Introduction

Semistate models based on differential-algebraic equations (DAEs) are nowadays extensively used in circuit simulation programs. Modelling approaches based on node tableau analysis (NTA), augmented node analysis (ANA) or modified node analysis (MNA) set up the circuit equations in differential-algebraic form [GF99a, GF99b].

Within the differential-algebraic framework, the *index* of the circuit model becomes a standard measure for the numerical problems faced during computer simulation. So far, most index characterizations [ET00, Rei98, Ria06, RT05, Tis99, Tis03] are based on passivity assumptions, which amount to positive definiteness hypothesis in the conductance matrix and, in some cases, in reactance matrices.

In this direction, the goal of the present work is to introduce graph-theoretic index characterizations relaxing passivity assumptions to algebraic conditions on some circuit trees. The attention is focused on circuits without controlled sources and accommodating only certain coupled devices: previous index characterizations for passive circuits within this scope can be derived as particular instances of the results here discussed.

Section 2 presents some background on circuit modelling. We then analyse index-1 configurations in Sect. 3: assuming that there is no coupling among resistors, we show in Theorem 1 that index-1 in augmented models turns out to be equivalent to the absence of both V–C loops and I–L cutsets together with a non-zero sum for the conductance products in so-called *proper* trees.

Index equivalences are known that allow one to extend our results to general Node Tableau Analysis models [RT05].

As indicated in Sect. 4, these procedures can be applied to characterize index 1 MNA configurations, although the quasilinear structure of these models involve more technicalities and, for the sake of brevity, details are not included here. Networks without coupling restrictions, index 2 configurations, and circuits with controlled sources stay within the scope of future research.

2 Differential-Algebraic Circuit Models

Mathematical models of lumped electrical circuits usually take the form of a differential-algebraic equation, including:

– A topological component, characterized by (incidence) matrices describing the network
– A dynamical component, linked to reactive elements: inductors and capacitors
– A physical component, derived from the constitutive relationships of the network elements.

In this communication, special attention is paid to the formulation known as ANA. ANA models are simpler than NTA formulations and can be regarded as an intermediate step to arrive both at MNA and at a state space equation. The corresponding DAE, which can be derived from general NTA [RT05] by eliminating several branch and current variables, reads:

$$C(v_C)v_C' = i_C, \tag{1a}$$

$$L(i_L)i_L' = A_{\mathcal{L}}^T e, \tag{1b}$$

$$0 = A_{\mathcal{R}}\gamma(A_{\mathcal{R}}^T e) + A_C i_C + A_{\mathcal{V}}i_{\mathcal{V}} + A_{\mathcal{L}}i_L + A_{\mathcal{I}}i_S(t), \tag{1c}$$

$$0 = A_C^T e - v_C, \tag{1d}$$

$$0 = A_{\mathcal{V}}^T e - v_S(t), \tag{1e}$$

where $A = (A_C\ A_{\mathcal{L}}\ A_{\mathcal{R}}\ A_{\mathcal{V}}\ A_{\mathcal{I}})$ is the incidence matrix for the circuit split for its different elements, and γ stands for the $i - v$ characteristic of resistors. If the capacitance and inductance matrices $C(v_C)$, $L(i_L)$ are non-singular, then this model can be written in the semiexplicit form

$$u' = \varphi(u, v, t), \tag{2a}$$

$$0 = \psi(u, v, t), \tag{2b}$$

where the dynamic variables $u = (v_C, i_L)$ correspond to capacitor voltages and inductor currents, and the algebraic variables $v = (e, i_C, i_{\mathcal{V}})$ comprise node voltages, capacitor currents and voltage source currents. Equation (2a) comprises the dynamic relations (1a)–(1b), whereas (2b) stands for (1c)–(1e):

note that (1c) expresses Kirchhoff Current Law, whilst (1d)–(1e) correspond to Kirchhoff Voltage Law for capacitors and voltage sources, respectively. The excitation terms are defined by current sources $i_S(t)$ and voltage sources $v_S(t)$. Semiexplicit systems such as (2) are said to be index 1 if and only if the derivative $\psi_v(u, v, t)$ defines a non-singular matrix [BCP96].

3 Index 1 Configurations

Tree-like structures have a fundamental impact on the characterization of index 1 configurations. In the following, a *C–V loop* is a loop formed exclusively by capacitors and/or voltage sources; similarly, an *I–L cutset* is a cutset formed exclusively by inductors and/or current sources; *proper trees* are defined as C–R–V trees whose branches contain every voltage source, every capacitor and (possibly) some resistors.

Theorem 1. *Assume that $L(i_L)$ and $C(v_C)$ are non-singular and that there is no coupling among resistors. Then, system (1) is index 1 if and only if*

$T_{1,2}$ *There are neither C–V loops nor I–L cutsets and*
A_1 *The sum of conductance products in proper trees does not vanish.*

Proof. If $C(v_C)$ and $L(i_L)$ are non-singular, (1) is index 1 if and only if

$$J = \psi_v(u, v, t) = \begin{pmatrix} A_{\mathcal{R}} G A_{\mathcal{R}}^T & A_{\mathcal{CV}} \\ A_{\mathcal{CV}}^T & 0 \end{pmatrix} \tag{3}$$

is non singular, where $A_{\mathcal{CV}} = (A_C \ A_V)$ and $G(A_{\mathcal{R}}^T e) = \gamma'(A_{\mathcal{R}}^T e)$ has been written as G to simplify notation. To analyse this matrix we will perform the following factorization:

$$J = J_1 J_2 J_3 = \begin{pmatrix} A_{\mathcal{R}} & A_{\mathcal{CV}} & 0 \\ 0 & 0 & I_{\mathcal{CV}} \end{pmatrix} \begin{pmatrix} G & 0 & 0 \\ 0 & 0 & I_{\mathcal{CV}} \\ 0 & I_{\mathcal{CV}} & 0 \end{pmatrix} \begin{pmatrix} A_{\mathcal{R}}^T & 0 \\ A_{\mathcal{CV}}^T & 0 \\ 0 & I_{\mathcal{CV}} \end{pmatrix}, \tag{4}$$

where we have split the information derived from the topological description of the circuit, included in J_1 and J_3, and the relevant information coming from the constitutive relationships of the circuit elements (amounting in this case to matrix G), included in J_2.

Letting $n_e + 1$, n_C, n_R and n_V be respectively the number of nodes, capacitors, conductances and voltage sources in the circuit, set $n = n_e + n_C + n_V$, $m = n_R + 2(n_C + n_V)$. With this notation, the matrices J, J_1, J_2 and J_3 above have dimensions $n \times n$, $n \times m$, $m \times m$ and $m \times n$, respectively. Under the existence of a proper tree, the C–R–V subgraph connects all nodes in the circuit, which implies that $n_e \leq n_R + n_C + n_V$ and then $n \leq m$.

From (3) and (4) it can be proven necessary that there are neither C–V loops nor I–L cutsets; for if there is a C–V loop, then $A_{\mathcal{CV}}$ has not full column rank, and if there is an I–L cutset, then $(A_{\mathcal{R}} \; A_{CV})$ has not full row rank. These two conditions directly lead to the existence of at least one proper tree.

In order to delve into (4), we can now apply the Cauchy–Binet formula [HJ85] to J; this is a well-known algebraic property which expands the determinant of a matrix product into a sum of products of determinants, namely,

$$\det J_1 J_2 J_3 = \sum_{\alpha,\beta} \det J_1^{\omega,\alpha} \det J_2^{\alpha,\beta} \det J_3^{\beta,\omega}, \tag{5}$$

where the sum is taken over all index sets $\alpha, \beta \subseteq \{1, \ldots, m\}$ with cardinality n, and $\omega = \{1, \ldots, n\}$; $J_2^{\alpha,\beta}$ is the $n \times n$ submatrix of J_2 defined by the rows indexed by α and the columns indexed by β, and $J_1^{\omega,\alpha}$ (resp. $J_3^{\beta,\omega}$) is the submatrix of J_1 (resp. of J_3) including entries from all rows in J_1 (resp. all columns in J_3) and the columns indexed by α (resp. the rows indexed by β).

In the terms of expansion (5), it can easily be seen that some matrix blocks from (4) must be wholly or partially included for the corresponding determinant not to vanish. Actually, all non-zero terms in the formula can be shown to have the following structure:

$$\begin{vmatrix} A_{\tilde{\mathcal{R}}} & A_{CV} & 0 \\ 0 & 0 & I_{CV} \end{vmatrix} \begin{vmatrix} \tilde{G} & 0 & 0 \\ 0 & 0 & I_{CV} \\ 0 & I_{CV} & 0 \end{vmatrix} \begin{vmatrix} A_{\tilde{\mathcal{R}}}^T & 0 \\ A_{CV}^T & 0 \\ 0 & I_{CV} \end{vmatrix}, \tag{6}$$

where a tilde stands for a set of conductances of the original circuit. For $(A_{\tilde{\mathcal{R}}} \; A_{CV})$ in (6) to have full column rank, C–V–$\tilde{\text{R}}$ must be a tree, which is necessarily proper as it includes all capacitors and voltage sources.

If there is no coupling among conductances, then G is a diagonal matrix and the calculation of (6) leads to the following sum:

$$\det J = \pm \sum_{T \in \mathcal{T}_{\mathrm{p}}} \prod_{G_i \in T} G_i, \tag{7}$$

where $\mathcal{T}_{\mathrm{p}}$ is the set of proper trees in the circuit. $\square$

For circuits with no coupling among conductances, previous results such as Theorem 2 below can be derived from Theorem 1. Specifically, the definiteness assumption $\hat{A}_1$ in cases without conductive coupling can be seen as a particular instance of condition A_1 above. Note the broader scope of Theorem 1 provided by the full index-1 characterization implicit in the "if and only if" there.

Theorem 2 ([RT05]). *Assume that $L(i_L)$ and $C(v_C)$ are non-singular. Then, system (1) is index 1 if*

$T_{1,2}$ *There are neither C–V loops nor I–L cutsets and*
$\hat{A}_1$ *G is (positive or negative) definite.*

Indeed, if there are neither C–V loops nor I–L cutsets, then there must exist at least one proper tree within the circuit. Since all conductances are simultaneously positive or negative, all conductance products have the same sign and sum (7) is not null.

4 Concluding Remarks

As shown in Sect. 3, tree-like structures are central in circuit index characterizations. The case of index 1 in ANA has been examined in the present
communication. Employing an analogous procedure to the one detailed here,
necessary and sufficient conditions for index 0 in MNA can be obtained. A
similar, more complex process can also be used to characterize index 1 for
MNA; this involves the use of projectors and so called *normal trees*.

The inclusion of couplings, the analysis of index 2 configurations and the
study of models with controlled sources are future research lines.

References

[BCP96]   Brenan, K.E., Campbell, S.L., Petzold, L.R.: Numerical solution of initial-
          value problems in differential-algebraic equations. SIAM, Philadelphia
          (1996)
[ET00]   Estévez-Schwarz, D., Tischendorf, C.: Structural analysis of electric cir-
          cuits and consequences for MNA. Internat. J. Circuit Theory Appl., **28**,
          131–162 (2000)
[GF99a]   Günther, M., Feldmann, U.: CAD-based electric-circuit modeling in in-
          dustry. I: Mathematical structure and index of network equations. Surv.
          Math. Ind., **8**, 97–129 (1999)
[GF99b]   Günther, M., Feldmann, U.: CAD-based electric-circuit modeling in in-
          dustry. II: Impact of circuit configurations and parameters. Surv. Math.
          Ind., **8**, 131–157 (1999)
[HJ85]   Horn, R.A., Johnson, C.R.: Matrix analysis. Cambridge University Press,
          Cambridge (1985)
[Rei98]   Reiszig, G.: The index of the standard circuit equations of passive
          RLCTG-networks does not exceed 2. Proc. ISCAS'98, **3**, 419–422 (1998)
[Ria06]   Riaza R.: Time-domain properties of reactive dual circuits. Internat. J.
          Circuit Theory Appl., **34**, 317–340 (2006)
[RT05]   Riaza R., Torres-Ramírez, J.: Nonlinear circuit modelling via nodal
          methods. Internat. J. Circuit Theory Appl., **33**, 281–305 (2005)
[Tis99]   Tischendorf, C.: Topological index calculation of DAEs in circuit simula-
          tion. Surv. Math. Ind., **8**, 187–199 (1999)
[Tis03]   Tischendorf, C.: Coupled systems of differential algebraic and partial
          differential equations in circuit and device simulation. Modeling and
          numerical analysis. Habilitationsschrift, Humboldt-Univ, Berlin (2003)

Fingerprint Classification using Entropy Sensitive Tracing

Preda Mihăilescu[1], Krzysztof Mieloch[2], and Axel Munk[2]

[1] Mathematical Institute, University of Göttingen, Germany
   preda@uni-math.gwdg.de
[2] Institute for Mathematical Stochastics, University of Göttingen, Germany
   mieloch@math.uni-goettingen.de, munk@math.uni-goettingen.de

1 Introduction

Fingerprints are currently the leading approach to biometric recognition [1]. The reasons are multiple – we mention on the one hand the more than centennial tradition of fingerprint use for forensic purposes and on the other hand the existence of some well-established experience – based rules derived along the line. Fingerprints have a specific flow dynamics, which comes in quite distinct flow patterns – these help define *classes* of fingerprints. The flow pattern carries various singularities, named *minutiae* – most important are line endings and bifurcations.

This chapter treats the classification of fingerprints. In order to establish the identity of a person from a given fingerprint image, it is necessary to search large databases. Hence, the first step is to reduce the search field by assigning a given fingerprint into one of a small number of categories. Then the matching is performed only among fingerprints which belong to the class of the template.

Most of the classification schemas currently used worldwide are variants of Henry's classification scheme [1]. The six most common classes are: *tended arch, arch, left loop, right loop, whorl* and *twin loop*.

While finger matching is usually performed with the help of local features, the fingerprints classification is generally based on global features, such as skin ridge flow. For classifications, one may consider orientation field as a feature vector [2–4]. Different procedures for reduction of the dimension of the orientation field, such as e.g. Karhunen–Loève transformation [6, 7] were proposed. With the reduced vector, the statistical classifiers are involved [5,8]. Further approaches are found in [9–11].

2 Tracer

We developed an approach for fingeprint recognition which begins with the extraction of the essential fingerprint data, organised in semantic structures. We have described the notions and objects for this novel fingerprint data extraction denoted *entracing* in [12].

The extracted data contains not only minutiae information but also information about interminutiae connections (ridges) as complete connected lines. The flow of these lines is used in this chapter for classification.

3 Characteristic Lines

An expert analyst's experience suggested that every Henry class has certain windings of connected lines which are specific and do not occur in other classes. They are to be found typically around the centre of the finger (see Fig. 1). This suggests the fact that, using the entraced connected lines, one may have a quick search for class – characteristic lines.

Guided by these insight, the research questions which we begin to address in the present chapter are the following:

A. What simple metric characterisation can be used for defining distinctive, class dependent, and connected line flows?
B. How accurate is the statistical classification based on the given metric?
C. When applied to large databases, how close can the new method approach the state of the art performances. Or, can one approach the intrinsic limits of classification?[1]

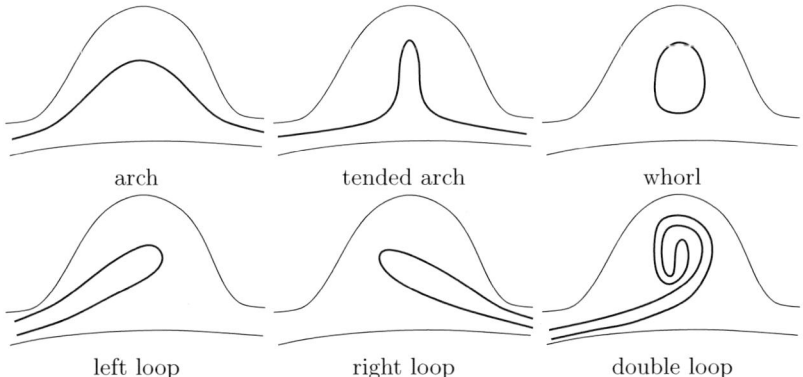

arch   tended arch   whorl

left loop   right loop   double loop

Fig. 1. Classes according to Henry's classification scheme

---

[1] It is well known that there are fingerprints which cannot be uniquely assigned to classes, while others cannot be assigned to any class, even by the human expert.

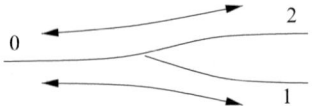

Fig. 2. Traversation of a bifurcation

3.1 Defining Characteristic Lines

As a result of tracing we get a usually disconnected graph $G = \{V, E\}$ where each vertex $v \in V$ has degree either 1 (ending or border point) or 3 (bifurcation). A characteristic line c is a path in the graph

$$c = \langle v_1, e_1, v_2, \ldots, e_{m-1}, v_m \rangle,$$

which connects two vertices v_1, v_m of degree 1 by edges and bifurcations (vertices of degree 3).

Let C be a set of all such paths with the additional condition that, in order to respect the natural flow, bifurcations are traversed only by following the larger angles (Fig. 2).

3.2 Characteristic Lines and Classes

Each edge in the graph $e \in E$ is a broken line and is defined, along with the vertex points $p(v_i) = p_i = (X_i, Y_i), i \in [1, n(e)]$, by a set of $n(e) - 1$ intermediate points, thus $e = \langle p_1 = (x_1, y_1), p_2, \ldots, p_{n(e)} \rangle$. The *metric features* used for classification are the total sum of orientation changes (TOC):

$$\Phi_t(c) = |(c, n(e))|$$

and the maximal orientation change (MOC):

$$\Phi_m(c) = \max_{k_1, k_2 \in [0, n(e)]} |\Phi(c, k_1) - \Phi(c, k_2)|,$$

where $\Phi(c, k)$ denotes the sum of arc changes in E:

$$\Phi(c, k) = \sum_{i=1}^{k} \angle (p_i - p_{i-1}, p_{i+1} - p_i)$$

The changes in orientation of a connected line around the centre of the finger appear to be a simple, class – distinctive value. For the main six Henry classes they vary around the following typical values[2]:

Class	Arch	Tended arch	Left and right loop	Whorl	Twin loop
TOC	0	0	π	2π	$\gg 2\pi$
MOC	$\frac{\pi}{3}$	π	π	2π	$\gg 2\pi$

[2] The distinction between left and right loops is done by considering the orientation/position of the start point.

3.3 Statistical Analysis

As mentioned in the previous section, the total orientation change $\Phi_t(\cdot)$ and the maximal orientation change $\Phi_m(\cdot)$ of a characteristic line are a good distinctive measure for classifying fingerprints.

In order to find representants for each class, in a training set of fingerprints whose classes are known, a density of those two features for all characteristic lines of all fingerprints in one class was estimated. Figure 3 illustrates computed densities of the maximal orientation change.

Consequently a fingerprint is classified according to the maximum likelihood of its feature vector against estimated densities from the training set.

On the test database (230 images from the dataset db02 in the FVC2002 database [1]) the classifier achieves an accuracy of 88.2%; the confusion matrix is reported in Table 1. The results compare well with state of the art publications and require very few computations.

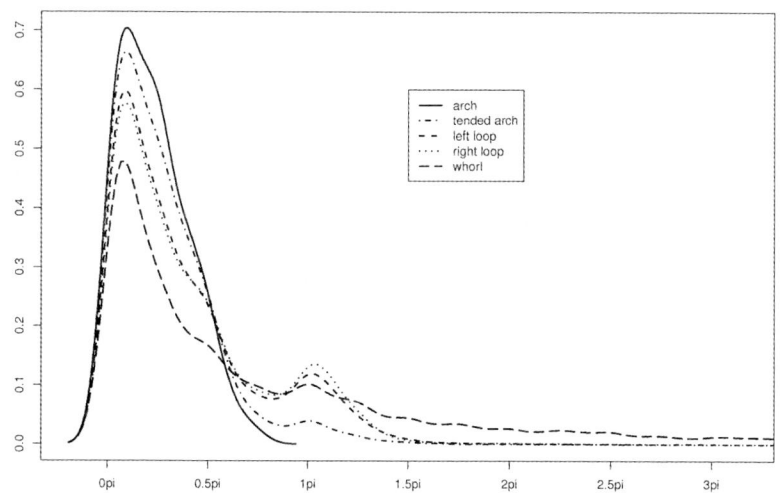

Fig. 3. Representants for all classes

Table 1. Confusion matrix

	A	TA	LL	RL	W	DL
A	36	3	1	0	0	0
TA	2	29	0	1	0	0
LL	1	1	37	0	0	1
RL	1	2	0	38	2	1
W	0	0	1	0	29	3
DL	0	0	2	1	4	34

4 Conclusions

We presented a new approach to fingerprint classification, which uses ridge data extracted by means of entropy sensitive tracing [12]. The approach goes in the direction of *characteristic lines* and is expected to have an appealing balance between the effective amount of information and operations used for class distinction and the accuracy of the classification. The chapters cover preliminary tests and statistics which encourage this hypothesis.

References

1. Davide Maltoni, Dario Maio, Anil K. Jain, Salil Prabhakar. Handbook of fingerprint recognition. Springer Verlag 2003.
2. Ballan, M., Sakarya, F. and Evans, B.: A Fingerprint classification technique using directional images. in 31st Asilomar conference on Signals, Systems and Computers, **1**, 101–104.
3. Mardia, K., Baczkowski, A., Feng, X. and Hainsworth. T.: Statistical Methods for Automatic Interpretation of Digitally Scanned Fingerprints. Pattern Recognition Letters, **18**, 1197–1203 (1997).
4. Kawagoe, M. and Tojo, A.: Fingerprint pattern classification. Pattern Recognition, **17**, 295–303 (1984).
5. Senior, A.: A hidden Markov model fingerprint classifier. In proc. of the 31st Asilomar conference on Signals, Systems and Computers, 306–310 (1997)
6. Cappelli, R., Maio, D., Maltoni, D.: Fingerprint classification based on Multi-space KL. In: Proc. of AutoID'99 (Workshop on Automatic Identification Advances Technologies), 117-120 (1999)
7. Cappelli, R., Maio, D., Maltoni, D.: Multi-space KL for pattern representation and classification. IEEE Transactions on Pattern Analysis and Machine Intelligence. (23), 977–996 (2001)
8. Hong, L., Wan, Y., Jain, A.: Fingerprint image enhancement: algorithm and performance evaluation. IEEE Trans. on pattern analysis and machine intelligence. **20**, 777–789 (1998)
9. Fritz, A.P., Green R.J.: Fingerprint classification using a hexagonal fast Fourier transform. Pattern recognition, **29**, 1587–1597 (1996)
10. Ehrhardt, J.C.: Hexagonal Fast Fourier Transform with rectangular output. IEEE Trans. on Signal Proceeding, **41**, 1469–1472 (1993)
11. Maio, D., Maltoni, D.: A structural approach to fingerprint classification. In: Proc. of the International Conference on Pattern Recognition, 578–592 (1996)
12. Mihăilescu, P., Mieloch, K., Munk, A.: Entropy sensitive fingerprint minutiae extraction, *submitted*.

An Invariant Domain Preserving MUSCL Scheme

Christophe Berthon

MAB, UMR 5466 LRC M03, Université Bordeaux 1, 351 cours de la libération, 33400 Talence, France
INRIA Futurs, Projet ScAlApplix, 351 cours de la libération, 33400 Talence, France
Christophe.Berthon@math.u-bordeaux1.fr

Summary. The second-order MUSCL schemes are considered in the present work. A new limitation procedure is detailed to enforce relevant robustness properties. The scheme is thus shown to preserve the invariant domain.

1 Introduction

In general, increasing the order of accuracy remains a delicate question when numerical finite volume methods are considered to approximate the solutions of hyperbolic systems. *Simpler* procedures have been introduced to increase the order of accuracy. One of the most popular has been proposed by van Leer [Lee79]; namely the MUSCL scheme. Our purpose is to analyze this scheme which has been used in many applications (for instance, see [Col90, DEO92]). For the sake of clarity in the presentation, we focus our attention on the Euler equations:

$$\begin{cases} \partial_t \rho + \partial_x \rho u = 0, \\ \partial_t \rho u + \partial_x (\rho u^2 + p) = 0, \\ \partial_t \rho E + \partial_x (\rho E + p) u = 0, \end{cases} \tag{1}$$

$$p = (\gamma - 1)\left(E - \rho \frac{u^2}{2}\right), \quad \gamma \in (1, 3]. \tag{2}$$

The admissible state space Ω is defined by

$$\Omega = \left\{ \mathbf{w} = (\rho, \rho u, \rho E) \in \mathbb{R}^3; \ \rho > 0, \ u \in \mathbb{R}, \ e(\mathbf{W}) = E - \frac{u^2}{2} > 0 \right\}.$$

A MUSCL scheme is considered to approximate the weak solutions of (1). For the sake of simplicity in the notations, we briefly recall the first-order conservative scheme based on piecewise constant approximation of the solution at time t^n, with $x_{i+\frac{1}{2}} = x_i + (x_{i+1} - x_i)/2$, where $(x_i)_{i\in\mathbb{Z}}$ denote the mesh nodes. To simplify the notations, the mesh is assumed to be uniform with size

$\Delta x = x_{i+1} - x_i$. To approximate the solution of (1) at time $t^{n+1} = t^n + \Delta t$, the sequence $(\mathbf{w}_i^{n+1})_{i \in \mathbb{Z}}$ is defined by the following conservative scheme [Tor99]:

$$\mathbf{w}_i^{n+1} = \mathbf{w}_i^n - \frac{\Delta t}{\Delta x}\left(\mathbf{F}(\mathbf{w}_i^n, \mathbf{w}_{i+1}^n) - \mathbf{F}(\mathbf{w}_{i-1}^n, \mathbf{w}_i^n)\right), \tag{3}$$

where $\mathbf{F} : \Omega \times \Omega \to \mathbb{R}^3$ denotes the numerical flux function assumed to be Lipschitz consistent. The time increment Δt is restricted according to the following CFL condition:

$$\frac{\Delta t}{\Delta x} \max_{i \in \mathbb{Z}} \left(|\lambda^{\pm}(\mathbf{w}_i^n, \mathbf{w}_{i+1}^n)|\right) \leq \frac{1}{2}, \tag{4}$$

where $\lambda^{\pm}(\mathbf{w}_i^n, \mathbf{w}_{i+1}^n)$ are the numerical velocity of the acoustic waves associated with the numerical flux function $\mathbf{F}(\mathbf{w}_i^n, \mathbf{w}_{i+1}^n)$.

In the sequel, the first-order scheme (3) is assumed to satisfy the numerical invariance of Ω: if $\rho_i^n > 0$ and $e_i^n > 0$ for all $i \in \mathbb{Z}$ then $\rho_i^{n+1} > 0$ and $e_i^{n+1} > 0$ for all $i \in \mathbb{Z}$.

The MUSCL schemes use a better reconstruction than a piecewise constant functions since piecewise linear functions are considered. In the cell $(x_{i-\frac{1}{2}}, x_{i+\frac{1}{2}})$, the inner approximations, located at $x_{i \pm \frac{1}{2}}$, are considered: $\mathbf{w}_i^{n,\pm} = \mathbf{w}^h(x_{i \pm \frac{1}{2}}, t^n)$. Let us note that we do not impose a conservative reconstruction since we do not enforce: $\mathbf{w}_i^n = \frac{1}{2}(\mathbf{w}_i^{n,-} + \mathbf{w}_i^{n,+})$.

2 Stability of the MUSCL Schemes

Arguing the above notations, the space second-order MUSCL scheme reads as follows:

$$\mathbf{w}_i^{n+1} = \mathbf{w}_i^n - \frac{\Delta t}{\Delta x}\left(\mathbf{F}(\mathbf{w}_i^{n,+}, \mathbf{w}_{i+1}^{n,-}) - \mathbf{F}(\mathbf{w}_{i-1}^{n,+}, \mathbf{w}_i^{n,-})\right), \tag{5}$$

where $\mathbf{F}$ is the numerical flux function introduced in (3).

Following an idea introduced by Perthame–Shu [PS96], the stability of the MUSCL scheme (5) is considered when involving a non-conservative gradient reconstruction. Of course, all the following results can be easily extended in the framework of a conservative gradient reconstruction. The reader is referred to [Bou04, KP94] where stability results are given when assuming a conservative reconstruction.

The stability of the scheme is based on the existence of a *ghost state*, denoted by $\mathbf{w}_i^{n,\star}$, defined as follows: $\alpha_i^- \mathbf{w}_i^{n,-} + \alpha_i^\star \mathbf{w}_i^{n,\star} + \alpha_i^+ \mathbf{w}_i^{n,+} = \mathbf{w}_i^n$, where the positive coefficients $\alpha_i^{\pm\star}$ are fixed in order to satisfy: $\alpha_i^- + \alpha_i^\star + \alpha_i^+ = 1$. In fact, a new mesh is thus defined where each cell $I_i = (x_{i-\frac{1}{2}}, x_{i+\frac{1}{2}})$ is split into three sub-cells: $I_i^- = (x_{i-\frac{1}{2}}, x_{i-\frac{1}{2}} + \alpha_i^- \Delta x)$, $I_i^\star = (x_{i-\frac{1}{2}} + \alpha_i^- \Delta x, x_{i-\frac{1}{2}} + (\alpha_i^- + \alpha_i^\star)\Delta x)$ and $I_i^+ = (x_{i-\frac{1}{2}} + (\alpha_i^- + \alpha_i^\star)\Delta x, x_{i+\frac{1}{2}})$. The vectors $\mathbf{w}_i^{n,\pm\star}$ are associated with the sub-cell $I_i^{\pm\star}$. The first-order scheme (3) is thus applied on the sub-mesh to obtain:

$$\mathbf{w}_i^{n+1,-} = \mathbf{w}_i^{n,-} - \frac{\Delta t}{\alpha_i^- \Delta x}\left(\mathbf{F}(\mathbf{w}_i^{n,-}, \mathbf{w}_i^{n,\star}) - \mathbf{F}(\mathbf{w}_{i-1}^{n,+}, \mathbf{w}_i^{n,-})\right),$$

$$\mathbf{w}_i^{n+1,\star} = \mathbf{w}_i^{n,\star} - \frac{\Delta t}{\alpha_i^\star \Delta x}\left(\mathbf{F}(\mathbf{w}_i^{n,\star}, \mathbf{w}_i^{n,+}) - \mathbf{F}(\mathbf{w}_i^{n,-}, \mathbf{w}_i^{n,\star})\right),$$

$$\mathbf{w}_i^{n+1,+} = \mathbf{w}_i^{n,+} - \frac{\Delta t}{\alpha_i^+ \Delta x}\left(\mathbf{F}(\mathbf{w}_i^{n,+}, \mathbf{w}_{i+1}^{n,-}) - \mathbf{F}(\mathbf{w}_i^{n,\star}, \mathbf{w}_i^{n,+})\right).$$

As a consequence, the approximate solution obtained by (5) rewrites as follows: $\mathbf{w}_i^{n+1} = \alpha_i^- \mathbf{w}_i^{n+1,-} + \alpha_i^\star \mathbf{w}_i^{n+1,\star} + \alpha_i^+ \mathbf{w}_i^{n+1,+}$. Since the states $\mathbf{w}_i^{n+1,\pm\star}$ result from a first-order scheme which satisfies the invariance of Ω, the following statement is obtained [Ber05] (see also [Ber06(2)] for a time and space second-order extension):

Theorem 1. *Assume that the first-order scheme (3) satisfies the invariance of Ω. Assume that $\mathbf{w}_i^n \in \Omega$ for all $i \in \mathbb{Z}$ and impose $\mathbf{w}_i^{n,\pm\star} \in \Omega$. Consider the following CFL condition:*

$$\frac{\Delta t}{\Delta x}\max_{i \in \mathbb{Z}}\left(|\lambda^\pm(\mathbf{w}_i^{n,-}, \mathbf{w}_i^{n,\star})|, |\lambda^\pm(\mathbf{w}_i^{n,\star}, \mathbf{w}_i^{n,+})|, |\lambda^\pm(\mathbf{w}_i^{n,+}, \mathbf{w}_{i+1}^{n,-})|\right)$$

$$\leq \frac{1}{2}\min_{i \in \mathbb{Z}}(\alpha_i^-, \alpha_i^\star, \alpha_i^+).$$

If $\mathbf{w}_i^n \in \Omega$ for all $i \in \mathbb{Z}$ then $\mathbf{w}_i^{n+1} \in \Omega$ for all $i \in \mathbb{Z}$.

Let us emphasize that the condition $\mathbf{w}_i^{n,\pm\star} \in \Omega$ turns out to be a new slope limitation. The gradient reconstruction must be modified according to the new conditions.

Now, we exhibit a non-conservative gradient reconstruction based on the primitive variables (ρ, u, p). In a cell I_i, the inner approximations located at $x_{i\pm\frac{1}{2}}$ write as follows:

$$\rho_i^{n,\pm} = \rho_i^n \pm \Delta\rho, \quad u_i^{n,\pm} = u_i^n \pm \Delta u, \quad p_i^{n,\pm} = p_i^n \pm \Delta p, \tag{6}$$

where $(\Delta\rho, \Delta u, \Delta p)$ denotes the increment vector which must satisfy the new limitations. The increments obtained involving some "usual" limitations (minmod, superbee, MC... [Tor99]) are denoted δX. These standard limiters are thus modified to enforce the condition: $\mathbf{w}_i^{n,\pm\star} \in \Omega$. To illustrate our purpose, let us consider the minmod limiter which writes as follows: $\delta X = \frac{1}{2}\mathrm{minmod}(X_i - X_{i-1}, X_{i+1} - X_i)$. Under the following assumption: $\alpha_i^- = \alpha_i^\star = \alpha_i^+ = \frac{1}{3}$, all the required conditions are satisfied with the choice

$$\Delta\rho = \rho_i^n \max\left(-1, \min(1, \frac{\delta\rho}{\rho_i^n})\right), \quad \Delta p = p_i^n \max\left(-1, \min(1, \frac{\delta p}{p_i^n})\right),$$

$$\Delta u = \text{sign}(\delta u) \sqrt{\min\left((\delta u)^2, \frac{p_i^n}{(\gamma-1)\rho_i^n\left(1+2\left(\frac{\Delta\rho}{\rho_i^n}\right)^2\right)}\right)}.$$

3 MUSCL Schemes for 2D Unstructured Meshes

The above stability results concerning the 1D MUSCL schemes are extended when considering 2D unstructured meshes [PQ94, KK05, PS96, Ber06]. Once again, the expected stability results come from the introduction of a relevant sub-mesh associated with relevant ghost states. Let us note from now on that the sub-mesh is an artefact useful in the proof but never informatically computed.

In a cell C_i with $\Lambda(i)$ adjacent cells $j(k)$, $1 \leq k \leq \Lambda(i)$, the edge which separates C_i and $C_{j(k)}$ is denoted $\Gamma_{ij(k)}$ where $\mathbf{n}_{ij(k)}$ is the associated outer unit normal. The state $\mathbf{w}_{ij(k)}$ denotes the second-order approximation of the solution in the cell C_i but located near the edge $\Gamma_{ij(k)}$. With these notations, the 2D MUSCL scheme reads as follows:

$$\mathbf{W}_i^{n+1} = \mathbf{W}_i^n - \frac{\Delta t}{|C_i|} \sum_{k=1}^{\Lambda(i)} |\Gamma_{ij(k)}| \phi(\mathbf{n}_{ij(k)}, \mathbf{w}_{ij(k)}, \mathbf{w}_{j(k)i}), \tag{7}$$

where $\phi(\mathbf{n}, \mathbf{w}_L, \mathbf{w}_R)$ denotes the numerical flux function.

The cell C_i is thus split into $\Lambda(i)+1$ sub-cells as displayed in Fig. 1. A ghost state $\mathbf{w}_i^\star$ is defined as follows: $\frac{|C_i^\star|}{|C_i|}\mathbf{w}_i^\star + \sum_{k=1}^{\Lambda(i)} \frac{|C_{ij(k)}|}{|C_i|}\mathbf{w}_{ij(k)} = \mathbf{w}_i^n$. We apply the same idea used to prove the stability of the 1D MUSCL scheme. To access such an issue, we introduce the following first-order scheme:

$$\mathbf{w}_i^{n+1} = \mathbf{w}_i^n - \frac{\Delta t}{|C_i|} \sum_{k=1}^{\Lambda(i)} |\Gamma_{ij(k)}| \phi(\mathbf{n}_{ij(k)}, \mathbf{w}_i, \mathbf{w}_{j(k)}),$$

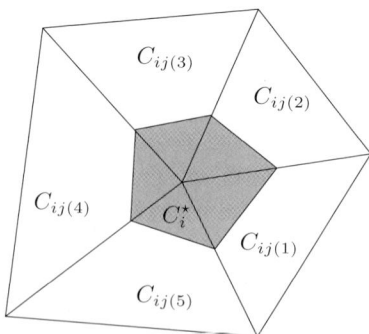

Fig. 1. Sub-cell decomposition of the cell C_i

This first-order scheme is considered to evolve in time the vectors $\mathbf{w}_i^\star$ and $\mathbf{w}_{ij(k)}$ stated on the sub-mesh. The following relation is easily obtained:

$$\mathbf{w}_i^{n+1} = \frac{|C_i^\star|}{|C_i|}\mathbf{w}_i^{n+1,\star} + \sum_{k=1}^{\Lambda(i)} \frac{|C_{ij(k)}|}{|C_i|}\mathbf{w}_{ij(k)}^{n+1}, \tag{8}$$

where $\mathbf{w}_i^{n+1}$ is given by (7). We thus obtain (see [Ber06])

Theorem 2. *Let $(\mathbf{w}_i^n)_{i\in\mathbb{Z}}$ be given by (7). Assume that the states $\mathbf{w}_i^n$, $\mathbf{w}_{ij}$ and $\mathbf{w}_i^\star$ belong to Ω. Assume that the sub-cells (C_{ij} or $C_i^\star$) are split into k elementary triangles T^k. Let us assume the following CFL restrictions: $\Delta t \frac{|\Gamma^k|}{|T^k|}\max|\lambda^\pm(\mathbf{n}_{\Gamma^k}, \mathbf{w}_{\Gamma^k}^+, \mathbf{w}_{\Gamma^k}^-)| \le 1$, where Γ^k is the edge of T^k which separates two sub-cells and where the states $\mathbf{w}_{\Gamma^k}^\pm$ define the value of the vector states in each side of this edge. If $\mathbf{w}_i^n \in \Omega$ for all $i \in \mathbb{Z}$ then $\mathbf{w}_i^{n+1} \in \Omega$ for all $i \in \mathbb{Z}$.*

From a numerical point of view, let us emphasize that the scheme is free from the definition of the sub-cells $C_i^\star$ and C_{ij}. For the sake of simplicity, we adopt the following condition: $\frac{|C_i^\star|}{|C_i|} = \frac{|C_{ij(k)}|}{|C_i|} = \frac{1}{\Lambda(i)+1}$.

References

[Ber05]   C. Berthon, Stability of the MUSCL schemes for the Euler equations, *Comm. Math. Sci.*, 3, pp. 133–158 (2005).

[Ber06]   C. Berthon, Robustness of MUSCL schemes for 2D unstructured meshes, *J. Comput. Phys.*, 218, pp. 495–509 (2006).

[Ber06(2)] C. Berthon, Why the MUSCL-Hancock scheme is L¹-stable, *Numer. Math.*, 104, pp. 27–46 (2006).

[Bou04]   F. Bouchut, Nonlinear stability of finite volume methods for hyperbolic conservation laws, and well-balanced schemes for sources, *Frontiers in Mathematics series*, Birkhäuser (2004).

[Col90]   P. Colella, Multidimensional upwind methods for hyperbolic conservation laws, *J. Comput. Phys.*, 87, pp. 171–200 (1990).

[DEO92]   L. J. Durlofsky, B. Engquist, S. Osher, Triangle based adaptive stencils for the solution of hyperbolic conservation laws, *J. Comput. Phys.*, 98, pp. 64–73 (1992).

[KK05]   B. Keen, S. Karni, A second order kinetic scheme for gas dynamics on arbitrary grids, *J. Comput. Phys.*, 205, pp. 108–130 (2005).

[KP94]   B. Khobalatte, B. Perthame, Maximum principle on the entropy and second-order kinetic schemes, *Math. Comput.*, 62, pp. 119–131 (1994).

[Lee79]   B. van Leer, Towards the ultimate conservative difference scheme. V. A second-order sequel to Godunov's method, *J. Comput. Phys.*, 32, pp. 101–136 (1979).

[PQ94]   B. Perthame, Y. Qiu, A variant of Van Leer's method for multidimensional systems of conservation laws, *J. Comput. Phys.*, 112, pp. 370–381 (1994).
[PS96]   B. Perthame, C.-W. Shu, On positivity preserving finite volume schemes for Euler equations, *Numer. Math.*, 73, pp. 119–130 (1996).
[Tor99] E. F. Toro, Riemann solvers and numerical methods for fluid dynamics. A practical introduction, Second edition. Springer-Verlag, Berlin (1999).

A Stable CE–SE Numerical Method for Time-Dependent Advection–Diffusion Equation

R. Company, E. Defez, L. Jódar, and E. Ponsoda

Instituto de Matemática Multidisciplinar, Universidad Politécnica de Valencia, Spain
{rcompany, edefez, ljodar, eponsoda}@imm.upv.es

Summary. In this chapter an efficient conservation element–solution element (CE–SE) to construct numerical solutions of time-dependent advection-diffusion equation initial value problems is presented. Stability conditions of the method are established in terms of data.

1 Introduction

Time-dependent advection–diffusion equation of the type

$$\left.\begin{array}{l} u_t(x,t) + a(t)u_x(x,t) - b(t)u_{xx}(x,t) = 0 \\[2mm] u(x,0) = f(x) \ ; \ b(t) \geq 0 \ ; \ (x,t) \in \mathbb{R} \times [0,+\infty[\end{array}\right\} \tag{1}$$

appears frequently in the modelization of physical and technological processes like, for example, the evaluation of the heating through radiations of microwaves when the properties of medium (dielectric properties, humidity, etc.) vary with time, [Met83, Poz90]. Equations of type (1) also appear in the study of the transmission of flows in industrial tubes [Chu84], conduction of heat in solids [Car95], etc. In order to solve (1) a modified version of conservation-elements and solution-elements (CE–SE) method was applied in [Def05]. The standard (CE–SE) method was used for the solution of conservation laws [Cha95], offering significant advantages with respect to other schemes like differences or finite-elements methods: the conservation of the flow is used as much in time as in space, and it also deals with independent variables and their derivative, calculating them simultaneous for each node. This (CE–SE) method has also been applied to the advection-diffusion equation with constant coefficients, [Wan99]. The modified method proposed in [Def05] preserve all these advantages.

For the sake of clarity in the presentation, the modified CE–SE scheme given in [Def05] can be summarized as follows. We take a mesh in $\mathbb{R} \times [0, +\infty[$, with points (x_j, t^n) given by $x_j = j\Delta x, t^n = n\Delta t$. We define the solution element $SE(j, n)$ like the rhombus centered in the mesh point (j, n) with diagonals Δx and Δt. For each solution element $SE(j, n)$, the approximate solution of (1) is defined by

$$U(x, t; j, n) = U_j^n + (U_x)_j^n \left[(x - x_j) - a(t^n)(t - t^n)\right], \quad \forall (x, t) \in SE(j, n),$$
(2)

where $(U)_j^n, (U_x)_j^n$ are undetermined constants. To obtain them, we define the vector

$$q(j, n) = \left[U_j^n, \frac{\Delta x}{4}(U_x)_j^n\right]^{\mathrm{T}}.$$
(3)

In [Def05] it is shown that $q(j, n)$ satisfies

$$q(j, n + 1) = Q_+(n + 1)\, Q_+(n + 1/2)\, q(j - 1, n)$$
$$+ \left[Q_+(n + 1)\, Q_-(n + 1/2) + Q_-(n + 1)\, Q_+(n + 1/2)\right] q(j, n)$$
$$+ Q_-(n + 1)\, Q_-(n + 1/2)\, q(j + 1, n), \quad \text{with}$$
(4)

$$Q_+(n) = \frac{1}{2}\begin{bmatrix} 1 + \nu_0^n & 1 - \nu^{n-1/2}\nu_2^{n-1/2} - \xi^n \\ \dfrac{-(1 - (\nu_0^n)^2)}{1 - \nu^n\nu_1^n + \xi^n} & \dfrac{-(1 - \nu_0^n)(1 - \nu^{n-1/2}\nu_2^{n-1/2} - \xi^n)}{1 - \nu^n\nu_1^n + \xi^n} \end{bmatrix},$$
(5)

$$Q_-(n) = \frac{1}{2}\begin{bmatrix} 1 - \nu_0^n & -(1 - \nu^{n-1/2}\nu_2^{n-1/2} - \xi^n) \\ \dfrac{1 - (\nu_0^n)^2}{1 - \nu^n\nu_1^n + \xi^n} & \dfrac{-(1 + \nu_0^n)(1 - \nu^{n-1/2}\nu_2^{n-1/2} - \xi^n)}{1 - \nu^n\nu_1^n + \xi^n} \end{bmatrix},$$
(6)

where the matrix coefficients are given by

$$\nu^n = a(t^n)\frac{\Delta t}{\Delta x} \qquad ; \, \nu^{n-1/2} = a(t^{n-1/2})\frac{\Delta t}{\Delta x};$$

$$\nu_1^n = <a>_1^n \frac{\Delta t}{\Delta x} \qquad ; \, \nu_2^{n-1/2} = <a>_2^{n-1/2} \frac{\Delta t}{\Delta x};$$

$$\nu_0^n = <a>_0^n \frac{\Delta t}{\Delta x} \qquad ; \qquad \xi^n = 4<b>_0^n \frac{\Delta t}{(\Delta x)^2};$$

$$<a>_0^n = \frac{2}{\Delta t}\int_{t^{n-1/2}}^{t^n} a(t)\mathrm{d}t \; ; \, <b>_0^n = \frac{2}{\Delta t}\int_{t^{n-1/2}}^{t^n} b(t)\mathrm{d}t;$$

$$<a>_2^{n-1/2} = \frac{8}{(\Delta t)^2}\int_{t^{n-1/2}}^{t^n} a(t)\left(t - t^{n-1/2}\right)\mathrm{d}t;$$

$$<a>_1^n = \frac{8}{(\Delta t)^2}\int_{t^{n-1/2}}^{t^n} a(t)\,(t^n - t)\,\mathrm{d}t\,.$$
(7)

The aim of this work is the study of the stability of this numerical scheme according to definition [Krö97, p.92].

Throughout this work, we will denote by $\| \, . \, \|$ the euclidean norm in $\mathbb{R}^2$. If $v \; : \; \mathbb{Z} \mapsto \mathbb{R}^2 \, / \, v(j) = [v_1(j), v_2(j)]^{\mathrm{T}}$, we will denote by $\| \, . \, \|_2$ the L^2-discrete norm with respect to the mesh Δx, which means $\|v\|_2^2 = \Delta x \sum_{j \in \mathbb{Z}} \left[v_1^2(j) + v_2^2(j) \right]$. The discrete Fourier transform of $v(j)$, denoted by $\widehat{v}(\theta)$, ant its inverse are defined, respectively, by

$$\widehat{v}(\theta) = \sum_{j \in \mathbb{Z}} \mathrm{e}^{-ij\theta} \left[v_1(j), v_2(j) \right]^{\mathrm{T}} \; ; \; \theta \in [-\pi, \pi] \; , \; v(j) = \int_{-\pi}^{\pi} \mathrm{e}^{ij\theta} \widehat{v}(\theta) \mathrm{d}\theta \; . \quad (8)$$

For $\left\{ \mathrm{e}^{-ij\theta} \right\}_{j \in \mathbb{Z}}$ representing an orthogonal system, the Parseval property gives

$$\|\widehat{v}\|_2^2 = \int_{-\pi}^{\pi} \sum_{j \in \mathbb{Z}} \left[(v_1(j))^2 + (v_2(j))^2 \right] \mathrm{d}\theta = \frac{2\pi}{\Delta x} \|v\|_2^2 \; . \quad (9)$$

For a matrix $A \in \mathbb{C}^{n \times n}$, we denote by $\rho(A)$ its spectral radius, defined by $\rho(A) = \max \left\{ |\lambda|; \lambda \in \sigma(A) \right\}$.

2 Stability Analysis

Given the numerical scheme summarized in the previous section, for vector (6) we define the sequence

$$q(n) \equiv \{q(j,n)\}_{j=-\infty}^{j=+\infty} = \{\ldots, q(-1,n), q(0,n), q(1,n), \ldots\} \; . \quad (10)$$

Let us assume that $q(n)$ satisfies

$$\|q(n)\|_2^2 = \sum_{j \in \mathbb{Z}} \|q(j,n)\|_2^2 \, \Delta x < \infty \; .$$

Taking the discrete Fourier transform in (10), one gets

$$q^\star(n, \theta) = \sum_{j \in \mathbb{Z}} q(j,n) \mathrm{e}^{-ij\theta} \; ; \; -\pi < \theta \leq \pi \; . \quad (11)$$

By (4) and (4), it follows that

$$q^\star(n+1, \theta) = \sum_{j \in \mathbb{Z}} q(j, n+1) \mathrm{e}^{-ij\theta}$$

$$= \sum_{j \in \mathbb{Z}} \left\{ Q_+(n+1) Q_+ \left(n+\frac{1}{2} \right) q(j-1, n) \right.$$

$$+ \left[Q_+(n+1) Q_-(n+\frac{1}{2}) + Q_-(n+1) Q_+ \left(n+\frac{1}{2} \right) \right] q(j, n)$$

$$\left. + Q_-(n+1) Q_- \left(n+\frac{1}{2} \right) q(j+1, n) \right\} \mathrm{e}^{-ij\theta} \; , \quad (12)$$

where matrices $\{Q_+(k), Q_-(k)\}$ are defined in (5) and (6), respectively. Then, one can rewrite (7) in the compact form

$$q^\star(n+1, \theta) = Q^{n+1}(\theta) Q^{n+\frac{1}{2}}(\theta) q^\star(n, \theta),$$

$$Q^k(\theta) = Q_+(k) e^{-i\frac{\theta}{2}} + Q_-(k) e^{i\frac{\theta}{2}} \; ; \; k = n+1, n+\frac{1}{2} \; . \qquad (13)$$

We will denote by the *amplification matrix* $Q^{n+1}(\theta) Q^{n+\frac{1}{2}}(\theta)$ introduced by (13). The following result gives a bound on the spectral radius for the amplification matrix.

Theorem 1. *Let $T > 0$ be fixed. If $|\nu(t)| = |a(t)| \dfrac{\Delta t}{\Delta x} < 1 \; ; \; \forall t \in [0, T]$, and*

$(\zeta^n + \Delta_-) > 0, \forall n > 0$ *semi-integer*, $n\Delta t < T$, $\Delta_- \equiv \dfrac{\nu^{n-\frac{1}{2}} \nu_1^{n-\frac{1}{2}} - \nu^n \nu_1^n}{2}$,

where $\nu^k, \nu_1^k, \zeta^n, \; k = n, n - \frac{1}{2}$; are defined by (1), then

$$\rho\left(Q^n(\theta)\right) \leq 1 \; ; \; \forall \theta \in [-\pi, \pi] \; .$$

Using the previous theorem, we can prove the following result on stability of the proposed scheme.

Theorem 2. *The numerical scheme is stable under conditions of Theorem 1.*

Proof. By (9) we can write for each $n \geq 0$

$$\|q(n+1)\|_2^2 = \frac{\Delta x}{2\pi} \|q^\star(n+1)\|_2^2 = \frac{\Delta x}{2\pi} \int_{-\pi}^{\pi} \|q^\star(n+1, \theta)\|^2 \, d\theta.$$

By (13) one gets

$$\|q(n+1)\|_2^2 \leq \frac{\Delta x}{2\pi} \int_{-\pi}^{\pi} \left\|Q^{n+1}(\theta)\right\|_2^2 \left\|Q^{n+\frac{1}{2}}(\theta)\right\|_2^2 \|q^\star(n, \theta)\|^2 \, d\theta \; . \qquad (14)$$

For matrix $Q^k(\theta)$, $0 \leq k \leq n+1$ and $\delta_{n+1} = \frac{(n+1)\Delta t}{2T} > 0$, there exists a norm [Ort72], denoted by $\| \cdot \|_{(k,\theta)}$, which fulfills

$$\left\|Q^k(\theta)\right\|_{(k,\theta)} \leq \rho\left(Q^k(\theta)\right) + \delta_{n+1} \leq 1 + \delta_{n+1} \; ,$$

and taking into account the equivalence of all matrix norms [Ort72], thus, there exists $M(k, \theta) > 0$, so that

$$\left\|Q^k(\theta)\right\|_2 \leq M(k, \theta)(1 + \delta_{n+1}) \; . \qquad (15)$$

Using (15) in expression (14), it follows that

$$\|q(n+1)\|_2^2 \leq \frac{\Delta x}{2\pi} \int_{-\pi}^{\pi} M^2(n+1, \theta) M^2\left(n + \frac{1}{2}, \theta\right)(1 + \delta_{n+1})^4 \|q^\star(n, \theta)\|^2 \, d\theta \; .$$

Taking $M(k) = \max\limits_{\theta \in [-\pi, \pi]} M(k, \theta)$, one gets

$$\|q(n+1)\|_2^2 \leq M^2(n+1) M^2(n + \tfrac{1}{2}) \left(1 + \delta_{n+1}\right)^4 \|q(n)\|_2^2 .$$

Iterating from $q(0)$ to $q(n)$, with $n\Delta t \leq T$, the above expression can be rewritten in the form

$$\|q(n+1)\|_2 \leq C(T) \left(1 + \delta_{n+1}\right)^{2(n+1)} \|q(0)\|_2 ,$$

$$C(T) = \max\limits_{n \leq \frac{T}{\Delta t}} M(n+1) M(0) \left(\prod_{i=1}^n M^2(i) \right).$$

Hence, it follows:

$$\left[\left(1 + \delta_{n+1}\right)^{n+1} \right]^2 \leq \left(e^{\delta_{n+1}} \right)^2 \ , \ \ \delta_{n+1} = \frac{(n+1)\Delta t}{2T},$$

and thus

$$\|q(n+1)\|_2 \leq C(T) e^{(n+1)\Delta t/T} \|q(0)\|_2 .$$

Taking into account the stability definition of [Krö97, p.92], it follows from the last expression that the proposed scheme is stable and the result is established. $\square$

References

[Car95]   Carslaw H.S., Jaeger J.C., Conduction of heat in solids, Oxford Univ. Press. 1995.

[Cha95]   Chang S.C. The method of Space-Time Conservation Element and Solution Element. A new Approach for solving the Navier Stokes and Euler Equations. *J. Comput. Phys.*, 119, pp. 295–324, 1995.

[Chu84]   Chua T.S., Dew P.M., The design of a variable step integrator for the simulation of gas transmission networks. *Int. J. Numer. Meths. in Engieneering.* Vol. 20, pp. 1797–1813, 1984.

[Def05]   Defez E., Company R., Ponsoda E., Jódar L., *Aplicación del método CE-SE a la ecuación de advección-difusión con coeficientes variables*, Congreso de Métodos Numéricos en Ingeniería (SEMNI), Granada, 4–7 de Julio 2005. Spain.

[Krö97]   Kröner D., Numerical Schemes for Conservation Laws. John Wiley, 1997.

[Met83]   Metaxas A.C., Meredith R.J. Industrial Microwave Heating. Peter Peregrinus, 1983.

[Ort72]   Ortega J.M., Numerical Analysis. A Second Course. Academic Press, New York, 1972.

[Poz90]   Pozar D.M., Microwave Engineering. Addison-Wesley, New York, 1990.

[Wan99]   Wang X.Y., Chow C.Y., Chang S.C., Application of the Space-time conservation Element and Solution Element method to one dimensional Advection-Difussion Problems. *NASA TM-1999-209068*, 1999.

A Random Euler Method for Solving Differential Equations with Uncertainties

J.C. Cortés, L. Jódar, and L. Villafuerte

Instituto de Matemática Multidisciplinar, Universidad Politécnica de Valencia Edificio 8G, 2, P.O. Box 22012, Valencia, Spain
{jccortes, ljodar}@imm.upv.es, lauvilal@doctor.upv.es

Summary. Industrial mathematical models often involve uncertainties in the data problem. In this chapter a vector random Euler method is proposed for constructing discrete mean square approximating processes of initial value problems for random differential equations with uncertainties. Convergence conditions and an illustrative example are included.

1 Introduction and Preliminaries

Random differential equations are powerful tools in order to model real problems with uncertainty [AB02] and [Soo73]. This chapter deals with the construction of a numerical method for systems of random differential equations of the form

$$\dot{\mathbf{X}}(t) = F(\mathbf{X}(t), t), \qquad t \in T = [t_0, t_1],$$
$$\mathbf{X}(t_0) = \mathbf{X}_0, \tag{1}$$

where $\mathbf{X}(t)$ and $F(\mathbf{X}(t), t)$ are m-dimensional vector stochastic processes and $\mathbf{X}_0$ is a m-dimensional random vector. We are interested in second-order random variables (2-r.v.'s) Y, having a density function f_Y, that is, Y satisfies

$$E\left[Y^2\right] = \int_{-\infty}^{\infty} y^2 f_Y(y) \mathrm{d}y < \infty,$$

where E denotes the expectation operator, and it allows us to introduce the Banach space L_2 of all the 2-r.v.'s endowed with the norm $\|Y\| = \sqrt{E[Y^2]}$, [Soo73, Chap 4]. Let X^j, $j = 1, ..., m$ be 2-r.v.'s, the m-dimensional second-order random vector is given by $\mathbf{X}^T = [X^1, ..., X^m]$. The space of all m-dimensional random vectors of second order(2-r.v.v.'s) with the norm

$$\|\mathbf{X}\|_m = \max_{j=1,...,m} \|X^j\|$$

is a Banach space and will be called the L_2^m-space. A stochastic process $X(t)$ defined on the probability space $(\Omega, \mathcal{F}, P)$ is called a second-order stochastic process (2-s.p.) if for each t, $X(t)$ is a 2-r.v. Let $\mathbf{X}(t)$, $t \in T$ be a second-order m-dimensional vector stochastic process (2-v.s.p.), hence the meaning of $\dot{\mathbf{X}}(t)$ is the mean square limit in L_2^m of the

$$\frac{\mathbf{X}(t + \Delta t) - \mathbf{X}(t)}{\Delta t} \qquad \text{as} \qquad \Delta \to 0.$$

Let $g : T \to L_2^m$ be a m.s. bounded function and let $h > 0$, then the m.s. modulus of continuity of g is the function

$$\omega(g, h) = \sup_{|t - t^*| \leq h} \|g(t) - g(t^*)\|_m, \qquad t, t^* \in T. \tag{2}$$

The function g is said to be m.s. uniformly continuous in T, if $\lim_{h \to 0} \omega(g, h) = 0$. Let $F(\mathbf{X}, t)$ be defined on $S \times T$ where S is a bounded set in L_2^m. We say that F is randomly bounded uniformly continuous in S, if

$$\lim_{h \to 0} \omega(F(\mathbf{X}, \cdot), h) = 0, \qquad \text{uniformly} \quad \text{for} \quad \mathbf{X} \in S. \tag{3}$$

Example 1. Consider the function $F(\mathbf{X}, t) = A(t)\mathbf{X} + \mathbf{C}(t)$, $0 \leq t \leq t_1$, where $\mathbf{X}^T = \begin{bmatrix} X^1, X^2 \end{bmatrix}$, $\mathbf{C}(t)^T = [0, B(t)]$ and

$$A(t) = A = \begin{bmatrix} 0 & 1 \\ -\omega_0^2 & 0 \end{bmatrix}. \tag{4}$$

Note that

$$(A(t)\mathbf{X} + \mathbf{C}(t))^T = \begin{bmatrix} X^2, B(t) - X^1\omega_0^2 \end{bmatrix},$$
$$(F(\mathbf{X}, t) - F(\mathbf{X}, t^*))^T = [0, B(t) - B(t^*)]. \tag{5}$$

By expression (3.115) of [Soo73, p. 63] and (5) it follows that

$$\|F(\mathbf{X}, t) - F(\mathbf{X}, t^*)\|_m = \max\{\|0\|, \|B(t) - B(t^*)\|\} = |t - t^*|^{\frac{1}{2}},$$

hence $F(\mathbf{X}, t)$ is randomly bounded uniformly continuous.

2 On the Random Euler Method

The goal of the present section is to show the mean square convergence in the fixed station sense of the random Euler scheme defined by

$$\left. \begin{array}{l} \mathbf{X}_{n+1} = \mathbf{X}_n + hF(\mathbf{X}_n, t_n), \\ \mathbf{X}(t_0) = \mathbf{X}_0. \end{array} \right\} \qquad n \geq 0, \tag{6}$$

where $\mathbf{X}_n$ and $F(\mathbf{X}_n, t_n)$ are 2-r.v.v.'s, $h = t_n - t_{n-1}$, $t_n = t_0 + nh$ and $F : S \times T \to L_2^m$, $S \subset L_2^m$ such that

P1: $F(\mathbf{X}, t)$ is randomly bounded uniformly continuous,
P2: $F(\mathbf{X}, t)$ satisfies the m.s. Lipschitz condition

$$\|F(\mathbf{X}, t) - F(\mathbf{Y}, t)\|_m \le k(t)\|\mathbf{X} - \mathbf{Y}\|_m, \tag{7}$$

where $\int_{t_0}^{t_1} k(t)\mathrm{d}t < \infty$.

The following theorem gives conditions for mean square convergence of scheme (6). It can be proved using the random mean value theorem as in the scalar case, see [CJV07]. However, due to the limitation of space we omit the proof.

Theorem 1. *If F satisfies the conditions P1 and P2 then the random Euler scheme (6) is m.s. convergent.*

3 Numerical Results

In this section we present an application of the random Euler method.

Example 2. Here we consider the vector form of the response of a mass-spring linear oscillator to a Brownian motion $B(t)$, see [Soo73, p. 158]

$$\dot{\mathbf{X}}(t) = A(t)\mathbf{X}(t) + \mathbf{C}(t), \quad \mathbf{X}(0) = \mathbf{X}_0, \quad 0 \le t \le t_1, \tag{8}$$

where

$$\mathbf{X}(t) = \begin{bmatrix} X^1(t) \\ X^2(t) \end{bmatrix}, \ A(t) = A = \begin{bmatrix} 0 & 1 \\ -\omega_0^2 & 0 \end{bmatrix}, \ \mathbf{C}(t) = \begin{bmatrix} 0 \\ B(t) \end{bmatrix}, \ \mathbf{X}(0) = \begin{bmatrix} 0 \\ 0 \end{bmatrix}. \tag{9}$$

By [Soo73, p. 154], the unique m.s. solution of (8) is given by

$$\mathbf{X}(t) = \int_0^t e^{A(t-s)} \begin{bmatrix} 0 \\ B(s) \end{bmatrix} \mathrm{d}s, \tag{10}$$

where

$$e^{A(t-s)} = \begin{bmatrix} \cos(\omega_0(t-s)) & \frac{1}{\omega_0}\sin(\omega_0(t-s)) \\ -\omega_0\sin(\omega_0(t-s)) & \cos(\omega_0(t-s)) \end{bmatrix}. \tag{11}$$

Hence

$$\mathbf{X}(t) = \begin{bmatrix} \frac{1}{\omega_0}\int_0^t \sin(\omega_0(t-s)) B(s)\mathrm{d}s \\ \int_0^t \cos(\omega_0(t-s)) B(s)\mathrm{d}s \end{bmatrix}, \tag{12}$$

and by [Soo73, p. 104], its expectation

$$E[\mathbf{X}(t)] = \begin{bmatrix} \frac{1}{\omega_0}\int_0^t \sin(\omega_0(t-s)) E[B(s)]\mathrm{d}s \\ \int_0^t \cos(\omega_0(t-s)) E[B(s)]\mathrm{d}s \end{bmatrix} = \begin{bmatrix} 0 \\ 0 \end{bmatrix}.$$

The variance of $\mathbf{X}(t)$ is obtained from expressions (2.15) and (2.16) of [CJV06]. It has the form

$$V(t) = \text{Var}\left[\mathbf{X}(t)\right] = \text{E}[\mathbf{X}(t)\mathbf{X}^T(t)] = \begin{bmatrix} V_{11}(t) & V_{12}(t) \\ V_{21}(t) & V_{22}(t) \end{bmatrix},$$

where

$$V_{11}(t) = \frac{1}{\omega_0^2} \int_0^t \int_0^t \sin\left(\omega_0(t-s)\right) \sin\left(\omega_0(t-r)\right) \min(r,s) dr\, ds, \tag{13}$$

$$V_{22}(t) = \int_0^t \int_0^t \cos\left(\omega_0(t-s)\right) \cos\left(\omega_0(t-r)\right) \min(r,s) dr\, ds,$$

$$V_{12}(t) = V_{21}(t) = \frac{1}{\omega_0} \int_0^t \int_0^t \sin\left(\omega_0(t-s)\right) \cos\left(\omega_0(t-r)\right) \min(r,s) ds\, dr.$$

As $\mathbf{X}_0 = \mathbf{0}$, expression of the random Euler scheme in this case takes the form

$$\mathbf{X}_n = (I + hA)^n \mathbf{X}_0 + h\sum_{i=0}^{n-1}(I+hA)^{n-i-1}\mathbf{C}(t_i) = h\sum_{i=0}^{n-1}(I+hA)^{n-i-1}\mathbf{C}(t_i)$$

$$= \sum_{i=0}^{n-1}\left(\sum_{j=0}^{n-i-1}\binom{n-i-1}{j}A^j h^{j+1}\right)\begin{bmatrix} 0 \\ B(t_i) \end{bmatrix}.$$

Hence, as $\text{E}\left[B(t_i)\right] = 0$, one gets

$$\text{E}\left[\mathbf{X}_n\right] = 0. \tag{14}$$

Note that for the matrix A given by (9), it follows that

$$A^{2n} = \begin{bmatrix} (-\omega_0^2)^n & 0 \\ 0 & (-\omega_0^2)^n \end{bmatrix} = (-1)^n\omega_0^{2n}I, \quad n \geq 0, \tag{15}$$

$$A^{2n+1} = (-1)^n\omega_0^{2n}\begin{bmatrix} 0 & 1 \\ -\omega_0^2 & 0 \end{bmatrix} = (-1)^n\omega_0^{2n}A, \quad n \geq 0, \tag{16}$$

and

$$A\begin{bmatrix} 0 \\ B(t) \end{bmatrix} = \begin{bmatrix} B(t) \\ 0 \end{bmatrix}, \quad I\begin{bmatrix} 0 \\ B(t) \end{bmatrix} = \begin{bmatrix} 0 \\ B(t) \end{bmatrix},$$

then, from (14), (15) and (16) one gets

$$X_n = \sum_{i=0}^{n-1}B(t_i)\begin{bmatrix} c_{ni} \\ d_{ni} \end{bmatrix}, \tag{17}$$

where

$$c_{ni} = \sum_{j=0}^{\left[\frac{n-i-1}{2}\right]}(-1)^j\omega_0^{2j}h^{2(j+1)}\binom{n-i-1}{2j+1}, \tag{18}$$

Table 1. Numerical results of the variance with for example 5.1

Points	$\tilde{V}_{11}$	$V_{11}(t)$	$\tilde{V}_{22}$	$V_{22}(t)$	$\tilde{V}_{21}$	$V_{21}(t)$
$t = 0.15$	2.4×10^{-6}	3.7×10^{-6}	9.8×10^{-4}	1.1×10^{-3}	4.6×10^{-5}	6.3×10^{-5}
$t = 0.30$	1.3×10^{-4}	1.2×10^{-4}	8.3×10^{-3}	8.8×10^{-3}	8.6×10^{-4}	9.9×10^{-4}
$t = 0.55$	2.1×10^{-3}	2.4×10^{-3}	5.0×10^{-2}	5.2×10^{-2}	1.0×10^{-2}	1.0×10^{-2}
$t = 0.80$	1.4×10^{-2}	1.5×10^{-2}	1.4×10^{-2}	1.5×10^{-2}	4.4×10^{-2}	4.5×10^{-2}
$t = 1.00$	4.2×10^{-2}	4.4×10^{-2}	2.7×10^{-1}	2.7×10^{-1}	1.0×10^{-2}	1.0×10^{-2}

$$d_{ni} = \sum_{j=0}^{\left[\frac{n-i-1}{2}\right]} (-1)^j \omega_0^{2j} h^{2j+1} \binom{n-i-1}{2j}. \tag{19}$$

From (17)–(19) it follows that

$$\tilde{V} = \text{Var}\left[\mathbf{X_n}\right] = \sum_{i=0}^{n-1}\sum_{j=0}^{n-1} \left(\min\left(i,j\right)\right) h \begin{bmatrix} c_{ni}c_{nj} & c_{ni}d_{nj} \\ c_{nj}d_{ni} & d_{ni}d_{nj} \end{bmatrix} = \begin{bmatrix} \tilde{V}_{11} & \tilde{V}_{12} \\ \tilde{V}_{21} & \tilde{V}_{22} \end{bmatrix}.$$

In Table 1, we compare the variance of the theoretical solution $V(t)$ and the variance of the approximation random Euler method $\tilde{V}$ with the step $h = 1/80$ and $\omega_0 = 1$. It shows that variance of the theoretical solution and the numerical Euler values are closer as n increases. Furthermore, the approximation is better when the points are closer to $t = 0$.

References

[AB02]   Ayala, A.G., Barron, R.: El método de yuxtaposición de dominios en la solución numérica de ecuaciones diferenciales estocásticas. (in Spanish), Proc. Métodos Numéricos en Ingeniería y Ciencias Aplicadas, (E. Oñate et all eds.) CIMNE, Barcelona, 267-276, (2002).

[CJV06]   Cortés, J.C., Jdar, L., Villafuerte, L.: Mean square numerical solution of random differential equations: facts and possibilities. Comput. Math. Applic, (in print) (2006).

[CJV07]   Cortés, J.C., Jódar, L., Villafuerte, L.: Numerical solution of random differential equations: a mean square approach. Mathematical and Computer Modelling, (in print) (2006).

[Soo73]   Soong, T.T.: Random Differential Equations in Science and Engineering. Academic Press, New York, (1973).

Cubic-Matrix Splines and Second-Order Matrix Models

M.M. Tung, L. Soler, E. Defez, and A. Hervás

Instituto de Matemática Multidisciplinar, Universidad Politécnica de Valencia, Spain
{mtung, edefez, ahervas}@imm.upv.es

Summary. We discuss the direct use of cubic-matrix splines to obtain continuous approximations to the unique solution of matrix models of the type $Y''(x) = f(x, Y(x))$. For numerical illustration, an estimation of the approximation error, an algorithm for its implementation, and an example are given.

1 Introduction

Matrix initial value problems of the form:

$$\left.\begin{array}{c} Y''(x) = f(x, Y(x)) \\ \\ Y(a) = Y_0 \ , \ \ Y'(a) \ = \ Y_1 \end{array}\right\} \ a \le x \le b \ , \tag{1}$$

are frequently encountered in different fields of physics and engineering (see, e.g. [Zha02]). In the scalar case, numerical methods for the calculation of approximate solutions of (1) can be found in [Col93]. For matrix problems, linear multi-step matrix methods with constant steps have been studied in [Jod93]. Although in this case there exist a priori error bounds for these methods (expressed as function of the data problem), these error bounds are given in terms of an exponential which depends on the integration step h. Therefore, in practice, h will take too small values. Problems of the type (1) can be written as an extended first-order matrix problem. Such a standard approach, however, involves an increase of the computational cost caused by the increase of the problem dimension. Recently, cubic-matrix splines were used in the resolution of first-order matrix differential systems [Def06], obtaining approximations that, among other advantages, were of class C^1 in the interval $[a, b]$, and easy to compute producing an approximation error $O(h^4)$. The present work extends this powerful scheme to the solution of matrix problems of type (1). Throughout this work, we will adopt the notation for norms and matrix cubic splines as in [Def06] and common in matrix calculus. The chapter is organized as follows. Section 2 develops the proposed method. Finally, in Sect. 3, an example is presented.

2 Construction of the Method

Let us consider the initial value problem

$$\left.\begin{array}{c} Y''(x) = f(x, Y(x)) \\[2mm] Y(a) = Y_0 \ , \ Y'(a) \ = \ Y_1 \end{array}\right\} \ a \le x \le b \ , \qquad (2)$$

where $Y_0, Y_1, Y(t) \in \mathbb{C}^{r \times q}$, $f : [a, b] \times \mathbb{C}^{r \times q} \times \longmapsto \mathbb{C}^{r \times q}$, $f \in \mathcal{C}^0 (T)$, with

$$T \ = \ \{(x, Y) \ ; \ a \le x \le b \ , \ Y \in \mathbb{C}^{r \times q}\} \ , \qquad (3)$$

and f fulfills the global Lipschitz's condition

$$\|f(x, Y_1) \ - \ f(x, Y_2)\| \ \le L \|Y_1 - Y_2\| \ , \ a \le x \le b \ , Y_1, Y_2 \in \mathbb{C}^{r \times q} \ . \quad (4)$$

Let us also use the partition of the interval $[a, b]$ defined by

$$\Delta_{[a,b]} = \{a = x_0 < x_1 < \ldots < x_n = b\} \ , \ x_k = a + kh \ , \ k = 0, 1, \ldots, n \ , \ (5)$$

where $h = (b - a)/n$, n being a positive integer. We will construct in each subinterval $[a + kh, a + (k + 1)h]$ a matrix-cubic spline approximating the solution of problem (2). For the first interval $[a, a + h]$, we consider that the matrix-cubic spline is given by

$$S\big|_{[a,a+h]} (x) \ = \ Y(a) + Y'(a)(x-a) + \frac{1}{2!}Y''(a)(x-a)^2 + \frac{1}{3!}A_0(x-a)^3 \ , \ (6)$$

where $A_0 \in \mathbb{C}^{r \times q}$ is a matrix parameter to be determined. It is straightforward to check: $S\big|_{[a,a+h]} (a) \ = \ Y(a), S'\big|_{[a,a+h]} (a) \ = \ Y'(a), S''\big|_{[a,a+h]} (a) \ = \ Y''(a) \ = f(a, S\big|_{[a,a+h]} (a))$. Thus, (6) satisfies the equations of problem (2) at point $x = a$. To fully construct the matrix-cubic spline, we must still determine A_0. By imposing that (6) is a solution of problem (2) in $x = a + h$, we have:

$$S''\big|_{[a,a+h]} (a + h) \ = \ f\left(a + h, S\big|_{[a,a+h]} (a + h)\right) \ , \qquad (7)$$

and obtain from (7) the matrix equation with only one unknown matrix A_0:

$$A_0 \ = \ \frac{1}{h}\left[f\left(a + h, Y(a) + Y'(a)h + \frac{1}{2}Y''(a)h^2 + \frac{1}{6}A_0 h^3\right) - Y''(a)\right] \ . \ (8)$$

Assuming that the matrix equation (8) has only one solution A_0, the matrix-cubic spline is totally determined in the interval $[a, a + h]$. Now, in the next interval $[a + h, a + 2h]$, the matrix-cubic spline is defined by:

$$S\big|_{[a+h,a+2h]} (x) = S\big|_{[a,a+h]} (a+h) + S'\big|_{[a,a+h]} (a+h)(x - (a + h))$$

$$+ \frac{1}{2!}S''\big|_{[a,a+h]} (a+h)(x-(a+h))^2 + \frac{1}{3!}A_1(x-(a+h))^3, \ (9)$$

so that $S(x)$ is of class $\mathcal{C}^2([a, a+h] \cup [a+h, a+2h])$, and all of the coefficients of matrix-cubic spline $S_{|_{[a+h,a+2h]}}(x)$ are determined with the exception of $A_1 \in \mathbb{C}^{r \times q}$. By construction, matrix-cubic spline (9) satisfies the differential equation (2) in $x = a+h$. We can obtain A_1 by requiring that the differential equation (2) holds at point $x = a + 2h$:

$$S''_{|_{[a+h,a+2h]}}(a+2h) \;=\; f\left(a+2h, S_{|_{[a+h,a+2h]}}(a+2h)\right) .$$

Expanding, we obtain the matrix equation with only one unknown matrix A_1:

$$A_1 = \frac{1}{h}\left[f\left(a+2h, S_{|_{[a,a+h]}}(a+h)+S'_{|_{[a,a+h]}}(a+h)h \right.\right.$$
$$\left.\left. +\frac{1}{2}S''_{|_{[a,a+h]}}(a+h)h^2 +\frac{1}{6}A_1 h^3\right) - S''_{|_{[a,a+h]}}(a+h)\right]. \tag{10}$$

Let us assume that the matrix equation (10) has only one solution A_1. This way the spline is now totally determined in the interval $[a+h, a+2h]$. Iterating this process, let us construct the matrix-cubic spline taking $[a+(k-1)h, a+kh]$ as the last subinterval. For the next subinterval $[a+kh, a+(k+1)h]$, we define the corresponding matrix-cubic spline as

$$S_{|_{[a+kh,a+(k+1)h]}}(x) = \beta_k(x) + \frac{1}{3!}A_k(x-(a+kh))^3$$

$$\text{where } \; \beta_k(x) = \sum_{l=0}^{2} \frac{1}{l!} S^{(l)}_{|_{[a+(k-1)h,a+kh]}}(a+kh)(x-(a+kh))^l . \tag{11}$$

With this definition, it is $S(x) \in \mathcal{C}^2 \left(\bigcup_{j=0}^{k}[a+jh, a+(j+1)h] \right)$ which fulfills the differential equation (2) at point $x = a+kh$. As an additional requirement, we assume that $S(x)$ satisfies the differential equation (2) at the point $x = a+(k+1)h$, i.e.

$$S''_{|_{[a+kh,a+(k+1)h]}}(a+(k+1)h) \;=\; f\left(a+(k+1)h, S_{|_{[a+kh,a+(k+1)h]}}(a+(k+1)h)\right) .$$

Subsequent expansion of this equation with the unknown matrix A_k yields

$$A_k = \frac{1}{h}\left[f\left(a+(k+1)h, \beta_k(a+(k+1)h)+\frac{1}{6}A_k h^3\right) - \beta''_k(a+(k+1)h)\right] . \tag{12}$$

Note that this matrix equation (12) is analogous to equations (8) and (10), when $k=0$ and $k=1$, respectively. For a fixed h, we will consider the matrix function of matrix variable $g : \mathbb{C}^{r \times q} \mapsto \mathbb{C}^{r \times q}$ defined by

$$g(T) = \frac{1}{h}\left[f\left(a+(k+1)h, \beta_k(a+(k+1)h)+\frac{1}{6}Th^3\right) - \beta''_k(a+(k+1)h)\right] .$$

952   M.M. Tung et al.

Relation (12) holds if and only if $A_k = g(A_k)$, that is, if A_k is a fixed point for function $g(T)$. Applying the global Lipschitz's conditions (4), it follows that $\|g(T_1) - g(T_2)\| \leq \frac{Lh^2}{6} \|T_1 - T_2\|$. Taking $h < \sqrt{\frac{6}{L}}$, $g(T)$ yields a contractive matrix function, which guarantees that (12) has unique solutions A_k for $k = 0, 1, \ldots, n-1$. Hence, the matrix-cubic spline is now fully determined. Taking into account [Los67, Theorem 5], the following result has been established:

Theorem 1. *If $h < \sqrt{\frac{6}{L}}$, then the matrix-cubic spline $S(x)$ exists in each subinterval $[a + kh, a + (k + 1)h]$, $k = 0, 1, \ldots, n-1$, as defined by the previous construction. Furthermore, if $f \in C^1(T)$, then $\|Y(x) - S(x)\| = O(h^3)\ \forall x \in [a, b]$, where $Y(x)$ is the theoretical solution of system (2).*

Depending on the function f, matrix equations (8) and (12) can be solved explicitly or by using some iterative method [Ort72]. Summarizing, we have the following algorithm:

- Take $n > \dfrac{(b - a)\sqrt{L}}{\sqrt{6}}$, $h = (b - a)/n$ and $[a,\ b]$ defined by (5).
- Solve (8) and determine $S_{|_{[a,a+h]}}(x)$ defined by (6).
- For $k = 1$ to $n - 1$, solve (12). Determine $S_{|_{[a+kh,a+(k+1)h]}}(x)$ defined by (11).

3 Example

The problem

$$Y''(t) + AY(t) = 0 , \tag{13}$$

with $Y(0) = Y_0$, $Y'(0) = Y_1$, has the exact solution

$$Y(t) = \cos\left(\sqrt{A}t\right)Y_0 + \left(\sqrt{A}\right)^{-1}\sin\left(\sqrt{A}t\right)Y_1 ,$$

where $\sqrt{A}$ denotes any square root of a non-singular matrix A [Har05]. The principal drawback of this formal solution is the difficult computation of $\sqrt{A}$, $\cos\left(\sqrt{A}t\right)$ and $\sin\left(\sqrt{A}t\right)$. The proposed method avoids this drawback. We consider problem (13) where $A = \begin{pmatrix} 1 & 0 \\ 2 & 1 \end{pmatrix}$, $Y_0 = \begin{pmatrix} 0 & 0 \\ 0 & 0 \end{pmatrix}$, $Y_1 = \begin{pmatrix} 1 & 0 \\ 1 & 1 \end{pmatrix}$, $t \in [0, 1]$, whose exact solution is $Y(t) = \sin\left[\begin{pmatrix} 1 & 0 \\ 1 & 1 \end{pmatrix}t\right] = \begin{pmatrix} \sin(t) & 0 \\ t\cos(t) & \sin(t) \end{pmatrix}$. In this case $L \approx 2.82843$. By Theorem 1, we need to take $h < 1.45647$, so we choose $h = 0.1$ for example. The results are summarized in the following table, where the numerical estimates have been rounded to the fourth relevant digit. In each subinterval, we evaluated the difference between the estimates of our numerical approach and the exact solution. The maximum of these errors are indicated in the third column.

Interval	Approximation	Max. error
$[0, 0.1]$	$\begin{pmatrix} x-0.1664x^3 & 0 \\ x-0.4986x^3 & x-0.1664x^3 \end{pmatrix}$	1.0072×10^{-6}
$[0.1, 0.2]$	$\begin{pmatrix} 1.00005x-0.0005x^2-0.1647x^3 & 0 \\ 1.0002x-0.0025x^2-0.4903x^3 & 1.0001x-0.0005x^2-0.1647x^3 \end{pmatrix}$	6.3032×10^{-6}
$[0.2, 0.3]$	$\begin{pmatrix} 1.0005x-0.0025x^2-0.1614x^3 & 0 \\ -0.0001+1.0022x-0.0124x^2-0.4738x^3 & 1.0005x-0.0025x^2-0.1614x^3 \end{pmatrix}$	2.0059×10^{-5}
$[0.3, 0.4]$	$\begin{pmatrix} -0.0002+1.0018x-0.0069x^2-0.1565x^3 & 0 \\ -0.0008+1.0088x-0.0344x^2-0.4494x^3 & -0.0002+1.0018x-0.0069x^2-0.1565x^3 \end{pmatrix}$	4.6213×10^{-5}
$[0.4, 0.5]$	$\begin{pmatrix} -0.0006+1.0049x-0.0147x^2-0.1500x^3 & 0 \\ -0.0028+1.0242x-0.0728x^2-0.4174x^3 & -0.0006+1.0049x-0.0147x^2-0.1500x^3 \end{pmatrix}$	8.8359×10^{-5}
$[0.5, 0.6]$	$\begin{pmatrix} -0.0016+1.0109x-0.0266x^2-0.1420x^3 & 0 \\ -0.0077+1.0536x-0.1316x^2-0.3782x^3 & -0.0016+1.0109x-0.0266x^2-0.1420x^3 \end{pmatrix}$	1.4964×10^{-4}
$[0.6, 0.7]$	$\begin{pmatrix} -0.0036+1.0210x-0.0436x^2-0.1327x^3 & 0 \\ -0.0176+1.1030x-0.2140x^2-0.3324x^3 & -0.0036+1.0210x-0.0436x^2-0.1327x^3 \end{pmatrix}$	2.3267×10^{-4}
$[0.7, 0.8]$	$\begin{pmatrix} -0.0073+1.0368x-0.0661x^2-0.1219x^3 & 0 \\ -0.0354+1.1791x-0.3227x^2-0.2807x^3 & -0.0073+1.0368x-0.0661x^2-0.1219x^3 \end{pmatrix}$	3.3941×10^{-4}
$[0.8, 0.9]$	$\begin{pmatrix} -0.0134+1.0597x-0.0947x^2-0.1100x^3 & 0 \\ -0.0646+1.2885x-0.4595x^2-0.2237x^3 & -0.0134+1.0597x-0.0947x^2-0.1100x^3 \end{pmatrix}$	4.7114×10^{-4}
$[0.9, 1]$	$\begin{pmatrix} -0.0229+1.0914x-0.1299x^2-0.0970x^3 & 0 \\ -0.1093+1.4378x-0.6253x^2-0.1623x^3 & -0.0229+1.0914x-0.1299x^2-0.0970x^3 \end{pmatrix}$	6.2838×10^{-4}

References

[Col93] Coleman, J.P.: Numerical method for $y'' = f(x, y)$. *Int. Proc. First Int. Colloq. Num. Anal.*, D. Bainov and V. Covachev (eds.), VSP. Utrecht, The Netherlands, pp. 27–38, 1993.

[Def06] Defez, E., Soler, L., Hervás, A., Tung, M.M.: Numerical solutions of matrix differential models using cubic matrix splines II. *Comput. Math. Appl.*, in print.

[Har05] Hargreaves, G.I., Higham, N.J.: Efficient algorithms for the matrix cosine and sine. *Numerical Algorithms* 40, pp. 383–400, 2005.

[Jod93] Jódar, L., Morera, J.L., Villanueva, R.J.: Numerical multistep matrix methods for $Y'' = f(t, Y)$. *Appl. Math. Comput.* 59, pp. 257–274. 1993.

[Los67] Loscalzo, F.R., Talbot, T.D.: Spline function approximations for solutions of ordinary differential equations. *SIAM J. Numer. Anal.*, 4(3), pp. 433–445, 1967.

[Ort72] Ortega, J.M., Rheinboldt, W.C.: Iterative Solution of Nonlinear Equations in Several Variables. Academic Press, 1972.

[Zha02] Zhang, J.F.: Optimal control for mechanical vibration systems based on second-order matrix equations. *Mechanical Systems and Signal Processing* 16(1), pp. 61–67, 2002, and references therein.

Part IV

Color Plates

Color Plates

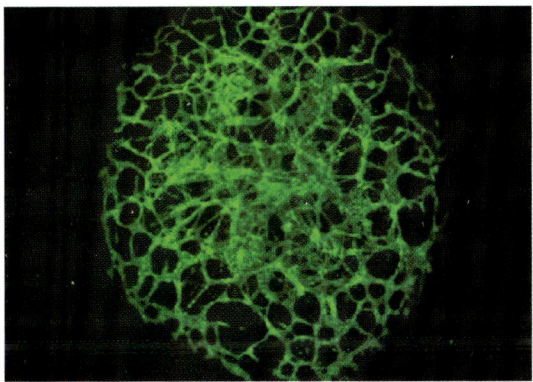

Fig. 1. (Capasso *et al*, **p. 6) Fig. 2.** Vascularization of an allantoid [Credit: Dejana et al. 2005]

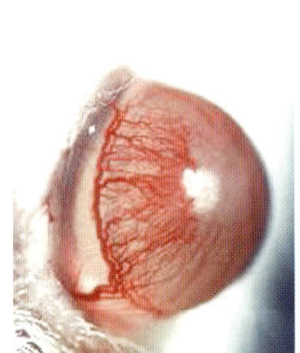 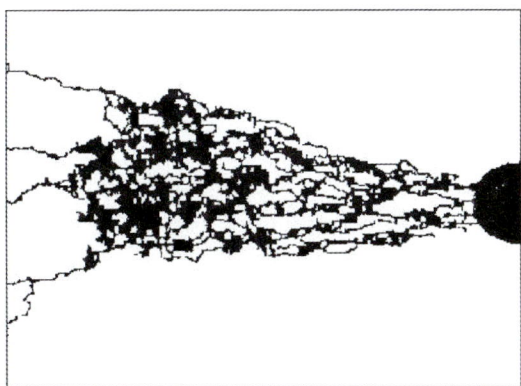

Fig. 2. (Capasso *et al*, **p. 7) Fig. 3.** *Left*: Angiogenesis on a rat cornea [Credit: Dejana et al. (2005). The white spot is a pellet implanted in the cornea containing an angiogenetic substance, emulating the effect of a tumor. *Right*: A simulation of an angiogenesis due to a localized tumor mass (*black* region on the *right*) (from [CA99])

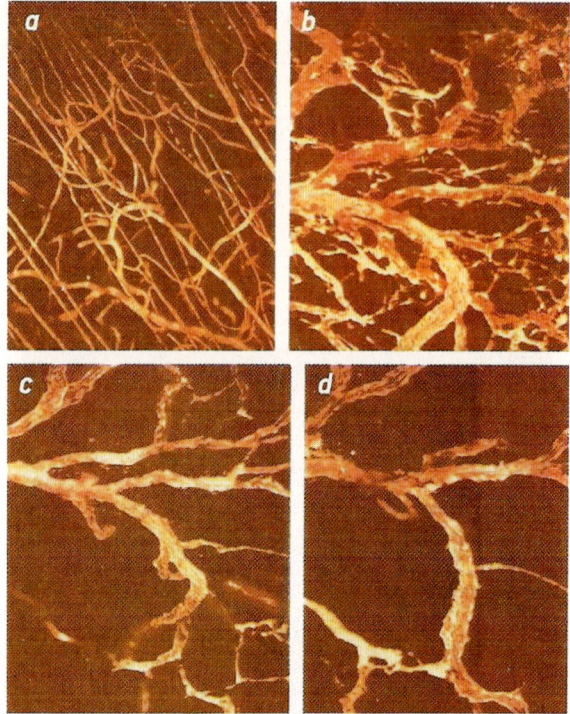

Fig. 3. (Capasso *et al*, p. 7) Fig. 4. Response of a vascular network to an anti-angiogenic treatment (from [JK01])

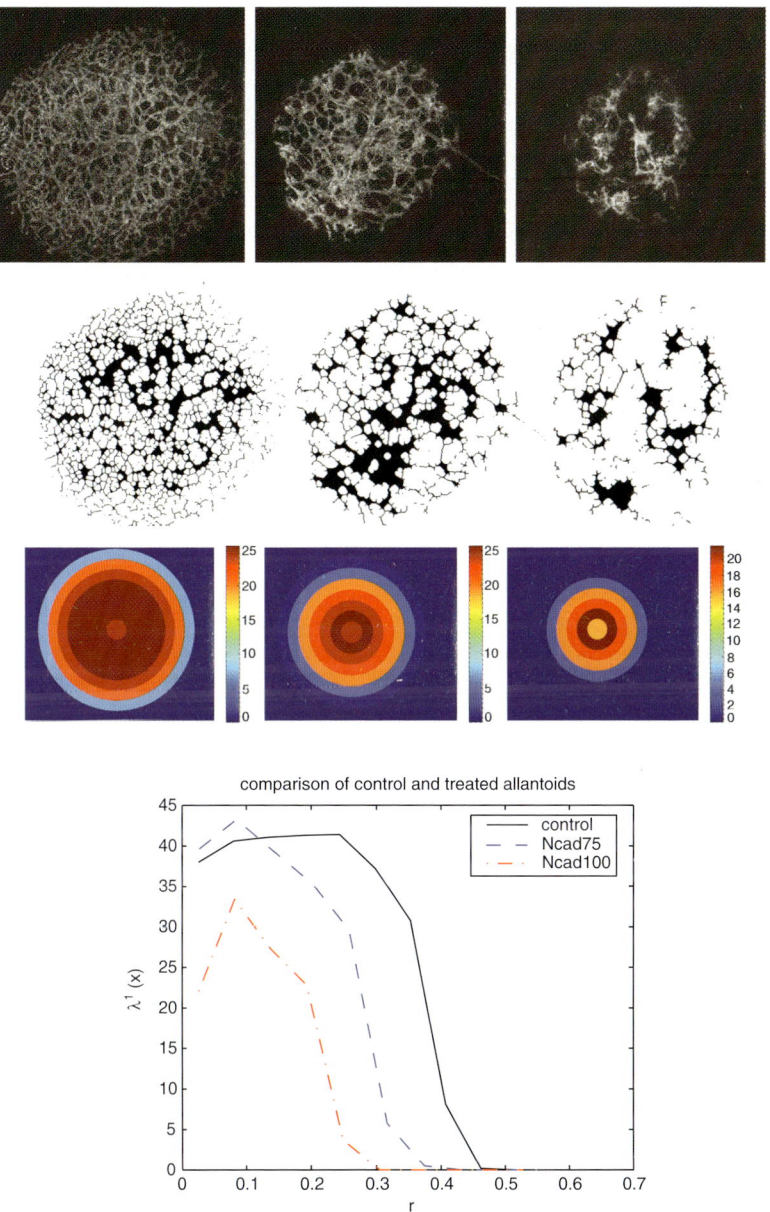

Fig. 4. (Capasso *et al*, p. 32) Fig. 18. Vascularization in allantoids. *First line*, from *left to right*: control experiment (untreated), treated with 0.75 mg of antiangiogenetic substance, treated with 1 mg of antiangiogenic substance. *Second line*: scheletonization of the upper images. *Third line*: 2D representation of the intensity estimate of the fibres in the skeletons; the space has been divided into ten spherical concentric shells. *Bottom line*: comparison of the radial estimates of the intensities of the three allantoids

Fig. 5. (Barrero *et al*, p. 39) Fig. 1. Cone, jet and spray in an electrospray; the electrosprayed liquid was methanol. The size of the charged droplets ranged between 380 and 720 nm, which are the wavelength of the blue and red radiation. As shown in the picture, droplets scatter the blue component avoiding its pass throughout the spray while the other components of the white light pass through the droplet cloud

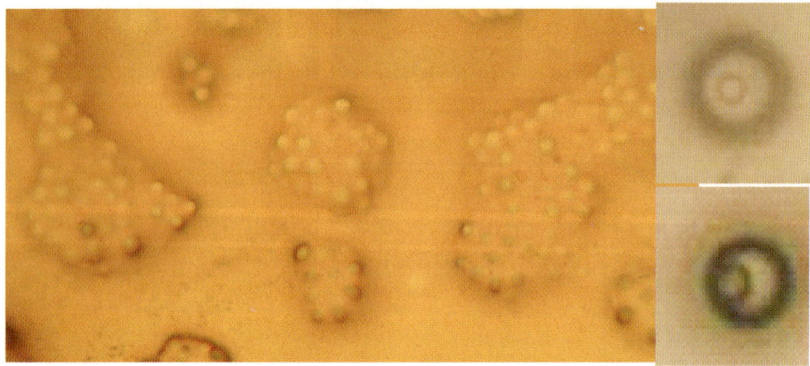

Fig. 6. (Barrero *et al*, p. 43) Fig. 5. Collection of near monodisperse capsules. Magnified views of two capsules formed under different parametrical conditions are also given in the two pictures on the *right*. In the *upper* one picture, the outer diameter is 10 μm, whereas the diameter of the capsule shown in the *lower* one is 8 μm

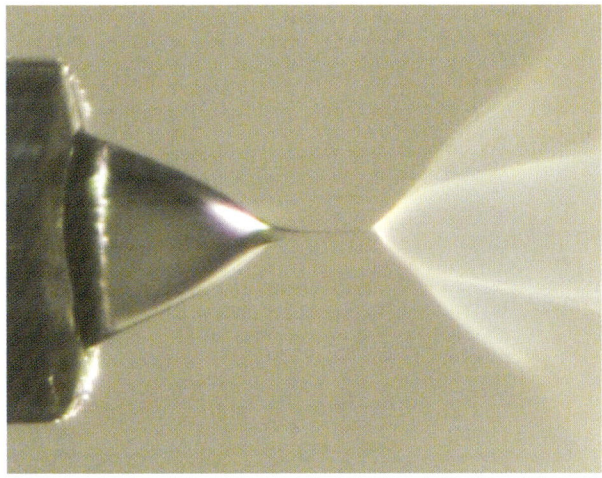

Fig. 7. (Barrero *et al*, p. 46) Fig. 8. Taylor cone of glycerol in a bath of hexane. The needle OD is 0.8 mm. The hydrosol in this case is formed by droplets of two different sizes: the main droplets, of $2\,\mu$m in diameter, and the satellite droplets, of about $0.8\,\mu$m in diameter

Fig. 8. (Bermúdez *et al*, p. 49) Fig. 1. Induction system

Fig. 9. (Bermúdez *et al*, p. 49) Fig. 2. Induction furnace

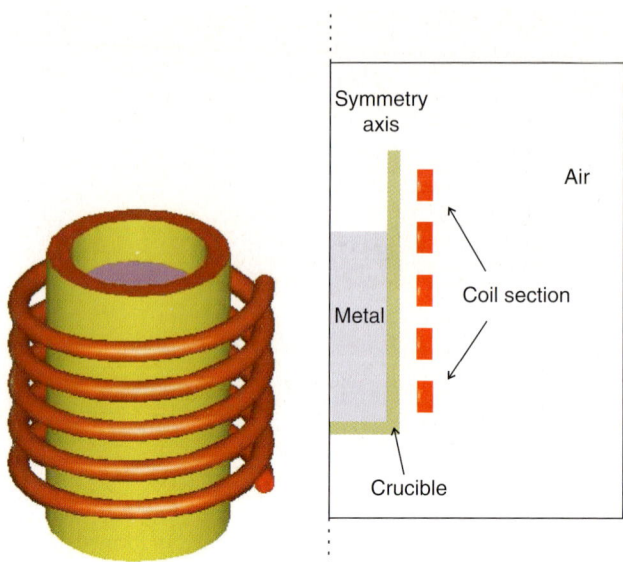

Fig. 10. (Bermúdez *et al*, p. 51) Fig. 3. Sketch of the induction furnace and diametral section

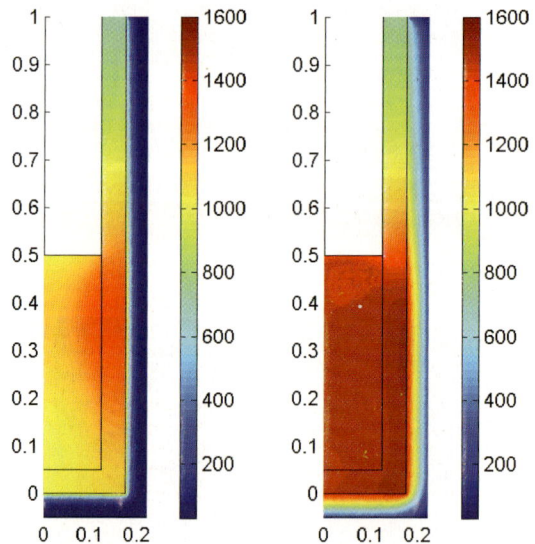

Fig. 11. (Bermúdez *et al*, p. 62) Fig. 10. Temperature field for $t = 30\,\text{min}$ (*left*) and $t = 180\,\text{min}$ (*right*)

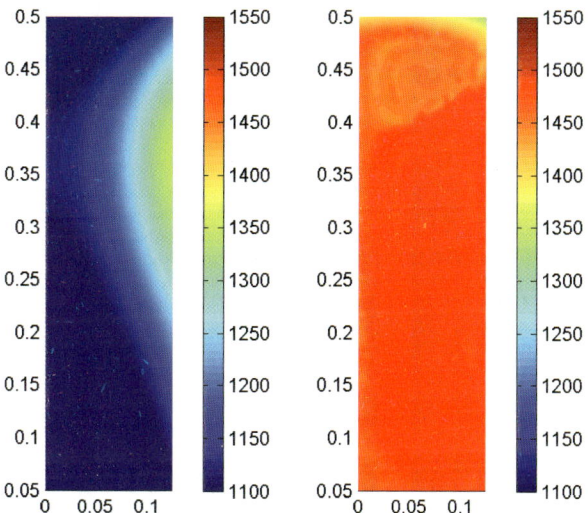

Fig. 12. (Bermúdez *et al*, p. 62) Fig. 11. Silicon temperature for $t = 30$ min (*left*) and $t = 180$ min (*right*)

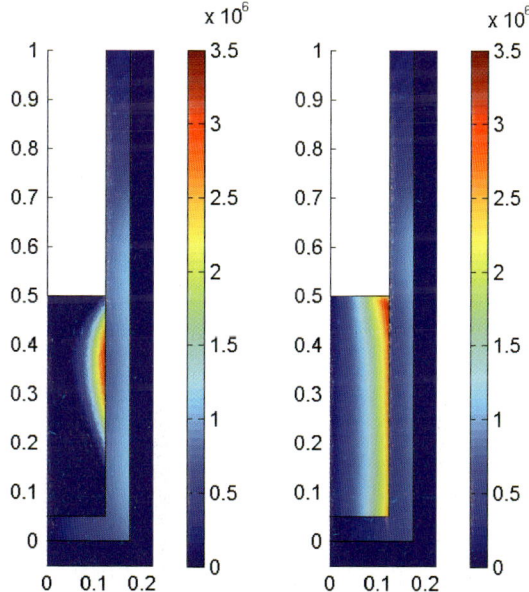

Fig. 13. (Bermúdez *et al*, p. 63) Fig. 12. Modulus of current density for $t = 30$ min (*left*) and $t = 180$ min (*right*)

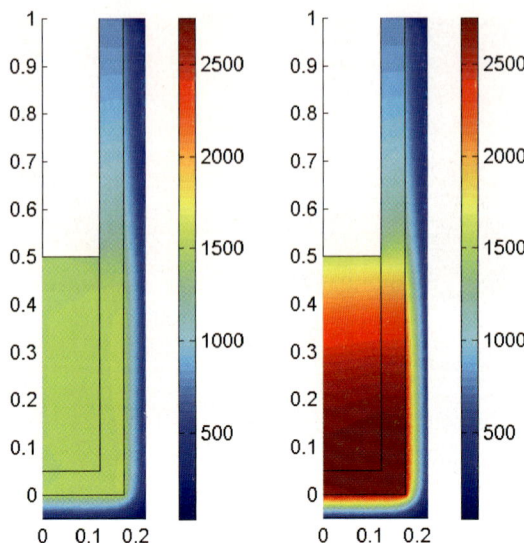

Fig. 14. (Bermúdez *et al*, p. 63) Fig. 13. Temperature with and without convection term ($t = 180\,\text{min}$)

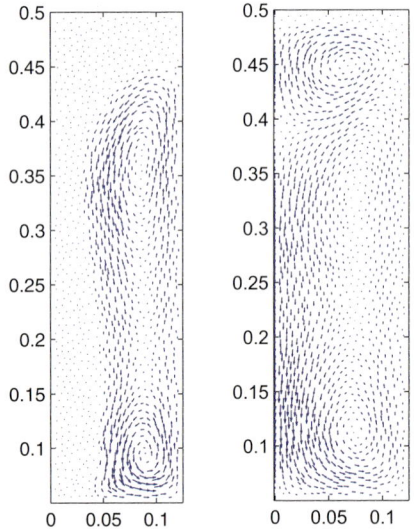

Fig. 15. (Bermúdez *et al*, p. 64) Fig. 15. Velocity field $t = 90$ min (*left*) and $t = 180\,\text{min}$ (*right*)

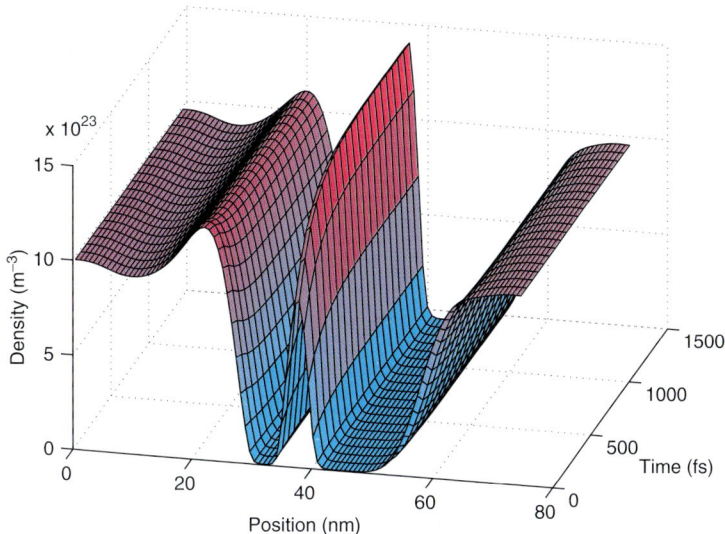

Fig. 16. (Degond *et al*, p. 116) Fig. 2. Evolution of the density from the peak (applied bias: 0.25 V) to the valley (applied bias: 0.31 V)

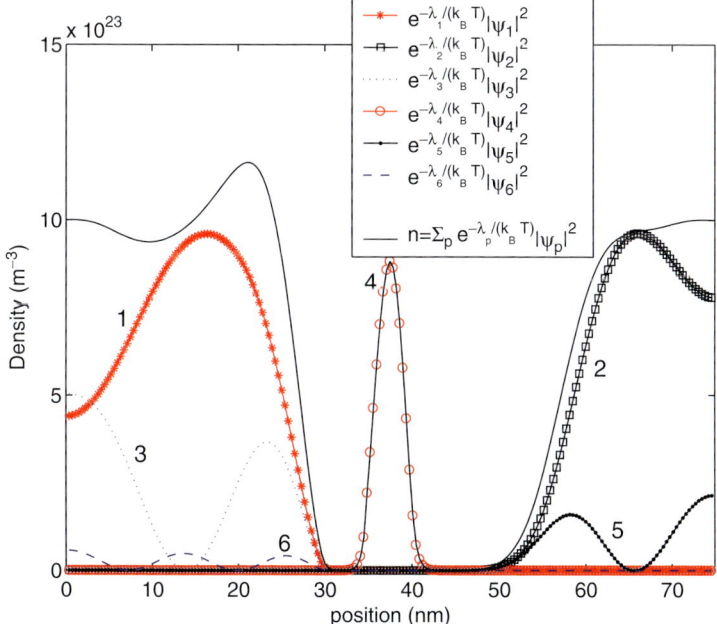

Fig. 17. (Degond *et al*, p. 117) Fig. 3. Density at the peak (applied bias: 0.25 V)

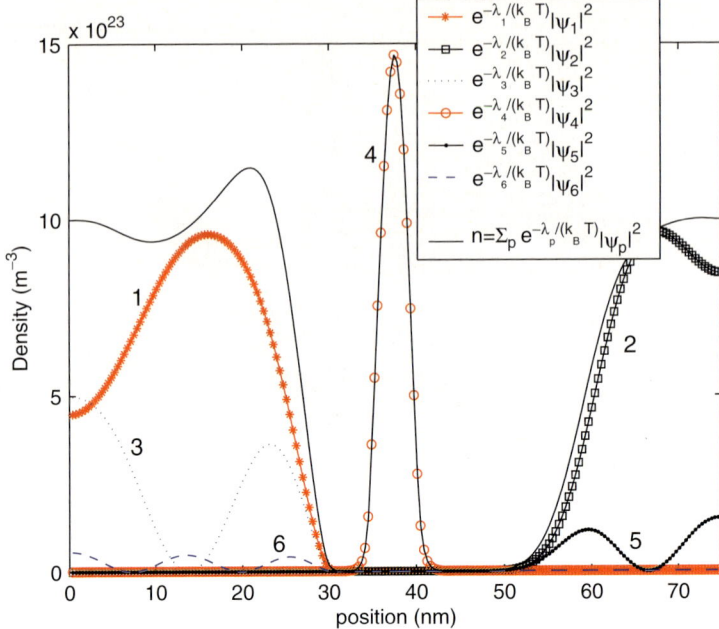

Fig. 18. (Degond _et al_, p. 118) Fig. 4. Density at the valley (applied bias: 0.31 V)

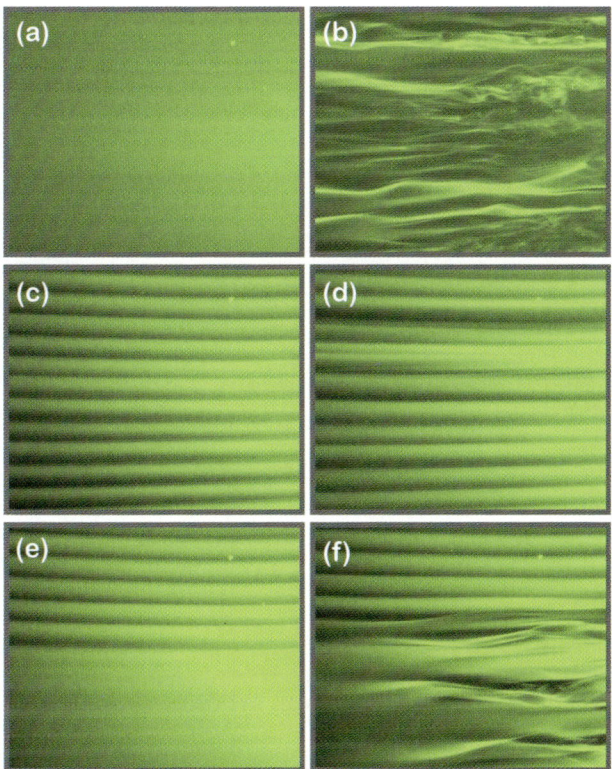

Fig. 19. (Cossu *et al*, p. 144) Fig. 5. Smoke flow visualizations from above with flow from left to right. (**a**) and (**b**) show the two-dimensional boundary layer, without streaks, with no excitation and with excitation of 201 mV, respectively. The flow in (**b**) is turbulent. (**c**) shows the streaky base flow with no excitation. In the presence of streaks with excitation of 450 mV (**d**), the flow remains laminar. (**e**) shows a half-streaky boundary layer obtained removing half the roughness elements and without forcing. With a forcing at 157 mV (**f**) the streaky part of the boundary layer remains laminar while the uncontrolled part undergoes transition (adapted from [FTBC06])

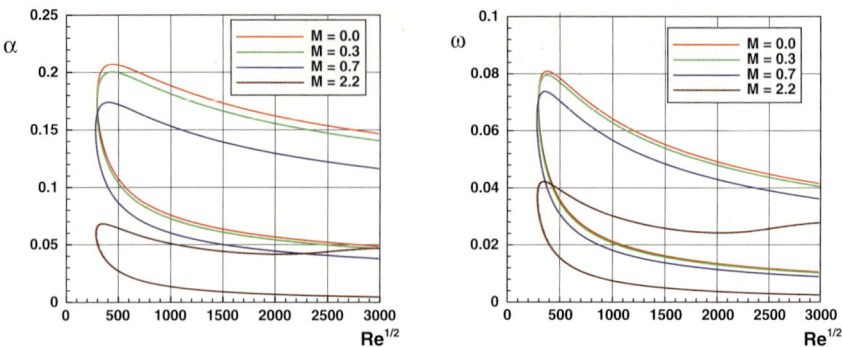

Fig. 20. (Martel *et al*, p. 161) **Fig. 1.** Neutral stability curves

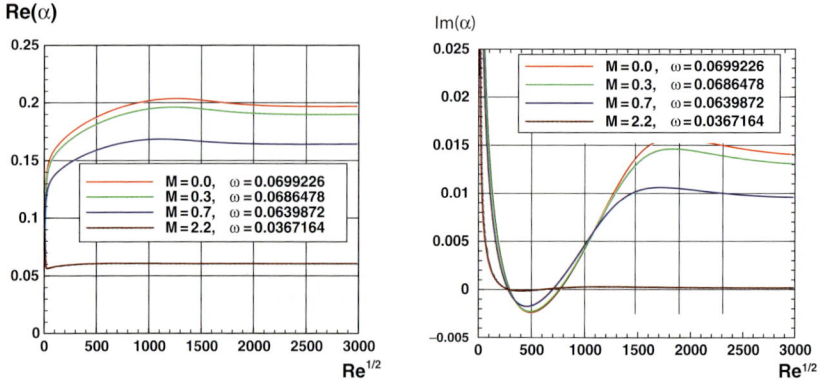

Fig. 21. (Martel *et al*, p. 161) **Fig. 2.** Wavenumber and damping rate vs. $R = \mathrm{Re}^{1/2}$ for fixed ω

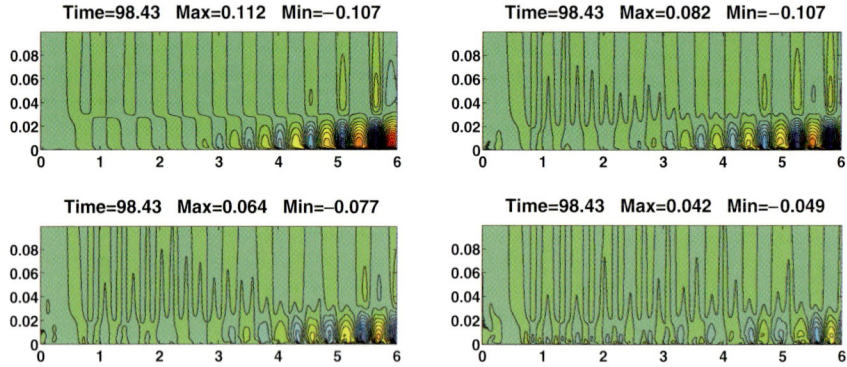

Fig. 22. (Martel *et al*, p. 166) **Fig. 6.** Streamwise perturbation velocity contours for the WT wave amplitudes corresponding to the dots marked in Fig. 5 ($a = 0, 0.0025, 0.004$ and 0.005 from *left* to *right* and *top* to *bottom*)

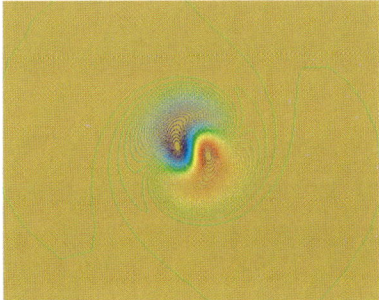

Fig. 23. (González, p. 200) Fig. 4. Eigenfunction $\hat{u}$ pertaining to the least-damped mode of a single Batchelor vortex at $Re = 100, \beta = 0.418$

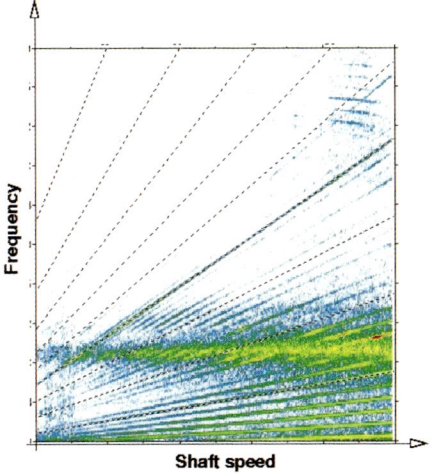

Fig. 24. (Corral _et al_, p. 216) Fig. 2. SG readings for a aerodynamically unstable LPT rotor blade

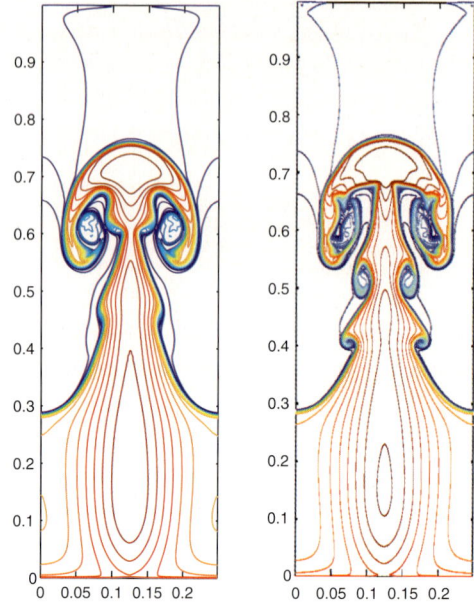

Fig. 25. (Serna, p. 234) Fig. 1. *Left*: MFF-WENO. *Right*: MFF-Weighted PowerENO

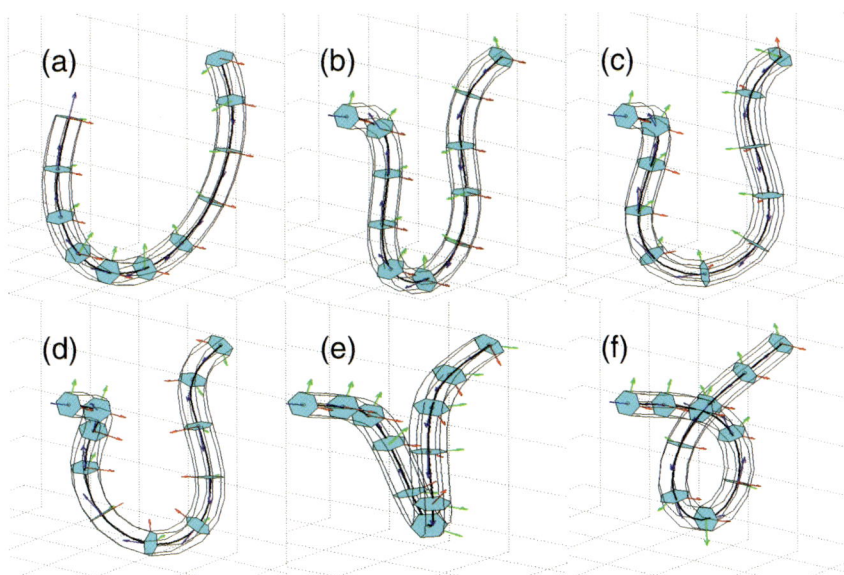

Fig. 26. (Linn _et al_, p. 252) Fig. 1. Sequential deformation of a discrete, hyperelastic Kirchhoff rod of symmetric cross section: (**a**) Starting from a circle segment, the tangents of the boundary frames are bent inward to produce (**b**) an (upside down) Ω-shaped deformation of the rod at zero twist. To demonstrate the effect of mutual coupling of bending and torsion in the discrete model, the boundary frame at $s = L$ is twisted counterclockwise by an angle of 2π while the other boundary frame at $s = 0$ is held fixed. The pictures (**c**)–(**f**) show snapshots of the deformation state taken at multiples of $\pi/2$. The overall deformation from (**a**) to (**f**) was split up into a sequence of 25 consecutive changes of the boundary conditions defined by the terminal frames of the rod. For a discretization of the cable into 10 segments, the simulation took 150 ms on 1 CPU of an AMD 2.2 GHz double processor PC, which amounts to an average computation time of 6 ms per step

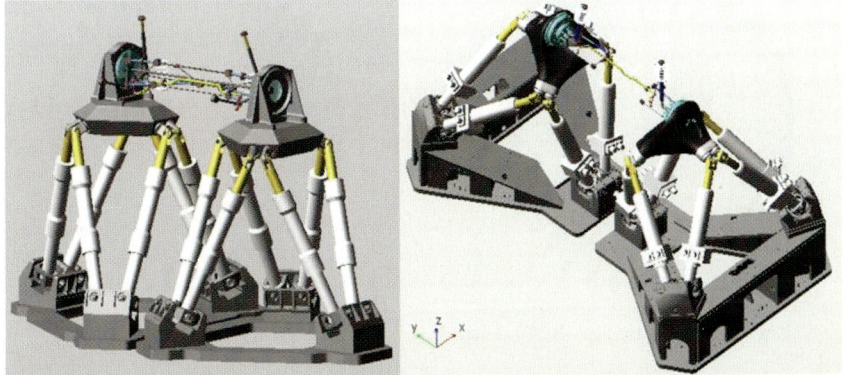

Fig. 27. (Speckert _et al_, p. 257) Fig. 2. Old (_left_) and new (_right_) design of the hexapod

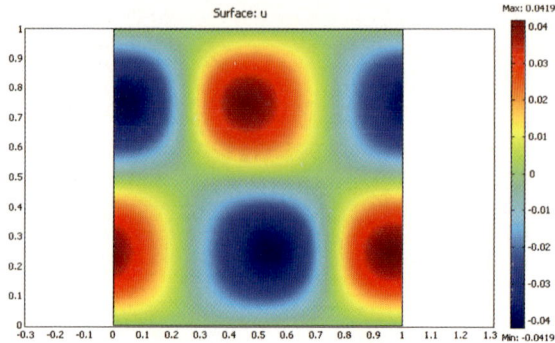

Fig. 28. (Svanstedt, p. 321) Fig. 1. Periodic cell solution (no convection)

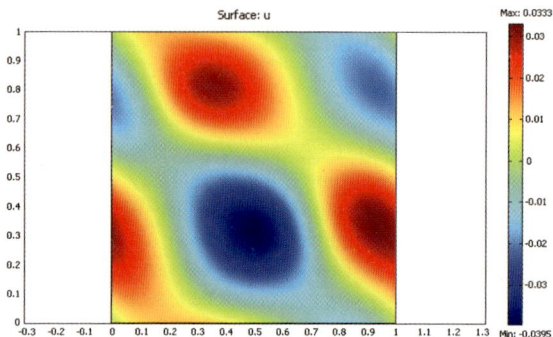

Fig. 29. Svanstedt, p. 321) Fig. 2. Periodic cell solution (large convection)

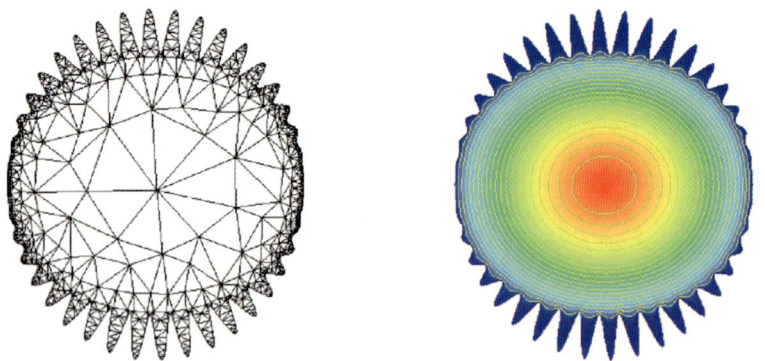

Fig. 30. (Neuss, p. 327) Fig. 1. Ω^ε with the initial mesh T_h^ε and u_h^ε

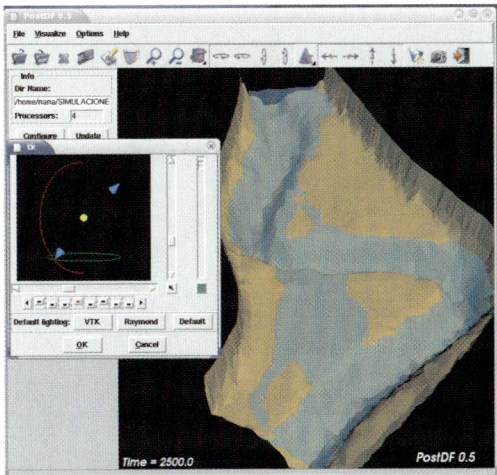

Fig. 31. (Castro *et al*, p. 359) Fig. 2. Screenshot of the post-processing tool

Fig. 32. (Castro *et al*, p. 360) Fig. 3. Mero river flood simulation

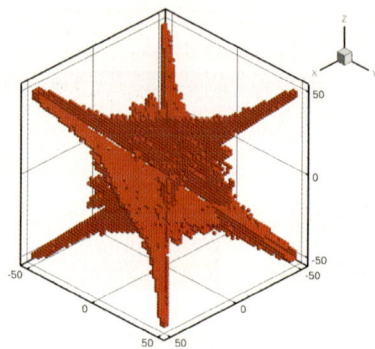

Fig. 33. (Ariza *et al*, p. 391) Fig. 2. Simulation cell containing one million Vanadium atoms. Slipped atoms after equilibration

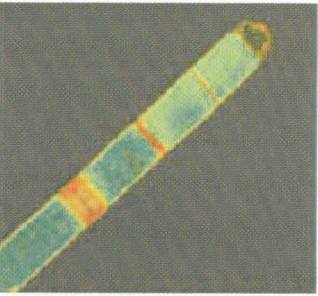

Fig. 34. (Jauho, p. 405) Fig. 1. A semiconductor nanowire grown at Lund University. The different colors indicate different materials, with different band-gaps. From [Sam03]

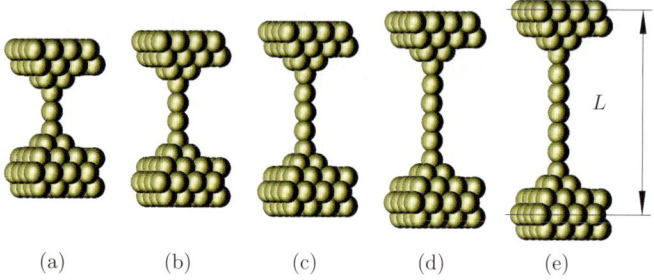

(a)   (b)   (c)   (d)   (e)

Fig. 35. (Jauho, p. 413) Fig. 8. Generic gold wire supercells containing 3–7 atoms bridging pyramidal bases connected to stacked Au(100) layers

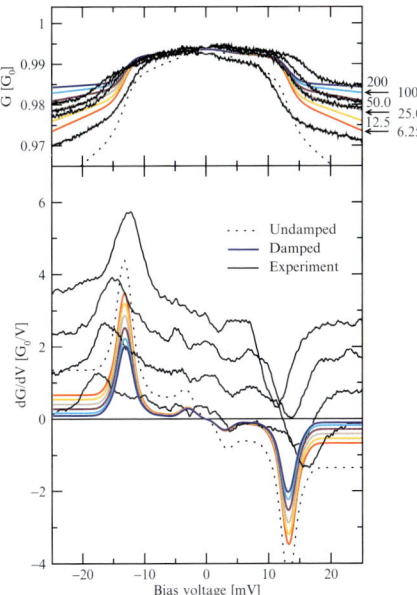

Fig. 36. (Jauho, p. 414) Fig. 9. Comparison between theory and experiment [AURV02] for the inelastic conductance of an atomic gold wire. The measured characteristics correspond to different states of strain of the wire (around 7 atoms long). The calculations are for the 7 atom wire at $L = 29.20$ Å. (Reproduced from [FPBJ06])

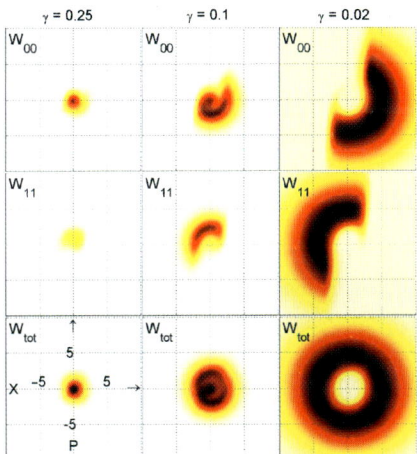

Fig. 37. (Donarini, p. 437) Fig. 3. Phase space picture of the tunnelling-to-shuttling transition. The respective rows show the Wigner distribution functions for the discharged (W_{00}), charged (W_{11}), and both (W_{tot}) states of the oscillator in the phase space. ($\Gamma = 0.05, \lambda = 1$)

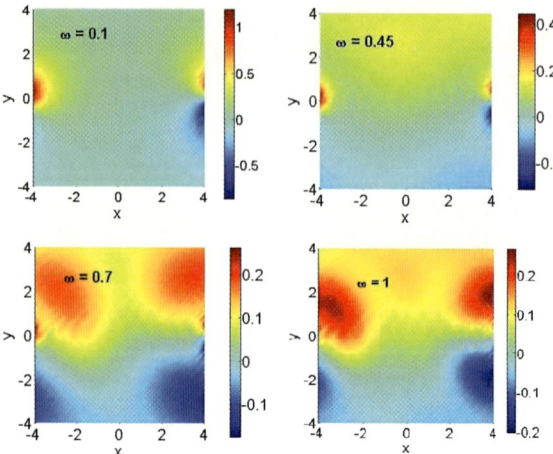

Fig. 38. (Cortijo *et al*, p. 492) Fig. 2. First order correction to the local density of states in a region around two pairs of heptagon–pentagon defects located out of the image for increasing values of the energy

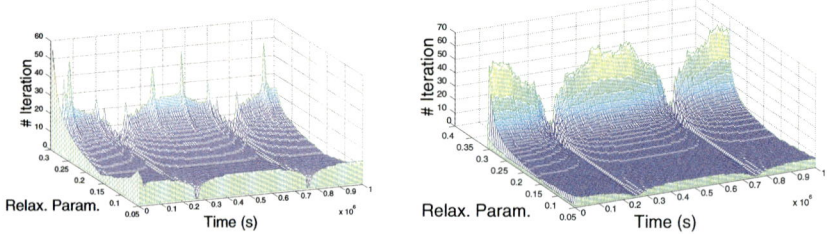

Fig. 39. (Alì *et al*, p. 503) Fig. 2. Sensitivity for Dirichlet b.c. (*left*) and Neumann b.c. (*right*)

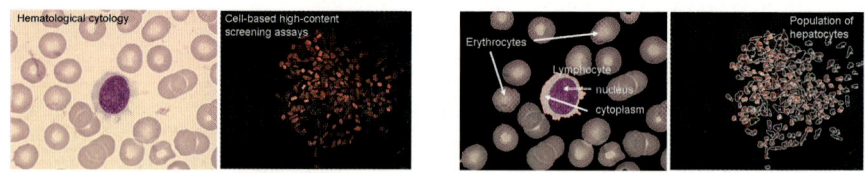

Fig. 40. (Angulo, p. 544) Fig. 1. Microscopic cell images (*left*), segmented cell shapes to be analysed (*right*)

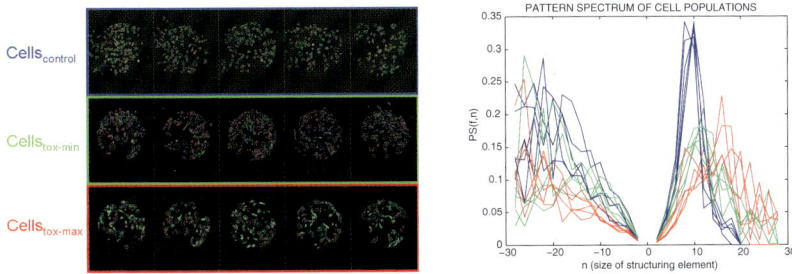

Fig. 41. (Angulo, p. 545) Fig. 2. *Left*, cell population based high content toxicity biosensor, three examples of toxic concentration. *Right*, pattern spectra, $PS(f, n)$, with openings (for size/shape description) and closing (for aggregation study) of size $n = -30$ to 30

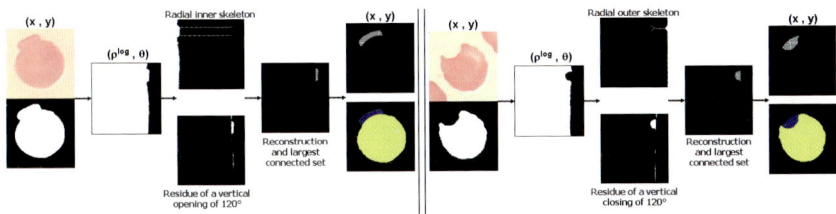

Fig. 42. (Angulo, p. 547) Fig. 3. Erythrocyte shape analysis: morphological algorithm for detecting extrusions (*left*) and intrusions (*right*)

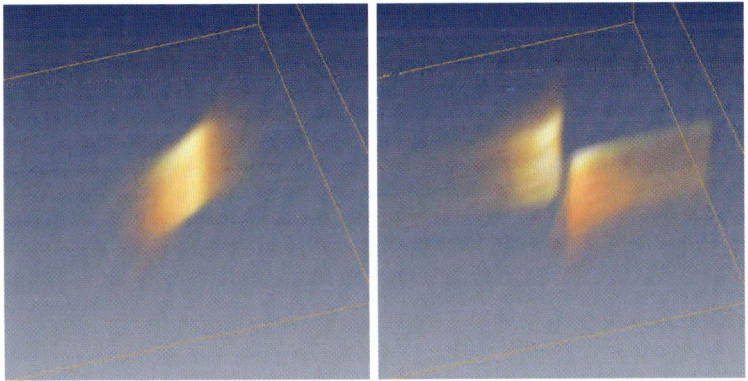

Fig. 43. (Kowar, p. 552) Fig. 3. Positive and negative part of a 3-D transducer pressure field with a pulse of 30 cycles. The reconstruction is performed with the loping Landweber–Kaczmarz method and the data is provided by GE Medical Systems Kretz Ultrasound

Fig. 44. (Morozova *et al*, p. 556) Fig. 2. 1D model with branching. Branching patterns for different values of parameters in the branching conditions

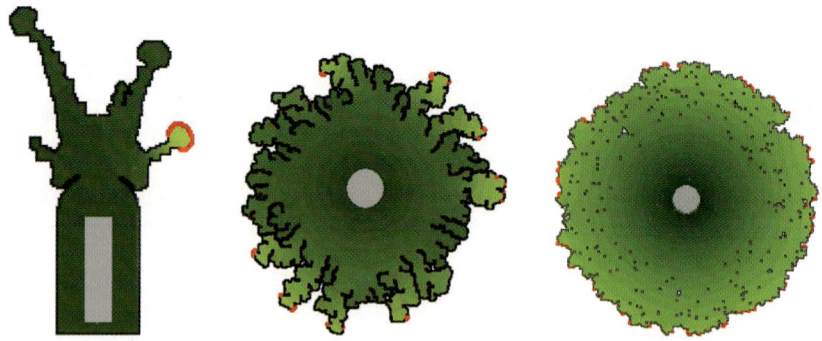

Fig. 45. (Morozova *et al*, p. 557) Fig. 3. 2D simulations. Nutrients are supplied through the internal rectangle (*left*) or through the internal circle: model without merging (*center*), model with merging (*right*)

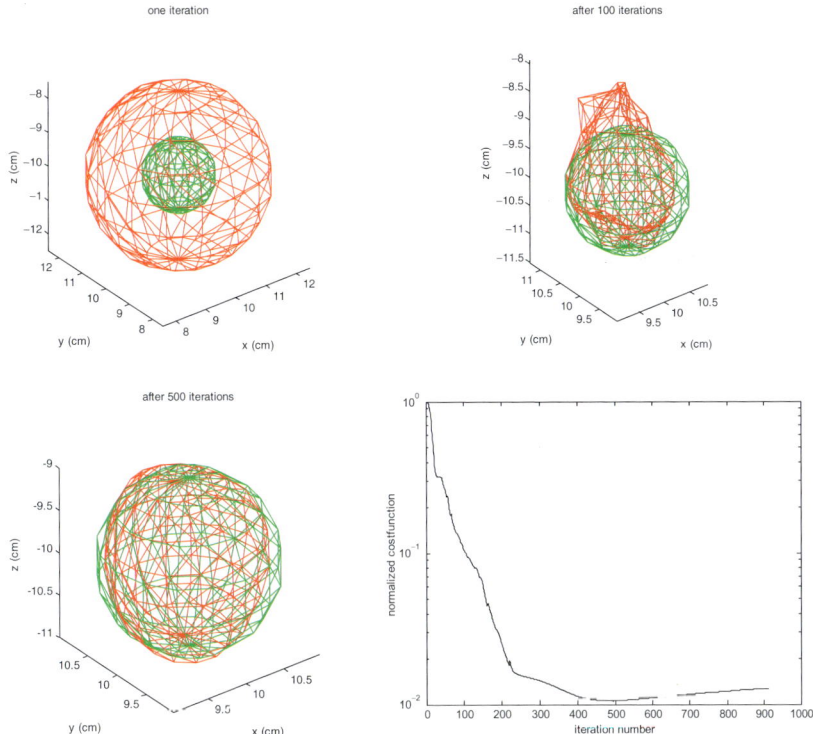

Fig. 46. (Dorn *et al*, p. 590) Fig. 1. Reconstruction of a small sphere. The true object is displayed in gray colour, and the reconstructed object by black colour in each iteration. *Top left*: after one iteration; *top right*: after 100 iterations; *bottom left*: after 500 iterations. The bottom right shows the evolution of the cost

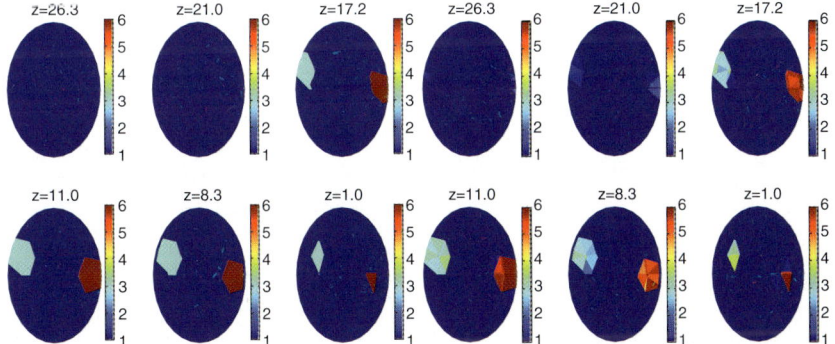

Fig. 47. (Polydorides, p. 612) Fig. 1. Planes of the simulated (*left*) and reconstructed (*right*) conductivity distributions

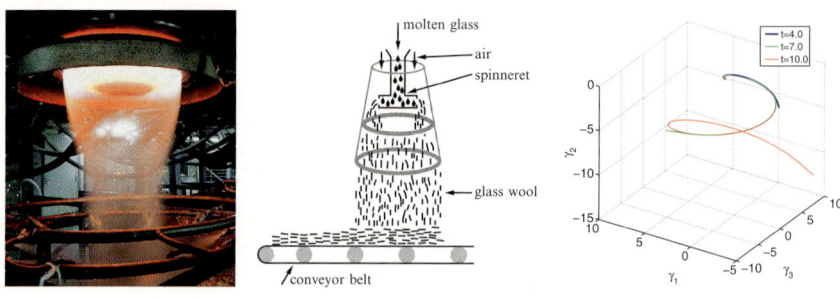

Fig. 48. (Panda *et al*, p. 686) Fig. 1. Glass wool production: plant, sketch, simulated fiber motion

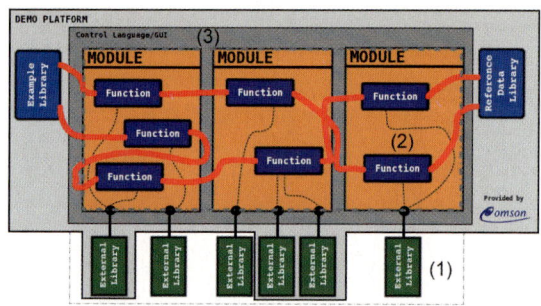

Fig. 49. (Alì *et al*, p. 737) Fig. 1. The demonstrator platform architecture

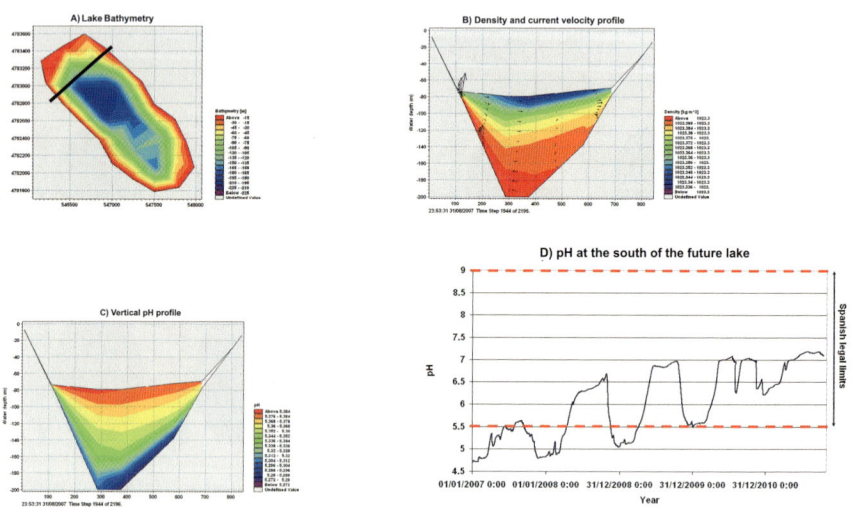

Fig. 50. (Bermúdez *et al*, p. 794) Fig. 1. Model results

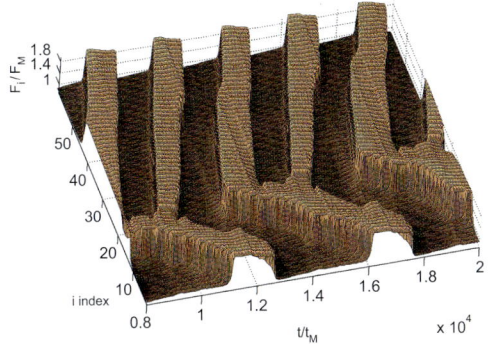

Fig. 51. (Arana *et al*, p. 816) Fig. 4. Electric field profiles during self-oscillations in a dc voltage biased 61-period undoped SL having a configuration as in Fig. 1. The voltage between the two ends of the SL is 1.765 V and the contact resistivities are 9.07 ($i = 0$) and 8.87 Ωcm ($i = N$)

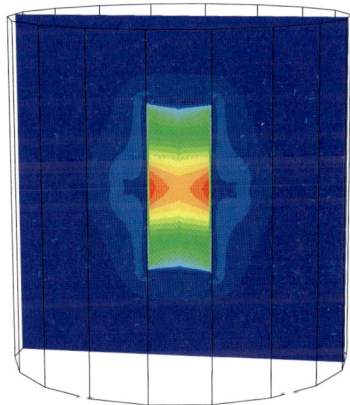

Fig. 52. (Bermúdez *et al*, p. 837) Fig. 4. Electromagnet. Intensity of the magnetic induction field, $|\boldsymbol{B}|$ in Ω

Part V

Contributor Index

List of Contributors

Printing: Krips bv, Meppel, The Netherlands
Binding: Stürtz, Würzburg, Germany